BIOLOGY

A GUIDE TO THE NATURAL WORLD

Technology Update

FIFTH EDITION

David Krogh

Boston   Columbus   Indianapolis   New York   San Francisco   Upper Saddle River
Amsterdam   Cape Town   Dubai   London   Madrid   Milan   Munich   Paris   Montréal   Toronto
Delhi   Mexico City   São Paulo   Sydney   Hong Kong   Seoul   Singapore   Taipei   Tokyo

Editor-in-Chief: *Beth Wilbur*
Senior Acquisitions Editor: *Star MacKenzie Burruto*
Executive Director of Development: *Deborah Gale*
Project Editor: *Frances Sink*
Media Producer: *Lee Ann Doctor*
Marketing Manager: *Amee Mosley*
Development Editor: *Debbie Hardin*
Art Development Editor: *Kim Quillin*
Art Editors: *Elisheva Marcus and Kelly Murphy*
Photo Editor: *Donna Kalal*
Photo Research: *Kristin Piljay*

Director of Production, Science: *Erin Gregg*
Managing Editor: *Mike Early*
Project Manager: *Camille Herrera, Jane Brundage, Shannon Tozier*
Production Service: *Progressive Publishing Alternatives*
Illustrations: *Imagineering Media Services, Inc.*
Text Design: *Gary Hespehneide*
Cover Design: *Riezebos Holzbaur Group*
Cover Production: *Fortuitous Publishing Services*
Manufacturing Buyer: *Christy Hall*
Cover Printer: *LSC Communications*
Printer and Binder: *LSC Communications*

Cover Photo Credits: chameleon © Edwin Giesbers/Nature Picture Library; Julia butterfly © Ralph Clevenger/Corbis; queen angelfish © Corbis/Stephen Frink; cheetah © Marvin Mattelson/Corbis; Ipheion 'Rolf Fiedler' © Mark Bolton/Corbis; Blue butterfly sitting on a yellow blossom © Fritz Rauschenbach/Corbis; Orange Peel Fungus © Frank Young/Corbis

Credits and acknowledgments borrowed from other sources and reproduced, with permission, in this textbook appear on p. C1.

Krogh, David, 1949-
 Biology : a guide to the natural world / David Krogh. — 5th ed.
   p. cm.
 Includes bibliographical references and index.
 ISBN-13: 978-0-321-61655-5 (alk. paper)
 ISBN-10: 0-321-61655-3 (alk. paper)
 1. Biology. I. Title.
 QH308.2.K76 2011
 570—dc22
                                                            2010027300

ISBN 10: 0-321-94676-6; ISBN 13: 978-0-321-94676-8 (Student edition)
ISBN 10: 0-321-94894-7; ISBN 13: 978-0-321-94894-6 (Professional copy)
ISBN 10: 0-321-94871-8; ISBN 13: 978-0-321-94871-7 (a la carte)

www.pearsonhighered.com

10 2020

BIOLOGY

A GUIDE TO THE NATURAL WORLD

Technology Update

About the Author

Author

DAVID KROGH has been writing about science for 27 years in newspapers, magazines, books, and for educational institutions. He is the author of *Smoking: The Artificial Passion,* an account of the pharmacological and cultural motivations behind the use of tobacco, which was nominated for the *Los Angeles Times* Book Prize in Science and Technology. In 1994, he began work on what would become *Biology: A Guide to the Natural World,* and in 1999 he completed its first edition. Since then, he has produced four more editions of *A Guide to the Natural World* along with a second textbook, *A Brief Guide to Biology.* He holds bachelor's degrees in journalism and history from the University of Missouri.

Art Development Editor

Kim Quillin received her B.A. in biology at Oberlin College and her Ph.D. in integrative biology from the University of California, Berkeley, and is now a lecturer in the Department of Biological Sciences at Salisbury University. Students and instructors alike have praised the illustration programs she has produced for three textbooks: David Krogh's *Biology: A Guide to the Natural World,* Colleen Belk and Virginia Borden's *Biology: Science for Life,* and Scott Freeman's *Biological Science.*

Preface

From the Author

Book titles may be the first thing any reader sees in a book, but they're often the last thing an author ponders. Not so with *Biology: A Guide to the Natural World*. The title arrived fairly early on, courtesy of the muse, and then stuck because it so aptly expresses what I think is special about this book.

Flip through these pages, and you'll see all the elements that students and teachers look for in any modern introductory textbook—rich, full-color art, an extensive study apparatus, and a full complement of digital learning tools. When you leaf slowly through the book and start to read a little of it, however, I think that something a little more subtle starts coming through. This second quality has to do with a sense of connection with students. The sensibility that I hope is apparent in *A Guide to the Natural World* is that there's a wonderful living world to be explored; that we who produced this book would like nothing better than to show this world to students; and that we want to take them on an instructive walk through this world, rather than a difficult march.

All the members of the teams who have produced the five editions of *A Guide to the Natural World* have worked with this idea in mind. We felt that we were taking students on a journey through the living world and that, rather like tour guides, we needed to be mindful of where students were at any given point. Would they remember this term from earlier in the chapter? Had we created enough of a bridge between one subject and the next? The idea was never to leave students with the feeling that they were wandering alone through terrain that lacked signposts. Rather, we aimed to give them the sense that they had a companion—this book—that would guide them through the subject of biology. *A Guide to the Natural World*, then, really is intended as a kind of guide, with its audience being students who are taking biology but not majoring in it.

Biology is complex, however, and if students are to understand it at anything beyond the most superficial level, details are necessary. It won't do to make what one faculty member called "magical leaps" over the difficult parts of complex subjects. Our goal was to make the difficult comprehensible, not to make it disappear altogether. Thus, the reader will find in this book fairly detailed accounts of such subjects as cellular respiration, photosynthesis, immune system function, and plant reproduction. It was in covering such topics that our concern for student comprehension was put to its greatest test. We like the way we handled these subjects and other key topics, however, and we hope readers will feel the same way.

What's New in the Fifth Edition?

The numerous changes made to the *Guide* for its fifth edition have added up to one global change in the book that longtime users may be able to perceive just by picking it up: It's shorter than it used to be. We had 791 numbered pages in the fourth edition, but this time around we have 725. The usual course for textbooks is to get larger as they get older, but we ended up going the other way in this edition after concluding that, in certain areas, we were providing more coverage than faculty thought was desirable. (You can see which areas have been trimmed by reading to the end of this section.) No book can perfectly meet the needs of all faculty with respect to content, but we've done our best to produce a book that has only as much content in it as most faculty say they need. Apart from this change, the fifth edition revision also includes:

- **A greatly revised lineup of essays.** Eighteen new essays have been written for this edition, most of them taking an applied slant. For example, on page 150, you'll find the essay "Using Photosynthesis to Fight Global Warming," which reviews some of the ways photosynthesis stands to be utilized as a tool societies can employ in their efforts to lessen climate change. Chapter 15's "Biotechnology Gets Personal" provides students with some idea of what they could expect to learn by having their own genomes scanned for signs of predispositions toward sickness or health (page 274). Then the essay "An Evolving Ability to Drink Milk" should apply in a personal way to every student in a classroom, as it shows how the forces of mutation and natural selection have worked with human culture to place all human adults alive today into one of two camps: those who can digest milk and those who can't (page 294).

- **An expanded set of review questions.** Faculty have made clear how important review questions are to student comprehension of biology, and we've responded by producing an edition that has more questions in it than any of its predecessors. The in-chapter "So Far" questions that made their debut in the fourth edition have been retained, while the end-of-chapter Multiple-Choice questions that existed up through the third edition have been brought back. The more detailed Brief Review and Applying Your Knowledge questions also appear at the end of each chapter and we're continuing to supply a host of review questions on the web.

- **Stand-alone chapters on plants, fungi, the nervous system, and the endocrine system.** In the *Guide's* fourth edition, plants and fungi were covered in a single chapter, and the same thing was true of the nervous and endocrine systems. In the fifth edition, each of these subjects gets its own chapter. One effect of this change is that the book's endocrine coverage has been significantly expanded, as you can see by turning to page 532.

- **A completely revised chapter on the immune system.** The *Guide's* coverage of the immune system has been revamped from start to finish and as such is right up-to-date. As you can see by turning to page 550, the chapter includes not only information on how the immune system works but information on two promising forms of immune therapy. This coverage is then complemented by two new essays that stand to be relevant to student lives: "Why is There No Vaccine for AIDS?" on page 566 and "Unfounded Fears about Vaccination" on page 568.

- **New coverage of applied topics.** Numerous applied topics are being covered for the first time in this edition, often in the essays mentioned earlier. Subjects taken up include the intertwining of science and big business in the development of new cancer drugs (page 12); the nature of the new targeted therapies for cancer (page 168); the pluses and minuses of making genetic information public (page 206); a research breakthrough in the fight against herpes (page 386); endorphin release as a possible motivation for the use of tanning beds (page 495); and the controversy about how much vitamin D Americans ought to be getting (page 602). We also have a greatly expanded table on methods of contraception, which you can see on page 642. Last but not least, the fourth edition of the *Guide* had a section on avian flu, but for reasons that may be obvious, the fifth edition has a section on swine flu (page 384).

- **New coverage of basic science topics.** The *Guide's* Chapter 14, on transcription, translation, and genetic regulation, is completely predictable in one way: Its section on transcription and translation never changes much while its section on genetic regulation never stops changing (page 253). Writing about genetic regulation in an age of RNA interference and alternatively spliced genes is like posting dispatches from the Lewis and Clark expedition: You're not sure where you're going to end up, but reports about the journey are fascinating. Genetic regulation is, however, only one of the basic science topics that's been greatly revised in the fifth edition. Chapter 15, on biotechnology, now has a long section on a topic whose name only came into wide circulation in the last couple of years: cell reprogramming. Material on both embryonic stem cells and induced pluripotent stem cells is included in this section, which starts on page 270. Chapter 19's coverage of the history of life on Earth has a greatly revised section on origin-of-life theories (page 342), and as you might expect the *Guide's* coverage of global warming and environmental issues in general has changed considerably since the fourth edition came out (page 704). In human evolution studies, a rough draft of the Neanderthal genome was completed in the last year, and the *Guide's* Chapter 20 includes coverage of that, along with an essay that explores the question of whether the discovery of cooking speeded up the evolution of human intelligence (page 371).

- **A different way of approaching animals.** For several editions, the *Guide* has covered the animal kingdom through the traditional means of taking students on a long walk through its major phyla. With this edition, we've taken a different tack, in accordance with advice we got from faculty. All the phyla are still covered, but in much less detail than before. What's been added is a look *across* the phyla at four subjects: reproduction, egg fertilization and protection, organs and circulation, and skeletons and molting. The end result is not only a different chapter, but a shorter one—something that was consistent with our goal of providing only as much coverage as faculty think is desirable.

- **A different way of approaching plants.** In the fourth edition of the *Guide*, plants were covered in three chapters: one in the book's evolution and diversity unit and two in the plant unit of the book, which was completely devoted to angiosperm anatomy and physiology. With this edition, some of the angiosperm A&P material has been moved to the plant diversity chapter, while the number of chapters in the plant unit has been reduced from two to one. Here again, we were following the advice of faculty, who told us we have been giving them more angiosperm A&P material than they needed. And here again the result was a reduction in the number of *Guide* pages.

- **The elimination of animal behavior.** The *Guide* has closed out nearly every one of its editions with an animal behavior chapter that was part of a larger ecology unit. Reviewing faculty liked this chapter and I certainly liked it, but it seemed to be used scarcely at all in non-majors classes and thus didn't fit in with our goal of producing a book that has only as much material in it as faculty recommend. So with the fifth edition of the *Guide*, we've eliminated animal behavior altogether.

The Technology Update

MasteringBiology has been greatly updated and expanded for this dynamic Technology Update. The Technology Update features margin callouts in the text, directing students and instructors to the robust MasteringBiology program.

There are over **30 new book-specific animations**. New topics include Population Growth, Behavioral Biology and the Principles of Evolution among others.

End of Chapter material is now available within MasteringBiology offering a greater number of assignable items for each chapter.

New Vocabulary Coaching activities allow students to connect terms and concepts in a logical fashion and build on previous knowledge.

New Learning Outcomes allow instructors to filter tagged activities by book specific and global outcomes to help focus student attention on core concepts and allow instructors to easily report on student progress.

New MasteringBiology Pre-Built Assignments are provided for each chapter. Instructors may choose among several time-saving options for auto-graded assignments that focus on current events, textbook reading quizzes, and coaching activities for the most complex topics.

The *Guide to the Natural World* Team

Textbooks are such enormous collective efforts that it's difficult for an author to thank all the people who've helped out, for the simple reason that the author hasn't *met* a good number of the people who've helped out. The best I can do, therefore, is name a few of the people whose assistance has been valuable to me personally as I've worked on the fifth edition of *A Guide to the Natural World*.

Developmental Editor Debbie Hardin gave the *Guide* the most thorough review it's had since it first came out. Every word, drawing, and photo in the book went under Debbie's microscope and the *Guide* is much better for having been analyzed in this way. The terrific artist and biologist Kim Quillin continued to produce nearly all the new and revised art for the book, working as usual with her winning combination of good cheer and devotion to clear scientific illustration. Among the people at Benjamin Cummings I'm indebted to are Production Supervisor Camille Herrera, who made sure all the pieces of the book moved forward together on time, and Design Manager Marilyn Perry, who gave the *Guide* its stylish fifth-edition look. Sylvia Rebert and her colleagues at Progressive Publishing Alternatives took thousands of pieces of *Guide* art and text and made a book out of them, doing so in a manner that allowed all parties to remain calm—no small feat, as anxiety about getting to the finish line on time is a built-in feature of producing a textbook. Overall direction for the *Guide*'s fifth edition came from a leader who provided a steady hand on the tiller, Star MacKenzie, whose devotion to producing a better book never wavered. Finally, I'm indebted to Project Editor Leata Holloway, who over a period of two years ably moved the *Guide* forward in a thousand different ways.

Apart from these publishing team members, more than 350 faculty have now carefully critiqued every word and image you see in the *Guide*. The names of reviewing faculty can be found beginning on page ix.

Finally, the test of any textbook is its effect in the classroom. As a consequence, for any textbook author, there is no such thing as too much feedback from those who stand at the front of the class. So, if you teach using the *Guide*, or if you've merely looked at it in some detail, here's an invitation: Write me and tell me what you think. Whether your reactions are positive or negative doesn't matter; the book stands to be improved by both kinds of responses. The more faculty I hear from, the better the book's sixth edition stands to be. My e-mail address is: rdkrogh@pacbell.net.

DAVID KROGH
Kensington, California

Student Supplements

Study Guide

0-321-68303-X / 978-0-321-68303-8

This tool is an effective and interactive study guide that helps students concentrate their study efforts. The Study Guide contains quizzes, word roots, key term reviews, and helpful summaries.

Study Card

0-321-68280-7 / 978-0-321-68280-2

This laminated quick-reference Study Card is developed specifically for *Biology: A Guide to the Natural World* and can be value-packaged with the book at no additional charge. The Study Card provides a brief, chapter-by-chapter review, including key figures, to help students review the most important topics in biology.

MasteringBiology Study Area

This comprehensive study resource has been revised to include new BioFlix™ animations on the toughest topics in biology and customized study plans for individual students. Also included are Web Links and references, Vocabulary Study Tools, Cumulative Tests, Web Animations with accompanying quizzes, Discovery Channel video clips, Tutoring Services, Lab Bench, Graph It! Activities, and MP3 lectures.

URL: http://www.masteringbiology.com/

Scientific American *Current Issues in Biology*

Volume 1: 0-8053-7507-4 / 978-0-8053-7507-7

Volume 2: 0-8053-7108-7 / 978-0-8053-7108-6

Volume 3: 0-8053-2146-2 / 978-0-8053-2146-3

Volume 4: 0-8053-3566-8 / 978-0-8053-3566-8

Volume 5: 0-321-54187-1 / 978-0-321-54187-1

Give your students the best of both worlds—accessible, dynamic, relevant articles from *Scientific American* magazine that present key issues in biology, paired with the authority, reliability, and clarity of Benjamin Cummings' non-majors biology texts. Articles include questions to help students check their comprehension and make connections to science and society. This resource is available at no additional charge when packaged with a new text.

Instructor Supplements

Instructor Resource DVD

0-321-68276-9 / 978-0-321-68276-5

The Instructor Resource CD-ROM includes PowerPoint® lectures that integrate figures and new BioFlix™ animations, BLAST animations, Discovery Channel® videos, all figures, art, and photos in JPEG format, label-edit PowerPoints, and over 200 Instructor Animations that accurately depict complex topics and dynamic processes described in the book. New, innovative Classroom Response System questions are offered for each chapter as a way to directly engage students in lectures.

Instructor Guide

0-321-68277-7 / 978-0-321-68277-2

This comprehensive guide provides chapter-by-chapter instructor and student objectives, chapter outlines, key terms, interactive activities with handouts, and answers to the Applying Your Knowledge critical thinking questions from the textbook. A separate section offers chapter-by-chapter media guides to aid teaching and learning introductory biology. The Instructor Guide is available in print and as a Microsoft® Word document

Transparency Acetates

0-321-72376-7 / 978-0-321-72376-5

Selected full-color acetates include illustrations and tables from the text.

Test Bank

0-321-72377-5 / 978-0-321-72377-2

All of the exam questions in the Test Bank have been peer reviewed, thus providing questions that set the standard for quality and accuracy. Test questions are ranked according to Bloom's taxonomy, and improved TestGen® software makes assembling tests easier. The Test Bank is also available in Course Management Systems and in Word® format on the Instructor Resource CD-ROM.

Blackboard

This open-access course management system contains preloaded content such as testing and assessment question pools. More content is also available in a premium version.

URL: http://www.aw-bc.com/blackboard

Reviewers

Fifth Edition Reviewers

Stephanie Aamodt, *Louisiana State University, Shreveport*

Stephen Ballard, *Hillyer College*

Cheryl Boice, *Lake City Community College*

Peggy Brickman, *University of Georgia*

Lisa G. Bryant, *Arkansas State University, Beebe*

Sara G. Carlson, *University of Akron*

Sanjoy Chakraborty, *New York City College of Technology, CUNY*

Gregory A. Dahlem, *Northern Kentucky University*

Kathleen DeCicco-Skinner, *American University*

Jean DeSaix, *University of North Carolina, Chapel Hill*

M. Omar Faison, *Virginia State Community College*

Elizabeth A. Fitch, *Motlow State Community College*

Tammy R. Gamza, *Arkansas State University, Beebe*

Anthony J. Gaudin, *Ivy Technical Community College*

George W. Gilchrist, *College of William and Mary*

Beverly Glover, *Western Oklahoma State University*

Jeanette Gore, *St. Petersburg College*

Kelly A. Hogan, *University of North Carolina, Chapel Hill*

Sue Hum-Musser, *Western Illinois University*

Diana E. Hurlbut, *Irvine Valley Community College*

Robert Iwan, *Inver Hills Community College*

Malcolm Ian Jenness, *New Mexico Institute of Mining and Technology*

Kyoungtae Kim, *Missouri State University*

Eliot Krause, *Seton Hall University*

Rukmani Kuppuswami, *Laredo Community College*

Dale Lambert, *Tarrant County Community College*

David Lambert, *Louisiana School for Math, Science, and the Arts*

Johnathan Lochamy, *Georgia Perimeter College*

Madeleine Love, *New River Community College*

Roger F. Martin, *Brigham Young University*

Bram Middeldorp, *Minneapolis Community and Technical College*

Martin M. Matute, *University of Arkansas*

Alice J. Monroe, *St. Petersburg College*

James Murphy, *Monroe Community College*

Kathryn Nette, *Cuyamaca College*

Rachel S. Potti, *Mountain View Community College*

Gregory Podgorski, *Utah State University*

Angela R. Porta, *Kean University*

Wendy M. Rappazzo, *Harford Community College*

Adam Rechs, *California State University, Sacramento*

Beverly Schieltz, *Wright State University*

Bo Sisnicki, *Piedmont Virginia Community College*

Colleen Smith Sisnicki, *Ashford University*

Marc A. Smith, *Sinclair Community College*

Meredith Somerville-Norris, *University of North Carolina, Charlotte*

Anthony Stancampiano, *Oklahoma City Community College*

Brian Sturtevant, *Oakland University*

Brent Todd, *Lake Land Community College*

Edward L. Vezey, *Oklahoma State University, Oklahoma City*

Robert L. Wallace, *Ripon College*

Jane M. Wattrus, *College of St. Scholastica*

Robert R. Wise, *University of Wisconsin, Oshkosh*

Past Edition Reviewers

Kim Anthony Aaronson, *Truman College*

Dawn Adams, *Baylor University*

Julie Adams, *Ohio Northern University*

John Alcock, *Arizona State University*

David L. Alles, *Western Washington University*

Sylvester Allred, *Northern Arizona University*

Gary Anderson, *University of California, Davis*

Marjay A. Anderson, *Howard University*

Michael F. Antolin, *Colorado State University*

Kerri Armstrong, *Community College of Philadelphia*

Mary Ashley, *University of Illinois, Chicago*

Bert Atsma, *Union County College*

Jessica Baack, *Montgomery College*

Kemuel Badger, *Ball State University*

James W. Bailey, *University of Tennessee*

Peter Bednekoff, *Eastern Michigan University*

Michael C. Bell, *Richland College*

William J. Bell, *University of Kansas*

Carla Bundrick Benejam, *California State University, Monterey Bay*

David Berrigan, *University of Washington*

Lois A. Bichler, *Stephens College*

Ian M. Bird, *University of Wisconsin, Madison*

A.W. Blackler, *Cornell University*

Andrew Blaustein, *Oregon State University*

Nancy Boury, *Iowa State University*

Robert S. Boyd, *Auburn University*

Bonnie L. Brenner, *Wilbur Wright College (City College of Chicago)*

Mimi Bres, *Prince George's Community College*

Leon Browder, *University of Calgary*

Arthur L. Buikema, *Virginia Polytechnic Institute and State University (Allegheny College)*

Steven K. Burian, *Southern Connecticut State University*

Janis K. Bush, *University of Texas, San Antonio*

Linda Butler, *University of Texas, Austin*

W. Barkley Butler, *Indiana University of Pennsylvania*

David Byres, *Florida Central Community College, South Campus*

James Cahill, *University of Alberta*

Chantae Calhoun, *Lawson State Community College*

Don Canfield, *University of Southern Denmark*

John Capeheart, *University of Houston-Downtown*

Kelly Sue Cartwright, *College of Lake County*

Marnie Chapman, *University of Alaska, Southeast Sitka*

Van D. Christman, *Ricks College*

Deborah C. Clark, *Middle Tennessee State University*

William S. Cohen, *University of Kentucky*

Karen A. Conzelman, *Glendale Community College*
Tricia Cooley, *Laredo Community College*
Richard Copping, *Lane College*
Lee Couch, *University of New Mexico*
Patricia Cox, *University of Tennessee, Knoxville*
John Crane, *Washington State University*
Robert Curry, *Villanova University*
Garry Davies, *University of Alaska, Anchorage*
Paula Dedmon, *Gaston College*
Miriam del Campo, *Miami-Dade Community College*
Brent DeMars, *Lakeland Community College*
Llewellyn Densmore, *Texas Technical University*
Jean Dickey, *Clemson University*
Christopher Dobson, *Front Range Community College*
Deborah Dodson, *Vincennes University*
Carolyn Doege, *Cy-Fair College*
Cathy Donald-Whitney, *Collin County Community College*
Matthew M. Douglas, *Grand Rapids Community College (University of Kansas)*
Lee C. Drickamer, *Southern Illinois University*
JodyLee Duek, *Pima College*
Albia Dugger, *Miami-Dade College*
Charles Duggins, Jr., *University of South Carolina*
Susan A. Dunford, *University of Cincinnati*
Douglas Eder, *Southern Illinois University*
Ron Edwards, *University of Florida*
Douglas J. Eernisse, *California State University, Fullerton*
Jamin Eisenbach, *Eastern Michigan University*
George Ellmore, *Tufts University*
Patrick E. Elvander, *University of California, Santa Cruz*
Michael Emsley, *George Mason University*
Elyce Ervin, *University of Toledo*
David W. Essar, *Winona State University*
Richard H. Falk, *University of California, Davis*
Michael Farabee, *Estrella Mountain Community College*
Rita Farrar, *Louisiana State University*
John Philip Fawley, *Westminster College*
Eugene J. Fenster, *Longview Community College*
Rebecca Ferrell, *Metropolitan State College at Denver*
Leanne H. Field, *University of Texas, Austin*
Teresa Fischer, *Indian River Community College*
Christine M. Foreman, *University of Toledo*
Carl S. Frankel, *Pennsylvania State University*
Ralph Fregosi, *University of Arizona*
Lawrence Friedman, *University of Missouri, St. Louis*

Carol Friesen, *Ball State University*
John L. Frola, *University of Akron*
Larry Fulton, *American River College*
Gail E. Gasparich, *Towson University*
Matt Geisler, *University of California, Riverside*
Mita Ghosh, *FCCJ*
Tejendra S. Gill, *University of Houston*
Claudette Giscombe, *University of Southern Indiana*
Carolyn Glaubensklee, *University of Southern Colorado*
Jack M. Goldberg, *University of California, Davis*
Elliott Goldstein, *Arizona State University*
Judith Goodenough, *University of Massachusetts, Amherst*
David M. Gordon, *Pittsburg State University*
Glenn A. Gorelick, *Citrus College*
Becky Graham, *The University of West Alabama*
Melvin H. Green, *University of California, San Diego*
G.A. Griffith, *South Suburban College*
Sherri Gross, *Ithaca College*
Edward Hale, *Ball State University*
Gail Hall, *Trinity College*
Linnea S. Hall, *California State University, Sacramento*
Madeline Hall, *Cleveland State University*
Kelly Hamilton, *Shoreline Community College*
James Hampton, *Salt Lake Community College*
Christopher Harendza, *Montgomery County Community College*
Steven C. Harris, *Clarion University*
Barbara Harvey, *Kirkwood Community College*
Steve Heard, *University of Iowa*
Walter Hewitson, *Bridgewater State College*
Jane Aloi Horlings, *Saddleback College*
Eva Horne, *Kansas State University*
Michael Hudecki, *State University of New York, Buffalo*
Michael Hudspeth, *Northern Illinois University*
Terry L. Hufford, *The George Washington University*
Carol A. Hurney, *James Madison University*
Catherine J. Hurlbut, *Florida Community College*
Andrea Huvard, *California Lutheran University*
Martin Ikkanda, *Los Angeles Pierce College*
Rose M. Isgrigg, *Ohio University*
Rebecca Jann, *Queens University of Charlotte*
Thomas W. Jurik, *Iowa State University*
Arnold J. Karpoff, *University of Louisville*
Karry Kazial, *SUNY Fredonia*
Anne Keddy-Hector, *Austin Community College*
Kathleen Keeler, *University of Nebraska*

Nancy Keene, *Pellissippi State Technical Community College*
Kevin M. Kelley, *California State University, Long Beach*
Jeanette J. Kiem, *Guilford Technical Community College*
Tom Knoedler, *Ohio State University, Lima Campus*
Don E. Krane, *Wright State University*
Jocelyn Krebs, *University of Alaska, Anchorage*
Kate Lajtha, *Oregon State University*
Kirkwood Land, *University of the Pacific*
Erika Ann Lawson, *Columbia College*
Mike Lawson, *Missouri Southern State College*
Brenda Leicht, *University of Iowa*
Harvey Liftin, *Broward Community College*
Lynne Lisenby, *Florida Community College at Jacksonville*
Ann S. Lumsden, *Florida State University*
Paul Lurquin, *Washington State University*
James Manser, *Harvey Mudd College*
Michael M. Martin, *University of Michigan, Ann Arbor*
Paul Mason, *Butte Community College*
Michel Masson, *Santa Barbara City College*
Andrea Mastro, *Pennsylvania State University*
Mary Victoria McDonald, *University of Central Arkansas*
Mary Colleen McNamara, *Albuquerque T-VI A Community College*
Hugh A. Miller, III, *East Tennessee University*
Leslie R. Miller, *Iowa State University*
Lee H. Mitchell, *Mount Hood Community College*
Robert Mitchell, *Community College of Philadelphia*
Jeremy Montague, *Barry University*
Scott M. Moody, *Ohio University*
Janice Moore, *Colorado State University*
Joseph Moore, *California State University*
Jorge A. Moreno, *University of Colorado*
Michael D. Morgan, *University of Wisconsin, Green Bay*
David Mork, *Saint Cloud State University*
Deborah A. Morris, *Portland State University*
Allison Morrison-Shetlar, *Georgia Southern University*
Richard Mortensen, *Albion College*
Christopher B. Mowry, *Berry College*
Michelle Murphy, *University of Notre Dame*
Courtney Murren, *University of Tennessee, Knoxville*
Richard L. Myers, *Missouri State University*
Royden Nakamura, *California Polytechnic State University*
Leann Naughton, *University of Wyoming*

Judy M. Nesmith, *University of Michigan, Dearborn*
Harry Nickla, *Creighton University*
James A. Nienow, *Valdosta State University*
Jane Noble-Harvey, *University of Delaware*
Paul Nolan, *Ithaca College*
Ray Ochs, *St. John's University*
Hunter O'Reilly, *University of Wisconsin, Milwaukee*
Marcy P. Osgood, *University of Michigan*
Andrea Ostrofsky, *University of Maine*
Onesimus Otieno, *Oakwood College*
Charles Owens, *King College*
Maya Patel, *Ithaca College*
Eckle L. Peabody, *Tulsa Community College*
Kathleen Pelkki, *Saginaw Valley State University*
Patricia A. Peroni, *Davidson College*
Rhoda E. Perozzi, *Virginia Commonwealth University*
Carolyn Peters, *Spoon River College*
John S. Peters, *College of Charleston*
Kim M. Peterson, *University of Alaska, Anchorage*
Raleigh K. Pettegrew, *Denison University*
Gary W. Pettibone, *State University of New York, College at Buffalo*
Holly C. Pinkart, *Central Washington University*
Barbara Pleasants, *Iowa State University*
John M. Pleasants, *Iowa State University*
Lynn Polasek, *Los Angeles Valley College*
F. Harvey Pough, *Arizona State University, West*
Don Pribor, *University of Toledo*
Michelle Priest, *University of Southern California*
Louis Primavera, *Hawaii Pacific University*
Paul Ramp, *Pellissippi State and Technical Community College*
Ameed Raoof, *University of Michigan Medical School*
Marceau Ratard, *Delgado Community College*
Regina Rector, *William Rainey Harper College*
Robert J. Reinsvold, *University of Northern Colorado*
Rick Relyea, *University of Pittsburgh*
Michael H. Renfroe, *James Madison University*
Dennis Richardson, *Quinnipiac University*
Todd Rimkus, *Marymount University*
Sonia J. Ringstrom, *Loyola University*
David A. Rintoul, *Kansas State University*
Darryl Ritter, *Okaloosa-Walton Community College*

Carolyn Roberson, *Roane State Community College*
Laurel Roberts, *University of Pittsburgh*
Jay Robinson, *San Antonio College*
Rodney A. Rogers, *Drake University*
William H. Rohrer, *Union County College*
Leslie Ann Roldan, *Massachusetts Institute of Technology*
Amanda Rosenzweig, *Delgado Community College*
Heidi Rottschafer, *University of Notre Dame*
John Rueter, *Portland State University*
Ron Ruppert, *Cuesta College*
Brian Sailer, *Sam Houston State University*
Nancy Sanders, *Northeast Missouri Sate University*
Gary Sarinsky, *City University of New York, Kingsborough Community College*
Julie Schroer, *Bismarck State College*
Steven Scott, *Merritt College*
Edna Seaman, *University of Massachusetts, Boston*
Ralph W. Seelke, *University of Wisconsin, Superior*
Prem P. Sehgal, *East Carolina University*
C. Thomas Settlemire, *Bowdoin College*
Robert Shetlar, *Georgia Southern University*
Cara Shillington, *Eastern Michigan University*
Mark A. Shotwell, *University of Slippery Rock*
Linda Simpson, *University of North Carolina, Charlotte*
Peter Slater, *University of St. Andrews, UK*
Ellen Smith, *Arizona State University, West*
Linda Smith-Staton, *Pellissippi State Technical Community College*
Philip J. Snider, *University of Houston*
Nancy G. Solomon, *Miami University*
Salvatore A. Sparace, *Clemson University*
Frederick W. Spiegel, *University of Arkansas*
Bryan Spohn, *Florida Community College at Jacksonville–Kent Campus*
Kathleen M. Steinert, *Bellevue Community College*
Allan R. Stevens, *Snow College*
Donald P. Streubel, *Idaho State University*
Gerald Summers, *University of Missouri, Columbia*
Christine Tachibana, *University of Washington*
Steven Tanner, *University of Missouri*
Todd Templeton, *Metropolitan Community College*
Rebekah J. Thomas, *Saint Leo University*
Joanne Tornow, *University of Southern Mississippi*

Todd T. Tracy, *Colorado State University*
Robin W. Tyser, *University of Wisconsin, LaCrosse*
Rani Vajravelu, *University of Central Florida*
Joseph W. Vanable, Jr., *Purdue University*
William A. Velhagen, Jr., *Longwood College*
Tanya Vickers, *University of Utah*
Janet Vigna, *Grand Valley State University*
Allan Hayes Vogel, *Chemeketa Community College*
Dennis Vrba, *North Iowa Area Community College*
Stephen Wagener, *Western Connecticut State University*
Jyoti R. Wagle, *Houston Community College*
John H. Wahlert, *Baruch College, The City University of New York*
Carol Wake, *South Dakota State University*
Timothy S. Wakefield, *John Brown University*
Charles Walcott, *Cornell University*
Gene Walton, *Tallahassee Community College*
Yunqiu Wang, *University of Miami*
Sarah Ward, *Colorado State University*
Jennifer M. Warner, *University of North Carolina, Charlotte*
Cheryl Watson, *Central Connecticut State University*
Richard Weinstein, *University of Tennessee, Knoxville*
Richard Barry Welch, *San Antonio College*
Jamie Welling, *South Suburban College*
Elizabeth Forston Wells, *George Washington University*
Mary Pat Wenderoth, *University of Washington*
Susan Whittemore, *Keene State University*
Allison Wiedemeier, *University of Missouri-Columbia*
Sandra Winicur, *Indiana University, South Bend*
William Wischusen, *Louisiana State University*
James Wise, *Hampton University*
Deborah Wisti-Peterson, *University of Washington*
Rachel Witcher, *University of Central Florida*
Mark A. Woelfe, *Vanderbilt University*
Lorne Wolfe, *Georgia Southern University*
Wade B. Worthen, *Furman University*
Robert Yost, *Indiana University Purdue University Indianapolis*
Calvin Young, *Fullerton College*
Martha C. Zúñiga, *University of California, Santa Cruz*
Victoria Zusman, *Miami-Dade Community College*

Instantly Transform Your Learning Experience with All-New Online Resources

Kevin Robertson, Ph.D.
Fire Ecology Specialist

◀ **Immerse yourself** in the world of biology with MasteringBiology, which now offers even more interactive activities including new Video Field Trips.

Achieve mastery ▶
of concepts through an adaptive system personalized to individual student needs

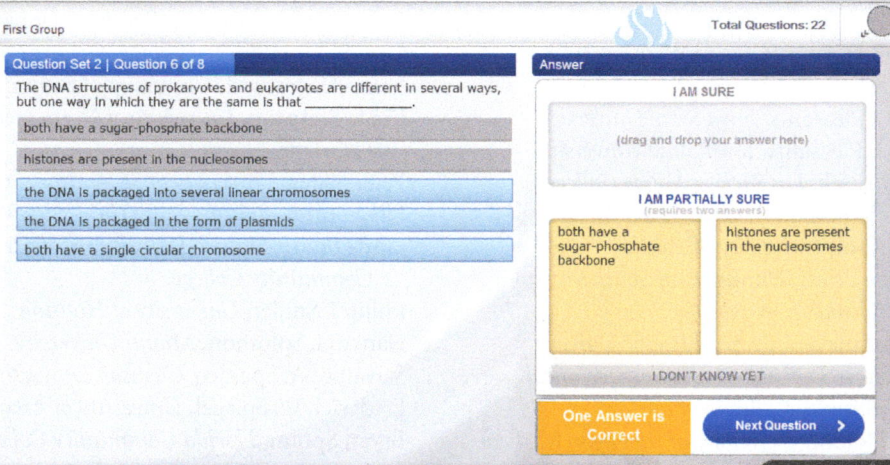

Use your mobile device ▶
to participate in class and assess your understanding of the material

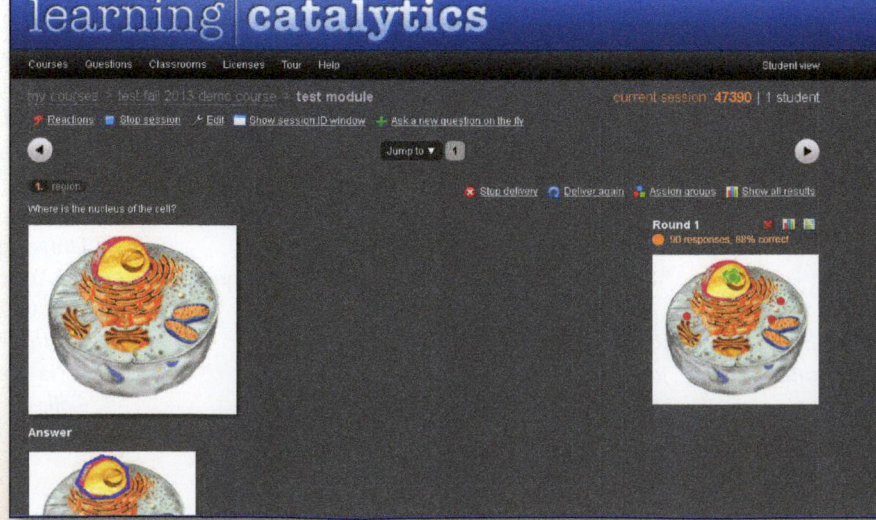

Enjoy Learning Biology with Fascinating and Effective Activities and Resources

New! Video Field Trips give you a fascinating behind-the-scenes experience of a prescribed fire burn, e-waste sites, solar energy, coal-fired power plants, wastewater treatment facilities, landfills, farms, and more. Each Video Field Trip includes assessment questions that may be assigned as homework.

New! Vocabulary Coaching activities allow you to connect terms and concepts in a logical fashion and build on your previous knowledge.

New! ABC News Videos cover a wide range of biological topics and show you how science is connected to your life.

SAVING THE WORLD'S OCEANS
ACID THREATENS LIFE UNDERSEAS

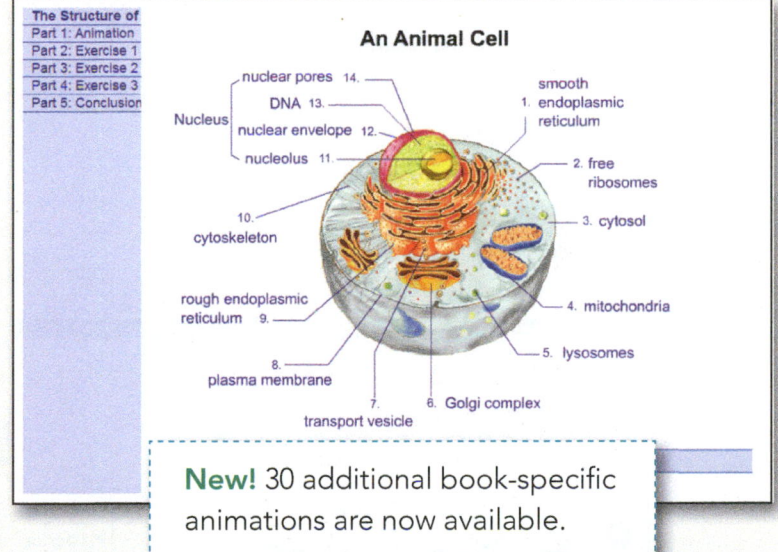

The Structure of
Part 1: Animation
Part 2: Exercise 1
Part 3: Exercise 2
Part 4: Exercise 3
Part 5: Conclusion

An Animal Cell

nuclear pores 14.
DNA 13.
Nucleus
nuclear envelope 12.
nucleolus 11.

smooth
1. endoplasmic
   reticulum

2. free
   ribosomes

10.
cytoskeleton

3. cytosol

rough endoplasmic
reticulum 9.

4. mitochondria

5. lysosomes

8.
plasma membrane

7.   6. Golgi complex
transport vesicle

New! 30 additional book-specific animations are now available.

MasteringBiology

Master Concepts and Improve Your Grade with a Personalized Study Plan

NEW! Dynamic Study Modules ▶ provide a comprehensive mobile learning experience that drives deep, contextual knowledge and understanding.

- You can indicate how confident you are of your answer, and the program will respond with appropriate follow-up questions.

- Full explanations and answers are embedded in the program for rapid learning and long-term retention.

- Progress charts provide you with immediate feedback for self-assessment.

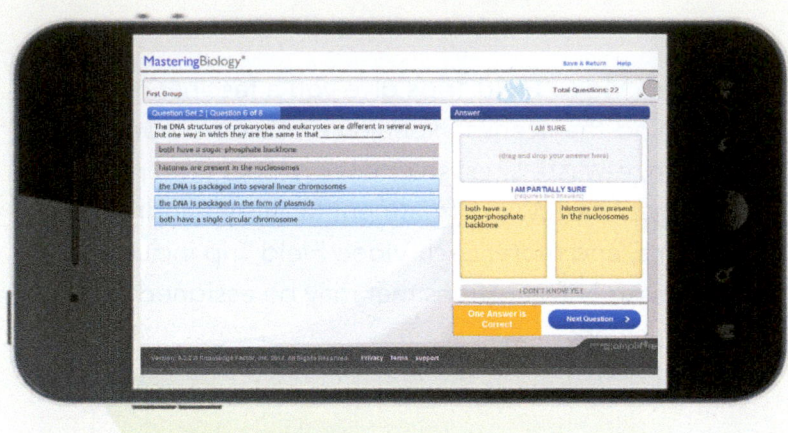

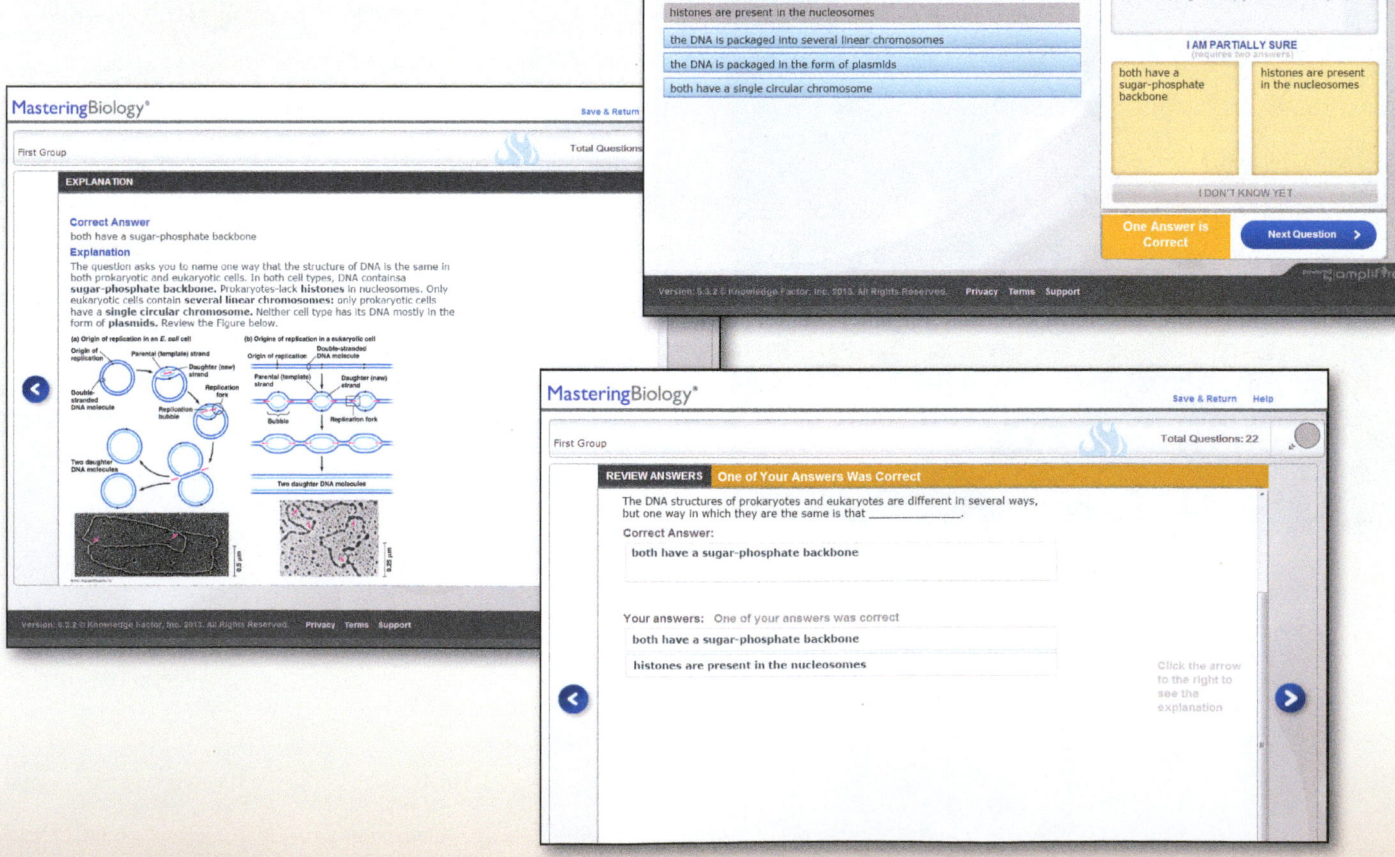

Communicate with Your Instructor and Signal Your Understanding in Real Time via Your Mobile Device During Class

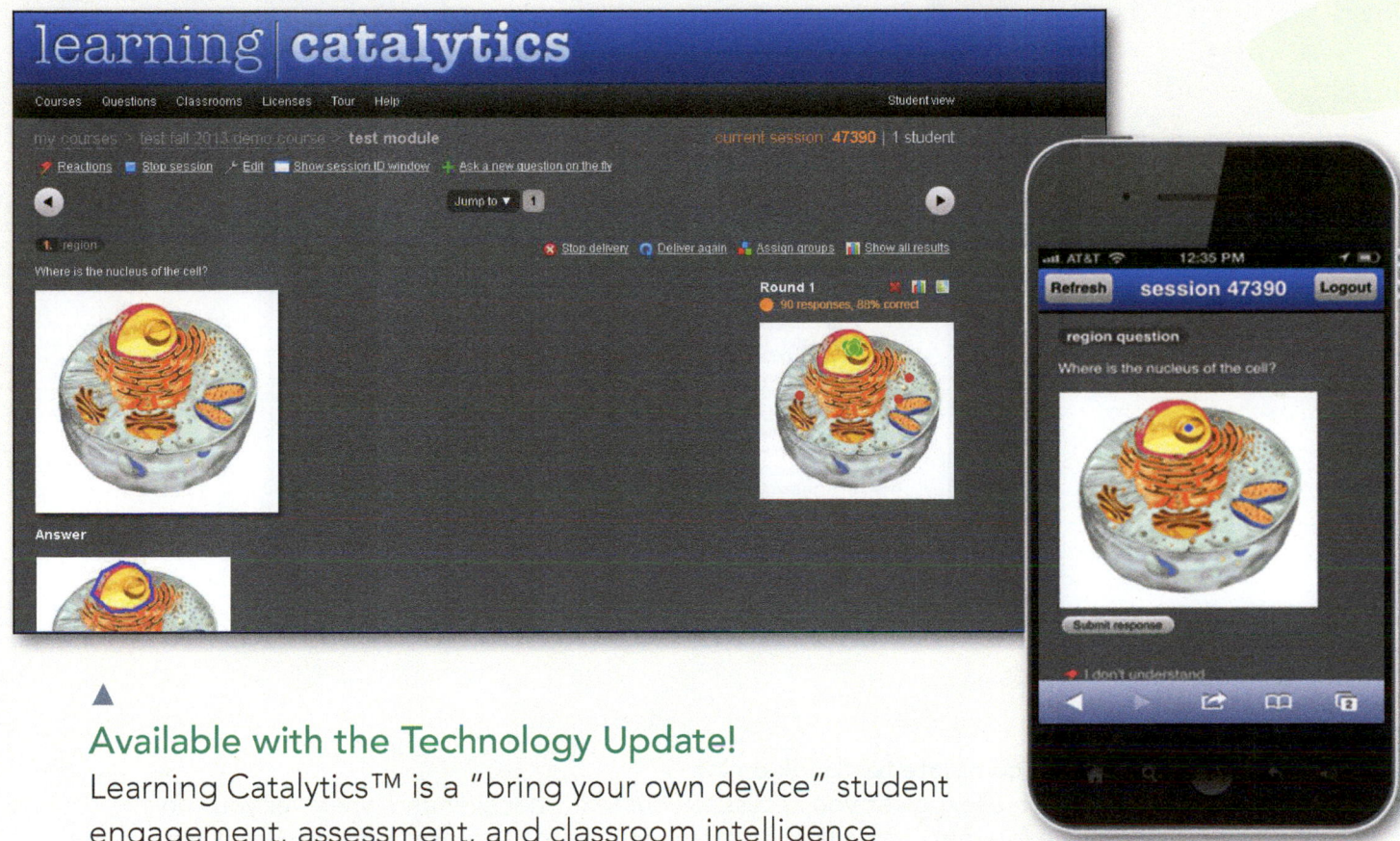

Available with the Technology Update!

Learning Catalytics™ is a "bring your own device" student engagement, assessment, and classroom intelligence system. This technology has grown out of 20 years of cutting edge research, innovation, and implementation of interactive teaching and peer instruction.

Review, Study, and Prepare Anytime You Choose by Using the Study Area

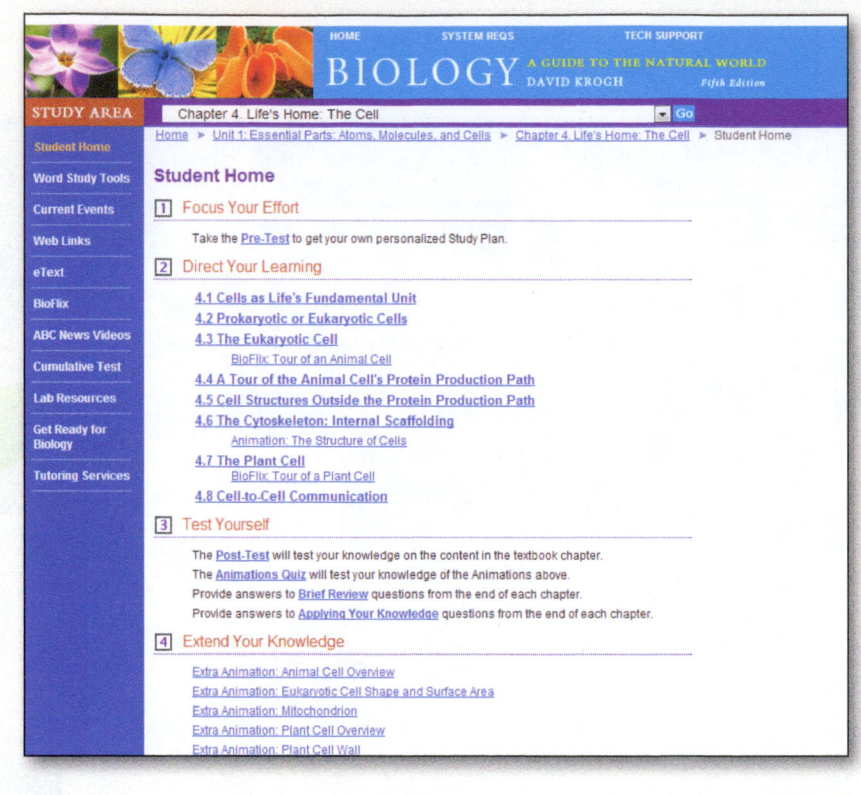

Access the complete textbook online

The Pearson eText gives you access to your text whenever and wherever you can access the internet, and with the new Pearson eText app existing subscribers who view their Pearson eText titles on a Mac or PC can also access their titles in a bookshelf on the iPad or an Android tablet either online or via download.

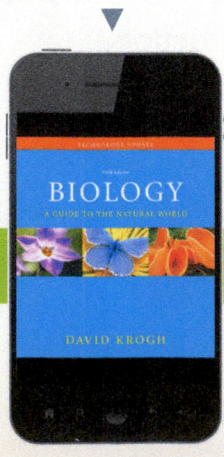
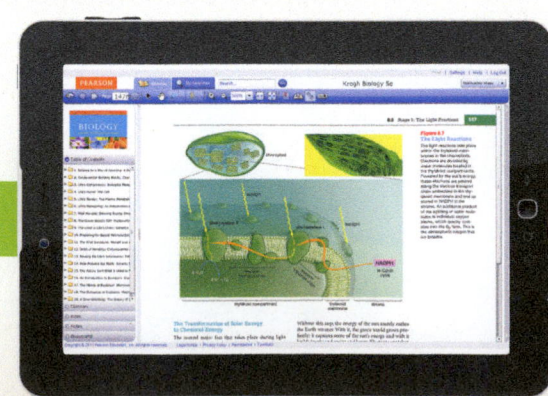

Let the Textbook and Online Resources Work Together to Guide You through the Material

Available on MasteringBiology®
Activities: Light Energy and Pigments; Properties of Light; Leaves: The Sites of Photosynthesis; Overview of Photosynthesis; The Sites of Photosynthesis; The Light Reactions; The Calvin Cycle; The Calvin Experiments; Photosynthesis in Dry Climates; **Bioflix:** Photosynthesis; **Current Events:** Storms Threaten Ozone Layer Over U.S., Study Says; Giants Who Scarfed Down Fast Food Feasts; **Video Field Trip:** Solar Energy; **Building Vocabulary; Questions:** Bioflix Quiz, Reading Questions, Test Bank

Energy from the sun helps produce the food that sustains nearly all living things.

MasteringBiology®

1. MasteringBiology Assignments
Activities: Properties of Light; Light Energy and Pigments; Leaves: The Sites of Photosynthesis; Overview of Photosynthesis; The Sites of Photosynthesis; The Light Reactions; The Calvin Cycle; The Calvin Experiments; Photosynthesis in Dry Climates; **Bioflix:** Photosynthesis; **Current Events:** Storms Threaten Ozone Layer Over U.S., Study Says; Giants Who Scarfed Down Fast Food Feasts; **Video Field Trip:** Solar Energy; **Building Vocabulary; Questions:** Bioflix Quiz, Reading Questions, Test Bank

2. eText
Read your book online, search, take notes, highlight text, and study more effectively.

3. The Study Area
Self-study tools to test comprehension.

◀ **The Technology Update**
features margin callouts in the text, directing students and instructors to a significantly more robust MasteringBiology program, introducing learning outcomes, book-specific animations, end-of-chapter material, new vocabulary activities, ABC videos, and more.

▼

Calvin Cycle

The Calvin Cycle

The Calvin Experiments

Bioflix: Photosynthesis

Figure 8.8
The Calvin Cycle
CO₂, ATP, and electrons
(contained in hydrogen

In the second stage of pho
captured in the light reactio
power a process b
from the atmosph
the resulting pro
high-energy carbo
let's begin by takin
left off. The short
where NADP⁺ has
energized electrons
There's also ATP i
the light reactions.

**Energized Su
a Cycle of Re**

The Light Reactions

8.3 Stage 1: The Lig Reactions

With these components and proce
trace the steps of the light reactions
first step is that solar energy, collecte
II's antennae molecules, arrives at th
This energy then gives a boost to a
reaction center in the two ways no
physically moves to another part of t
complex, the primary electron acc
pumped up the energy hill. With
reaction center chlorophyll has *lost*
loss leaves an energy "hole" in this
ing it an oxidizing agent. With the e
this imbalance, a special enzyme in t
splits water molecules that lie wit

Utilize the Engaging Content and Visuals in the Textbook

Author David Krogh guides you through the world of biology with his accessible and easy-to understand writing style. The book is a helpful companion that guides non-majors students on their voyage through biology by placing unfamiliar topics in context with everyday life.

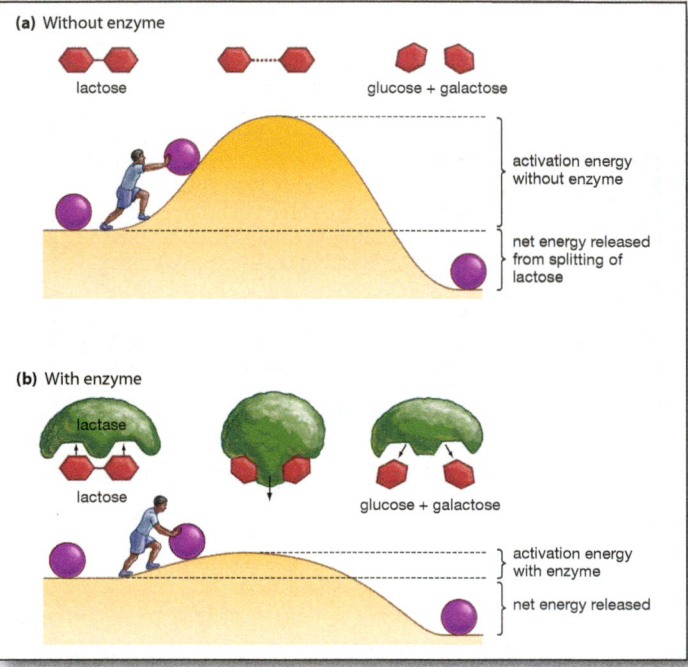

(a) Without enzyme

lactose

glucose + galactose

activation energy without enzyme

net energy released from splitting of lactose

(b) With enzyme

lactase

lactose

glucose + galactose

activation energy with enzyme

net energy released

Engaging Visuals

A strong illustration program guides you through structures and processes with clear three-dimensional detail; key information from the text is reinforced in the illustrations.

Essays

The Essays focus on social concerns and personal applications helping you apply what you are learning in class with your own experiences. Hot topics include:

- Using Photosynthesis to Fight Global Warming
- Biotechnology Gets Personal
- An Evolving Ability to Drink Milk

ESSAY

When the Cell Cycle Runs Amok: Cancer

The cell cycle reviewed in this chapter is, in one sense, a common natural process: Cells grow; they duplicate their chromosomes; these chromosomes separate; one cell divides into two. This goes on like clockwork, millions of times a second in each one of us. Then one day we learn that an aunt, a grandfather, or a friend is experiencing an *unrestrained* division of cells. Things aren't explained to us in this way, of course. We are simply told that someone we know has cancer.

At root, all cancers are failures of the cell cycle. Put another way, all cancers represent a failure of cells to limit their multiplication in the cell cycle. What is liver cancer, for example? It is a damaging multiplication of liver cells. First one, then two, then four, then eight liver cells move repeatedly through the cell cycle, and as their numbers increase, they destroy the liver's working tissues. Given that cancer manifests in this way, it's not surprising that a large portion of modern cancer research is *cell-cycle* research. The logic here is simple: To the extent that uncontrolled cell division can be stopped, cancer can be stopped.

In recent years, cell-cycle research has brought about a remarkable change in the kinds of treatments that are being developed for cancer. For decades, the basic idea behind cancer treatment was to kill as many cancer cells as possible, either by removing these cells from the body altogether (in surgery) or by destroying them with radiation or drugs. In contrast, nearly all the cancer drugs being developed today are intended not to kill cancerous cells. Instead, they are intended to disrupt the process by which these cells multiply. The idea is that, just as a light can be switched off by disrupting an electrical circuit, so cell division can be switched off by disrupting a *molecular* circuit—the chain of chemical reactions that keeps a cancerous cell moving constantly through cell division. The new breed of cancer treatments that are built on this idea are called targeted therapies.

To get a feel for what this approach means in practice, consider the first targeted therapy ever developed: a drug called Gleevec, which is aimed at combating a form of cancer, called chronic myeloid leukemia (CML), that results in an uncontrolled proliferation of certain types of white blood cells. In the 1990s, the researcher most responsible for Gleevec's use in CML treatment, Oregon Health and Sciences University Professor

Brian Druker, began to focus on the process by which CML cells come to their out-of-control state. Druker knew that CML begins in a single white blood cell when two of its chromosomes undergo an unusual form of segment-swapping with one another. When a segment from chromosome A fuses with a segment of chromosome B in this process, the result is that information from both chromosomes combines to create a mutant gene—a gene that does not exist in any normal white blood cell. Of course, this mutant gene (dubbed *BCR-ABL*) then results in a protein that is capable of carrying out some task in the white blood cell. Unfortunately, what the BCR-ABL protein does is flip a molecular switch; through it, the white blood cell starts multiplying more rapidly, even as it and its daughter cells start undergoing further destabilizing changes that ultimately result in full-blown leukemia.

Druker's key insight into stopping CML was to look very carefully at the BCR-ABL protein, which you can see represented in **Figure 1**. He realized that BCR-ABL is powered into activity following its binding with the energy-transfer molecule ATP. So, he wondered, what would happen if the ATP binding site on the BCR-ABL protein became unavailable—what would happen if this binding site became *occupied* by another molecule? The substance he eventually hit upon that takes on this occupying role

Figure 9.11
Cytokinesis in Animals
Cytokinesis in animal cells begins with an indentation of the cell surface, a cleavage furrow, shown here in a dividing frog egg.

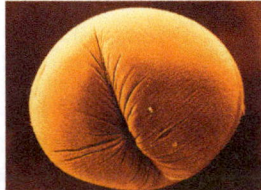

through the tightening of a cellular waistband that is composed of two sets of protein filaments working together. These filaments—the same type that allow your muscles to contract—form a *contractile ring* that narrows along the cellular equator (**Figure 9.11**). An indentation of the cell's surface (called a cleavage furrow) results from the ring's contraction; consequently, the fibers in the mitotic spindle are pushed closer and closer together, eventually forming one thick pole that is destined to break.

The dividing cell now assumes an hourglass shape; as the contractile ring continues to pinch in, one cell becomes two by means of something you looked at in Chapter 4: membrane fusion. The membranes on each

Instructor Resource DVD (IR-DVD)
978-0-321-68276-5 • 0-321-68276-9

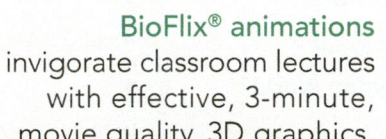

Everything you need for lectures is in one place. Enhanced menus make locating and assessing the digital resources for each chapter easy. All resources found on the IR-DVD are also available through the Instructor Resources Area of MasteringBiology.®

BioFlix® animations invigorate classroom lectures with effective, 3-minute, movie quality, 3D graphics. ▶

◀ These brief 3–5 minute **ABC News** video clips cover topics from fighting cancer to antibiotic resistance and introduced species.

▲ **More than 200 extra animations** are scientifically clear, accurate, and available in PowerPoint.

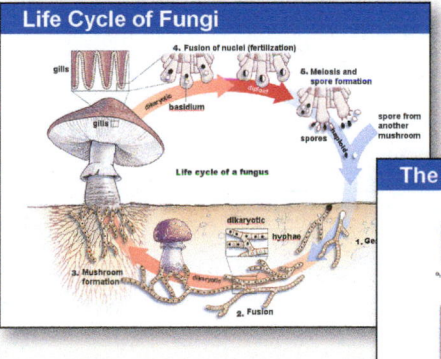

◀ Customizable **PowerPoint® Lectures** serve as lecture outlines and are included for each textbook chapter.

Expanded Test Bank questions are available as downloadable Microsoft® Word files and in TestGen.®

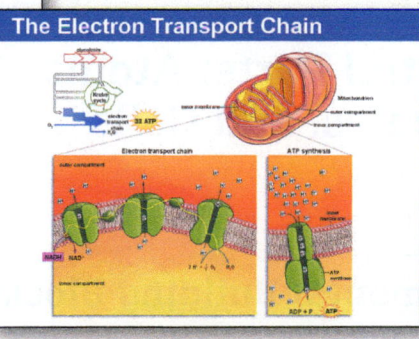

All of the **art, stepped-art, tables, and photos** from the book are available with customizable labels and are also available in PowerPoint.

▼

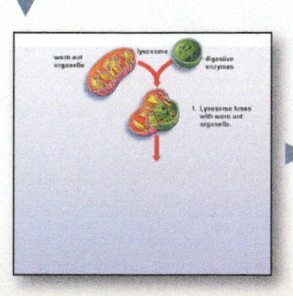 ▶ 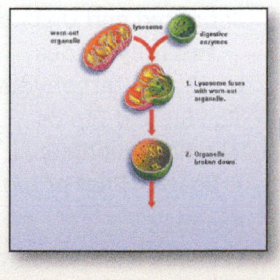 ▶ 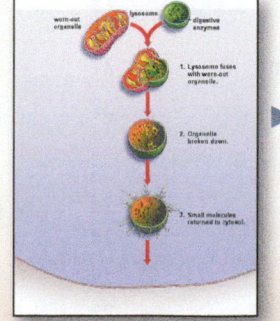 ▶ 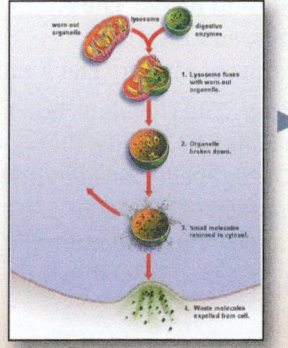 ▶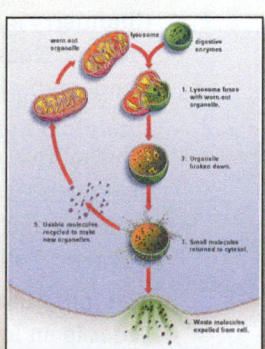

Contents

CHAPTER 4

Life's Home:
The Cell   64

CHAPTER 5

Life's Border:
The Plasma Membrane   92

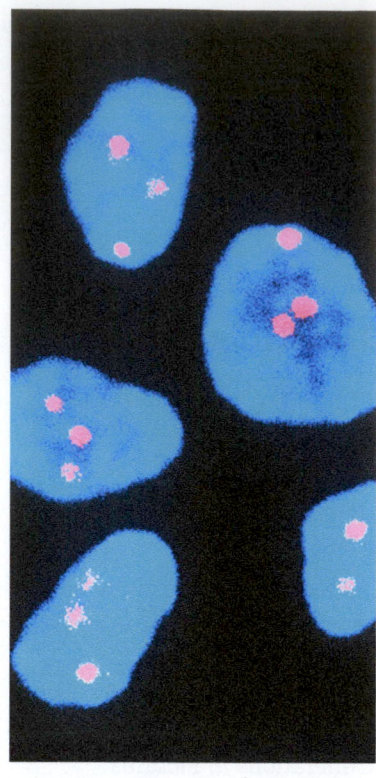

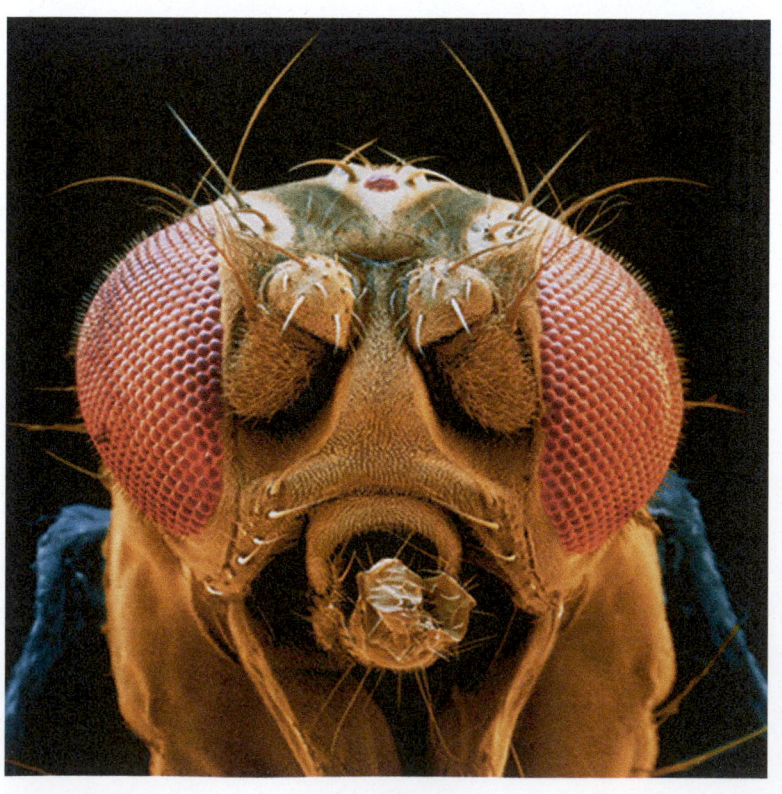

CHAPTER 15

The Future Isn't What It Used to Be: Biotechnology

Unit 4

Life's Organizing Principle: Evolution and the Diversity of Life

CHAPTER 16

An Introduction to Evolution:
Charles Darwin, Evolutionary Thought, and the Evidence for Evolution

CHAPTER 17

The Means of Evolution:
Microevolution

CHAPTER 18

The Outcomes of Evolution:
Macroevolution

CHAPTER 19

A Slow Unfolding:
The History of Life on Earth

CHAPTER 20

Arriving Late, Traveling Far:
The Evolution of Human Beings

CHAPTER 21

Viruses, Bacteria, Archaea, and Protists: The Diversity of Life 1

CHAPTER 22

Fungi:
The Diversity of Life 2

CHAPTER 23

Animals:
The Diversity of Life 3

CHAPTER 24

Plants:
The Diversity of Life 4

Unit 5
A Bounty That Feeds Us All: Plants

CHAPTER 25

The Angiosperms:
Form and Function in Flowering Plants

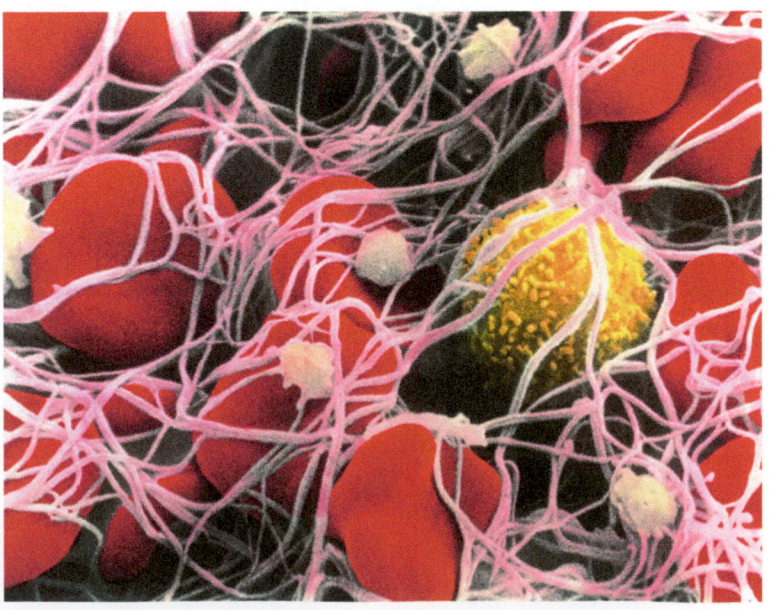

Unit 7

The Living World as a Whole: Ecology

CHAPTER **35**

An Interactive Living World 2:
Communities in Ecology 670

CHAPTER **36**

An Interactive Living World 3:
Ecosystems and Biomes 690

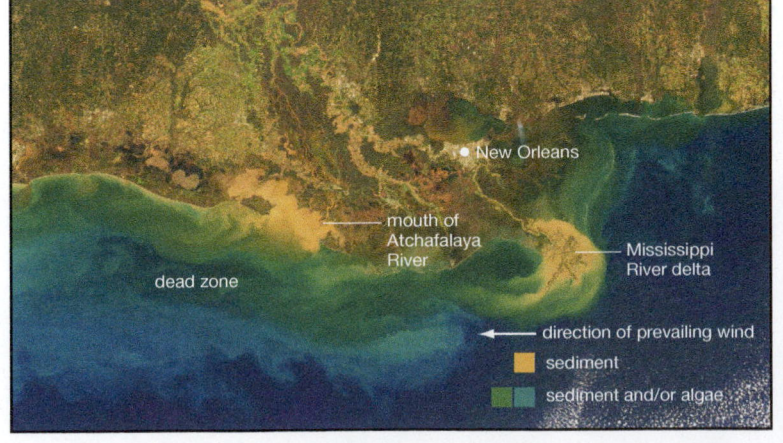

New Orleans
mouth of Atchafalaya River
Mississippi River delta
dead zone
→ direction of prevailing wind
■ sediment
■ sediment and/or algae

Science as a Way of Learning:

A Guide to the Natural World

Science has great impact on our lives now and stands to have greater impact on them in the future. Science is both a body of knowledge and a means of acquiring knowledge. Biology, a branch of science, is the study of life.

ESSAY

Available on MasteringBiology®

Activities: Metric System Review; Scientific Method; The Levels of Life Card Game; What Is Life? **GraphIt!:** An Introduction to Graphing; **Current Events:** Genome Detectives Solve a Hospital's Deadly Outbreak; **Building Vocabulary; Questions:** Reading Questions, Test Bank

Measurement plays a part in almost every scientific investigation. Here, an ecologist in Borneo measures a rafflesia plant (*Rafflesia keithii*), whose flower is believed to be the largest in the world.

In September 2008, the editors of *Nature* magazine noted an anniversary that had just occurred: Exactly 10 years earlier, the first employee of Google walked into the suburban garage that was serving as Google's headquarters at the time. In the intervening years, of course, Google became the largest distributor of information on the planet—something that prompted the *Nature* editors to pose a question to a group of cutting-edge science and technology analysts: What technology, currently at the stage Google was at in 1998, will have changed our world in the *next* 10 years as much as Google did in the past 10?

Of the six sets of predictions that *Nature* received, the most mind-bending came from Ian Pearson, an Englishman who makes his living consulting about the future. Pearson foresees a mash-up of our computer worlds and our everyday physical worlds that will come to us via a digital tool called the video visor, which may end up looking something like a set of wrap-around ski goggles spiked by a small wireless antenna. Each visor would be outfitted with a global positioning system and a high-speed Internet connection, thus giving its wearer a constant stream of information about his or her surroundings.

"So as you're walking down a busy city street," Pearson says, "you will be able to see reviews of shops and restaurants, adverts for services, other people who have similar interests to you, or whatever." All this will be customizable according to your interests. On a given day, you might be interested only in Chinese restaurants or perhaps in people you pass who play bridge (and who likewise are seeking out bridge players themselves). But the fun wouldn't stop there. As Pearson has written elsewhere, "Building appearance will depend on the taste of the person looking at it. So will the appearance of other people. People, buildings, and objects will emit a digital aura that contains alternative digital appearances as well as lots of marketing information. . . . You may well broadcast a different image according to the viewer. So your portable will act as a digital filter, deciding what you want to see and how you want to see it, and a personal advertiser, marketing you to the rest of the world."

And if you're thinking that the real stumbling block to this is that nobody would want to walk around with a set of ski goggles on, well, think how ordinary it now seems to see a person on the street talking out loud into a cell phone headset when only a few years ago such behavior seemed a little bizarre.

Metric System Review

1.1 How Does Science Impact the Everyday World?

The future that Pearson foresees may not come to pass, of course, but the thing that makes his vision so riveting is that we can imagine such leaps taking place in the near future because we've seen so many like them taking place in the recent past. The touch screen on an iPhone manages to seem like magic to us, even though it's fully *functional* magic, ready to enlarge a picture or get us to a website. A satellite image of our own house may cause us to shake our heads in disbelief that such a thing is possible, and yet there's our house, right in front of us on a computer screen.

We may categorize most of this change under the umbrella heading of *technology*, but when we look deeply enough into the roots of any technological innovation what we generally find is a contribution made by the endeavor we call science. The fundamental discovery that brought about modern electronics, including the computer, was research done on materials science that culminated with the invention of the transistor at Bell Labs in 1947. And most of the software that runs our electronic devices has its roots in university-based computer science research. What has happened in the years since World War II is that the fruits of this kind of research have been made ever more personal—moving from the laboratory, to businesses, and finally into our homes—even as the pace of technological innovation has been gaining speed over time.

To look at these events in connection with computer technology, average Americans heard about computers for 30 years following the invention of the transistor, but not until the mid-1980s were they *using* computers right on their desktops. Once this revolution in circuits got going, however, things came in a torrent. CD players were newfangled devices in the mid-1980s; now, people are downloading MP3 files into their portable global positioning systems. Then again, we need not go back to the 1980s to see how fast technological change is coming. On a typical day, do you send an e-mail to someone, talk on a cell phone, or visit a website? As late as the mid-1990s, the average person would not have done any of these things (**Figure 1.1**).

A similar story has taken place in connection with the raft of new products and processes that go under the heading of biotechnology. The fundamental breakthrough that brought about biotech was the description of the DNA molecule in 1953 by James Watson and Francis Crick. Some 25 years elapsed, however, between this discovery and the appearance of the first genetically engineered medicine (human insulin). But now? Eighty percent of U.S.-grown corn is genetically modified, criminals are regularly convicted (and innocent people freed) through use of DNA "fingerprints," cloned goats can provide us with human medicines, and in the near future you may be able to get a copy of your own personal genome—the inventory of all your DNA information—for less than $1,000, whereas in 2006 the cost of getting this information stood at more than $1 million.

It may go without saying that these various innovations and discoveries entail value judgments on the part of society. Research involving embryonic stem cells seems to have great potential to cure or alleviate the suffering caused by such conditions as spinal cord injuries and Parkinson's disease. Yet this research is dependent on the removal of cells from embryos that would otherwise be discarded by fertility clinics—a procedure that kills the embryos. After looking at both sides of this issue, California voters decided in 2004 that, over the ensuing 10 years, they would spend $3 billion of state money on embryonic stem cell research. They did so, however, only after *national* political leaders came to a very different conclusion about this research. From 2001 to 2009, federal funding of it was severely restricted by order of President George Bush, who believed that the killing of human embryos amounted to a taking of human life. The Bush funding restrictions lasted only as long as the Bush presidency, however. The change in voter sentiment that brought Barack Obama to office in 2009 resulted in a directive from him that loosened federal limitations on stem cell research. Of course, stem cells are not the only science-driven issue that society is grappling with. Think of genetically modified foods, the dizzying possibility of a human clone, and suspects who are listed on arrest warrants not by their names but by their DNA profiles.

Note that all these issues have been brought to society by science. We might say that science has presented society with *options*, about which society then makes decisions. But things also work the other way around—society brings issues to science. Why? To get advice. Since scientists are in the business of investigating nature, they function, in effect, as society's eyes and ears on the natural world. Thus, if we want to know whether Earth really is warming, whether we're in danger from an errant comet, or what the likelihood is of swine flu spreading, then it's not politicians, economists, or business executives who are going to tell us; it is scientists. In line with this, scarcely a week goes by without scientists rendering judgments on vital issues that touch on nature's processes. Do polar bears need protection as an endangered species? Does the practice of thinning out forests reduce the risk of catastrophic forest fires? What are the risk factors for diabetes? In all these instances, governments and average citizens look to

Figure 1.1
Then and Now

(a) A technician enters data into the world's first programmable computer, run initially in 1948. Called "Baby," the computer was more than 6 feet tall and almost 16 feet wide but had a total memory of only 128 bytes.

(b) In contrast, this MSI touch-screen computer, displayed at the 2010 Las Vegas Consumer Electronics Show, was two feet across, less than three inches thick, and had a total memory of 640 billion bytes.

(a)

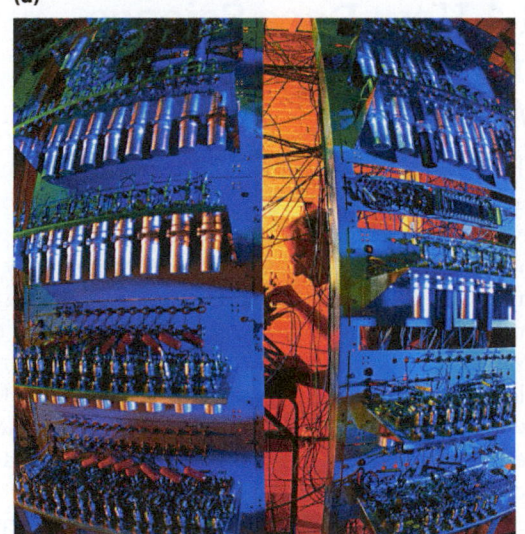

(b)

science to provide answers (after which it's up to governments and average citizens to act on the advice they get).

If we take a step back from all this, the message is that science and technology are now woven more tightly into the fabric of society than ever before. Accordingly, to fully participate in the workforce, to make informed choices at the ballot box, or simply to make routine decisions, the average person must now be more scientifically literate than ever before. To get a sense of how this plays out in everyday life, let's look at some biology-related news that came to Americans through one magazine (*Time*) during one short period (June through October 2008). In the process, you'll see how learning something about science can provide a foundation for understanding the kind of science-related news that comes our way every day.

How Safe Are Vaccines? *Time* asked in its June 2 cover story, which noted that a small but growing number of children in the United States are not being vaccinated against common childhood diseases because of their parents' belief that vaccination serves to cause illness, rather than prevent it (**Figure 1.2**).

Figure 1.2
Science in the News

The importance of science to everyday life is reflected in the large number of news stories that focus on science. Pictured are several magazines whose cover stories concerned science-related issues.

Time didn't have the space to fill in its readers on how vaccination works or what kinds of illnesses it's aimed at stopping, but readers of this book can learn about both topics by turning to the *Guide*'s Chapter 29, on the human immune system, which begins on page 550.

It's Not Just Genetics, said *Time* in its story of June 12 on the reasons for the national epidemic of childhood obesity. Factors other than genes, such as poverty and place of residence, also play a part in predisposing children to being obese. But how common is it for genes to work in concert with "environmental" factors such as these to produce a given physical condition? And what is it about being overweight that's so hazardous to human health? Readers interested in the first question can turn to Chapter 11's primer on inheritance, which begins on page 190, while readers interested in the second question should see "Why Fat Matters and Why Exercise May Matter More" in Chapter 26 on page 492.

Mopping Up the CO_2 Deluge, in *Time*'s July 3 issue, was about ways to combat global warming by getting rid of the excess carbon dioxide that human beings have put into the atmosphere over the last couple of hundred years. One of these proposed "geoengineering" techniques is to fertilize the world's oceans with iron, thus allowing algae to flourish on a scale large enough to soak up significant amounts of atmospheric CO_2. Both the algae and the carbon within them would then sink to the bottom of the ocean when the algae die—or so the proposal goes. But why do algae soak up CO_2? To find out, see Chapter 8, on photosynthesis, beginning on page 140. And what are the root causes of global warming? Chapter 36 provides some answers, beginning on page 704.

He Won His Battle with Cancer, in *Time*'s September 4 edition, was on the glacially slow progress the United States is making in its fight against cancer. "Overall, the death-rate from cancer dropped just 5 percent from 1950 to 2005," *Time* noted, while during the same period, "deaths from heart disease dropped 64 percent." Though there are many different kinds of cancer, all of them are the result of a breakdown in a basic biological process: cell division, which you can read about in Chapter 9, beginning on page 156. And what causes these breakdowns? Mutations to DNA, which you can read about in Chapter 13, beginning on page 232.

The Forgotten Plague, in *Time*'s October 2 edition, was about the resurgence of a disease that at one time seemed to be all but vanquished: tuberculosis, which has recently appeared in forms that are resistant to all the tuberculosis-fighting antibiotics we have. But what is an antibiotic, and why have many of these drugs begun losing their effectiveness against diseases such as tuberculosis? To find out, see "Bacteria and Human Disease" in Chapter 21, beginning on page 389.

1.2 What Is Science?

Having looked a little at the impact that science has, it might be helpful now to consider the question of what science *is*. The point here is to give you some sense of the underpinnings of science—to review something about the how and why of it before you begin looking at the nature of one of its disciplines, biology.

Science as a Body of Knowledge

Science is in one sense a process—a *way* of learning. In this respect, it is an activity carried out under certain rules, which we'll get to shortly. **Science** is also a body of knowledge about the natural world. It is a collection of unified insights about nature, the evidence for which is an array of facts. The unified insights of science are commonly referred to as *theories*.

It is unfortunate but true that the word *theory* means one thing in everyday speech and something almost completely different in scientific communication. In everyday speech, a theory can be little more than a hunch. It is an unproven idea that may or may not have any supportive evidence. In science, meanwhile, a **theory** is a general set of principles, supported by evidence, that explains some aspect of nature. There is, for example, a Big Bang theory of the universe. It is a general set of principles that explains how our universe began and then developed over time. Among its principles are that a cataclysmic explosion occurred about 13.7 billion years ago, and that after it, matter first developed in the form of gases, which then coalesced into the stars we can see all around us. There are numerous facts supporting these principles, such as the current size of the universe and its average temperature.

As you might imagine, with any theory this grand some *pieces* of it are in dispute; some facts don't fit the theory, and scientists disagree about how to interpret this piece of information or that. On the whole, though, these insights have withstood the questioning of critics and stand as a scientific theory.

The Importance of Theories

Far from being a hunch, a scientific theory actually is a much more valued entity than is a scientific fact because the theory has an *explanatory* power, while a fact generally is an isolated piece of information. That the universe is about 13.7 billion years old is a wonderfully interesting fact, but it explains little in comparison with the Big Bang theory. Facts are important; theories could not be supported or refuted without them. But science is first and foremost in the theory-building business, not the fact-finding business.

Science as a Process: Arriving at Scientific Insights

So how does a body of facts and theories come about? What is the process of scientific investigation, in other words? When science is viewed as a process, it could be defined as a means of coming to understand the natural world through observation and the testing of hypotheses. This process is referred to as the **scientific method** (**Figure 1.3**). The starting state for scientific inquiry is always *observation*: A piece of the natural world is observed to work in a certain way. Then follows the *question,* which broadly speaking is one of three types: a "what" question, a "why" question, or a "how" question. Biologists have asked, for example: What are genes made of? Why does the number of species decrease as we move from the equator to the poles? How does the brain make sense of visual images?

Following the formulation of the question, various hypotheses are proposed that might answer it. A **hypothesis** is a tentative, testable explanation for an observed phenomenon. In almost any scientific question, several hypotheses are proposed to account for the same observation. Which one is correct? Most fre-

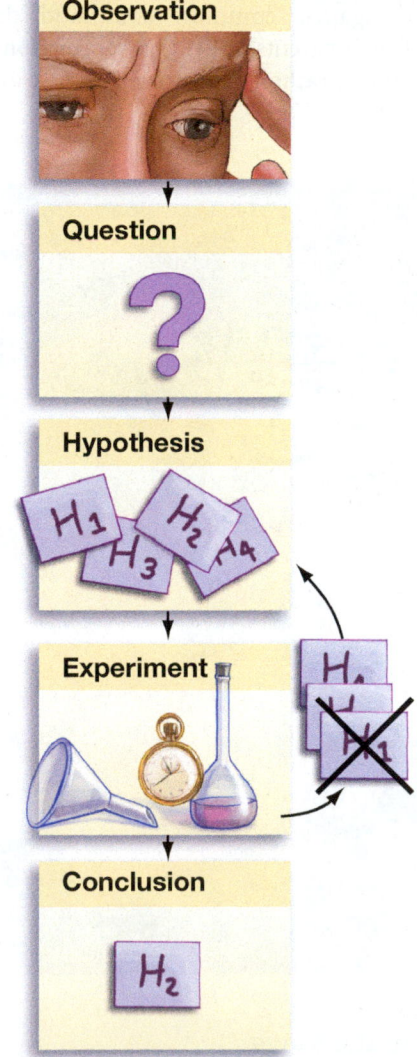

Figure 1.3
Scientific Method
The scientific method enables us to answer questions by testing hypotheses.

Scientific Method

GraphIt!: An Introduction to Graphing

quently in science, the answer is provided by a series of experiments, which are controlled tests of the question at hand. Scientists don't regard all hypotheses as equally worthy of undergoing experimental test. By the time scientists arrive at the experimental stage, they usually have an idea of which is the most promising hypothesis among the contenders and then proceed to put that hypothesis to the test. Let's see how this worked in an example from history.

The Scientific Method at Work: Pasteur and Spontaneous Generation

Does life regularly arise from anything *but* life, or can it be created "spontaneously," through the coming together of basic chemicals? The latter idea had wide acceptance from the time of the ancient Romans, and

as late as the nineteenth century it was championed by some of the leading scientists of the day. So, how could the issue be decided? The famous French chemist and medical researcher Louis Pasteur formulated a hypothesis to address this question (**Figure 1.4**). He believed that many purported examples of life arising spontaneously were simply instances of airborne microscopic organisms landing on a suitable substance and then multiplying in such profusion that they could be seen. Life came from life, in other words, not from spontaneous generation. But how could this be demonstrated? In 1860, Pasteur sterilized a meat broth in glass flasks by heating it, while at the same time heating the glass *necks* of the flasks, after which he bent the necks into a "swan" or S shape. The ends of the flasks remained open to the air, but

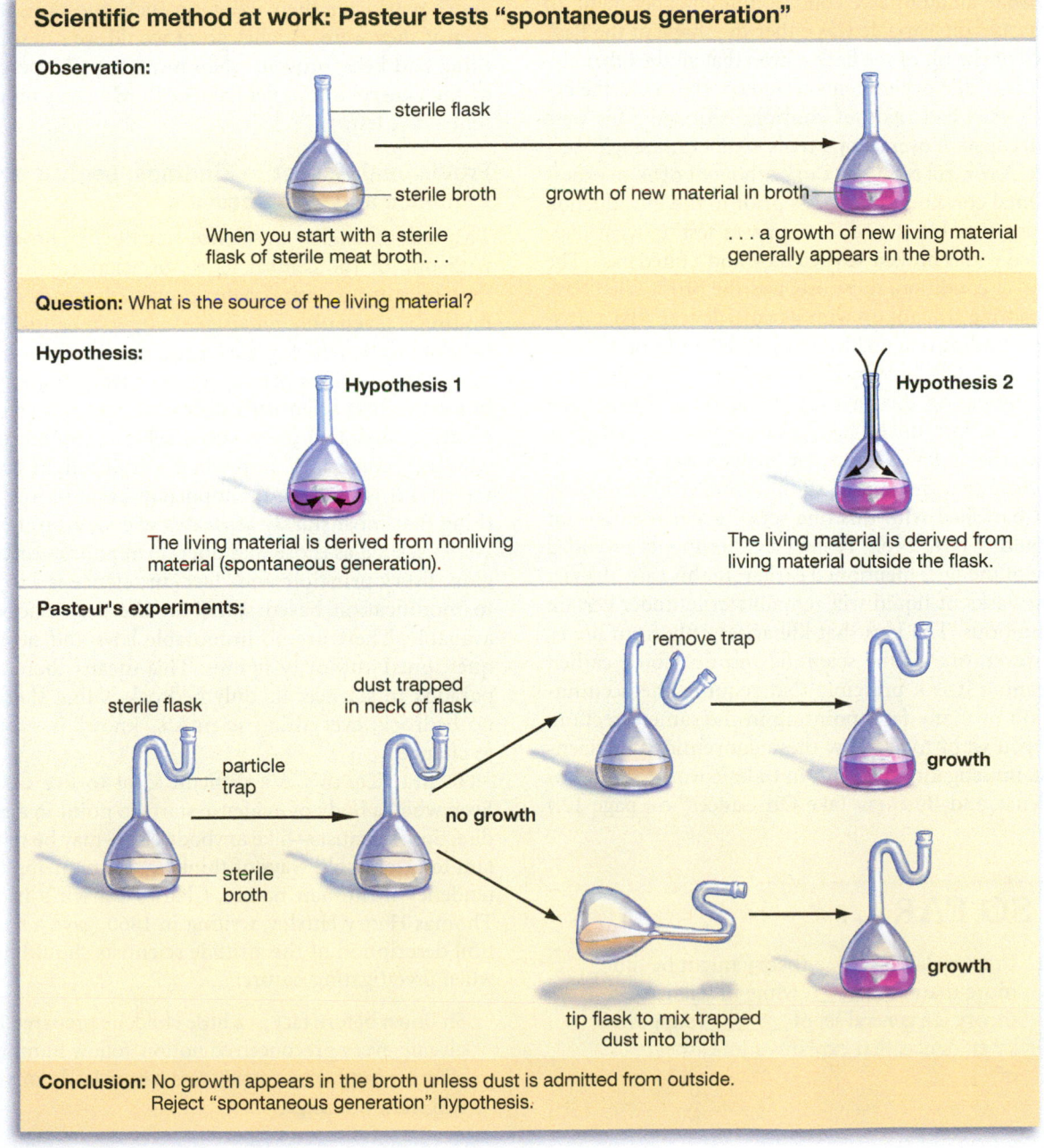

Scientific method at work: Pasteur tests "spontaneous generation"

Observation:

sterile flask

sterile broth

growth of new material in broth

When you start with a sterile flask of sterile meat broth . . .

. . . a growth of new living material generally appears in the broth.

Question: What is the source of the living material?

Hypothesis:

Hypothesis 1

The living material is derived from nonliving material (spontaneous generation).

Hypothesis 2

The living material is derived from living material outside the flask.

Pasteur's experiments:

sterile flask

particle trap

sterile broth

dust trapped in neck of flask

no growth

remove trap

growth

tip flask to mix trapped dust into broth

growth

Conclusion: No growth appears in the broth unless dust is admitted from outside. Reject "spontaneous generation" hypothesis.

Figure 1.4
Pasteur's Experiments and the Scientific Method
Nineteenth-century observation made clear that life would appear in a medium, such as broth, that had been sterilized, but what was the source of this life? One hypothesis was that it arose through *spontaneous generation,* meaning it formed from the simple chemicals in the broth. Conversely, Pasteur hypothesized that it originated from airborne microorganisms. He designed an experiment that offered evidence for this hypothesis. The device he used was an S-shaped flask, which enabled air to enter the flask freely while trapping all particles (including invisible microorganisms) in a bend in the neck.

inside the flasks there was not a sign of life. Why? The broth remained sterile because microbe-bearing dust particles got trapped in the bend of the flask's neck. When Pasteur broke the neck off before the bend, however, the flask soon had a riot of bacterial life growing within it. In another test, Pasteur tilted the flask so that the broth *touched* the bend in the neck, a change that likewise got the microbes growing.

Elements in Pasteur's Experiments

Now, note what was at work here. Pasteur had a preconceived notion of what the truth was, and he designed experiments to test his hypothesis. Critically, he performed the same set of steps several times in the experiments, keeping all elements the same each time—except for one. The nutrient broth was the same in each test; it was heated the same amount of time and in the same kind of flask. What *changed* each time was one critical **variable:** an adjustable condition in an experiment. In this case, the variable was either the shape of the flask neck or the tilt of the flask. Given that all the other elements of the experiments were kept the same, the experiments had rigorous controls: All conditions were held constant over several trials except for a single variable. A control *condition* can be thought of as an experimental condition that exists prior to the introduction of any variables tested. Pasteur was testing what happened with a broken-necked flask and a tilted flask. The control condition, therefore, was the broth-filled flask left sitting straight up with its particle trap intact. Pasteur's finding that no life grew in this condition is interesting but tells little by itself. We learn something only by comparing this finding to the result Pasteur got when he introduced his variables: that life *did* grow when the neck was broken or the flask was tilted.

Note that the idea of spontaneous generation was not banished with this one set of experiments—nor should it have been. Pasteur's experiments provided one of the *facts* mentioned earlier; in this case, the fact that flasks of liquid will remain sterile under certain conditions. The idea that life arises only from life is, however, one of the scientific *theories* noted earlier, meaning it is a principle that requires the accumulation of many facts pointing in the same direction. (If you want to see how these conventions of scientific investigation play out in today's world, see "How Science and Business Take On Cancer" on page 12.)

SO FAR . . .

1. In everyday speech, a theory might be little more than a _____, while in science a theory is a general set of _____, supported by evidence, that explains some aspect of _____.

2. In science, a tentative, testable explanation for an observed phenomenon is referred to as a _____.

3. All properly executed scientific experiments must have rigorous controls, meaning that all aspects of the experiment must be held _____ except for the condition, known as a _____, which is being tested for.

When Is a Theory Proven?

At what point does a theory become proven? An irony of the orderly undertaking called science is that there's nothing orderly about this transition. No scientific supreme court exists to make a decision. Scientists aren't polled for their views on such questions, and even if they were, at what point would we say something had been proven: when more than 50 percent of the experts in the field assent to it? When no dissenters are left?

Provisional Assent to Findings: Legitimate Evidence and Hypotheses

This lack of finality in science actually fits, however, with one of the central tenets of science, which is that nothing is ever finally proven. Instead, every finding is given only *provisional* assent, meaning it is believed to be true for now, pending the addition of new evidence. This principle is so deeply embedded in science that scientists rarely have reason to think about it (just as drivers would seldom contemplate why they come to a stop when a traffic light turns red). Yet it is profoundly important because it is the thing that most starkly separates science from belief systems, such as those that operate in politics or religion. Every principle and "fact" in science is subject to modification based solely on the best evidence available. There are no immutable laws and no unquestioned authority figures. This means there is a paradox in science: Its only bedrock is that there is no bedrock; everything scientists "know" is subject to change.

In practice, this is a difficult ideal to live up to. Even when a body of evidence starts to point in a new direction, scientists—like anybody else—may be reluctant to give up old ways of thinking. Recognizing this tendency in human nature, Charles Darwin's friend Thomas Henry Huxley, writing in 1860, gave a beautiful description of the attitude scientists should have when investigating nature:

> Sit down before fact as a little child, be prepared to give up every preconceived notion, follow humbly wherever and whatever abysses nature leads, or you will learn nothing.

This principle of science's openness to revision is one of three important scientific principles having to do with the scientific process. Here are all three:

- Every assertion regarding the natural world is subject to challenge and revision based solely on evidence.

- Any scientific hypothesis or claim must be *falsifiable,* meaning open to negation through scientific inquiry. The assertion that "UFOs are visiting Earth" does not rise to the level of a scientific claim because there is no way to prove that this is *not* so.

- Scientific inquiry concerns itself only with natural explanations for natural phenomena. Put another way, *supernatural* explanations for the workings of nature lie outside the realm of science and thus cannot be examined through the scientific process.

At first blush, this third principle may seem a little strange. If science is about the testing of hypotheses through open inquiry, then when carrying out such inquiry, why can't scientists test all possible explanations—including supernatural explanations? For example, if we find (as we do) that the Hawaiian Islands have a wildly uneven representation of living things—there are 500 species of one kind of fly on the islands but not a single native species of reptile—then why can't scientists investigate a claim such as this: Hawaii has the mix of creatures it does because an intelligent designer arranged things in this way?

The primary answer to this question is that an intelligent designer would, by definition, have abilities that lie outside those in the natural world. Any entity that could arrange species on the Hawaiian Islands as it saw fit would be free to violate all the principles that underlie science: the laws of physics and chemistry, to say nothing of biology. And if this is the case, then scientists would have little more ability than, say, artists to investigate intelligent design in Hawaii because scientific principles need not apply to the question at hand.

SO FAR...

1. Every finding in science is subject to _____ based on the accumulation of new _____.

2. Any scientific claim must be falsifiable, meaning open to _____ through means of scientific inquiry.

3. Science concerns itself only with _____ explanations for natural phenomena, as opposed to _____ explanations for these phenomena.

1.3 The Nature of Biology

Let's shift now from an overview of science to a more narrow focus on **biology,** which can be defined as the study of life. But what is life? It may surprise you to learn that there is no standard short answer to this question. Indeed, the only agreement among scholars seems to be that there is not, and perhaps cannot be, a short answer to this question. The main impediment to such a definition is that any quality common to all living things is likely to exist in some non-living things as well. Some living things may "move under their own power," but so does the wind. Living things may grow, but crystals and fire do the same thing. Therefore, biologists generally define life in terms of a group of characteristics possessed by living things. Looked at together, these characteristics are sufficient to separate the living world from the non-living. We can say that living things:

- Can assimilate (take in) and use energy
- Can respond to their environment
- Can maintain a relatively constant internal environment
- Possess an inherited information base, encoded in DNA, that allows them to function
- Can reproduce through use of the information encoded in DNA
- Are composed of one or more cells
- Evolved from other living things
- Are highly organized compared to inanimate objects

Every one of these qualities exists in all the varieties of Earth's living things. The simplest bacterium needs an energy source no less than any human being. Our energy source is the food that's familiar to us; the bacterium's might be the remains of vegetation in the soil. The bacterium responds to its environment, just as we do. You would take action if you smelled gas in your house; the bacterium would move away if it encountered something it regarded as noxious. Humans maintain a relatively stable internal environment by, for example, sweating when they are hot. When the bacterium's external environment becomes too hot, it has certain genes that will switch on to keep it functioning. Both humans and bacteria use the molecule DNA as a repository of the information necessary to allow them to live. Bacteria and human beings both reproduce—bacteria through simple cell division, human beings through the use of two kinds of reproductive cells (egg and sperm). A bacterium is a single-celled life form, while humans are a 10-trillion-celled life form. Bacteria and humans both evolved from complex living things and ultimately share a single common ancestor.

There are some exceptions to these "universals." For example, the overwhelming majority of honeybees

The Levels of Life Card Game

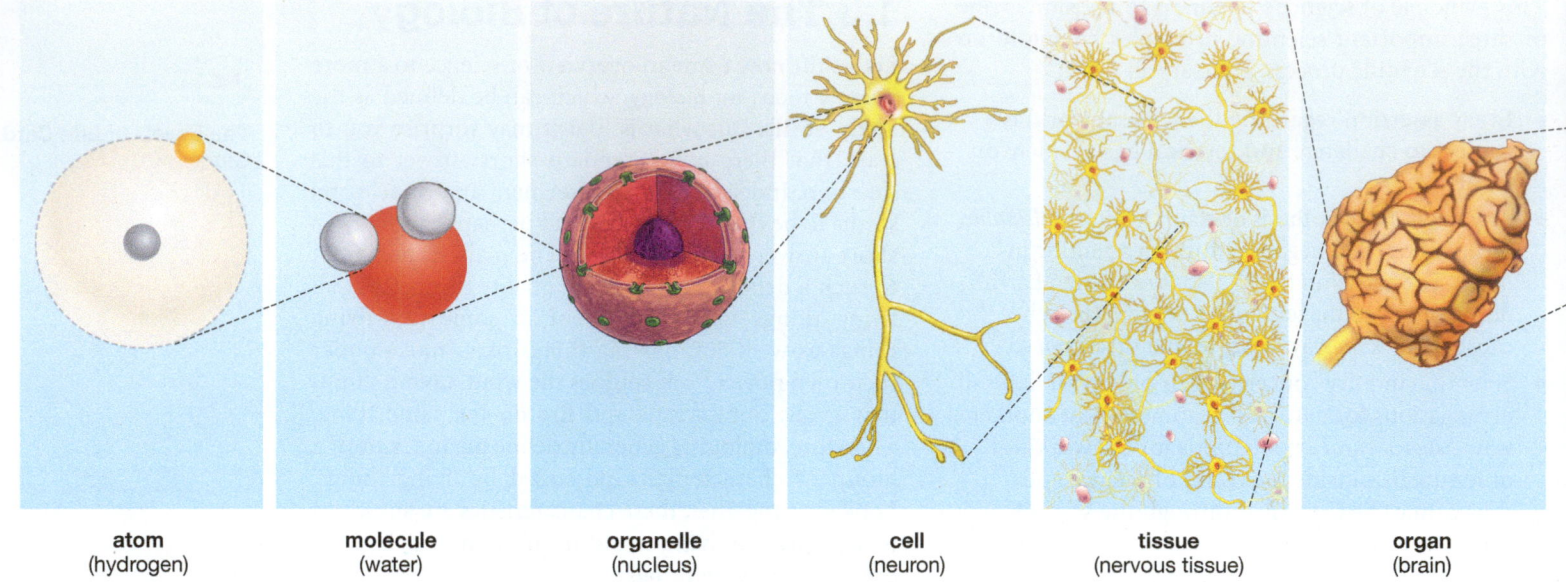

| atom (hydrogen) | molecule (water) | organelle (nucleus) | cell (neuron) | tissue (nervous tissue) | organ (brain) |

Figure 1.5
Levels of Organization in Living Things

What Is Life?

and ants are sterile females; they can't reproduce, but no one would doubt that they're alive. In general, however, if something is living, it has all these qualities.

Life Is Highly Organized in a Hierarchical Manner

One item on the list of qualities requires a little more explanation. Living things are indeed highly organized compared to inanimate objects. But to put a finer point on this, living things are organized in a "hierarchical" manner, meaning one in which lower levels of organization are progressively integrated to make up higher levels. The main levels in this hierarchy could be compared to the organization of a business. In a corporation, there may be several individuals making up an office, several offices making up a department, several departments making up a division, and so forth. In life, there is one set of organized "building blocks" making up another, as you can see in **Figure 1.5**.

When we begin to think about life in terms of the integration of these many structural levels, we can see that it is not just "highly" organized. Nothing else comes *close* to it in organizational complexity. The sun is a large thing, but it is an uncomplicated thing compared to even the simplest organism. Consider that you have about 10 trillion cells in your body and that, with some exceptions, each type of cell has in it a complement of DNA that is made up of chemical building blocks. How many building blocks? Three billion of them. Now, you probably know that most cells divide regularly: one cell becoming two, the two becoming four, and so on. Each time this happens, each of the 3 billion DNA building blocks must be faithfully *copied* so that both cells resulting from cell division have their own complete copy of DNA. And this copying of the molecule—before anything is

actually done with it—is carried out in almost all the varieties of the 10 trillion cells we have. Complex indeed. Let's see what life's levels of organization are.

Levels of Organization in Living Things

The building blocks of matter, called *atoms,* lie at the base of life's organizational structure (see Chapter 2 for more about them). Atoms come together to form *molecules,* which are entities consisting of a defined number of atoms that exist in a defined spatial relationship to one another. A molecule of water is one atom of oxygen bonded to two atoms of hydrogen, with these atoms arranged in a precise way. Molecules in turn form *organelles,* meaning "tiny organs," in a cell. Each of your cells has, for example, a structure called a nucleus that contains the cell's primary complement of DNA. Such an organelle is not just a collection of molecules that exist close to one another. It is a highly organized structure, as you can tell just from looking at the rendering of it in the sea lion in Figure 1.5.

Atoms, molecules, and organelles are all component parts of life, but they are not themselves living things. At the next step up the organizational chain, however, we reach the entities that are living. These are *cells,* the fundamental units of life; the simplest, smallest entities that carry out all of life's basic processes. Indeed, if we ask where life exists *outside* cells, the answer most experts would give is: nowhere at all. Every product of life—the material that makes up our bones, the wood that helps make up a tree—comes from cells, and every process that enables life is initiated by cells.

Large organisms such as the sea lion in Figure 1.5 then go on to have *tissues,* meaning *collections* of cells that serve a common function. The sea lion has, for example, collections of the nerve cells called neurons,

an system
(ous system)

organism
(sea lion)

population
(colony)

community
(giant kelp forest)

ecosystem
(Southern California coast)

biosphere
(Earth)

all of which serve the common function of transmitting electrical signals. Each collection of these cells is referred to as neural tissue. Several kinds of tissues can then come together to form a functioning unit known as an *organ*. The sea lion's brain, for example, is made up not only of neurons but of cells called *glia* that help support neurons. Several organs and related tissues then can be integrated into an *organ system*. The sea lion's brain, its spinal cord, and all the nerves that extend from these organs constitute the sea lion's nervous system. An assemblage of cells, tissues, organs, and organ systems can then form a multicelled *organism* such as the sea lion. (However, back down at the cell level, a one-celled bacterium is also an organism; it's just not one that has organs, tissues, and so forth.)

From this point out, life's levels of organization all involve many organisms living together. Members of a single type of living thing (a species), living together in a defined area, make up what is known as a *population*. When you look at *all* the kinds of living things in a given area, you are looking at a *community*. When you consider the members of a community and the non-living elements with which they interact (such as climate and water), the result is an *ecosystem*. Finally, all the communities of Earth—and the physical environment with which they interact—make up the *biosphere*.

Life's Spectacular Diversity

Life at the level of populations, communities, and ecosystems is tremendously complex for the simple reason that life at these levels is tremendously diverse. We can get an intuitive sense of this diversity by just thinking about the stunning *forms* that life comes in: whales, bats, trees, toadstools, algae. But life's diversity also exists in forms that are less visible to us. A single gram of fertile midwestern soil—an amount of soil

about the size of a quarter of a teaspoon—is likely to contain 10,000 different *species* of bacteria, along with an assortment of roundworms, fungi, insects, and perhaps plant root fragments. When we look at the living world as a whole, we find that the lowest estimate for the total number of species in it is about 4 million, while higher-end estimates come in at 10 to 15 million, and the highest of them all is 100 million. We really have no idea which of these estimates best approximates the truth, however. In all the time that scientists have been looking at the living world, they simply have been unable to catalogue its vast diversity.

1.4 Special Qualities of Biology

Biology traces its origins to the ancient Greeks. In the work of such Greeks as Hippocrates and Galen, we can find the origins of modern medical science. In the work of Aristotle and others, we can find the origins of "natural history," which led to what we think of today as mainstream biology and the larger category of the **life sciences:** a set of disciplines that focus on varying aspects of the living world. Apart from biology, the life sciences include such areas of study as veterinary medicine and forestry.

Despite its ancient origins, biology is, in a sense, a much younger science than, say, physics, which is one of the physical sciences, meaning the natural sciences not concerned with life. Western Europe's revolution in the physical sciences probably can be dated from the sixteenth century, when Nicholas Copernicus published his work *On the Revolution of Heavenly Spheres,* which demonstrated that Earth moves around the sun. Meanwhile, biology did not come into its own as a science until the *nineteenth* century. Prior to the

THE PROCESS OF SCIENCE

How Science and Business Take On Cancer

At one time, getting a diagnosis for the form of cancer known as chronic myeloid leukemia, or CML, was tantamount to receiving a death sentence, because up through the late 1990s this disease of white blood cells was inevitably fatal. On average, three to five years elapsed between the date patients were diagnosed with CML and the date they died from it. Today, however, 85 to 90 percent of CML patients are alive five years after diagnosis, and it seems likely that the vast majority of these patients will be alive a great deal longer. The reason their future looks bright is that CML has been transformed. Whereas once it was a deadly disease, it now has taken on the qualities of a chronic, manageable disorder—a condition that can be controlled in the manner of, say, diabetes or asthma.

What has made the difference with CML is a single drug, called Gleevec, that was approved for medical use by the U.S. Food and Drug Administration in 2001. Gleevec is extremely expensive—it would not be unusual for a single patient's supply to cost $30,000 per year—and it is not a cure for CML, because patients must keep taking it for as long as they live. But for most of the 5,000 Americans diagnosed with CML each year, it has meant the difference between life and death.

For a single drug to have made this kind of difference in the treatment of a particular cancer is almost without precedent in cancer therapy. But Gleevec is remarkable in several other respects as well. To an unusual degree, it resulted from the vision and persistence of a single researcher, Brian Druker, who since 1993 has been at the Oregon Health and Science University in Portland, Oregon (**Figure 1**). Apart from this, Gleevec is now seen as the drug that pointed the way toward the future in cancer research, in that it was the first drug to target what might be called the circuitry of a particular cancer—the chemical pathway by which a given form of cancer gets started. (For more on this aspect of Gleevec, see "When the Cell Cycle Runs Amok" on page 168.) Despite all this, however, there are elements of Gleevec's history that could be found in connection with any cancer drug. In some ways, its development is a textbook example of how science and big business work together in tackling the daunting challenge of cancer.

Gleevec's story began in the 1980s, when Druker, then fresh out of medical school at the University of California, San Diego, was carrying out a type of research that underlies the development of any cancer treatment. This is so-called basic research, which is aimed at coming to a better understanding of either cancer in general or of a particular cancer—how does it get started, how does it change over time, what might its weak points be? Such research can be funded by private foundations, but in the United States today, the largest single funder of this research by far is the federal government—specifically a branch of the federal government known as the National Institutes of Health (NIH), which is made up of 27 separate institutes and centers. In 2008, NIH's National Cancer Institute provided researchers across the country with about $4.9 billion in cancer research funding. This large federal investment in cancer research, however, does not extend to developing the cancer *drugs* that end up being administered by physicians. In nearly every case, the research on such drugs is funded by pharmaceutical companies, which are almost invariably *large* companies because of the costs involved: Some experts put the current cost of bringing a single drug to market at an average of $1.2 billion.

Gleevec was a product of an intertwining of both this "applied," pharmaceutical research and basic, NIH-funded research. In the 1980s, Druker was carrying out basic research on a family of proteins, called the tyrosine kinases, that promote the process by which human cells divide—one cell becoming two, two becoming four, and so on. At the same time, the Swiss pharmaceutical giant Ciba-Geigy (which later became Novartis) was likewise working with tyrosine kinases and came to Druker's lab, beginning in 1986, to seek out expertise on the effectiveness of a set of drugs that were aimed at *inhibiting* the activity of tyrosine kinases. The idea was that drugs that could disrupt the activity of these kinases could be used to stop the uncontrolled division of cells that is a hallmark of all forms of cancer.

Figure 1
Cancer Research Pioneer
Cancer researcher and physician Brian Druker of the Oregon Health and Science University.

By 1990, Druker was focusing strictly on CML and had come to believe it might be arrested by means of disabling a mutant protein, called BCR-ABL, that becomes active in the white blood cells of all CML patients. The challenge was to find a compound that could carry out this disabling task. When, in 1993, Druker began to search for such a compound, he contacted Ciba-Geigy biochemist Nicholas Lydon, who sent him several tyrosine kinase inhibitors developed in his lab, one of which went on to show promise in laboratory

that serve as the make-or-break test for all human medicines: clinical trials.

No drug can be prescribed for use in the United States without having first been approved by the U.S. Food and Drug Administration. And to gain FDA approval, every drug must be put through a series of rigorous tests on human beings that demonstrate, first, that the drug is safe and, second, that it has a positive effect on the condition it is intended to combat. The gold standard for demonstrating both things to the FDA is the

of CML patients who had not responded to the standard therapy of the time, which was a form of the drug interferon. This trial, which went on for a little less than a year, was intended only to measure the safety of Gleevec, but a strange thing happened during this test: Gleevec demonstrated an astounding *effectiveness* among the patients. White blood cell counts returned to normal in 53 of 54 patients who received high doses of Gleevec—a 98 percent response rate in a type of trial that would have been counted as a success had it achieved a 10 percent response rate. With the Phase I results in, Novartis moved quickly into Phase II trials, and the results there were equally stellar. So effective was Gleevec in Phase II that it gained FDA approval as a prescription drug before its Phase III trials were ever completed. Three years elapsed from the time a volunteer first took a dose of Gleevec to the time it was approved for use in CML patients. By way of comparison, the FDA normally takes between five and eight years to approve a drug.

Gleevec is now seen as the drug that pointed the way toward the future in cancer research.

tests that Druker and Lydon reported on together. This was the compound that became Gleevec. The problem was that Ciba-Geigy's successor, Novartis, had little interest in proceeding with *human* tests of this substance—that is, with tests of it in CML patients. One of the problems with such testing was that no one had ever given a tyrosine kinase inhibitor to a human being before, and the assumption of many experts was that doing so could be dangerous. Apart from this, however, proceeding with the development of Gleevec seemed to make little sense to Novartis as a business proposition. The 5,000 Americans who are diagnosed with CML each year represent a small number of cancer patients compared to, say, the 46,000 Americans who are diagnosed with kidney cancer or the 215,000 who are diagnosed with lung cancer. Thus, even if Gleevec worked—which was far from certain following the lab tests of it—demand for it was bound to be small, while the costs of developing it were bound to be large. As a simple matter of economics, then, Novartis had little incentive to find out whether it would be effective as a cancer treatment. Eventually, however, Druker convinced the company to see it through the expensive set of evaluations

clinical trial, which is a set of tests that typically operate in three stages: a Phase I trial that seeks to demonstrate nothing more than the safety of the drug, as measured by its effect among a few dozen patients; a Phase II trial that expands the patient-base to several hundred volunteers and that is looking for both safety and effectiveness; and a Phase III trial that often enrolls thousands of patients at dozens of clinical settings that may be located around the world. These trials are models of the scientific method in that they must adhere to two watchwords: they must be *controlled* and they must be *randomized*. They are controlled in that the drug must demonstrate its effectiveness *compared to* another therapy—generally the therapy that is standard for the disease in question at the time. And they are randomized in that patients will randomly be assigned to get either the new drug or the standard therapy. The costs for all three phases of a clinical trial are borne entirely by the pharmaceutical firm developing the drug in question, and these costs are considerable; a given company could easily spend $400 million testing a drug that goes through Phase III trials.

Novartis began its Phase I trial of Gleevec in June 1998 with a small group

Today, Gleevec is used not just for CML, but for nine other forms of cancer as well; as a result, it is now administered to some 200,000 cancer patients worldwide. Druker receives no income from any sales of Gleevec, as he began his work with it only after it had been patented as a potential medicine. His achievements in connection with it have, however, been widely recognized by both cancer patients and cancer researchers. In 2009, he and Nicholas Lydon were named winners of the Lasker-DeBakey Clinical Medical Research Award—sometimes called the American Nobel Prize—for their work in developing Gleevec. The two shared the award with Charles Sawyers of Memorial Sloan-Kettering Cancer Center in New York, who was instrumental in developing compounds that are effective in treating CML patients whose cancers have become resistant to Gleevec.

1800s, biology was almost purely "descriptive," meaning that the naturalists whom we would today call biologists largely confined themselves to describing living things—what kinds there were, where they lived, what features they had, and so forth. Beginning in about the 1820s, however, biologists began to formulate biological theories as that term was defined earlier. They began to postulate that all life exists within cells, that life comes only from life, that life is passed on through small packets of information that we now call genes, and so on. To put this another way, biologists in the nineteenth century began describing the *rules* of the living world, whereas before they were largely describing *forms* in the living world.

This change moved biology closer to the same scientific footing as physics, but biology was then, and remains now, a very different kind of science from any of the physical sciences, with physics a clear case in point. One reason for this difference is that the component parts of physics are uniform and far fewer than is the case in biology. Physics deals with only 92 stable elements, such as hydrogen and gold, and to a first approximation, if you've seen one electron, you've seen them all.

Meanwhile, in biology, if you've seen one species, you've seen just that—one species. Each species is at least marginally different from another, and many are greatly dissimilar. Moreover, each species has all the organizational levels of elements in physics *and more*. (They not only have electrons and atoms, but organelles, cells, tissues, and so on.) Biology is concerned with the rules that govern all species, and

you've seen that there are some biological "universals." However, when cancer researchers are looking for the principles that underlie cell division, they are likely to be looking at only one of two main kinds of cells; when ecologists are looking at what causes dry grassland to turn into desert, their findings are likely to have little relevance to the rain forest. Put simply, the living world is tremendously diverse compared to the non-living world, and such diversity means that biology is concerned less with universal rules than is the case in the physical sciences.

Evolution: Biology's Chief Unifying Principle

Almost all biologists would agree that the most important thread that runs through biology is **evolution**: the gradual modification of populations of living things over time, with this modification sometimes resulting in the development of new species. Evolution is central to biology because every living thing has been shaped by it. (There are no known exceptions to this universal.) Given this, the explanatory power of evolution is immense. Why do peacocks have their finery, or frogs their coloration, or trees their height (**Figure 1.6**)? All these things stand as wonders of nature's diversity, but with knowledge of evolution they are wonders of diversity that *make sense*. For example, why do so many unrelated stinging insects look alike? Evolutionary principles suggest that they *evolved* to look alike because of the general protection this provides from predators. Think of yourself for a moment as a bee predator.

(a) A peacock displaying his plumage

(b) A red-eyed tree frog from Central America

(c) Pine trees in the South Pacific

Figure 1.6
Evolution Has Shaped the Living World

(a) Leafcutter bee

(b) Paper wasp

Figure 1.7
Similar Enough to Yield a Benefit
These are two of the many stinging insects that have the black-and-yellow striped coloration that warns away predators.

Having once gotten stung, would you annoy any roundish insect that had a black-and-yellow striped coloration? You probably learned your lesson about this in connection with one species, but many species of insects are now protected from you simply by virtue of the coloration they share (**Figure 1.7**). Thus, there were reproductive benefits to individual insects that, through genetic chance, happened to get a slightly more striped coloration: They left more offspring because they were bothered less by predators. Over time, entire populations moved in this direction. They evolved, in other words.

The means by which living things can evolve is a topic this book takes up beginning in Chapter 16. For now, just keep in mind that a consideration of evolution is never far from most biological observations. So strong is evolution's explanatory power that in uncovering something new about, say, a sequence of DNA or the life cycle of a given organism, one of the first things a biologist will ask is: What evolutionary advantage can this have provided?

On to Chemistry

How do you get a handle on something as diverse and complex as life? One way is to start small, with life's building blocks, and then see how these units come together to make up complete organisms. We'll be taking this approach over the next several chapters—starting small and working our way up, you might say. We'll actually be starting *very* small in this tour, with the building blocks called atoms. For us, the central question concerning these tiny entities will be: How do they come together to make up life's larger molecules? As you'll see, the short answer to this question is: They obey the laws of chemistry.

SO FAR . . .

1. Arrange the following levels of organization in living things, going from least inclusive to most inclusive: cell, community, molecule, organ, ecosystem, organism.

2. The most important principle in biology, the theory of evolution, concerns the gradual _____ of populations of living things over time, sometimes resulting in the development of new _____.

3. Biology can be defined as the study of _____.

Summary

1.1 How Does Science Impact the Everyday World?

- Science is increasingly important to society, which often turns to scientists to answer questions about health, the environment, and other domains of life. (p. 3)

1.2 What Is Science?

- Science is a body of knowledge: a collection of unified insights about nature. Science is also a way of learning: a process of coming to understand the natural world through observation and experimentation. (p. 6)

- The unified insights of science are known as theories. A theory is a general set of principles, supported by evidence, that explains some aspect of nature. (p. 6)

- Science uses the scientific method, in which an observation leads to a question about the natural world. Then comes a hypothesis—a tentative, testable explanation that most often will be tested through a series of experiments. (p. 6)

- In science, every assertion is subject to challenge and revision; scientific claims must be falsifiable, meaning open to negation through scientific inquiry; and scientific inquiry is limited to investigating natural explanations for natural phenomena. (p. 9)

1.3 The Nature of Biology

- Biology is the study of life. Life is defined by a group of characteristics: Living things can assimilate energy, respond to their environment, maintain a relatively constant internal environment, and reproduce. In addition, they possess an inherited information base, encoded in DNA, that allows them to function; they are composed of one or more cells; they are evolved from other living things; and they are highly organized compared to inanimate objects. (p. 9)

- Life is organized in a hierarchical manner in increasing complexity: from atoms to molecules, organelles, cells, tissues, organs, organ systems, organisms, populations, communities, ecosystems, and the biosphere. (p. 10)

1.4 Special Qualities of Biology

- The living world is exceedingly diverse, and as such, biology focuses less on universal rules than is the case in the physical sciences. (p. 14)

- Biology's chief unifying principle is evolution, defined as the gradual modification of populations of living things over time, with this modification sometimes resulting in the development of new species. Evolution helps explain the forms and processes seen in living things. (p. 14)

Key Terms

biology   9

evolution   14

hypothesis   6

life sciences   11

science   6

scientific method   6

theory   6

variable   8

Understanding the Basics

Multiple-Choice Questions

(Answers are in the back of the book.)

1. Which of the following statements best describes the nature of a scientific hypothesis?
   a. A hypothesis is an idea that is widely accepted as a description of objective reality by a majority of scientists.
   b. A hypothesis must stand alone and not be based on prior knowledge.
   c. A hypothesis is a tentative explanation that can be tested, usually through experimentation.
   d. A hypothesis must deal with an aspect of the natural world never dealt with before.
   e. A hypothesis, when accepted, becomes a scientific law.

2. A theory, as defined in scientific discourse, is:
   a. an established fact about the natural world, such as the distance from Earth to the sun.
   b. a long-accepted belief about the natural world.
   c. a concept that is in doubt among most scientists.
   d. a set of principles, supported by evidence, that explains some aspect of nature.
   e. an initial assumption about how some aspect of nature works.

3. Pasteur's experiments on spontaneous generation made correct use of a variable in that Pasteur:
   a. varied the bacteria he employed with each experiment.
   b. used statistics to prove his hypothesis.
   c. observed the bacteria as they were growing in the flasks.
   d. held all conditions constant in each test except one.
   e. was willing to vary to the extent necessary from the standard hypotheses of his day.

4. Which of the following characteristics are true of all scientific claims? (Select all that apply.)
   a. They are capable of negation through further scientific inquiry.
   b. They can be negated by expert opinion.
   c. They are natural explanations for either natural or supernatural phenomena.

d. They stand or fall solely on the basis of evidence.

e. They are regarded as provisional, pending the addition of new evidence.

5. Evolution is a central, unifying theme in biology because:
   a. it is not a falsifiable hypothesis.
   b. humans have evolved from ancestors we share with present-day monkeys.
   c. the enormously diverse forms of life on Earth have all been shaped by it.
   d. it has occurred in the past, even though it no longer operates today.
   e. almost all biologists believe in it.

6. Biologists generally define life in terms of a group of characteristics possessed by living things. Which of the following is *not* a characteristic of living things?
   a. All living things possess an inherited information base, encoded in DNA, that allows them to function.
   b. All living things can respond to their environment.
   c. All living things can maintain a relatively constant internal environment.
   d. All living things evolved from other living things.
   e. All living things are composed of more than one cell.

Brief Review

(Answers are in the back of the book.)

1. What is science? In what ways is science different from a belief system such as religious faith?

2. What is a controlled experiment?

3. How did Louis Pasteur cast doubt on the idea of spontaneous generation?

4. Describe the defining features of life as we know it on Earth.

5. Living things are organized in a hierarchical manner. List all the levels of the biological hierarchy that you can.

Applying Your Knowledge

1. Would you agree that it is valuable for a nation to have a citizenry that is reasonably well versed in science? Give reasons for your answer. Would you say this need has become especially urgent in the last two decades? If so, why?

2. Although science cannot investigate the supposed workings of the supernatural, the scientific method can be used to investigate some *claims* that the supernatural has been at work. Can an astrologer know something about you just by knowing the place and time of your birth? A scientific test of this question, carried out in the 1980s, involved providing a group of 30 top-level astrologers with nothing but birth information for a group of people and then seeing whether the astrologers could predict anything about the personalities of these people (as measured by a standard personality "inventory"). The result was that the astrologers did no better than chance in trying to predict personality. Many of the standard tools of science were at work in this test of astrology—for example, controls (the test was the same for each astrologer) and statistical analysis (used to see whether the astrologers did better than chance). Using this test as a case in point, are all claims of supernatural effects open to scientific investigation? Can you think of other claims that could be investigated or any that could not?

3. If you were sent on an interplanetary mission to investigate the presence of life on Mars, what would you look for? Would you explore the land and the atmosphere? Imagine you discover an entity you suspect is a living being. Realizing that life elsewhere in the universe may not be organized by the same rules as on Earth, which of the features of life on Earth, if any, would you insist that the entity display before you would declare it living?

MasteringBiology®

1. MasteringBiology Assignments
Activities: Metric System Review; Scientific Method; The Levels of Life Card Game; What Is Life? **GraphIt!**: An Introduction to Graphing; **Current Events**: Genome Detectives Solve a Hospital's Deadly Outbreak; **Building Vocabulary; Questions**: Reading Questions, Test Bank

2. eText
Read your book online, search, take notes, highlight text, and study more effectively.

3. The Study Area
Self-study tools to test comprehension.

2

Fundamental Building Blocks:
Chemistry, Water, and pH

Life is carried on through chains of chemical reactions. The qualities of water have shaped life. The degree to which substances are acidic or basic has strong effects on living things.

Available on MasteringBiology®

Activities: Structure of the Atomic Nucleus; Structure of Atoms, Elements, Isotopes; Covalent Bonds; Electron Arrangement; Nonpolar and Polar Molecules; Ionic Bonds; Hydrogen Bonds; Chemical Bonding; Dissociation of Water Molecules; Cohesion of Water; The Polarity of Water; Acids, Bases, and pH; **Chemistry Review:** Atoms and Molecules; Reactions & Equilibrium; Water; Acids, Bases, & pH; **Current Events:** A Gold Rush in the Abyss; Citizens' Testing Finds 20 Hot Spots Around Tokyo; Moonlighting as a Conjurer of Chemicals; **Building Vocabulary; Questions:** Reading Questions, Test Bank

Life began as a series of chemical reactions in water, and water remains indispensable to all living things today. Here, a red-eyed tree frog (*Agalychnis callidryas*) sits under a plant leaf during a rainstorm.

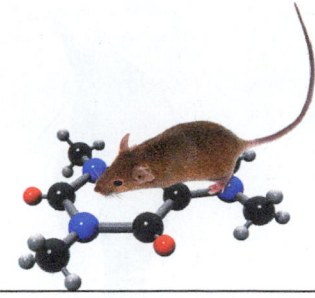

Cities are made of buildings and buildings are made of bricks; bricks are made of earth, and earth is made of ...? To answer this question, in this chapter we will look at what the material world is made of. Biology is our subject, but to fully understand it, you need to learn a little about what underlies biology. You need to learn a little about what *biology* is made of, in a sense. To do this, you need to understand some of the basics in another field: chemistry.

How is chemistry relevant to biology? Well, the average person probably is aware that living things are made up of individual units called cells, but beyond this bit of knowledge, reality fades and a kind of fantasy takes over. In it, the cells that populate people, plants, or birds carry on their activities under the direction of their own low-level consciousness. A cell *decides* to move, it *decides* to divide, and so on. Not so. By the time our story is finished, many chapters from now, it will be clear that the cells that make up complex living things do what they do as the result of a chain of chemical reactions. Repulsion and bonding, latching on and re-forming, depositing and breaking down—what makes people, plants, and birds function at this cellular level is *chemistry*. Given this, the basic principles of chemistry are important to biology—important enough that we'll spend most of this chapter reviewing them. Once this is done, we'll touch on one of life's most important substances, water. Then we'll finish the chapter with a look at pH, a chemistry-related concept that has to do with how acidic or "basic" watery solutions are.

2.1 Chemistry's Building Block: The Atom

What is chemistry concerned with? Look around you. Do you see a table, light from a lamp, a patch of night or daytime sky? Everything that exists can be viewed as falling into one of two categories: matter or energy. You will learn something about energy in this chapter, but we are most concerned here with *matter and its transformations,* which is the subject of chemistry. Matter can be defined as anything that takes up space

and has **mass.** This latter term is a measure of the *quantity* of matter in any given object. How much space does an object occupy—how much volume, to put it another way—and how *dense* is the matter within that space? These are the things that define mass. For our purposes, we may think of mass as equivalent to weight, although physics makes a distinction between these two things.

As we behold matter all around us, it is natural to ask, What is its nature? A child sees a grain of sand, pounds it with a rock, sees the smaller bits that result, and wonders: What is this stuff like at the end of these divisions? Not surprisingly, adults also have wondered about this question—for centuries. About 2,400 years ago, the Greek philosopher Plato accepted the notion that all matter is made up of four primary substances: earth, air, fire, and water. A near-contemporary of his, Democritus, believed that these substances were in turn made up of smaller units that were both invisible and in*divi*sible—they could not be broken down further. He called these units *atoms* (**Figure 2.1**, on the next page).

Well, let's give at least one cheer for Democritus because he had it partly right. Centuries of painstaking work between his time and ours has confirmed that matter is indeed composed of tiny pieces of matter, which we still call atoms, but these atoms are not indivisible, as Democritus thought. Rather, they are themselves composed of constituent parts. A superficial account of *all* the parts scientists have discovered to date would go on for pages and still be incomplete. Physicists are continually slamming together parts of atoms with ever-greater force in an effort to determine what else there may be at the heart of matter. This is what the machines called "atom smashers" do. (The physicists who run them could be compared to people who, in trying to find out what parts a watch has, throw it on the ground and record the way its

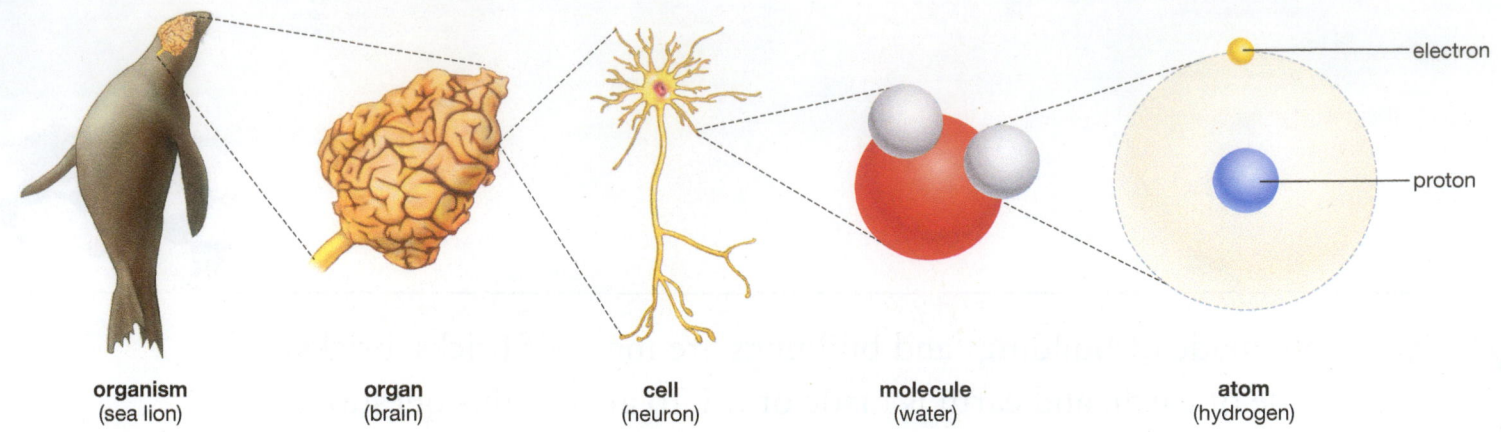

organism
(sea lion)

organ
(brain)

cell
(neuron)

molecule
(water)

electron

proton

atom
(hydrogen)

Figure 2.1
The Building Blocks of Life
Viewing this sea lion at increasing levels of magnification, we eventually arrive at the building block of all matter, the atom.
The atom selected here, from among a multitude that make up the sea lion, is a single hydrogen atom, composed of one
proton and one electron.

various mechanisms fly out on impact.) This is interesting stuff, but it is purely the business of physics, with little relation to biology. We are not concerned here with what's at the very end of matter's divisions. We do care a good deal, however, about what's nearly at the end of them.

Protons, Neutrons, and Electrons

Structure of the Atomic
Nucleus

For our purposes, there are three important constituent parts of an atom: **protons, neutrons,** and **electrons.** These three parts exist in a spatial arrangement that is always the same, regardless of what atom we're dealing with. Protons and neutrons are packed tightly together in a core (the atom's **nucleus**), and electrons move around this core some distance away (**Figure 2.2**). The one variation on this theme is the substance hydrogen, the lightest of all the kinds of matter we will run into. Hydrogen has no neutrons but rather only one proton in its nucleus and one electron in motion around it.

These three "subatomic" particles have mind-bending sizes and proportions. As the chemist P. W.

Atkins has pointed out, an atom is so small that 100 million carbon atoms would lie end to end in a line of carbon about this long: _____ (3 centimeters). Things are just as disorienting when we consider the size of an atom as a whole relative to the nucleus. The whole atom, with electrons at its edge, is 100,000 times bigger than the nucleus. So, if you were to draw a model of an atom *to scale* and began by sketching a nucleus of, say, half an inch, you'd have to draw some of its electrons more than three-quarters of a mile away.

Although the nucleus accounts for little of the space an atom takes up, it accounts for almost all of the mass an atom has. So negligible are electrons in this regard, in fact, that all of the mass (or weight) of an atom is considered to reside with the protons and neutrons of the nucleus.

The components of atoms have another quality that interests us: electrical charge. Protons are positively charged, and electrons are negatively charged. Neutrons—as their name implies—have no charge; they are electrically neutral. Because all these particles do not exist separately but *combine* to form an atom, as a whole the atom may be electrically neutral as well. The negative charge of the electrons balances out the positive charge of the protons. Why? Because in this state the number of protons an atom has is exactly equal to the number of electrons it has (although we'll see a different, "ionic" state later in this chapter). In contrast, the number of neutrons an atom has can vary in relation to the other two particles.

With this picture of atoms in mind, we can begin to answer the question that has been handed down to us through history: What is matter? We certainly have a common sense answer to this question. Matter is any substance that exists in our everyday experience. For example, the iron that goes into cars is matter. But what is it that differentiates this iron from, say, gold? The answer is that an iron atom has 26 protons in its nucleus, while a gold atom has 79.

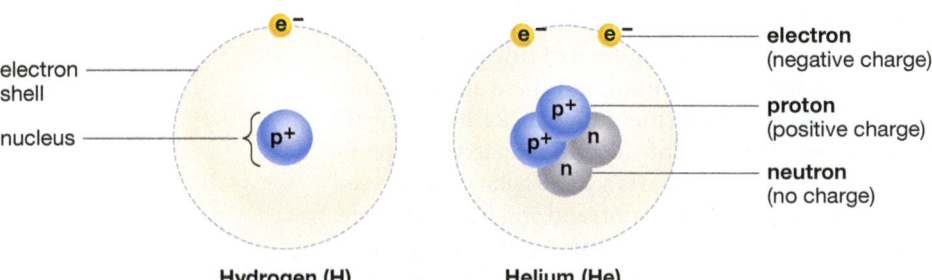

electron
shell

nucleus

e⁻

p⁺

Hydrogen (H)

e⁻   e⁻

p⁺
p⁺ n
n

Helium (He)

electron
(negative charge)

proton
(positive charge)

neutron
(no charge)

Figure 2.2
Representations of Atoms
One conceptualization of two separate atoms, hydrogen and helium. The model is not drawn to
scale; if it were, the electrons would be perhaps a third of a mile away from the nuclei. The model
also is simplified, giving the appearance that electrons exist in track-like orbits around an atom's
nucleus. In fact, electrons spend time in volumes of space that have several different shapes.

Fundamental Forms of Matter: The Element

Gold is an **element**—a substance that is "pure" in that it cannot be reduced to any simpler set of component substances through chemical processes. And the thing that defines each element is the number of protons it has in its nucleus. A solid-gold bar, then, represents a huge collection of identical atoms, each of which has 79 protons in its nucleus (**Figure 2.3**). In making gold jewelry, an artist may combine gold with another metal such as silver or copper to form an alloy that is stronger than pure gold, but the gold atoms are still present, all retaining their 79-proton nuclei.

Given what you've just read about protons, neutrons, and electrons, you may wonder why gold—or any other element—cannot be reduced to any simpler set of component substances. Aren't protons and neutrons components of atoms? Yes, but they are not component *substances* because they cannot exist by themselves as matter. Rather, protons and neutrons must *combine* with each other to make up atoms.

Assigning Numbers to the Elements

In the same way that buildings can be defined by a location and thus have a street number assigned to them, elements, which are defined by protons in their nuclei, have an **atomic number** assigned to them. Scientists have constructed the atomic numbering system so that it goes from the smallest number of protons to the largest. So, hydrogen, which has only one proton in its nucleus, has the atomic number 1. The next element, helium, has two protons, so it is assigned the atomic number 2. Continuing on this scale all the way through the elements found in nature, we end with uranium, which has an atomic number of 92. (You can see all of these elements laid out for you in a periodic table at the back of the book, in Appendix 2, on page AP3.)

Given this view of the nature of matter, we are now in a position to answer the question posed at the beginning of the chapter: What is a handful of earth—or anything else—made of? The answer is: one or more elements. If you look at **Figure 2.4**, you can see the most important elements that go into making up both Earth's crust and human beings.

Figure 2.3
Pure Gold
Gold is an element because it cannot be reduced to any simpler set of substances through chemical means. Each gold bar is made up of a vast collection of identical atoms—those with 79 protons in their nuclei.

Chemistry Review:
Reactions and Equilibrium

Earth's crust

other 8%
oxygen (O) 50%
silicon (Si) 26%
aluminum (Al) 8%
calcium (Ca) 3%
iron (Fe) 5%

Human body

7% other
10% hydrogen (H)
65% oxygen (O)
18% carbon (C)

Figure 2.4
Constituent Elements
The major chemical elements found in Earth's crust (including the oceans and the atmosphere) and in the human body.

Structure of Atoms, Elements, Isotopes

Isotopes

All this seems like a nice, tidy way to identify elements—one element, one atomic number, based on number of protons—except that we're leaving out something. Recall that atoms also have neutrons in their nuclei; that these neutrons add weight to the atom; and that the number of neutrons can vary independently of the number of protons. What this means is that, in thinking about an element in terms of its weight, we have to take neutrons into account. Furthermore, because the number of neutrons in an element's nucleus may vary, we can have various *forms* of elements, called **isotopes.** Most people have heard of one example of an isotope, whether or not they recognize it as such. The element carbon has six protons, giving it an atomic number of 6. In its most common form, it also has six neutrons. However, a relatively small amount of carbon exists in a form that has *eight* neutrons. Well, the element is still carbon, and in this form the number of its protons and neutrons equals 14, so the *isotope* is carbon-14 (which is used in determining the age of objects ranging from pyramids to plants, as you can see in "Finding the Iceman's Age in an Isotope" on page 26).

Most elements have several isotopes. Hydrogen, for example, which usually has one proton and one electron, also exists in two other forms: deuterium, which has one proton, one electron, and one neutron; and tritium, which has one proton, one electron, and two neutrons (**Figure 2.5**).

The Importance of Electrons

In our account so far of the subatomic trio, we have had much to say about protons and neutrons but little to say about electrons. This was necessary because we needed to discuss the nature of matter, but in a sense, you can regard what has been set forth to this point as so much stage-setting because what's most important in biology is the way elements *combine* with other elements. And in this combining, it is the outermost electrons that play a critical role. Just as you come into contact with the world through what lies at your surface—your eyes, your ears, your hands—so atoms link with one another through what lies at their periphery.

2.2 Chemical Bonding: The Covalent Bond

The process of chemical combination and rearrangement is called **chemical bonding,** and for us it represents the heart of the story in chemistry. When the outermost electrons of two atoms come into contact, it becomes possible for these electrons to reshuffle themselves in a way that allows the atoms to become attached to one another. This can take place in two ways: One atom can *give up* one or more electrons to another, or one atom can *share* one or more electrons with another atom. Giving up electrons is called *ionic bonding;* sharing electrons is called *covalent bonding.* A third type of bonding, which we'll discuss shortly, is also important for our purposes; it's known as hydrogen bonding.

Energy Always Seeks Its Lowest State

Atoms that undertake bonding with one another do so because they are in a more stable state after the bonding than before it. A frequently used phrase is helpful in understanding this kind of stability: Energy always seeks its lowest state. Imagine a boulder perched precariously on a hill. A mere shove might send it rolling toward its lower energy state—at the

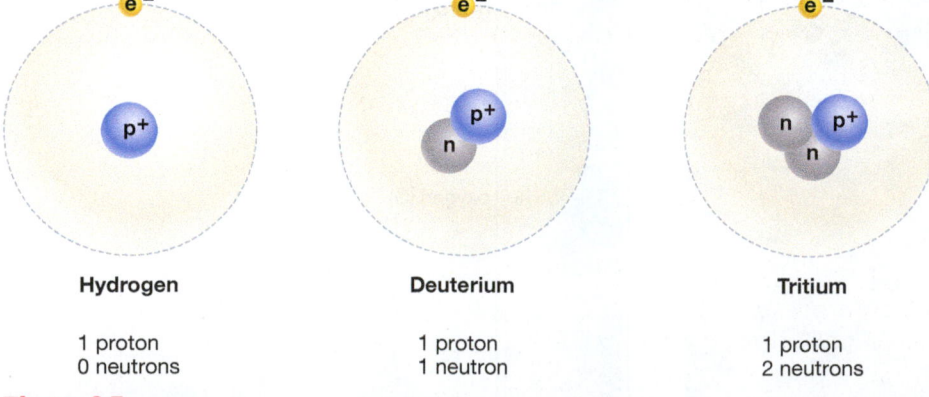

Figure 2.5
Same Element, Different Forms
Pictured are three isotopes of hydrogen. Like all isotopes, they differ in their number of neutrons.

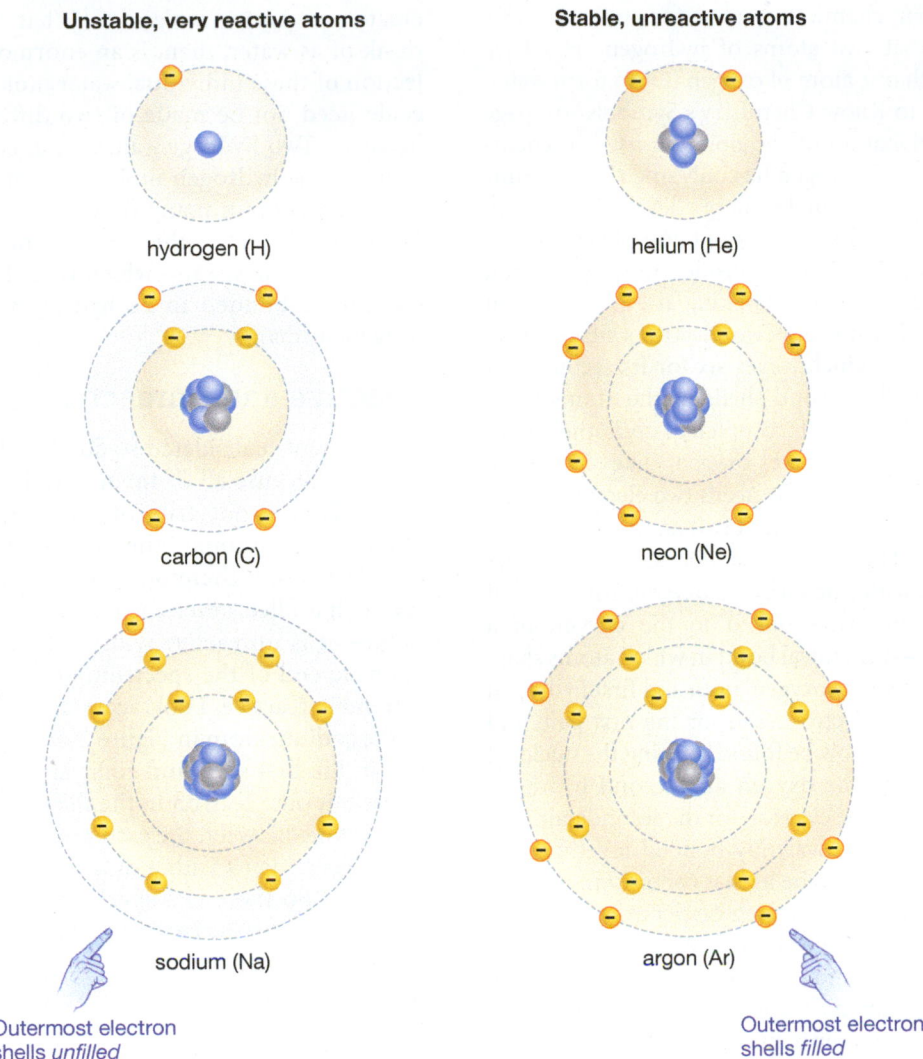

Unstable, very reactive atoms

hydrogen (H)

carbon (C)

sodium (Na)

Outermost electron shells *unfilled*

Stable, unreactive atoms

helium (He)

neon (Ne)

argon (Ar)

Outermost electron shells *filled*

Figure 2.6
Electron Configurations in Some Representative Elements

The concentric rings represent energy levels or "shells" of the elements, and the dots on the rings represent electrons. Hydrogen has but a single shell and a single electron within it, while carbon has two shells with a total of six electrons in them. Helium, neon, and argon have filled outer shells and are thus unreactive. Hydrogen, carbon, and sodium do not have filled outer shells and are thus reactive—they readily combine with other elements.

bottom of the hill. It would not then roll *up* the hill, either spontaneously or with a light shove, because it is now existing in a lower energy state than it did before—one that is clearly more stable than its former precarious perch. With electrons, the energy is not gravitational but electrical. Atoms bond with one another to the extent that doing so moves them to a lower, more stable energy state. The critical thing for our purposes is that atoms move to this more stable state by filling what is known as their *outer shells*.

Seeking a Full Outer Shell: Covalent Bonding

What are the outer shells? Electrons reside in certain well-defined "energy levels" outside the nuclei of atoms. The number of these energy levels varies depending on the element in question. Here, we only need to note the practical effect of these levels on bonding: *Two* electrons are required to fill the first energy level (or shell) of any given atom, but *eight* are

required to fill all the levels thereafter in most of the elements that make up the living world. If you look at the electron configurations in **Figure 2.6**, you can see that two elements pictured there—hydrogen and helium—have so few electrons in orbit around them that they have nothing *but* a first energy level, while the other elements pictured have two or three energy levels. This means that hydrogen and helium each require only two electrons in orbit around their nuclei to have filled outer shells, but that the other elements pictured require eight electrons to complete their outer electron shells.

Chemical Bonding in One Instance: Water

To see how chemical bonding works in connection with this concept of filled outer shells, take a look at the bonding that occurs with the constituent parts of one of the most simple (and important) substances on Earth, water. In so doing, you'll see one of the kinds of bonding we talked about—covalent bonding.

Covalent Bonds

The familiar chemical symbol for water is H_2O. This means that two atoms of hydrogen (H_2) have combined with one atom of oxygen (O) to form water. (See "Getting to Know Chemistry's Symbols" on page 29 for an explanation of the notation used in chemistry.) Recall that hydrogen has only one electron running around in its single energy level. Also recall, however, that this first level is not completed until it has *two* electrons in it. Next, consider the oxygen atom, which has eight electrons. Looking at **Figure 2.7**, you can see what this means: Two electrons fill oxygen's first energy level, which leaves six for its second. But remember that the second shells of the atoms we're concerned with are not completed until they hold eight electrons. Therefore oxygen, like hydrogen, would welcome a partner. It needs two electrons to fill its outer shell—something that two *atoms* of hydrogen could provide. The outcome is a bonding of two hydrogen atoms with one atom of oxygen. And each of the hydrogen atoms is linked to the oxygen in a **covalent bond**—a chemical bond in which atoms share pairs of electrons. The oxygen atom and first hydrogen atom donate one electron each for the first pair, and these electrons can now be found orbiting the nuclei of both atoms. Then the oxygen and second hydrogen atom each donate one electron for the second pair. The result? Three atoms covalently bonded together, and all of them "satisfied" to be in that condition.

Note that when this pairing of electrons happens, no matter has been gained or lost. We started with two atoms of hydrogen and one atom of oxygen, and we finish that way. The difference is that these atoms are now bonded. This points up an important principle known as the **law of conservation of mass,** which states that matter is neither created nor destroyed in a chemical reaction.

What Is a Molecule?

When two or more atoms combine in this kind of covalent reaction, the result is a **molecule:** an entity consisting of a defined number of atoms covalently bonded together. Here, one atom of oxygen has covalently bonded with two atoms of hydrogen to

create *one water molecule.* (What we commonly think of as water, then, is an enormous, linked collection of these individual water molecules.) A molecule need not be made of two different elements, however. Two hydrogen atoms can covalently bond to form one hydrogen molecule. Conversely, a molecule can contain many different elements bonded together. Consider the sucrose molecule—better known as table sugar—which is $C_{12}H_{22}O_{11}$ (12 carbon atoms bonded to 22 hydrogen atoms and 11 oxygen atoms).

Reactive and Unreactive Elements

The elements considered so far all welcome bonding partners because all of them have incomplete outer shells. This is not true of all elements, however. There is, for example, the helium atom, which has two electrons. It therefore *comes equipped,* we might say, with a filled outer shell. As such, it is extremely stable—it is unreactive with other elements. At the opposite end of the spectrum are elements that are extremely reactive. Look again at the representation of the sodium atom in Figure 2.6. It has 11 electrons: two in the first shell and eight in the second, which leaves but one electron in the third shell—a very unstable state. Between the extremes of sodium and helium are elements with a range of outer (or *valence*) electrons. So there is a spectrum of stability in the chemical elements, based on the number of outer-shell electrons each element has—from one to eight, with one being the most reactive and eight being the least reactive.

Covalent Polar and Nonpolar Bonding

Not all covalent bonds are created alike. When two hydrogen atoms come together, the result is a hydrogen molecule (H_2). Now, in the hydrogen molecule, the electrons are shared *equally.* That is, the two electrons the hydrogen atoms share are equally attracted to each hydrogen atom. This is not the case, however, with the water molecule.

Figure 2.7
Covalent Bonding

(a) A covalent bond is formed when two atoms share one or more pairs of electrons. The starting state in this reaction is two hydrogen atoms and one oxygen atom. Note that none of these atoms has outer-shell stability—each hydrogen atom would need one more electron to achieve stability, while the oxygen atom would need two more.

(b) In bonding, all the atoms achieve this stability; in this case, two pairs of electrons are shared—one pair between one hydrogen atom and the oxygen atom and the other pair between the second hydrogen and the oxygen. The result is the creation of a water molecule.

MB

Chemistry Review: Atoms and Molecules

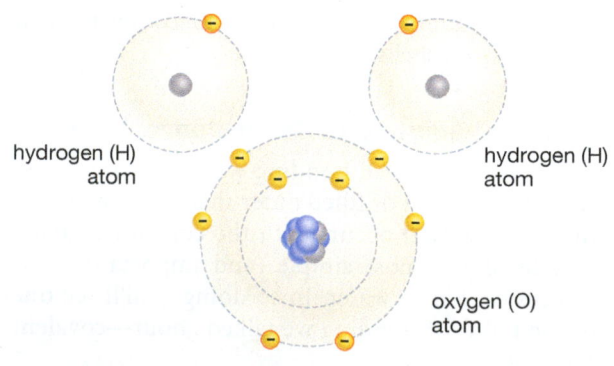

(a) Two hydrogen atoms and one oxygen atom

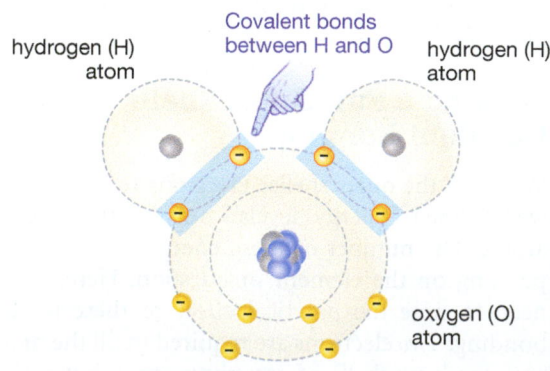

(b) One water molecule

Look at the representation of the water molecule in **Figure 2.8a**. As it turns out, the oxygen atom has greater power to attract electrons to itself than do the hydrogen atoms. The term for measuring this kind of pull is **electronegativity.** Because the oxygen atom has more electronegativity than do the hydrogen atoms, it tends to pull the shared electrons away from the hydrogen and toward itself. When this happens, the molecule takes on a **polarity,** or a difference in electrical charge at one end as opposed to the other. Because electrons are negatively charged and because they can be found closer to the oxygen nucleus, the oxygen end of the molecule becomes slightly negatively charged, while the hydrogen regions become slightly positively charged. We still have a covalent bond, but it is a specific type: a **polar covalent bond.** Conversely, with the hydrogen molecule—where electrons are shared equally—we have a **nonpolar covalent bond.**

To grasp the importance of this, consider the water molecule with its positive and negative regions. What's going to happen when it comes into contact with *other* polar molecules? The oppositely charged parts of the molecules will attract, and the similarly charged parts will repel. It's like having a bar magnet and trying to bring its positive end into contact with the positive end of another magnet; left on its own, the second magnet just flips around, so that positive is now linked to negative. In the same way, molecules flip around in relation to their polarity.

It is possible for atoms with different electronegativity to link and still have the resulting molecule be nonpolar. Water is polar because the atom with more electronegativity (the oxygen) lies to one *side* of the two hydrogen atoms. Meanwhile, in the methane molecule shown, four hydrogen atoms are arranged in a symmetrical way around a central, and more electronegative, carbon atom (**Figure 2.8b**). In this arrangement, the differing charges balance each other out, leaving methane with no positive or negative end—meaning it is nonpolar. (Keep this quality of methane in mind, as it will be important in

the consideration of why oil and water don't mix.) In summary, some molecules are polar while others are nonpolar, and this difference has significant consequences for chemical bonding.

SO FAR . . .

1. Atoms tend to bond with one another to the extent that they do not have a _____ outer shell, which means having _____ outer-shell electrons for most elements in living things, but only _____ outer-shell electrons for the simple elements hydrogen and helium.

2. When two atoms share one or more pairs of electrons, they have become linked through a _____ bond.

3. In some of these bonds, one of the linked atoms may exert a greater pull on the shared electrons. When this happens, the result can be a _____ bond, a bond in which differing regions of the resulting molecule take on a difference in _____ in one region, as compared to another.

2.3 The Ionic Bond

So we've gone from nonpolar covalent bonding, in which electrons are shared equally, to polar covalent bonding, in which electrons are pulled to one side of the resulting molecule. What if we carried this just one step further and had instances when the electronegativity differences between two atoms were so extreme that electrons were pulled *off* one atom, only to latch on to the atom that was attracting them? This is what happens in our second type of

 Nonpolar and Polar Molecules

 Electron Arrangement

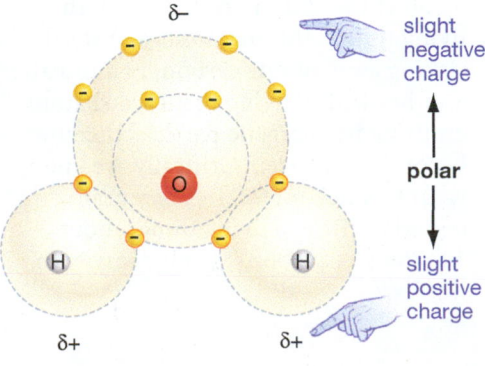

(a) Polar water molecule

slight negative charge

polar

slight positive charge

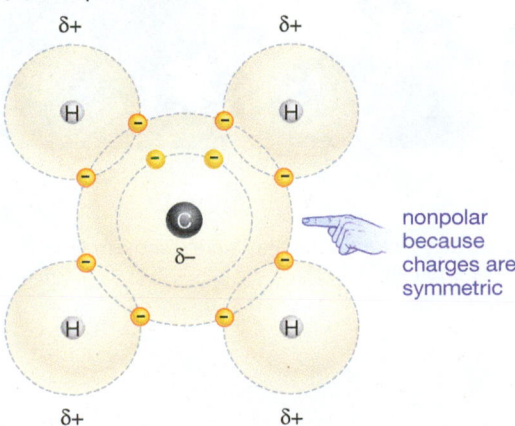

(b) Nonpolar methane molecule

nonpolar because charges are symmetric

Figure 2.8
Polar and Nonpolar Covalent Bonding

(a) In the water molecule, the oxygen atom exerts greater attraction on the shared electrons than do the hydrogen atoms. Thus, the electrons are shifted toward the oxygen atom, giving the oxygen atom a partial negative charge (because electrons are negatively charged) and the hydrogen atoms a partial positive charge. ("Partial" here is indicated by the Greek symbol δ.)

(b) In the methane molecule, the carbon atom is more electronegative than the hydrogen atoms, but the methane molecule as a whole is nonpolar because its hydrogen atoms are arranged symmetrically around the central carbon atom, meaning the partial charges that exist balance each other out.

ESSAY

Finding the Iceman's Age in an Isotope

In September 1991, two German hikers walking in a mountain pass that spanned the border between Austria and Italy were startled to see a human body protruding from a section of ice at the bottom of some rock outcroppings (**Figure 1**). The couple's discovery was initially investigated by Austrian police, who at first wondered whether the body might be that of a music teacher who had disappeared from the Italian city of Verona in 1939. In the days that followed, investigators removed the corpse from the ice and then helicoptered it to a research institute at the University of Innsbruck in Austria. There, an expert on prehistoric cultures looked at both the body and some articles found with it and estimated that the frozen corpse was not 50 years old, but more like 4,000 years old. The man in the ice—dubbed Oetzi for the Otzal section of the Alps in which he was found—had made a

Carbon-14 dating has taken us from the realm of guessing about dates in the distant past to the realm of knowing about them.

fatal journey over the mountains long before the countries of Austria or Italy ever existed. But exactly when had Oetzi lived? The 4,000-year figure was just a rough guess, based mostly on the articles found with the body. To get a more precise date, scientists employed a kind of atomic clock—one that begins ticking in the remains of all living things from the moment of their death. As it turns out, this clock runs on a special variety of the element carbon.

All living things take in carbon each day in order to live, and the pathway by which they obtain this carbon is well understood. Plants take in carbon dioxide from the atmosphere and incorporate the carbon from it into their tissues; then animals eat plants, thus incorporating the carbon into *their* tissues. The vast majority of the carbon atoms that move through this system are carbon-12 atoms, but a very small proportion of them are carbon-14 atoms, which are constantly being produced in the atmosphere. Each carbon-14 atom is unstable, however; each is fated to emit an energetic particle from its nucleus and thus be transformed into an atom of ordinary nitrogen. This process goes under a name that may sound familiar: radioactive decay. Carbon-14 is a radioactive isotope of carbon.

Several features of carbon-14's decay are critical in allowing it to serve as an atomic clock. First, its decay takes a long time. Take any sample of carbon-14 atoms, and half of them will decay into nitrogen in 5,730 years; of the carbon atoms that remain in the sample, another half will decay into nitrogen in the *next* 5,730 years, and so on (**Figure 2**). In living things, this manifests as a steady ticking away of one carbon-14 atom after another within living tissue, each atom emitting its energetic particle, thereby becoming nitrogen. Critically, we know what this tick rate is: In every gram of carbon from a living thing, about 14 atoms of carbon-14 will decay each

Figure 1
A Startling Find
The remains of the "iceman" Oetzi as they looked when they were discovered in the mountainous border region of Austria and Italy in 1991.

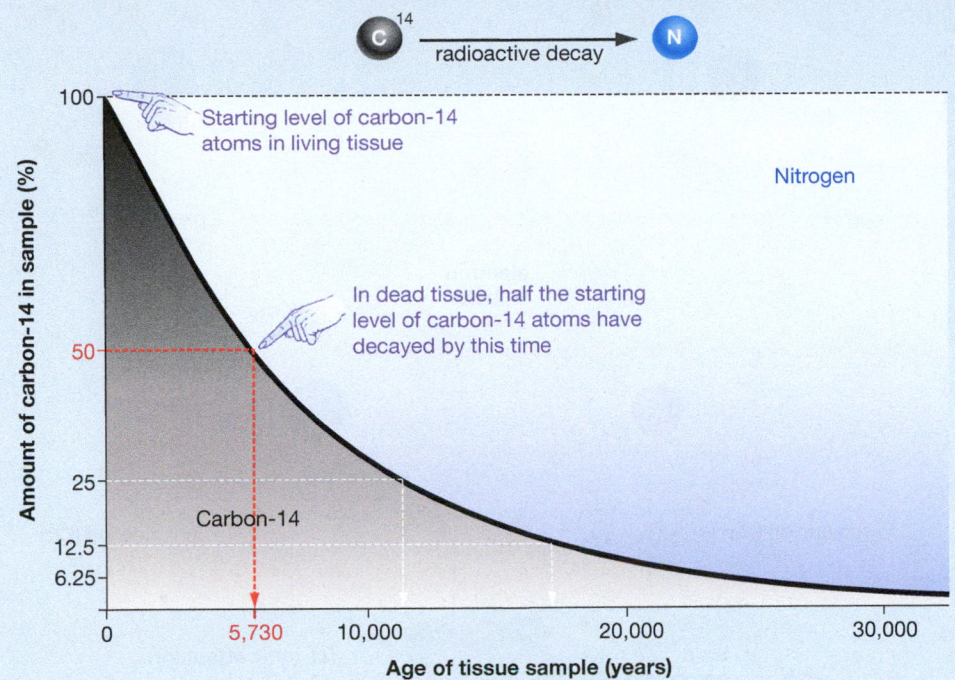

Figure 2
Using Carbon as a Clock
Carbon-14 decays into nitrogen at such a steady rate that scientists can use it to calculate the approximate age of almost any object 50,000 years or younger that was composed originally of living tissue.

minute. Just as important, we can *count* these ticks; the emission of an energetic particle from one carbon-14 atom as it is transforming into nitrogen will register as one tick on a radioactive counter.

With all this in mind, here's how carbon-14 dating worked in the case of Oetzi. On the day this hardy traveler died, we know he had eaten some grain and some deer meat, meaning the movement of carbon atoms through plants and animals and then through him was proceeding as usual. The critical thing, however, comes with Oetzi's death, for it's at this point that *carbon stopped flowing into him.* He quit taking in carbon from the outside world, including carbon-14. With this, no newly formed carbon-14 atoms entered him, but all of the carbon-14 atoms *already* present in his tissues were

decaying away at their usual rate. On the day he died, 14 ticks per minute would have taken place for every gram of carbon in his body. But go, say, 2,800 years out from his death and the number of ticks drops to 10.5 per minute. Why? Because by this time, 25 percent of the carbon-14 atoms in his body had become nitrogen and therefore were no longer emitting energetic particles. All that was necessary to determine the approximate date of his death, therefore, was to get a sample of carbon from his tissues and count the number of carbon-14 ticks being emitted from it. The lower the number of ticks, the longer ago he lived. Two labs performed a modern variation on this type of analysis, and they both came to the same conclusion: there is a 95.4 percent chance that Oetzi lived sometime between 5,111 and 5,381 years ago.

If you compare these relatively precise dates with the rough estimate that Oetzi lived about 4,000 years ago, you can see what carbon-14 dating has done for human knowledge: It has taken us from the realm of guessing about dates in the distant past to the realm of knowing about them. Critically, it's not just human or even just animal tissues that can be dated. Wood found in an ancient Egyptian pyramid once was a part of a living tree, so it can be dated; cloth that made up a medieval tapestry once was a part of a living plant, so it can be dated. Trees knocked down by an ancient glacier can give us an approximate date for the glacier's advance. Given this wide utility, it's not surprising that tens of thousands of samples of once-living tissue go through carbon-14 analysis each year at laboratories around the world. The primary limitation on the technique is that it doesn't work on items that are more than about 50,000 years old; at that point, the carbon-14 in any sample has run out.

For many years, carbon-14 dating was done through the counting method outlined above; scientists had to get gram-sized samples of tissue and then count up particle emissions over several days. More recently, this method has been replaced by a technique called accelerator mass spectrometry, which not only can directly count the number of carbon-14 atoms in a sample, but can do so on samples that are only thousandths of a gram in size. The basic principle behind carbon-14 dating has remained unchanged, however, from the time chemist Willard Libby first developed it at the University of Chicago in the 1940s. Then as now, the idea was to find out how many carbon-14 atoms remain in a sample of once-living tissue. In 1960, Libby received the ultimate in thanks for a job well done in science when he was awarded the Nobel Prize in chemistry for giving the world the gift of carbon-14 dating.

Figure 2.9
Ionic Bonding

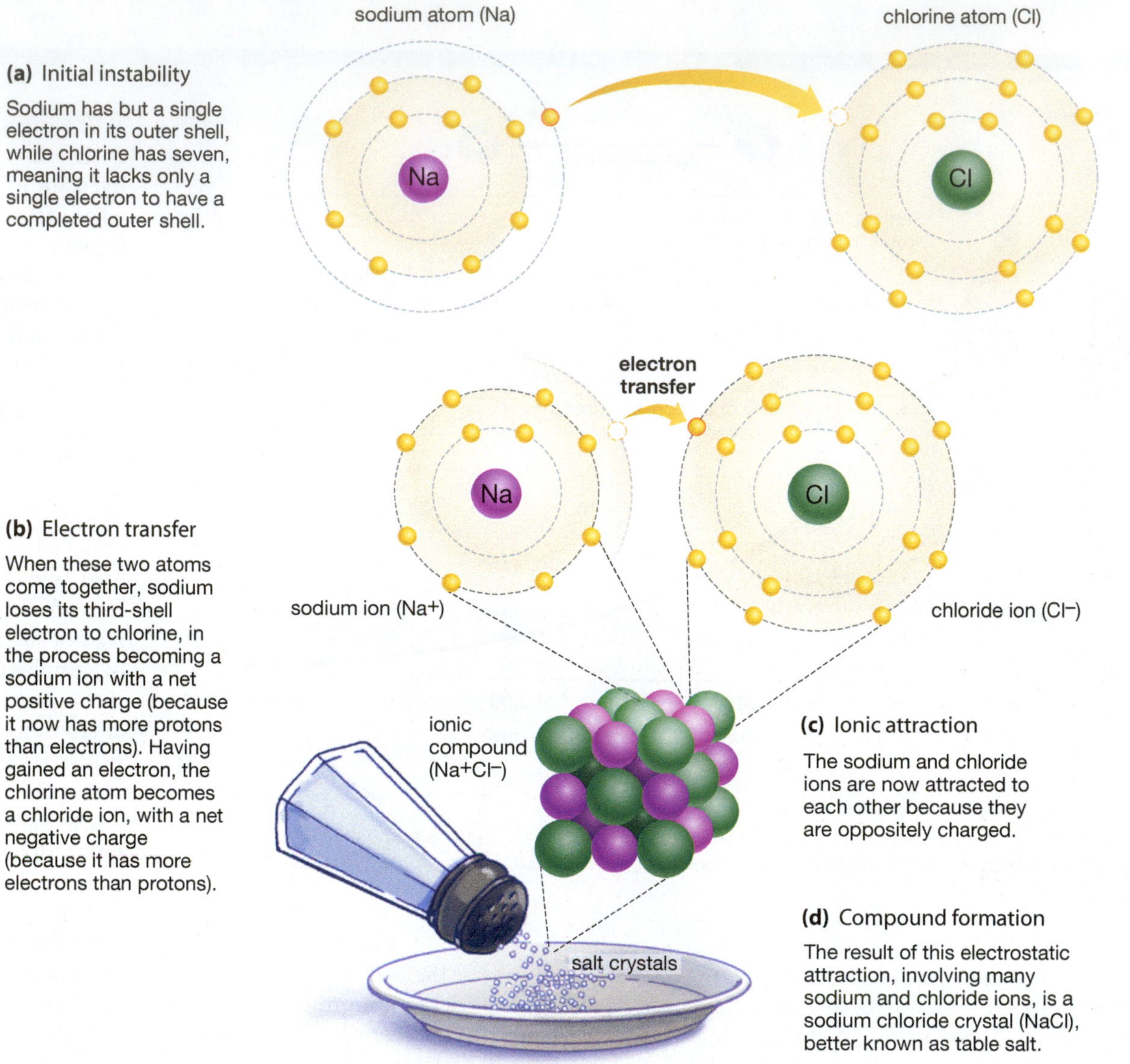

sodium atom (Na)   chlorine atom (Cl)

(a) Initial instability

Sodium has but a single electron in its outer shell, while chlorine has seven, meaning it lacks only a single electron to have a completed outer shell.

electron transfer

(b) Electron transfer

When these two atoms come together, sodium loses its third-shell electron to chlorine, in the process becoming a sodium ion with a net positive charge (because it now has more protons than electrons). Having gained an electron, the chlorine atom becomes a chloride ion, with a net negative charge (because it has more electrons than protons).

sodium ion (Na+)

chloride ion (Cl−)

ionic compound (Na+Cl−)

(c) Ionic attraction

The sodium and chloride ions are now attracted to each other because they are oppositely charged.

salt crystals

(d) Compound formation

The result of this electrostatic attraction, involving many sodium and chloride ions, is a sodium chloride crystal (NaCl), better known as table salt.

Ionic Bonds

bonding, ionic bonding. The classic illustration of this type of bonding involves the sodium we looked at earlier and the element chlorine. Recall that sodium has 11 electrons, meaning that there is a lone electron flying around in its outer, third electron shell. Chlorine, in contrast, has 17 electrons, meaning it has 7 electrons in the third shell. Remember that 8 usually is the magic number for outer-shell stability. Sodium could get to this number by *losing* one electron, while chlorine could get to it by *gaining* one electron. That's just how this encounter occurs: Sodium does in fact lose its one electron, chlorine gains it, and both parties become stable in the process (**Figure 2.9**).

What Is an Ion?

But this story has a postscript. Having lost an electron (with its negative charge), sodium (Na) then takes on an overall *positive* charge. Having gained an electron,

chlorine (Cl) takes on a negative charge. Each is then said to be an **ion**—a charged atom or, to put it another way, an atom whose number of electrons differs from its number of protons. We denote the ionized forms of these atoms like this: Na^+, Cl^-. Were an atom to gain or lose more electrons, a number would be put before the charge sign. For example, to show that the magnesium atom has lost two electrons, we would write Mg^{2+}.

Note that we now have two ions, Na^+ and Cl^-, with differing charges in proximity to one another. They are therefore attracted to one another through an *electrostatic attraction* and have therefore become bonded through **ionic bonding:** a chemical bonding in which two or more ions are linked by virtue of their opposite charge. This hardly ever happens with just two ions, of course. Many billions of them are bonded in this way, up, down, and sideways from each other, thus forming the kind of solid, crystal structure you can see in **Figure 2.9c**. What's produced through this

Getting to Know Chemistry's Symbols

One of the "languages of science" is the system of symbols that chemistry uses. Somewhat intimidating at first viewing, this system soon comes to serve its intended purpose of conveying a lot of information quickly.

Our starting place is that each chemical element has its own symbol, so that hydrogen is represented by H, carbon by C, and platinum by Pt. When we begin to combine these elements into molecules, it is necessary to specify how *many* atoms of each element are part of the molecule. If we have two atoms of oxygen together—which is the way oxygen is usually packaged in our atmosphere—we have the molecule O_2. Three molecules of O_2 is written as $3O_2$. This kind of notation is known as a **molecular formula.** It is helpful in stipulating the makeup of molecules, from the simple, such as oxygen, to the complex, such as chlorophyll, which is notated $C_{55}H_{72}MgN_4O_5$.

Anyone who sees a molecular formula learns a lot about which atoms are in a molecule but nothing about the way the atoms are *arranged* in relation to one another. (Look at chlorophyll's formula. Is there a line of 55 carbon atoms followed by 72 hydrogen atoms? From the molecular formula, how could you tell?) To convey this ordering information, chemists and biologists use **structural formulas**—two-dimensional representations of a given molecule. Methane (CH_4) is a very simple molecule composed, as the molecular formula shows, of one atom of carbon and four of hydrogen. In a structural formula, these constituent parts are conceptualized like this:

$$H - \underset{\underset{H}{\displaystyle |}}{\overset{\overset{H}{\displaystyle |}}{C}} - H$$

methane

Note that there is a single line between each hydrogen atom and the central carbon atom. This signifies that the bond between any of the hydrogen atoms and carbon is a single bond: Each line represents *one* pair of electrons being shared. Thus:

$$H : \overset{H}{\underset{H}{C}} : H \text{——electron pairs}$$

There can also be double bonds and triple bonds. When carbon dioxide forms, there are two oxygen atoms. Each shares two pairs of electrons with a lone carbon atom. Here's how we would notate the double bond in a carbon dioxide (CO_2) molecule:

$$O = C = O$$
carbon dioxide

Although structural (or "skeletal") formulas can tell us a good deal about the ordering of atoms in a molecule, they tell us little about the *three-dimensional* arrangements of atoms. For this, we rely on two other kinds of representations, the **ball-and-stick model** and the **space-filling model.** An ammonia molecule (NH_3) is pictured next in both forms, with the molecular and structural formulas added to show the progression:

$$NH_3 \qquad\qquad H - \overset{..}{\underset{\underset{H}{\displaystyle |}}{N}} - H$$

molecular   structural
formula   formula

ball-and-stick   space-filling
model   model

Note that the ball-and-stick model gives us a better idea of molecular angles of the atoms, but that the space-filling model gives us a better idea of the relative size of these atoms and how one actually hugs the other. (The dots above the "N" in the structural formula represent an unshared pair of electrons.)

Finally, it is useful to have a way to notate the "before" and "after" stages of a chemical reaction. This is done by employing a simple arrow, as when carbon reacts with hydrogen to form methane: $C + 4H \rightarrow CH_4$. The carbon and hydrogen atoms on the left are **reactants,** the arrow means "yields," and the methane molecule on the right is the **product** of the reaction. Here's a graphic representation of what is happening:

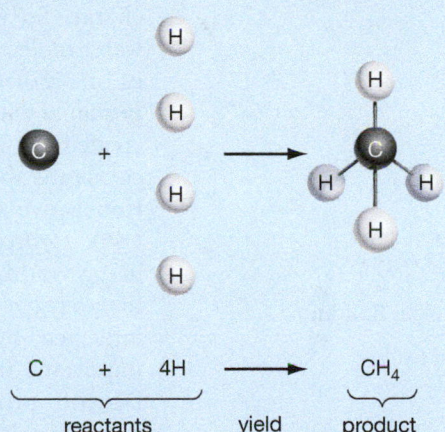

$$\underbrace{C \quad + \quad 4H}_{\text{reactants}} \quad \underrightarrow{}_{\text{yield}} \quad \underbrace{CH_4}_{\text{product}}$$

An important thing to keep in mind about notation goes back to the law of conservation of mass: Matter is neither created nor destroyed in a chemical reaction. It follows that reactions such as the one above must be *balanced:* We must have the same number of atoms when the reaction is finished (on the right) as when it started (on the left). So we couldn't have $C_2 + 4H \rightarrow CH_4$. This is an unbalanced reaction. We started out with two atoms of carbon on the left but somehow ended up with one atom of carbon on the right. Nature doesn't play that way.

process is an **ionic compound:** a collection of the atoms of two or more elements that have become linked through ionic bonding.

Note that this compound is not a *molecule* as H_2O is. It does not have a defined number of atoms—as in H_2O's two atoms of hydrogen and one of oxygen—and the many atoms that do make it up are linked through ionic bonds, rather than covalent bonds. The particular ionic compound pictured in Figure 2.9 actually is familiar. Sodium and chlorine combine to create sodium chloride, which is better known as table salt. The notation for it should properly be written Na^+Cl^-, but it is usually denoted just as NaCl.

2.4 The Hydrogen Bond

We now turn to a final variant on bonding called *hydrogen bonding.* Recall that in any water molecule, the stronger electronegativity of the oxygen atom pulls the electrons shared with the hydrogen atoms toward the oxygen nucleus, giving the oxygen end of the molecule a partial negative charge and the hydrogen end of the molecule a partial positive charge. So what happens when you place several water molecules together? A positive hydrogen atom of one molecule is weakly attracted to the negative, *unshared* electrons of its oxygen neighbor. Thus is created the **hydrogen bond,** which links an already covalently bonded hydrogen atom with an electronegative atom (in this case, with oxygen; **Figure 2.10**). Hydrogen bonding is a linkage that, in the living world, nearly always pairs hydrogen with either oxygen or nitrogen. Hydrogen bonds, usually indicated by a dotted line in illustrations, are important in many of the molecules of life—in DNA, in proteins, and elsewhere. These bonds are, however, relatively weak compared to either covalent or ionic bonds. In collections of water molecules, the result is a set of bonds that are constantly breaking and then re-forming. Each water molecule can be bonded to four others, but no individual set of bonds lasts for long. Nevertheless, hydrogen bonding manages to strongly shape the qualities of water, as you'll soon see.

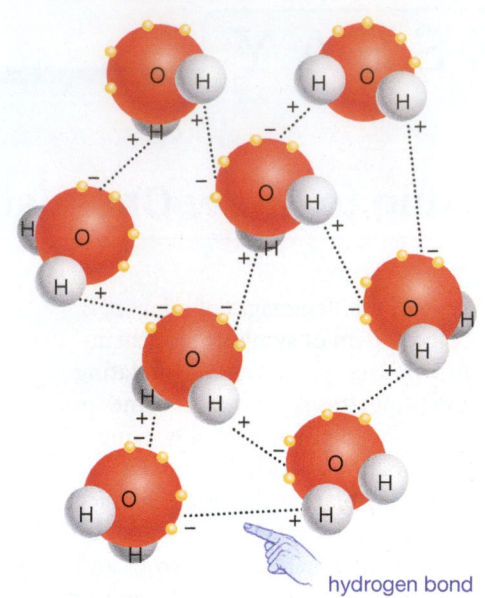

hydrogen bond

Figure 2.10
Hydrogen Bonding

The hydrogen bond, in this case between water molecules, is indicated by the dotted line. It exists because of the attraction between hydrogen atoms, with their partial positive charge, and the unshared electrons of the oxygen atom, with their partial negative charge.

2.5 Three-Dimensional Shape in Molecules

We now need to make more explicit what has been noted only by implication in our diagrams of water molecules: that molecules and ionic compounds have a three-dimensional shape. It is useful to depict them as two-dimensional chains, rings, and such, but in real life a molecule is as three-dimensional as a sculpture. A fair number of shapes are possible; atoms may be lined up in a row, in triangles, or in pyramid shapes. As an example, look at the water and methane molecules in **Figure 2.11**. You can see that in water there is a definite spatial configuration: its hydrogen atoms are splayed out from its oxygen atom at an angle of 104.5°. In methane, meanwhile, the hydrogen atoms have an angle of 109.5° between them.

Figure 2.11
Three-Dimensional Representations of Molecules

(a) In the case of water, there is an angle of 104.5° between hydrogen atoms.

(b) Methane is a molecule with an angle of 109.5° between hydrogen atoms.

(a) Water (H_2O)

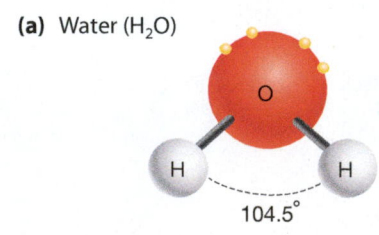

104.5°

(b) Methane (CH_4)

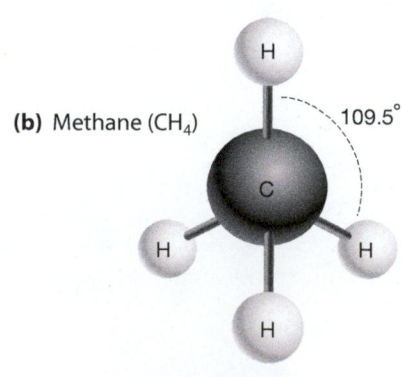

109.5°

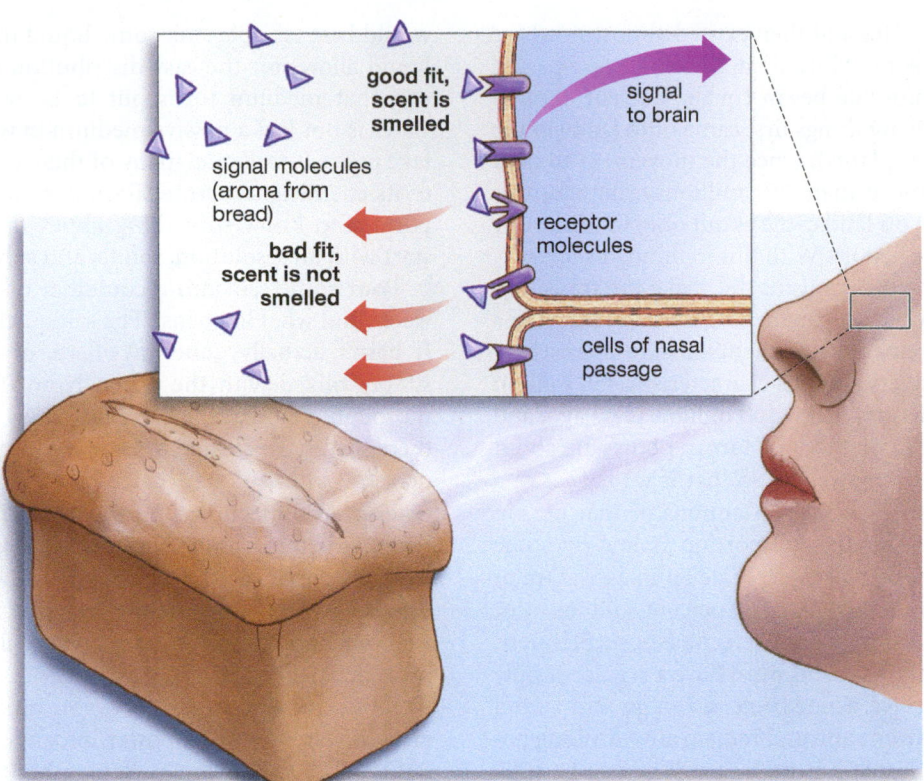

Figure 2.12
The Importance of Molecular Shape
Gas molecules wafting off bread (the triangles) bind with specific receptors on the surface of the cell, thus acting as signaling molecules that set a cellular process in motion. For this binding to take place, the gas and nasal receptor molecules must fit together.

Why does molecular shape matter? It is critical in enabling biological molecules to carry out their activities. This is because molecular shape determines the capacity of molecules to latch on to or "bind" with one another. When, for example, you smell the aroma of fresh-baked bread, gas molecules wafting off the bread bind with receptor molecules in your nasal passages, thus sending a message to the brain about the presence of bread. It is the precise shape of the gas molecules and nasal receptor molecules that allows them to bind with one another. Look at **Figure 2.12** to see how this works. If you look at **Figure 2.13**, then you can get an idea of how large some biological molecules are relative to the simple molecules considered so far.

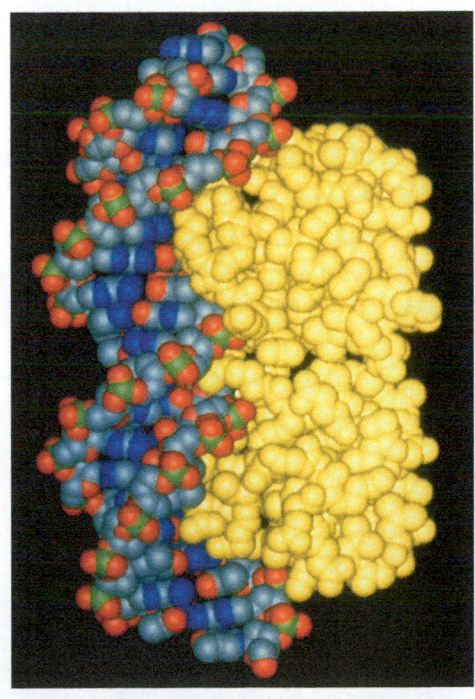

Figure 2.13
Complex Binding
A computer model of some real-life molecular binding. In this case, a protein (the yellow atoms) binds to a length of DNA. This space-filling model provides some idea of the enormous number of atoms and the complicated shapes that make up some of the molecules employed by living things.

SO FAR...

1. An ionic bond occurs when one atom _____ one or more _____ to another atom and the resulting ions become attached because of their differing _____.

2. A hydrogen bond links an already covalently bonded hydrogen atom with a more _____ atom. In biology, such bonds almost always occur between hydrogen and either _____ or _____.

3. Molecular shape is important in biology because it determines the ability that molecules have to _____ with one another.

2.6 Water and Life

Having looked at some of the ways in which atoms form the world's substances, we now look in detail at just *one* substance, water. Why such emphasis on water? Because of its great importance to our central subject, which is life.

How are life and water related? Well, consider how the two things have been intertwined over time. Life got started in water and then existed almost nowhere *but* water for eons. More than 3 billion years passed between the time life began (in the ancient oceans) and the time living things first came onto land (in the form of ancient plants). Since the movement to land occurred no more than 450 million years ago, this means that life on land came about only in the last 11 percent of life's history. With this in mind, it's not surprising that when organisms did make the transition to land, they could only manage this feat by carrying a watery environment with them—inside themselves. The plant pioneers had this characteristic 450 million years ago, dinosaurs had it 100 million years ago, and human beings have it today. Human bodies are about 66 percent water by weight, so that if we have, say, a 128-pound person, about 85 pounds of that person will be water. And this proportion is low by some standards: Among most vertebrate animals, the water proportion is more like 70 or 80 percent, and for most plant tissues, the figure is 90 percent. Not surprisingly, then, almost all organisms must have a regular supply of water to survive. Some bacteria can go into a kind of suspended animation and remain alive for long periods without water, but no living thing can be fully *functional* without water. To reproduce, to move, to obtain nutrients—to carry out most of life's basic processes, in other words—all living things must have water. Given this, if we ask how important water is to life, the answer is that the two things are inseparable.

Water Is a Major Player in Many of Life's Processes

But what gives water this status? Recall the notion we touched on earlier of life being, at one level, a series of chemical reactions. It makes sense that these reactions would best take place in some liquid medium—since liquid allows for the easy distribution of materials—and that medium turns out to be water. However, water is not just a passive medium in which reactions take place. It *facilitates* many of these reactions thanks to its chemical structure. To understand this, it's important to know something about three terms that start with an *s*: solution, solute, and solvent.

Pour some salt into a container of water, stir the water, and what happens? The salt quickly disappears. It hasn't actually gone anywhere, of course; it has simply mixed with the water. Now, if it has mixed thoroughly, so that there are no lumps of salt here or there, you have created a **solution**—a mixture of two or more kinds of molecules, atoms, or ions that is *homogenous,* meaning uniform throughout. The salt is what's being dissolved, so it is the **solute.** The water is doing the dissolving, so it is the **solvent.** And with this, we get to the point about water: it's a terrific solvent, which is to say it has a great ability to dissolve other substances.

If you look at **Figure 2.14**, you can get a detailed view of water's solvent power in connection with the salt we just talked about. Remember that water is a polar molecule, which is to say a molecule that has differing electric charges at one end as opposed to the other. Attracted by the polar nature of the water molecule, the sodium and chloride ions that make up a salt crystal separate from the crystal—and from each other. Each ion is then surrounded by several water molecules. These units keep the sodium and chloride ions from getting back together. In other words, they keep the ions evenly dispersed throughout the water, which is what makes this a solution. Water works as a solvent here because the ionic compound sodium chloride carries an electrical charge. What *generally*

Dissociation of Water
Molecules

Figure 2.14
Water's Power as a Solvent

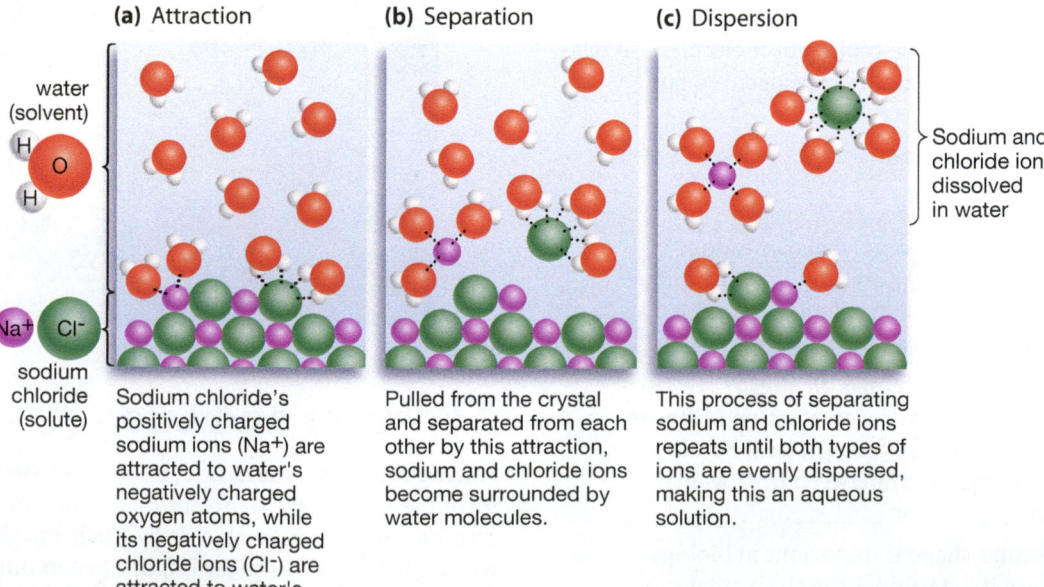

(a) Attraction   **(b)** Separation   **(c)** Dispersion

water
(solvent)

H
O
H

Na+ Cl⁻

sodium
chloride
(solute)

Sodium chloride's positively charged sodium ions (Na+) are attracted to water's negatively charged oxygen atoms, while its negatively charged chloride ions (Cl⁻) are attracted to water's positively charged hydrogen atoms.

Pulled from the crystal and separated from each other by this attraction, sodium and chloride ions become surrounded by water molecules.

This process of separating sodium and chloride ions repeats until both types of ions are evenly dispersed, making this an aqueous solution.

Sodium and chloride ions dissolved in water

makes water work as a solvent, however, is its ability to form, with other molecules, the hydrogen bonds we talked about earlier.

Water's Unusual Properties

When we note water's ability to act as a solvent, we're actually not giving water its due. Water is not just *a* solvent: Over the range of substances, nothing can match it as a solvent. It can dissolve more compounds in greater amounts than can any other liquid.

But solvency power is merely the beginning of water's abilities. It is a multitalented performer. And it achieves this status because it is ... odd. Compared to other molecules, water is like some zany eccentric whose powers stem precisely from its eccentricity. Consider the fact that ice floats on water. This is so because the solid form of H_2O is less dense than the liquid form—a strange reversal of nature's normal pattern. Things work this way because of water's hydrogen bonding. You may remember that, in a collection of water molecules, the hydrogen bonds between individual molecules are constantly breaking and then re-forming. This allows the molecules to pack relatively closely together, as you can see in **Figure 2.15**. Start to move water toward freezing, however, and things change; now the water molecules are able to form the maximum number of hydrogen bonds with each other, and the effect of this is that each water molecule sits *farther apart* from its neighbor. The more space there is between molecules of any substance, the less dense that substance is. In this case, the result is that ice is less dense than liquid water.

This may seem like some minor, quirky quality, but it actually has the effect of making possible life as we know it. Ice on the surface of water insulates the water beneath it from the freezing surface temperatures and wind above, creating a warmer environment for organisms such as fish. If ice *sank,* on the other hand, then the entire body of water would freeze solid at colder latitudes, creating an environment in which few living things could survive for long.

Moderating Temperature: Specific Heat

Water serves as an insulator not only when it is frozen, but also when it is liquid or gas. This has to do with a quality called **specific heat:** the amount of energy required to raise the temperature of a substance by 1°Celsius. As it turns out, water has a high specific heat. Put a gram container of drinking alcohol (ethyl alcohol) side by side with one of water, heat them both, and it will take almost twice as much energy to raise the temperature of the water 1°C as it will the alcohol. Having absorbed this much heat, however, water then has the capacity to *release* it when the environment around it is colder than the water itself. The result? Water acts as a great heat buffer for Earth.

Chemistry Review: Water

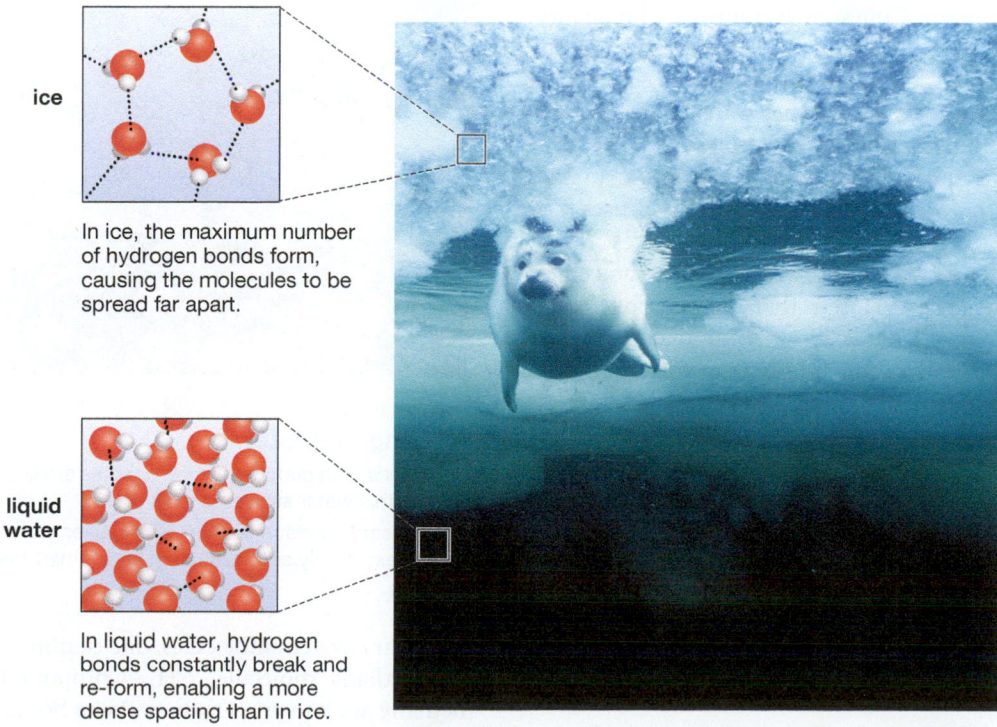

ice

In ice, the maximum number of hydrogen bonds form, causing the molecules to be spread far apart.

liquid water

In liquid water, hydrogen bonds constantly break and re-form, enabling a more dense spacing than in ice.

Figure 2.15
Life Made Possible under the Ice

Water is unusual in that its solid form (ice) is less dense than its liquid form. This means that ice floats, and this in turn means that life can flourish in cold-weather aquatic environments. Pictured is a harp seal swimming under the ice in Canada's Gulf of St. Lawrence.

The oceans absorb tremendous amounts of radiant energy from the sun only to release this heat when the temperature of the air above the ocean gets colder. Without this buffering, temperature on Earth would be less stable. People who have spent a day and a night in a desert can attest to this effect. The searing heat of the desert day radiates off the desert floor; but at night, with little water vapor in the air to capture this heat, the desert cools dramatically. In the same way, our *internal* temperature can remain much more stable because the water that makes up so much of us is able to first absorb and then release great amounts of heat. The sweat created in exercise has considerable cooling power because each drop of perspiration carries with it a great deal of heat.

Water derives these powers once again from its hydrogen bonds. Weak and shifting though individual hydrogen bonds may be, collectively they have great strength. Heat is the motion of molecules. To get molecules moving, though, chemical bonds must first be broken. Because water has a formidable set of hydrogen bonds, it takes a lot of energy to break the bonds and get its molecules moving.

Cohesion and Surface Tension

Cohesion of Water

This tendency of water molecules to stay together is called *cohesion,* and it is very important in the plant world; thanks to this quality, evaporation can work with a process called osmosis to draw water from a plant's roots all the way to its leaves and out into the air as water vapor. Cohesion also imparts the quality of *surface tension* to water in places where water meets air. Water molecules below the surface are equally attracted in all directions to other water molecules. *At the surface, however, water molecules have no such attraction to the air above them*—they are pulled down and to the side, but not up. This causes the "beading" that water droplets do on surfaces. More important for our purposes, surface water molecules pack together more closely than do interior molecules, allowing all kinds of small animals to move across the surface of water, rather than sinking into it. Note the familiar water strider in **Figure 2.16**.

Having looked at the qualities of water, you may be tempted to think how *uncanny* it is that water has all these characteristics that are conducive to life. But remember that life went on solely in water for billions of years before living things ever came onto land. Thus, water has fundamentally conditioned life as we know it. Being surprised about water's life-enhancing qualities is like being surprised that a stage is a good place to put on a play.

Hydrophobic and Hydrophilic Molecules

With its great complexity, life requires molecules that *cannot* be dissolved by water. Such is the case with

(a) Walking on water

(b) Beading up

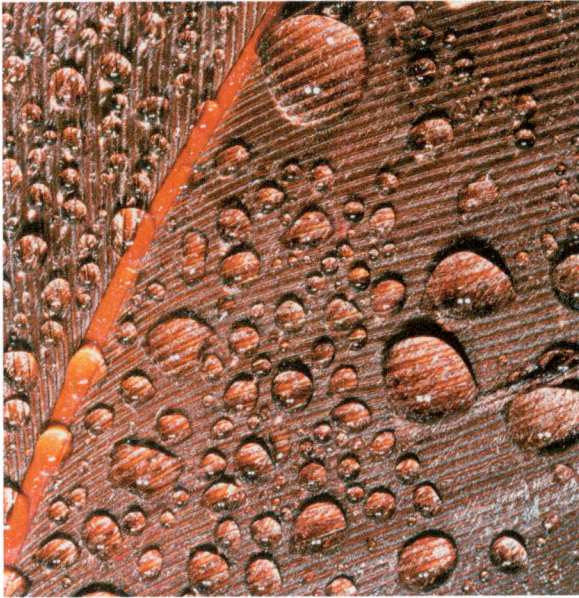

Figure 2.16
Walking on Water

(a) Water's high surface tension enables small animals, such as this water strider, to walk on it.

(b) This same surface tension also causes water to bead up on waxy or oily surfaces, such as this bird feather.

nonpolar covalent molecules, one example of which is the methane molecule, CH_4. Compounds such as methane are known as **hydrocarbons** because they are made up solely of hydrogen and carbon atoms. Petroleum products are more complex hydrocarbons than methane, but they also are nonpolar, and you can see a vivid demonstration of their lack of water solubility in oil spills. Oil doesn't dissolve in water because the

Figure 2.17
Oil and Water Do Not Mix
When there is an oil spill in the ocean, the oil stays concentrated even as it spreads because oil and water do not form chemical bonds with each other. Here, trawlers are using a boom to clean up after an oil spill in Great Britain.

oil carries almost no electrical charge that water can bond with; thus water has no way to separate one oil molecule from another (**Figure 2.17**).

The ability of molecules to form bonds with water has a couple of important names attached to it. Compounds that will interact with water—such as the sodium chloride considered earlier—are known as **hydrophilic** ("water loving"), while compounds that do not interact with water, such as oil, are known as **hydrophobic** ("water fearing"). Both terms are misleading in that no substance has any emotional relationship with water. *Hydrophobic* is particularly off the mark because water does not repel hydrophobic molecules; instead, the strong bonds that water molecules form with each other cause them to form circles around concentrations of hydrophobic molecules, as if they had lassoed them. (You can see this at home anytime you pour some cooking oil into a pan of water.)

The importance of hydrophobic molecules can be illustrated in part by the common milk carton. Why is the milk carton important? Because it can keep milk separate from everything else. We living organisms need some kind of "carton" that can separate the world outside of us from ourselves. Likewise, organisms have great use *within* themselves for compartments that can be sealed off to one degree or another. If water broke down every molecule of life it came in contact with, then it would break down all these divisions of living systems. Molecules do not have to be completely hydrophilic or hydrophobic. Indeed, as you'll see, a number of important molecules have both hydrophilic and hydrophobic portions.

The Polarity of Water

SO FAR . . .

1. Pour some salt into a container of water and stir until the salt is no longer visible. In this situation, the salt is the _____, the water is the _____, and the uniform mixture of the two substances is a _____.

2. Thanks to the _____ linking them, water molecules have great _____, meaning a tendency to stay together, and a high _____, with the result that it takes a relatively large amount of energy to raise their temperature.

3. Oil is an example of a substance that will not readily bond with water and thus is _____, while sodium chloride is an example of a substance that is _____ because it will readily bond with water.

2.7 Acids and Bases

When considering the question of the water-based or "aqueous" solutions so common in nature, an important concept is that of acids, bases, and the pH scale used to measure their levels.

We all have had experience with acids and bases, whether we've called them by these names or not. Acidic substances tend to be a little more familiar: lemon juice, vinegar, tomatoes. Substances that are strongly acidic have a well-deserved reputation for being dangerous: The word *acid* is often used to mean something that can sear human flesh. It might seem to follow that bases are benign, but ammonia is a strong base, as are many oven cleaners. The safe zone for living tissue in general lies with substances that are neither strongly acidic nor strongly basic. Science has developed a way of measuring the degree to which something is acidic or basic—the pH scale. So widespread is pH usage that it pops up from time to time in television advertising ("It's pH-balanced!").

Acids, Bases, and pH

Acids Yield Hydrogen Ions in Solution, Bases Accept Them

The *H* in pH stands for hydrogen, while the *p* can be thought of as standing for power. Thus, we get "hydrogen power," which describes what lies at the root of pH. An **acid** is any substance that *yields hydrogen ions* when put in aqueous solution. A **base** is any substance that *accepts* hydrogen ions in aqueous solution.

How might this yielding or accepting come about? Recall first that an ion is a charged atom, and that atoms become charged through the gain or loss of one or more electrons. Because electrons carry a negative charge, the loss of an electron leaves an atom with a net positive charge. Also recall that the hydrogen atom amounts to one central proton and one electron that circles around it. A hydrogen ion, then, is a lone proton that has lost its electron—a positively charged ion whose symbol is H^+.

Now, suppose you put an acid—hydrochloric acid (HCl)—into some water. What happens is that HCl *dissociates* or breaks apart into its ionic components, H^+ and Cl^-. The HCl has therefore yielded a hydrogen ion (H^+). Now, with a greater concentration of hydrogen ions in it, the water is more acidic than it was (**Figure 2.18**).

What about bases? There is a compound called sodium hydroxide (NaOH)—better known as lye—that, when poured into water, dissociates into Na^+ (sodium) and OH^- (hydroxide) ions. The place to look in this case is the OH^- ions. Negatively charged as they are, they would readily bond with positively charged H^+ ions. In other words, they would *accept* H^+ ions in solution, which is the definition of a base. This accepting of the H^+ ions makes the solution more basic—or to look at it another way, less acidic. Thus, acids and bases are something like the two ends of a teeter-totter: When one goes up, the other comes down. As you might have guessed, in the right proportions they can balance each other out perfectly. Look at Figure 2.18 to see how this would play out with the solutions you've looked at so far. Mixing them, the collection of H^+ and OH^- ions now comes

Figure 2.18
Hydrogen Ions and pH

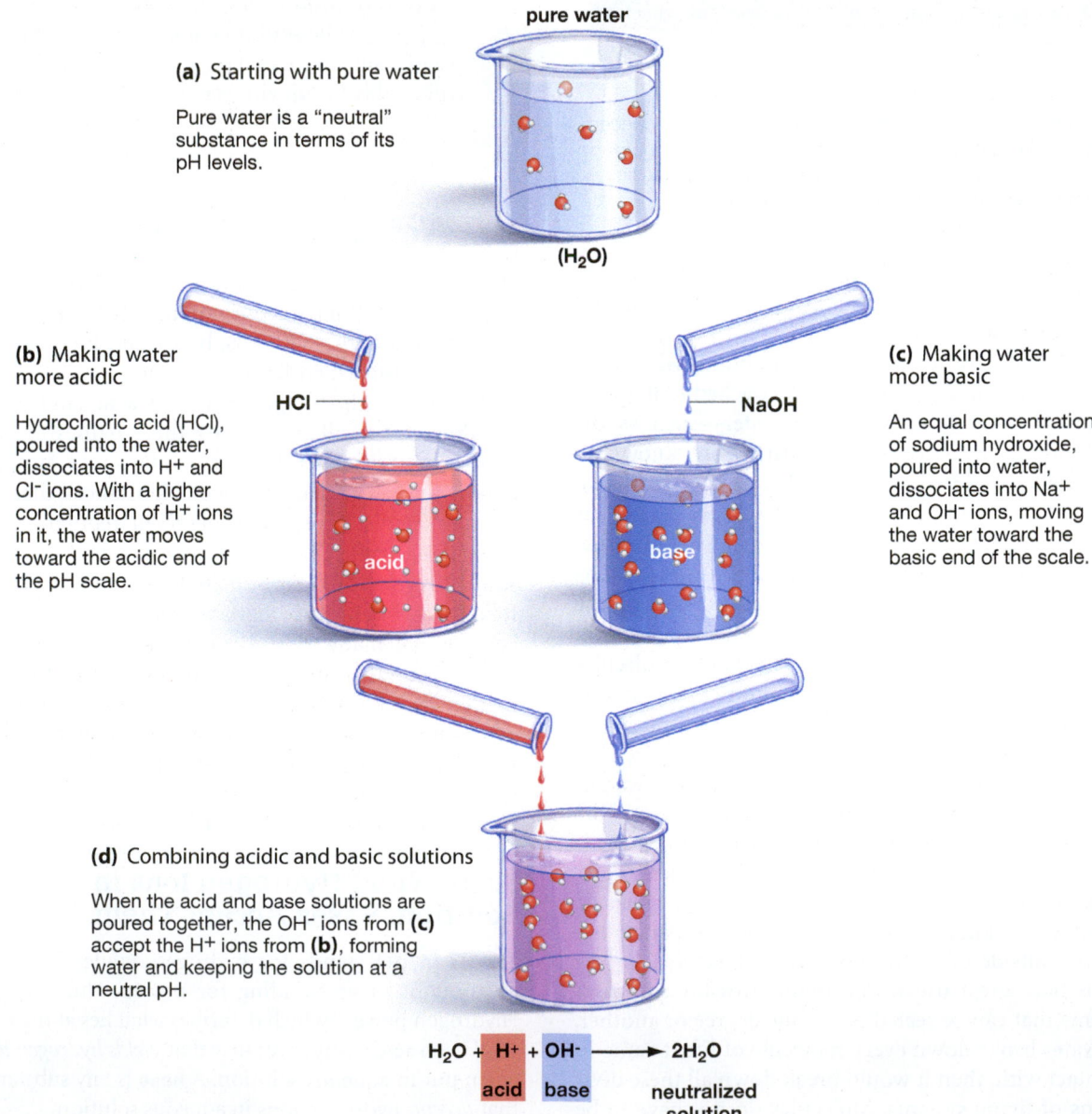

pure water

(a) Starting with pure water

Pure water is a "neutral" substance in terms of its pH levels.

(H_2O)

(b) Making water more acidic

Hydrochloric acid (HCl), poured into the water, dissociates into H^+ and Cl^- ions. With a higher concentration of H^+ ions in it, the water moves toward the acidic end of the pH scale.

HCl

acid

(c) Making water more basic

An equal concentration of sodium hydroxide, poured into water, dissociates into Na^+ and OH^- ions, moving the water toward the basic end of the scale.

NaOH

base

(d) Combining acidic and basic solutions

When the acid and base solutions are poured together, the OH^- ions from **(c)** accept the H^+ ions from **(b)**, forming water and keeping the solution at a neutral pH.

$$H_2O + H^+ + OH^- \longrightarrow 2H_2O$$

acid base

neutralized solution

together in water, and for each pair of ions that interacts, the result is:

$$H^+ + OH^- \rightarrow H_2O$$

Water, which is neutral on the pH scale. The acid and the base have perfectly balanced one another out. The OH^- ion, generally referred to as the **hydroxide ion,** is important because compounds that yield them in quantity are strongly basic and can be used to move solutions from the acidic toward the basic.

Ranking Substances on the pH Scale

Look at **Figure 2.19,** on the next page, for an idea of how acidic or basic some common substances are. Following from the notion of what pH amounts to, it's clear that battery acid, for example, is strongly acidic because when it dissociates in solution it yields a large number of hydrogen ions. As you move on to lemon juice and then tomatoes, however, the acids are weaker, which is to say they are substances that yield fewer H^+ ions in solution. By the time you get to human blood, you've arrived at substances that *accept* hydrogen ions.

The net effect of all this yielding and accepting of hydrogen ions is the concentration of H^+ ions in solution. Through this concentration, the notion of pH can be *quantified*—can have numbers attached to it. What is employed here is the **pH scale,** a scale used in measuring the relative acidity of a substance. Look at Figure 2.19 again, this time in connection with the numbers on the right of the scale. You can see that 0 on the scale is the most acidic, while 14 is the most basic. It's important to note that the pH scale is *logarithmic:* A substance with a pH of 9 is 10 times as basic as a substance with a pH of 8 and 100 times as basic as a substance with a pH of 7.

Some pH Terminology

Now here are some notes on pH terminology:

- As a solution becomes more basic, its pH *rises.* Thus, the higher the pH, the more basic the solution; the lower the pH, the more acidic the solution. Oven cleaner is said to have a high pH, while lemon juice has a low pH.

- Given that a hydrogen ion amounts to a single proton, it is also correct to say that an acid is something that yields *protons* in solution, while a base is something that accepts protons. This is how you will often hear hydrogen ions discussed in biology.

- A solution that is basic is also referred to as an **alkaline** solution.

Why Does pH Matter?

So, why do we care about pH? The brief answer is because living things are sensitive to its levels in many ways. There is, for example, a class of proteins called enzymes that you'll be reading about in Chapter 3. Enzymes are chemical tools that must retain a specific shape to function. However, if you put an enzyme in a solution with a pH that is too acidic, the enzyme loses its shape. Why? The charged nature of the acidic solution starts breaking down the enzyme's hydrogen bonds. (Remember, lots of positively charged protons are floating around in an acidic solution.) Likewise, cell membranes—which to some degree perform the milk carton function noted earlier—can start to break down if pH levels begin to go outside normal limits. Membranes and enzymes are so fundamental to life that when they are interfered with, death can result. It is not surprising, then, that many organisms have developed so-called acid-base **buffering systems,** meaning physiological systems that function to keep pH within normal limits. What are these limits? The usual range for living things is about 6–8, with the pH of the human cell being about 7 and that of the blood in our arteries about 7.4. However, some parts of the body have special pH requirements. The interior of your stomach, for example, can have a pH as low as 1—an extremely acidic environment that not only helps break down food but that also kills most bacteria that ride in on the food. Even a stomach can become too acidic, however (perhaps because of what we've eaten), in which case people may turn to the *pharmaceutical* pH buffers commonly known as antacids. These substances do just what their name implies: They raise pH levels in the stomach, thus helping to alleviate the symptoms of "heartburn."

Beyond its effects on individuals, pH can affect entire communities of living things. You may have heard of acid rain as a phenomenon that affects forests, streams, and lakes. Well, **acid rain** is just what it sounds like: rain whose pH level has been skewed toward the acidic side of the pH scale, with air pollution being the cause of this shift. As you can imagine, when rain itself has an altered pH, there can be widespread trouble for living things. Starting in the 1970s, this began to be the case in many environments worldwide, and this remains true today. (In the United States, the problem has been most severe in an area stretching from Michigan eastward through Pennsylvania and New York State.) Air pollution regulations have lessened the severity of acid rain in recent years, but it remains a long-term problem that requires constant monitoring.

On to Biological Molecules

In this chapter, you've learned a fair amount about some of nature's most fundamental building blocks:

Chemsitry Review: Acids, Bases, & pH

Figure 2.19

Common Substances and the pH Scale

Chemists use units called moles per liter to measure the concentration of substances in solution. The pH scale, derived from this framework, measures the concentration of hydrogen ions per liter of solution. The most acidic substances on the scale have the greatest concentration of hydrogen ions, while the most basic (or alkaline) substances have the lowest concentration of hydrogen ions. The scale is logarithmic, so that wine, for example, is 10 times as acidic as tomatoes and 100 times as acidic as black coffee.

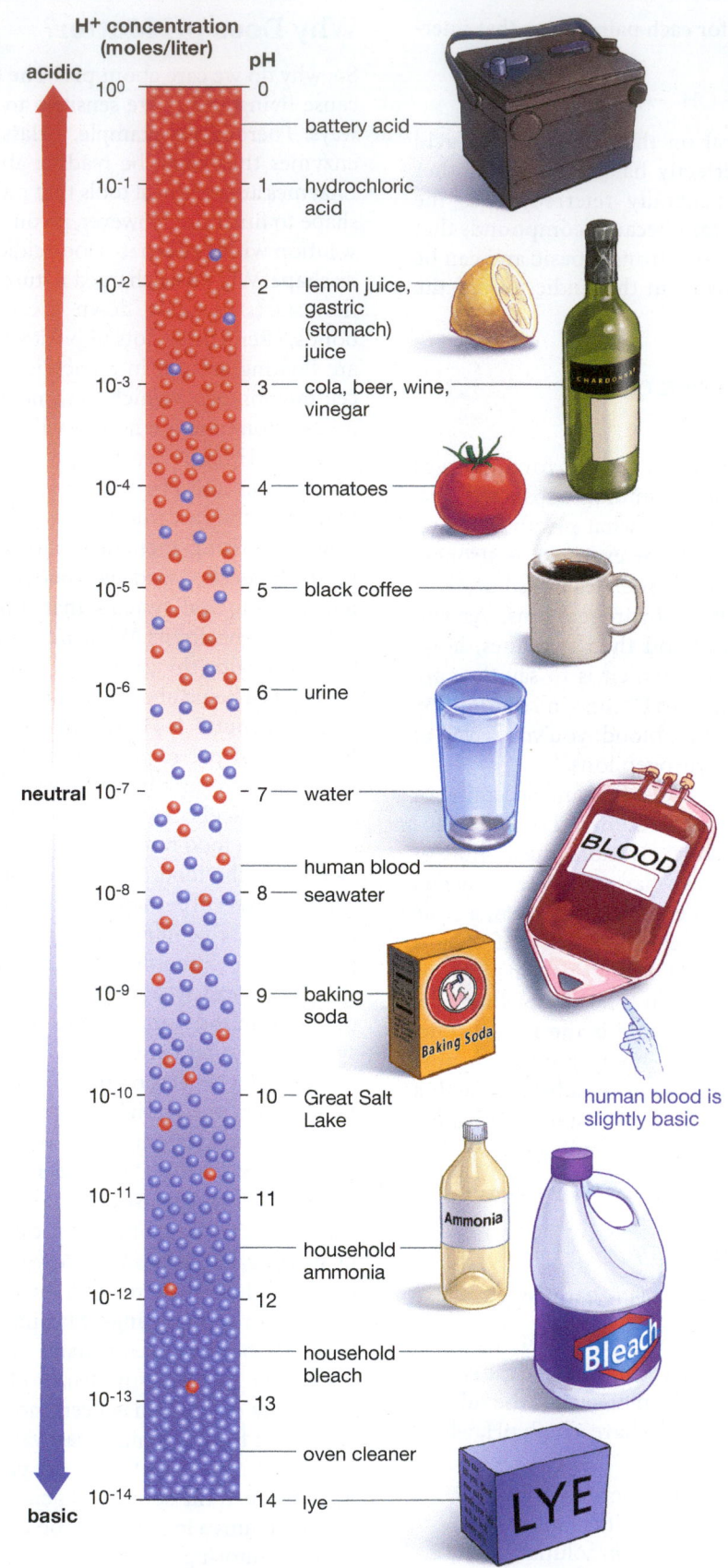

H⁺ concentration (moles/liter)

	pH	
acidic 10^0	0	
10^{-1}	1	battery acid
		hydrochloric acid
10^{-2}	2	lemon juice, gastric (stomach) juice
10^{-3}	3	cola, beer, wine, vinegar
10^{-4}	4	tomatoes
10^{-5}	5	black coffee
10^{-6}	6	urine
neutral 10^{-7}	7	water
		human blood
10^{-8}	8	seawater
10^{-9}	9	baking soda
10^{-10}	10	Great Salt Lake
10^{-11}	11	
		household ammonia
10^{-12}	12	
		household bleach
10^{-13}	13	
		oven cleaner
basic 10^{-14}	14	lye

human blood is slightly basic

atoms, ions, the simple substance water, and a special ion (the H^+ ion) whose levels govern pH. Now it's time to see how these building blocks fit together to create some of the larger-scale component parts of life. These component parts usually are referred to as *biological molecules*. Everyone has heard of at least some biological molecules. Carbohydrates are one variety of them, for example, and it's easy to name foods that we think of as carbohydrates. But what *is* a carbohydrate? What is the fundamental nature of this component part of life? To sharpen the point, how does a carbohydrate differ from a protein? In Chapter 3, you'll find out as we explore the types of molecules that make up the living world.

SO FAR . . .

1. An acid is any substance that _____ hydrogen _____ when put into an aqueous solution, while a base is any substance that _____ them.

2. The neutral point on the pH scale is _____, while the most acidic point is _____, and the most basic or alkaline point is _____.

3. Living things tend to have internal pH environments that hover around _____.

CHAPTER 2 REVIEW

Summary

2.1 Chemistry's Building Block: The Atom

- The fundamental unit of matter is the atom, whose most important constituent parts are protons, neutrons, and electrons. Protons and neutrons exist in the atom's nucleus; electrons move at some distance around the nucleus. Protons are positively charged; electrons are negatively charged; neutrons carry no charge. (p. 20)

- An element is any substance that cannot be reduced to a simpler set of constituent substances through chemical means. Each element is defined by the number of protons in its nucleus. (p. 21)

- The number of neutrons in an atom can vary independently of the number of protons. Thus, a single element can exist in various forms, called isotopes. (p. 22)

2.2 Chemical Bonding: The Covalent Bond

- Atoms can link to one another in the process of chemical bonding. One form

this bonding can take is covalent bonding, in which atoms share one or more electrons. (p. 22)

- Chemical bonding comes about as atoms "seek" their lowest energy state. An atom achieves this state when it has a filled outer electron shell. (p. 22)

- A molecule is an entity consisting of a defined number of atoms covalently bonded together. (p. 24)

- Atoms of different elements differ in their power to attract electrons. The term for measuring this power is electronegativity. Through electronegativity, a molecule can take on a polarity, meaning a difference in electrical charge at one end compared to the other. Covalent chemical bonds can be polar or nonpolar. (p. 25)

2.3 The Ionic Bond

- Two atoms will undergo a process of ionization when the electronegativity differences between them are great enough that one atom loses one or more electrons to the other. This process creates ions: atoms whose number of electrons differs from

their number of protons. The charge differences that result from ionization can produce an electrostatic attraction between ions. This attraction is an ionic bond, and the product of this bonding is an ionic compound: a collection of the atoms of two or more elements that have become linked through ionic bonding. (p. 25)

2.4 The Hydrogen Bond

- Hydrogen bonding links an already covalently bonded hydrogen atom with an electronegative atom. (p. 30)

2.5 Three-Dimensional Shape in Molecules

- Three-dimensional molecular shape determines the capacity molecules have to bind with one another. (p. 30)

2.6 Water and Life

- No living thing can be fully functional without a steady supply of water. Water's chemical properties allow it to facilitate many of the chemical reactions that take place in living things. (p. 31)

- Water has several other qualities that have affected life on Earth, among them cohesion and high specific heat. **(p. 33)**

- Some molecules do not interact with water and are called hydrophobic. Molecules or ions that do interact with water, by virtue of being charged or polar, are called hydrophilic. **(p. 34)**

2.7 Acids and Bases

- An acid is any substance that yields hydrogen ions when put in an aqueous (water) solution. A base is any substance that accepts hydrogen ions in solution. A base added to an acidic solution makes that solution less acidic, while an acid added to a basic solution makes that solution less basic. **(p. 35)**

- The concentration of hydrogen ions a solution has determines how basic or acidic that solution is, as measured on the pH scale (running from 0 to 14, with 0 most acidic, 14 most basic, and 7 neutral). **(p. 37)**

- The pH scale is logarithmic; a substance with a pH of 9 is 10 times as basic as a substance with a pH of 8. Living things function best in a near-neutral pH. **(p. 37)**

Key Terms

acid 35
acid rain 37
alkaline 37
atomic number 21
ball-and-stick model 29
base 35
buffering system 37
chemical bonding 22
covalent bond 24
electron 20
electronegativity 25
element 21
hydrocarbon 34
hydrogen bond 30
hydrophilic 35
hydrophobic 35
hydroxide ion 37
ion 28
ionic bonding 28

ionic compound 30
isotope 22
law of conservation of mass 24
mass 19
molecular formula 29
molecule 24
neutron 20
nonpolar covalent bond 25
nucleus 20
pH scale 37
polar covalent bond 25
polarity 25
product 29
proton 20
reactant 29
solute 32
solution 32
solvent 32
space-filling model 29
specific heat 33
structural formula 29

Understanding the Basics

Multiple-Choice Questions

(Answers are in the back of the book.)

1. Carbon is an element with an atomic number of 6. Based on this information, which of the following statements is true? (More than one may be true.)
   a. Carbon can be broken down into simpler component substances.
   b. Carbon cannot be broken down into simpler component substances.
   c. Each carbon atom will always have 6 neutrons.
   d. Each carbon atom will always have 6 protons.
   e. Number of protons + number of electrons = 6

2. Neon has an atomic number of 10 and thus has eight electrons in its second energy level. Thus neon (select all that apply):
   a. has a strong tendency to form covalent bonds.
   b. has a filled outer shell.
   c. has no tendency to form covalent bonds.

   d. is polar.
   e. all of these

3. Oxygen and hydrogen differ in their electronegativity. Thus:
   a. They can share electrons, but unequally.
   b. Sometimes oxygen takes electrons completely away from hydrogen.
   c. They can share electrons equally.
   d. Hydrogen is attracted to oxygen but does not bond with it.
   e. They have the same number of protons.

4. A molecule that does not have a net electrical charge at one end as opposed to the other is:
   a. an isotope.
   b. a polar molecule.
   c. a reactant.
   d. a nonpolar molecule.
   e. a solvent.

5. You add sugar to your coffee, and the sugar dissolves. Thus the coffee is the _____ and the sugar is the _____.
   a. solute; solvent
   b. solvent; solute
   c. polar covalent bond; nonpolar covalent bond
   d. nonpolar covalent bond; polar covalent bond
   e. ionic bond; hydrogen bond

6. Near an ocean or other large body of water, air temperatures do not vary as much with the seasons as they do in the middle of a continent. This tendency of water to resist changes in temperature is the result of water's:
   a. high density.
   b. low density.
   c. being a good solvent.
   d. low specific heat.
   e. high specific heat.

7. Janine has dry skin, so she uses body oil every morning. The oil seals in some of the water on her skin, so that it doesn't get as dry. This is possible because oils:
   a. are hydrophilic.
   b. are rare in nature.
   c. have a high specific heat.
   d. are more dense than water.
   e. are hydrophobic.

8. Some plants live in bogs in which the pH is about 2. Thus these plants are able to survive in a(n) _____ external environment.
   a. basic
   b. buffered
   c. acidic
   d. neutral
   e. alkaline

Brief Review

(Answers are in the back of the book.)

1. As with most elements, carbon comes in several forms, one of which is carbon-14. What are these forms called, and how does one differ from the other?

2. Compare the size of an atom with the size of its nucleus. Where are the electrons? In light of this, what makes up most of the volume of an atom?

3. Draw a line and label one end "complete + or − charge" and the other end "no charge" to indicate the charges on the molecules or ions after bonding has occurred. Along the line, indicate where polar covalent bonds, nonpolar covalent bonds, and ionic bonds should be placed.

4. Why are atoms unlikely to react when they have their outer shell filled with electrons?

5. Give two examples of the ways in which water acts as a heat buffer.

6. Why do living things need to keep such a tight control on their internal pH?

Applying Your Knowledge

1. In the Middle Ages, alchemists labored to turn common materials such as iron into precious metals such as gold. If you could journey back in time, how could you convince an alchemist that iron cannot be changed into gold?

2. Why does a balloon filled with helium float? Hydrogen can make balloons float, but it is not used for this purpose today because it is flammable. Based on chemical principles reviewed in the chapter, can you see why helium is not flammable? (*Hint:* Think what you are adding to a fire when you blow on it.)

3. How is water's importance to life reflected in the normal range of internal pH values that living things are likely to have?

Life's Components:
Biological Molecules

Four kinds of carbon-based molecules form the living world.

Available on MasteringBiology®

Activities: The Chemistry of Carbon; Diversity of Carbon-Based Molecules; Isomers; Functional Groups; Carbohydrates; Lipids; Proteins; Protein Structure; Protein Functions; Heritable Information: DNA; Nucleic Acid Structure; **Chemistry Review:** Organic Molecules; Carbohydrates; Lipids; Proteins; Nucleic Acids; **Current Events:** Asked to Get Slim, Cheese Resists; The Fat Trap; From Ancient Giants, Finding New Life to Help the Planet; **Building Vocabulary; Questions:** Reading Questions, Test Bank

Living things such as this giant redwood tree (*Sequoia sempervirens*) can grow to enormous sizes thanks to the strength of materials they themselves produce.

We are stardust . . .

Billion-year-old carbon

       from "Woodstock," by Joni Mitchell

Songwriters usually specialize in emotional truth, so it's no small feat that, in just a couple of lines, Joni Mitchell could sum up the literal truth of what we are made of. Admittedly, her billion-year-old figure is off—Earth actually is about 4.6 billion years old, and its carbon is older yet. Even so, she was right about the essentials. Are we stardust? To a large extent,

yes. All the heavy elements in our bodies—calcium, iron, carbon—were first cooked up in the fiery interior of stars and then dispersed into space when these stars exploded. Over time, these elements became a part of planets such as the Earth.

But surely we're more than carbon, right? Well, by weight we are more oxygen than anything else, mostly because oxygen is a principal element in water, which makes up so much of our bodies. Even so, it's carbon that really is life's element. How so? Well, consider the way in which living things amount to *concentrations* of carbon when compared to the makeup of the non-living world around them. As the environmental scientist Vaclav Smil has noted, Earth's crust is about 0.05 percent carbon, and Earth's atmosphere is about 0.01 percent carbon. Yet Earth's *organisms* are 18 percent carbon, even when water is taken into account, and nearly 50 percent carbon when water is factored out.

Given such figures, it's not surprising that, just as money must flow through an economy for it to operate, so carbon must flow through the living world for it to operate. First, plants and other photosynthesizers take in carbon from the atmosphere—over 100 billion tons of it per year, in the form of atmospheric CO_2. Then the plants, and all the organisms that eat them, use this carbon to make up their tissues and to power their activities. Finally, all these organisms die, and as their remains are broken down, some of the carbon in them is released into the atmosphere. But this is not a one-way trip; sooner or later, this same carbon is likely to move back into the living world, to be made part of a tree or a mushroom or a human being once again.

3.1 Carbon's Place in the Living World

So, why does carbon have this special place in life? The answer can be summed up in one word: linkage. Recall from Chapter 2 that carbon has four outer-shell,

or "valence" electrons but that most elements need *eight* outer-shell electrons for maximum stability. This means that carbon achieves maximum stability by linking up with four more electrons, which in turn means that carbon's bonding capacity is great. It can form very large and complex molecules, which is exactly what is needed in connection with the tremendous complexity of living things. Moreover, the bonds that carbon creates are covalent—it is *sharing* electrons with other atoms—which means its linkages are more stable than those formed by elements that bond ionically. Given the natural forces that can buffet life, such as ultraviolet radiation and heat, this stability is an important quality first in getting life going and then in keeping it going. So important is carbon that it forms one whole area of chemistry—**organic chemistry,** a branch of chemistry devoted to the study of molecules that have carbon as their central element (**Figure 3.1**, on the next page).

How does carbon link up with itself or other atoms? Last chapter, you had a look at a model of a very simple carbon-based molecule, methane. Here it is again:

$$
\begin{array}{c}
\text{H} \\
| \\
\text{H} - \text{C} - \text{H} \\
| \\
\text{H}
\end{array}
$$

Note that the only elements present are hydrogen and carbon, which makes this molecule a **hydrocarbon.** Now, observe a slightly more complex hydrocarbon, the familiar gas propane (C_3H_8).

$$
\begin{array}{c}
\text{H} \quad \text{H} \quad \text{H} \\
| \quad\;\; | \quad\;\; | \\
\text{H} - \text{C} - \text{C} - \text{C} - \text{H} \\
| \quad\;\; | \quad\;\; | \\
\text{H} \quad \text{H} \quad \text{H}
\end{array}
$$

You can see that, instead of being surrounded by four hydrogen atoms, the carbons here link with other carbons as well as with the hydrogens. This process of

 MB

The Chemistry of Carbon

Figure 3.1
Pure Carbon, Mostly Carbon

A diamond is pure carbon and the hardest natural material known. Surrounding the diamond is a lump of coal, which has a high carbon content because it is made up mostly of the remains of carbon-rich, ancient vegetation.

Diversity of Carbon-Based Molecules

extension keeps going with still more complex hydrocarbons. For obvious reasons, the preceding configuration is known as a *straight-chain* carbon molecule. But carbon has more tricks up its sleeve than simple straight-line extensions. The next hydrocarbon up the line is butane, the fuel in cigarette lighters. It can be just another straight-chain extension, looking like this:

But it can also look like this:

Isomers

These two forms of butane are known as *isomers*—molecules that have the same chemical formulas but differ in the spatial arrangement of their atoms.

Carbon can also form rings. Here is the structure of benzene (C_6H_6), which is found in petroleum products:

Note the three sets of double lines in the molecule. Recall that this means there are double *bonds* between these atoms; the atoms involved share *two* pairs of electrons.

It just so happens that all the carbon-based molecules introduced so far are hydrocarbons, but this is

Chemistry Review: Organic Molecules

the exception rather than the rule. Here are two representations of a molecule mentioned already—glucose, better known as blood sugar:

So here are some added oxygen atoms. Notice how the second model, below the first, with its heavy line down at the bottom, looks a little different than the benzene ring? Such a model is meant to give you a slightly more realistic picture of the actual arrangement of atoms when a carbon ring is present. Think of the ring as lying at an angle to the paper, with the heavy line closer to you than the line in the back; the H and OH groups then lie above and below the plane of the ring, as shown. You may also notice that when things get this complicated, space starts getting a little tight for writing in the letters that stand for the various atoms that are a part of the molecule. Because most rings are formed predominantly of carbon, it is common to dispense with the Cs when a structural formula is presented. Here is a third representation of the glucose model, written in this stripped-down form:

You can see that carbon is *assumed* to exist at each bond juncture in the ring; if some other element occupies one of these points, it is explicitly noted. Often the solitary Hs that are attached to carbons are left out as well.

Such models as these, as you'll recall from the last chapter, don't show the actual three-dimensional form of a molecule. To do so, ball-and-stick or space-filling models are required. Examples of these models are shown in **Figure 3.2**. With them, you can start to see how carbon is literally central to the molecules of the living world. Note how the carbon atoms form the core of the glucose molecule, with the hydrogen and oxygen atoms on the periphery.

3.2 Functional Groups

Having seen how carbon-based molecules start to become specialized through the addition of other kinds of atoms, it's worth noting that such atoms often are affixed in *groups*. Each of these units is a **functional group:** a group of atoms that confers a special property on a carbon-based molecule. Just as an attachment on a power tool might make the difference between a drill and a screwdriver, so a functional group can make the difference between one molecule and another. Look here at the formula for the hydrocarbon ethane, which is a flammable gas:

$$
\begin{array}{c}
\quad\; H \quad H \\
\quad\; | \quad\;\; | \\
H - C - C - H \\
\quad\; | \quad\;\; | \\
\quad\; H \quad H
\end{array}
$$

Now let's substitute a functional group, called the hydroxyl or –OH group, for the rightmost hydrogen atom in ethane and thus get:

$$
\begin{array}{c}
\quad\; H \quad H \\
\quad\; | \quad\;\; | \\
H - C - C - OH \\
\quad\; | \quad\;\; | \\
\quad\; H \quad H
\end{array}
$$

With this, we no longer have a gas; we have a liquid—the drinkable liquid ethyl alcohol. Anytime an –OH group is added to a hydrocarbon chain, some sort of alcohol is formed, be it the isopropyl alcohol used to disinfect cuts, the methanol used to power turbine engines, or some other variety. Thus, the –OH group has a generalized function in connection with hydrocarbons and is therefore referred to as a *functional group.* Now note another effect of this group. The ethane we started with was *nonpolar*—there was no difference in charge at one end of the ethane molecule as opposed to the other. But the addition of the group changed this; the oxygen atom it contains has such strong electronegativity that ethyl alcohol does have a polarity, meaning it can bond with other charged or polar molecules, including water. This exemplifies a general characteristic of functional groups, which is that they often impart an electrical charge, or at least a polarity, onto molecules, which makes a big difference in their bonding capacity. The –OH group serves this polarizing function not only when it is added to hydrocarbons, but also when it is part of other molecules as well. If you look at **Table 3.1**, you can see some of the functional groups that are most important in living

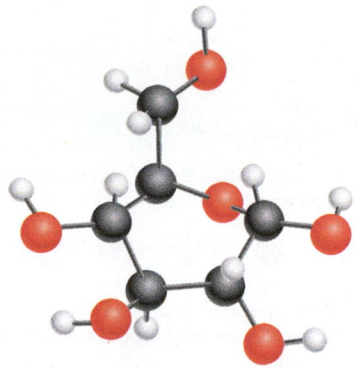   

ball-and-stick model of glucose   space-filling model of glucose

Figure 3.2
Ball-and-Stick and Space-Filling Models of Glucose
The simple sugar glucose is the single most important energy source for our bodies. Because it has several –OH groups, it is highly hydrophilic and thus readily breaks down in water. Note how the carbon atoms (in black) form the core of the molecule, with oxygen (in red) and hydrogen (in white) at the periphery.

things. You'll be seeing all of these groups as the chapter progresses: the carboxyl group when we look at fats (or lipids), the amino group when we go over proteins, and the phosphate group when we review both lipids and DNA.

Functional Groups

> ## SO FAR ...
>
> 1. The element _____ is central to the molecules that make up the living world. This element comes to its special status because of the great power it has to form stable _____.
>
> 2. A functional group is a group of _____ that confers a special property on a _____ molecule.

Table 3.1

Functional Groups

Group	Structural Formula	Found in
Carboxyl (–COOH)	$-C{\Large\langle}^{O}_{OH}$	fatty acids, amino acids
Hydroxyl (–OH)	—OH	alcohols, carbohydrates
Amino (–NH₂)	$-N{\Large\langle}^{H}_{H}$	amino acids
Phosphate (–PO₄)	$-O-\overset{\overset{O}{\|\|}}{\underset{\underset{O^-}{\|}}{P}}-O^-$	DNA, ATP

Table 3.2

Monomers, Polymers

If the monomer is . . .	The polymer is . . .
A monosaccharide (for example, glucose, fructose) 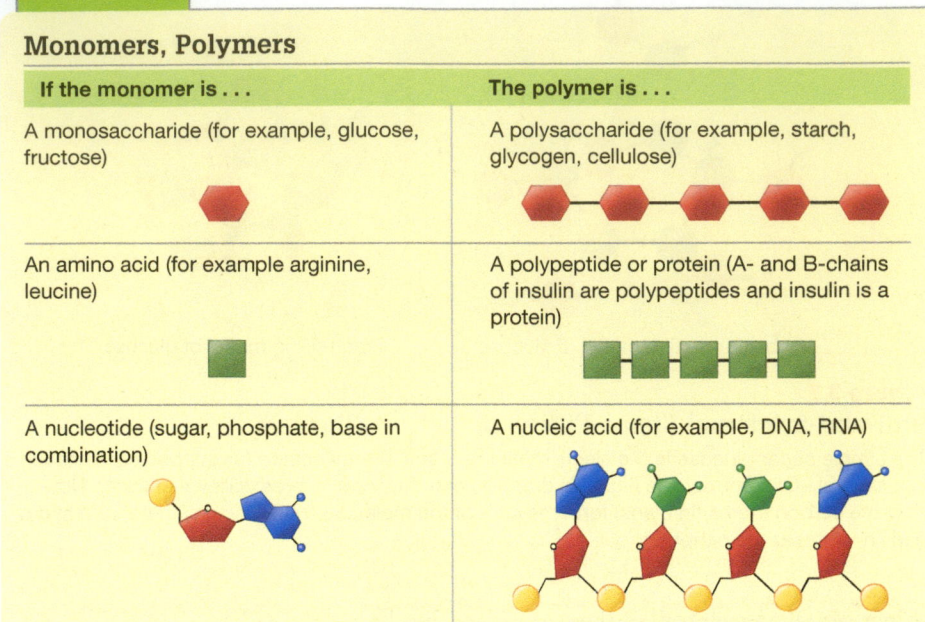	A polysaccharide (for example, starch, glycogen, cellulose)
An amino acid (for example arginine, leucine)	A polypeptide or protein (A- and B-chains of insulin are polypeptides and insulin is a protein)
A nucleotide (sugar, phosphate, base in combination)	A nucleic acid (for example, DNA, RNA)

3.3 Carbohydrates

With this overview of carbon structures under your belt, you're ready to begin looking at some of the classes of carbon-based molecules—the molecules of living things. You'll explore four groupings of these organic molecules: carbohydrates, lipids, proteins, and nucleic acids.

As you get into this section, it will be a great help to keep in mind that *complex* organic molecules often are made from *simpler* molecules. Many of the molecules you'll be reading about have a building-blocks quality to them. Take a simple sugar, or monosaccharide, such as glucose, put it together with another monosaccharide (fructose), and you have a larger *di*saccharide called sucrose (better known as table sugar). Put *many* monosac-

Figure 3.3
Carbohydrates in Foods

Breads, cereals, and pasta make up a significant proportion of our diets. These foods are all rich in carbohydrates, one of the four main types of biological molecules.

charide units together, and you have a polysaccharide, such as starch. The starch is an example of a **polymer**—a large molecule made up of many similar or identical subunits. Meanwhile, the glucose is an example of a **monomer**—a small molecule that can be combined with other similar or identical molecules to make a polymer. Look at **Table 3.2** for examples of both.

Carbohydrates: From Simple Sugars to Cellulose

Happily, for purposes of memory, the elements in the first molecules we'll look at, carbohydrates, are all hinted at in the name: **Carbohydrates** are organic molecules that always contain carbon, oxygen, and hydrogen and that in many instances contain nothing *but* carbon, oxygen, and hydrogen. Furthermore, they usually contain exactly twice as many hydrogen atoms as oxygen atoms. For example, the carbohydrate glucose ($C_6H_{12}O_6$) contains 12 atoms of hydrogen and 6 of oxygen. Most people think of carbohydrates purely in terms of foods such as breads and pasta (**Figure 3.3**), but as you'll see, carbohydrates have more roles than this in nature.

The building blocks of the carbohydrates are the **monosaccharides** mentioned earlier. These are the monomers of carbohydrates, collectively referred to as **simple sugars.** You've already seen several views of one of these monomers, glucose. Glucose has a use in and of itself: Much of the food we eat is broken down into it, at which point it becomes our most important energy source. Glucose can also bond with other monosaccharides, however, to form more complex carbohydrates. Let's take a look at how this happens. If you look at **Figure 3.4**, you can see an example of two glucose molecules bonding to create the disaccharide called maltose.

Several things are worth noting here. In maltose, as you can see, the link that joins the two glucose monomers is a single oxygen atom linked to carbons of each of the glucose units. To get this, two atoms of hydrogen and one atom of oxygen are split off (on the left side of the equation) from the original glucose molecules. What becomes of these three atoms? Look at the far right-hand side of the reaction and see the $+ H_2O$. The products of this reaction are a molecule of maltose and a molecule of water.

Now note the arrows in the middle of the reaction. You've been used to seeing a single, rightward-pointing arrow, but here the arrows go both ways. This means that this is a reversible reaction—it can proceed in either of the two directions. Maltose can be split apart to yield two *glucose* molecules. Finally, note the existence of the –OH functional group in both the glucose and maltose molecules.

Kinds of Simple Carbohydrates

You have thus far seen examples of one of the simplest carbohydrates, the monosaccharide glucose, and one

glucose + glucose = maltose + water

CH₂OH ... CH₂OH ... CH₂OH ... CH₂OH ... + H₂O

Figure 3.4
Carbohydrates Follow a Building-Blocks Model
In this example, two units of the monosaccharide (or simple sugar) glucose link to form the disaccharide maltose. In addition to maltose, the reaction yields water.

slightly more complex carbohydrate, a disaccharide called maltose. There are many kinds of mono- and disaccharides, however. Among the monosaccharides are, for example, fructose and deoxyribose. Among disaccharides, there are sucrose and lactose. At the risk of pointing out the obvious, note that all these sugars have –ose at the end of their name: If it's an –ose, it's a sugar (**Figure 3.5**).

Complex Carbohydrates

If you go another couple of bumps up in complexity from disaccharides, you get to the polymers of carbohydrates, the **polysaccharides.** The *poly-* in polysaccharide means "many," while *saccharide* means "sugars," and the term is apt. In the polysaccharide molecule cellulose, for example, there may be 10,000 glucose units linked with one another. The basic unit here is the six-carbon monosaccharide glucose $C_6H_{12}O_6$ from which chains of glucose units are built. The complexity of these molecules gives them their alternate name, *complex carbohydrates.* Four different types of complex carbohydrates interest us: starch, glycogen, cellulose, and chitin (**Table 3.3**, on the next page).

Starch is a complex carbohydrate found in plants; it exists in the form of such foods as potatoes, rice, carrots, and corn. In plants, these starches serve as the main form of carbohydrate *storage,* sometimes as seeds (rice and wheat grains) or sometimes as roots (carrots or beets) (Table 3.3).

Another complex carbohydrate, **glycogen,** serves as the primary form of carbohydrate storage in animals. Thus, glycogen does for animals what starch does for plants. For this reason, glycogen is sometimes called animal starch. The starches or sugars we eat are broken down eventually into glucose, at which point some of this glucose may be used immediately as an energy source. Some may not be needed right away, however, in which case it's moved into muscle and liver cells to be stored as the more complex carbohydrate glycogen (Table 3.3).

Cellulose is a structural, complex carbohydrate produced by plants and other organisms. Despite this in-

nocuous-sounding function, cellulose is important because it makes up so much of the natural world. It is easily the most abundant carbohydrate on Earth: Trees, cotton, leaves, and grasses are largely made of it. The dry-weight of a single white oak tree, for example, is about 4,200 pounds, of which 1,200 pounds might be cellulose. When you think of how many trees there are in the world—to say nothing of the uncountable acres of grass, cotton, and so forth—you begin to get an idea of how much cellulose there is on Earth. When this carbohydrate is enmeshed with a hardening substance called lignin, the result is a set of cell walls that can hold up giant redwood trees. Because cellulose is so dense and rigid, it is not surprising that human beings cannot digest it. This statement may make you wonder about grass-eating *cows,* but in fact cows have cellulose digested for them by special bacteria in their digestive tract. Cellulose is important to humans because it is our major source of insoluble fiber, which helps move foods through the digestive tract. Because cellulose exists in the cell walls of plants, we can get our fiber from such foods as whole grains.

Carbohydrates

Figure 3.5
Sugars Come in Many Forms

Sucrose or table sugar comes to us from sugarcane or sugar beets, and glucose is found in corn syrup. Fructose comes to us in sweet fruits and in high-fructose corn syrup, which often is used to sweeten soft drinks.

Our fourth complex carbohydrate, **chitin,** forms the external skeleton of the arthropods—all insects, spiders, and crustaceans, for example. In these animals, chitin plays a structural role similar to that of cellulose in plants: It gives shape and strength to the structure of the organism (Table 3.3).

SO FAR...

1. Biological molecules often have a building-blocks quality to them, with smaller _____ linked together to form larger _____.

2. Carbohydrates are organic molecules that always contain _____, _____, and _____. The building blocks of carbohydrates are the simple sugars or _____, which link to form _____.

3. The four most important varieties of complex carbohydrates in the living world are _____, which serves as a carbohydrate storage molecule in plants; _____, which serves this same function in animals; _____, which makes up a large portion of plant stems and leaves; and _____, which forms the external skeleton of all arthropods.

Chemistry Review: Carbohydrates

Table 3.3

Four Examples of Complex Carbohydrates

	Starch	Glycogen	Cellulose	Chitin
Structure				
Function	Serves as a form of carbohydrate storage in many plants	Serves as a form of carbohydrate storage in animals	Provides structural support for plants and other organisms	Makes up a large portion of the outer "skin" or cuticle of arthropods
Example	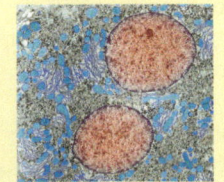 Starch granules within cells of a raw potato slice	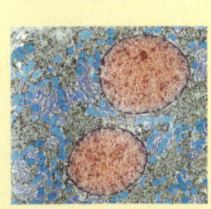 Glycogen granules (black dots) within a liver cell	 Cellulose fibers within the cell wall of a marine algae cell	 The chitinous cuticle of a tick

3.4 Lipids

We turn now to our second major group of biological molecules, the **lipids**, a class of molecules whose defining characteristic is that they do not readily dissolve in water. It turns out that lipids are made of the same elements as carbohydrates—carbon, hydrogen, and oxygen. But lipids have much more hydrogen, relative to oxygen, than do the carbohydrates. We're all familiar with some lipids; they exist as fats, oils, cholesterol, and as hormones such as testosterone and estrogen. Unlike the other biological molecules you'll be studying, however, a lipid is not a polymer composed of component-part monomers; no single structural unit is common to all lipids. Thus, the one characteristic shared by pure lipids is that they do not readily dissolve or "break down" in water. Remember a discussion last chapter about the need living things have for internal compartments that are sealed off from one another? Well, thanks to their insolubility, lipids are able to serve this function. In addition, lipids have considerable powers to store energy and to provide insulation (think of the abundant fat on a polar bear).

The Glyceride Lipids

The most common kind of lipid, the glyceride, can be thought of as a molecule in two parts. The first part is a "head" composed of the alcohol glycerol:

The second part is one or more fatty acids; an example can be seen in **Figure 3.6**. This is stearic acid, one of the fatty acids in animal fat. As you can see, it amounts to a long chain of hydrogen and carbon atoms—its hydrocarbon portion—that terminates with a COOH or "carboxyl" functional group on the left. Now, bringing the carboxyl part of this fatty acid chain together with the –OH group of glycerol is what makes a glyceride. But you can see that the glycerol has *three* –OH groups on the right. Thus there is, you might say, docking space on glycerol for three fatty acids, and in the synthesis of many glycerides, this

linkage takes place: Three fatty acids link with glycerol to form a triglyceride, which is the most important dietary form of lipid. **Figure 3.7** shows how it works schematically.

The Rs on the right end of the COOH group stand for whatever the hydrocarbon chain is of that particular fatty acid. To get a better idea of the shape of a completed triglyceride, look at the space-filling model in **Figure 3.8**, on the next page. This particular triglyceride, called tristearin, has three stearic fatty acids that stem like tines on a fork from the glycerol. This is only one possibility among many and is exceptional, rather than usual, in that all three fatty acids are the same. Among the dozens of fatty acids that exist, several *different* kinds of fatty acids generally will hook up in glycerol's three –OH "slots" to form a triglyceride. Actual fat products, such as butter, are composed of different proportions of various fatty acids.

We've seen three fatty acids linking with glycerol to form *triglycerides*, but monoglycerides (one fatty acid joined to glycerol) and diglycerides (two fatty acids joined with glycerol) can be formed as well. Triglycerides are the most important of the glycerides, however, because they constitute about 90 percent of the lipid weight in foods.

With all this under your belt, you're ready for a couple of definitions. A **triglyceride** is a lipid molecule formed from three fatty acids bonded to glycerol. A **fatty acid** is a molecule found in many lipids that is composed of a hydrocarbon chain bonded to a carboxyl group.

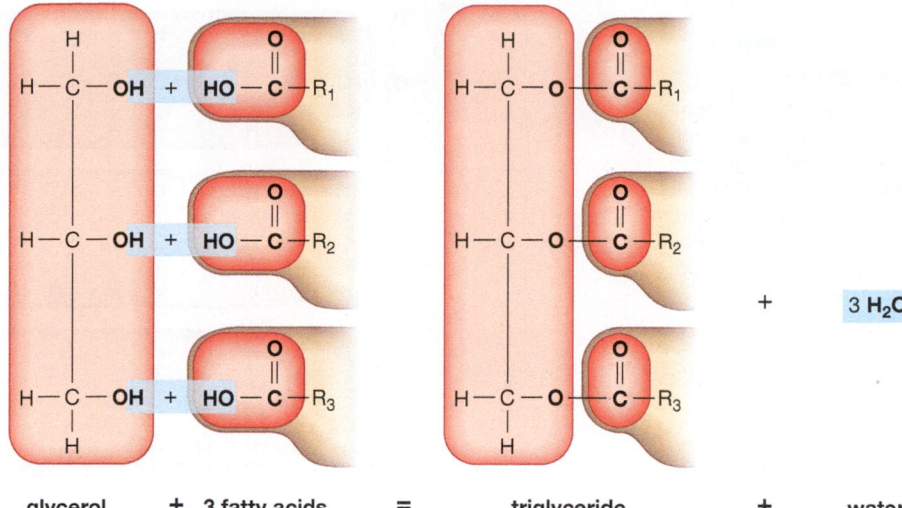

glycerol　+　3 fatty acids　=　triglyceride　+　water

Figure 3.7
Formation of a Triglyceride

Lipids

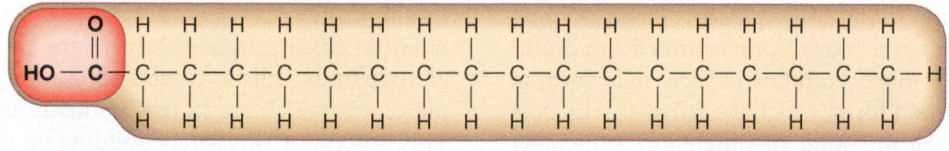

Figure 3.6
Structural Formula for Stearic Acid

Figure 3.8
The Triglyceride Tristearin
This lipid molecule is composed of three stearic fatty acids, stemming rightward from the glycerol –OH "head." Tristearin is found in both beef fat and the cocoa butter that helps make up chocolate.

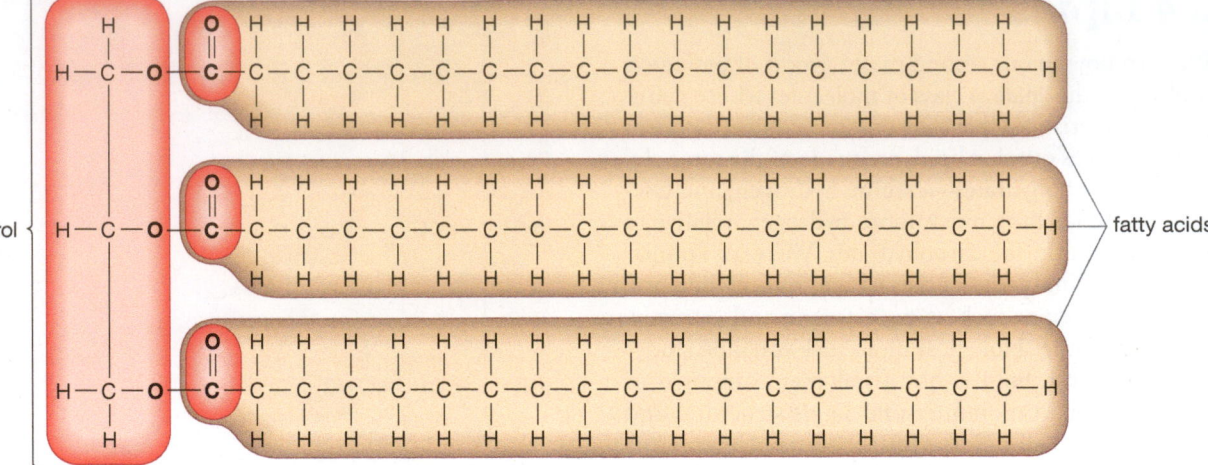

Chemistry Review: Lipids

Saturated and Unsaturated Fatty Acids in Triglycerides

If you now look at **Figure 3.9**, you can see three different fatty acids: palmitic, oleic, and linoleic. There is an obvious difference among them in that the oleic and linoleic hydrocarbon chains are "bent" compared to the straight-line palmitic chain. On closer inspection, you can see that the palmitic acid has an unbroken line of single bonds linking its carbon atoms. Meanwhile, the oleic acid has one double bond between the carbons in its chain, and the linoleic acid has two. Furthermore, note that these double bonds exist precisely where the "kinks" appear in these molecules.

What these variations describe are the differences between three *kinds* of fatty acids. First, there is a **saturated fatty acid**—a fatty acid with no double bonds between the carbon atoms of its hydrocarbon chain. Then, there is a **monounsaturated fatty acid** (a fatty acid with one double bond between carbon atoms) and a **polyunsaturated fatty acid** (a fatty acid with two or more double bonds between carbon atoms). What a saturated fatty acid is saturated *with* is hydrogen atoms. A given monounsaturated fatty acid could theoretically have its lone carbon-carbon double bond replaced with a *single* bond between the carbons. This change would entail the addition of two more hydrogen atoms at the bond sites. Once this happened, this fatty acid would contain the maximum number of hydrogen atoms possible, meaning it would be a saturated fatty acid.

This may not sound like much of a difference, but it has several important consequences. First, at room temperature, as you move from saturated to unsaturated, you also move from fats in their solid form to fats in their liquid form, which we call **oils.** Why does saturation make this kind of difference? Remember that saturated fatty acids have a straight-chain form; as such, they can pack together tightly in a triglyceride,

like so many boards in a lumberyard. In contrast, unsaturated fatty acids stick out at varying angles to one another; thus, they have a disorder that generally makes them liquid at room temperature. In short, the "kinks" in unsaturated fatty acids have consequences.

The distinction between saturated and unsaturated fatty acids also has another important effect, this one related to human health. Think of it this way: To the degree that *fatty acids* are saturated, the *fats* they make up will be saturated. And consumption of saturated fats has long been linked with heart disease—saturated fats raise total blood cholesterol levels, and high cholesterol levels are a risk factor for heart disease. The picture on dietary fats in general is more complex, however; some are neutral with respect to health and others actually are beneficial. You can read more about this in "From Trans Fats to Omega-3s" on page 54.

Lipid and Carbohydrate Energy Storage and Use

Lipids and the carbohydrates we looked at have something in common in connection with energy use on the one hand and energy storage on the other. To be used for energy, the fats we eat must first be broken down into their glycerol and fatty acid components. To be stored away, however, these component parts must be combined, thus producing triglycerides. Carbohydrates, meanwhile, may come to us in their complex starch form, but to be used for energy expenditure, they must first be broken down into their simple carbohydrate building blocks, usually meaning glucose (**Figure 3.10**, on page 52). Should we not need glucose immediately for energy expenditure, glucose units will be combined into larger glycogen molecules, which are stored in muscle and liver cells. This process of alternately building up molecules for energy storage or breaking them down for energy expenditure is a major task of living things.

(a) Palmitic acid

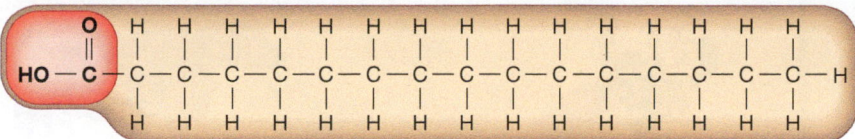

saturated (no double bonds)

(b) Oleic acid

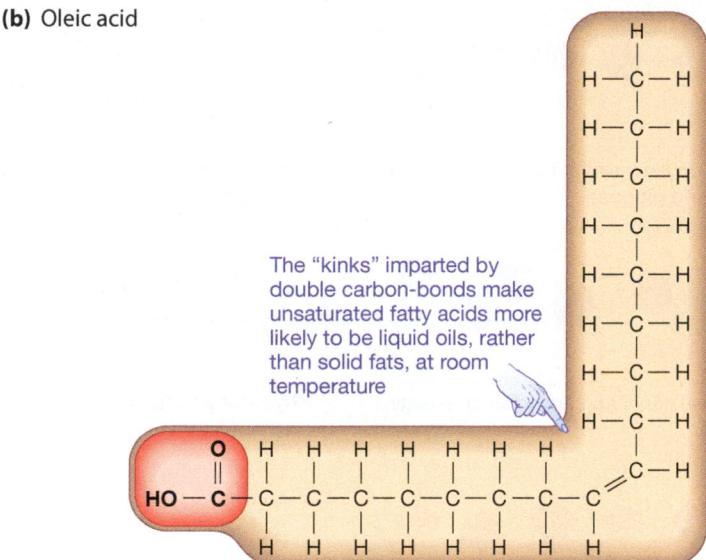

The "kinks" imparted by double carbon-bonds make unsaturated fatty acids more likely to be liquid oils, rather than solid fats, at room temperature

monounsaturated (one double bond)

(c) Linoleic acid

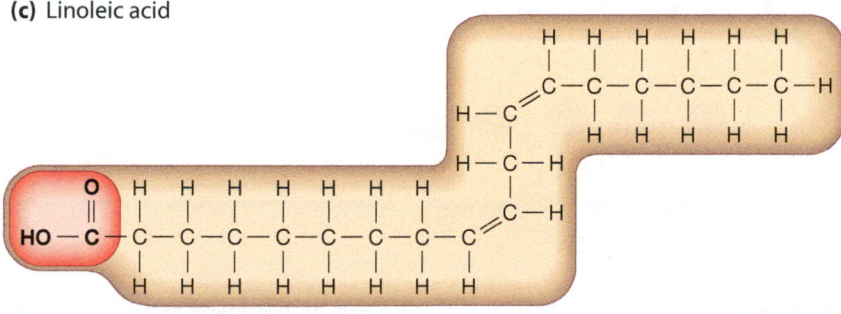

polyunsaturated (more than one double bond)

Figure 3.9

Saturated and Unsaturated Fatty Acids

The degree to which fatty acid hydrocarbon chains are "saturated" with hydrogen atoms has consequences for both the form these lipids take and for human health.

(a) The hydrocarbon "tail" in palmitic acid is formed by an unbroken line of carbons, each with a single bond to the next, making this a saturated fatty acid.

(b) In oleic acid, a double bond exists at one point between two carbon atoms, making this a monounsaturated fatty acid.

(c) The carbons in linoleic acid have double bonds in two locations, making this a polyunsaturated fatty acid.

Figure 3.10
Storage and Use of Carbohydrates and Lipids
Carbohydrates and lipids generally are stored in one form but used in another.

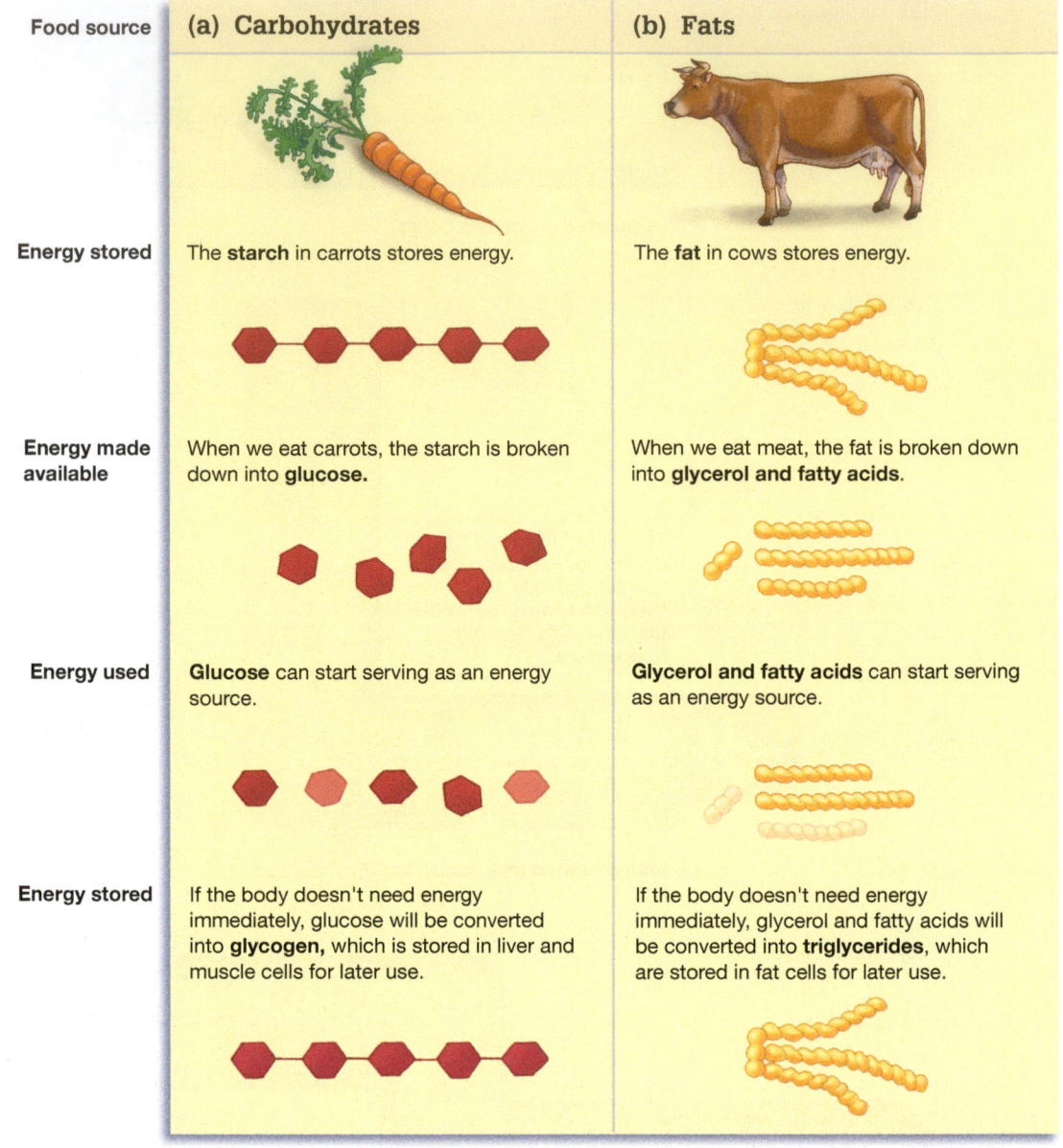

	(a) Carbohydrates	(b) Fats
Food source		
Energy stored	The **starch** in carrots stores energy.	The **fat** in cows stores energy.
Energy made available	When we eat carrots, the starch is broken down into **glucose.**	When we eat meat, the fat is broken down into **glycerol and fatty acids**.
Energy used	**Glucose** can start serving as an energy source.	**Glycerol and fatty acids** can start serving as an energy source.
Energy stored	If the body doesn't need energy immediately, glucose will be converted into **glycogen,** which is stored in liver and muscle cells for later use.	If the body doesn't need energy immediately, glycerol and fatty acids will be converted into **triglycerides**, which are stored in fat cells for later use.

The Steroid Lipids

Our second variety of lipid molecules, the **steroids,** can be defined as a class of lipids that have, as a central element in their structure, four carbon rings. What separates one steroid from another are the various side chains that can be attached to these rings (**Figure 3.11**). When you see how different steroids are structurally from the triglycerides, you can understand why the monomers-to-polymers framework doesn't apply to lipids.

Among the most well-known steroids are cholesterol and two of the steroid hormones, testosterone and estrogen. Like fats in general, cholesterol has a bad reputation, but also like fats in general, it serves good purposes, too. **Cholesterol** is a steroid molecule that forms part of the outer membrane of all animal cells and that acts as a precursor for many other steroids.

One of the steroids formed from it is the principal "male" hormone testosterone; another is the principal "female" hormone estrogen. (Both hormones actually are produced in both sexes, though in differing amounts.) The term *steroids* by itself undoubtedly rings a bell because the phrase "on steroids" has come to mean artificially bulked up or supercharged. In this common usage, steroids refers to manufactured drugs that are close chemical cousins of one variety of natural steroids, the muscle-building steroid hormones.

The Phospholipids

Our next class of lipids, phospholipids, has something of the same makeup as triglycerides in that a phospholipid has a glycerol head with fatty acids attached to it. But where triglycerides have three fatty acids stemming from the glycerol, phospholipids have only two (**Figure 3.12a**). Linking up with glycerol's third –OH group is a charged **phosphate group,** which is a phosphorus atom surrounded by four oxygen atoms. Thus, we can define a **phospholipid** as a charged lipid molecule composed of two fatty acids, glycerol, and a phosphate group.

The combination of a phosphate group and fatty acid tails is extremely important because it gives phospholipids a dual nature: Being hydrocarbons, the fatty acid tails are hydrophobic, but the phosphate head is hydro*philic*—it will readily bond with water—because it is charged. In Figure 3.12b, you can see the effect of this structure in solution. Imagine a phospholipid as a marker buoy in deep water. No matter how you push the hydrocarbon tail around, it's going to end up waving free *out* of the water, while the head is going to be submerged in it. You will learn more about these molecules later, when you study cells; for now, just note that the material on the periphery of cells—the outer membrane of a cell—is largely made of phospholipids. Living things need the kind of partitions described earlier, and these partitions are composed to a significant extent of phospholipids.

A Fourth Class of Lipids Is Wax

Ever wonder why water rolls off a duck's back? It's because the duck's feathers are coated by another variety of lipid: a wax. Likewise, beehives are made largely of wax (beeswax, of course), and an apple can

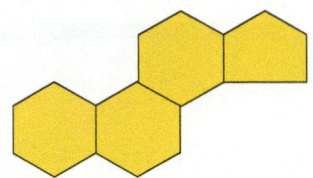

(a) Four-ring steroid structure

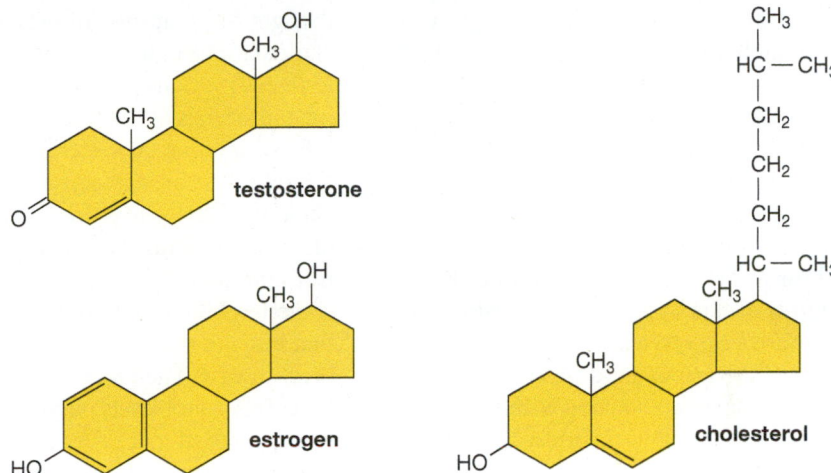

(b) Side chains make each steroid unique

testosterone

estrogen

cholesterol

Figure 3.11
Structure of Steroids
(a) The basic unit of steroids, four interlocked carbon rings.
(b) Types of steroids, each differentiated from the other by the side chains that extend from the four-ring skeleton.

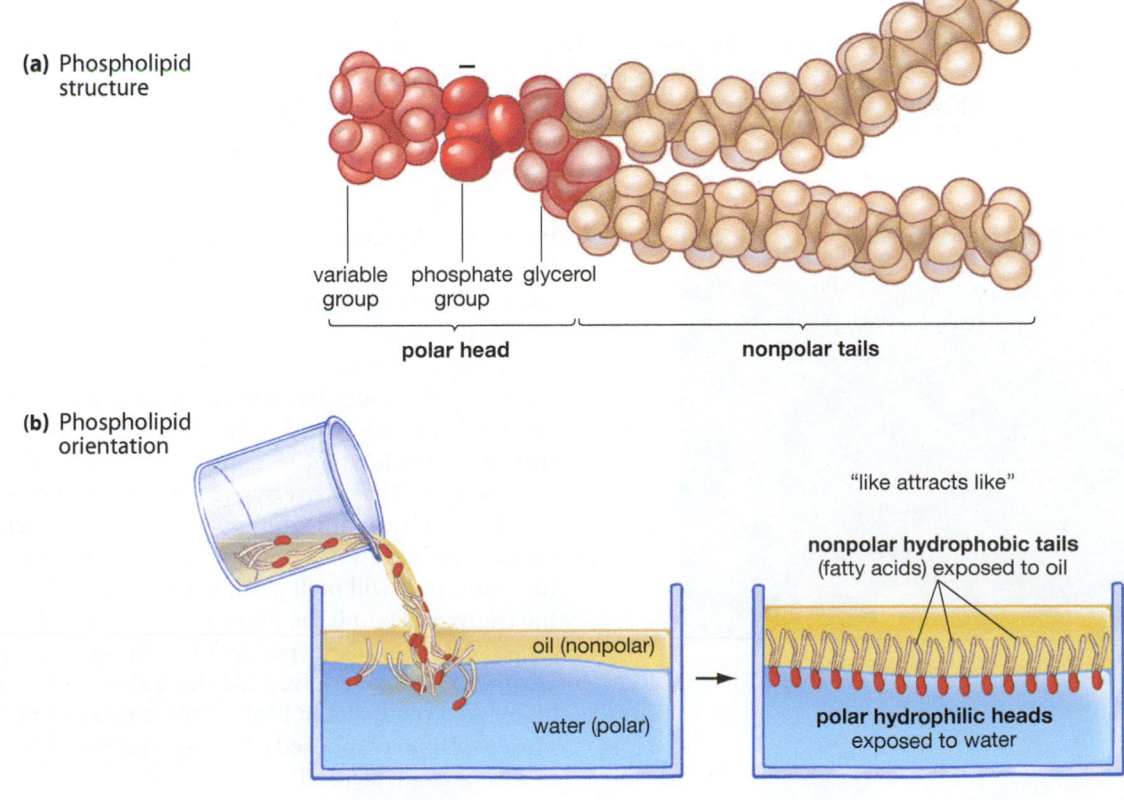

(a) Phospholipid structure

variable group · phosphate group · glycerol
polar head   **nonpolar tails**

(b) Phospholipid orientation

"like attracts like"

nonpolar hydrophobic tails
(fatty acids) exposed to oil

oil (nonpolar)

water (polar)

polar hydrophilic heads
exposed to water

Figure 3.12
A Dual-Natured Molecule

(a) Phospholipids are composed of two long fatty acid "tails" attached to a "head" containing a phosphate group (which carries a negative charge) and another variable group (which often carries a charge).

(b) Because the head is polarized, it can bond with water and thus will remain submerged in it; the tails have no such bonding capability.

ESSAY

From Trans Fats to Omega-3s: Fats and Health

Saturated and unsaturated fats, trans fats, omega-3's. What a confusing situation. Average Americans might understandably throw up their hands, assume that all fats are unhealthy, and try to reduce their intake of every variety. But a closer look at the various kinds of fats reveals that, while some are indeed bad for the heart, others don't affect its health one way or another, and still others are actually good for it. Thus it's possible to rate various kinds of fats based on their health effects and it's easy to learn which fats are contained in various foods. Here is a hierarchy of dietary fats starting with the most healthful and going to the most harmful:

- Polyunsaturated fats containing omega-3 fatty acids
- Monounsaturated fats and polyunsaturated fats that do not contain omega-3s
- Saturated fats
- Trans fats

To understand this ranking, it's important to understand the effects these various fats have on two substances produced in the body: low-density lipoproteins, or LDLs,

and high-density lipoproteins, or HDLs. The LDLs (or "bad cholesterol") can be thought of as capsules of protein that carry cholesterol from the liver and small intestines *to* various places in the body, including the arteries of the heart. Meanwhile, HDLs (or "good cholesterol") can be thought of as protein capsules that carry cholesterol to the liver *from* the heart and other tissues, thus clearing this cholesterol from the system. It is the lodging of LDL-cholesterol molecules in the heart's arteries that initiates the most prevalent form of heart disease. Therefore, any food that raises LDL cholesterol levels ought to be avoided, while any food that raises HDL cholesterol levels ought to be sought out. So where do the various forms of dietary fat rank on this and other measures of health? Here is the rundown on all of them, looking at them again from most healthy to least healthy.

Polyunsaturated Fats Containing Omega-3s

Standing at the top of the "good fats" list are fats that contain omega-3 fatty acids. Omega-3s are one variety of polyunsatu-

rated fatty acid—fatty acids that have two or more carbon-carbon double bonds somewhere in their hydrocarbon chains. Now, the far end of the hydrocarbon chain (away from the glycerol head) is called the *omega end*. When a carbon double bond is located between the third and fourth carbons up from the omega end, the result is an omega-3 fatty acid. Polyunsaturated fats containing these fatty acids raise HDL levels, guard against blood clot formation, reduce fat levels generally in the bloodstream, and reduce the growth of the fatty deposits that clog heart arteries. Omega-3 fatty acids are found most abundantly in certain kinds of fatty fish, including salmon, albacore tuna, mackerel, lake trout, and sardines; and in lesser amounts in plant-based foods such as canola oil, walnuts and the soybean-based product tofu.

Monounsaturated Fats and Other Polyunsaturated Fats

Monounsaturated fats, which are mostly found in oils, leave both LDL and HDL levels unchanged. However, one particular source of monounsaturated fats may actually work to prevent heart disease. Several studies have shown that residents of Mediterranean countries have significantly lower levels of heart disease than people in other parts of the developed world. What these residents have in common, in terms of

Figure 3.13
A Waxy Outer Covering

Wax can form a tight seal, which is just what plants need to help them keep water in while keeping microbial invaders out. Pictured is a cross section of a Haworthia leaf, which like all leaves has a waxy outer covering called a cuticle, visible as the layer of material running across the top of the leaf. Cuticles on leaves are pock-marked with microscopic pores, called stomata, that can open to allow the plant to take in the carbon dioxide it needs to perform photosynthesis.

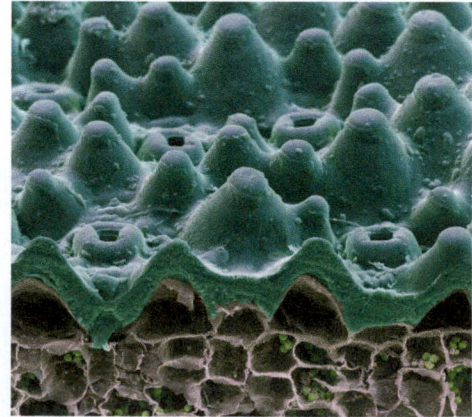

be polished because its outer surface has a wax coating that takes on a shine when the apple is rubbed. As these examples show, waxes often serve in nature to seal one thing off from another—the duck from water and the apple from the world around it. Waxes are extremely widespread in plants; almost all plant surfaces exposed to air will have a thin covering (called a cuticle) that is composed largely of wax (**Figure 3.13**). These coverings conserve water even as they protect the plant from outside invaders, such as fungi. Waxes are solid yet pliable at lower temperatures, but they will melt at higher temperatures. Like the triglycerides and the phospholipids we looked at, waxes are composed partly of fatty acid chains. However, whereas triglycerides have three fatty acids linked to a small alcohol (glycerol), a **wax** can be defined as a lipid composed of a single fatty acid linked to a long-chain alcohol.

diet, is consumption of large amounts of the monounsaturated fats in olive oil. In 2005, researchers uncovered a reason that olive oil may be beneficial: It contains a naturally occurring substance (called oleocanthal) that stands to work against heart disease by cutting down on the inflammation that helps it get going. Other sources of monounsaturated fats are peanut oil and avocados.

When we look at polyunsaturated fats other than those that contain omega-3s, we find that they leave LDL levels unchanged, although they do slightly lower the "good" HDL cholesterol levels. Even so, most experts regard them, at worst, as neutral to health. This type of polyunsaturated fat can be found in safflower and corn oil.

Saturated Fats

Nothing in your diet will raise LDL levels as much as saturated fats do. Most studies have found that each 1 percent increase in these fats leads to a 2 percent increase in LDL cholesterol levels. Given an effect such as this, the American Heart Association recommends limiting calories from saturated fats to less than 7 percent of total daily calories. The usual source for saturated fats is animal fat—primarily fatty meat and dairy products—but they can also be found in a few plant products, such as coconut oil and cocoa butter.

Trans Fats

Trans fats stand apart from all the other fats we've looked at in that they are produced not by nature but by an industrial process, called *hydrogenation,* in which hydrogen is bubbled into mono- or polyunsaturated oils. Food manufacturers do this for several reasons: It turns these oils into fats at room temperature, it often gives them a creamy texture that consumers like, and it increases the shelf life of the foods that result. The problem is that hydrogenation changes the chemical structure of fatty acids in ways that make them unhealthful. In most instances, this process has the effect of changing the *orientation* of the hydrogen atoms that already exist at carbon-carbon double bonds. Normally these pairs of hydrogens are on the same side, like this:

cis form (hydrogen atoms on same side of chain)

With so-called partial hydrogenation, they can end up on the opposite or *trans* sides, like this:

trans form (hydrogen atoms on opposite sides of chain)

Thus, we get the term *trans* fatty acids. The trans fats that result from this change are found in a dwindling, but still significant, number of packaged and fast foods: cookies, French fries, cakes, crackers, doughnuts, popcorn, and candy, to name a few. Trans fats earned their reputation as the least healthy fats by having a twin set of bad effects that not even saturated fats can manage: They raise LDL levels while lowering HDL levels. In addition, they appear to boost fat levels in general in the blood while impairing the ability of blood vessels to open or "dilate." As such, the American Heart Association recommends limiting trans fat intake to less than 1 percent of total daily calories.

How can you know how much trans fat you are consuming? For packaged foods, it's easy. Beginning in 2006, the federal Food and Drug Administration (FDA) required food producers to list trans fats separately on the Nutrition Facts labels seen on all packaged products. Check these labels to find out the amount of trans fats you're getting. Chances are, however, that you won't be getting much in the future. The FDA labeling requirement produced a surge of activity among packaged producers to rid their products of trans fats. Hence, what you'll often see on the Nutrition Facts labels these days is "Trans fat 0 g," meaning zero grams of trans fat.

SO FAR . . .

1. The defining characteristic of lipids is that they do not _____ in _____.

2. The fats in foods are composed mostly of triglycerides, which are formed from three _____ bonded to a _____. A saturated fatty acid is one that has no double bonds between the _____ of its _____ chain.

3. Steroids are a class of lipids whose central structure includes four _____. Examples include the reproductive hormones _____ and _____. Another class of lipids is important in making up the outer membrane of cells; these are the _____, composed of glycerol, two fatty acids, and a _____ group.

3.5 Proteins

Living things must accomplish a great number of tasks just to get through a day, and the diverse biological molecules you've been looking at allow this to happen. You've seen carbohydrates do some things, and you've seen lipids do some more, but in the range of tasks that molecules accomplish, proteins reign supreme. Witness the fact that almost every chemical reaction that takes place in living things is hastened—or, in practical terms, *enabled*—by a particular kind of protein called an enzyme. These molecules function in nature like some vast group of tools, each taking on a specific chemical task. Accordingly, an animal cell might contain up to 4,000 different types of enzymes.

We might marvel at proteins solely because of what enzymes can do, but the amazing thing is that

Proteins

Chemistry Review:
Proteins

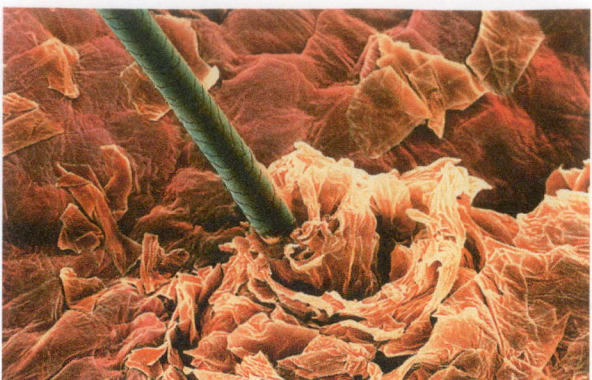

Figure 3.14
Made of Protein

A human hair, shown here emerging from a follicle in the skin, is composed mostly of the protein keratin. Each hair is composed of many groups of keratin strands that wrap around each other, giving hair its structure, which is flexible and much longer than it is wide.

Protein Structure

enzymes are only one class of proteins. Proteins also form the scaffolding, or structure, of a good deal of tissue; they're active in transporting molecules from one site to another; they allow muscles to contract and cells to move; some hormones are made from them. If you factor out water, they account for about half the weight of the average cell. In short, it's hard to overestimate the importance of these molecules (**Figure 3.14**). **Table 3.4** lists some of the different kinds of proteins.

Proteins Are Made from Chains of Amino Acids

Proteins are prime examples of the building-block type of molecule described earlier. The monomers in this case are called amino acids. String a minimum number of them together in a chain—some say 10, some say 30—and you have a **polypeptide,** defined as a series of amino acids linked in linear fashion. When the polypeptide chain *folds up* in a specific three-dimensional manner, you have a **protein,** defined as a large, folded chain of amino acids. As a practical matter, proteins are likely to be made of hundreds of amino acids strung together and folded up. As you'll see, it's not unusual for two or more polypeptide chains to be part of a single protein.

Figure 3.15a gives the fundamental structural unit for amino acids, followed by a couple of examples. You can see carbon at the center of this unit, an amino functional group off to its left, and a carboxyl group to its right. What differentiates one amino acid from another is the group of atoms that occupies the R or "side-chain" position. In **Figure 3.15b**, you can see examples of the actual amino acids tyrosine and glutamine with their different occupants of the R position.

(a) What all amino acids have in common is an amino group and a carboxyl group attached to a central carbon.

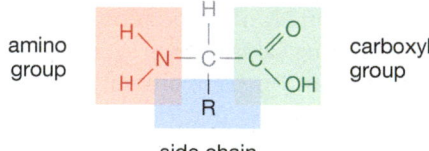

(b) What makes the 20 amino acids unique are the side-chains attached to the central carbon.

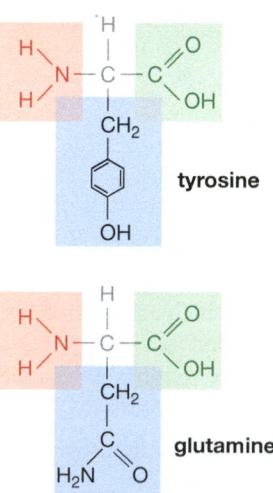

Figure 3.15
Structure of Amino Acids

(a) Basic amino acid structure.

(b) Two examples of actual amino acids, tyrosine and glutamine.

Table 3.4		
Types of Proteins		
Type	**Role**	**Examples**
Enzymes	Quicken chemical reactions	Sucrase: Positions sucrose (table sugar) in such a way that it can be broken down into component parts of glucose and fructose.
Hormones	Chemical messengers	Growth hormone: Stimulates growth of bones
Transport	Move other molecules	Hemoglobin: Transports oxygen through blood
Contractile	Movement	Myosin and actin: Allow muscles to contract
Protective	Healing; defense against invader	Fibrinogen: Stops bleeding Antibodies: Combat microbial invaders
Structural	Mechanical support	Keratin: Hair, Collagen: Cartilage
Storage	Stores nutrients	Ovalbumin: Egg white, used as nutrient for embryos
Toxins	Defense, predation	Bacterial diphtheria toxin
Communication	Cell signaling	Glycoprotein: Receptors on cell surface

A Group of Only 20 Amino Acids Is the Basis for All Proteins in Living Things

Although only two examples are shown here, it is a group of 20 amino acids that is the basis for all proteins in living organisms. The thousands of proteins that exist can be made from a mere 20 amino acids because these amino acids can be strung together in different *orders*. Substitute an alanine here for a glutamine there, and you've got a different protein. In this, amino acids commonly are compared to letters of the alphabet. In English, substituting one letter can take us from *bat* to *hat*. In the natural world, 20 amino acids can be put together in different order to create a multitude of proteins, each with a different function. (A list of all 20 primary amino acids can be found in Chapter 14, on page 245.)

The stringing together of amino acids happens in a regular way: The carboxyl group of one amino acid joins to the amino group of another. Look at **Figure 3.16** to see how three amino acids come together.

Shape Is Critical to the Functioning of All Proteins

As a protein's amino acids are strung together in sequence, all the kinds of chemical forces discussed in Chapter 2 begin to work on them. With this, the chain begins to twist, turn, and fold into its unique three-dimensional shape. It turns out that, in the functioning of proteins, this shape, or protein conformation, is utterly crucial. Here's an example of why. If you look at the top illustration in **Figure 3.17**, on the left you can see a computerized model of a protein found on the surface of the influenza virus. If you look on the right, you will see a model of an immune system protein, called an antibody, that fights foreign invaders such as influenza viruses. Now, how does the antibody interact with a virus? It binds to it as shown in the lower illustration, and this process is made possible by the exact fit of antibody to invading virus. Look at the shape of the two molecules to see how closely they conform to one another. The binding of an antibody to an invader sounds important enough, but recall that proteins are the chemical enablers called enzymes, and they're transport molecules, and so forth. In all these functions, shape is critical. The architect Frank Lloyd Wright had a famous dictum about his designs: "Form and function are one." The same thing can be said of proteins.

There Are Four Levels of Protein Structure

So, what forms do proteins take? Well, the answer to this depends on what vantage point you adopt in looking at them, as there are four levels of structure

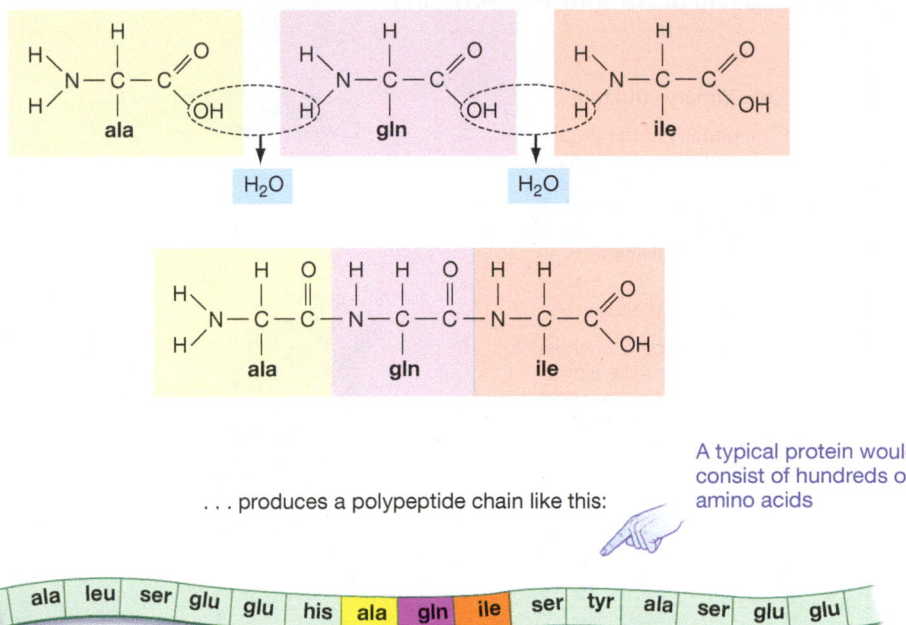

The linkage of several amino acids . . .

. . . produces a polypeptide chain like this:

A typical protein would consist of hundreds of amino acids

Figure 3.16
Beginnings of a Protein
Amino acids join together to form polypeptide chains, which fold up to become proteins. The linking of amino acids yields water as a by-product. In this figure, alanine (ala) first joins with glutamine (gln), which then is linked to isoleucine (ile).

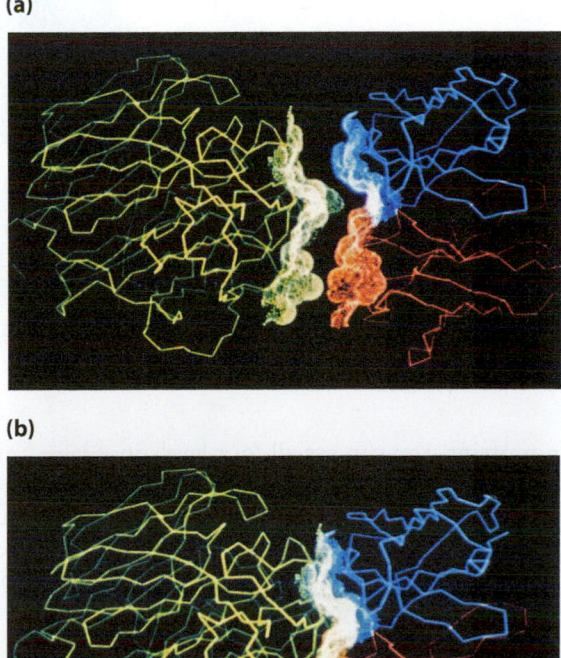

(a)

(b)

Figure 3.17
Hand-in-Glove Fit
Molecular shape is a critical element in the functioning of most proteins.

(a) The computer-model on the left is of a molecule found on the surface of an influenza virus. On the right is a model of a portion of human protein, called an antibody, that attacks invaders such as viruses.

(b) The antibody helps disable the virus by binding with it, as shown. The antibody is able to carry out this binding because it has a shape that is complementary to that of the virus molecule.

Four Levels of Structure in Proteins

(a) Primary structure

The primary structure of any protein is simply its sequence of amino acids. This sequence determines everything else about the protein's final shape.

(b) Secondary structure

Structural motifs, such as the corkscrew-like alpha helix, beta pleated sheets, and the less organized "random coils" are parts of many polypeptide chains, forming their secondary structure.

(c) Tertiary structure

These motifs may persist through a set of larger-scale turns that make up the tertiary structure of the molecule.

(d) Quaternary structure

Several polypeptide chains may be linked together in a given protein, in this case hemoglobin, with their configuration forming its quaternary structure.

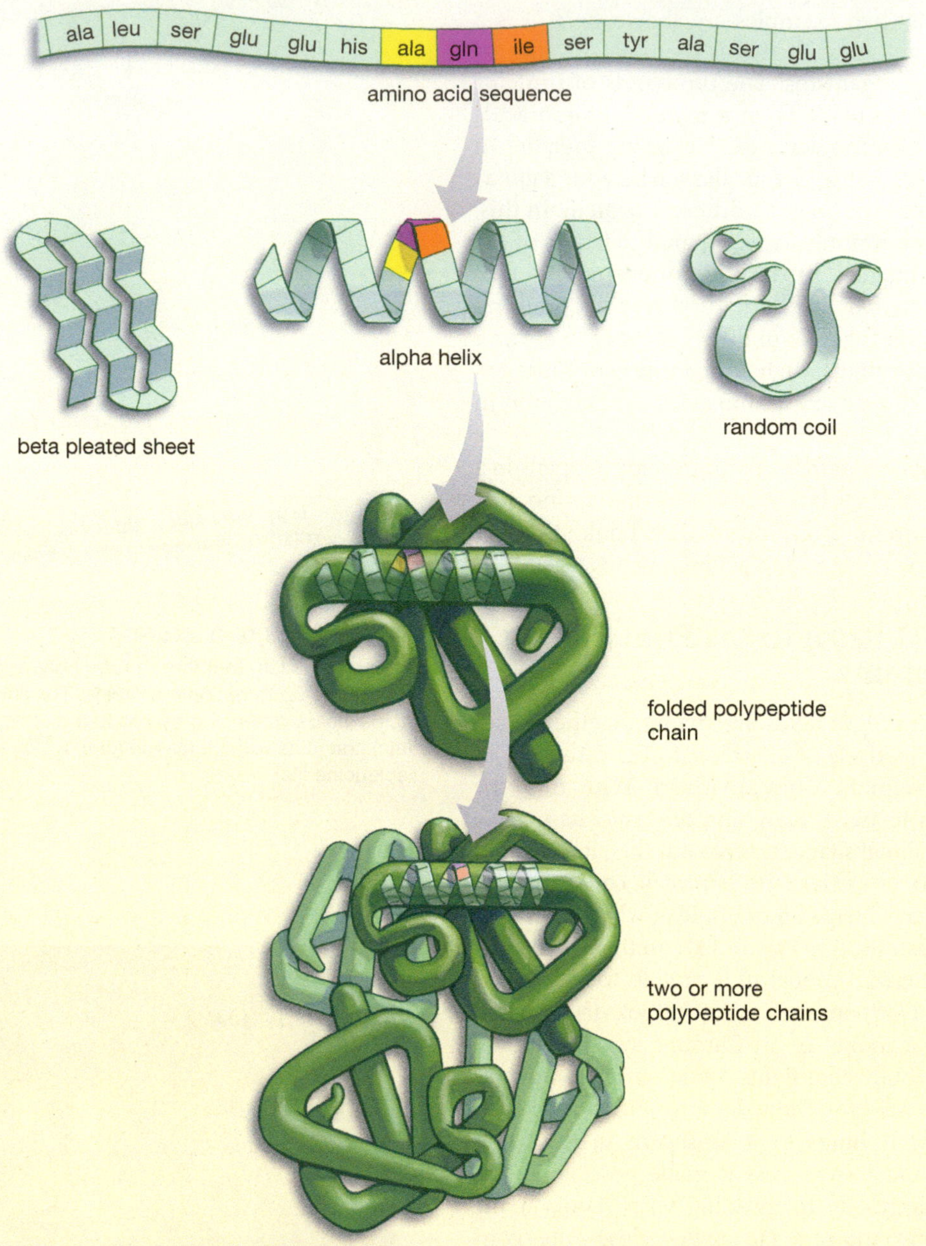

| ala | leu | ser | glu | glu | his | ala | gln | ile | ser | tyr | ala | ser | glu | glu |

amino acid sequence

beta pleated sheet

alpha helix

random coil

folded polypeptide chain

two or more polypeptide chains

Figure 3.18
Four Levels of Structure in Proteins

in proteins. You can see all four levels in **Figure 3.18**. The first of these, the **primary structure** of a protein, is simply its sequence of amino acids. Everything about the final shape of a protein is dictated by this sequence. Electrochemical attraction and repulsion forces act on this structure, and the result is the folded-up protein.

As it turns out, when these forces begin to operate on the amino acid sequence, a couple of common shapes begin to emerge in the **secondary structure** of proteins, defined as the structure that proteins assume after folding up. The *alpha helix*, a common secondary structure of proteins, has a shape much like a corkscrew. Another common secondary structure in proteins, the *beta pleated sheet*, takes a form like the folds of an accordion. Proteins can be made almost entirely of alpha helices. This is the case with hair, nails, horns, and the like. Likewise, proteins can be made entirely of beta pleated sheets. The most familiar example of this is silk, in which the beta sheets lie pancake-style on one another. Often, however, alpha helices and beta pleated sheets form what we might think of as

Protein Functions

design motifs within a larger protein structure; they periodically give way to the less-regular segments called random coils. The larger-scale three-dimensional shape that a protein takes is its **tertiary structure.** The way in which *two or more* polypeptide chains come together to form a protein results in that protein's **quaternary structure.**

Proteins Can Come Undone

As noted, proteins fold up into a precise conformation in order to function. However, proteins can *lose* their shape and thus their functionality. We noted this last chapter in connection with pH and enzymes. In the wrong pH environment, an enzyme can unfold, losing its tertiary structure and thus losing its ability to accelerate a chemical process. Alcohol works as a disinfectant on skin because it *denatures* or alters the shape of the proteins of bacteria.

Lipoproteins and Glycoproteins

Some molecules in living things are hybrids or combinations of the various types of molecules you've been looking at. **Lipoproteins,** as their name implies, are molecules that are a combination of lipids and proteins. Active in transporting fats throughout the body, lipoproteins are transport molecules that amount to a capsule of protein surrounding a globule of fat. Almost everyone has heard of two varieties of lipoproteins: high-density lipoproteins and low-density lipoproteins, also known as HDLs and LDLs, which are important components of our diet. What makes them high or low in density is the ratio of protein to lipid in them; lipid is less dense than protein, so a low-density lipoprotein has a relatively large amount of lipid compared to protein. The LDLs have acquired a reputation as dietary villains because they carry cholesterol *to* outlying tissues, including the coronary arteries of the heart, where this cholesterol may come to reside, thickening eventually into "plaques" that can block coronary arteries and bring about a heart attack. The HDLs, meanwhile, are regarded as the cavalry; they carry cholesterol *away* from outlying cells to the liver. A high proportion of HDLs in relation to cholesterol is predictive of keeping a healthy heart.

    Glycoproteins are combinations of proteins and *carbohydrates.* Where do we find these molecules? One place is the surface of cells, which usually are peppered with a profusion of antenna-like structures, called *receptors,* that allow a cell to receive signals from outside itself. A typical receptor is likely to be composed of both protein and carbohydrate, with the protein forming the "stem" of the receptor and the carbohydrate forming side chains that extend from the stem and serve as the actual binding sites for signaling molecules that may pass by. Some hormones are glycoproteins, along with many other proteins released from cells.

3.6 Nucleic Acids

Nucleic acids are the last major class of biological molecule we'll look at. The building blocks for them, called nucleotides, can be important molecules in and of themselves, as you'll see in Chapter 6 when we look at an energy-transferring nucleotide called ATP (adenosine triphosphate). Here, however, we'll look primarily at just one nucleic acid, which has the distinction of being perhaps the most famous biological molecule of them all.

DNA: Information for the Construction of Proteins

You've learned that proteins perform a large number of biological functions and that one class of proteins, the enzymes, may be represented with up to 4,000 types in a single animal cell. If you had a factory that turned out 4,000 different kinds of tools, you would obviously need some direction on how each of these tools was to be manufactured: This part of the tool goes here first, and then that goes there, and so on. There is a molecule that in essence provides this kind of information for the construction of proteins. It is **DNA,** or **deoxyribonucleic acid:** the primary information-bearing molecule of life, composed of two linked chains of nucleotides. How many nucleotides? In human beings, about 3 billion of them are strung together to form the DNA that comes packaged in the units called chromosomes.

    The information contained in the DNA molecule is much like the information contained in a cookbook, only what the DNA "recipes" call for are precisely ordered chains of amino acids—those that go into making up different proteins. "Start with an alanine, then add a cysteine, then a tyrosine . . .," and so on for hundreds of steps, the DNA-encoded instructions will say, after which, through a series of steps, a protein becomes synthesized and then gets busy on some task. A player in this series of steps is another nucleic acid, **ribonucleic acid** or **RNA,** a molecule composed of nucleotides that is active in the synthesis of proteins. The functions of RNA include ferrying the DNA-encoded instructions to a kind of workbench in the cell, called a ribosome, where proteins are put together. A different sort of RNA actually helps make up this workbench as well, as you'll see next chapter.

Heritable Information: DNA

Figure 3.19
Nucleotides Are the Building Blocks of DNA

(a) The basic unit of the DNA molecule is the nucleotide, a molecule in three parts: a sugar (deoxyribose), a phosphate group, and a nitrogen-containing base.

(b) A computer-generated space-filling model of DNA.

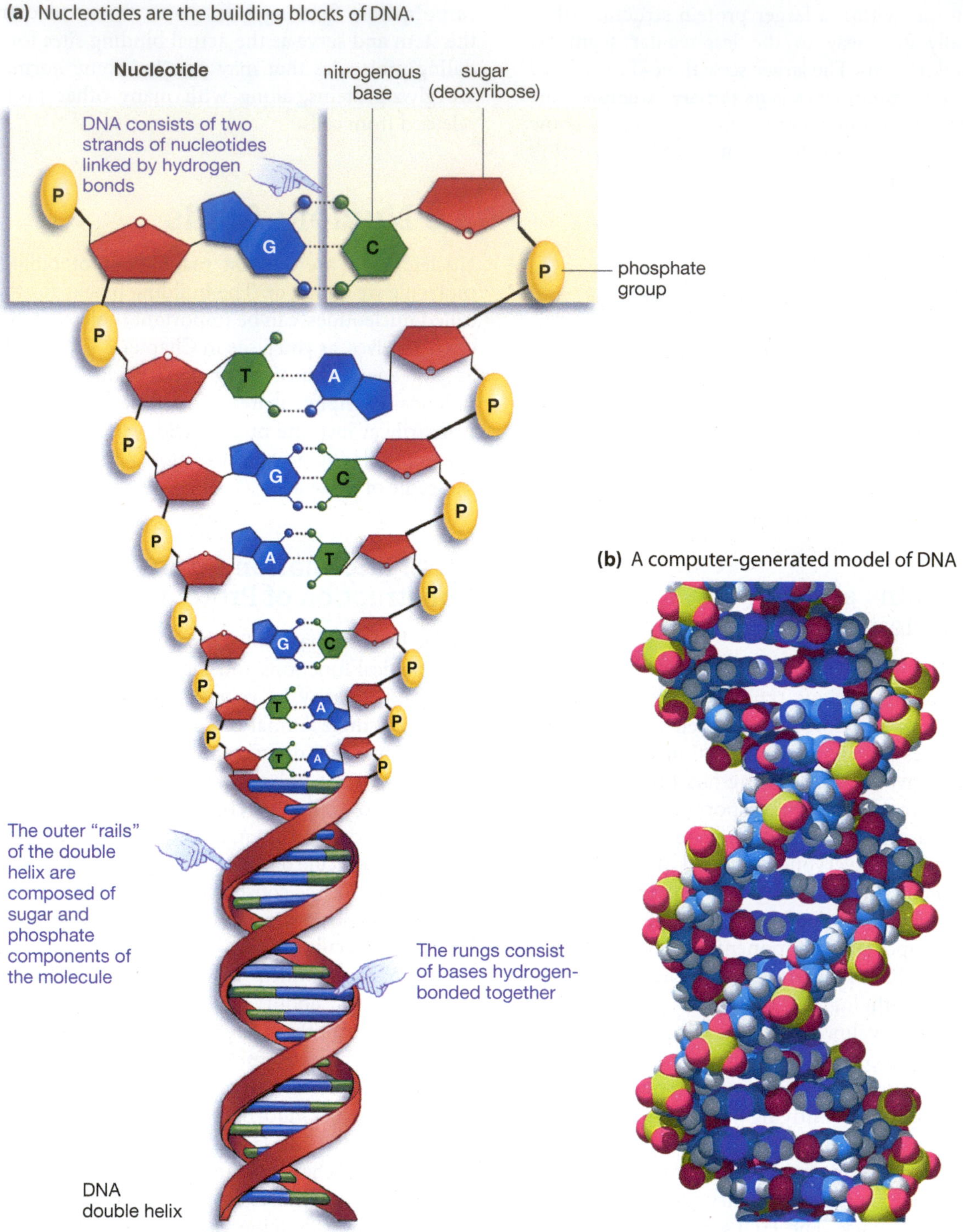

(a) Nucleotides are the building blocks of DNA.

Nucleotide   nitrogenous base   sugar (deoxyribose)

DNA consists of two strands of nucleotides linked by hydrogen bonds

G   C

phosphate group

T   A

G   C

A   T

G   C

T   A

T   A

The outer "rails" of the double helix are composed of sugar and phosphate components of the molecule

The rungs consist of bases hydrogen-bonded together

DNA double helix

(b) A computer-generated model of DNA

The Structural Unit of DNA Is the Nucleotide

Look at **Figure 3.19**, and you'll see the structure of the monomers that make up the nucleic acid polymer of DNA. These are **nucleotides,** each of which is a molecule in three parts: a phosphate group, a sugar (deoxyribose), and one of four nitrogen-containing bases: adenine (A), guanine (G), cytosine (C), or thymine (T). The sugar and phosphate components of the nucleotides link to one another to form the

outer "rails" of the DNA molecule, while the bases point toward the molecule's interior. Two chains of nucleotides are linked, via hydrogen bonds, to form DNA's famous double helix.

On to Cells

Before getting to the story of this elegant DNA molecule, you need to learn a little bit about the territory in which it does its work. What you'll be looking at is a profusion of jostling, roiling, ceaselessly working

Nucleic Acid Structure

Table 3.5

Summary Table of Biological Molecules

Type of Molecule	Subgroups	Examples and Roles
Carbohydrates	Monosaccharides	Glucose: energy source
	Disaccharides	Sucrose: energy source
	Polysaccharides	Glycogen: storage form of glucose
		Starch: carbohydrate storage in plants; used by animals in nutrition
		Cellulose: plant cell walls, structure; fiber in animal digestion
		Chitin: external skeleton of arthropods
Lipids	Triglycerides: 3 fatty acids and glycerol	Fats, oils (butter, corn oil): food, energy, storage, insulation
	Fatty acids: components of triglycerides	Stearic acid: food, energy sources
	Steroids: four-ring structure	Cholesterol: fat digestion, hormone precursor, cell membrane component
	Phospholipids: polar head, nonpolar tails	Cell membrane structure
Proteins	Enzymes: chemically active	Sucrase: breaks down sugar
	Structural	Keratin: hair
	Lipoproteins: protein-lipid molecule	HDLs, LDLs: transport of lipids
	Glycoproteins: protein-sugar molecule	Cell surface receptors
Nucleic acids	Deoxyribonucleic acid (DNA)	DNA contains information for the production of proteins
	Ribonucleic acid (RNA)	One variety of RNA carries DNA's information to the sites of protein production, the ribosomes; another variety of RNA helps make up ribosomes.

chemical factories that make up all living things. These factories are called cells, the subject of Chapter 4. Look at **Table 3.5** and you'll find a summary of all the types of molecules reviewed in this chapter.

SO FAR...

1. Each protein is made from a chain of _____ that folds up into a specific shape that allows the protein to take on its working role. The primary structure of a protein is simply its sequence of _____, while its secondary structure is the shape the protein assumes after having _____.

2. HDL and LDL are acronyms for high-density and low-density _____. These molecules are combinations of _____ and _____.

3. DNA and RNA are nucleic acids, which are polymers composed of the monomers called _____, each of which is in turn composed of a _____, a _____, and one of four _____.

MB

Chemistry Review: Nucleic Acids

CHAPTER

3 REVIEW

Summary

3.1 Carbon's Place in the Living World

- Carbon is central to life because most biological molecules are built on a carbon framework. Carbon's outer shell has four of the eight outer-shell electrons necessary for maximum stability in most elements. Carbon atoms are thus able to form stable, covalent bonds with a wide variety of atoms, a capacity that allows carbon to serve as the central element in the large, complex molecules that make up living things. (p. 43)

3.2 Functional Groups

- Groups of atoms known as functional groups can confer special properties on carbon-based molecules, thus altering the molecule's function. (p. 45)

3.3 Carbohydrates

- Carbohydrates are formed from the building blocks or monomers of simple sugars. These monomers can be linked to form larger carbohydrate polymers (polysaccharides, or complex carbohydrates). Four polysaccharides are critical in the living world: starch, the nutrient storage form of carbohydrates in plants; glycogen, the nutrient storage form of carbohydrates in animals; cellulose, a rigid structural carbohydrate produced by plants and other organisms; and chitin, a tough carbohydrate that forms the external skeleton of arthropods. (p. 46)

3.4 Lipids

- The defining characteristic of all lipids is that they do not readily dissolve in water. No one structural element is common to all lipids. (p. 49)

- Among the most important lipids are: triglycerides, composed of a glyceride and three fatty acids; steroids, which have a core of four carbon rings; phospholipids, composed of two fatty acids, glycerol, and a charged phosphate group; and waxes, which are composed of a single fatty acid linked to a long-chain alcohol. (p. 49)

3.5 Proteins

- Proteins are a diverse group of biological molecules composed of the monomers called amino acids. Sequences of amino acids are strung together to produce polypeptide chains, which then fold up into working proteins. Important groups of proteins include enzymes, which hasten chemical reactions, and structural proteins. (p. 55)

- The primary structure of a protein is its amino acid sequence; this sequence determines a protein's secondary structure—the form a protein assumes after having folded up. The tertiary structure is the larger-scale three-dimensional shape that a protein assumes; the way two or more polypeptide chains come together to form a protein results in that protein's quaternary structure. The activities of proteins are determined by their final folded shapes. (p. 57)

- Lipoproteins are combinations of lipids and proteins. (p. 59)

- Glycoproteins are combinations of carbohydrates and proteins. (p. 59)

3.6 Nucleic Acids

- Nucleic acids are polymers composed of nucleotides. The nucleic acid DNA (deoxyribonucleic acid) is composed of nucleotides that contain a sugar (deoxyribose), a phosphate group, and one of four nitrogen-containing bases. DNA is a repository of genetic information. The sequence of its bases contains the information for the production of the huge array of proteins produced by living things. (p. 60)

Key Terms

carbohydrate   46
cellulose   47
chitin   48
cholesterol   52
deoxyribonucleic acid (DNA)   59
fatty acid   49
functional group   45
glycogen   47
glycoprotein   59
hydrocarbon   43
lipid   49
lipoprotein   59
monomer   46
monosaccharide   46
monounsaturated fatty acid   50
nucleotide   60
oil   50
organic chemistry   43
phosphate group   53
phospholipid   53
polymer   46
polypeptide   56
polysaccharide   47
polyunsaturated fatty acid   50
primary structure   58
protein   56
quaternary structure   59
ribonucleic acid (RNA)   59
saturated fatty acid   50
secondary structure   58
simple sugar   46
starch   47
steroid   52
tertiary structure   59
triglyceride   49
wax   54

Understanding the Basics

Multiple-Choice Questions

(Answers are in the back of the book.)

1. Carbon is able to serve as life's central element because it:
   a. is able to bond with hydrogen atoms to form hydrocarbon chains.
   b. will readily bond with water molecules.
   c. has four outer-shell electrons and thus can form stable bonds with many other elements.
   d. has six outer-shell electrons and thus can form stable bonds with many other elements.
   e. forms strong ionic bonds with many other elements.

2. Functional groups often confer the general property of _____ on _____ molecules.
   a. hydrophilia; water-based
   b. hydrophobia; water-based
   c. stability; carbon-based
   d. stability; unstable
   e. polarity; carbon-based

3. Glucose and the amino acid tyrosine are members of different classes of biological molecules, but they are alike in that both are _____ that go into making up _____.
   a. monomers; polymers
   b. monosaccharides; polysaccharides
   c. polysaccharides; monosaccharides
   d. polymers; monomers
   e. nucleotides; DNA

4. When you eat starch such as spaghetti, an enzyme in your mouth breaks it down to maltose. Eventually, the maltose enters your small intestine, where it is broken down to glucose, which you can absorb into your bloodstream. The starch is a(n) _____, the maltose is a _____, and the glucose is a(n) _____.
   a. protein; dipeptide; amino acid
   b. monosaccharide; disaccharide; polysaccharide
   c. triglyceride; fatty acid; glycerol
   d. amino acid; dipeptide; protein
   e. polysaccharide; disaccharide;, monosaccharide

5. Which of these is not a difference between saturated and unsaturated fats?
   a. Saturated fats are more likely to be solid at room temperature; unsaturated fats are more likely to be liquid.
   b. Saturated fats are a type of cholesterol or steroid, whereas unsaturated fats are a triglyceride.
   c. Saturated fats have fatty acids that pack closely together, owing to their straight-line construction. Unsaturated fats have fatty acids arranged at varying angles to one another, because of their double bonds.
   d. Saturated fats are composed of fatty acids that have no double bonds in their chemical structure. Unsaturated fats are composed of fatty acids that have one or more double bonds in their chemical structure.
   e. All of the above are differences between saturated and unsaturated fats.

6. The myoglobin protein, which carries oxygen in muscle cells, has only the first three levels of protein structure. In other words, it lacks a quaternary level. From this you can conclude that myoglobin:
   a. is made of nucleic acids.
   b. is made of only one polypeptide chain.
   c. lacks hydrogen bonds.
   d. is not helical or pleated.
   e. is a fiber.

7. You received your genetic information from your parents in the form of DNA. This DNA carried the instructions for making:
   a. carbohydrates such as glycogen.
   b. fatty acids.
   c. phospholipids for making the membranes of your cells.
   d. proteins.
   e. all of the above.

8. John is lactose intolerant. The -ose ending indicates that John cannot digest a certain:
   a. sugar.
   b. polysaccharide.
   c. protein.
   d. steroid.
   e. enzyme.

Brief Review

(Answers are in the back of the book.)

1. Why is it accurate to look at life as "carbon based," and what quality gives carbon this special role?

2. Both low-density lipoproteins (LDLs) and high-density lipoproteins (HDLs) are involved with carrying fats through the bloodstream. If your LDL count is unusually high, should you be concerned? What if your HDL count is high? Why are they different?

3. What are steroids, as that term is defined in biochemistry? What are steroids in the popular sense of the term?

4. List as many functions of proteins as possible. Why are proteins able to do so many different types of jobs?

5. Why is it accurate to think of DNA as an information-bearing molecule?

Applying Your Knowledge

1. Scientists who look for life outside Earth have a favorite candidate for an element that could take on the role that carbon does here on Earth. The element is silicon, which has 14 electrons, compared to carbon's 6. Based on the rules of chemical bonding you learned about last chapter, why should silicon's electron number make it a candidate for a central biological element?

MasteringBiology®

1. MasteringBiology Assignments
Activities: The Chemistry of Carbon; Diversity of Carbon-Based Molecules; Isomers; Functional Groups; Carbohydrates; Lipids; Proteins; Protein Structure; Protein Functions; Heritable Information: DNA; Nucleic Acid Structure; **Chemistry Review:** Organic Molecules; Carbohydrates; Lipids; Proteins; Nucleic Acids; **Current Events:** Asked to Get Slim, Cheese Resists; The Fat Trap; From Ancient Giants, Finding New Life to Help the Planet; **Building Vocabulary; Questions:** Reading Questions, Test Bank

2. eText
Read your book online, search, take notes, highlight text, and study more effectively.

3. The Study Area
Self-study tools to test comprehension.

Life's Home:
The Cell

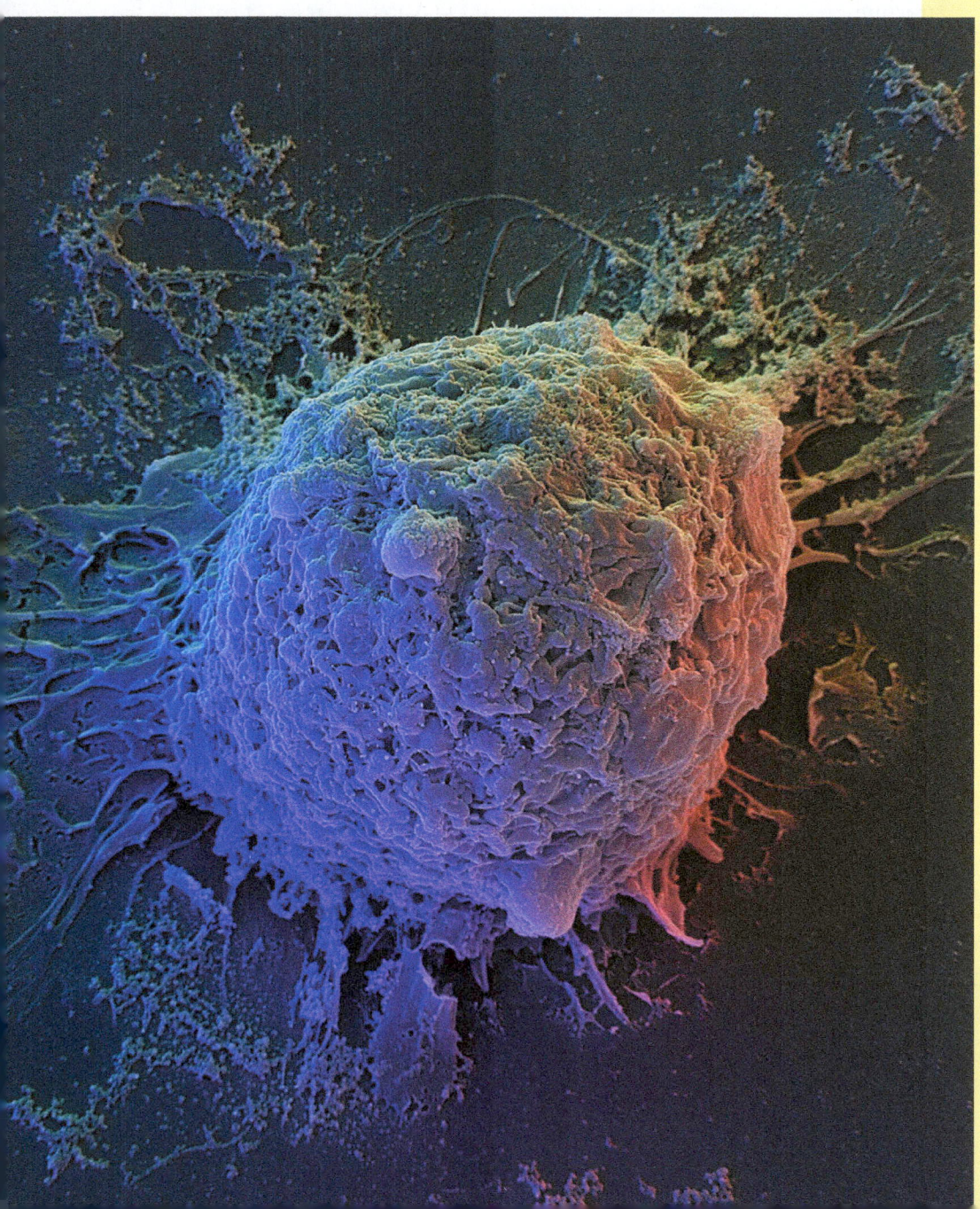

Cells are the fundamental units of life.
Pictured is an artificially colored human
embryonic cell.

*All life exists within cells. These tiny
entities can be compared to factories
whose products maintain life.*

Available on MasteringBiology®

Activities: Form Fits Function: Cells; Prokaryotic
Cell Structure and Function; Review: Animal Cell
Structure and Function; Role of the Nucleus and
Ribosomes in Protein Synthesis; The Endomembrane
System; The Structure of Cells; Cilia and Flagella;
Endosymbiosis; Cell Junctions; **Bioflix:** Tour of an
Animal Cell; Tour of a Plant Cell; **Current Events:**
A First: Organs Tailor-Made with Body's Own Cells;
Peering Over the Fortress That Is the Mighty Cell;
Building Vocabulary; Questions: Bioflix Quiz,
Reading Questions, Test Bank

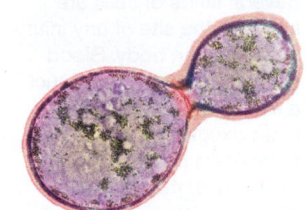

The United States Senate has 100 members in it, and all of them do occasionally gather to consider given issues. But if you want to know how the Senate actually gets something done, you must look elsewhere. The business of the nation is simply too complex for all Senate members to deal with all issues. Instead, issues are dealt with one at a time within the specialized working units of the Senate known as committees.

Something similar to this goes on in the natural world. If we look at birds, trees, or people and ask how any of these living things actually gets something done, the answer again is through working units—in this case the working units known as cells.

4.1 Cells as Life's Fundamental Unit

Muffins are in the oven, and you are in the living room. Gas molecules from the baking muffins waft into the living room, and some of them happen to make their way to your nose, there to travel a short distance to your upper nasal cavity and land on a set of ceaselessly waving hair-like projections called cilia. These actually are the extensions of some specialized nerve cells. If enough gas molecules bind with enough of the cilia, an impulse is passed along (through other nerve cells) to trigger not only the *sensation* of smell but also the *association* of this smell with muffins you have smelled in the past. How do we know muffins are in the oven? Through cells. How do we move our hands or read this page? Through cells. Life's working units are cells, and in our amazingly complex natural world, there is great specialization in them (**Figure 4.1**, on the next page). The nerve cells in human beings are specialists in transmitting nerve signals, red blood cells are specialists in transporting oxygen, and muscle cells are specialists in contracting.

Running like a thread through this diversity, however, is the unity of cellular life. Every form of life either is a single cell or is composed of cells. The one possible exception to this is viruses, but even they must use the machinery of cells to reproduce. There is unity, too, in the way cells come about: Every cell comes *from* a cell. Human beings are incapable of producing cells from scratch in the laboratory, and so far as we can tell, nature has fashioned cells from simple molecules only once—back when life on Earth got started. The fact that all cells come from cells means that each cell in your body is a link in a cellular chain that stretches back more than 3.5 billion years.

Form Fits Function: Cells

Life's Fundamental Unit

(a) Human red and white blood cells inside a blood vessel. The large, dark ovals that can be seen in the background are flat cells that form the interior lining of the blood vessel.

(b) Cells of the fungus brewer's yeast, which is used to make wine, beer, and bread.

(c) Cells of a spinach leaf, with the leaf seen in cross section. Two layers of outer or "epidermal" cells can be seen at top and bottom; most of the cells in between perform photosynthesis.

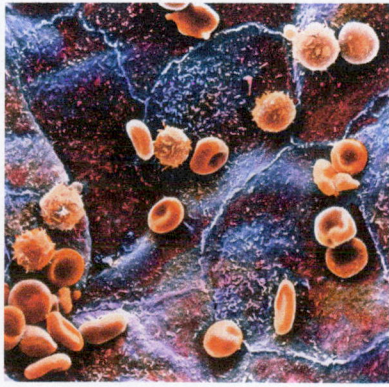

(a)

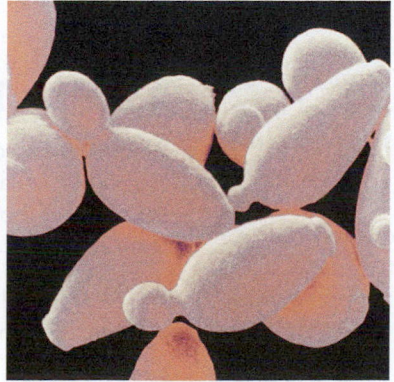

(b)

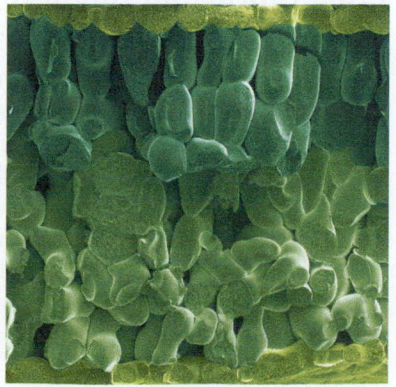

(c)

Figure 4.1
Cells Specialize

Several kinds of cells are active at the site of any injury to the human body. Blood circulation increases at such a site, with the result that more oxygen-bearing red blood cells pass through it. Meanwhile, the body's immune system will launch an attack against microbes that entered the injured site. The yellow cells visible at the center of the photo are immune system cells called leukocytes that will take part in this attack. The site of any injury also will be sealed off partly through the activity of a group of cell fragments called platelets, visible in the photo as the smaller pink objects.

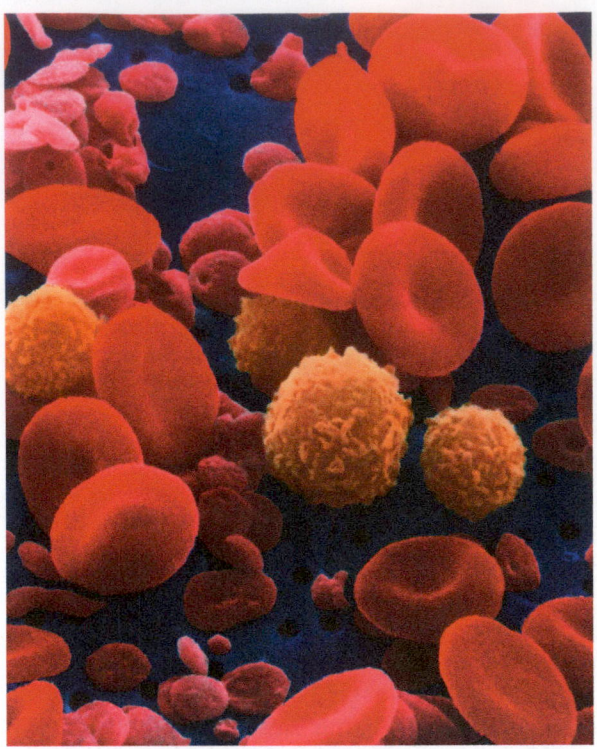

Prokaryotic Cell Structure and Function

4.2 Prokaryotic and Eukaryotic Cells

So, what are these tiny working units we call cells? It follows, from the variety of cells that exist, that there is no such thing as a "typical" cell. There are, however, certain *categories* of cells that are important, and the two most important of these are prokaryotic and eukaryotic cells. Every cell that exists is one or the other, and this simple either-or quality extends to the organisms that fall into these camps. All prokaryotic cells either are bacteria or another microscopic form of life known as archaea. Setting bacteria and archaea aside, *all other cells* are eukaryotic. This means all the cells in plants, in animals, in fungi—all the cells in every living thing except in the bacteria and the archaea.

Prokaryotic and Eukaryotic Differences

The name *eukaryote* comes from the Greek *eu,* meaning "true," and *karyon,* meaning "nucleus," while *prokaryote* means "before nucleus." These terms describe the most critical distinction between the two cell types. **Eukaryotic cells** are cells whose primary complement of DNA is enclosed within a nucleus. **Prokaryotic cells** are cells whose DNA is not enclosed within a nucleus. To complete this circle, a **nucleus** is a membrane-lined compartment that encloses the primary complement of DNA in eukaryotic cells.

While having a nucleus is the most important difference between eukaryotic and prokaryotic cells, it is not the only difference. It may seem sensible to think of two single-celled creatures—one a prokaryotic bacterium, the other a eukaryotic amoeba—as very similar, but the distance between them as life-forms is immense. Human beings and chimpanzees are nearly identical in comparison. Eukaryotic cells tend to be much larger than their prokaryotic counterparts; indeed, thousands of bacteria could easily fit into an average eukaryotic cell (see "The Size of Cells" on page 68). Eukaryotes are often multicelled organisms, while prokaryotes are always single-celled (**Figure 4.2**). Beyond this, eukaryotic cells are *compartmentalized* to a far greater degree than is the case with prokaryotes. The nucleus in eukaryotic cells turns out to be only one variety of **organelle:** a highly organized structure, internal to a cell, that serves some specialized function. Eukaryotic cells contain several different kinds of these "tiny organs." There are organelles called mitochondria, for example, that transform energy from food, and organelles called lysosomes that recycle the raw materials of the cell. In prokaryotic cells, meanwhile, there is only a single type of organelle—a kind of workbench for producing proteins we'll look at later.

If we look at eukaryotes that are familiar to us, such as trees, mushrooms, and horses, it's easy to see that there is a fantastic diversity in the *forms* of eukaryotes compared to the strictly single-celled prokaryotes. It does not follow from this, however, that prokaryotes are uniform; on the contrary, they actually differ greatly from one another with respect to, say, the ways they obtain their nutrition and the environments in which they live. And they are extremely successful, if success is defined as inhabiting these environments in great numbers. As the biologists Lynn Margulis and Karlene Schwartz have observed, more bacteria are living in your mouth right now than the number of people who have ever existed. But for the range of biological structures and processes we want to look at, eukaryotic cells are more diverse and hence are the initial focus of our study. (For an account of the first person ever to behold the micro-world of cells, see "First Sightings: Anton van Leeuwenhoek" on page 88.)

4.3 The Eukaryotic Cell

What are the constituent parts of eukaryotic cells? Here are five larger structures that in turn are composed of component parts. The first two have been mentioned already (**Figure 4.3**):

- The cell's nucleus
- Other organelles (which lie outside the nucleus)

Figure 4.2
Prokaryotic and Eukaryotic Cells Compared

Prokaryotic cells	Eukaryotic cells
DNA spread through much of cell	within membrane-bound nucleus
Size much smaller	much larger
Organization always single-celled	often multicellular
Organelles only one type of organelle	many types of organelles

- The **cytosol,** a protein-rich, jelly-like fluid in which the cell's organelles are immersed
- The *cytoskeleton,* a kind of internal scaffolding consisting of three sorts of protein fibers; some of these can be likened to tent poles, others to monorails
- The **plasma membrane,** the outer lining of the cell

In any discussion of a cell, you are likely to hear the term **cytoplasm,** which simply means the *region* of the cell inside the plasma membrane but outside the nucleus. The cytoplasm is different from the cyto*sol*. If

you removed all the structures of the cytoplasm—meaning the organelles and the cytoskeleton—what would be left is the cytosol, which is mostly water. This does not mean that the cytosol is simply a passive medium for the other structures, but it is not an organized structure in the way the organelles are. Almost all the organelles you'll be seeing are encased in their own membranes, just as the whole cell is encased in its plasma membrane. What are membranes? Until you get the formal definition next chapter, think of a membrane as the flexible, chemically active outer

Figure 4.3
Eukaryotic Cell

This cutaway view of a eukaryotic cell displays the elements that nearly all such cells possess.

Components of eukaryotic cells

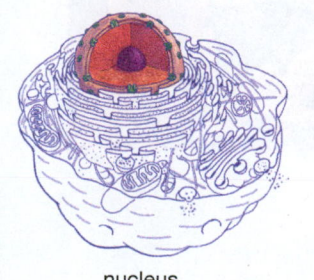

nucleus

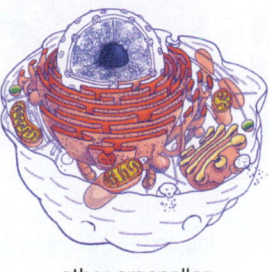

other organelles

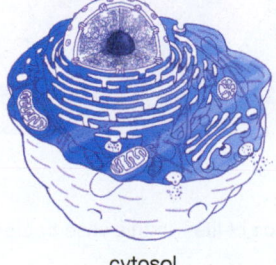

cytosol

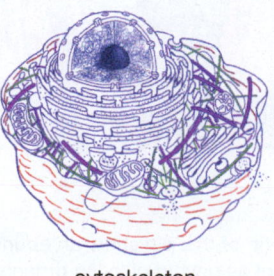

cytoskeleton

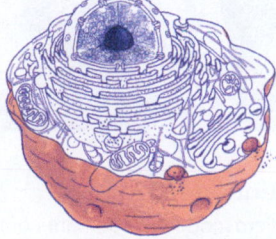

plasma membrane

ESSAY

The Size of Cells

For several chapters, you've been going over atoms, ions, molecules, and such—things small enough to be invisible to the unaided eye. For the most part, you've had to imagine what these entities look like, simply because most of them are so small that such images as we have of them don't tend to be very informative. If you flip through the pages of this chapter and those to come, however, you begin to see a fair number of actual photographs. They don't have the same quality as summer vacation snapshots, but they are recognizable as pictures, or more properly as **micrographs,** meaning pictures taken with the aid of a microscope. Micrographs enable us to see surprising things; for example, hundreds of bacteria on the tip of a pin (**Figure 1**). So, with the cells that are introduced in this chapter, there has obviously been a bump up in size into a world that is more easily visible with the help of various kinds of technology.

In taking stock of the micro-world, two units of measure are particularly valuable. The smaller of them is the **nanometer,** abbreviated as nm; the larger is the **micrometer,** abbreviated μm. (The unfamiliar-looking first letter is the Greek symbol for a small *m.* Scientists are not trying to be obscure in using it; another unit of the metric system, the millimeter, lays claim to the mm abbreviation.) What these abbreviations stand for is a billionth of a meter (nm) and a millionth of a meter (μm). A meter equals about 39.6 inches, or just over a yard, which gives you some starting sense of physical reality in understanding the rest of these sizes.

Now look at **Figure 2** to see what size various objects are. Atoms are at the bottom of the scale, at about a tenth of a nanometer. Something less than a 10-fold increase gets you to the size of the protein building blocks, called amino acids, that you looked at in Chapter 3. Another

10-fold-plus increase and you've reached the upper limit on proteins.

You have to go better than 10 times larger than this, however, before you arrive at the size of something that is actually *living* as opposed to something that is a component part of life. The living entities here are cells, specifically bacterial cells that measure 200–300 nm. This extreme on the small side of cells has a counterpart on the large side with the single cells we call chicken or ostrich eggs and with certain nerve cells that can stretch out to a meter in length. In general, however, we just cross into the micrometer range with the smaller bacterial cells: about 1–10 μm. The cell size for most plants and animals falls in a range that is a little less than 10 times larger than this, about 10–30 μm. At 10 μm perhaps a billion cells could fit into the tip of your finger. And how about in your whole body? A standard estimate of the actual number of human cells is 10 trillion.

So, Why So Small?

Having learned how small cells are, you might then well ask: *Why* are they so small? As noted, cells are small chemical factories,

(a) Bacteria on a pin, magnified x 85

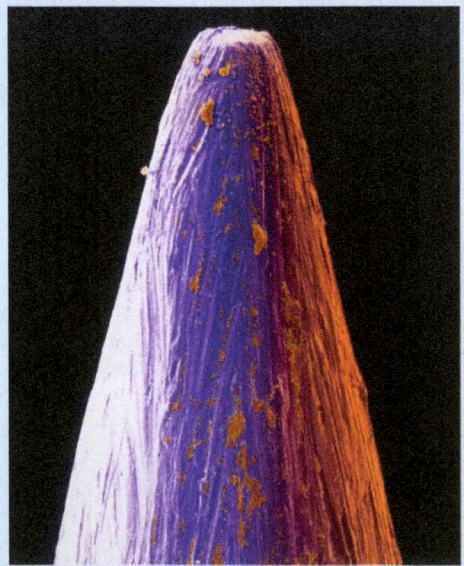

(b) Magnified x 425

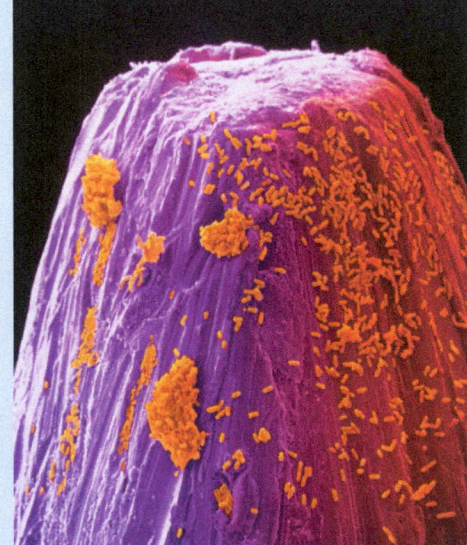

(c) Magnified x 2100

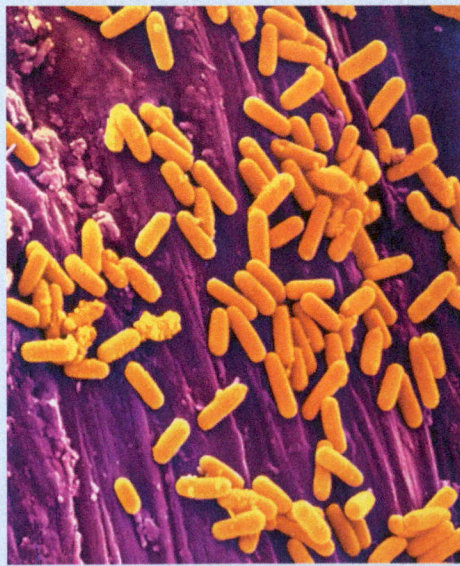

Figure 1
Hidden Life
Microscope enlargements of the tip of this pin show an abundance of life—in this case bacteria—thriving on an object that we normally think of as devoid of living organisms.

and just like any factory, they are constantly shipping things in and out. The primary size-limiting factor for cells is having enough surface area to export and import all that they need.

This constraint comes about because of a fundamental mathematical principle: As the surface area of an object increases, its *volume* increases even more. Say you have a cube 1 inch long on each of its sides. Its surface area is 6 square inches (Length × Width × Number of sides), while its volume is 1 cubic inch (Length × Width × Height). Now, say you increase the side dimension to 8 inches. The surface area goes from 6 to 384 square inches, but the *volume* goes from 1 to 512 cubic inches. At a 1-inch dimension, there were *six* square inches of surface area for every cubic inch of volume; now, there are only *three-quarters* of an inch of surface for every cubic inch of volume. Beyond a certain volume, then, a cell simply would not have enough surface area to import and export all the materials it needs. This effectively sets an upper limit on how big cells can be and helps explain why most of the cells in an elephant are no bigger than those in an ant, although the elephant does have more cells than the ant. **Figure 3** shows you some size comparisons in the micro-world.

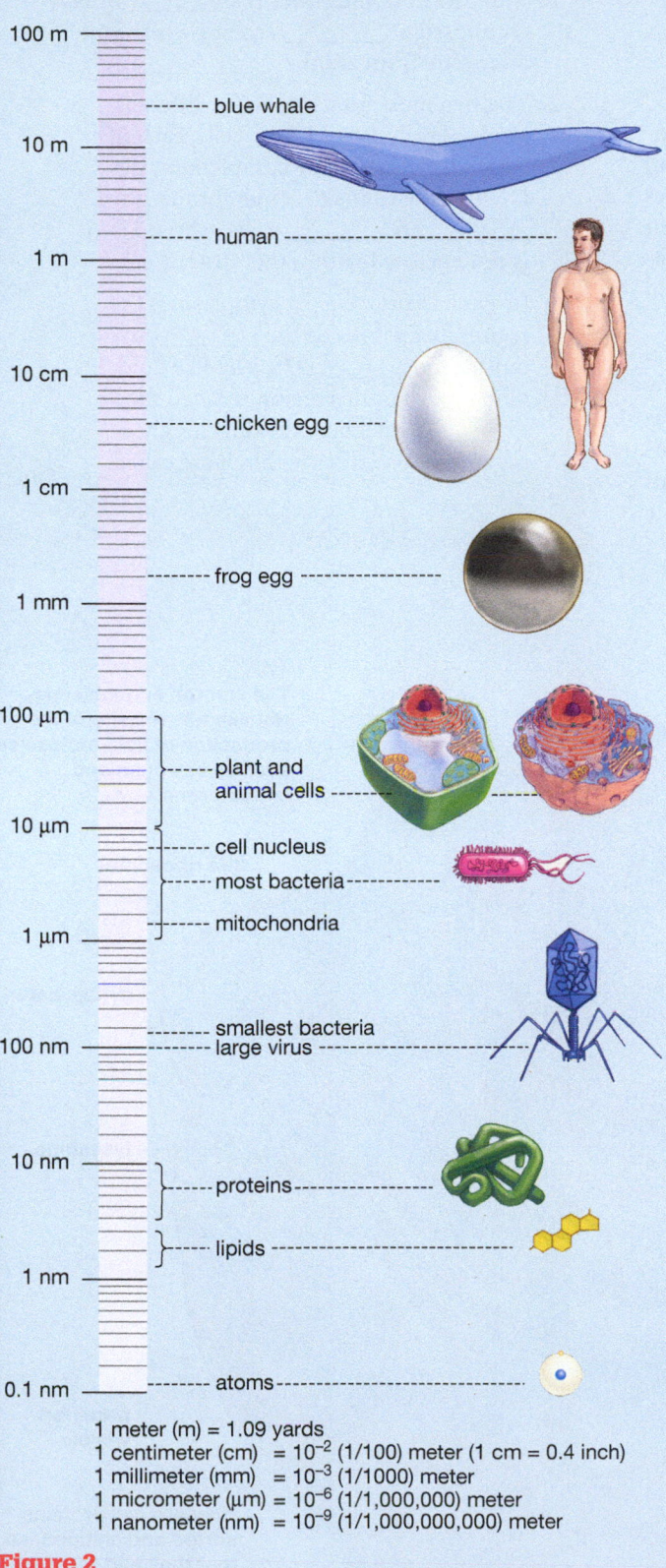

1 meter (m) = 1.09 yards
1 centimeter (cm) = 10^{-2} (1/100) meter (1 cm = 0.4 inch)
1 millimeter (mm) = 10^{-3} (1/1000) meter
1 micrometer (μm) = 10^{-6} (1/1,000,000) meter
1 nanometer (nm) = 10^{-9} (1/1,000,000,000) meter

Figure 2
Little and Big
The sizes of some selected objects in the natural world.

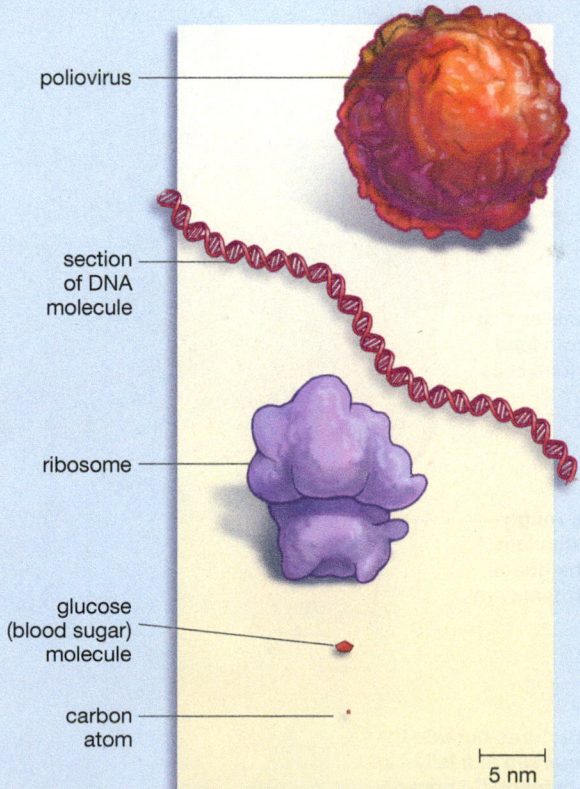

Figure 3
Small Is a Relative Thing
The sizes and shapes of five natural-world entities, each magnified a million times.

lining of a cell or of one of its compartments. In the balance of this chapter, we'll explore all of the constituent parts listed above except for the plasma membrane, which is so special it gets its own chapter.

The Animal Cell

To get a sense of what eukaryotic cells are like, it's convenient to look at two types of them: animal cells and plant cells. These cell types have more similarities than they do differences, but they are different enough that it will be helpful to look at them separately. We'll start by examining animal cells and then look at how plant cells differ from them.

Insofar as we can characterize that elusive creature, the "typical" animal cell (**Figure 4.4**), it is roughly spherical, surrounded by, and linked to, cells of similar type, immersed in a watery fluid, and about 25 micrometers (μm) in diameter, meaning that about 30 of them could fit side by side within the period at the end of this sentence.

Bioflix: Tour of an Animal Cell

Figure 4.4
The Animal Cell

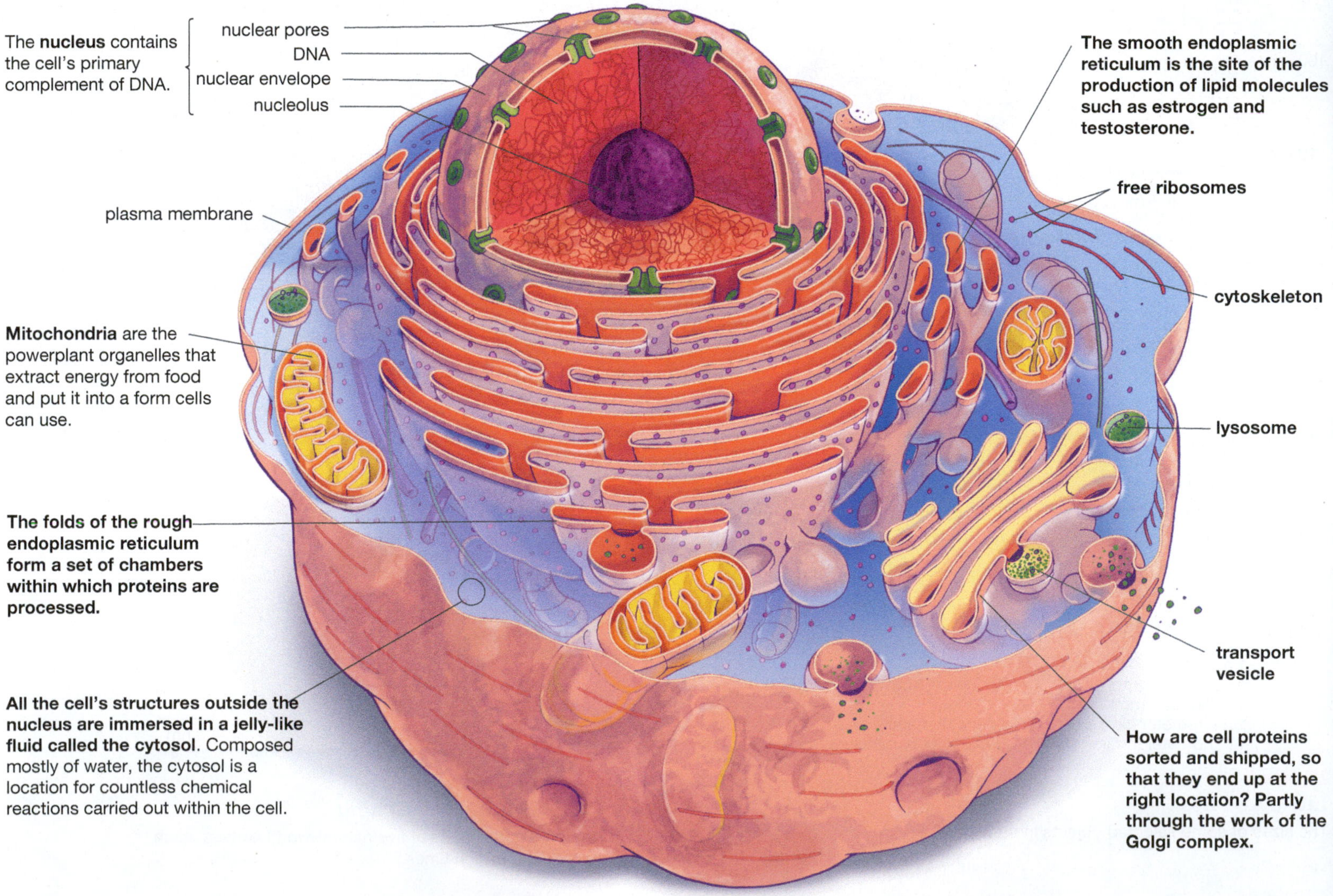

SO FAR . . .

1. Every form of life either is a _____ or is composed of _____, each of which can arise only from another _____.

2. The two most fundamentally different kinds of cells are _____ cells, each of which has its primary complement of DNA enclosed inside a membrane-lined _____; and _____ cells, whose DNA is not enclosed within this structure.

3. In a eukaryotic cell, the cytoplasm is the region that lies inside the _____ but outside the _____. The jelly-like fluid filling much of this region is called the _____, while the individual highly organized structures within it are called _____.

The **nucleus** contains the cell's primary complement of DNA.

nuclear pores
DNA
nuclear envelope
nucleolus

plasma membrane

Mitochondria are the powerplant organelles that extract energy from food and put it into a form cells can use.

The folds of the **rough endoplasmic reticulum** form a set of chambers within which proteins are processed.

All the cell's structures outside the nucleus are immersed in a jelly-like fluid called the **cytosol**. Composed mostly of water, the cytosol is a location for countless chemical reactions carried out within the cell.

The **smooth endoplasmic reticulum** is the site of the production of lipid molecules such as estrogen and testosterone.

free ribosomes

cytoskeleton

lysosome

transport vesicle

How are cell proteins sorted and shipped, so that they end up at the right location? Partly through the work of the **Golgi complex**.

4.4 A Tour of the Animal Cell's Protein Production Path

You are now going to take an extended tour of the animal cell, which can be thought of as a living factory. Much of the first part of your trip will be spent tracing the way this factory puts together a product—a protein—for export outside itself. Just as a new employee might tour an assembly line as a means of learning about factory equipment, so you are going to follow the path of protein production as a means of learning about cell equipment.

Figure 4.5 shows the path you'll be taking, from nucleus to the outer edge of the cell. Don't be bothered by the unfamiliar terms in Figure 4.5 because they'll all be explained in the text. The important thing is that, before we begin our tour, you have some sense of the path that protein production takes.

Beginning in the Control Center: The Nucleus

As noted in Chapter 3, proteins are critical working molecules in living things, and DNA contains the information for putting proteins together from a starting set of the building blocks called amino acids. Our entire complement of DNA is like a cookbook that contains individual protein recipes, which we call genes. A given gene's chemical sequence in effect says, "Connect this amino acid to this one, then add this one, then this one . . . ," the final result being the specifications for a completed protein. In the eukaryotic cell, as you've seen, DNA is largely confined within a nucleus bound by a membrane. If you look at **Figure 4.6** on the next page, you can see the nature of this membrane. It is the **nuclear envelope:** the double membrane that lines the nucleus in eukaryotic cells.

There comes a point in the life of most cells when they divide, one cell becoming two. Because (with a few exceptions) all cells must possess the set of instructions that are contained in DNA, it follows that when a cell divides, its original complement of DNA must *double,* so that both cells that result from the cell division can have their own DNA. The nucleus, then, is not just the site where DNA exists; it is the site where new DNA is put together, or "synthesized," for this doubling.

Messenger RNA

At the end of Chapter 3, you saw that, while DNA exists in a cell's nucleus, the proteins that DNA codes for are put together in a cell's cytoplasm. Thus, the information contained in a length of DNA has to be transported from the nucleus to the cytoplasm. How does this happen? DNA's instructions are copied onto a second long-chain molecule, called RNA, and *it* ferries the DNA information out of the nucleus to the site of protein synthesis. RNA actually comes in several forms, but the form that DNA's instructions are copied onto is the well-named *messenger RNA* or mRNA for short. Given that mRNA goes to the cytoplasm, it must, of course, have some way of getting out of the nucleus; its exit points turn out to be thousands of channels that stud the surface of the nuclear envelope—the *nuclear pores* that you can see in Figure 4.6.

mRNA Moves Out of the Nucleus

Imagine shrinking in size so that the nucleus seems about as big as a house, with you standing outside it. What you would see in protein synthesis is lengths of

(MB)

Review: Animal Cell Structure and Function

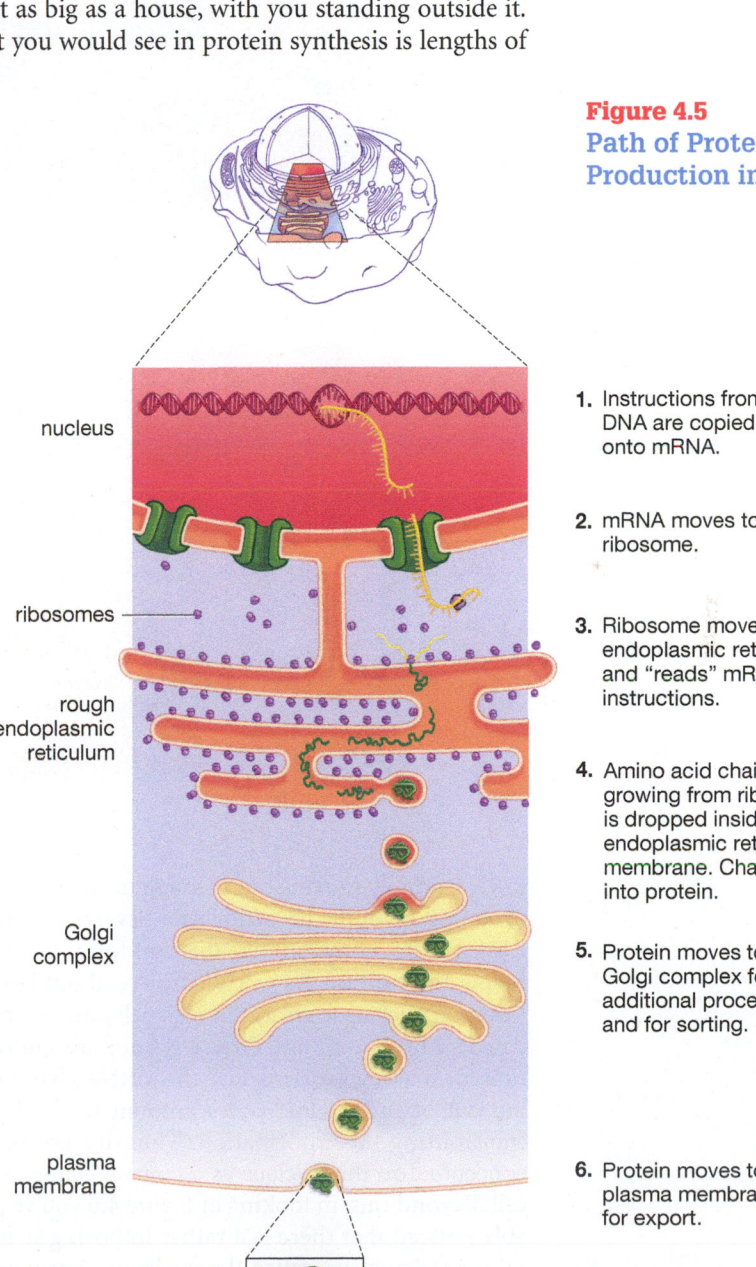

Figure 4.5
Path of Protein Production in Cells

nucleus

ribosomes

rough endoplasmic reticulum

Golgi complex

plasma membrane

1. Instructions from DNA are copied onto mRNA.

2. mRNA moves to ribosome.

3. Ribosome moves to endoplasmic reticulum and "reads" mRNA instructions.

4. Amino acid chain growing from ribosome is dropped inside endoplasmic reticulum membrane. Chain folds into protein.

5. Protein moves to Golgi complex for additional processing and for sorting.

6. Protein moves to plasma membrane for export.

Figure 4.6
The Cell's Nucleus

The DNA of eukaryotic cells is sequestered inside a compartment, the nucleus, which is lined by a double membrane known as the nuclear envelope. Compounds pass into and out of the nucleus through a series of microscopic channels called nuclear pores. Protein production depends on the information encoded in DNA's sequence of chemical building blocks. This information is copied onto a length of messenger RNA (mRNA), which then exits from the nucleus through a nuclear pore. (Micrograph: × 4,400)

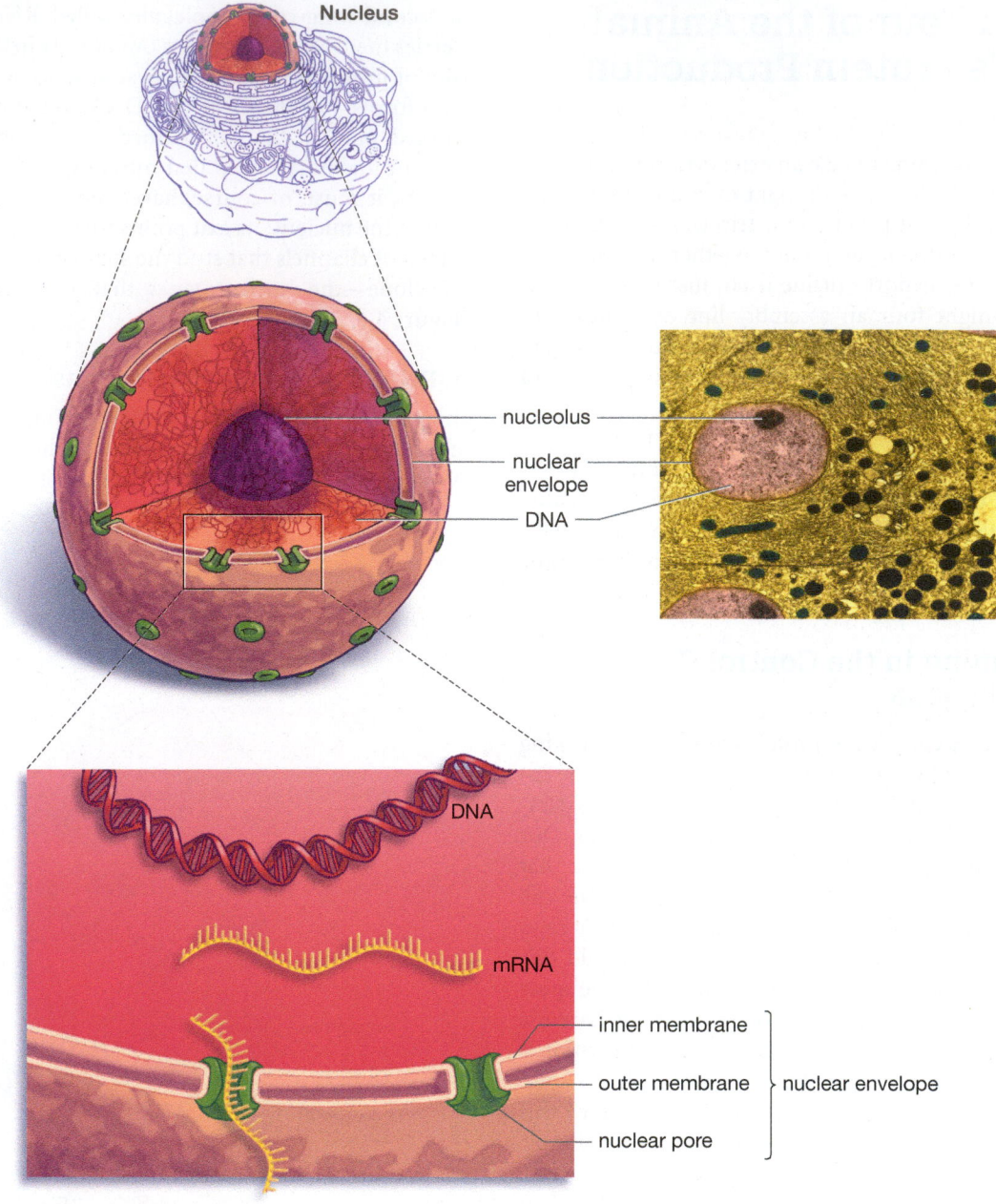

Nucleus

nucleolus

nuclear envelope

DNA

DNA

mRNA

inner membrane

outer membrane   nuclear envelope

nuclear pore

mRNA rapidly moving out through nuclear pores and dropping off into the cytoplasm. Several varieties of RNA actually come out like this, but for now, let's just follow the trail of the mRNA. You're about ready to leave the nucleus to continue your cell tour, but before you do, note two things. First, DNA contains information for making proteins, and the mRNA chains coming out of the nuclear pores amount to a means of transmitting this information. Thus, it's not hard to conceptualize the nucleus as a control center for the cell. Beyond this, in looking at Figure 4.6 you've probably noticed that there is a rather imposing structure *within* the nucleus, called the nucleolus. For now, just *hold that thought;* the mRNA chains have come out of the nuclear pores and are making a short trip to another part of the cell.

Ribosomes

Small structures called ribosomes are the destinations for our lengths of mRNA. For now, we can define the **ribosome** as an organelle that serves as the site of protein synthesis in the cell. If an mRNA chain amounts to a set of instructions for putting a protein together, a ribosome amounts to a machine that carries out these instructions. Each ribosome acts as a site through which an mRNA chain is run; as this happens, the ribosome "reads" the chemical sequence in the mRNA chain and starts putting together amino acids in the order specified by the mRNA sequence. The result is a chain of amino acids that grows from the ribosome—a chain that eventually will fold up into the molecule we call a protein (**Figure 4.7**). You may remember that the kind of protein we are tracking will eventually be exported out of the cell altogether. The synthesis of this kind of protein actually stops when only a very short sequence of the amino acid (or "polypeptide") chain has grown from the ribosome. Why? "Export" polypeptide chains need to be processed within other structures in the cell before they can become fully functional proteins. The first step in this process is for the ribosome, and its associated cargo, to migrate a short distance in the cell and then attach to another cell structure.

The Rough Endoplasmic Reticulum

If you look again at Figure 4.4, you can see that, though the nucleus cuts a roughly spherical figure out of the cell, there is in essence a folded-up continuation of the nuclear envelope on one side. This mass of membrane has a name that is a mouthful: the **rough endoplasmic reticulum**. This structure can be defined as a network of membranes that aids in the processing of proteins in eukaryotic cells. The rough endoplasmic reticulum is rough because it is studded with ribosomes; it is endoplasmic because it lies within (*endo*) the cyto*plasm*; and it is a reticulum because it is a net-

work, which is what *reticulum* means in Latin. Understandably, it is generally referred to as the rough ER or the RER.

Our ribosome, bound with its attached mRNA and polypeptide chains, will migrate to the rough ER and dock on its outside face, thus joining a multitude of other ribosomes that have done the same thing. As noted, the polypeptide chain that is output from the ribosome is in essence an unfinished protein that needs to go through more processing before it can be exported. The first step in this processing leads only to the other side of the ER wall in which the ribosome is embedded. As the ribosome goes on with its work, the polypeptide chain it is producing drops into chambers inside the rough ER.

If you look at Figure 4.7, you can see that the entire rough ER takes the shape of a set of flattened sacs. The membrane that the rough ER is composed of forms the periphery of these sacs. Inside are the internal spaces of the rough ER. As polypeptide chains enter the internal spaces, they first fold up into their protein shapes, as you saw in Chapter 3. Beyond this, most proteins exported from cells have sugar side chains added to them here. Quality control of the production line is in operation in the rough ER as well. Polypeptide chains that have

Role of the Nucleus and Ribosomes in Protein Synthesis

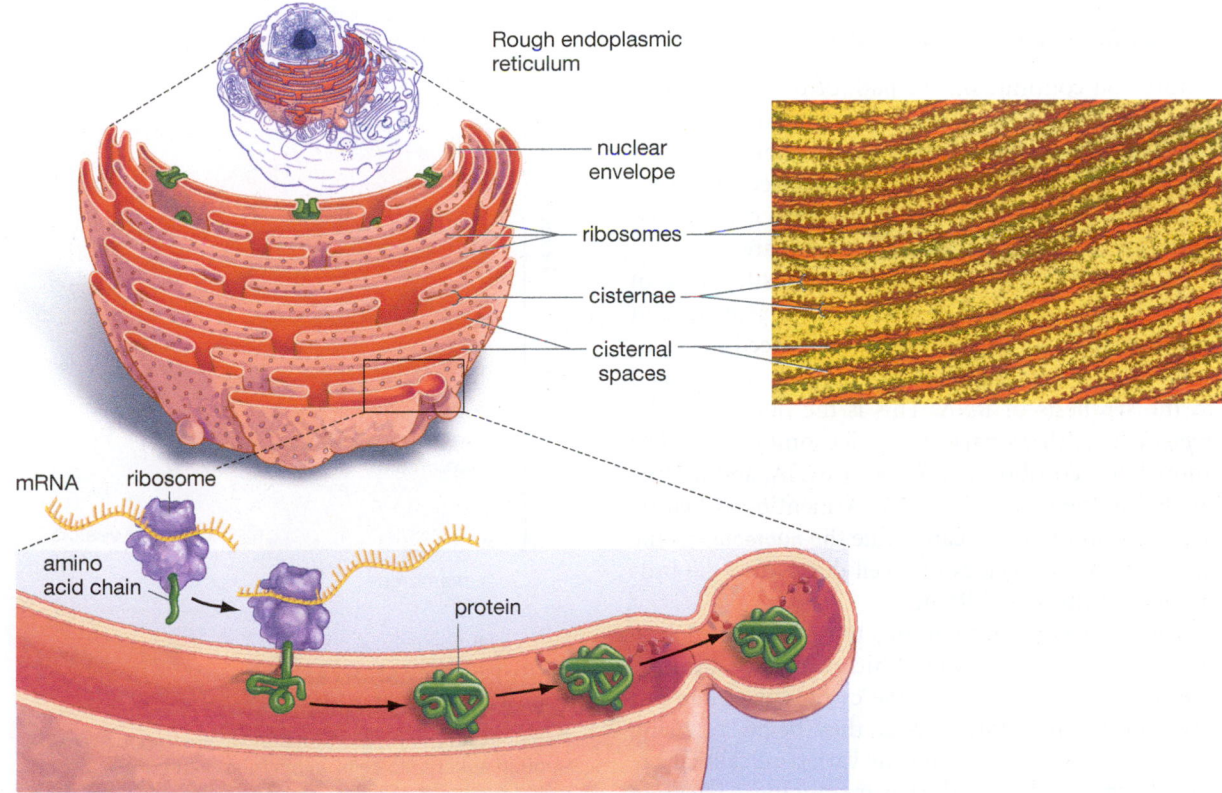

1. mRNA docks on ribosome. Amino acid chain production begins.
2. Ribosome docks on ER. As it is completed, amino acid chain moves into ER's internal space.
3. Amino acid chain folds up making a protein.
4. Sugar side chains added to protein.
5. Vesicle formed to house protein while in transport.

Figure 4.7
Proteins Taking Shape: Ribosomes and the Rough Endoplasmic Reticulum

The steps of the lower figure begin with a length of messenger RNA (mRNA) that has migrated from the nucleus of a cell.

faulty sequences are detected in the rough ER internal space and ejected out of the protein production line altogether, after which they are broken down into their component parts, which will be recycled. Other proteins will pass the quality control tests, however, and will then move out of the rough ER for more processing.

Several Locations for Ribosomes

All of the mRNA that comes out of the nuclear pores goes to ribosomes, but only some of these ribosomes end up migrating to the rough ER. A multitude of ribosomes will remain "free ribosomes," which is to say ribosomes that remain free-standing, in the cytosol. What makes the difference? Remember how, in the ribosome we looked at, only a small stretch of the polypeptide chain it was producing emerged before the ribosome first halted its work and then migrated to the rough ER? That small stretch of the chain contained a chemical signal that said, in effect, "rough ER processing needed." The result was the ribosome's move to the rough ER. In general, these rough ER-linked ribosomes produce proteins that will reside in the cell's membranes or that will be exported out of the cell altogether (making them "secretory" proteins). Meanwhile, free ribosomes tend to produce proteins that will be used within the cell's cytoplasm or nucleus.

A Pause for the Nucleolus

Before you continue on the path of protein processing, think back a bit to the discussion about the nucleus, when you were asked to hold the thought about the large structure *inside* the nucleus called the nucleolus. This is the point where its story can be told because now you know what ribosomes are.

It turns out that ribosomes are mostly made of RNA. (They are made of a mixture of proteins and *ribo*nucleic acid.) So great is the cell's need for ribosomes that a special section of the nucleus is devoted to the synthesis of RNA. This is the nucleolus. The *type* of RNA that's part of the ribosomes is, fittingly enough, called ribosomal RNA or rRNA, and it's one of the multiple varieties of RNA mentioned before. With this in mind, we can define the **nucleolus** as the area within the nucleus of a cell devoted to the production of ribosomal RNA.

Ribosomes are brought only to an unfinished state within the nucleolus, after which they pass through the nuclear pores and into the cytoplasm; when put together in final form there, they begin receiving mRNA chains. They are the one variety of cellular organelle that is not lined by a membrane. (They are also the one variety of organelle that prokaryotic cells have.) The cell's traffic in ribosomes is considerable; with millions of them in existence, lots are going to be wearing out all the time. Perhaps a thousand need to be replaced every minute.

The Endomembrane System

Elegant Transportation: Transport Vesicles

The proteins that have been processed within the rough ER need to move out of it and to their "downstream" destinations before being exported. But how do proteins move from one location to another within the cell? Recall that the rough ER and the nuclear membrane amount to one long, convoluted membrane. And, as just noted, all the organelles in the cell except ribosomes are "membrane-bound," meaning they have membranes at their periphery. Each of these membranes has its own chemical structure, but collectively they have an amazing ability to work together: A piece of one membrane can *bud off,* as the term goes, carrying inside it some of our proteins-in-process. Moving through the cytosol, this tiny sphere of membrane can then *fuse* with another membrane-bound organelle, releasing its protein cargo in the process. You can see how this works in **Figure 4.8**, which offers a closer view of

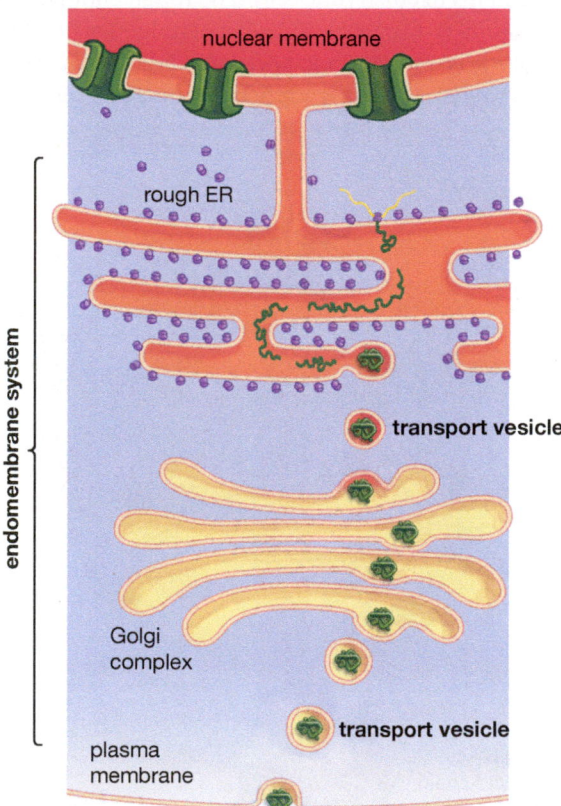

Figure 4.8
Cellular Transportation:
The Endomembrane System

Proteins and other cellular materials are transported through eukaryotic cells within membrane-lined spheres called transport vesicles. These vesicles can emerge, soap bubble-like, from one membrane-lined organelle, travel through the cell's cytosol, and then fuse with a second membrane-lined organelle, unloading their cargo in the process. The cell's interactive network of transport vesicles and membrane-lined organelles is referred to as its endomembrane system.

one part of the protein-production path. The membrane-lined spheres that move within this network, carrying proteins and other molecules, are called **transport vesicles.** The network itself is known as the **endomembrane system:** an interactive group of membrane-lined organelles and transport vesicles within eukaryotic cells.

This system gives cells a remarkable capability. One minute a piece of membrane may be an integral part of, say, the rough ER; the next, it is separating off as a spheroid vesicle and moving through the cytosol, carrying proteins within it. (These vesicles are not moving under their own power, however. As you'll see later, they are propelled along the "monorails" of the cell's internal skeleton.) Many different kinds of proteins are being processed at any one time in the rough ER sacs, but most of those under construction are initially bound for the same place—the Golgi complex.

Downstream from the Rough ER: The Golgi Complex

Once a transport vesicle, bearing proteins, has budded off from the rough ER, it then moves through the cytosol to fuse with the membrane of another organelle, one first noticed by Italian biologist Camillo Golgi at the beginning of the twentieth century. The **Golgi complex** is a network of membranes that processes and distributes proteins that come to it from the rough ER. Some side chains of sugar may be trimmed from proteins here, or phosphate groups may be added, but the Golgi complex does something else as well. Recall that some proteins in this production line are bound for export outside the cell, while others will end up being used within various membranes in the cell. It follows that proteins have to be *sorted and shipped* appropriately, and the Golgi does just this, acting as a kind of distribution center. Chemical "tags" that are part of the proteins often allow for this routing. Remember how, in the rough ER, carbohydrate side chains might be attached to a newly formed protein? Often, these side chains serve as the routing tags; other times, a section of the protein's amino acid sequence will serve this function.

The Golgi is similar to the ER in that it amounts to a series of connected membranous sacs with internal spaces (**Figure 4.9**). Proteins arrive at the Golgi housed in transport vesicles that fuse with the Golgi "face" nearest the rough ER, at which point the vesicles release their protein cargo into the Golgi internal spaces for processing. Once processed, proteins of the sort we are following eventually bud off from the outside face of the Golgi, now housed in their final transport vesicles.

From the Golgi to the Surface

For secretory proteins, the journey that began with the copying of DNA information onto mRNA is almost over. Once a vesicle buds off from the Golgi, all that remains is for it to make its way through the cytosol to the plasma membrane at the outer reaches of

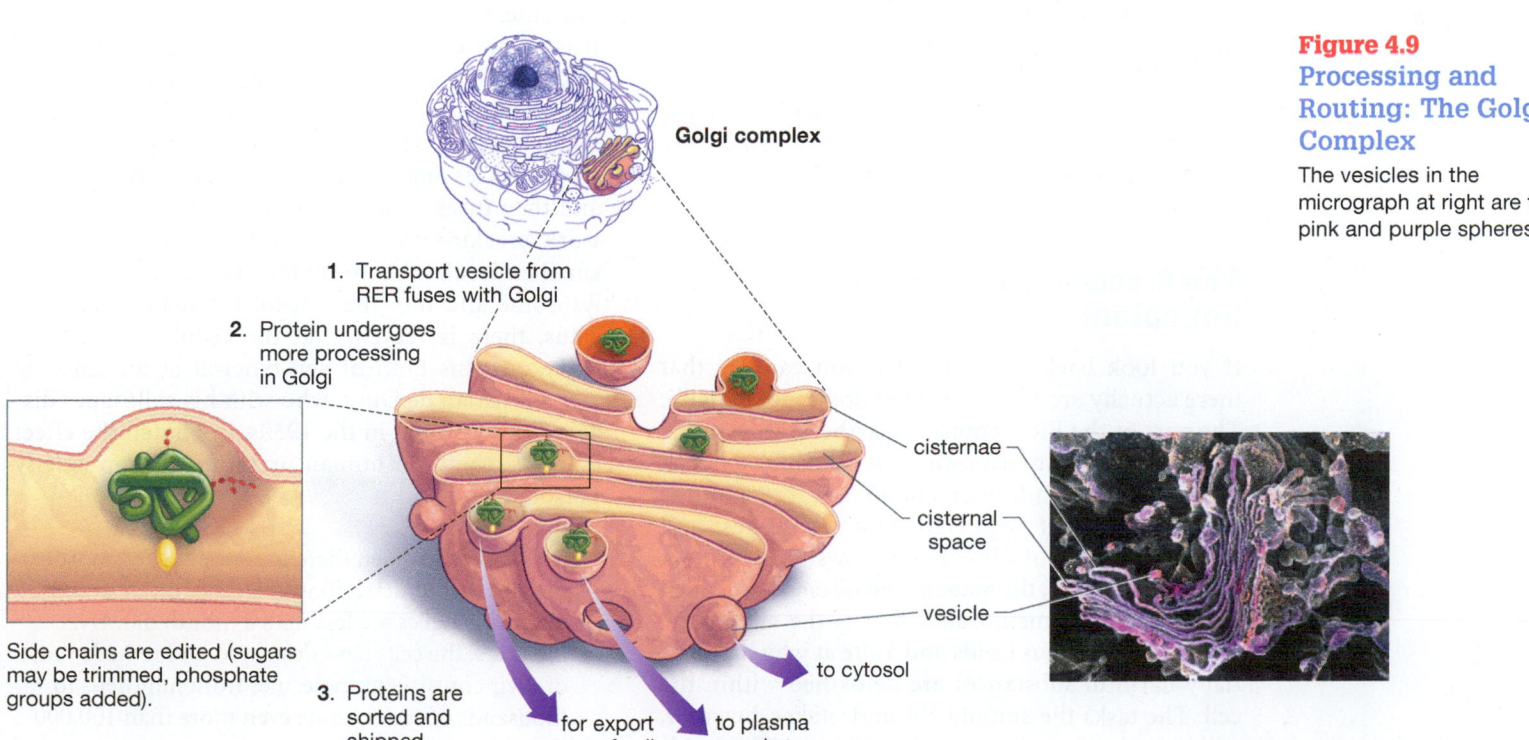

Golgi complex

1. Transport vesicle from RER fuses with Golgi

2. Protein undergoes more processing in Golgi

Side chains are edited (sugars may be trimmed, phosphate groups added).

3. Proteins are sorted and shipped...

for export out of cell

to plasma membrane

to cytosol

cisternae

cisternal space

vesicle

**Figure 4.9
Processing and Routing: The Golgi Complex**

The vesicles in the micrograph at right are the pink and purple spheres.

the cell. There, the vesicle fuses with the plasma membrane, and the protein is ejected into the extracellular world. With it, one finished product of the cellular factory has rolled out the door.

SO FAR . . .

1. The cell's nucleus holds the cell's primary complement of _____, which contains information for the production of _____.

2. Proteins of the sort followed in the text are put together within tiny organelles called _____ that lie outside the nucleus. During their production, these proteins fold up upon entering the _____, where they undergo editing, after which they move to the _____, where they undergo further editing and are sorted for distribution.

3. Proteins and other materials move through the cell within membrane-lined spheres called _____, which, together with membrane-lined organelles, make up the _____ system.

4.5 Cell Structures Outside the Protein Production Path

A functioning cell engages in more activities than the protein synthesis and shipment just reviewed for the simple reason that cells do a lot more than produce proteins.

The Smooth Endoplasmic Reticulum

If you look back to Figure 4.4, you can see that there actually are *two* kinds of endoplasmic reticuli. The part of the ER membrane, farther out from the nucleus, that has no ribosomes is called the smooth endoplasmic reticulum or smooth ER. It's "smooth" because it is not peppered with ribosomes, and this very quality means it is not a site of protein synthesis. Instead, the **smooth endoplasmic reticulum** is a network of membranes that is the site of the synthesis of various lipids and a site at which potentially harmful substances are detoxified within the cell. The tasks the smooth ER undertakes, however, will vary in accordance with cell type. The lipids we normally think of as "fats" are put together and stored in the smooth ER of liver and fat cells, while the "steroid" lipids reviewed last chapter—testosterone and estrogen—are put together in the smooth ER of the ovaries and testes. The detoxification of potentially harmful substances, such as alcohol, takes place largely in the smooth ER of liver cells.

Tiny Acid Vats: Lysosomes and Cellular Recycling

Any factory must be able to get rid of some old materials while recycling others. A factory also needs new materials, brought in from the outside, that probably will have to undergo some processing before use. A single organelle in the animal cell aids in doing all these things. It is the **lysosome,** an organelle found in animal cells that digests worn-out cellular materials and foreign materials that enter the cell. Several hundred of these membrane-bound organelles may exist in any given cell. You could think of them as sealed-off acid vats that take in large molecules, break them down, and then return the resulting smaller molecules to the cytosol. What they cannot return, they retain inside themselves or expel outside the cell. They carry out this work not only on molecules entering the cell from the outside (say, invading bacteria) but also on materials that exist inside the cell—on worn-out organelle parts, for example (**Figure 4.10**).

A given lysosome may be filled with as many as 40 different enzymes that can break larger molecules into their component parts—an enzymatic array that allows each lysosome to break down most of what comes its way. A lysosome generally gets hold of its macromolecule prey through the work of a larger network of membrane-lined vesicles. One of these vesicles engulfs, say, a worn-out organelle part and then fuses with a lysosome, which then goes to work breaking the organelle fragment down. The small molecules that result then pass freely out of the lysosome and into the cytosol for reuse elsewhere. Thus, there is recycling at the cellular level. Cells carry out this kind of self-renewal at an amazing rate. Christian de Duve, who with his colleagues discovered lysosomes in the 1950s, has noted the effect of this activity on human brain cells. In an elderly person, he noted,

> such cells have been there for decades. Yet, most of their mitochondria, ribosomes, membranes, and other organelles are less than a month old. Over the years, the cells have destroyed and remade most of their constituent molecules from hundreds to thousands of times, some even more than 100,000 times.

Molecular house cleaning of this sort operates continually within all eukaryotic cells, but the *pace* of this activity changes in accordance with an organism's nutritional state. The organelles and proteins of a person who is starving will be dismantled at a much higher rate than is the case in a person who has had plenty to eat. Indeed, when nutrition is running low, perfectly *functional* cellular components seem to be taken apart in this process, after which the resulting molecules are used to meet the energy needs of the cell—the cellular equivalent of throwing furniture into a fireplace as a means of keeping warm. On the other side of this coin, we now have evidence indicating that an *underactive* molecular recycling system plays a part in producing Alzheimer's disease. In this case, faulty protein fragments are allowed to build up in brain cells, rather than being degraded through lysosomal activity. The result is a set of brain cells that end up dying because of the proteins that have accumulated in them.

Extracting Energy from Food: Mitochondria

Just as there is no such thing as a free lunch, there is no such thing as free lysosome activity, or ribosomal action, or protein export. There is a price to be paid for all these things, and it is called energy expenditure. The fuel for this energy is contained in the food that cells ingest. But the energy in this food has to be extracted and transferred to a molecule that can easily dispense it as needed—a molecule called ATP (adenosine triphosphate). And where does this energy extraction and transfer take place? Mostly in the tiny "power plant" organelles of the cell, called mitochondria. Just as a city's power plant takes in coal and turns out electricity, so mitochondria take in food and turn out ATP. Thus, **mitochondria** can be defined as organelles that are the primary sites of energy conversion in eukaryotic cells. While cells that don't use much energy might only have one or two mitochondria, an energy-ravenous liver cell might have a thousand. Most of the heat in our bodies is generated within mitochondria, and almost all the food we eat is ultimately consumed in them.

Figure 4.11, on the next page, gives you an idea of the structure of a mitochondrion. Note that there is a continuous outer membrane enclosing an inner membrane that has a series of folds in it. The effect of these infoldings is to give mitochondria a larger internal surface area for carrying out their energy transformation activities.

Few details about mitochondria are included here because much of Chapter 7 is devoted to them. Suffice it to say that to carry out their work, mitochondria need not only food but oxygen. (Ever

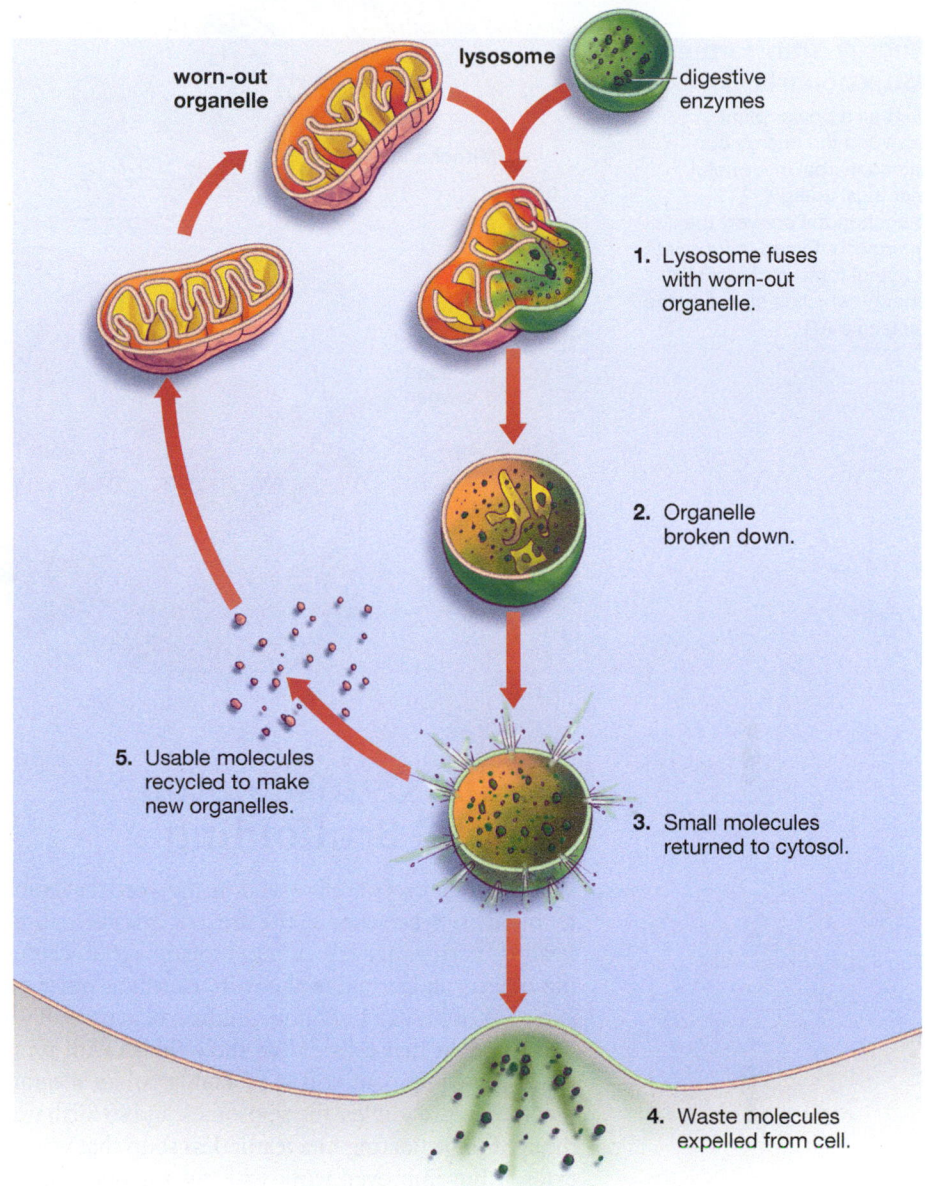

Figure 4.10
Cellular Recycling: Lysosomes
Lysosomes are membrane-bound organelles that contain potent enzymes capable of digesting large molecules, such as worn-out organelles. The useful parts of such organelles will be returned to the cytosol and used elsewhere—a form of cellular recycling.

worn-out organelle

lysosome

digestive enzymes

1. Lysosome fuses with worn-out organelle.

2. Organelle broken down.

5. Usable molecules recycled to make new organelles.

3. Small molecules returned to cytosol.

4. Waste molecules expelled from cell.

wonder why you need to breathe?) The *products* of mitochondrial activity, meanwhile, are the ATP we talked about along with water and carbon dioxide. (Ever wonder where the carbon dioxide you exhale comes from?) Another thing to note about mitochondria is that they are the descendants of resident aliens. They started out as bacterial cells that invaded eukaryotic precursor cells more than 1.5 billion years ago only to end up becoming part of these larger cells (see "The Stranger Within," on page 80).

Figure 4.11
Energy Transformers: Mitochondria
Just as a power plant converts the energy contained in coal into useful electrical energy, mitochondria convert the energy contained in food into a useful form of chemical energy, which is stored in the molecule ATP.

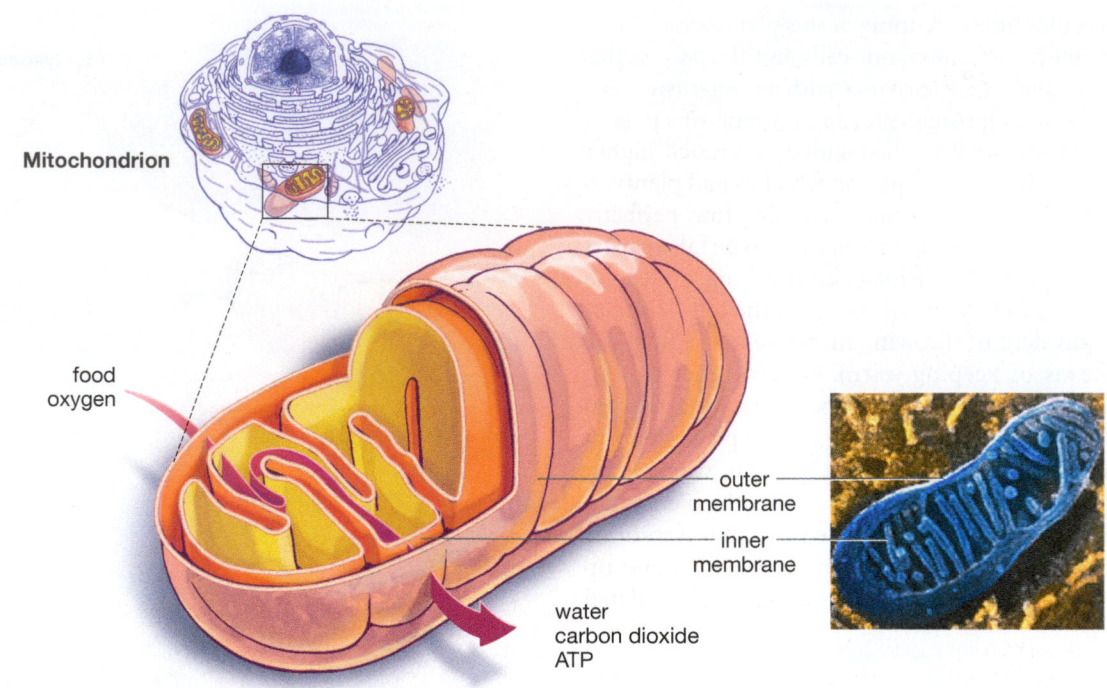

Mitochondrion

food
oxygen

outer membrane
inner membrane

water
carbon dioxide
ATP

MB

The Structure of Cells

4.6 The Cytoskeleton: Internal Scaffolding

When you see a set of cells in action, the word that comes to mind is *hyperactive*. Cells are not passive entities. Jostling, narrowing, expanding, moving about, capturing objects and bringing them in, expelling other objects: This is life as a bubbling cauldron of activity. It was once thought that cells did all these things with pretty much the equipment you've looked at so far, meaning that the Golgi complex, the ribosomes, and so forth were thought to be floating in a featureless soup that was the cytosol. But improved work with microscopes showed that there is, in fact, a complex forest within the cytoplasm. It is the **cytoskeleton,** a network of protein filaments that functions in cell structure, cell movement, and the transport of materials within the cell. This network is found in its full form only in eukaryotic cells, but in the last few years scientists have confirmed that bacteria have several types of cytoskeletal fibers as well.

Some of the cytoskeleton's fibers are permanent and relatively static, but many are moving, and some are assembled or disassembled very rapidly. The cytoskeleton usually is divided into three component parts. Ordered by size, going from smallest to largest in diameter, these are microfilaments, intermediate filaments, and microtubules (**Figure 4.12**). Let's take a look at some of the characteristics of each.

Microfilaments

The most slender of the cytoskeletal fibers, **microfilaments** are made of the protein *actin* and serve as a support or "structural" filament in almost all eukaryotic cells. Microfilaments can also help cells move

or capture prey, essentially by growing very rapidly at one end—in the direction of the movement or extension—while decomposing rapidly at the other end. You can see a vivid example of microfilament-aided cell extension in **Figure 4.13**.

Intermediate Filaments

Intermediate filaments are filaments of the cytoskeleton intermediate in diameter between microfilaments and microtubules. These in-between-sized proteins are the most permanent of the cytoskeletal elements, perhaps coming closest to our everyday notion of what a skeleton is like. They stabilize the positions of the nucleus and other organelles within the cell.

Microtubules

Microtubules are the largest of the cytoskeletal filaments, taking the form of tubes composed of the protein tubulin. Microtubules play a structural role in the cell; in fact, theirs seems to be the preeminent structural role in the sense of determining the shape of the cell. But they take on several other tasks in addition. They serve, for example, as the monorails of the cell's internal skeleton discussed earlier. Recall that protein-laden vesicles move from one organelle to another in the cell. These spheres are moving along the "rails" of microtubules, while sitting atop the "engine" of one of the so-called motor proteins (**Figure 4.14** on page 80). How does this work? One family of motor proteins serves to move vesicles toward a cell's periphery while a second family of proteins serves to move vesicles toward a cell's center. If you've ever "walked" a fork across a flat surface, you understand the kind of motor-protein movement that's pictured in Figure 4.14. One of the

(a) Microfilaments (in red)   **(b)** Intermediate filaments   **(c)** Microtubules

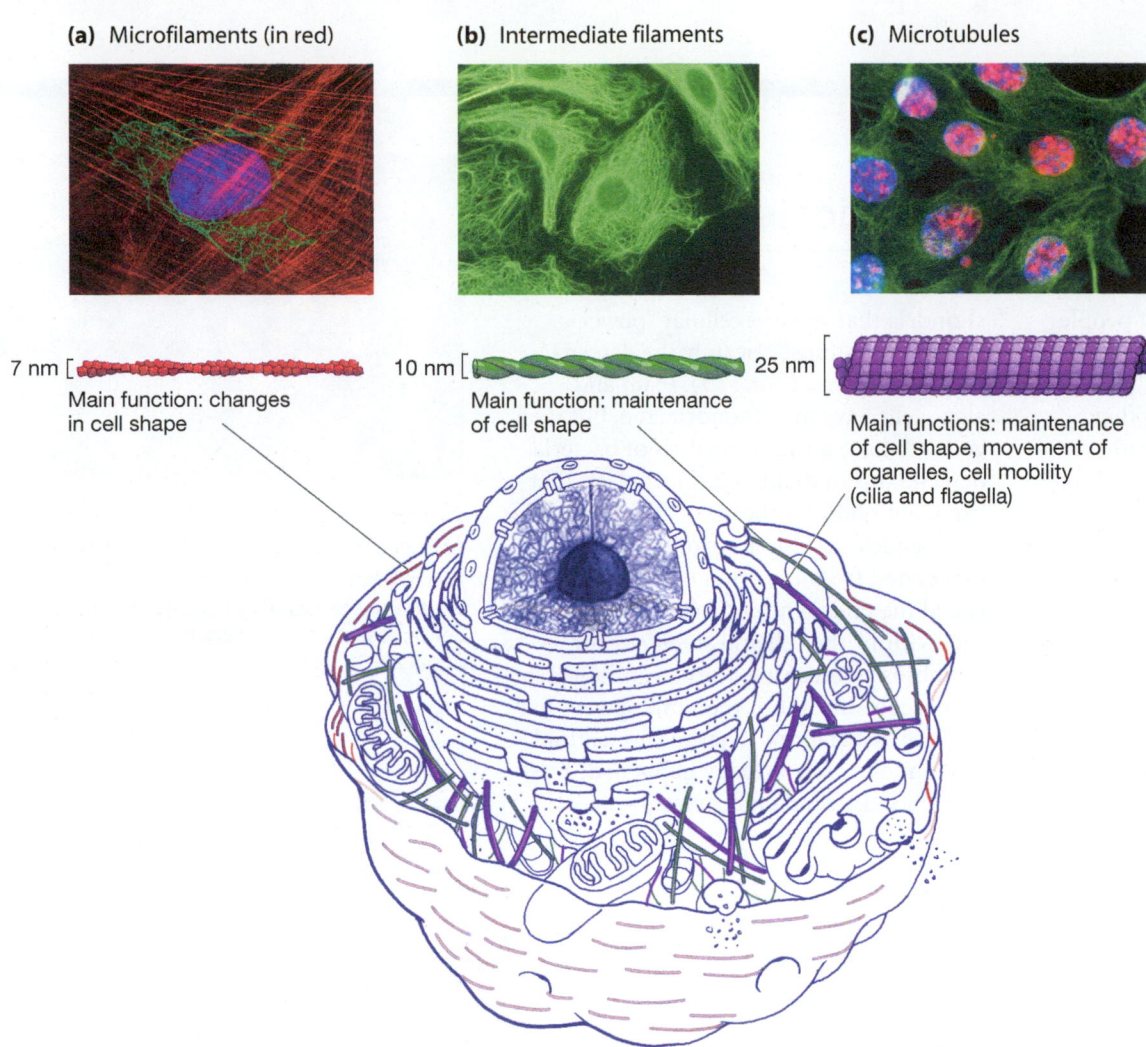

7 nm

Main function: changes in cell shape

10 nm

Main function: maintenance of cell shape

25 nm

Main functions: maintenance of cell shape, movement of organelles, cell mobility (cilia and flagella)

Figure 4.12

Structure and Movement: The Cytoskeleton

Three types of fibers form the inner scaffolding or cytoskeleton of the eukaryotic cell: microfilaments, intermediate filaments, and microtubules.

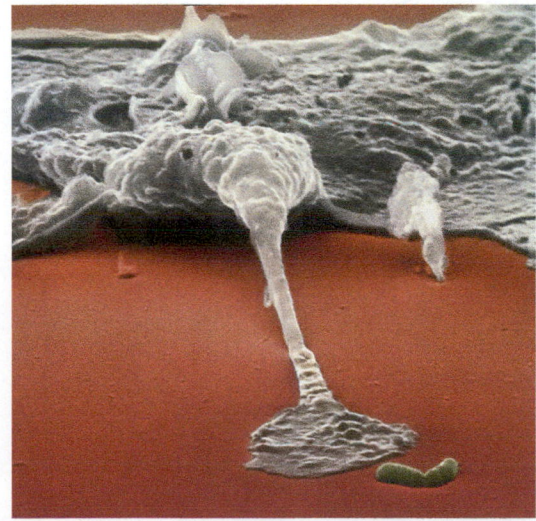

Figure 4.13

Microfilaments in Action

Certain cells can move or capture prey by sending out extensions of themselves called pseudopodia ("false feet"). It is the rapid construction of actin microfilaments—in the direction of the extension—that makes this possible. Here, a type of blood-borne guard cell called a macrophage is about to use a pseudopodium it is constructing to capture a green bacterium.

protein's "feet" detaches from the microtubule and swings around in front of the other foot, after which the process is reversed. Each of these steps is 8 nanometers long, and each is powered by a release of energy from an ATP molecule.

Cell Extensions Made of Microtubules: Cilia and Flagella

Microtubules also form the underlying structure for two kinds of cell extensions, cilia and flagella. **Cilia** are microtubular extensions of cells that take the form of a large number of active, hair-like growths stemming from them. The function of cilia is simple: Move back and forth very rapidly, perhaps 10 to 40 times per second. The effect of this movement can be either to propel a cell or to move material *around* a cell. Cilia are extremely common among single-celled organisms and in some of the cells of simple animals (sponges, jellyfish). You saw an example of cilia in humans earlier in connection with our sense of smell. Our lungs also are lined with cilia whose job it is to sweep the lungs clean of whatever foreign matter has been inhaled. These cilia, like most others, all beat at once in the same direction, acting like rowers in a crew.

Cilia and Flagella

ESSAY

The Stranger Within: Endosymbiosis

If pressed to think about it, most people would probably agree that there's something slightly creepy about being made up of cells. Jostling and dividing as they are, working and dying within us without so much as asking permission, they give us the feeling that the unitary *self* that we so cherish actually amounts to an unruly collection of creatures within.

How sobering it is, then, to realize that we almost certainly are composite beings in more ways than one. Each of our *cells* has within it the vestiges of other living things: the descendants of bacteria that long ago invaded our ancestors' cells, only to take up residence there. In animal cells, the mito-

chondria that serve as cellular "powerhouses" are almost certainly the descendants of bacteria (**Figure 1**). Plant and algae cells have mitochondria, too, but then go on to have an additional set of bacterial descendants in them: the chloroplasts that carry out photosynthesis.

The idea that these structures are descended from free-standing bacteria is called the endosymbiotic theory—*endo* for "within" the cell and *symbiotic* for "symbiosis," meaning a situation in which two organisms not of the same species live in close association. What makes us think that endosymbiosis really happened? First, mitochondria and chloroplasts—which,

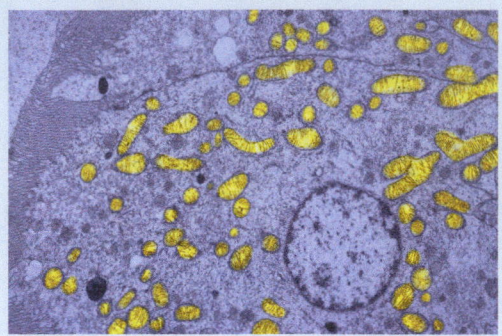

Figure 1

Once Invaders, Now Organelles

The mitochondria within this cell, colored yellow, are the descendants of bacteria that invaded a set of host eukaryotic cells billions of years ago. Over evolutionary time, these bacteria were transformed into energy-converting organelles that today are found in almost all eukaryotic cells.

Figure 4.14

Several Functions for Microtubules

(a) They are the "rails" on which vesicles move through the cell, carried along by "motor proteins."

(b) They exist outside the cell in the form of cilia, which are profuse collections of hair-like projections that beat rapidly, forming currents that can propel a cell or move material around it.

(c) They are also found outside cells in the form of flagella. The flagellum on this sperm cell is enabling it to seek entry into an egg.

Endosymbiosis

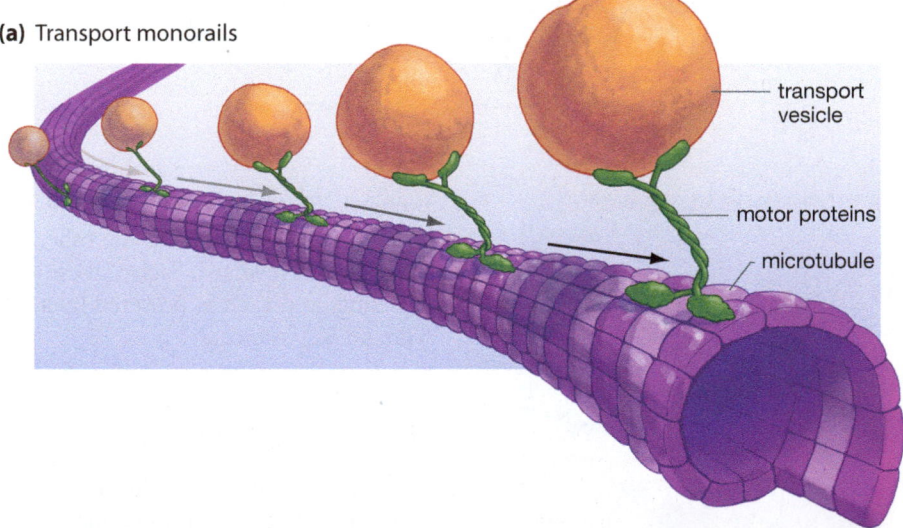

(a) Transport monorails

transport vesicle

motor proteins

microtubule

(b) Cilia

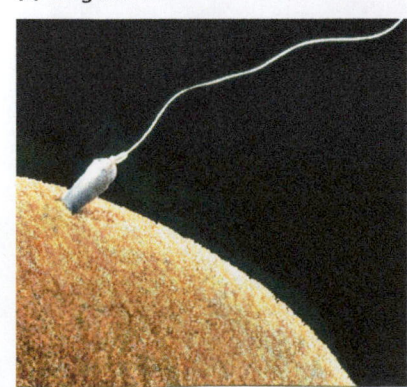

(c) Flagellum

remember, are *organelles* within eukaryotic cells—have many of the characteristics of free-standing *cells,* specifically free-standing bacterial cells. They have their own ribosomes and their own DNA, both of a bacterial type, and they reproduce through a bacteria-like division, partly under their own genetic control. Moreover, the sequencing of mitochondrial DNA indicates that all mitochondria are descended from a single species of bacteria whose closest modern relatives make a living by invading other cells. (As it happens, the single closest living relative of the ancient mitochondrial invaders is *Rickettsia prowazekii,* the cause of modern typhus.)

Mitochondria appear to have completed the transition to endosymbiotic living somewhere between 2.2 and 1.5 billion years ago—so far in the past that the receiving or "host" cells for their initial invasion were a kind of early version of today's eukaryotic cells. The reason this invasion turned into a long-lasting merger was straightforward: Both the bacterial and eukaryotic cells benefited from the new arrangement, though in different ways. At the time, oxygen was making up more and more of Earth's atmosphere, and the mitochondrial ancestors were able to use or "metabolize" oxygen in the extraction of energy from food, while the early eukaryotes they invaded were fairly intolerant of oxygen. Once the two kinds of organisms began living together, the host eukaryote provided food to the bacterium, and the bacterium allowed the host to thrive in an oxygenated world. Then, over time, through a process of internal gene transfer, the bacteria made a transition from organisms that were living inside a host to organelles that were simply part of the host.

This union of the two types of cells was an important event in the evolution of living things in that it scaled up the energy available to eukaryotes. This is so because oxygen-aided energy transfer is terrifically *efficient* compared to energy transfer that does not use oxygen. In our own bodies, the oxygen-aided or "aerobic" energy transfer that takes place in our mitochondria allows us to extract perhaps 15 times as much energy from each bite of an apple than would be the case with energy transfer that doesn't use oxygen. Large organisms require large amounts of energy, and the only way they can get such quantities is by metabolizing oxygen. Thus, the sheer size of the living world is to some degree a product of a microbial merger that began as a microbial invasion billions of years ago.

Cilia grow from eukaryotic cells in great profusion, but it is a different story with **flagella**—the relatively long, tail-like extensions of some cells that function in cell movement (Figure 4.14). It is sometimes the case that several flagella will sprout from a given cell, but often there is but a single flagellum. Only one kind of animal cell is flagellated, and it scarcely needs an introduction: A sperm is a single cell that whips its flagellum in a corkscrew motion to get to an unfertilized egg.

In Summary: Structures in the Animal Cell

In your tour of the cell so far, you've pictured a cell as a factory, one that synthesizes proteins in a "production line" that starts with DNA in the nucleus and then goes to the ribosomes (via mRNA), to the rough ER, to the Golgi complex, and finally to the protein's destination (the plasma membrane, export, and so on). You've also seen that cells have other structures, such as lysosomes for digestion and recycling, mitochondria for energy extraction, the smooth endoplasmic reticulum for lipid synthesis, and the cytoskeleton for structure and movement. If you look at **Figure 4.15**, you can see, in metaphorical form, a "map" of these component parts within the cell. **Table 4.1**, on the next page, lists cellular elements found in plant and animal cells as well as some elements found only in plant cells.

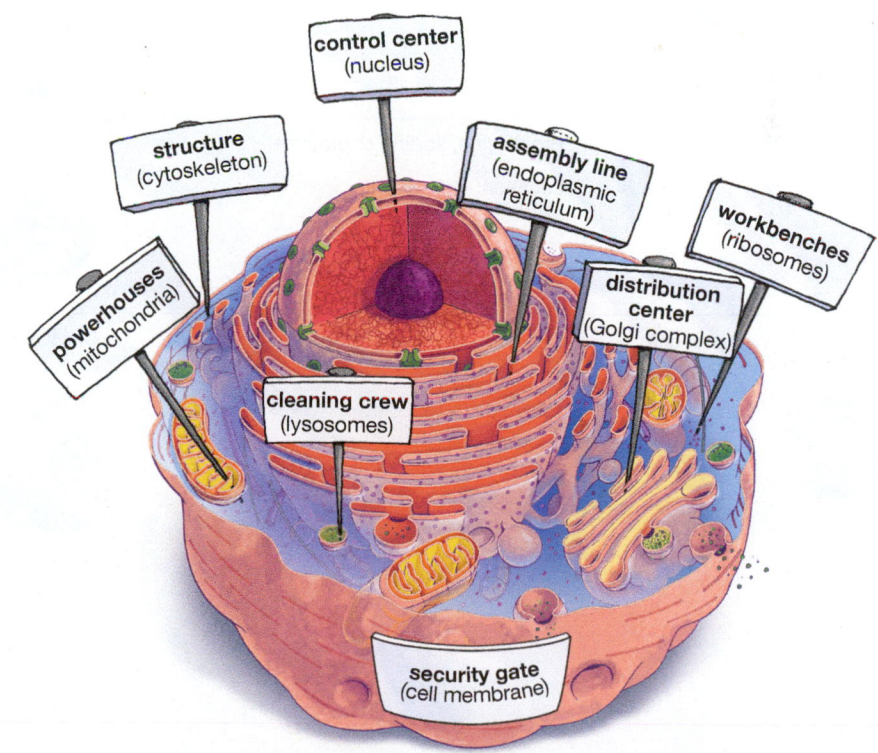

Figure 4.15
The Cell as a Factory
This comparison may help you to remember some roles of the different parts of the cell.

Table 4.1

Structures in Plant and Animal Cells

Name	Function and Location	Name	Function and Location
Nucleus	Site of most of the cell's DNA Location: Inside nuclear envelope	Mitochondria	Transform energy from food Location: Cytoplasm
Nucleolus	Synthesis of ribosomal RNA Location: Nucleus	Rough endoplasmic reticulum	Protein processing Location: Cytoplasm
Ribosomes	Sites of protein synthesis Location: Rough ER, Free-standing in cytoplasm	Smooth endoplasmic reticulum	Lipid synthesis, storage; detoxification of harmful substances Location: Cytoplasm
Cytoskeleton	Maintains cell shape, facilitates cell movement and movement of materials within cell Location: Cytoplasm	Vesicles	Transport of proteins and other cellular materials Location: Cytoplasm
Cytosol	Protein-rich fluid in which organelles and cytoskeleton are immersed Location: Cytoplasm	Central vacuole (in plant cells only) 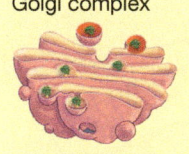	Nutrient storage, cell pressure maintenance, pH balance Location: Cytoplasm
Golgi complex	Processing, sorting of proteins Location: Cytoplasm	Chloroplasts (in plant cells only) 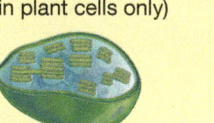	Photosynthesis Location: Cytoplasm
Lysosomes (in animal cells only) 	Digestion of imported materials and cell's own used materials Location: Cytoplasm	Cell walls (in plant cells only) 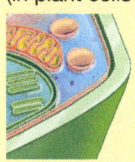	Limit water uptake; maintain cell membrane shape, protect from outside influences Location: Outside plasma membrane

SO FAR...

1. Lipids, including fats, are put together in the organelle called the _____, while worn-out cellular materials are broken down and recycled in the organelles called _____.

2. The organelles called mitochondria transform the _____ from food into a usable chemical form, contained in a molecule called _____.

3. The cell's internal scaffolding, made up of different types of protein fibers, is called the _____.

4.7 The Plant Cell

The animal cell you just looked at has lots in common with the cells that you'd see in any eukaryotic living thing. If you looked at the cells that make up plants or fungi, for example, you'd see that they have a nucleus, ribosomes, mitochondria, and so forth. But, as you can imagine, the cells of a fungus have to be *somewhat* different from those of an animal since fungi and animals are such different types of organisms. To give you an idea of some of the kinds of differences found in the cells of life's various kingdoms, we'll look briefly now at the plant cell, which differs from an animal cell in several basic ways.

A quick look at **Figure 4.16** will confirm for you the first part of this story—how structurally similar plant and animal cells are. As you see, a plant cell has a nucleus (with a nucleolus), the smooth and rough ERs, a cytoskeleton—most of the things you've just gone over in animal cells. Indeed, there is only one struc-

ture present in the animal cells you've looked at that plant cells don't have: the lysosome. What jumps out at you when you look at plant cells is not what they lack compared to animal cells, but what they *have* that animal cells do not. As you can see in **Figure 4.17** on the next page, these additions are:

- A thick cell wall
- Structures called chloroplasts
- A large structure called a central vacuole

The Central Vacuole

The central vacuole pictured in Figure 4.17 is so prominent that it appears to be a kind of organelle continent surrounded by a mere moat of cytosol. And, indeed, in a mature plant cell, one or two central vacuoles may comprise 90 percent of cell volume. Although animal cells can have vacuoles, the imposing central vacuole in plants is different. For a start, it is composed mostly of water, which demonstrates just how watery plant cells are. A typical animal cell may be 70 percent water, but for plant cells the water proportion is likely to be 90 to 98 percent.

The watery environment of the central vacuole contains hundreds of other substances. Many of these are nutrients; others are waste products. There are also hydrogen ions, pumped in to keep the cell's cytoplasm at a near-neutral pH. The sheer number of larger molecules or "solutes" in the central vacuole spurs the movement of *water* into it through an effect called osmosis (which we'll be looking at next chapter). As a result of this influx of water, the central vacuole expands, pushing up against the cell's cytosol and giving the plant cell as a whole a healthy internal pressure. Given this task and the others, we can define the **central vacuole** as a large, watery plant organelle whose functions include the maintenance of cell pressure, the storage of cell nutrients, and the retention and degradation of

Tour of a Plant Cell

Figure 4.16
Common Structures in Animal and Plant Cells

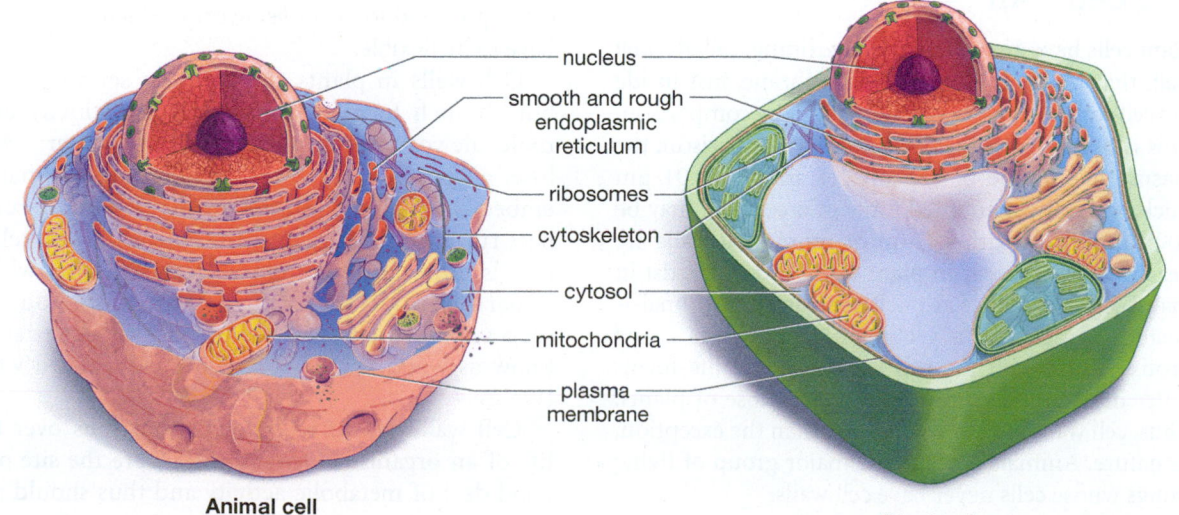

nucleus

smooth and rough endoplasmic reticulum

ribosomes

cytoskeleton

cytosol

mitochondria

plasma membrane

Animal cell

Plant cell

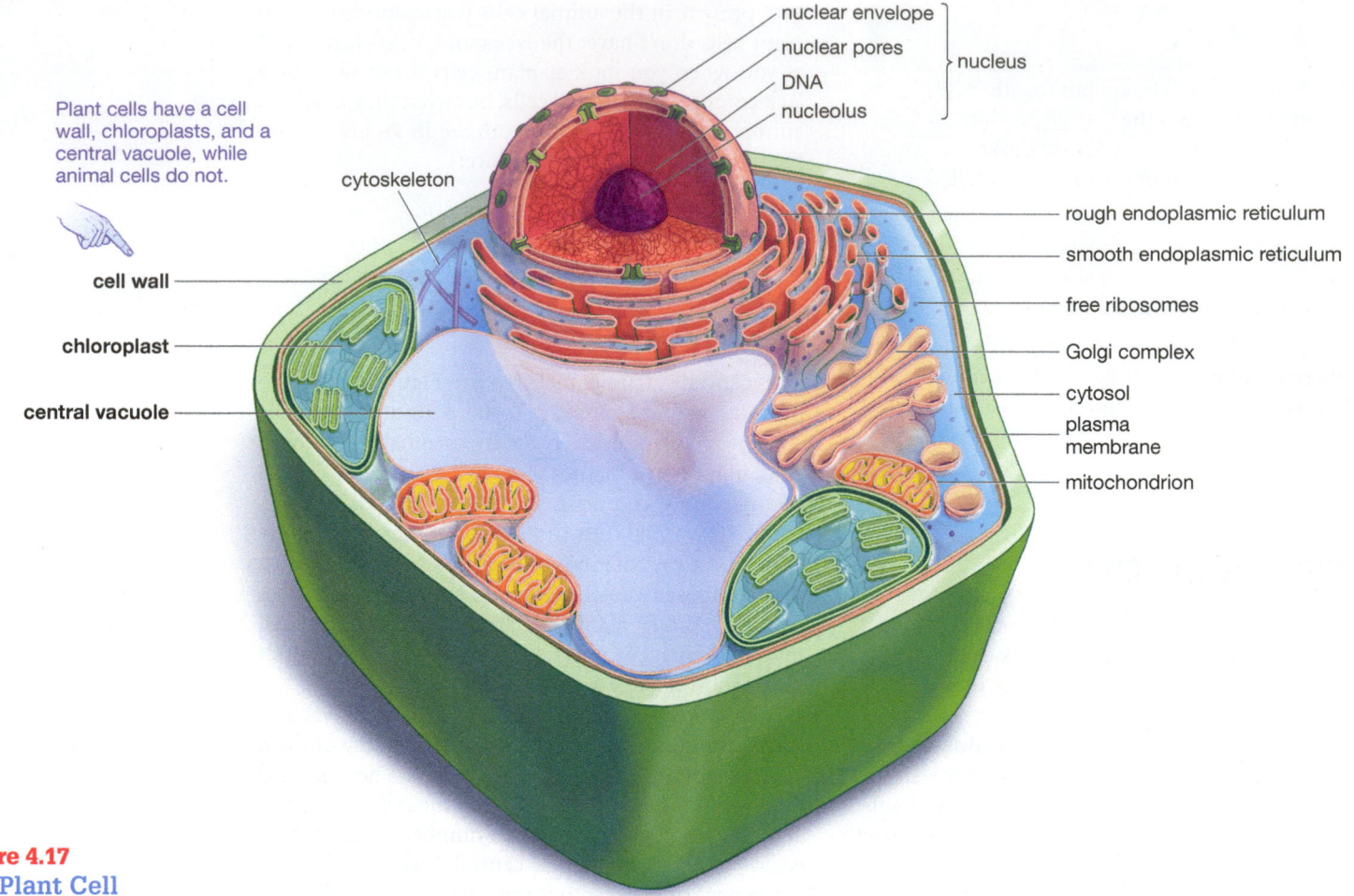

Plant cells have a cell wall, chloroplasts, and a central vacuole, while animal cells do not.

Figure 4.17
The Plant Cell

cell waste products. Despite the vacuole's size, however, it is the other two structures—the cell wall and the chloroplasts—that interest us most as they tell us the most about differences among the cells of various types of living things. Let's have a look at these two structures.

The Cell Wall

Plant cells have an outer protective lining, called a **cell wall,** that makes their plasma membrane, just inside the cell wall, look rather thin and frail by comparison. This is because it *is* thin and frail by comparison; the plasma membrane of a plant cell may be 0.01 μm thick, while the combined units of a cell wall may be 700 times this width—7 μm or more. Cell walls are nearly always present in plant cells. And they exist in many organisms that are neither plant nor animal— bacteria, fungi, and a group of living things called protists—although the cell walls of these life-forms differ in chemical composition from those of plants. Thus, cell walls are the rule, rather than the exception, in nature. Animals are the one major group of living things whose cells never have cell walls.

What do cell walls do for plant cells? They provide them with structural strength, put a limit on their absorption of water (as you'll see in the next chapter), and generally protect plants from harmful outside influences. So, if they're so useful, why don't *animal* cells have them? Cell walls make for a rather rigid, inflexible organism—like plants, which are fixed in one place. Animals, meanwhile, need to be mobile, both to catch prey and to avoid *being* prey, which means they have to be flexible.

Cell walls in plants can come in several forms, but all such forms will be composed chiefly of a molecule you were introduced to last chapter: cellulose, a complex sugar or "polysaccharide" that is embedded within cell walls in the way reinforcing bars run through concrete. In some cell walls, cellulose is joined by a compound called lignin, which imparts considerable structural strength. You can see a vivid demonstration of this in the material we know as wood, which is largely made of cell walls (**Figure 4.18a**).

Cell walls can serve different functions over the life of an organism. Generally, they are the site of a good deal of metabolic activity and thus should not

(a)

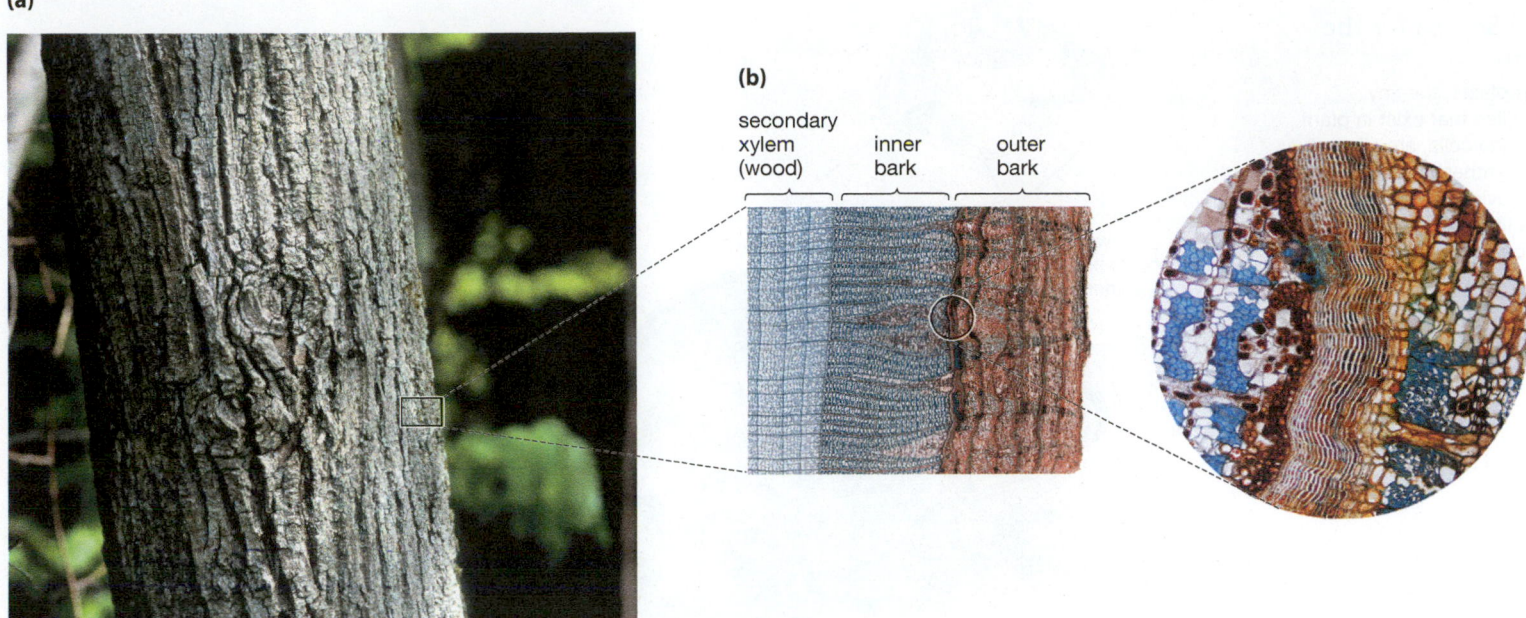

(b)

secondary xylem (wood)   inner bark   outer bark

Figure 4.18
Great Strength from Small Things
(a) Trees can reach great heights because of the strength of their secondary xylem—better known as wood—which is largely made of cell walls. Pictured is an American basswood tree (*Tilia americana*).
(b) Cell walls play important roles in living cells, but they also are valuable as the strong, remaining components of dead cells. The outer bark of the basswood tree, seen in the blow-up at right, is composed mostly of dead cells whose cell walls are infused with a waxy substance that acts to protect the tree from invading insects and microorganisms.

be considered mere barriers. On the other hand, the outer portion of tree bark consists mostly of *dead* cell walls, which clearly are serving a barrier function (**Figure 4.18b**).

Chloroplasts

Everyone knows that a major difference between plants and animals is that, while animals must get their food from outside themselves, plants make their own food through the process of photosynthesis. Where does photosynthesis take place in plants? Inside the organelles known as chloroplasts. It's hard to overstate the importance of these tiny, oblong structures, given that they are the food factories for most of the living world. Lions may eat zebras, but zebras eat grass, and grass is essentially produced inside chloroplasts. Thanks to their ability to harness the sun's power, these membrane-laden organelles can start with nothing more than water, a few minerals, and atmospheric carbon dioxide and end up turning out a sugar that builds an entire plant. A by-product of this operation—a kind of refuse of photosynthesis—is the oxygen that sustains most of Earth's organisms (**Figure 4.19** on the next page). **Chloroplasts** can be defined as organelles that are the sites of photosynthesis in algae cells, as well as plant cells. No other organisms possess these specialized organelles, although

certain bacteria are able to perform photosynthesis without them. In plants, chloroplasts are especially abundant in the cells of leaves. A cell that lies toward a leaf's interior might contain hundreds of these organelles, each of which is capable of performing photosynthesis on its own.

Also note that, as with mitochondria, chloroplasts are the descendants of resident aliens. They originally were a form of free-standing bacteria that were capable of performing photosynthesis. Ancient eukaryotic cells then ingested these bacteria and, over time, the two organisms came to live as one, with the bacteria eventually being transformed into mere organelles within the larger eukaryotic cells. Thus, the green world of trees and plants and algae—this massive edifice that sustains nearly all animals—ultimately was powered into being through the activity of some of the smallest living things we know of.

4.8 Cell-to-Cell Communication

Most of what you've seen so far has made the cell seem like an isolated entity, but this is not the case. Single-celled organisms can exist as separate entities, along with certain plant or animal cells (red blood

Figure 4.19
Food Source for the World

Chloroplasts, the tiny organelles that exist in plant and algae cells, are sites of photosynthesis—the process that provides food for most of the living world. (Micrograph: × 13,000)

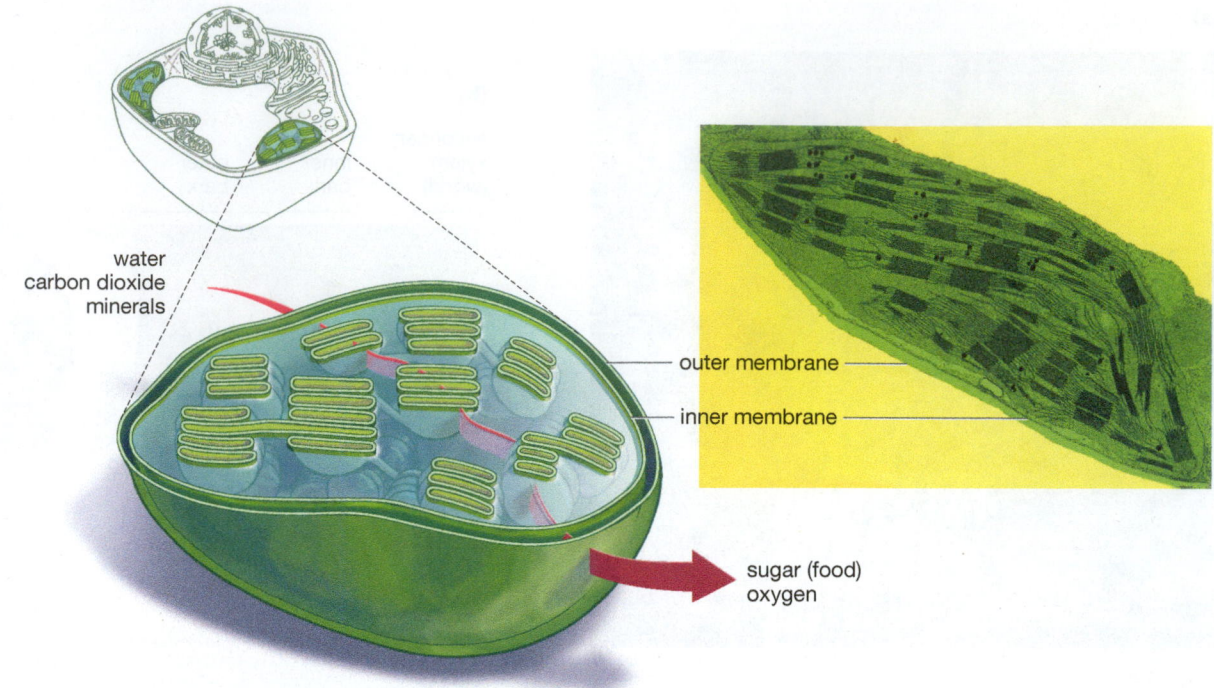

water
carbon dioxide
minerals

outer membrane

inner membrane

sugar (food)
oxygen

Cell Junctions

cells, for example), but most plant and animal cells are linked together in organized collections referred to as *tissues.* Not surprisingly, these assemblages of cells—whether plant or animal—have the ability to communicate with one another.

Communication among Plant Cells

Having noted the thickness of something like the cell walls in plants, you might wonder how one plant cell could interact with another. Communication between plant cells takes place quite readily, however, through a series of tiny channels in the plant cell wall called **plasmodesmata** (singular, plasmodesma). The structure of these channels is such that the cytoplasm of one plant cell is continuous with that of another—so much so that the cytoplasm of an entire plant can be properly looked at as one continuous whole (**Figure 4.20a**). The structure of plasmodesmata is more complex than that of a simple opening, but the basic idea is of a channel-like linkage between two plant cells.

Communication among Animal Cells

There are no plasmodesmata in animal cells, but there are three other kinds of *cell junctions,* or linkages, one of which serves to facilitate cell communication. It is called

a gap junction, and it consists of clusters of protein structures that shoot through the plasma membrane of a cell from one side to the other, forming a kind of tube. When these tubes line up in adjacent cells, the result is a channel for passage of small molecules and electrical signals (**Figure 4.20b**). Thus, we can define a **gap junction** as a protein assemblage that forms a communication channel between adjacent animal cells. Note that animal gap junctions and plant plasmodesmata are very different kinds of channels. Plasmodesmata can be thought of as permanent channels between plant cells, whereas gap junctions open only as necessary.

On to the Periphery

Having looked at what is inside the cell, you've arrived at the cell's periphery, the plasma membrane, which is where you'll be staying for a while. It may at first seem strange to devote a whole chapter to an outer boundary. How much attention would you pay, after all, to a factory's wall as opposed to its contents? The answer is: A lot, if that wall could facilitate communication with the outer world, continually renew itself, and let some things in while keeping others out. Such is the case with the plasma membrane, a slender lining that manages to make one of the most fundamental distinctions on Earth: Inside, life goes on; outside it does not. We'll look at its story in the chapter that's coming up.

Plant tissues

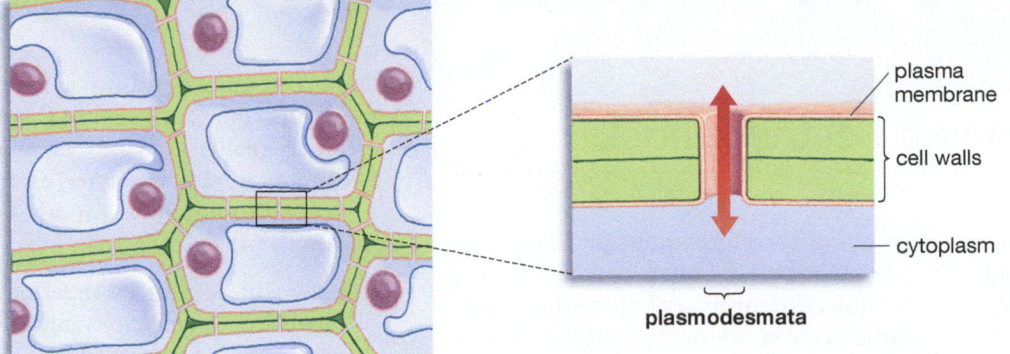

(a) **Plasmodesmata**

In plants, a series of tiny pores between plant cells, the plasmodesmata, allow for the movement of materials among cells. Thanks to the plasmodesmata channels, the cytoplasm of one cell is continuous with the cytoplasm of the next; the plant as a whole can be thought of as having a single complement of continuous cytoplasm.

Animal tissues

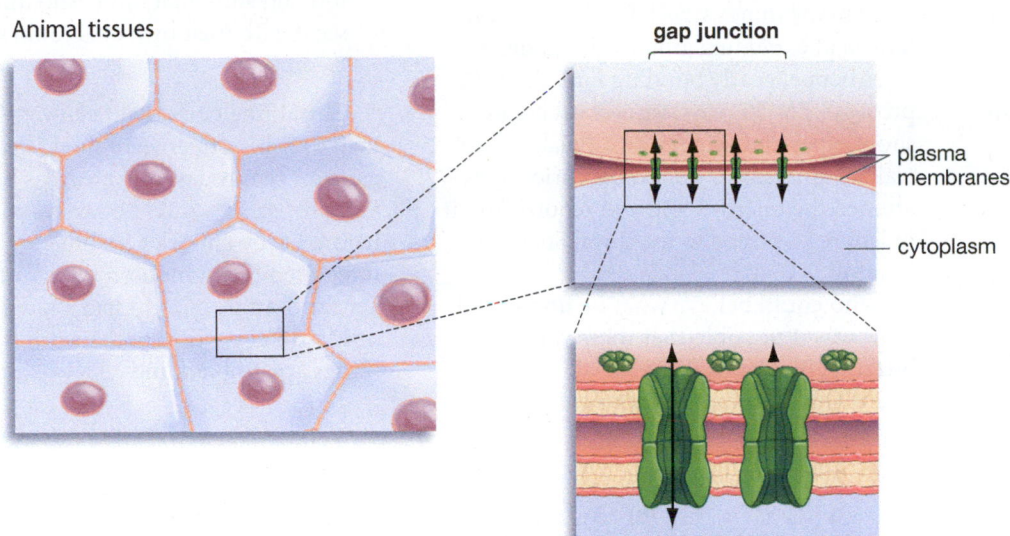

(b) **Gap junctions**

In animals, protein assemblies come into alignment with one another, forming communication channels between cells. A cluster of many such assemblies—perhaps several hundred—is called a gap junction.

Figure 4.20
Cell Communication

SO FAR . . .

1. Plant cells have two organelles and one structure that animal cells do not have. The organelles are _____, which are the sites of photosynthesis, and a large _____, which performs multiple functions, including the storage of nutrients and the maintenance of internal cell pressure. The structure is the _____, which provides protection and limits the plant cell's uptake of water.

2. Animal cells have only a single organelle that plant cells do not have. This is the _____, which breaks down and recycles materials in the cell.

3. In plants, the channels that facilitate cell-to-cell communication are called _____, while in animals, the tube-like protein structures that facilitate such communication are called _____.

THE PROCESS OF SCIENCE

First Sightings: Anton van Leeuwenhoek

To the list of explorers that includes Columbus and Balboa we could add the name Anton van Leeuwenhoek. This unassuming Dutchman was the great early voyager into another world, the micro-world.

It was not until the seventeenth century that human beings realized that things as small as cells existed. Leeuwenhoek began to report in the 1670s on what he saw with the aid of a device invented at the end of the 1500s—the microscope (**Figure 1**). One of Leeuwenhoek's contemporaries, Englishman Robert Hooke, coined the term *cell* after viewing a slice of cork under a simple microscope, but Hooke's purpose was to reveal the detailed structure of familiar, small objects, such as the flea. Leeuwenhoek, by contrast, revealed the *existence* of creatures unimagined until his time. Moreover, he carried out this work in the most extraordinary fashion: Laboring alone in the small town of Delft with palm-sized magnifiers he himself had created, looking at anything that struck his fancy. (And many things struck his fancy; he once looked at exploding gunpowder under a microscope, nearly blinding himself in the process.) For 50 years, while working as shopkeeper and minor city official, this untrained amateur of boundless curiosity examined the micro-world and reported on it in letters he posted to the Royal Society in London.

Who could believe what he uncovered? How was it possible that there was a buzzing, blooming universe of "animalcules" (little animals) whose existence had been completely unsuspected? Prior to his work, no one thought that any creature smaller than a worm could exist within the human body. Yet, examining scrapings from his own mouth, Leeuwenhoek tells us:

> I saw, with as great a wonderment as ever before, an inconceivably great number of little animalcules, and in so unbelievably small a quantity of the foresaid stuff, that those who didn't see it with their own eyes could scarce credit it.

The animalcules that Leeuwenhoek beheld over his career were single-celled organisms that today are known as bacteria and protists. It would be two hundred years before Leeuwenhoek's findings were fully integrated into a modern theory of cells. Yet the man from Delft had shown that only a small portion of the world's living things are visible things.

(a) What Leeuwenhoek could see

(b) Leeuwenhoek with two of his microscopes

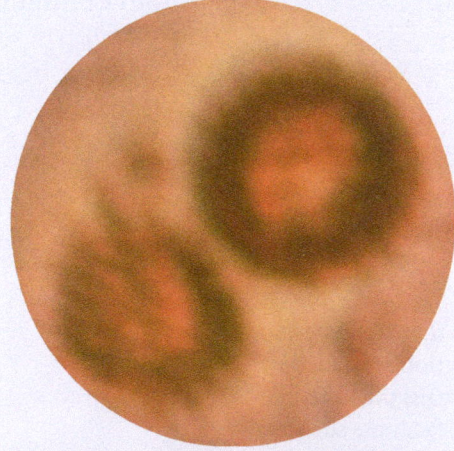

Figure 1

(a) What Leeuwenhoek Could See
Biologist Brian Ford used an actual 300-year-old Leeuwenhoek microscope to capture this image of spiny spores from a truffle. (Micrograph: × 600)

(b) Leeuwenhoek's Microscopes Were Handheld and Paddle-Shaped
Leeuwenhoek revealing the micro-world to Queen Catherine of England, wife of King Charles II (1630–1685).

Summary

4.1 Cells as Life's Fundamental Unit

- With the possible exception of viruses, every living thing is a cell or is composed of cells and all cells come into existence only through the activity of other cells. (p. 65)

4.2 Prokaryotic and Eukaryotic Cells

- All cells are either prokaryotic (bacteria or archaea) or eukaryotic (all other cells). Eukaryotic cells store most of their DNA in their cell nucleus while prokaryotic cells do not have a nucleus. Eukaryotic cells tend to be much larger than prokaryotic cells and they have more organelles. Many eukaryotes are multicelled organisms while all prokaryotes are single-celled. (p. 66)

4.3 The Eukaryotic Cell

- The five principal components of the eukaryotic cell are the nucleus, other organelles, the cytosol, the cytoskeleton, and the plasma membrane. Organelles are highly organized structures within the cell that carry out specialized functions. The cytosol is the jelly-like fluid outside the nucleus in which most organelles are immersed. The cytoplasm is the region of the cell inside the plasma membrane but outside the nucleus. The cytoskeleton is a network of protein filaments that have diverse functions. The plasma membrane is the outer lining of the cell. (p. 66)

4.4 A Tour of the Animal Cell's Protein Production Path

- Information for constructing proteins is contained in the DNA located in the cell nucleus. This information is copied onto a length of messenger RNA (mRNA) that departs the nucleus through its nuclear pores and goes to ribosomes. Many ribosomes that receive mRNA chains then migrate to and embed in a membrane network called the rough endoplasmic reticulum (RER). The polypeptide chains produced by the ribosomal "reading" of the mRNA sequences are dropped from ribosomes into the internal spaces of the RER, where they undergo folding and editing. (p. 71)

- Materials move from one structure to another in the cell via the endomembrane system, in which a piece of membrane, with proteins inside, can bud off from one organelle as a transport vesicle that moves through the cell and then fuses with another membrane-lined structure. (p. 74)

- Once protein processing is finished in the RER, proteins undergoing processing move to the Golgi complex where they are processed further and marked for shipment to appropriate cellular locations. (p. 75)

4.5 Cell Structures Outside the Protein Production Path

- The smooth ER is a network of membranes that functions to synthesize lipids and to detoxify harmful substances. Lysosomes are organelles that break down worn-out cellular structures or foreign materials that come into the cell and return their components to the cytoplasm

for re-use. Mitochondria are organelles that extract energy from food and transfer this energy to an energy-dispensing molecule, ATP. (p. 76)

4.6 The Cytoskeleton: Internal Scaffolding

- The cytoskeleton provides the cell with structure, facilitates the movement of materials inside the cell, and facilitates cell movement. (p. 78)

- From smallest to largest in diameter, the three principal types of cytoskeleton elements are microfilaments, intermediate filaments, and microtubules. Microfilaments help the cell move and capture prey by forming rapidly in the direction of movement. Intermediate filaments provide support and structure to the cell. Microtubules provide structure to cells and facilitate the movement of materials inside the cell by serving as transport "rails." (p. 78)

- Cilia and flagella are extensions of cells composed of microtubules. Cilia extend from cells in great numbers, serving to either move the cell or to move material around the cell. In contrast, one—or at most a few—flagella extend from cells that have them and serve to facilitate cell movement. (p. 79)

4.7 The Plant Cell

- Plant cells have two organelles not found in animal cells—chloroplasts and central vacuoles—and have one structure not found in animal cells, the cell wall. Conversely, plant cells lack one organelle found in animal cells, the lysosome. The cell wall gives the plant structural strength and helps regulate the intake of water; the central vacuole stores nutrients, degrades waste products, and helps maintain internal cell pressure; and chloroplasts are the sites of photosynthesis. **(p. 83)**

4.8 Cell-to-Cell Communication

- Plant cells have plasmodesmata: channels that are always open and that hence make the cytoplasm of one plant cell continuous with that of another. Adjacent animal cells have gap junctions that are composed of protein assemblages that open only as necessary, allowing the movement of small molecules and electrical signals between cells. **(p. 85)**

Key Terms

cell wall 84

central vacuole 83

chloroplast 85

cilia 79

cytoplasm 67

cytoskeleton 78

cytosol 67

endomembrane system 75

eukaryotic cell 66

flagella 81

gap junction 86

Golgi complex 75

intermediate filament 78

lysosome 76

microfilament 78

micrograph 68

micrometer 68

microtubule 78

mitochondria 77

nanometer 68

nuclear envelope 71

nucleolus 74

nucleus 66

organelle 66

plasma membrane 67

plasmodesmata 86

prokaryotic cell 66

ribosome 72

rough endoplasmic reticulum 73

smooth endoplasmic reticulum 76

transport vesicle 75

Understanding the Basics

Multiple-Choice Questions

(Answers are in the back of the book.)

1. Jerome has strep throat, a bacterial infection. The cause of the infection is:
   a. the growth of a virus.
   b. the presence of archaea.
   c. eukaryotic cells dividing in his throat.
   d. organelles that take control of his organs—in this case, his throat.
   e. prokaryotic cells.

2. Where would you expect to find a cytoskeleton?
   a. primarily inside the nucleus
   b. as the internal structure of a mitochondrion
   c. between bacterial cells
   d. throughout the cytosol
   e. as an outer coat on an insect

3. Suppose you found a cell that had many mitochondria, a nucleus, and a cell wall made of cellulose. You would assume that you had found:
   a. a plant cell.
   b. an animal cell.
   c. a bacterial cell.
   d. one of these three types, but you will not know which type without further investigation.
   e. either a plant cell or an animal cell, but not a bacterial cell.

4. Cells in the pancreas manufacture large amounts of protein. Which of these would you expect to find a large amount of, or number of, in pancreatic cells?
   a. lysosomes
   b. rough endoplasmic reticulum
   c. smooth endoplasmic reticulum
   d. chloroplasts
   e. plasma membrane

5. Suppose that a cell could be seen to lack microtubule filaments. Which of these would be the most likely effect of this condition?
   a. an inability to get energy out of food
   b. an inability to manufacture proteins
   c. an inability to break down worn-out organelles
   d. an inability of the skeletal muscles to contract
   e. an inability to move proteins from one part of the cell to another

6. Heart muscle cells have a number of gap junctions connecting them to their adjoining cells. From this you can conclude that heart muscle cells:
   a. exchange nuclei very frequently.
   b. have plasmodesmata.
   c. move vacuoles from cell to cell.
   d. communicate frequently.
   e. lack the ability to divide.

7. Which is the correct ranking of these "small" things, from smallest to largest? (Use typical sizes.)
   a. animal cells<atoms<bacteria< proteins<amino acids
   b. atoms<amino acids<proteins< bacteria<animal cells
   c. atoms<proteins<amino acids< animal cells<bacteria
   d. bacteria<atoms<amino acids< proteins<animal cells
   e. bacteria<animal cells<atoms <amino acids<proteins

8. Cells need large amounts of ribosomal RNA to make proteins. The ribosomal RNA is made in a specialized structure known as _____, which is found in _____.
   a. a chloroplast; the cytosol
   b. a ribosome; the cytosol
   c. the endoplasmic reticulum; the nucleus
   d. the nuclear envelope; the nucleus
   e. the nucleolus; the nucleus

Brief Review

(Answers are in the back of the book.)

1. Why are cells necessarily so small? Why couldn't a brain cell, for example, be large enough to be seen with the naked eye?

2. What are some of the differences between eukaryotic cells and prokaryotic cells?

3. Insulin, a hormone that reduces blood-sugar levels in the body, is a protein that is exported from special cells in the pancreas. Describe the set of steps that leads from a length of insulin-coding DNA in the nucleus to the release of insulin in the bloodstream.

4. What are cilia and flagella? What are some of the roles they play in the human body?

5. Consider what plants would be like if they had no plasmodesmata. Explain why this would be a problem.

6. What is a gap junction, and what is its function?

Applying Your Knowledge

1. The text notes that animals could not function as they do if their cells were encased in the thick cell walls that surround plant cells. What about the reverse case? Why would plants be unable to function if their cells lacked cell walls?

2. The earliest organisms we have evidence of were all single-celled. Why do you think life had to "start small" like this?

3. Why do membranes figure so prominently in eukaryotic cells? What essential function do they serve?

MasteringBiology®

1. MasteringBiology Assignments

Activities: Form Fits Function: Cells; Prokaryotic Cell Structure and Function; Review: Animal Cell Structure and Function; Role of the Nucleus and Ribosomes in Protein Synthesis; The Endomembrane System; The Structure of Cells; Cilia and Flagella; Endosymbiosis; Cell Junctions; **Bioflix:** Tour of an Animal Cell; Tour of a Plant Cell; **Current Events:** A First: Organs Tailor-Made with Body's Own Cells; Peering Over the Fortress That Is the Mighty Cell; **Building Vocabulary; Questions:** Bioflix Quiz, Reading Questions, Test Bank

2. eText

Read your book online, search, take notes, highlight text, and study more effectively.

3. The Study Area

Self-study tools to test comprehension.

Life's Border:
The Plasma Membrane

The outer lining of cells is in a sense the outer border of life. In its roles as protector, gatekeeper, and message carrier, this lining is indispensable to life.

Available on MasteringBiology®

Activities: Membrane Structure; Selective Permeability of Membranes; Diffusion; Osmosis and Water Balance in Cells; Plasma Membranes and Diffusion; Active Transport; Facilitated Diffusion; Exocytosis and Endocytosis; **Bioflix:** Membrane Transport; **Building Vocabulary; Questions:** Bioflix Quiz, Reading Questions, Test Bank

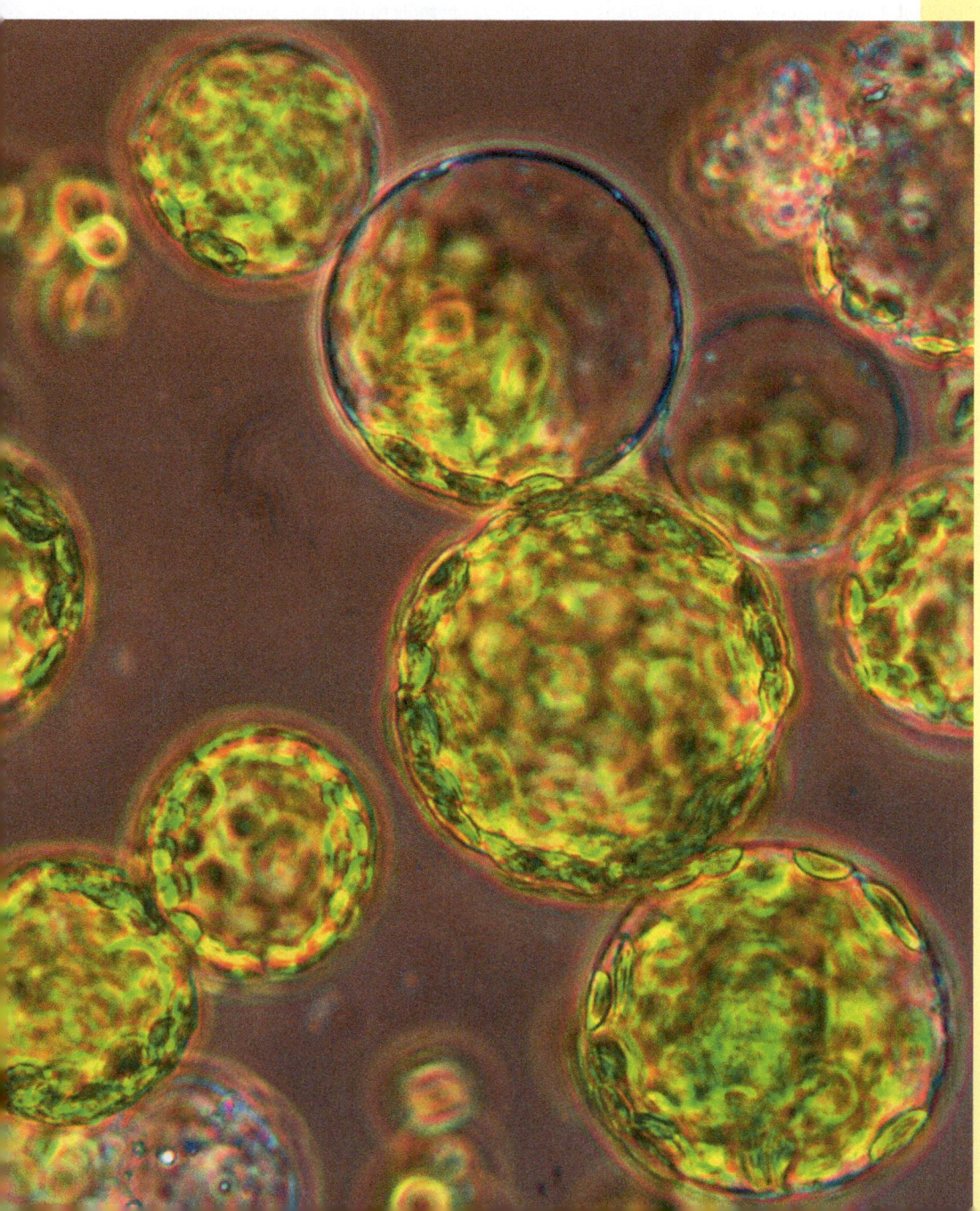

Every cell has, at its periphery, a plasma membrane that serves as the interface between the cell and the world outside it.

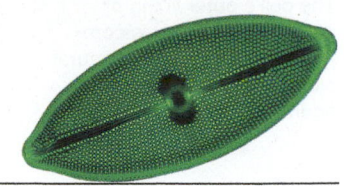

Things have steadily been getting better for the estimated 30,000 Americans who suffer from the disease cystic fibrosis (CF). At one time, 80 percent of the victims of this illness didn't make it to their first birthday. But with each passing year, small advances in the treatment of CF have led to small increases in life expectancy. Even so, today's average CF

patient lives only to age 37. How does CF bring about this shortening of lives? Although its effects are many, its lethal effects usually have to do with the lungs. More specifically, CF causes a dangerous layer of mucus to accumulate in the passageways of the lungs. The lungs of healthy people are lined by a layer of mucus, too. But in their case, the layer is thin and wet—it's pliable enough, we might say, that it can be regularly swept away by the hair-like cilia that function like tiny brooms in the lung passageways. Cystic fibrosis patients, conversely, have a mucus layer that is thicker and drier and that becomes a site for repeated bacterial infections. In time, these infections can result in the destruction of the lung passageways.

So, what accounts for the difference between the lungs of a healthy person and those of a person suffering from CF? Because of a faulty gene, the substance chloride cannot be transported in sufficient quantity from the inside of a CF patient's lung cells to the outside of them (where the mucus builds up). Healthy individuals have a protein that acts as a channel for chloride—a kind of tunnel that spans the cell's outer membrane. In contrast, the cells of CF patients have either defective chloride channels or none at all. The result is a lack of chloride outside their cells, which gets the mucus buildup going.

As noted last chapter, life goes on only inside cells. The fact that the cell's outer, or "plasma," membrane is out on life's edge may prompt the thought that it's *merely* at the edge, as if the real action is taking place deep within the cell. But consider, as in CF, the effect of having a *defective* plasma membrane. The focus in this chapter is to learn more about this important cell lining.

5.1 The Nature of the Plasma Membrane

So, what is the nature of the plasma membrane? First, it's worth noting that it's very much like the membranes described in Chapter 4. Much of what follows about the plasma membrane could also be said of the various membranes that are internal to the cell—those that line lysosomes or that make up the Golgi complex, for example. There are some important differences between the plasma and other membranes, however, because the plasma membrane is constantly interacting with the world outside the cell, whereas internal membranes are not.

When we start to consider the qualities of the plasma membrane, we find that, even in the micro-world of cells, it is an extremely *thin* entity. If you stacked 10,000 of them on top of one another, their combined depth would be about that of a sheet of paper. Next, the plasma membrane has a fluid and somewhat fatty makeup. If we looked for a counterpart to it in our everyday world, a soap bubble might come close, meaning this membrane is very flexible. Yet it is stable enough to stay together despite being constantly re-formed, thanks to the endless movement of materials in and out of it. If you imagine half the surface of your skin remaking itself every 30 minutes or so, you begin to get the picture about how dynamic this membrane is. Let's look now at the most important component parts of it. All these components are illustrated in **Figure 5.1** on the next page. They are:

- **The phospholipid bilayer.** Just as bread is made mostly from flour, the plasma membrane is made

Membrane Structure

Figure 5.1
The Plasma Membrane

The cell's outer lining or plasma membrane is composed of four main structural elements, as shown in the figure.

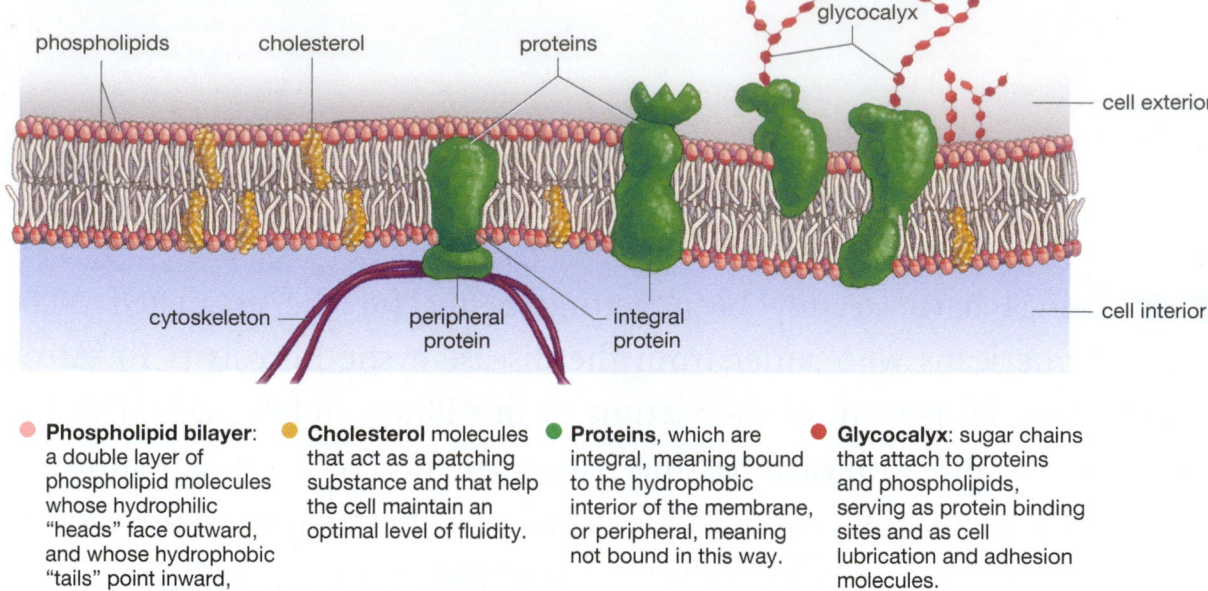

phospholipids   cholesterol   proteins   glycocalyx   cell exterior

cytoskeleton   peripheral protein   integral protein   cell interior

- **Phospholipid bilayer:** a double layer of phospholipid molecules whose hydrophilic "heads" face outward, and whose hydrophobic "tails" point inward, toward each other.

- **Cholesterol** molecules that act as a patching substance and that help the cell maintain an optimal level of fluidity.

- **Proteins,** which are integral, meaning bound to the hydrophobic interior of the membrane, or peripheral, meaning not bound in this way.

- **Glycocalyx:** sugar chains that attach to proteins and phospholipids, serving as protein binding sites and as cell lubrication and adhesion molecules.

mostly from two layers of phospholipid molecules, with "tails" that mingle together, but "heads" that point in opposite directions—to the interior of the cell in one direction and to the world outside the cell in the other.

- **Cholesterol.** Like mortar between bricks, cholesterol acts as a "patching" material for the membrane in animal cells; it also keeps the membrane at an optimal level of fluidity.

- **Proteins.** These molecules shoot through the membrane and lie on either side of it, serving as support structures, signaling antennas, identification markers, and cellular passageways.

- **Glycocalyx.** This is the collective term for a set of carbohydrate chains that form a layer on the outside of the membrane, functioning in cell lubrication, adhesion, and signaling.

Now let's look at each of these components in greater detail.

First Component: The Phospholipid Bilayer

You may recall the discussion in Chapter 3 of phospholipids—molecules that have two long fatty acid chains linked to a phosphate-bearing group (**Figure 5.2**). You may also recall that, because fatty acid chains are lipids, they are hydrophobic or "water fearing," meaning they will not bond with water. Conversely, phosphate groups are charged, hydro*philic* molecules, meaning they are "water loving" and will readily bond with water. When such phosphate and lipid components are put together, the resulting phospholipid is a molecule with a dual nature. Its phosphate "head" will seek water, but its fatty acid "tail" will avoid it. Drop a

group of phospholipids in water, and they will arrange themselves into the form you see in Figure 5.2: two *layers* of phospholipids sandwiched together, the tails of each layer pointing inward (thus avoiding water) and the heads pointing outward (thus bonding with the watery environment that lies both inside and outside the cell). With this in mind, the **phospholipid bilayer** can be defined as a chief component of the plasma membrane, composed of two layers of phospholipids, arranged with their fatty acid chains pointing toward each other.

This phospholipid structure has a couple of important consequences for the plasma membrane. First, it helps give the membrane its fluid nature—the bilayer's fatty acid tails have a chemical makeup that is more like that of oil than shortening. Second, because these are two hydrophobic layers of tails, the only substances that can pass through them with ease are other hydrophobic substances or a few very small molecules that aren't hydrophobic. This stems from a rule of thumb in chemistry: Like dissolves like. Fats dissolve in fats, charged molecules dissolve in charged molecules. Thus, various steroid hormones (which, remember, are lipids) gain fairly easy entry into the cell, as do fatty acids, because these substances will dissolve in the lipid bilayer. Conversely, hydrophilic substances—ions, amino acids, all kinds of polar molecules—will not dissolve in the membrane and thus cannot make it past the fatty acid chains without help. What kind of help? Stay tuned.

Second Component: Cholesterol

Molecules of the lipid material cholesterol nestle between phospholipid molecules throughout the plasma membrane in animal cells, performing two

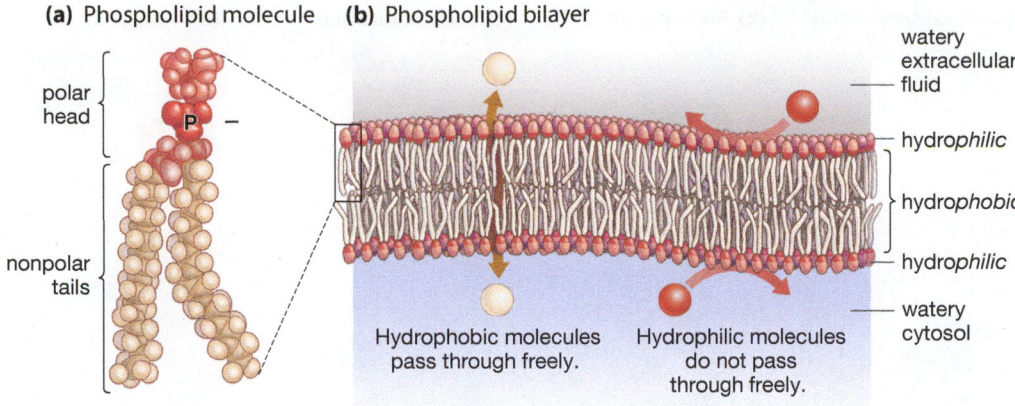

(a) Phospholipid molecule   **(b)** Phospholipid bilayer

polar head

nonpolar tails

watery extracellular fluid

hydro*philic*

hydro*phobic*

hydro*philic*

watery cytosol

Hydrophobic molecules pass through freely.

Hydrophilic molecules do not pass through freely.

Figure 5.2
Dual-Natured Lining

(a) The essential building block of the cell's plasma membrane is the phospholipid molecule, which has two elements: a hydrophilic "head," containing a charged phosphate group that bonds with water; and hydrophobic "tails," composed of fatty acids that do not bond with water.

(b) Two layers of phospholipids sandwich together to form the plasma membrane.

Selective Permeability of Membranes

functions. First, they act as a kind of patching substance on the bilayer, keeping some small molecules from getting through. Second, they help keep the membrane at an optimum level of fluidity. Without cholesterol, the plasma membrane would react to low temperatures the same way butter does: It would change from a flexible substance to a hardened one—a dysfunctional condition for a cell. The cholesterol molecules nestled between the phospholipid's fatty acid tails keep this from happening by keeping these chains from packing too closely together. When temperatures rise, however, the carbon rings of the cholesterol molecule interfere with the movement of the phospholipid's fatty acid tails, thus keeping the membrane from becoming too fluid.

Third Component: Proteins

Embedded within, or lying on, the phospholipid bilayer is a third major group of membrane components, *membrane proteins,* of which there are two principal types, integral and peripheral. **Integral proteins** are plasma membrane proteins that are bound to the membrane's hydrophobic interior. In some cases, these proteins span the entire membrane—popping out on either side of it—while in other cases, they extend only partway in. **Peripheral proteins** are plasma membrane proteins that lie on either side of the membrane but are not bound to its hydrophobic interior. They are usually attached to integral proteins at the surface of the membrane.

It is hard to overstate the importance of proteins to the plasma membrane. By itself, the lipid bilayer has a handful of capabilities. With the addition of proteins, it is turned into a structure with a huge array of capabilities. Here is a selected set of the roles that membrane proteins play.

Structural Support

Peripheral proteins that lie on the interior or "cytoplasmic" side of the membrane often are attached to elements of the cell's cytoskeleton. Thus anchored,

they help to stabilize various parts of the cell and play a part in giving animal cells their characteristic shape (**Figure 5.3** on the next page). (Why not plant cells? Because their shape is determined by the cell wall.)

Recognition

Much like military sentries, some cells need to know the answer to a critical question: Who goes there, friend or foe? These sentries are immune system cells, which circulate through the body in a never-ending search for cells that have been infected by viruses or other microbial invaders. The way immune cells can tell a friend (a normal cell) from a foe (an infected cell) is that almost all cells in the body constantly take part in a kind of molecular show-and-tell: They put on display, within proteins that extend from their plasma membranes, fragments of proteins they've recently been producing. Immune system cells have receptors that can *bind with* the resulting protein complexes. In most cases, the message that will be conveyed by this binding is, "Normal cell here; pass on by." But a cell that's been infected with a virus will be putting a set of *viral* protein fragments on display within one of its recognition proteins, thus conveying the message, "I've been infected; begin an immune system attack."

Communication

Cells communicate with one another in various ways. Signals can be sent from one cell to a neighboring cell, and communication over longer distances is possible through the chemical messengers known as hormones. These signals are likely to be channeled through a **receptor protein**—a plasma membrane protein that binds with a signaling molecule.

Receptor proteins—usually just called receptors—have binding sites on them with a shape that is so specific they generally will bind only to a single type of signaling molecule. To take one example of this, the hormone insulin is a signaling molecule. The message it conveys to cells is: "Glucose levels are rising in the bloodstream; construct some channels in your plasma membranes so that you can take some of this glucose

Figure 5.3
Roles of Membrane Proteins
Plasma membrane proteins serve four principal functions, as shown in the figure.

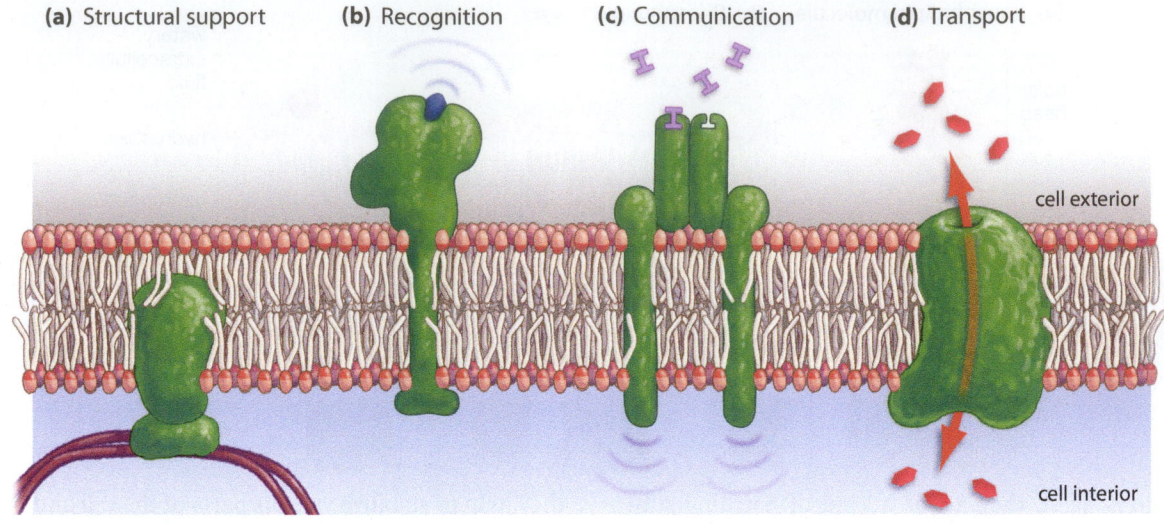

(a) Structural support   **(b)** Recognition   **(c)** Communication   **(d)** Transport

cell exterior

cell interior

Membrane proteins can provide structural support, often when attached to parts of the cell's scaffolding or "cytoskeleton."

Protein fragments held within recognition proteins can serve to identify the cell as "normal" or "infected" to immune system cells.

Receptor proteins, protruding out from the plasma membrane, can be the point of contact for signals sent to the cell via traveling molecules, such as hormones.

Proteins can serve as channels through which materials can pass in and out of the cell.

in." Now, how does insulin transmit this message? It binds with an insulin receptor protein whose binding site lies outside the cell, and this binding causes a chemical change on the portion of the protein that lies *inside* the cell. The result is a cascade of cellular reactions that puts in motion the construction of the channels. Note that, through this means, a signaling molecule need not come into the cell to have an effect; it merely needs to bind with a receptor protein that spans the plasma membrane.

Transport

You have seen that most materials cannot simply pass through the cell's plasma membrane. Yet, cells need many of these very materials. How do they get in? Different kinds of integral proteins take on a transport task—they act as channels through which materials pass both into cells and out of them. In some cases, these proteins form what might be called passive passageways—microtunnels that simply provide a hydrophilic opening through a hydrophobic environment. In other cases, however, these protein channels are actively involved in moving materials through the membrane. We'll review all the forms of protein-aided transport shortly. For now, just note the existence of **transport proteins:** proteins that facilitate the movement of molecules or ions from one side of the plasma membrane to the other.

Fourth Component: The Glycocalyx

If you look at Figure 5.1 and focus on the extracellular face of the plasma membrane, you can see, protruding from proteins and phospholipids alike,

a number of short, branched extensions. These are simple carbohydrate chains, meaning sugar chains. It turns out that these chains serve as the actual binding sites for many signaling molecules, including insulin. (Remember in Chapter 4 how many of the proteins that were constructed had sugar side chains added to them? Well, now you can see what some of these chains do.) Sugar chains also serve to lubricate cells and allow them to stick to other cells by acting as an adhesion layer. Collectively, all these chains form the **glycocalyx:** an outer layer of the plasma membrane composed of short carbohydrate chains that attach to membrane proteins and phospholipid molecules.

The Fluid-Mosaic Membrane Model

With this review of component parts under your belt, you're ready for a definition of the **plasma membrane.** It is a membrane, forming the outer boundary of many cells, composed of a phospholipid bilayer that is interspersed with proteins and cholesterol and coated on its exterior face with carbohydrate chains.

Taking a step back, the general message is that the plasma membrane is a loose, lipid structure peppered with proteins and coated with sugars. A better image, however, might be of a *sea* of lipids that is peppered with proteins, some of which are fixed in place but some of which have a great deal of freedom to move sideways or "laterally" across the membrane, something like icebergs on an ocean. When we overlay this quality of movement on the basic membrane structure just reviewed, the result is the contemporary view of the plasma membrane, the **fluid-mosaic model:** a conceptualization of the plasma membrane as a fluid, phospholipid bilayer that has within it a mosaic of both stationary and mobile proteins.

SO FAR . . .

1. The plasma membrane of an animal cell has four principal components. These are the _____, which is composed of two layers of molecules with fatty acid tails that point toward each other; the lipid molecule _____, which serves as a patching and fluidity control substance; many _____, which serve in such diverse roles as passageways and communication molecules; and the _____, which is a collective term for a set of carbohydrate chains attached to the outside of the membrane.

2. The phospholipid molecules of the plasma membrane are dual natured. The "heads" of these molecules are _____, meaning they will readily bond with _____. Meanwhile, the tails point inward, toward each other, because they are _____, meaning they will not bond with _____.

3. Plasma membrane proteins serve in the role of _____ by binding with signaling molecules that come from outside the cell and in the role of _____ by acting as channels for substances moving into and out of the cell.

5.2 Diffusion, Gradients, and Osmosis

It may be obvious that one of the main roles of the plasma membrane is to let in substances that are needed within the cell while keeping out substances that aren't needed or that actually are harmful. Having learned something about the membrane's makeup, you will now see how it carries out this task of passage and blockage.

It is necessary to begin this story, however, not with the cell but with the more general notion of how substances go from places where they are more concentrated to places where they are less concentrated. Anyone who has put some food dye in water has a sense of how this works: A few drops of, say, red dye will tumble into the liquid and start dispersing; eventually, the result is a solution that is uniformly more reddish than plain water (**Figure 5.4**). The question we'll look at is why this takes place.

Random Movement and Even Distribution

All molecules are constantly in motion, and this motion is random. Because liquid molecules are bonded together so weakly, they can *act* on this motion—they can slide past each other and thus go from one location to another. (Meanwhile, the molecules that make up solids are so tightly bonded together that their motion amounts simply to vibrating in place.) Because liquid molecules move randomly, they will naturally move from any initial *ordered* state—in our example,

(a) Dye is dropped in.   **(b)** Diffusion begins.   **(c)** Dye is evenly distributed.

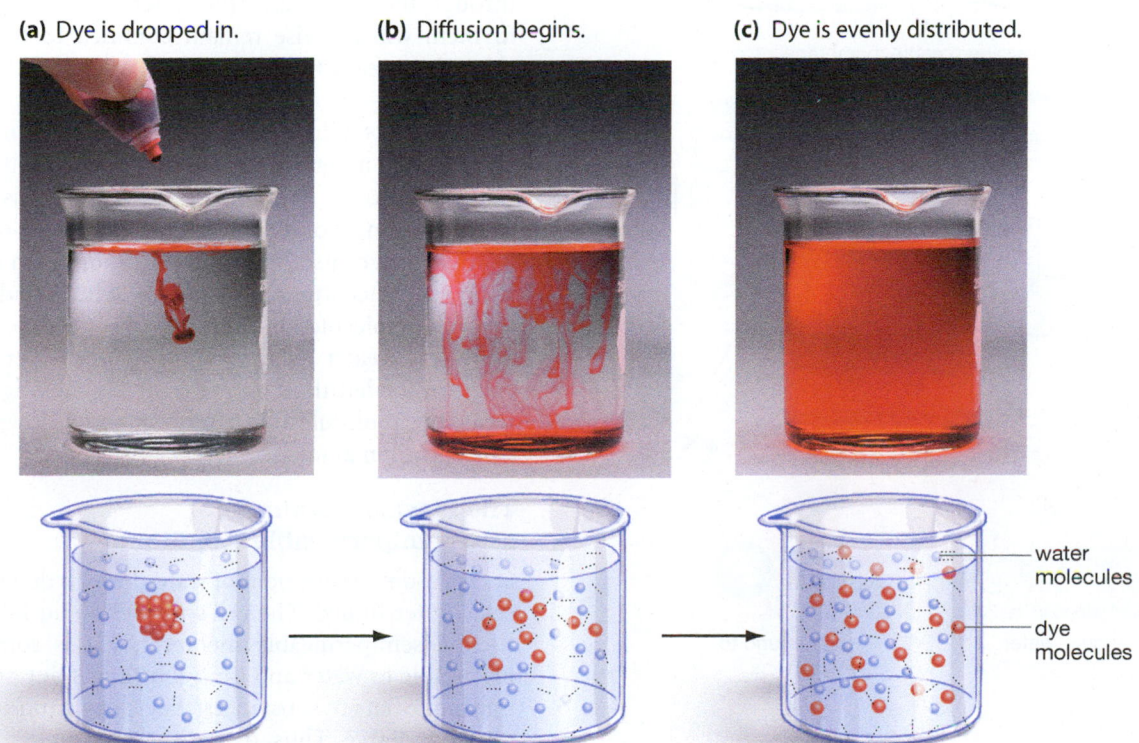

water molecules

dye molecules

Figure 5.4
From Concentrated to Dispersed

Diffusion is the movement of molecules or ions from areas of their greater concentration to areas of their lesser concentration. In this sequence of photos and diagrams, a few drops of red dye, added to a beaker of water, are at first heavily concentrated in one area but then begin to diffuse, eventually becoming evenly distributed throughout the solution.

Diffusion

Osmosis and Water Balance in Cells

their concentrated state when first dropped in the container—to their most disordered state, meaning evenly distributed throughout a given volume.

What is at work here is **diffusion**: the movement of molecules or ions from a region of their higher concentration to a region of their lower concentration. This notion carries with it the concept of a gradient. A **concentration gradient** is defined as the difference between the highest and lowest concentration of a solute within a given medium. In our example, the solute is the dye, and the medium is water. As with bicycles coasting down a grade, the natural tendency for any solute is to move down its concentration gradient, from higher concentration to lower. Our dye did just that. Bikes and solutes can move *up* a grade or gradient, but there is a price to be paid, and that price is the expenditure of energy. (On its own, would red dye ever come back to its concentrated state in the container? No. Some work would have to be performed for this to happen, which means that some energy would have to be expended.)

Diffusion through Membranes

So far, you have looked at molecules diffusing in an undivided container. But the subject here is divisions—those provided by the plasma membrane. Let us consider, therefore, what happens to a solution in a container that is divided by a membrane. If that membrane is *permeable* to both water and the solute—that is, if both water and the solute can freely pass through it—and the solute lies only on one side of the membrane, then the predictable happens: The solute moves down its concentration gradient, diffusing right through the membrane, eventually becoming evenly distributed on both sides of it.

Now let's imagine a *semi*permeable membrane, one that water can freely move through but that solutes cannot. **Figure 5.5** shows you what happens if more solute—in this case, salt—is put on the right side of the membrane than the left. Water flows through the membrane both ways, but *more* water flows into the right chamber, which has a greater concentration of solutes in it. The result is that the solution on the right side rises to a higher level than the one on the left.

This seems strange on first viewing, as if gravity were taking a vacation on the right side of the container. But what has been demonstrated here is **osmosis**: the net movement of water across a semipermeable membrane from an area of lower solute concentration to an area of higher solute concentration. Why should this occur? In the case of a solute like salt, water molecules will surround and bond with the sodium and chloride ions that salt separates into in solution. Because these solutes are not free to pass through the membrane, the water molecules bound to them will likewise remain confined to the right side. This means that more "free" water will exist on the left side—water that is free to act on the natural tendency of liquid molecules to remain in motion. The result is a net movement of water into the right side (see Figure 5.5b). Looked at one way, this is just another example of diffusion—of a substance moving from an area of its higher concentration to an area of its lower concentration. Once the solute binds with the water molecules on the right side of the container, there is a greater concentration of free water molecules on the left than on the right. By moving to the right, the molecules are simply moving down their concentration gradient.

The Plasma Membrane as a Semipermeable Membrane

So, what does movement of water have to do with the cell's outer lining? The cell's phospholipid bilayer is itself a semipermeable membrane. It is somewhat permeable to water and lipid substances but not permeable to larger substances that carry a positive or negative charge. Thus, osmosis can take place across

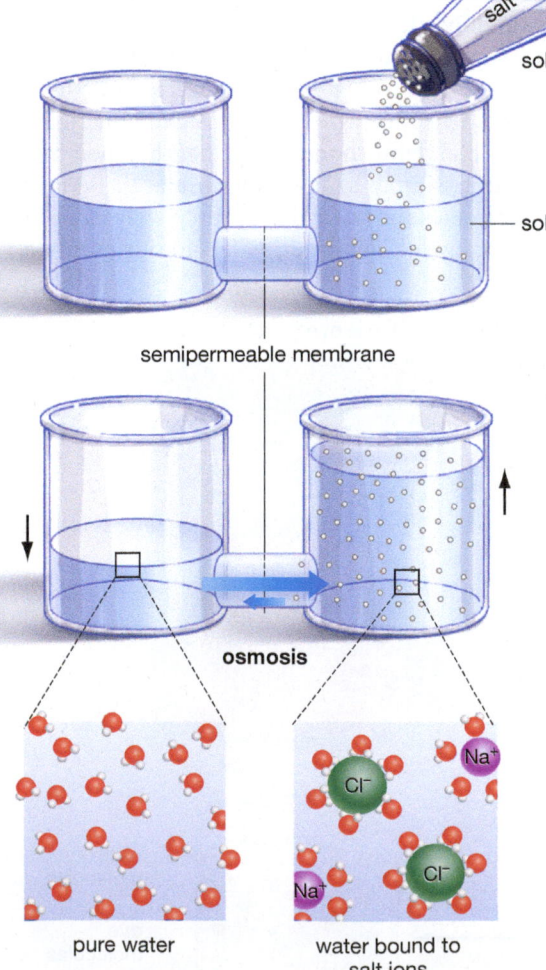

(a) An aqueous solution divided by a semipermeable membrane has a solute —in this case, salt— poured into its right chamber.

salt

solute

solvent

semipermeable membrane

(b) As a result, though water continues to flow in both directions through the membrane, there is a net movement of water toward the side with the greater concentration of solutes in it.

osmosis

(c) Why does this occur? Water molecules that are bonded to the sodium (Na⁺) and chloride (Cl⁻) ions that make up salt are not free to pass through the membrane to the left chamber of the container.

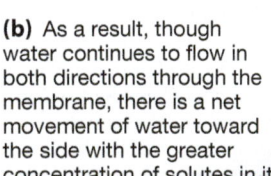

pure water

water bound to salt ions

Figure 5.5
Osmosis in Action

the plasma membrane—indeed, it does all the time. It is the primary means by which plants *get* water, and it is a player in all sorts of routine metabolic processes in animals. In humans, for example, much of the fluid portion of blood is at one point driven out of the blood vessels called capillaries by the force of blood pressure. How does this fluid get back in? Osmosis. Proteins that remain in the capillaries after the fluid has moved out function like the salt in the container: They bring about an osmosis that pulls the fluid back into the blood vessels.

Or think about a more extreme example. Why are we always told that someone who is stranded at sea should never drink salt water? Because doing so would provide an enormous concentration of sodium chloride ions in the watery fluid that lies *outside* the cells. This is no different in principle than dumping more solutes into the right side of the container in Figure 5.5. The result would be water flowing out of cells, dehydration of the cells, and in extreme cases, death. (What actually kills people in this situation is a shrinkage of brain cells due to the dehydration.)

This is an example of an *osmotic imbalance,* which is a situation in which osmotic pressure is either drying out cells or flooding them. In the case of flooding,

plant cells have a great advantage over animal cells—their cell walls. Remember from Chapter 4 that one function of the cell wall is to regulate water uptake? Well, now we can see why this is so valuable. Animal cells, which do not have a cell wall, can expand until they burst when water comes in. Plant cells, conversely, will expand only until their membranes push up against the cell wall with some force, setting up a pressure level that keeps more water from coming in. Such tight quarters actually are an optimal condition for plants. A nice, crisp celery stick is one that has achieved this kind of *turgid* state, while a droopy celery stick that has lost this quality has cells that are *flaccid;* flowers or leaves in the latter condition are *wilted* (**Figure 5.6**).

Osmosis and Cell Environments

Is a given cell likely to lose water to its surroundings, gain water from them, or have a balanced flow back and forth with them? Any of these things is possible, depending on what the solute concentration is outside the cell as opposed to inside. Three terms are helpful in describing the various conditions that can exist. A fluid that has a higher concentration of solutes than another is said to be a **hypertonic**

Plasma Membranes
and Diffusion

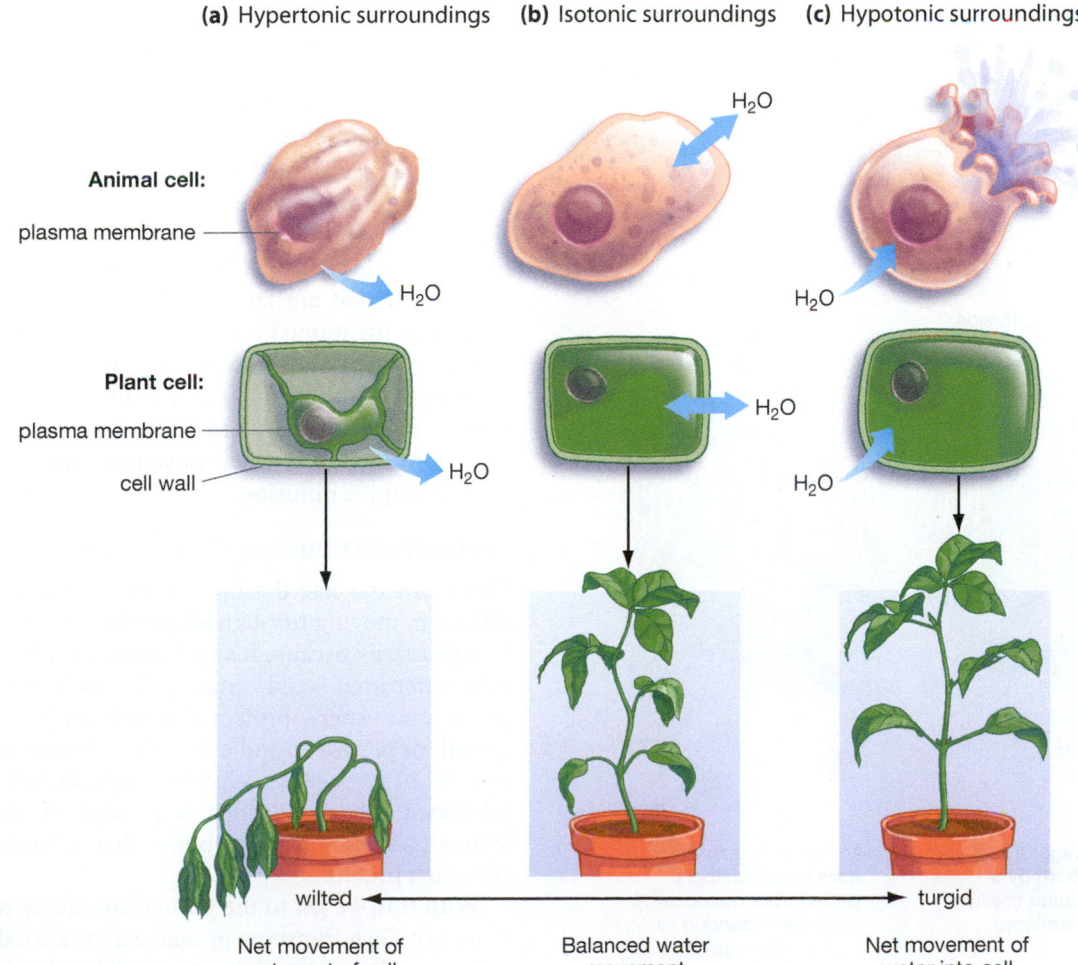

(a) Hypertonic surroundings   **(b)** Isotonic surroundings   **(c)** Hypotonic surroundings

Animal cell:

plasma membrane

H_2O

H_2O

H_2O

H_2O

Plant cell:

plasma membrane

cell wall

H_2O

H_2O

H_2O

wilted   turgid

Net movement of
water out of cell

Balanced water
movement

Net movement of
water into cell

Figure 5.6
Osmosis in Cells

(a) When solutes (such as salt) exist in greater concentration outside the cell than inside, water moves out of the cell by osmosis, and the cell shrinks.

(b) When solute concentrations inside and outside the cell are balanced, water movement into and out of the cell is balanced as well.

(c) When solutes exist in greater concentration inside the cell than outside, water moves into the cell by osmosis.

solution. If a cell's surroundings are hypertonic to the cell's cytoplasm, water will flow out of the cell. Two solutions that have equal concentrations of solutes are said to be **isotonic.** If one of these solutions is the cell's cytoplasm and the other is the fluid surrounding the cell, water flow will be balanced between cell and surroundings. Finally, a fluid that has a lower concentration of solutes than another is a **hypotonic solution.** If a cell's surroundings are hypotonic to the cell's cytoplasm, water will flow into the cell. Figure 5.6 gives examples of what can happen to cells in all three types of environments.

(MB)

Active Transport

Figure 5.7
Transport through
the Plasma Membrane

SO FAR . . .

1. Diffusion, the movement of molecules or ions from a region of their _____ to a region of their _____ takes place spontaneously in liquids, meaning it takes place without the input of any _____.

2. Osmosis is the net movement of water across a _____ membrane, from an area of lower _____ to an area of higher _____.

3. If a cell is surrounded by a fluid that is hypertonic to the cell's cytoplasm, water will flow _____ the cell. If a cell is surrounded by a fluid that is hypotonic to the cell's cytoplasm, water will flow _____ the cell.

5.3 Moving Smaller Substances In and Out

This excursion into the land of diffusion and osmosis has prepared us to start looking at the ways materials actually move into and out of the cell, across the plasma membrane. The big picture here is that some molecules are able to cross with no more assistance than is provided by diffusion. Other molecules require diffusion and the protein channels we talked about, and still others require channels and the expenditure of energy to get across. Different terms are applied to these different kinds of transport. **Active transport** is any movement of molecules or ions across a cell membrane that requires the expenditure of energy. **Passive transport** is any movement of molecules or ions across a cell membrane that does not require the expenditure of energy (**Figure 5.7**). Our review of transport begins with two varieties of passive transport.

Passive Transport

Simple Diffusion

Gases such as oxygen and carbon dioxide are such small molecules that they need only move down their concentration gradients to pass into or out of the cell. Having been delivered by blood capillaries to an area just outside the cell, oxygen exists there in greater concentration than it does inside the cell. Moving down its concentration gradient, it diffuses through the plasma membrane and emerges on the other side. This is an example of **simple diffusion,** which is diffusion through a cell membrane that does not require a special protein channel. Molecules that are larger than oxygen but that are fat soluble, such as the steroid hormones mentioned before, also move into the cell in this manner because of their ability to dissolve in the bilayer of the membrane. Carbon dioxide, which is formed *in* the cell as a result of the transfer of energy from food, has a net movement *out* of the cell through simple diffusion.

Facilitated Diffusion: Help from Proteins

Water can traverse the lipid bilayer through simple diffusion, moving through *despite* its hydrophilic nature primarily because it's such a small molecule. Yet, with water itself—and certainly with polar molecules larger than water—protein channels begin to be required for passage in and out of the cell. With this, the method of passage is no longer *simple* diffusion; it is **facilitated diffusion,** which is passage of materials through the plasma membrane that is aided by a transport protein.

With this, we get to the protein specificity referred to earlier. Each transport protein acts as a conduit for only one substance or at most a small family of related

Passive transport

simple diffusion

facilitated diffusion

Active transport

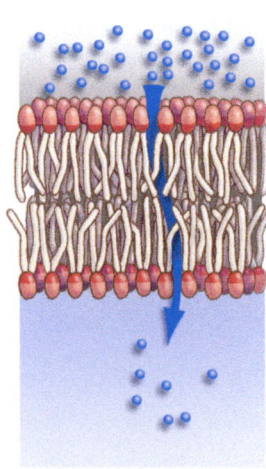

Materials move down their concentration gradient through the phospholipid bilayer.

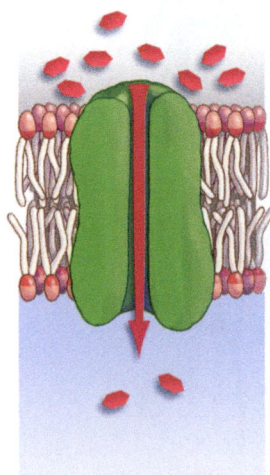

The passage of materials is aided both by a concentration gradient and by a transport protein.

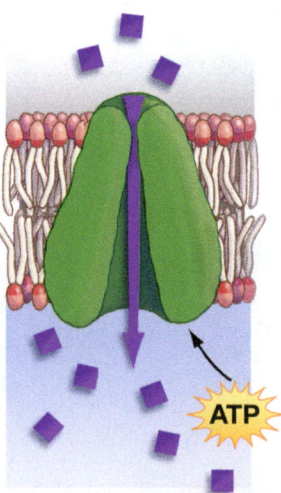

ATP

Molecules again move through a transport protein, but now energy must be expended to move them against their concentration gradient.

substances. Transport through these proteins begins with the kind of binding explained earlier. A circulating molecule of glucose, for example, latches onto the binding site of a glucose transport protein, as you can see in **Figure 5.8**. This binding causes the protein to change its shape, thus allowing the glucose molecule to pass through the membrane. In addition to glucose, other hydrophilic molecules, such as amino acids, move through the plasma membrane in this way. Note that this facilitated diffusion does not require the expenditure of energy because it has a concentration gradient working in its favor. Glucose exists in greater concentration outside most cells than inside them. It is moving down its concentration gradient, then, by passing into the cell through a protein channel. So facilitated diffusion has something in common with simple diffusion. *Both* processes are driven by concentration gradients, meaning that neither requires metabolic energy. Thus, both are specific examples of passive transport.

Although the glucose in our example moved only one way in facilitated diffusion (into the cell), in general transport proteins are a channel for movement either way through the membrane. All that is required is a concentration gradient and binding of the material with the transport protein.

Active Transport

If passive transport—either in simple or facilitated diffusion—were the only means of membrane passage available, cells would be totally dependent on concentration gradients. If, for example, molecules of a given amino acid come to exist in the same concentration on both sides of the plasma membrane, these amino acid travelers will continue to be transported, but they will be flowing out of the cell at the same rate they are flowing in. For cells, the problem is that some solutes are *needed* in greater concentration, say, outside the cell. So, how do cells deal with a situation like this? They use pumps: They expend energy and pump the needed materials out of themselves. This is active transport at work. The energy for this transport usually is supplied by a molecule we've had occasion to note before, ATP (adenosine triphosphate). One of the most important and best-studied of the ATP-driven transport mechanisms is called the sodium-potassium pump.

The Sodium-Potassium Pump

The sodium-potassium pump allows the cell to maintain high concentrations of potassium ions (K^+) inside the cell and sodium ions (Na^+) outside the cell. Imagine for a moment being able to take a snapshot of a cell at a moment in time that would allow us to see these two ion concentrations. With lots of Na^+ ions outside the cell, the result is predictable: These ions leak slowly into the cell through transport proteins as they follow not only their concentration gradient but an electrochemical gradient—these positively charged Na^+ ions are attracted to a net *negative* charge that tends to exist inside cells. But our fantasy snapshot of this scene would also reveal something else: An abundance of protruding proteins that are

Bioflix: Membrane Transport

Facilitated Diffusion

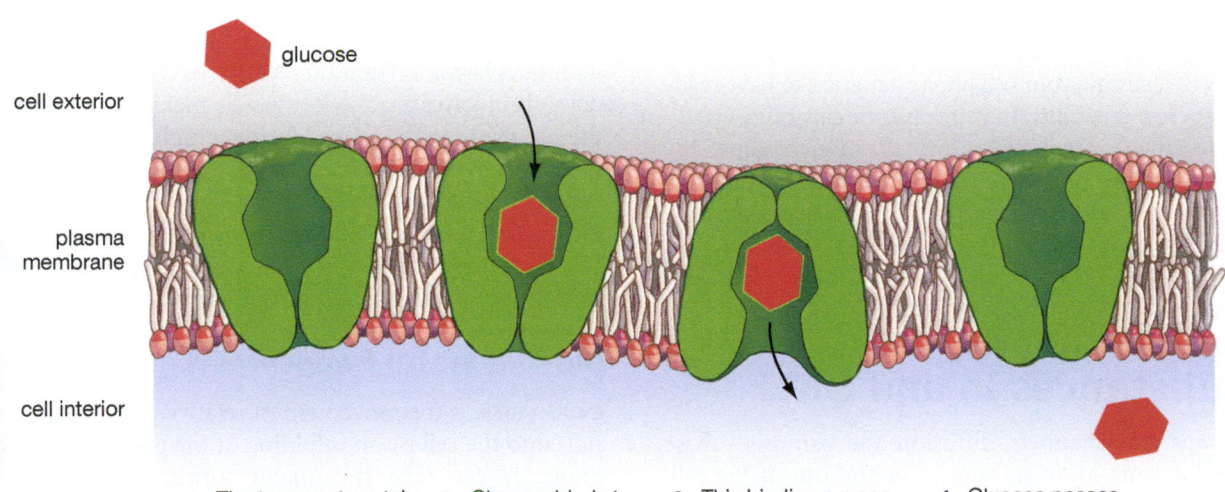

| 1. The transport protein has a binding site for glucose that is open to the outside of the cell. | 2. Glucose binds to the binding site. | 3. This binding causes the protein to change shape, exposing glucose to the inside of the cell. | 4. Glucose passes into the cell and the protein returns to its original shape. |

Figure 5.8
Facilitated Diffusion

Owing to its size and hydrophilic nature, a molecule such as glucose cannot move into a cell through simple diffusion, even when it has a concentration gradient working in its favor. Its movement must be facilitated by a transport protein, such as the one pictured in the figure. This movement of glucose is passive transport, however, in that no expenditure of cellular energy is required to bring it about.

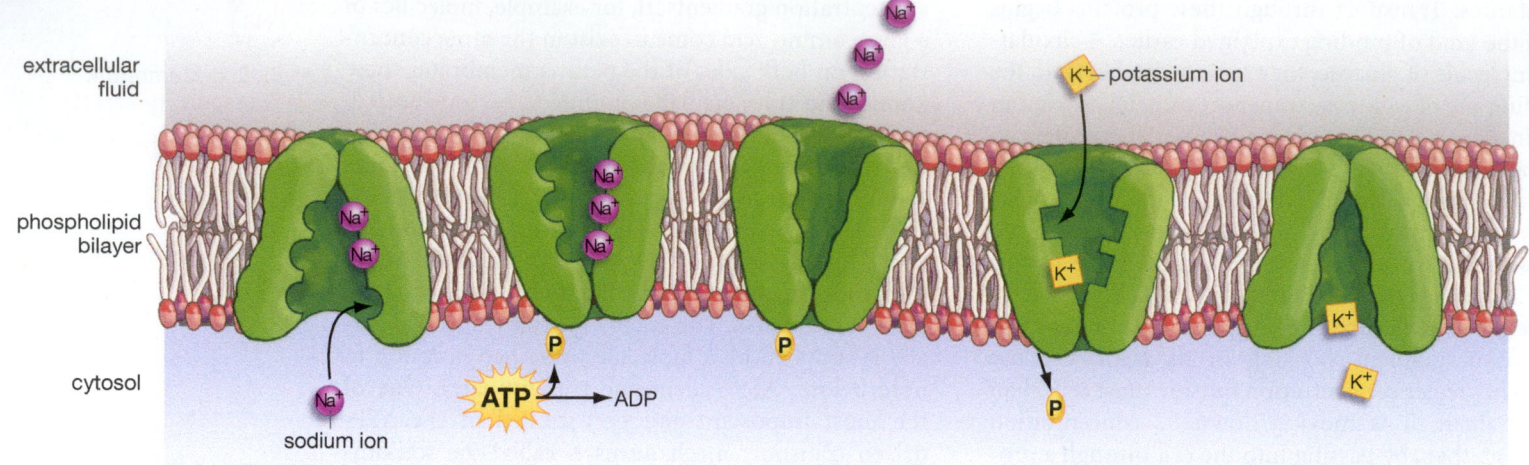

extracellular
fluid

phospholipid
bilayer

cytosol

K⁺ — potassium ion

ATP → ADP

sodium ion

1. Three sodium ions (Na⁺) located within the cell's cytoplasm bind with a transport protein.

2. The energy molecule ATP gives up an energetic phosphate group to the transport protein.

3. This binding causes the protein to open its channel to the extra-cellular fluid; to lose its Na⁺ binding sites, thus releasing the ions into the fluid; and to create binding sites for potassium ions (K⁺).

4. Two K⁺ move into the protein's K⁺ binding sites, which brings about the release of the phosphate group.

5. This loss returns the protein to its original shape, releasing the K⁺ into the cytoplasm and readying the protein for binding with another set of Na⁺.

Figure 5.9
The Sodium-Potassium Pump

Exocytosis and
Endocytosis

busy pumping these Na⁺ ions the *other way*, acting like so many bilge pumps moving seawater that has leaked into a ship back out into the ocean. The result is maintenance of a high concentration of Na⁺ ions outside the cell (and K⁺ ions inside). If you look at **Figure 5.9**, you can see how this works.

As may be obvious, this is an energy-intensive operation. As such, you may wonder: Why would a cell spend so much energy to maintain this imbalance in sodium and potassium ions? Well, nerve cells undertake this task for the same reason you undertake the task of charging your cell phone batteries: to have ready access to a capability. In your case, the capability is talking on the phone. In the nerve cell's case, the capability is being able to rapidly take in a large number of Na⁺ ions as a means of transmitting a nerve signal—something you can read more about in Chapter 27.

5.4 Moving Larger Substances In and Out

Pumps and channels, diffusion and osmosis—these mechanisms move substances in and out of the cell, but as it turns out they only move relatively *small* substances. Remember an earlier discussion about the immune system cells that check to see if a given cell is friend or foe? Well, if an immune sentry finds a bacterial foe, it may have to ingest this *whole cell*. As you might guess, a cell cannot do this by employing little channels or pumps. You learned a bit in Chapter 4 about what does happen in these cases; now, you'll be looking at it in somewhat greater detail.

The methods in question here are *endocytosis*, which brings materials into the cell, and *exocytosis*, which sends them out. What these mechanisms have in common is their use of vesicles—the membrane-lined enclosures you learned about last chapter that alternately bud off from membranes or fuse with them.

Movement Out: Exocytosis

Exocytosis is defined as the movement of materials out of the cell through a fusion of a transport vesicle with the plasma membrane. As **Figure 5.10** shows, exocytosis involves a transport vesicle making its way to the plasma membrane and fusing with it, whereupon the vesicle's contents are released into the extracellular fluid. You observed in Chapter 4 that cells use exocytosis when they are exporting proteins. For single-celled creatures, waste products may also be released into the extracellular fluid through this process.

Movement In: Endocytosis

Endocytosis is the movement of relatively large materials into the cell by an infolding of the plasma membrane. It can take two major forms: pinocytosis, which brings in molecules that are larger than those we've looked at so far, and phagocytosis, which brings in much larger molecules yet, as you'll see.

Pinocytosis

There actually are several varieties of pinocytosis, but we'll only look at one representative form here, called clathrin-mediated endocytosis or CME. This means of bringing materials into cells was once known as

(a) Exocytosis

(b) Micrograph of exocytosis

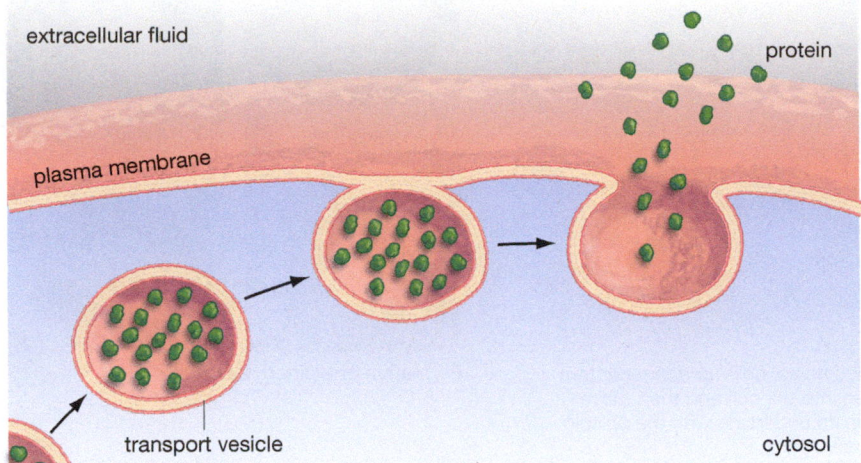

extracellular fluid

protein

plasma membrane

transport vesicle

cytosol

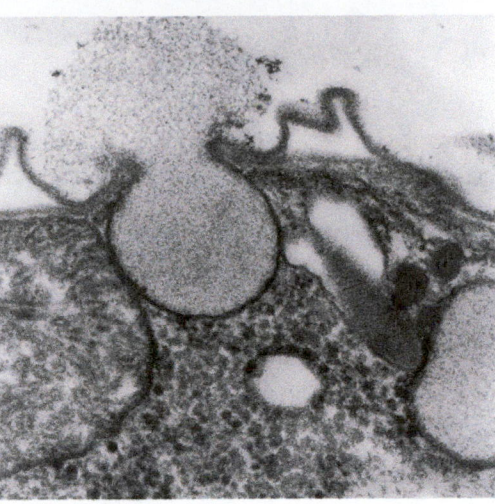

Figure 5.10
Movement Out of the Cell
(a) In exocytosis, a transport vesicle—perhaps loaded with proteins or waste products—moves to the plasma membrane and fuses with it. This section of the membrane then opens, and the contents of the former vesicle are released to the extracellular fluid.
(b) Micrograph of material expelled from the cell through exocytosis.

receptor-mediated endocytosis, but it turns out that almost all forms of endocytosis employ receptors. The receptors active in CME first bind with molecules of the material that's brought into the cell. Then, while the receptors are holding onto their molecular cargo, they move laterally across the membrane and congregate in a depression referred to as a coated pit. What the pit is coated *with*, on its underside, is the chemically active protein clathrin (hence the term clathrin-mediated endocytosis). Eventually, the pit deepens (or "invaginates") and pinches off, creating the familiar vesicle moving into the cell. CME is very important in getting nutrients and other substances into cells. It is the way cholesterol gets into our cells, for example. Alas, it is also the way certain invaders get into our cells: Polio and flu viruses are among the molecules that can bind to human CME receptors, thus giving them access to our cells. Taking a step back, in CME, we can see an example of **pinocytosis:** the movement of relatively large materials into a cell by means of the creation of transport vesicles that are produced through an invagination of the plasma membrane (**Figure 5.11** on the next page).

Phagocytosis

Phagocytosis literally means "cell eating," and the term can be apt since this is the means by which many one-celled creatures take in food. It's also the means, mentioned earlier, by which human immune system cells ingest whole bacteria. Given functions like these, it will come as no surprise to learn that phagocytosis is bringing *large* materials into the cell. Indeed, the vesicles used in it might be 10 times the size of the vesicles used in pinocytosis.

Phagocytosis begins when the cell sends out extensions of its plasma membrane called *pseudopodia* ("false feet"). These surround the food and fuse their ends together. What was once outside is now inside, encased in a vesicle and moving toward the cell's interior. Once this happens, lysosomes in the cell move to the vesicle, fuse with it, and begin breaking down the material in it. **Phagocytosis** can be defined as the movement of large materials into a cell by means of wrapping extensions of the plasma membrane around the materials and fusing the extensions together. While phagocytosis is seen fairly often in one-celled organisms, there are only three types of cells in the human body that are capable of carrying it out, all of them immune system cells. While these cells sometimes serve in the role of warrior by ingesting whole invaders, they also have a less-exalted function, which is that of janitor. They clean up debris within the body—fragments of the body's own cells or larger molecules—by ingesting these materials, after which they are broken down into their component parts.

On to Energy

Your tour of the cell and its plasma membrane is now complete. What's coming up next is something called bioenergetics. In our story of biology so far, there have been continuing vague references to proteins that are called enzymes, to a molecule called ATP, and to energy expenditures. This has been rather like a description of an elephant that avoids the subject of its trunk. We will now dive into the subject of bioenergetics, however, which describes forces that are so basic that the entire living world is shaped by them.

(a) Pinocytosis

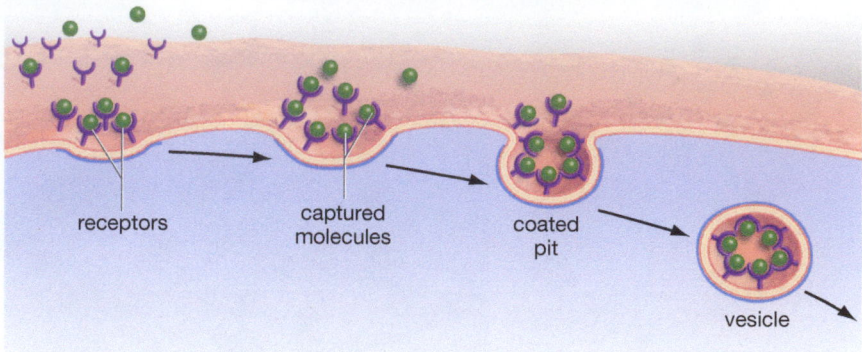
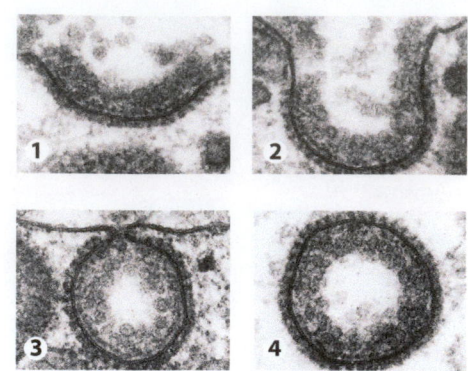

Formation of a pinocytosis vesicle.

In this form of pinocytosis, called clathrin-mediated endocytosis, cell-surface receptors bind to individual molecules of the substance to be taken into the cell and then move laterally across the plasma membrane to a pit, coated on its underside with the protein clathrin, that will become a vesicle that moves into the cell.

(b) Phagocytosis

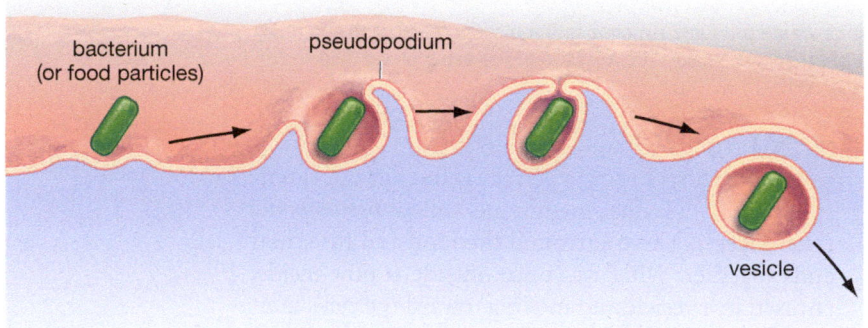

In phagocytosis, food particles—or perhaps whole organisms—are taken in by means of "false feet" or pseudopodia that surround the material. Pseudopodia then fuse together, forming a vesicle that moves into the cell's interior with its catch enclosed.

A human immune system cell called a macrophage (colored blue) uses phagocytosis to ingest an invading yeast cell.

Figure 5.11
Two Ways to Get Relatively Large Materials into the Cell

SO FAR . . .

1. Smaller materials move into and out of the cell through one means that requires the expenditure of energy, _____, and one means that does not require the expenditure of energy, _____. In turn, there are two varieties of the latter means of transport: simple diffusion, meaning diffusion through a cell membrane that does not require a special _____; and facilitated diffusion, meaning diffusion that does require a _____.

2. Active transport allows for the net movement of molecules into or out of a cell even when there is a _____ working against this movement.

3. Exocytosis is the movement of materials out of the cell through the fusion of a _____ with the _____. The movement of relatively large materials into the cell takes two forms: _____, in which entire cells or other large materials may be ingested through use of plasma membrane extensions called pseudopodia; and _____, in which somewhat smaller materials are brought in through an infolding or "invagination" of the membrane to create a vesicle that then moves inside the cell.

C H A P T E R

5 REVIEW

Summary

5.1 The Nature of the Plasma Membrane

- The plasma membrane is a thin, fluid entity that manages to be very flexible and yet is stable enough to stay together despite constant movement of materials in and out of it. The four principal components of the plasma membrane in animal cells are a phospholipid bilayer, molecules of cholesterol interspersed within the bilayer, proteins that are embedded in or that lie on the bilayer, and short carbohydrate chains on the cell surface, collectively called the glycocalyx, that function in cell adhesion and as binding sites. (p. 93)

- Phospholipids are composed of two fatty acid chains linked to a charged phosphate group. The fatty acid chains are hydrophobic; the phosphate group is hydrophilic. Phospholipids arrange themselves into bilayers, with the fatty acid "tails" of each layer pointing inward and the phosphate "heads" pointing outward. (p. 94)

- The cholesterol molecules that, in animal cells, are interspersed between phospholipid molecules in the plasma membrane act as patching that helps keep some small molecules from moving through the membrane and keeps the membrane at an optimal level of fluidity. (p. 94)

- Membrane proteins function in structural support; in cell identification, by serving as external recognition proteins that interact with immune system cells; in communication, by serving as external receptors for signaling molecules; and in transport, by providing channels for the movement of compounds into and out of the cell. (p. 95)

- The fluid-mosaic model views the plasma membrane as a fluid, phospholipid bilayer that has a mosaic of proteins either fixed within it or capable of moving laterally across it. (p. 96)

5.2 Diffusion, Gradients, and Osmosis

- Diffusion is the movement of molecules or ions from a region of their higher concentration to a region of their lower concentration. A concentration gradient defines the difference between the highest and lowest concentrations of a solute within a given medium. Through diffusion, compounds naturally move from higher to lower concentrations, meaning down their concentration gradients. Energy must be expended, however, to move compounds against their concentration gradients. (p. 97)

- A semipermeable membrane allows some compounds to pass through freely while blocking passage of others. Osmosis is the net movement of water across a semipermeable membrane from an area of lower solute concentration to an area of higher solute concentration. (p. 98)

- Osmosis is responsible for much of the movement of fluids into and out of cells. Osmotic imbalances can cause cells either to dry out from losing too much water or, in the case of animal cells, to break from taking too much water in. Plant cells generally do not have the latter problem because their cell walls limit their uptake of water. (p. 98)

- Cells will gain or lose water relative to their surroundings in accordance with what the solute concentration is inside the cell as opposed to outside it. A cell will lose water to a surrounding solution that is hypertonic to the cytoplasmic fluid; a cell will gain water when the surrounding solution is hypotonic. Water flow will be balanced between the cell and its surroundings when the surrounding fluid and the cytoplasmic fluid are isotonic to each other. (p. 99)

5.3 Moving Smaller Substances In and Out

- Some compounds are able to cross the plasma membrane strictly through diffusion; others require diffusion and special protein channels; still others require protein channels and the expenditure of cellular energy. Active transport is any movement of molecules or ions across a cell membrane that requires the expenditure of energy. Passive transport is any movement of molecules or ions across a cell membrane that does not require the expenditure of energy. (p. 100)

- Simple diffusion and facilitated diffusion are two forms of passive transport. For either form to bring about a net movement of materials into or out of a cell, a concentration gradient must exist. Facilitated diffusion requires both a concentration gradient and a protein channel. (p. 100)

- To maintain differing concentrations of substances on opposite sides of the plasma membrane, cells employ chemical pumps, through which they expend energy to move compounds across the plasma membrane against their concentration and electrochemical gradients. This is active transport. (p. 101)

5.4 Moving Larger Substances In and Out

- Larger materials are brought into the cell through endocytosis and moved out through exocytosis. Both mechanisms employ vesicles. In exocytosis, a transport vesicle moves from the interior of the cell to the plasma membrane and fuses with it, at which point the contents of the vesicle are released to the environment outside the cell. (p. 102)

- There are two principal forms of endocytosis: pinocytosis, the movement of moderate-sized molecules into a cell by means of the creation of transport vesicles produced through an infolding of a portion of the plasma membrane; and phagocytosis, in which certain cells use pseudopodia to first surround and then engulf whole cells, fragments of them, or other large organic materials. As in pinocytosis, such materials are brought into the cell inside vesicles that bud off from the plasma membrane. **(p. 102)**

Key Terms

active transport   100

concentration gradient   98

diffusion   98

endocytosis   102

exocytosis   102

facilitated diffusion   100

fluid-mosaic model   96

glycocalyx   96

hypertonic solution   99

hypotonic solution   100

integral protein   95

isotonic solution   100

osmosis   98

passive transport   100

peripheral protein   95

phagocytosis   103

phospholipid bilayer   94

pinocytosis   103

plasma membrane   96

receptor protein   95

simple diffusion   100

transport protein   96

Understanding the Basics

Multiple-Choice Questions

(Answers are in the back of the book.)

1. Which of the following groupings describes the four principal components of the plasma membrane?
   a. Phospholipid bilayer, proteins, DNA, ribosomes
   b. Phospholipid bilayer, ribosomes, glycocalyx, proteins
   c. Phospholipid bilayer, proteins, cholesterol, glycocalyx
   d. Phospholipid bilayer, lipoproteins, glycocalyx, cholesterol
   e. Phospholipid bilayer, DNA, proteins, cholesterol

2. Because of the fatty acid chains that help make up the phospholipid bilayer, the only substances that can pass through the plasma membrane purely by means of diffusion are:
   a. small proteins.
   b. amino acids and lipids.
   c. small carbohydrates.
   d. lipids or very small polar molecules.
   e. lipids or polar molecules of any size.

3. True or false: You eat spaghetti for dinner and digest the starch in it to glucose. The glucose is now in high concentration in the cells of your small intestine, relative to its concentration in your bloodstream. Predictably, with the help of a transport protein, the glucose diffuses across the plasma membrane of your intestinal cells, into your bloodstream.
   a. True.
   b. False, because molecules cannot move across membranes by diffusion.
   c. False, because diffusion moves materials from low concentration to high concentration.
   d. False, because molecules are too large to move across membranes.
   e. False, because energy is required for diffusion.

4. In osmosis, water moves across a semipermeable membrane:
   a. wherever energy pulls it.
   b. toward an area of lower solute concentration.
   c. toward an area of higher solute concentration.
   d. independently of solutes.
   e. any of the above could be true in different situations.

5. Tyrone gave Lakeisha a rose for the anniversary of their first date. Now the rose has wilted, because:
   a. the cells have taken too much water in and burst.
   b. the cell walls have broken down.
   c. the plasma membrane has lost the proteins that keep water in.
   d. the phospholipid bilayer has broken down.
   e. the plant's cells have insufficient water in them to push against their cell walls.

6. You get a cut on your finger and some bacteria enter. Your immune system cells kill off the invaders by ingesting them. This is an example of:
   a. pinocytosis.
   b. phagocytosis.
   c. receptor-mediated endocytosis.
   d. exocytosis.
   e. active transport.

7. According to the fluid-mosaic model of membrane structure, the membrane is made up of a:
   a. continuous base of proteins with phospholipids in it that have fairly free lateral movement.
   b. bilayer of phospholipids with proteins interspersed in it that have fairly free lateral movement.
   c. bilayer of protein with layers of phospholipids on the outside and inside.
   d. bilayer of phospholipid with layers of proteins on the outside and inside.
   e. bilayer of phospholipids sandwiched between two bilayers of proteins.

8. Endocytosis and exocytosis are similar in that both involve the use of:
   a. vesicles that carry materials into and out of the cell.
   b. the Golgi apparatus to package materials.
   c. diffusion to carry food and wastes across the plasma membrane.
   d. starch to supply large amounts of energy.
   e. the rough ER to manufacture proteins.

Brief Review

(Answers are in the back of the book.)

1. Imagine a receptor on the surface of one of your muscle cells that responds only to glucose. Plenty of insulin molecules are passing by this receptor, but it does not respond to them. Why not?

2. You often hear how dangerous it is for athletes to take steroids to "bulk up" and how many problems steroids cause in various parts of the body. Why do these drugs affect all parts of the body—not just the muscles?

3. Explain why the phospholipid "heads" of the plasma membrane are always pointed toward the cytosol and extracellular fluid, whereas the "tails" are always oriented toward the middle of the membrane.

4. Describe the form of endocytosis, called clathrin-mediated endocytosis, that is reviewed in the text.

5. As you saw in this chapter, all cellular membranes are similar in structure. However, what differences are likely to exist between plasma membranes and the membranes that exist in the interior of the cell?

Applying Your Knowledge

1. All life-forms on Earth arguably exist either as single cells or as collections of cells. All such cells have a plasma membrane at their periphery (although many cells have cell walls outside their plasma membrane). Why do you think the plasma membrane exists in all living things? Why aren't there living things that don't have a membrane at their periphery?

2. Because materials move from high concentration to low concentration via simple diffusion, and because they move the same way via facilitated diffusion, why do cells make proteins to carry out facilitated diffusion? Isn't this just a waste of energy?

3. Place a number of marbles on one end of a cafeteria tray. Shake the tray gently a few times, keeping it level. What happens to the marbles? Start over and shake faster the next time. What happens now? Explain how this is like chemical diffusion.

MasteringBiology®

1. MasteringBiology Assignments
Activities: Membrane Structure; Selective Permeability of Membranes; Diffusion; Osmosis and Water Balance in Cells; Plasma Membranes and Diffusion; Active Transport; Facilitated Diffusion; Exocytosis and Endocytosis; **Bioflix:** Membrane Transport; **Building Vocabulary; Questions:** Bioflix Quiz, Reading Questions, Test Bank

2. eText
Read your book online, search, take notes, highlight text, and study more effectively.

3. The Study Area
Self-study tools to test comprehension.

Life's Mainspring:
An Introduction to Energy

Energy flows through life in a never-ending pattern of acquisition, storage, and use.

Available on MasteringBiology®

Activities: Energy Transformations; Energy and Biology; Chemical Reactions and ATP; The Structure of ATP; How Enzymes Work; Enzymes; **Chemistry Review:** Enzymes and Pathways; **Building Vocabulary; Questions:** Reading Questions, Test Bank

All living things must have a constant supply of energy. Here, a golden stone-fly (*Hesperoperla pacifica*) expends considerable energy getting from one place to another near the banks of Oregon's Metolius River.

When my daughter Tessa was little, she had a teddy bear with a music box in it that let us listen to "Brahms's Lullaby." All I had to do was wind up the key that attached to the music box. After a few turns of the wrist, the melody would cycle through several times, then slow for a last few bars, then stop. This process could be repeated as many times as I wound the music box. As we listened, a mechanism in the music box kept the melody coming out at an even tempo. Even the slowing at the end seemed just right for a lullaby, since the audience was a sleepy child.

6.1 Energy Is Central to Life

The forces that determined how Tessa's teddy bear worked also turn out to condition life because, to function, both the music box and life must be supplied with the same thing: energy. Just as the music from the box depended on energy from an "outside hand" (mine) to continue, so almost all life on Earth depends on energy from an outside source: the sun, whose daily showering of energetic rays lets the green world of plants and algae flourish. These organisms then pass their bounty along to animals and other living things in the form of food to eat and oxygen to breathe.

Beyond this, the sun and the music box have something else in common: Their energy always runs downhill, from more concentrated to less. The spring in the music box spontaneously unwinds, but it will not rewind itself. The sun sends out light and heat, but this energy does not spontaneously come together to form another sun. The music from the music box eventually must come to a stop, and in a similar way, even the sun will not shine forever. With all its grandeur, its energy is winding down, its brilliance dispersing. The difference is that once the sun's showering of energy comes to a stop, nothing will restart it.

While the sun's gift lasts, however, living things can transform its energy and use it to do things like sprout leaves and swim through oceans. Such complex activities make it imperative, though, that living things be able to *control* the energy they capture. The spring within the music box unwinds at a measured rate, spinning out the melody at just the right pace; similarly, living things have elaborate mechanisms in place that allow them to make the most of the energy they receive.

Principles such as these are important enough to biology that this chapter and the two that follow are devoted to the subject of energy and life. In this chapter, you'll learn some basics about energy and some energy-saving molecules called enzymes. Then, in Chapter 7, you'll see how living things "harvest" energy, meaning how they extract energy from food. Finally, in Chapter 8, you'll look at the ways that plants, algae, and some bacteria capture the sun's energy in the process called photosynthesis. We'll start now by looking at the energy basics.

6.2 The Nature of Energy

What is energy? It is commonly defined as the capacity to do work, but this definition merely passes the question mark along to the word *work*. Given this, here's one possible definition of **energy**: the capacity to bring about movement against an opposing force. The thing that makes energy so tricky as a concept is that, although we can measure it with great precision, and we can experience its effects, we cannot grasp it or see it. We can see the water that drives a waterwheel, but who has seen the energy the water contains? (**Figure 6.1** on the next page.)

The Forms of Energy

Energy can be thought of as coming in different forms, many of them familiar. Mechanical energy is captured in the wound-up spring of the musical teddy bear. Chemical energy is the energy held in the chemical bonds of, for example, a peach or a lump of coal.

Figure 6.1
Where's the Energy?
A waterwheel turns as a result of energy supplied by moving water. Though the movement of the water is plainly visible, the energy contained in the water is not.

Energy Transformations

One other way to conceptualize energy is as potential and kinetic. **Potential energy** is stored energy: the rock perched precariously at the top of the hill; the charged ions kept on one side of a cell membrane. Conversely, **kinetic energy** is energy in motion, as with the rock tumbling down the hill or charged ions rushing in through a protein channel. If we define energy as the capacity to bring about movement against an opposing force, we can see how kinetic energy fits into this scheme. A rock that tumbles off a cliff and into a lake would cause the water in the lake to splash upward—against the force of gravity.

The Study of Energy: Thermodynamics

Given how important energy is to life, it makes sense that there is a branch of biology, called *bioenergetics*, that links biology with a scientific discipline known as **thermodynamics,** the study of energy. Although thermodynamics is a mathematical and rather abstract discipline, research into it was prompted originally by a very down-to-earth goal: the desire to build a better steam engine. British inventors had harnessed steam power in the eighteenth century, and this discovery had the awesome impact of sparking the Industrial Revolution. The word *revolution* is fitting here because before the advent of steam power, people used the stored energy in, say, wood or coal solely to produce *heat*. The steam engine was a device that allowed them to use heat to perform *work* on a scale that was previously impossible. Heat could be channeled, it was discovered, to drive an engine, which in turn could power any number of industrial processes.

With this development, new questions confronted scientists: How exactly did heat drive a steam engine?

Was there a finite quantity of energy, such that it could all be used up? Research carried out in England, Germany, and France during the nineteenth century yielded the answers to these questions, and the result was the development of some insights known as the laws of thermodynamics.

The First Law of Thermodynamics: The Transformation of Energy

The **first law of thermodynamics** states that energy is never created or destroyed but is only transformed. The sun's energy is not used up by green plants; rather, some of this energy is *converted* by the plants into chemical form.

At first blush, this is a tricky concept to grasp. We might ask: How can something as seemingly unenergetic as, say, a peach contain energy? Well, setting aside the water in it, a peach is mostly composed of carbohydrates and, as you may recall from Chapter 3, carbohydrates are molecules made of carbon, oxygen, and hydrogen atoms that are linked together in very specific ways. You could think of the sun's rays as having moved the electrons in these atoms up an energy hill, at which point they were locked into place, at their higher level, through the process of chemical bonding. At this point, a given electron is like a rock perched at the top of a hill. Just as a rock can fall down the hill, releasing energy as it goes, so the electrons in the peach can fall down an energy hill, releasing energy as they go.

A critical point here, however, is that the tree that sprouted this peach could put only *some* of the solar energy it received into chemical form in a peach. As it turns out, a plant cannot convert all of the energy it receives into carbohydrates. Likewise, a car engine actually converts only a fraction of the energy contained in gasoline into the movement of a piston. What happens to the energy that is not stored in carbohydrates or transformed into motion? It can't be destroyed; that's clear from the first law. What happens is that it is converted into the form of energy known as *heat*. For every energy "transaction" that takes place, at least some of the original energy is converted into heat. This is true for every transaction in every energy system: trees producing peaches, electric utilities burning coal, and your brain decoding this sentence. The way this concept is usually phrased, however, is that some of the original energy is *lost* to heat. At first glance, this may seem like an unfair knock on heat, which after all has intrinsic value. Why, then, is energy "lost" to heat?

Well, consider what might be called the before and after of an energy transaction involving coal. In the "before," a lump of coal is a very ordered object containing carbon atoms that exist in a precise spatial relationship to one another. Thus bonded, these atoms are not free to move. However, bringing a flame to the lump of coal causes the coal's carbon atoms to react with oxygen. Chemical bonds are broken through this process, and the energy stored in these bonds is released as heat. But heat is the random motion of molecules, and these molecules have no tendency to stay concentrated. Thus, it's clear that we have gone from the *ordered, concentrated* energy in the chemical bonds of coal to the *disordered, dispersed* energy of heat.

The Second Law of Thermodynamics: The Natural Tendency toward Disorder

And this gets us to the critical thing. Energy transformations will spontaneously run *only* from greater order to lesser order. Some of the heat produced in a steam engine is useful; it's driving a piston. Some of it is not, however—it's simply dissipated. But *all* of it amounts to a form of energy that is more dispersed than the chemical bonds in coal and that will quickly disperse further. Once this happens, there is no chance that it will spontaneously convert into anything *but* heat. We expect to see a lump of coal burn to ashes and to feel the hot air that results disperse, but we do not expect to see the ashes re-form into a lump of coal or to feel the hot air concentrate itself again.

In yielding energy, matter goes from a more-ordered state to a less-ordered state. The scientific principle that speaks to this is the **second law of thermodynamics:** Energy transfer always results in a greater amount of disorder in the universe. In connection with this law, there is a term you will hear— *entropy,* which is a measure of the amount of disorder in a system. The greater the entropy, the greater the disorder. **Figure 6.2** summarizes the principles of the first and second laws of thermodynamics. **Figure 6.3** shows you how much energy is lost to heat in several energy systems.

The Consequences of Thermodynamics

All of this may seem like just so much … *hot air,* but it turns out that few things so profoundly condition the universe we live in as these thermodynamic rules. Indeed, the relentless increase in entropy that the second law entails may dictate nothing less than the fate of the universe.

Allow yourself, for a second, to think of the universe as consisting solely of the sun, the Earth, and the space between them. The sun is a concentration of hydrogen that is undergoing a nuclear reaction, thus releasing light. Living things on Earth manage to capture buckets of the solar energy in this light through the activities of plants and the other photosynthesizers. But every time one of these buckets is poured into another—every time solar energy from the sun is transformed into energy stored in plants,

for example—there is some spillage into heat. And once heat is generated, there is no spontaneous way it can make its way back into the orderly form of sunlight or carbohydrates. With every firing of a car's cylinder or contraction of a muscle, then, some amount of energy is lost to heat.

Against this, life as we know it is made possible by ordered concentrations of energy. The sun can sustain life, but dispersed heat will not. One day, our sun will have wound down like the spring in a music box, bringing to an end the energy source for life on Earth.

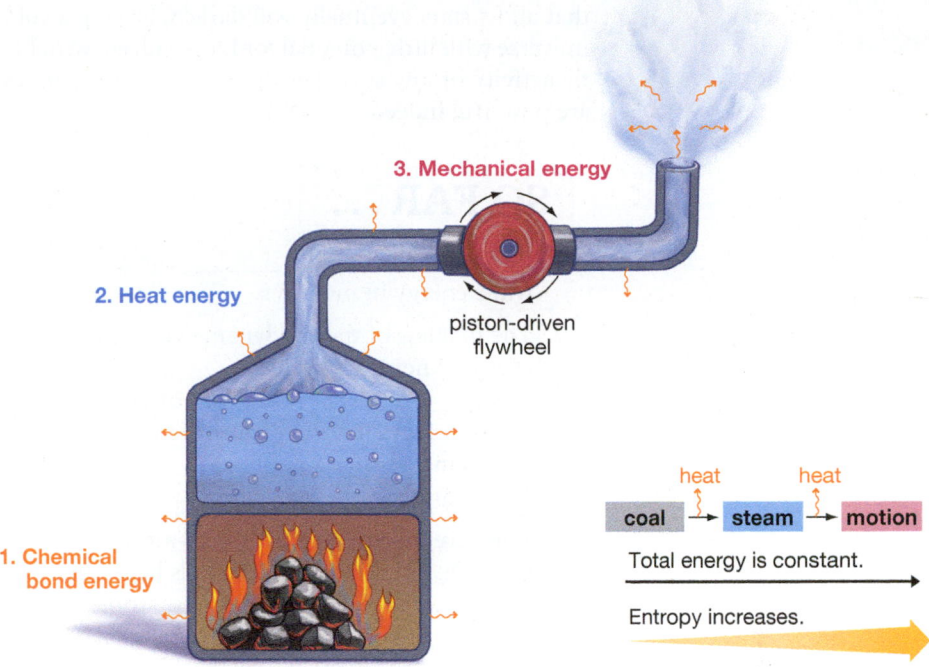

Figure 6.2
Energy Transformed

In a steam engine, energy locked up in the chemical bonds of coal is transformed into heat energy and mechanical energy. No energy is lost in this process, but energy is transformed from a more-ordered form (the chemical bonds of coal) to a less-ordered form (heat). Thus, the amount of disorder—or entropy—increases in the transaction.

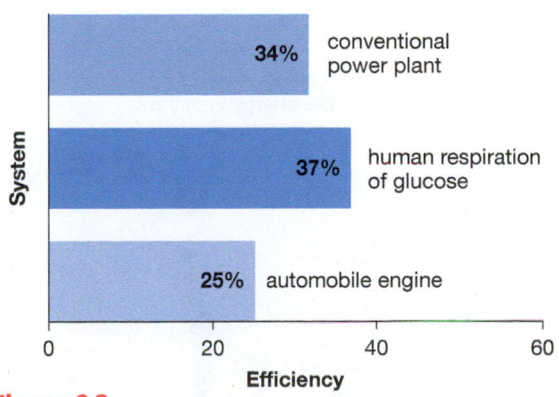

Figure 6.3
Energy Efficiency

The efficiency of several energy systems, as measured by the proportion of energy they receive relative to what they then make available to perform work. In measuring the efficiency of the car engine, the question is: How much of the energy contained in the chemical bonds of gasoline is converted by the car into the kinetic energy of wheel movement? In each system, most of the energy not available for work is lost to heat.

One possible scenario for the universe as a whole is that all its stars eventually will darken, leaving a cold universe with little potential for life—indeed, with little activity of any sort. The laws of thermodynamics are powerful indeed.

SO FAR ...

1. Energy that is stored is _____ energy, while energy in motion is _____ energy.

2. The first law of thermodynamics says that energy is never _____ or _____ but is only _____. The second law of thermodynamics says that energy transfer always results in a greater amount of _____ in the universe.

3. Whenever energy is transferred—for example, from the chemical bonds in gasoline to the movement of a car's pistons—some of the original quantity of energy is lost to _____.

6.3 How Is Energy Used by Living Things?

A sprouting plant is constantly building itself up; it is making larger, more complex molecules (proteins, starches) from smaller, simpler ones (amino acids, simple sugars). This increasing organization stands in contrast to the spontaneous course of things—breakdown and disorder—and thus may seem to be a violation of the second law, but it is not. In the context of the universe as a whole, living things are contributors to its entropy. Any given biological activity, such as building a leaf, generates heat and thus increases the disorder in the universe as a whole. But living things can bring about *local* increases in order (in themselves) by using the energy that comes ultimately from the sun.

Up and Down the Great Energy Hill

The building up of complex molecules from simpler ones is an example of work, and in it, we begin to see how living things use energy. A starchy carbohydrate is a more complex thing than the simple sugars it is made of. It is a more *ordered* arrangement of atoms than is a group of individual simple sugars. As such, the second law tells us, it should *take* energy to make a starch from simple sugars, and this is indeed the case. Going in the opposite direction, the breakdown of a starchy carbohydrate into simple sugars is an action favored by the second law because it brings about *less* order. Such a process therefore should not take energy. Indeed, this process should *release* energy, and this too is the case.

Breakdown operations, such as starches breaking down into sugars, are examples of **exergonic reactions:** reactions in which the starting set of molecules (the reactants) contains more energy than the final set of molecules (the products). Meanwhile, buildup operations are examples of **endergonic reactions:** reactions in which the products contain more energy than the reactants. This is what happens when, for example, simple glucose molecules are brought together to form glycogen, which is a storage form of carbohydrates (**Figure 6.4**).

Figure 6.4
Energy Stored and Released

It takes energy to build up a more complex molecule (in this case, glycogen) from simpler molecules (in this case, glucose units). Such a buildup is thus an endergonic or uphill reaction. Conversely, energy is released in reactions in which more complex molecules are broken down into simpler ones. Such reactions are exergonic or downhill reactions.

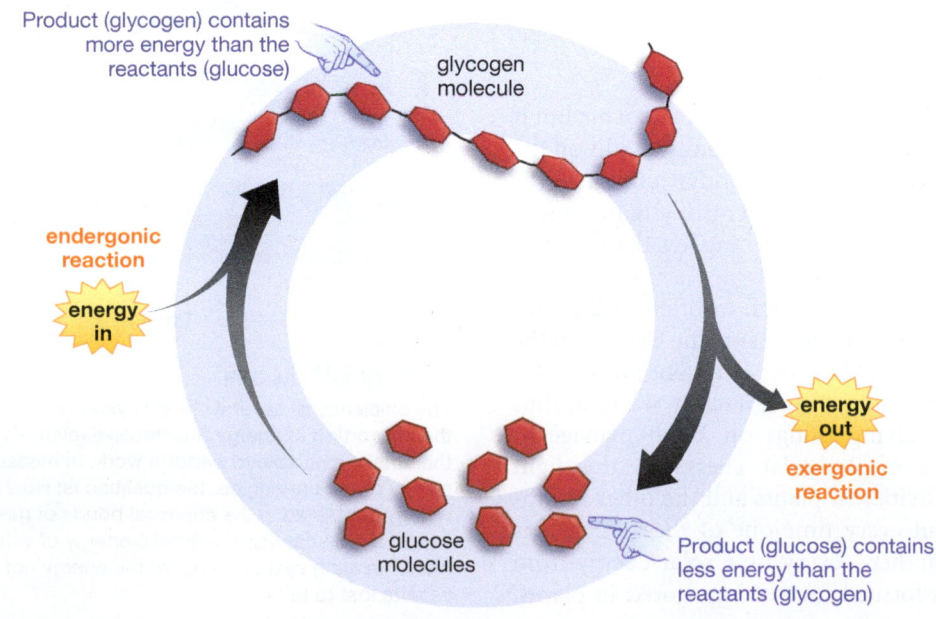

Product (glycogen) contains more energy than the reactants (glucose)

glycogen molecule

endergonic reaction

energy in

glucose molecules

energy out

exergonic reaction

Product (glucose) contains less energy than the reactants (glycogen)

Coupled Reactions

With this as background, we arrive at an important insight: Endergonic and exergonic reactions are linked in living things. Why should this be so? If you want to get something uphill—into a higher energy state—you have to power that process, which means *using* some energy. Just as it takes energy to get a bicycle uphill, so it takes energy to power the endergonic reaction that takes us from glucose molecules to glycogen. Now, where is this energy going to come from? It must come from an energy-releasing *exergonic* reaction. The result is a **coupled reaction:** a chemical reaction in which an exergonic reaction powers an endergonic reaction.

With this, we reach a critical point about how living things operate: through an endless series of coupled reactions. Want to move a muscle? Two kinds of long, slender muscle protein filaments must slide past each other for this to happen. For this to work, a huge number of tiny, pivoting extensions on one kind of muscle filament must first latch onto and then pull the other kind of filament (**Figure 6.5**). Now, how are these extensions going to "cock" themselves and get ready for a pull? This cocking is an endergonic reaction—it is a movement up the energy hill. Once our muscle extensions are cocked, they will be in a higher energy state, just as the bar on a mousetrap is in a higher energetic state after it's pulled back. Following what was just said, an endergonic reaction such as this must be powered by an exergonic reaction—energy must be released from somewhere for those muscle extensions to be moved into their cocked state. Now, where does the energy for this come from? To follow the whole trail, this energy came originally from the sun, after which it was captured by a photosynthesizing organism—a peach tree, for example. Then, say

we ate a peach from the tree; now, the energy that was locked up in the peach is locked up in us in the form of glucose molecules. So, does the energy to pull back the muscle extensions go directly from the glucose to the extensions? No. One more energy transfer must be made first. The energy in the glucose must be transferred into a molecule called ATP.

6.4 The Energy Dispenser: ATP

It is ATP that will release the energy that allows the muscle extensions to be cocked. Only with the energy dispensed by ATP can our muscles contract. You could think of ATP as a kind of "middleman"—it receives the energy from food and then doles it out for activities such as muscle contraction. Why do we need an intermediate molecule of this sort? Well, think of it this way. Power plants have considerable use for coal because, as we've seen, coal can yield energy. But it may go without saying that most people don't use coal's energy *directly* in their homes. Instead, power plants must convert the energy in coal into a form that people *can* use in their homes—electricity. In this same way, the energy contained in food must be converted into a form the human body can use, and this form is overwhelmingly the molecule known as ATP, which can accept relatively large amounts of energy from a food source and then proceed to pass on this energy to a wide range of the molecules found in the body. The upshot is that every time a muscle contracts or a nerve signal fires, ATP is doling out the energy for these operations. Every time a sperm cell whips its tail, ATP is supplying the energy. And what goes for energy use in human beings goes for energy use in all forms of life.

Energy and Biology

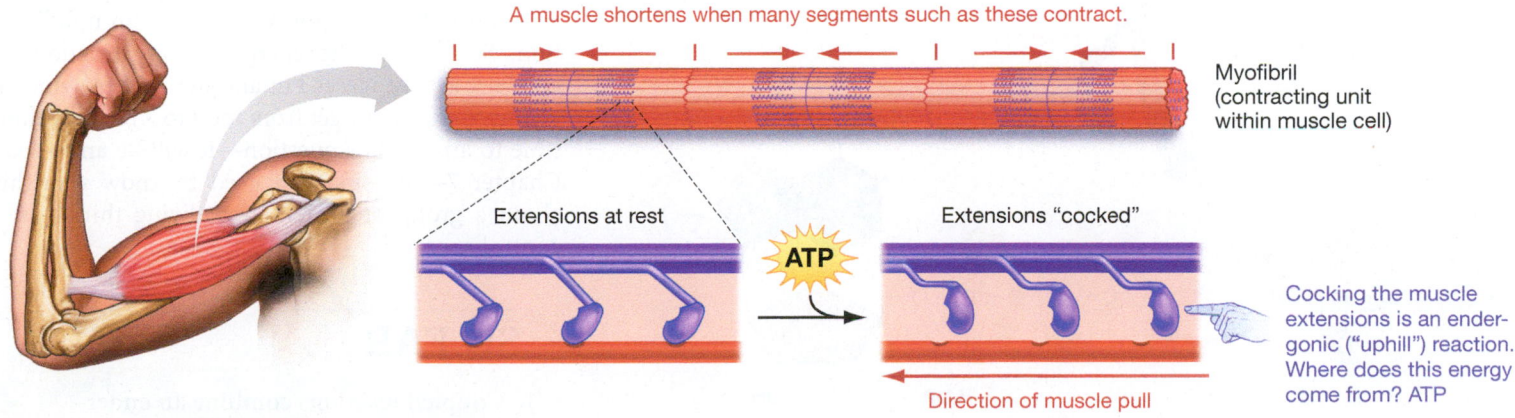

A muscle shortens when many segments such as these contract.

Myofibril (contracting unit within muscle cell)

Extensions at rest

ATP

Extensions "cocked"

Cocking the muscle extensions is an endergonic ("uphill") reaction. Where does this energy come from? ATP

Direction of muscle pull

Figure 6.5
Powering a Coupled Reaction
For any muscle to contract, one set of muscle filaments must pull on a second set, thus shortening the muscle segment that both are part of. To accomplish this, extensions of the active filaments must be put in a "cocked" position. Energetically, this is a movement up the energy hill and thus is an energy-using or endergonic reaction that must be powered by an energy-releasing or exergonic reaction. Together, the two reactions constitute a single reaction—a coupled reaction. This coupled reaction, like so many in living things, is powered by energy released from the molecule ATP.

Chemical Reactions
and ATP

The Structure of ATP

The simplest bacterium is as dependent on ATP as we are. In sum, **ATP** can be defined as the most important energy transfer molecule in living things.

So, what's the structure of this molecule? To answer this question, let's start by thinking about where ATP's full name, **adenosine triphosphate,** comes from. ATP has a sugar (called ribose) and a nitrogen-containing base (called adenine) that together make up a compound called adenosine, as you can see in **Figure 6.6**. Phosphate groups are also linked to adenosine—three phosphate groups, which gives us adenosine *tri*phosphate. The place to look for energy storage in ATP is in the phosphate groups. Notice that they all are negatively charged. This is the key to understanding why ATP serves as such an effective energy transfer molecule. Remember that like charges repel each other, and in this case three phosphate groups are doing just that. Energy was required to put these phosphate groups together as they are, in this relatively unstable state. This linkage represents a move *up* the energy hill. With ATP, it is as though someone squeezed a jack-in-the-box back into the box and shut

the lid. It should be no surprise that when the lid is opened, a good deal of energy will be released.

If you look now at **Figure 6.7**, you can see the details of how ATP does its job in the muscle contraction we just touched on. The essence of the operation is that a muscle-filament extension is moved from a 45-degree angle to a "cocked" *90*-degree angle thanks to the energy input from ATP—a position from which the extension can now pull the muscle filament to the left. The result, when carried out by a huge number of extensions and filaments, is contraction of a muscle.

The ATP/ADP Cycle

There is another side to this ATP-driven reaction, however. Note that once ATP loses its outer phosphate group, it is not ATP anymore, but two molecules—one called ADP and another labeled P (for phosphate). ATP works by losing (or "donating") a phosphate group in this way, but once this happens, it becomes a molecule with only *two* phosphate groups: adenosine *di*phosphate or ADP. However, this is a temporary condition. Once an ADP molecule is finished with a reaction, it is free to have a third phosphate group added to it again. This is just what happens: ADP returns to being ATP, which is capable of providing energy for yet another reaction. This shuttling from ATP to ADP and back, illustrated in Figure 6.6, takes place constantly in cells.

Between Food and ATP

The essence of our story so far is that, in animals, energy that comes originally from the sun is first stored in the food we eat and is then transferred to ATP, which then dispenses the energy to power our daily activities. But if you think about it, there's a piece missing here. How does that third phosphate group get onto ADP, making it ATP? This is an uphill reaction. So, where's the energy-*yielding* reaction that allows it to take place? Put another way, by which reactions does energy get from food to ATP? It is almost time to answer this question—it will be answered in Chapter 7—but first you need to know something about a group of proteins that living things use to control energy.

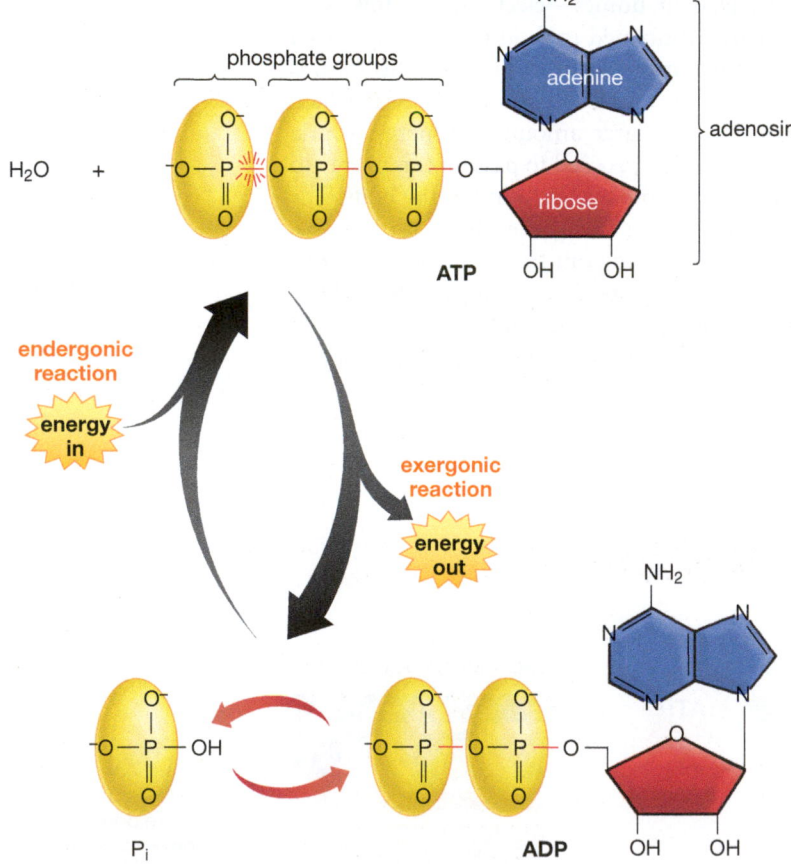

Figure 6.6
Life's Most Important Energy Transfer Molecule
ATP stores energy in the form of chemical bonds between its phosphate groups. When the bond between the second and outermost phosphate group is broken, the outermost phosphate separates from ATP, and energy is released. This separation transforms ATP into adenosine diphosphate or ADP, which then goes on to pick up another phosphate group, becoming ATP again.

SO FAR . . .

1. Coupled reactions combine an endergonic reaction, in which the _____ have more energy than the _____, and an exergonic reaction, in which the _____ have more energy than the _____.

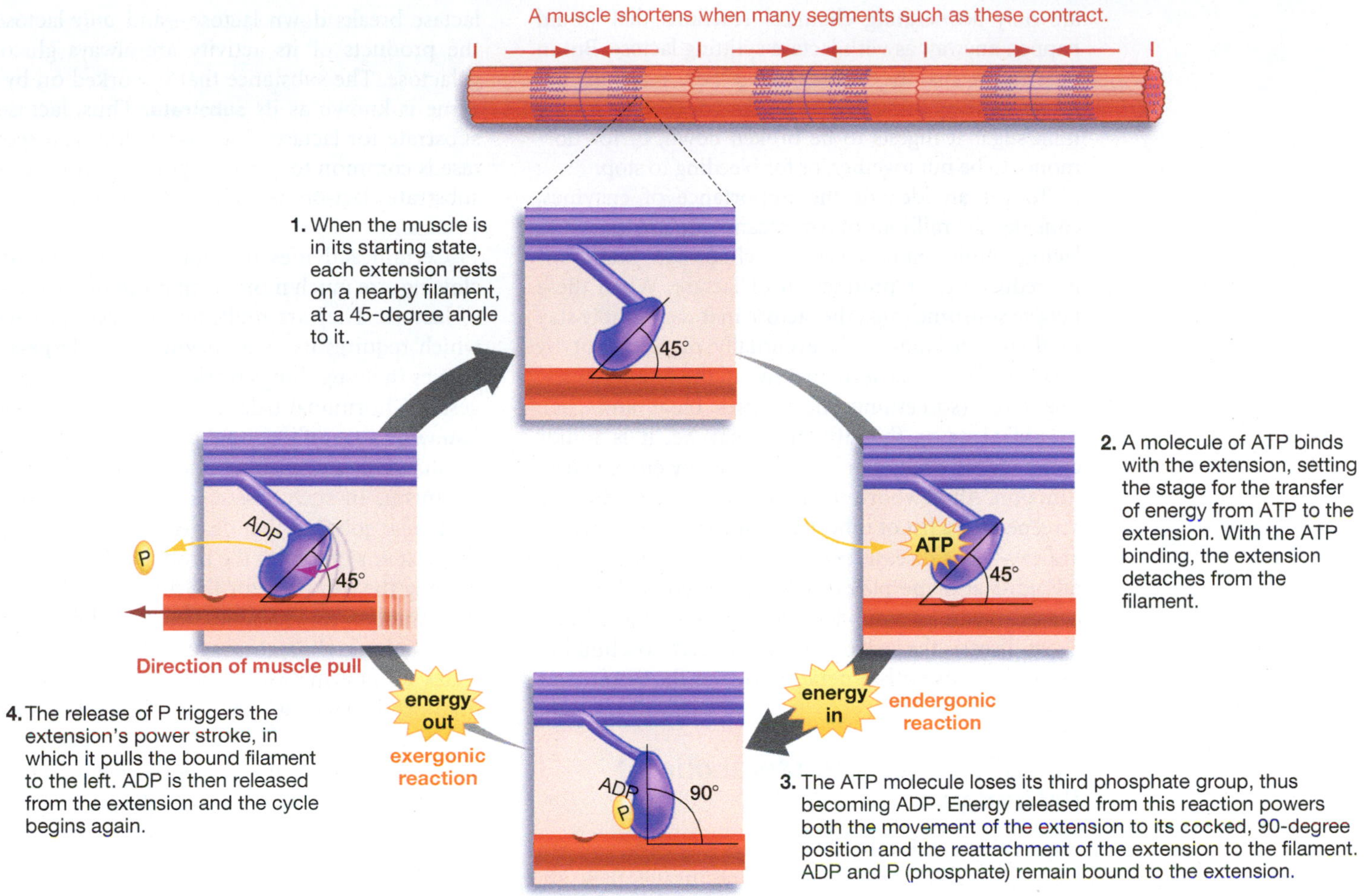

A muscle shortens when many segments such as these contract.

1. When the muscle is in its starting state, each extension rests on a nearby filament, at a 45-degree angle to it.

45°

2. A molecule of ATP binds with the extension, setting the stage for the transfer of energy from ATP to the extension. With the ATP binding, the extension detaches from the filament.

ATP

45°

ADP

P

Direction of muscle pull

4. The release of P triggers the extension's power stroke, in which it pulls the bound filament to the left. ADP is then released from the extension and the cycle begins again.

45°

energy out

exergonic reaction

energy in

endergonic reaction

ADP

P

90°

3. The ATP molecule loses its third phosphate group, thus becoming ADP. Energy released from this reaction powers both the movement of the extension to its cocked, 90-degree position and the reattachment of the extension to the filament. ADP and P (phosphate) remain bound to the extension.

Figure 6.7
Muscle Contraction, Powered by ATP

2. The energy for most coupled reactions in the body is supplied by the molecule _____, in particular by the splitting off of this molecule's third or outer _____.

3. After powering a reaction, ATP becomes the molecule _____, which, with the addition of another _____, becomes ATP once more.

6.5 Efficient Energy Use in Living Things: Enzymes

Lactose—better known as milk sugar—is composed of two simple sugars, glucose and galactose, that are linked together by a single atom of oxygen. It follows from what you've seen so far that lactose will split into its glucose and galactose parts without any energy requirement. Indeed, a little energy will be released in this process because chemical bonds have been broken and two smaller molecules have been produced. Such a splitting is therefore a downhill reaction.

If we actually took a small amount of lactose, however, and put it in some plain water, it would certainly be hours—it might be days—before any of the lactose molecules would break down into glucose and galactose. How can this be? Millions of people around the world drink milk, and it doesn't seem to pile up in them.

Accelerating Reactions

The secret is that when lactose is metabolized in living things, it isn't being split up in plain water. Something else is present that speeds up this process immensely, accelerating it perhaps a billionfold. That something is an **enzyme,** a type of protein that accelerates a chemical reaction. In the case of lactose, the particular enzyme is called lactase, but it is merely one of thousands of enzymes known to exist. Each of these compounds works on some chemical process, but not all are involved in splitting molecules. Some enzymes combine molecules; some rearrange them.

Given their numbers, it may be apparent that enzymes are involved in a lot of different activities, but this doesn't begin to give them the recognition they deserve. Enzymes facilitate nearly every chemical process that takes place in living things. No organism could survive without them. Technically, these compounds

Enzymes

How Enzymes Work

are only accelerating chemical reactions that would happen anyway, as with lactase splitting lactose. But in practical terms, they are *enabling* these reactions because no living thing could wait days or months for the milk sugar it ingests to be broken down, or for hormones to be put together, or for bleeding to stop.

To get an idea of the importance of enzymes, consider the millions of Americans who are known as lactose intolerant because, after childhood, their bodies reduced their production of lactase. When these people consume milk, the lactose in it *does* simply stay in their intestines. It is eventually digested not by them but by the bacteria that live in the human digestive tract (something that causes bloat and gas). Troublesome as this affliction may be, it is minor compared to other conditions caused by enzyme dysfunction. All newborn babies in the United States are screened for a list of inherited diseases, three of which are Tay-Sachs disease, phenylketonuria, and congenital adrenal hyperplasia. Each of these diseases is caused by an enzyme deficiency. In the case of Tay-Sachs disease, the lack of an enzyme leads to a buildup of fat deposits within nerve cells, and the result usually is death by age 4.

Specific Tasks and Metabolic Pathways

Chemistry Review: Enzymes and Pathways

From the sheer number of enzymes that exist, you can probably get a sense of how specifically they are matched to given tasks. Although some enzymes can work on groups of similar substances, a specific enzyme usually facilitates a specific reaction: It works on one or perhaps two molecules and no others. Thus,

lactase breaks down lactose—and *only* lactose—and the products of its activity are always glucose and galactose. The substance that is worked on by an enzyme is known as its **substrate.** Thus, lactose is the substrate for lactase. The *-ase* ending you see in lactase is common to most enzymes. As you can see, the substrate (lactose) is identified in the enzyme's name (lactase).

Certain activities in living things, such as blood clotting, are much more complex than the breakdown of lactose. They are multistep processes, each step of which requires its own enzyme. Most large-scale activities in living things work this way: leaf growth, digestion, hormonal balance. The result is a **metabolic pathway:** a set of enzymatically controlled steps that results in the completion of a product or process in an organism. In such a sequence, each enzyme does a particular job and then leaves the succeeding task to the next enzyme, with the product of one reaction becoming the substrate for the next (**Figure 6.8**). This is the way that, for example, cholesterol gets produced: One cholesterol precursor leads to another through a series of enzymatically enabled reactions, with the final product being cholesterol itself. The sum of all the chemical reactions that a cell or larger organism carries out is known as its **metabolism.** To put things simply, enzymes are active in all facets of the metabolism of all living things.

6.6 Enzymes and the Activation Barrier

All this information about enzymes is well and good, you may be saying at this point, but what does it have to do with energy? The answer is that in carrying out their tasks, enzymes are in the business of lowering the amount of energy needed to get chemical reactions going. And this means that the reactions can get going faster.

An analogy may be helpful here. Consider the now-familiar rock perched at the top of a hill. Situated in this way, it has a good deal of potential energy. If it rolled down the hill, it would release this potential energy as kinetic energy. But here is the critical question: What would be required to get this rock going? It is perched at the top of the hill because it is *stable* at the top of the hill. To get it going would require *additional* energy in the form of, say, a push.

Now consider the lactose molecule. It is "perched" like the rock, in a sense, because it lies "uphill" from the glucose and galactose it can break into. But just as the rock is stable because of its position in the ground, so the lactose is stable because it has a strong set of chemical bonds holding its atoms in a fixed position. Now, some energy is going to be required to break those bonds, thus getting the reaction going. But is there anything that can *lower* the amount of energy

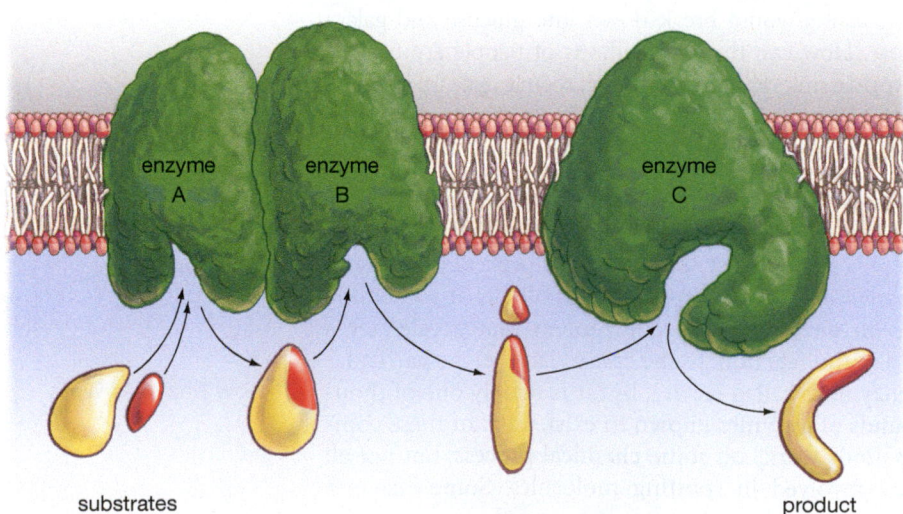

Figure 6.8

Metabolic Pathway: Sequence of Enzyme Action

Most processes in living organisms are carried out through a metabolic pathway—a sequential set of enzymatically controlled reactions in which the product of one reaction serves as the substrate for the next. Enzymes perform specific tasks on specific substrates. In this example, enzyme A combines two substrates, enzyme B removes part of its substrate, and enzyme C changes the shape of its substrate.

required to set this process in motion? The answer is yes—an enzyme. If you look at **Figure 6.9**, you can see this in schematic terms. Imagine that you had to push a boulder up the hill in (a) to get it to roll down the other side. Now imagine that you only had to push a boulder up the hill in (b) to get the same result. In which situation would you be able to get your desired result *faster*? What is at work here is lowering of **activation energy:** the energy required to initiate a chemical reaction.

How Do Enzymes Work?

How do enzymes carry out this task of lowering activation energy? As you'll see shortly, they bind to their substrates and in so doing make these substances more vulnerable to chemical alteration. The amazing thing is that they do this without being permanently altered themselves. Enzymes are **catalysts**—substances that retain their original chemical composition while bringing about a change in a substrate. At the end of its splitting of the lactose molecule, lactase has exactly the same chemical structure as it did prior to this operation. It is thus free to pick up another lactose molecule and split *it*. And all this takes place in a flash: The fastest-working enzymes can carry out 100,000 chemical transformations per second.

Enzymes generally take the form of globular or ball-like proteins with a shape that includes a kind of pocket into which the substrate fits. If you look at **Figure 6.10**, you can see a space-filling model of one enzyme, called hexokinase. You can also see, buried there in the middle, its substrate—glucose, which is better known as blood sugar. As with all proteins, enzymes are made up of amino acids, but only a few of the hundreds of amino acids in an enzyme are typically involved in actually binding with the substrate. Five or six would be common. These amino acids help form the substrate pocket, known as the **active site:** the portion of an enzyme that binds with a substrate, thus helping transform it.

In some cases, the participants at the active site include, along with amino acids, one or more accessory molecules. One variety of these molecules is the **coenzymes:** molecules other than amino acids that facilitate the work of enzymes by binding with them. If you've ever wondered what vitamins do, a big part of the answer is: function as coenzymes. All the B vitamins, for example, serve as coenzymes that sit in the active site of an enzyme and provide an added chemical attraction or repulsion that allows the enzyme to do its job.

An Enzyme in Action: Chymotrypsin

The details of how enzymes carry out their work are complex, and they vary according to enzyme. Let's consider in a general way, however, the activity of a much-studied enzyme called chymotrypsin.

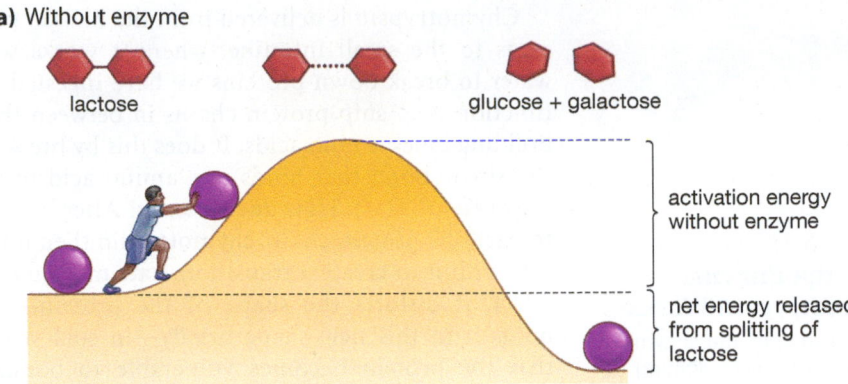

(a) Without enzyme

lactose   glucose + galactose

activation energy without enzyme

net energy released from splitting of lactose

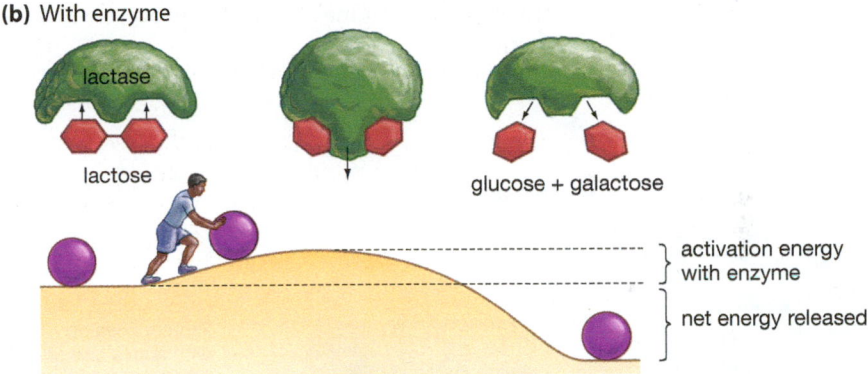

(b) With enzyme

lactase

lactose   glucose + galactose

activation energy with enzyme

net energy released

Figure 6.9
Enzymes Accelerate Chemical Reactions
Without an enzyme, the amount of energy necessary to split lactose is high. In the presence of the enzyme lactase, however, a low activation energy is sufficient to get the process started.

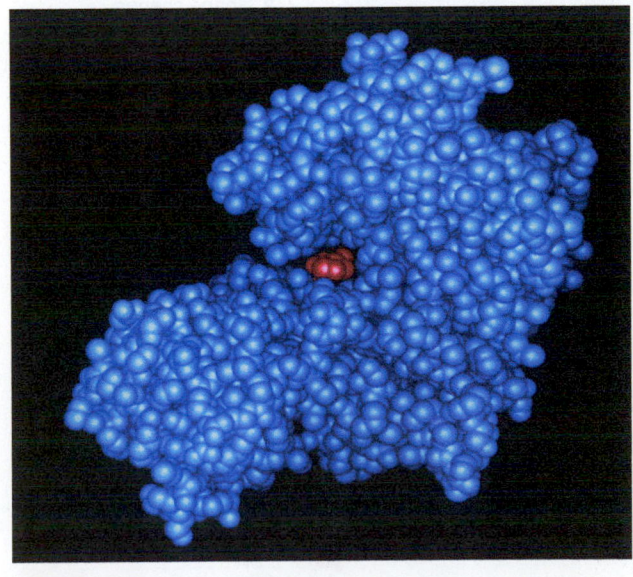

Glucose molecule (shown in red) binding to active site of hexokinase enzyme

Figure 6.10
Shape Is Important in Enzymes
Substrates generally fit into small grooves or pockets in enzymes. The location at which the enzyme binds the substrate is known as the enzyme's active site. Pictured is a computer-generated model of a glucose molecule (in red) binding to the active site of an enzyme called hexokinase (in blue).

Chymotrypsin is delivered from the human pancreas to the small intestine, where it works with water to break down proteins we have ingested. Its function is to snip protein chains in between their building block amino acids. It does this by breaking the single bond that binds one amino acid to the next (**Figure 6.11**). How does it work? After binding to part of a protein chain, chymotrypsin then interacts with it to create a transition-state molecule. In effect, it distorts the shape of the protein—and holds it in this new shape briefly—in such a way that the protein becomes vulnerable to bonding with ionized water molecules. This allows carbon and nitrogen atoms to latch on to new partners, and the protein chain is cut in two. Chymotrypsin then returns to its original form and proceeds to a new reaction.

Figure 6.11
How the Enzyme Chymotrypsin Works
In this example, the enzyme chymotrypsin is facilitating the breakdown of a protein by changing the protein's shape. In the absence of chymotrypsin, this process would take a billion times longer.

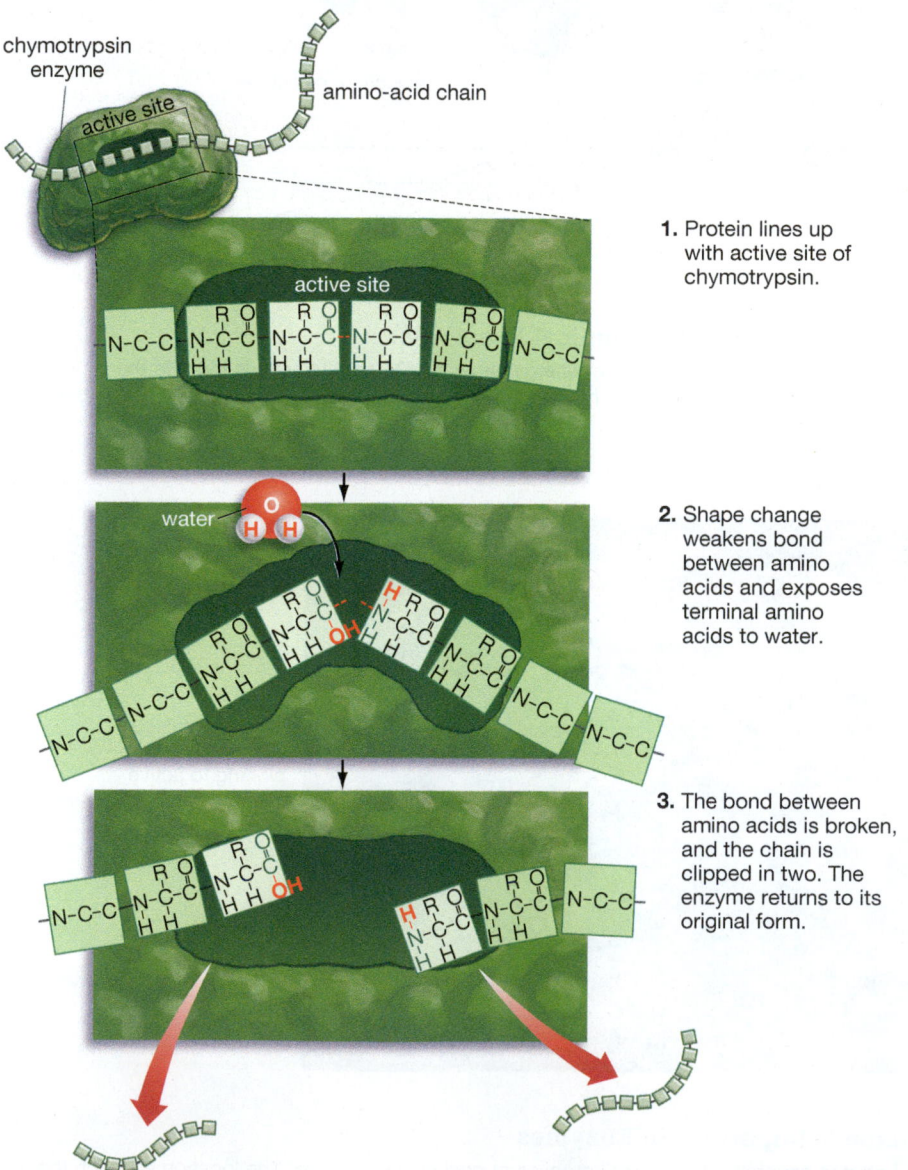

chymotrypsin enzyme

amino-acid chain

active site

1. Protein lines up with active site of chymotrypsin.

active site

water

2. Shape change weakens bond between amino acids and exposes terminal amino acids to water.

3. The bond between amino acids is broken, and the chain is clipped in two. The enzyme returns to its original form.

6.7 Regulating Enzymatic Activity

The activity of enzymes doesn't proceed in a monotonous, clock-like way, but instead is regulated through several different means. To put this another way, several factors influence the amount of "product" an enzyme turns out. One of these factors, logically, is the amount of substrate in the enzyme's vicinity. If there's no substrate to work on, no product can be turned out.

The work of enzymes can, however, be reduced in other ways. For example, some molecules that are not the normal substrate for an enzyme can nevertheless bind to the enzyme's active site. This happens all the time in natural ways in our bodies, but to sharpen the point, consider the well-known drug Lipitor, which millions of people take to lower their cholesterol levels. How does this drug work? Recall that there is a metabolic pathway by which cholesterol is produced—one cholesterol precursor after another is modified in a set of reactions, each one of which is facilitated by an enzyme. But one of these enzymes has an active site that *Lipitor* can fit into. When this happens, this enzyme is occupied—it can't bind with the cholesterol precursor that is its normal substrate. And this effect is dose dependent: The greater the number of enzyme molecules that are occupied by Lipitor, the fewer the number of enzyme molecules that can help produce cholesterol. Thus, the ultimate effect of Lipitor is to lower the amount of cholesterol produced in the body. What we see with Lipitor is an example of **competitive inhibition**: a reduction in the activity of an enzyme by means of a compound other than the enzyme's usual substrate binding with it in its active site.

Allosteric Regulation of Enzymes

Living things routinely regulate enzyme activity by another means. Earlier, you saw a model of an enzyme called hexokinase, which works on the substrate glucose. The product of this reaction is something called glucose-6-phosphate. Although this is an enzymatic product, it can bind to hexokinase at a site *other* than its active site and in so doing change the shape of hexokinase. This change renders hexokinase less able to bind with its substrate, glucose, meaning hexokinase's work is temporarily reduced. If you look at **Figure 6.12**, you can see how this process works schematically.

This is but one example of the process known as **allosteric regulation**—the regulation of an enzyme's activity by means of a molecule binding to a site on the enzyme other than its active site. Many variations on this theme are possible. In the case of hexokinase, its own product (glucose-6-phosphate) bound with it and cut down on its activity, but molecules other than the enzymatic product can bind with an enzyme at one of its additional binding sites. Furthermore, the effect of such binding need not be negative; allosteric binding can increase the activity of enzymes as well as reduce it.

The importance of such regulation is that enzymes are not simply fated to turn out product as long as there is substrate to work on. If a cell has too much of a product, allosteric control can cut down on it; if there is too little, its concentration can be increased.

On to Energy Harvesting

Our story of energy concluded in this chapter with a look at enzymes and energy. Next chapter, however, we'll pick up the story of energy with a more detailed look at a process touched on earlier—the transfer of energy from food to ATP. The question we will tackle is this: What are the steps by which our bodies extract energy from food and then channel this energy into pushing phosphate groups onto ADP, thus making it ATP? Put another way, how does our primary energy-dispensing molecule become energized?

SO FAR ...

1. An enzyme is a type of _____ that _____ a chemical reaction.

2. The substance that an enzyme works on is known as its _____. A metabolic pathway is a linked series of enzymatically controlled reactions in which the _____ of one reaction becomes the _____ for the next.

3. Activation energy is the energy required to _____ a chemical reaction. Enzymes work by _____ activation energy.

(a) Substrate becomes product.

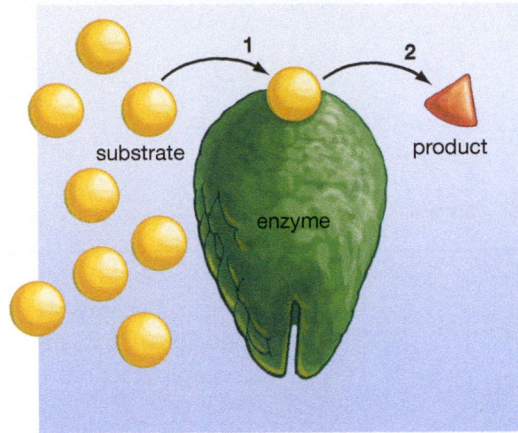

1. Substrate binds to enzyme.

2. Enzyme transforms substrate to product.

(b) Product feeds back on enzyme.

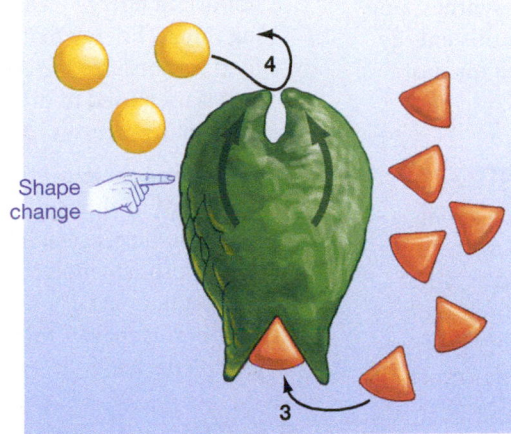

3. Product binds to a *different* site on the enzyme, causing the enzyme to change shape.

4. The new shape of the enzyme prevents it from binding to any more substrate.

Figure 6.12
Allosteric Regulation of Enzymes
The frequency of chemical reactions can be controlled by the binding of a molecule to an enzyme at a site other than its active site. In this instance, the enzyme's own product binds with the enzyme in this way, thereby reducing the enzyme's activity.

CHAPTER

6 REVIEW

Summary

6.1 Energy Is Central to Life

- All living things require energy. The sun is the ultimate source of energy for most living things. The sun's energy is captured on Earth by photosynthesizing organisms, which then pass this energy on to other organisms in the form of food. (p. 109)

6.2 The Nature of Energy

- Energy is the capacity to bring about movement against an opposing force. Energy can be conceptualized as either potential energy, meaning stored energy; or kinetic energy, meaning energy in motion. The study of energy is known as thermodynamics. (p. 109)

- The first law of thermodynamics states that energy is never created or destroyed, but is only transformed; the second law of thermodynamics states that energy transfer will always result in a greater amount of disorder in the universe. (p. 110)

- In every energy transaction, some energy will be lost to heat, the most disordered form of energy. Entropy is a measure of the amount of disorder in a system. (p. 111)

6.3 How Is Energy Used by Living Things?

- Living things can bring about local increases in order (in themselves) through their metabolic processes, but it takes energy to do this. (p. 112)

- Energy in living things is stored in endergonic reactions in which the products of the reaction contain more energy than the starting substances (or reactants). Energy is released in exergonic reactions, in which the reactants contain more energy than the products. Endergonic and exergonic reactions are linked in coupled reactions

in which an energy-yielding exergonic reaction powers an energy-requiring endergonic reaction. (p. 112)

6.4 The Energy Dispenser: ATP

- Adenosine triphosphate (ATP) is the most important energy transfer molecule in living things. Energy that is extracted from food is transferred to ATP, and the energy in it is then used to drive a vast array of metabolic processes. (p. 113)

- ATP drives chemical reactions by donating one of its three phosphate groups to them. In the process, it becomes the two-phosphate molecule adenosine diphosphate (ADP). To again become ATP, it must have a third phosphate group attached to it. (p. 114)

6.5 Efficient Energy Use in Living Things: Enzymes

- An enzyme is a type of protein that accelerates the rate at which a chemical reaction takes place in an organism. Nearly every chemical process that takes place in living things is facilitated by an enzyme. (p. 115)

- The substance that an enzyme helps transform through chemical reaction is called its substrate. Many activities in living things are controlled by metabolic pathways, in which a series of reactions is undertaken in sequence, each facilitated by its own enzyme. In such a series, the product of one reaction becomes the substrate for the next. The sum of all the chemical reactions that a cell or larger living thing carries out is its metabolism. (p. 116)

6.6 Enzymes and the Activation Barrier

- Enzymes work by lowering activation energy, meaning the energy required to initiate a chemical reaction. Enzymes are catalysts: They bring about a change in

their substrates without being chemically altered themselves. (p. 116)

- Enzymes generally take the form of globular proteins whose shape includes a pocket, called the active site, into which the enzyme's substrate fits. Only a few of the hundreds of amino acids that typically make up an enzyme will be involved in substrate binding. In some cases, substrate binding is facilitated by coenzymes. (p. 117)

6.7 Regulating Enzymatic Activity

- Enzyme activity can be controlled by competitive inhibition (a reduction in the activity of an enzyme by means of a compound other than the enzyme's usual substrate binding with the enzyme in its active site) or allosteric regulation (a molecule binding with the enzyme at a site other than its active site). Through such means, enzyme activity can be finely tuned in accordance with cellular needs. (p. 118)

Key Terms

Understanding the Basics

Multiple-Choice Questions
(Answers are in the back of the book.)

1. According to the first law of thermodynamics, energy:
   a. is never lost or gained, but is only transformed.
   b. always requires an ultimate source such as the sun.
   c. can never be gained, but can be lost.
   d. can never really be harnessed.
   e. can never be transformed.

2. Each time there is a chemical reaction, some energy is exchanged. According to the second law of thermodynamics, with each exchange:
   a. some energy is lost, but other energy is created.
   b. some energy must come from the sun.
   c. some energy is transformed into heat.
   d. energy is gained for future use.
   e. some energy is permanently destroyed.

3. ATP stores energy in the form of:
   a. mechanical energy.
   b. heat.
   c. complex carbohydrates.
   d. chemical bonds.
   e. amino acids.

4. People who are lactose intolerant lack a compound called lactase in their digestive tract. You know that this compound, lactase, is probably a(n):
   a. hormone, because it causes a change in the person.
   b. coenzyme, because it helps the enzyme, lactose, do its job.
   c. vitamin, because it is essential to health.
   d. substrate, because it is involved with food.
   e. enzyme, because its name ends in -ase.

5. Glycolysis is a metabolic pathway that helps living things extract energy from food. From this we know that glycolysis:
   a. consists of a series of chemical reactions.
   b. uses a number of enzymes.
   c. involves the modification of a series of substrates.
   d. proceeds by means of each enzyme leaving a succeeding reaction to a different enzyme.
   e. all of the above

6. Each molecule of ATP stores a good deal of energy because:
   a. ATP is such a large protein.
   b. ATP captures energy directly from the sun.
   c. ATP's third phosphate group exists in a highly energetic state.
   d. a long metabolic pathway is required for the production of ATP.

7. Enzymes:
   a. accelerate specific chemical reactions.
   b. are not chemically altered by binding with a substrate.
   c. lower the activation energy of specific chemical reactions.
   d. all of the above
   e. a and c only

8. The active site of enzyme A is occupied by a molecule other than its substrate. Enzyme B has had its activity reduced by means of a molecule binding to it at a site other than its active site. What is taking place in connection with enzyme A and then enzyme B?
   a. a reaction acceleration, a coupled reaction
   b. a lowering of activation energy, ATP use
   c. allosteric regulation, competitive inhibition
   d. allosteric regulation, coenzyme participation
   e. competitive inhibition, allosteric regulation

Brief Review
(Answers are in the back of the book.)

1. A plant leaf is more ordered and complex than are carbon dioxide and water. In photosynthesis, plants take small, relatively disordered carbon dioxide, water, and mineral molecules and make complex, ordered sugars out of them. How is it possible for this to happen without violating the second law of thermodynamics?

2. The text notes that, in connection with muscle contraction, extensions of one kind of muscle filament are "cocked" in an uphill reaction, and that ATP plays a part in this. Explain how this is a coupled reaction.

3. Where does the energy come from that is stored and released by ATP? In what way is ATP an energy transfer molecule?

4. The average person thinks of enzymes only as substances in their digestive tract that break down food. How would you explain the role of enzymes to your little sister, who has heard about digestion in grade-school biology?

5. What is meant by allosteric regulation of an enzyme?

Applying Your Knowledge

1. In light of the laws of thermodynamics, explain why you get so much hotter when you run than when you are sitting still.

2. Which is easier to have: a messy, disordered room or a neat, ordered room? Is the second law of thermodynamics at work in this situation? What is the price of bringing order to a room?

3. Physicists say that it is impossible to create a perpetual motion machine—a machine whose own activity keeps it running perpetually. What energy principle precludes the possibility of a perpetual motion machine?

MasteringBiology®

1. MasteringBiology Assignments
Activities: Energy Transformations; Energy and Biology; Chemical Reactions and ATP; The Structure of ATP; How Enzymes Work; Enzymes; **Chemistry Review:** Enzymes and Pathways; **Building Vocabulary; Questions:** Reading Questions, Test Bank

2. eText
Read your book online, search, take notes, highlight text, and study more effectively.

3. The Study Area
Self-study tools to test comprehension.

Vital Harvest:

Deriving Energy from Food

To remain alive, living things must put the energy they acquire into a form they can use—ATP.

Available on MasteringBiology®

Activities: Build a Chemical Cycling System; Overview of Cellular Respiration; Glycolysis; Fermentation; The Citric Acid Cycle; Electron Transport; **Bioflix:** Cellular Respiration; **Current Events:** Still Counting Calories? Your Weight Loss Plan Might Be Outdated; **Building Vocabulary; Questions:** Bioflix Quiz, Reading Questions, Test Bank

How do living things extract energy from food? Through the process of cellular respiration. Here, a snowcap hummingbird (*Microchera albocoronata*) feeds from a flower in a Costa Rican rain forest.

In May of 1996, the world's attention was riveted on news coming out of Nepal, in Southeast Asia. Disaster had struck several groups of mountain climbers who had been attempting to scale the world's highest peak, Mt. Everest. A vicious storm had taken the members of two Everest expeditions by surprise as they were moving down from Everest's summit, toward the safety of lower altitudes. Within 36 hours, five climbers had lost their lives, and another was so badly frostbitten that one of his hands had to be amputated.

One of the expeditions was led by a highly respected guide from New Zealand, Rob Hall. Courageously remaining near the summit in an effort to bring down one of his clients (who had collapsed), Hall ended up spending a night in a howling storm on Everest without a tent. When the sun rose the next day, the factors working against him included cold and wind, which he might have encountered in lots of inhospitable places on Earth. Also bedeviling him, however, was something that human beings rarely encounter: a lack of oxygen. At his altitude—sitting in the snow at about 28,500 feet—the oxygen he took in with each breath was little more than a third of the amount he would have inhaled with each breath at sea level. Technology might have come to his aid, as he had two oxygen bottles with him, but his intake valve for them had become clogged with ice.

Down at Everest's lower altitudes, other climbers—some of them Hall's friends—had to endure the agony of talking to him by two-way radio while not being able to reach him physically. All of them knew that a lack of oxygen was draining not only his muscles but his mind. Here is Hall in one of his radio transmissions asking about another guide on the ill-fated expedition: "Harold was with me last night, but he doesn't seem to be with me now. Was Harold with me? Can you tell me that?" In the end, the mountain without mercy claimed Rob Hall. He never got up from the place where he spent his night on Everest.

The tragedy of the 1996 Everest expeditions drives home a point and raises a paradox: No one needs to be reminded that we need to breathe in order to live, and most people are aware that oxygen is the most important thing that comes in with breath. That said, of the next 100 people you meet, how many could tell you what oxygen is doing to sustain life? Put another way, why do we need to breathe?

The short answer is that breathing and oxygen are in the energy transfer business. They are part of a system that allows us to extract, from food, energy that is then used to put together the "energy-dispensing" molecule, ATP. As you'll recall from Chapter 6, the body uses ATP (adenosine triphosphate) to power activities that range from muscle contraction to thinking to cell repair. Living things need large amounts of ATP to live, and organisms such as ourselves use oxygen to produce most of our ATP. If we don't get enough oxygen to keep making ATP, our bodies and minds start failing.

Exhausted Near the Summit

Two members of the ill-fated expeditions that climbed Mount Everest in May 1996 approach the mountain's peak.

Oxygen is not required for all the "harvesting" we do of the energy contained in food. When you lift a box, the short burst of energy that is required comes largely from a different sort of energy harvesting, which you'll learn more about shortly. But even here, the essential *product* of energy harvesting is ATP. Keep that in mind, and you won't get lost in the byways of energy transfer. How does the body produce enough ATP to allow us to read a book or climb a mountain? In this chapter, you'll find out.

7.1 Energizing ATP

Given the importance of ATP to this story, let us begin with a brief review of how this molecule works. Recall from Chapter 6 that each ATP molecule has three phosphate groups and that ATP powers a given reaction by losing the outermost of these phosphate groups. In this process, ATP is transformed into a molecule with *two* phosphate groups, adenosine diphosphate, or ADP. To return to its more "energized" ATP state, it must have a third phosphate group attached again. This is, however, a trip *up* the energy hill because the product (ATP) contains more energy than the reactants (ADP + a phosphate group). Thus, getting that third phosphate group onto ADP requires energy. It is like preparing a spring-loaded mousetrap for action. It takes energy to pull a mousetrap bar back, and the same is true of putting a third phosphate group onto ATP (**Figure 7.1**). Where does the energy come from? For animals such as ourselves, it comes from food; energy that is extracted from food powers the phosphate group up the energy hill—and literally onto ADP.

In tracking the story of *how* energy goes from food to ATP, we will use as an example one particular food molecule, glucose, to see how energy is harvested from it. Although the details here are complex, the essential story is simple. Electrons derived from glucose, which is high in energy, will be running

downhill; they will be channeled off, a few at a time, and their downhill drop will power the *uphill* push needed to attach a phosphate group onto ADP. The glucose-derived electrons will be transferred to several intermediate molecules in their downhill journey, but the *final* molecule that will receive them at the bottom of the energy hill is oxygen. We need to breathe because we need oxygen to serve as this final electron acceptor.

As you are following this process, you may well wonder why so many steps are required to transfer energy from food to ATP. Why isn't there just one step in energy transfer—food to ATP? It turns out that the gradual transfer of energy, through many steps, allows the body to make the most of the energy it receives. Think of it this way. You could have water drop 1,000 feet straight off a cliff onto a waterwheel below. In this case, the wheel would simply be pulverized by the water crashing onto it. In the process, the potential energy that had been contained in the water at the top of the cliff would be transformed almost entirely into useless heat. Conversely, you could have a stream that ran briskly downhill for 1,000 vertical feet with waterwheels spaced periodically to channel the energy off a little at a time, thereby conserving the energy for work. The steps you are going to read about amount to this kind of controlled channeling off of energy—the energy that's contained in food.

Redox Reactions

The basis for electron transfers down the energy hill is straightforward: Some substances more strongly attract electrons than do others. A substance that loses one or more electrons to another is said to have undergone **oxidation.** We hear the related word *oxidized* all the time—when paint on an outdoor surface has become dulled, for example, or when metal rusts (**Figure 7.2**). Meanwhile, the substance that *gains* electrons in this reaction is said to have undergone **reduction.** (This seems about as logical as saying a

MB

Build a Chemical Cycling System

Figure 7.1
Storing and Releasing Energy

Adenosine triphosphate (ATP) is the most important energy-dispensing molecule in our bodies. The energy it contains is used to power everything from muscle contraction to thinking.

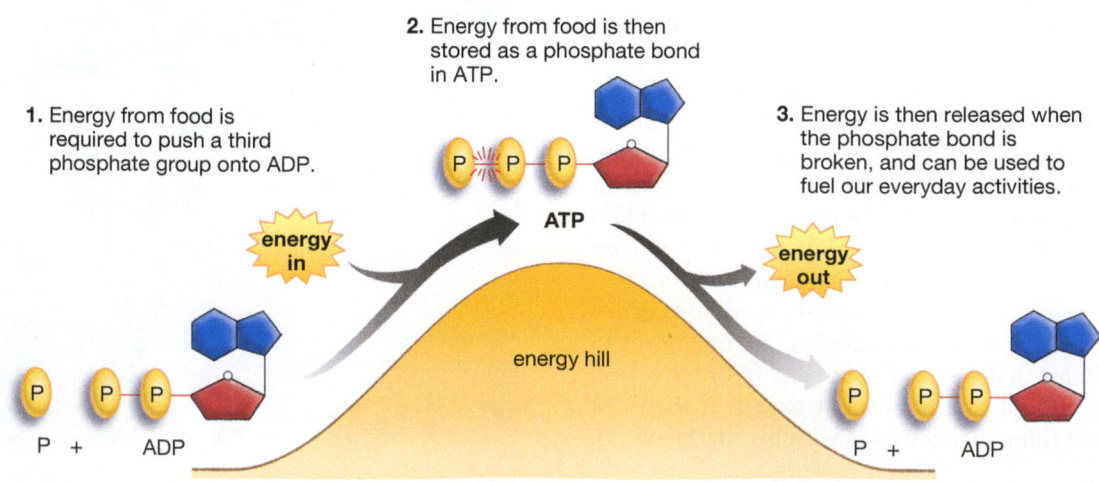

1. Energy from food is required to push a third phosphate group onto ADP.

2. Energy from food is then stored as a phosphate bond in ATP.

3. Energy is then released when the phosphate bond is broken, and can be used to fuel our everyday activities.

energy in

ATP

energy out

energy hill

P   +   ADP

P   +   ADP

Figure 7.2
An Effect of Oxidation
Rust has formed on a hook atop a corroded steel sheet.

country lost a war by winning. The way to think about it is that, because electrons carry a negative charge, any substance that gains electrons has had a reduction in its *positive charge*.)

In cells, oxidation and reduction never occur independently. If one substance is oxidized, another must be reduced. It's like a teeter-totter; if one side goes down, the other must come up. The combined operation is known as a reduction-oxidation reaction, or simply a **redox reaction:** the process by which electrons are transferred from one molecule to another. Critically, the substance oxidized in a redox reaction has its electrons traveling energetically downhill.

Many Molecules Can Oxidize Other Molecules

The term *oxidation* might give you the idea that oxygen must be involved in any redox reaction, but this is not the case. Any compound that serves to pull electrons from another is a so-called oxidizing agent. In living things, a large number of molecules are involved in energy transfer, and each has a certain tendency to gain or lose electrons relative to the others.

This is how electrons can be passed down the energy hill: The starting, energetic molecule of glucose is oxidized by another molecule, which in turn is oxidized by the *next* molecule down the hill. The whole thing might be thought of as a kind of downhill electron bucket brigade. Molecules that serve to transfer electrons from one molecule to another in ATP formation are known as **electron carriers.** The thing that makes their role a little complicated is that many of the electrons they accept are bound up originally in hydrogen atoms. You may remember from Chapter 3 that a hydrogen atom amounts to one proton and one electron. In transferring a hydrogen atom, then, a molecule is transferring a single electron (bound to a proton), which means a redox reaction has taken place.

Redox through Intermediates: NAD

The most important electron carrier in energy transfer is a molecule known as **nicotinamide adenine dinucleotide,** or **NAD,** which can be thought of as a city cab that can exist in two states: loaded with passengers or empty. And like a cab, it can switch very easily between those two states. The passengers that NAD picks up and drops off are electrons (**Figure 7.3**).

The "empty" state that NAD comes in is ionic: NAD^+. Remembering the discussion about ions, you

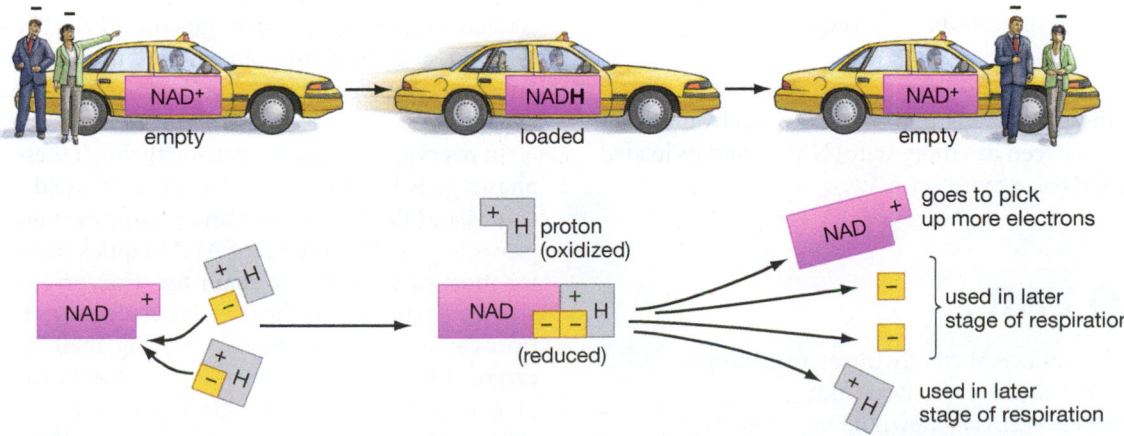

empty   loaded   empty

goes to pick
up more electrons

used in later
stage of respiration

used in later
stage of respiration

1. NAD^+ within a cell, along with two hydrogen atoms that are part of the food that is supplying energy for the body.

2. NAD^+ is reduced to NAD by accepting an electron from a hydrogen atom. It also picks up another hydrogen atom to become NADH.

3. NADH carries the electrons to a later stage of respiration then drops them off, becoming oxidized to its original form, NAD^+.

Figure 7.3
The Electron Carrier NAD^+

In its unloaded form (NAD^+) and its loaded form (NADH), this molecule is a critical player in energy transfer, picking up energetic electrons from food and transferring them to later stages of respiration.

can see that NAD$^+$ is positively charged, meaning that it has fewer electrons than protons. In a redox reaction, what NAD$^+$ does is pick up, in effect, one hydrogen atom (an electron and a proton) and one solo electron (from a second hydrogen atom). The isolated electron that NAD$^+$ picks up turns it from positively charged to neutral: NAD$^+$ → NAD; the whole hydrogen atom takes it from NAD to NADH. Keeping an eye on redox reactions here, NAD$^+$ has become NADH by oxidizing a substance—by accepting electrons from it.

So much for half of NAD's role: picking up passengers. *Now,* as NADH, it is loaded with these passengers. It can proceed down the energy hill to donate them to molecules that have a greater potential to accept electrons than it does. Having dropped its passengers off with such a molecule, it returns to being the empty NAD$^+$ and is ready for another pickup. Through this process, NADH transfers energy from one molecule to another.

How Does NAD Do Its Job?

Bioflix: Cellular Respiration

The molecule that NAD$^+$ is oxidizing is the starting glucose molecule (or actually derivatives of it). To carry out its oxidizing role, NAD$^+$ obviously needs to be brought together with the glucose derivatives. What have you been looking at that brings substances together in this way? Enzymes! One of the things you'll be seeing, therefore, is NAD$^+$ and glucose derivatives brought together by enzymes, at which point NAD$^+$ accepts electrons from these derivatives, later to hand them off to another molecule. Electron transfer through intermediate molecules such as NAD$^+$ provides the energy for most of the ATP produced. Thus, when looking at the diagrams in this chapter that outline respiration, you'll see:

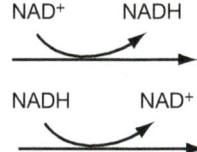

What these drawings indicate is the electron carrier shifting between its empty state (NAD$^+$) and its loaded state (NADH) or vice versa.

SO FAR...

1. The essence of energy transfer in animals is that electrons derived from _____ run energetically downhill to power the uphill process by which a third _____ is put onto the body's energy-dispensing molecule _____.

2. A substance that loses one or more electrons to another is said to have undergone _____, while the substance that received the electrons in this reaction has undergone _____. The entire, coupled reaction is called a _____ reaction.

3. The electrons derived from food are transferred to the later stages of energy harvesting by molecules known as _____, the most important of which is one called _____ in its "empty" form and _____ in its "loaded" or reduced form.

7.2 The Three Stages of Cellular Respiration

All the energy players are finally in place: glucose, redox reactions, electron carriers, enzymes, and the rest. In terms of a molecular formula, the big picture on respiration looks like this:

$$C_6H_{12}O_6 + 6O_2 + 36ADP + 36P \rightarrow$$
$$6CO_2 + 6H_2O + 36ATP$$

The starting molecule on the first row is glucose, which is food, storing chemical energy. You can also see, next to it, the oxygen ($6O_2$) that is needed as the final electron acceptor. Then there is ADP, the two-phosphate molecule; and inorganic phosphate molecules (the 36P), which are pushed on ADP to make it ATP. The *products* of this reaction, on the second row, are carbon dioxide ($6CO_2$) and water ($6H_2O$) as by-products and energetic ATP molecules. How *much* ATP does the breakdown of one molecule of glucose yield? The answer is contained in the formula: a maximum of 36 molecules.

Although many separate steps are involved in actually getting energy from glucose to ATP, respiration can be divided into three main phases. These are known as *glycolysis,* the *Krebs cycle,* and the *electron transport chain.*

In overview, energy harvesting through these three phases goes like this: We get some ATP yield in glycolysis and the Krebs cycle, and glycolysis is useful in providing small amounts of ATP in quick bursts. But for most organisms, the main function of glycolysis and the Krebs cycle is the transfer of electrons to electron carriers such as NAD$^+$. Bearing their electron cargo, these carriers then move to the third phase of energy harvesting, the electron transport chain (ETC). Here, the electron carriers are themselves oxidized, and the resulting movement of their electrons through the ETC provides the main harvest of ATP. Indeed, about 32 of the approximately 36 molecules of ATP that are netted per glucose molecule are

(a) In metaphorical terms

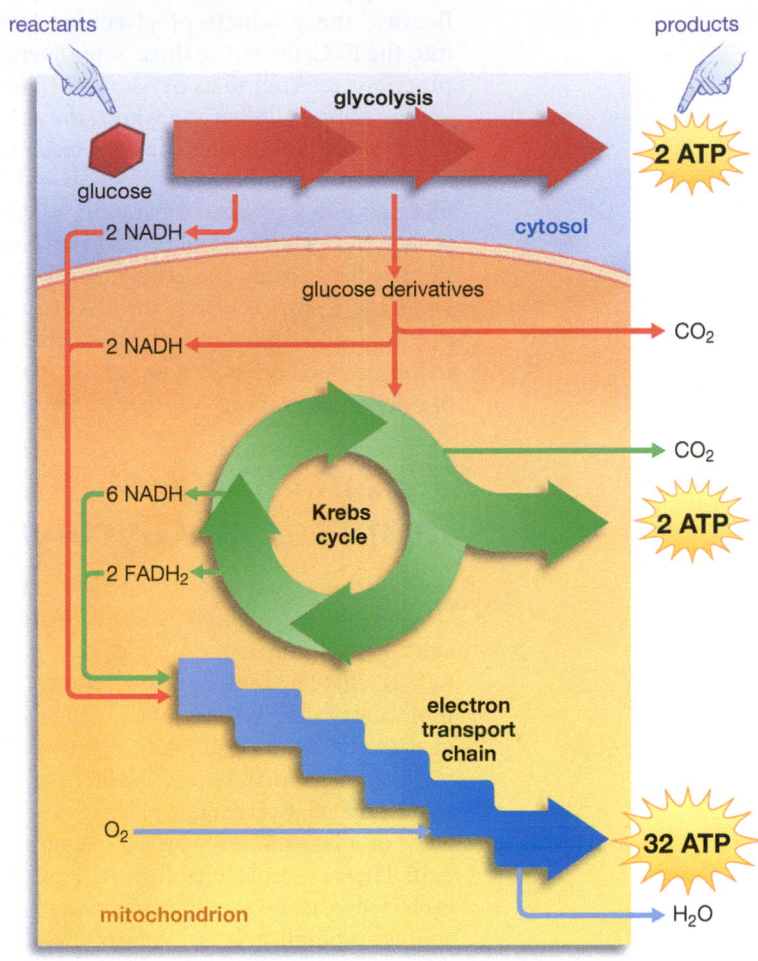

(b) In schematic terms

Just as the video games in some arcades can use only tokens (rather than money) to make them function, so our bodies can use only ATP (rather than food) as a direct source of energy. The energy contained in food—glucose in the example—is transferred to ATP in three major steps: glycolysis, the Krebs cycle, and the electron transport chain. Though glycolysis and the Krebs cycle contribute only small amounts of ATP directly, they also contribute electrons (on the left of the token machine) that help bring about the large yield of ATP in the electron transport chain. Our energy-transfer mechanisms are not quite as efficient as the arcade machine makes them appear. At each stage of the conversion process, some of the original energy contained in the glucose is lost to heat.

As with the arcade machine, the starting point in this example is a single molecule of glucose, which again yields ATP in three major sets of steps: glycolysis, the Krebs cycle, and the electron transport chain (ETC). These steps can yield a maximum of about 36 molecules of ATP: 2 in glycolysis, 2 in the Krebs cycle, and 32 in the ETC. As noted, however, glycolysis and the Krebs cycle also yield electrons that move to the ETC, aiding in its ATP production. These electrons get to the ETC via the electron carriers NADH and $FADH_2$, shown on the left. Oxygen is consumed in energy harvesting, while water and carbon dioxide are produced in it. Glycolysis takes place in the cytosol of the cell, but the Krebs cycle and the ETC take place in cellular organelles, called mitochondria, that lie within the cytosol.

obtained in the ETC. If you look at **Figure 7.4** you can see two representations of the three stages of energy extraction.

Glycolysis: First to Evolve, Less Efficient

The division among the three energy-harvesting phases is in one sense a division of evolutionary history. For most organisms, energy harvesting's first phase, glycolysis, could be likened to a nineteenth-century steam engine working side by side with a state-of-the-art electric turbine. By itself, glycolysis produces only two ATP molecules per glucose molecule, which is pretty

small in comparison to the 36 ATP molecules likely to come from all three phases. And glycolysis doesn't even yield many electrons compared to energy harvesting's second phase, the Krebs (or citric acid) cycle. Yet, glycolysis takes place in all living things—at the very least, as the first phase in energy harvesting—and for certain primitive one-celled organisms it can be the *only* phase in energy harvesting.

All of this points to glycolysis as a more ancient form of energy harvesting than the other two phases. It existed first, and then at some later time the processes we now call the Krebs cycle and the ETC developed during the course of evolution. The critical factor in this change was the use of oxygen in energy

Figure 7.4

Overview of Energy Harvesting

Overview of Cellular Respiration

harvesting—specifically the use of oxygen in the ETC. Because the products of glycolysis feed ultimately into the ETC, the entire three-stage energy-harvesting process is referred to as oxygen-dependent or *aerobic* energy transfer. Taken as a whole, the aerobic harvesting of energy is known as **cellular respiration.**

The separation between glycolysis and the other two phases is also a physical separation. In eukaryotic cells such as ours, glycolysis takes place in the watery material that lies outside the cell's nucleus—the cytosol—while the Krebs cycle and the ETC take place within tiny organelles, called mitochondria, that are immersed in the cytosol. Let's start now to look at all three phases of cellular respiration.

7.3 First Stage of Respiration: Glycolysis

Glycolysis

Glycolysis, the first stage of energy harvesting, means "sugar splitting," which is appropriate because our starting molecule is the simple sugar glucose. Briefly, here is what happens during glycolysis. First, this one molecule of glucose has to be prepared, in a sense, for energy release. ATP actually has to be used, rather than synthesized, in two of the first steps of glycolysis, so that the relatively stable glucose can be put into the form of a less-stable sugar. This sugar is then split in half. The two molecules formed have three carbons each (whereas the starting glucose had six). Once this split is accomplished, the steps of glycolysis take place in duplicate; it is like taking one long piece of cloth, dividing it in two, and then proceeding to do the same things to both of the resulting pieces. **Figure 7.5** sets forth the individual steps of glycolysis.

Now, what are the results of these steps? Glycolysis accomplishes three valuable things in energy harvesting: It yields two ATP molecules, it yields two energized molecules of NADH, and it results in two molecules of pyruvic acid. These pyruvic acid molecules are the derivatives of the original glucose molecule. Now *they* are the molecules that will be oxidized—that will have electrons removed from them—in the next stage of energy harvesting. Keeping our eye on the ball here, we have some net ATP through glycolysis, some traveling electrons that will help make more ATP in the ETC, and a derivative of the original glucose molecule (pyruvic acid) that will now move to the next stage of energy harvesting, where more energy will be extracted from it.

Before continuing on to this next stage, you may wish to read more about what might be called the consequences of glycolysis. Even oxygen-using organisms such as ourselves rely on glycolysis when short bursts of energy are needed. And, as noted, for some organisms glycolysis is the end of the line in energy harvesting. You can read about this in "When Energy Harvesting Ends at Glycolysis" on page 130. Glycolysis

fits into a special place in providing energy for human exercise; you can read about this in "Energy and Exercise" on page 136.

SO FAR . . .

1. The three main stages of oxygen-dependent or aerobic energy harvesting are _____, the _____ cycle, and the _____. The first of these takes place in the cell's _____, while the second and third take place in the cellular organelles called _____.

2. While the first two of these stages do yield some ATP, their most important contribution to energy harvesting is the channeling off of the _____ used to bring about the large harvest of ATP in _____.

3. Glycolysis nets two of energy harvesting's end-product molecules, _____, it yields two "loaded" molecules of the electron carrier _____, and it results in two molecules of the substrate _____, which are the downstream derivatives of the starting molecule of glucose.

7.4 Second Stage of Respiration: The Krebs Cycle

Since glycolysis yielded only 2 of the 36 molecules of ATP that eventually are harvested, the glucose derivative that resulted from glycolysis, pyruvic acid, still clearly has a lot of energy left in it. Some of this energy will be harvested in the second stage of cellular respiration, the **Krebs cycle.** This stage is named for the German-English biochemist Hans Krebs, who in the 1930s used the flight muscles of pigeons to find out how aerobic respiration works. Because the first product of the Krebs cycle is citric acid, the cycle is sometimes referred to as the **citric acid cycle.** The ATP yield in this cycle is once again a paltry 2 molecules, but the Krebs cycle serves to set the stage for the big harvest of ATP in the ETC by stripping our molecules of pyruvic acid of their energetic electrons and sending them to the ETC, as you'll see.

Site of Action Moves from the Cytosol to the Mitochondria

Glycolysis yielded its two molecules of pyruvic acid in the cytosol, but the site of energy harvesting quickly shifts, with the pyruvic acid, to the new site of energy transfer, the mitochondria. These organelles are where

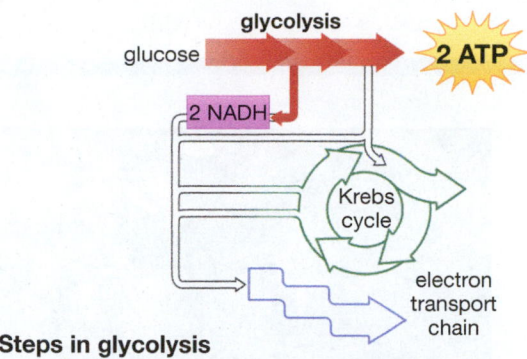

Steps in glycolysis

1. Delivered by the bloodstream, glucose enters a cell and immediately has a phosphate group from ATP attached to it. Because this process, called phosphorylation, attaches the phosphate to the sixth carbon of glucose, it now goes under the name glucose-6-phosphate. Note that one molecule of ATP has been *used* in this step.

   ATP ledger now reads: -1 ATP

2. Glucose 6-phosphate is rearranged to become a molecule called fructose-6-phosphate.

3. Another molecule of ATP is used to add a second phosphate to fructose-6-phosphate, which now becomes fructose-1,6-diphosphate.

   ATP ledger now reads: -2 ATP

4. The single, six-carbon sugar fructose-1,6-diphosphate now becomes two molecules of a 3-carbon sugar, glyceraldehyde-3-phosphate, each with a phosphate group attached. From here on out, glycolysis happens in duplicate: What happens to one of the glyceraldehyde molecules happens to the other.

5. An enzyme brings together glyceraldehyde-3-phosphate, the electron carrier NAD⁺, and a phosphate group. The glyceraldehyde-3-phosphate molecule is oxidized by NAD⁺, which in its new form, NADH, moves to the electron transport chain bearing its electron cargo. The oxidation of NAD⁺ is energetic enough that it allows the phosphate group to become attached to the main molecule, now called 1,3-diphosphoglyceric acid. Because everything is happening in duplicate, two NADH molecules are produced.

6. 1,3-diphosphoglyceric acid loses one of its phosphate groups, thus becoming 3-phosphoglyceric acid. The reaction is energetic enough to push this phosphate group onto an ADP molecule, yielding ATP. Because of duplication, two ATP are produced.

   ATP ledger now reads: 0 ATP

7. In two reactions, 3-phosphoglyceric acid becomes phosphoenolpyruvic acid, which generates more ATP as it transfers its phosphate group to ADP. Two more ATP molecules are produced. The phosphate transfer turns phosphoenolpyruvic acid into pyruvic acid—the derivative of the original glucose that now will enter the Krebs cycle.

   ATP ledger now reads: +2 ATP

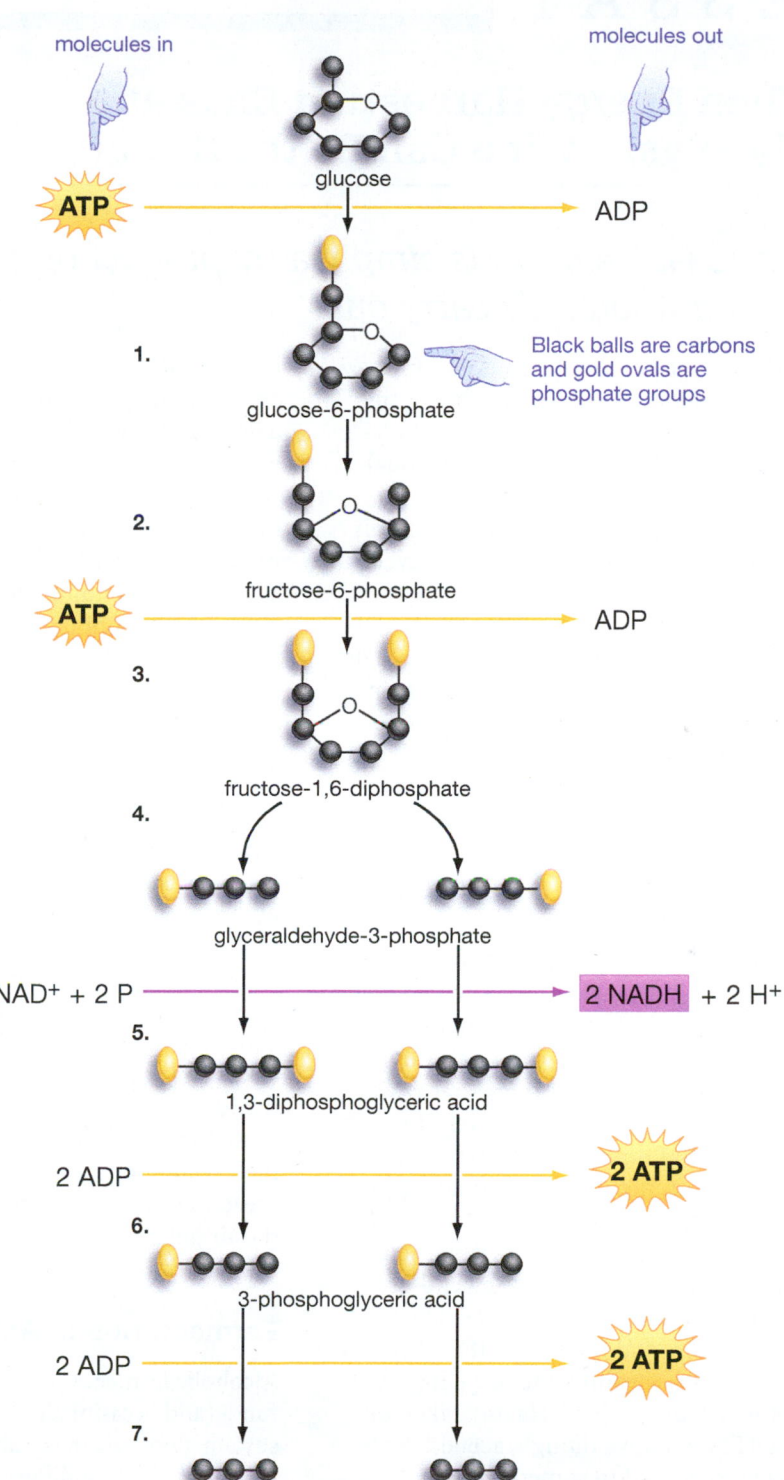

Figure 7.5
Glycolysis

In glycolysis, the single glucose molecule is transformed in a series of steps into two molecules of a substance called pyruvic acid. These two molecules then move on to the next stage of cellular respiration (the Krebs cycle). Glycolysis also produces two molecules of electron-carrying NADH, which move to the electron transport chain. Although two molecules of ATP are used in the early stages of glycolysis, four are produced in the later stages, for a net production of two ATP.

ESSAY

When Energy Harvesting Ends at Glycolysis, Wine Can Be the Result

For yeasts, alcohol is simply a by-product of the fermentation they carry out.

For some bacteria—and even sometimes for people—energy harvesting ends with the set of steps called glycolysis rather than proceeding on through the Krebs cycle and oxygen-dependent electron transport chain (ETC). When this happens, organisms need a way to recycle their energy transfer molecules in a way that doesn't depend on oxygen. The NADH they produced in glycolysis needs to lose its added electrons and become NAD^+ again, so it can be reused in glycolysis. The solution to this dilemma is a kind of electron dumping, the products of which turn out to be some of the most familiar substances in the world.

Fermentation in Yeast

Yeasts provide a good example of the way in which this works because yeasts are busy single-celled fungi that can live by glycolysis alone, or that can go through aerobic respiration. Say, then, that yeasts are working away on sugar, going through glycolysis, but doing so in an oxygenless environment. Recall that the final "substrate" product of glycolysis is two molecules of a substance called pyruvic acid. In yeasts, the pyruvic acid they end up with is converted to a molecule called acetaldehyde, and *it* takes on the electrons from NADH, meaning the recycling problem is now solved. Having taken on NADH's electrons, though, acetaldehyde now is converted to something very familiar: ethanol, better known as drinking alcohol.

Human beings *put* yeasts in environments in which this will happen, of course,

because this is how we make wine and beer. Imagine yeast cells inside a dark, airless wine cask, working away on burgundy grape juice, harvesting energy from the grape juice sugars, but in so doing turning out pyruvic acid, which is turned into alcohol. This continues until the alcohol content of the wine reaches a level (about 14 percent) at which the yeast can no longer survive in it. This whole process is known as *alcoholic fermentation,* the process by which yeasts produce alcohol as a by-product of glycolysis they perform in an oxygenless environment.

For yeasts, alcohol is simply a waste by-product of the glycolysis they employ in *anaerobic* energy conversion, meaning conversion in an oxygenless environment. The same is true of the carbon dioxide that also results from their conversion of pyruvic acid. Here again, humans have found great use for the refuse of nature. Put the right yeast into dough, and its continuing fermentation produces the CO_2 that causes bread to rise and become "light" because of the air holes now in it. (And the alcohol? It evaporates.)

Fermentation in Animals

Alcoholic fermentation is the process that fungi (and occasionally plants) employ to sustain glycolysis in the absence of oxygen. But animals take a different tack, because in them the product of glycolysis, pyruvic acid, accepts the electrons from NADH. In this process, pyruvic acid is turned into another substance: *lactic acid.* Thus, the

Yeast Inside, Performing Fermentation
This man is inspecting wine that is fermenting in a barrel. The alcohol in the wine is a by-product of the glycolysis carried out by yeast in an oxygenless environment.

kind of fermentation that animals (and certain bacteria) carry out is called *lactate fermentation.* Have you ever experienced the muscle "burn" that comes with, for example, climbing several flights of stairs? If so, you have experienced a buildup of lactic acid in your muscles.

But why should this happen? Why should glycolysis ever end in lactate fermentation for us, when, unlike yeast, we always have access to oxygen? The problem turns out not to be a matter of oxygen access, but a matter of oxygen *delivery.* During a quick burst of energy use, oxygen cannot be delivered into our muscle cells fast enough to accommodate the big increase in our energy expenditure. When a muscle's capacity for aerobic energy transfer has reached its limit, it turns to glycolysis and lactate fermentation to supply more ATP. Note, however, that glycolysis and lactate fermentation are limited, short-term means of supplying energy in animals such as ourselves. These processes do not supply enough energy to sustain us for long; their role is to supply enough energy to meet our needs until such time as oxygen-aided energy transfer can ramp itself up.

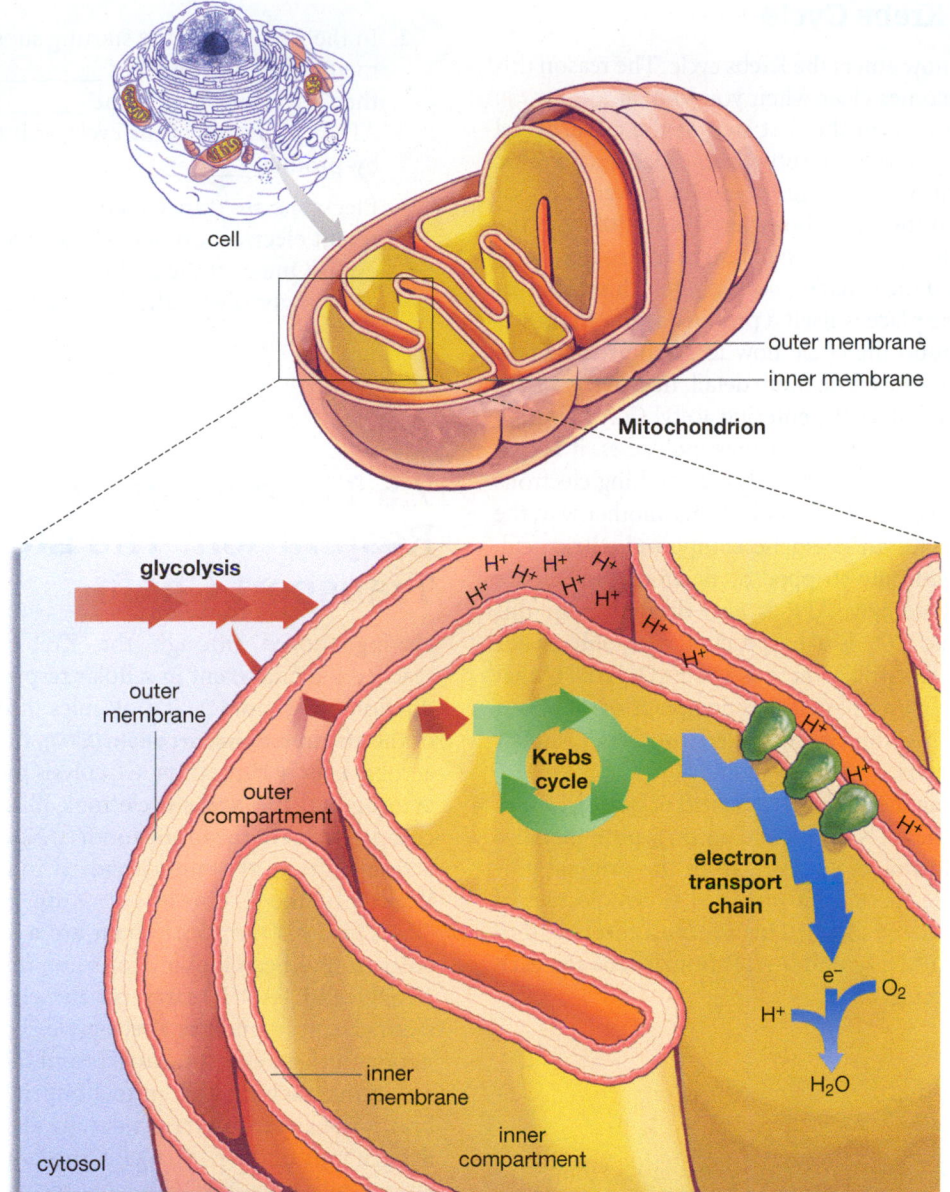

Figure 7.6
Energy Transfer in the Mitochondria
Mitochondria are organelles, or "tiny organs," that exist within cells. They are the location for the second and third sets of steps in cellular respiration, the Krebs cycle and the electron transport chain, respectively.

all the action takes place from here on out, so now would be a good time to look at **Figure 7.6** and examine their structure. You can see that mitochondria have both an inner and an outer membrane. The Krebs cycle reactions take place to the interior of the inner membrane—in an area known as the inner compartment—while the reactions of the ETC take place *within* the inner membrane.

Between Glycolysis and the Krebs Cycle, an Intermediate Step

There actually is a transition step in respiration between glycolysis and the Krebs cycle. In it, the three-carbon pyruvic acid molecule combines with a substance called coenzyme A, thus forming the substrate acetyl coen-

zyme A (acetyl CoA). One outcome of this reaction is that acetyl CoA enters the Krebs cycle as the derivative of the original glucose molecule. There are, however, two other products of this reaction. One is a carbon dioxide molecule that, like others you'll be seeing in the Krebs cycle, eventually diffuses out of the cell and into the bloodstream. This is conceptually an important point because *this* is the source of the carbon dioxide that we exhale with each breath. This CO_2 is officially a by-product of cellular respiration, but we could think of it as a waste product—a kind of exhaust that comes from the burning of food. The third product of this reaction is one more molecule of NADH, which finds its way to the ETC. This reaction takes place twice for each glucose molecule, as you can see in **Figure 7.7**, on the next page.

Fermentation

The Citric Acid Cycle

Into the Krebs Cycle

Acetyl CoA now enters the Krebs cycle. The reason this is a *cycle* becomes clear when you look at **Figure 7.8**. You can see that, in the first step of the cycle, acetyl CoA combines with a substance called oxaloacetic acid to produce citric acid. If you look over at about 11 o'clock on the circle, however, you can see that the last step in the cycle is the *synthesis* of oxaloacetic acid. Thus, a substance that's necessary for this chain of events to take place is itself a product of the chain.

Stroll around the circle now to see how the Krebs cycle functions in a little more detail. In essence, what's happening is that, as the entering acetyl CoA molecule is transformed into these various molecules, it is oxidized by electron carriers, with the resulting electrons moving on to the ETC. To look at this another way, the cycle starts with an energetic compound, citric acid, that is methodically stripped of electrons (and hence energy). Accordingly, ATP is also derived from this process, while CO_2 is a by-product. The only major player that's unfamiliar here is the FAD–$FADH_2$ that can be seen taking part in a redox reaction at about 8 o'clock. It is simply another electron carrier, similar to NAD^+/NADH.

In counting the Krebs cycle's yield of both ATP and electron carriers (NADHs and so on), note that *two* turns around this cycle result from the original glucose molecule (because it provided two molecules of the starting acetyl CoA). This means a total yield of 6 NADH, 2 $FADH_2$, and 2 ATP from Krebs for each glucose molecule.

Figure 7.7
Transition between Glycolysis and the Krebs Cycle

The pyruvic acid product of glycolysis does not enter directly into the Krebs cycle. Rather, it must first be transformed into acetyl coenzyme A. The consequences of this reaction are the production of CO_2, which diffuses into the bloodstream, and the production of an NADH molecule, which continues onto the electron transport chain.

SO FAR . . .

1. In aerobic respiration, the pyruvic acid that is the product of glycolysis is converted to the molecule _____, which is the substrate that is transferred to the Krebs cycle.

2. In the Krebs cycle, the starting substrate is methodically stripped of _____, which then are transferred to the _____. Some ATP is produced in the cycle, as is the by-product _____.

3. Electrons are transferred to two different electron carriers in the Krebs cycle. One of these is the _____ carrier seen in glycolysis; the other is the carrier _____.

7.5 Third Stage of Respiration: The Electron Transport Chain

Having moved through the Krebs cycle, you've reached the final event in cellular respiration: the production of 34 more ATP molecules through the work of the **electron transport chain (ETC)**, the third stage of aerobic energy harvesting. Glycolysis took place in the cytosol, and the Krebs cycle took place in the inner compartment of the mitochondria. Now, however, the action shifts to the mitochondrial inner membrane, the site of the ETC (**Figure 7.9**, on page 134). The "links" in this transport chain are a series of molecules. You've been looking for some time now at how NADH and $FADH_2$ carry off the electrons derived from glycolysis and the Krebs cycle. This is the destination of these electron carriers and their cargo.

Upon reaching the mitochondrial inner membrane, the electron carriers donate electrons and hydrogen ions to the ETC. Again, this is a trip down the energy hill, only *NADH* is now the molecule whose electrons exist at a higher energy level. Thus, when NADH runs into the appropriate enzyme in the ETC, NADH is oxidized, donating its electrons and proton to the ETC. This process then is simply repeated down the whole ETC, each carrier donating electrons to the next electron carrier in line. A carrier is reduced by receiving these electrons; donating them to the next carrier, it then returns to an oxidized state. Each carrier thus alternates between oxidized and reduced states.

Looking at Figure 7.9, you can see the carriers in the ETC: There are three large enzyme complexes with two smaller mobile molecules that link them. When NADH arrives at the inner membrane, it bumps into the ETC's first carrier and donates electrons to it; this carrier then donates these electrons to the next carrier, and so on down the line to the last electron acceptor, which is oxygen in the inner compartment (note the "$\left(\frac{1}{2}O_2\right)$," which will be explained shortly).

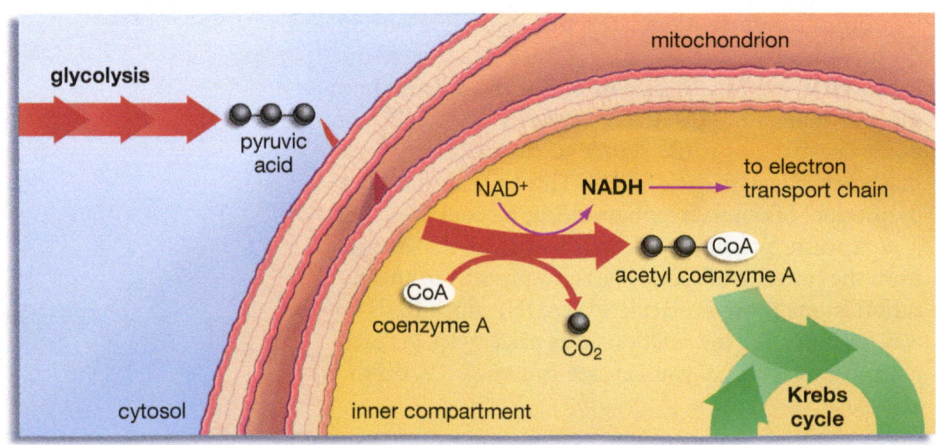

glycolysis

pyruvic acid

mitochondrion

NAD^+

NADH → to electron transport chain

CoA

acetyl coenzyme A

CoA

coenzyme A

CO_2

Krebs cycle

cytosol

inner compartment

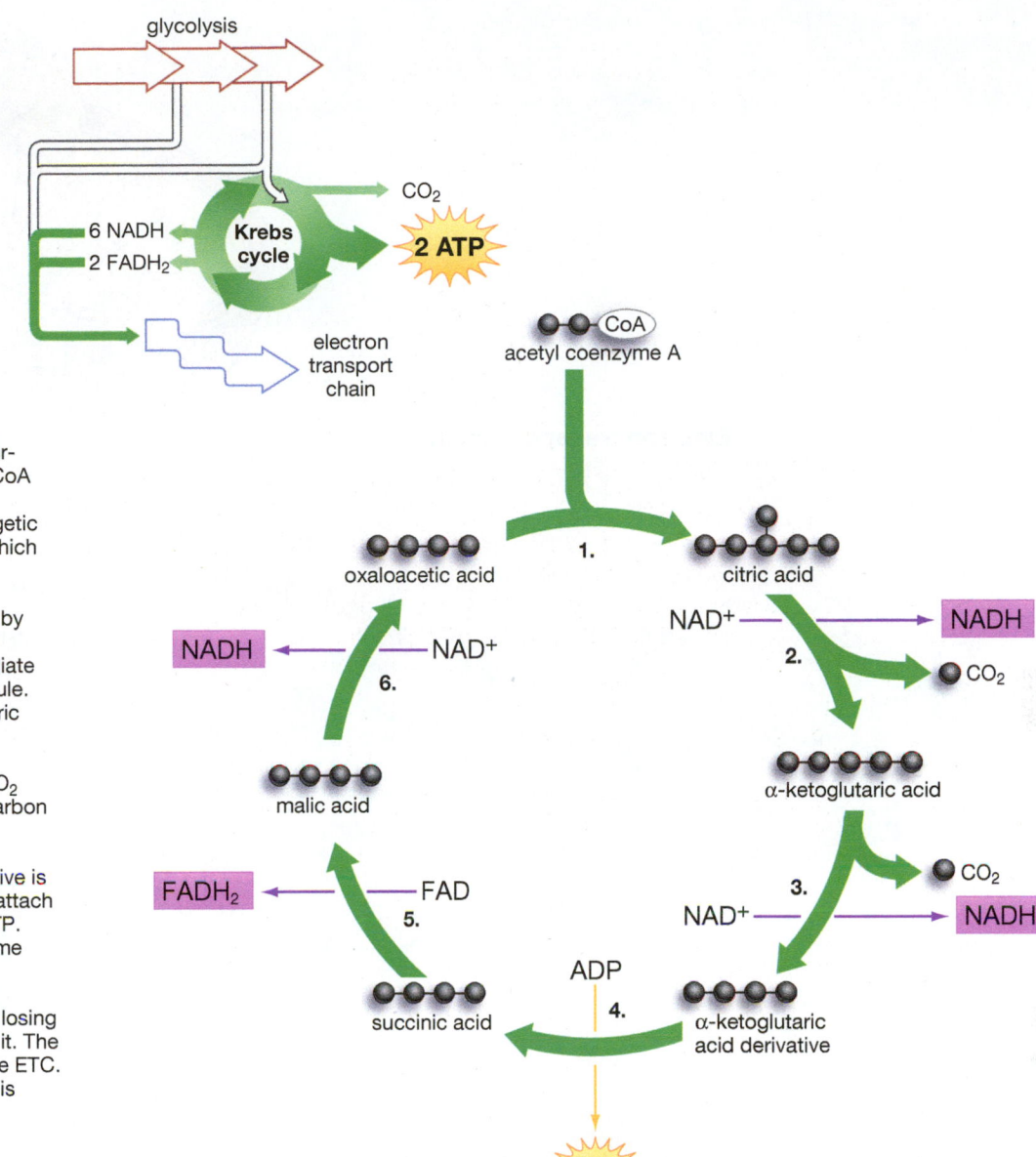

Steps in the Krebs cycle

1. Acetyl CoA combines with the four-carbon oxaloacetic acid and the CoA fragment separates from this compound. The result is the energetic six-carbon molecule citric acid, which will now be oxidized.

2. A citric acid derivative is oxidized by NAD^+; the resulting NADH carries electrons to the ETC. An intermediate molecule then loses a CO_2 molecule. Citric acid is now alpha-ketoglutaric acid.

3. Alpha-ketoglutaric acid loses a CO_2 molecule and the resulting four-carbon molecule is oxidized by NAD^+.

4. An alpha-ketoglutaric acid derivative is split, releasing enough energy to attach a phosphate to ADP, making it ATP. Alpha-ketoglutaric acid has become succinic acid.

5. Succinic acid is oxidized by FAD, losing two complete hydrogen atoms to it. The resulting $FADH_2$ then moves to the ETC. In a series of steps, succinic acid is transformed into malic acid.

6. Malic acid is oxidized by NAD^+. This step transforms malic acid to oxaloacetic acid—the molecule that enters the first step of the Krebs cycle.

Figure 7.8
The Krebs Cycle
The Krebs cycle is the major source of electrons that are transported to the electron transport chain by the carriers NADH and $FADH_2$. For each molecule of acetyl coenzyme A that enters the cycle, three NADH molecules, one $FADH_2$, and one ATP are produced. Because two acetyl CoA molecules enter per molecule of glucose, the yield per glucose is six NADH, two $FADH_2$, and two ATP.

Where's the ATP?

At this point, you may be tapping your foot, waiting for this promised harvest of ATP to appear. So far, all that's taken place is a transfer of electrons in the ETC. Well, note what happens in the first ETC enzyme complex. The movement of electrons through it releases enough energy to power the movement of *hydrogen ions* (H^+ ions) through the complex, pushing them from the inner compartment into the outer compartment. (The fall of the electrons causes the enzyme complex to change its shape in a way that facilitates the passage of H^+ ions through it and into the outer compartment.) By the time the electrons have completed their movement through the ETC, this pumping of H^+ ions will occur with the two other large enzyme complexes. Critically, these H^+ ions are being pumped *against* their concentration and electrical gradients. Put another way, the ions are being pumped up the energy hill with energy supplied by the *downhill* fall of electrons through the ETC.

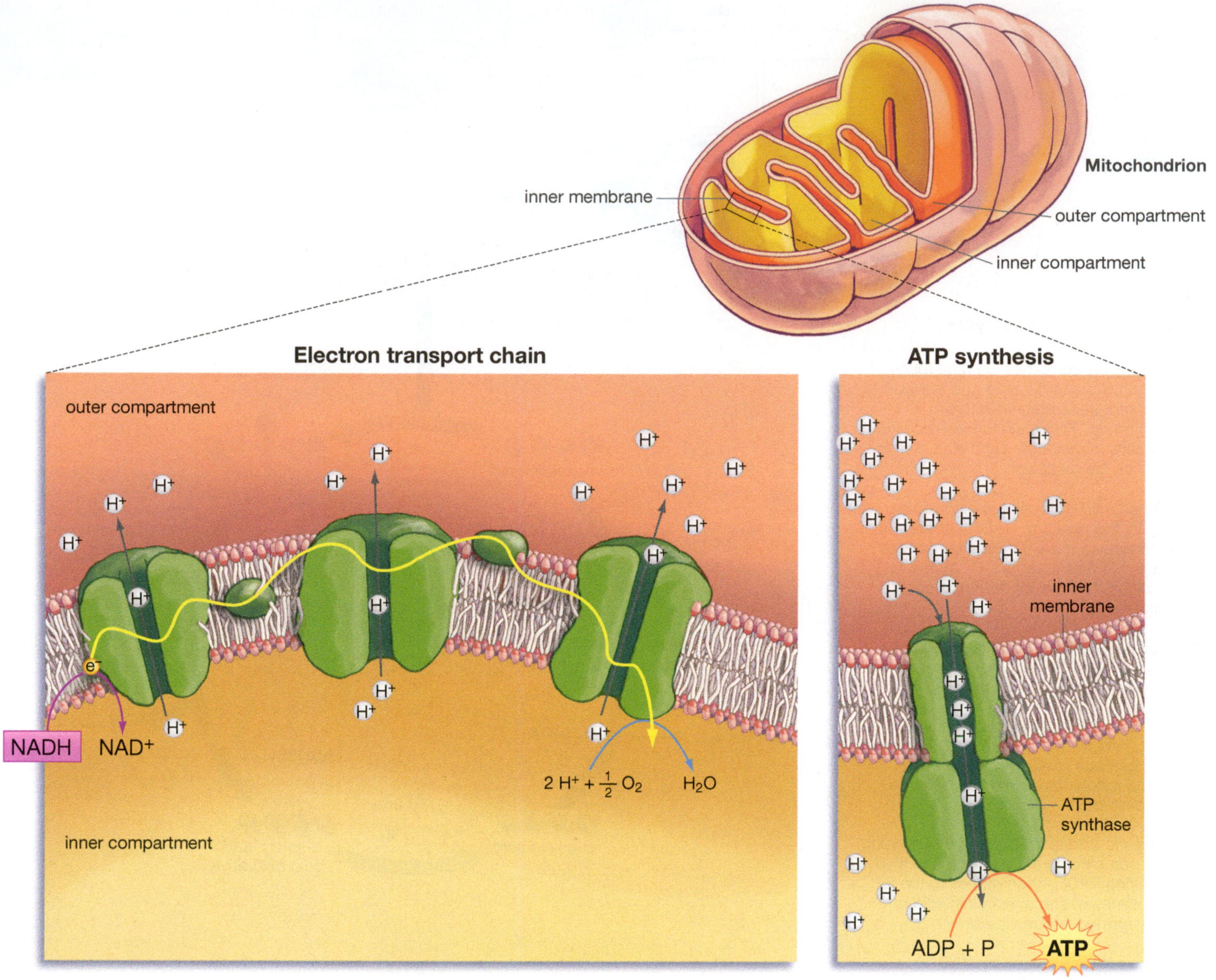

Mitochondrion

inner membrane

outer compartment

inner compartment

Electron transport chain

outer compartment

H+

NADH   NAD+

e⁻

$2\,H^+ + \frac{1}{2}\,O_2$ H_2O

inner compartment

ATP synthesis

inner membrane

ATP synthase

ADP + P   **ATP**

Figure 7.9
The Electron Transport Chain (ETC)

The movement of electrons through the ETC powers the process that provides the bulk of the ATP yield in respiration. The electrons carried by NADH and FADH₂ are released into the ETC and transported along its chain of molecules. The movement of electrons along the chain releases enough energy to power the pumping of hydrogen ions (H⁺) across the membrane into the outer compartment of the mitochondrion. It is the subsequent energetic "fall" of the H⁺ ions back into the inner compartment that drives the synthesis of ATP molecules by the enzyme ATP synthase.

Electron Transport

Here's where the ATP is produced at last. The H⁺ ions that have been pumped into the outer compartment now move back down their concentration and energy gradients, into the inner compartment, through a special enzyme called ATP synthase. This remarkable enzyme is driven by the H⁺ ions flowing through it, which cause part of the enzyme to rotate (as fast as 100 revolutions per second). We could think of it as being something like a waterwheel, except that its movement is driven not by water, but by the movement of H⁺ ions that are passing through it, on their way back into the inner com-

partment. It is this energetic spinning that puts a third phosphate group (P) onto ADP, thus making ATP. **ATP synthase** can thus be defined as an enzyme that functions in cellular respiration by bringing together ADP and inorganic phosphate molecules to produce ATP.

The basic mechanism here—the pumping of hydrogen ions powering the synthesis of ATP—was first proposed in 1961 by British biochemist Peter Mitchell, who won the Nobel Prize in 1978 for his contribution to bioenergetics. Lest we pass by it too quickly, note that in this mechanism, we can see the essence of aerobic respiration. In the links of the ETC, the downhill drop of electrons derived from food has powered the uphill synthesis of the molecule that supplies the energy for the vast majority of our activities.

The Reduction of Oxygen, the Production of Water

The story of the breakdown of glucose—the saga of *one molecule's* transformation!—is nearly complete. The only unfinished business lies at the end of the ETC. As noted earlier, oxygen is the final acceptor of the electrons that move through the ETC. In the mitochondrial inner compartment, a single oxygen atom $\left(\frac{1}{2}O_2\right)$ accepts two electrons from the last enzyme complex in the ETC and two H^+ ions. These then come together to produce H_2O: water, which is like CO_2 in that it's a by-product of cellular respiration.

This process for dealing with the electrons that made their way through respiration may sound like some minor housekeeping detail, but it's not. The electrons that are donated *to* the ETC by NADH and $FADH_2$ must be taken *from* the ETC somehow. To understand why, imagine a factory that makes, say, refrigerators. Now imagine that these refrigerators move down an assembly line on metal platforms, but that the platforms are suddenly allowed to start piling up at the end of the assembly line. Once this pile grows large enough, nothing can come off the end of the assembly line—indeed, the whole thing grinds to a halt. Just so, the body has an assembly line that results in both a product (ATP) and some materials that are used in manufacturing it (our de-energized electrons). Should these electrons not be taken away by oxygen, they would remain in place in the ETC—in its last enzyme complex. Should this happen, the complex would cease to accept electrons from the *previous* complex up the chain, and this process would continue until the whole assembly line would be "backed up" all the way to glycolysis. Some organisms (such as yeast) can get by with just the small amount of ATP that glycolysis provides, but this is not the case for large organisms such as ourselves. We must have the larger amount of energy provided by aerobic energy transfer. And ground zero for a *lack* of energy transfer in us is the brain, which has cells that are voracious users of energy. Five minutes without oxygen is enough to put a human being into the condition known as a persistent vegetative state. We need to breathe because the energy assembly line will only keep moving if the electrons that are part of it are taken away by oxygen.

Bountiful Harvest: ATP Accounting

Let's think for a moment about what might be called the ATP accounting in our story. In the ETC,

each NADH molecule will be responsible for the production of three molecules of ATP. And how many NADH are coming? Ten: two produced in glycolysis, two from pyruvic acid conversion, and six from the Krebs cycle. Each $FADH_2$ carrier joins the ETC a little later than does NADH and as such is responsible for roughly two ATP. And how many $FADH_2$ are coming? Two, both of them produced in the Krebs cycle. When we put the $FADH_2$ and NADH output together, we find that 34 ATP are produced in the ETC as a result of the electrons brought by these carriers. Given this, think about what a mighty energy extractor the ETC is: 34 ATP compared to two in glycolysis. Imagine one car that got 34 miles to the gallon compared to another that got two. Our final accounting on ATP yield is the 34 from the ETC, plus the two from glycolysis, plus two from the Krebs cycle, yielding 38 in all. From this, we need to subtract two ATP that are spent transporting the NADH produced in glycolysis into the mitochondria. The final result is thus a maximum of 36 net ATP per molecule of glucose from all the steps of aerobic respiration.

7.6 Other Foods, Other Respiratory Pathways

By looking at aerobic respiration as it applies to glucose, which served as our starting food molecule, you've been able to go through the whole respiratory chain, but there are many kinds of nutrients besides glucose—for example, fats, proteins, and other sugars. All these can provide energy by being oxidized within the chain of reactions you've just gone over. However, none of them proceeds through the chain in exactly the same way as glucose. Instead, various nutrients can be channeled into the cell respiration chain in different ways.

To give one example of how this channeling of nutrients works, consider the possible fates of triglycerides. As you learned in Chapter 3, triglycerides are fats made of a three-carbon glycerol "head" and three long hydrocarbon chains that look like tines on a fork. Imagine that a given cell needs energy from fat. The question is, how do triglycerides go through aerobic respiration, as opposed to glucose?

The first thing that happens is that a triglyceride molecule is split by enzymes into its constituent parts of fatty acids and glycerol. Glycerol is then converted into glyceraldehyde phosphate. This formidable name may not ring a bell, but if you look back at step 4 in Figure 7.5, you'll see that it's one of the downstream derivatives of glucose in glycolysis. To put this another way, glycerol does not first get converted to glucose and then march through all of the steps of glycolysis. Instead, it joins the glycolytic pathway several steps down from glucose and

ESSAY

Energy and Exercise

One minute we're sleeping, the next we're looking for our running shoes, and 10 minutes after that we're bounding down the road trying to cover three miles at a faster pace than yesterday. Coming to the steep hill at the end of our run, we go all out—charging to the top of it, slowing only after we start going downhill, breathless and spent. How can the human body cope with a range of energy demands that runs from sleeping to sprinting? Its secret is having a kind of energy trio at its service, with each member of the group specializing in delivering energy in particular situations.

The essence of any kind of exercise is the contraction of muscles, and the only energy molecule that can power the contraction of skeletal muscle is ATP. You have seen that the human body uses a three-stage process to produce ATP: glycolysis, the Krebs cycle, and the electron transport chain (ETC). As a whole, this system is dependent on oxygen, but glycolysis is not directly dependent on it. For brief periods

of time, we can produce a good deal of the ATP we need through a glycolysis that is decoupled from the Krebs cycle and the ETC. Oxygen-independent glycolysis can thus be thought of as the first member of the body's energy trio, while *aerobic* or oxygen-using energy transfer can be thought of as the second member. Our cells also have a capacity to *store* small amounts of ATP, however, and they can also stockpile another molecule, phosphocreatine (PCr), that acts as a reservoir of phosphate groups that can be used to produce ATP. This stored ATP/PCr is the third member of the energy ensemble. Let's now see how all three players would work together in the course of a bike ride, starting out at a fairly brisk pace.

With the first turns of the pedals, bike riders bump up their body's energy demands tremendously. In their cycling, they could easily burn up 10 times the calories they would while in a resting state. (As a point of comparison, in doing housework a person typically burns up

about three times the calories he or she does while at rest.)

So, how do muscle cells accommodate such large increases in their need for ATP? In the first few seconds, the small reservoirs of ATP/PCr take the lead role, with a proportionately smaller contribution from glycolysis and a smaller contribution yet from aerobic metabolism. These differences reflect the time it takes for each of these systems to get going. You've seen how many steps there are in glycolysis and how many *more* steps there are in the Krebs cycle and ETC. By comparison, the output of ATP from stored ATP/PCr is instantaneous. Remember, though, that the ATP/PCr reservoir is small. A person who had to depend solely on it for a full-ahead sprint would have only enough energy to last about six seconds. On a longer bike ride, glycolysis and ATP/PCr are contributing roughly equal amounts of ATP within 10 seconds. Within 30 seconds, glycolysis has greatly eclipsed PCr as an energy provider.

At about this same 30-second mark, the third player, aerobic respiration, is supplying only about 20 percent of the ATP, but its contribution is growing fast. In **Figure 1**, you can see how the aerobic contribution to ATP yield grows over the

Figure 7.10
Many Respiratory Pathways

Glucose is not the only starting material for cellular respiration. Other carbohydrates, proteins, and fats can also be used as fuel for cellular respiration. These reactants enter the process at different stages.

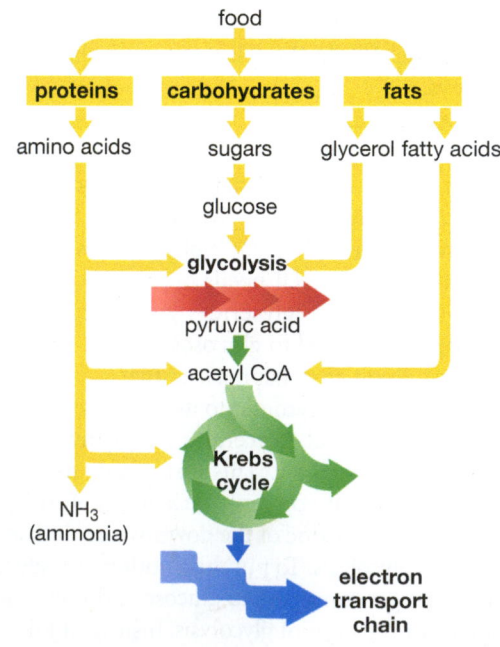

is then converted to pyruvic acid. Then, as pyruvic acid, it goes through the Krebs cycle and ETC, producing ATP. And the fatty acids? They are converted to acetyl CoA, which you may remember is the substrate that enters the Krebs cycle, yielding energy by becoming oxidized there. In **Figure 7.10**, you can see a summary of the way foods are broken down through various respiratory pathways.

On to Photosynthesis

This long walk through cellular respiration has illustrated how living things harvest energy from food. Recall, however, that almost all living things have one source to thank for this food: the sun's energy, which plants capture and then lock up in a sugar they produce. This capture of energy takes place through a process that has a beautiful symmetry with the cellular respiration you've just looked at. The process in question is called photosynthesis.

course of a long, strenuous workout. At the start, oxygen-dependent respiration is supplying less than 10 percent of our ATP, but 10 minutes into the workout it is delivering 85 percent.

You might think that this would be the final division of ATP contributions among the three players, but the amounts they supply can vary over time, depending on the intensity of the activity. **Figure 2**

shows how this would work in an actual professional-level bicycle race. Going along at a steady state in "pack" riding, you can see the overriding contribution of aerobic metabolism. In the final sprint, however, stores of ATP/PCr—replenished during the pack riding—are fully used, with the contribution of glycolysis increasing as well. In stretches of difficult hill climbing, meanwhile, glycolysis is fully used.

Of course, lots of sports have no aerobic component to them. The energy for a sport like football is supplied almost entirely by stored ATP/PCr. Even in a competition as long as a 400-meter sprint—about once around a regulation track—about 70 percent of the energy comes from a combination of ATP/PCr and glycolysis.

Duration of maximal exercise

	Seconds			Minutes					
	10	30	60	2	4	10	30	60	120
Percent anaerobic	90	80	70	50	35	15	5	2	1
Percent aerobic	10	20	30	50	65	85	95	98	99

Figure 1
Different Contributions over Time

Relative contributions of anaerobic and aerobic respiration to exercise during the duration of a workout. "Anaerobic" here means a combination of glycolysis and stored ATP/PCr release.

(Adapted from P. O. Astrand and K. Rodahl, *Textbook of Work Physiology.* New York: McGraw Hill Book Company, © 1977.)

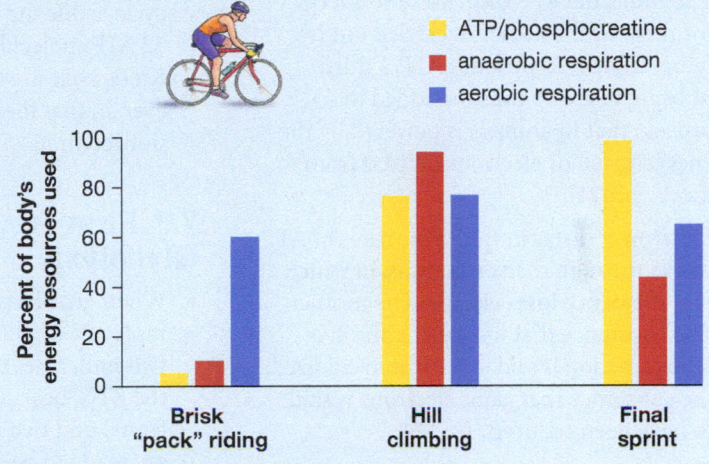

Figure 2
Different Activities, Different Energy Stores

Energy-harvesting processes used at different stages in a bicycle race.

(Adapted from J. T. Kearney, "Training the Olympic Athletes." *Scientific American,* June 1996, p. 54.)

SO FAR . . .

1. In the electron transport chain (ETC), the electron carriers NADH and FADH$_2$ _____ electrons to a series of enzyme complexes located within the inner _____ of the mitochondria. The energetic downhill drop of this electron transfer powers the movement of _____ into the outer compartment of the mitochondria.

2. The physical movement of _____ back into the inner compartment through the enzyme _____ is a downhill energetic drop that powers the process by which _____ are added to ADP, thus yielding _____.

3. The electrons picked up at various stages of energy harvesting by NADH and FADH$_2$ must be accepted by _____ at the end of the ETC if energy transfer is to continue.

CHAPTER

7

REVIEW

Summary

7.1 Energizing ATP

- The molecule ATP dispenses the energy for most of the activities carried out by living things. To produce ATP, a third phosphate group must be added to ADP, a process that in animals is powered by the energetic fall of electrons derived from food. (p. 124)

- Electron transfer in the production of ATP works through redox reactions, in which one substance loses electrons to another. The substance that loses electrons in a redox reaction is said to have been oxidized; the substance that gains electrons is said to have been reduced. (p. 124)

- Electrons are carried between one part of the energy-harvesting process and another by electron carriers, the most important of which is NAD. In its "empty" state, this molecule exists as NAD$^+$. Through a redox reaction, it picks up one hydrogen atom and another single electron from food, thus becoming NADH, a form it will retain until it drops off its energetic electrons (and a proton) in a later stage of the energy-harvesting process. (p. 125)

7.2 The Three Stages of Cellular Respiration

- The three principal stages of harvesting energy from food are glycolysis, the Krebs cycle, and the electron transport chain (ETC). Collectively, this three-stage, oxygen-dependent process is known as cellular respiration. For most organisms, glycolysis serves as a primary means of energy extraction only in certain limited situations, but it must take place for the Krebs cycle and ETC processes to go forward. (p. 126)

- Glycolysis is carried out in the cell's cytosol while the Krebs cycle and the ETC take place in the organelles called mitochondria, which lie within the cytosol. Glycolysis yields 2 net molecules of ATP per molecule of glucose, as does the Krebs cycle, while the ETC yields a maximum of 32 ATP molecules. Glycolysis and the Krebs cycle are critical to the ETC, however, in that they yield electrons that are utilized in it. (p. 126)

7.3 First Stage of Respiration: Glycolysis

- When glycolysis begins with a single molecule of glucose, the products are two molecules of NADH (which move to the ETC, bearing their energetic electrons) and two molecules of ATP (which are ready to be used). Glycolysis also produces two molecules of pyruvic acid—the derivatives of the original glucose molecule—which move on to the Krebs cycle. (p. 128)

7.4 Second Stage of Respiration: The Krebs Cycle

- In the transition between glycolysis and the Krebs cycle, each pyruvic acid molecule that was produced in glycolysis combines with coenzyme A, thus forming acetyl CoA, which enters the Krebs cycle. Carbon dioxide (which diffuses to the bloodstream) is another product of this reaction, as is one more molecule of NADH (which moves to the ETC). (p. 131)

- In the Krebs (or citric acid) cycle, the derivatives of the original glucose molecule are oxidized, thus yielding more energetic electrons that are transported by NADH and the electron carrier FADH$_2$ to the ETC. The net energy yield of the Krebs cycle is six molecules of NADH, two molecules of FADH$_2$, and two molecules of ATP per molecule of glucose. (p. 132)

7.5 Third Stage of Respiration: The Electron Transport Chain

- The ETC is a series of molecules located within the mitochondrial inner membrane. On reaching the ETC, NADH and FADH$_2$ are oxidized by molecules in the chain. Each carrier in the chain is then reduced by accepting electrons from the carrier that came before it. The last electron acceptor in the ETC is oxygen. (p. 132)

- The movement of electrons through the ETC releases enough energy to power the movement of hydrogen ions through the three ETC protein complexes, moving them from the mitochondrion's inner compartment to its outer compartment. The movement of these ions down their concentration and charge gradients, back into the inner compartment through an enzyme called ATP synthase, drives the synthesis of ATP from ADP and phosphate. In the inner compartment, oxygen accepts hydrogen ions and electrons from the ETC, forming water. If oxygen is not present to accept the ETC electrons, the energy-harvesting process downstream from glycolysis comes to a halt. (p. 133)

7.6 Other Foods, Other Respiratory Pathways

- Different nutrients and their derivatives can be channeled through different pathways in cellular respiration in accordance with the needs of an organism. (p. 135)

Key Terms

Understanding the Basics

Multiple-Choice Questions

(Answers are in the back of the book.)

1. You need energy to think, to keep your heart beating, to play a sport, and to study this book. This energy is directly dispensed by _____, which is produced in the process of cellular respiration.
   a. enzymes
   b. ATP
   c. NAD⁺
   d. vitamins
   e. proteins

2. Energy transfer in living things works through redox reactions, in which one substance is _____ by another substance, thereby _____.
   a. oxidized ... losing electrons to it
   b. transported ... becoming more energetic
   c. digested ... becoming more energetic
   d. reduced ... losing electrons to it
   e. oxidized ... gaining electrons from it

3. _____ and _____ are important not so much for the ATP produced in them, but for their _____.
   a. Glycolysis ... the Krebs cycle ... yield of electrons transported to the ETC
   b. Glycolysis ... the ETC ... yield of electrons transported to the cytosol
   c. Redox reactions ... fatty acid breakdown ... yield of calories
   d. The Krebs cycle ... the ETC ... numerous redox reactions
   e. The Krebs cycle ... the ETC ... fatty acid breakdown

4. The carbon dioxide we exhale with each breath is produced in _____ and in _____. This CO_2 can be thought of as a(n) _____ of cellular respiration.
   a. glycolysis ... the ETC ... waste product
   b. glycolysis ... the ETC ... ultimate product

   c. the formation of acetyl CoA ... the Krebs cycle ... waste product
   d. glycolysis ... the Krebs cycle ... intermediate product
   e. Krebs cycle ... the reduction of water ... intermediate product

5. At most, how many net molecules of ATP can be produced per glucose molecule in cellular respiration?
   a. 2
   b. 8
   c. 24
   d. 36
   e. 72

6. Where does the Krebs cycle occur in this model of a mitochondrion?

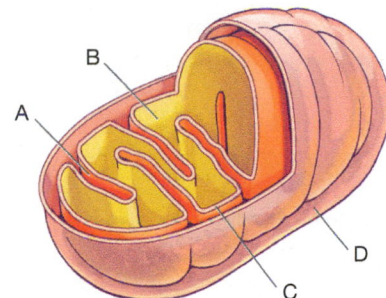

   a. A
   b. B
   c. C
   d. D
   e. none of these

7. When NADH passes its electrons to the ETC, it is
   a. electrolyzed.
   b. polarized.
   c. reduced.
   d. oxidized.
   e. deformed.

8. We need to breathe because we need
   a. both atmospheric nitrogen and the oxygen for energy transformation.
   b. oxygen to donate electrons to NAD+.
   c. nitrogen to donate phosphate groups to oxygen.

   d. oxygen to act as the final acceptor of electrons in the ETC.
   e. oxygen to donate phosphate groups to ADP, making it ATP.

Brief Review

(Answers are in the back of the book.)

1. Living things need ATP to power most of the processes that go on within them. In essence, how does cellular respiration yield this ATP?

2. Where do you expect people would have more mitochondria: their skin cells or their muscle cells? Why?

3. In the ETC, what is physically moved across the inner membrane—and energetically moved to a higher state—to drive the attachment of phosphate groups to ADP?

4. Where does the energy come from that powers this movement across the inner membrane in the ETC?

5. Most of the ATP we use is produced in the ETC. So, when we have to run across the street to avoid traffic, does the ATP we are using come primarily from the ETC?

Applying Your Knowledge

1. In Madeleine L'Engle's children's novel *A Wrinkle in Time,* the mitochondria in one of the characters start to die. Describe what would happen to people who lost their mitochondria and explain why it would happen.

2. Most of the heat in the human body is generated within mitochondria. Why should this be?

3. The food we eat is broken down in our digestive tract, but where are most of the calories in this food actually "burned"?

CHAPTER

8

The Green World's Gift:

Photosynthesis

Plants use the sun's energy to make their own food—and ours.

Available on MasteringBiology®
Activities: Light Energy and Pigments; Properties of Light; Leaves: The Sites of Photosynthesis; Overview of Photosynthesis; The Sites of Photosynthesis; The Light Reactions; The Calvin Cycle; The Calvin Experiments; Photosynthesis in Dry Climates; **Bioflix:** Photosynthesis; **Current Events:** Storms Threaten Ozone Layer Over U.S., Study Says; Giants Who Scarfed Down Fast Food Feasts; **Video Field Trip:** Solar Energy; **Building Vocabulary; Questions:** Bioflix Quiz, Reading Questions, Test Bank

Energy from the sun helps produce the food that sustains nearly all living things.

On a summer's evening, out by a pond as the sun is going down, nature tunes up her symphony. The crickets bring their chirping to the fore, while frogs chime in with their own countermelody. Silently, something else is changing as the last light fades: The green world is shutting down. Microscopic pores that dot the undersides of leaves are closing

up, ceasing to be openings for the carbon dioxide that flows in and the water vapor that flows out during the day. The green world is alive at night, but it is only fully active during the day, when a special talent that it has is set in motion by the warm rays of the sun.

8.1 Photosynthesis and Energy

The green world's special talent is called photosynthesis, and the wonder of it can perhaps best be appreciated by imagining what life would be like if we humans possessed it. The bone and muscle within us—and the energy to make them work—come from the foods we eat: carbohydrates, fats, and proteins. Imagine, though, having the ability to flourish simply by spending time in the sun and having access to three things: water, small amounts of minerals, and carbon dioxide. In this condition, we would not *eat* carbohydrates in the usual forms (bread, sugar, potatoes). Rather, we would transform the simple gas carbon dioxide *into* carbohydrates using nothing more than water, minerals, and sunlight. Carbohydrates produced in this way are as useful as any food; they can be stored away as starches, used to provide ATP for energy needs, or transformed to make proteins. This, then, is what plants do: *They make their own food.* They use the energy provided by the sun to take a simple gas, carbon dioxide, join it to a carbohydrate (a sugar), and then *energize* that sugar, thus transforming it into food. In this activity, they are joined by a variety of other photosynthesizing organisms (**Figure 8.1**).

From Plants, a Great Bounty for Animals

The process of photosynthesis is not only good for plants; it's good for other living things, including

animals, since most of the living world ultimately depends on photosynthesis for food. Try naming something you eat that isn't a plant or doesn't itself eat plants to survive, and you will come up with a very short list indeed.

Food, however, is only half the bounty that plants provide. A by-product of photosynthesis—a kind of castoff from the work of plants—is oxygen. As part of photosynthesis, plants break water molecules apart. In doing so, they use electrons and protons from H_2O, but they leave behind oxygen molecules (O_2). This is the oxygen that exists in our atmosphere; it is the oxygen that we breathe in every minute of every day or suffer the mortal consequences.

With this, you can step back and begin to view the largest picture of them all with respect to energy flow in living things. It is a great pathway in which energy comes from the sun and then, in photosynthesis, is stored in plants in the chemical bonds of the carbohydrates they produce. Thus stored—in such forms as wheat grains or grass leaves—these carbohydrates

Light Energy and Pigments

Figure 8.1
Three Types of Photosynthesizers

Photosynthesis is carried out not only by familiar plants, such as these sunflowers **(a)**, but also by algae, such as this giant kelp **(b)**, and by some bacteria, such as these cyanobacteria **(c)**.

(a) Sunflowers (plants)

(b) Giant kelp (algae)

(c) Cyanobacteria (bacteria)

Figure 8.2
Energy flows through the living world in two great phases. The first of these is photosynthesis, in which the sun's energy is captured by plants and then stored within them. The second is cellular respiration, in which this stored energy is extracted and then transferred to ATP, which in turn powers most of the activities that organisms undertake.

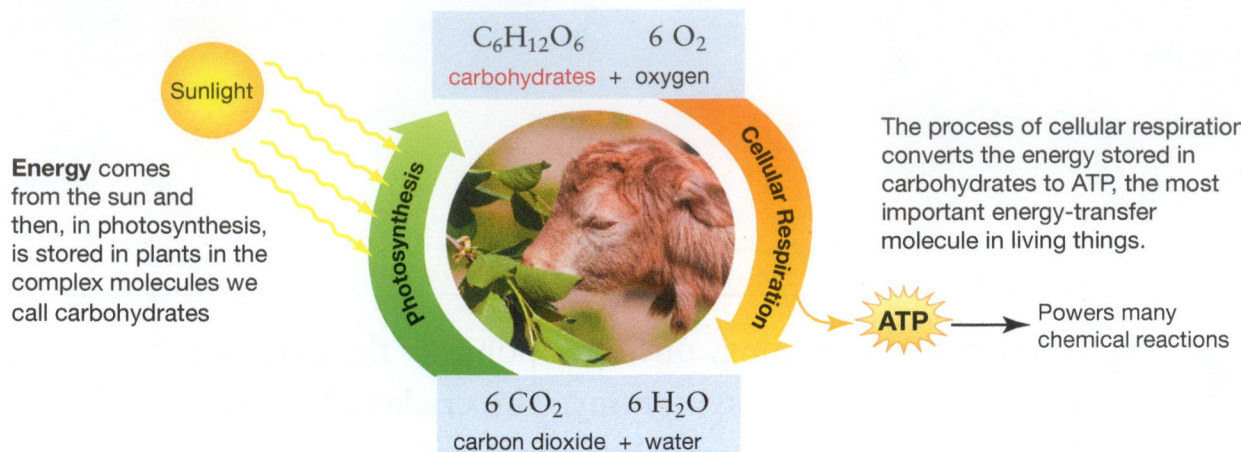

Energy comes from the sun and then, in photosynthesis, is stored in plants in the complex molecules we call carbohydrates

The process of cellular respiration converts the energy stored in carbohydrates to ATP, the most important energy-transfer molecule in living things.

ATP → Powers many chemical reactions

Properties of Light

Figure 8.3
The Electromagnetic Spectrum
Sunlight is composed of rays of many different wavelengths, but we can see only those in the visible light range (represented by the color spectrum). When the light of the sun shines on a green leaf, it is primarily the sun's green rays that are reflected, while frequencies in other ranges are absorbed; frequencies in the red and blue ranges drive photosynthesis most strongly.

can be broken down and used, either by the plants themselves or by the organisms that eat them. The breakdown of this food ends with the "cellular respiration" covered last chapter that provides ATP. If you look at **Figure 8.2**, you can see this whole process displayed in graphic form.

Note that photosynthesis and cellular respiration are trips up and down the energy hill that became so familiar in Chapter 7. In respiration, the trip was down the hill, meaning from more stored energy (in food) to less, as the food was broken down to produce ATP. Photosynthesis is a trip up the energy hill. Here, electrons are removed from water, boosted to a more energetic state by the power of sunlight, and then brought together with a sugar and carbon dioxide, resulting in an energy-rich sugar—food in the form of a carbohydrate. The story of how this carbohydrate production is accomplished is the subject of this chapter. To encapsulate this story in a definition, **photosynthesis** is the process by which certain groups

of organisms capture energy from sunlight and convert this solar energy into chemical energy that is initially stored in a carbohydrate.

We could think of this process in terms of a kind of recipe—a formula that shows us what "ingredients" go into photosynthesis and what products come out.

$$sunlight\ energy$$
$$6CO_2 + 6H_2O \rightarrow C_6H_{12}O_6 + 6O_2$$

Over on the left, you can see six molecules of carbon dioxide ($6CO_2$) and six molecules of water ($6H_2O$) as starting ingredients (or "reactants"), and then on the right is a molecule of glucose ($C_6H_{12}O_6$) and six molecules of oxygen ($6O_2$) as products. Intervening in the middle, and driving the whole process, is the energy contained in the rays of the sun.

8.2 The Components of Photosynthesis

We can think of photosynthesis as starting with the absorption of sunlight by leaves. *Absorption* here means just what it sounds like: Light is taken in by the leaves. Actually, leaves capture only a *portion* of the light that falls on them. The sunlight that makes it to the Earth's surface is composed of a spectrum of energetic rays (measured by their "wavelengths") that range from very short ultraviolet rays, through visible light rays, to the longer and less-energetic infrared rays. (In **Figure 8.3** you can see that these wavelengths are, in turn, part of a larger range of electromagnetic radiation.) Photosynthesis is driven by part of the *visible light* spectrum—mainly by blue and red light of certain wavelengths. Tune a laser to emit blue light of a certain wavelength, shine this beam on a plant, and the plant will absorb a great deal of this light. Tune the laser to emit *green* light, on the other hand, and the plant will reflect most of this light. This is why plants

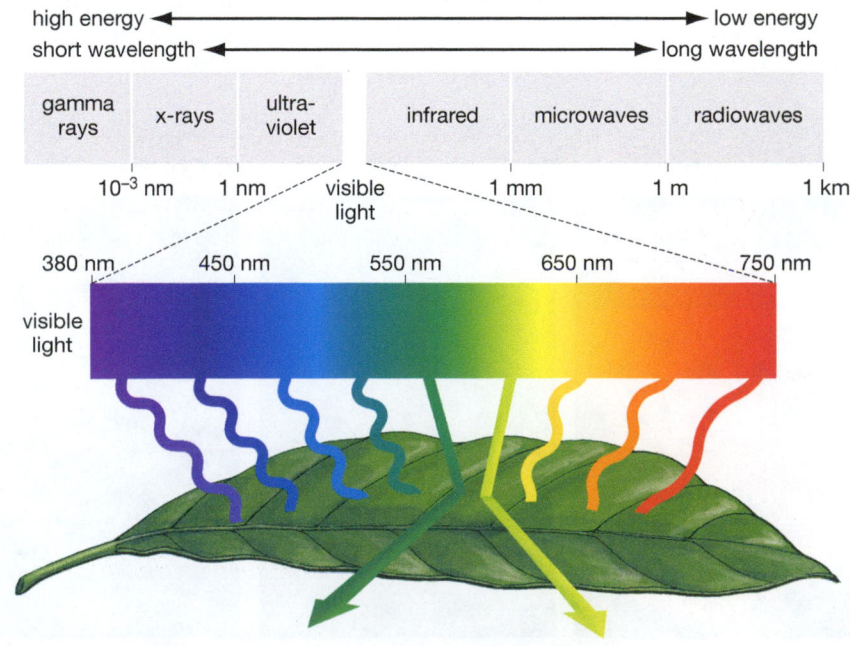

high energy ←————————→ low energy
short wavelength ←————————→ long wavelength

gamma rays	x-rays	ultra-violet		infrared	microwaves	radiowaves

10^{-3} nm   1 nm   visible light   1 mm   1 m   1 km

380 nm   450 nm   550 nm   650 nm   750 nm

visible light

are green: They strongly reflect the green portion of the visible light spectrum.

Where Does Photosynthesis Occur?

So, where is this light absorption occurring within plants? To answer this question, you need to know something about the primary playing fields of photosynthesis: leaves.

If you look at the idealized leaf in **Figure 8.4**, you can see that it can be likened to a kind of cellular sandwich, with one layer of outer (or epidermal) cells at the top, another layer of epidermal cells at the bottom, and several layers of mesophyll cells in between. Take time now to read through Figure 8.4's drawings and captions; as you do, you'll go from the leaf's layers of cells, then into the cells themselves, and then into the tiny structures *within* the cells—the organelles called chloroplasts—that are the actual sites of photosynthesis.

Here is some detail on the structures seen in Figure 8.4. The **stomata** seen in the figure are the microscopic pores, mentioned at the start of the chapter, that open during the day to let carbon dioxide pass

Leaves: The Sites of Photosynthesis

Figure 8.4
Site of Photosynthesis

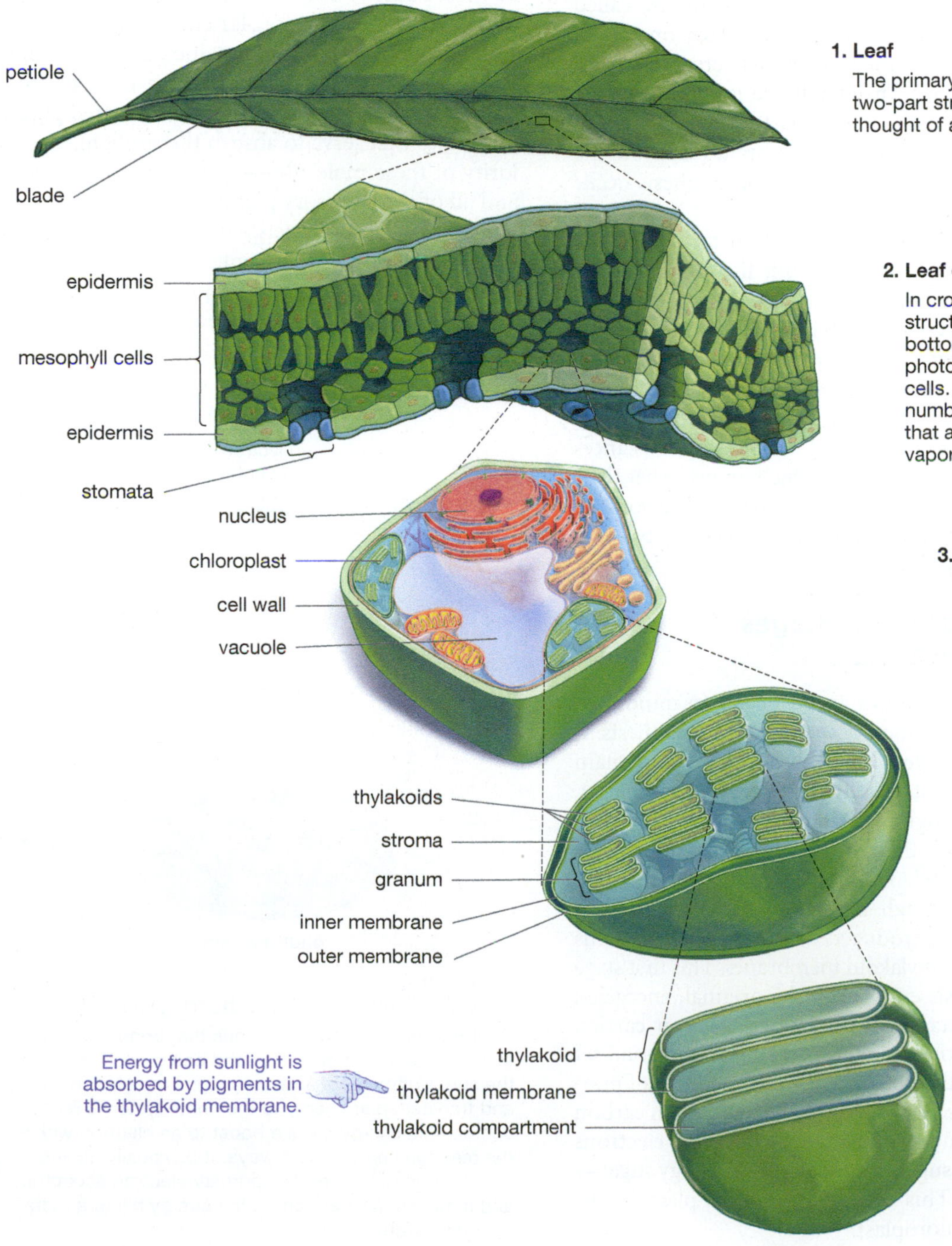

1. Leaf

The primary site of photosynthesis in plants, leaves have a two-part structure: a petiole (or stalk) and a blade (normally thought of as the leaf).

petiole
blade

2. Leaf cross section

In cross section, leaves have a sandwich-like structure, with epidermal layers at top and bottom and mesophyll cells in between. Most photosynthesis is performed within mesophyll cells. Leaf epidermis is pocked with a large number of microscopic openings, called stomata, that allow carbon dioxide to pass in and water vapor to pass out.

epidermis
mesophyll cells
epidermis
stomata

3. Mesophyll cell

A single mesophyll cell within a leaf contains all the component parts of plant cells in general, including the organelles—called chloroplasts—that are the actual sites of photosynthesis.

nucleus
chloroplast
cell wall
vacuole

4. Chloroplast

Each chloroplast has an outer membrane at its periphery; then an inner membrane; then a liquid material, called the stroma, that has immersed within it a network of membranes, the thylakoids. These thylakoids sometimes stack on one another to create a granum.

thylakoids
stroma
granum
inner membrane
outer membrane

5. Granum

Electrons used in photosynthesis will come from water contained in the thylakoid compartment, and all the steps of photosynthesis will take place either within the thylakoid membrane, or in the stroma that surrounds the thylakoids.

Energy from sunlight is absorbed by pigments in the thylakoid membrane.

thylakoid
thylakoid membrane
thylakoid compartment

Overview of
Photosynthesis

The Sites of
Photosynthesis

into leaves and water vapor pass out of them. There is no shortage of stomata for this work; a given square centimeter of plant leaf may contain from a thousand to a hundred thousand of them.

Going down a couple of levels, **chloroplasts** can be defined as the organelles within plant and algae cells that are the sites of photosynthesis. Anywhere from one to several hundred of these oblong structures might exist within a given leaf cell, each one capable of carrying out photosynthesis on its own. As Figure 8.4 shows, chloroplasts present another example of something you've seen before—membranes within membranes. The chloroplast has outer and inner membranes at its periphery. Inside them, in the interior of the chloroplast, there is a network of chloroplast membranes, active in photosynthesis, called **thylakoids.** (These thylakoids often stack on top of one another like so many pancakes, creating structures called *grana*.) Thylakoids are immersed in the liquid material of the chloroplast, the **stroma.** Keep your eye on the thylakoid membranes and on the stroma because all the steps of photosynthesis occur in one of these two places.

The thylakoid membranes are, in fact, the place where photosynthesis starts, with the absorption of sunlight we just talked about. Any compound that strongly absorbs certain visible wavelengths of sunlight is called a *pigment*. Thylakoid membranes contain a pigment, called **chlorophyll a,** that can be defined as the primary pigment active in plant photosynthesis. Chlorophyll *a* is then aided by several substances known as *accessory pigments*, which do just what their name implies: They aid chlorophyll *a* in absorbing energetic rays from the sun, after which they pass the absorbed energy along.

The Two Essential Stages in Photosynthesis

Keeping these structures and concepts in mind, you can now begin tracing the steps of photosynthesis. It is helpful to divide photosynthesis into two main stages. In the first stage (the *photo* of photosynthesis), the power of sunlight will do two things—strip water of electrons and then boost these electrons to a higher energy level. Thus boosted, the electrons are passed along through a series of electron carriers similar to those introduced in Chapter 7. All of this takes place in the thylakoid membranes. This first stage of photosynthesis ends when the original, energized electrons get attached to a mobile electron carrier, called $NADP^+$, that transports them to the second set of reactions. In this second stage (the *synthesis* of photosynthesis), the electrons come together with carbon dioxide and a sugar. The attachment of the electrons and CO_2 to this sugar produces a high-energy sugar—meaning food. This second stage takes place in the stroma of the chloroplast.

The steps in the first stage of photosynthesis obviously depend directly on sunlight. Thus, these steps are sometimes referred to as the *light reactions*. Meanwhile, the second set of steps is referred to as the *Calvin cycle,* in honor of Melvin Calvin, the scientist who first revealed their details. You will be following both sets of steps as they occur in time—that is, from the collection of sunlight to the synthesis of carbohydrates.

The Working Units of the Light Reactions

Looking first at the light reactions, we see that they take place within a set of working units. Each of these units is a **photosystem**—an organized complex of molecules within a thylakoid membrane that, in photosynthesis, collects solar energy and transforms it into chemical energy. If you look at **Figure 8.5**, you can see the components that make up a photosystem. First, there is a group of a few hundred pigment molecules that serve to absorb the sunlight. The majority of these molecules—most of the chlorophyll *a* and all of the accessory pigments—serve only as "antennae" that absorb energy from the sun and pass it on. At the center of the antennae system, however, there is a **reaction center**—a pair of special chlorophyll *a* molecules and associated compounds that first receive the solar energy from photosystem pigments

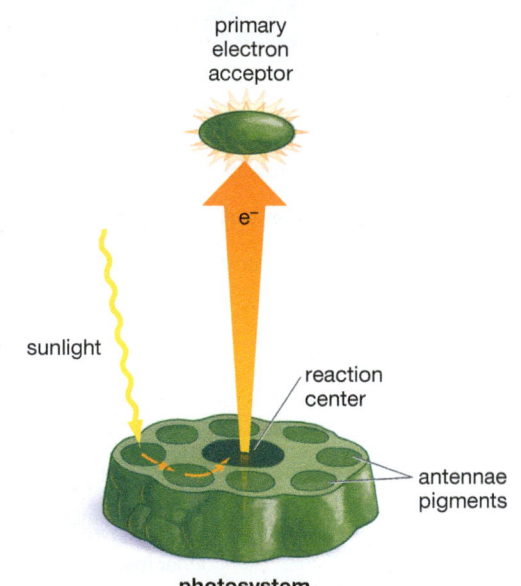

Figure 8.5

The Working Units of Photosynthesis

Photosystems are multipart units that bring together electrons derived from water and energy derived from the sun. Hundreds of antennae pigments absorb sunlight and transfer solar energy to the photosystem's reaction center. This energy gives a boost to an electron within the reaction center in two ways: It physically moves the electron to the center's primary electron acceptor, and it moves the electron up the energy hill to a more energetic state.

and then transform this solar energy into chemical energy. As you can see in Figure 8.5, the reaction center includes a molecule that could be thought of as the first recipient of the absorbed solar energy, the *primary electron acceptor*.

You could conceptualize photosynthesis as beginning when sunlight is absorbed by any of the hundreds of antennae pigment molecules in a photosystem unit. The absorbed energy is then passed on to the pair of chlorophyll *a* molecules in a reaction center. As a result, electrons from this pair are "moved" in a couple of ways, one physical and one metaphorical. Physically, these electrons are transferred to the primary electron acceptor. You also could think of the initial "distance traveled" by the electrons in a metaphorical way, however: as a movement up the energy hill. The energy from sunlight is pumping the electrons up the hill in photosynthesis.

Energy Transfer through Redox Reactions

How can electrons be transferred from one molecule to another? Recall from Chapter 7 that some substances have a tendency to *lose* electrons to other substances, and that any time one substance loses (or "donates") electrons to another, it is said to have been oxidized. Meanwhile, the substance that gained elec-

trons is said to have been reduced. These two reactions always happen together; if one substance is oxidized, another must be reduced. This kind of reduction-oxidation reaction has a shorter name: a *redox reaction*.

Many of the photosynthetic steps you will be seeing are redox reactions. Indeed, you have already seen one. The movement of electrons within the reaction center—from the chlorophyll *a* molecules to the primary electron acceptor—is a redox reaction. The chlorophyll molecules were oxidized by the primary electron acceptor, meaning they lost electrons to it. What happens in the light reactions is a continuation of this energy transfer: Electrons are passed on by means of one electron carrier oxidizing another.

As electrons physically move through the light reactions in this way, they are also moving both up and down the energy hill—they are becoming first more and then less energetic. If you look at **Figure 8.6**, you can trace their movement up and down the energy hill as they make their journey through the electron transport chain you can see there. With energy supplied by the sun, electrons are pumped up a couple of formidable energy gradients (in photosystems II and I) only to come partway back down them as they "seek" their lowest energy state. Critically, they are releasing energy as they fall—just as a boulder releases energy by rolling downhill.

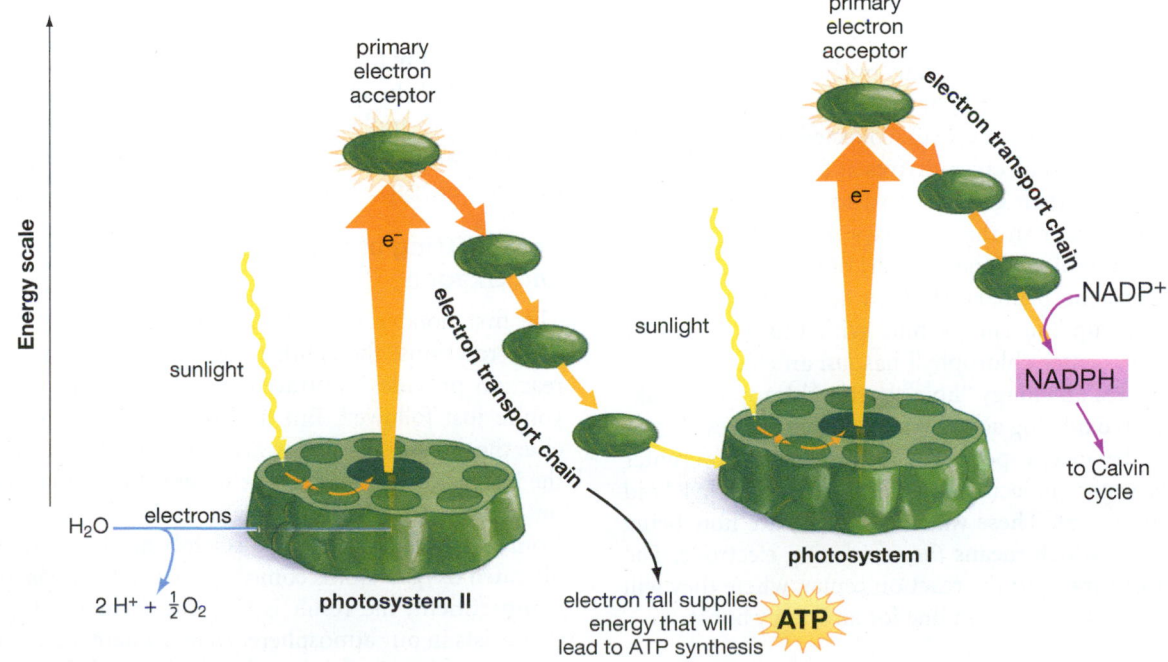

Figure 8.6
Collecting Solar Energy, Boosting Electrons

The solar energy gathered by photosystems II and I is used to energize electrons that exist within the photosystems' reaction centers. Energetically, the electrons are boosted up, and then move down, two energy hills. Physically, they first move to primary electron acceptors and then down electron transport chains until they are at last taken up by NADP+ to form NADPH. The NADPH molecules then transfer the electrons to the Calvin cycle, in which they are used to make sugars.

SO FAR . . .

1. In photosynthesis, the power of the sun is used to boost the energy of _____ that are derived from water and bring them together with both a low-energy sugar and _____ from the atmosphere, thus producing an energy-rich _____, meaning food.

2. In plants, photosynthesis takes place within the organelles called _____, which are found most abundantly in certain cells that make up the interior portions of _____. Pigments within the membranes of these organelles, in particular a pigment called _____, serve to absorb sunlight, which provides the energy that drives photosynthesis.

3. Photosynthesis can be divided into two major sets of steps. In the first set, known as the _____, absorbed solar energy moves electrons derived from water in two ways: (a) physically to a molecule called a _____ and (b) metaphorically to a higher _____.

8.3 Stage 1: The Light Reactions

The Light Reactions

With these components and processes in mind, let's trace the steps of the light reactions (Figure 8.6). The first step is that solar energy, collected by photosystem II's antennae molecules, arrives at the reaction center. This energy then gives a boost to an electron in the reaction center in the two ways noted: The electron *physically* moves to another part of the reaction center complex, the primary electron acceptor, and is also pumped up the energy hill. With this activity, the reaction center chlorophyll has *lost* an electron. That loss leaves an energy "hole" in this chlorophyll, making it an oxidizing agent. With the energy provided by this imbalance, a special enzyme in the reaction center splits water molecules that lie within the thylakoid compartment. These water molecules are now being oxidized, which means *they* are losing electrons. The electrons travel to the reaction center, where they will be the next electrons in line for an energy boost.

A Chain of Redox Reactions and Another Boost from the Sun

By following Figure 8.6 in a general way, you can see what happens next to electrons that are boosted in photosystem II. After arriving at the primary electron acceptor, they fall back down the energy hill as they are transferred through a series of electron transport chain molecules, each oxidizing its predecessor. At the bottom of this hill, they arrive at photosystem I, which includes a slightly different kind of reaction center. This center is also receiving solar energy and uses it to boost electrons to a higher energy state. From this energetic state, electrons are transferred down the second energy hill—until they are received by the electron carrier $NADP^+$, which is very much like the electron carrier NAD^+ reviewed in Chapter 7. In accepting electrons, $NADP^+$ becomes reduced to NADPH, an electron carrier that ferries the electrons into the next stage of photosynthesis. This second stage is the Calvin cycle, which will yield the high-energy sugar that is the essential product of photosynthesis.

The Physical Movement of Electrons in the Light Reactions

In this movement up and down the energy hill, the electrons have moved physically through the chloroplast. As **Figure 8.7** shows, they started out in the water of the thylakoid compartment, moved into and then through the thylakoid membrane—handed off at each step by the various electron transport molecules—and then finally ended up in the stroma, attached to NADPH.

What Makes the Light Reactions So Important?

The steps just reviewed are the essence of the light reactions. To make sure you don't miss the forest for the trees, let's be clear about two momentous things that have taken place within these steps: the splitting of water and the transformation of solar energy to chemical energy.

The Splitting of Water: Electrons and Oxygen

The first notable event takes place near the start of these reactions: the splitting of water. Critically, this reaction provides the traveling electrons whose path you've just followed. But it also provides something else: the oxygen that today accounts for 21 percent of the Earth's atmosphere. In the water splitting that goes on in photosynthesis, hydrogen atoms are removed from H_2O, while the oxygen is left behind. When two liberated oxygen atoms come together in the thylakoid compartment, the result is O_2—the form of oxygen that exists in our atmosphere. *Here* is where we get the substance that is the breath of life itself for most species, all of it contributed by living things. Given its importance, it is no small irony that oxygen is, in another sense, a kind of green-world refuse—leftovers from the process by which plants strip water of the electrons they use to carry out photosynthesis.

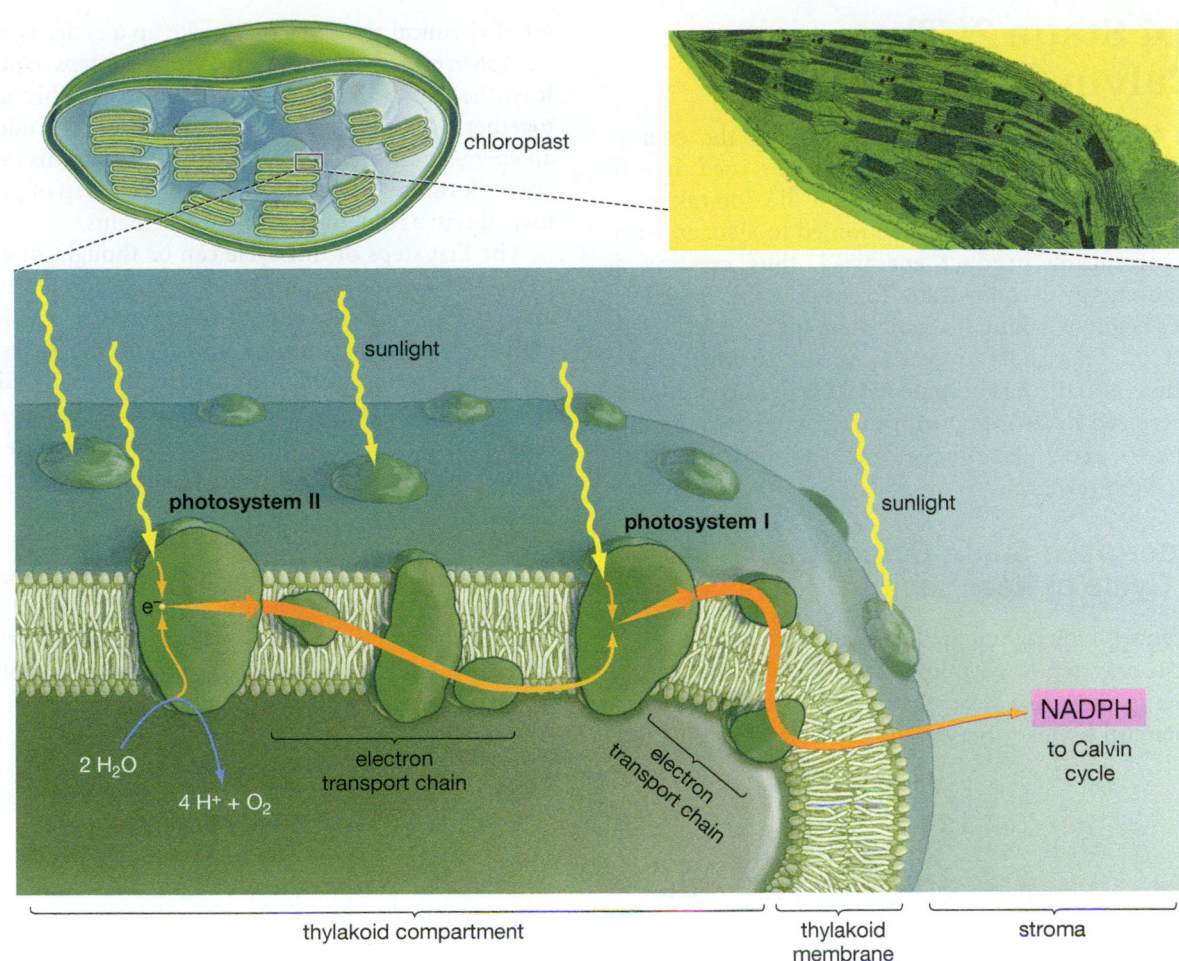

chloroplast

sunlight

photosystem II

photosystem I

sunlight

e⁻

NADPH

to Calvin cycle

2 H₂O

4 H⁺ + O₂

electron transport chain

electron transport chain

thylakoid compartment

thylakoid membrane

stroma

Figure 8.7
The Light Reactions
The light reactions take place within the thylakoid membranes in the chloroplasts. Electrons are donated by water molecules located in the thylakoid compartments. Powered by the sun's energy, these electrons are passed along the electron transport chain embedded in the thylakoid membrane and end up stored in NADPH in the stroma. An additional product of the splitting of water molecules is individual oxygen atoms, which quickly combine into the O_2 form. This is the atmospheric oxygen that we breathe.

The Transformation of Solar Energy to Chemical Energy

The second major feat that takes place during light reactions is the transformation of solar energy to chemical energy. When the energy of sunlight moves to a chlorophyll molecule in a reaction center, it boosts an electron there from what is known as a ground state to an excited state. Thus far, this has been described as a movement up a metaphorical energy hill. In physical terms, however, this electron is moving farther out from the nucleus of an atom. To appreciate what's special about photosynthesis, consider what the *common* fates are for such excited electrons. One is for the electrons to drop back down to the original (and more stable) state, in the process releasing as *heat* the energy they have absorbed. This is the process by which black objects get hot on sunny days. Another possible fate is for falling electrons to release part of their energy as light in the process known as fluorescence.

In photosynthesis, however, the energized electrons are *transferred to a different molecule*—the initial electron acceptors of photosystems II and I. They don't fall back to their ground state, releasing relatively useless heat or light in the process. They are passed on in a redox reaction. This is the bridge to life as we know it.

Without this step, the energy of the sun merely makes the Earth warmer. With it, the green world grows profusely; it captures some of the sun's energy and with it builds trunks and grains and leaves. Electrons are taken from water, boosted to a higher energy state by the sun, and then transferred to molecules that have the ability to *store* the energy they've received, rather than letting it dissipate as heat. When looked at globally, this set of chemical reactions has enormous consequences. By one reckoning, photosynthesis produces up to 155 billion tons of material each year—about 25 tons for every person on Earth.

Production of ATP

It's also important to note that a primary function of the fall of electrons between photosystems II and I is a release of energy that is used for the production of ATP. Put together through a process similar to the one you looked at last chapter, this ATP will be used to power the reactions that are coming up in the second stage of photosynthesis. Thus, the sun's power is providing energy that is put into *two* forms in the light reactions: ATP and the energetic electrons that are channeled into NADPH. Looking at the big picture, you can see that the *photo* of photosynthesis is an energy-capturing operation.

8.4 Stage 2: The Calvin Cycle

The Calvin Cycle

The Calvin Experiments

Bioflix: Photosynthesis

Figure 8.8
The Calvin Cycle
CO_2, ATP, and electrons (contained in hydrogen atoms) from NADPH are the input into the Calvin cycle, while a sugar (G3P) is its output.

In the second stage of photosynthesis, the energy captured in the light reactions will be *used*. It will power a process by which carbon dioxide taken in from the atmosphere will be joined to a sugar, with the resulting product energized, thus creating a high-energy carbohydrate. To see how this happens, let's begin by taking stock of where the light reactions left off. The short answer is, in the stroma. That's where $NADP^+$ has become NADPH by accepting the energized electrons pouring out from photosystem I. There's also ATP in the stroma, another product of the light reactions.

Energized Sugar Comes from a Cycle of Reactions

Even with the right ingredients, however, it takes more than one step to go from carbon dioxide and a low-energy sugar to a sugar that has enough energy in its chemical bonds to serve as food. What's required is a set of chemical reactions that make up a cycle. This is the **Calvin cycle,** or the C_3 cycle, the set of steps in photosynthesis in which energetic electrons are brought together with carbon dioxide and a sugar to produce an energetic carbohydrate. The highlights of this cycle are set forth in **Figure 8.8**. This is the *synthesis* of photosynthesis: a bringing together of elements.

The first steps of this cycle can be thought of as a process of **fixation**—of a gas being incorporated into an organic molecule. Specifically, carbon dioxide, which comes in through the stomata on the leaves of the plant, is being fixed into the starting sugar, which is called RuBP. (From the time plants are embryos, they have small amounts of this sugar in them as a legacy from their parent plants.) This is a terrifically important step for it is the way life builds itself up with materials that lie *outside* of life. Note that carbon is being taken in from the atmosphere and then made part of a living thing—a plant.

The next reactions in the cycle are the energizing steps of the process (which you can see in section 2 of the cycle). Here is where our low-energy sugar receives the energetic products of the light reactions.

1. **Carbon fixation.** An enzyme called rubisco brings together three molecules of CO_2 with three molecules of the sugar RuBP. In this reaction, one carbon from each CO_2 molecule is being added to the five-carbon RuBP, and this is being done three times. The three resulting six-carbon molecules are immediately split into six three-carbon molecules named 3-PGA (3-phosphoglyceric acid).

2. **Energizing the sugar.** In two separate reactions, six ATP molecules react with six 3-PGA, in each case transferring a phosphate onto the 3-PGA. The six 3-PGA derivatives oxidize (gain electrons from) six NADPH molecules; in so doing, they are transformed into the energy-rich sugar G3P (glyceraldehyde 3-phosphate).

3. **Exit of product.** One molecule of G3P exits as the output of the Calvin cycle. This molecule, the product of photosynthesis, can be used for energy or transformed into materials that make up the plant.

4. **Regeneration of RuBP.** In several reactions, five molecules of G3P are transformed into three molecules of RuBP, which enter the cycle.

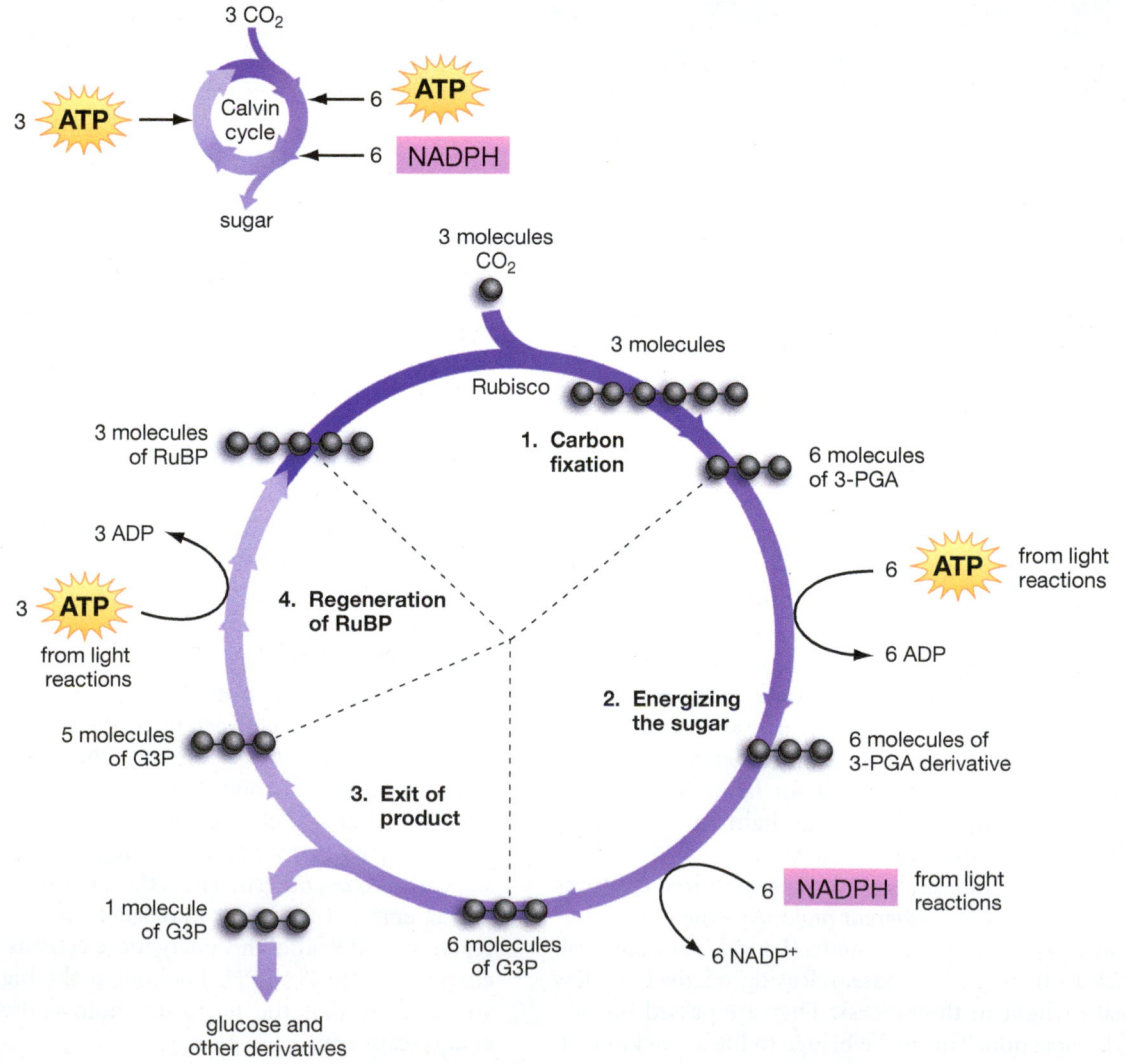

The RuBP derivatives first interact with ATP and then receive energetic electrons from NADPH—the electrons that came originally from water and that were boosted up the energy hills by sunlight. With this, a relatively low-energy sugar has been energized—it has been moved up the energy hill—from which position it can now serve as food. The energized sugar that is produced, called G3P, is the essential product of photosynthesis. One molecule of it is netted with each turn of the Calvin cycle, as you can see in section 3. The remainder of the cycle can be thought of as a preparation of molecules for another trip around the cycle. In section 4, several reactions are carried out that result in the formation of more RuBP, which then enters the cycle again.

The Ultimate Product of Photosynthesis

The importance of photosynthesis becomes clear only with an understanding of what G3P can give rise to once it exits the Calvin cycle. This sugar has something in common with steel coming out of a factory in that it can be turned into many things. Put two molecules of G3P together, and you get the more familiar six-carbon sugar glucose. In turn, many molecules of glucose can come together to form the large storage molecules known as starches, familiar to us in such forms as potatoes or wheat grains. Beyond this, sugar is used to make proteins, which are then used as structural components of the plant or as enzymes. And sugar can be broken down, of course, to provide more of the ATP that powers a plant's activities. Given all this, if we ask what the *ultimate* product of photosynthesis is, the answer turns out to be: *the whole plant*. If you look at **Figure 8.9**, you can see a summary of the process of photosynthesis.

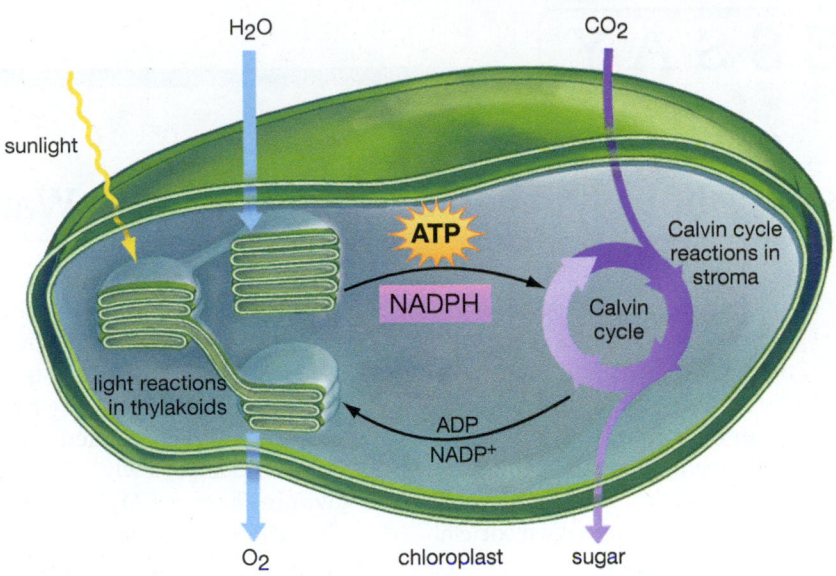

Figure 8.9
Summary of Photosynthesis in the Chloroplasts of Plant Cells
In the light reactions, which take place in the thylakoids of chloroplasts, solar energy is converted to chemical energy that is stored temporarily in the molecules ATP and NADPH. Water is required for these reactions and oxygen is a by-product. In the Calvin cycle, which takes place in the stroma of chloroplasts, the energy stored in ATP and NADPH is used to produce a high-energy sugar (G3P). Carbon dioxide is required for the cycle's reactions.

SO FAR . . .

1. The energy acquired in the light reactions takes two forms: energetic _____ stored in the carrier _____ and energy stored in the energy transfer molecule _____. An important by-product of the light reactions is the ____ produced when water is split.

2. The energy from the light reactions is used to power the second major set of steps in photosynthesis, the _____ cycle.

3. In this cycle, an energy-poor _____ is brought together with carbon dioxide from the _____, and the resulting product is energized with _____ derived from the light reactions. This process is powered by the ____ produced in the light reactions.

8.5 Photorespiration and the C₄ Pathway

You have now gone through the basic process of photosynthesis. It's necessary, however, to say something about what might be called a glitch in this process and then to note a mechanism that evolved within one group of plants that allows them to avoid the problem.

The glitch is called *photorespiration*, and it takes place within the Calvin cycle you just went over. Recall that this cycle begins with carbon dioxide (from the atmosphere) being joined to a low-energy sugar (from the plant). What is it that brings these two compounds together? Readers of the last few chapters will not be surprised to learn that it is an enzyme—a massive one called rubisco. Unfortunately, this molecular worker turns out to be something like the giant in "Jack and the Beanstalk": extremely powerful but slow moving and prone to making errors.

Extremely powerful? Rubisco is the bridge between the living world and the non-living world. It is the molecule that allows plants to capture, from the atmosphere, the carbon atoms that are the building blocks for all living things. One of the problems with the way it works, however, is its speed—or lack of it. Whereas an average enzyme can catalyze 1,000 reactions per second, rubisco can manage only three.

ESSAY

Using Photosynthesis to Fight Global Warming

How important can one of nature's processes be to one of society's issues? Really important. Just ask the scientists, economists, and political leaders from Boston to Bangalore who today are struggling to get a better handle on the process of photosynthesis as a means of combating the critical societal issue of global warming.

How are the two things related? Remember that global warming's root cause is a buildup of so-called greenhouse gases in the atmosphere, with the most important of these gases being carbon dioxide (CO_2). Now recall that plants and other photosynthesizing organisms take CO_2 *out* of the atmosphere and use it to build themselves up. Not all of the carbon they take in serves this function; indeed, much of the carbon they capture will quickly be returned to the atmosphere when the plants burn up some of their food stores, thus producing CO_2 as a waste product. Even allowing for this, however, trees and other photosynthesizers constitute what is known as a carbon sink: On balance they *remove* CO_2 from the atmosphere. Meanwhile, entities such as coal-fired power plants are so-called carbon sources—they *add* CO_2 to the atmosphere.

Given this, what happens when you cut down trees on a large scale, often burning them in the bargain? You get what you can see in **Figure 1**: an enormous quantity of human-generated greenhouse gases, second in size only to the quantity of gases society generates by burning fossil fuels. Trees that are merely cut down will start to decay, slowly releasing the carbon within them as CO_2. Trees that are burned, meanwhile, will release this CO_2 immediately. And of course trees that cease to exist will also cease to serve as carbon sinks. This part of the problem is less pronounced in the tropics, where mature trees are thought to return about as much CO_2 to the atmosphere as they take from it. But in so-called temperate and boreal forests—think of the eastern United States and north-central Canada—trees serve as significant carbon sinks, locking away an additional 1.5 billion tons of carbon each year, according to a recent analysis.

Given all this, one way to reduce global warming is, of course, to reduce global deforestation. This is one of the reasons we hear so much about tropical rainforests: They lock up an enormous amount of carbon and, among all the world's forests, they are being cut and burned at the highest rate. In 2005 alone, Brazil's Amazon

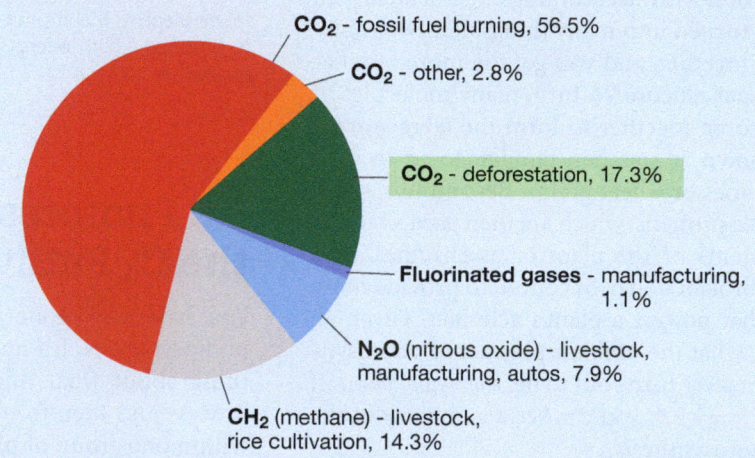

CO$_2$ - fossil fuel burning, 56.5%

CO$_2$ - other, 2.8%

CO$_2$ - deforestation, 17.3%

Fluorinated gases - manufacturing, 1.1%

N$_2$O (nitrous oxide) - livestock, manufacturing, autos, 7.9%

CH$_2$ (methane) - livestock, rice cultivation, 14.3%

Figure 1
Big Contributor to Global Warming

Deforestation produces about 17 percent of all human-generated greenhouse gases. Only the burning of fossil fuels puts a greater amount of CO_2 into the atmosphere.

(Adapted from *Climate Change 2007: Synthesis Report*, Intergovernmental Panel on Climate Change, Figure 2.1.)

This snail's pace has consequences. Because rubisco works so slowly, every plant must have a lot of it. As a result, it is easily the world's most abundant protein; at least 11 pounds of it exist for every person on Earth.

Slow pace is one thing, but it turns out that rubisco has a problem in even binding with the right substance: It can bind to *oxygen* as well as to carbon dioxide; indeed, it does this all the time. In C$_3$ photosynthesis, rubisco may fix up to one molecule of O$_2$ for every three of the carbon dioxide it incorporates. Worse, when this happens, some of the CO_2

that *has* been fixed by a plant is released by it. It is as if a carpenter building a house took boards that were stored inside and began throwing them out the window.

Photorespiration is especially likely to take place when temperatures rise, because heat prompts the stomata on leaves to close to preserve water. The effect of closed stomata, however, is that both water and *oxygen* are kept inside plant cells—remember that plants produce oxygen during photosynthesis—while CO_2 from the atmosphere is kept out of them. The result, understandably, is that rubisco binds

rainforest lost 7,200 square miles of forest—an area about the size of Massachusetts—through the logging practice known as clear-cutting.

Harmful as this kind of deforestation is, even halting it altogether wouldn't be enough to stop global warming. Instead, society is going to have to find many ways to cut down on carbon sources while channeling material into carbon sinks. Most of the ideas we hear about today have to do with sources—specifically with reducing society's use of fossil fuels. But scientists and public policy makers are also thinking about carbon sinks, which essentially means they are thinking about photosynthesis. They are asking, for example, whether it would be possible to increase photosynthesis worldwide by focusing on the portion of it carried out on the world's oceans. As it turns out, ocean-dwelling algae (and certain types of bacteria that float alongside them) undertake slightly more than half of the world's photosynthesis already. But algae often cannot flourish in the open ocean because these waters are deficient in iron, which is a basic nutrient. One idea to combat global warming, therefore, has been to fertilize the oceans with iron in the way that agricultural fields are fertilized with nitrogen. With this addition to the ocean environment, the idea goes, the algae would flourish and then would sink to the bottom of the ocean as they die, carrying with them the carbon they've taken in while alive.

This idea has come in for some sharp criticism; among other things, it appears doubtful that a significant proportion of algae in a so-called algal bloom actually sink after they die—something that would have to happen if the carbon these organisms capture is to be kept out of the atmosphere. But this idea is representative of the kind of proposals that have gotten increased attention as the world scrambles to figure out what it's going to do about global warming.

A related set of ideas has to do not with increasing photosynthesis per se, but rather with locking up the *products* of photosynthesis so that they don't decay, thus releasing CO$_2$. For example, a project currently underway in California's San Joaquin Delta, northeast of San Francisco, seeks to demonstrate the potential of "carbon-capture" farming, which in this case would mean paying farmers to "grow" marshlands of the sort that flourished throughout the delta before it was drained for farming, back in the nineteenth century. The idea is to slightly submerge the land once again and then grow marsh grasses and cattails on it. Such plants can quickly build up into a peat soil that has the benefit of *not* decaying because, as it accumulates in depth—by as much as 4 inches per year—it will be kept submerged. A pilot project has shown that this process can lock up or "sequester" as much as 25 metric tons of CO$_2$ per acre per year. Project directors calculate that if this savings were applied to all 321,000 acres of suitable land in the California delta, the effect would be the equivalent of transforming all of California's SUV's into gas-sipping small hybrids.

A question that follows from this, of course, is how delta farmers are going to profit from producing carbon sinks rather than agricultural crops, but the answer actually is not hard to envision. One of the general strategies to combat global warming is the development of government-run cap-and-trade systems, in which industries that emit CO$_2$ in the course of their business (such as electric utilities) would be allowed to offset these emissions by purchasing carbon "credits" from businesses whose work reduces atmospheric CO$_2$—in this case, California farmers.

To make a dent in global warming, carbon-capture farming would, of course, have to be ramped up to an enormous scale. But such farming is just one of the carbon-capture ideas that have been put on the table. How about burying old or dead trees below ground, so that they won't decay? How about putting photosynthesizing bacteria right at the source of coal-plant emissions, where they would soak up CO$_2$ and other greenhouse gases? Both these ideas have been proposed by serious scientists, and the latter idea currently is undergoing pilot-project testing. Only time will tell whether either proposal will work, of course, but note that both involve photosynthesis. This process of nature is by far the most powerful mechanism ever developed for first capturing and then locking up atmospheric carbon. Given this, it makes sense that a world that has an atmospheric carbon problem would be looking to photosynthesis for answers.

---

frequently with oxygen. The problem with this is that oxygen cannot serve as a building block for plant growth; only carbon can do that. Thus, when rubisco fails to bind with CO$_2$, plant growth is undercut. With all this in mind, here is a definition of **photorespiration**: a process in which the enzyme rubisco reduces carbon fixation in photosynthesis by binding with oxygen instead of with carbon dioxide. Meanwhile, **rubisco** can be defined as an enzyme that allows organisms to incorporate atmospheric carbon dioxide into their own sugars during the process of photosynthesis.

Evolution, however, produced an adaptation that allows some warm-climate plant species to lessen the photorespiration problem. It is a mechanism that uses a special carbon-fixing "front-end" and then a carbon-dioxide shuttle into the Calvin cycle. The so-called C$_4$ plants that make use of this mechanism don't initially use rubisco to fix carbon dioxide. Rather, they employ a different enzyme that will not bind with oxygen. The immediate product of this reaction is a four-carbon molecule (hence the C$_4$ cycle's name). A derivative of that molecule moves to a special group of cells and releases the CO$_2$ that will

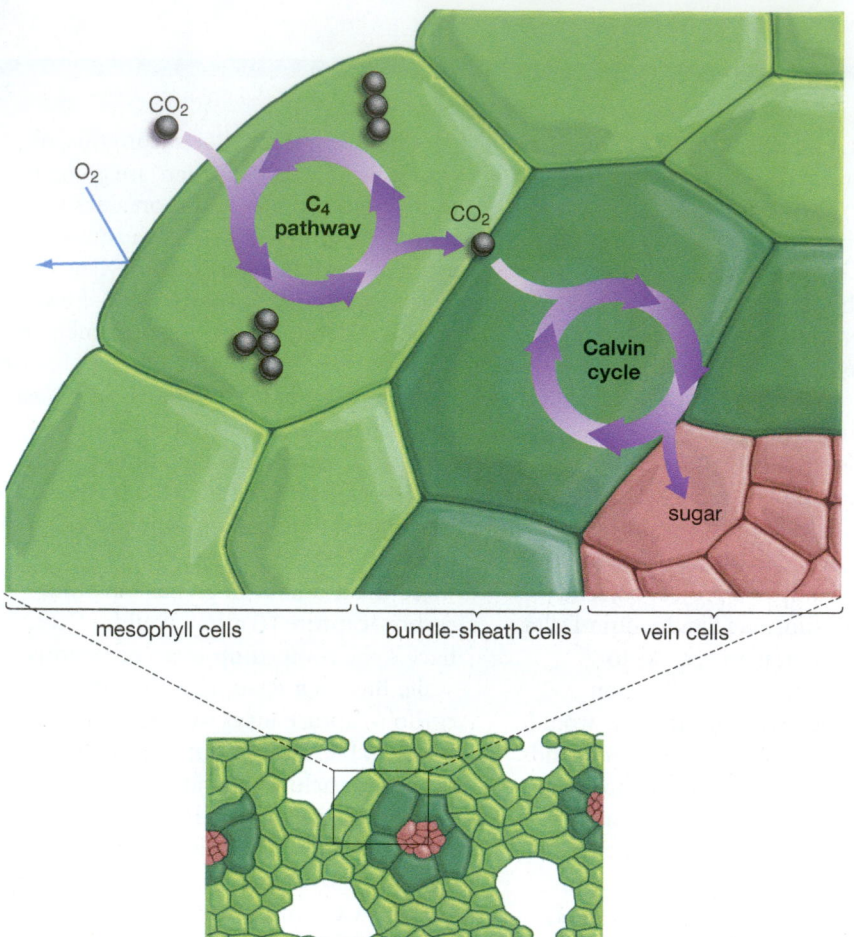

mesophyll cells   bundle-sheath cells   vein cells

8.6 CAM Photosynthesis

When plants live in climates that are not just warm but *dry*, a large part of their survival comes down to retaining water. Photosynthesis, however, works against water retention because, as you have seen, when CO_2 can pass in, water vapor can pass out. Evolution produced a third form of photosynthesis, however, that minimizes this problem for the dry-weather plants known as succulents. These plants *close* their stomata during the day and then open them at night. They then carry out C_4 metabolism at night, but only up to a point. They fix carbon dioxide into an initial four-carbon molecule and then stand pat. The CO_2 stays "banked" in them, awaiting the energy of the next day's sun, which will power the production of the ATP they need to carry out the Calvin cycle. This is **CAM photosynthesis**, defined as a form of photosynthesis, undertaken by plants in hot, dry climates, in which carbon fixation takes place at night and the Calvin cycle occurs during the day. This process gets the job done for cactus, pineapple, orchid, and some mint family plants, among others. As you may have guessed, CAM is an acronym; it stands for crassulacean acid metabolism. The succulent plant family Crassulaceae (which includes

Figure 8.10
C_4 Photosynthesis

Photorespiration occurs when the C_3 enzyme rubisco binds to oxygen rather than to carbon dioxide (thereby undercutting the production of sugar). C_4 plants contain a different enzyme in their mesophyll cells that binds to carbon dioxide but not to oxygen. Then the carbon dioxide is escorted into special bundle-sheath cells, where rubisco can bind to it, thus initiating the Calvin cycle.

then bind to rubisco and go through the Calvin cycle. These special cells are wrapped around leaf veins and are hence named the bundle-sheath cells (**Figure 8.10**). What the C_4 system provides—even at times when little CO_2 may be coming in through the stomata—is a relatively high concentration of CO_2 where it's needed: in the cells where the Calvin cycle takes place. We can thus define **C_4 photosynthesis** as a form of photosynthesis in which carbon dioxide is first fixed to a four-carbon molecule and then transferred to special cells in which the Calvin cycle is undertaken, bundle-sheath cells.

Lest all this seem like a purely theoretical issue, consider the fact that an international consortium of plant researchers is hard at work today trying to develop a C_4 version of the world's most important food crop, rice (**Figure 8.11**). The perceived need for such a plant is straightforward: Asia's population is expected to grow by 50 percent in the next 40–50 years, but the region's rice yields haven't changed appreciably in the last 30 years. Researchers calculate that these yields could be boosted by up to 50 percent if they could alter rice in such a way that it could carry out C_4 photosynthesis, rather than the C_3 variety it undertakes now. The goal is to develop a C_4 plant through a combination of bioengineering and old-fashioned plant breeding.

Figure 8.11
Producing a Better Rice Plant

A consortium called the International Rice Institute, headquartered in the Philippines, is leading the effort to develop a C_4 rice plant as a means of increasing rice production worldwide. In the photo, an IRI worker is inspecting one of the institute's experimental varieties of rice.

jade plants) was the first group of plants in which CAM metabolism was discovered. The three methods of photosynthesis we've looked at—C₃, C₄, and CAM—are summarized in **Table 8.1**.

On to Genetics

In the last three chapters, we've been looking at energy as it flows from the sun to plants and then on to the rest of the living world. In this journey, energy acts as a great enabler of life—as the essential force that allows life to exist in the first place and that, on a daily basis, allows flowers to open and birds to fly. Having learned something about how energy underlies life, however, it's now time to see what life does with its energy. Initially, we'll be interested in an activity that is undertaken on a grand scale within the confines of life: the amazing ability living things have of producing more living things. Two parents produce a son, but what exactly did the parents pass on that resulted in the color of this child's hair or the shape of his nose? And what are the chances that he might inherit a physical malady that afflicted his grandfather but that did not affect either of his parents? There's a branch of biology that seeks to answer questions such as these; it's called genetics, and it's the next stop in our tour of the living world.

SO FAR . . .

1. The enzyme rubisco brings together the _____ (from the plant) and _____ (from the atmosphere), creating the product that moves through the Calvin cycle. Unfortunately, rubisco also has an affinity for binding with _____, a process known as _____.

2. An evolutionary adaptation to this process, found in plants that thrive in _____ climates, is _____ photosynthesis, in which carbon dioxide is first fixed to a four-carbon molecule and then transferred into cells that carry out the Calvin cycle.

3. Plants that exist in climates that are both warm and dry employ a variation on photosynthesis, _____ photosynthesis, in which carbon dioxide is brought into the plant only _____, while the Calvin cycle is carried out only _____.

Photosynthesis in Dry Climates

Table 8.1

Three Modes of Photosynthesis

	C₃	C₄	CAM
Used by:	Majority of plants	Corn, sugarcane (warm environments)	Cactus, pineapple, orchid (dry environments)
Benefits:	Efficient use of ATP	Less photorespiration	Less water loss
Problems:	Photorespiration	Uses up more ATP	Uses up more ATP; hard to "bank" enough CO₂
How It Works			

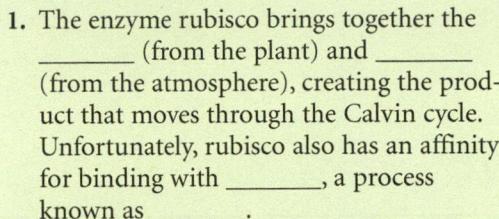

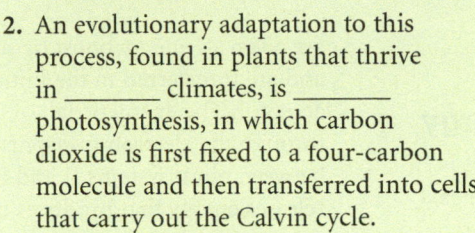

CHAPTER

8 REVIEW

Summary

8.1 Photosynthesis and Energy

- The sugar produced in photosynthesis is the source of food for most living things. Photosynthesis also is responsible for the atmospheric oxygen used by living things in cellular respiration. **(p. 141)**

8.2 The Components of Photosynthesis

- In plants and algae, photosynthesis takes place in the organelles called chloroplasts, which exist in abundance in the mesophyll cells of plant leaves. The energy for photosynthesis comes mostly from blue and red wavelengths of visible sunlight that are absorbed by pigments in the chloroplasts. Plant leaves contain microscopic pores called stomata that can open and close, letting carbon dioxide in and water vapor out. **(p. 142)**

- In the first stage of photosynthesis (the light reactions), electrons derived from water are energetically boosted by the power of sunlight. These electrons are passed through a series of electron carriers, ending up as part of NADPH, which carries them to the second stage of photosynthesis (the Calvin cycle), in which electrons are brought together with carbon dioxide and sugar to produce a high-energy sugar. **(p. 144)**

8.3 Stage 1: The Light Reactions

- In the light reactions, photosynthesis works by absorbing and transmitting solar energy through a pair of molecular complexes, photosystems II and I. **(p. 146)**

- In these reactions, water is split, yielding both the electrons that move through the light reactions and the oxygen that organisms breathe. The derived electrons

are given an energy boost by solar energy and are transferred to the initial electron acceptor. The energetic fall of electrons through the electron transport chain between photosystems II and I also releases energy that produces the ATP that powers the second stage of photosynthesis. **(p. 146)**

8.4 Stage 2: The Calvin Cycle

- In the Calvin or C_3 cycle, carbon dioxide is brought together with the sugar RuBP. The resulting compound is energized with the addition of electrons supplied by the light reactions. The result is the essential product of photosynthesis, the high-energy sugar G3P, which can be used for energy or for plant growth. **(p. 148)**

8.5 Photorespiration and the C_4 Pathway

- In C_3 plants, the enzyme rubisco frequently binds with oxygen rather than with carbon dioxide—a process called photorespiration that undercuts photosynthesis and that occurs with greater frequency in warm-weather climates. **(p. 149)**

- In some warm-weather plants, a form of photosynthesis called C_4 photosynthesis evolved that employs an enzyme that binds with carbon dioxide but not with oxygen. The carbon dioxide is then shuttled to special bundle-sheath cells and released, after which it moves into the Calvin cycle. **(p. 151)**

8.6 CAM Photosynthesis

- Many dry-weather plants employ CAM photosynthesis, in which the plant's stomata open only at night, letting in and fixing CO_2, which is then "banked" until sunrise, when the sun will supply the energy needed to power the Calvin cycle. **(p. 152)**

Key Terms

C_4 photosynthesis   152
Calvin cycle   148
CAM photosynthesis   152
chlorophyll *a*   144
chloroplast   144
fixation   148
photorespiration   151
photosynthesis   142
photosystem   144
reaction center   144
rubisco   151
stomata   143
stroma   144
thylakoid   144

Understanding the Basics

Multiple-Choice Questions

(Answers are in the back of the book.)

1. In photosynthesis, energy from the sun is being _____ while in the cellular respiration looked at in Chapter 7, energy is being _____ .
   a. turned into light ... turned solely into heat
   b. entirely lost to heat ... stored as ADP
   c. stored in carbohydrates ... released from storage in food
   d. absorbed as green light ... stored in carbohydrates
   e. spent as carbohydrates ... stored as carbohydrates

2. Stomata are the _____ on leaves that serve not only to let water vapor out but to let _____ in.
   a. proteins ... oxygen
   b. cilia ... oxygen
   c. outer cell layers ... sunlight

d. stems ... carbon dioxide

e. pores ... carbon dioxide

3. Two important products of the light reactions in photosynthesis are:

   a. electrons, oxygen.

   b. electrons, carbon dioxide.

   c. chloroplasts, carbon dioxide.

   d. rubisco, oxygen.

   e. rubisco, carbon dioxide.

4. Suppose you become a biologist. On a bright summer day, while studying the rate of photosynthesis in a desert plant, you discover that its stomata are closed, yet it is producing ATP. This plant must be carrying out:

   a. photorespiration.

   b. the C_3 cycle.

   c. the C_4 pathway.

   d. CAM photosynthesis.

   e. CO_2 uptake.

5. Most plant photosynthesis occurs in the _____ of leaves.

   a. upper epidermal cells

   b. lower epidermal cells

   c. petioles

   d. stomata

   e. mesophyll cells

6. The essential product of photosynthesis is:

   a. ATP.

   b. the enzyme rubisco.

   c. the sugar G3P.

   d. plant structural proteins.

   e. carbon dioxide

7. The electrons boosted to a higher energy level in photosystem II are derived from the splitting of _____, which results in the release of the by-product _____.

   a. carbon dioxide ... ATP

   b. water ... oxygen

c. NADPH ... carbon dioxide

d. ATP ... carbon dioxide

e. rubisco ... oxygen

8. The enzyme rubisco is important because:

   a. it aids in splitting water molecules, thus yielding oxygen.

   b. it powers the light reactions through use of ATP.

   c. it brings together carbon dioxide from the atmosphere with the plant's own sugar.

   d. it is photosystem II's primary electron acceptor.

   e. it is photosystem I's primary electron acceptor.

Brief Review

(**Answers are in the back of the book.**)

1. Since plants capture energy through photosynthesis, do they need to carry out cellular respiration as well? Why or why not?

2. Sunlight reaching Earth is composed of a spectrum of energetic rays. Describe the rays that make up this spectrum. Which portion of the spectrum drives photosynthesis?

3. What are the component parts of a photosystem?

4. Which plant would you expect to grow more rapidly in a warm-weather climate—a C_3 plant or a C_4 plant? Explain your answer.

5. The text notes that the fall of electrons between photosystems II and I provides energy used to produce ATP. What is this ATP used for?

Applying Your Knowledge

1. Rubisco has been called the most important protein on Earth. Why should it be regarded as so important?

2. In the mid-seventeenth century, Dutch scientist Jan Baptista van Helmont struggled to understand where plants get the material that allows them to grow larger. He planted a willow tree of a given weight in a fixed quantity of soil and then, five years later, weighed both the willow and the soil. He found that the tree had gained 164 pounds during the period but that the soil had lost very little weight. His conclusion was that the willow's weight gain came solely from the water it absorbed through its roots. If you could be transported back to van Helmont's time, what would you tell him about the source of the willow's weight gain?

MasteringBiology®

1. MasteringBiology Assignments

Activities: Properties of Light; Light Energy and Pigments; Leaves: The Sites of Photosynthesis; Overview of Photosynthesis; The Sites of Photosynthesis; The Light Reactions; The Calvin Cycle; The Calvin Experiments; Photosynthesis in Dry Climates; **Bioflix:** Photosynthesis; **Current Events:** Storms Threaten Ozone Layer Over U.S., Study Says; Giants Who Scarfed Down Fast Food Feasts; **Video Field Trip:** Solar Energy; **Building Vocabulary; Questions:** Bioflix Quiz, Reading Questions, Test Bank

2. eText

Read your book online, search, take notes, highlight text, and study more effectively.

3. The Study Area

Self-study tools to test comprehension.

The Links in Life's Chain:

Genetics and Cell Division

Life depends on information that is stored in DNA. This information is duplicated and then apportioned every time a cell divides.

Available on MasteringBiology®

Activities: The Cell Cycle; Mitosis and Cytokinesis Animation; Causes of Cancer; Cell Division for Bacteria; **Bioflix:** Mitosis; **Current Events:** Cancer's Secrets Come Into Sharper Focus; **ABC Video:** Henrietta Lacks' Cells; The Effect of Exercise on Cells; **Chemistry Review:** Nucleic Acids; **Building Vocabulary; Questions:** Bioflix Quiz, Reading Questions, Test Bank

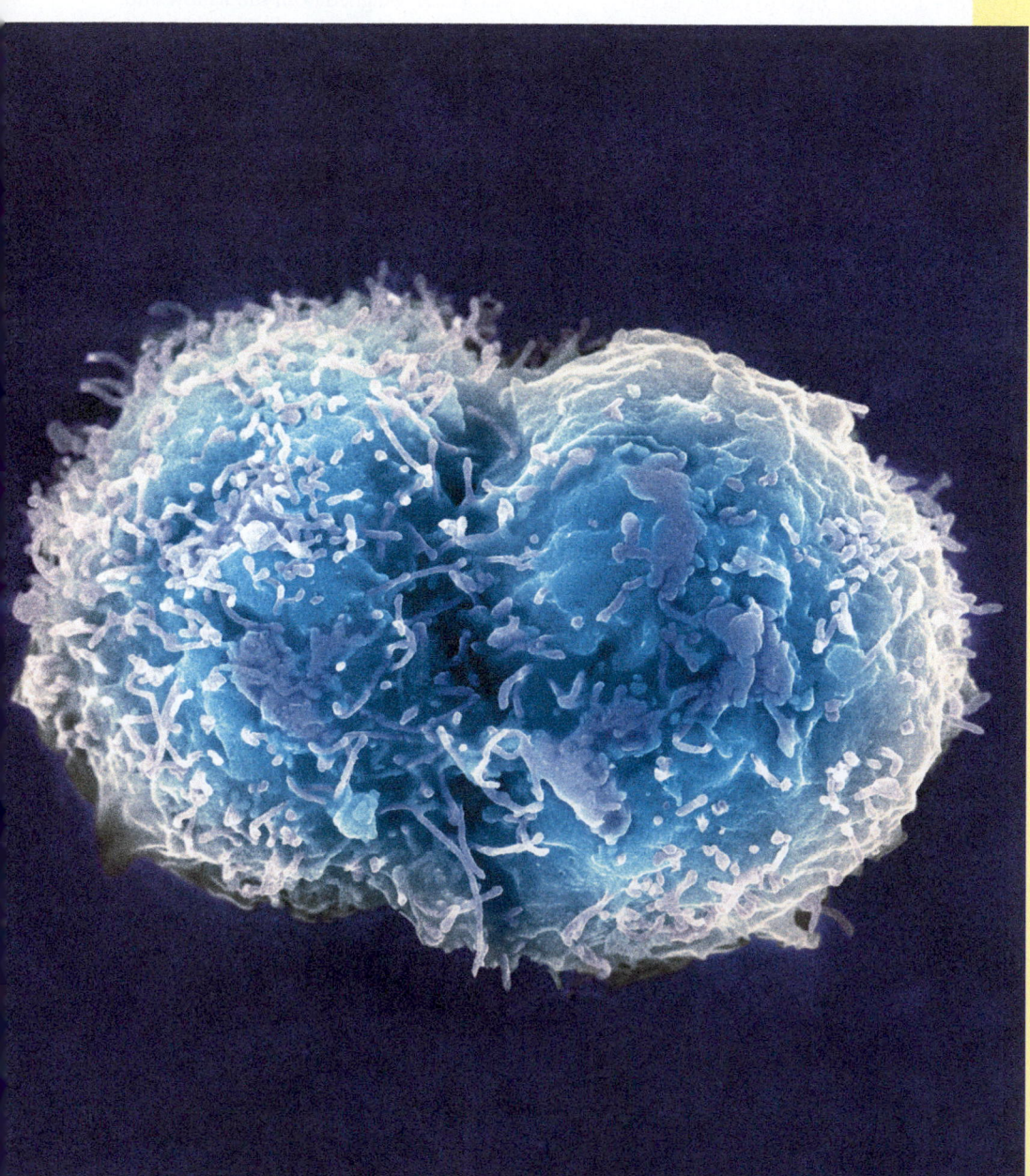

How do living things grow and reproduce? Through cell division. Here, one human cell is about to become two through this tightly choreographed process.

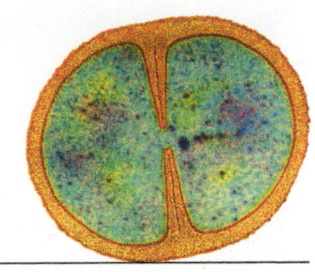

Everyone knows how one generation of living things gives rise to another, at least in a general sense. But the details of how this works are not apparent just by looking at Earth's creatures. A brief sexual encounter takes place, and the next thing we know there are new kittens, chicks, or babies. Only the encounter and the newborns are obvious; what comes in between is mysterious.

The problem, of course, is that most of the component parts of reproduction are hidden away and, in any event, are so small that the process is completely invisible to the naked eye. Given this, it's no wonder that for centuries human beings wondered what it was that one generation passed on, so that the next generation could be formed.

A similar kind of puzzlement existed over a related question: What is the mechanism that controls the development of a living thing as it goes from being a microscopic cell to being a human baby in full cry? For that matter, how is it that the adult body is able to build muscle or repair a minor wound? What, in short, is the mechanism that controls day-to-day physical functioning?

how living things develop and then function each day, the answer is: largely through the information provided by the genes in the genome.

To understand what this latter role means in practice, let's think about what happens when you eat something—say, a candy bar. The result of doing this is that you'll get a surge of simple sugar coursing through your veins; this is blood-sugar that needs to move into your cells to satisfy their energy needs. Now, this sugar will move into these cells largely through the action of the protein insulin, which helps create channels in cells through which nutrients can pass. But as you can imagine, none of us has a huge supply of ready-made insulin stored within us, just waiting for our blood-sugar surges. Instead, this protein has to be produced *as needed*—whenever we have a surplus of

Figure 9.1
Passing on Genetic Information
Half of each person's genome comes from the father and half from the mother.

9.1 An Introduction to Genetics

It took roughly a century of hard-won scientific advances for these questions to be answered. In the end, it turned out that a single substance is central to the reproduction of living things, their development from single cells, and their day-to-day functioning. This substance is deoxyribonucleic acid, a long, vanishingly thin molecule known everywhere today by its acronym, DNA.

DNA contains what is known as an organism's **genome:** the complete collection of that organism's genetic information. This information exists in units called genes that lie along DNA's famous double helix.

But how is it that this one molecule can undertake all the tasks just described? Well, if we ask what it is that humans *pass on* in reproduction, the answer turns out to be: half of a father's genome and half of a mother's genome, with both these halves coming together in a fertilized egg to produce a whole, new genome (**Figure 9.1**). With respect to the question of

blood-sugar in circulation. And the way this insulin gets produced is that the body's protein-producing machinery will go to a strand of DNA to retrieve one tiny bit of the information that is stored along it: the information on how to put together insulin.

This story deepens with recognition of the fact that insulin is only one protein among thousands that will be put together in a typical cell. And as it turns out, the information for the production of *all* of these proteins is contained in DNA. To grasp the importance of this, it's helpful to remember something first noted in Chapter 3: that proteins are the living world's most active and versatile molecules, serving not only as structural building blocks of life but as the chemically active workers that carry out most of life's essential tasks. Thus, there are proteins that form cartilage and hair; there are communication proteins that help form the receptors on cells; there are hormonal proteins and transport proteins. And then there are enzymes, the catalyst proteins that speed up—or in practical terms, *enable*—chemical reactions in living things. Without these proteins, almost nothing could get done in the body; with them, a dazzling set of chemical reactions takes place every second in every living thing. In sum, it's difficult to overstate how important proteins are to life, and the way proteins get produced is through the information contained in DNA.

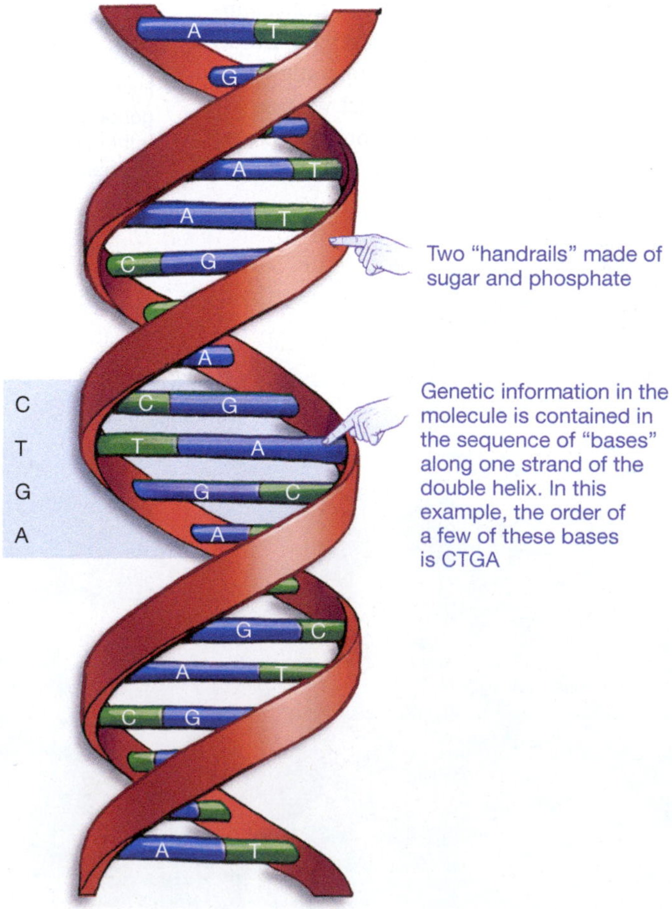

But what is the actual linkage here? That is, how does the information in DNA bring about the production of a protein? In Chapter 14, you'll get the details of this process, but for now let's just look at the essentials of the story. Imagine you are a telegraph operator in the Old West, and your "key" starts tapping frantically. You leap forward in your chair, listening for the sound of short and long bursts on the key, and you begin writing down the letters these sounds stand for. Here's the message you get:

.–.. ––– –.–. –.–   – ....– ..–. .

which means:

L   o   c   k   the   s   a   f   e

You walk over and do just that. There is thus a code here, Morse code, that has resulted in some action. In DNA there is likewise a code, but instead of a code specified in short or long sounds, it is a code specified in chemical substances. There are four of these substances, and they lie along the double helix. They are the "bases" adenine, thymine, guanine, and cytosine, abbreviated A, T, G, C, respectively.

A particular sequence of these bases—usually thousands of letters long—contains information that can be acted on, just as the Morse code did. Only the action here is the production of a particular protein. One series of A's, T's, C's, and G's contains the information for the production of one protein, but a *different* sequence of A's, T's, G's, and C's specifies a different protein. These separate sequences of bases are separate genes.

You saw in Chapter 3 that proteins are composed of building blocks called amino acids. The base sequence that makes up a gene, then, is like a message that says, "Give me this amino acid, now this one, then this one, . . . " and so on for hundreds of amino acids until a protein is created that folds up and gets busy on some task. This task can be one small part of *forming* a human being—as in embryonic development—or of *maintaining* one, as in an adult. One generation gives rise to another by passing on this entire complement of information.

The Architecture of DNA

It is the architecture of the DNA molecule that makes this protein production process possible. Look at the general scheme of it pictured in **Figure 9.2**. As you can see, DNA looks something like an open, spiral staircase, the handrails of which are a repeating series of sugar (deoxyribose) and phosphate molecules. Meanwhile, the messages you've just been reading about are contained in the steps of the staircase in the order in which the bases A, T, C, and G are placed along the helix. As you can see, these bases extend inward from both of the DNA rails. The order in which the bases occur along the rails is extremely varied. It is this order that "encodes" the information for the production of proteins.

Chemistry Review: Nucleic Acids

Figure 9.2
Information-Bearing DNA Molecule

Two "handrails" made of sugar and phosphate

Genetic information in the molecule is contained in the sequence of "bases" along one strand of the double helix. In this example, the order of a few of these bases is CTGA

The Path of Protein Synthesis

Look now at **Figure 9.3** and see the path that is followed in protein production or "synthesis." The DNA you have been reading about is contained in the nucleus of the cell. The first step in synthesis is that a stretch of it unwinds there, and its message—the order of a string of bases—is copied onto a molecule called messenger RNA (mRNA). This length of mRNA then exits from the cell nucleus. Its destination is a molecular workbench in the cell's cytoplasm, a structure called a ribosome. It is here that both message (the mRNA sequence) and raw materials (amino acids) come together to make the product (a protein). The mRNA sequence is "read" within the ribosome, and as this happens a chain of amino acids is put together in the ribosome in the order specified by the mRNA sequence. When the chain is finished and folded up, a protein has come into existence.

Genetics as Information Management

The story you have been reading about is that of **genetics,** meaning the study of physical inheritance among living things. In reflecting on this story, the first thing to note is that it concerns the storage, duplication, and transfer of *information*. True, it is information encoded in chemical form, but that shouldn't bother us. Information comes in lots of forms; there are the familiar words on a page, but there are also bar graphs, smoke signals, or a baby's smile.

To grasp the fact of genetics as molecular information, think back to Chapter 6, where you looked at a particular protein, an enzyme called lactase that helps break up milk sugar (lactose) into its components of glucose and galactose. In line with what you've seen so far, there is a gene that prompts the production of lactase. This gene's message is copied onto a length of mRNA, which then migrates to a ribosome in the cytoplasm. Lactase is put together there, amino acid by amino acid, and once synthesized, it folds up and gets busy clipping lactose molecules.

Now, consider the role of the gene in this process. It didn't do the lactose clipping; it didn't even migrate to the cytoplasm to be involved in the synthesis of lactase. In this case, its role was more like that of a … cookbook recipe. It specified this amino acid, then that amino acid, and eventually lactase was produced.

From One Gene to a Collection

Now, in considering not just one gene, but rather the entire *collection* of genes in a living thing—its genome, in other words—it is easy to see that we are dealing with a vast *library* of information. The size of such a collection can vary, but to give you some idea, the human genome is estimated to include 20,000–25,000 genes. We generally think that information comes to us after birth and from outside the body, but in fact we are born with a volume of information that has been amassed, edited, and passed down to us over 3.8 billion years of evolution.

The amazing thing about such collections is that there is not just one copy of them in living things; there may be trillions. Most cells within an organism contain a complete copy of that organism's genome. A given type of cell puts to use (or "expresses") only some parts of this genome, while another type of cell uses other parts; but this is merely a way of saying that different genes are active in different cells. This is what distinguishes a liver cell, say, from a bone cell. These cells do different things and, accordingly, need to synthesize different proteins. But *both* kinds of cells have within them the whole genomic library; they merely put different parts of it to use.

Cells duplicate, of course. One cell divides to become two, the two divide to become four, and so on. Given this, if each cell in the body is to have its own copy of the genome, then each time a cell duplicates, the genome that lies within that cell must duplicate as well, prior to the actual cell division. This is just what happens, in a process we'll look at now.

Figure 9.3
The Path of Protein Synthesis

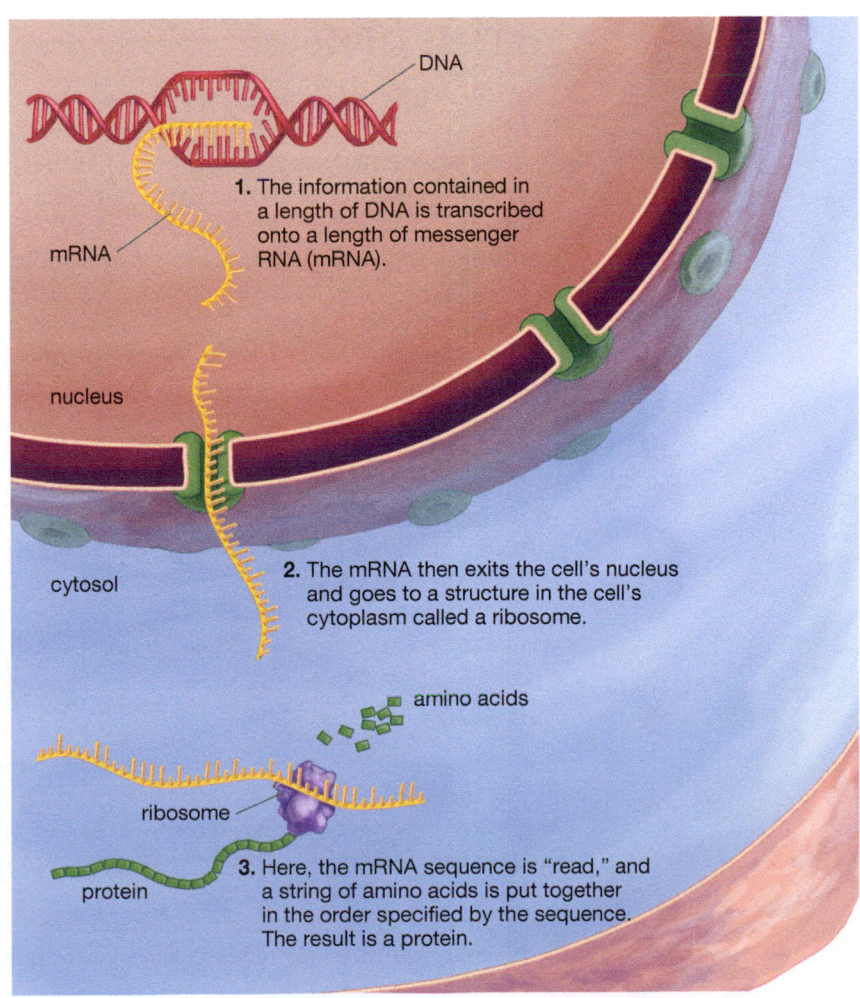

1. The information contained in a length of DNA is transcribed onto a length of messenger RNA (mRNA).

DNA

mRNA

nucleus

cytosol

2. The mRNA then exits the cell's nucleus and goes to a structure in the cell's cytoplasm called a ribosome.

amino acids

ribosome

protein

3. Here, the mRNA sequence is "read," and a string of amino acids is put together in the order specified by the sequence. The result is a protein.

SO FAR . . .

1. Genes contain information for the production of _____. The building blocks of genes are a set of four _____ whose symbols are _____, _____, _____, and _____.

2. Proteins are made up of a set of building blocks called _____. In any given protein, the order of these building blocks is specified by the order of the _____ that make up a _____.

3. Proteins are produced by means of information encoded in DNA being copied onto a molecule called _____, which then ferries the information to a structure called a _____. At this structure, _____ are linked to one another in the order specified by the _____, thus resulting in a _____.

9.2 An Introduction to Cell Division

The Cell Cycle

How does a baby grow, or a plant develop, or a wound heal? Always through cell division. As noted in Chapter 4, with the possible exception of viruses, life exists only inside cells. Further, we know that cells come only from other cells. And the *way* cells come from cells is by dividing. To understand the continuity in life, then, we need to have some understanding of how cells divide.

The sheer numbers involved in this process are mind-boggling. In your body, as many as 25 million cell divisions are taking place each second. It is likely that every one of these divisions serves to maintain your body in some way, yet there are instances in which cells divide excessively. How does the disease we call cancer take hold? Always through the *unrestrained* division of cells. As you can see in "When the Cell Cycle Runs Amok" on page 168, cancer is intimately related to basic processes of cellular division.

Why Do Cells Divide?

But why should cells divide at all? That is, why don't they just stay as they are? One answer is that cells die and need to be replaced. Beyond this, cells can only grow so large before they become too large—past a certain size, they can't function. (See "The Size of Cells" on page 68.) The solution to this dilemma is to divide and start small again. Moreover, more cells may

be needed at one point in life than in another. To take one example, the material that makes up human bones is secreted by cells. A growing teenager needs to add more of this bone material than does a mature adult. So, where is the teenager going to get this additional material? From additional cells—produced through cell division. None of this is to say that *every* cell divides. Most human brain cells are formed in the first three months of embryonic existence and then live for decades, with relatively few of them ever dividing again. Much the same is true of the leaf cells in plants; they divide only when the leaf is very small, then grow for a time, and then function at this mature size for as long as the leaf lives. At the other extreme, cells located in human bone marrow never *stop* dividing as they produce red blood cells (each one of which only lives for about two months). The output here is staggering: About 180 million new red blood cells are produced inside us each minute through cell division.

"Cell division" actually may be a somewhat misleading term if it is taken to mean a simple separation of cellular material, akin to a candy bar being split between two friends. Indeed, a splitting does occur in cell division, but certain parts of the cell must *duplicate* before this happens. Then, the duplicated material is parceled out with fine precision, half of it going to one "daughter" cell and half to the other.

What's being duplicated and apportioned is DNA. Because a cell's full complement of DNA contains such critical information, it would not do for a cell to be left with 50 or 75 percent of this information; rather, it needs the whole thing—no more and no less.

With this in mind, here is the big picture on cell division. First, there is a duplication of DNA; then there is the movement of two precisely matched quantities of DNA to opposite sides of the "parent" cell; finally, there is the splitting of the parent cell into two daughter cells. The duplication of DNA is known as *replication,* the apportioning of it into two identical quantities is known as *mitosis,* and the splitting of the cellular material is known as *cytokinesis* (**Figure 9.4**). The goal for the remainder of this chapter is to learn a little bit about replication and a good deal more about mitosis and cytokinesis.

The Replication of DNA

So how does the first part of cell division work—the DNA replication? If you look at **Figure 9.5**, you can see a simplified representation of the DNA molecule shown in Figure 9.2. Notice that the strands of the double helix have started to unwind, which is an initial step in DNA replication (Figure 9.5, step 1). Each of the two resulting single strands then serves as a template or pattern on which a new strand is created

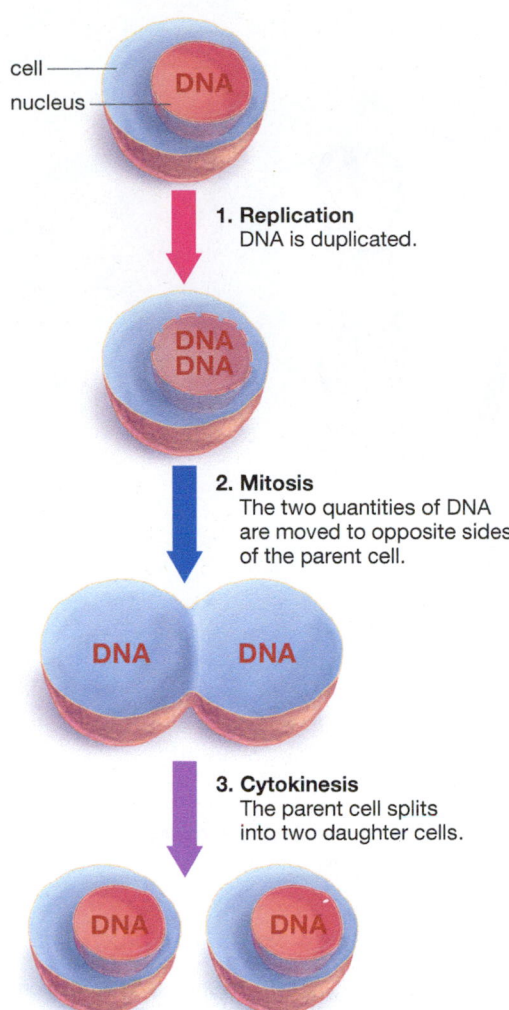

Figure 9.4
Overview of Cell Division

cell
nucleus
DNA

1. Replication
DNA is duplicated.

DNA
DNA

2. Mitosis
The two quantities of DNA
are moved to opposite sides
of the parent cell.

DNA   DNA

3. Cytokinesis
The parent cell splits
into two daughter cells.

DNA   DNA

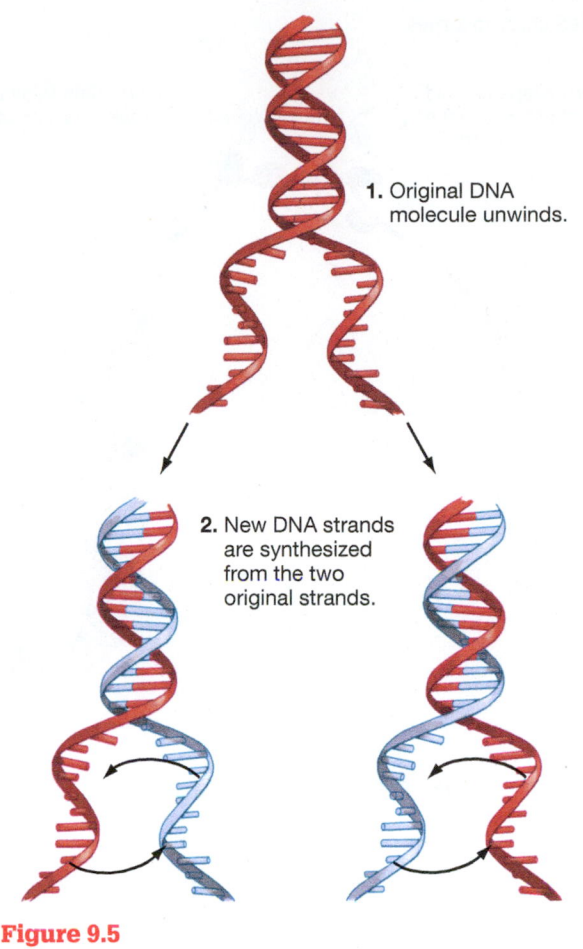

1. Original DNA
molecule unwinds.

2. New DNA strands
are synthesized
from the two
original strands.

Figure 9.5
DNA Replication

(1) A DNA molecule unwinds.

(2) Each of the single strands of the original molecule (in
red) serves as a template or pattern for the creation of a
second DNA strand (in blue).

(Figure 9.5, step 2). Earlier, DNA structure was compared to a spiral staircase. Now, think of that staircase as splitting right down the middle of its steps
(where the A and T or the G and C bases meet).
Then, a new *half*-staircase is created that bonds with
one of the old DNA strands; a second half-staircase
is likewise created that bonds with the *second* original strand. (These new strands are the blue ones in
Figure 9.5, step 2.) How does this happen? Free-
floating DNA bases will bond, one base at a time,
with bases on the original strand. This process continues down the line until two strands have been
created, each composed of one "old" half-strand
and one "new" half-strand. This replication process,
which doubles the cell's DNA content, takes place in
advance of the cell dividing.

In Chapter 13, you will metaphorically walk along
the rungs of the double helix to grasp the details of
this doubling of DNA strands. Right now, however,
you'll be considering DNA on a larger scale, as it
comes packaged.

9.3 DNA in Chromosomes

You have thus far conceptualized each cell's DNA as
spiraling out in *one* long double helix, but it does not,
in fact, take this form. Rather, the DNA in each cell
comes divided and packaged into individual units of
DNA called **chromosomes.** Different organisms have
different numbers of chromosomes; human cells
have 46, for example, while onion cells have 16.
The chromosomes of eukaryotic organisms, such as
ourselves, amount to DNA "packages" in several
senses. They consist not only of DNA itself but of
proteins around which DNA is wrapped. The result
is **chromatin:** a molecular complex, composed of
DNA and associated proteins, that makes up the
chromosomes of eukaryotic organisms. Thus, a number of chromosomes, composed of chromatin, exist
in a cell's nucleus, detached but in close physical
proximity to one another, and the *collection* of chromosomes makes up nearly the entire complement of
a cell's DNA.

(a) DNA is packaged in units called chromosomes

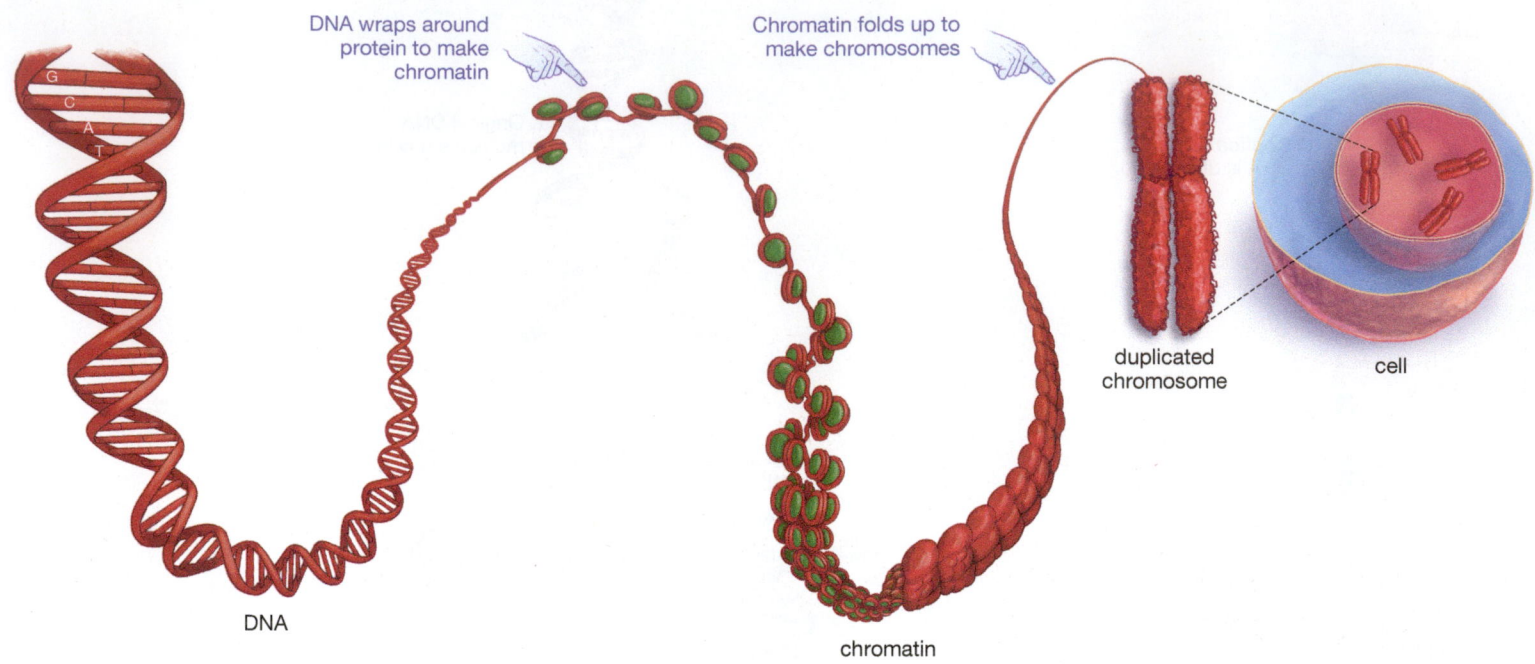

DNA wraps around protein to make chromatin

Chromatin folds up to make chromosomes

DNA

chromatin

duplicated chromosome

cell

(b) DNA replication at two levels

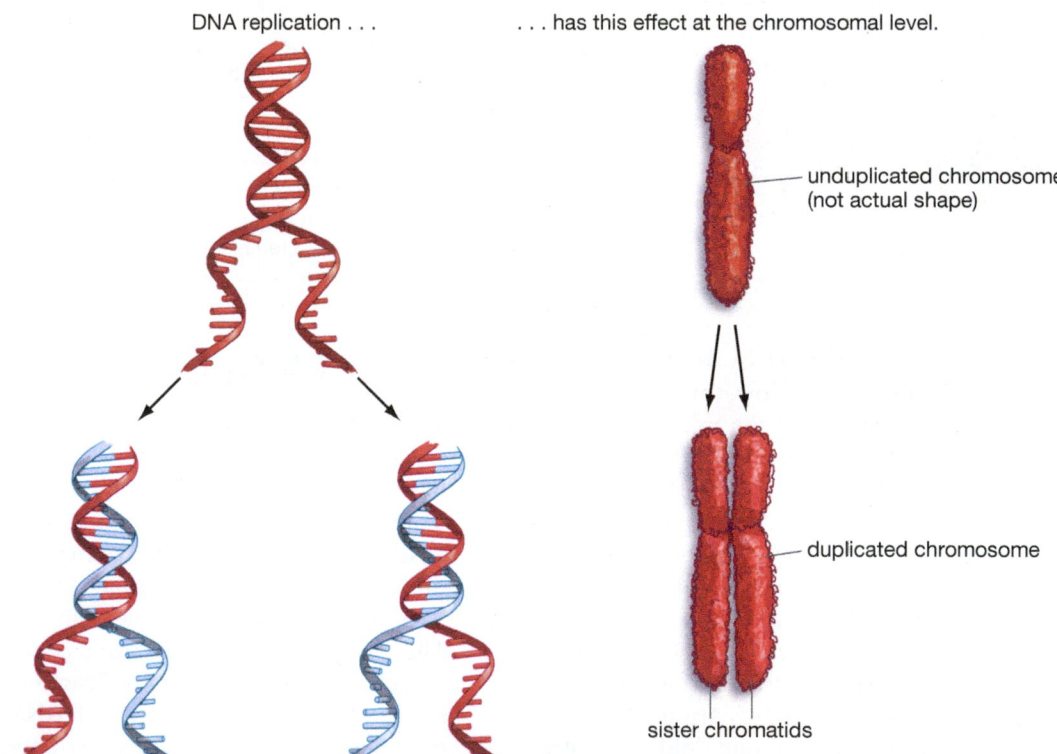

DNA replication . . .

. . . has this effect at the chromosomal level.

unduplicated chromosome (not actual shape)

duplicated chromosome

sister chromatids

Figure 9.6
Chromosomes and DNA Replication

(a) The DNA molecule is bound up with proteins, with the resulting combination known as chromatin. This chromatin then makes up structures called chromosomes.

(b) When DNA replicates, the result is two copies of the original DNA molecule. At the chromosomal level, the result of this replication is a single chromosome in duplicated state. The chromosome is composed of two sister chromatids—the two copies of the original DNA molecule. An unduplicated chromosome doesn't actually have the well-defined shape of the chromosome in the figure. It's shown this way merely for purposes of comparison with the duplicated chromosome.

If you look at **Figure 9.6a**, you can see how to think about the double helix being packaged into chromosomes. **Figure 9.6b** shows you how to relate the "staircase-splitting" DNA replication just described to what this means at the chromosomal level. When we say that DNA is replicating, this is another way of saying that the chromosomes that DNA helps make up are *duplicating*. Once that has happened, the result is chromosomes in "duplicated state," which means individual chromosomes, each of which is made up of two sister chromatids. Each chromatid is one of the newly replicated DNA double helices plus its associated proteins, as you can see in Figure 9.6b. More formally, a **chromatid** is one of the two identical strands of chromatin that make up a chromosome in its duplicated state. (It probably goes without saying that despite their name, sister chromatids are neither siblings nor females.)

Matched Pairs of Chromosomes

Chromosomes are detached from one another, but that does not mean they are completely different from one another. In eukaryotes, in fact, chromosomes tend to come in *pairs* that are close, but not exact, matches. The 46 chromosomes we have come to us as 23 chromosomes from *each parent*. Critically, with one exception that you'll soon get to, these are 23 matched pairs of chromosomes, each chromosome from the mother matching with one from the father.

What is a "matched pair" of chromosomes? Well, we have a chromosome 1 that we inherit from our mother, and it contains a set of genes very similar to those that lie on the chromosome 1 we inherit from our father. And the way the genes on these two chromosomes are similar is that they code for the same types of proteins. For example, human chromosome 1 contains a gene that codes for a protein, called F5, that helps blood to clot following an injury. For our purposes, the important thing is that the *F5* gene exists on *both* the maternal and paternal copies of chromosome 1. Indeed, if we were to look at all the roughly 4,200 genes found on chromosome 1 we would find this same symmetry: A given gene found on *pa*ternal chromosome 1 will almost certainly have a counterpart on *ma*ternal chromosome 1. This pattern then continues for the genes found on chromosome 2, chromosome 3, and so forth. It's easy to see, then, that the paternal and maternal copies of a given chromosome are the same in function; any two chromosomes that share this quality are said to be **homologous chromosomes.**

It's important to note, however, that homologous chromosomes are not *exactly* alike, because there can be variations on the genes found on the members of any homologous pair. If a given paternal chromosome has a gene that codes for, say, hair color, then it's a safe bet that the matching maternal chromosome will likewise have a gene that codes for hair color. But a gene on the paternal chromosome may help code for red hair, while the gene on the matching maternal chromosome may help code for blond hair. Nevertheless, both chromosomes have genes that code for hair color, as well as for thousands of other traits, and as such are said to be homologous.

X and Y Chromosomes

The one exception to the matched-pairs rule in human chromosomes is the so-called sex chromosomes of males. Human females have the 23 pairs of homolo-

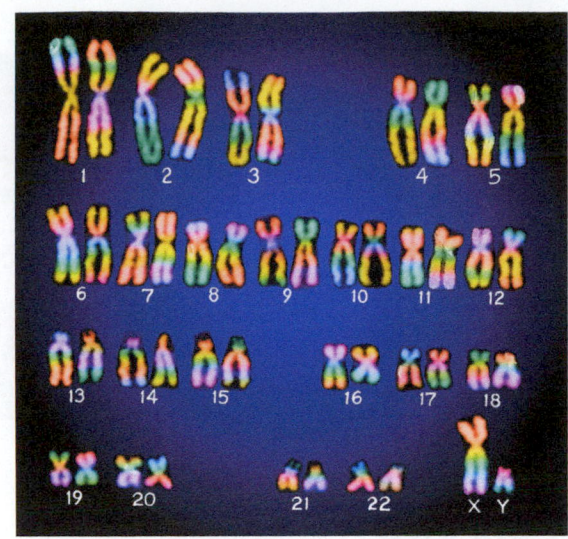

Figure 9.7
A Karyotype Displays a Full Set of Chromosomes
One member of each chromosome pair comes from the individual's father and the other member from the mother. Each paired set of chromosomes is said to be "homologous," meaning the same in function. (The two chromosomes over the number 1 are a homologous pair, as are the two over number 2, and so forth.)

gous chromosomes mentioned before; 22 of these are pairs of *autosomes,* or non-sex chromosomes, but one is a pair of homologous X chromosomes that, when present, means a growing embryo will be a female. Males, on the other hand, have the 22 pairs of homologous autosomes and one X chromosome; but then they also have one *Y* chromosome that, when present, means a growing embryo will be a male.

This whole scheme can be set forth in a visual display called a **karyotype,** meaning a pictorial arrangement of a complete set of human chromosomes. If you look at **Figure 9.7**, you can see an example of a karyotype, with its 46 chromosomes arranged in order, 22 of them matched pairs. In this case, one pair, at lower right, has one X chromosome and one Y chromosome, meaning this is a male.

The chromosomes in Figure 9.7 are in the duplicated state noted earlier, each of them being composed of two sister chromatids. For the sake of clarity, it would be nice to see a picture of some *un*duplicated chromosomes, but we have no pictures that are the counterpart to Figure 9.7. This is so because chromosomes take on an easily discernible shape only after they duplicate, prior to cell division. As cell division approaches, the DNA–protein complex changes its form; it tightens up, condensing mightily to produce well-defined chromosomes (**Figure 9.8b**, on the next page). After cell division is complete, the cell's chromosomes return to their relatively formless state, shown in **Figure 9.8a**.

Why this change in form? This condensing before cell division has the same effect as you taking all your scattered belongings and packing them into boxes just before moving from one apartment to another. Remember that DNA is packing up to leave as well, in this case for life in a successor cell. Were it not to tighten into its duplicated chromosomal form, its elongated fibers would get tangled up in the move.

Figure 9.8
DNA Can Be Arranged in Two Ways

(a) Replication occurs when the chromosomes are in a relatively formless state. In this state, the chromatin that makes up the chromosomes has yet to condense into its compact arrangement.

(b) Mitosis occurs after chromatin condenses into the easily discernible shapes of duplicated chromosomes. The micrograph is of color-enhanced duplicated human chromosomes.

(a) DNA in uncondensed form

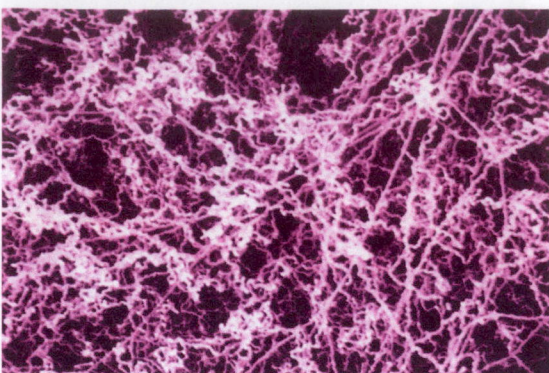

(b) DNA condensed into duplicated chromosomes

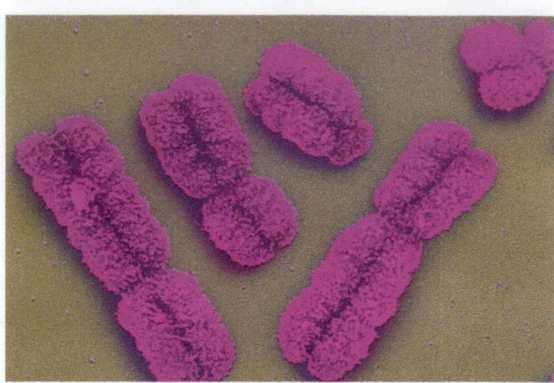

SO FAR . . .

1. For any cell to divide, the DNA it contains must first double or _____ so that both of the daughter cells that are produced will have their own complement of DNA. To look at this process another way, prior to cell division each of the chromosomes in a cell must _____.

2. Chromosomes are composed of both DNA and its associated _____. In duplicated state, each chromosome is composed of two identical _____.

3. In human beings (and many other organisms) chromosomes come in pairs that are said to be _____, meaning the same in function. The pairs rule is seen in women, who have a pair of chromosomes that confer female gender, the _____ chromosomes. Meanwhile, men have one _____ chromosome (like women) but one _____ chromosome, which confers the male gender.

Chromosome Duplication as a Part of Cell Division

With the chromosomes duplicated, you are now ready to look at the process by which they will split up. Before following this path, though, recognize that this separation occurs within the larger process that we are following: the division of the cell as a whole. As you saw in Chapter 5, there are many more things inside a cell than its complement of chromosomes. There are mitochondria, lysosomes, ribosomes, and so forth and the cytosol in which they are immersed. Together, these things lie in the cell's cytoplasm—the area *outside* the nucleus, as opposed to chromosomes, which lie *inside* the nucleus. With cell division, half of the cytoplasmic material goes to one daughter cell and

half to the other. It is helpful, then, to conceptualize cell division as having two separable components. **Mitosis** is the separation of a cell's duplicated chromosomes prior to cytokinesis. **Cytokinesis** is the physical separation of one cell into two daughter cells. These two processes are part of the big picture of cell division sketched earlier, which now can be stated in a slightly different way. First, chromosomes duplicate; then chromosomes separate and move to opposite sides of the parent cell (mitosis); then the parent cell splits into two daughter cells (cytokinesis).

The Cell Cycle

If you look at **Figure 9.9**, you can see how all these processes are linked over time. The figure illustrates the **cell cycle:** the repeating pattern of growth, genetic duplication, and division seen in most cells. There are two main phases in this cycle. First, there is **interphase:** that portion of the cell cycle in which the cell simultaneously carries out its work and, in preparation for division, duplicates its chromosomes. Second, there is **mitotic phase** (or M phase): that portion of the cell cycle that includes both mitosis and cytokinesis.

Interphase and M phase are in turn subdivided into smaller phases. In interphase, first there is G_1, standing for "gap-one." What goes on here are normal cell operations and cell growth. The cell then enters the S or "synthesis" phase, which is the synthesis of DNA, resulting in the duplication of the chromosomes. When this ends, interphase's gap-two or G_2 period begins, during which there is more cytoplasmic growth and a preparation for cell division. Because DNA synthesis is stopped in G_2 as well as G_1, the word *gap* in G_1 and G_2 can be thought of as a gap in DNA synthesis.

The length of the cell cycle varies greatly from one type of cell to another. In a typical animal cell, the total cell-cycle length averages about 24 hours. Within this cycle, mitotic phase takes up only about 30 minutes. Within interphase, an animal cell spends roughly 12 hours in G_1, 6 hours in S, and 6 hours in G_2. Some cells divide only rarely, while others—such as most

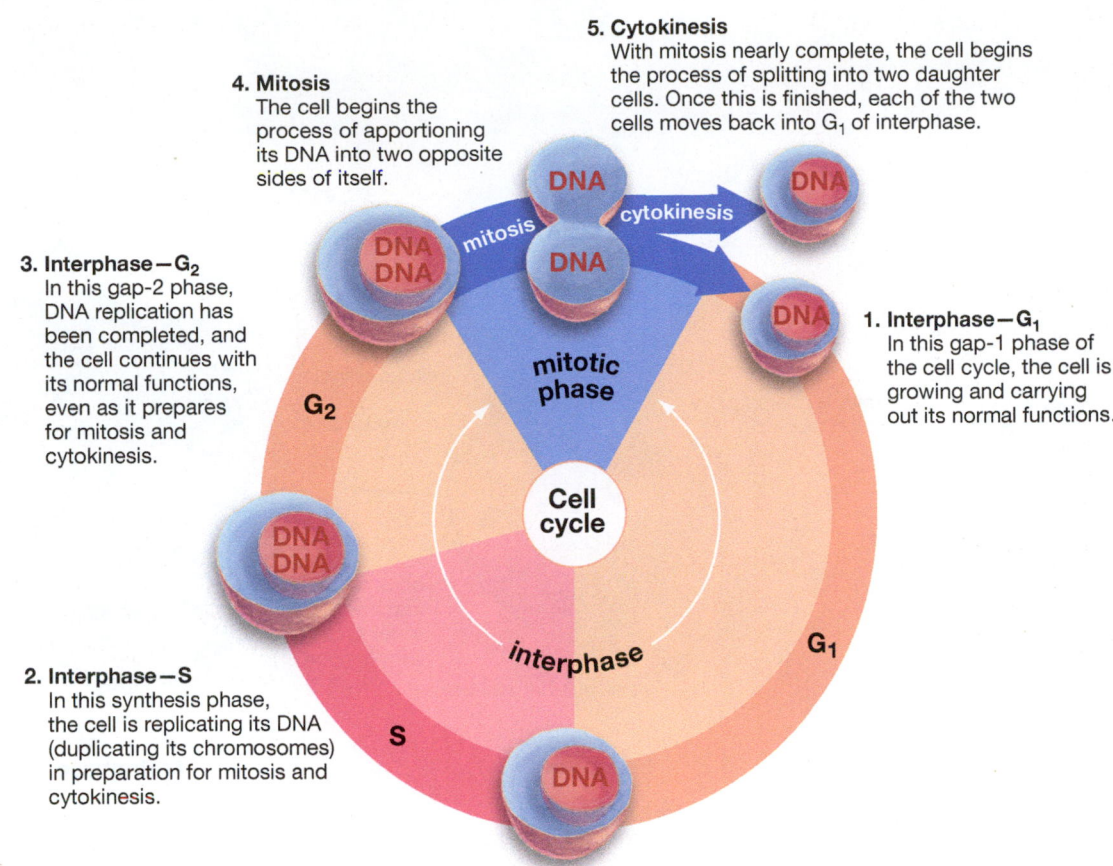

5. **Cytokinesis**
With mitosis nearly complete, the cell begins the process of splitting into two daughter cells. Once this is finished, each of the two cells moves back into G₁ of interphase.

4. **Mitosis**
The cell begins the process of apportioning its DNA into two opposite sides of itself.

3. **Interphase—G₂**
In this gap-2 phase, DNA replication has been completed, and the cell continues with its normal functions, even as it prepares for mitosis and cytokinesis.

1. **Interphase—G₁**
In this gap-1 phase of the cell cycle, the cell is growing and carrying out its normal functions.

2. **Interphase—S**
In this synthesis phase, the cell is replicating its DNA (duplicating its chromosomes) in preparation for mitosis and cytokinesis.

Figure 9.9
The Cell Cycle
The three main stages of cell division—DNA replication, mitosis, and cytokinesis—can be seen here in the context of a complete cell cycle. This cycle traditionally is divided into two main phases, interphase and mitotic (or M) phase, which are in turn divided into other phases.

nerve cells—don't divide at all. We could think of these cells as existing in an extended G₁ phase. But to distinguish them from cells that are simply moving through G₁, they are said to be in G₀ phase.

9.4 Mitosis and Cytokinesis

It is the second main phase of the cell cycle, mitotic phase, that is the focus of the rest of this chapter. Now the concern is not with how a cell manages its general functions, or with how it replicates its DNA, but with how it *divides*—how it carries out both mitosis and cytokinesis. Mitosis turns out to have four phases: prophase, metaphase, anaphase, and telophase. We'll now look at these phases in order, after which we'll review cytokinesis.

The Phases of Mitosis

Prophase

The beginning of mitosis marks the end of the cell's interphase (**Figure 9.10**, on the next page). The starting state is that the cell's DNA, which was replicated during the S phase, has just begun to pack itself into well-defined chromosomes. When we can finally *see* such chromosomes with a microscope, we can say that mitosis has begun, with prophase. Now the packing job continues; before it is done, the DNA will have coiled and condensed into one eight-thousandth of its

S-phase length. When this is over in human beings, there are 46 well-defined chromosomes in the nucleus in duplicated state (meaning they are composed of 92 chromatids). Meanwhile, the nuclear envelope—the double membrane surrounding the nucleus—begins to break up.

While this is going on, big changes are taking place outside the nucleus. Recall from Chapter 4 the structures called **microtubules**: protein fibers that are part of the cell's cytoskeleton or internal fiber network. In some of their roles, microtubules can be compared to tent poles that can shorten or lengthen as need be. Throughout mitosis, microtubules stretch the cell as a whole and physically move the cell's chromosomes around.

In an interphase cell there exists, just outside the nucleus, a **centrosome**: a cellular structure that acts as an organizing center for the assembly of microtubules. This centrosome duplicates, so that now there are two microtubule organizing centers. Now, in prophase, these two centrosomes start to move apart. Where are they going? To the poles. From here on out, it is convenient to think of mitosis and cytokinesis in terms of a global metaphor: The centrosomes migrate to the cellular poles, while the chromosomes first align, and then separate, along a cellular equator called the metaphase plate. Unlike microtubules or the centrosome, the **metaphase plate** is not a physical structure but instead is a plane located midway between

Bioflix: Mitosis

Mitosis and Cytokinesis Animation

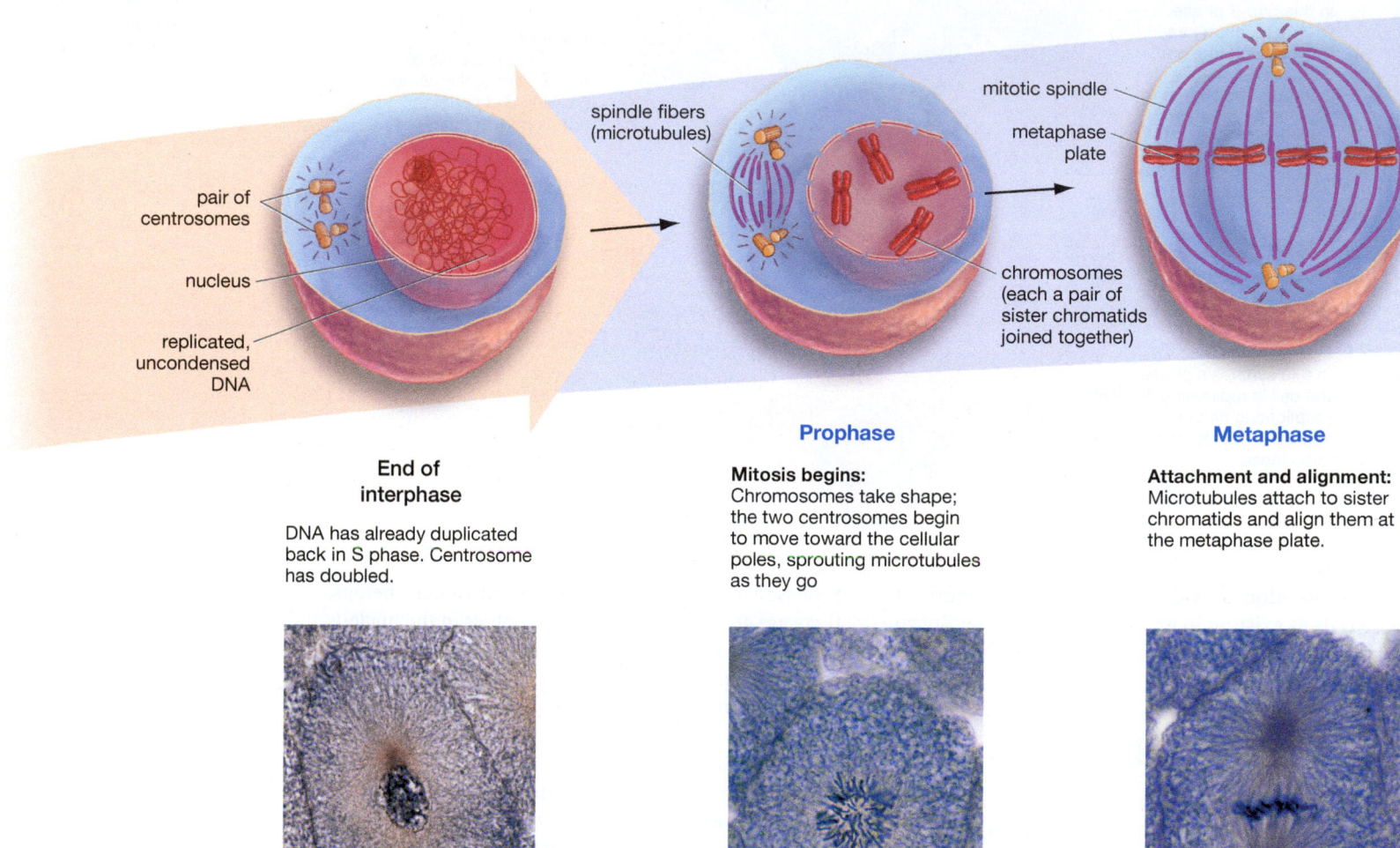

pair of centrosomes

nucleus

replicated, uncondensed DNA

spindle fibers (microtubules)

chromosomes (each a pair of sister chromatids joined together)

mitotic spindle

metaphase plate

Prophase

Metaphase

End of interphase

DNA has already duplicated back in S phase. Centrosome has doubled.

Mitosis begins: Chromosomes take shape; the two centrosomes begin to move toward the cellular poles, sprouting microtubules as they go

Attachment and alignment: Microtubules attach to sister chromatids and align them at the metaphase plate.

Figure 9.10
Mitosis and Cytokinesis

The micrographs at the bottom of the figure are pictures of mitosis and cytokinesis in a whitefish embryo.

the poles of a dividing cell. As the centrosomes move apart, they begin sprouting microtubules in all directions. One variety of them forms a football-shaped cage around the nuclear material, while a second variety attaches to the chromosomes themselves. Taken together, the microtubules active in cell division are known as the **mitotic spindle.**

Metaphase

By the time metaphase begins, the nuclear envelope has disappeared completely, and the microtubules

that were growing toward the chromosomes now *attach* to them. Through a lively back-and-forth movement, the microtubules align the chromosomes at the equator. With this, each chromatid now faces the pole *opposite* that of its sister chromatid, and each chromatid is attached to its respective pole by perhaps 30 microtubules.

Anaphase

At last the genetic material divides. As you may have guessed, this is a parting of sisters. The sister

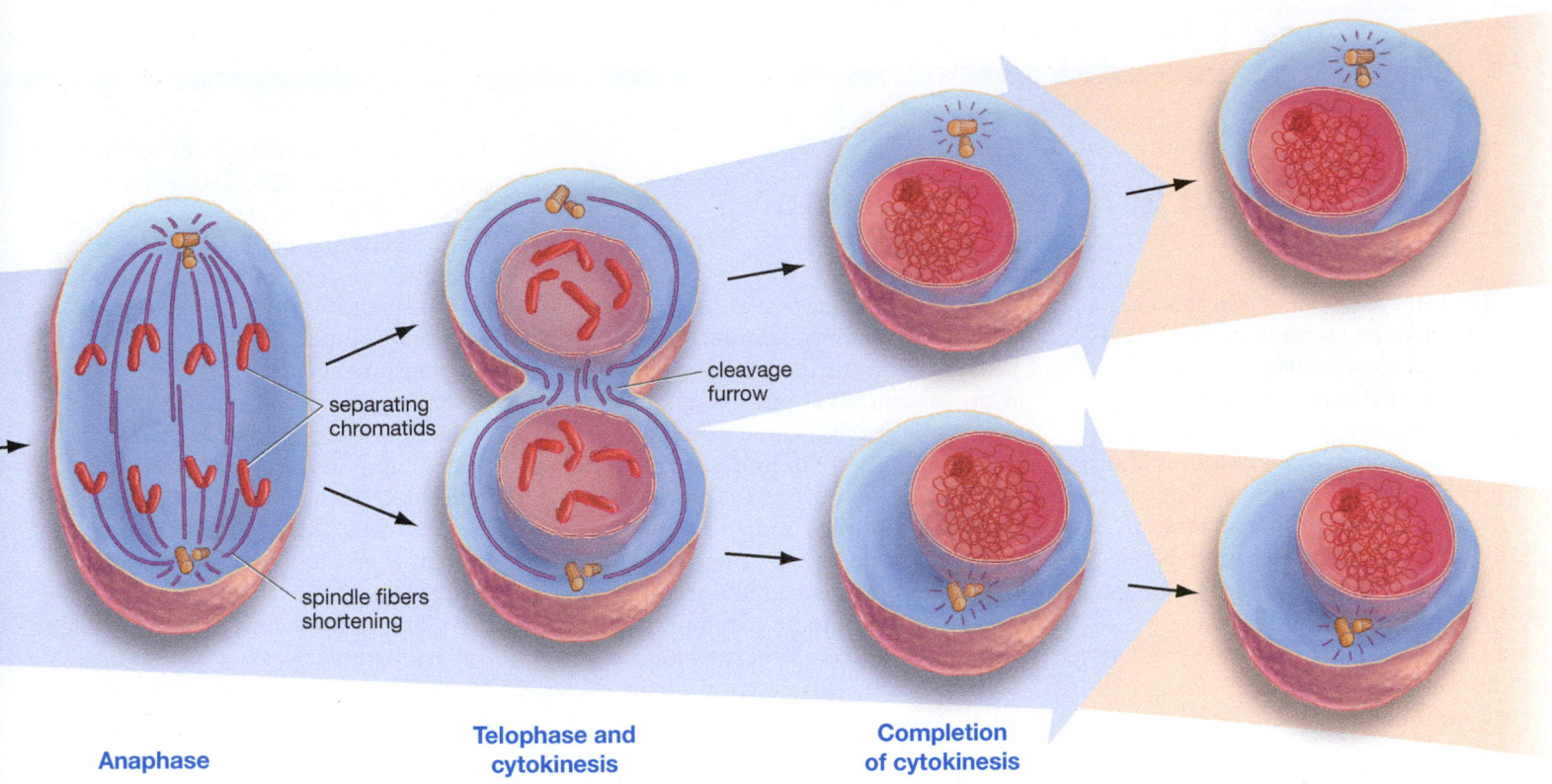

separating chromatids

cleavage furrow

spindle fibers shortening

Anaphase

Separation:
Sister chromatids are moved to opposite poles in the cell, each chromatid now becoming a full-fledged chromosome.

Telophase and cytokinesis

Exit from mitosis:
Chromosomes decondense; nuclear envelopes form around the two separate complements of chromosomes. Cleavage furrow begins to form.

Completion of cytokinesis

One cell becomes two:
The cell membrane pinches together completely; membranes on either side fuse together, creating two cells.

Beginning of interphase

These two cells now enter the G_1 phase of interphase.

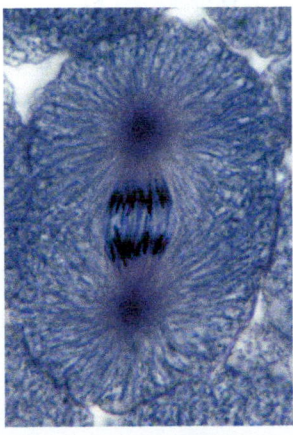

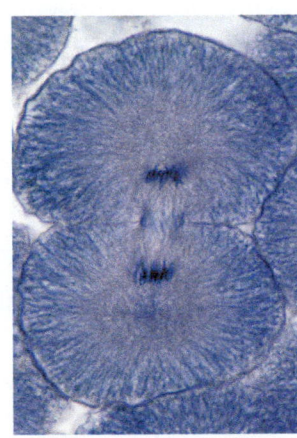

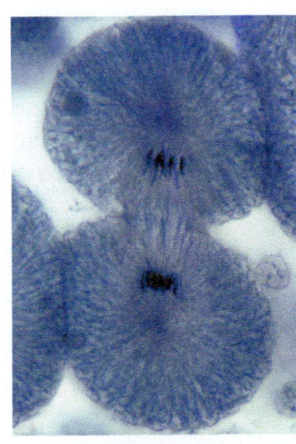

chromatids are pulled apart, each now becoming a full-fledged chromosome. All 46 chromatid pairs divide at the same time, and each member of a chromatid pair moves toward its respective pole, pulled by a shortening of the microtubules to which it is attached.

Telophase

Telophase represents a return to things as they were before mitosis started. The newly independent chromosomes, having arrived at their respective poles,

now unwind and lose their clearly defined shape. New nuclear membranes are forming. When this work is complete, there are two finished daughter nuclei lying in one elongating cell. Even as this is going on, though, something else is taking place that will result in this one cell becoming two.

Cytokinesis

Cytokinesis actually began back in anaphase and is well under way by the time of telophase. It works

ESSAY

When the Cell Cycle Runs Amok: Cancer

The cell cycle reviewed in this chapter is, in one sense, a common natural process: Cells grow; they duplicate their chromosomes; these chromosomes separate; one cell divides into two. This goes on like clockwork, millions of times a second in each one of us. Then one day we learn that an aunt, a grandfather, or a friend is experiencing an *unrestrained* division of cells. Things aren't explained to us in this way, of course. We are simply told that someone we know has cancer.

At root, all cancers are failures of the cell cycle. Put another way, all cancers represent a failure of cells to limit their multiplication in the cell cycle. What is liver cancer, for example? It is a damaging multiplication of liver cells. First one, then two, then four, then eight liver cells move repeatedly through the cell cycle, and as their numbers increase, they destroy the liver's working tissues. Given that cancer manifests in this way, it's not surprising that a large portion of modern cancer research is *cell-cycle* research. The logic here is simple: To the extent that uncontrolled cell division can be stopped, cancer can be stopped.

In recent years, cell-cycle research has brought about a remarkable change in the kinds of treatments that are being developed for cancer. For decades, the basic idea behind cancer treatment was to kill as many cancer cells as possible, either by removing these cells from the body altogether (in surgery) or by destroying them with radiation or drugs. In contrast, nearly all the cancer drugs being developed today are intended not to kill cancerous cells. Instead, they are intended to disrupt the process by which these cells multiply. The idea is that, just as a light can be switched off by disrupting an electrical circuit, so cell division can be switched off by disrupting a *molecular* circuit—the chain of chemical reactions that keeps a cancerous cell moving constantly through cell division. The new breed of cancer treatments that are built on this idea are called targeted therapies.

To get a feel for what this approach means in practice, consider the first targeted therapy ever developed: a drug called Gleevec, which is aimed at combating a form of cancer, called chronic myeloid leukemia (CML), that results in an uncontrolled proliferation of certain types of white blood cells. In the 1990s, the researcher most responsible for Gleevec's use in CML treatment, Oregon Health and Sciences University Professor Brian Druker, began to focus on the process by which CML cells come to their out-of-control state. Druker knew that CML begins in a single white blood cell when two of its chromosomes undergo an unusual form of segment-swapping with one another. When a segment from chromosome A fuses with a segment of chromosome B in this process, the result is that information from both chromosomes combines to create a mutant gene—a gene that does not exist in any normal white blood cell. Of course, this mutant gene (dubbed *BCR-ABL*) then results in a protein that is capable of carrying out some task in the white blood cell. Unfortunately, what the BCR-ABL protein does is flip a molecular switch; through it, the white blood cell starts multiplying more rapidly, even as it and its daughter cells start undergoing further destabilizing changes that ultimately result in full-blown leukemia.

Druker's key insight into stopping CML was to look very carefully at the BCR-ABL protein, which you can see represented in **Figure 1**. He realized that BCR-ABL is powered into activity following its binding with the energy-transfer molecule ATP. So, he wondered, what would happen if the ATP binding site on the BCR-ABL protein became unavailable—what would happen if this binding site became *occupied* by another molecule? The substance he eventually hit upon that takes on this occupying role

Figure 9.11
Cytokinesis in Animals

Cytokinesis in animal cells begins with an indentation of the cell surface, a cleavage furrow, shown here in a dividing frog egg.

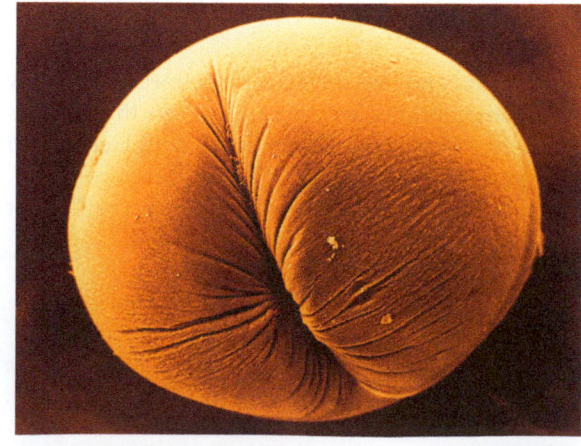

through the tightening of a cellular waistband that is composed of two sets of protein filaments working together. These filaments—the same type that allow your muscles to contract—form a *contractile ring* that narrows along the cellular equator (**Figure 9.11**). An indentation of the cell's surface (called a cleavage furrow) results from the ring's contraction; consequently, the fibers in the mitotic spindle are pushed closer and closer together, eventually forming one thick pole that is destined to break.

The dividing cell now assumes an hourglass shape; as the contractile ring continues to pinch in, one cell becomes two by means of something you looked at in Chapter 4: membrane fusion. The membranes on each

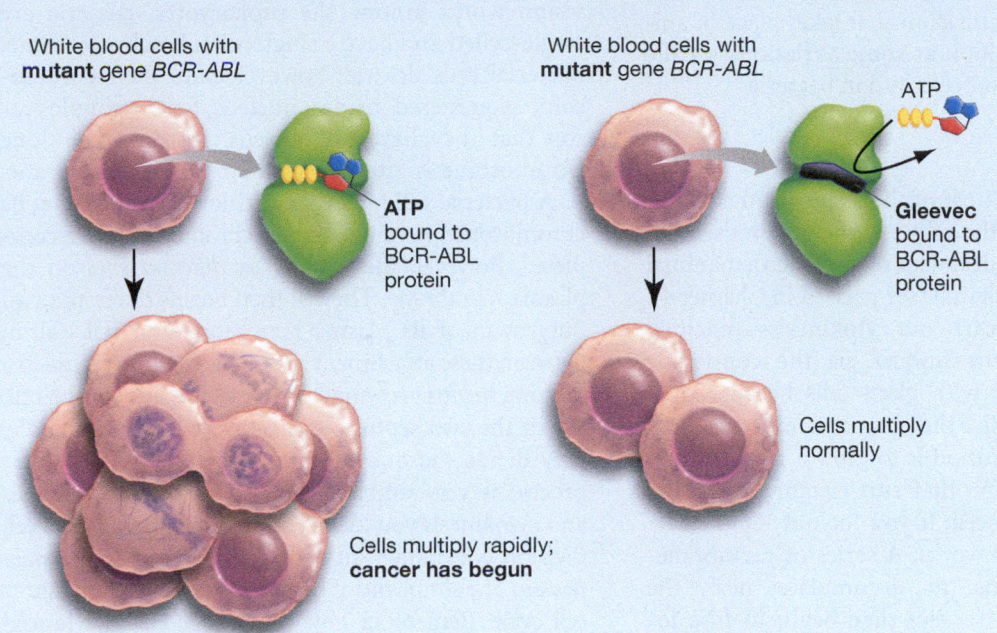

Figure 1

Blocking Cancer's Progress

The drug Gleevec is a so-called targeted therapy for cancer in that it blocks a molecular signaling pathway that sustains a particular cancer, chronic myeloid leukemia, which affects white blood cells. Gleevec binds to a site on the mutant protein BCR-ABL, thus preventing the energy-dispensing molecule ATP from binding to the site. With this, BCR-ABL is deprived of the energy it needs to carry out its reactions and can no longer keep the cell moving through the process of cell division.

was the drug that became Gleevec, which had been developed in the 1990s by the pharmaceutical firm Ciba-Geigy. (For more on the development of Gleevec, see "How Science and Business Take on Cancer," on page 12).

As it turned out, the effect of Gleevec was revolutionary—in two ways. First, this drug turned CML from an invariably deadly form of cancer into a chronic, manageable illness. Eight-five to 90 percent of CML patients now survive at least five years and it appears that the vast majority of these patients will never die from CML. Second, Gleevec was proof-positive of the potential of targeted therapy. It showed that disrupting the circuitry of a given form of cancer could upend that

cancer altogether. Since Gleevec's debut, two other notable targeted therapy drugs have been brought to market: Herceptin, which works against certain forms of breast cancer; and Avastin, which is aimed at denying cancer tumors the blood supply they need to keep growing. Neither drug has had the dramatic effect of Gleevec, but it would have been unreasonable to expect every targeted therapy drug to yield Gleevec's startling results.

A broader measure of Gleevec's impact, however, comes not from looking at cancer drugs currently in use, but from looking at cancer drugs currently in development. Almost all the drugs that are in the research pipelines of pharmaceutical firms today are forms of targeted therapy. Rather than being intended to kill cancer cells outright, these drugs are intended to disrupt the specific molecular pathways of particular cancers. A change such as this may ultimately result in a completely different way of thinking about cancer. Whereas cancer researchers have traditionally defined different forms of cancer by their anatomical sites of origin—liver cancer, lung cancer, and so forth—they are today defining cancers more and more by the molecular pathways that sustain them. Only time will tell what will come of this sweeping change in the fight against cancer, but the results so far have been promising.

half of the hourglass circle toward each other and then fuse. With this, the two cells become separate. Mitosis and cytokinesis are over, and the two daughter cells slip back into the relative quiet of interphase.

SO FAR . . .

1. Cell division has two separable components: Mitosis is the movement of a cell's _____ into opposite parts of the cell, while cytokinesis is the physical separation of one cell into two _____.

2. Mitosis and cytokinesis take place in the relatively short _____ phase of the cell cycle. Meanwhile, a typical cell spends most of its time in the cycle's _____, carrying out its normal function and duplicating its chromosomes in preparation for division.

3. In mitosis, the apportioning of DNA comes about through the separation of _____, each of which then becomes a fully functional _____ in one of the newly formed daughter cells.

Causes of Cancer

Cell Division for Bacteria

9.5 Cell Division in Plants and Bacteria

Having looked at cell division as it takes place in animal cells, we'll now look at some variations on the process, as seen in plant cells and in bacteria.

Plant Cells

Plant cells carry out most of the steps of mitosis just as animal cells do. In the splitting of cytokinesis, however, plant cells must deal with something that animal cells don't have: the cell wall (see page 84 in Chapter 4). The way animal cells carry out cytokinesis—pinching the plasma membrane inward via the contractile ring—wouldn't work with plant cells because their thick cell wall lies *outside* the plasma membrane.

The plant cell's solution is to grow a new cell wall and plasma membrane that run roughly down the middle of the parent cell. If you look at **Figure 9.12**, you can see how this works. A series of membrane-lined vesicles begins to accumulate near the metaphase plate; the vesicles then begin to fuse together; eventually, they will form a flat *cell plate* that runs from one side of the parent cell to the other. The membrane portion of the cell plate then fuses with the original plasma membrane of the parent cell, and the result is two new adjacent plasma membranes that split the parent cell in two. Meanwhile, the material enclosed *inside* the cell-plate membrane is the foundation for the new cell-wall segments of the daughter cells.

Prokaryotes: Bacteria and Archaea

It's also important to recognize that some cells have fundamentally different ways of replicating. The cells of prokaryotes—the bacteria and archaea you looked at in Chapter 4—are a case in point.

If you look at **Figure 9.13**, you can see how cell division works among the prokaryotes. Bacteria are single-celled and have a single, circular chromosome. Bacterial cells do not, however, have their chromosome sequestered inside a nucleus, for the simple reason that they have no nucleus. Instead, their lone chromosome is attached to their plasma membrane.

A bacterial cell begins its division by duplicating its chromosome. As the daughter chromosomes are completed, however, they attach to *different sites* on the plasma membrane. The cell then begins developing an outgrowth of its plasma membrane and cell wall in between these attachment sites. This outgrowth, called a septum, begins growing from opposite sides of the cell. When the two septum extensions join in the middle, they divide the one cell into two (**Figure 9.14**). This process is very different from the eukaryotic mitosis and cytokinesis you looked at. As such, prokaryotic cell division goes under an entirely different name: **binary fission.** The simplicity of binary fission makes for a short cell cycle. Remember how the animal cells we looked at took about 24 hours to go through a cycle? Bacteria can complete their cycle in as little as 20 minutes.

On to Meiosis

The cell division reviewed in this chapter concerns "somatic" cells—the kind that form bone, muscle, nerve, and many other sorts of tissue. Given a distribution this wide, you might well ask: What kind of cells are *not* somatic? The answer is only one kind—the kind that forms the basis for each succeeding generation of sexually reproducing living things. The division of these cells, through a process known as meiosis, is the subject of Chapter 10.

Figure 9.12
Cytokinesis in Plants

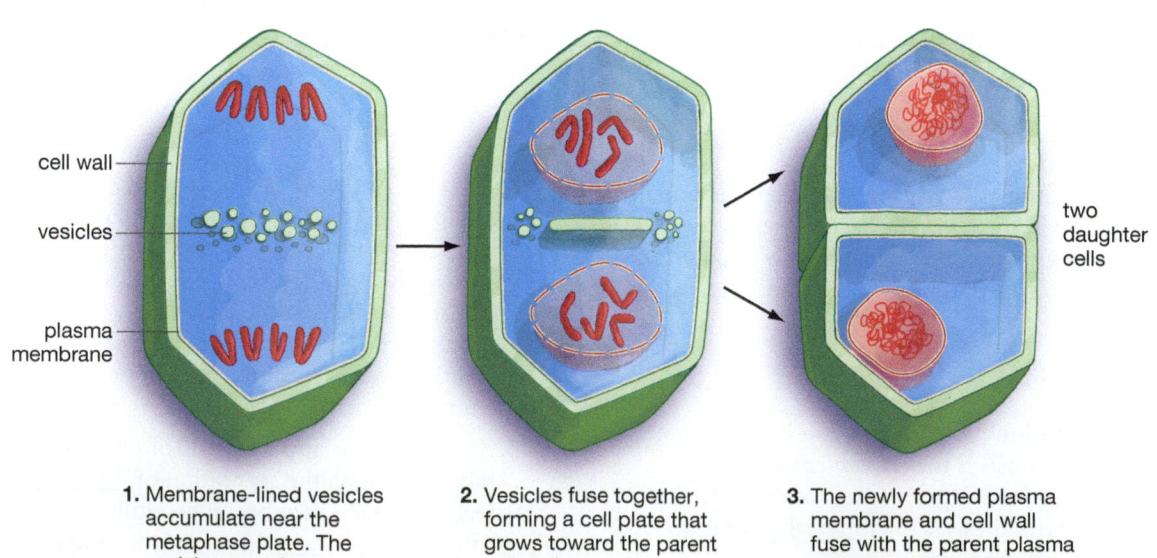

cell wall

vesicles

plasma membrane

two daughter cells

1. Membrane-lined vesicles accumulate near the metaphase plate. The vesicles contain precursors to the cell wall.

2. Vesicles fuse together, forming a cell plate that grows toward the parent cell wall.

3. The newly formed plasma membrane and cell wall fuse with the parent plasma membrane and cell wall, forming two distinct daughter cells.

Figure 9.13
Binary Fission in Bacteria

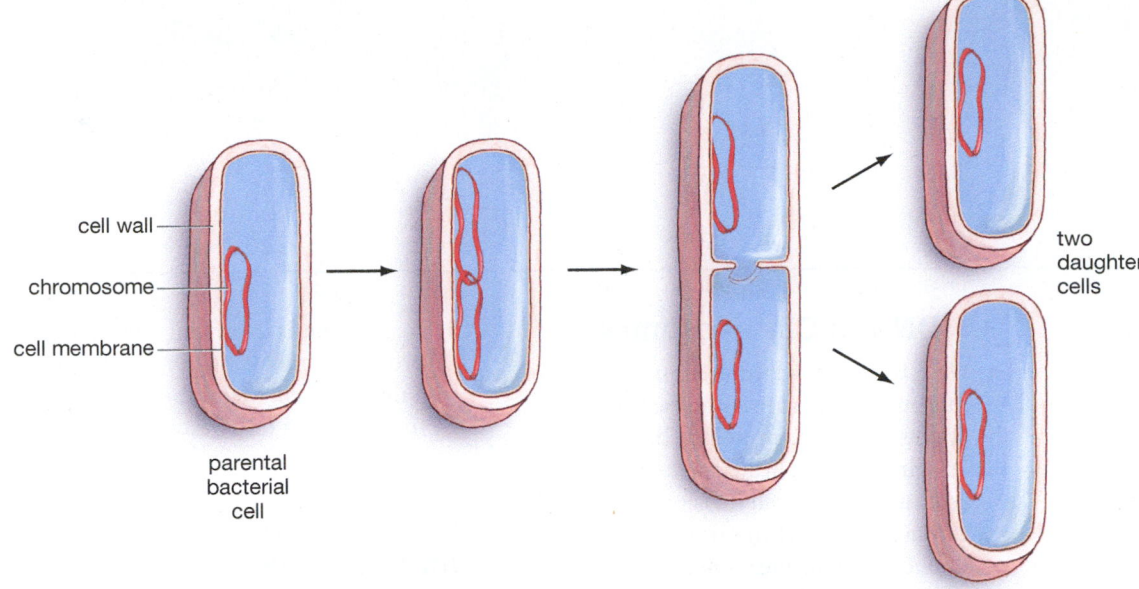

cell wall

chromosome

cell membrane

parental bacterial cell

two daughter cells

1. Bacterial cell starts with a single, circular chromosome attached to its plasma membrane.

2. The chromosome replicates and the daughter chromosomes attach to different sites on the plasma membrane.

3. The cell membrane and wall grow an extension between the attachment points of the two chromosomes.

4. The cell wall and membrane join together in the middle, resulting in two new cells.

SO FAR . . .

1. Owing to their cell walls, plant cells cannot carry out _____ in the same way that animal cells do. Thus, plant cells build a new _____ and _____ that run roughly down the middle of the parent cell, creating two daughter cells that then separate.

2. Bacterial cells are prokaryotes, meaning their cells have no _____. In cell division, a bacterial cell's single chromosome duplicates and the resulting daughter chromosomes attach to different sites on the _____, after which the cell divides in two. This process is known as _____.

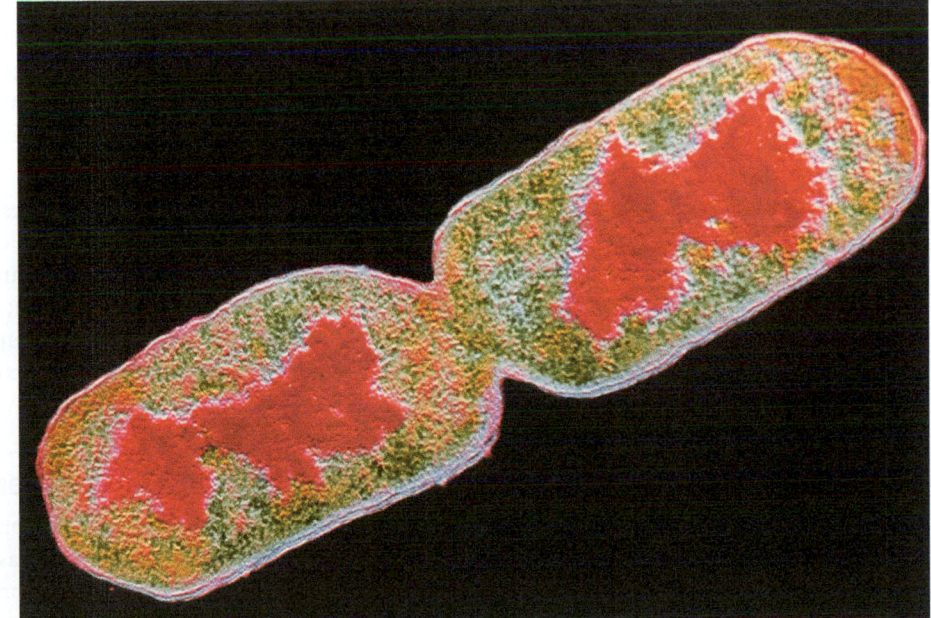

Figure 9.14
Cell Division for a Resident of the Human Gut
An *Escherichia coli* (*E. coli*) bacterium in a late phase of cell division.

9 REVIEW

Summary

9.1 An Introduction to Genetics

- DNA contains the information for the production of proteins, which carry out a wide variety of tasks in living things. **(p. 157)**

- The information in DNA is encoded in chemical substances called bases, laid out along the DNA double helix in four varieties: adenine (A), thymine (T), guanine (G), and cytosine (C). One series of bases contains information for the production of one protein; a different series of bases specifies a different protein. Each series of protein-specifying bases is known as a gene. **(p. 158)**

- In protein synthesis, a sequence of DNA bases is copied onto the molecule messenger RNA (mRNA). This molecule moves out of the cell's nucleus to a ribosome, located in the cell's cytoplasm, where the mRNA sequence is brought together with amino acids. As the mRNA sequence is "read" within the ribosome, a chain of amino acids is linked together in the ribosome in the order specified by the mRNA sequence. The result is a chain of amino acids that folds into a protein. **(p. 159)**

9.2 An Introduction to Cell Division

- Cell division includes the duplication of DNA (replication); the apportioning of the copied DNA into two quantities in a parent cell (mitosis); and the physical splitting of this parent cell into two daughter cells (cytokinesis). In DNA replication, the two strands of the double helix unwind, after which each single strand serves as a template for construction of a second, complementary strand. **(p. 160)**

9.3 DNA in Chromosomes

- DNA comes packaged in units called chromosomes, which are composed of DNA and its associated proteins. Chromosomes exist in an unduplicated state until DNA replicates, before cell division. DNA replication results in chromosomes that are in duplicated state, meaning one chromosome composed of two identical sister chromatids. **(p. 161)**

- Chromosomes in human beings (and many other species) come in matched pairs, with one member of each pair inherited from the mother and father, respectively. Such homologous chromosomes have closely matched sets of genes, although many of these are not identical. Human beings have 46 chromosomes—22 matched pairs plus either a matched pair of X chromosomes (in females) or an X and a Y chromosome (in males). **(p. 163)**

- The cell cycle has two main phases: (1) interphase, in which the cell carries out its work, grows, and duplicates its chromosomes in preparation for division; and (2) mitotic phase, in which the duplicated chromosomes separate (mitosis) and the cell splits in two (cytokinesis). **(p. 164)**

9.4 Mitosis and Cytokinesis

- There are four stages in mitosis: prophase, metaphase, anaphase, and telophase. Duplicated chromosomes line up along the metaphase plate, with sister chromatids on opposite sides of the plate. Attached to fibers called microtubules, the sister chromatids are pulled apart, to opposite poles of the parent cell. Once cell division is complete, a pair of sister chromatids that once formed a single chromosome will reside in separate daughter cells, with each sister chromatid now functioning as a full-fledged chromosome. **(p. 165)**

- In cytokinesis in animal cells, a ring of protein filaments tightens at the middle of a dividing cell. Membranes being pinched together then fuse, resulting in two daughter cells. **(p. 167)**

9.5 Cell Division in Plants and Bacteria

- In cytokinesis in plant cells, the cells grow new cell walls and plasma membranes near the metaphase plate, thus dividing the parent cell into two. **(p. 170)**

- In cell division in prokaryotes, cells double their single, circular chromosome, with the two resulting chromosomes attaching to different sites on the plasma membrane. Then an outgrowth of plasma membrane and cell wall, called a septum, begins growing from opposite sides of the cell, between the two chromosomes. When the two septum extensions join, they divide the one cell into two. **(p. 170)**

Key Terms

Understanding the Basics

Multiple-Choice Questions

(Answers are in the back of the book.)

1. Proteins serve in living things as:
   a. hormones.
   b. enzymes.
   c. cell receptors.
   d. structural building blocks.
   e. all of the above.

2. The four bases found along the DNA double-helix are:
   a. adenine, thymine, guanine, and cytosol.
   b. adenine, thyroxine, glucose, and cytosine.
   c. adrenaline, thymine, glucosamine, and uracil.
   d. adenine, thymine, guanine, and cytosine.
   e. none of the above.

3. A gene can be described as:
   a. a protein that contains enough information to carry out a task.
   b. a hormone that prompts some action inside a cell.
   c. a series of DNA bases that transfer energy within a cell.
   d. a series of DNA bases that contain information for production of a protein.
   e. a protein that delivers information to a ribosome.

4. The human genome contains:
   a. 23 pairs of chromosomes.
   b. 48 chromosomes.
   c. as many as 25,000 genes.
   d. b and c, above.
   e. a and c, above.

5. Prophase has begun when:
   a. well-defined chromosomes are distinguishable.
   b. a cleavage furrow has formed.
   c. chromosomes have aligned along the metaphase plate.
   d. chromosomes have duplicated.
   e. sister chromatids have separated.

6. If a cell is not dividing, and has not yet replicated its DNA (duplicated its chromosomes), what phase of the cell cycle would it be in?
   a. gap-two (G_2) phase
   b. synthesis (S) phase
   c. gap-one (G1) phase
   d. prophase
   e. mitotic phase

7. The average length of the cell cycle in animal cells is about:
   a. 16 hours.
   b. 24 hours.
   c. 30 minutes.
   d. 2 hours.
   e. 12 hours.

8. The cell cycle is linked to cancer because cancer always entails:
   a. too much time spent in S phase.
   b. a slow-down in mitosis.
   c. unrestrained cell division.
   d. a misalignment of chromosomes.
   e. not enough time spent in M phase.

Brief Review

(Answers are in the back of the book.)

1. In what way does DNA help form and then maintain our bodies?

2. The drawing below shows a cell in a phase of mitosis. What phase is shown? How many chromatids are present in this stage? What is the total number of chromosomes each daughter cell will have?

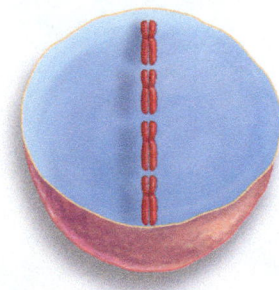

3. How do mitosis and cytokinesis differ?

4. Name the four phases of mitosis and describe the major events occurring in each phase.

5. Compared to animal cells, what additional cell structure must plant cells deal with when they are undergoing cytokinesis? Given this difference, how do plant cells carry out cytokinesis?

6. In what way are the 23 pairs of human chromosomes "matched" pairs of chromosomes?

7. The protein synthesis pathway—DNA → mRNA → protein—is a very important concept in biology. Within this framework:
   a. What are the information-bearing "building blocks" of DNA?
   b. Where is DNA located in the cell?
   c. What does the "m" in mRNA stand for?
   d. Using this knowledge, what is the function of mRNA, and where is it used?
   e. What are the building blocks of proteins?
   f. Where does protein formation occur?

Applying Your Knowledge

1. Why is it accurate to think of each human being as the owner of a library of ancient information?

2. People who are born with the condition called Down syndrome generally have three copies of a certain chromosome (chromosome 21), rather than the usual two copies. What can you deduce from this regarding the proteins produced by a person with Down syndrome?

3. The tools of computer science are today being applied to the genomes of living things thanks to the happy coincidence that the information in both computers and genomes exists in digital form. In computers, digital information is expressed as a series of 0's and 1's. What is the digital form that a genome's information comes in?

MasteringBiology®

1. MasteringBiology Assignments
Activities: The Cell Cycle; Mitosis and Cytokinesis Animation; Causes of Cancer; Cell Division for Bacteria; **Bioflix:** Mitosis; **Current Events:** Cancer's Secrets Come Into Sharper Focus; **ABC Video:** Henrietta Lacks' Cells; The Effect of Exercise on Cells; **Chemistry Review:** Nucleic Acids; **Building**

Vocabulary; Questions: Bioflix Quiz, Reading Questions, Test Bank

2. eText
Read your book online, search, take notes, highlight text, and study more effectively.

3. The Study Area
Self-study tools to test comprehension.

10

Preparing for Sexual Reproduction:

Meiosis

One special kind of cell division yields the sex cells that give rise to succeeding generations and serves as the foundation for much of the diversity seen in the living world.

Available on MasteringBiology®

Activities: Meiosis Animation; Origins of Genetic Variation; Asexual and Sexual Life Cycles; **Bioflix:** Meiosis; **Building Vocabulary; Questions:** Bioflix Quiz, Reading Questions, Test Bank

Animals, fungi, plants. How did the living world become so diverse? In part, through the process of meiosis.

A real-life couple—let's call them Jack Fennington and Jill Kent—combine their last names when they get married and thus become Jack and Jill Fennington-Kent. Now, what would happen if the Fennington-Kent's children were to continue in this tradition? Their daughter, Susie, might marry, say, Ralph Reeson-Dodd, which would make her Susie Fennington-Kent-Reeson-Dodd. If *Susie's* daughter, Alicia, were to keep this up, she might be fated to become Alicia Fennington-Kent-Reeson-Dodd-Garcia-Lee-Minderbinder-Green, and so on.

This growing chain of names illustrates a kind of problem that living things have managed to address in carrying out reproduction. *Reproduction* in this context does not mean the mitotic cell division you looked at in Chapter 9. Reproduction here means *sexual* reproduction, in which specialized reproductive cells come together to produce offspring. The cells described in Chapter 9—the ones that undergo mitosis—are known as *somatic* cells, which in animals means all the cells in the organism except for one type. What type is this? Sex cells; cells called *gametes*. These are reproductive cells, known more commonly as eggs and sperm.

Eggs and sperm are not commonly thought of as cells, but that's just what they are, complete with plasma membranes, nuclei, chromosomes, and all the rest (though sperm eventually lose much of this material). Eggs and sperm are special cells, however, in that they are not destined to divide like somatic cells and thus make more eggs and sperm. Rather, they will *fuse*—sperm fertilizing egg—to make a zygote, which can develop into a whole organism. It is sperm and egg that take part in sexual reproduction and that thus bear some relation to the Fennington-Kent problem.

Here's how it plays out in humans. Human somatic cells, as you have seen, have 23 pairs of chromosomes, or 46 chromosomes in all. If an egg and sperm each brought 46 chromosomes to their *union,* the result would be a zygote with 92 chromosomes. The next generation down would presumably have 184 chromosomes, the next after that 368, and so on. This would be about as functional as having Fennington-Kent-Reeson-Dodd-Garcia-Lee-Minderbinder-Green as a last name.

10.1 An Overview of Meiosis

How do chromosomes avoid the problem of doubling with each generation? In sexual reproduction, chromosome union is preceded by chromosome *reduction.* The reduction comes in the cells that give rise to sperm and egg. When these cells divide, the result is sperm or egg cells that have only *half* the usual, somatic number of chromosomes. Human sperm or eggs, in other words, have only 23 chromosomes in them. (Not the 23 *pairs* of chromosomes that somatic cells have, mind you; 23 chromosomes.) Each 23-chromosome sperm can then unite with a 23-chromosome egg to produce a 46-chromosome zygote that can develop into a new human being. In each generation, then, there is first a halving of chromosome number (when egg and sperm cells are produced), followed by a coming together of these two halves (when sperm and egg unite). This basic pattern holds true for sexual reproduction in all eukaryotes.

Now here are a few terms that will be helpful in understanding this process. The kind of cell division that results in the halving of chromosome number is called *meiosis,* in contrast with *mitosis,* which you looked at in Chapter 9. In mitosis, there is no halving of chromosome number; just a duplication of chromosomes and then a division of these chromosomes into each of two daughter cells.

In animals, cells that are produced in meiosis—thus getting the half-number of chromosomes—are always reproductive cells, the term for which is **gametes,** as we've seen. Such reduced-chromosome cells are said to be in the haploid state, the term *haploid* meaning "single number." When egg and sperm unite, however, it marks a return to the *diploid,* or "double-number" state of cellular existence, meaning 46 chromosomes in human beings. (More formally, **haploid** means possessing a single set of chromosomes, while **diploid** means possessing two sets of chromosomes.) **Meiosis** can thus be defined as a process in which a single diploid cell divides to produce four haploid reproductive cells. Diploid cells are also sometimes referred to as **2n** cells (the "2" here standing for a doubled number of chromosomes), while haploid cells are said to be **n.** Thus, eggs and sperm are n, while the diploid cells that give rise to them are 2n.

In what follows, we will be looking at meiosis as it occurs in human beings; toward the end of the chapter, we'll go over some variations on reproduction in other kinds of organisms.

10.2 The Steps in Meiosis

Meiosis Animation

How does the chromosomal halving we've been talking about take place? Let's go over the process of meiosis and see. **Figure 10.1** shows meiosis, in a stripped-down form, as compared to the mitosis described in Chapter 9. Two essential differences between the two processes can be seen in the figure. First, you may remember that the formula for mitosis was: duplicate once, divide once. Duplicate the chromosomes once, then divide the original cell once. With meiosis, on the other hand, the formula is: duplicate once, divide *twice.* Meiosis includes one chromosome duplication followed by two cellular divisions, which means the process produces four cells (instead of the two that come about in mitosis). The other big difference you can see is that the first of the meiotic divisions is a separation of homologous chromo*somes*, not the separation of chroma*tids* that you saw in mitosis. The subsequent meiotic division is, however, just like mitotic division: It separates the chromatids that make up the homologous chromosomes.

But what is this term *homologous*? You may remember from Chapter 9 that homologous chromosomes are chromosomes that are the same in function (and hence size). These chromosomes do similar things, which is to say they have genes on them that code for similar proteins. We have a chromosome 1 that we inherit from our mother, and it is homologous to the chromosome 1 we inherit from our father. Likewise, we inherit a chromosome 2 from our mother that is homologous with chromosome 2 from our father, and so forth for 23 pairs. The lone exception to this pairs rule is that the X and Y chromosomes males have are not homologous with one another.

The steps of meiosis are separated into two multi-step stages, called meiosis I and meiosis II. The big picture here is that, in meiosis I, homologous chromosomes are positioned close together, on opposite sides of the cellular equator called the metaphase plate. Then these chromosomes will move apart into opposite parts of the cell—parts that will become different daughter cells. In meiosis II, the chroma*tids* of these now-separated chromosomes will, in turn, sepa-

rate into different daughter cells. Let's walk through the steps of meiosis now, as illustrated in **Figure 10.2** on pages 178–179. These steps will reveal a process that not only divides up genetic material but provides for a great deal of the diversity that exists among living things.

Meiosis I

Prophase I

Meiosis I begins, in prophase I, with the same appearance of 46 identifiable chromosomes that you observed in mitosis. The first big difference between meiosis and mitosis also appears in prophase I. It is that, in meiosis, *homologous chromosomes pair up.* In mitosis, 46 individual chromosomes condensed, but did not pair up in any way. In meiosis, however, each pair of homologous chromosomes link up as they condense. Maternal chromosome 5, for example, aligns with *pa*ternal chromosome 5, maternal chromosome 6 with paternal 6, and so on. When two homologous chromosomes have paired up in this way, they are called a **tetrad** (which you can see illustrated in Figure 10.3).

A critical bit of part-swapping then takes place between the non-sister chromatids of the paired chromosomes. This process is called *crossing over* (or *recombination*), and you'll be looking at it in detail later. Once this crossing over has finished, the homologous chromosomes begin to separate from one another, although they remain in alignment.

Metaphase I

Still existing as tetrads, the homologous chromosomes, attached to microtubules, are moved to the metaphase plate, via a lengthening and shortening of the microtubules; in this step of meiosis, the maternal member of a given pair lies on one side of the plate and the paternal member on the other. A critical point about this, however, is that the alignment adopted by any one pair of chromosomes bears no relation to the alignment adopted by any other pair. It may be, for example, that paternal chromosome 5 will line up on what we might call side A of the metaphase plate; if so, then maternal chromosome 5 ends up on side B. Shift to chromosome 6, however, and things could just as easily be reversed: The *maternal* chromosome might end up on side A and the paternal on side B. It is thus a throw of the dice regarding which side of the plate a given chromosome lines up on. More important, it is this chance event that determines which *cell* a chromosome joins because each side of the plate becomes a separate cell once cell division is finished (something you can see illustrated in Figure 10.4). As you will see, this randomness is critical in the shuffling of genetic material.

Anaphase I

In anaphase I, the paired, homologous chromosomes now begin to move away from each other, toward their respective poles, pulled through the disassembly of the

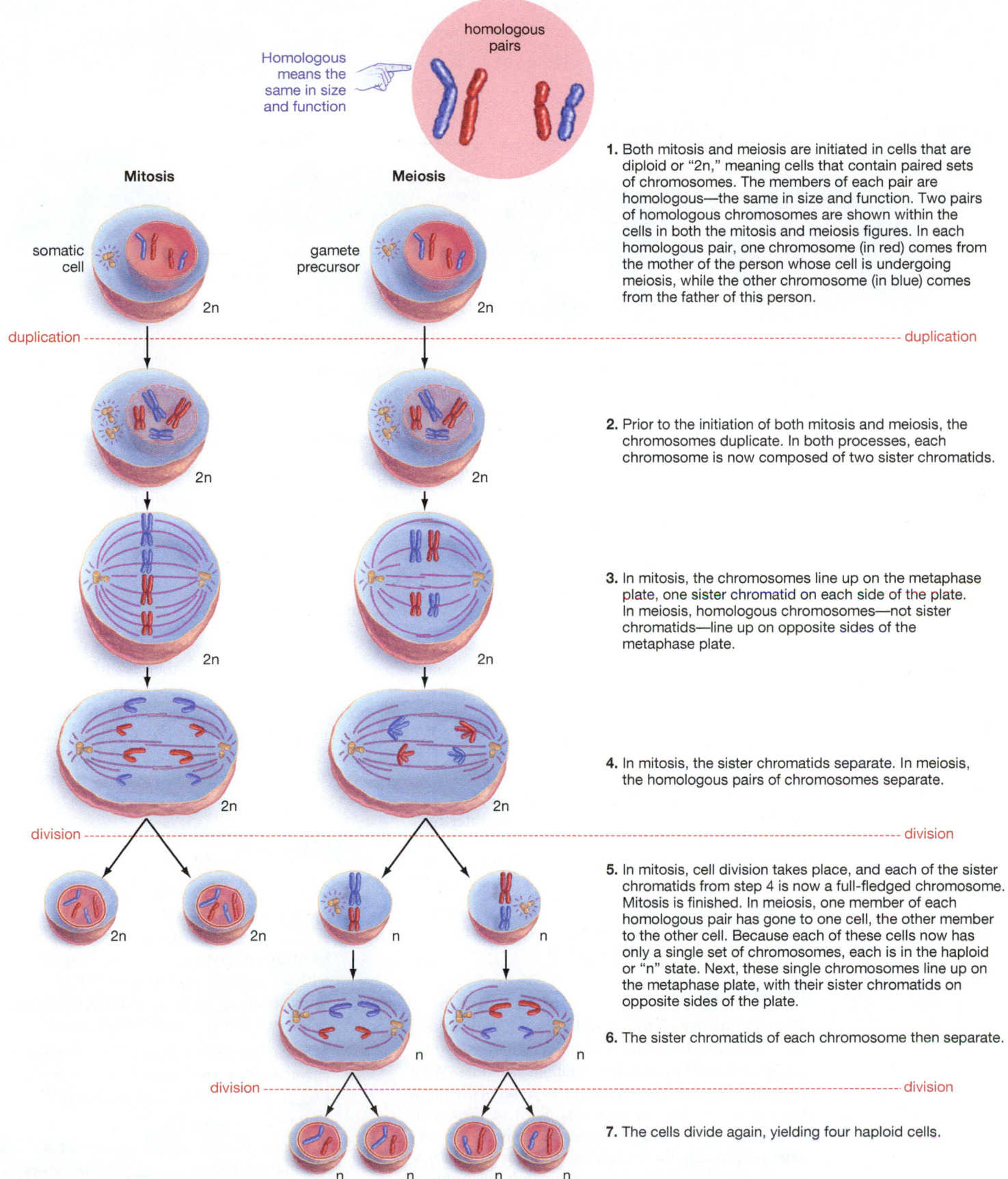

1. Both mitosis and meiosis are initiated in cells that are diploid or "2n," meaning cells that contain paired sets of chromosomes. The members of each pair are homologous—the same in size and function. Two pairs of homologous chromosomes are shown within the cells in both the mitosis and meiosis figures. In each homologous pair, one chromosome (in red) comes from the mother of the person whose cell is undergoing meiosis, while the other chromosome (in blue) comes from the father of this person.

2. Prior to the initiation of both mitosis and meiosis, the chromosomes duplicate. In both processes, each chromosome is now composed of two sister chromatids.

3. In mitosis, the chromosomes line up on the metaphase plate, one sister chromatid on each side of the plate. In meiosis, homologous chromosomes—not sister chromatids—line up on opposite sides of the metaphase plate.

4. In mitosis, the sister chromatids separate. In meiosis, the homologous pairs of chromosomes separate.

5. In mitosis, cell division takes place, and each of the sister chromatids from step 4 is now a full-fledged chromosome. Mitosis is finished. In meiosis, one member of each homologous pair has gone to one cell, the other member to the other cell. Because each of these cells now has only a single set of chromosomes, each is in the haploid or "n" state. Next, these single chromosomes line up on the metaphase plate, with their sister chromatids on opposite sides of the plate.

6. The sister chromatids of each chromosome then separate.

7. The cells divide again, yielding four haploid cells.

Figure 10.1
Meiosis Compared to Mitosis

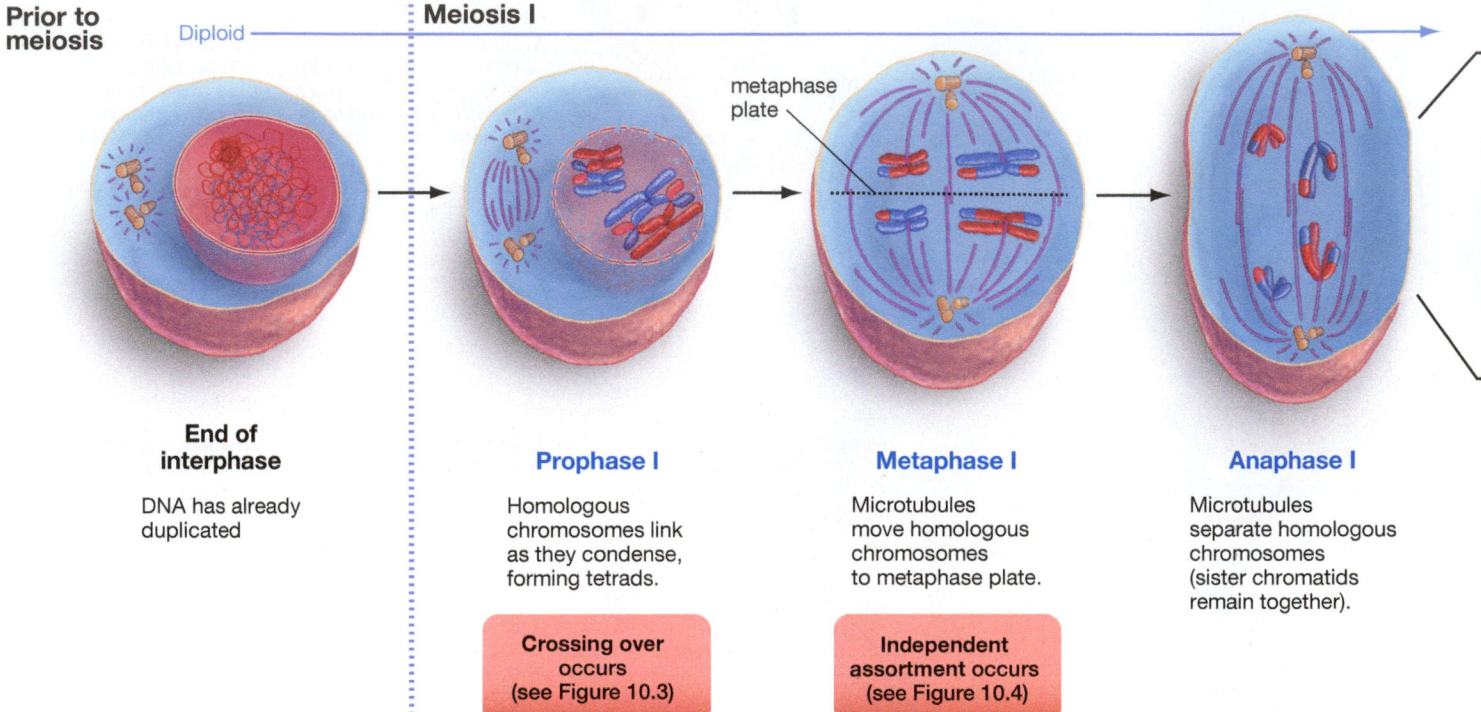

Prior to meiosis

Diploid

Meiosis I

metaphase plate

End of interphase

DNA has already duplicated

Prophase I

Homologous chromosomes link as they condense, forming tetrads.

Crossing over occurs (see Figure 10.3)

Metaphase I

Microtubules move homologous chromosomes to metaphase plate.

Independent assortment occurs (see Figure 10.4)

Anaphase I

Microtubules separate homologous chromosomes (sister chromatids remain together).

Figure 10.2
The Steps of Meiosis

Bioflix: Meiosis

microtubule spindles they are attached to. Although homologous chromosomes have now separated, the sister chromatids that make them up are still together.

Telophase I

With chromosome movement toward the poles completed, the original cell now undergoes cytokinesis, dividing into two completely separate daughter cells. With this, there are two haploid cells, whereas in the beginning there was one diploid cell. (Why are the cells now haploid? Meiosis I moved one set of chromosomes to one cell and the second set to another cell. Therefore, each cell now has only one set of chromosomes, meaning each is haploid.)

Meiosis II

There is little pause for interphase between meiosis I and meiosis II; little time for the daughter cells to regroup and carry on regular metabolic activities. What happens instead is another division—or set of divisions, because there are now two cells. Each of these cells now has 23 duplicated chromosomes in it. After a very brief prophase II, these chromosomes line up at a new metaphase plate, with sister chromatids on opposite sides of each plate (metaphase II in the figure). What happens next is the same thing that happened in mitosis. The *sister chromatids* now separate, moving toward

opposite poles; with this, they assume the role of full-fledged chromosomes (anaphase II). Once this separation is completed, cytokinesis occurs. Where once there were two cells, now there are four (telophase II). The difference between this process and mitosis is that each of these cells has 23 chromosomes in it instead of 46.

SO FAR . . .

1. Meiosis is the process by which the reproductive cells commonly known as _____ and _____ are formed. This takes place by means of one ____ cell, meaning a cell with a paired set of chromosomes, giving rise to four _____ cells, meaning cells that each have a single set of chromosomes.

2. Meiosis I involves a separation not of sister chromatids, as in mitosis, but a separation of _____. Meiosis II involves a separation of _____.

3. In human beings, meiosis starts with a single cell that contains _____ chromosomes and ends with four cells, each of which contains _____ chromosomes.

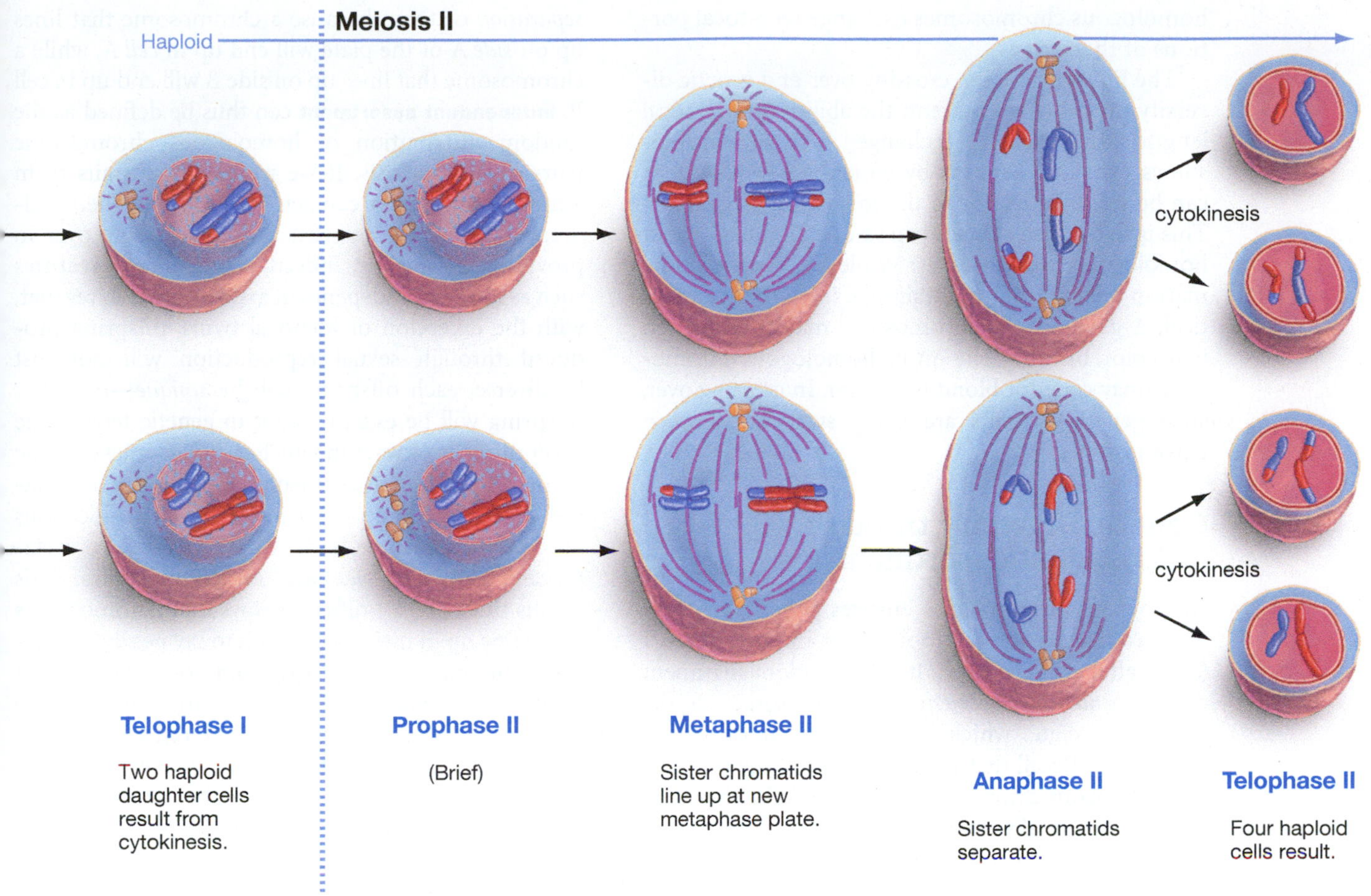

Meiosis II

Haploid →

Telophase I

Two haploid
daughter cells
result from
cytokinesis.

Prophase II

(Brief)

Metaphase II

Sister chromatids
line up at new
metaphase plate.

cytokinesis

Anaphase II

Sister chromatids
separate.

Telophase II

Four haploid
cells result.

10.3 The Significance of Meiosis

You have now examined the mechanics of meiosis, but what are the effects of this process? First, the "Fennington-Kent" problem noted at the beginning of the chapter has been solved. There will be no 92-chromosome zygotes because egg and sperm will each bring only 23 chromosomes to their union, thus yielding a combined 46 chromosomes in human somatic cells. Just as notable, however, are two kinds of diversity that meiosis brings about. The first of these is genetic diversity in offspring; the second is a large-scale diversity in the natural world.

Genetic Diversity through Crossing Over

What is genetic diversity? Recall that, back in prophase I of meiosis, the homologous chromosome pairs aligned for a time and then engaged in something called crossing over (or recombination). In this, there is an exchange of *parts* of chromosomes prior to their lining up at the metaphase plate.

Look at **Figure 10.3** to see how this important process works. In the first stage of meiosis I, there is a physical breaking of non-sister chromatids and

then a "reunion" of these reciprocal chromosomal sections onto new chromatid partners. With this, what had been a "maternal" chromosome now has a portion of the paternal chromosome within it, and vice versa. Formally, then, **crossing over** can be defined as a process, occurring during meiosis, in which

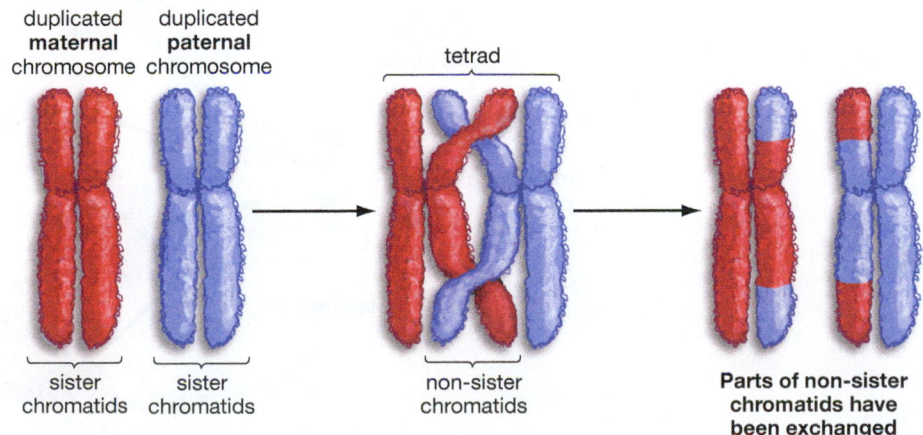

duplicated **maternal** chromosome   duplicated **paternal** chromosome

tetrad

sister chromatids   sister chromatids   non-sister chromatids   **Parts of non-sister chromatids have been exchanged**

Figure 10.3
Crossing Over and Genetic Diversity

In this process, which occurs in prophase I of meiosis, two homologous chromosomes join with each other, thus forming a tetrad, after which non-sister chromatids from each chromosome exchange reciprocal portions of themselves. In this way, genetic material from maternal and paternal sources is shuffled during the formation of gametes.

homologous chromosomes exchange reciprocal portions of themselves.

The linkage between crossing over and genetic diversity probably is apparent; the ability of reciprocal lengths of DNA to be exchanged between chromosomes provides a means by which the genetic deck can be reshuffled prior to the formation of gametes. This is so because of what you saw in Chapter 9 about homologous chromosomes: While the genetic information on any two will be similar, it will not be identical. A gene on one chromosome may code for red hair color, but the gene on its homologous chromosome may code for blond hair color. In crossing over, such genetic variants are being swapped between chromosomes.

Genetic Diversity through Independent Assortment

Origins of Genetic Variation

Another way that meiosis ensures genetic diversity is a crucial event that took place in metaphase I (right after crossing over). It is the random alignment or *independent assortment* of chromosomes at the metaphase plate, which you can see illustrated in **Figure 10.4**. Recall that the alignment of any one pair of homologous chromosomes along the plate bears no relation to the alignment of any other homologous pair. If you looked at, say, chromosome 5, the paternal member of the pair might line up on "side A" of the metaphase plate, meaning maternal chromosome 5 will be on side B. Shift to chromosome 6, however, and things could just as easily be reversed. This random alignment of chromosomes then leads to a random

separation of them because a chromosome that lines up on *side* A of the plate will end up in *cell* A, while a chromosome that lines up on side B will end up in cell B. **Independent assortment** can thus be defined as the random distribution of homologous chromosome pairs during meiosis. If we wonder how traits from a mother and father can get "mixed up" in their children, independent assortment joins crossing over in providing an answer. Indeed, for complex creatures such as humans, independent assortment ensures that, with the exception of identical twins, offspring produced through sexual reproduction will not just be diverse; each offspring will be *unique*—no other offspring will be exactly like it in genetic terms. The sheer number of ways in which chromosomes can line up in independent assortment makes this a mathematical certainty. Think about it: We have two locations for any chromosome (side A or side B of the metaphase plate), and we have 23 chromosomes. This means there are 2^{23} different ways that chromosomes could line up in meiosis, which is to say nearly 8.4 million different ways. The egg from your mother that produced you may have been, say, variation 1,527,000 in this range, while the egg that produced your brother or sister may have been variation 4,573,000.

Why does this matter? Remember that there are genes on these chromosomes, and that the genes on any given paternal chromosome may differ somewhat from those on its matched maternal chromosome. Which chromosome the offspring gets thus makes a difference—in eye color, height, or any of thousands of other characteristics. Given this, each of us is very much affected by the way paternal and

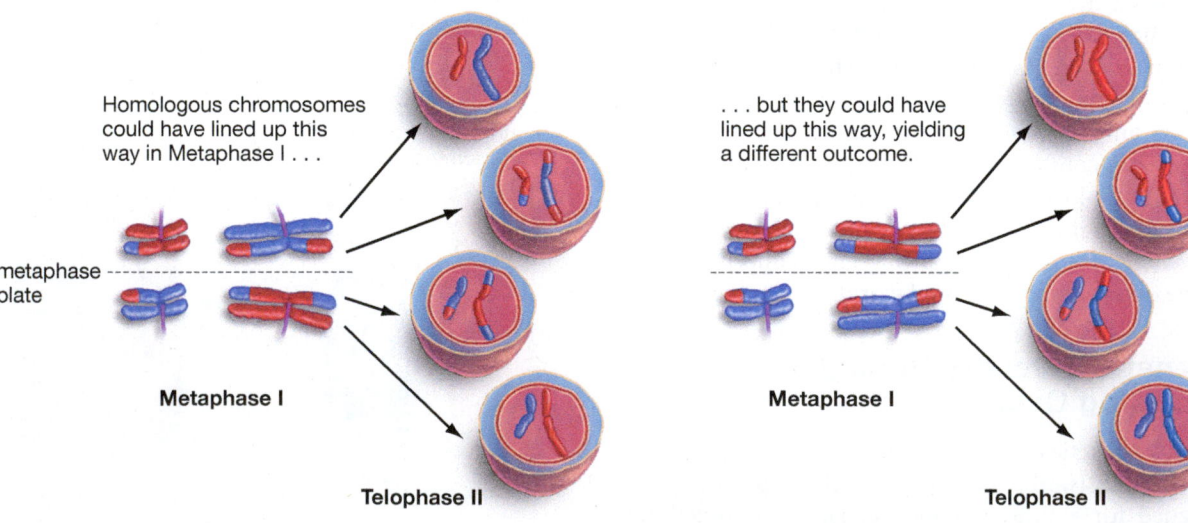

Homologous chromosomes could have lined up this way in Metaphase I . . .

. . . but they could have lined up this way, yielding a different outcome.

metaphase plate

Metaphase I

Telophase II

Metaphase I

Telophase II

Figure 10.4

Independent Assortment and Genetic Diversity

In this process, which occurs in metaphase I of meiosis, homologous pairs of chromosomes align on opposite sides of a cell's metaphase plate. Critically, the alignment that each pair assumes at the plate—maternal chromosome on one side, paternal on the other—is a random event. The alignment that any one pair assumes does not affect the alignment of any other pair. This independent assortment of homologous chromosomes is a second means by which meiosis produces genetic diversity in gametes—and hence in offspring.

maternal chromosomes happened to line up in two very special meioses: those that produced the egg cell and the sperm cell that gave rise to us.

When we add independent assortment to crossing over, it's easy to see why separate children from the same parents can come out looking so different—from each other as well as from the parents themselves (**Figure 10.5**). Overall, meiosis generates genetic *diversity,* while its counterpart, mitosis, does not. Mitosis makes genetically exact copies of cells; it *retains* the qualities that cells have from one generation to the next. Meiosis mixes genetic elements each time it produces reproductive cells and thus brings about genetic variation in succeeding generations.

Diversity in the Living World

This genetic variation has an importance way beyond that of making siblings look different from each other. It turns out that this diversity is responsible, in significant part, for the fantastically diverse natural world around us. Think of it: two-foot ferns, 100-yard redwood trees, bats, bees, whales, mushrooms. How did the natural world come to be such a remarkably diverse place? The short answer is that evolution is spurred on by the *differences* in offspring that meiosis brings about. Let's see why.

Three frog siblings survive to reproductive age from a clutch of eggs laid by a female and fertilized by a male. Thanks to crossing over and independent assortment, one of these frogs may end up having, say, a darker coloration than all the other frogs of its species in its area. In the formation of the eggs and sperm that led to this frog, chromosome parts were swapped (in crossing over), and chromosomes lined up (in independent assortment) in such a way that this frog was the darkest frog possible among all the frogs in its population. Further, it happens that, because of shifting rain patterns in this frog's habitat, this dark coloration is starting to be useful for evading predators. This frog will thus survive to have more offspring of its own than do other frogs around it. By being selected for survival in this way—by undergoing what is called natural selection—this frog may have started its population down the road to having a darker coloration.

Now, it's easy to see that coloration is only one of a countless number of traits that could be selected for in this way. Taller height in a tree? Diving stamina in a dolphin? Meiosis and sexual reproduction would lead the way to these traits by producing, in one generation after another, *this* tree that grew a little taller or *that* dolphin that dove a little deeper. Just as General Motors provides an array of models for consumers to choose from each year, meiosis and sexual reproduction provide an array of models for *nature* to choose from each generation.

To put a finer point on this, consider an alternative to meiosis and sexual reproduction. Bacteria are single-celled organisms that possess a single, circular chromosome. As you saw in Chapter 9, bacteria divide through the process of binary fission, one bacterium becoming two in the course of cell division. As it turns out, that's it for bacteria as far as reproduction goes. There is no chromosomal part swapping or independent assortment of chromosomes in reproduction. And there is no fusion of two gametes from separate organisms, as occurs with eggs and sperm. Instead, there is just an exact duplication of the bacterial chromosome and a splitting of one bacterial cell into two. In general, then, every "daughter" bacterial cell is a genetic replica of its parent cell—a clone of it. What this means is that differences among bacteria are not a built-in feature of their reproduction. Instead, such differences as come to pass are dependent on their DNA mutating (which is rare) or on a form of gene transfer they are capable of (which is intermittent). Natural selection simply does not have as much to choose from in asexually reproducing organisms such as bacteria. Where is the equivalent of our darker-colored frog in a group of bacterial cells descended in a line? How could an equivalent difference come about, given that each bacterium is a clone of its predecessor?

What evidence do we have that sexual reproduction makes a practical difference? Well, consider that bacteria and their fellow single-celled organisms, the archaea, came into existence perhaps 3.8 billion years ago. Then, for 1.8 billion years after that, the living world consisted of nothing *but* bacteria and archaea—it took that long for anything else to evolve from them. We are not sure when sexual reproduction got going, but we have some evidence that it was taking place by 1.2 billion years ago. So, in the last 1.2 billion years, all the fish and birds and trees and fungi and flowers that we see before us developed through

Figure 10.5
Ensuring Variety

The shuffling of genetic material that occurs during meiosis is the primary reason that children look different from their parents and from each other.

sexual reproduction, whereas in the 1.8 billion years after bacteria and archaea got going, they changed very little in physical form. Although other factors contributed to the ramping up of diversity on Earth, meiosis and sexual reproduction were key players in it. The British geneticist J. B. S. Haldane is said to have remarked that "Evolution could roll on fairly efficiently without sex but such a world would be a dull one to live in." How did the living world come to be such a diverse place? In part through the very small-scale process you've been looking at, meiosis.

10.4 Meiosis and Sex Outcome

Let us think now about what meiosis means in relation to passing on sex to offspring. How do humans, in particular, come to be male or female? In humans, there is one exception to the rule that chromosomes come in homologous pairs. Human females do indeed have 23 pairs of matched chromosomes, including 22 pairs of "autosomes" and one matched pair of **sex chromosomes,** meaning the chromosomes that determine what sex an individual will be. The sex chromosomes in females are called X chromosomes, and each female possesses two of them. In males, conversely, there are 22 autosome pairs, one X sex chromosome, and then one Y sex chromosome. It is this Y chromosome that leads to the male sex. (**Figure 10.6** illustrates the differences in the X and Y sizes.)

In meiosis I in a female, the female's two X chromosomes, being homologous, line up together at the metaphase plate. Then, these chromosomes separate, each of them going to different cells (**Figure 10.7**). In males, the non-homologous X and Y chromosomes line up as if they were homologues. Then one resulting

cell gets an X chromosome, while the other gets the Y chromosome.

Sex determination is then simple. Each of the eggs produced by a female bears a single X chromosome. If this egg is fertilized by a sperm bearing an X chromosome, the resulting child will be a female. If, however, the egg is fertilized by a sperm bearing a Y chromosome, the child will be a male.

Because females don't have Y chromosomes to pass on, it follows that the Y chromosome that any male has must come from his father. Meanwhile, the single X chromosome that any male has must come from his mother. Females, conversely, carry one X chromosome from their mother and one from their father.

SO FAR . . .

1. Two features of meiosis ensure that each generation of sexually reproducing organisms will differ from its parents genetically. In one of these features, crossing over, _____ exchange reciprocal portions of themselves. In the other feature, independent assortment, there is a random distribution of _____ at the metaphase plate.

2. The reason that crossing over and independent assortment ensure genetic diversity is that any two members of a pair of homologous chromosomes _____ in the genetic information they carry.

3. In human beings, any fertilized egg that carries two _____ chromosomes will be a female; fertilized eggs destined to become males, meanwhile, carry one _____ chromosome (donated by the mother) and one _____ chromosome (donated by the father).

Figure 10.6
The X and the Y

The human X chromosome can be seen on the left, while the human Y chromosome is on the right. The difference in size between them is in part a reflection of the difference in the number of genes each of them contains. The Y chromosome only has only 78, while the X has about 1,500.

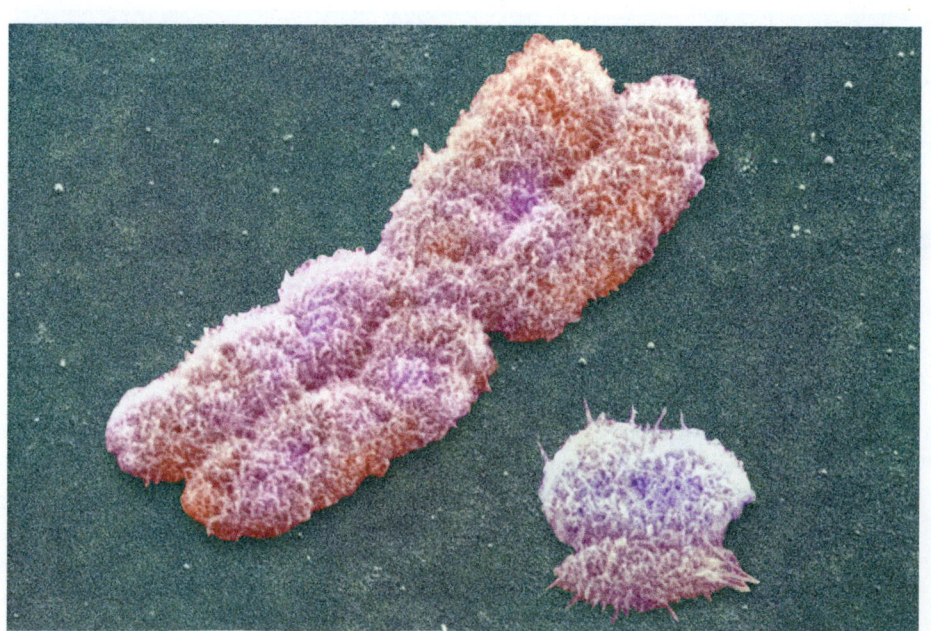

10.5 Gamete Formation in Humans

You've seen that, through meiosis, a single diploid cell gives rise to four haploid gametes. But what are the diploid cells that go through meiosis? As you might expect, there is a whole sequence of actions involved in the process by which egg and sperm are produced—a process known as *gamete formation*. In humans, the starting female cells in gamete formation are known as **oogonia,** while the starting male cells are **spermatogonia.** These are diploid cells that give rise to two other sets of diploid cells, the **primary oocytes** and **primary spermatocytes.** These, then, are the cells that undergo meiosis, yielding haploid egg and sperm cells.

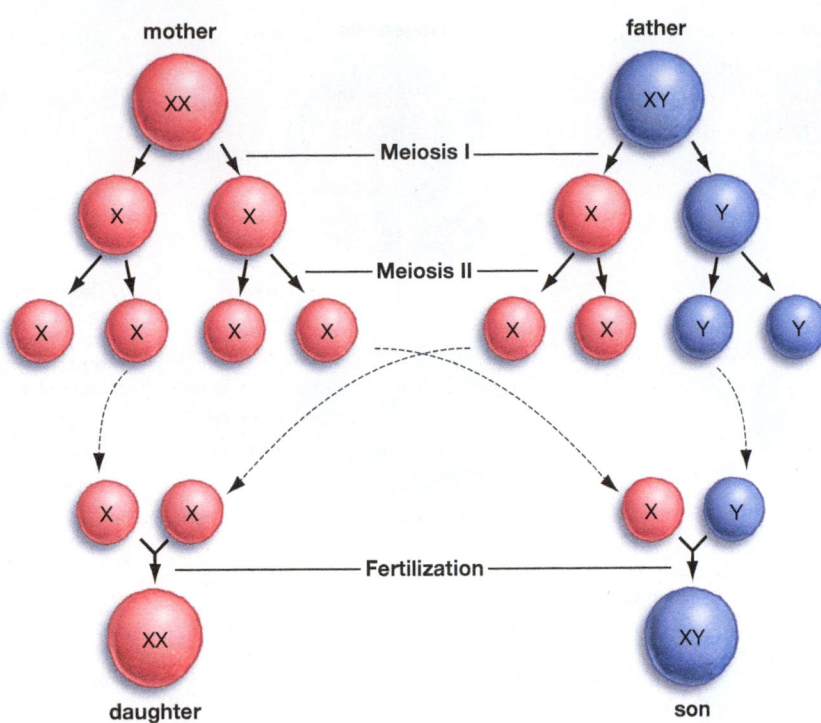

1. Early in meiosis I, the mother's two X chromosomes line up at the metaphase plate (XX). Meanwhile, in the father's meiosis, his X and Y chromosomes line up (XY).

2. The X-X and X-Y pairs then separate into different cells.

3. The chromatids that made up the duplicated chromosomes separate, yielding individual eggs and sperm.

4. Should an X-bearing sperm from the male fertilize the egg, the child will be a girl; should a Y-bearing sperm from the male fertilize the egg, the child will be a boy.

Figure 10.7
Sex Outcome in Human Reproduction

Sperm Formation

The starting cells in male gamete formation, the spermatogonia, exist in the male testes. Each time a spermatogonium divides, it gives rise to two cells, of course, but there's a twist to this. One of the cells that's produced will be a primary spermatocyte that will in turn give rise to a sperm cell. But the *other* cell will be another spermatogonium. Thus, spermatogonia are sperm-producing "factories" that generate not only sperm but more factories as well (**Figure 10.8**, on the next page). It is this self-generating ability that qualifies spermatogonia as "stem" cells and that allows males to *keep* producing sperm throughout their lives, beginning at puberty.

The primary spermatocytes that come from spermatogonia go through meiosis I and thus produce the haploid *secondary spermatocytes*. Meiosis II then ensues, and the result is four haploid *spermatids*. These are not yet the familiar, tadpole-like sperm; those come only after spermatids develop further, a process that takes about three weeks. (You can read more about this in Chapter 33, beginning on page 639.) There is thus an assembly-line quality to sperm production. Once a male reaches puberty, he starts producing sperm in this manner, with each sperm moving right through the entire process in a matter of weeks, and the process continuing until the male dies. As you'll see, female egg production takes a much different route.

What is the nature of the sperm produced? A given sperm cell must, of course, fully develop its best-known feature, its tail-like flagellum (which in the human body is found *only* on sperm cells). It turns out that each flagellum is composed of the microtubules you have been

seeing so much of in cell division. Moving the flagellum in a whip-like motion, microtubules serve to propel the sperm on their journey (**Figure 10.9**, on page 185). Because this journey is difficult—only a tiny fraction of the sperm that start it will complete it—and because sperm contribute nothing but their set of chromosomes to a fertilized egg, sperm have no use for most of the cellular organelles reviewed back in Chapter 4: lysosomes, the Golgi complex, and so forth. In consequence, sperm lose all these structures and more as they develop. By the time a given sperm is mature, it amounts to little more than a haploid set of chromosomes in front, a collection of "engines" (mitochondria) in the middle, and a propeller (the flagellum) in the back. About 250 million sperm are produced within a male each day, and about that number will be released with each ejaculation. Sperm that are not released will age and eventually are destroyed by the body's immune system.

Egg Formation: Stem Cells in Females?

The need of male sperm to become stripped-down, mobile DNA packages stands in contrast to the needs of female ova or eggs, which must have rich resources of nutrients and building materials to sustain life once they have been fertilized. This difference in function is reflected in size: The volume of an egg is 200,000 times the volume of a sperm. (Figure 10.9 allows you to get an idea of this difference.) Accordingly, eggs are created in a process that maximizes cell content, at least for the fraction of eggs that move fully through development.

Egg formation begins with the cells of the female called the *oogonia*. These cells are very different from

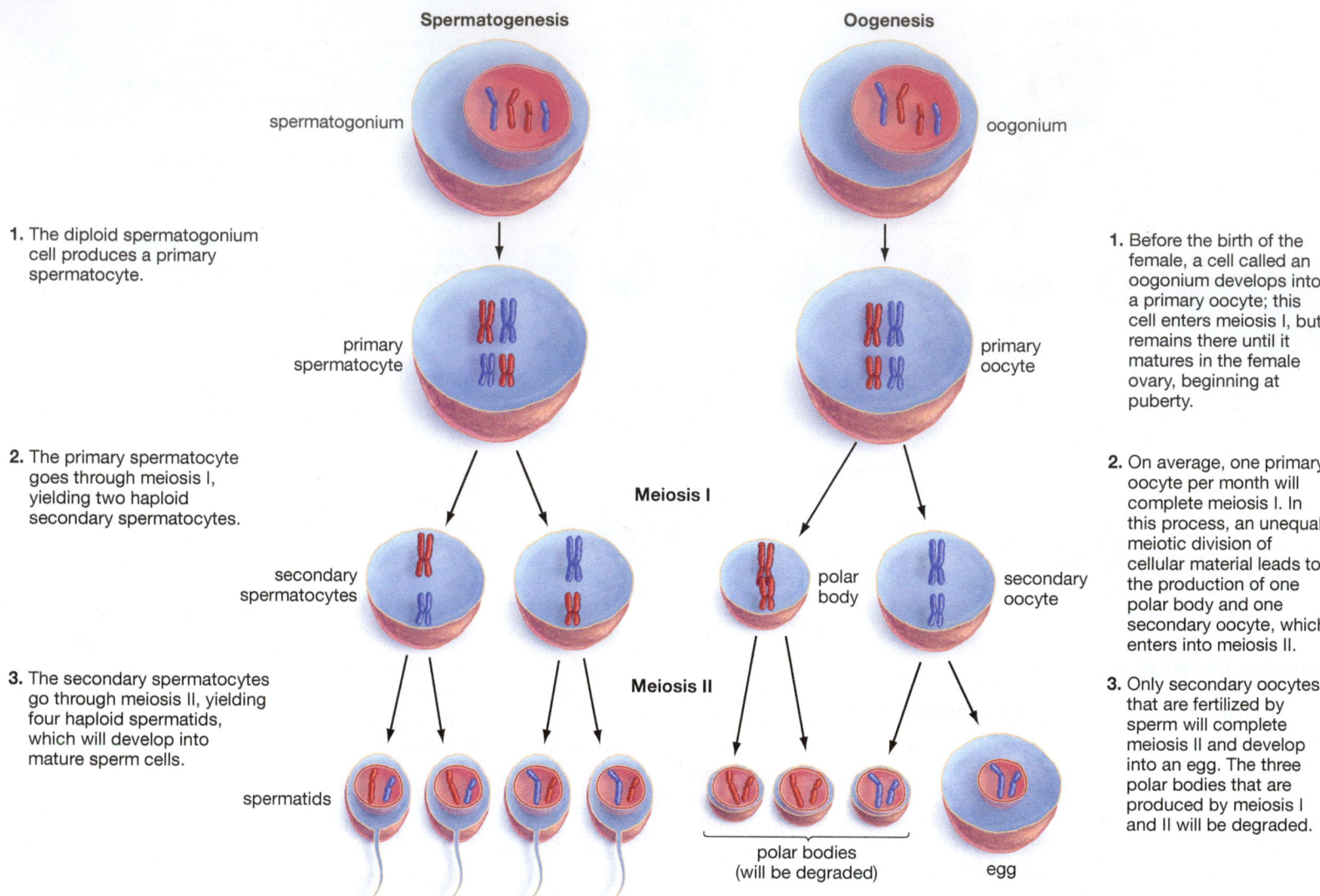

Spermatogenesis

Oogenesis

spermatogonium

oogonium

1. The diploid spermatogonium cell produces a primary spermatocyte.

primary spermatocyte

primary oocyte

1. Before the birth of the female, a cell called an oogonium develops into a primary oocyte; this cell enters meiosis I, but remains there until it matures in the female ovary, beginning at puberty.

2. The primary spermatocyte goes through meiosis I, yielding two haploid secondary spermatocytes.

Meiosis I

secondary spermatocytes

polar body

secondary oocyte

2. On average, one primary oocyte per month will complete meiosis I. In this process, an unequal meiotic division of cellular material leads to the production of one polar body and one secondary oocyte, which enters into meiosis II.

3. The secondary spermatocytes go through meiosis II, yielding four haploid spermatids, which will develop into mature sperm cells.

Meiosis II

spermatids

polar bodies (will be degraded)

egg

3. Only secondary oocytes that are fertilized by sperm will complete meiosis II and develop into an egg. The three polar bodies that are produced by meiosis I and II will be degraded.

Figure 10.8
Sperm and Egg Formation in Humans

their counterpart cells in males (the spermatogonia). How? They are produced in large numbers only up to the seventh month of fetal life, and it's possible that none are created after birth. Put another way, these cells seem to be produced only while a female is a fetus. Further, of the millions produced, most die before the female is born. Some survive to become the generation of cells, the primary oocytes, that enter meiosis.

Under this view, as you can see, adult females have no reproductive stem cells—cells that have the ability to keep on producing more of themselves in the way that spermatogonia do. Oogonia appear to exist only up until birth, and until quite recently, scientists were confident that no other variety of reproductive stem cell existed in adult females. This in turn was thought to be the primary reason that women lose the ability to reproduce in middle age: They run out of eggs, in part because they have no stem cells that are capable of giving rise to new eggs.

All this seemed well understood until 2004, when researchers from Harvard claimed that, in adult female

mice, they had detected the presence of reproductive stem cells. This report rocked the world of *human* reproductive research because of its implications. Imagine if such cells could be found or coaxed into development in human females. Fertility might be prolonged for older women or enhanced for women of any age. At present, therefore, the reproductive research community has an important question before it: Do adult female mammals possess reproductive stem cells? Despite much research in the last few years, no one can say with certainty whether they do or not. Only more research will provide an answer.

Egg Formation: From Oogonia to Eggs

Setting aside the question of reproductive stem cells, the path by which most eggs are produced in women is well understood. It starts with the oogonia we just talked about—the cells that are produced largely, or perhaps entirely, prior to the birth of the female.

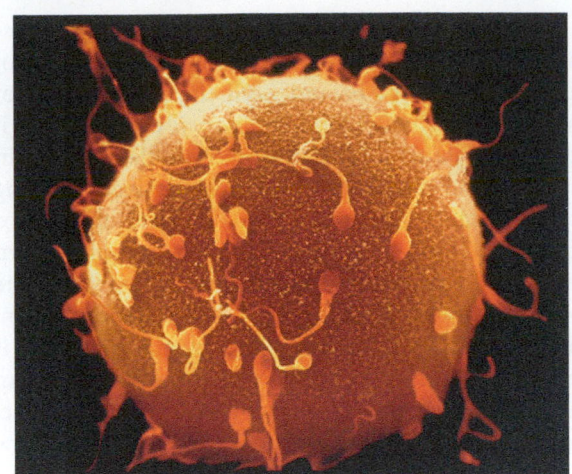

Figure 10.9
The Products of Gamete Formation
A human egg surrounded by much smaller human sperm.

These cells give rise to the next generation of cells, the primary oocytes, which are the cells that enter into meiosis I (Figure 10.8). Amazingly enough, meiosis I is where these oocytes *stay*—for years. Imagine a collection of oocytes in the female ovaries (which are walnut-sized structures just right and left of center in the female pelvic area). Once puberty begins, each oocyte is theoretically capable of maturing into an egg, but an average of only *one* oocyte per month actually starts this maturation process. Which oocyte is this? The one released from the ovary each month in the process called ovulation. All the other oocytes in the ovaries halt their meiosis in prophase of meiosis I. Their chromosomes are duplicated, and they have completed crossing over, but they have not yet lined up at the metaphase plate. And this is where things stay for most of the oocytes a female possesses. Only the one oocyte per month that is expelled from the ovary will complete meiosis I and thus enter into meiosis II. Even this meiosis is "arrested," however, until this oocyte is fertilized by a sperm. It is the union with a sperm that prompts the completion of meiosis II. Thus, an oocyte that came into being when a female was a fetus might remain suspended in the cell cycle for years or even decades. Of the millions of oocytes originally created, only a few hundred will mature even through metaphase of meiosis I during a woman's lifetime.

How do certain oocytes get "selected" to mature through metaphase I—how do they get selected to be ovulated, in other words? This occurs in connection with an intricate interplay that involves their surrounding tissue, hormones, and the happenstance of timing. You'll read more about this process in Chapter 33.

One Egg, Several Polar Bodies

For the few oocytes selected, the process of going through meiosis I and II means proceeding from one diploid cell to four haploid cells. In this process, however, almost all the cytoplasm from the original diploid cell will be shunted to *one* of the four daughter cells, thus preferentially building up its stores of nutrients and other cytoplasmic material. This happens in two rounds: meiosis I and meiosis II. When cytokinesis comes in meiosis I, one daughter cell gets almost all the cytoplasm, while the second daughter cell gets almost none. This same thing happens in meiosis II. The result? One richly endowed haploid oocyte and two, or perhaps three, very small haploid **polar bodies:** non-functional cells produced during meiosis in females (Figure 10.8). (Only two polar bodies and the oocyte will be the end result if the polar body produced in meiosis I does not continue through meiosis II, which often happens.) The oocyte has a chance at taking part in producing offspring, but the polar bodies are bound for oblivion rather than a shot at immortality. They will eventually degrade into their constituent substances.

10.6 Life Cycles: Humans and Other Organisms

The process of the formation of the human gametes called egg and sperm is followed, as you have seen, by the union of these gametes. In humans, the oocyte that is expelled from the ovary makes a slow trip down the female's uterine (or Fallopian) tube, during which time it may encounter a traveling sperm. If it does, the result will be a fertilized egg (or zygote), which then may develop into a whole human being. One of the things that will eventually occur in this new human being is the development of another group of eggs or sperm through the process of meiosis. Thus, do we see a **life cycle:** the repeating series of steps that occur in the reproduction of an organism. You have looked at the human life cycle, but it's important to recognize that it amounts to only one life cycle among many that exist.

Not All Reproduction Is Sexual

As noted earlier, human reproduction is **sexual reproduction:** the union of two reproductive cells to create a new organism. In many other species, there is **asexual reproduction** or reproduction that does not involve sex. Asexual reproduction actually is seen in a vast array of organisms. In fact, mammals are the only grouping of living things that never carries out asexual reproduction. Admittedly, asexual reproduction is rare in creatures as complex as birds, fish, and reptiles, but it does happen. In the American southwestern desert, for example, there are a few species of whiptail lizard that are entirely female

Figure 10.10
All-Female Species
Pictured is a whiptail lizard, *Aspidoscelis uniparens,* that reproduces solely through partheno-genesis. Because the eggs these lizards produce are not fertilized by males, they contain only one set of chromosomes, passed on by the mother. Hatchlings develop from these haploid eggs, however, and as a consequence the hatchlings themselves are haploid and genetically identical to the mother. The end result is a species that is entirely female.

Asexual and Sexual Life Cycles

(**Figure 10.10**). The female whiptail lays haploid eggs, but they are never fertilized. Instead, they develop into haploid clones of her through regular mitotic cell division. This method of reproduction (called *parthenogenesis*) also results in "drone" male honey-bees. Those eggs that a queen bee chooses not to fertilize (with stored-up sperm) develop into drones through parthenogenesis.

Other forms of asexual reproduction also are possible. In fact, you've already been introduced to one of them. Recall that a bacterial cell engages in a form of asexual reproduction, called *binary fission,* in which it duplicates its single chromosome and then divides in two. Beyond this, anyone who has ever seen a cutting taken from a plant knows that a snipped-off stem or branch will, when properly planted, sprout its own roots and grow into an independent plant, complete with the ability to take part in sexual reproduction. This process is called *vegetative reproduction,* and it has a counterpart in the animal world. Cut the arm and part of a central disk from a sea star, for example, and, through *regeneration,* this severed limb will grow into a whole, multi-armed sea star (**Figure 10.11**).

As we go through the branching tree of life, from more complex creatures to less complex ones, asexual reproduction becomes more common. By the time we reach bacteria, it is the only form of reproduction that exists. Note that all forms of asexual reproduction produce offspring that are genetically identical to their parents. There is no mixing and matching of chromosomes from different parents in these asexual processes; just a cloning of one genetically identical individual from its parent.

Variations in Sexual Reproduction

Within the framework of sexual reproduction, as you can imagine, there are many variations. For example, separate gametes need not come from separate organisms. Some animals, such as tapeworms, are hermaphrodites, meaning they have both male and female reproductive parts. In some cases, hermaphrodites

Figure 10.11
Regenerating the Whole from a Part
A sea star off the coast of Papua New Guinea has regenerated a body and other arms from a piece of one arm and part of its central disk.

fertilize themselves, but generally they are fertilized by another member of their species. Most plants contain both male and female parts, and some plants "self-fertilize"—their sperm can fertilize their eggs—as with a famous pea plant you will be looking at in Chapter 11.

On to Patterns of Inheritance

Our tour of cell division over the last couple of chapters has yielded a look not only at how one cell becomes two, but at how chromosomes operate within this process. Given the paired nature of chromosomes and their precisely ordered activity during meiosis, it stands to reason that there would be some predictability or pattern to the passing on of traits. This is the case, as it turns out. And the person who first recognized this pattern was a monk who worked in nineteenth-century Europe in an obscurity he did not deserve.

SO FAR ...

1. Spermatogonia are regarded as _____ cells because they can produce not only a specialized variety of cell (the spermatocyte) but also more spermatogonia. Meanwhile, it is possible that the female counterpart to spermatogonia, the oogonia, are produced in females only _____.

2. The function of eggs as rich sources of nutrients and cellular materials results in oogenesis producing not four cells but one viable cell and several _____.

3. Each organism produced through asexual reproduction is genetically _____ to its parent.

CHAPTER 10 REVIEW

Summary

10.1 An Overview of Meiosis

- In humans, nearly all cells have paired sets of chromosomes, meaning these cells are diploid. In meiosis, a single diploid cell divides to produce four haploid cells—cells that contain a single set of chromosomes. (p. 175)

- The haploid cells produced through meiosis are called gametes, which are the reproductive cells of many organisms. Female gametes are eggs; male gametes are sperm. When the haploid sperm and haploid egg fuse, a diploid fertilized egg is produced, setting into development a new generation of organism. (p. 175)

10.2 The Steps in Meiosis

- There are two primary stages to meiosis. In meiosis I, chromosome duplication is followed by a pairing of homologous

chromosomes with one another, during which time they exchange reciprocal sections of themselves. Homologous chromosome pairs then line up on opposite sides of the metaphase plate. These homologous pairs are then separated, each of them becoming part of a separate daughter cell. In meiosis II, the sister chromatids of the duplicated chromosomes are separated into separate daughter cells. (p. 176)

10.3 The Significance of Meiosis

- Meiosis generates genetic diversity among offspring in two ways. In prophase I of meiosis, homologous chromosomes pair with each other and, in the process called crossing over, exchange reciprocal segments of themselves. Then in metaphase I of meiosis, there is an independent assortment of maternal and paternal

chromosome pairs on either side of the metaphase plate. This chance alignment determines which daughter cell each chromosome will end up in. (p. 179)

- Thanks to the differences it ensures among offspring, meiosis is responsible to a significant extent for the diversity of life-forms seen today in the living world. By contrast, asexual reproduction, as seen in bacteria and other organisms, produces organisms that are exact genetic copies of the parent. (p. 181)

10.4 Meiosis and Sex Outcome

- Human females have 23 matched pairs of chromosomes—22 pairs of autosomes and one pair of sex-determining X chromosomes. Human males have 22 autosome pairs, one X chromosome, and one sex-determining Y chromosome. Each egg that a female produces has a single X chromosome in it. Each sperm that a male produces has either an X or a

Y chromosome within it. If a sperm with a Y chromosome fertilizes an egg, the offspring will be male. If a sperm with an X chromosome fertilizes the egg, the offspring will be female. **(p. 182)**

10.5 Gamete Formation in Humans

- Male gamete formation begins with cells called spermatogonia, which are stem cells in that they give rise not only to diploid primary spermatocytes but also to more spermatogonia. Primary spermatocytes then go through meiosis, producing haploid secondary spermatocytes, which in turn give rise to the spermatids that develop into mature sperm cells. **(p. 182)**

- Female gamete formation begins with cells called oogonia, most or all of which are produced prior to the birth of the female. These give rise to primary oocytes—the cells that will go through meiosis I, producing haploid oocytes. Only the primary oocyte released each month during ovulation completes meiosis I. Only those ovulated oocytes that are fertilized by sperm complete meiosis II. Only one of the four cells produced in meiosis will receive enough cytoplasmic material during cell division to be able to develop into a haploid egg; the other cells are destined to become non-functional polar bodies. **(p. 184)**

10.6 Life Cycles: Humans and Other Organisms

- Asexual reproduction can take several forms. Bacteria reproduce through binary fission, in which a given bacterial cell replicates its single chromosome and then divides in two. Plants can engage in vegetative reproduction; other organisms, such as worms and sea stars, can carry out regeneration. **(p. 185)**

Key Terms

Understanding the Basics

Multiple-Choice Questions

(Answers are in the back of the book.)

1. With respect to chromosomes, meiosis involves:
   a. no duplications, 1 reduction.
   b. 1 duplication, 2 reductions.
   c. 2 duplications, 1 reductions.
   d. 2 duplications, 2 reductions.
   e. none of the above.

2. Independent assortment occurs during:
   a. prophase I.
   b. anaphase II.
   c. metaphase I.
   d. metaphase II.
   e. both c and d.

3. Which of the following statements is not true regarding egg formation in humans?
   a. Oogonia are not produced until the eleventh or twelfth year.
   b. At the completion of division, one oocyte and two or three polar bodies exist.
   c. In both meiosis I and II, cytoplasmic material will be preferentially shunted to one cell.
   d. Oocyte maturation occurs in approximately a 28-day cycle.
   e. Each egg contains 23 chromosomes.

4. What are the two sources of genetic diversity in meiosis?
   a. binary fission and regeneration
   b. crossing over and independent assortment
   c. oogenesis and crossing over
   d. spermatogenesis and oogenesis
   e. vegetative reproduction and regeneration

5. Meiosis and sexual reproduction have fostered diversity in the natural world because they:
   a. are so accurate in copying genetic information.
   b. are nearly certain to produce differences among offspring.
   c. are such ancient processes.
   d. produce genetic replicas, one generation after another.
   e. are used by such a wide range of organisms.

6. Eggs are relatively large because they _____; sperm are relatively small because they _____.
   a. must contain spare sets of chromosomes … function through mobility
   b. must contain room for the growing zygote … have only one set of chromosomes
   c. must contain room for the sperm … have a short life-span
   d. must contain nutrients for the zygote … function through mobility
   e. exist in a harsh environment … exist in a nurturing environment

7. In prophase I of meiosis:
   a. maternal chromosome 1 joins with maternal chromosome 2.
   b. maternal chromosomes exchange genetic material with other maternal chromosomes.
   c. paternal chromosomes exchange genetic material with other paternal chromosomes.
   d. homologous chromosomes exchange genetic material.
   e. Both c and d are correct.

8. At the completion of meiosis I, each human cell going through the process contains:
   a. 23 chromosomes, 46 chromatids.
   b. 46 chromosomes, 46 chromatids.
   c. 46 chromosomes, 92 chromatids.
   d. 23 chromosomes, 23 chromatids.
   e. none of the above.

9. Asexual reproduction is _____ and always results in _____.
   a. found in only a few types of organisms … offspring that are different from parents
   b. found in all types of organisms … offspring that are sometimes genetically identical to parents

c. reproduction that results from a fusion of two reproductive cells … twice as many offspring as parents

d. found more commonly in simple organisms than in complex ones … offspring that are genetically identical to parents

e. found in only a few mammals and birds … offspring that are different from parents

Brief Review

(Answers are in the back of the book.)

1. Describe the essential differences between mitosis and meiosis.

2. Does chromosome reduction take place during meiosis I or II?

3. Why are men able to keep producing sperm throughout their lifetime, from puberty forward?

4. How does cytokinesis in the production of the eggs that females produce differ from cytokinesis in regular mitotic cell division?

5. Define and distinguish between somatic cells and gametes.

6. A diploid organism has four chromosomes. Name the phases of mitosis or meiosis shown in the following figures.

7. In what ways does meiosis ensure genetic diversity in offspring?

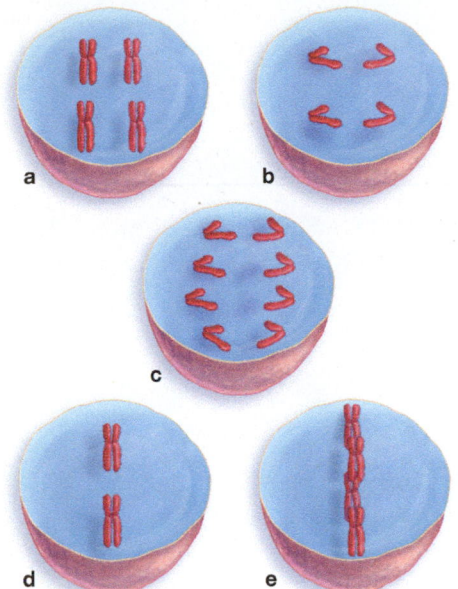

8. When human egg and sperm come together, how many chromosomes does each kind of gamete have? What is the genetic makeup of the resulting zygote?

Applying Your Knowledge

1. Ultimately, is it the paternal or maternal gamete that delivers the sex-determining chromosome? Describe the process.

2. Mammals have now been cloned by scientists (think of Dolly the sheep). Suppose that half the world's human births were achieved through cloning. How would our world be different?

3. All the bananas we eat come from trees that are produced through "cuttings"—a stem from an existing tree is planted in the ground, resulting in a new tree. Thus, each tree is a clone of another. Why would growers find it advantageous to produce an enormous series of identical clones, rather than using trees that reproduce sexually?

The First Geneticist:
Mendel and His Discoveries

A straightforward set of rules governs the way living things pass on many of their traits. Gregor Mendel was the first person to understand what those rules are.

Available on MasteringBiology®

Activities: Dihybrid Cross; Gregor's Garden; Variations on Mendel; **Building Vocabulary; Questions:** Reading Questions; Test Bank

By experimenting with generations of garden peas such as these, Gregor Mendel became the first person to understand some of the basic rules of genetics.

In the Czech Republic, in the oldest part of the city of Brno, stands the former Monastery of St. Thomas. Now a research center, St. Thomas has a small, fenced-in garden that sits just outside its historic living quarters. At one end of the garden stands an ivy-covered tablet, inscribed with a few simple words: "Prelate Gregor Mendel made his experiments for his law

here." Visitors to the garden can cast their eyes upward and see the rooms that Gregor Johann Mendel lived in while carrying out his work in the years between 1856 and 1863 (**Figure 11.1**). Mendel labored during this time on a species of common peas, *Pisum sativum,* using tweezers and an artist's paintbrush as his tools. Like many scientific discoverers, Mendel looked at ordinary objects; and as with many scientists, his work was tedious and exacting. Nevertheless, the distance from what Mendel did to what Mendel learned is the distance from the ordinary to the historic.

Coming to adulthood in the mid-nineteenth century, Mendel knew nothing of the elements of genetics we have reviewed so far. Chromosomes and the genes that lie along them had not even been discovered when he was working with his peas; little knowledge existed of meiosis or mitosis, to say nothing of DNA. And as it happened, Mendel himself played no direct role in discovering any of these things. Yet, this unassuming monk, the son of eastern European peasant farmers, generally is accorded the title of the father of genetics. What contribution earned him this honor?

Figure 11.1
Austrian Monk and Naturalist Gregor Mendel

11.1 Mendel and the Black Box

Mendel's achievement was to comprehend what was going on inside what might be called the "black box" of genetics without ever being able to look inside that box himself. Science is filled with so-called black-box problems, in which researchers know what goes *into* a given process and what comes *out.* It is what is going on in between—in the black box—that is a mystery. In the case of genetics, what lies inside the black box is DNA, chromosomes, meiosis, and so forth—all the component parts of genetics, in other words, and the way they work together. Because of the timing of his birth, Mendel had no knowledge of any of these things. Until Mendel, however, nobody had looked carefully at even the starting and ending points of this black-box problem: at what went in and what came out. Mendel's original pea plants represented the input side to the black box of genetics, while the offspring he got from breeding these plants represented the output. By looking carefully at generations of parents and offspring—at both sides of the box, in a sense—Mendel was able to infer something about what had to be going on within. As you will see, his main inferences were correct: (1) that the basic units of genetics are material elements; (2) that these elements come in pairs; (3) that these elements (today called genes) can retain their character through many generations; and (4) that gene pairs *separate* during the formation of gametes.

These insights may sound familiar because all of them were approached from another direction in Chapters 9 and 10. Here are the lessons you went over then:

1. Genes are material elements—lengths of DNA.

2. In human beings, genes come in pairs, residing in pairs of homologous chromosomes.

3. Chromosomes make copies of themselves, thus giving the genes that lie along them the ability to be passed on intact through generations.

4. In meiosis, homologous chromosomes line up next to each other and then separate, with each member of a pair ending up in a different egg or sperm cell.

By observing generations of pea plants and applying mathematics to his observations, Mendel inferred that something like this had to be happening in reproduction. Because Mendel was the first to perceive a set of principles that govern inheritance, we date our knowledge of genetics from him.

11.2 The Experimental Subjects: *Pisum sativum*

Beginning work in his monastery garden after a period of scientific study in Vienna, Mendel picked the pea plant *Pisum sativum* as the species on which he would carry out his experiments. If you look at **Figure 11.2**, you can see something of the life cycle of this garden pea. Note that what we think of as peas in a pod are seeds in this plant's ovary. Each of these seeds begins as an unfertilized egg, just as a human baby begins as a maternal egg that is unfertilized. Sperm-bearing pollen, landing on the plant's stigma, then set in motion the fertilization of the eggs.

Importantly, *Pisum* plants can *self*-pollinate. The anthers of a given flower release pollen grains that land on that flower's stigma. Each seed that develops

Figure 11.2
Life Cycle of the Pea Plant

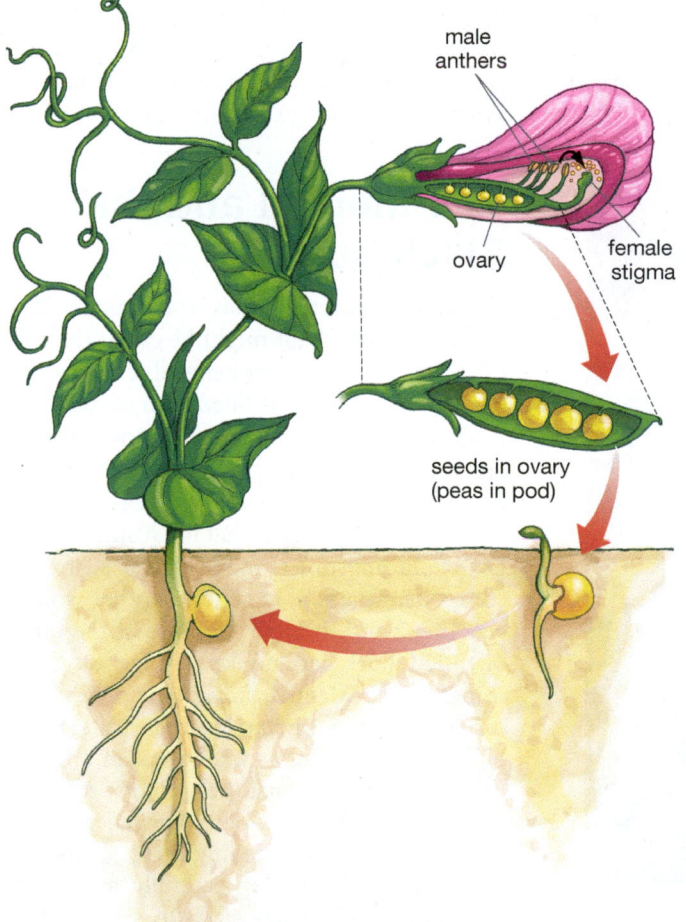

male anthers

ovary

female stigma

seeds in ovary (peas in pod)

1. **Self-pollination.** Pollen from the male anthers of a plant falls on the female stigma of that same plant.

2. **Fertilization.** Sperm from the pollen fertilize the plant's eggs, which lie inside the ovary. These eggs will develop into seeds in the ovary (peas in a pod), which represent a new plant generation. Each seed is fertilized separately.

3. **Germination.** Each seed can be planted and grown into a separate plant.

4. **Development.** Seedlings develop into mature seed plants, capable of producing their own offspring.

from the resulting fertilizations can then be planted in the ground and give rise to a new generation of plant. The way to think of the seeds in a pod is as multiple offspring from multiple fertilizations—one pollen grain fertilizes this seed, another pollen grain fertilizes another. This is why, as you'll see, seeds that are in the same pod can have different characteristics.

Although the pea plants can self-pollinate, they can also **cross-pollinate**—one plant can pollinate another. Indeed, Mendel could accomplish such cross-pollination through his own efforts, by going to work with his tweezers and paintbrushes, as shown in **Figure 11.3**.

In directing pollination, Mendel could control for certain attributes in his pea plants—qualities now referred to as *characters*. If you look at **Table 11.1**, on the next page, you can see that Mendel looked at seven characters in all—such things as stem length, seed color, and seed shape. Note that each of these characters comes in two varieties. There are yellow or green seeds, for example, and purple or white flowers. Such character variations are known as *traits*. Mendel referred to

each of these variations as being either a "dominant" or a "recessive" trait, as the table shows. For now, you can think of a recessive trait as one that tends to remain hidden in certain generations of pea plants. A dominant trait, meanwhile, can be thought of as a trait that tends to appear in those same generations. You'll get a formal definition of dominant and recessive later.

Phenotype and Genotype

Now, let's put these traits in another context. Taken together, traits represent what are called **phenotypes.** Broadly speaking, a phenotype is a physiological feature, bodily characteristic, or behavior of an organism. In the context of our look at *Pisum sativum*, phenotype means the plant's visible physical characteristics. Purple flowers are one phenotype, white flowers another; yellow seeds are one phenotype, green seeds another. Meanwhile, any phenotype is in significant part determined by an organism's underlying **genotype,** meaning its genetic makeup.

Figure 11.3
How to Cross-Pollinate Pea Plants

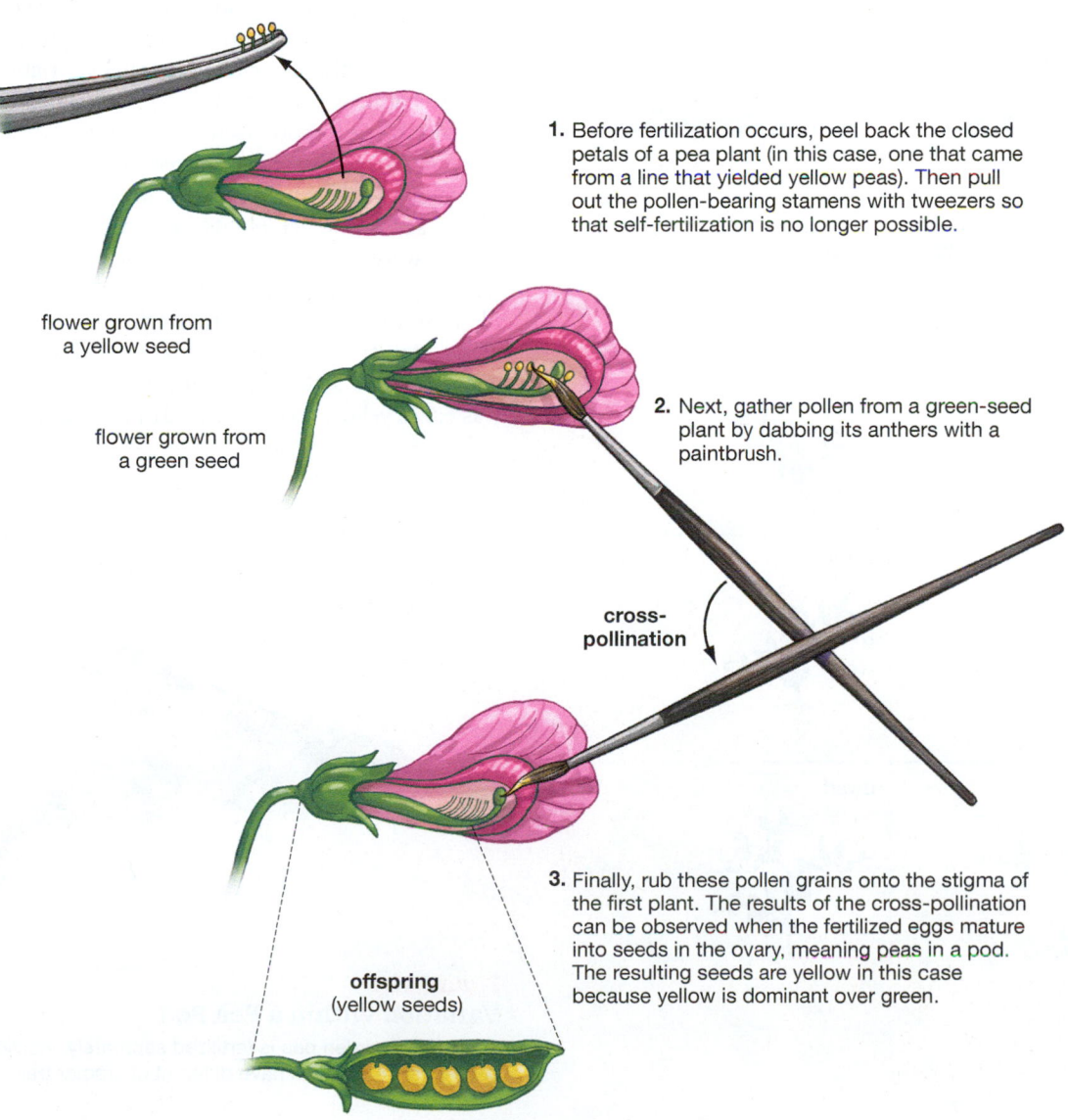

flower grown from a yellow seed

flower grown from a green seed

cross-pollination

offspring (yellow seeds)

1. Before fertilization occurs, peel back the closed petals of a pea plant (in this case, one that came from a line that yielded yellow peas). Then pull out the pollen-bearing stamens with tweezers so that self-fertilization is no longer possible.

2. Next, gather pollen from a green-seed plant by dabbing its anthers with a paintbrush.

3. Finally, rub these pollen grains onto the stigma of the first plant. The results of the cross-pollination can be observed when the fertilized eggs mature into seeds in the ovary, meaning peas in a pod. The resulting seeds are yellow in this case because yellow is dominant over green.

11.3 Starting the Experiments: Yellow and Green Peas

Mendel began by making sure that his starting plants "bred true" for the phenotypes under study. That is, he assured himself that, for example, all of his purple-flowered plants would produce nothing but generations of purple offspring if they self-pollinated.

Table 11.1

Pea-Plant Characters Studied by Mendel

Character Studied	Dominant Trait	Recessive Trait
Seed shape	smooth	wrinkled
Seed color	yellow	green
Pod shape	inflated	wrinkled
Pod color	green	yellow
Flower color	purple	white
Flower position	on stem	at tip
Stem length	tall	dwarf

Parental, F_1, and F_2 Generations

The **parental generation** in such experiments is often referred to in shorthand as the **P** generation. Meanwhile, the offspring of the parental generation are known as the **first filial generation,** the word *filial* indicating a son or a daughter. The shorthand for first filial is F_1. The F_x form can be used for any succeeding generation. You will be seeing a lot of the second filial, or F_2 generation, for example.

So, how did Mendel proceed with his plants? Let's start with just one example. Mendel took, for his P generation, some plants from a line that bred true for yellow seeds and other plants from a line that bred true for green seeds. Then he took pollen from one variety of the plants and fertilized the other variety, as seen in Figure 11.3. The plants fertilized in this way then produced their own seeds—every one of which was yellow. (Remember that although these seeds were developing within the pods of the parental generation, they represented the new generation—the F_1 generation.) Getting all yellow seeds was an interesting result in itself. Because each seed in a pea-plant pod is fertilized separately, it can be the case that a given pod will contain both yellow and green seeds. (Other pea characters can be variable within a pod as well, as you can see in **Figure 11.4**.) Yet, all the seeds in this generation were yellow, indicating that yellow was a dominant trait and green a recessive one.

Having viewed these results, Mendel then planted his F_1 seeds and let plants grow from them. These plants were then allowed to *self*-pollinate—sperm from a given plant fertilized the eggs from that same plant, so that instead of "crossing" one plant with another, Mendel was crossing a plant with itself. What he got from this self-pollination, in the F_2 generation, was 6,022 yellow seeds and 2,001 green seeds.

Some of these peas have a smooth texture, while others are wrinkled

Figure 11.4
Variation within a Pea Pod

Since each garden pea is fertilized separately, individual peas within a pod can have different character traits.

The Power of Counting

In counting the seeds, Mendel was taking a giant step forward. Why? Experimenters before Mendel had gotten the kind of results he had. What they did not do, however, was undertake a careful *counting* of their results and then analyze the results in terms of proportions.

Looking at the specific results, two things stand out. First, green seeds disappeared in F_1, but came back in F_2. Recall that there was not a green seed to be found in the F_1 generation pods, but in F_2, there are 2,001 of them. Second, green seeds came back in F_2 as *one-fourth* of the seeds as a whole. The F_2 generation isn't divided 50/50, half yellow and half green; instead, there are roughly 3 yellow seeds for every 1 green seed—or to put it another way, a 3:1 ratio of yellow to green seeds. As it turned out, Mendel got the same result with each of the seven characters he was studying, as you can see in **Table 11.2**. Note that the 3:1 ratio is a proportion of dominant to recessive traits—more yellow seeds than green, more purple flowers than white. It was *solely* the dominant traits that appeared in the F_1 generation, but in the F_2 generation the dominants are simply appearing in greater proportion than the recessives.

Interpreting the F_1 and F_2 Results

What did Mendel learn from these results? For one thing, he saw that inheritance for his peas was not a matter of the "blending" of their characteristics. For example, the flower colors purple and white were retained as just that. No intermediate phenotypes—say, lavender flowers—resulted from the cross in any generation. This finding ran contrary to the notion, popular in Mendel's time, that two given traits would blend into a homogeneous third entity, as coffee and milk will blend together.

Beyond this finding, it was apparent that, dominance notwithstanding, plants could retain the *potential* for recessive phenotypes, even though those phenotypes might not appear in a given generation. Mendel's F_1 plants had no green seeds, but in F_2 the green seeds were back. It was reasonable to assume, therefore, that the yellow-seed F_1 plants retained a green-seed *element*, which got expressed only in the F_2 generation. Finally, because his phenotypes came in pairs, it was reasonable for Mendel to hypothesize that the elements likewise came in pairs.

If you think Mendel's *pairs of elements* sound suspiciously like the "pairs of genes" on homologous chromosomes we've seen before, you are right. However, it's time to start referring to these "matched pairs of genes" in scientific terminology. It is more accurate to think of matched pairs of genes as alternative forms of a single gene. The proper name for an alternative form of a gene is an **allele.** Thus, the pea

plant had a single gene for seed color that came in two alleles—one of which coded for yellow seeds, the other of which coded for green seeds. These alleles resided on separate homologous chromosomes.

Table 11.2		
Ratios of Dominant to Recessive in Mendel's Plants		
Dominant Trait	**Recessive Trait**	**Ratio of Dominant to Recessive in F2 Generation**
Smooth seed	Wrinkled seed	2.96:1 (5,474 smooth, 1,850 wrinkled)
Yellow seed	Green seed	3.01:1 (6,022 yellow, 2,001 green)
Inflated pod	Wrinkled pod	2.95: 1 (822 inflated, 299 wrinkled)
Green pod	Yellow pod	2.82:1 (428 green, 152 yellow)
Purple flower	White flower	3.14:1 (705 purple, 224 white)
Flower on stem	Flower at tip	3.14:1 (651 along stem, 207 at tip)
Tall stem	Dwarf stem	2.84:1 (787 tall plants, 277 dwarfs)
Average ratio, all traits	**3:1**	

11.4 Another Generation

Having seen recessive phenotypes reappear in F_2, Mendel decided to extend his breeding. In his next experiment, he took a set of F_2 generation seeds—all of them yellow—planted them, and let the plants they grew into self-pollinate. Then, he looked at the color of the F_3 seeds contained in the pods of these plants. What he found was that, of 519 plants grown, 166 yielded only yellow seeds, while the remaining 353 plants had a mixture of yellow and green seeds in the pods, in the familiar 3:1 ratio. But when he similarly took F_2 green seeds, planted them, and let them self-pollinate, these plants produced nothing but green seeds.

What could have caused these results? The answer was the underlying elements Mendel had hit upon, meaning the alleles for the color gene. When he planted his yellow seeds, some of them yielded all-yellow offspring, but others yielded mixed (green and yellow) offspring. He reasoned that there must be *two kinds* of yellow plants: those that were "pure" yellow (the 166 that yielded solely yellow seeds) and those that were "mixed" yellow (the 353 that produced both yellow and green seeds). Pure yellow had nothing but yellow alleles within and would produce nothing but yellow seeds when self-pollinated. Meanwhile, the mixed yellow would have one yellow and one green allele within and thus would produce pods that contained seeds of both colors. Finally, the green seeds had nothing but green alleles and thus produced nothing but green seeds.

(a) P generation crosses

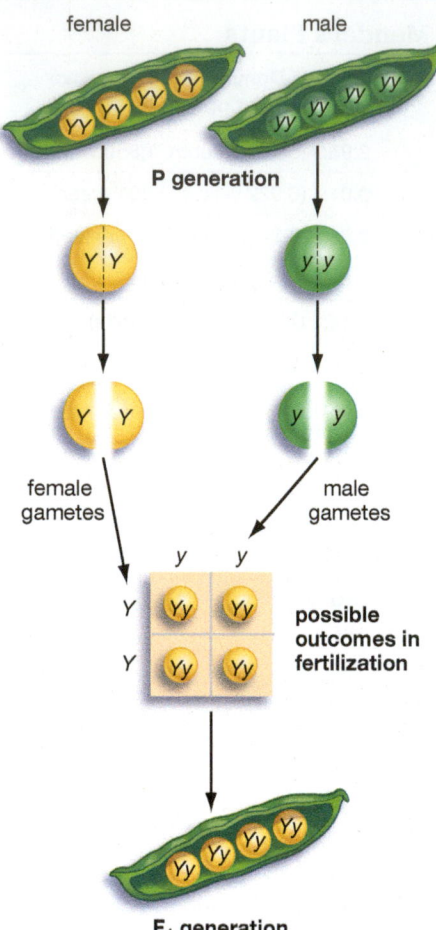

female   male

P generation

female gametes   male gametes

possible outcomes in fertilization

F₁ generation

1. Female gametes are being provided by a plant that has the dominant, yellow alleles (*YY*); male gametes are being provided by a plant that has the recessive, green alleles (*yy*).

2. The cells of the pea plants that give rise to gametes start to go through meiosis.

3. The two alleles for pea color, which lie on separate homologous chromosomes, separate in meiosis, yielding gametes that each bear a single allele for seed color. In the female, each gamete bears a *Y* allele; in the male, each bears a *y* allele.

4. The Punnett square shows the possible combinations that can result when the male and female gametes come together in the moment of fertilization. (If you have trouble reading the Punnett square, see Figure 11.5b). The single possible outcome in this fertilization is a mixed genotype, *Yy*.

5. Because *Y* (yellow) is dominant over *y* (green), the result is that all the offspring in the F₁ generation are yellow because they all contain a *Y* allele.

(b) How to read a Punnett square

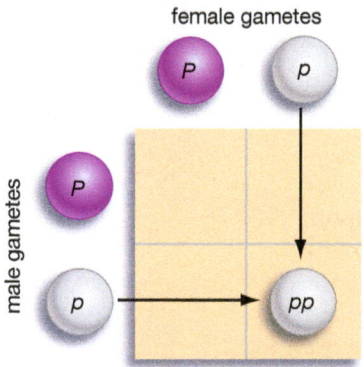

female gametes

male gametes

female gametes

male gametes

1. A *p* gamete from the male combines with a *p* gamete from the female to produce an offspring of *pp* genotype (and white color).

2. A *p* gamete from the male combines with a *P* gamete from the female to produce an offspring of *Pp* genotype (and purple color).

Figure 11.5
Visualizing Mendel's Results

(a) Mendel's P Generation Crosses
(b) Examples of Punnett Squares

SO FAR . . .

1. A physical characteristic, such as the height of a person or the color of one of Mendel's peas, represents a _____ in these organisms, which is in large part based on their genetic makeup or _____.

2. In Mendel's experiments, the first generation of plants that he got, from crossing parental generation plants, exhibited only _____ traits, while the next generation of plants, which resulted from self-pollination of plants, exhibited both _____ and _____ traits, in a 3:1 ratio.

3. Genes come in alternative forms known as _____, which reside on separate _____.

Mendel's Generations in Pictures

Let's now make these points clearer by reviewing all three generations of experiments schematically.

The F₁ Generation

To start out, it will be helpful to introduce a common, shorthand method of denoting alleles. Let us refer to dominant and recessive alleles by uppercase and lowercase letters, with uppercase used for dominant types and lowercase used for recessive types. Thus, a "pure" yellow-seeded plant, having *two* yellow alleles, would be symbolized as *YY*. Meanwhile, pure green-seeded plants are *yy*, while mixed seeds are *Yy*. Let us say, just as an example, that female gametes are being supplied by a plant that is *YY*, while male gametes are coming from a plant that is *yy*. In meiosis, as you've seen, homologous chromosomes separate. This happens in the pea plants, meaning that each of these plants will contribute *one member* of its gene pair to its respective gametes—the *YY* female contributing a *Y* gamete, and the *yy* male contributing a *y* gamete. When these gametes fuse in the moment of fertilization, the result is a *Yy* hybrid in the F₁ generation. If you look at **Figure 11.5**, you will be introduced to a time-honored way to represent such outcomes, the Punnett square, and see Mendel's F₁ results symbolized as well.

Note that, because *Y* is dominant, all the seeds in the F₁ offspring pods will have a yellow *phenotype*, even though every one of these seeds contains a mixed *genotype*. (Every one is *Yy*.) It takes only one *Y* allele for a seed to be yellow; for a seed to be green, then, it must have two *y* alleles.

The F₂ Generation

Next come the F₁ crosses that yielded the F₂ generation. The starting point here is the F₁ seeds, which are all of the mixed *Yy* type, as you can see in **Figure 11.6**.

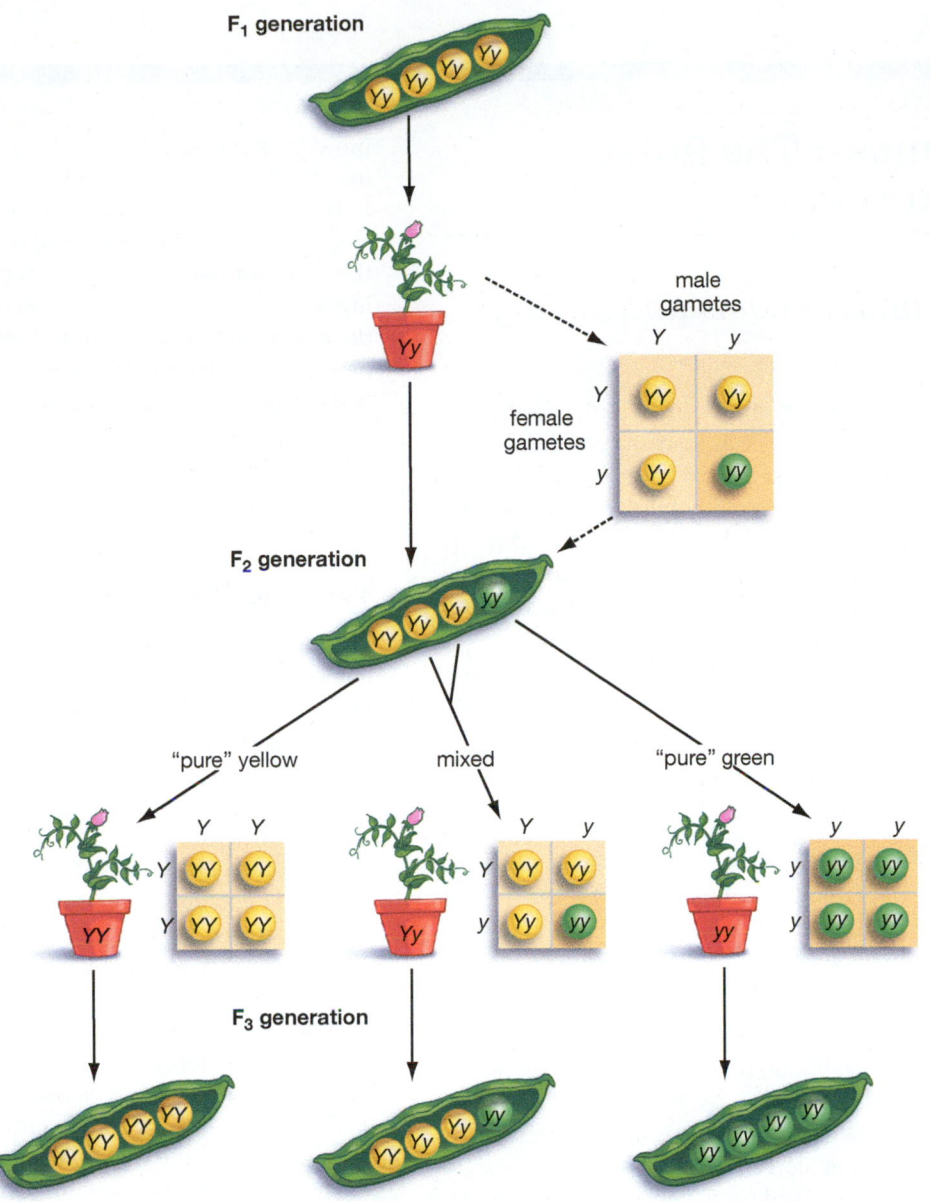

F₁ generation

male gametes

female gametes

F₂ generation

"pure" yellow   mixed   "pure" green

F₃ generation

1. **From the F₁ to the F₂ generation**

   The starting point is the F₁ generation, a set of seeds that all have the *Yy* genotype. These seeds are planted and the plants go through meiosis, yielding the gametes shown around the Punnett square. When these gametes come together in self-fertilization, the possibilities include *YY* and *yy* combinations, as well as the *Yy* combination seen in the F₁ generation. The existence of *yy* individuals is the reason green seeds reappear in the F₂ generation. Because *Y* is dominant, the green phenotype could not appear in seeds that had even a single *Y* allele.

2. **From the F₂ to the F₃ generation**

   With three starting genotypes (*YY, Yy, yy*) the F₂ generation yields plants that have these three genotypes, though there are more F₂ plants of "mixed" genotype than of either "pure" genotype.

Figure 11.6
From the F₁ to the F₃ Generation

Meiosis then occurs in the gamete precursors, which results in a separation of these alleles: Half the gametes now contain *Y* alleles and the other half *y*. You can see from the figure why this cross can now give us back the green-seed phenotype, on average in a 1:3 proportion with yellow seeds. Note that there are twice as many mixed-genotype seeds (*Yy*) as either variety of pure seed (*YY* or *yy*). Important mathematical principles underlie this outcome. You can read more about them in "Proportions and Their Causes" on page 198.

The F₃ Generation

Finally, in Figure 11.6, you can see why Mendel got the results he did when he went from F₂ to F₃. Recall that when Mendel planted and self-fertilized his *yellow* F₂ seeds, he got 519 plants, of which 166 produced nothing but yellow seeds. Now you can see why: These were not only phenotypically yellow; they were "pure" yellow in terms of genotype (*YY*). Mendel also got 353

plants that produced both yellow and green seeds. This is because the seeds these plants grew from were of mixed genotype (*Yy*); in reproduction, their offspring would be of both pure (*YY, yy*) and mixed (*Yy*) genotype. Finally, when Mendel planted and self-fertilized *green* F₂ seeds, all the plants that resulted from these seeds bred true for green seeds because all of them began as *yy*. **Figure 11.7** illustrates how the three genotypes discussed here yield only the two phenotypes of yellow or green.

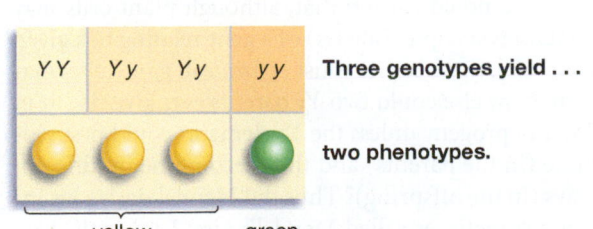

Three genotypes yield . . .

two phenotypes.

yellow   green

Figure 11.7
Three Genotypes Yield Two Phenotypes

The two alleles for seed color (*Y* = yellow and *y* = green) can result in three genotypes (*YY, Yy, yy*), but these can yield only two phenotypes (yellow and green).

ESSAY

Proportions and Their Causes: The Rules of Multiplication and Addition

What are the odds of tossing up two coins and having both come up heads?

To be a good songwriter, a person needs to be both a good musical composer and a good lyricist. Of all the people who decide to take up songwriting, say that 1 in 10 is a good composer and that 1 in 10 is a good lyricist. Now, what is the probability that a person who takes up songwriting will be *both* a good composer and a good lyricist? The answer to this question turns out to have relevance to all kinds of questions in our world, including the results that Gregor Mendel got.

When Mendel examined the F_2 pods that contained the F_3 seeds, he found that there were many more pods that had mixed seed colors within (yellow and green) than pods with seeds that were either all yellow or all green. It was the F_1 meiosis that set the stage for this outcome, as you can see here:

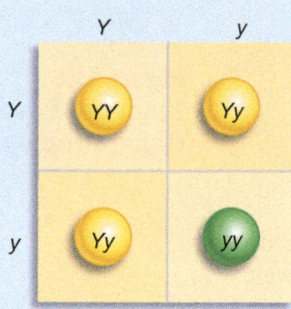

The Punnett square provides an intuitive visual sense of why things turned out

the way they did. Meiosis in the F_1 plants produced twice as many mixed F_2 seeds (*Yy*) as either type of "pure" seeds (*YY* or *yy*). It's even more helpful, however, to understand a principle that underlies this outcome. A simple coin-tossing example can illustrate what is at work. Suppose that you are going to throw two coins in the air—a nickel and a dime. What is the likelihood that *both* coins will come up heads? There is only one way to get this result:

1. Nickel = heads; dime = heads

Now, consider tossing the two coins and specifying a *mixed* result. There are *two* ways that this condition can be satisfied:

1. Nickel = heads; dime = tails
2. Dime = heads; nickel = tails

The seeds are in the same situation. What is the likelihood of getting a "pure" yellow seed? As you can see by looking again at the Punnett square, there is only one way to get this: female *Y*, male *Y*. Now consider specifying a mixed result. There are two ways to get this: female *y*, male *Y*; and female *Y*, male *y*.

The Rule of Multiplication

To state a principle that can be applied to any situation, let's look at this concept in a

more formal way. What are the odds of tossing up one coin and getting heads? Fifty/fifty, right? This means a 50 percent chance, which equals 0.5. Now, what are the odds of tossing up two coins and having both come up heads? A principle called the **rule of multiplication** comes into play here. It states that the probability of any two events happening is the product of their respective probabilities. In this case, the probability for each head coming up is 0.5, so the equation is $0.5 \times 0.5 = 0.25$, or a 25 percent probability of two heads.

The Rule of Addition

Now, what is the probability of getting one coin that comes up heads and one that comes up tails—when either of them can be heads or tails? When an outcome can occur in two or more *different* ways, as is the case here, the probability of this happening is the *sum* of the respective probabilities. This principle is known as the **rule of addition.** The probability of nickel = heads, dime = tails is the same 0.25 probability you saw in the rule of multiplication, but the probability of nickel = tails, dime = heads is also 0.25. The probability that *either* of these outcomes will take place is thus $0.25 + 0.25 = 0.50$, or 50 percent.

All this gives us a way to figure the songwriter probability—or any set of independent probabilities. Remember the stipulation that 1 in 10 aspiring songwriters is a good composer and 1 in 10 is a good lyricist. Thus, under the rule of multiplication, $0.10 \times 0.10 = 0.01$, meaning 1 in 100 is likely to be good at both composing and lyric writing.

The Law of Segregation

For inheritance to work this way, Mendel saw something we noted earlier: that, although plant cells may contain two copies (alleles) of a gene relating to a given character, these copies must *separate* in gamete formation. How else could two *Yy* parents ever give rise to *yy* (or *YY*) progeny unless the *Yy* elements could first separate (in the parents) and then recombine in different ways (in the offspring)? Thus did Mendel derive his insight, sometimes called Mendel's First Law or the **Law**

of Segregation: Differing characters in organisms result from two genetic elements (alleles) that separate in gamete formation, such that each gamete gets only one of the two alleles. As noted, the physical basis for this law, which Mendel knew nothing of, is the separation of homologous chromosomes during meiosis.

Homozygous and Heterozygous Conditions

It's time to add a couple more terms to the concepts you have been considering. There is scientific terminology for the genotypically "pure" and "mixed" organisms noted earlier. An organism that has two identical alleles of a gene for a given character is said to be **homozygous** for that character (as with *YY* or *yy*). An organism that has differing alleles for a character is said to be **heterozygous** for that character (as with *Yy*). You often see homozygous used in combination with the terms dominant and recessive. For example, a *yy* plant is a *homozygous recessive*, while a *YY* is a *homozygous dominant* (**Figure 11.8**). Knowledge of these terms puts you in a position to understand the formal definitions of dominant and recessive promised earlier. **Dominant** is a term used to designate an allele that is expressed in the heterozygous condition. In a heterozygous pea plant (*Yy*), the yellow allele (*Y*) is expressed, meaning it is dominant over the green allele (*y*). **Recessive** is a term used to designate an allele that is not expressed in the heterozygous condition. The green allele (*y*) is recessive because it is not expressed when it exists heterozygously with the yellow allele (*Yy*).

11.5 Crosses Involving Two Characters

Thus far, the subject has been how pea plants come to differ in *one* of their characters, seed color. When people breed organisms for a single difference such as this, looking to see how the offspring will come out, the procedure is known as a **monohybrid cross.** Mendel, however, went on to ask: What happens if you breed organisms for *two* characters? What happens, in other words, if you undertake what is known as **dihybrid cross**? The question Mendel was seeking to answer in carrying out his dihybrid crosses was: Do given characters travel together in heredity, or do they travel separately? If you get one character passed on to a plant, do you necessarily get the other?

Crosses for Seed Color and Seed Shape

One dihybrid cross that Mendel performed involved, for its first character, the yellow and green seeds you've become so familiar with. In addition, this cross involved a second character of these seeds: their shape, which can be smooth or wrinkled (as you saw in Figure 11.4). It's clear that yellow color is dominant to green, and as Table 11.1 shows, smooth seed shape is dominant to wrinkled. We have been denoting the alleles for seed color as *Y* for yellow and *y* for green. In the same fashion, let us now denote seed-shape alleles as *S* for smooth and *s* for wrinkled.

You can see, in **Figure 11.9a** on the next page, what the phenotypic outcomes were for these crosses from the P to F₁ generations. In the F₁ generation, all the seeds were smooth and yellow. When the F₁s were self-

pollinated, however, the F₂ phenotypes that resulted came out as 315 smooth yellow seeds; 108 smooth green seeds; 101 wrinkled yellow seeds; and 32 wrinkled green seeds. These numbers work out to a 9:3:3:1 ratio, meaning 9 parts smooth yellow; 3 parts smooth green; 3 parts wrinkled yellow; and 1 part wrinkled green.

A Hidden, Underlying Ratio

For Mendel, these imposing figures represented another opportunity to perceive something about how the black box of genetics operated. He realized early on that this ratio might be a composite, hiding an underlying reality for each of the *single* characters he was studying. Think for a second, as Mendel did, of only one of these characters, the color of the seeds. Here, the result is:

315 (smooth) yellow seeds	108 (smooth) green seeds
101 (wrinkled) yellow	32 (wrinkled) green
416 yellow seeds	140 green seeds

This equals a yellow:green ratio of about 3:1. In other words, the familiar 3:1 ratio in the F₂ generation still holds. It held as well when Mendel looked only at seed *shape*. This suggested something to Mendel that turned out to be another of his major insights: that characters—in this case, seed shape and seed color—are transmitted *independently* of one another. When these two characters were being crossed at the same time, the results for each were the same as when they were crossed as single characters in the

Dihybrid Cross

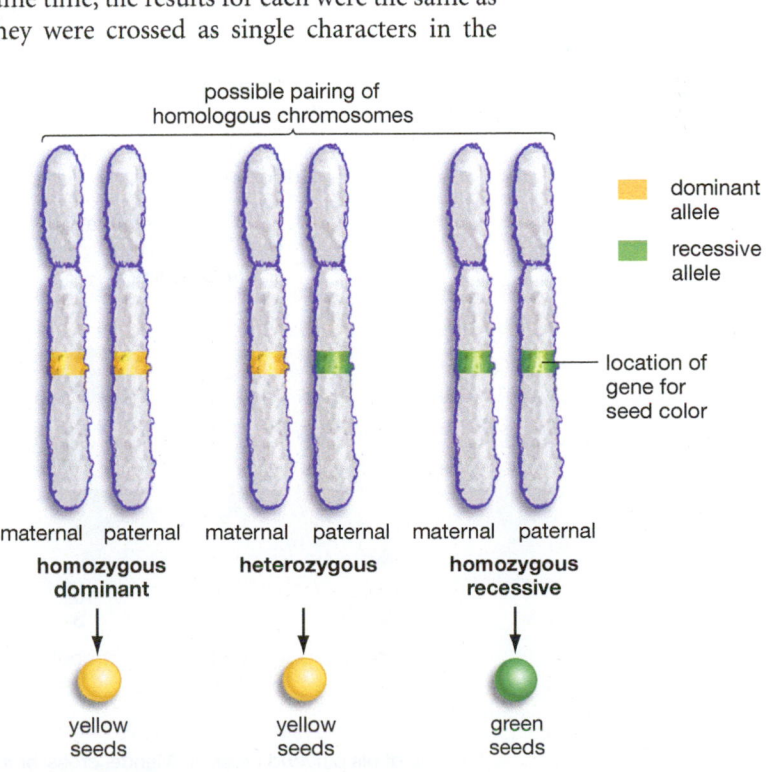

Figure 11.8
Chromosomes and Phenotypes
The figure shows how alleles on chromosomes yielded the pea-color phenotypes that Mendel observed.

earlier experiments. Thus, one character's transmission did not appear to affect the other's.

Understanding the 9:3:3:1 Ratio

Still, what's the basis of the formidable 9:3:3:1 ratio that Mendel got in his dihybrid cross, with its mixture of wrinkled yellows, smooth greens, and so forth? This is simply a more complex example of the kind of outcomes you saw earlier with the aid of the Punnett square. If you look at the square in **Figure 11.9b** and start adding up internal squares *by phenotype*, Mendel's 9:3:3:1 ratio starts making sense. You can see, for example, where the 1 in the ratio comes from: There is only 1 part wrinkled green seeds because it is only the *sy* (pollen) and *sy* (egg) combination that yields this phenotype. Likewise, you will get three parts wrinkled yellow seeds by adding up the three possible male and female gamete combinations that could bring this about.

The Law of Independent Assortment

This outcome could only come about, however, if Mendel's fundamental insight about dihybrid crosses was correct: that characters were being transmitted independently of one another. If one character had affected another's transmission, these ratios would have been very different. This insight of Mendel's is now known as Mendel's Second Law, or the **Law of Independent Assortment**. It states that during gamete formation, gene pairs assort independently of one another.

Independent Assortment and Chromosomes

Now recall that the underlying physical basis for this law was set forth in Chapter 10: In meiosis, pairs of homologous chromosomes *assort independently* from one another at the metaphase plate. It may be, for example, that paternal chromosome 5 will line up on side A of

Gregor's Garden

Figure 11.9
Phenotype Ratios in a Dihybrid Cross

(a) Results of Mendel's dihybrid cross

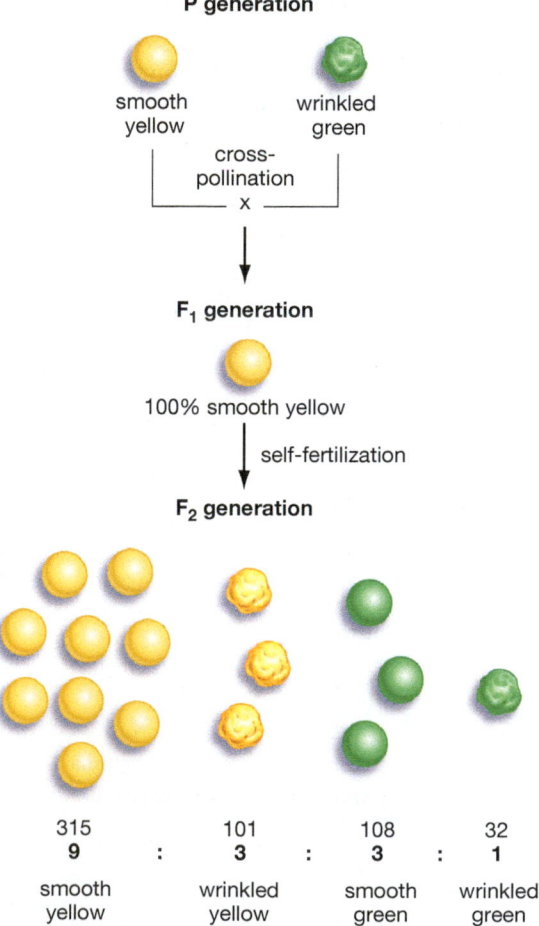

In one of his dihybrid crosses, Mendel cross-bred plants that had smooth yellow seeds with those that had green wrinkled seeds. The result was a generation of plants that all had smooth yellow seeds. When these plants self-fertilized, the result was an F₂ generation that had the phenotypes shown in a 9:3:3:1 ratio.

(b) Why Mendel got these results

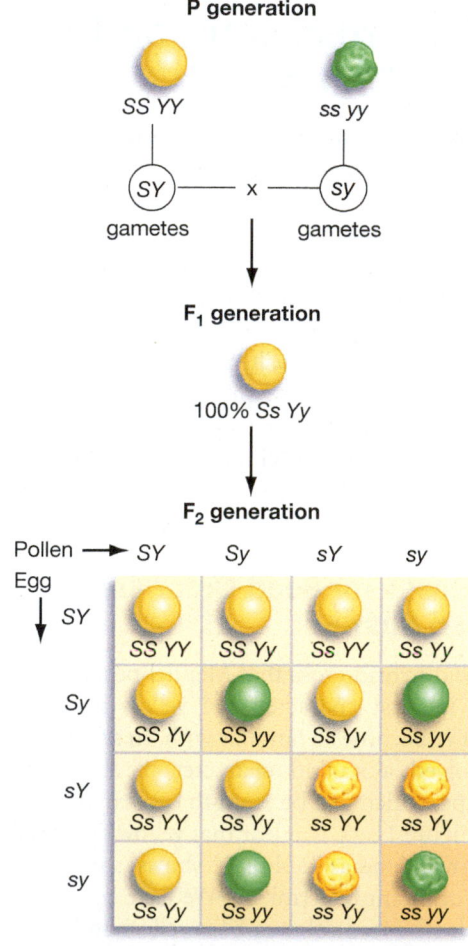

The Punnett square demonstrates why Mendel got the 9:3:3:1 phenotypic ratio in his dihybrid cross. Nine combinations yield smooth yellow seeds, 3 yield smooth green seeds, 3 yield wrinkled yellow seeds, while only 1 results in a wrinkled green seed.

the metaphase plate; if so, then maternal chromosome 5 ends up on side B. For chromosome 6, however, things could just as easily be reversed: The *maternal* chromosome might end up on side A and the paternal on side B (Figure 10.4, page 180). The genes for Mendel's seed color exist on the plant's chromosome 1, while the genes for seed shape exist on its chromosome 7. Because these are separate chromosomes, they assort independently at the metaphase plate, meaning they are passed on independently to future generations.

11.6 Reception of Mendel's Ideas

When Mendel finished his experiments, he delivered two lectures on them to a local scientific society in 1865. It would be nice to report that the society members' jaws dropped open once Mendel let them in on how inheritance worked in the living world, but no such thing happened. His findings were ultimately published in a scientific journal that was distributed in Germany, Austria, the United States, and England. Mendel even took it upon himself to get 40 reprints of his paper, thereafter sending some out to various scientists in an effort to spark some exchange on his findings—all to no avail. His work sank nearly without a trace, finally to be rediscovered in 1900, 16 years after his death.

This poor early reception notwithstanding, Mendel's insights have stood the test of time. For certain kinds of phenotypes, his rules of inheritance are extremely reliable. To this day, biologists will begin an observation by noting that "Mendelian rules tell us that . . . " or "Mendelian inheritance operates here." Even allowing that he had intellectual predecessors, the fact is that Mendel did not just add to the discipline of genetics—he founded it.

SO FAR . . .

1. Mendel's law of segregation states that differing traits in organisms result from two differing _____ that _____ in gamete formation, such that each gamete gets only one of the two.

2. Homozygous refers to an organism that has two _____ alleles for a character, while heterozygous refers to an organism that has two _____ alleles for the character.

3. An allele that is expressed in the heterozygous condition is said to be _____, while an allele that is not expressed in the heterozygous condition is said to be _____.

11.7 Incomplete Dominance and Codominance

Mendel knew as well as anyone that the rules he discovered did not apply in all instances of inheritance. In his peas, crossing white- and purple-flowered plants resulted in an F_1 generation in which all flowers were purple. In some other species, however, the results are different. Crossing a true-breeding red-flowered *snapdragon* with a true-breeding white-flowered snapdragon produces neither red- nor white-flowered F_1 snapdragons, but *pink* snapdragons (**Figure 11.10** on the next page). This might seem to indicate that inheritance can indeed work through the blending of genetic traits—a notion that Mendel's work had quashed. Blending is not taking place, however. When breeding is continued through the F_2 generation, red and white flowers come back, along with the pink, in a ratio of 1 part red, 2 parts pink, 1 part white. This 1:2:1 ratio demonstrates that red and white alleles have not irretrievably blended into pink but rather have come together in F_1, only to separate out again in F_2. Still, how can the pink F_1 snapdragons be explained?

Variations on Mendel

Incomplete Dominance

First, think about what genes do: They contain information regarding the production of proteins. In the case of colors, such proteins can bring about the formation of pigments. This is the case with the snapdragons; they have a gene for color, and one of the alleles of this gene brings about the production of red pigment. Meanwhile, the second allele of this gene is nonfunctional—it brings about the production of no pigment at all. Two of the red alleles, then, yield a red color, while one red allele produces only enough pigment to yield a pink color. Neither the red nor the white allele, therefore, is completely dominant; each is thus said to have **incomplete dominance,** meaning a genetic condition in which the heterozygote phenotype is intermediate between either of the homozygous phenotypes.

Codominance

The notion of genes as protein-producing entities can help us steer through another variation on Mendelian inheritance, this one having great importance for variety in the living world. When people speak of blood "types," what they mean is types of "glycolipids"—lipid molecules with carbohydrate chains attached—that extend from the surface of red blood cells. These surface extensions come in many different varieties, the two most important of which are designated A and B.

Blood types are completely under genetic control, with a single gene that lies on chromosome 9 determining what blood type a person will have—A, B,

AB, or O. There are two copies of this gene in each individual because there are two copies of chromosome 9 in each individual—one inherited from the mother, the other inherited from the father. Thus, there exists the possibility for *alleles*, or gene variants, in each individual. One of these alleles codes for a protein that modifies surface extensions into the A form, while the other codes for a protein that modifies the extensions into the type B form.

Now, a single person could have two alleles that code for type A extensions, or a person could have both alleles coding for type B extensions. In the first case, a person would have type A blood and in the second type B blood. However, a person could have one allele that codes for the type A extensions *and* one that codes for type B, in which case that person would have type AB blood. Finally, a person could have two alleles that are inactive—that code for neither extension—and would thus have type O blood, as **Table 11.3** shows.

What's new in this case is, first, another aspect of the idea of dominance in alleles. In the blood types, *both* the A and B alleles produce proteins that modify the cell-surface extensions; each allele thus has an independent effect on phenotype. People who are heterozygous for these alleles thus do not get a phenotype that, like pink flowers, lies halfway between A and B. They get both the A and B versions of the molecules on the surface of their blood cells. Alleles that have this kind of effect result in **codominance:** a condition in which two alleles of a given gene have different phenotypic effects, with both effects manifesting in organisms that are heterozygous for the gene.

Figure 11.10
How Red and White Yield Pink: Incomplete Dominance in Snapdragons

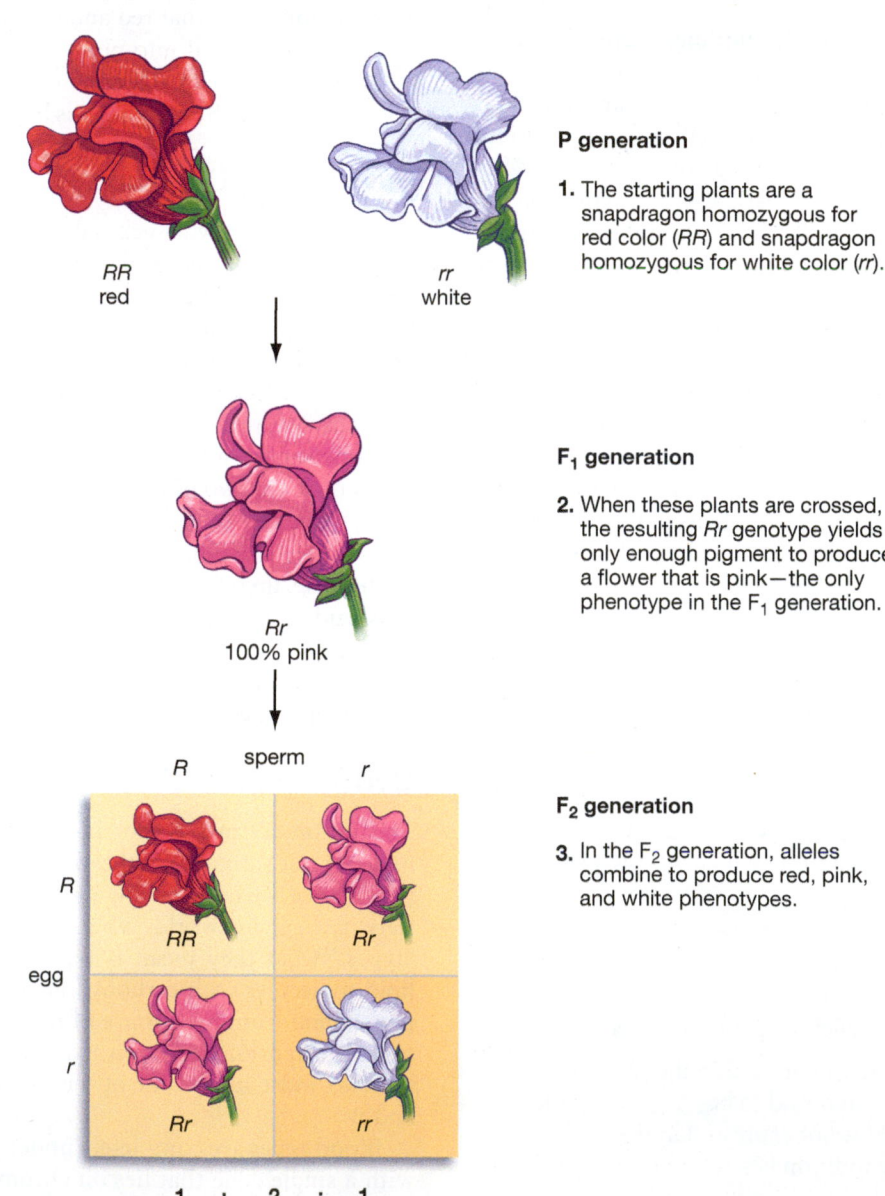

RR
red

rr
white

P generation

1. The starting plants are a snapdragon homozygous for red color (*RR*) and snapdragon homozygous for white color (*rr*).

Rr
100% pink

F₁ generation

2. When these plants are crossed, the resulting *Rr* genotype yields only enough pigment to produce a flower that is pink—the only phenotype in the F₁ generation.

sperm

R   *r*

egg

R

R *RR*

Rr

r

Rr

rr

F₂ generation

3. In the F₂ generation, alleles combine to produce red, pink, and white phenotypes.

1 : 2 : 1
red   pink   white

Table 11.3

Human Blood Types

This blood type (phenotype) . . .	. . . has these surface glycolipids . . .	. . . and is produced by these genotypes
A		AA or AO
B		BB or BO
AB		AB
O	(no surface glycolipids)	OO

The familiar ABO human blood-typing system refers to glycolipid molecules that extend from the surface of red blood cells. People whose blood is "type A" have A extensions on their blood cells. It is also possible to have only B extensions (and be type B); to have both A and B extensions (and be type AB); or to have none of these extensions (and be type O). Note that a person whose genotype is AO is phenotypically type A; likewise, a person whose genotype is BO is phenotypically type B.

Dominant and Recessive Alleles

All this gets us to an important point, which is the relationship between dominant and recessive alleles. In common speech, the word *dominance* is used all the time to mean "bringing another under submission by means of superior power." Alleles, however, are not locked in a power struggle with one another. Alleles contain information for the production of proteins. Such proteins (or lack of them) can create dominant phenotypes (smooth peas), incompletely dominant phenotypes (pink snapdragons), or codominant phenotypes (AB blood). "Dominance," then, just refers to the ways in which a given allele is expressed. As you've seen, a recessive allele is not *eliminated* by being paired with a dominant allele—it simply isn't expressed in a given generation. When this generation gives rise to another, the recessive allele will be passed on intact, at which point it may be expressed. When the pea plant's recessive green allele was passed on, for example, it was expressed when brought together with another green allele. This has relevance for traits more familiar to us. People seem to have an idea that human blond hair, to give one example, eventually will cease to exist because the alleles for it are recessive to those for dark hair. The alleles for blond hair are, however, not disappearing. To the extent that alleles for it come together in reproduction, there will always be blond-haired people, as you can see in **Figure 11.11**.

SO FAR ...

1. Incomplete dominance, such as that of the pink snapdragon, exists when a _____ phenotype is intermediate between that of either of the _____ phenotypes.

2. Codominance exists when alleles have _____ effects on a phenotype, with both effects manifesting in organisms that are _____ for a codominant gene.

3. A recessive allele is not eliminated when paired with a dominant allele in a single individual; it simply is not _____. It may have phenotypic effects in a later generation, however, if paired with _____.

11.8 Multiple Alleles and Polygenic Inheritance

ABO blood types have one more lesson for us. In Mendel's seed shapes, we saw two phenotypic variants: smooth and wrinkled. In blood types, we saw four: A, B, AB, and O. As you know, no one person can have more than *two* alleles for a given gene. As a result, a person may be, say, A and B, but he or she cannot be A, B, and O. It is thus only in a *population* of humans that the full range of ABO alleles can be found. And therein lies the lesson: In a population, alleles can come not just in two, but in many variants.

When three or more alleles of the same gene exist in a population, they are known as **multiple alleles.** Such alleles are important in and of themselves for they can bring diversity to traits that are governed by single genes (such as human blood types). Their more general significance comes, however, in connection with traits that are controlled not by one gene, but by many. Recall that seed texture in Mendel's peas was governed by a single gene (which had smooth and wrinkled allelic variants). This situation is, however, the exception rather than the rule in the living world. Most traits are governed by many genes. The human traits of height, weight, eye color, and skin color are each controlled by several genes. In the wider living world, the color of a wheat grain, the length of an ear of corn, and the amount of milk a cow gives are all controlled by many genes acting together. The term for such genetic

Figure 11.11
Alleles Are Retained Across Generations
The fraternal Hodgson twins, Kian on the left and Remee on the right, were born one minute apart in 2005 to English parents Kylie Hodgson (at far left) and Remi Horder. Both parents are biracial, from mixed-race, black-white unions, a circumstance that set the stage for the unusual genetic outcome in their twins. Kian happened to inherit alleles, passed on from both of her black grandparents, that produced her darker skin, black hair, and brown eyes, while Remee inherited alleles, passed on from both of her white grandparents, that led to lighter skin, blond hair, and blue eyes.

influence is **polygenic inheritance**, meaning the inheritance of a genetic character that is determined by the interaction of multiple genes, with each gene having a small additive effect on the character.

When we have many genes contributing some small increment to a trait, the result is what you see in **Figure 11.12**. Human beings don't come in two heights, or three or four; they display what is known as a "continuous variation" in height, each person being just barely taller or shorter than the next. Continuous variation also holds true for human skin color. Human beings don't really have "black" or "white" skin. Instead, they have skin that comes in a range of hues, in which one color shades imperceptibly into another. Likewise, trees of a given species come in a range of heights, and the beaks of birds come in a range of lengths. Polygenic inheritance creates this fine-grained diversity.

Now, note something else about Figure 11.12a. Most of the students fall in the middle range of heights in the group. Put another way, there are fewer tall or short students than there are students of medium height. This height distribution means that the group as a whole takes on a kind of shape—a bell shape. This is, in fact, the famous **bell curve,** meaning a distribution of values that is symmetrical around the average. Most biological traits manifest in this way. Look, for example, at the beak depths in a group of Darwin's finches in Figure 11.12b. Graphs like this just confirm what we know intuitively—that traits in living things tend to cluster around the middle of a range of values, rather than at the extreme ends of the range.

In polygenic inheritance, with its many genes and alleles, gene interactions are so complex that predictions about phenotype are a matter of *probability*, not certainty. Human height is largely under genetic control, and there are means of trying to predict the height of children based on the height of their parents. (The starting point generally is to add together the height of the parents and divide the result by two.) All that such predictions can do, however, is specify that if parents are of a given height, then there is a certain probability that their children will fall into a given range of heights. Think how different this is from the situation with Mendel's peas. There, you could confidently predict that a given cross would yield, say, all green peas.

11.9 Genes and Environment

Scientists work awfully hard at trying to determine the probabilities for certain polygenic outcomes—those related to disease. In carrying out this work, they are essentially doing what Mendel did: observing traits in a parental generation and then observing traits in the parents' offspring. Such work is the basis for the warnings we sometimes hear about how much a person's risk for a particular cancer goes up if that person has a relative who has contracted the disease. (To learn more about how science is working to predict disease susceptibility on the basis of genotype, see "Scanning DNA for Disease Markers" on page 206.) With polygenic illnesses such as cancer, however, it is difficult to

(a) Continuous variation in human height

(b) The bell curve

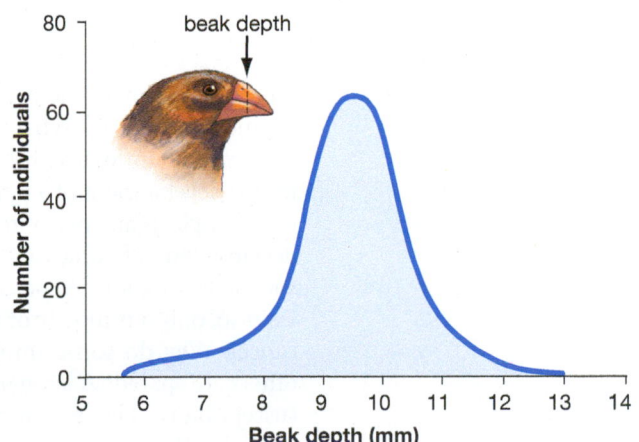

Figure 11.12
Continuous Variation and the Bell Curve

(a) Human heights exist in a range, with no fixed increments between heights of individuals. Such "continuous variation" results from polygenic inheritance, in which each of several genes contributes a small additive effect to a character. University of Connecticut students in the picture have been arranged by height to show how continuous variation works in this one human trait. Note that the group as a whole takes on the shape of a bell. The students' heights are distributed in a pattern that creates a bell curve.

(b) A bell curve can also be seen in this graph, which plots the average beak depth of a population of Darwin's finches from the Galapagos Islands. Note how most of the finches had beak depths that fell within a middle range of values for the trait.

ESSAY

Scanning Genomes for Disease Markers

Should an individual's genetic information ever become public information?

Since genes underlie the traits seen in healthy people—their height, their skin color, and so forth—it's not surprising that they also can play a part in producing the traits seen in unhealthy people: weak hearts, failing eyesight, bad kidneys. Scientists have been trying to decipher the linkage between genetic makeup and disease since the early twentieth century, but from that time until very recently they had to deal with a critical missing piece of information: They didn't know what the genetic makeup of any human being was. This state of affairs changed dramatically in 2003 with the completion of the Human Genome Project, which revealed the order of the 3 billion A's, T's, G's, and C's that make up the full complement of human DNA. The hope at the time was that, with this information in hand, scientists would be able to look at the genomes of, say, cancer patients and pinpoint the ways in which their genetic

makeup had made them susceptible to their disease.

This kind of detective work has borne some fruit in recent years, but the yield has not been nearly what scientists had hoped for. Now, however, a Harvard biologist named George Church has proposed tackling the problem by taking what might be called a Wikipedia approach to genes and disease. His idea is to sequence the genomes of 100,000 average citizens, get detailed physical profiles of each of these people, and then make all this information freely available on the Web. Straightforward as this idea may sound, in the time since it was announced, it has sparked a vigorous debate on a societal issue: Should an individual's genetic information ever become public information?

The scientific impetus behind Church's Personal Genome Project (PGP) is that, even with the human genome sequence in

hand, medical researchers today have inadequate information to use when looking for linkages between genetic makeup and disease. At present, when researchers want to look for such connections, they get DNA samples from disease victims and then sequence a number of target regions in their genome—that is, regions that seem to have a reasonable chance of playing a part in their illness. Then they look for single-letter differences between these regions and the same regions in a comparable group of healthy individuals. In such a case, one small part of the result might look like so:

Standard genome: AATC**G**CGGAC
Patient's genome: AATC**T**CGGAC

The difference between the two genomes, called a single nucleotide polymorphism, or SNP, is likely to be one of many such differences between patients and healthy individuals. Because major diseases are almost always *polygenic* diseases—that is, afflictions in which many genes play a role—the question then becomes: With what frequency do given *sets* of SNPs show up in patients when compared to healthy individuals? The problem thus far has been that almost no clear associations have emerged between SNP patterns and major diseases; scientists have

separate the genetic factors from what are called "environmental" factors. What's an environmental factor? Any external influence that is favorable or unfavorable for the development of a trait in an organism. Smoking, for example, is an environmental factor that influences development of lung cancer. Indeed, smoking is responsible for about 90 percent of all lung cancer cases. Even so, only a minority of smokers ever contract lung cancer. Why do some smokers get this disease while others are spared? A large part of the answer is genetic susceptibility: The genetic makeup of some individuals makes them more susceptible to the environmental influence of smoking. Thus, we can see genes and environment working together to create the outcome of lung cancer. As it turns out, this is almost always the case in nature. For polygenic traits—which is to say, most traits—it is rare for genes to be the sole determining factor.

A clear example of how genes and environment work together can be seen in **Figure 11.13**, which shows a picture of the most popular species of hydrangea flower, *Hydrangea macrophylla*. Any gardener could start with a blue hydrangea, like the one in the picture on the right, take a "cutting" from it, do a little preparatory work, plant the cutting in the ground, and get a whole new hydrangea plant as a result. For our purposes, the important point is that this second plant is a clone of the first. It is not produced by mixing eggs and sperm; it is an exact genetic replica of the first plant. It can, however, be a plant with *pink* flowers instead of blue if some garden lime is mixed into its surrounding soil. It's possible, therefore, to have two hydrangea plants with the exact same genotype but with very different phenotypes—pink or blue flowers. The message here is one of genetic limitation. Genotype specifies

been unable to say that a given set of SNP differences is related in any meaningful way to a given condition. To get a feel for what this means in practice, consider a recent test for this kind of inquiry that involved not disease but the character of human height, which is known to be 80 to 90 percent under genetic control. After looking at SNP patterns in more than 30,000 individuals, scientists then looked at how tall each individual was and found that no set of SNPs among them could account for more than 5 percent of the variance seen in their heights. Thus, while we know that genes have a huge effect on height, we don't know *where* this control lies within the genome.

Enter George Church, who believes that the answer to this issue may lie in giving scientists around the world access to two kinds of information that they don't have now: DNA sequence data on a large number of complete human genomes—rather than just individual regions of the genome—and physical profiles of the people whose genomes have been sequenced. In short, Church's Personal Genome Project is aimed at providing researchers not just with genotypic information on a huge group of people, but with *phenotypic* information on them as well—their height, weight, blood pres-

sure, cholesterol levels, allergies they suffer from, their mood states, even their television-watching habits. Once all this information is up on the Web, researchers around the world will have an enormous database to tap into in looking for connections between genetic makeup and disease.

Despite this potential, the initial phases of PGP created quite a stir because, as may be apparent, the project is a large-scale experiment in turning private information into public information. At present, only two people in the world have made their genomes public (scientists James Watson and Craig Venter). Now Church is proposing to go public with the genotypes and phenotypes of 100,000 people. These volunteers will be given numerical identities on the PGP website, but the project is careful to warn prospective participants that no level of confidentiality can be assured for any of them over the long run. So, suppose that a SNP pattern is discovered for, say, Alzheimer's disease and that 1,000 PGP volunteers share in this pattern. At present, federal law forbids employers and health insurance companies from discriminating on the basis of genetic makeup, but life insurance and long-term care insurance companies are not

restricted in this way. So, wouldn't it be reasonable to assume that these companies would begin trying to figure out if anyone who comes to them seeking insurance is one of the 1,000 PGP volunteers who shares in the Alzheimer's SNP pattern? Beyond this, what about the close relatives of PGP volunteers—their siblings, parents, and children? How much genetic privacy can they have if a genetic profile very similar to theirs has been made public on the Web?

Church is the first to admit that, because something like PGP has never been tried, no one is sure how the privacy issues it raises will play out. As such, he and his team have taken a cautious approach in the project's early phases. PGP's first 10 volunteers were required to have a high level of knowledge regarding genetics and they were encouraged to consult with close relatives before taking part. Even when the project ramps up to a larger scale, volunteers will be required to demonstrate a fair amount of familiarity with genetics. Despite misgivings about such privacy concerns, however, the PGP experiment has, you might say, opened for business. Anyone who wants to be part of it can learn more by going to personalgenomes.org.

Figure 11.13
Flower Color Is Affected by Environment

Both the blue hydrangeas on the right and pink hydrangeas on the left are from the same species (*Hydrangea macrophylla*), yet they have different coloration because of different chemical conditions in their soil. (The soil the blue plants are in is more acidic than the soil the pink plants are in.) This is but one example of how genes and environment work together to produce the traits we see in living things.

only so much about phenotype. Protein products that bring about a blue plant in one environment will bring about a pink plant in another. Across the natural world, things generally work this way. Most genes do not come with an unvarying ability to bring about an effect. Rather, genes and environment work together to create the phenotypes we see in living things.

On to the Chromosome

This completes the survey of Mendel's work and the variations on his ideas of inheritance. As you have seen, Mendel's findings lay dormant for some decades in the nineteenth century. What sent scientists scurrying back to his reports were experiments on those tiny entities you have looked at in detail before: chromosomes, the subject of the next chapter.

SO FAR . . .

1. The usual case in nature is that a given character in an organism is influenced by _____ genes. To put this another way, the character is produced through _____ inheritance. True or false: This form of inheritance tends to produce fixed increments of difference between organisms.

2. When we measure the traits found in living things, we find that they tend to be distributed in a form known as the _____, which can be defined as a range of values that is symmetrical around the _____.

3. The hydrangea example in the book shows that the same genes can produce different phenotypes depending on _____ influences that act on an organism.

CHAPTER 11 REVIEW

Summary

11.1 Mendel and the Black Box

- Gregor Mendel was the first person to comprehend the most basic principles of genetics. (p. 191)

11.2 The Experimental Subjects: *Pisum sativum*

- Mendel experimented by breeding pea plants (*Pisum sativum*) that possessed a set of seven attributes called characters, each of which came in two forms, referred to as traits, one of them dominant, the other recessive. In his experiments, he bred generations of pea plants and observed the frequencies with which given traits showed up in each generation. (p. 192)

- A phenotype is any physiological feature, bodily characteristic, or behavior of an or-

ganism. Phenotypes in any organism are determined in significant part by that organism's genotype, meaning its genetic makeup. (p. 193)

11.3 Starting the Experiments: Yellow and Green Peas

- Mendel realized that it was possible for organisms to have identical phenotypes, such as yellow seeds, but differing underlying genotypes. (p. 195)

- Mendel discovered that the basic units of genetics are material elements that come in pairs. These elements, today called genes, come in alternative forms called alleles. One allele for a given gene resides on one chromosome, while the other allele for the gene resides on a second chromosome that is homologous to the first. Mendel demonstrated that genes retain their character through many generations rather than being "blended" together. (p. 195)

11.4 Another Generation

- Mendel realized that alleles separate during the formation of gametes. Although Mendel did not know it, the physical basis for this is that the alleles he was observing resided on homologous chromosomes, which always separate in meiosis. An organism that has two identical alleles for a given character is said to be homozygous for that character. An organism that has differing alleles for a character is said to be heterozygous for that character. Dominant means expressed in the heterozygous condition. Recessive means not expressed in the heterozygous condition. (p. 195)

11.5 Crosses Involving Two Characters

- Mendel observed that the genes for the characters he studied were passed on independently of one another. This was so because the genes for these characters resided on separate, non-homologous chromosomes. The physical basis for what he found is the independent assortment of chromosome pairs during meiosis. (p. 199)

11.6 Reception of Mendel's Ideas

- During Mendel's lifetime, almost no one recognized the significance of his discoveries. (p. 201)

11.7 Incomplete Dominance and Codominance

- Incomplete dominance operates when neither allele for a given gene is completely dominant, with the result that a heterozygous genotype will yield an intermediate phenotype. (p. 201)

- In some instances, differing alleles of the same gene will have independent effects in a single organism. In such a situation, neither allele is dominant; each has a separate phenotypic effect, meaning that each is said to be codominant. (p. 201)

11.8 Multiple Alleles and Polygenic Inheritance

- While in many species an individual can have no more than two alleles for a given gene, within a population there can be many allelic variants of a single gene. Most traits in living things are governed not by one gene but by many, with each of these genes often having several allelic variants. Inheritance of such traits is known as polygenic inheritance, a situation in which each gene has a small additive effect on a trait. (p. 204)

- Polygenic inheritance tends to produce continuous variation in phenotypes, in which there are no fixed increments of difference between individuals. (p. 204)

- The traits produced in polygenic inheritance tend to manifest in bell-curve distributions, in which most individuals display middle-range, rather than extreme, trait values. Gene interactions and gene–environment interactions are so complex in polygenic inheritance that predictions about phenotypes are a matter of probability. (p. 205)

11.9 Genes and Environment

- The effects of genes can vary greatly in accordance with the environment in which the genes are expressed. (p. 205)

Key Terms

allele   195

bell curve   205

codominance   202

cross-pollinate   193

dihybrid cross   199

dominant   199

first filial generation (F_1)   194

genotype   193

heterozygous   199

homozygous   199

incomplete dominance   201

Law of Independent Assortment   200

Law of Segregation   198

monohybrid cross   199

multiple alleles   204

parental generation (P)   194

phenotype   193

polygenic inheritance   205

recessive   199

rule of addition   198

rule of multiplication   198

Understanding the Basics

Multiple-Choice Questions

(Answers are in the back of the book.)

1. The first filial generation, or F_1 results when:
   a. two P generation organisms are crossed.
   b. the progeny of a P generation is crossed.
   c. a P generation is crossed with an F_2 organism.
   d. an F_3 organism is crossed with an F_2 organism.
   e. both a and d.

2. When Mendel crossed pure-bred yellow peas with pure-bred green peas, the F_1 generation contained:
   a. all yellow peas.
   b. all green peas.
   c. green and yellow peas.
   d. green, wrinkled peas.
   e. both b and d.

3. If a cross is made between a pure-breeding green, smooth (*yySS*) plant and a pure-breeding yellow, wrinkled (*YYss*) plant, what is the result?
   a. all green, wrinkled peas
   b. all yellow, wrinkled peas
   c. all yellow, smooth peas
   d. yellow wrinkled; green wrinkled
   e. none of the above

4. Mendel's first law states that:
   a. during the formation of gametes, gene pairs demonstrate independent assortment.
   b. during the formation of gametes, genetic elements (alleles) segregate from each other.
   c. the probability of any two events happening is the product of their respective probabilities.
   d. if an organism is pure-breeding for any trait, it will display incomplete dominance.
   e. Mendel did not write any laws regarding genetics.

5. A trait whose phenotype in the heterozygous condition is an intermediate of the two homozygous conditions (that is, heterozygous pink flowers, homozygous red flowers, homozygous white flowers) demonstrates:
   a. polygenic inheritance.
   b. incomplete dominance.
   c. intermediate dominance.
   d. codominance.
   e. multiple alleles.

6. Mendel's second law states that:
   a. no organism can have more than two alleles for a given character.
   b. no two organisms can ever be genetically identical.
   c. organisms will tend to be heterozygous for most traits, rather than homozygous.
   d. the way one gene pair assorts during gamete formation will affect the way other gene pairs assort.
   e. gene pairs assort independently of one another during gamete formation.

7. Which of the following is true about polygenic inheritance? (More than one answer may be correct.)
   a. It is the rule rather than the exception in living things.
   b. Its outcomes are always predictable.
   c. It is never seen in human beings.
   d. It produces phenotypes that grade smoothly into one another.
   e. It always produces heterozygous individuals.

8. Which of the following is true regarding the effects of genes and environment on phenotypic variation?
   a. Gene expression will always be greatest in the native environment.
   b. Gene expression is not affected by a change in environment.
   c. Phenotypes are based only on environmental influences.
   d. Environment may easily affect genotype, not phenotype.
   e. The phenotypes of genetically identical organisms may vary in different environments.

Brief Review

(Answers are in the back of the book.)

1. What are the differences between:
   a. phenotype and genotype?
   b. dominant and recessive?
   c. codominance and incomplete dominance?

2. True or false: When Mendel conducted his genetic research, neither DNA nor chromosomes had been discovered.

3. How many alleles is an individual human being likely to possess for any given gene? Name the source of each allele.

4. In a case of Mendelian genetics where one gene affects one trait, how many different genotypes may bring about a dominant phenotype (for example, yellow pea color)?

5. If a heterozygous yellow-seed pea plant is crossed with a homozygous green-seed pea plant, approximately what percentage of the progeny peas would be expected to be yellow?

6. What term is now given to Mendel's element of inheritance?

7. Name and describe the two laws that are credited to Mendel and his research.

8. Using the symbols A, a, B, and b, what Mendelian crosses would be expected to result in the following phenotypic ratios?
   a. 1:1
   b. 3:1
   c. 1:1:1:1
   d. 9:3:3:1

Applying Your Knowledge

1. Stands of aspen trees often are a series of genetically identical individuals, with each succeeding tree growing from the severed shoot of another tree. Using what you've learned of genetics in this chapter, would you expect one aspen tree in a stand to differ greatly from another in its phenotype? Would you expect each to look exactly like the next in terms of phenotype?

2. The ABO blood-typing system mentioned in the text has great practical significance because the human immune system will attack cells displaying blood-cell surface molecules that it recognizes as "foreign"—that is, molecules that a person does not have on his or her own blood cells. Given this, what is the "universal donor" blood type? What is the "universal recipient" type?

3. If, as the text states, the effects of genes can be expected to vary in accordance with the environment in which these genes work, would you expect this phenomenon to be applicable to human genes, such as those that help produce such traits as height, weight, or introversion and extroversion?

Genetics Problems

(Answers are in the back of the book.)

One of Mendel's many monohybrid (single-character) crosses was between a true-breeding (homozygous) parent bearing purple flowers and a true-breeding (homozygous) parent bearing white flowers. All the offspring had purple flowers. When these offspring were allowed to self-fertilize, Mendel observed that their offspring occurred in the ratio of about three purple-flowered plants to one white-flowered plant. The allele deter-mining purple flowers is symbolized P, and the allele determining white flowers is symbolized p. Use this information to answer the following questions:

1. What are the genotypes of the parents?
   a. PP; pp
   b. PP; Pp
   c. Pp; Pp
   d. pp; pp

2. Which generation has only purple-flowered plants?
   a. P
   b. F_1
   c. F_2
   d. F_3

3. Which generation tells you which of the traits is dominant?
   a. P
   b. F_1
   c. F_2
   d. F_3

4. How many phenotypes are present in the F_2 generation?
   a. 1
   b. 2
   c. 3

5. How many different genotypes are present in the F_2 generation?
   a. 1
   b. 2
   c. 3

6. Mendel allowed the white-flowered F_2 plants to self-fertilize. What proportion of these F_2 plants produced only white-flowered offspring?
   a. none
   b. $\frac{1}{4}$
   c. $\frac{1}{3}$
   d. $\frac{3}{4}$
   e. all

7. Mendel allowed the purple-flowered F_2 plants to self-fertilize. What proportion of these F_2 plants produced only purple-flowered offspring?
   a. none
   b. $\frac{1}{4}$
   c. $\frac{1}{3}$
   d. $\frac{3}{4}$
   e. all

8. If a heterozygous purple-flowered pea plant is crossed with a homozygous white-flowered plant, what proportion of offspring will be white-flowered?
   a. none
   b. $\frac{1}{4}$
   c. $\frac{1}{2}$
   d. $\frac{3}{4}$
   e. all

9. Given what you know about the alleles that determine purple and white color in pea flowers, why is it possible to have a purple-flowered heterozygote but not a white-flowered heterozygote?

10. In considering flower color in pea plants, we are dealing with _____ gene(s) and _____ allele(s).
    a. 1; 1
    b. 1; 2
    c. 1; 4
    d. 2; 1
    e. 2; 2

11. The next three questions require you to apply rules of probability.
    a. In a toss of two coins, what is the chance of obtaining two heads?
       (1) 0
       (2) 12.5 percent
       (3) 25 percent
       (4) 50 percent
       (5) 75 percent
    b. What rule of probability was used to determine the chance of obtaining two heads?
    c. Now, use this rule to solve the following problem. In a cross between two purple-flowered heterozygotes, 75 percent of their offspring are expected to be purple flowered. In a cross between two heterozygotes with yellow-colored seeds, 25 percent of their offspring are expected to be green. If two purple-flowered, yellow-seeded individuals, each heterozygous for both genes, are crossed, what proportion of their offspring is expected to be purple-flowered with green seeds?

12. By observing flower color in snapdragons, is it possible to unambiguously determine the genotype? Is the same true for flower color in peas? Why or why not?

13. If the alleles for snapdragon flower color were codominant instead of incompletely dominant, a possible phenotype for a heterozygous individual would be:
    a. pure red flowers.
    b. pure pink flowers.
    c. pure white flowers.
    d. a mosaic pattern of red and white regions on flowers.

14. If two individuals with AB blood type marry, what proportion of their children are expected to have AB blood type?
    a. none
    b. 0.25
    c. 0.50
    d. 0.75
    e. all

Units of Heredity:

Chromosomes and Inheritance

Chromosomes are critical players in the process by which living things pass on traits. Some of our worst diseases result from chromosomes that have failed to function properly.

Available on MasteringBiology®

Activities: X-Linked Recessive Traits; Some Human Genetic Disorders; Polyploid Plants; Morgan's Flies; **ABC News Videos:** The Genetics of Short Sleepers; **Current Events:** From Californians' DNA, a Giant Genome Project; In Choosing a Sperm Donor, a Roll of the Genetic Dice; **Building Vocabulary; Questions:** Reading Questions; Test Bank

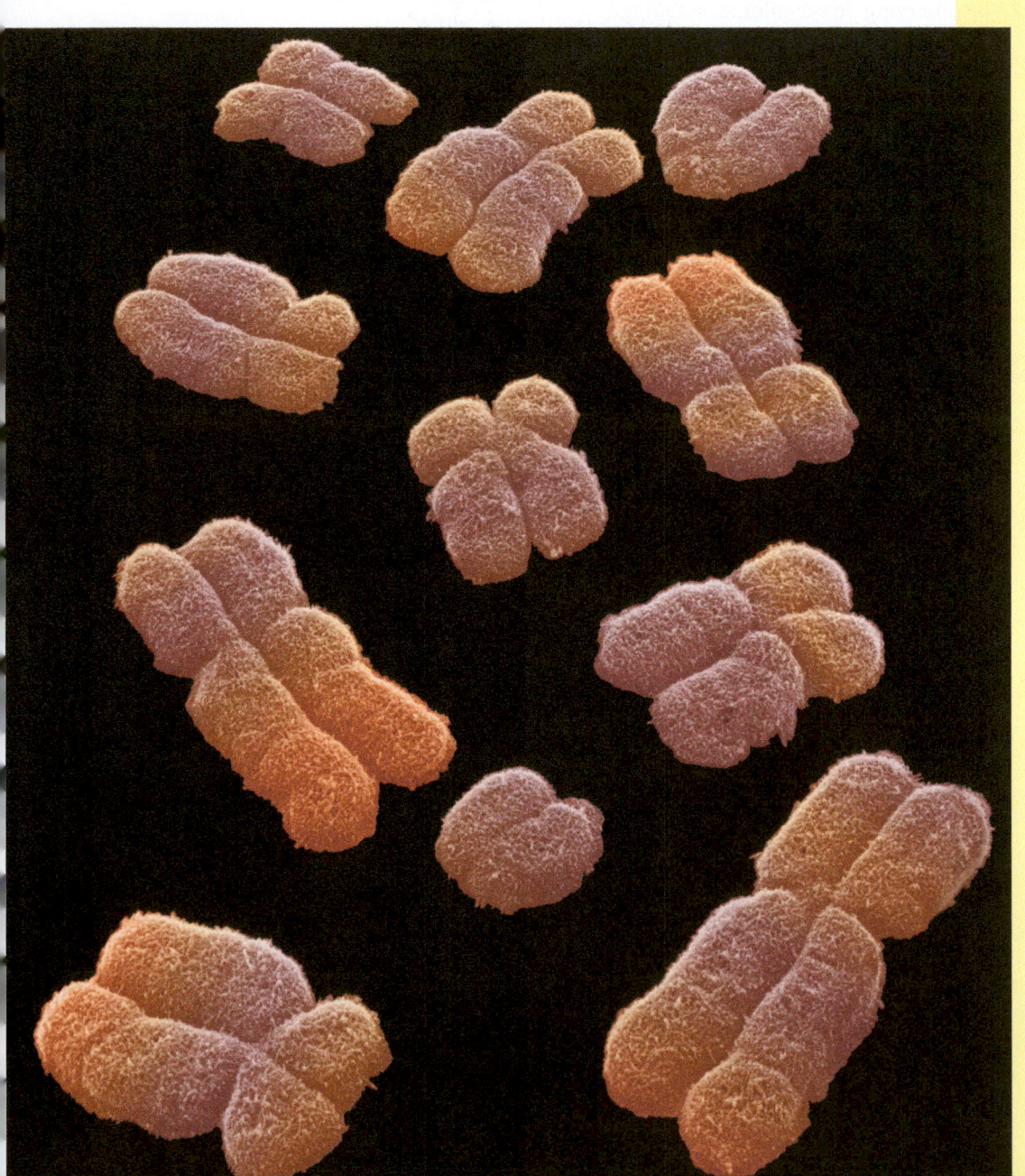

Chromosomes are units of heredity composed of DNA and proteins. Pictured is a partial set of human chromosomes in their well-defined duplicated state.

The condition known as Down syndrome is almost always caused by the inheritance of an extra chromosome. But how can a parent who does not have an extra chromosome pass one on to a son or daughter? The gene that causes sickle-cell anemia also seems to provide some protection against malaria. How can it have both effects? The chromosomal makeup of women offers them a kind of protection against being color-blind. Why should this be so?

From these questions, it may be apparent that chromosomes and the genes on them are centrally involved in issues of human health. You saw last chapter that chromosomal activity lies at the root of several of the genetic principles that Gregor Mendel uncovered. Mendel's "pairs of elements," which separate in reproduction, had as their physical basis the pairs of homologous chromosomes—one inherited from the male, one from the female—that first pair up and then separate from one another in meiosis. The differing traits that Mendel investigated so thoroughly, such as yellow or green peas in a pod, had as their physical basis the pairs of *alleles* or differing forms of a gene that lie on homologous chromosomes.

But there are aspects of chromosomal functioning that Mendel's work scarcely touched on. For one, there is the issue of sex chromosomes. As noted in Chapter 10, the sex of human beings is determined by their chromosomal makeup. Human females have two X chromosomes, while human males have one X and one Y chromosome. Apart from determining sex, what part do these special chromosomes play in heredity? Next, how do the terms *dominant* and *recessive* that you looked at last chapter play out with respect to human disease? And how can genetically based diseases be traced through generations of a family? Finally, you saw in Chapter 10 that a critical step in meiosis is the *crossing over* (or recombination) that occurs early in it. In this process, homologous chromosomes first intertwine and then swap pieces of themselves, after which they separate in meiosis. But what happens when chromosomes fail to properly carry out this part swapping or the separation that follows? The short answer is that such mistakes are responsible for a number of the afflictions that affect human beings, including Down syndrome.

So, in this chapter you will look at sex chromosomes, dominant and recessive disease genes, tracing diseases through generations, and diseases caused by chromosomal mistakes. All the disorders that will be reviewed are transmitted through so-called Mendelian inheritance, meaning they are all disorders in which a single gene or chromosome means the difference between sickness and health. In contrast, many widespread diseases, such as heart disease and cancer, are polygenic—they are based on many genes working in concert, with a given set of genes only predisposing a person toward sickness or health, rather than ensuring it. Let's start now with sex and heredity.

12.1 X-Linked Inheritance in Humans

Minor cuts are simply an irritation to most of us, but for people suffering from the condition known as hemophilia, such cuts can be life-threatening. Hemophilia is a failure of blood to clot properly. A group of proteins interact to make blood clot, but about 80 percent of hemophiliacs lack a functioning version of just one of these proteins (called factor VIII). Genes contain the information for the production of proteins, of course, so at root hemophilia is a genetic disease—a faulty gene is causing the faulty factor VIII protein. This same process is at work in a couple of other well-known afflictions: the disease called Duchenne muscular dystrophy and a far less serious condition, red-green color blindness. These disorders have more in common than being genetically based, however. All of them are known as *X-linked* disorders because the genes that cause them lie on the X chromosome. As it turns out, X-linked disorders claim more male victims than female. Why should this be so? A look at color blindness can provide the answer.

Figure 12.1
Typical Test for Red-Green Color Blindness
If you do not see two numbers (3 and 8) inside the large circle, you may have this recessive trait.

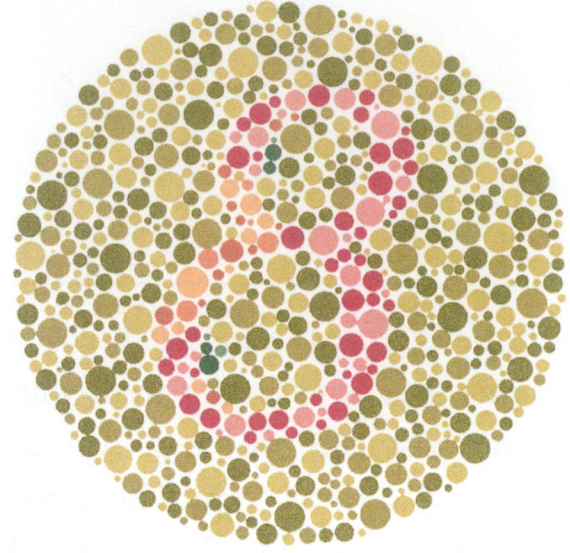

X-Linked Recessive Traits

The Genetics of Color Vision

Despite its name, "color blindness" almost never means a *total* inability to perceive color. Rather, it most often means an inability to distinguish among certain shades of red and green (**Figure 12.1**). To understand how color vision can malfunction in this way, it's important to understand how it commonly works. When light enters our eyes, it travels to the back of them and lands on a group of cells (called cones) that have within them molecules called pigments. A pigment is simply a substance that strongly absorbs a given color of light. The pigments in cone cells turn out to be molecules composed of two main parts, one of which is a protein called opsin. Because genes code for proteins, it's easy to see how color blindness is genetically based: We have genes that are coding for the protein portions of our light-absorbing pigments. More specifically, we

have a gene that codes for blue-absorbing pigment, and it lies on chromosome 7; then we have genes that code for both red- and green-absorbing pigments, and they lie very close to one another on the X chromosome. A red-green color-blind person, then, is one in whom these X-located genes fail to code for the proper pigment proteins.

Alleles and Recessive Disorders

In the normal case, as you've seen, a given gene comes in two variant forms, or *alleles,* in each person. One allele for a gene might reside on, say, the chromosome 1 we inherit from our mother; if so, the second allele for this gene will reside on the chromosome 1 we inherit from our father. There is therefore a kind of backup system in human genetics: Genes come in pairs of alleles that lie on separate, homologous chromosomes. If one allele for a given trait is defective, there will usually exist a second, functional allele for that same trait on a homologous chromosome.

This genetic structure usually works fine when a condition is a so-called **recessive disorder,** meaning a genetic disorder that will not exist in the presence of a functional allele. In such a condition, a person does not have to have *two* functional alleles; a single "good" allele will do. Red-green color blindness turns out to be a recessive disorder: A person who has even one functional allele for red pigment and one functional allele for green pigment will not be color-blind. So, why do men suffer more from this condition than women? Because men have no backup: They have only one allele for red pigment and one allele for green pigment because they have only one *X chromosome* (and one Y). Meanwhile, women have two X chromosomes. If the pigment alleles are defective on one of their chromosomes, the alleles on their other X chromosome will still provide them with full color vision.

Given this, think about the interesting way that color blindness is passed along. Say there is a mother who is not color-blind herself but who is heterozygous for the trait. That is, she has functional color alleles on one of her X chromosomes but dysfunctional alleles on her other X chromosome. Should her son happen to inherit this second X chromosome, he will be color-blind because the X chromosome he got from his mother is his *only* X chromosome (**Figure 12.2**). Meanwhile, a daughter who inherited this chromosome would likely be protected by her second X chromosome—the one she got from her father. Not surprisingly, then, more males are color-blind than females. About 8 percent of the male population has some degree of color blindness, while for females the figure is about half a percent. Women enjoy a similar protection in connection with other X-linked disorders, such as hemophilia and muscular dystrophy. (For more on the way hemophilia can be passed down through generations, see Figure 12.6 on page 219.)

Figure 12.2
The Value of Having Two X Chromosomes

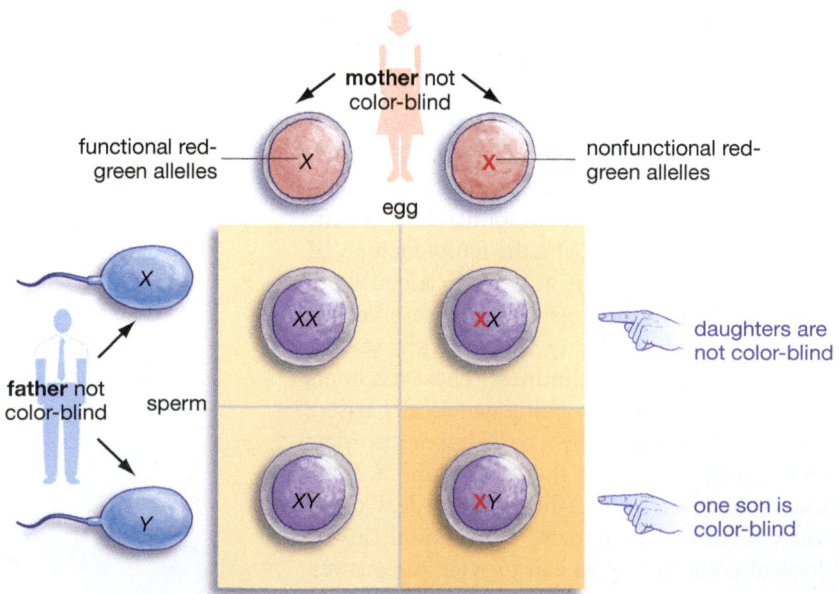

mother not color-blind

functional red-green alleles

nonfunctional red-green alleles

egg

father not color-blind

sperm

daughters are not color-blind

one son is color-blind

12.2 Autosomal Genetic Disorders

Recessive Disorders

The X and Y chromosomes are, of course, only two of the chromosomes in the human collection, and any chromosome can have a malfunctioning gene on it. The chromosomes other than the X and Y chromosomes are called *autosomes*. A recessive dysfunction related to an autosome is thus known as an **autosomal recessive disorder.** (For a list of all the disorders you'll be looking at and more, see **Table 12.1**.)

Sickle-Cell Anemia

A well-known example of an autosomal recessive disorder is sickle-cell anemia, which affects populations derived from several areas on the globe, including Africa. In the United States it is, of course, most widely known as a disease affecting African Americans. The "sickle" in the name comes from the curved shape that is taken on by the red blood cells of its victims. Red blood cells carry oxygen to all parts of the body; in their normal shape, they look a little like doughnuts with incomplete holes. When they take on a sickle shape, however, red cells clog up capillaries, resulting in decreased oxygen supplies to various parts of the body (**Figure 12.3**). In the United States, the average life expectancy for men with the condition is 42 years; for women, it is 48 years.

The question is, what causes a red blood cell to take on this lethal, sickled shape? There is a protein, called hemoglobin, that carries the oxygen within red blood cells. The vast majority of people in the world have one form of hemoglobin, called hemoglobin A, but sickle-cell anemia sufferers have another form of this protein, hemoglobin S, which coalesces into crystals that distort the cell.

Now, how is sickle-cell anemia passed from parents to offspring? Because it is a recessive condition, any child who gets it must inherit *two* alleles for it—one allele from the mother, one from the father, each of them coding for the hemoglobin S protein. This means, of course, that both parents must themselves have at least one hemoglobin S allele; both parents must be at least heterozygous for the trait. In this situation, the laws governing inheritance of this disorder are simply those of Mendel's monohybrid cross: A child of these two parents has a 25 percent chance of inheriting the disorder. You can look at the Punnett square in **Figure 12.4a** on the next page to see how this works out.

It's important to note the status of the parents in this example. They are heterozygous for a recessive debilitation, meaning they do not suffer from the condition themselves because they are protected by their single "good" allele. Each of them is, however, a **carrier** for the condition: a person who does not suffer from a recessive genetic debilitation but who carries an allele for it that can be passed along to offspring. In this respect, they are just like the mother in the color blindness example.

Table 12.1

Selected Examples of Human Genetic Disorders

Type	Name of Condition	Effects
X-linked recessive disorders	Hemophilia	Faulty blood clotting
	Duchenne muscular dystrophy	Wasting of muscles
	Red-green color blindness	Inability to distinguish shades of red from green
Autosomal recessive disorders	Albinism	No pigmentation in skin
	Sickle-cell anemia	Anemia, intense pain, infections
	Cystic fibrosis	Impaired lung function, lung infections
	Phenylketonuria	Mental retardation
	Tay-Sachs disease	Nervous system degeneration in infants
	Werner syndrome	Premature aging
Autosomal dominant disorders	Huntington disease	Brain tissue degeneration
	Marfan syndrome	Ruptured blood vessels
	Polydactyly	Extra fingers or toes
Aberrations in chromosome number	Down syndrome	Mental retardation, shortened life span
	Turner syndrome	Sterility, short stature
	Klinefelter syndrome	Dysfunctional testicles, feminized features
Aberrations in chromosome structure	Cri-du-chat syndrome	Mental retardation, malformed larynx
	Fragile-X syndrome	Mental retardation, facial deformities

Some Human Genetic Disorders

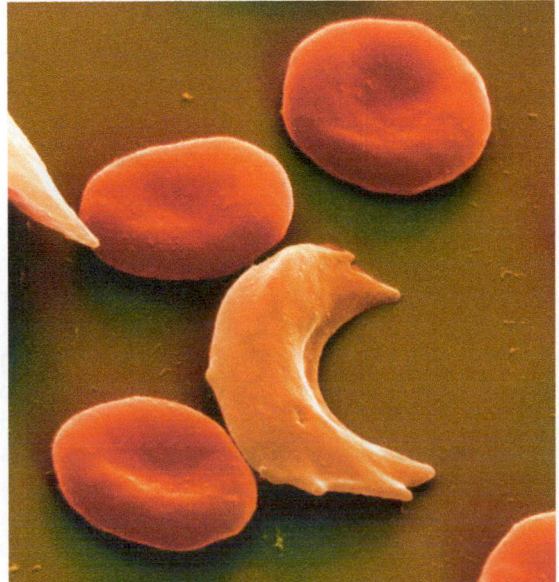

Figure 12.3
Healthy Cells, Sickled Cell
Three normal red blood cells flank a single sickled cell.

Figure 12.4
Transmission of Recessive and Dominant Disorders

(a) Sickle-cell anemia: transmission of a recessive disorder.

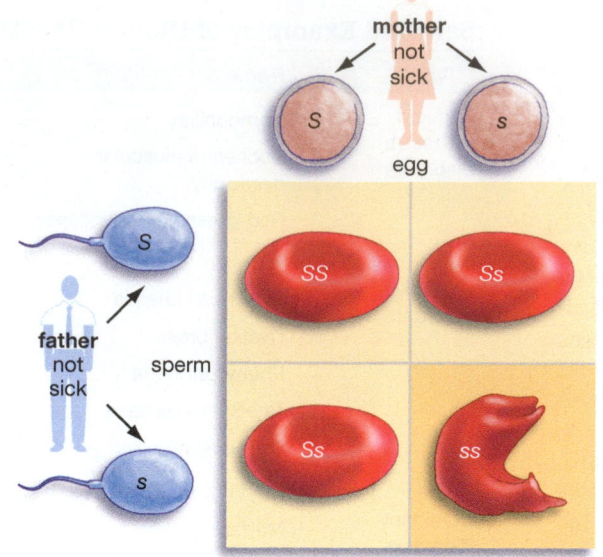

Sickle-cell anemia is a recessive autosomal disorder; both the mother and father must carry at least one allele for the trait in order for a son or a daughter to be a sickle-cell victim. When both parents have one sickle-cell allele, there is a 25% chance that any given offspring will inherit the condition.

25% probability of inheriting the disorder

(b) Huntington disease: transmission of a dominant disorder.

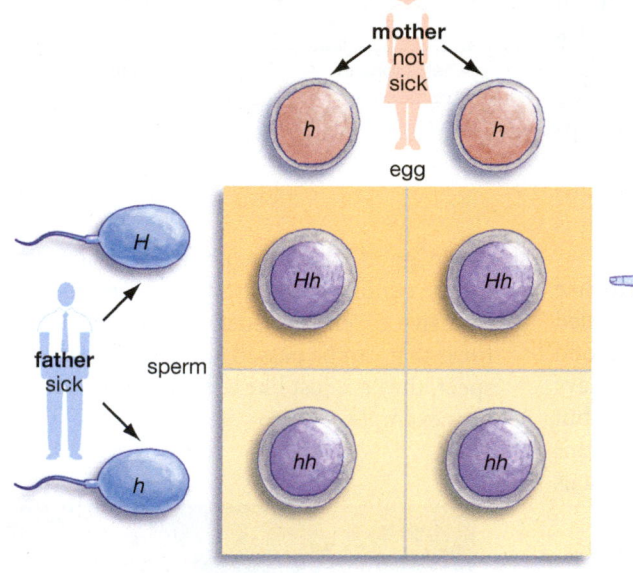

 50% probability of inheriting the disorder

In Huntington disease, if only a single parent has a Huntington allele there is a 50% chance that a son or daughter will inherit the condition.

Malaria Protection: Hemoglobin S as a Useful Protein

Sickle-cell anemia offers another lesson in connection with the notion of hemoglobin S as a "faulty" protein. It is in fact quite a functional protein in certain circumstances, namely, when it appears heterozygously (with hemoglobin A) in people who live in regions where malaria is common. The most severe form of malaria is caused by a single-celled parasite that is transmitted into human beings by the *Anopheles* mosquito. Traveling through the bloodstream, these parasites invade and ultimately destroy red blood cells. For reasons we still don't understand, having hemoglobin

S in one's system makes red blood cells resistant to invasion. Thus, it is likely that hemoglobin S became as widespread as it is because of its value in resisting malaria. The price of this resistance, however, is that some offspring are not heterozygous for hemoglobin S. They are homozygous for it, meaning they have sickle-cell anemia.

Dominant Disorders

Although sickle-cell anemia is an autosomal disorder and red-green color blindness an X-linked disorder, both are recessive disorders. That is, a person with even a single properly functioning allele will not suffer from them. However, there are also **dominant disorders:** genetic conditions in which a single faulty allele can cause damage, even when a second, functional allele exists. This leads to the concept of an **autosomal dominant disorder,** simply meaning a dominant genetic disorder caused by a faulty allele that lies on an autosomal chromosome. There is, for example, an autosomal dominant disorder called Huntington disease. Affecting about 30,000 Americans, it results in both mental impairment and uncontrollable spastic movements called *chorea*. Perhaps its most famous victim was the American folksinger Woody Guthrie. Like most Huntington sufferers, Guthrie did not begin to show symptoms of the disease until well into adulthood—*after* he had children, who then may have had the disease passed along to them. You can look at the Punnett square in **Figure 12.4b** to see how the inheritance pattern of an autosomal dominant disease such as Huntington differs from that of an autosomal recessive illness. Because a parent need only pass on a single Huntington allele for a son or daughter to suffer from the condition, the chances of any given offspring getting the disease from a single affected parent are one out of two.

SO FAR . . .

1. A recessive genetic disorder is one that will not exist in the presence of a _____, while a dominant disorder is one in which a _____ can cause trouble, even in the presence of a _____.

2. True or false: A woman who has only one X chromosome containing dysfunctional alleles for red-green color vision is not color-blind.

3. To inherit an autosomal recessive disorder, both parents of an offspring must be at least _____ for the condition, meaning each parent must have at least _____.

12.3 Tracking Traits with Pedigrees

Confronted with a medical condition, such as Huntington, that is running through a family, scientists find it helpful to construct a medical **pedigree,** defined as a familial history intended to track genetic conditions. Normally set forth as diagrams, medical pedigrees do more than give a family history of disease. They can be used to ascertain whether a condition is dominant or recessive, and X-linked or autosomal. This can help establish probabilities for *future* inheritance of the condition—something that can be very helpful for couples thinking of having a child.

If you look at **Figure 12.5**, you can see a simple pedigree for albinism: a lack of skin pigmentation, which is known to be an autosomal recessive condition. In the figure, you can see some of the standard symbols used in pedigrees. A circle is used for a female and a square for a male. Parents are indicated by a horizontal line connecting a male and a female, while a vertical line between the parents leads to a lower row that denotes the parents' offspring. This second row—a horizontal line of siblings—has an order to it: oldest child on the left, youngest on the right. A circle or square that is filled in indicates a family member who has the condition (in this case, albinism), while a symbol that is half filled in indicates a person known to be heterozygous for a recessive condition (a carrier). One of the strengths of a pedigree is that it can sometimes tell researchers which persons are heterozygous carriers of a recessive condition. Remember that a carrier does not display *symptoms* of the condition. So, how can a pedigree reveal this?

Take a look at Figure 12.5. The only thing that would be apparent from actually seeing any of the people in the pedigree is that two of them—a female in generation II and a male in generation III—have the condition of albinism. But a little knowledge of Mendelian genetics also allows some other deductions. If the condition had been dominant, then it would have manifested itself in at least one of the parents on the left in generation I. Because this was not the case, it is fair to deduce that this is a recessive

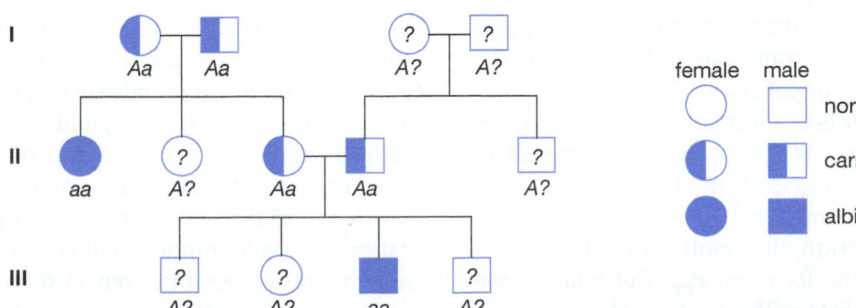

Figure 12.5
A Hypothetical Pedigree for Albinism through Three Generations

female male

○ ☐ normal

◑ ◩ carrier

● ■ albino

condition. The fact that it is recessive, however, means that *both* parents in generation I had to be carriers for the allele—both have to be heterozygous for the condition, which is why their symbols can be half shaded in. Things are less clear with the parents on the right in generation I. One of their sons did not manifest the condition himself but went on to have a son who did. From this, we know that the son in generation II had to have been heterozygous for the condition. But, from which parent did this son get his albinism allele? Either, or both, of his parents in generation I could have been heterozygous for the condition and yet between them, only have passed along a single albinism allele.

This mixture of certainty and uncertainty leads to the genotype labeling you see in the figure, with *A* representing the dominant "normal" allele and lowercase *a* representing the recessive albinism allele. We know, for example, that both parents on the left in generation I had to be *Aa,* but all we can say for sure about the parents on the right in the same generation is that at least one of them was a carrier. As you can imagine, pedigrees can get much more elaborate than this one. To see how a trait can be tracked through many generations, look at **Figure 12.6**.

12.4 Aberrations in Chromosomal Sets: Polyploidy

Polyploid Plants

All of the maladies you've looked at so far have resulted from dysfunctional genes that exist among a standard array of chromosomes. In humans, this array is 22 pairs of autosomes and either an XX combination (in females) or an XY combination (in males). Meanwhile, Mendel's peas had seven pairs of chromosomes, and the common fruit fly has four pairs. Whatever the *number* of chromosomes, note the similarity in all these species: Their chromosomes come in pairs, or to put it another way, they all have two *sets* of chromosomes, meaning the organisms that possess them are diploid.

In many situations, however, organisms end up with more than two sets of chromosomes. This is called **polyploidy,** a condition in which one or more entire sets of chromosomes has been added to the genome of a diploid organism.

You may wonder how a human being could end up with more than two sets of chromosomes. After all, in meiosis, sperm and egg end up with just one set of chromosomes in them. When they later fuse, in the moment of conception, the result is then two sets of chromosomes in the fertilized egg. But what if *two* sperm manage to fuse with one egg? The result is a fertilized egg with three sets of chromosomes, and the

outcome is polyploidy. Normally, an egg lets in only one sperm, and it blocks the others hovering about with a kind of chemical gate-slamming mechanism. (For more on this, see page 644 in Chapter 33.) But occasionally, more than one sperm manages to make it in. Human polyploidy can also get started before fertilization, when meiosis malfunctions and gives a single egg or sperm two sets of chromosomes, meaning that there will be three sets in total when egg and sperm fuse.

Polyploidy actually is tolerated well in some organisms. Indeed, it has come about countless times in plant species and generally gives rise to perfectly robust plants. In human beings, however, polyploidy is a disaster. Perhaps only 1 percent of human embryos with the condition will survive to birth, and none of these babies lives long. Concerns about chromosome number in living persons, then, center not on addition or deletion of whole sets of chromosomes, but rather on the gain or loss of *individual* chromosomes.

12.5 Incorrect Chromosome Number: Aneuploidy

The condition called **aneuploidy** is one in which an organism has either more or fewer chromosomes than normally exist in its species' full set. This usually means that an organism has gained or lost a *single* chromosome, although more complex patterns are possible, as you'll see. Among human genetic malfunctions, aneuploidy is unusual in that it occurs quite commonly and yet goes largely unrecognized. This is so because aneuploidy occurs most often not in fully formed human beings but in embryos. A would-be mother may know only that she is having a hard time getting pregnant. What she may not know is that she actually *has* been pregnant—perhaps several times—but that aneuploidy has doomed the embryo in each case. Doomed it in what way? The most common outcome of aneuploidy is a pregnancy that ends in miscarriage. Many of these miscarriages happen so early in a pregnancy that a woman never realizes she was pregnant; the microscopic embryo may simply be part of material that is discharged, perhaps with the woman's next menstrual period. Even when a miscarriage is recognized as such, tests can seldom be run that can identify aneuploidy as the cause. But by undertaking careful examinations of thousands of pregnancies, scientists have concluded that aneuploidy is responsible for about 30 percent of miscarriages. And how common is miscarriage? In the best study conducted to date, scientists found that it takes place in 31 percent of pregnancies. A small proportion of embryos do manage to survive aneuploidy, but even in these instances there are consequences. Down syndrome is one of the outcomes of aneuploidy.

Three Lessons from a Royal Pedigree

the labels in black are referred to in the text

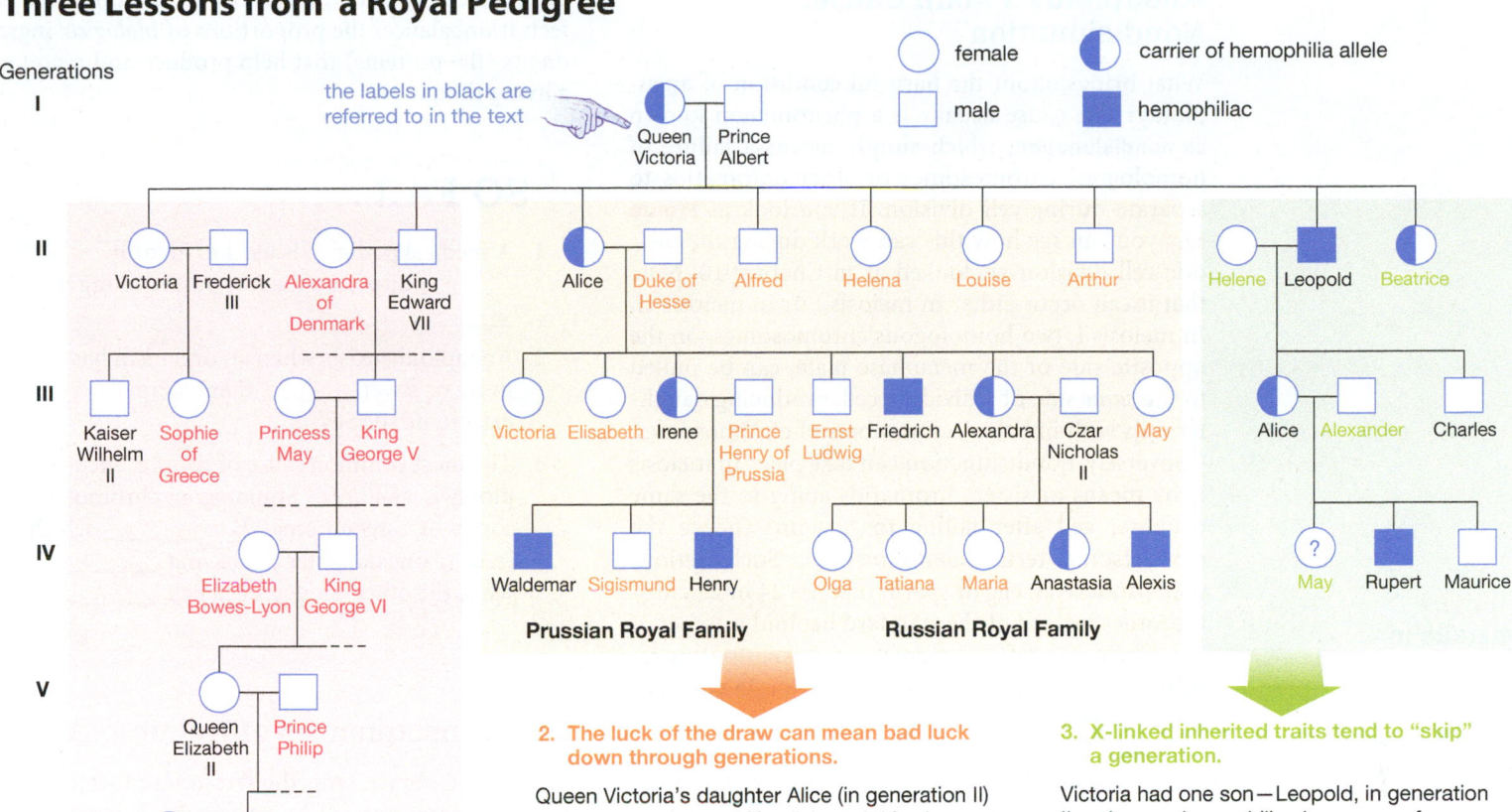

Legend:
- ○ female
- ◑ carrier of hemophilia allele
- □ male
- ■ hemophiliac

Generations I–VII

British Royal Family

Prussian Royal Family   Russian Royal Family

1. The luck of the draw can mean good luck down through generations.

The hemophilia allele almost certainly arose as a mutation in either Queen Victoria or in one of her parents. Whichever the case, the Queen was undoubtedly a carrier for the allele. For Britain's royal family, the key event was that King Edward VII (toward the left in generation II) did not inherit the hemophilia allele from his mother, but instead inherited her second "good" allele. With this, the British Royal House—right down through today's princes William and Harry—avoided hemophilia, barring its introduction through marriage.

2. The luck of the draw can mean bad luck down through generations.

Queen Victoria's daughter Alice (in generation II) did inherit the hemophilia allele and this chance event had momentous consequences. Though she herself was only a carrier, she had one son, Friedrich, who was hemophiliac (dying at age 3); and two daughters who, as carriers, went on to pass the allele along. One of these daughters, Alexandra, became the czarina of Russia after marrying Czar Nicholas II. The couple had a male child, Alexis, who inherited his mother's hemophilia allele—now passed down through at least three generations—and was thus a hemophiliac. So precarious was Alexis' condition that his mother turned to a semiliterate monk, Rasputin, as a means of trying to care for him. Anger with Rasputin among the Russian populace may have played a part in bringing down the Russian Royal House in the revolutions of 1917. Neither Alexis nor his sisters lived long enough to have children, since they and their parents were murdered by communist revolutionaries in 1918. We know the genetic status of Alexis' sisters, however, thanks to a 2009 analysis carried out on the remains of members of the royal family. Among the four sisters, only one was a carrier for the hemophilia allele—most likely Anastasia, to judge by anthropological evidence.

3. X-linked inherited traits tend to "skip" a generation.

Victoria had one son—Leopold, in generation II—who was hemophiliac by reason of inheriting his mother's faulty allele. Note, however, that among Leopold's offspring, there are no hemophiliacs. The next hemophiliac is found one generation further down, in Rupert, in generation IV. Why should things work this way? Remember that the gene for hemophilia is on the X chromosome and that every male must get his lone X chromosome from his mother. Thus, Leopold's son, Charles, was not at risk of being hemophiliac because Charles' single X chromosome came from his mother. However, Leopold's daughter, Alice, carried two X chromosomes—one "good" chromosome from her mother and the faulty X chromosome from Leopold. She then passed on Leopold's chromosome to one of her sons (Rupert), though not to the other (Maurice). The luck of the draw was on display again.

Figure 12.6

A Pedigree for Hemophilia in Europe's Royal Families

Hemophilia is a rare, recessive genetic disorder, but it managed to affect a large portion of Europe's royal family members in the nineteenth and twentieth centuries after being passed on to them from England's Queen Victoria. As the main text notes, hemophilia usually is caused by a particular variant form of a gene (an allele) that leads to a faulty form of a blood-clotting protein called Factor VIII. In the case of Queen Victoria and her descendants, however, a mutated allele that leads to a closely related protein, called Factor IX, was responsible. Since the genes for factors VIII and IX reside on the X chromosome, hemophilia is a condition that is much more likely to affect men than women. The pedigree shown here for hemophilia among Europe's royal families has been greatly simplified; in most of the British royal family, for example, only the heirs to the throne and their spouses are shown.

Aneuploidy's Main Cause: Nondisjunction

What brings about the harmful condition of aneuploidy? The cause usually is a phenomenon known as **nondisjunction,** which simply means a failure of homologous chromosomes or sister chromatids to separate during cell division. If you look at **Figure 12.7** you can see how this can work during the meiotic cell division we looked at in Chapter 10. Note that it can occur either in meiosis I or in meiosis II. In meiosis I, two homologous chromosomes, on the opposite side of the metaphase plate, can be pulled to the *same* side of a dividing cell, producing daughter cells with imbalanced numbers of chromosomes. Conversely, nondisjunction can take place in meiosis II by means of sister chromatids going to the same daughter cell after failing to "disjoin" (hence the cumbersome term *nondisjunction*). Such actions then produce an egg or sperm that has 24 or 22 chromosomes instead of the standard haploid number of 23; after union with a normal egg or sperm, this results in an embryo that has either 47 or 45 chromosomes instead of the standard 46.

The effect of this is that the embryo has a genetic imbalance—one that perhaps can be made clear by an analogy. Imagine that while baking chocolate chip cookies, you decided to increase the amount of flour called for in the recipe. This change might be of little consequence if you increased all the other ingredients in the same proportion. If, however, you increased nothing but the flour, the final product would be hard and flavorless because the ingredients in it would be unbalanced. An added chromosome has this same effect: It unbalances the proportions of *biological* ingredients (the proteins) that help produce and maintain a living thing.

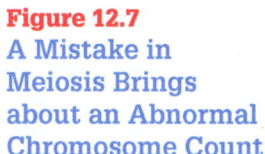

Figure 12.7
A Mistake in Meiosis Brings about an Abnormal Chromosome Count
Nondisjunction can occur in either meiosis I or meiosis II, when either chromosomes or chromatids fail to separate properly. When it occurs in meiosis I, 100 percent of the resulting gametes will be abnormal; when it takes place in meiosis II, only 50 percent will be abnormal.

SO FAR . . .

1. A medical pedigree is used to identify _____ disorders that are passed along in _____.

2. Aneuploidy exists when an organism has more or fewer _____ than normally exist in its species' _____.

3. The most common cause of meiotic aneuploidy is a failure of homologous chromosomes or sister chromatids to _____, which leads to one daughter cell having _____, while the other daughter cell has _____.

The Consequences of Aneuploidy

As noted, embryos with the wrong number of chromosomes are not likely to survive. Setting aside meiotic aneuploidies that affect the X and Y chromosomes, *all* such aneuploidies result in miscarriage except for those that give an embryo an additional chromosome 13, 18, or 21. It is the gain of an additional chromosome 21 that causes the most well-known outcome of aneuploidy in human beings.

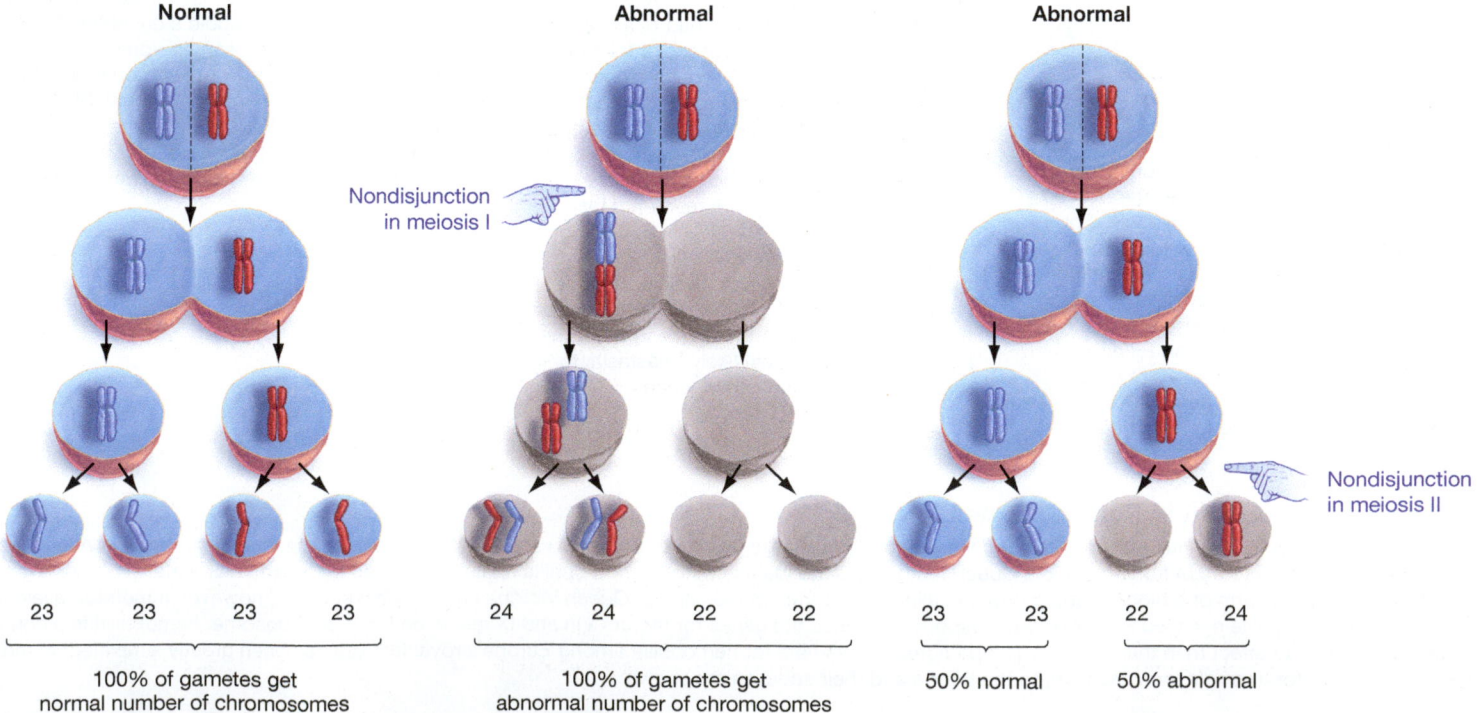

| Normal | Abnormal | Abnormal |

Nondisjunction in meiosis I

Nondisjunction in meiosis II

| 23 | 23 | 23 | 23 | | 24 | 24 | 22 | 22 | | 23 | 23 | 22 | 24 |

100% of gametes get normal number of chromosomes

100% of gametes get abnormal number of chromosomes

50% normal   50% abnormal

Down syndrome is, in some 95 percent of cases, a condition in which a person has three copies of chromosome 21 rather than the standard two. Ninety percent of these cases come about because of nondisjunction in egg formation, with the remaining 10 percent caused by nondisjunction in sperm formation. Seen in about 0.1 percent of all live births, Down syndrome results in an array of effects: smallish, oval heads; IQs that are well below normal; infertility in males; short stature; and reduced life span in both sexes.

It is well known that when women pass the age of about 35, their risk of giving birth to a Down syndrome child increases dramatically (**Figure 12.8**). Nevertheless, most Down syndrome children are born to mothers under the age of 35 for the simple reason that younger mothers give birth to most children. As such, in 2007 the American College of Obstetricians and Gynecologists recommended that all pregnant women, regardless of age, undertake Down syndrome screening, whereas previously the group recommended such testing only for women 35 or older.

Scientists are not certain why a mother's age should figure so prominently in Down syndrome, although several hypotheses have been put forward to account for this. Most scientists agree that one piece of the puzzle is something you saw in Chapter 10: In egg development, the process of meiosis generally stretches on for decades. Any given egg that develops in a woman starts meiosis I before the woman is born, but may not complete this meiosis until she is, say, in her thirties. In the intervening decades, this egg's cellular machinery has aged, perhaps to the point that it can no longer separate its chromosomes or chromatids properly.

Beginning in the late 1960s, it became possible to test developing fetuses for genetic abnormalities. In recent years, however, reproductive technology has moved beyond testing. Prospective parents can now choose, from a group of embryos they have created, embryos that do not have any recognizable genetic defects. You can read more about this in "PGD: Screening for a Healthy Child" on page 222.

Abnormal Numbers of Sex Chromosomes

Aneuploidy can affect not only autosomes but sex chromosomes as well, meaning the two X chromosomes that females have and the X and Y chromosomes that males have. Embryos often survive sex chromosome aneuploidies, but the result, once again, is usually debilitations. An example is Turner syndrome, which produces people who are phenotypically female, but who have only one X chromosome. Such females, then, have only 45 chromosomes, rather than the usual 46. Their state is sometimes referred to as XO, the "O" signifying the missing X chromosome. The absence of a second sex chromosome causes a range of afflictions. Females with Turner syndrome have ovaries that don't develop properly (which causes sterility), they are generally short, and they often have brown spots (called nevi) over their bodies.

(a) Maternal age and Down syndrome risk

Mother's age	Chances of giving birth to a child with Down syndrome
20	1 in 1925
25	1 in 1205
30	1 in 885
35	1 in 365
40	1 in 110
45	1 in 32

(b) Karyotype of a person with Down syndrome

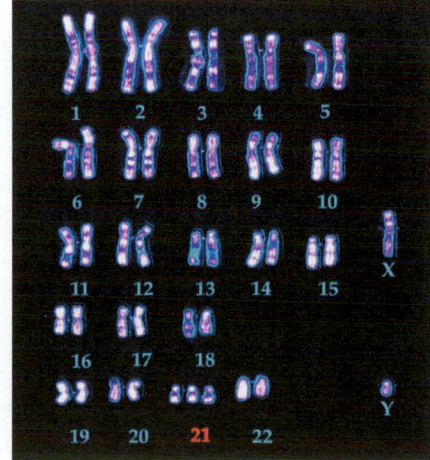

Figure 12.8
Down Syndrome

(a) The risk of giving birth to a Down syndrome child increases dramatically past maternal age 35. Despite this, most Down syndrome children are born to women under the age of 35 because younger women give birth to most children.

(b) A karyotype is a visual display of an entire set of chromosomes. Pictured is a karyotype of a person with Down syndrome. Note the three chromosomes above the number 21. The karyotype shows that this person is a male, as indicated by the Y chromosome.

ESSAY

PGD: Screening for a Healthy Child

"Is the baby OK?" In a life full of questions, it's possible that none carries more weight than this one. In most cases, the answer will be reassuring, but in some it will be devastating. Small wonder, then, that scientists constantly are trying to perfect ways to know in advance whether a newborn will be healthy. You have seen that a large number of debilitating human conditions have genetic causes. Further, you know that nearly every kind of cell in a person's body contains a complete copy of that person's genome, or set of genes. Therefore, to check on someone's genetic well-being, all that is necessary is to have access to a small collection of that person's cells. Such cells exist, of course, in embryos as well as in newborn babies and adults. Thus, it's not hard to see what the procedure would be for prenatal genetic testing: Gather embryonic cells and examine their DNA and chromosomes.

This kind of testing has been going on, in ever-more-refined forms, since the late 1960s. From that time until recently, however, most prospective parents who got bad news from a genetic test had two choices: abort the fetus or bring a child with an incurable condition into the world. Now, an ever-growing number of reproductive clinics are providing a third choice. Parents who are willing to undertake so-called in vitro fertilization—fertilization of the mother's eggs in a laboratory—can have each one of the early-stage embryos that results from this process tested for genetic trouble. Of this initial group of embryos, only those found to be genetically healthy are candidates to

be inserted back into the mother. The hope is that at least one of them will implant in the mother's uterus and result in a child.

Right now, some clinics allow parents to use PGD to choose the sex of their child.

This process, called preimplantation genetic diagnosis or PGD, can be used by couples who have no known genetic risk factors but who are simply having a hard time conceiving a child. As the main text notes, a significant proportion of human embryos have genetic defects (such as aneuploidy). Moreover, some couples may be particularly prone to producing such embryos, which generally are lost in miscarriages. PGD allows couples to screen their embryos for genetic problems before implantation in the uterus, thus greatly upping their chances of bringing a child to term.

PGD can also be used, however, in connection with couples who are *likely* to produce a child with genetic defects. In an extreme example, imagine a prospective father who knows he is fated to fall victim to Huntington disease because he has been found to carry the defective Huntington allele. Since Huntington is a dominant disorder, his odds of passing the disease along to one of his children are one in two. With PGD, only those embryos without the harmful allele would be selected for implantation.

Of course, more conventional means of genetic screening, such as amniocentesis, can provide information about conditions

such as Huntington, but there is a difference. In amniocentesis, cells are obtained from an embryo that has been developing in a mother's uterus for a minimum of 14 weeks. By the time the results of amniocentesis are in, more than four months of a nine-month pregnancy may have elapsed. Most prospective parents would make a distinction between aborting a four-month-old fetus on the one hand or not implanting a four-day-old embryo on the other. Of course, not everyone feels this way. For some people, all embryos represent human life; therefore, producing an embryo and then not using it amounts to taking a human life. In PGD, most embryos have two fates: Those that are not implanted are eventually destroyed. (However, a small number of "embryo adoptions" are now taking place, in which preserved embryos from fertility clinics are, with the biological parents' consent, implanted in an adoptive mother and then brought to term.)

Beyond this issue, PGD also raises another ethical question: What limits should be placed on choosing from among embryos? Right now, some clinics allow parents to use PGD to choose the *sex* of their child. If this seems disturbing, consider that, with our expanding knowledge of genetics, it may be possible in the future to select for a tall or muscular child. Prospective parents who would never abort a fetus to get such an outcome might not object to choosing from among embryos to get it, particularly since more embryos must be produced than will be used. Parents may think: Since a selection process is going to take place anyway (for healthy genes), why not select for things like height as well?

Such questions become less abstract with a little knowledge of how PGD works. Many of its steps are those that occur with any in vitro fertilization (IVF). The starting point for IVF is to administer to the prospective mother a group of hormones that stimulate maturation of a large group of eggs in her ovaries. (These eggs are the oocytes you looked at in Chapter 10.) Anywhere from a week to two weeks later, a physician uses a needle attached to an ultrasound guide to remove the mature eggs from the ovaries. Although numbers vary, it would not be uncommon for 10 or 12 eggs to be recovered in this process. The husband then provides sperm to fertilize the eggs, with this fertilization taking place in a laboratory.

Three days after fertilization, each egg has divided into an eight-cell embryo. If you look at **Figure 1a**, you can see what comes next: Using a tiny hollow tube called a pipette, a physician gently suctions one of these cells away from the others. (Embryonic cells are so undifferentiated at this point that the loss of any one of them does not interfere with development of the others into a child.) The cell that is removed provides the DNA for testing. If you look at **Figure 1b**, you can see the results of one of the tests. Recall that children born with Down syndrome generally have a third copy of chromosome 21. It is possible to *test* for a third chromosome 21 by preparing a length of DNA whose "bases" will latch onto the unique set of DNA bases that exist on chromosome 21. Moreover, these lengths of "homing" DNA are fluorescent—they light up. The result is what you see in the figure: The prepared DNA has latched onto three chromosome 21's in these embryonic cells; they have revealed that, had this embryo been implanted in the mother, the result would have been a child with Down syndrome.

By use of this test and others, embryos produced through IVF are ranked according to their health. Parents then choose from among them, selecting perhaps three or four that will be transferred back into the mother by a physician.

(a)

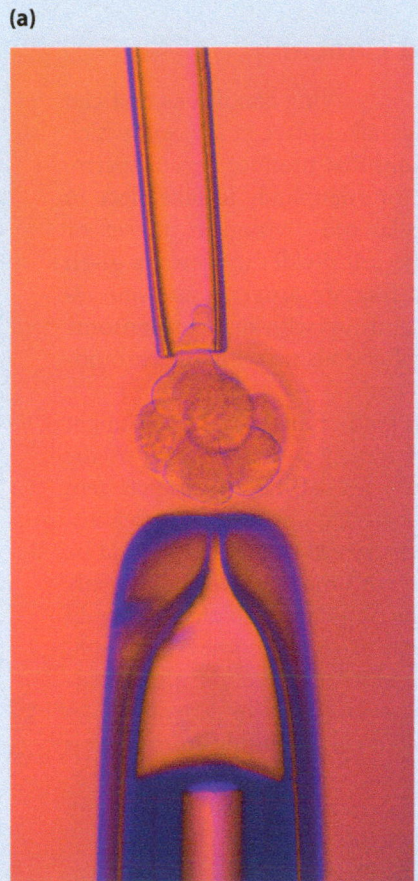

(b)

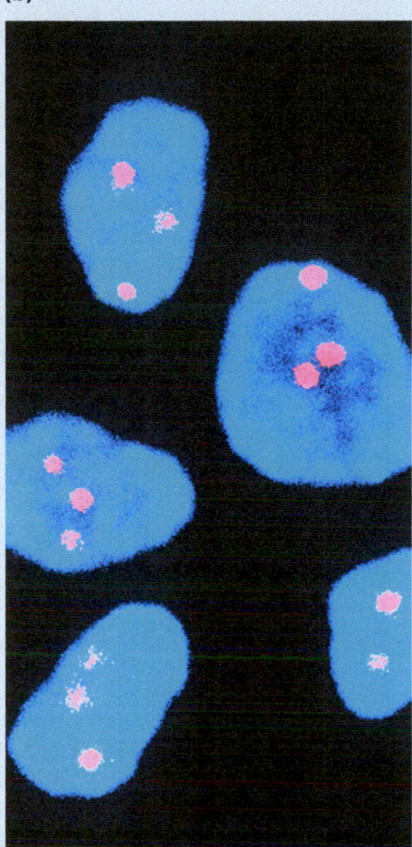

Figure 1

(a) Removing a Cell for Testing

A human embryo at the eight-cell stage (center) being manipulated to have one of its cells removed for preimplantation genetic diagnosis. A pipette at left holds the embryo, while a smaller pipette at right gently suctions off one cell from it. The cell's genetic material is then checked for abnormalities. If this embryo is determined to be normal, it may be inserted into the mother's uterus, where it can develop into a baby.

(b) Down Syndrome Diagnosed

Genetic testing of this embryo has revealed that each of its cells has three copies of chromosome 21, indicated by the pink areas in the nuclei of the cells. Such an embryo would develop into a child with Down syndrome. The pink areas represent fluorescently labeled DNA that has bound or "hybridized" with regions of DNA specific to chromosome 21.

Turner syndrome results from a loss of a chromosome, but all manner of sex chromosome *additions* can also take place. There are, for example, XXY men who, while phenotypically male in most respects, tend to have a number of feminine features: some breast development, a more feminine figure, and lack of facial hair. When coupled with other characteristics (such as tall stature and dysfunctional testicles), the result is a condition called Klinefelter syndrome.

Aneuploidy and Cancer

We've thus far looked at varying conditions that can result when aneuploidy takes place in meiosis, which is to say the cell divisions that produce eggs or sperm. But it's also possible for aneuploidy to take place in *mitosis,* meaning regular cell division in nonsex or "somatic" cells. As one cell divides into two, a given chromatid can fail to migrate properly, and one resulting daughter cell ends up with one chromosome too many, while the other ends up with one chromosome too few.

Now, note that this kind of aneuploid event might take place in a child, a teenager, or an adult. If so, the only cells affected would be the "line" of cells that stem from the original two cells that underwent aneuploidy. Put another way, not every cell in the body would be affected, as is the case with meiotic aneuploidy; only the daughter cells of the original aneuploid cells would have the wrong number of chromosomes. You might think: Well, this is bound to be less serious than aneuploidy resulting from meiosis; and in general, this is the case. Most aneuploid cells resulting from mitosis will die, but a few can survive. Indeed, mitotic aneuploidy is often seen in a destructive kind of cell found in the body, the cancer cell.

Researchers have known about aneuploidy in cancer cells for decades. Today, it's clear that almost all cancer cells have undergone aneuploidy—that is, almost all cancerous cells have the wrong number of chromosomes in them. But for many years, it was assumed that such aneuploidies were the *result* of a cancer in progress rather than the *cause* of it. In recent years, however, this idea has been turned on its head. Some prominent cancer researchers believe that aneuploidy is a cause of cancer. To be sure, this idea differs from the leading hypothesis about cancer: that it results from a series of mutations to individual genes. The problem with this idea is that, despite decades of research, no one has yet identified a series of mutations that will predictably cause any of the most serious forms of cancer. Meanwhile, it has been demonstrated that, at the least, aneuploidy can occur before the very earliest stages of some forms of cancer. The question is whether it is the triggering event in these cancers (**Figure 12.9**). Only more research will answer this question. For now, don't be surprised if you start see-

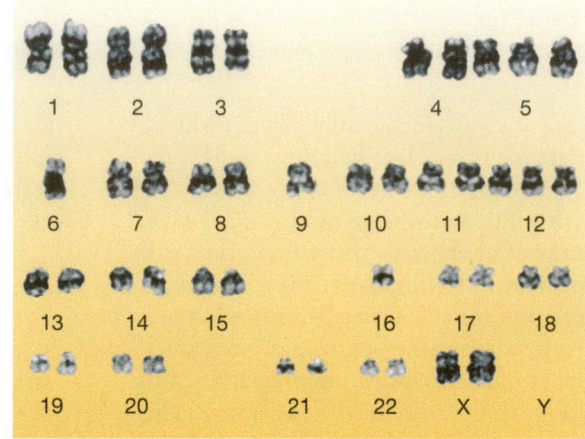

Figure 12.9
Aneuploidy before Cancer

In this karyotype of a cell taken from a young girl, aneuploidy can be seen to exist in a complex form. Note that the cell has the standard two copies of chromosomes 1 through 4, but then three copies of chromosome 5, only one copy of chromosome 6, and so on. This aneuploidy was present in the girl during embryonic development, meaning it was present prior to the time she developed a form of childhood cancer (called an embryonal rhabdomyosarcoma). Evidence of this sort has convinced some researchers that aneuploidy can be a cause of cancer.
(Karyotype courtesy of Dr. Sandra Hanks and Dr. Nazneen Rahman.)

ing newspaper stories about aneuploidy and cancer. We're certain that the cellular mistake of aneuploidy has devastating consequences every day. The question is whether cancer is one of them.

12.6 Structural Aberrations in Chromosomes

Aberrations can occur *within* a given chromosome, sometimes resulting from interactions *between* chromosomes. A frequent cause of such change is that pieces of chromosomes can break off from the main chromosomal body. Such a chromosomal fragment may then be lost to further genetic activity, or it may rejoin a chromosome—either the one it came from or another—sometimes to harmful effect. These structural aberrations may come about spontaneously, but they may also be caused by exposure to such agents as radiation, viruses, and chemicals.

Deletions

A chromosomal **deletion** occurs when a chromosome fragment breaks off and then does not rejoin any chromosome. Such an event might take place during meiosis in a healthy parent, who then could pass along a complement of 23 chromosomes, one chromosome of which would be missing a segment. If the deleted piece were large enough, the zygote

(a) Cri-du-chat syndrome

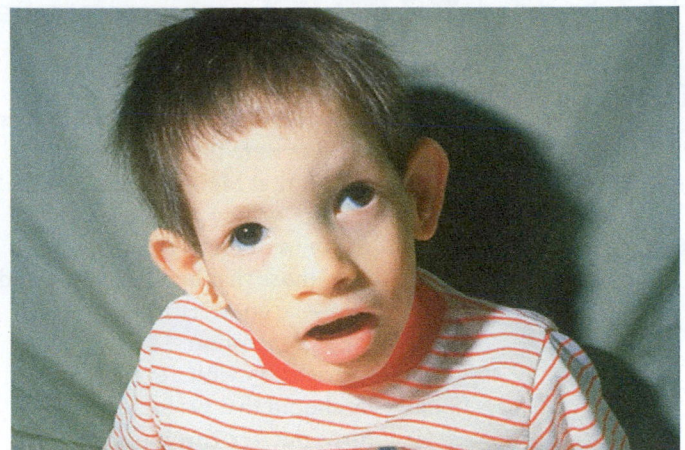

(b) Cri-du-chat karyotype

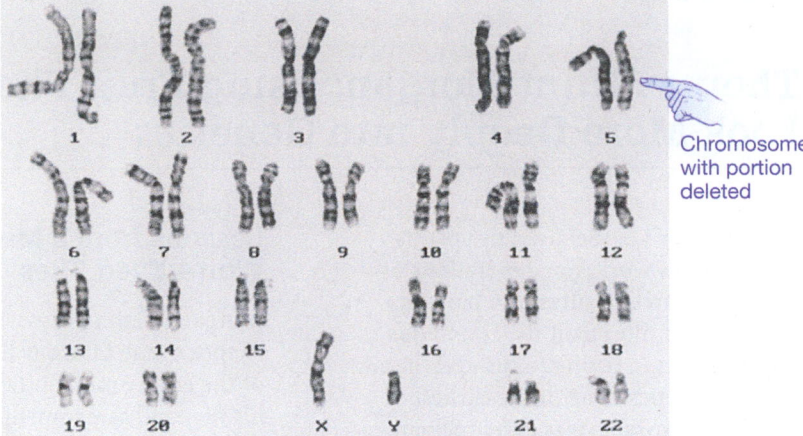

Chromosome with portion deleted

Figure 12.10
Chromosomal Deletion
(a) A 5-year-old boy affected by cri-du-chat syndrome, which results from a chromosomal deletion. Note the small head and low-set ears.
(b) Karyotype of a person with cri-du-chat syndrome. Note, in chromosome 5, that the chromosome on the right is shorter than the chromosome on the left because of loss of a chromosomal section.

would likely not survive. You can see how critical even a portion of a chromosome is by looking at children who suffer from a rare condition called *cri-du-chat* ("cry of the cat") syndrome, which results from the deletion of the far end of the "short arm" of chromosome 5. If you look at **Figure 12.10**, you can see a chromosome set that shows you this deletion. Children born with the cri-du-chat condition exhibit a host of maladies, among them mental retardation and an improperly constructed larynx that early in life produces sounds akin to those of a cat.

Inversions and Translocations

When a chromosome fragment rejoins the chromosome it came from, it may do so with its orientation "flipped," so that the fragment's chemical sequence is out of order. This is an **inversion:** a chromosomal abnormality that comes about when a chromosomal fragment that rejoins a chromosome does so with an inverted orientation (**Figure 12.11**). A **translocation** is a chromosomal abnormality that occurs when two chromosomes that are not homologous exchange pieces, leaving both with improper gene sequences, which can have phenotypic effects.

Duplications

There is, of course, a perfectly normal process of chromosomal part swapping—crossing over. But consider what might happen if two homologous chromosomes exchanged *unequal* pieces of themselves in crossing over. One would lose genetic material, while the other would gain it. Because these are homologous (meaning paired) chromosomes, the fragment added to the latter chromosome would duplicate some of the material it already has. This is but one of the ways in which genetic duplication can take place. Such duplication is something of a double-edged sword. For individuals, it

Figure 12.11
Chromosomal Structural Changes

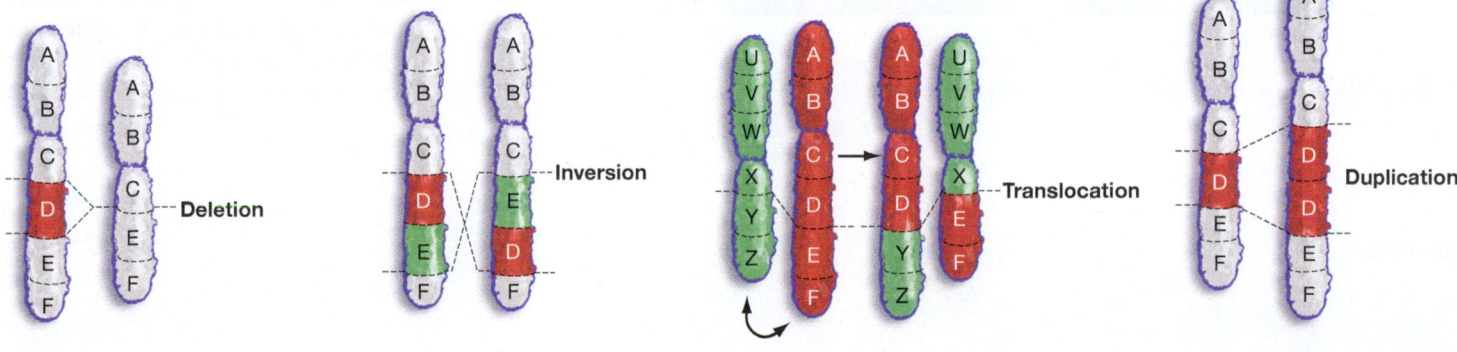

Deletion

Inversion

Translocation

Duplication

THE PROCESS OF SCIENCE

Thomas Hunt Morgan: Using Fruit Flies to Look More Deeply into Genetics

Gregor Mendel used pea plants in his work, but generations of researchers after him have employed a common fruit fly, *Drosophila melanogaster,* in trying to unravel the secrets of genetics. One of the earliest *Drosophila* workers was a Kentuckian named Thomas Hunt Morgan, who greatly deepened our understanding of heredity with work he began in 1908 in his lab at Columbia University.

The *Drosophila* fly has a lot going for it as an experimental subject: It is small, it has unusually large chromosomes, and it has what is known as a short generation time. Whereas Mendel got one or two generations of pea plants a year, Morgan got one generation of *Drosophila* every 12 days or so. In practical terms, this meant that Morgan's group had to wait at most a month to see the results of their experiments.

Lessons from a Mutation: White-Eyed Flies

Morgan's path to discovery began with a chance event. One day he looked into one of the empty milk bottles in which he kept his flies and saw something strange: Among numerous normal or "wild-type" *Drosophila* with red eyes was a single male

The importance of Morgan's insight was that he had linked a particular trait to a particular chromosome, which was a first.

with white eyes (**Figure 1**). Morgan correctly assumed that the variation he saw had come about because of a spontaneous genetic change or "mutation" in the fly. He then crossed his white-eyed male with females that he knew bred true for red eyes

and got all red-eyed flies in the F₁ generation. No surprise there; it simply looked as though red eyes were the same type of dominant trait that Mendel had seen in, for example, his yellow peas. Indeed, when Morgan went on to breed the F_1 generation—those that had mixed red and white alleles—he got a 3:1 red-to-white ratio in the F_2 generation. The strange thing was that every white-eyed fly was a *male.* Why should that be?

As it turns out, sex chromosomes exist in *Drosophila* as they do in human beings: Females have two X chromosomes, while males have one X and one Y. Doing further experiments on these flies, Morgan eventually decided that the gene for eye color had to lie on a *particular* chromosome: the X chromosome. Why? Any fly that was white-eyed had to have only white-eyed alleles (because white was recessive). Normally, of course, this would mean *two* white-eyed alleles, but there was an exception: Males could be white-eyed if they had but a single white allele—because they have only a *single X chromosome.* You can look at **Figure 2** to see how this plays out schematically. The importance of Morgan's insight was not that he had grasped something about eye color in the fly. It was that he had linked a particular trait to a particular chromosome, which was a first.

When Genes Separate: Recombination

Continuing with *Drosophila* breeding work, Morgan and his colleagues soon were finding all kinds of mutations in the flies. There was, for example, a mutation

(a) Eye of red-eyed *Drosophila*

(b) Eye of white-eyed *Drosophila*

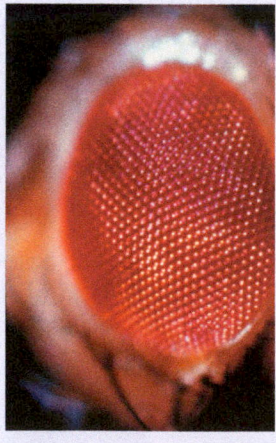 

Figure 1
The Mutation Morgan Saw

for "miniature" wings, as opposed to full ones. This trait followed the rules described earlier for the white-eyed mutation and accordingly was deemed to be an X-linked characteristic. Not surprisingly, the white-eyed and miniature-winged mutations tended to be transmitted *together* when both of these mutations were bred for because the genes for both of them lay on the same chromosome. The question was, though, did genes on the same chromosome *always* travel together? Surprisingly, the answer was no. For example, miniature wings were usually passed on together with white eyes, but not always. Sometimes, one trait would appear in an offspring, but the other would not. Since genes for these traits were on one chromosome, how did they get separated from one another?

A paper published in Europe in 1909 gave Morgan a clue. Viewing cells under a microscope, F. A. Janssens had observed chromosomes entwining with each other during an early phase of meiosis. Given this and his own data, Morgan made a creative leap. He suggested that chromosomes can swap parts with each other during their meiotic intertwining. Sound at all familiar? Remember, in the Chapter 10 review of meiosis, the discussion of the phenomenon known as crossing over? Here, with Morgan, is the insight that led to this now-established tenet of science. Appropriately enough, insights such as these won Morgan a Nobel Prize in 1934.

Figure 2
Why All Morgan's White-Eyed F₂'s Were Males

The F₂ female flies that inherited one white-eyed X-chromosome allele also inherited a second red-eyed X allele, making them red-eyed. However, the males inheriting the white-eyed allele had no second X chromosome. Instead, they inherited a Y chromosome—which has no genes on it for eye color—leaving them white-eyed.

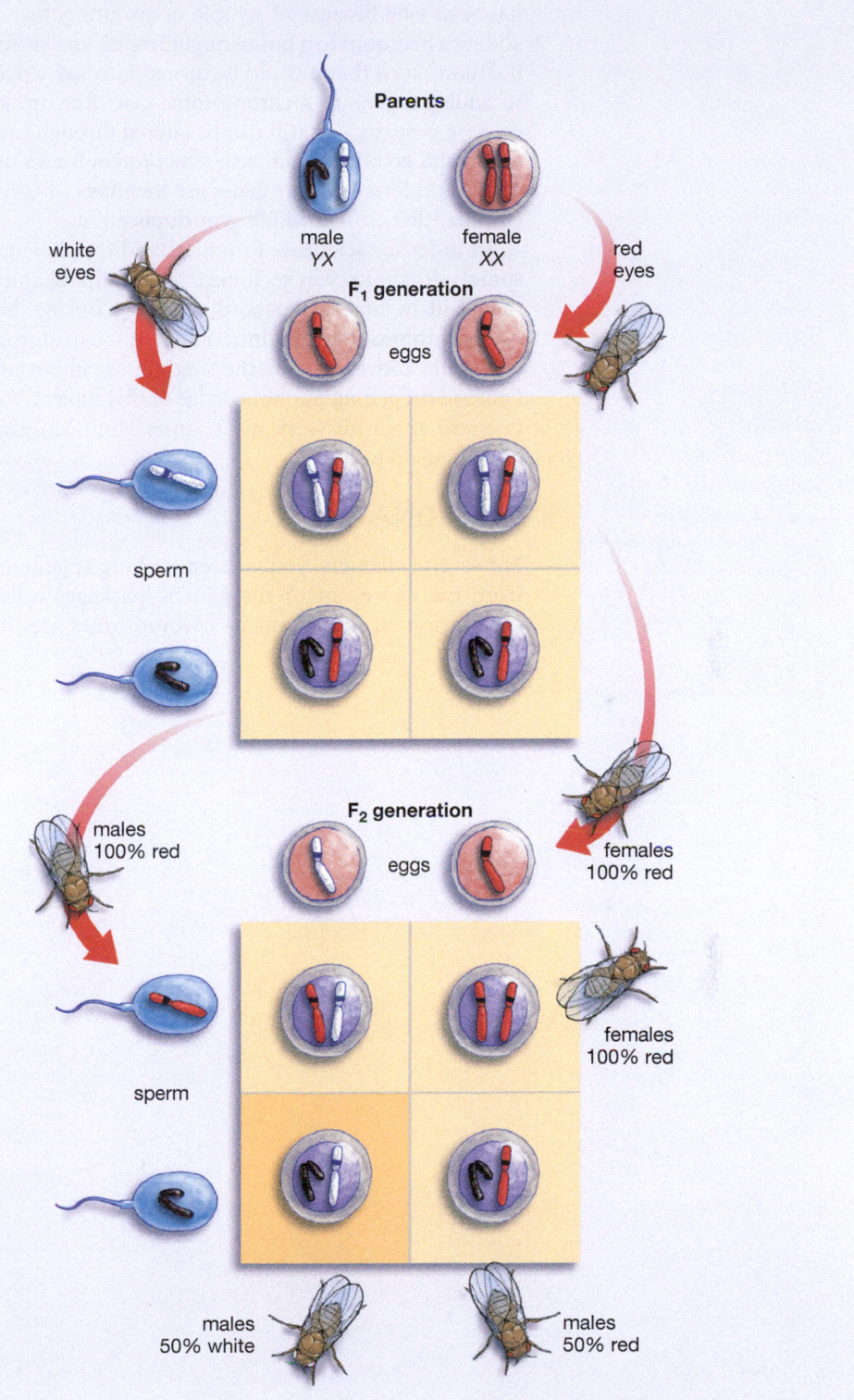

Parents

white eyes   male YX   female XX   red eyes

F₁ generation

eggs

sperm

males 100% red   F₂ generation   eggs   females 100% red

sperm   females 100% red

males 50% white   males 50% red

may have no effect at all, although large duplications can have negative effects, such as mental retardation. For the living world as a whole, however, duplication has been vital because of its role in evolution. Just as adding a bedroom to a house might free up an existing bedroom, such that it could be turned into, say, a den, so adding a gene to a chromosome can "free up" an existing gene, such that it can be altered through mutation and go on to produce a new protein for an organism. At least half our genes are members of "gene families" that arose through gene duplication.

In order for scientists to understand the chromosomal afflictions you've looked at in this chapter, they had to first understand the basic linkage between chromosomes and inheritance. A scientist from Columbia University was the single most important figure in deepening our knowledge of this subject. You can read about his work in "Thomas Hunt Morgan" beginning on page 226.

On to DNA

For several chapters, you've been looking at genetics from the viewpoint of the genetic packages called chromosomes. Important as chromosomes are, it's probably been apparent that the real genetic control lies with the units that help make them up: the DNA sequences we call genes. As you've seen, it is genes that make pea-plant flowers purple or white—and that give a person Huntington disease or not. It's time to explore just what these genes are and how they operate.

SO FAR ...

1. Which of the following can be a result of aneuploidy: (a) miscarriage, (b) Down syndrome, (c) Turner syndrome?

2. The aneuploidy that exists in cancerous cells develops during the process of _____, rather than _____, and thus is not seen in every cell in the body, but only in a _____ of cells.

3. Chromosomal aberrations include deletions, in which a person is missing a _____ of a _____.

12 REVIEW

Summary

12.1 X-Linked Inheritance in Humans

- X-linked genetic conditions stem from dysfunctional alleles located on the X chromosome. Men are more likely to suffer from these conditions because men have only a single X chromosome. A woman with a dysfunctional allele on one of her X chromosomes usually will be protected by a second functional allele that lies on her second X chromosome. **(p. 213)**

- Recessive genetic conditions are those that will not exist in the presence of a single functional allele. Persons who do not suffer from recessive genetic conditions may still possess an allele for them, which they can pass on to their offspring. Such carriers are heterozygous for the condition—the alleles they have for the trait differ, one being functional, the other being dysfunctional. **(p. 214)**

12.2 Autosomal Genetic Disorders

- Autosomal genetic disorders involve neither the X nor the Y chromosome. Any person who suffers from a recessive autosomal disorder must be homozygous for the alleles that cause the disorder. **(p. 215)**

- Dominant disorders are those in which a single allele can bring about the condition, regardless of whether a person also has a normal allele. **(p. 216)**

12.3 Tracking Traits with Pedigrees

- In tracking inherited diseases, scientists often construct medical pedigrees, which are genetic familial histories that normally take the form of diagrams. **(p. 217)**

12.4 Aberrations in Chromosomal Sets: Polyploidy

- Many species have diploid or paired sets of chromosomes. In human beings, this means 46 chromosomes in all: 22 pairs of autosomes and either an XX chromosome pair (for females) or an XY pair (for males). The state of having more than two sets of chromosomes is called polyploidy. Many plants are polyploid, but the condition is fatal for human beings. **(p. 218)**

12.5 Incorrect Chromosome Number: Aneuploidy

- In aneuploidy, an organism has either more or fewer chromosomes than normally exist in its species' full set. Aneuploidy is responsible for a large proportion of the miscarriages that occur in human pregnancies. A small proportion of embryos survive aneuploidy, but the children who result from these embryos are born with such conditions as Down syndrome. **(p. 218)**

- The cause of meiotic aneuploidy usually is nondisjunction, in which homologous chromosomes or sister chromatids fail to separate correctly, leading to eggs or sperm that have one too many or one too few chromosomes. **(p. 220)**

- Aneuploidy can come about in regular cell division (mitosis) as well as in meiosis. A number of cancer researchers believe that mitotic aneuploidy can be a cause of cancer rather than an effect of it. **(p. 224)**

12.6 Structural Aberrations in Chromosomes

- Harmful aberrations can occur within chromosomes, with many coming about because of mistakes in chromosomal interactions. Chromosomal aberrations include deletions, inversions, translocations, and duplications. **(p. 224)**

Key Terms

Understanding the Basics

Multiple-Choice Questions

(Answers are in the back of the book.)

1. Examples of sex-linked genetic diseases include:
   a. sickle-cell anemia, red-green color blindness, and polyploidy.
   b. Duchenne muscular dystrophy, Down syndrome, and red-green color blindness.
   c. hemophilia, red-green color blindness, and Duchenne muscular dystrophy.
   d. Huntington disease, hemophilia, and Klinefelter syndrome.
   e. Down syndrome, cri-du-chat, and hemophilia.

2. Healthy individuals least likely to suffer from malaria are those who:
   a. are heterozygous for hemoglobin (one hemoglobin A, one hemoglobin S).
   b. are homozygous for hemoglobin S alleles.

c. are hemophiliacs.

d. are homozygous for hemoglobin A alleles.

e. both a and c.

3. Inheritance of an autosomal dominant disorder differs from inheritance of an autosomal recessive disorder in that:

a. a dominant disorder may be passed on only if both parents are affected.

b. a dominant disorder is evident only if the offspring is homozygous for the allele.

c. a dominant disorder is more often seen in females.

d. a dominant disorder may be passed on even if only one parent is affected.

e. both b and d.

4. An individual having 44 autosomes and one X chromosome would be classified as:

a. polyploid.

b. aneuploid.

c. having Klinefelter syndrome.

d. having Turner syndrome.

e. both b and d.

5. A human embryo with 69 chromosomes would:

a. die in the womb or shortly after birth.

b. be considered polyploid.

c. have fewer problems than a plant with the same condition.

d. both a and b.

e. all of the above

6. A carrier is a person who:

a. suffers from a recessive disorder.

b. is heterozygous for a recessive disorder.

c. is homozygous for a recessive disorder.

d. does not suffer from a recessive disorder but possesses an allele for it.

e. b and d.

7. Which of the following statements is *not* true regarding Down syndrome?

a. Affected individuals usually have three copies of chromosome 21.

b. The condition is usually caused by nondisjunction in egg formation.

c. Most Down syndrome children are born to women 35 years of age or older.

d. Approximately 0.1 percent of all live births are children with Down syndrome.

e. The physical characteristics associated with the disorder include mental retardation, male infertility, short stature, and reduced life span.

8. Which of the following diseases usually is not evident until well into adulthood?

a. Huntington disease

b. Down syndrome

c. cri-du-chat

d. Turner syndrome

e. none of the above

Brief Review

(Answers are in the back of the book.)

1. In sex-linked diseases, which gender is more frequently affected—male or female? Why?

2. How can the serious genetic defect known as aneuploidy be both widespread and relatively unrecognized?

3. The text notes the possibility that some forms of human cancer may result from aneuploidy. What is the fundamental difference between an aneuploidy that might lead to cancer and one that results in Down syndrome?

4. Compare and contrast the four chromosome structural aberrations discussed in this chapter: deletions, inversions, translocations, and duplications.

5. Given the following situations, answer accordingly:

a. A human cell with 47 chromosomes has probably undergone _____.

b. Two sperm manage to fuse with a single egg, producing an embryo. The result would be _____.

6. Why would a person with Klinefelter syndrome exhibit both male and female features?

7. Certain genotypes may be advantageous in specific situations. Give an example of a heterozygous genotype that is advantageous over either homozygous genotype. Why is this the case?

Applying Your Knowledge

1. In the United States, 8 percent of African Americans are carriers for the sickle-cell allele, while in central Africa the figure is 20 percent. What could account for this difference?

2. The text notes that the technique known as preimplantation genetic diagnosis (PGD) allows prospective parents to screen embryos not only for serious genetic defects such as Down syndrome but for traits that may be regarded as more or less desirable, such as male or female gender. What limits, if any, should be placed on parents' ability to choose from among embryos for traits such as this?

3. Thinking back to what you learned in Chapter 10 about how sperm are produced, why is it that the cells that give rise to sperm undergo nondisjunction less frequently than the cells that give rise to eggs?

Genetics Problems

(Answers are in the back of the book.)

1. A man who is a carrier for sickle-cell anemia, a recessive genetic disease, marries a woman who is not a carrier. What proportion of their children is expected to be afflicted with sickle-cell anemia?

2. Hemophilia is an X-linked recessive disease that prevents blood clotting. If a woman who is a carrier for hemophilia marries a man who is not a carrier, what proportion of their sons are expected to have hemophilia? What is the chance that their first child will have hemophilia? (Assume that the sex of this child is unknown until birth.)

Consider the following pedigree for a human autosomal trait. (Note that in pedigrees, generations are indicated by Roman numerals and individuals within a generation are numbered from left to right with Arabic numerals.)

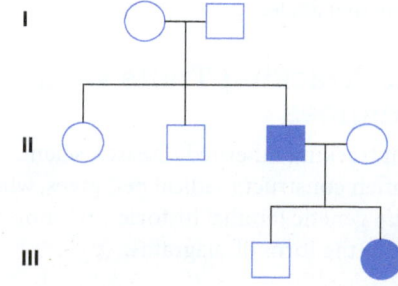

3. Is the allele that determines this trait dominant or recessive? Explain your reasoning.

4. Using symbols *A* and *a* for the dominant and recessive alleles, respectively, what is the genotype of the affected male in the second generation (individual II-3)?

5. What is the genotype of this individual's mate?

6. If female III-2 marries a heterozygous man, what proportion of their children is expected to have this trait?

7. Consider the following pedigree, which assumes a fantasy gene for ear shape. In answering the questions below about it, use the symbol *P* for the regular allele or *p* for a pointy-shaped allele if this is an autosomal trait; or X^+ for the regular allele or X^p for the pointy allele if this is an X-linked trait.

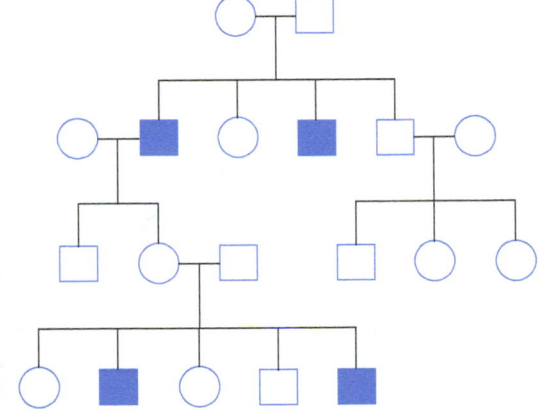

This pedigree indicates the trait is most likely inherited as an:
a. autosomal dominant.
b. autosomal recessive.
c. X-linked dominant.
d. X-linked recessive.

8. What is the genotype of individual I-1?

9. If a woman who is a carrier for this trait married individual III-1, what proportion of their children is expected to show this trait? What proportion of their sons is expected to show this trait?

10. Construct a pedigree to answer the following question: Red-green color blindness is an X-linked recessive trait. A woman who has a color-blind mother and a father with normal color vision marries a man with normal vision. This couple has a son. What is the chance that the son is color-blind?

11. Neurofibromatosis is a genetic disorder associated with an allele of one autosomal gene. Individuals with neurofibromatosis have uneven skin pigmentation and skin tumors. A man with neurofibromatosis marries a woman who does not carry the allele for this disorder. The couple has five children, three of whom have neurofibromatosis. The most likely explanation for this outcome is that neurofibromatosis is a _____ trait and that the man is _____.
a. recessive; homozygous recessive
b. recessive; heterozygous
c. dominant; homozygous dominant
d. dominant; heterozygous
e. a and b are both possible explanations.

12. Hyperphosphatemia, a disease that causes a form of rickets (abnormal bone growth and development), is inherited as an X-linked dominant trait. A woman who is heterozygous for the disease allele marries a man who has a normal allele. What proportion of their sons will have hyperphosphatemia? What proportion of their daughters will have hyperphosphatemia? If hyperphosphatemia were an X-linked recessive trait, how would your answer differ?

Passing on Life's Information:

DNA Structure and Replication

DNA serves as a storehouse of information for living things. Scientists could not understand how DNA's information could be passed on from one generation to the next until they understood how all its component parts fit together.

Available on Mastering Biology®

Activities: DNA; One-Gene-One-Enzyme Hypothesis; DNA and RNA Structure; DNA Double Helix; DNA Replication, An Overview; Mutations; **Bioflix:** DNA Replication; **Current Events:** Bits of Mystery DNA, Far From "Junk", Play Crucial Role; New Drug Trial Seeks to Stop Alzheimer's Before it Starts; The Cancer Sleeper Cell; **Building Vocabulary; Questions:** Bioflix Quiz; Reading Questions; Test Bank

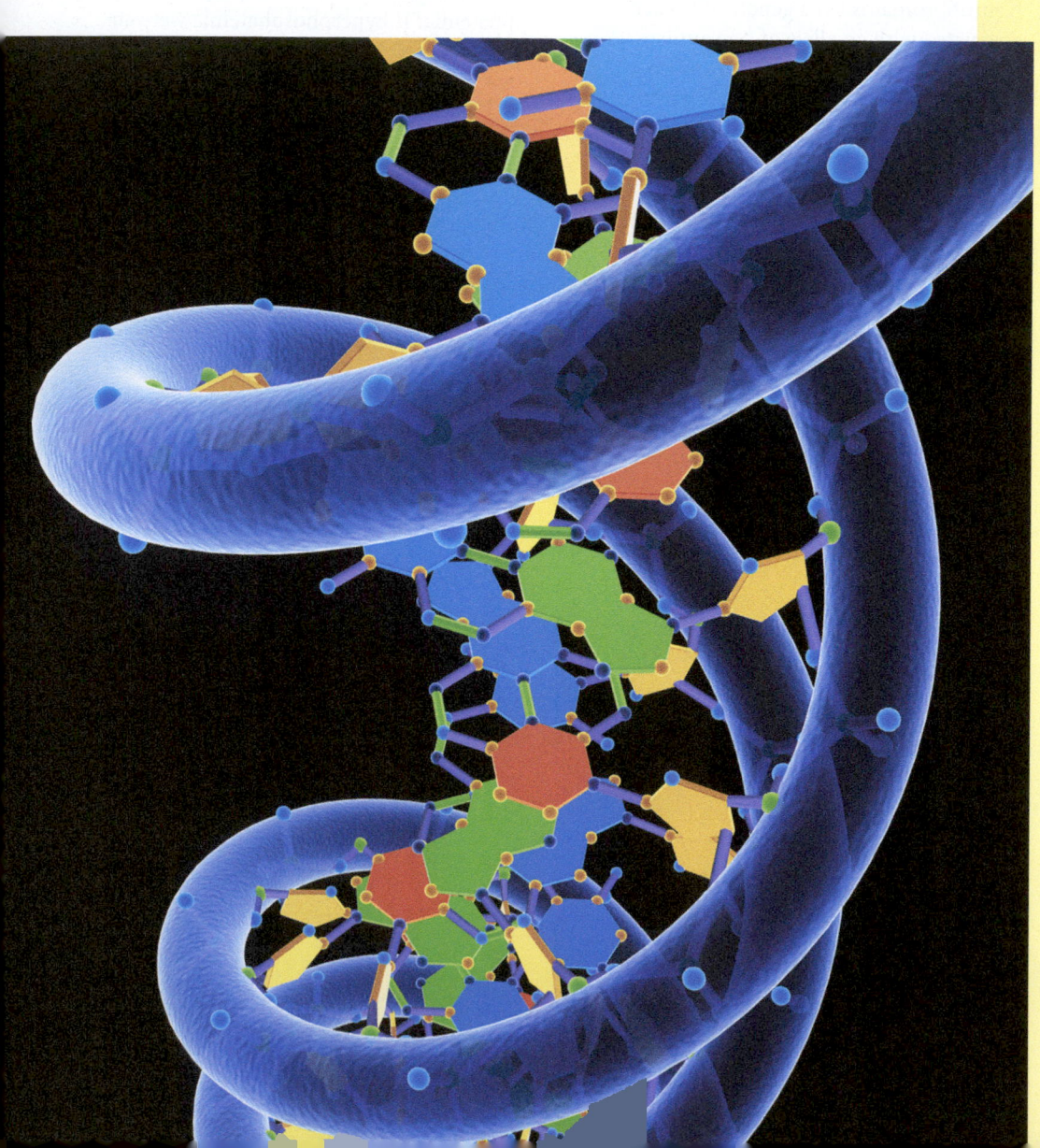

Life is able to build on itself thanks to the structure of DNA, shown here in a computer-generated model.

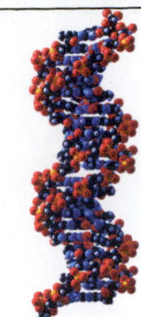

"In research the front line is almost always in a fog," wrote Francis Crick in his 1988 memoir, *What Mad Pursuit.* Crick had first-hand knowledge of this. In 1953, with James Watson, he discovered the structure of DNA, an effort that brought him to the front line of research, which is where he remained to the end of his long, productive life. Crick's

book makes clear what a tremendous difference there is between learning something about nature and discovering something about it. In learning, dozens of voices stand ready to instruct; there are books, lectures, videos, CDs—a world of well-organized information. The person who seeks to discover something about nature, meanwhile, is confronted with silence. Nature goes on: Cells divide, chromosomes condense, birds migrate. But *why* do these things operate the way they do? All the researcher has as a guide are the things themselves, working away. The trick is to devise some test (called an experiment) that can make nature's routine operations yield information. In this process of teasing out truth, knowledge is gained in very small increments. Scientists are like the first people down a darkened maze of tunnels; they must make their way exceedingly slowly, feeling each square inch of wall space as they go, after which they can only leave lights on behind them (in the form of their scientific papers, books, and so forth).

13.1 The Form and Function of Genes

To get a feel for the process of discovery, consider the state of genetics early in the twentieth century. By 1920, it had been demonstrated beyond any doubt that genetic information resided on chromosomes. Within a decade or so, by looking at some abnormally large fly chromosomes, scientists could observe in great detail such chromosomal processes as crossing over. By viewing the banding patterns on these chromosomes, they could even identify the rough location of some genes.

Yet, what was a gene? What was the physical nature of this unit of heredity, and how did it work? No one knew. Because genes are much smaller than the chromosomes

on which they reside, there was no hope of simply viewing one under a microscope. In observing chromosomes, scientists were only "looking" at genes in the way that any of us is "looking" at lunar rocks by glancing up at the moon. Through the 1920s and 1930s, then, genes could be described only in vague, functional terms, as in: A gene is an entity that lies along a chromosome and brings about a phenotypic trait in an organism (for example, the green or yellow peas of Mendel's experiments).

Things were to clear up somewhat in the ensuing years. The central achievement in genetics in the late 1930s and early 1940s was a more concrete description of what genes do: They bring about the production of proteins. The central achievement in genetics from the mid-1940s to the early 1950s was strong evidence indicating what genes are composed of: deoxyribonucleic acid, or DNA for short.

DNA Structure and the Rise of Molecular Biology

By the early 1950s, a key question became *how* DNA carried out its genetic function. To find out, scientists had to piece together its exact chemical structure. This investigation turned out to be a watershed event in biology. Just as opening a mechanical watch and observing how its parts fit together would allow a person to understand how the watch works, so deciphering the structure of DNA allowed biologists to understand how genetics works at its most fundamental level. How could genetic information be passed on from one generation to the next? How could DNA be a versatile enough molecule to specify the makeup of thousands of proteins? The structure of the DNA molecule suggested an answer to these questions, as you'll see.

Apart from what this investigation uncovered, the inquiry itself stood as a symbol of a new era in biology.

DNA

When Gregor Mendel did his experiments with peas, he was looking at whole organisms. When T. H. Morgan was doing his work with *Drosophila* flies, he was interested in whole chromosomes, which he knew to be composed of several types of molecules. In the early 1950s, however, the search turned to a single molecule. How was DNA structured, and how did this structure allow it to carry out its various functions? Biological research of this sort has grown ever more important in the decades since the 1950s. It is today known as **molecular biology:** the investigation of life at the level of its individual molecules.

13.2 Watson and Crick: The Double Helix

James Watson and Francis Crick may not be instantly recognized scientific names in the way that, say, Albert Einstein or Louis Pasteur are, but there's a certain public awareness that these two researchers did something important in connection with DNA (**Figure 13.1**). What they did was present to the world, in 1953, the structure of DNA. Atom by atom, bond by bond, this is how DNA fits together, they said in unveiling DNA's now-famous configuration, the double helix.

The two were a seemingly unlikely pair to make an epochal scientific discovery. Watson, an American,

was a 23-year-old who was scarcely more than a year out of graduate school, and Crick, an Englishman, was a 35-year-old just then working on his doctorate when the two met in the fall of 1951 at Cambridge University in England. Ending up together by coincidence at the same laboratory, they realized in short order their mutual interest in the structure of DNA, and to the neglect of projects they were supposed to be working on, they began several rounds of model building and brainstorming that resulted in their breakthrough.

Watson and Crick were greatly aided in their investigation by the work of others. Although the DNA molecule was too small to be seen by even the most powerful microscopes of the time, something about its structure could be inferred from a technique called X-ray diffraction. In this process, a purified form of a molecule is bombarded with X rays. The way these rays scatter on impact then reveals something about the structure of the molecule.

If you look at **Figure 13.2**, you can see the results of some X-ray diffraction. As you can imagine, it takes a highly trained observer to be able to deduce anything about the structure of a molecule from such an image. Fortunately for Watson and Crick, such a person was working just up the road from them. She was Rosalind Franklin (**Figure 13.3**), a researcher at King's College in London and one of the handful of individuals then

Figure 13.1
Young and Famous
James Watson, on the left, and Francis Crick, with a model of the DNA double helix, shortly after they published their paper on the molecule's structure.

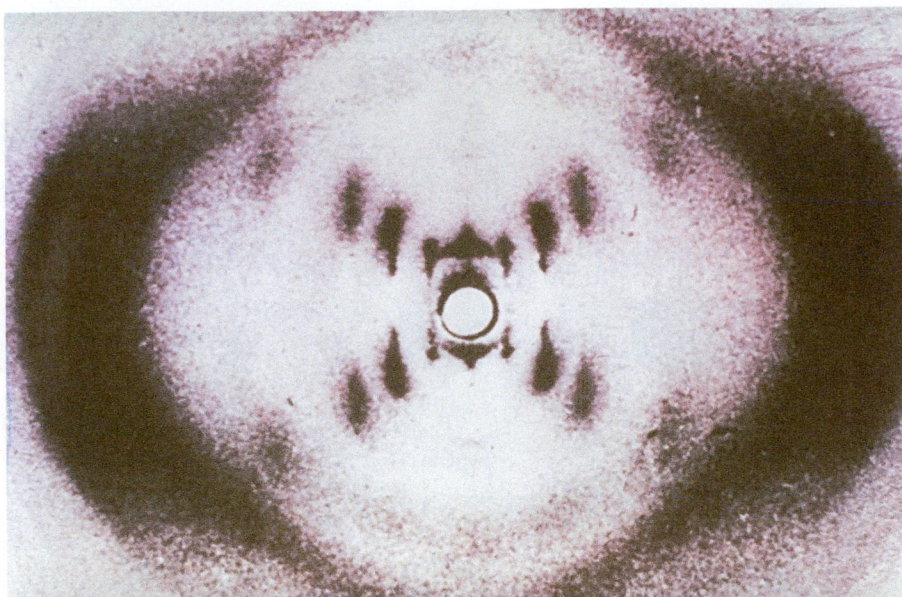

Figure 13.2
Imaging DNA
One of Franklin's X-ray diffraction images of DNA. The "cross" formed of dark spots indicated the molecule had a helical structure.

skilled in performing X-ray diffraction on DNA. She and her colleague, Maurice Wilkins (**Figure 13.4**), were themselves working on the structure of DNA at the time, as were other researchers in America. Thus did Watson, at least, regard the search for DNA structure to be a race between several teams, a fact that concentrated his efforts wonderfully. In 1962, Watson, Crick, and Wilkins were awarded the Nobel Prize in Medicine or Physiology for their work on DNA. Rosalind Franklin died of cancer in 1958 at the age of 37. Nobel Prizes are not awarded posthumously; it's unknown what would have happened had she lived.

Figure 13.3
DNA Investigator
Rosalind Franklin, whose work in X-ray diffraction was important in revealing the structure of the DNA molecule.

Figure 13.4
Nobel Laureate
Maurice Wilkins, who with Watson and Crick was awarded the Nobel Prize for research that revealed DNA's structure.

Figure 13.5
The Structure of DNA

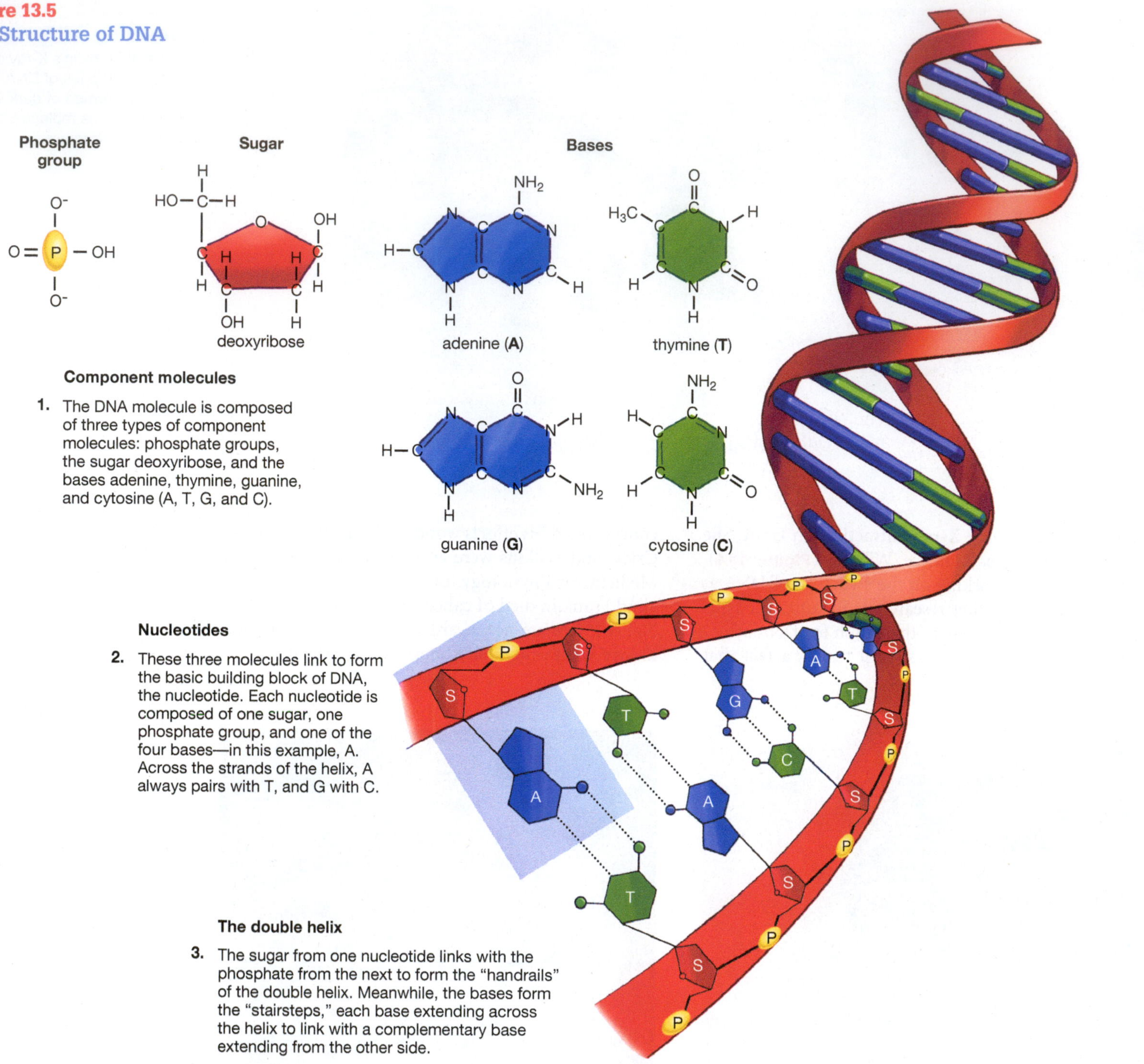

Phosphate group

Sugar

Bases

adenine (**A**)

thymine (**T**)

deoxyribose

guanine (**G**)

cytosine (**C**)

Component molecules

1. The DNA molecule is composed of three types of component molecules: phosphate groups, the sugar deoxyribose, and the bases adenine, thymine, guanine, and cytosine (A, T, G, and C).

Nucleotides

2. These three molecules link to form the basic building block of DNA, the nucleotide. Each nucleotide is composed of one sugar, one phosphate group, and one of the four bases—in this example, A. Across the strands of the helix, A always pairs with T, and G with C.

The double helix

3. The sugar from one nucleotide links with the phosphate from the next to form the "handrails" of the double helix. Meanwhile, the bases form the "stairsteps," each base extending across the helix to link with a complementary base extending from the other side.

DNA and RNA Structure

DNA Double Helix

13.3 The Components of DNA and Their Arrangement

A good way to understand the structure of DNA is to look at the building blocks that make it up. First, there is a phosphate group and, second, a sugar called *deoxyribose*, both of which can be seen in (**Figure 13.5**). Third, there are four possible DNA "bases": adenine, guanine, thymine, and cytosine—A, G, T, and C,

for short. When you link together a phosphate group, a deoxyribose molecule, and one of the four bases, you get the basic building block of DNA, the *nucleotide*, one of which is highlighted in the larger DNA molecule in the figure.

When you look at this larger figure as a whole, you can see that the double helix looks something like a spiral staircase. By focusing on its exterior "handrails," you can see how sugar and phosphate fit together to form a kind of chain whose links are ordered as: sugar-phosphate-sugar-phosphate (symbolized in the figure by S and P). The "steps" lying between these handrails are composed of DNA's bases. As the figure shows, each of the bases amounts to a *half*-stairstep, if you will. Each extends inward from one handrail of the double helix and is then joined to a base extending inward from the other handrail of the DNA molecule. (The bases are linked via hydrogen bonds, symbolized by the dotted lines in the figure.)

The figure shows some examples of one base being paired with another. At the bottom, for example, an A base on the left is paired with a T base on the right. Go two sets of nucleotides up, and a G on the left is being paired with a C on the right. This turns out to be one of the fundamental rules about DNA structure. If you viewed a billion DNA *base pairs*, as they're called, you would find the same thing: A always pairing with T, and G always pairing with C, across the helix. (Any two bases that can pair together in this way are said to be *complementary*; thus, A is complementary to T, and G is complementary to C. Likewise, you can have complementary chains of DNA.)

The Structure of DNA Gives Away the Secret of Replication

It was this structure of DNA that suggested the answer to one of the great questions of genetics: How is genetic information passed on? To put this another way, how does a cell make a copy of its own DNA? As you've seen, a full complement of our DNA is contained in nearly every cell in our body. Yet cells divide, and each pair of daughter cells contains exactly the same complement of DNA as did the parent cell they came from. This means that genetic information is passed on by means of DNA being copied, with one copy of this molecule ending up in one daughter cell and a second copy in the other. Indeed, such copying extends to the formation of egg and sperm cells, meaning this is the way genetic information is passed on from one *generation* to the next. But how could this work?

The structure Watson and Crick discovered suggested a way. We've observed that A must always pair with T, and G with C, across the two strands of the double helix. This rule, Watson and Crick saw, meant that each single strand of DNA could serve as a *template* for the synthesis of a new single strand (**Figure 13.6**). Each A on an old strand would specify the place for a T on the new, each G on the old a place for a C on the new, and so forth. All that was required was for the two old strands to separate—splitting the stairsteps right down the middle—and for new strands to be put together that were complementary to the old.

DNA Structure and Protein Production

The second thing DNA structure suggested was a partial answer to another great question of genetics: How

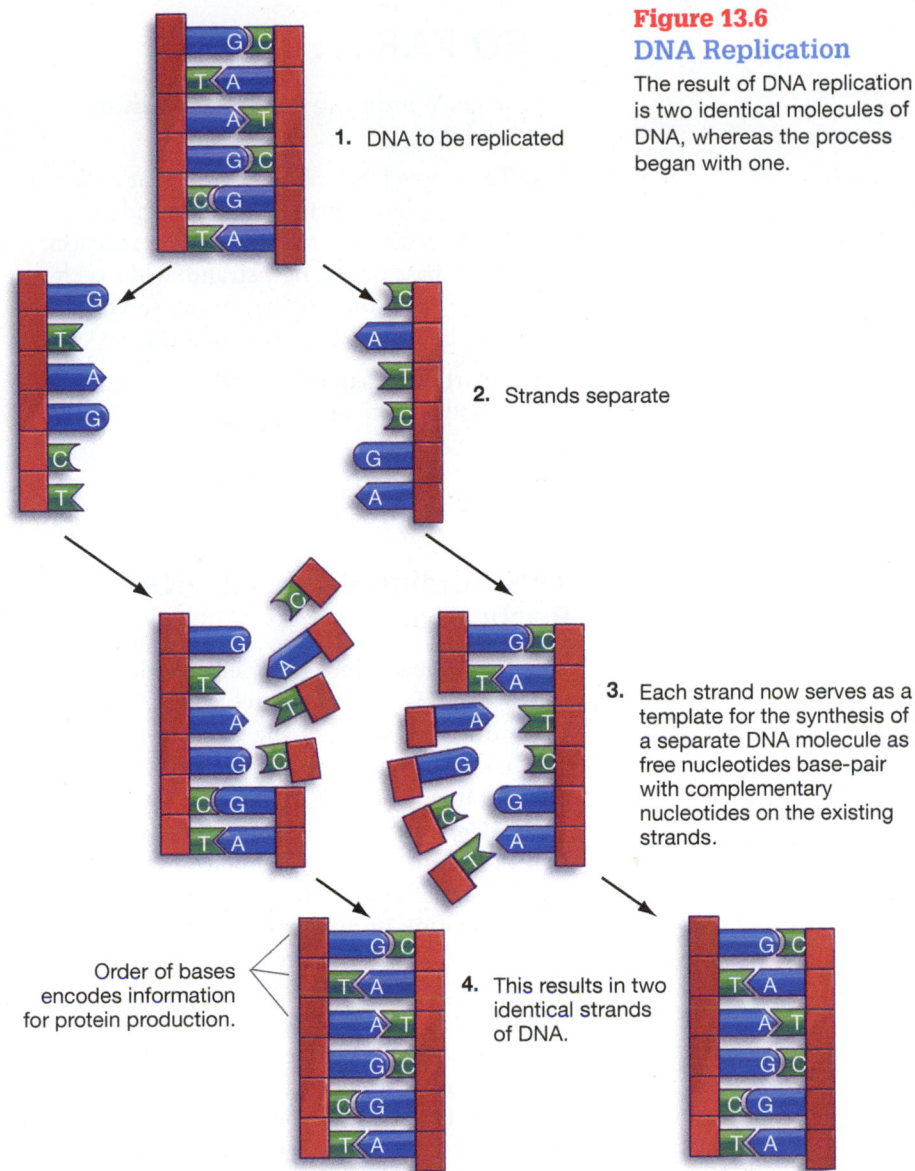

Figure 13.6
DNA Replication
The result of DNA replication is two identical molecules of DNA, whereas the process began with one.

1. DNA to be replicated

2. Strands separate

3. Each strand now serves as a template for the synthesis of a separate DNA molecule as free nucleotides base-pair with complementary nucleotides on the existing strands.

Order of bases encodes information for protein production.

4. This results in two identical strands of DNA.

could DNA be versatile enough to specify the dazzling array of proteins that all living things produce? The handrails of DNA's double helix are monotonous—phosphate, sugar, phosphate, sugar. But the bases can be laid out along these handrails in an extremely varied manner. A has to pair with T and G with C *across* the helix; but *along* the handrails, the bases can come in any order. Look at the top drawing in Figure 13.6 to see the order this hypothetical group of bases comes in; starting from the bottom right, A-G-C-T-A-C. Strung together by the thousands, such bases can specify a particular protein, just as in Morse code a series of short and long clicks can specify a particular word. And thanks to the variability of DNA's bases along the handrails, an enormous number of proteins can be specified. Next chapter, we'll look at the process by which cells start with the information contained in DNA and end up with completed proteins.

DNA Replication, An Overview

SO FAR...

1. James Watson and Francis Crick discovered the _____.

2. The key to DNA replication is that each _____ base on an original strand of DNA serves as a template for the addition of a T base on the new strand, while each _____ on the original strand serves as a template for the addition of a G base.

3. A particular sequence of DNA _____ can specify a particular _____.

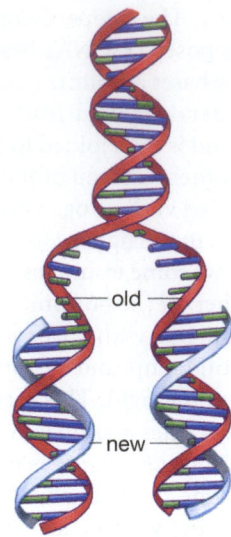

Figure 13.7
How Life Builds on Itself
Each newly synthesized DNA molecule is a combination of the old and the new. An existing DNA molecule unwinds, and each of the resulting single strands (the old, in red) serves as a template for a complementary strand that will be formed through base pairing (the new, in blue).

The Building Blocks of DNA Replication

Bioflix: DNA Replication

You have already looked briefly at the process by which DNA is copied or "replicated," but we now need to review some of the details of this process. The basic steps here are straightforward. The essential building block of DNA is the unit called the **nucleotide,** which can be seen in the highlighted portion of Figure 13.5. Every nucleotide contains one sugar and one phosphate group. What differentiates one nucleotide from another is the base that is attached to it; in this case the base is A, but as you can see, others have T, C, or G.

Figure 13.6 shows the process of DNA replication in overview. Note that the first thing that happens is that the joined strands of the double helix unwind, separating from one another. The nucleotides on each of the single strands are then paired with free-floating nucleotides that line up in new, complementary strands. Because *both* strands of the original double helix are being paired with new strands, the end product is two double helices, where before there was one. This is important because each of these strands is destined to serve in separate cells—the daughter cells that are produced from a single parent cell during the process of cell division.

You can see in Figure 13.6 that the addition of nucleotides moves in differing directions on the two strands. This is so because of something you can see in Figure 13.5: The two strands of the double helix have opposite orientations. Note that the sugars of the right-hand strand, for example, are upside down relative to the sugars of the left-hand strand. (If the strands were people, we would think of them lying head to *foot*, relative to each other.) Because new nucleotides can be added to only one end of a DNA strand, the nucleotides are added in different directions.

Something Old, Something New

It's worth drawing a little finer point on one aspect of this process: Each resulting double helix is a combination of the old and the new. Each has one "parental" strand of DNA and one newly synthesized complementary strand (**Figure 13.7**). This combination is conceptually important because this is how life builds on itself. We think of life as a process in which the old gives rise to the new, but here we can see how this is accomplished. The old *specifies* the new, even as it is joining with it, in a newly synthesized strand of DNA. If we ask why life has such remarkable continuity—why it is that one generation of living thing tends gives rise to another that is so much like it—we find that this process is largely responsible. Through it, what is old will be passed on largely intact to what is new.

As noted, each instance of DNA replication comes in the service of producing something new: the two daughter cells that arise from a single parent cell during the process of cell division. As such, each pair of newly formed double helices is destined to part company. As part of the mitosis we looked at in Chapter 9, each of the new strands of DNA is moved to opposite sides of a dividing parent cell. Once this cell splits into two, the separation of the helices is complete; each strand now resides in a separate daughter cell and each now serves as an independently functioning segment of DNA. At the chromosomal level, each newly replicated helix is, when joined by appropriate proteins, one of the sister chromatids you've seen so much of in the last few chapters.

Enzymes Facilitate Replication

The DNA replication that sets all this in motion may sound simple in overview, but in fact its details are enormously complicated. As you might imagine, such a process could not proceed without enzymes to catalyze it. To name just two groups of them, there are enzymes called helicases that unwind the double helix, separating its two strands to make the bases on them available for base pairing. There is also another group of enzymes, collectively known as **DNA polymerases,** that move along each strand of the double helix, joining together nucleotides as they are added—one by one—to form the new, complementary strands of DNA.

Editing Out Mistakes

The base pairing that takes place in replication goes on millions of times in a given stretch of DNA: Free-standing A's are aligned with complementary T's, while C's are aligned with G's. When the entire human genome is considered, 3.2 *billion* base pairs must be aligned in this way, not just once, but twice. (The human genome is 3.2 billion base pairs long, but each human cell contains two copies of each chromosome.) The amazing thing about this copying is how few mismatches are produced in it by the time it's completed. The error rate in human replication—the rate at which the *wrong* bases have been brought together—might be only one in every billion bases by the end of replication. Yet *during* replication, such a mistake might be made once in every 100,000 bases. Obviously, to start with a given number of mistakes but to end up with far fewer, the cell's genetic machinery has to be capable of correcting its errors.

This happens partly through the services of the versatile DNA polymerases, which are able to perform a kind of DNA editing: They remove a mismatched nucleotide and replace it with a proper one. Interestingly enough, this happens through a kind of backspacing. Normally, the DNA polymerases are moving along a DNA chain, linking together recently arrived nucleotides. When an error is detected, however, they stop, move "backward," remove the incorrect nucleotide, put in the correct one, and then move forward again.

13.4 Mutations

With this consideration of alterations in DNA structure, you have arrived at an important concept: that of a **mutation,** which can be defined as a permanent alteration of a DNA base sequence. Such alterations come about because the cell's various DNA error-correcting mechanisms are not foolproof; they do not correct all the mistakes that occur. For the balance of this chapter, we'll focus on these genetic mistakes, which have a powerful effect both on human health and on evolution.

DNA's makeup can be altered in many ways. A slight change in the chemical form of a base might, for example, cause a G to link up across the helix with a T (instead of with its normal partner, C), as you can see in **Figure 13.8**. Then, in a subsequent round of DNA replication, the cell might "repair" this error in such a way that a permanent mistake is introduced: An A-T pair now exists, whereas the original sequence had a G-C pair. Permanent mistakes like these are called **point mutations,** meaning a mutation of a single base pair in the genome. Such mutations stand in contrast to the kind of whole-chromosome aberrations pre-

One-Gene-One-Enzyme Hypothesis

Mutations

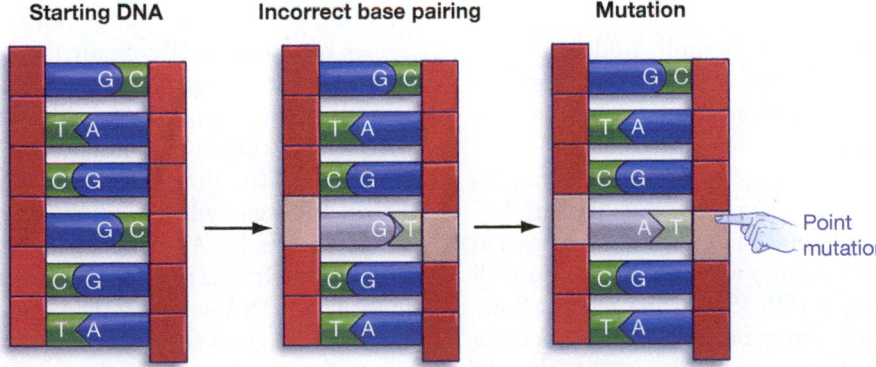

Figure 13.8
One Route to a Mutation

1. In replicating a cell's DNA, mistakes are sometimes made, such that one base can be paired with another base that is not complementary to it (G with T in this case).

2. The next time a cell replicates its DNA, the replication repair mechanism may "fix" this error in such a way that a permanent alteration in the DNA sequence results. The original G will be replaced, instead of the wrongly added T. The result is an A-T base pair, whereas the cell started with a G-C base pair.

sented in Chapter 12, although these also qualify as mutations under some definitions.

In what way are mutations "permanent"? Think of how DNA replication works. Before any cell can divide, it must first make a copy of its complement of DNA. Should this DNA contain an uncorrected mistake, it too will be copied—again and again, actually, with each succeeding cell division. Most mutations have no noticeable effect on an organism, and as you'll see, mutations are vital to the process of evolution. But the concept of mutation is quite rightly fearful to us because, in relatively rare instances, mutations can have disastrous effects.

Cancer and Huntington Mutations

A cancerous growth is a line of cells that has undergone a special kind of mutation—one that causes the affected cells to proliferate wildly. As an example, the skin cancer known as melanoma causes skin cells called melanocytes to start dividing very rapidly. For our purposes, the question is: How does this process get going? In 2002, British scientists looked at numerous lines of melanoma cells that had been taken from cancer patients. They found that 59 percent of these cells contained a mutation in a gene called *BRAF*. This gene, like all others, amounts to a sequence of DNA bases—a sequence of C's, T's, A's, and G's. When the British scientists looked at how the *BRAF* gene had actually mutated, they found that in 92 percent of cases, there had been a single change: an A had been substituted for a T at *BRAF*'s 1,796th base. What they found, in other words, was a point mutation, nearly 2,000 bases within the sequence of bases that makes up the *BRAF* gene. And the effect of this mutation? It produces a protein that is a slightly altered version of the normal BRAF protein. This protein is altered enough, however, that it keeps melanocytes moving through the cell-division cycle, causing them to multiply wildly.

Cell proliferation is not the only kind of trouble mutations can bring about. Chapter 12 introduced Huntington disease, which causes spastic movements, severe dementia, and ultimately death among its victims. In all people, there is a gene, called *IT15*, that has within it a number of repeats of a particular "triplet" of bases, CAG. People with 34 or fewer CAG repeats in their *IT15* gene are fine; they will suffer no illness at all from the gene. People with 35–39 CAG repeats, however, may develop Huntington, although this is not a certainty. Meanwhile, people who have more than 40 repeats definitely will develop the disease, and the more repeats they have, the younger they will be when its symptoms first appear. Unlike the case with cancer, however, the Huntington mutation does not cause cells to multiply wildly. Instead, the mutated *IT15* gene creates a faulty protein called huntingtin, which cannot be broken down by nerve cells and ends up

building up inside them, eventually killing them. Because Huntington is caused by a repeating group of three nucleotides, it is referred to as a "trinucleotide repeat" disease. At least eight other diseases of the nervous system fall into this category.

Heritable and Nonheritable Mutations

Though Huntington disease and melanoma are both caused by mutations, there is an important distinction to be made between them. Most mutations come about in the body's **somatic cells,** which is to say cells that do not become eggs or sperm. This is the case with the mutations that bring about melanoma. Conversely, some mutations arise in **germ-line cells,** meaning the cells that do become eggs or sperm. This is the case with the Huntington mutation. The important point here is that germ-line cell mutations are *heritable*—they can be passed on from one generation to the next, in the ways reviewed in Chapter 12. In contrast, although a line of melanoma cells may be quite harmful, it is separate from the line of cells that gives rise to eggs or sperm. For this reason, melanoma cannot be passed on from one generation to the next.

What Causes Mutations?

A critical question, of course, is what causes DNA to mutate? One answer is so-called environmental insults. The chemicals in cigarette smoke are powerful *mutagens,* meaning substances that can mutate DNA. So is the ultraviolet light that comes from the sun. To look at the latter example in a little more detail, ultraviolet light is a form of radiation that can link adjacent T's together in a single strand of DNA; sometimes it even causes both strands of the DNA helix to break. When this latter damage takes place in a cell, one of three things can happen. First, enzymes can successfully repair the damage. Second, the cell will recognize that this damage *cannot* be repaired, in which case it may either stop dividing (in the process known as cellular senescence) or commit suicide (in the process known as apoptosis). Third, the cell will undertake none of these responses to its DNA alteration but will simply keep on dividing. In this instance, the erroneous "spelling" in the cell's DNA will be copied over and over as this original cell gives rise to two cells, which then give rise to four, and so on. In short, a mutation is being passed on in a line of cells.

Not all mutations are caused by environmental influences. Mutations happen simply as random, spontaneous events. The collision of water molecules with DNA can remove nucleotide bases or alter them in such a way that their base-pairing properties are changed. The very process of eating and breathing produces so-called free radicals that can damage DNA. And the DNA replication machinery itself may introduce errors, irrespective of any outside influences.

Once such errors occur, it is understandable that some of them will go uncorrected simply because of the size of the replication operation. As noted, every time a human cell divides, 3.2 billion bases have to be copied twice. And some 25 million cells are dividing each second in human beings. The surprise, therefore, is not that mistakes happen, but that so *few* of them happen, and that fewer still become permanent. (Recall the one-in-a-billion error rate in DNA replication that was observed earlier.) After learning about the kinds of effects mutations can have on organisms, however, you can understand why this error rate is so low: Life as we know it could not exist with a high error rate, at least in the portions of genome that code for proteins. Remember that within 3.2 billion A's, T's, G's, and C's, the substitution of a single A for a single T can help bring about melanoma. How could life exist if such mistakes were common?

Mutations and Evolutionary Adaptation

While it's true that, for individuals, mutations can have a negative effect, for the living world as a whole mutations have been vitally important because of a role they play in evolution. It turns out that germ-line mutations are the only means by which completely new genetic information can be added to a species' genome. Without mutations, organisms can shuffle *existing* genetic information in myriad ways. (Think of meiosis, with its chromosomal part swapping and random alignment of chromosome pairs). But no amount of genetic recombination could have produced, for example, the eyes that some living things possess. To go from no eyes to eyes, there had to have been some mutations along the line—some accidental reorderings of DNA sequences that either produced entirely new proteins or that altered the production of existing proteins in beneficial ways. Modifications such as these are vital to living things given their struggle to get along in environments that are constantly changing. (Temperatures shift; streams dry up.) When environments change, species either change with them or become extinct. And the primary way major changes come about in species is through mutations. You'll be looking at this topic again in the evolution unit of this book. For now, however, isn't it interesting to ponder the fact that the living world adapts partly through its mistakes?

On to Protein Production

Back in Chapter 9, you saw that since antiquity, human beings have speculated about how one generation of living things can give rise to another—about *how* it is that life goes on. The detailed answer to this question turned out to be the something-old, something-new quality of DNA replication: parental DNA strands serve as templates for new strands, with the result that the qualities of one generation can be passed along to the next. Splendid as this function is, it is only one of the two great tasks carried out by our genetic machinery. You have just reviewed the process by which genetic information is replicated; now we'll look at the process by which this information is used. How can a stretch of DNA bring about the production of a protein? You'll see in the chapter coming up.

SO FAR . . .

1. In DNA replication, both of the strands of the double helix serve as templates for _____.

2. The _____ are a group of enzymes that add nucleotides to a replicating DNA chain.

3. A permanent alteration in a DNA base sequence is known as a _____.

CHAPTER

13 REVIEW

Summary

13.1 The Form and Function of Genes

- In trying to decipher the structure of DNA, scientists were performing work in molecular biology, defined as the investigation of life at the level of its individual molecules. (p. 233)

13.2 Watson and Crick: The Double Helix

- James Watson and Francis Crick discovered the chemical structure of DNA in 1953, which ushered in a new era in biology because it allowed researchers to understand some of the most fundamental processes in genetics. (p. 234)

- Watson and Crick's research was aided by the work of others, including Rosalind Franklin, who was using X-ray diffraction to learn about DNA's structure. (p. 234)

13.3 The Components of DNA and Their Arrangement

- The DNA molecule is composed of building blocks called nucleotides, each of which consists of one sugar (deoxyribose), one phosphate group, and one of four bases: adenine, guanine, thymine, or cytosine (A, G, T, or C). The sugar and phosphate groups are linked in a chain that forms the "handrails" of the DNA double helix. Bases then extend inward and are joined to each other in the middle by hydrogen bonds. A always pairs with T across the helix; G always pairs with C. (p. 236)

- DNA is copied by means of each strand of DNA serving as a template for the synthesis of a new, complementary strand. The DNA double helix first divides down the middle. Each A on an original strand then specifies a place for a T in a new strand, while each G specifies a place for a C, and so forth. Because both strands of the original double helix are being paired with new strands in this process, the end product is two double helices. (p. 237)

- DNA can encode the information for the huge number of proteins used by living things because the sequence of bases along DNA's handrails can be laid out in an extremely varied manner. A collection of bases in one order encodes the information for one protein, while a different sequence of bases encodes the information for a different protein. (p. 237)

- Each double helix produced in replication is a combination of one parental strand of DNA and one newly synthesized complementary strand. This is how life builds on itself. Each instance of DNA replication occurs prior to the division of one parent cell into two daughter cells. When this division is complete, each of the newly formed helices will reside in separate daughter cells. (p. 238)

- Enzymes called DNA polymerases move along the double helix in DNA replication, bonding together new nucleotides in complementary DNA strands. (p. 239)

13.4 Mutations

- The error rate in DNA replication is very low, partly because repair enzymes are able to correct replication mistakes. When such mistakes are made and then not corrected, the result is a mutation: a permanent alteration of a DNA base sequence. (p. 239)

- Most mutations have no noticeable effect on an organism, but when mutations do have an effect, it is generally negative. Cancers result from a line of cells that have undergone types of mutations that cause them to proliferate wildly. (p. 240)

- Mutations that come about in the body's germ-line cells, meaning cells that become eggs or sperm, can be passed from one generation to another. Most mutations, however, come about in the body's somatic cells, which are cells that do not give rise to eggs or sperm, and these mutations cannot be passed on. (p. 240)

- Mutagens, such as cigarette smoke or ultraviolet light, can mutate DNA. Spontaneous events that are brought about by normal cellular processes can also lead to mutations. (p. 240)

- Mutations are important to evolution because they are the only means through which completely new genetic information can be added to a species' genome. Mutations can, in rare instances, either produce new proteins or alter the production of existing proteins in ways that are beneficial. (p. 241)

Key Terms

DNA polymerase   239
germ-line cell   240
molecular biology   234
mutation   239
nucleotide   238
point mutation   239
somatic cell   240

Understanding the Basics

Multiple-Choice Questions

(Answers are in the back of the book.)

1. James Watson and Francis Crick:
   a. proved that DNA contains information for the production of proteins.

b. elucidated the nature of genetic mutations.

c. discovered DNA.

d. proved that chromosomes were made partly of DNA.

e. discovered the structure of DNA.

2. The type of biological study that was exemplified through the discovery of DNA structure is called:
   a. detailed biology.
   b. component biology.
   c. microscopic biology.
   d. elemental biology.
   e. molecular biology.

3. The components of the DNA handrails are _____ and _____.
   a. nucleotides, phosphates
   b. sugars, nucleotides
   c. sugars, bases
   d. sugars, phosphates
   e. none of the above

4. Which of the following combinations is an example of a nucleotide?
   a. sugar + phosphate + adenine
   b. sugar + phosphate + guanine
   c. sugar + phosphate + thymine
   d. sugar + phosphate + cytosine
   e. all of the above

5. Which of the following DNA sequences contains a mismatched base pair?
   a. TCAA
   a. AGTT
   b. CAGC
   b. GTCG
   c. GATA
   c. CTCT
   d. AATT
   d. TTAA
   e. GACG
   e. CTGC

6. DNA polymerases work to:
   a. unwind the double helix during replication.
   b. cause mutations in DNA.
   c. join together nucleotides in a growing, complementary DNA strand.

d. correct errors in DNA base sequences during replication.

e. both c and d.

7. Watson and Crick made their momentous discovery about DNA in _____.
   a. 1943
   b. 1953
   c. 1921
   d. 1973
   e. 1962

8. How are the base pairs of the DNA double helix linked together?
   a. phosphate bonds
   b. hydrogen bonds
   c. carbon-carbon bonds
   d. sugar bonds
   e. basic bonds

Brief Review

(Answers are in the back of the book.)

1. How does the phrase "something old, something new" describe the method of DNA replication employed by the cell?

2. What two fundamental questions of genetics did the discovery of the structure of DNA help answer?

3. True or False: DNA polymerases can correct errors made in DNA replication.

4. What is a mutation? Are all mutations harmful?

5. What is the nature of the Huntington disease mutation, and what is its effect at a cellular level?

Applying Your Knowledge

1. Given the following DNA sequence, determine the complementary strand that would be added in replication:

   ATTGCATGATAGCC

2. Would you expect cancer to arise more often in types of cells that divide frequently (such as skin cells) or in types of cells that divide rarely or not at all (such as nerve cells in the brain)? Explain your reasoning.

3. The Huntington disease referred to in the main text sends all of its victims into dementia, physical immobility, and ultimately death, but usually does not start doing so until the person is at least 35. Imagine yourself as the adolescent child of a parent who has started to manifest Huntington symptoms. Your chances of inheriting this incurable condition are one out of two, but you have thus far shown no symptoms of it. The question is: Do you get yourself tested to see if you inherited the harmful Huntington allele? Do you find out for sure whether you will, or will not, have this illness, or do you live with uncertainty? Assuming you've arrived at an answer, consider it now in the context of the conclusion that most people in this situation come to: Ninety-five percent of such people decline to be tested.

MasteringBiology®

1. MasteringBiology Assignments

Activities: DNA; One-Gene-One-Enzyme Hypothesis; DNA and RNA Structure; DNA Double Helix; DNA Replication, An Overview; Mutations; **Bioflix:** DNA Replication; **Current Events:** Bits of Mystery DNA, Far From "Junk", Play Crucial Role; New Drug Trial Seeks to Stop Alzheimer's Before it Starts; The Cancer Sleeper Cell;

Building Vocabulary; Questions: Bioflix Quiz; Reading Questions; Test Bank

2. eText

Read your book online, search, take notes, highlight text, and study more effectively.

3. The Study Area

Self-study tools to test comprehension.

How Proteins Are Made:

Genetic Transcription, Translation, and Regulation

Like workers using a set of blueprints to construct a house, cells use the information contained in DNA to construct proteins. Each cell can finely tune its production of proteins in accordance with its needs.

Available on Mastering Biology®
Activities: Overview of Protein Synthesis; Transcription; Translation; Control of Transcription; Human Genome Sequencing; RNA Processing
Bioflix: Protein Synthesis; **Chemistry Review:** Nucleic Acids; **Current Events:** The Search for Genes Leads to Unexpected Places; **Building Vocabulary; Questions:** Bioflix Quiz; Reading Questions; Test Bank

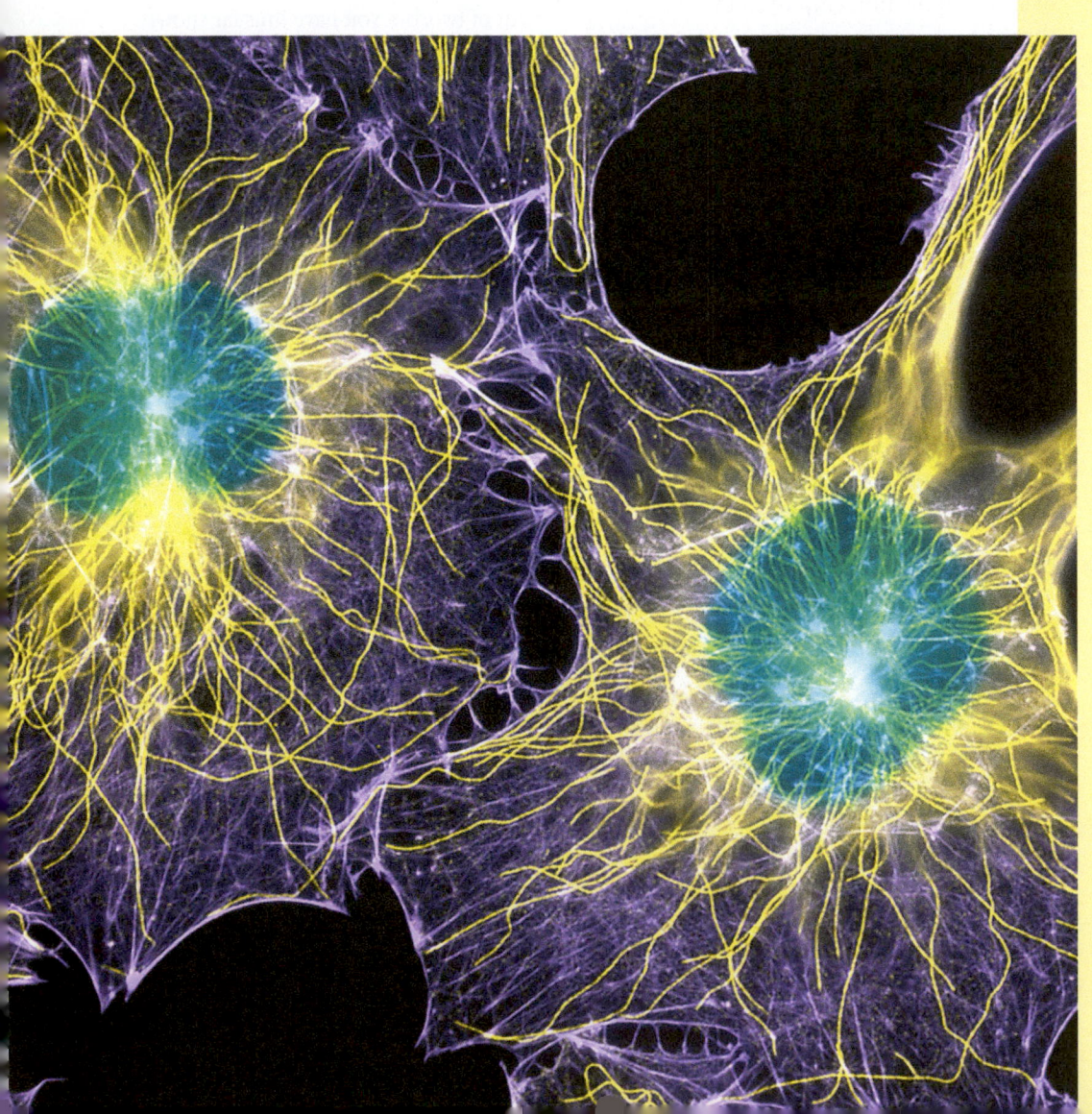

Two kinds of structural proteins have been stained to be visible in these mouse cells: actin (in blue) and microtubules (in yellow).

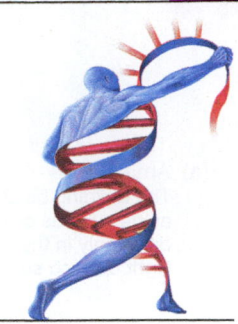

The scientific advance that was reviewed in Chapter 13, the discovery of the structure of DNA in 1953, bears some comparison to the birth of Napoleon: Both events had momentous consequences, but not for a while. Only slowly did the discoveries come that, in time, were seen as hinging on James Watson and Francis Crick's work. So significant were these advances, however, that the years between 1953 and 1966 are now regarded by some scholars as a kind of golden age of genetics—a time when basic findings about this field came fast and furious, such that by the mid-1960s scientists had at last grasped the essentials of how the genetic machinery works.

These discoveries were made in connection with both of the two great tasks of genetics: on the one hand, the copying or "replication" of DNA that you reviewed in Chapter 13; on the other, the means by which DNA's genetic information brings about the production of proteins. The first task has to do with how genetic information is passed on, the second with how genetic information is used. This second task is the subject of this chapter.

14.1 The Structure of Proteins

Because you'll be dealing extensively with proteins here, now may be a good time to review some basic information about them. As you saw in Chapter 3, proteins fit into the "building blocks" model of biological molecules. The blocks in this case are amino acids. String a number of these together, and you have a **polypeptide** chain, which then folds up in a specific three-dimensional manner, resulting in a protein. Proteins are likely to be made of hundreds of amino acids strung together, often in several linked polypeptide chains.

Though there are hundreds of thousands of different proteins, each one of them is put together from a starting set of a mere 20 amino acids. If you wonder how such diversity can proceed from such simplicity, think of the English language, which has thousands of words but only 26 letters in its alphabet. It is the *order* in which letters occur that determines whether, for example, "cat" or "act" is spelled out; just so, it is the order of amino acids that determines which protein is synthesized.

If you look at **Figure 14.1a** on the next page, you can see the chemical structure of two amino acids, glycine (gly) and isoleucine (ile). If you look at **Figure 14.1b**, you can see that these two amino acids are the first two that occur in one of the two polypeptide chains that make up the unusually small protein we call insulin. **Figure 14.1c** then shows a three-dimensional representation of insulin—the folded-up form this protein assumes before taking on its function of moving blood sugar into cells. A list of all 20 primary amino acids and their three-letter abbreviations can be found in **Table 14.1**.

For our purposes, the question is: How do the chains of amino acids that make up a protein come into being? How do gly, ile, val, and so forth come to be strung together in a specific order to create this protein? You know from what you've studied so far that genes can be thought of as "recipes" for proteins, and that these recipes are set forth as a series of chemical "bases"—the A's, T's, C's, and G's that lie along a DNA strand. How, then, do we get from *this* series of DNA bases to *that* series of amino acids—to a protein? Let's find out.

14.2 Protein Synthesis in Overview

In overview, the process of protein synthesis can be described fairly simply. In eukaryotic organisms such as ourselves, the DNA just referred to is contained in the nucleus of the cell. The first step is that a stretch of DNA unwinds there, and its message—the order of a string of A's, T's, C's, and G's—is copied onto a molecule called messenger RNA (mRNA). This

Table 14.1	
Amino Acids	
Alanine	ala
Arginine	arg
Asparagine	asn
Aspartic Acid	asp
Cysteine	cys
Glutamic Acid	glu
Glutamine	gln
Glycine	gly
Histidine	his
Isoleucine	ile
Leucine	leu
Lysine	lys
Methionine	met
Phenylalanine	phe
Proline	pro
Serine	ser
Threonine	thr
Tryptophan	trp
Tyrosine	tyr
Valine	val

glycine **(gly)**

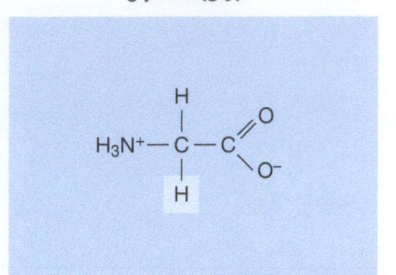

isoleucine **(ile)**

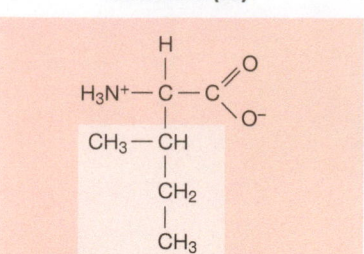

(a) Amino acids
The building blocks of proteins are amino acids such as glycine and isoleucine, which differ only in their side-chain composition (light colored squares).

(b) Polypeptide chain
These amino acids are strung together to form polypeptide chains. Pictured is one of the two polypeptide chains that make up the unusually small protein insulin.

(c) Protein
Polypeptide chains function as proteins only when folded into their proper three-dimensional shape, as shown here for insulin. Note the position of the glycine and isoleucine amino acids in one of the insulin polypeptide chains (colored light green).

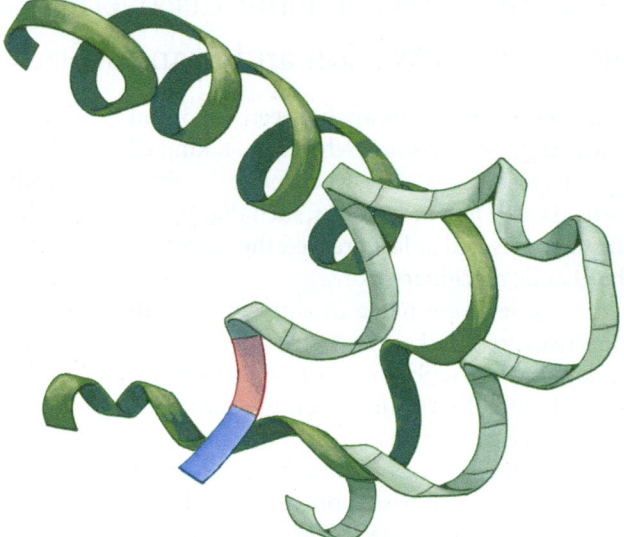

Figure 14.1
The Structure of Proteins

MB

Overview of Protein
Synthesis

length of mRNA then exits from the cell nucleus (**Figure 14.2**).

The destination of this mRNA is a molecular workbench in the cell's cytoplasm, a structure called a ribosome. More formally, a **ribosome** is an organelle, located in the cell's cytoplasm, that is the site of protein synthesis. At the ribosome, both the recipe (the mRNA sequence) and the raw materials (amino acids) come together to make the product (a protein). As the mRNA sequence is "read" within the ribosome, something grows from it: a chain of amino acids that have been linked together in the ribosome in the order specified by the mRNA sequence. When the chain is finished and folded up, a protein has come into existence. And how do amino acids get to the ribosomes? They are brought there by a second type of RNA, transfer RNA (tRNA).

As may be apparent from this account, protein synthesis divides neatly into two sets of steps. The first set is called **transcription**: the process by which the genetic information encoded in DNA is copied onto

messenger RNA. The second set is called **translation**: the process by which information encoded in messenger RNA is used to assemble a protein at a ribosome. Let's now look in more detail now at both of these processes, starting with transcription.

SO FAR . . .

1. Though there are hundreds of thousands of proteins active in living things, all of them are put together from a starting set of 20 of the building blocks known as _____.

2. The information for building proteins that is encoded in DNA is passed on first to _____, which then migrates to an organelle called a _____, where the protein is put together.

14.3 A Closer Look at Transcription

From what's been reviewed so far, you can see that a key player in transcription (and translation) is RNA, whose full name is ribonucleic acid. As it turns out, RNA is structurally very similar to DNA. For one thing, RNA has the sugar and phosphate "handrail" components you saw in DNA last chapter. Then, in both molecules, this two-part structure is joined to a third element, a base, as you can see in **Figure 14.3**.

There are, however, differences between DNA and RNA. One is that RNA usually is single stranded, whereas DNA is structured in two strands that form its famous double helix. Beyond this, recall that any given DNA building block or "nucleotide" has one of four bases: adenine, guanine, cytosine, or thymine (A, G, C, or T). RNA uses the first three of these, but then uses uracil (U) instead of the thymine (T) found in DNA.

Passing on the Message: Base Pairing Again

Given the chemical similarity between DNA and RNA, it's not hard to see how DNA's genetic message can be passed on to messenger RNA: Base pairing is at work again. Recall from Chapter 13 how DNA is replicated. The double helix is unwound, after which bases along the now-single DNA strands are paired up with complementary DNA bases. Every T on a single DNA strand is

Transcription

1. In transcription, a section of DNA unwinds and nucleotides on it form base pairs with nucleotides of messenger RNA, creating an mRNA chain.

2. This segment of mRNA then leaves the cell nucleus, headed for a ribosome in the cell's cytoplasm, where translation takes place.

Translation

3. Joining the mRNA chain at the ribosome are amino acids, brought there by transfer RNA molecules. The length of messenger RNA is then "read" within the ribosome. The result? A chain of amino acids is linked together in the order specified by the mRNA sequence.

4. When the chain is finished and folded up, a protein has come into existence.

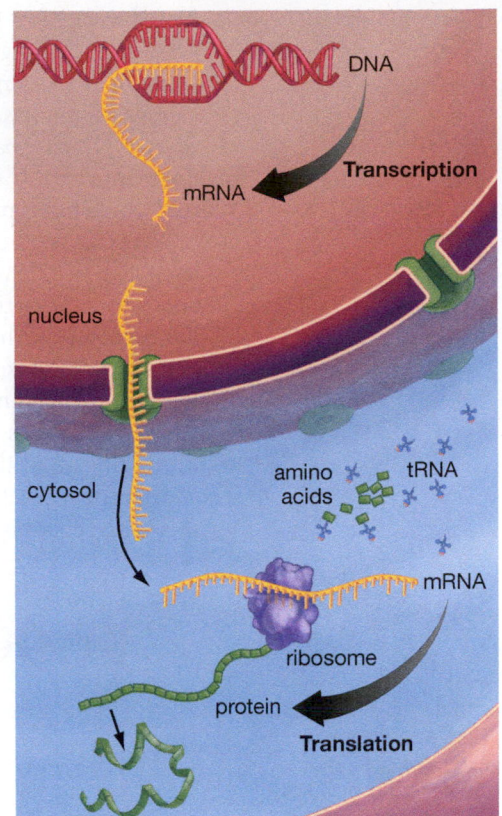

Figure 14.2
The Two Major Stages of Protein Synthesis

(a) Comparison of RNA and DNA nucleotides

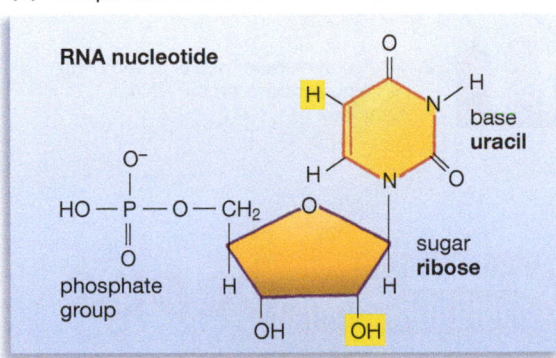

(b) Comparison of RNA and DNA three-dimensional structure

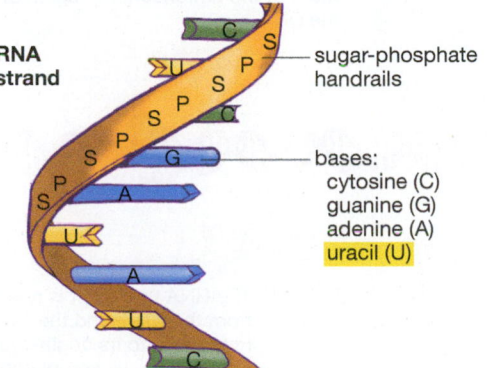

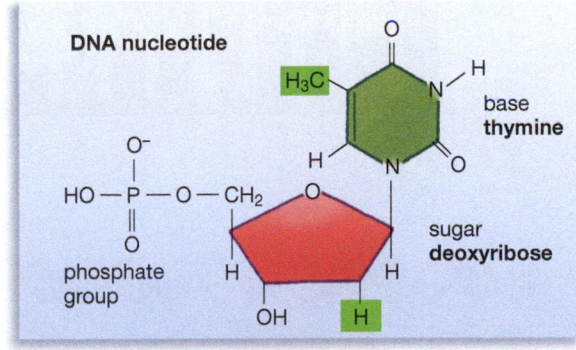

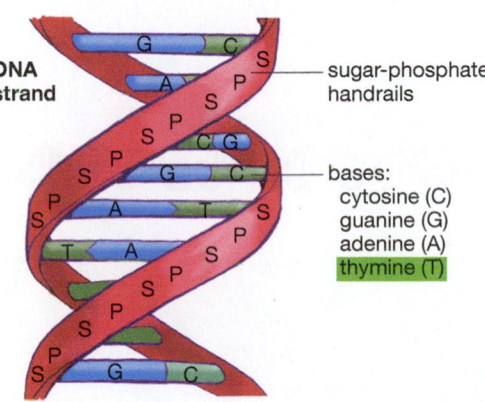

Figure 14.3
RNA and DNA Compared

(a) Each building-block nucleotide of both DNA and RNA is composed of a phosphate group, a sugar—ribose in RNA, deoxyribose in DNA—and one of four bases. RNA and DNA both use the bases adenine, guanine, and cytosine (A, G, and C), but RNA uses the base uracil (U) instead of the thymine (T) that DNA uses.

(b) Both RNA and DNA amount to linked chains of these nucleotides. The "handrails" of both RNA and DNA are composed of the sugar molecule of one nucleotide linked to the phosphate molecule of the next nucleotide (thus the S-P-S-P labeling). The "stairsteps" stemming from the handrails are formed by the bases of the nucleotides.

Chemistry Review:
Nucleic Acids

paired with a free-floating A, and every C is linked with a G, thus yielding a complementary DNA *chain.* Because of RNA's similarity to DNA, the bases RNA has can also form base pairs with DNA. The twist to this RNA–DNA base pairing is that each A on a DNA strand links up with a U on the RNA strand, instead of the T that would be A's partner in DNA-to-DNA base pairing.

You can see how transcription works by looking at **Figure 14.4**. It will come as no surprise to you that an enzyme is critically involved in this process. **RNA polymerase,** as this complex of enzymes is known, actually undertakes two critical tasks: It unwinds the DNA sequence and then strings together a series of RNA nucleotides that is complementary to it, thus producing an initial RNA chain.

In organisms such as ourselves (the eukaryotes), this initial chain is not actually the finished messenger

RNA we've been talking about. Instead, it is an RNA chain called a *primary transcript* that must undergo some modification or "editing" before it qualifies as messenger RNA. You can read more about this editing in section 14.5, on genetic regulation. For now, we'll set it aside and follow the fate of a completed mRNA chain. Before leaving the subject of mRNA, however, here's a more formal definition of it: **Messenger RNA (mRNA)** is a type of RNA that encodes, and carries to ribosomes, information for the synthesis of proteins.

A Triplet Code

Thus far, we have been looking at what might be called the flow of genetic information (from DNA to mRNA). A separate issue in protein synthesis, however, has to do with the linkage between DNA and

Figure 14.4
Transcription Works through Base Pairing
Thanks to their chemical similarity, DNA and RNA can engage in base pairing, and this base pairing is how RNA transcripts are synthesized.

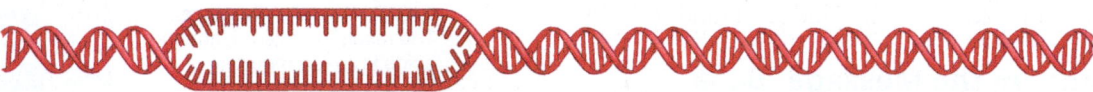

1. RNA polymerase unwinds a region of the DNA double helix.

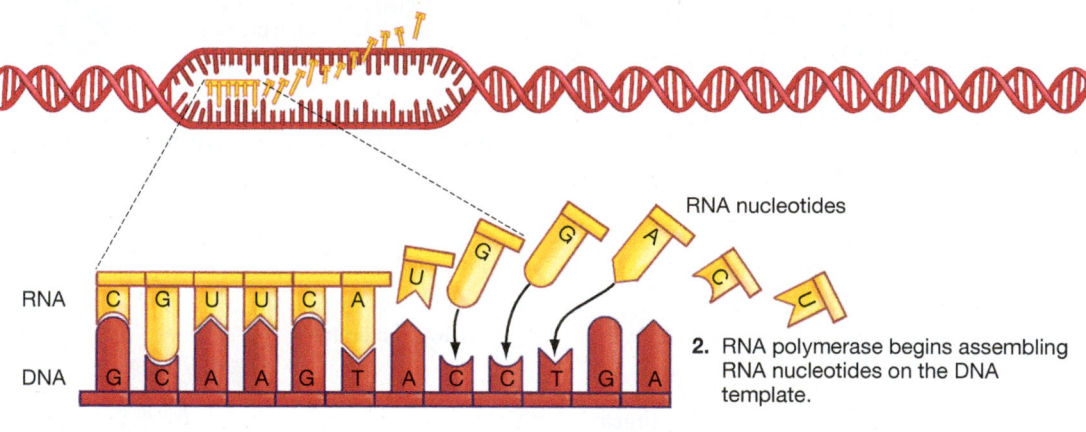

RNA nucleotides

2. RNA polymerase begins assembling RNA nucleotides on the DNA template.

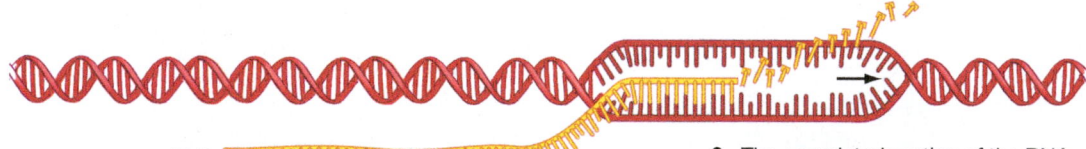

3. The completed portion of the RNA transcript separates from the DNA. Meanwhile, RNA polymerase unwinds more of the untranscribed region of the DNA.

4. The RNA transcript is released from the DNA, and the DNA is rewound into its original form. Transcription is completed.

RNA on the one hand and amino acids on the other. The question here is: How many DNA bases does it take to code for an amino acid? As **Figure 14.5** shows, the answer is three. Each three bases in a DNA sequence pair with three mRNA bases, but each group of three mRNA bases then codes for a *single* amino acid. Each coding triplet of mRNA bases is known, appropriately enough, as a **codon.**

A question that follows is this: If you have a given three mRNA bases, *which* amino acid do they specify? In Morse code, the sounds . — . (short-long-short) code for the letter *R*. But if we know that a given mRNA codon has the base sequence UCC, what *amino acid* does this code for? Today, we know that the answer is serine (ser), but at one time this was not clear. Indeed, it took years for scientists to figure out all the linkages between codons and amino acids. When this work was completed, the result was the **genetic code,** which can be defined as the inventory of linkages between nucleotide triplets and the amino acids they code for. If you look at "The Importance of the Genetic Code" on page 254, you can learn more about the significance of this inventory.

14.4 A Closer Look at Translation

With a length of RNA having first been transcribed from a length of DNA and having then moved to a ribosome as mRNA, many of the players are in place for translation—the second stage of protein production.

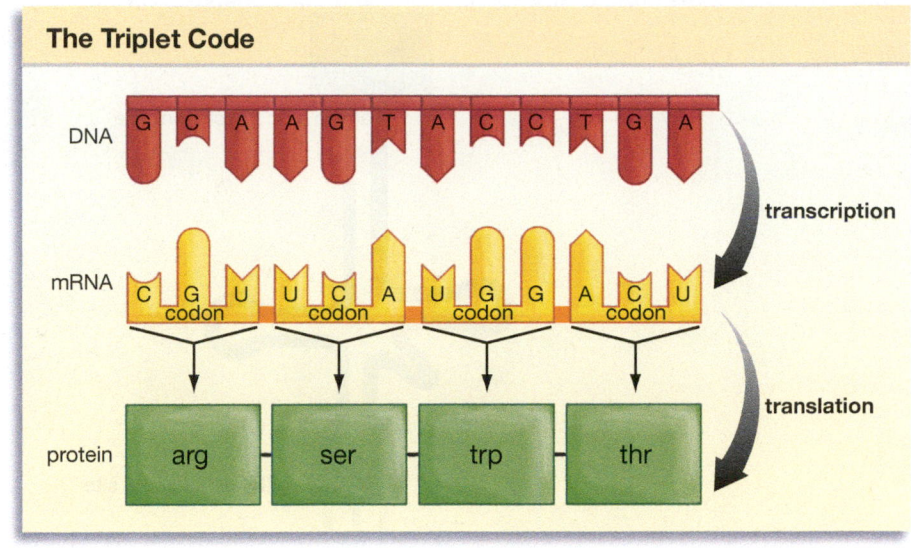

The Triplet Code

What's also needed, however, are the building blocks of proteins, amino acids. As noted, they are brought to ribosomes by a second form of RNA, called transfer RNA (or tRNA).

The Nature of tRNA

If you look at **Figure 14.6**, you can see why tRNA is aptly placed within the translation phase of protein synthesis. When, in everyday speech, we think of a *translator,* we think of someone who can communicate in *two* languages. Transfer RNA effectively does

Figure 14.5
Triplet Code

Each triplet of DNA bases codes for a triplet of mRNA bases (a codon), but it takes a complete codon to code for a single amino acid.

MB

Translation

Figure 14.6
Bridging Molecule

Transfer RNA (tRNA) molecules link up with amino acids on the one hand and mRNA codons on the other, thus forming a chemical bridge between the two kinds of molecules in protein synthesis. They also transfer amino acids to ribosomes, as shown in the steps of the figure.

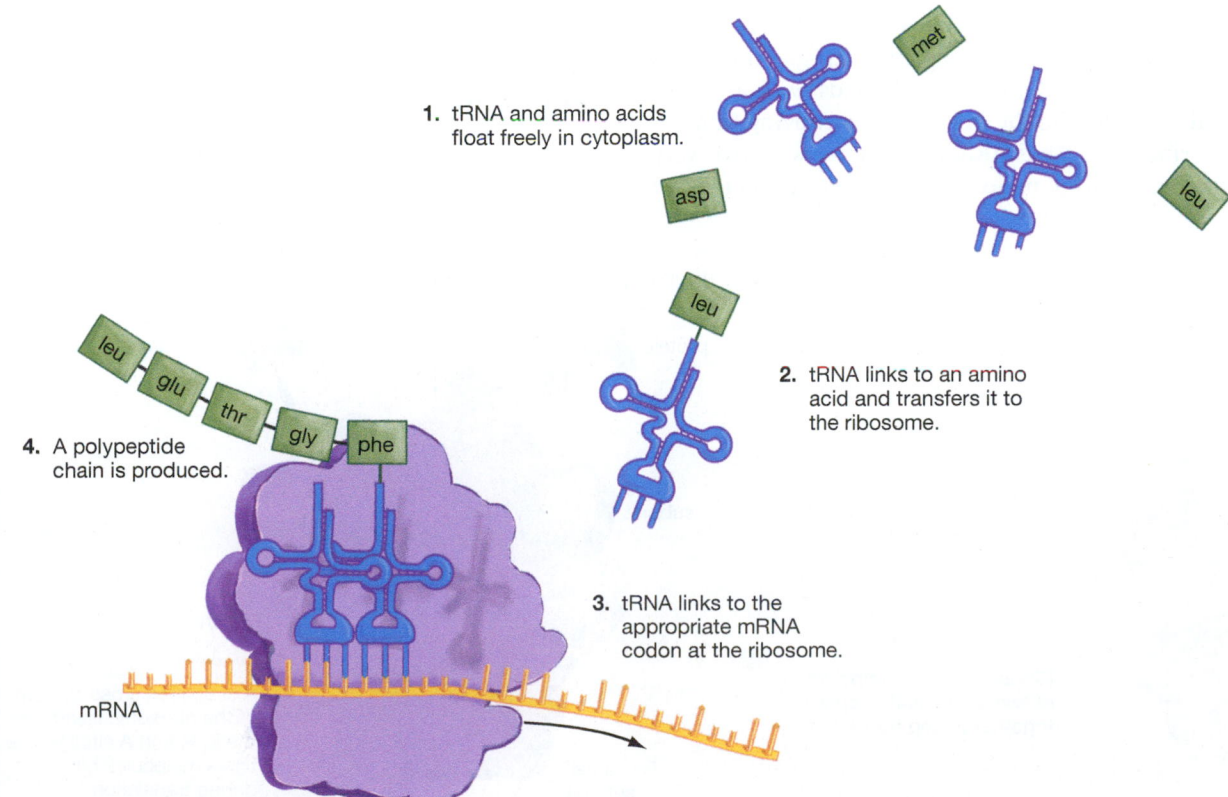

1. tRNA and amino acids float freely in cytoplasm.

2. tRNA links to an amino acid and transfers it to the ribosome.

3. tRNA links to the appropriate mRNA codon at the ribosome.

4. A polypeptide chain is produced.

mRNA

ribosome

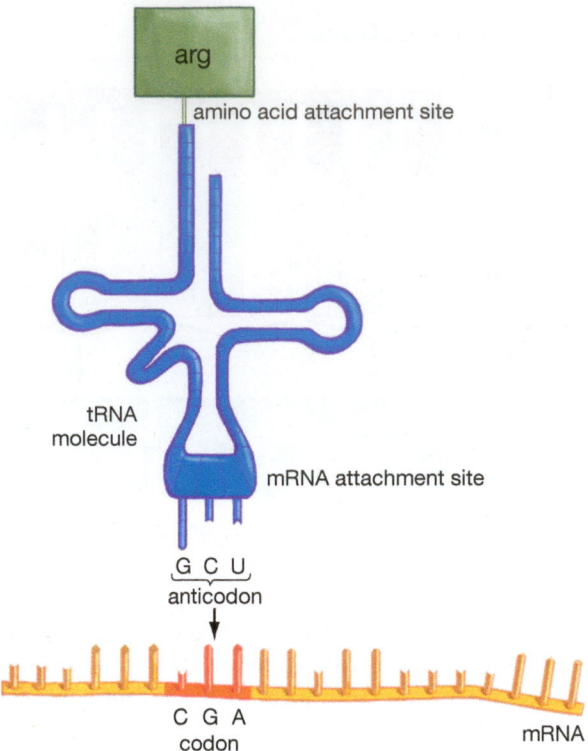

Figure 14.7
The Structure of Transfer RNA
In this two-dimensional model of transfer RNA, one end of the tRNA molecule can be seen binding to a specific amino acid (arg), while the other end binds to its counterpart mRNA codon (CGA).

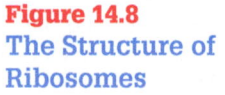
Figure 14.8
The Structure of Ribosomes

hand, nucleic acids on the other—thereby serving as a translator between them. **Transfer RNA (tRNA)** can be defined as a form of RNA that, in protein synthesis, binds with amino acids, transfers them to ribosomes, and then binds with messenger RNA.

Figure 14.7 provides a more detailed look at how transfer RNA carries out this function. You can see that it binds with an mRNA codon by means of three bases it possesses, called an **anticodon.** At its other end (the 12 o'clock position in the figure), there is an attachment site for the amino acid. The linkage between amino acids and tRNA molecules is specific enough that any tRNA that binds to an mRNA codon will be carrying an appropriate amino acid.

The Structure of Ribosomes

You've now been introduced to almost all the players that will be active in protein synthesis at the ribosome. But what is the structure of the ribosome itself? If you look at **Figure 14.8**, you can begin to get an idea. Note that ribosomes are composed of two "subunits"—one larger than the other—both made of a mixture of proteins and yet another type of RNA, **ribosomal RNA (rRNA).** The two subunits may float apart from one another in the cytoplasm until prompted to come together by the process of translation. You can see that when the subunits have been joined, there exist three binding sites (which are convenient to think of as slots), one of them an "E" site, the next a "P" site, and the third an "A" site. You'll be looking at their roles shortly. **Table 14.2** sets forth the three types of RNA reviewed so far that are active in protein synthesis.

The Steps of Translation

this. One end of each tRNA molecule links with a specific *amino* acid, which it finds floating free in the cytoplasm. Then, transferring this amino acid to the ribosome, this tRNA molecule binds with a *nucleic* acid—a codon on the mRNA that is moving through the ribosome. Thus, tRNA is binding with two very different kinds of molecules—amino acids on the one

With all the players introduced, let's see how translation works. To keep things simple, we'll follow the

(a) Large and small ribosomal units

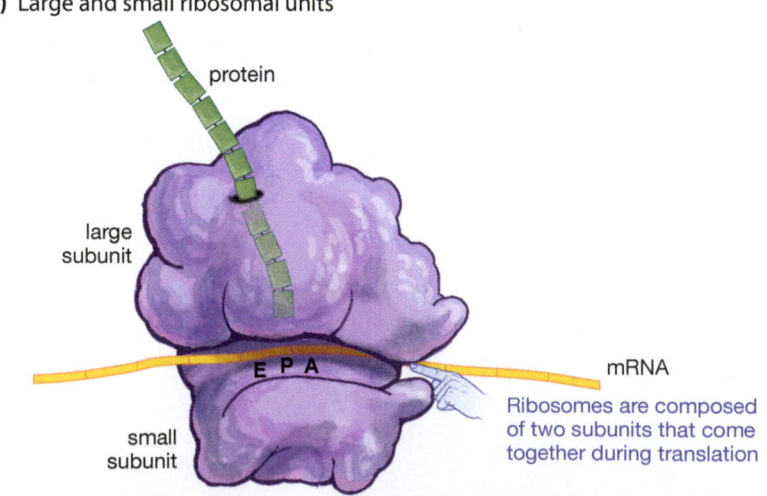

protein

large
subunit

small
subunit

E P A

mRNA

Ribosomes are composed of two subunits that come together during translation

(b) Binding sites in the ribosome

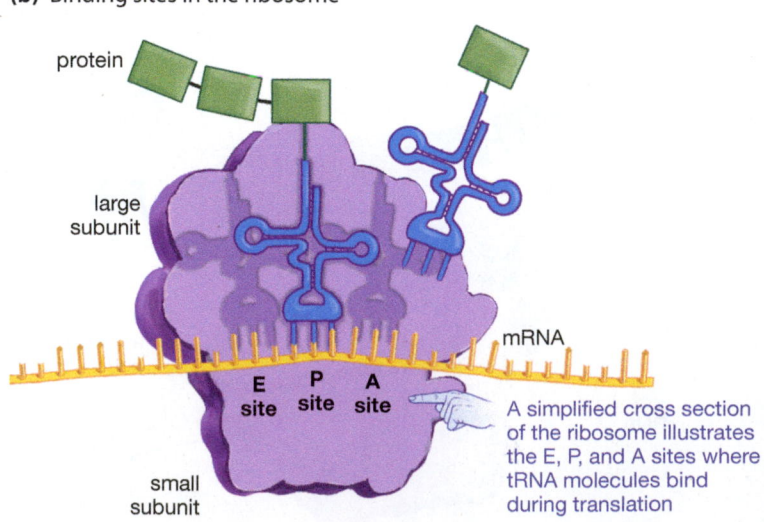

protein

large
subunit

small
subunit

E P A
site site site

mRNA

A simplified cross section of the ribosome illustrates the E, P, and A sites where tRNA molecules bind during translation

process as it occurs in prokaryotes. Our starting point is an mRNA transcript that is ready to begin binding with a ribosome. Meanwhile, nearby tRNA molecules have linked to their appropriate amino acids.

mRNA Binds to Ribosome, First tRNA Arrives

In this first step, the mRNA chain arrives at the ribosome and binds to the ribosome's small subunit (**Figure 14.9**). The mRNA codon AUG is the usual "start" codon for a polypeptide chain. Next, a tRNA molecule with the appropriate anticodon sequence (UAC) binds to this AUG codon. This tRNA arrives bearing its appropriate amino acid, which is methionine (met).

Polypeptide Chain Is Elongated

Following this, the large ribosomal subunit becomes part of the ribosome, providing the ribosome's A, P, and E binding sites (**Figure 14.10** on the next page). Next, amino acids will start to be joined together in a chain. As you can see, this elongation process begins with a second incoming tRNA molecule binding to an mRNA codon in the A site. Because it's a CUG codon that has moved into the site, a tRNA with a GAC anticodon binds to it. This tRNA comes bearing the amino acid leucine (leu). Now here's how we get the chain: The met amino acid attached to the tRNA in the P site bonds with the leu amino acid attached to the tRNA in the A site. In this process, the bond is broken between met and its original tRNA.

Once this occurs, a kind of molecular musical chairs ensues: The ribosome effectively shifts one codon to the right. With this, the tRNA that had been in the P site is relocated to the E site. What does the E stand for? Exit. The tRNA in this site bears no amino acid now; soon, it will be ejected from the ribosome altogether. Meanwhile, the tRNA that had been in the A site moves to the P site, and a new codon shifts into the now-vacated A site. You can see

what comes next: A tRNA binds with this new codon in the A site. Soon, the growing polypeptide chain will bind with this tRNA's amino acid, and the process will continue.

Termination of the Growing Chain

There are three separate codons that don't code for any amino acid but that instead act as stop signals for polypeptide synthesis. Any time one of these "termination" codons moves into the ribosome's A site, it doesn't bind with an incoming tRNA but instead brings about a severing of the linkage between the P-site tRNA and the polypeptide chain. Indeed, the whole translation apparatus comes apart at this

Bioflix: Protein Synthesis

Table 14.2

Types of RNA

Type of RNA	Function	Functions in
Messenger RNA (mRNA)	Carries DNA sequence information to ribosomes	Nucleus, migrates to ribosomes in cytoplasm
Transfer RNA (tRNA)	Provides linkage between mRNA and amino acids; transfers amino acids to ribosomes	Cytoplasm
Ribosomal RNA (rRNA)	Structural component of ribosomes	Cytoplasm

---

First step of translation

1. A messenger RNA transcript binds to the small subunit of a ribosome as the first transfer RNA is arriving. The mRNA codon AUG is the "start" sequence for most polypeptide chains. The tRNA, with its methionine (met) amino acid attached, then binds this AUG codon.

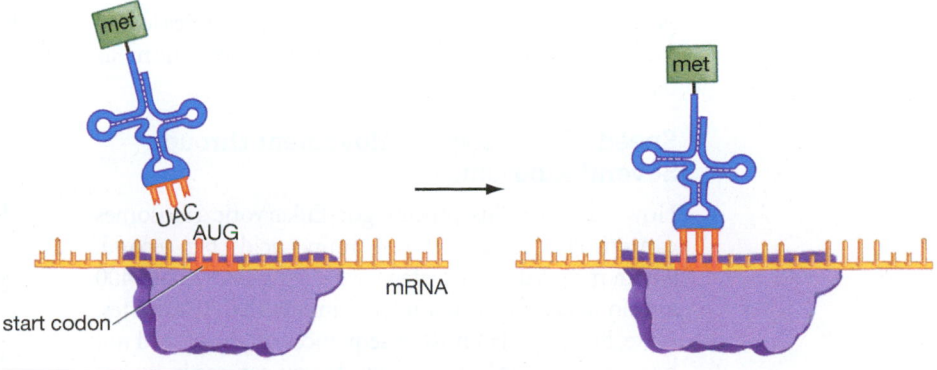

Figure 14.9
How Translation Begins
In this initial step in translation, a messenger RNA chain binds to the small unit of a ribosome, setting the stage for the arrival of the first transfer RNA, which binds with its complementary codon in the mRNA chain.

Translation continues

2. The large ribosomal subunit joins the ribosome, as a second tRNA arrives, bearing a leucine (leu) amino acid. The second tRNA binds to the mRNA chain, within the ribosome's A site.

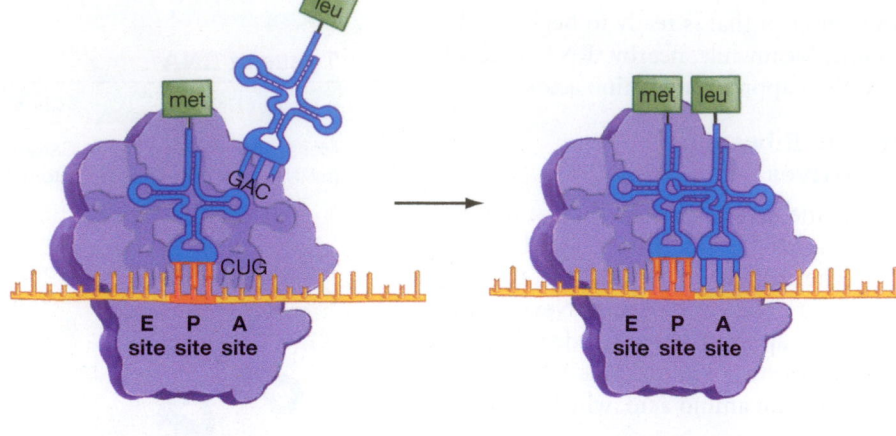

3. A bond is formed between the newly arrived leu amino acid and the met amino acid, thus forming a polypeptide chain. The ribosome now effectively shifts one codon to the right, relocating the original P site tRNA to the E site, the A site tRNA to the P site, and moving a new mRNA codon into the A site.

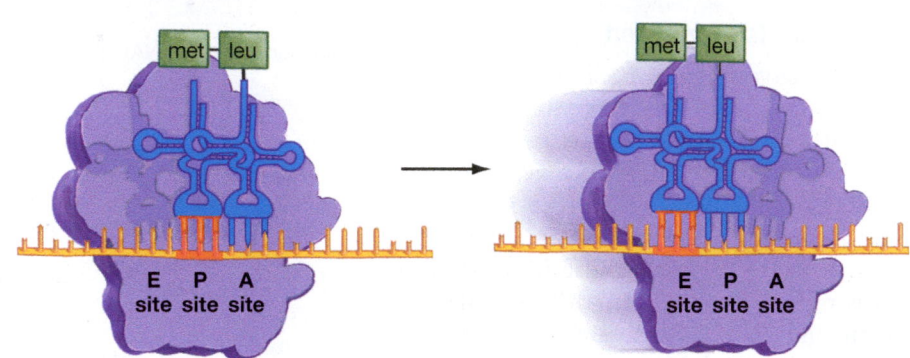

4. The E site tRNA leaves the ribosome, even as a new tRNA binds with the A site mRNA codon, and the process of elongation continues.

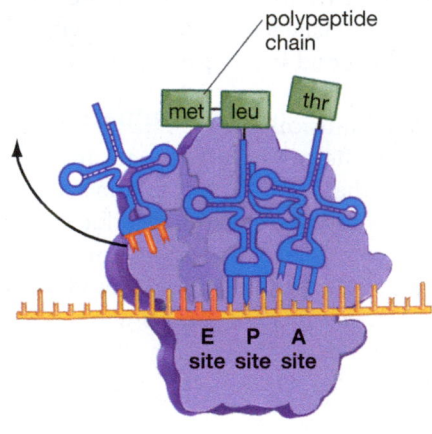

Figure 14.10
Next Steps of Translation

point, with the polypeptide chain being released to fold up and be processed as a protein. Translation has been completed.

Speed of the Process; Movement through Several Ribosomes

How fast does this process go? Eukaryotic ribosomes can string together up to 4 amino acids per second, which means that an average-sized protein of about 400 amino acids can be completed in less than 2 minutes. Note, however, that mRNA sequences often are read not by one ribosome but by many. As you can see in **Figure 14.11**, several ribosomes—perhaps scores of them—might move over a given mRNA transcript, with the re-sult that identical polypeptide chains grow out of each

ribosome. This greatly increases the number of proteins that can be put together in a given period of time.

SO FAR . . .

1. DNA passes on its information to messen-ger RNA by means of _____.

2. Each three _____ bases code for three _____ bases, which code for one _____.

3. A transfer RNA molecule attaches to an _____ on one end and a _____ on the other.

(a)

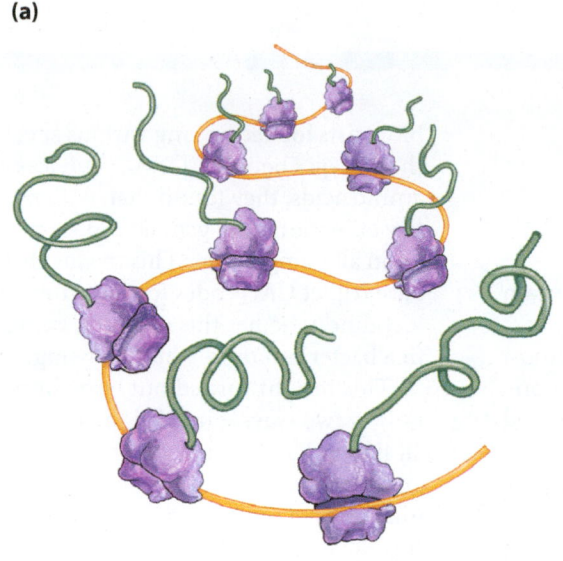

(b)

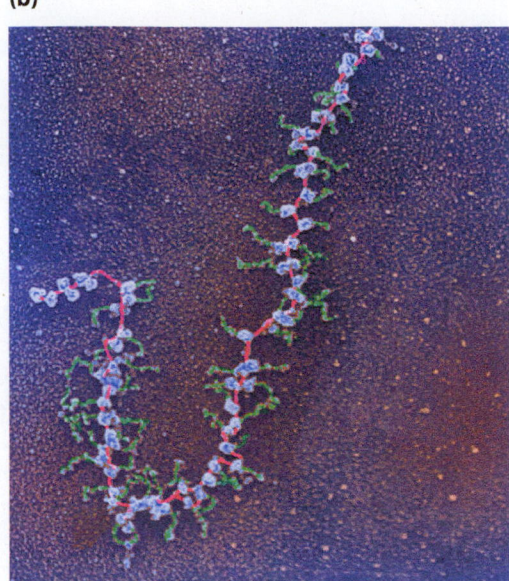

Figure 14.11
Mass Production

(a) An mRNA transcript can be translated by many ribosomes at once, resulting in the production of many copies of the same protein.

(b) A micrograph of this process in operation.

14.5 Genetic Regulation

From what you've read so far, you might have gotten the impression that transcription never ends for genes—that all genes are constantly being transcribed, thus bringing about a constant production of the proteins they code for. But a genome that worked like this would be like a restaurant that ceaselessly turned out every dish on its menu, no matter how many customers it had. Life is complex, which means that organisms need to finely tune their production of proteins. To look at but one example of this, after we eat a meal, one of the proteins that comes into play is a hormonal protein, called CCK, that prompts our pancreas to release enzymes that break down the food we've consumed. Now, in a person who hasn't eaten in awhile, the gene for the CCK protein is transcribed only at very low levels. After a person consumes a meal high in proteins and fats, however, the story changes. Now transcription of the CCK gene proceeds at high levels, with the result that much greater quantities of the CCK protein start being produced. In short, the activity of the CCK gene is *regulated*—something that is true of all genes. The protein production you've been reading about operates through a system of genetic regulation that turns out to be very important. Indeed, cutting-edge research conducted over the past decade has begun to suggest that it is the regulation of genes, rather than genes themselves, that has been the most important factor in making human beings as complex as they are and in producing the fantastic diversity we see today in the living world. We'll now go over some of the highlights of genetic regulation, but before we do, we need to look a little more closely at the entity that gets regulated, the **human genome,** defined here as the full complement of DNA found in the nucleus of each human cell.

Our Genome Contains More than Genes

At one time, scientists thought that genes in the human genome sat right beside each other like pearls on a string. Beginning in the 1970s, however, scientists began to realize that the genome contains large segments of DNA that do not code for proteins. Indeed, over time it became apparent that, in all eukaryotic organisms, it is the *coding* sections of the genome that are few and far between. Completion of the Human Genome Project in 2003 revealed that less than 2 percent of the human genome codes for proteins. To gain some perspective on this number, consider that although the DNA in any of our cells would stretch to about 6 feet in length if uncoiled, less than 1 inch of this length is protein-coding sequence.

So, what is the rest of this DNA doing in our genome? When scientists first began to investigate this question, they thought that most non-coding DNA served no function at all. This seemed reasonable since, at the time, long stretches of human DNA were known to have come about through a kind of genetic invasion: During the course of evolution, *viruses* were able to insert their DNA into our genome through a cut-and-paste operation. This would have resulted only in some small additions to the genome but for the fact that some inserted viral sequences could make copies of themselves—copies that *also* got spliced into the genome, along with some internally derived sequences that had this same capability. The ultimate effect of this was big: Some 45 percent of the DNA in our genome seems to have been produced through this copy-and-splice process. Scientists in the early days of DNA sequencing were aware of the origins of this DNA and thus reasonably assumed it amounted to "junk DNA," which is to say DNA that

 Control of Transcription

 Human Genome Sequencing

ESSAY

The Importance of the Genetic Code

Like scholars trying to extract meaning from an ancient written language, biologists in the early 1960s were trying to extract meaning from an ancient molecular language. They knew that triplets of mRNA bases were coding for amino acids, but what was the linkage? If you had a given RNA triplet, *which* amino acid did it code for? Although it took them the better part of a decade to find out, eventually they succeeded. If you look at **Figure 1**, you can see a summation of their work—the genetic code in its entirety. Let's look a little closer at this code to understand its nature and significance.

One of the notable things about the code is that not every mRNA triplet in it codes for an amino acid. Note, in Figure 1, that three mRNA codons specify "stop" codes that bring an end to the synthesis of a polypeptide chain. One triplet (AUG) specifies a special amino acid, methionine (met), which serves as the starting amino acid for most polypeptide chains. Note also that the genetic code is redundant. The amino acid phenylalanine (phe), for example, is coded for not only by the UUU triplet but by UUC as well. Indeed, almost all the amino acids are coded for by more than one mRNA codon. Leucine (leu) and serine (ser) are coded for by six.

What is the importance of knowing this code? For one thing, it gives scientists a powerful tool. If they know one "end" of a genetic chain, they are instantly given insight into the other. Knowing something about a DNA sequence means knowing something about a protein sequence—and vice versa. If you know, for example, that a protein has a tryptophan (trp) amino acid at one point in its sequence, you know that a UGG triplet has to exist at a corresponding point in the *messenger RNA* sequence for this protein. This then tells you to look for an ACC sequence at this point in the DNA sequence for the protein.

Apart from providing these kinds of practical linkages, the cracking of the genetic code also had the effect of revealing how unified life is at the genetic level. As biologists looked among various species at the linkages between DNA sequences and amino acids, they found that, with only a few exceptions, the genetic code is universal in all living things. This means that the base triplet CAC codes for the amino acid histidine, whether this coding is going on in a bacterium or in a human being.

This insight turned out to be important in two ways. First, it is evidence that all life on Earth is derived from a single ancestor. How else can we explain all of life's diverse organisms sharing this very specific code? Such a complex molecular linkage—this triplet equals that amino acid—is extremely unlikely to have evolved *more* than once. What seems likely is that it came to be employed in an ancient common ancestor, after which it was passed on to all of this organism's descendants—every creature that has subsequently lived on Earth. Note, then, that this code has been passed on from one generation to the next, and from one evolving species to the next, for billions of years. We humans share in an informational linkage that stretches back billions of years and runs through all the contemporary living world.

Apart from this, the universality of the genetic code has a very practical consequence. It means that genes from one organism can function in another. This has both good and bad consequences for human beings. First the bad: Viruses that cause diseases ranging from colds to AIDS can "hijack" the human cellular machinery precisely because their genes function in human cells. (The human cellular machinery will put together proteins whether these proteins are called for by human or viral DNA.) Now the good news: Using biotechnology processes you'll be learning about in Chapter 15, human beings can today use viruses and bacteria to manufacture all kinds of products, including medicines such as human insulin and human growth hormone. In these cases, human genes are being put to work inside microorganisms. This transferability of genes from one species to another has, as its basis, the universality of the genetic code.

second base

first base	U	C	A	G	third base
U	UUU phe	UCU ser	UAU tyr	UGU cys	U
	UUC	UCC	UAC	UGC	C
	UUA leu	UCA	UAA Stop	UGA Stop	A
	UUG	UCG	UAG Stop	UGG trp	G
C	CUU leu	CCU pro	CAU his	CGU arg	U
	CUC	CCC	CAC	CGC	C
	CUA	CCA	CAA gln	CGA	A
	CUG	CCG	CAG	CGG	G
A	AUU ile	ACU thr	AAU asn	AGU ser	U
	AUC	ACC	AAC	AGC	C
	AUA	ACA	AAA lys	AGA arg	A
	AUG met (Start)	ACG	AAG	AGG	G
G	GUU val	GCU ala	GAU asp	GGU gly	U
	GUC	GCC	GAC	GGC	C
	GUA	GCA	GAA glu	GGA	A
	GUG	GCG	GAG	GGG	G

Figure 1
The Genetic Code Dictionary
If we know what a given mRNA codon is, how can we find out what amino acid it codes for? This dictionary of the genetic code offers a way. Say you want to confirm that the codon CGU codes for the amino acid arginine (arg). Looking that up here, C is the first base (go to the C row along the "first base" line), G is the second base (go to the G column under the "second base" line), and U is the third (go to the codon parallel with the U in the "third base" line).

did neither harm nor good to human beings but that simply resided in the human genome like an uninvited house guest.

A good deal of non-coding DNA may yet turn out to be junk DNA in this sense, but as we'll see, we now have reason to wonder whether any of the 3.2 billion base-pairs in our genome deserves this label. What has been clear for decades, however, is that some of our non-coding DNA does serve a set of vital functions—a set of regulatory functions that we'll now turn to.

Promoters, Enhancers, and Proteins That Bind Them

When organisms develop as embryos, there are genes in them that control the development of their mid-body or "thoracic" structures, such as the vertebrae that make up their backbones. One of these genes, called *Hoxc8*, is nearly identical in creatures as diverse as chickens, mice, and snakes. Yet, as you can see in **Figure 14.12**, a chicken has 7 vertebrae, while a mouse has 13. So the same developmental gene is at work in these two animals, yet it is yielding very different outcomes in them. If we ask what accounts for this, the answer is that the mouse *Hoxc8* is being *transcribed* more than the chicken *Hoxc8*. More transcription means the mouse has more of the protein that *Hoxc8* codes for; this in turn means a broader distribution of this protein in the mouse embryo, and the result is more vertebrae in the mouse.

But why would *Hoxc8* get transcribed more in a mouse than a chicken? You may recall that for any gene to be transcribed, an enzyme complex called RNA polymerase has to go down the line on the gene's sequence, pairing up its DNA bases with complementary RNA bases. But this is not an all-or-nothing proposition. RNA polymerase is *more or less* likely to get busy transcribing a gene depending on how it is chemically positioned on a DNA sequence that lies just "upstream" from the gene. This is the "promoter" sequence that you can see in the figure—our first example of non-coding DNA. For transcription to take place in any gene, a multipart protein complex must bind with the promoter sequence to help RNA polymerase align with it.

The question then becomes: What factors stand to facilitate or impede this alignment? If you look just upstream from the promoter in the figure, you can see a facilitating factor: a sequence called an enhancer. This, too, is a sequence of DNA, but like the promoter, it does not code for anything. Its sole role is to serve as a binding site for proteins that help get RNA polymerase positioned at the promoter. Now, remember that the *Hoxc8* gene is nearly identical in lots of animals. But this is not true of the *Hoxc8* enhancer; its sequence differs markedly from chickens, to mice, to snakes. Accordingly, the mouse's enhancer binds with a different set of aligning proteins than does the chicken's. And these mouse proteins more strongly induce transcription than do the chicken's. The result is more *Hoxc8* transcription in the mouse and ultimately more vertebrae in it.

So, note what's at work here. First, there are two stretches of DNA—the promoter and enhancer

Figure 14.12
Genetic Regulation in Action

(a) For transcription of any gene to begin, RNA polymerase must be aligned at a sequence of DNA called the promoter, which lies just upstream from a given gene—in this case the *Hoxc8* gene. A multipart protein complex, seen here surrounding RNA polymerase, is active in getting a base level of transcription going. Another DNA sequence, called an enhancer, can facilitate transcription by acting as a binding site for other proteins that can interact with the promoter-site protein complex.

(b) The enhancer in the mouse has a different DNA sequence than that of the enhancer in the chicken—a sequence that binds a different set of proteins. These proteins interact with the promoter sequence proteins in such a way as to better align RNA polymerase at the promoter. The result is increased transcription of the *Hoxc8* gene in the mouse.

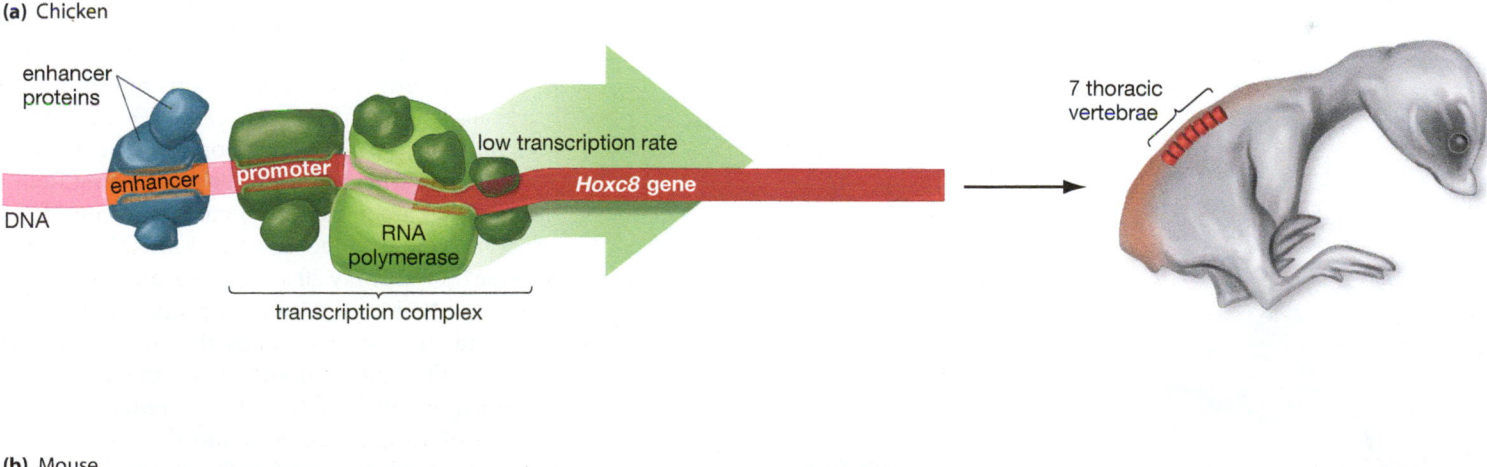

(a) Chicken

enhancer proteins

enhancer

promoter

DNA

RNA polymerase

transcription complex

low transcription rate

Hoxc8 gene

7 thoracic vertebrae

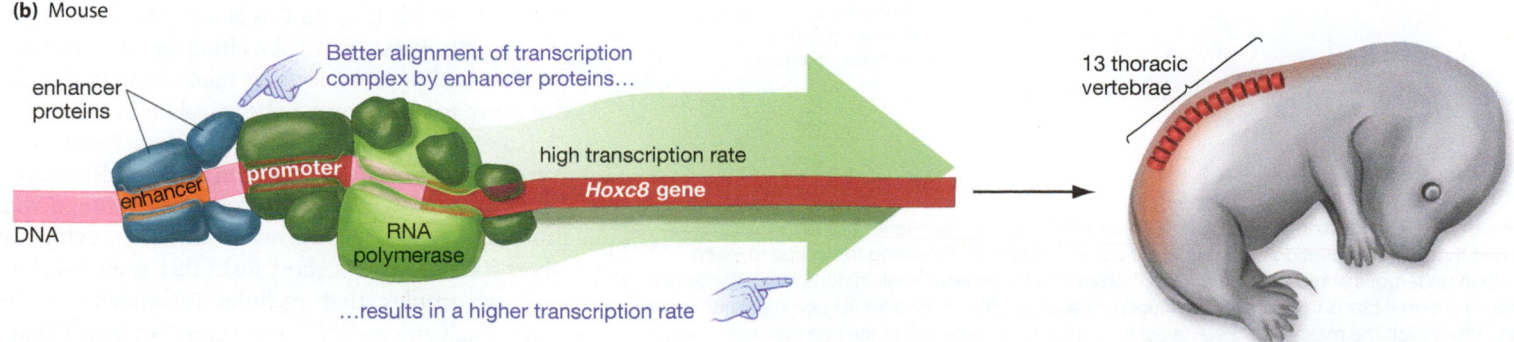

(b) Mouse

Better alignment of transcription complex by enhancer proteins…

enhancer proteins

enhancer

promoter

DNA

RNA polymerase

high transcription rate

Hoxc8 gene

…results in a higher transcription rate

13 thoracic vertebrae

RNA Processing

sequences—that do not code for protein but that do serve a regulatory function (facilitating transcription by RNA polymerase). Second, note what it is that's binding to these regulatory sequences: proteins. One multipart protein complex binds to the promoter sequence (getting a base-level of transcription going), while another complex binds to the enhancer sequence (increasing the rate of transcription). Now, where did these proteins come from? They were produced through transcription and translation, just like any other protein. Thus, DNA codes for proteins that then *feed back* on DNA itself, controlling its transcription. (Appropriately enough, these DNA-binding proteins are called transcription factors.) With this,

you can begin to see what lies at the root of genetic regulation: The whole system is *self*-regulating. DNA brings forth proteins and some of these proteins then go on to control the transcription of DNA.

Finally, notice that it is not necessary to have a change in the sequence of a gene to have a change in the effect of that gene. The mouse and the chicken have the same *Hoxc8* gene; where they differ is in their segments of regulatory DNA for this gene. Over evolutionary time, mutations worked to make the *Hoxc8* regulatory segments different in the two species. So when we think of mutation working hand-in-hand with evolution to channel species in different directions, it's not necessarily a mutation of *genes* that we're talking about; it may be a mutation of regulatory DNA. Indeed, most of the scientists who look at evolution by studying the embryonic development of animals have become convinced that regulatory mutations have been far more important than gene mutations in bringing about the differences we see among animals. Why? Because mutations to regulatory sequences allow an organism to *fine-tune* its use of a protein (as with *Hoxc8*), whereas mutations to protein coding sequences are likely to result in the *elimination* of a protein— often a disastrous event for an organism, since so many proteins serve a vital function.

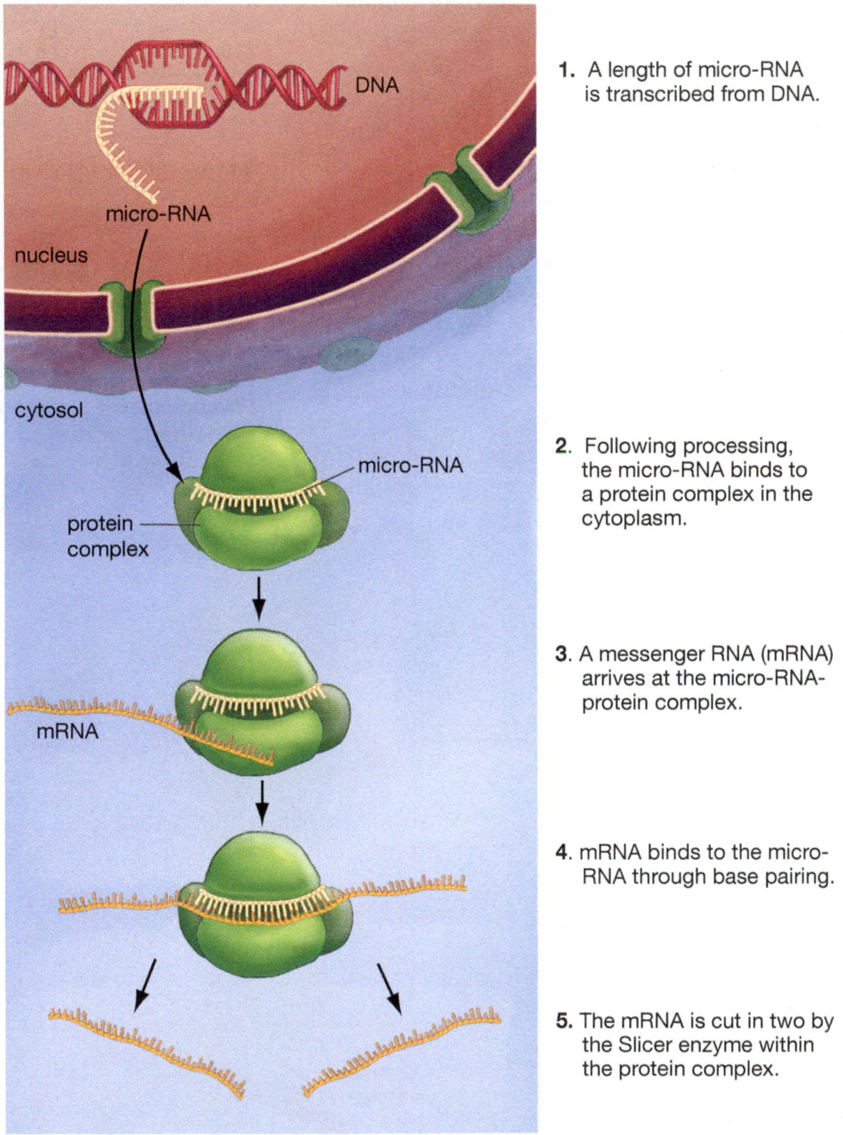

A Second Form of Regulation: Micro-RNAs

The message "DNA makes RNA makes protein" has come through loud and clear in this chapter, but this message actually needs a little modification. DNA does indeed code for RNA, but not all RNA then goes on to code for protein. As it turns out, some of the RNA that results from transcription ends up serving a bricks-and-mortar function: Once it's produced, it goes on to help make up ribosomes or transfer RNA. But transcription also produces non-coding RNA sequences that are *regulatory*. A collective name for one variety of these sequences is *micro-RNAs*—"micro" because they are only 20 to 22 bases long. So far, about 500 human micro-RNA sequences have been discovered, and one estimate holds that twice that many may actually exist. Almost all of the micro-RNAs identified to date have the effect of reducing the production of specific proteins, and the most common means by which they do this is well known: They interfere with *messenger* RNAs, either by targeting them for destruction or by locking them away before they can move to a ribosome to be translated.

How would this work? If you look at **Figure 14.13**, you can see one example. A length of micro-RNA comes into being through transcription, goes through some processing, and then moves to the cell's cytoplasm, where it's cleaved to a short form that is enclosed in a protein complex that includes an enzyme named Slicer. Each micro-RNA has a base sequence that is

1. A length of micro-RNA is transcribed from DNA.

2. Following processing, the micro-RNA binds to a protein complex in the cytoplasm.

3. A messenger RNA (mRNA) arrives at the micro-RNA-protein complex.

4. mRNA binds to the micro-RNA through base pairing.

5. The mRNA is cut in two by the Slicer enzyme within the protein complex.

Figure 14.13

Micro-RNAs and Gene-Silencing

Genes can be silenced through the activity of micro-RNAs—short sequences of RNA that are produced through transcription but that do not code for proteins. Following transcription, each micro-RNA undergoes some processing and then moves to a special protein complex in the cell's cytoplasm. There it binds to a recently produced messenger RNA transcript via complementary base pairing, after which the messenger RNA is cut in two by an enzyme within the complex called Slicer.

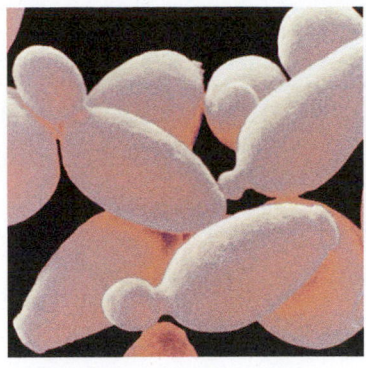

Saccharomyces cerevisiae
(baker's yeast)

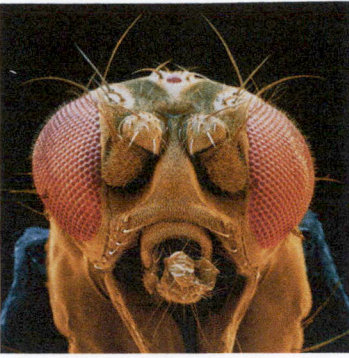

Drosophila melanogaster
(fruit fly)

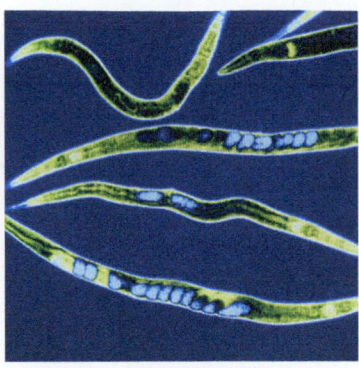

Caenorhabditis elegans
(roundworm)

Arabidopsis thaliana
(mustard plant)

Estimated number of genes:

6,034   13,061   19,099   25,000

complementary to a messenger RNA that has recently been produced through transcription. As you might guess, base pairing brings the micro-RNA and mRNA sequences together, at which point Slicer cuts the messenger RNA in two. The result? Decreased production of the protein that the messenger RNA coded for.

Now, it may seem strange that our genome would first produce messenger RNAs that it then goes on to destroy, but in some instances it may be cost-effective to "silence" genes after they have been transcribed rather than to stop their transcription altogether. If we ask how important this gene silencing is, the answer seems to be: very important. In animal studies conducted to date, scientists have found micro-RNAs operating in every organ system they have examined—in the eyes, the lungs, the liver, and the pancreas, for example. And in 2007, several teams of scientists used genetic engineering to produce groups of mice that *lacked* some of their normal micro-RNAs. The result was one group of mice that could not fight off infections and another group that died young because they developed holes in their hearts. You might think that any part of the genetic operation that is this critical would have been well understood for decades. In fact, however, the existence of micro-RNAs was completely unknown to scientists until the late 1990s. Since that time, so many roles have been discovered for them that some scientists have become convinced they are active in regulating all the major metabolic processes in the body.

A Third Form of Regulation: Alternative Splicing

When the first stage of the Human Genome Project was completed in 2000, biologists got a surprise: The human genome was shown to contain far fewer genes than the 100,000 that had been the standard estimate for years. The teams that sequenced the genome first came up with an estimate of about 30,000 human genes and then, three years later, produced the estimate that still stands today—between 20,000 and 25,000 genes. These figures actually were surprising in two ways: We learned from them not only that we have fewer genes than previously imagined but that some rather humble organisms have nearly as many genes as we do. You can see some comparisons in **Figure 14.14**. Note that we humans—with our vision, speech, and 10 trillion cells—have, at most, 6,000 more genes than the microscopic roundworm *Caenorhabditis elegans,* which is eyeless, speechless, and made up of 959 cells. So how are we able to do so much more with our 25,000 genes than the roundworm does with its 19,000? Part of the explanation lies in a third form of genetic regulation.

Remember how, back in this chapter's section on transcription, we noted that the initial length of RNA that is transcribed from DNA—the primary transcript—has to undergo some editing before it becomes messenger RNA? **Figure 14.15** presents the classic view of how this editing takes place. As you can

Figure 14.14
Not as Much Difference as We Thought

Genome sequencing has revealed that the human genome probably contains about 20,000–25,000 genes. This is, at most, about 6,000 more genes than the tiny roundworm *C. elegans* has and about 12,000 more than the *Drosophila* fruit fly has.

Figure 14.15
Editing Out Non-coding Sequences

In eukaryotes, lengths of DNA that are transcribed contain sequences that code for amino acids as well as sequences that do not code for them. Both kinds of sequences are copied onto RNA primary transcripts. Once this is done, however, the non-coding sequences (called introns) are removed by enzymes through the editing process shown, while the coding sequences (called exons) are spliced back together. The result is a messenger RNA molecule.

exon 1   intron   exon 2   intron   exon 3   primary transcript

enzyme

enzymes cut out the introns

messenger RNA

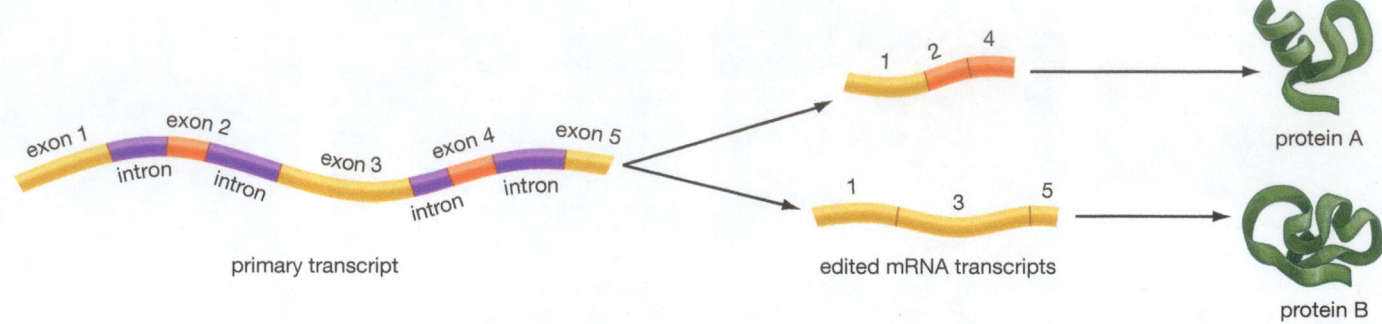

Figure 14.16
Alternative Splicing

In the figure, a single primary transcript, containing five exons or protein-coding portions, has been transcribed from a single gene. The human genetic machinery can then splice these exons together in alternative ways to yield different proteins—A or B in the figure.

see, the primary transcript has some sequences cut out of it, after which the sequences that remain are spliced back together. The sequences that are cut out are called introns (for *in*tervening sequences), while the sequences that are retained are called exons (because most of them are *ex*pressed as proteins). It is the exons that code for protein, in other words, while the introns are not coding for protein. Given the exons' seemingly more important role, you might think that a gene would mostly be made up of them, but the reverse is true: Introns account for more than 90 percent of the length of the average human gene.

Note that, in the classic view, all the introns are edited out, while all the exons are retained. But if you look at **Figure 14.16**, you can see that many genes are more versatile than this in that their primary transcripts can be edited in different ways. Each of these alternative edits then yields a different messenger RNA, and each of these mRNAs can yield a different protein.

What's at work here is another form of genetic regulation, this one known as **alternative splicing,** a process in which a single primary transcript can be edited in different ways to yield multiple messenger RNAs. This is one of the ways we do so much with our relatively small number of genes. Thanks to alternative splicing, our 20,000–25,000 genes don't produce 20,000–25,000 proteins; they produce more than 90,000 proteins. Under one estimate, 94 percent of human genes undergo alternative splicing, whereas for the *C. elegans* roundworm, the figure is about 10 percent.

The Importance of Genetic Regulation

The three forms of genetic regulation we've looked at may at first just seem to be a fine-tuning of the basic transcription and translation process, but we have reason to believe that regulation is much more important than this. To look at the examples we've just gone over, what gives mice more vertebrae than chickens? What has been the most important genetic factor in producing the various forms of animals seen in the world? What is a critical factor in making human beings more complex than roundworms? In every case, the answer is genetic regulation.

To extend this notion, what makes for complexity when we look across the living world at *many* different kinds of organisms? **Figure 14.17** gives you some idea. What seems to go hand-in-hand with complexity is the proportion of an organism's genome that does *not* code for proteins—the proportion of its genome that potentially is involved in genetic regulation. And here, we humans really do shine. Remember how more than 98 percent of our DNA is non-coding? This is the highest percentage we know of in any organism.

The obvious implication of all this is that the gene—defined here as a sequence of DNA that codes for protein—seems to be just one player in the genetic operation rather than the central player in it. The reason it's necessary to ask whether this is true, however, is that many of the elements of regulation we've just gone over sit right at the cutting edge of contemporary genetics research. The basic process of protein production we looked at earlier—DNA makes RNA makes protein—has been understood

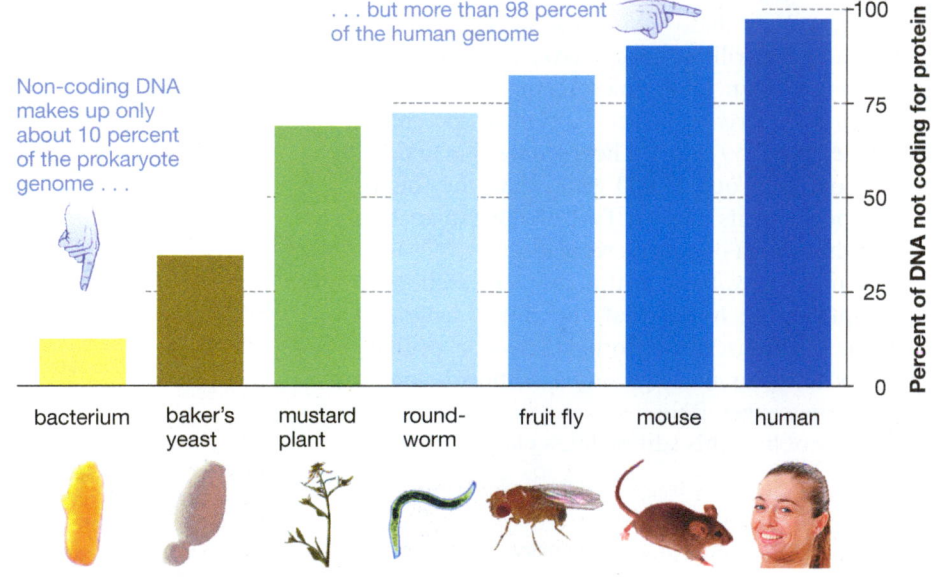

Figure 14.17
What Makes for Complexity in Living Things?

The seven categories of living things in the figure are arranged from least complex on the left to most complex on the right. Note that complexity in these organisms increases in accordance with the proportion of DNA in their genomes that does not code for proteins.

(Adapted with permission from "Increasing Biological Complexity Is Positively Correlated with the Relative Genome-wide Expansion of Non-Protein-Coding DNA Sequences," Ryan J. Taft and John S. Mattick, *Genome Biology*, 5:1, 23, 2003.)

for decades, at least in outline. We are only beginning to comprehend, however, the nature of micro-RNAs and some of the non-coding portions of the genome. There are researchers who feel that these component parts of genetics are important enough that we have, in essence, *mis*understood the genetic operation for decades. While the jury is still out on this issue, what is clear is that recent discoveries about regulation and non-coding DNA are forcing biologists to ask fundamental questions about genetics. How does our genetic machinery really work? The hope is that research will soon be providing some answers.

14.6 Genetics and Life

The 20,000–25,000 genes that human beings have are contained in a genome that is about 3.2 billion base pairs long. To make this large number a little less abstract, consider the following exercise. If we took the base sequence of the human genome and simply arrayed the single-letter symbols for the bases on a printed page, going like this:

AATCCGTTTGGAGAAACGGCCCTATTG
GCAGCAAGGCTCTCGGGTCGTCAACG
CGTATTAAACATATTTCAAGGCTCTA ...

it would take about 1,000 telephone books, each of them 1,000 pages long, merely to record it all. We're talking, then, about an unbroken series of these base symbols that goes on for a million pages. That is one measure of the size of the human genome. Simply maintaining this genome—making sure that it is passed on from one cell and generation to another—requires that all of these 3.2 billion bases be faithfully copied each time a cell divides. (They have to be copied twice, actually, because we have two copies of each chromosome.) When we start thinking about *using* this genome—about producing proteins from it—all the layers of genetic regulation just sketched out come into play. Genes help bring about proteins; some of these proteins feed back on other genes and control their production of other proteins; RNA sequences get spliced in different ways, yielding different proteins, and so forth.

The obvious message here is that the human genetic operation is unimaginably complex. But it turns out that humans are only marginally special in this regard. If we were to walk through the genetic operation of even a humble bacterium, you would see that its complexity is stunning, too. What's important to recognize is that this complexity is necessary because it *enables life*. Living things must obtain energy, they must grow, they must respond constantly to their environment, they must reproduce, and they must coordinate all these activities. Only an incredibly sophisticated system of information storage and use could allow a self-contained entity to do all these things. And genetics is exactly such a system. With each twitch of a muscle or swig of a soft drink, information repositories called genes are opened or closed within us, setting in motion the delivery of the tiny working entities called proteins—or stopping delivery of them, if that's what is called for.

Most of the universe is made up of relatively simple, inanimate objects. But we living things are different. We get to inherit and then use an enormous library of information that was amassed by our ancestors over a period of 3.8 billion years. And this endowment has made us complex, capable, and conscious to a greater degree than anything else we know of in the universe. Looked at one way, you might say that all living things are born rich.

Biotechnology Is Next

If each gene in a genome is something like a book in a library, think what the effect would be if organisms could, in effect, check books out of each other's libraries. Actually, you don't have to imagine what this would be like; it's happening right now with great frequency. Soybeans in the fields of Indiana and Iowa have the genes of bacteria spliced into their genomes. And human genes have been spliced into the genomes of bacteria so that these one-celled creatures can turn out human medicines. Moreover, we are now able to read individual letters in our own genetic library with an efficiency that lets us catch criminals on the basis of flakes of dandruff they leave at crime scenes. This may sound fanciful, but it's not. It's biotechnology, the subject of Chapter 15.

SO FAR ...

1. A central feature of genetic regulation is that proteins called transcription factors bind to non-coding _____, thus regulating the _____ of genes.

2. Small lengths of RNA called _____ have the effect of "silencing" genes in the genome by means of targeting their _____ for destruction or otherwise interfering with them.

3. In eukaryotic organisms such as ourselves, the RNA chains, called primary transcripts, that code for proteins must undergo _____ on their way to becoming _____. The segments of the primary transcripts that are removed in this process are called _____, while the segments that are retained are called _____.

CHAPTER

14 REVIEW

Summary

14.1 The Structure of Proteins

- Proteins are composed of building blocks called amino acids. A string of amino acids is called a polypeptide chain, which when folded into a three-dimensional shape is a protein. Hundreds of thousands of proteins are put together from a starting set of 20 amino acids. The order in which the amino acids are linked in a polypeptide chain determines which protein will be produced. Proteins often are composed of two or more linked polypeptide chains. **(p. 245)**

14.2 Protein Synthesis in Overview

- There are two principal stages in protein synthesis: transcription, during which the information encoded in DNA is copied onto a length of messenger RNA (mRNA); and translation, during which amino acids brought to a ribosome by transfer RNA (tRNA) molecules are linked together within the ribosome in the order specified by the mRNA sequence. **(p. 245)**

14.3 A Closer Look at Transcription

- The information in DNA is transferred to mRNA through complementary base pairing. The enzyme RNA polymerase unwinds the DNA sequence to be transcribed and then strings together a chain of RNA nucleotides that is complementary to it. In all eukaryotes, the initial RNA chain transcribed from a DNA sequence is the primary transcript, which must undergo editing before becoming an mRNA chain. **(p. 247)**

- Each three coding bases of DNA pair with three RNA bases, but each group of three

mRNA bases then codes for a single amino acid. Each triplet of mRNA bases that codes for an amino acid is called a codon. The genetic code is the inventory of linkages between base triplets and the amino acids they code for. **(p. 248)**

14.4 A Closer Look at Translation

- Transfer RNA serves as a bridging molecule in protein synthesis thanks to its ability to bind with both amino acids and nucleic acids (in the form of mRNA). A given tRNA molecule binds with a specific amino acid in the cell's cytoplasm and then transfers that amino acid to a ribosome in which an mRNA transcript is being "read." There, another portion of the tRNA molecule, an anticodon, binds with the appropriate codon in the mRNA chain. **(p. 249)**

- Ribosomes, the "workbenches" of protein synthesis, are composed of proteins and ribosomal RNA (rRNA). Each bears A, P, and E binding sites, which tRNA molecules occupy, thus facilitating the synthesis of a polypeptide chain. **(p. 250)**

- In translation, a succession of tRNA molecules arrive at a ribosome, bound to their appropriate amino acids, and then bind to their appropriate codon in the mRNA transcript. As this proceeds, a succession of amino acids is linked together into a polypeptide chain. **(p. 251)**

14.5 Genetic Regulation

- Protein production is carefully regulated in living things in that genes are transcribed in accordance with the needs of an organism. **(p. 253)**

- Less than 2 percent of the DNA in the human genome codes for proteins. Some non-coding segments of DNA may

be "junk" that never had a function, but scientists are increasingly skeptical that any human DNA fits this description. **(p. 253)**

- All gene transcription requires that RNA polymerase be properly aligned at a non-coding promoter DNA sequence that lies just "upstream" from a gene sequence. There often is also a non-coding enhancer sequence that lies at some distance from the promoter sequence. Proteins called transcription factors bind to both the promoter and enhancer sequences, thus facilitating the alignment of RNA polymerase at the promoter. Transcription factors are produced through normal transcription and translation. **(p. 255)**

- DNA codes for several different forms of RNA, but of these, only mRNA then goes on to code for proteins. DNA also codes for several varieties of regulatory RNA, one of which is micro-RNA, most of the known sequences of which reduce the production of particular proteins, often by targeting their mRNAs for destruction. **(p. 256)**

- In all eukaryotes, the RNA primary transcript undergoes editing in which some sequences (called introns) are cut out and the remaining sequences (called exons) are spliced back together. The result is a completed mRNA chain. Introns do not code for protein, but most exons do. **(p. 257)**

- Some relatively simple organisms have nearly as many genes as human beings do. Human beings are much more complex than these organisms, however, thanks in part to alternative splicing, in which a primary transcript is edited in different ways, thus yielding different mRNA chains, which in turn result in different proteins. **(p. 257)**

- Many researchers believe that the regulation of genes is the most important factor in bringing about the differences among organisms. There seems to be little relation between the number of genes an organism has and its complexity. The complexity of organisms does seem to correlate with the proportion of their DNA that does not code for protein. (p. 258)

14.6 Genetics and Life

- Life is made possible by the fantastic ability of genetic systems to store, use, and pass on information. (p. 259)

Key Terms

alternative splicing 258

anticodon 250

codon 249

genetic code 249

human genome 253

messenger RNA (mRNA) 248

polypeptide 245

ribosomal RNA (rRNA) 250

ribosome 246

RNA polymerase 248

transcription 246

transfer RNA (tRNA) 250

translation 246

Understanding the Basics

Multiple-Choice Questions

(Answers are in the back of the book.)

1. A chain of _____ results in a polypeptide chain that folds up into a _____.
   a. RNA bases … transcript
   b. amino acids … gene
   c. proteins … ribosome
   d. DNA bases … gene
   e. amino acids … protein

2. The genetic code is an inventory of which _____ specify(ies) which _____.
   a. three mRNA bases … amino acid
   b. mRNA base … amino acid
   c. amino acids … proteins
   d. polypeptide chain … protein
   e. three DNA bases … three mRNA bases

3. The "t" of tRNA stands for:
   a. tripartite
   b. teleporting
   c. transfer
   d. tracking
   e. transcribing

4. Which of the following statements is true regarding ribosomes?
   a. They are composed of two separate units, each made of rRNA combined with protein.
   b. They are used to translate the mRNA sequence into a protein composed of amino acids.
   c. They have three different tRNA binding sites
   d. all of the above
   e. a and b only

5. Why can transfer RNA be referred to as a "bridging" molecule?
   a. It forms a chemical bridge between DNA and messenger RNA.
   b. It creates the linkage between the base pairs of the double helix.
   c. It links species through evolution.
   d. It links ribosomes together.
   e. It binds to both amino acids and messenger RNA.

6. A typical eukaryote protein might be _____ amino acids long and could be produced in less than _____ in translation.
   a. 50 … 5 minutes
   b. 50,000 … 5 hours
   c. 5 … 5 seconds
   d. 400 … 2 minutes
   e. 4,000 … 4 minutes

7. In one form of genetic regulation, proteins called transcription factors bind to _____, thus facilitating the work of _____.
   a. ribosomes … transfer RNA
   b. messenger RNA … ribosomes
   c. non-coding DNA sequences … RNA polymerase
   d. coding DNA sequences … RNA polymerase
   e. transfer RNA … ribosomes

8. An intron is a sequence of _____ RNA that is _____ a primary transcript, while an exon is a sequence of _____ RNA that is _____ a primary transcript.
   a. non-coding … cut out of … coding … left in
   b. transfer … cut out of … non-coding … left in
   c. coding … left in … non-coding … cut out of
   d. transfer … left in … transfer … cut out of
   e. non-coding … left in … coding … left in

Brief Review

(Answers are in the back of the book.)

1. What two great tasks are carried out by the genetic machinery of living things?

2. What are the three types of RNA that are used by a cell any time a protein is produced? What do their abbreviations stand for? What are their functions?

3. Define transcription and translation.

4. During transcription, DNA neither unwinds itself nor pairs its own bases with complementary RNA bases. How are these tasks accomplished?

5. How many DNA bases does it take to code for an RNA codon? How many amino acids does an RNA codon code for?

6. Describe one means by which the genetic operation is self-regulating.

7. What recent research results have given scientists reason to doubt the importance of genes relative to the importance of the regulation of genes?

Applying Your Knowledge

1. Given the following sequence of mRNA, what would the resulting amino acid sequence be?

AUGAAACGGGGACCAAUGGAUAACUAA

MasteringBiology®

1. MasteringBiology Assignments
Activities: Overview of Protein Synthesis; Transcription; Translation; Control of Transcription; Human Genome Sequencing; RNA Processing; **Bioflix:** Protein Synthesis; **Chemistry Review:** Nucleic Acids; **Current Events:** The Search for Genes Leads to Unexpected Places; **Building Vocabulary;**

Questions: Bioflix Quiz; Reading Questions; Test Bank

2. eText
Read your book online, search, take notes, highlight text, and study more effectively.

3. The Study Area
Self-study tools to test comprehension.

CHAPTER 15

The Future Isn't What It Used to Be:
Biotechnology

By manipulating the information contained in DNA, scientists are raising hopes—and some fears—on a grand scale.

Available on MasteringBiology®
Activities: Copying DNA through PCR; **ABC News Video:** Genetically Altered Salmon; **Building Vocabulary; Questions:** Reading Questions; Test Bank

Is it possible to produce an artificial spider's thread? Scientists have done it by splicing genes into bacteria—just one of the marvels made possible through biotechnology.

The little mouse that can be seen in **Figure 15.1** looked typical enough to anyone who saw it after it was born in 2008. It was active, it fed normally, and it even went on to father a litter of healthy offspring. Admittedly, it was produced through cloning, but that in itself wasn't particularly surprising—since the mid-1990s, lots of animals have been produced through cloning. This mouse was different, however, in that it was a clone of a mouse that had been dead for 16 years. To be sure, the body of this "donor" mouse had been kept in a deep freeze for all those years, but outside of this, nothing special had been done to preserve its cells or its DNA. Teruhiko Wakayama and his colleagues in Kobe, Japan, simply thawed out the frozen mouse, isolated some of its DNA, and used a double-round of cloning procedures to produce the mouse in the photo and three more like it.

As you might imagine, the point of this work was not to give the world a few more mice. Instead, Wakayama and his colleagues wanted to know whether it was possible to use tissue that had been frozen to "resurrect" members of a given species. The potential value of this work stemmed from the fact that a growing number of Earth's animals—the Siberian tiger, the white rhino, the mountain gorilla—are teetering on the brink of extinction. Given this, Wakayama reasoned, society may wish to develop a means of bringing such species back *after* all their living members are gone. This kind of resurrection would only be thought of as a last-ditch means of preserving a species—as a procedure that might be employed should all other conservation efforts fail. But in a world where numerous species now have a tenuous grasp on existence, research such as this ultimately could provide society with one last card to play.

As it happened, the very month Wakayama's results were published, society was abuzz with talk of another kind of animal resurrection, this time of a species that *has* gone extinct—a long time ago. Researchers from Pennsylvania State University succeeded in sequencing nearly 3 billion bases of DNA extracted from a Siberian wooly mammoth that died nearly 20,000 years ago. And the scientists who accomplished this feat believe they'll eventually be able to get an accurate reading of the entire 4 billion bases of woolly mammoth DNA. It was a short step from this to wondering whether mammoths could be brought back to life through a combination of genetic engineering and cloning. As it turns out, though, there is a long list of hurdles standing in the way of achieving this resurrection—no one is going to be producing a live woolly mammoth anytime soon. Yet Pennsylvania State researcher Stephan C. Schuster now believes this feat is at least theoretically possible, whereas in the period before his team's sequencing accomplishment, he regarded the idea as "science fiction."

But isn't this the way with science and living organisms these days—things that once were pure fantasy now have actually been done or at least seem conceivable? In one instance after another, scientists working in the field known as biotechnology have been upsetting our notions of what is possible and what is impossible. We may assume that all animals must have both abiological mother and father, but the cloned mice noted earlier arguably had neither. We may assume that humans and sheep are fundamentally separate entities, but scientists have been adding human "stem" cells to sheep fetuses, thus producing sheep that have mostly human liver tissues. Even the simple bacterium routinely has human genes spliced into it, which allows it to turn out human proteins for medical use. Meanwhile, if you have

Figure 15.1
A Live Mouse from Dead Tissue

The brown mouse in the photo is one of four cloned in 2008 from a mouse that died 16 years earlier. In the intervening years, the cloned mouse's tissues were kept frozen.

eaten anything made from soybeans recently, you probably have eaten food from a plant that was bioengineered. And if a criminal in your town leaves so much as six or seven skin cells on the steering wheel of a car, chances are police will be able to get a DNA profile of him, which would be bad news for him if a DNA profile of him already exists in a police database.

Taken together, achievements such as these amount to a revolution; but like all revolutions, this one seems to be inspiring both hope and fear in equal measure. For many people, the idea of bringing an animal species back from the dead doesn't mean we're living in a world that is better controlled; it means we're living in a world that is out of control. Forget about the "promise" of biotechnology, this view goes; just put the genie back in the bottle. On the other hand, who wouldn't want better medicines or cheaper foods?

In thinking about such questions, it's helpful to view biotechnology not as a monolithic entity, but rather as a collection of separate *capabilities*. There is the capability to clone, the capability to splice genes into organisms, and so forth. The purpose of this chapter is to provide you with a basic understanding of some of these capabilities. Once you see what's at work in them, you'll be in a better position to make judgments about the degree to which biotechnology holds promise or peril.

15.1 What Is Biotechnology?

The term *biotechnology* is familiar to most of us, but what does it really mean? **Biotechnology** can be defined as the use of technology to control biological processes as a means of meeting societal needs. One biological process that can be controlled is the way a cell copies its DNA every time it divides. In the 1980s, a California scientist realized there was a way to use technology to exploit this natural copying, such that a tiny DNA sample could rapidly be copied over and over, thus yielding a relatively large DNA sample. One result of this *polymerase chain reaction* or PCR process is the ability police have to get DNA "matches" in criminal cases on the basis of tiny flecks of human tissue. Like other biotech processes we'll be looking at, PCR is a marriage of both biology and technology. By the time you finish this chapter, you'll see why biotechnology is such an appropriate term.

Because biotechnology is so varied, any account of it has to be limited to a few examples chosen from among thousands that exist. In this chapter, we will consider four aspects of biotechnology:

- Transgenic biotechnology—the splicing of DNA from one species into another
- Reproductive cloning—the production of mammals through cloning
- Cell reprogramming—the production of needed cell types through the reprogramming of genomes

- Forensic biotechnology—the use of biotechnology to establish identities (of criminals, of crime victims, and so forth)

Let's start now by reviewing an area of biotechnology in which genes are moved from one organism to another.

15.2 Transgenic Biotechnology

Human beings grow to their full height under the influence of human growth hormone (HGH), which normally is secreted by the human pituitary gland. HGH's role in promoting growth is, of course, most important during childhood and adolescence. A faulty pituitary gland can greatly reduce the amount of HGH young people have in their system, leaving them abnormally short. For years, the only way to get HGH was to laboriously extract it from the pituitaries of dead human beings, a practice that not only yielded too little HGH to go around but that also turned out to be unsafe.

Enter biotechnology, which in the mid-1980s produced synthetic HGH in the following way. Using collections of human cells, the gene for HGH was snipped out of the human genome and inserted into the *Escherichia coli* bacterium. Each bacterium that took on the HGH gene began transcribing and then translating this gene, which is to say turning out a small quantity of HGH protein. These bacteria were then grown in vats by the billions. The result? Collectible quantities of HGH, clinically indistinguishable from that produced in human pituitary glands, manufactured by a biotech firm, and shipped to pharmacies worldwide.

Note that in this process, a gene was taken from one species (a human being) and spliced into another species (a bacterium). With this, the bacterium became a **transgenic organism:** an organism whose genome has stably incorporated one or more genes from another species. This same phenomenon is repeated time and time again in biotechnology. The goats you can see in **Figure 15.2** are transgenic—they have incorporated a human gene into their genome. But how is this possible? How do you cut DNA out of one genome and paste it into another? To get an answer, we'll take a walking tour of some basic biotech processes. As a starting point, imagine that, as with human growth hormone, the goal is to produce quantities of a hypothetical human protein through the use of a living organism.

A Biotech Tool: Restriction Enzymes

In the early 1970s, it became possible for scientists to cut genomes at particular places, with the discovery of **restriction enzymes.** These are enzymes, occurring naturally in bacteria, that are used in biotechnology to cut DNA into desired fragments. (In nature, bacteria use them to cut up the DNA of invading viruses.) In isolating restriction enzymes, scientists found that many of them had a wonderful property: They didn't

Figure 15.2
Transgenic Animals
A firm called GTC Biotherapeutics has specialized in splicing human DNA into the genomes of goats. The result is goat's milk that contains therapeutic human proteins, among them a protein called antithrombin that prevents the formation of blood clots.

cule encounters another GGATCC sequence, it also makes cuts, which effectively is like making a second cut in a piece of rope, giving us rope *fragments.* *Bam*HI's recognition sequence may be GGATCC, but another restriction enzyme will have a different recognition sequence and will make its cuts between a different pair of bases. Indeed, several thousand restriction enzymes have been identified so far, which cut in hundreds of different places

Note that with *Bam*HI each of the resulting DNA fragments has one strand that protrudes. Restriction enzymes that make this kind of cut are particularly valuable for they produce "sticky ends" of DNA, so named because they have the potential to *stick to* other complementary DNA sequences. In the fragment on the lower left in Figure 15.3, for example, the protruding sequence CTAG could now easily form a base pair with any piece of DNA whose sequence is the complementary GATC. So great is the need for restriction enzymes that they can now be ordered from biochemical suppliers, much as a person might order a set of socket wrenches from a hardware store.

Another Tool of Biotech: Plasmids

From our look at manufacturing human growth hormone, recall that the human gene for HGH was inserted into *E. coli* bacteria, which then started turning out quantities of this protein. The question is, how did this human gene get into a bacterium? Several methods of transfer are available, but for now let's focus on a specific kind of DNA delivery vehicle. As it turns out, bacteria have small DNA-bearing units that lie *outside* their single chromosome. These are the **plasmids,** extra-chromosomal rings of bacterial DNA that can be as little as 1,000 base pairs in length

just cut DNA randomly; they cut it at very specific places. Here is how it works with an actual restriction enzyme called *Bam*HI. The two strands of DNA's double helix are complementary, as we've seen, and they also happen to run in opposite directions. Thus, the sequence GGATCC would look like **Figure 15.3** if we were viewing both strands of the helix.

Now, the *Bam*HI restriction enzyme will move along the double helix, leaving the DNA alone until it comes to this series of six bases, known as its recognition sequence, and here it will make identical cuts on both strands of the DNA molecule, always between adjacent G nucleotides. When another *Bam*HI mole-

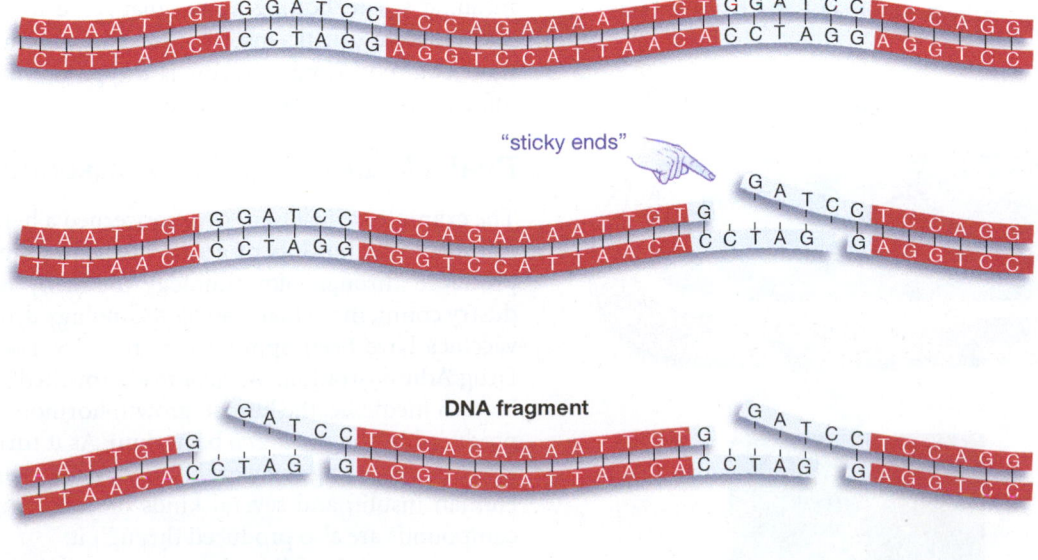

1. A portion of a DNA strand, highlighted here, has the recognition sequence GGATCC.

2. A restriction enzyme moves along the DNA strand until it reaches the recognition sequence and makes a cut between adjacent G nucleotides.

3. A second restriction enzyme makes another cut in the strand at the same recognition sequence, resulting in a DNA fragment.

Figure 15.3
The Work of Restriction Enzymes

(**Figure 15.4**). Plasmids can replicate independently of the bacterial chromosome, but just as important for biotech's purposes, they can *move into* bacterial cells.

How do plasmids do this? Bacteria are capable of taking up DNA from their surroundings, after which this DNA will function—that is, code for proteins—inside the bacterial cells. Appropriately enough, this process is known as **transformation:** a cell's incorporation of genetic material from outside its boundary. Some bacterial cells are naturally adept at transformation, while others, such as *E. coli,* can be induced to perform it by means of chemical treatment. Critically, plasmid DNA can be taken in via transformation and continue to function, as plasmid DNA, inside the bacteria.

Using Biotech's Tools: Getting Human Genes into Plasmids

At this point, you know about a couple of tools in the biotech tool kit: restriction enzymes and the transformation process involving plasmids. Let's now see how they work together. As you can see in **Figure 15.5**, the process starts with a gene of interest in the human genome. The first step is to use restriction enzymes on this human DNA. Knowing, say, the starting and ending sequence of the gene of interest, a restriction enzyme is selected that allows part of the genome to be cut into a manageable fragment including this sequence of interest, preferably in a sticky-ended form.

Here's where the beauty of restriction enzymes really comes into play. If the same restriction enzyme is now used on the DNA of isolated *plasmids,* the result is complementary sticky ends of plasmid and human DNA. In other words, these segments of human and plasmid DNA, through sticky-ended base pairing, fit together like puzzle pieces.

When the DNA fragments are mixed with the "cut" plasmids, that's just what happens: Human and plasmid DNA form base pairs, and the human DNA is

incorporated into the plasmid circle. With this, a segment of DNA that was once part of the human genome has now been *re-combined* with a different stretch of DNA (the plasmid sequence). This process gives us the term **recombinant DNA,** defined as two or more segments of DNA that have been combined by humans into a sequence that does not exist in nature.

Getting the Plasmids Back inside Cells, Turning out Protein

To this point, what has been done is to produce a collection of independent plasmids having a human gene as part of their makeup. Remember, though, the goal is to turn out quantities of whatever protein the human gene is coding for. To do this requires a vast quantity of plasmids working away, which means working away *back inside* bacterial cells, and it's here that transformation comes into play. If the plasmids are put into a medium containing specially treated bacterial cells, a few of these cells will take up plasmids through transformation (at which point these cells have become transgenic, as noted earlier). Once this happens, the plasmids start *replicating* along with the bacterial cells themselves. As one bacterial cell becomes two, two become four, and so on, the plasmids are replicating away as well (Figure 15.5). And as the cell count reaches into the billions, collectible quantities of the protein begin to be turned out via instructions from the human gene inserted into the plasmid DNA.

A Plasmid Is One Kind of Cloning Vector

In this example, plasmids were the vehicle that served to take on some foreign DNA and then to ferry it into working bacterial cells. Note, however, that plasmids are only one vehicle among several that can be used. Collectively, such vehicles are known as **cloning vectors,** meaning self-replicating agents that serve to transfer and replicate genetic material. Next to plasmids, the most common cloning vector is a type of virus that infects bacteria—a bacteriophage.

Real-World Transgenic Biotechnology

The example we just went over concerned a hypothetical medicine, but lots of actual medicines are being produced through biotechnology today. By one industry count, more than 250 biotechnology drugs and vaccines have been approved by the U.S. Food and Drug Administration. We've already touched on one biotech medicine: the human growth hormone that is produced inside the *E. coli* bacterium. As it turns out, *E. coli* is something of a workhorse in medical biotech. Human insulin and several kinds of cancer-fighting compounds are also produced through it.

E. coli and its fellow bacteria are not the only transgenic "bio-factories" used in medical biotech-

Figure 15.4
Transfer Agent

Plasmids are small rings of bacterial DNA that are not a part of the bacterial chromosome. They can exist outside bacterial cells and then be taken up by these cells. This artificially colored micrograph shows a type of plasmid, from *E. coli* bacteria, that is commonly used in genetic engineering.

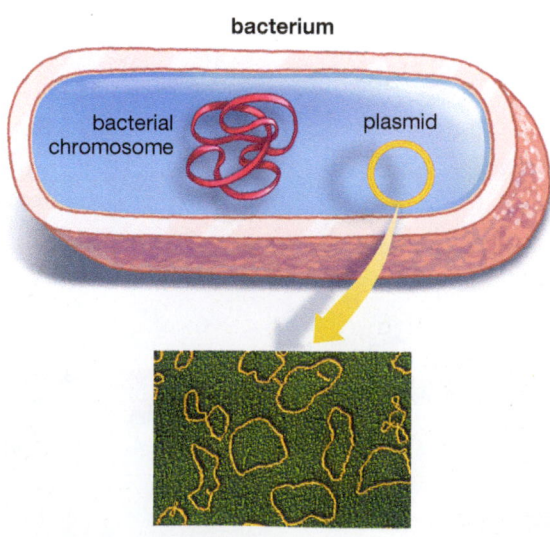

Figure 15.5
How to Use Bacteria to Produce a Needed Human Protein

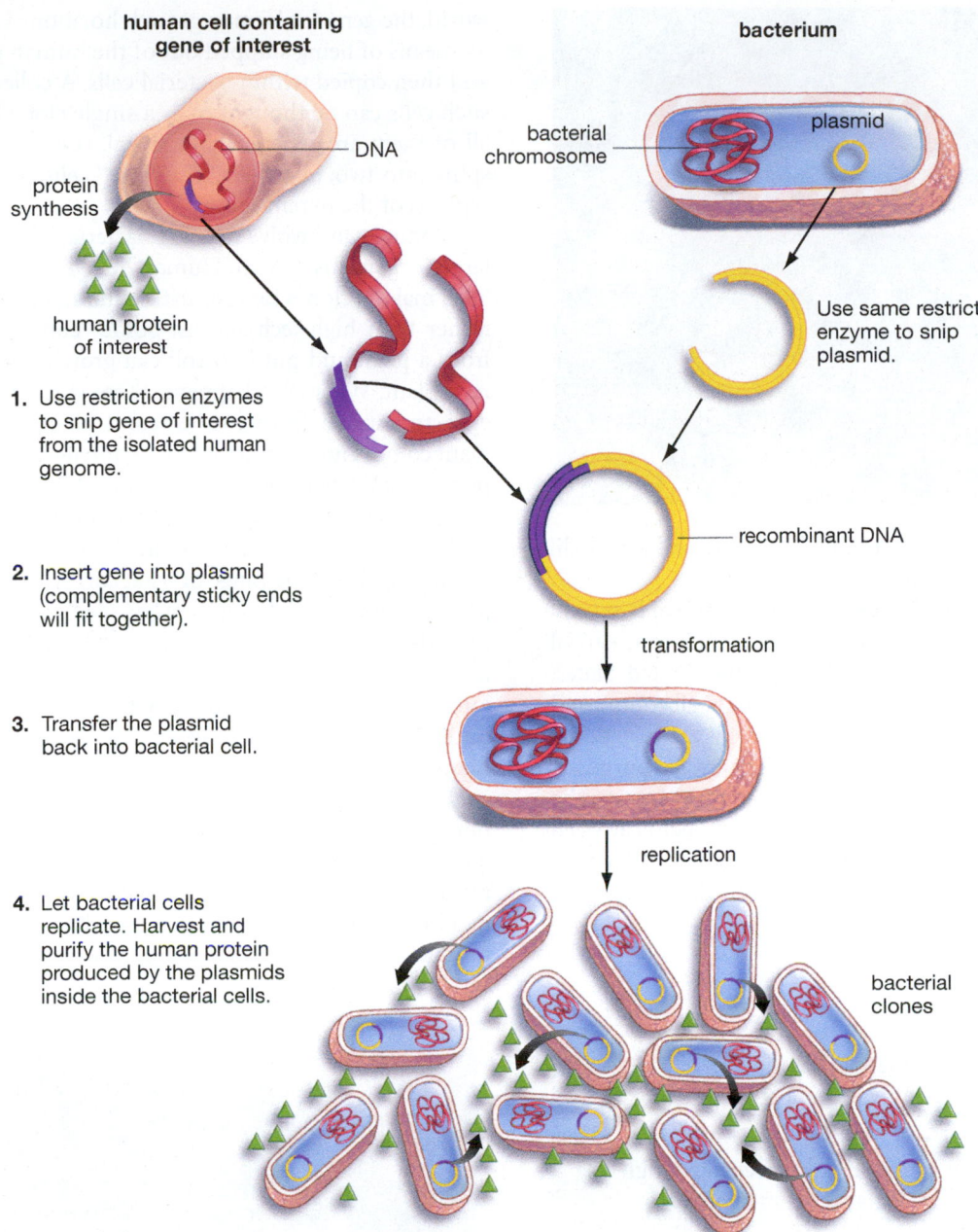

human cell containing gene of interest

DNA

protein synthesis

human protein of interest

bacterium

bacterial chromosome

plasmid

Use same restriction enzyme to snip plasmid.

1. Use restriction enzymes to snip gene of interest from the isolated human genome.

2. Insert gene into plasmid (complementary sticky ends will fit together).

recombinant DNA

transformation

3. Transfer the plasmid back into bacterial cell.

replication

4. Let bacterial cells replicate. Harvest and purify the human protein produced by the plasmids inside the bacterial cells.

bacterial clones

nology, however. Yeast, hamster cells, and even mammals are used in this way as well. Goats of the type pictured in Figure 15.2 produce several kinds of human proteins in their milk, including one called antithrombin that prevents the formation of blood clots. You might wonder why large, expensive goats would be used to turn out proteins when small, inexpensive bacteria are available. The answer is that lactating goats naturally produce a tremendous amount of protein. A single goat can produce as much antithrombin in a year as can be derived from 90,000 human blood donations. You also might wonder how a human gene can be spliced into a goat genome. The answer is, through the splicing of a human gene into a single goat cell and then a cloning of an adult goat from that cell.

Genetically Modified Food Crops

Animals are not the only kind of transgenic organisms. You may have heard of genetically modified, or "GM," food crops. It won't surprise you to hear that all of these crops are transgenic. If you look at **Figure 15.6** on the next page, you can see a transgenic rice that has been dubbed "golden rice" because of its ability to produce its own beta carotene, which the human body converts to vitamin A. Conventional rice does not contain beta carotene, and this fact has health consequences. The World Health Organization estimates that half a million children go blind each year from vitamin A deficiency and that one to two million children die from it. The hope is that golden rice can help alleviate this problem (though this is a hope that has existed for a long time now). Golden rice actually is transgenic in more ways

Figure 15.6
A More Nutritious Rice
Grains of the genetically engineered "golden rice" stand out next to grains of ordinary rice. The rice has a golden color because it is able to produce its own beta carotene, which the human body converts into vitamin A.

than one: Genes from both a bacterium and a daffodil have been spliced into it.

Though the public seems scarcely aware of it, several other transgenic food crops are in full commercial production right now. Indeed, in the United States, 92 percent of all soybeans, 86 percent of all cotton, and 80 percent of all corn crops currently are genetically modified, though only for a couple of limited purposes. One of these is to allow the crops to resist insects; the other is to allow them to tolerate the chemical weed killers known as herbicides. So, why is biotechnology's role in agriculture so limited? Stay tuned.

SO FAR ...

1. A transgenic organism is one whose genome has stably incorporated one or more _____ from _____.

2. DNA is cut into useful fragments through use of the molecular scissors known as _____.

3. The plasmids noted in the text had _____ DNA spliced into them as a first step in using them to turn out a human _____.

15.3 Reproductive Cloning

The concept of *cloning* has come up a couple of times so far. Thanks mostly to movies, this word has taken on some sinister implications, as if biological cloning were inherently a Frankenstein-like procedure. But it need be no more threatening than the process of making copies of a single gene. To **clone** simply means "to make an exact genetic copy of." An individual clone is one of these exact genetic copies. Thus, in the biotech

world, the gene for human growth hormone is cloned by means of being snipped out of the human genome and then copied within bacterial cells. A collection of such cells can be thought of as a single clone because all of them are genetically identical. (One bacterium splits into two, and both "daughter" cells are genetic replicas of the parent cell.)

Cloning can involve not just bacteria, however, but larger organisms as well. Human beings actually have been making clones for centuries, although by low-tech rather than high-tech means. Some "cuttings" taken from a plant and put into soil can grow into a whole new plant. When this happens, there is no mixing of eggs and sperm from two different plants. The new plant comes entirely from one individual—the original plant—and thus meets the definition of a clone in that it is an exact genetic copy of another entity.

All this said, in recent years biotechnology has greatly expanded the range of what can be cloned. The potential for such cloning appears to be very broad indeed—something that has inspired both great optimism and great concern. Biotechnology has now joined with various reproductive technologies to produce clones not just of genes or cells or plants but of mammals, with the starting cells for these mammals coming from adult mammals. Though we've mentioned a couple of these cloned animals already, the most famous one by far is Dolly the sheep, cloned by researcher Ian Wilmut and his colleagues in Scotland in 1997 (**Figure 15.7**). Dolly and similar animals are products of **reproductive cloning:** cloning intended to

Figure 15.7
Revolutionary Sheep
Dolly, the first mammal ever cloned from an adult mammal, is shown here as a young sheep with her surrogate mother.

produce adult mammals of a defined genotype. Powerful in its own right, this procedure gains added potential when combined with the basic biotech processes you've been reviewing.

Reproductive Cloning: How Dolly Was Cloned

Dolly, who died in 2003, was a clone as that term is defined: She was, to a first approximation, an exact genetic replica—in this case, of another sheep. Here's how she was produced (**Figure 15.8**). A cell was taken from the udder of a six-year-old adult sheep and cultured in the laboratory, meaning that the original cell divided into many "daughter" cells. While this was going on, researchers took an *egg* from a second sheep and removed its nucleus, meaning they removed all its nuclear DNA. Then they placed the udder cell (which had DNA) next to the egg cell (which did not have DNA) and applied a small electric current to the egg. This had two effects: It caused the two cells to fuse into one, and it mimicked the stimulation normally provided when a *sperm* cell fuses with an egg. With this, the udder cell DNA began to be reprogrammed. As a result of this reprogramming, the fused cell started to develop as an embryo. (Though the egg cell had its DNA removed, it still contained all kinds of egg-cell proteins whose normal function is to trigger development of an embryo. It was these factors still in the egg that reprogrammed the donor-cell DNA.) After the embryo had developed to a certain point, the researchers implanted it in a third sheep, which served as a surrogate mother. The result, 21 weeks later, was the lamb Dolly, who went on to give birth to two sets of her own lambs.

Every cell in Dolly had DNA in its nucleus that was an exact copy of the DNA in the six-year-old donor sheep. Thus, Dolly was a clone of that sheep. There was no second parent contributing chromosomes to Dolly, no mixing of genetic material at the moment of conception—just a copying of one individual from another. The procedure through which she was cloned is known today as **somatic cell nuclear transfer (SCNT)**: a means of cloning mammals through fusion of one somatic (non-sex) cell with an egg cell whose nucleus has been removed (an "enucleated" egg cell).

Cloning and Recombinant DNA

Once Dolly was born, science was off to the races in terms of producing clones of mammals. To date, horses, mules, cows, pigs, cats, mice, goats, and dogs are among the animals that have been cloned, almost all of them through SCNT. But to what end? That is, what use do scientists have for these animals? Actually, we've already talked about one use. Remember the cloned goats we looked at earlier—the transgenic ones able to produce human proteins in their milk? They are intended to serve as living factories for

human medicines. (Indeed, this is the role Ian Wilmut and his colleagues envisioned for cloned animals before they ever produced Dolly.) Along other lines, scientists have now produced cloned pigs whose tissues have been engineered to contain high levels of heart-healthy omega-3 fats.

In both these examples, we can see something that is true of reproductive cloning in general: So far, it's

Figure 15.8
How Dolly Was Cloned

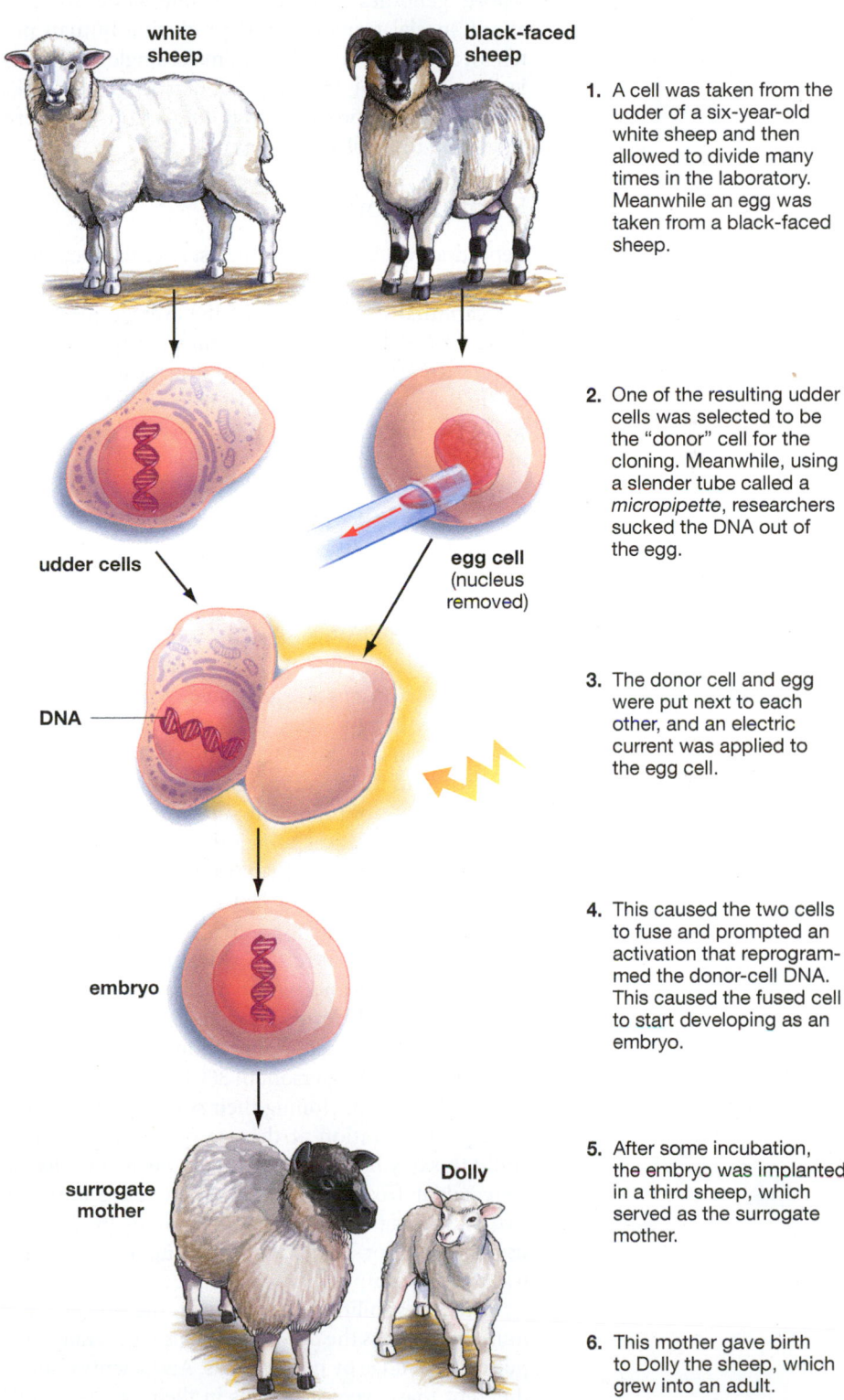

1. A cell was taken from the udder of a six-year-old white sheep and then allowed to divide many times in the laboratory. Meanwhile an egg was taken from a black-faced sheep.

2. One of the resulting udder cells was selected to be the "donor" cell for the cloning. Meanwhile, using a slender tube called a *micropipette*, researchers sucked the DNA out of the egg.

3. The donor cell and egg were put next to each other, and an electric current was applied to the egg cell.

4. This caused the two cells to fuse and prompted an activation that reprogrammed the donor-cell DNA. This caused the fused cell to start developing as an embryo.

5. After some incubation, the embryo was implanted in a third sheep, which served as the surrogate mother.

6. This mother gave birth to Dolly the sheep, which grew into an adult.

envisioned not so much as a means of producing genetic copies of normal animals; rather, it's seen as a means of producing genetically *altered* animals that can serve some special purpose. It is true that for a hefty fee (at least $140,000 at present), you can get a clone of your favorite dog. And, as noted earlier, scientists are looking into cloning as a means of preserving endangered species. But by and large, cloning today is thought of as a means of producing animals whose genomes have been manipulated to bring about special traits (such as producing human proteins). Scientists know how to make single cells genetically capable of bringing about such traits. What cloning does is provide a way to go from an altered single cell to a full-grown animal.

Human Cloning

Reproductive cloning may interest scientists because of its potential to yield products such as medicines, but it has captured the attention of the average person because it raises the possibility of *human* cloning. Despite some grandiose claims to the contrary, we have no reason to believe that a human being has ever been cloned so far. But this does not mean that the feat is technically impossible or that it will never be attempted.

The prospect of a human clone is so dizzying that it's worthwhile to think about what such a person would represent in biological terms. He or she would be a genetic replica of the person who provided the donor cell with the DNA in it. One helpful way to think of this person's biological status is in terms of a more familiar concept: that of an identical twin. As it happens, identical twins also are genetic replicas of one another. Early in a human pregnancy, separate cells from a single embryo can start forming as two separate embryos, and the result is identical twins. Cloning is a *different* means of producing genetically identical individuals, but the critical point is that a donor and a clone would be no more alike than two identical twins.

The parallels between twins and clones end there, however, because the donor-DNA cell for a clone can come from so many different sources. It could, for example, come from a person of any age—an 80-year-old, an 8-year-old; it could even come from a fetus. Indeed, the donor cell used in cloning need not even come from a living person. In 2001, an American couple sought help in cloning their son, who had died in a hospital operation at the age of 10 months. How could this boy be cloned? Through the use of cells that were taken from him while he was alive and then frozen. It was these cells that the parents proposed to use as the donor-DNA cells in cloning the boy (which was never attempted).

Given possibilities such as these, the prospect of a human clone has the power to stun us—generally to repulse us, to judge by news reports. And scientists are no different than average citizens in their aversion to the idea of human cloning. But does this mean that no scientist in the world would be willing to undertake the procedure? Actually, several have already announced that they are willing to carry it out, despite the fact that cloning today remains a largely *experimental* procedure that often results in animals that have serious physical defects. As Ian Wilmut has written with cloning expert Rudolf Jaenisch:

> Newborn clones often display respiratory distress and circulatory problems, the most common causes of neonatal death. Even apparently healthy survivors may suffer from immune dysfunction, or kidney or brain malformation, which can contribute to death later.

With possibilities such as these in mind, would anyone actually attempt to clone a human being? Only time will tell.

SO FAR ...

1. To "clone" means to make an exact _____.

2. Reproductive cloning is cloning intended to produce _____ of a defined genotype.

3. Dolly was a clone of the sheep who _____.

15.4 Cell Reprogramming

In one way, human cells are like human beings: They start off with the potential to be lots of things, but eventually most of them commit themselves to being just one thing. This course of events takes place during human "development," meaning the process by which an embryo becomes a fully formed human being. Early in this process, cells need to be adaptable enough to go down different developmental pathways. Why? Because the embryo is developing in so many basic ways. For example, it needs to create both muscle cells and bone cells. Fortunately, there is a type of embryonic cell that is flexible enough to give rise to either type of cell. As time goes on, however, more and more tissues become fully specialized, which means there is less and less need for cells that are developmentally flexible. By the time humans reach adulthood, specialization has almost completely won out over flexibility. The vast majority of cells in our bodies have undergone what is known as *commitment*: a developmental process that results in cells whose roles are completely determined. Almost all your muscle cells, for example, are committed in that they can be nothing but muscle cells and can give rise to nothing but muscle cells.

The problem with this sequence of events is that the tissues of fully developed human beings can become damaged. They break down or become diseased; they are injured in car accidents or fires. And many damaged tissues will not regenerate themselves. A person whose spinal cord is damaged does not generate a new spinal cord. Thus, for medicine, the question for decades has been: Could we use developmental processes to generate new tissues? One obvious way to go about this would be to harness the power of the cells that, in nature, generate new kinds of tissue—early embryonic cells. For most of the past 10 years, one variety of these cells, *embryonic stem cells* (ESCs), seemed to hold the most promise to generate needed human tissues. All this changed in 2007, however, with the announcement by two teams of scientists that they had been able to transform ordinary adult skin cells into cells that have all the developmental power of ESCs. These induced pluripotent stem cells (iPS cells) thus join ESCs in having the potential to generate needed human tissues.

As we'll see, the commonality between ESCs and iPS cells is that they represent two forms of *cell reprogramming*: ESCs are being reprogrammed to yield desired cell types and ordinary adult cells are being reprogrammed to yield iPS cells that then go on to yield desired cell types. The challenge now before scientists is to use both forms of reprogramming to generate large quantities of very specialized cells—those that can take on specific tasks after being introduced into diseased or damaged human bodies. Let's take a look now at some of the ways scientists are attempting to do this, first in connection with ESCs.

Embryonic Stem Cells

A **stem cell** can be defined as any cell that can give rise to more cells of its kind, along with at least one variety of specialized cell. At different stages of life, human bodies possess different kinds of stem cells, but we'll limit our consideration here to a single variety of stem cell—one that in nature exists only in the first few weeks of human life.

When human sperm and egg come together, they create a fertilized egg (an embryo) that has begun to undergo cell division. If we go forward about five days from the first division, the embryo has developed into a hollowed-out, fluid-filled ball of 100–150 cells—a **blastocyst.** One section of the blastocyst becomes the placenta that will help nurture a growing embryo, while another section, called the inner cell mass, constitutes the embryo itself (**Figure 15.9**). It is the cells of this latter section that interest us, for these are **embryonic stem cells (ESCs),** which can be defined as cells from the blastocyst stage of a human embryo that are capable of giving rise to all the types of cells in the adult body.

Given a developmental power this extensive, you might wonder why the medical potential of ESCs wasn't realized years ago. The answer is that it was only in 1998 that scientist James Thomson of the University of Wisconsin, Madison, succeeded in *culturing* ESCs—that is, in getting them to go through cell division outside of embryos, in laboratory containers. Only with this advance could scientists begin to work with ESCs, channeling groups of them down different developmental pathways—toward becoming nerve or heart cells, for example.

So far, the high-water mark of this channeling has been the development of cells that can produce a type of nervous system tissue, called myelin, that normally wraps around cells called neurons. In victims of spinal cord injuries, this myelin wrapping often is destroyed at the site of injury, and there is no natural way for it to be regenerated. The good news for spinal cord patients is that, in 2005, researchers in California coaxed human ESCs into becoming myelin-producing nerve cells. Better yet, when these cells were injected into the spinal cords of rats whose spinal cords had been damaged, the new cells started producing a myelin covering around spinal cord neurons in the damaged area. The result? Rats who had been paraplegic—who had lost control of their hind legs—regained enough function in them to do a kind of impaired walking, rather than dragging themselves along with their front legs.

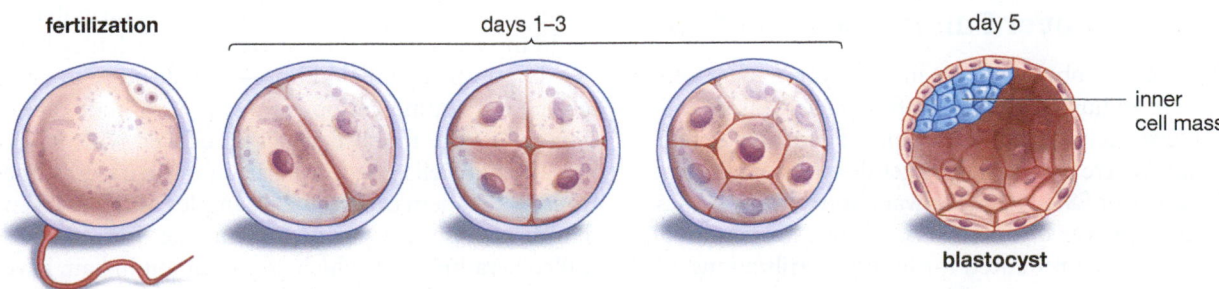

fertilization   days 1–3   day 5   inner cell mass

blastocyst

Figure 15.9
Cells with a Great Ability to Specialize
Human development starts with the fertilization of egg by sperm, after which the resulting embryo proceeds through several days of cell division. By day 5, it has developed into a hollow ball of 100–150 cells called a blastocyst. In nature, one segment of the blastocyst, the inner cell mass, develops into the baby.

At the moment, the pharmaceutical company that financed this research is preparing to proceed with the first human trials of this regeneration procedure.

In ESC work of this sort, the cell reprogramming noted earlier takes place when scientists coax ESCs down a desired developmental pathway. Without this guidance, ESCs would only produce more of themselves. By adding the right chemical compounds to a culture of ESCs, however, scientists get more ESCs along with cells of the desired type. The ability of ESCs to generate more of themselves is important in itself, however, for it is this ability that allows ESCs to *keep on* producing specialized cells in large numbers. In the myelin-regeneration experiments noted above, scientists had to inject a pea-size volume of about 1.5 million nerve cells into the spinal cord of each rat—a volume that would have been impossible to produce had each ESC been generating nothing but nerve cells.

This regenerative capacity of ESCs is then linked to a second ability they have, which is their power to produce many different *types* of cells (something that's not surprising, given that, in nature, they end up giving rise to every kind of cell in the body). Thus far, researchers have managed to coax ESCs into producing nerve cells, heart cells, immune cells, blood cells, and more.

Given strengths such as these, you might think the future of ESCs in medicine would be assured, but this is not the case. For one thing, a number of basic questions about these cells have to be answered before they can be used in connection with most human illnesses. Once cells derived from ESCs have been implanted, can we be sure they won't then differentiate further—into other types of specialized cells? Can implanted cells be counted on to stay where they are needed, rather than "migrating" to other areas? Serious as such questions are, they may actually end up playing a minor role in deciding whether ESCs have a future in medicine. A bigger threat to them at the moment is that their role might be usurped by a different kind of stem cell—one whose discovery was driven in part by a moral issue surrounding ESCs.

The Fight over Embryonic Stem Cells

To state the obvious, human embryonic stem cells come from human embryos—from collections of cells that have the potential to become fully formed human beings. Where do researchers get the embryos that are the source of ESCs? Almost always from fertility clinics, which each year discard thousands of excess embryos that have been produced for in vitro fertilizations. (A couple undergoing fertility treatments generally will produce more embryos than they need to bring about a pregnancy. With the couple's consent, excess embryos that would have been discarded are used instead in scientific research.) When scientists extract stem cells from these early stage embryos, however, the embryos are destroyed, and this is a matter of great concern to many people. For them, the destruction of an embryo is a destruction of human life. In contrast, many people feel that an embryo in the blastocyst stage represents only a *potential* human life. And others hold that, no matter what one's view of this issue, society has a greater responsibility toward impaired children and adults than it does toward embryos.

Deeply held beliefs such as these eventually resulted in a national political decision: In 2001, then-President George Bush decided that the federal government—the country's largest single source of biological research funds—would severely restrict funding for research on embryonic stem cells. (Only those genetically identical "lines" of stem cells in existence at the time the president made his decision were eligible for federal funding.) Apart from limiting funding for ESC research, this decision greatly disrupted ongoing ESC research. Single research labs that got both private and federal funding had to maintain clear lines of separation—often a physical separation—between their federally funded and privately funded research operations. After 2001, then, the scientific community had a great impetus to develop stem cells that were not derived from human embryos. Eventually, two groups of scientists succeeded in doing exactly this.

Induced Pluripotent Stem Cells

In December 2007, scientific teams led by Shinya Yamanaka of Kyoto University in Japan and James Thomson of the University of Wisconsin, Madison, announced that, by inserting just four developmental genes into ordinary adult skin cells, they had been able to produce cells that have all the generative potential of ESCs (**Figure 15.10**). These are the **induced pluripotent stem cells (iPS cells)** noted earlier: cells that have been induced into a state of pluripotency through the introduction of genes from outside their genome. (A "pluripotent" cell is simply one that is capable of giving rise to all the types of cells in the body. Thus, both iPS cells and ESCs are pluripotent.)

In the years since iPS cells were unveiled, they have lived up to their initial billing—they do seem to have all the developmental power of ESCs without carrying any of the moral baggage associated with ESCs. Because iPS cells are derived from ordinary cells, the sources for them can be small samples of tissue taken harmlessly from a person of any age. Indeed, cells called keratinocytes, which are abundant in hair, have turned out to be particularly easy to reprogram into iPS cells, which means that iPS cell lines can be initiated by means of plucking out a single human hair.

The ordinary origins of iPS cells may end up giving them another substantial advantage over ESCs: an ability to help produce *personalized* lines of specialized cells. To understand this, think first about any specialized cells that might be produced through ESCs. Having been derived originally from an embryo, and then placed into the body of a patient, such cells are likely to be regarded as invading *foreign* cells by the patient's immune system, meaning these cells would set off an immune system attack. Thus, transplanted cells derived from ESCs could be "rejected" by a body in the same way a transplanted kidney can be. Conversely, iPS cells that give rise to specialized cells could come originally from ... patients themselves! Patients could donate some of their own cells; these cells would be reprogrammed into iPS cells; and the iPS cells would go on to yield the specialized cells that would be implanted. In short, patients would be able to generate their own therapeutic cells.

Patient-specific cells of this sort would require such an enormous research expenditure for each patient served that no one thinks they will become a reality anytime soon. But a kind of half-way measure might be possible. In organ transplantation today, doctors try to match donated organs with organ recipients in accordance with whether both organ and recipient have an immune system compatibility. Specifically, doctors look to see if the cells of both the organ and the recipient have, on their surface, similar sets of immune system proteins—proteins called HLA markers. With iPS cells, it may become possible to produce *banks* of cells that are specialized not just for cell type but for HLA type as well. The result would be collections of readily available cells that could be counted on not to set off severe immune reactions.

Disease treatments such as these lie off in the future, but iPS cells are proving invaluable to medical scientists right now as a means of *learning* about human disease. Consider, for example, an inherited disease called spinal muscular atrophy (SMA) that can cause death in children. As SMA researcher Michael Sendtner wrote recently in the journal *Nature,* prior to the advent of iPS cells, anyone who studied SMA ran up against two powerful obstacles. First, this disease exists only in humans, which means it can't be studied in lab animals such as mice. Second, the cells that SMA most strongly affects—cells called spinal motor neurons—cannot be isolated and cultured. But as other researchers have found, it's possible to take common cells called fibroblasts from an SMA patient, reprogram those cells into iPS cells, and then coax the iPS cells into producing stable lines of the needed spinal motor neurons. Critically, each of these neurons has the genetic makeup of the SMA victim—meaning, the genetic makeup that causes SMA. Since spinal motor neurons are developing from iPS cells in this procedure, it becomes possible to trace the way SMA develops *over time* in the neurons. Beyond this, it becomes possible to see how the neurons

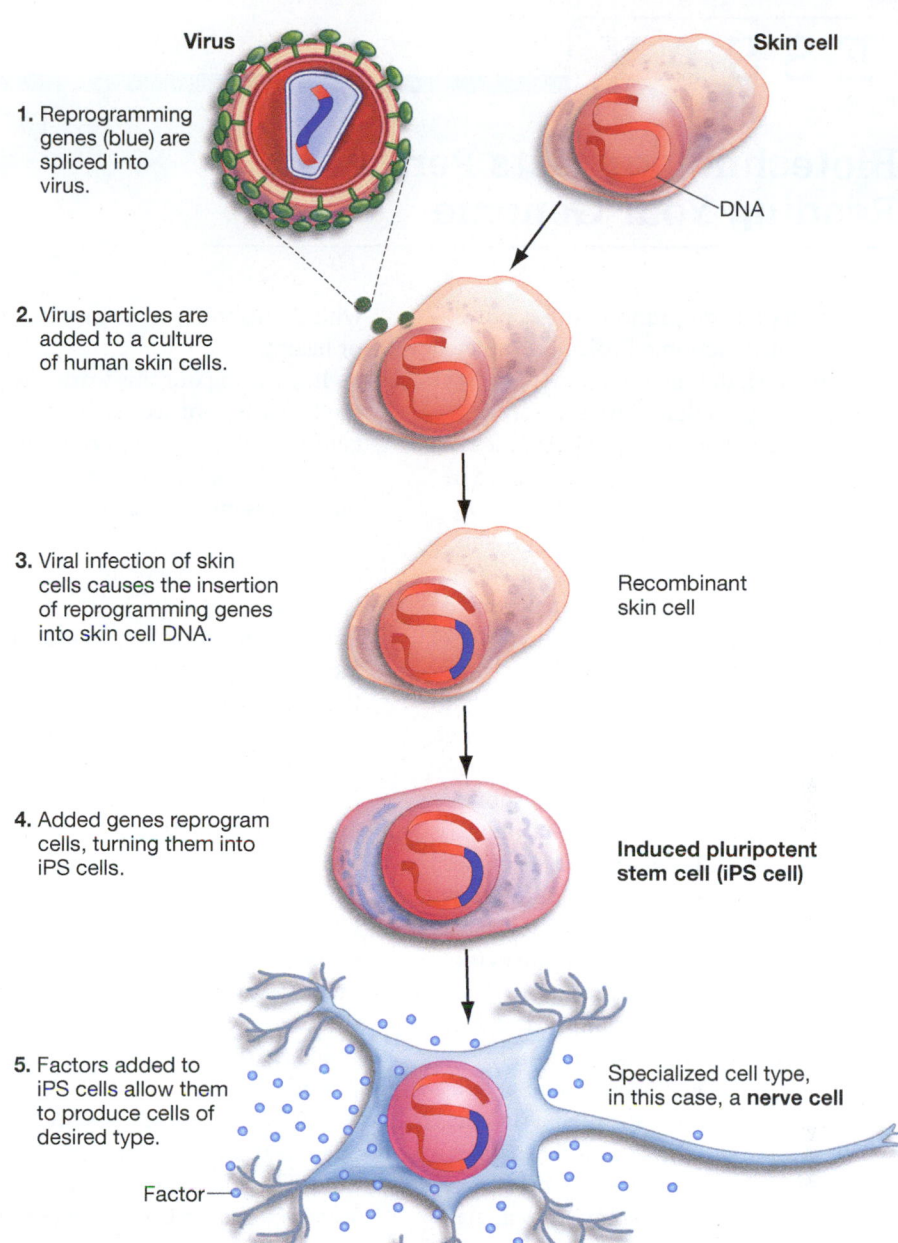

Figure 15.10
A New Kind of Stem Cell
How induced pluripotent stem cells are produced and used.

respond to various drugs that are applied to them—drugs that may eventually serve as treatments for the disease. In short, iPS cells have made it possible to produce what some researchers have called "diseases in a dish." So promising is this avenue of research that scores of disease-specific iPS cell lines are now in development.

Research advances such as these have provided scientists with a great sense of optimism about iPS cells. To be sure, any treatments these cells might provide are years away—at present, scientists are merely investigating the details of their developmental power

ESSAY

Biotechnology Gets Personal: Reading Your Genome

Though it was grand in impact, the Human Genome Project, completed in 2003, did have a name that was a bit misleading, in that there is no single, standard human genome. Instead, each of us has a personal genome—a sequence of 3.2 billion A's, T's, G's, and C's that is unique to us (except for those of us who have an identical twin). If we could line up the genomes of any two unrelated individuals, we'd find about three million differences between them, which is to say three million places at which their bases do not line up. Scientists have long known that genomic variation leads to variation in disease susceptibility—one person responds to a modest weight gain by becoming diabetic while another does not; one person develops lung cancer from smoking while another does not. What's different today is that, thanks to modern DNA sequencing and analysis, scientists have a hope of *identifying* the genetic variations that leave people susceptible to these conditions. Such work is aimed at ushering in an era of personal genomics, meaning a time in which meaningful information can be extracted from each person's genome.

The personal genomics era may actually be with us now, in the sense that scientists have already identified some genetic variations that modestly raise or lower a person's chances of contracting diseases such as Type 2 diabetes and rheumatoid arthritis. Indeed, the budding ability of scientists to make associations of this sort has sparked the development of a *commercial* aspect to personal genomics. At present, for a fee of as little as $399, commercial genotyping firms will obtain a sample of your DNA, analyze it, and provide you with the most up-to-date information they have on how any genetic variants you possess correlate with predispositions toward sickness or health.

In carrying out this work, the genotyping firms are not sequencing the entire genomes of any of their customers—that task would be far too expensive to carry out at present. Instead, they are scanning customers' genomes for what are called single-nucleotide polymorphisms, or SNPs (pronounced "snips"). At a given point in the genome, for example, individual A may have the sequence CCGTC while individual B has the sequence CCGTA. The final letter in these sequences constitutes a SNP, assuming that the least common variant of this letter is found in at least 1 percent of the population. About 10 million SNPs are thought to exist in the human genome. Some of these occur within sequences of DNA that code for proteins and thus stand to be able to *alter* proteins, but most lie outside of protein-coding regions. Even these latter SNPs, however, can serve as so-called markers for disease and can thus provide valuable information. Suppose, for example, that researchers find that a particular group of eight SNPs occurs at a higher frequency in the genomes of asthma sufferers than in the genomes of individuals without asthma. It's not necessary to know whether these SNPs are doing anything to cause asthma; all that's necessary to know is that a person's odds of being asthmatic are raised by possessing this particular group of SNPs.

The challenge to commercial genotyping firms is to stay current on what science has revealed about associations such as these, after which they can inform customers about their *personal* associations. How does a given customer's SNP lineup for, say, Parkinson's disease compare with a lineup seen in Parkinson's victims?

The ability of private firms to provide clients with this kind of information is a testament to how far personal genomics has come in recent years. Nevertheless, a key question for consumers today is whether they stand to get much *useful* information from an analysis of their SNP patterns. If there's a consensus among experts about this today, it seems to be that consumers don't stand to benefit greatly from peering inside their own genomes in this way, for the simple reason that science has not yet learned enough about the links between SNPs and disease. Consider, for example, what Harvard psychologist Steven Pinker learned in 2008 when he had his genome analyzed by a commercial genotyping firm. As he wrote in the *New York Times Magazine*:

The two biggest pieces of news I got about my disease risks were a 12.6 percent chance of getting prostate cancer before I turn 80 compared with the average risk for white men of 17.8 percent, and a 26.8 percent chance of getting Type 2 diabetes compared with the average risk of 21.9 percent. Most of the other outcomes involved even smaller departures from the norm. For a blessedly average person like me, it is completely unclear what to do with these odds. A one-in-four chance of developing diabetes should make any prudent person watch his weight and other risk factors. But then so should a one-in-five chance.

This said, personal genomics is a rapidly changing field, and what is true today may not be true tomorrow. Scientific work aimed at identifying the genetic variations that predispose people to disease or health is proceeding at a rapid pace. Should this work bear fruit, then commercial genotyping operations would presumably have customers flocking to them, for the simple reason that they'll have so much useful information to offer.

ACGGCTAAAATCGCTTAATTGGCCGCGCTATCGATGCTAGGGTAGCCTAGTTAAAAGATAGATCGAGATTA

while looking for any safety risks they might pose. Thus far, however, iPS cells have shown a remarkable set of strengths. In one test after another, they have put the power of cell reprogramming on full display.

SO FAR ...

1. Embryonic stem cells (ESCs) are derived from embryos when they are in the blastocyst stage, which occurs about (a) 5 days (b) 5 weeks (c) 5 months following conception.

2. Induced pluripotent stem cells (iPS cells) are like ESCs in that they seemingly can give rise to _____ human cell types, yet they are unlike ESCs in that they are derived from _____ cells.

3. iPS cells could in theory solve the problem of the rejection of transplanted tissue because, through such cells, transplanted tissue might be generated from a patient's _____.

15.5 Forensic Biotechnology

After New York's World Trade Center (WTC) was destroyed on September 11, 2001, public officials had to undertake the grim task of trying to identify all the victims of the attack to provide some sense of closure to the victims' loved ones. In the weeks and months following the attack, therefore, officials collected samples of DNA not only from WTC victims but from the toothbrushes and razors of persons who were missing since the tragedy and hence *presumed* to be victims. They also collected DNA from relatives of these presumed victims. With this DNA database in hand, workers then made an effort to match one kind of DNA with another—to match the DNA found on a toothbrush, for example, with the DNA from a WTC victim.

High-profile cases such as these are the most visible manifestation of a huge change that has come about in the way society identifies persons—criminals, biological fathers, and victims, most notably. In the criminal arena, the most highly valued type of physical evidence today is not fingerprints; it is tiny bits of human tissue from which DNA can be extracted. But how do police use DNA to get a "match" in a criminal case? Let's find out.

The essence of "forensic DNA typing," as it is known, is to get two sets of physical samples—first, a sample left by a crime's perpetrator in such forms as blood or semen; and second, a blood sample from a suspect. By comparing the DNA patterns of both samples, forensic scientists can establish odds of whether the suspect was present at the crime scene. If he or she was, but had no good reason to be there, this is strong evidence that the suspect committed the crime.

Another way of gathering DNA evidence is to use a victim's blood sample and test it against, for example, blood found on a suspect's clothing.

The Use of PCR

When forensic DNA typing first came into use in the 1980s, a significant limitation on it was that relatively *large* samples of human tissue had to be obtained to get enough DNA to work with; blood samples, for example, had to be about the size of a quarter. Eventually, however, this problem was all but eliminated by something called the **polymerase chain reaction (PCR):** a technique for quickly producing many copies of a specific segment of DNA. Thanks to PCR, if a criminal so much as licks an envelope and leaves it behind, police can retrieve enough DNA to get a match.

The PCR process is very simple in outline. Four kinds of materials are mixed together to carry it out. The first of these is a quantity of DNA itself. Then, there is a collection of DNA nucleotides and DNA polymerase (which, remember, goes down the line on single-stranded DNA, affixing nucleotides to available bases). Finally, there are two DNA "primer" sequences— short sequences of single-stranded DNA that act as signals to DNA polymerase, saying "start adding nucleotides here."

The first step in the process is that the starting quantity of DNA is heated until the two strands of DNA's double helix separate, resulting in two single strands of DNA (**Figure 15.11**, on the next page). As the heated DNA mixture cools, primers will attach to each of the now-separate strands of DNA. Then the DNA polymerases go down the line, starting from the primers, linking nucleotides to the template DNA and thus producing strands that are complementary to the original strands. The result is two double-stranded lengths of DNA, both identical to the original double strand. In short, the DNA sample has been doubled in one copying "cycle." Then the entire process of heating and cooling is repeated. Since each copying cycle takes only 1–3 minutes, by the time 90 minutes have passed, millions of copies can be created. Moreover, thanks to the specificity of primers, the sequence that is copied can be limited to a small sequence of interest within the larger genome. So useful is PCR that it is employed in countless facets of biological research. We are focusing here on one specific use for PCR, but it has so many applications that in 1993, its inventor, Kary Mullis, was awarded the Nobel Prize in Chemistry for his achievement.

Finding Individual Patterns

Once PCR has done its job, the police have enough DNA to begin looking for the individual patterns within it—one set of patterns found in the crime-scene DNA, one set of patterns seen in the suspect's DNA. But what are these patterns that each of us

Copying DNA through PCR

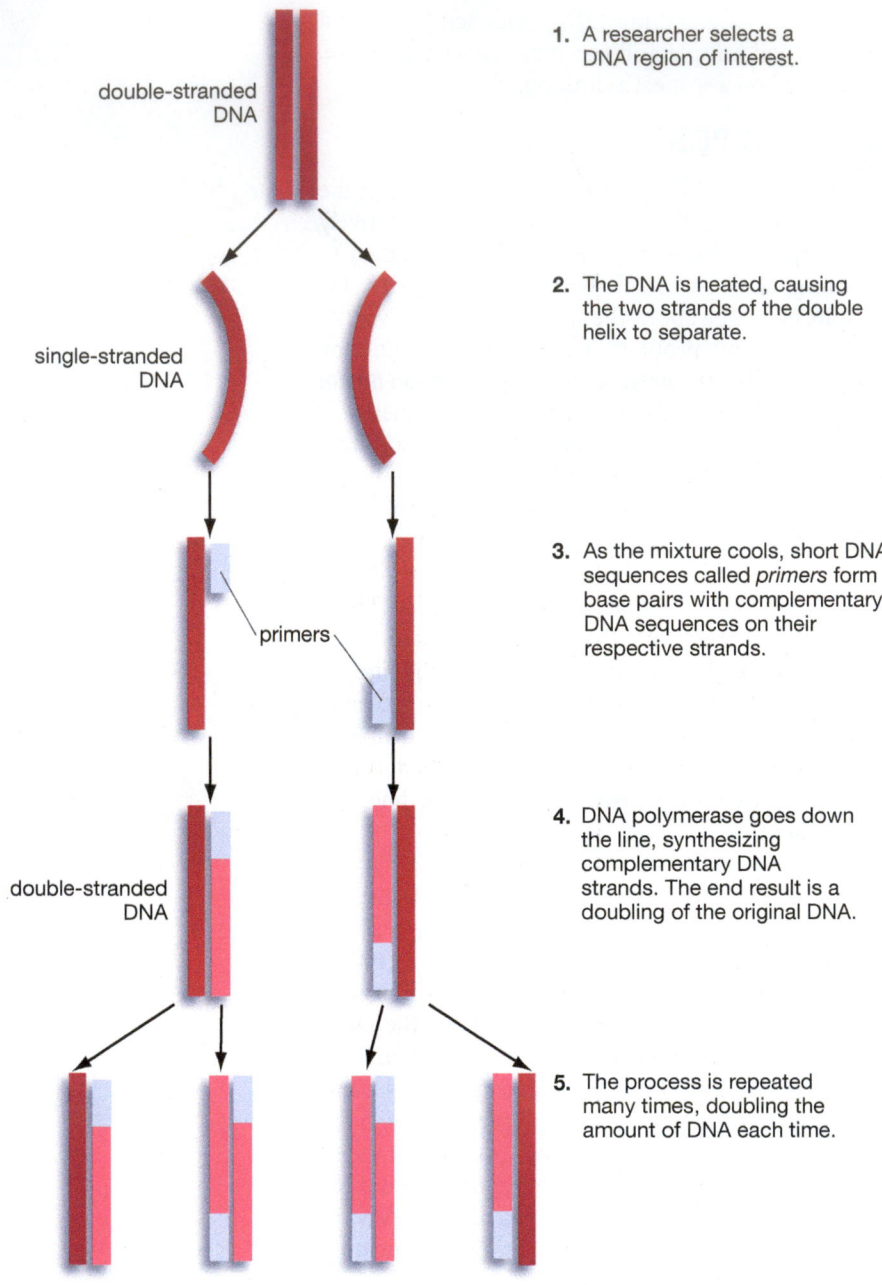

1. A researcher selects a DNA region of interest.

double-stranded DNA

2. The DNA is heated, causing the two strands of the double helix to separate.

single-stranded DNA

3. As the mixture cools, short DNA sequences called *primers* form base pairs with complementary DNA sequences on their respective strands.

primers

4. DNA polymerase goes down the line, synthesizing complementary DNA strands. The end result is a doubling of the original DNA.

double-stranded DNA

5. The process is repeated many times, doubling the amount of DNA each time.

Figure 15.11

DNA Copying Machine

The polymerase chain reaction (PCR) makes copies of a given length of DNA very quickly.

Now, the chances are small that two unrelated individuals will have an identical number of repeats at even one location in the genome. But modern forensic DNA typing looks at STR repeats at 13 different locations. Imagine yourself, then, as a forensic investigator who is reviewing both DNA from a crime scene and DNA from a suspect in the crime. At genomic location 1, the STR patterns are:

Crime scene: TCAT TCAT TCAT TCAT
  Suspect: TCAT TCAT TCAT TCAT

So, you have a match at this one location. If the STR patterns continue to line up this way at the other 12 locations, then what once was just probable becomes certain. Your suspect was certainly at the crime scene because the odds of any two unrelated people having identical STR patterns at 13 locations are astronomically small.

It's important to recognize that this kind of DNA evidence can be used to prove innocence as well as guilt. DNA testing regularly exonerates not only suspects but people who have been *convicted* of crimes. In the past 20 years, more than 200 prisoners who were jailed for various offenses have been freed on the basis of DNA testing. If you look at the essay "Reading DNA Profiles," you can see the means by which STR patterns are visualized in today's criminal cases.

Beyond matching a known suspect to a crime scene or victim, forensic DNA typing has another power. Increasingly, it is allowing police to make so-called *cold hits*. All 50 states now require criminals convicted of violent crimes to provide DNA samples. Capitalizing on this, the FBI has put together a national database of the DNA profiles of these criminals. Given this, imagine a situation in which a crime is committed, and no suspect was identified, but in which a DNA sample was retrieved. A DNA profile will be produced from this sample—a profile that can be matched against all those that are stored in a DNA database. If the profile from the crime scene matches a criminal's stored profile, that in itself is sufficient evidence to make an arrest. And this is happening all the time—cold hits result in hundreds of arrests across the nation each year.

carries? Human genomes are filled with short sequences of DNA that are repeated over and over, from 3 to 50 times, like this:

TCATTCATTCATTCATTCAT

This TCAT example is not hypothetical. It is an actual short tandem repeat (or STR) that police departments use in today's DNA typing. The usefulness of STRs is that, at a given location in the genome, one person will have one number of tandem repeats, while another person is likely to have a different number of repeats, like so:

Individual 1: TCAT TCAT TCAT TCAT
Individual 2: TCAT TCAT TCAT TCAT TCAT TCAT

SO FAR . . .

1. PCR has been described as a copying machine for _____.

2. In forensic biotechnology, the value of PCR is that it allows the establishment of identities based on _____ samples of DNA.

3. The DNA of every individual contains groups of _____ whose number can be used to establish the identity of that individual.

Reading DNA Profiles

What does a forensic DNA profile look like? Law enforcement agencies use the kind of imaging you can see below. The upper profile was produced from DNA taken from a crime scene, while the lower profile was produced from DNA obtained from a suspect in the crime. The short tandem repeat, or STR, patterns that differ from one person to another are represented by the peaks that run across the rows in both figures. Each peak, or set of two peaks, represents one location in the genome—a location whose STR pattern was examined. In the upper figure's first row, the peak labeled 16 indicates that the person who left DNA at the crime scene had 16 repeats at this location. The

nearby 17 then indicates that this person had 17 repeats at this same location, but on his second, homologous chromosome. Meanwhile, the 24 that can be seen farther to the right represents the number of repeats at a different location in the genome. In this case, the person had 24 repeats on both chromosomes—he was homozygous for repeats at this location.

The scale running across the top of the figure indicates the number of DNA base pairs in the lengths of DNA that were examined at each location. Thus, the 16 repeats at the first peak were found in a section of DNA that was about 125 base pairs long. The height of each peak has no bearing on identity but simply reflects the

quantity of DNA that was derived, through the PCR process, for each location. The X and Y peaks on the second row represent a section of the genome that differs between men and women. Had a woman's DNA been found at the crime scene, there would have been no Y peak.

A quick look at both the upper and lower figures reveals that the STR patterns in them match up perfectly. The obvious implication is that the suspect was at the crime scene. Indeed, the odds that the crime-scene DNA came from anyone *other* than the suspect are infinitesimally small. What are the chances that a person selected at random from a general population would have the same STR profile as the one found in the crime-scene DNA? The odds of this occurring range from 1 in 4.6 trillion for the U.S. Caucasian population to 1 in 36 trillion for the U.S. southwestern Hispanic population.

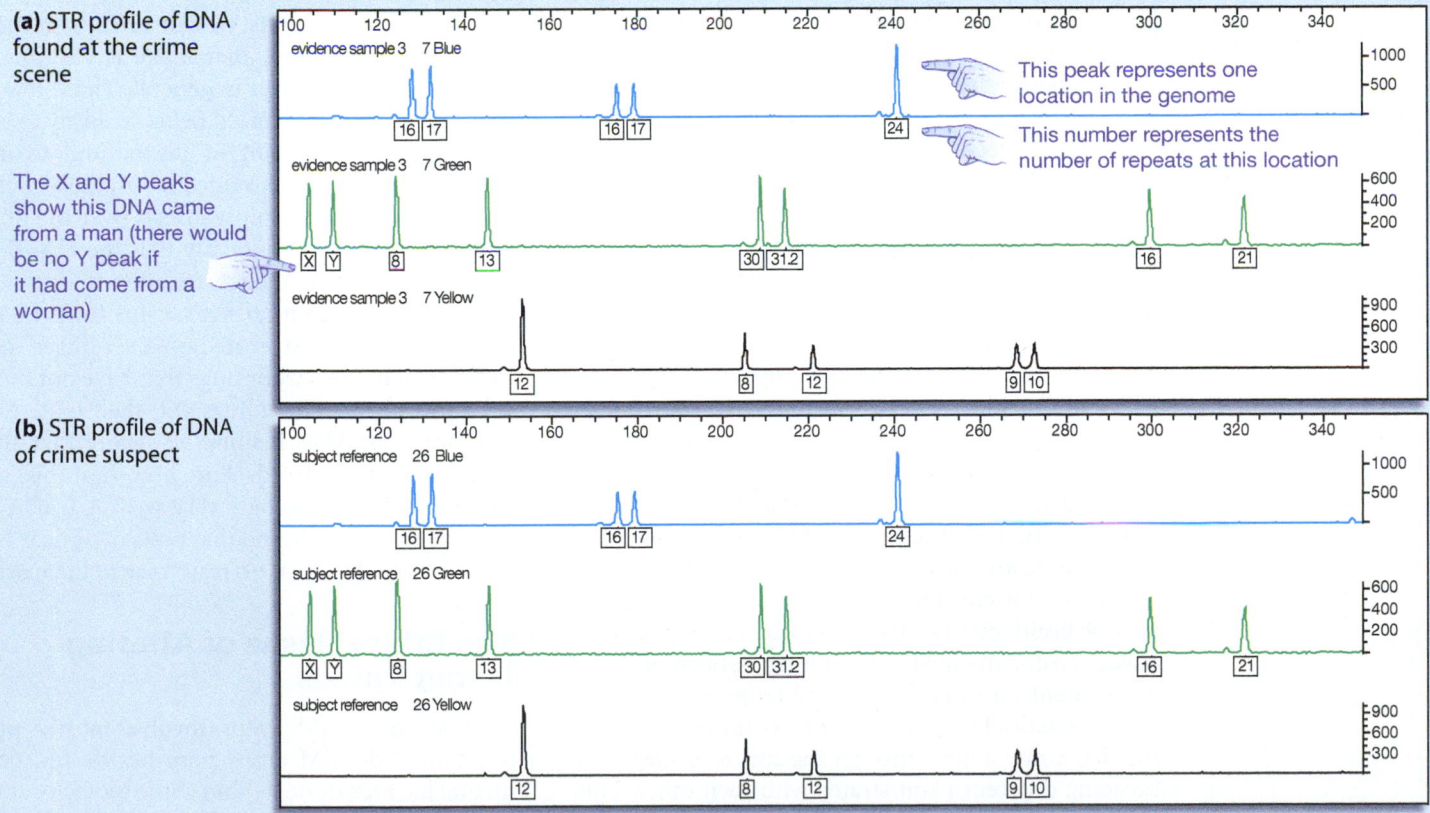

(a) STR profile of DNA found at the crime scene

(b) STR profile of DNA of crime suspect

15.6 Controversies in Biotechnology

Genetically Modified Crops

You may recall the biotech crops we looked at earlier—the ones that are transgenic, in that they have one or more genes from another species spliced into them. Worldwide, tens of millions of acres have been planted with these genetically modified, or GM, crops. It turns out, however, that almost all this acreage is devoted to just *four* crops: cotton, corn, soybeans, and canola. Moreover, GM crops have not been modified to make their ultimate products—food or clothing—preferable to consumers in some way. Instead, all four GM crops have been modified to be preferable to *farmers,* either by being resistant to pests or tolerant of herbicides. (One of the latter crops is "Roundup-ready" soybeans, whose seeds contain a bacterial gene that allows them to tolerate the herbicide Roundup.) You might think that the list of GM crops will certainly be getting longer in the future, but this may not be the case. Consider that in 1999 about 120 GM fruits and vegetables were being field-tested in the United States, but that by 2003 this number had dropped to about 20.

So, why are there so few GM crops, either in the fields or in the works? A large part of the answer can be summed up in two words: consumer resistance. Worldwide, a significant number of consumers have become persuaded that biotech foods are dangerous, and the effect of this has been a continued shelving of plans for biotech crops. In 2004, for example, the Monsanto Corporation gave up on a program it had started seven years earlier aimed at producing GM wheat. The company worried that few American or Canadian farmers would plant the wheat because they feared resistance to it from European and Japanese consumers. Such resistance, the farmers thought, might effectively put a "do not import" label on all U.S. and Canadian wheat, whether genetically modified or not. This same chain of reasoning is operating in connection with other GM foods.

And what is the basis for consumer fears about GM foods? Nothing that has occurred so far. In 2004, the National Academy of Sciences published a report whose central conclusion is still valid: "To date, no adverse health effects attributed to genetic engineering have been documented in the human population." The report went on to note that food crops actually have been "genetically modified" for centuries, although this has come about through the low-tech means of breeding different plant strains with each other. This form of genetic mixing, which is relatively uncontrolled, stands in contrast to the biotech variety, in which a small number of identifiable genes are inserted into (or removed from) a plant's genome. Yet it is biotech crops that have encountered resistance as "frankenfoods," while plants produced through conventional means can be changed without notice.

Resistance to GM crops is based on more than the fear of their health effects, however. The other great concern about them has to do with environmental effects. If you look at **Figure 15.12**, you can see a case in point in this battle. Pictured are two sets of cotton plants, one set genetically modified, the other not. The modified plants contain genes from a bacterium, called *Bacillus thuringiensis,* that is found naturally in the soil and that produces proteins toxic to a number of insects. Collectively, these proteins constitute a natural insecticide known as Bt, which has been sprayed on crops for years, mostly by "organic" farmers, meaning those who do not use human-made fertilizers, pesticides, or herbicides. Enter biotechnology firms, which in the 1990s spliced Bt genes into crop plants, with the result that these plants—corn, cotton, and potatoes—now produce their own insecticide. The results have in some ways been an environmentalist's dream. In one survey conducted in the American Southeast, farmers who planted Bt cotton reduced the amount of chemical insecticides they applied to their fields by 72 percent. They did this, moreover, while increasing cotton yields by more than 11 percent.

Despite such benefits, we can also see, in the use of Bt seeds, the qualities that make environmentalists uneasy about GM crops in general. Those few insects that survive in Bt-enhanced fields are likely to give rise to populations resistant to the natural toxin. This raises the prospect of a valued natural insecticide losing its effectiveness against insects over the long run. So alarming is this possibility that the U.S. Environmental Protection Agency requires that at least 20 percent of any given farmer's crops must be non-Bt plants. The idea is to create non-Bt "refuges" near the Bt fields that will harbor bugs that have not built up a resistance to Bt. These bugs will then mate with Bt-resistant bugs, thus helping to ensure that Bt resistance does not spread. And how has this strategy worked so far? Seemingly very well. A USDA survey completed in 2003 found that pests in or near Bt fields had developed little or no resistance to the insecticide.

The Ethical Issue of Altering Living Things

Disputes over GM crops involve mostly practical questions—do GM crops pose health and environmental hazards or not?—but some biotech controversies are essentially ethical in nature. We've touched on one of these ethical controversies already, which is the

Figure 15.12
Built-in Resistance to Pests
The cotton plants on the left have had genes spliced into them from the bacterium *Bacillus thuringiensis.* These genes code for proteins that function as a pesticide. Meanwhile, the cotton plants on the right are not transgenic in this way—they have not had genes from another organism spliced into them. Both stands of cotton were under equal attack from insects, but the Bt-enhanced plants fared much better.

use of human embryonic stem cells to produce needed human tissues. A more general ethical objection to biotech research, however, has to do with the question of altering living things—of altering nature, if you like.

Of course, human beings have been altering nature for centuries through such means as the plant breeding we talked about. But for an example that is a little more relevant, consider one of our favorite pets, the dog. The fantastic variety in dog breeds that we see today has mostly come about in the past 400 years, and this has happened almost entirely as a result of the human practice of selectively breeding one kind of dog with another. So is this inherently unethical? If your answer is no, then what about producing a sheep whose liver tissue is mostly made of human cells? Is there some ethical line that is crossed when we go from mixing different breeds of an animal to mixing different species of animal? (The animal that results from this mixing is "chimeric" in scientific parlance.) If the answer to this is no, then what about injecting immature human brain cells into the brain of a monkey? This has been done already, as a means of conducting Parkinson's disease research, and the monkeys have, to outside observers' eyes, remained "all monkey." But is there a point at which this becomes unethical? Perhaps the point at which a monkey might develop the first glimmer of human consciousness?

The possibilities that biotech has opened up for chimeric animals drives home the point that we have entered a new era in altering nature. Given biotech's power, however, it isn't necessary to consider chimeras to confront the kind of "altering" dilemmas that might eventually be upon us. Suppose you could alter a chicken's own genes so that you could create an animal that could lay eggs but that had no eyes or legs. (This would make these chickens easier to control, which would in turn bring the price of eggs down.) Suppose further that these animals suffered no apparent discomfort as a result of this manipulation. If you could do such a thing, would you?

On to Evolution

In touring the strands of DNA's double helix, you've had many occasions to see how this microscopic molecule can profoundly affect our macroscopic world. DNA replicates, it mutates, it is passed on from one generation to the next. And some organisms will have more *success* in passing on their DNA—in reproducing, in other words—than will others. This last fact has been critical in bringing about the huge variety of life-forms that Earth houses, from bacteria to bats to trees. Over billions of years, genetics has interacted with Earth's myriad environments to produce the grand story of life. That story is called evolution, and it is the subject of the next four chapters.

15 REVIEW

Summary

15.1 What Is Biotechnology?

- Biotechnology is the use of technology to control biological processes as a means of meeting societal needs. **(p. 264)**

15.2 Transgenic Biotechnology

- A transgenic organism is one whose genome has stably incorporated one or more genes from another species. **(p. 264)**

- Restriction enzymes are proteins derived from bacteria that can cut DNA in specific places. Plasmids are small, extra-chromosomal rings of bacterial DNA that can exist outside of bacterial cells and that can move into these cells through the process of transformation. **(p. 264)**

- Human DNA can be inserted into plasmid rings. Scientists use the same restriction enzyme to cut both the human DNA of interest and the plasmids. Complementary "sticky ends" of the fragmented human and plasmid DNA will then bind together, splicing the human DNA into the plasmid, producing recombinant DNA. **(p. 266)**

- Plasmids with human DNA spliced into them can then be taken up into bacterial cells through transformation. As these cells replicate, the plasmid DNA inside them replicates as well. Because the plasmids are producing the protein coded for by the human DNA that has been spliced into them, the result is production of a quantity of the human protein of interest. **(p. 266)**

- Transgenic organisms that are used to produce a number of medicines and vaccines include bacteria, yeast, hamster cells, and mammals such as goats. Transgenic food crops are planted in abundance today in the United States. **(p. 266)**

15.3 Reproductive Cloning

- A clone is a genetically identical copy of a biological entity. Genes can be cloned, as can cells, and clones of plants exist in the form of

plants grown from "cuttings." Reproductive cloning is the process of making adult clones of mammals of a defined genotype. Today in reproductive cloning of mammals, an egg cell has its nucleus removed and is fused with an adult cell containing a nucleus and, therefore, DNA. The fused cell then starts to develop as an embryo and is implanted in a surrogate mother. This process is called somatic cell nuclear transfer (SCNT). **(p. 268)**

- Reproductive cloning can work in tandem with various recombinant DNA processes to produce adult mammals possessing desired traits. A cell can be made transgenic for such a trait and then used as the donor DNA cell in producing an adult mammal with the trait. **(p. 269)**

- No human being has been cloned to date, but the possibility of such cloning cannot be ruled out. A human clone would be genetically identical to the person who provided the donor DNA cell. **(p. 270)**

15.4 Cell Reprogramming

- Two promising methods now exist for generating human cells that are needed to treat victims of accident or disease: production through embryonic stem cells and production through induced pluripotent stem cells. Both methods employ the reprogramming of cells to yield desired cell types. **(p. 270)**

- Stem cells are cells that have a continuing capacity to produce more cells of their own type, along with at least one type of specialized daughter cell. One source of human stem cells is the blastocyst. Cells from the inner cell mass of the blastocyst are embryonic stem cells (ESCs): cells that can give rise to all the different cell types in the adult human body. Scientists harvest ESCs from blastocysts and then induce the ESCs to produce needed cell types. **(p. 271)**

- In 2007, two research teams developed a type of human stem cell that is not derived from an embryo: the induced pluripotent stem cell (iPS cell), which was first produced by means of splicing four develop-

mental genes into the genomes of ordinary adult skin cells. iPS cells appear to have all the developmental power of ESCs; they hold promise of reducing problems of tissue rejection in medical transplantation procedures; and they are being widely used as a means of studying human disease. **(p. 272)**

15.5 Forensic Biotechnology

- Identities of criminals, biological fathers, and disaster victims often are established through the use of forensic DNA typing. **(p. 275)**

- The polymerase chain reaction is a technique for quickly producing many copies of a segment of DNA. It is useful in situations, such as crime investigations, in which a large amount of DNA is needed for analysis, yet the starting quantity of DNA is small. **(p. 275)**

- Forensic DNA typing usually works through comparisons of short tandem repeat patterns that are found in all human genomes. **(p. 275)**

15.6 Controversies in Biotechnology

- Opponents of genetically modified crops are concerned about their effect on human health and the environment. There is no evidence so far that these crops have had detrimental effects, but consumer resistance to the crops has sharply limited both the types being planted and being developed. **(p. 278)**

- Some biotech controversies are essentially ethical in nature. One general controversy has to do with the question of what level of constraint society ought to impose on the modification of living things. **(p. 278)**

Key Terms

biotechnology 264

blastocyst 271

clone 268

Understanding the Basics

Multiple-Choice Questions

(Answers are in the back of the book.)

1. How is biotechnology defined?
   a. the study of technical biology   **b.** the use of biology to control technological processes as a means of meeting societal needs   **c.** the use of technology to control biological processes as a means of meeting societal needs   **d.** the technological process of cloning genes   **e.** the commercialization of biological research results

2. Researchers who work with DNA are able to cut it into desired lengths through the use of:
   a. transformation.   **b.** cloning.
   c. RNA.   **d.** restriction enzymes.
   e. DNA polymerase.

3. In transgenic biotechnology involving bacteria, which sequence of events would be followed in the process of harvesting a protein of interest?
   a. Replicate host bacterial cells; isolate gene that codes for protein; splice gene into plasmids; harvest protein of interest; transform plasmids into bacteria.
   b. Transform plasmids into bacteria; isolate gene that codes for protein; replicate host bacterial cells; splice gene into plasmids; harvest protein of interest.
   c. Harvest protein of interest; isolate gene that codes for protein; splice gene into plasmids; replicate host bacterial cells; transform plasmids into bacteria.
   d. Isolate gene that codes for protein; splice gene into plasmids; replicate host bacterial cells; transform plasmids into bacteria; harvest protein of interest.
   e. Isolate gene that codes for protein; splice gene into plasmids; transform plasmids into bacteria; replicate host bacterial cells; harvest protein of interest.

4. Which of the following is an example of a transgenic organism?
   a. *E. coli* bacterium with a human gene inserted   **b.** Dolly the sheep   **c.** rice with bacterium and daffodil genes inserted   **d.** all of the above   **e.** a and c only

5. What is a clone?
   a. an exact genetic copy of a gene or an entire living thing   **b.** a cell that can take up plasmids   **c.** a segment of DNA of a specified length   **d.** an offspring produced by test-tube fertilization   **e.** a segment of DNA that can prime the polymerase chain reaction

6. Dolly the sheep was cloned by means of:
   a. applying an electric current to an egg and sperm in a laboratory.   **b.** taking an embryo from one sheep and implanting it in another.   **c.** taking a cell from an adult sheep and fusing it with a sheep egg that had its DNA removed.   **d.** inserting new genes into an existing mammary cell.
   e. getting sperm to fuse with egg outside a living sheep.

7. As a means of producing large quantities of cells needed for the regeneration of human tissues, medical scientists are utilizing:
   a. somatic cell nuclear transfer, plasmids, and gene splicing.   **b.** gene splicing, polymerase chain reaction, and transgenic organisms.   **c.** embryonic stem cells and induced pluripotent stem cells.
   d. somatic cell nuclear transfer, embryonic stem cells, and induced pluripotent stem cells.   **e.** gene splicing, cloned cells, and transgenic organisms.

8. Criminal investigators who want to make copies of a DNA sample from a crime do so through:
   a. the polymerase chain reaction.

b. transformation.   **c.** cloning vectors.
d. DNA sequencing.   **e.** iPS cells.

Brief Review

(Answers are in the back of the book.)

1. Why is it useful to have organisms such as bacteria turning out human proteins?

2. Why are restriction enzymes that cut DNA to produce sticky ends useful? If the following piece of DNA were cut with *Bam*HI, what would the fragment sequences be?

   ATCGGATCCTCCG
   TAGCCTAGGAGGC

3. Why did Dolly the sheep end up looking like the sheep that "donated" the udder cell rather than the sheep that donated the egg?

4. Why is the polymerase chain reaction (PCR) so valuable?

5. Name two advantages that induced pluripotent stem cells (iPS cells) seem to have relative to embryonic stem cells (ESCs).

Applying Your Knowledge

1. One of the motives put forth for human cloning is that people want to replace children or other loved ones who have died. To what extent could a clone of a loved one be a replacement for that person? If the technique had then been available, should doctors in the nineteenth century have preserved the DNA of Abraham Lincoln for cloning?

2. Should society demand that there be no risks to genetically modified foods before it allows them to be developed? Does any significant technology have no risks associated with it?

3. What limits, if any, should be put on the ability of human beings to modify the genomes of other living things? Should human beings be free to carry out any modification that does not harm an organism?

MasteringBiology®

1. MasteringBiology Assignments
Activities: Copying DNA through PCR; **ABC News Video:** Genetically Altered Salmon; **Building Vocabulary; Questions:** Reading Questions; Test Bank

2. eText
Read your book online, search, take notes, highlight text, and study more effectively.

3. The Study Area
Self-study tools to test comprehension.

An Introduction to Evolution:

Charles Darwin, Evolutionary Thought, and the Evidence for Evolution

Charles Darwin is the person most responsible for deepening our understanding of how life evolved on Earth. Evidence uncovered from his time to ours has confirmed his insights.

Available on MasteringBiology®
GraphIt!: Species Area Effect and Island Biogeography; **ABC News Video:** Exploring Evolution in the Solomon Islands; Protecting the Galapagos Islands; **Current Events:** Evolution Right Under Our Noses; **Building Vocabulary**; **Questions:** Reading Questions; Test Bank

The theory of evolution explains countless features of the natural world, including the dazzling plumage of the peacock (*Pavo cristatus*).

Standing in a coastal redwood grove in California, we see trees stretching hundreds of feet into the air and wonder: Why are they so tall? Watching a nature program on television, we observe the exquisite plumage of the male peacock but are puzzled that such a *cumbersome* display could exist on a creature that has to be wary of predators. Leafing through a book, we discover that the color vision of honeybees is most sensitive to the colors that exist in the very flowering plants they pollinate. We wonder: Can this be an accident?

For as long as humans have existed, people have asked questions like these. Yet for most of human history, even the most informed individuals had no way to make sense of the variations that nature presented. Why does the redwood have its height or the peacock its plumage? Through the middle of the nineteenth century, the answer was likely to be either a shrug of the shoulders or a statement such as: "The creator gave unique forms to living things for reasons we can't understand."

Jumping ahead to our own time, modern-day biologists would give different answers to these questions. Their replies would go something like this:

- Redwoods (and trees in general) are so tall because, over millions of years, they have been in competition with one another for the resource of sunlight. Individual trees that had the genetic capacity to grow tall got the sunlight, flourished, and thus left relatively many offspring, which made for taller trees in successive generations.

- Extravagant plumage exists on male peacocks because it is attractive to *female* peacocks when they are choosing mating partners. Such finery is indeed cumbersome and carries a price (in lack of mobility), but this price is outweighed by the value of the feathers. Over time, the peacock ancestors with more attractive plumage mated more often and thus sired more offspring than peacocks with less-impressive plumage. This made for successive generations of peacocks with more elaborate plumage.

- It is not an accident that honeybee eyesight is most sensitive to the colors that exist in the flowering plants they pollinate. Both bees and flowering plants flourish because of their interactions with one another—bees get food, and flowers get pollinated. It is likely that flower color, or bee vision, or both, became modified over time in ways that maximized this interaction.

Comparing contemporary answers such as these to that single nineteenth-century answer, you might well ask: What is it that came *in between,* so that we no longer must simply stand in puzzled wonderment before the diversity of nature?

16.1 Evolution and Its Core Principles

What intervened was a set of intellectual breakthroughs in the mid-nineteenth century that today go under the heading of the *theory of evolution.* Two principles lie at the core of this theory.

Common Descent with Modification

One of these principles has been labeled **common descent with modification.** It holds that particular groups, or species, of living things can undergo modification in successive generations, with such change sometimes resulting in the formation of new, separate species—one species separating into two, these two then separating further. If you had a diagram of this "branching" of species and then looked at it from top to bottom, what you'd see is a reduction of all these branches into one, meaning that all living things on Earth ultimately are descended from a single, ancient ancestor. Evolution itself can have several definitions. In this chapter, we will think of it as synonymous with common descent with modification. In Chapter 17, you'll get a more technical definition of it.

Natural Selection

Common descent with modification is joined to a second evolutionary principle, **natural selection:** The process through which traits that confer a reproductive advantage to individual organisms grow more common in populations of organisms over successive generations. In the redwood example, the genetically based trait of increased tallness conferred a reproductive advantage: It allowed taller trees to capture more sunlight and hence to leave more offspring than shorter trees. The offspring of these taller trees would likewise tend to inherit the genetic capacity to grow tall, meaning they, too, would leave more offspring than other trees of *their* generation. In this way, the trait of tallness was becoming increasingly common in this population of trees from one generation to the next. Put another way, this population of trees was *evolving* in the direction of greater tallness by means of a process in which taller trees were selected to leave more offspring than shorter trees. As you will see, of all the processes that shape evolution, none is as important as natural selection.

The Importance of Evolution as a Concept

Evolution is not just important to biology; it is *central* to it. If biology were a house, evolution's principles would be the mortar binding its stones together. Why? Because all life on Earth has been shaped by evolution's key principles: natural selection and common descent with modification. This is easy to see once you think about any given organism. Take your pick: a fish, a tree, a human being? *All* of them evolved from ancestors, all of them ultimately evolved from a common ancestor, and natural selection was the primary process that shaped their evolution.

Given this, the theory of evolution provides a means for us to understand nature in all its buzzing, blooming complexity. It allows us to understand not just things that are familiar to us, such as the forms of the redwoods and peacocks, but all manner of new natural phenomena that come our way. When physicians observe that strains of infectious bacteria are becoming resistant to antibiotics, they don't have to ask: How could this happen? General evolutionary principles tell them that organisms evolve, that bacteria are capable of evolving very quickly (because they can produce many generations a day), and that antibiotic resistance is simply an evolutionary adaptation to the human use of antibiotics.

Evolution Affects Human Perspectives Regarding Life

So far-reaching is the theory of evolution that its importance stretches beyond the domain of biology and into the realm of basic human assumptions about the world. Once persuasive evidence for the theory had been amassed, it meant that human beings could no longer view themselves as something *separate* from all other living things. Common descent with modification tells us that we are descended from ancestors, just like other organisms. Thus, we sit not on a pedestal, viewing the rest of nature beneath us, but rather on one ordinary branch of an immense evolutionary tree.

Beyond this, evolution meant the end of the idea of a *fixed* living world in which, for example, birds have always been birds and whales have always been whales. Birds are the descendants of dinosaurs, while whales are the descendants of medium-sized mammals that walked on land (**Figure 16.1**). Life-forms only appear to be fixed to us because human life is so short relative to the time frames in which species undergo change.

Finally, in evolution human beings are confronted with the fact that, through natural selection, life has been shaped by a process that has no mind and no goals. Like a river that digs a canyon, natural selection shapes, but it does not design; it has no more "intentions" than does the wind. Trees that received more sunlight left more offspring, and thus trees grew taller over time. This is not a "decision" on the part of nature; it is an outcome of an impersonal process. And if evolution has no consciousness, it follows that it can have no morals. This force that has so powerfully shaped the natural world is neither cruel nor kind, but simply indifferent.

Figure 16.1
From Land to Water
Whales, such as this breaching humpback, are warm-blooded, air-breathing mammals. How can we explain such classic mammalian traits in creatures who live solely in water? Whales are the descendants of four-legged land-dwelling mammals who made a gradual transition from land to water over a very long period—about eight million years, beginning about 50 million years ago. Their closest living relative is the hippopotamus.

SO FAR . . .

1. The principle of common descent with modification holds that groups of living things can undergo modifications that,

over time, can lead to the formation of new _____ and that all living things on Earth are descended from a _____.

2. The principle of natural selection holds that some traits will confer a _____ to individual organisms and that these traits will accordingly become more _____ over time in a population of organisms.

3. Evolutionary theory supports the idea that humans are _____ other organisms and thus are not _____ from other living things.

16.2 Charles Darwin and the Theory of Evolution

How did such a far-reaching theory come into existence? Thinking back to this book's unit on genetics, you may recall that the basic principles of that field were developed over the course of about a century by a large number of scientists. In contrast, a single person—nineteenth-century British naturalist Charles Darwin—is credited with bringing together the essentials of the theory of evolution (**Figure 16.2**).

Darwin's Contribution

Darwin's contribution was two-fold. First, he developed existing ideas about descent with modification while providing a large body of evidence in support of them. Second, he was the first person to perceive natural selection as the primary driving force behind descent with modification.

To be sure, the study of evolution has been refined and expanded since Darwin. And it should be noted that before Darwin had ever published anything on his theory, a contemporary of his, fellow Englishman Alfred Russel Wallace, independently arrived at his insight about the importance of natural selection to evolution. (This makes Wallace the co-discoverer of this principle.) Yet, there is universal agreement that it is Charles Darwin who deserves primary credit for providing us with the core ideas that exist in evolutionary biology even today.

Darwin's Journey of Discovery

Darwin was born on February 12, 1809, in the country town of Shrewsbury, England. He was the son of a prosperous physician, Robert Darwin, and his wife, Susannah Wedgwood Darwin, who died when Charles was 8. Young Charles seemed destined to follow in his father's footsteps as a doctor, being sent away to the University of Edinburgh at age 16 for medical training. But he found medical school boring, and his medical

career came to a halt when, in the days before anesthesia, he found it unbearable to watch surgery being performed on children. His father then decided that he should study for the ministry. At the age of 20, Darwin set off for Cambridge University to spend 3 years that he later recalled as "the most joyous in my happy life." Darwin's happiness came in part from the fact that theology at Cambridge took a backseat to what had been his true passion since childhood: the study of nature. From his early years, he had collected rock, animal, and plant specimens and was an avid reader of nature books. His studies at Cambridge did yield a divinity degree, but they also gave Darwin a solid background in what we would today call life science and Earth science.

An Offer to Join the Voyage of the Beagle

Darwin's training, and the contacts he made at school, came together in one of the most fateful first-job

Figure 16.2
Scientist of Great Insight

Charles Darwin, late in his life.

(Painting by John Collier, 1883. London, National Portrait Gallery. ©Archiv/Photo Researchers.)

Figure 16.3

What the *Beagle* Looked Like

A reconstruction of the *HMS Beagle,* sailing off the coast of Tierra del Fuego in South America.

Thus did Darwin spend time on a research vessel—almost 5 years in all—beginning in England two days after Christmas 1831 and ending back there in October 1836. Just 22 when he left, he was prone to seasickness; he had to share a 10-by-15-foot room with two other officers; he was not a traveler by nature; and the journey was dangerous (three of the *Beagle*'s officers died of illness during it). Yet Darwin was happy because of the work he was doing: looking, listening, collecting, and thinking about it all. Here he is writing about a part of coastal Brazil:

> A most paradoxical mixture of sound and silence pervades the shady parts of the wood. The noise from the insects is so loud that it may be heard even in a vessel anchored several hundred feet from the shore; yet within the recesses of the forest a universal silence appears to reign. To a person fond of natural history such a day as this brings with it a deeper pleasure than he can ever hope to experience again.

offers ever extended to a recent college graduate. One of Darwin's Cambridge professors arranged to have him be the resident naturalist aboard the *HMS Beagle,* a ship that was to undertake a survey of coastal areas around the world (**Figure 16.3**). In **Figure 16.4** you can trace the *Beagle*'s journey. In addition to numerous stops on the east coast of South America (the ship's primary survey site), the *Beagle* also stopped briefly at the remote Galapagos Islands, about 600 miles west of Ecuador—a visit you'll learn more about later.

16.3 Evolutionary Thinking before Darwin

Darwin beheld these sights and sounds during a period in the nineteenth century in which change was in the air with respect to ideas about life on Earth.

Charles Lyell and Geology

The single book Darwin took with him when he boarded the *Beagle* was the first volume of Charles Lyell's *Principles of Geology,* published in 1830 and bearing a message that even Lyell's fellow scientists

Figure 16.4

Journey into History

The main mission of *HMS Beagle*'s 5-year voyage of 1831–1836 was to chart some of the commercially promising waters of South America. Darwin observed nature and collected specimens throughout the ship's journey, but for purposes of the theory of evolution, the ship's most important stops came in 1835 on the Galapagos Islands, west of South America.

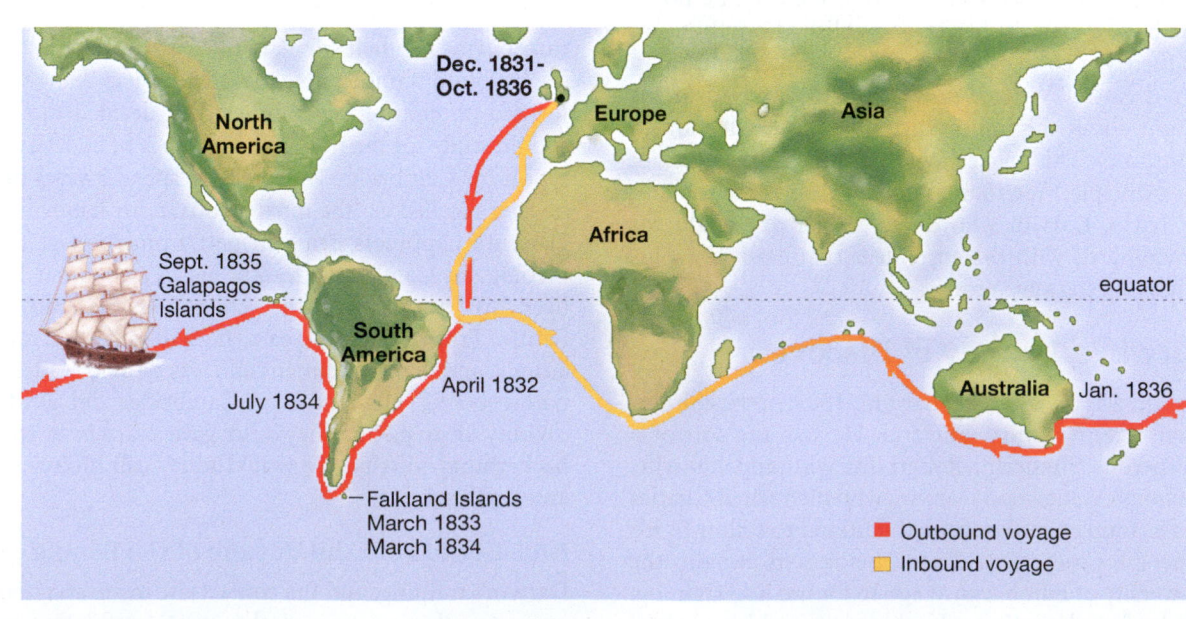

Figure 16.5
Earth's Life-Forms Change over Time

Animals known as ammonites swam in Earth's oceans from about 400 million years ago to 65 million years ago. Pictured are ammonite fossils dating from 150 million years ago that were unearthed in sediments that are now part of Great Britain's coastline. Ammonites became extinct at the time Earth's dinosaurs did, but they are closely related to such living animals as octopus and squid.

found hard to accept: that geological forces *still operating* could account for the changes geologists could see in the Earth's surface. (As Darwin put it, "that long lines of inland cliffs had been formed, the great valleys excavated, by the agencies which we still see at work.") Under this view, Earth had not been put into final form at a moment of creation, but rather was steadily undergoing change. If such a thing were possible for the Earth itself, why not for the creatures that lived on it (**Figure 16.5**)?

Jean-Baptiste de Lamarck and Evolution

An idea along this line *had* been proposed, by the French naturalist Jean-Baptiste de Lamarck, in a book published in 1809. Lamarck believed that organisms changed form over generations through what has been termed the inheritance of acquired characteristics. Ducks or frogs did not originally have webbed feet, he said, but in the act of swimming they stretched out their toes to move more rapidly through the water. In time, membranes grew between their toes, effectively becoming webbing; critically, this characteristic was passed along to their offspring (**Figure 16.6**). Over time, he believed, an animal would acquire enough changes that one species would diverge into two, with this branching extending all the way to human beings. In his work, Lamarck lent support to a *means* of evolution that we know today is false. (Animals don't acquire and then pass along traits such as webbed feet in accordance with their activities.) Yet, note what Lamarck

got right: that organisms can evolve; that one kind of organism can be ancestral to a different kind of organism.

Georges Cuvier and Extinction

Another French scientist, Georges Cuvier, in examining the fossil-laden rocks of the Paris Basin early in the nineteenth century, provided conclusive evidence of the extinction of species on Earth. This was a radical notion at the time because many Christians believed that the creator would never allow one of his creatures to perish. (One amateur fossil collector who believed this was Thomas Jefferson.) This much about extinction Cuvier got right. However, seeing in his rock layers seeming "breaks" in the sequence of animal forms—a layer of simple forms would be followed by a layer of more complex forms as he went from older to newer layers—he held that there had been a series of catastrophes, such as floods, that wiped out life in given areas, after which the creator had carried out *new* acts of creation, bringing more complex life-forms into being with each act.

If you were to survey a broad swath of scientific thought leading up to Darwin, you'd repeatedly find what you've seen with Lamarck and Cuvier: a given scientist getting things partly right and partly wrong (which is almost always the way in science). The result was a rich mix of scientific findings and fanciful speculation that existed in Darwin's world as he boarded the *Beagle*. He himself believed at the time that species were fixed entities; that they did not change over time.

16.4 Darwin's Insights Following the *Beagle*'s Voyage

Darwin observed and collected wherever he went, but the most important stop on his journey took place nearly 4 years into it, when the *Beagle* stopped for a

Figure 16.6
The Duck's Feet

Jean-Baptiste de Lamarck proposed that new species evolve through a branching evolution, which is correct. He also proposed, however, that one of the forces driving this evolution is the inheritance of acquired characteristics, which does not take place in the form he imagined. Under Lamarck's view, common at the time, a duck got its webbed feet because of activities it carried out during its lifetime, and this change was then passed on to its offspring.

Figure 16.7
The Galapagos Islands

Galapagos is Spanish for *tortoise.* A part of Ecuador, the islands are still an important site of evolutionary research. Here a giant tortoise (*Geochelone nigra*) moves through the Alcedo Volcano area on the Galapagos island of Isabela.

scant five weeks at the remote series of volcanic outcroppings called the Galapagos Islands (**Figure 16.7**). There, in the dry landscape, amid broken pieces of black lava, he saw strange iguanas and tortoises and mockingbirds that varied from one island to another. He shot and preserved several of these mockingbird "varieties" for transport back home, along with a number of small birds that he took to be blackbirds, wrens, warblers, and finches.

Perceiving Common Descent with Modification

Arriving back in England in October of 1836, Darwin soon donated a good deal of his *Beagle* collection to

the Zoological Society of London. If there was a single most important flash of insight for him regarding descent with modification, it came when one of the society's bird experts gave him an initial report on the birds he had collected on the Galapagos. Three of the mockingbirds were not just the "varieties" that Darwin had thought; they were separate species, as judged by the standards of the day. Beyond that, the small birds he had believed to be blackbirds, finches, wrens, and warblers were all finches—separate species of finches, each of them found nowhere but the Galapagos (**Figure 16.8**).

With this, Darwin began to see a pattern: The Galapagos finches were related to an ancestral species that could be found on the mainland of South America, hundreds of miles to the east. Members of that ancestral species had come by air to the Galapagos and then, fanning out to separate islands, had diverged over time into separate species. Thus it was with Galapagos tortoises and iguanas and cactus plants as well; they had common mainland ancestors, but on these islands had diverged into separate species. Darwin began to perceive the infinite branching that exists in evolution.

Perceiving Natural Selection

But what drives this branching? In England and set on a career as gentleman naturalist, Darwin married and settled in London for a time before moving to a small village south of London, where he would live out his days (**Figure 16.9**). Two years after his homecoming, inspiration on evolution's key process came to him not from looking at nature, but from reading a book on human population and food supply—*An Essay on the Principle of Population,* by T. R. Malthus. As Darwin later wrote in his autobiography:

Figure 16.8
Three of Darwin's Finches

Thirteen species of finch evolved on the Galapagos Islands, all of them descendants of a single species of finch native to South America. Shown, left to right, are *Geospiza fuliginosa, G. fortis,* and *G. magnirostris.*

Figure 16.9
Where *On the Origin of Species* Was Written
In this study in his house in rural England, Charles Darwin conducted scientific research and wrote his groundbreaking work, *On the Origin of Species by Means of Natural Selection*.

I happened to read for amusement Malthus on Population, and being well prepared to appreciate the struggle for existence which everywhere goes on . . . it at once struck me that under these circumstances favourable variations would tend to be preserved and unfavourable ones to be destroyed.

Thus did natural selection occur to Darwin as the driving force behind evolution. Organisms that had "favorable variations" were preserved (they lived and left many offspring), while those that had unfavorable variations were destroyed (they left fewer offspring or none at all). As a result, over generations, organisms evolved in the direction of the favorable variations.

You might think that, with two major insights in place—common descent with modification and its link to natural selection—Darwin would have rushed to inform the world of them. In fact, more than 20 years were to elapse between his reading of Malthus and the 1859 publication of his great book, *On the Origin of Species by Means of Natural Selection*. In between, though incapacitated much of the time with a mysterious illness, Darwin published on geology, bred pigeons, and spent 8 years studying the variations in barnacles—activities that each had relevance to his theory and that yet constituted a kind of holding pattern for him as he contemplated informing the world of a theory that he knew would be controversial. Darwin finally got to work on what he thought would be his "big species book" in 1856.

16.5 Alfred Russel Wallace

Two years after Darwin began this labor, half a world away and unbeknownst to him, a fellow English naturalist lay in a malaria-induced delirium on an Indonesian island. Alfred Russel Wallace made a living collecting bird and butterfly specimens from the then-exotic lands of South America and Southeast Asia, selling his finds to museums and collectors. He had thought long and hard (and even published) about how species originate. Now, shivering with malarial fever in a hut in the tropical heat, with the question on his mind again, Wallace came on the very insight that had come to Darwin 20 years earlier: Natural selection is the process that shapes evolution. Recovering from his illness, Wallace wrote out his ideas over the next few days and sent them to a scientific hero of his in England—Charles Darwin! He asked Darwin to read his manuscript and submit it to a journal if Darwin thought it worthy.

Darwin was stunned as he faced the prospect of another man being the first to bring to the world an insight he thought was his alone. Nevertheless, he would not allow himself to be underhanded with the younger scientist. He informed some of his scientist friends about his plight, and (without informing Wallace) they arranged for both Wallace's paper and some of Darwin's letters, sketching out his ideas, to be presented at a meeting of a scientific society in London on July 1, 1858. The readings before the scientific society turned out to have little immediate effect, but they did prod Darwin into working on what would become *On the Origin of Species*, published some 16 months later. This event had a great effect. Indeed, it set off a thunderclap whose reverberations can still be heard today.

16.6 Darwin: Accepted, Doubted, and Vindicated

Sparking both scorn and praise, Darwin's ideas were fiercely debated in the years after 1859. At least within scientific circles, however, it did not take long for common descent with modification to be accepted. Fifteen years or so after *On the Origin of Species* was published, almost all naturalists had become convinced of it, and it's not hard to see why. With evolution as a framework, so many things that had previously seemed curious, or even bizarre, now made sense.

For example, years before *Origin* was published, scientists had known that, at a certain point in their

Figure 16.10
Different Embryos, Same Structure

(Adapted from M. K. Richardson, 1997.)

Pharyngeal slits exist in these five vertebrate animals . . .

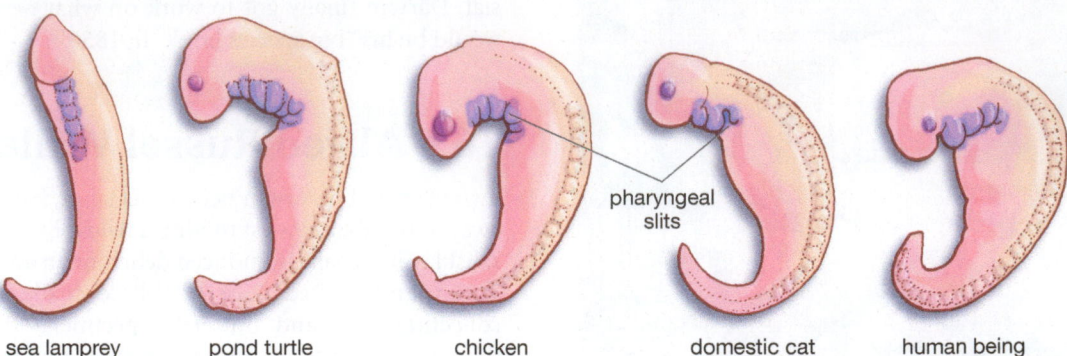

pharyngeal slits

sea lamprey   pond turtle   chicken   domestic cat   human being

. . . evidence that all five evolved from a common ancestor.

embryonic development, organisms as diverse as fish, chickens, and humans all have structures known as pharyngeal slits (**Figure 16.10**). In fish, these structures go on to be *gill* slits; in humans, they develop into the eustachian tubes, among other things. The question was: Why would land-dwelling and sea-dwelling animals share this common structure in embryonic life? For that matter, why would organisms as different as a human and a chicken share this structure? With evolution, the answer became clear. All of them shared a common vertebrate *ancestor,* which had the slits. The vertebrate ancestral line had branched out into various species, yet elements of the common ancestor persisted in these different species in their embryonic stage.

You might think that this would leave open the question of why organisms should differ more as adults than as embryos, but Darwin's theory made sense of this, too. Remember that natural selection is the primary process that brings about the differences that we see among species. Now consider one trait that natural selection stands to operate on in all animal species—the trait of movement or "locomotion." If you think about it, *adult* animals must deal with the issue of movement all the time, but embryonic animals never deal with it (because they don't move). Hence, for millions of years, natural selection was channeling the adult forms of vertebrate species in different ways with respect to their means of movement (fins, legs, wings). But this channeling was going on *only* in the adult forms. Likewise, natural selection was only affecting adults with respect to such traits as mating and the means of obtaining food. (Embryos don't mate, and food is delivered to them.) Hence, natural selection was driving differences in the adult forms of animals more than it was driving differences in embryonic forms. The result? Early embryos of various species could retain the ancestral vertebrate form more than adults could.

In bringing together countless loose ends such as this, evolution became the mortar that unified the study of the living world. (It's interesting to think of

biologists from this period as readers who start going over a detective novel a *second* time, saying, "Of course!" and "Why didn't I see that?") Given this, scientists had little trouble accepting the *fact* of evolution—that is, the occurrence of common descent with modification. But the primary process that Darwin said was driving evolution, natural selection, was not embraced so quickly. Indeed, the twentieth century would be almost halfway over before it was generally accepted.

The Controversy over Natural Selection

The essential stumbling block for the theory of evolution by natural selection was this: It asserted that traits can vary in ways that confer reproductive advantages on given individuals and that these variations can be passed on from one generation to the next. With the trees considered earlier, the trait was height, and the variation that conferred a reproductive advantage was greater height. In a given environment, therefore, added height represents a *difference* that allows some members of the species to leave more offspring than others, leading eventually to a taller species over all. But in the middle of the nineteenth century, no one could imagine how such differences could reliably be passed down over many generations. Even scientists of the time could not understand this because they had no grasp of the field that we today call genetics.

Coming to an Understanding of Genetics

Thinking back to the genetics section of this book, recall, from Chapter 11, that a trait such as tree height is likely to be governed by many genes working in tandem, with each of these genes coding for a protein that has some small, additive effect on height. Indeed, such multi-gene, or "polygenic," inheritance is at work with most of the traits in living things. (See "Multiple Alleles and Polygenic Inheritance" in Chapter 11, page

204.) A tree's "height genes" will be shuffled in countless ways when tree egg and sperm are formed and when they fuse in fertilization. As a result, gene combinations that yield particularly tall trees will come together in certain instances. In addition, outright mutations—alterations in genetic information—can take place, showing up as new physical traits, one of them being a taller tree. Most important for our purposes, whatever the genetic information is, it *persists*; it is passed on largely intact from one generation to another in the small informational packets we call genes. The result is small physical differences that can be reliably passed on from one generation to the next.

Darwin could not have imagined a better mechanism for carrying out natural selection as he had envisioned it. The problem was that neither he nor anyone else in his time knew that things worked like this. Most nineteenth-century scientists believed that inheritance worked through a "blending" process akin to mixing different-colored paints. If you blend red and blue paint, for example, you get something that lies halfway between them, purple. Under such a system, differences in traits would not be preserved over time; they would be averaged into oblivion. Blending would work to make groups of organisms the same, not to allow differences among them. How could natural selection work if there were no persistent differences to select from?

The vindication of natural selection required, first, a demonstration (by Gregor Mendel) that inheritance doesn't work by blending. Rather, it works through the transmission of the stable genetic units we now call genes. Then, in the early part of the twentieth century, came the demonstration that, through polygenic inheritance, genes could account for very small physical differences that arise within a given population. Trees come in a *range* of heights, with the differences in this range governed by the combinations of individual genes described earlier.

Darwin Triumphant: The Modern Synthesis

Advances in genetics such as this were later joined to other types of evolutionary research, such that in the period roughly from 1937 to 1950 there took place what is known as the **modern synthesis**: the convergence of several lines of biological research into a unified evolutionary theory. Taxonomists reported on how species are distributed throughout the world. Mathematical geneticists provided guidance on how evolution *could* work, given the shifting around of genes among organisms. Paleontologists studied evolution in the fossil record. Critically, these different types of scientists were involved in a process of mutual education with one another; the findings of geneticists, for example, were now known to taxonomists, and vice versa. The upshot was a greatly deepened understanding of how evolution

works—an understanding that provided conclusive evidence for Darwin's assertions about the importance of natural selection. The architect of evolutionary theory had been vindicated nearly 100 years after publishing his great work. (For another example of natural selection's importance, see "An Evolving Ability to Drink Milk" on page 294.)

SO FAR...

1. True or false: Charles Darwin was the first person to perceive common descent with modification.

2. The co-discoverer of natural selection was _____.

3. Following the publication of *On the Origin of Species,* the evolutionary principle of _____ was quickly accepted by most scientists, but the importance of _____ would not be accepted by most authorities until midway through the twentieth century.

16.7 Opposition to the Theory of Evolution

So, why are Darwin's essential insights so widely believed to be correct? We'll now review some of the evidence for these insights, but first a word about why a look at this evidence is desirable. When you studied cells in this book, you did not review the "evidence for cells." Nor did you go over the "evidence for energy transfer" when studying cellular respiration, nor the "evidence for lipids" when studying biological molecules. What makes evolution different from these topics?

The False Notion of a Scientific Controversy

Evolution has an unusual status among major biological subdisciplines in that, almost alone among them, its findings are regularly challenged as being unproven or simply wrong. For anyone unfamiliar with biology, these attacks make it appear that the theory of evolution is not a body of knowledge, solidly grounded in evidence, but rather a kind of scientific guess about the history of life on Earth.

What Is a Theory?

Several factors are critical in allowing the appearance of a "scientific debate" on the validity of the theory of evolution, when in fact none exists. The first of these has to do with a simple misunderstanding regarding terminology. You saw in Chapter 1 that, to the average person, the word *theory* implies an idea that certainly is unproven and that may be pure speculation. In science, however, a theory is a general set of principles, supported by evidence, that explains some aspect of the natural world. Accordingly, we have Isaac Newton's theory of gravitation, Albert Einstein's general theory of relativity, and many others. Lots of principles go into making up the theory of evolution, some of them on surer footing than others. Over time, however, some of these principles have achieved the status of established fact—we are as sure of them as we are sure that the Earth is round. Among these principles are the fact that evolution has indeed taken place and the fact that this has occurred over billions of years.

The Nature of Historical Evidence

A second factor that provides an opening for the opponents of evolution is the nature of the evidence for it. If you were called on to provide the "evidence for cells" referred to earlier, a look through a microscope at some actual cells might be enough to stop this "debate" in its tracks. Conversely, one of evolution's most important manifestations—radical transformations of life-forms—can never be observed in this way because these transformations take place over vast expanses of time. Asking to "see" the equivalent of a dinosaur evolving into a bird before you'll believe it is like asking to see the European colonization of North America before you'll acknowledge it.

Evolution is taking place right now, all around us, but the evidence for evolution is historical evidence to a degree that is not true of, say, the study of genetics or of photosynthesis. Any ancient historical record is fragmentary, and evolution's historical record is ancient indeed. The fossils that scientists analyze are what remain from hundreds of millions of years of weathering and decay. Such an incomplete record leaves room for a great deal of interpretation among scientists—far more than is the case in purely experimental science. This interpretation has to do, however, with the *details* of evolution, not with its core principles. It has to do with what group descended from what other, with the rate at which evolution has proceeded, or with the role that pure chance has played in evolution. It does not have to do with *whether* evolution has occurred.

16.8 The Evidence for Evolution

So, what makes scientists so sure about the core principles of evolution? In any search for truth, we are more comfortable when lines of evidence are internally consistent and then go on to agree with *other* lines of evidence. The evidence for evolution satisfies both criteria. If we were to find even a single glaring inconsistency within or between lines of evidence, the whole body of evolutionary theory would be called into question. But no such inconsistencies have turned up yet, and scientists have been looking for them for 150 years.

Radiometric Dating

One claim of evolutionary theory, as you have seen, is that evolution proceeds at a leisurely pace, with billions of years having elapsed between the appearance of life and the present. Yet how do we know that Earth isn't, say, 46,000 years old, as opposed to the 4.6 billion that scientists believe it to be? The conceptual bedrock on which we have determined the age of the Earth and its organisms is **radiometric dating,** a technique for determining the age of objects by measuring the decay of the radioactive elements they contain. As volcanic rocks are formed, various elements in their surroundings are incorporated into them. Some of these elements are radioactive, meaning they emit energetic rays or particles and "decay" in the process.

With such decay, one element can be transformed into another, the most famous example being uranium-238, which becomes lead-206 through a long series of transformations. The critical thing is that this transformation proceeds at a fixed *rate*; it is as steady as the most accurate clock imaginable. It takes 4.5 billion years for half a given amount of uranium-238 to decay into lead-206 (hence the term *half-life*). When such a transformation takes place in a cooled rock, the original or "parent" element is trapped within the rock, as is the "daughter" element, and the atoms of both elements can be counted. Therefore, comparing the proportion of a parent element to the daughter element in a rock sample provides a date for the rock with a fair amount of precision. There are now more than 40 different radiometric dating techniques used, each of them employing a different radioactive isotope. Some of these isotopes have very long half-lives (as with uranium-238), but some have relatively short half-lives. (See Chapter 2, page 26, for information on an isotope with a very short half-life, carbon-14.) Given the range of such "radiometric clocks," scientists have been able to date objects from nearly the formation of the Earth to the present, although there are some gaps in the picture.

Fossils

Today, one line of evidence for evolution is the similarity of fossil types by sedimentary layers. Looking at the same geologic layers of sediment worldwide, scientists find similar types of fossils, with a general movement toward more complex organisms as they go up through the newer strata. Chapter 1 noted that scientific claims must be *falsifiable*—they must be open to being proved false with the discovery of new

Figure 16.11
Ancient Organism, Now Extinct
Shown is a fossilized trilobite, a kind of arthropod that became extinct long before the animals known as primates came to exist. This fossilized trilobite dates from about 370 million years ago and was found in present-day Morocco.

evidence. The fossil record presents a falsifiable claim. For example, creatures called trilobites (**Figure 16.11**) had a long run on Earth, existing in the ancient oceans from about 500 million years ago until some 245 million years ago, when they became extinct. By contrast, the evolutionary lineage that humans are part of, the primates, *began* no more than 81 million years ago. If scientists were to find, in a single fossil bed, fossils of trilobites existing side by side with those of early primates, our whole notion of evolutionary sequences would be called into question. No such incompatible pairing has happened, however, and the strong betting is it won't. Given this, the line of fossil evidence is internally consistent—but there's more. When we compare fossil placement with the dates we

get from radiometric dating, we get excellent agreement *between* these two lines of evidence. We don't find trilobites embedded in sediments that turn out to be 81 million years old.

Morphology and Vestigial Characters

Morphology is the study of the physical forms that organisms take. In comparative morphology, some classic evidence for evolution is seen in the similar forelimb structures found in a very diverse group of mammals—in a whale, a cat, a bat, and a gorilla, as seen in **Figure 16.12**. Look at what exists in each case: one upper bone, joined to two intermediate bones, joined to five digits. Evolutionary biologists postulate that the four mammals evolved from a common ancestor, adapting this 1-2-5 structure over time in accordance with their environments. Such features are said to be **homologous,** meaning the same in structure owing to inheritance from a common ancestor.

Along similar lines, if you look in detailed human anatomy textbooks, you'll see listings for the muscles of the human head. One of these, for example, is a muscle that retracts the tongue (the hyoglossus) and another is a muscle that helps us swallow (the aryepiglottic). In any such list, however, you'll also find the auricular muscles—three of them, which lie on each side of the head. And what do these muscles do? *They wiggle our ears.* That's it; no other natural use has ever been found for them. And, of course, in most people, they are useless for even this. Now, muscles are tissues that are specialized to do one thing—contract—and relative to fat tissue, they burn up a lot of energy. Given this, we have to ask: Why do we humans have auricular muscles? If we imagine that our bodies were designed, then wouldn't it be strange

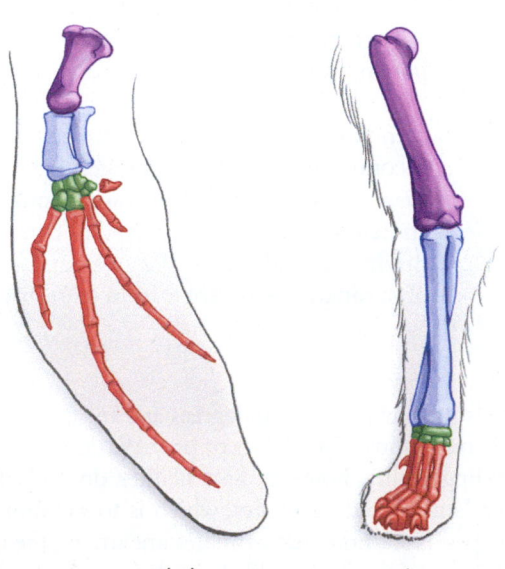

whale   cat

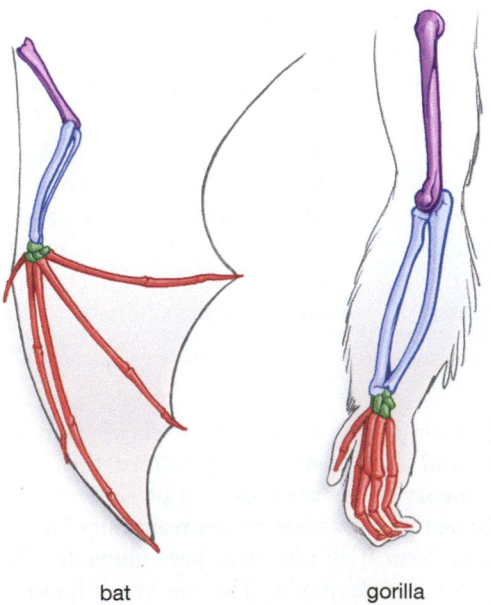

bat   gorilla

Figure 16.12
Four Animals, One Forelimb Structure

Whales, cats, bats, and gorillas are all descendants of a common ancestor. As a result, the bones in the forelimbs of these diverse organisms are very similar despite wide differences in function. Four sets of homologous bones are color-coded for comparison. Note that in each case there is one upper bone, joined to two intermediate bones, joined to five digits.

ESSAY

An Evolving Ability to Drink Milk

Anyone who thinks that the effects of evolution lie strictly off in the past might consider asking themselves the following question: How would you feel about having a glass of milk right now? If the answer is you'd feel fine about it, then that puts you in the company of hundreds of millions of adults worldwide who would feel the same way. Meanwhile, for nearly 100 percent of adults in China, and for nearly that percentage of adults in parts of Africa, the idea of drinking a glass of milk is about as appealing as the idea of acquiring a low-grade food poisoning, because both things can have the same effect: abdominal pain, bloating, and gas. Within the United States, at least 30 million adults are averse to milk: Some 75 percent of adult African Americans and 90 percent of adult Asian Americans cannot digest it. Conversely, no more than 5 percent of adults of Northern European descent have a problem with it.

So, why should human beings have such differing reactions to a healthy food? The answer is that, for tens of thousands of years, *no* adult human beings appear to have had an ability to digest milk. Then, in the last 7,000 years, a human cultural change—the initiation of animal herding—worked hand-in-hand with the evolutionary forces of mutation and natural selection to produce select groups of people who could digest milk throughout their lives. It's possible that all of the people in the world today who can digest milk as adults are descended from members of these original groups.

Human beings of any age who digest milk do so by producing an enzyme, called lactase, that breaks down the milk sugar lactose when it enters the small intestines.

Virtually all human *infants* can break down the lactose in milk—something that's not surprising, given that through most of human history, all human infants were nurtured on milk from their mothers. When early childhood ends, however, a majority of the world's people stop producing the lactase enzyme. The gene for lactase, called *LCT*, ceases to be transcribed in them, which means it effectively shuts down. Individuals in whom this takes place become lactose-intolerant; any lactose that enters their small intestines simply stays in the digestive tract for a time, drawing in water while feeding bacteria—conditions that lead to bloat, gas, and discomfort. The fact that the *LCT* gene shuts down in most people actually is not surprising, however, since for most of human history an *active LCT* gene would have been useless in adults. This is so because, for the first 190,000 years of human history, humans had no source of milk other than mothers' milk. The default position for the human lactase gene was to shut down after early childhood because *after* early childhood most humans consumed no milk at all.

In some parts of the world, however, the human nutritional environment began to change considerably beginning about 10,000 years ago, when people first domesticated herd animals. The cattle, sheep, goats, and camels that human beings brought under control would eventually serve as rich sources of milk in the human diet. Of course, such milk initially had a limited value because of adult lactose intolerance, but this was about to change. Research conducted in the last 10 years has shown that the ability of adults to digest

milk arose separately in at least three places—Northern Europe, the Middle East, and Eastern Africa—and that this occurred through the same mechanism in each case: an accidental mutation to human DNA followed by rounds of natural selection that spread the ability to digest milk through ever-expanding proportions of the human population.

In 2002, Leena Peltonen and her colleagues in Helsinki, Finland, lifted the curtain on the Northern European part of this story. An analysis Peltonen conducted of nine extended Finnish families revealed a mutation to the DNA of some family members that correlated perfectly with their ability to digest milk. Exactly 13,910 bases "upstream" from the lactase *LCT* gene, the C base found in most human beings mutated into a T base in the ancestors of these Northern Europeans. This small change made all the difference, for it allowed the binding of a protein to the DNA of these Europeans that kept their transcription of the lactase gene going, even after they passed childhood. Thus, anyone who inherited this "T_13910" mutation effectively inherited a lifelong ability to digest milk.

The discovery of this European mutation deepened a question that University of Maryland, College Park, researcher Sarah Tishkoff had been asking herself since the late 1990s: What accounts for the "lactase persistence" seen in the cattle-herding tribal populations of East Africa? (**Figure 1**) Sequencing of the DNA of some lactase-persistent East Africans made clear that they didn't share the T_13910 mutation seen in Europeans. To find out what did account for their ability to digest lactose, Tishkoff and her colleagues spent long stretches of time over a period of 3 years driving around East Africa in a Land Rover, obtaining DNA samples from 470 Africans

for a designer to put in place these sets of costly, specialized, and useless structures? Against this, evolutionary theory makes sense of their presence.

What we think of as our ears are really our outer ears, or "pinna." Almost all mammals have pinna, but these usually are *mobile* pinna. The common house cat

swivels its pinna up to 180 degrees as a means of homing in on sounds of predators or prey. We don't have this swiveling ability, however, and neither do any of the other "anthropoid" primates, which is to say monkeys and apes. Now, note that primates are among the latest evolving mammals. So, a likely explanation for human

Figure 1

From a Human Cultural Practice, an Evolving Human Ability

A Masai tribesman in Kenya tends to a herd of cattle. The diet of most Masai consists largely of meat and milk derived from the cows and goats they raise. The East African region that is home to the Masai is one of three in the world in which human adults first acquired the ability to digest milk.

while testing them for their ability to digest lactose. In 2007, Tishkoff let the world in on what this work revealed: The ancestors of Africans in Kenya and Tanzania developed a mutation to their DNA that, while different from the European mutation, nevertheless had the same effect; it drove lifelong transcription of the *LCT* gene. One year after this discovery was announced, the Peltonen group in Finland added a third chapter to the human lactase story. DNA testing of modern-day Saudi Arabians revealed that their ancestors had experienced two mutations—completely separate from the European and African mutations—that now are widespread in the Middle East and that account for the ability of many Middle Eastern residents to digest milk in adulthood.

Modern DNA analysis is so powerful that it allows researchers to assign rough dates to the appearance of mutations such

as these. Peltonen's group estimated that one of the two Middle Eastern lactase mutations first appeared about 4,100 years ago, though this event could have taken place about 2,100 years to either side of this date. Meanwhile, the African mutation is thought to have appeared sometime between 3,000 and 7,000 years ago, and the Northern European mutation is thought to have arisen about 7,000 years ago. Taking a step back from these dates, note, first, how well they correlate with the initiation of animal herding 10,000 years ago. Second, notice how *recently* lactase persistence arose within the context of the 200,000 years that human beings have existed as a species. If human existence were an hour long, the ability of any adult humans to digest milk would have come about only in the last two minutes of that hour. We might marvel at the fact that such a large share of humanity has become lactase-persistent in this

relatively short period of time, but a little reflection on the mechanism that brought this change about, natural selection, makes clear that a rapid spread of lactase persistence was all but inevitable.

Imagine two girls born, say, 6,000 years ago in what is now Finland. Life in this region is precarious for everyone; the winters are long and food is scarce even in warm months. The Northern European lactase mutation appeared about 1,000 years prior to the birth of these girls, which means that only some Northern Europeans now possess this genetic variant. Imagine that one of the girls we're considering has inherited this variant but the other hasn't. This means that from the age of, say, 4, one of these girls can obtain nourishment from cow or goat's milk while the other cannot. Given this, which of these girls is more likely to *survive* into adulthood? The answer, of course, is the girl with the lactase mutation. The milk that she consumes will provide her, all year round, with precious calories and calcium. In the environment in which she's growing up, this girl has been *selected* to survive by virtue of her genetic makeup. And with this survival comes an increased likelihood that she will go on to have children of her own—children who themselves will have a reproductive advantage, assuming that they too have the lactase-persistence mutation.

When this process is repeated in one generation after another, the result is the rapid spread of a trait through a population. Today in parts of Northern Europe, 98 percent of the population is believed to be lactase-persistent. In 7,000 years, then, lactase persistence evolved from a 0 percent prevalence to a 98 percent prevalence in this one part of the world. How strong a force is evolution through natural selection? Well, how would you feel about having a glass of milk?

auricular muscles is that they are *vestiges* of our mammalian evolutionary history. Such muscles had a real function in our mammalian ancestors: They moved the pinna of these animals. In the anthropoid primates, however, moveable pinna became less useful for some reason (the three-dimensional vision we have?) and

over time these muscles fell into their current, functionless state.

Auricular muscles are one example among many of a **vestigial character,** meaning a structure in an organism whose original function has been lost during the course of evolution. The ostrich is a bird and it has wings, but it cannot fly. Ostrich wings do have several modern-day functions; they help balance the bird as it runs, for example. But, like muscles, wings are structures *specialized* for something—lifting animals off the ground. So is it reasonable to assume that a designer fashioned wings for the ostrich just to help it with balance? To paraphrase chemist Douglas Theobald, this would be like designing a computer keyboard that is meant to be used in hammering tacks. Evolution has a different explanation: The ostrich came from birds that could fly, but its own line *evolved* into a flightless condition. Looked at one way, the wings of the ostrich and the auricular muscles of human beings are like calling cards left by our ancestors; they tell us where we came from.

Evidence from Gene Modification

In the genetics unit, you saw that every living thing on Earth employs DNA and uses an almost identical "genetic code" (*this* triplet of DNA bases specifies *that* amino acid). At the very least, this means there is a unity running through all earthly life; it is also consistent with the idea that all life on Earth ultimately had a single starting point—the single common ancestor mentioned earlier.

In recent decades, molecular biology has also provided another check on what scientists have long believed to be true about evolutionary relationships— about which species are more closely related and about how long it's been since they've shared a common ancestor. All genes amount to a sequence of A, T, G, and C bases along DNA's double helix. And there are some genes that do very similar things in different organisms. There is, for example, a gene called *hedgehog* that helps regulate embryonic development in the *Drosophila* fruit fly and a gene called *sonic hedgehog* that helps regulate embryonic development in mice. Though the *hedgehog* and *sonic hedgehog* genes are similar, we would not expect them to be identical. This is so because gene base sequences change over time through the process of mutation. If the *rate* of this change is constant, then mutations are like a molecular clock that's ticking: With the passage of a given amount of time, you get a set number of mutations. Given this, the longer it has been since two organisms shared a common ancestor, the greater the number of base differences we should see in the sequence of genes like *hedgehog* and *sonic hedgehog*.

To better understand this last point, it may be helpful to think in terms of something more familiar: natural language. The European settlers of Australia were predominantly English. Once the Australians and the English became fundamentally separate, however, their manner of speaking began to evolve independently. For example, we could think of the word *outback* as a "mutation" in Australian speech.

Meanwhile, England would be developing its own "mutations," meaning words not used in Australia. And the longer the Australians went on being separate from the English, the greater the number of such differences we would see between the two languages. Just so, the longer it has been since two species went down separate evolutionary lines, the greater the number of base differences we would see in similar genes they have, as measured by the different mutations each has acquired.

This gene-modification hypothesis has been put to the test in connection with a gene that codes for an enzyme called cytochrome *c* oxidase, which exists in organisms as different as humans, moths, and yeasts. Going into this experiment, evolutionary theory predicted that there should be fewer DNA base-pair differences between the cytochrome *c* oxidase genes of, say, a human and a duck than between a human and a moth because all the *other* evidence we had told us that humans and ducks share a more recent common ancestor than do humans and moths. In **Figure 16.13** you can see that evolutionary theory was fully borne out by the DNA sequencing. There are 17 sequence differences between a human and a duck but 36 between a human and a moth. Indeed, all the differences between the species pictured fall into line with evolutionary theory. Thus, we have another confirmation for evolution between lines of evidence—between DNA sequencing *and* the fossil record *and* radiometric dating *and* comparative morphology.

Experimental Evidence

Finally, much evidence for evolution has been provided in recent years by experiment and observation. At first glance, this may seem rather improbable because, as you've seen, evolutionary change can often be perceived only over long stretches of time. Scientists have devised clever ways to catch evolution in the act, however.

Consider the experiments of John Endler, who believed that the male guppies he was studying were being pulled in two directions by natural selection. On the one hand, those that were larger and had brighter coloration were chosen more often by females for mating. On the other, these very characteristics made the males more vulnerable to predators. Endler saw that he could test natural selection by putting some guppies in a predator-free environment. In only a few generations, brighter coloration and larger tails evolved in the males. When Endler then reintroduced predators into this population, things went the other way: Over several generations, smaller tails and less-brilliant colors evolved in the males. Both of these outcomes are precisely what evolutionary theory would predict. In both instances, the physical

MB

GraphIt!: Species Area Effect and Island Biogeography

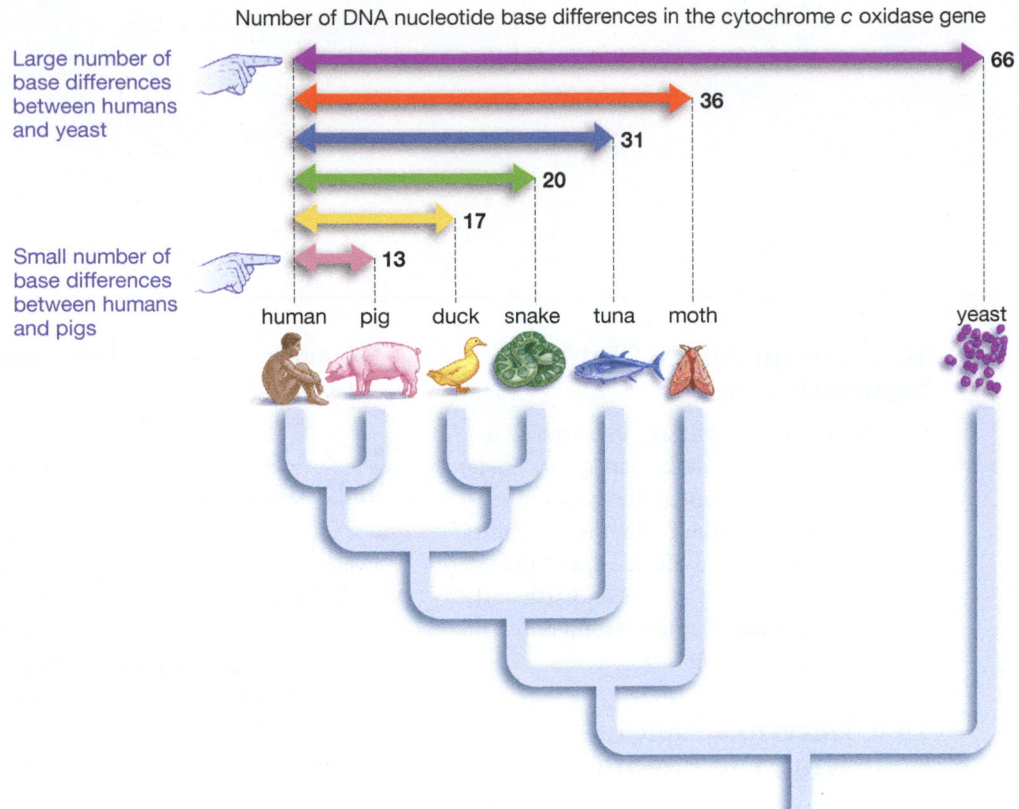

Number of DNA nucleotide base differences in the cytochrome *c* oxidase gene

Large number of base differences between humans and yeast

Small number of base differences between humans and pigs

human   pig   duck   snake   tuna   moth   yeast

66
36
31
20
17
13

Figure 16.13
Using Molecules to Track Evolution

Diverse organisms—such as yeast, moths, and pigs—all have a gene that codes for an enzyme called cytochrome *c* oxidase. Over time, however, the cytochrome *c* oxidase genes have undergone mutations that have altered the sequence of their DNA "building blocks," called bases. The longer it has been since any two species shared a common ancestor, the more differences there should be in their cytochrome *c* oxidase bases. There are 13 differences between the bases found in the human cytochrome *c* oxidase gene and the one found in pigs, but there are 66 differences between the human and yeast genes.

(Data from Whitfield, Philip. 1993. From *So Simple a Beginning: The Book of Evolution*. New York: Macmillan: Maxwell Macmillan International.)

characteristics of the guppies were modified in directions that maximized their reproductive success—more mates with the brighter colors in the predator-free environment, and a longer life (and hence more reproduction) with the drab coloration in the predator-laden environment.

On to How Evolution Works

Given the abundance of evidence for evolution, biologists long ago stopped asking whether it occurred. For decades the really interesting questions have been: Through what means has evolution proceeded? At what pace? In what direction, if any? These are the questions we'll look at in the next three chapters.

SO FAR...

1. The age of the Earth has been established through the use of _____, in which scientists test for the _____ of parent and daughter elements in a given sample of rock or other material.

2. _____ features are features in different species that are the same in structure owing to inheritance from a _____.

3. The auricular muscles of human beings and the wings of ostriches are examples of _____ characters.

16 REVIEW

Summary

16.1 Evolution and Its Core Principles

- A key principle within the theory of evolution is common descent with modification: the process by which species of living things can undergo modification over time, with such change sometimes resulting in the formation of new species. All species have descended from other species and all living things descend ultimately from a single, common ancestor. (p. 283)

- A second key principle in the theory of evolution is natural selection: the process through which traits that confer a reproductive advantage to individual organisms grow more common in populations of organisms over successive generations. (p. 284)

- With an understanding of the theory of evolution, human beings have become aware that (1) we are descended from other varieties of living things and that (2) species are not fixed entities, but instead are constantly undergoing modification. (p. 284)

16.2 Charles Darwin and the Theory of Evolution

- Charles Darwin deserves primary credit for the theory of evolution. He developed existing ideas about descent with modification while providing much evidence in support of them. And he was the first to perceive natural selection as the primary process that drives evolution. Darwin's insights were inspired by the research he carried out during a 5-year voyage he took around the world on the ship *HMS Beagle*, beginning in 1831. (p. 285)

16.3 Evolutionary Thinking before Darwin

- Some of Darwin's ideas can be traced to the work of Lyell, de Lamarck, and Cuvier, who respectively noted the dynamic geological nature of the Earth, the possibility of descent with modification, and the extinction of some species on Earth and the appearance of others within different time frames. (p. 286)

16.4 Darwin's Insights Following the *Beagle*'s Voyage

- Darwin understood descent with modification for several years before he comprehended that natural selection was the most important process driving it. (p. 287)

16.5 Alfred Russel Wallace

- English naturalist Alfred Russel Wallace is the co-discoverer of natural selection as the principal process underlying evolution. (p. 289)

16.6 Darwin: Accepted, Doubted, and Vindicated

- Descent with modification was accepted by most scientists not long after publication of Darwin's *On the Origin of Species by Means of Natural Selection* in 1859 because the principle explained so many facets of the living world. (p. 289)

- That natural selection is the most important process underlying evolution was not generally accepted until the middle of the twentieth century, when the modern synthesis in the theory of evolution brought together lines of evidence from genetics, the fossil record, and the distribution of organisms throughout the world. (p. 290)

16.7 Opposition to the Theory of Evolution

- One factor leading to the appearance of a scientific debate over evolution is confusion about the meaning of the word *theory*, which in everyday speech can mean an idea that is pure speculation but that in science means a general set of principles, supported by evidence, that explains some aspect of the natural world. (p. 291)

16.8 The Evidence for Evolution

- Six lines of evidence that are consistent with the theory of evolution are (1) radiometric dating, (2) the fossil record, (3) the common occurrence of homologous physical structures in different organisms, (4) the existence of vestigial characters in various organisms, (5) variations found in the DNA sequences of various organisms, and (6) experimental demonstrations of evolution. (p. 292)

Key Terms

common descent with modification   283

homologous   293

modern synthesis   291

morphology   293

natural selection   284

radiometric dating   292

vestigial character   295

Understanding the Basics

Multiple-Choice Questions

(Answers are in the back of the book.)

1. In science, a theory is:
   a. an untested hypothesis.
   b. a general set of principles, supported by evidence, that explains some aspect of the natural world.
   c. speculation about the natural world, based on general knowledge of a field.
   d. an observation.
   e. the first idea proposed to explain some aspect of the natural world.

2. Which of the following are central ideas in the theory of evolution by natural selection? (List all that apply.)
   a. Organisms vary in traits that affect their reproduction.
   b. Descent with modification occurs over generations.
   c. Evolution has a goal.
   d. Traits that confer a reproductive advantage will become more common in populations over time.
   e. Traits that are acquired during the life of an individual contribute to evolution.

3. Which of the following observations provide evidence for evolution? (List all that apply.)
   a. Monkey and trilobite fossils are never found in the same fossil beds.
   b. Athletic training can produce an increase in muscle mass.
   c. DNA sequences of genes shared by various species vary in accordance with predictions about how closely related those species are.
   d. The genetic code is nearly universal.
   e. Species whose adult forms look very different may have similar features in embryonic life.

4. During the formulation of his theory of evolution by natural selection, Darwin brought together ideas and results from several disciplines. Match the person with the phrase that describes his work.

   a. Lyell   Evidence of extinction of species
   b. Wallace   One form of organism can evolve into another
   c. Lamarck   Natural selection as a mechanism underlying evolution
   d. Cuvier   Geological forces observable today caused changes in Earth

5. Some of the following structures are homologous with each other. Which ones are they, and which two do not belong to this group?
   a. whale flipper
   b. insect leg
   c. bat wing
   d. cat forelimb
   e. octopus tentacles

6. Discoveries in the field of _____ provided key evidence needed for natural selection to be widely accepted as the mechanism of evolution (select one).
   a. radiometric dating
   b. fossils
   c. genetics
   d. DNA sequencing
   e. medicine

7. Important implications of the theory of evolution by natural selection include (select all that apply):
   a. All organisms, including humans, are descended from a common ancestor.
   b. The biological world is constantly evolving.
   c. Nature designs things intentionally.
   d. Traits found in organisms are the outcomes of a goal-less process.
   e. Humans occupy one ordinary branch on an enormous tree of life.

8. John Endler performed an experiment involving guppies in which he demonstrated evolution driven by natural selection that worked through a predator–prey relationship. Which of the following are true statements about the results of his experiments? (Select all that are correct.)
   a. Males with longer tails and brighter coloration were more frequently eaten by predator fish.
   b. Females preferred to mate with dull-colored males.
   c. When males were put in a predator-free environment, brighter coloration and larger tails evolved in them.
   d. In the predator-free environment, the average brightness of females increased.
   e. When predators were reintroduced, duller coloration and smaller tails evolved in the males.

Brief Review

(Answers are in the back of the book.)

1. What is homology, and why does it provide evidence for evolution?

2. How does the evidence presented in Figure 16.13 on page 297 support the conclusion that humans are more closely related to pigs than to yeast?

3. Describe two examples in which agreement among different lines of evidence provides evidence for evolution.

4. What is one observation that Darwin made in the Galapagos that influenced his thinking about evolution, and why was it important? Would he have been as likely to formulate his theory had he not had the opportunity to sail on the *Beagle*? Why or why not?

5. What major tenet of Darwin's theory of evolution was accepted by almost all scientists within Darwin's lifetime? What major tenet of his theory did not gain acceptance until midway through the twentieth century?

Applying Your Knowledge

1. Using evolutionary principles, explain why large ears might be expected to evolve in a terrestrial plant eater such as a rabbit or deer but not in an aquatic mammal such as a seal.

MasteringBiology®

1. MasteringBiology Assignments
GraphIt!: Species Area Effect and Island Biogeography; **ABC News Video:** Exploring Evolution in the Solomon Islands; Protecting the Galapagos Islands; **Current Events:** Evolution Right Under Our Noses; **Building Vocabulary;** Questions: Reading Questions; Test Bank

2. eText
Read your book online, search, take notes, highlight text, and study more effectively.

3. The Study Area
Self-study tools to test comprehension.

The Means of Evolution:
Microevolution

Evolution at its most fundamental level is called microevolution.

Available on MasteringBiology®

Activities: Evolution and Genetics; The Bottleneck Effect; Three Modes of Selection; **Bioflix:** Mechanisms of Evolution; **ABC News Videos:** Herbicide-resistant Pigweed; **Current Events:** New Fossils Indicate Early Branching of Human Family Tree; A Colorful Way to Watch Evolution in Nebraska's Sand Dunes; Mammoth Hemoglobin Offers More Clues to Its Arctic Evolution; **Building Vocabulary; Questions:** Bioflix Quiz, Reading Questions, Test Bank

What is it that evolves? The essential unit that does so is the population. Pictured is part of a migrating population of snow geese (*Anser caerulescens*).

When people hear the word *evolution,* what generally comes to mind is the grand sweep of evolution: the story of how microorganisms were the first life-forms; of how dinosaurs came and went; of how human beings arose as newcomers on the planet. It is quite a story, to be sure, but to fully appreciate it, the place to look initially is not at evolution's outcomes,

but rather at its processes. *How* do generations of organisms become modified over time? That's the subject of this chapter. When you understand it, you'll be in a position to comprehend evolution on its grander scale.

17.1 What Is It That Evolves?

In approaching this topic, the first question to be answered is: What is it that evolves? If you think about it, it's pretty clear that it is not individual organisms. A tree may inherit a mix of genes that makes it slightly taller than other trees, but if this tree is considered in isolation, it is simply one slightly taller tree. If it were to die without leaving any offspring, nothing could be said to have evolved because no persistent quality (added height) has been passed on to any group of organisms.

In Chapter 16 you were introduced to the idea of a **species,** which can briefly be defined as a group of organisms that can successfully interbreed with one another in nature but that don't successfully interbreed with members of other such groups. You might think that it is species—such as horses or American elm trees—that evolve. Indeed, scientists often speak this way when they refer, for example, to a species of amphibious mammal having evolved into today's whales. But this is a kind of shorthand whose inaccuracy becomes apparent when species are considered as they live in the real world.

Populations Evolve

Think of a hypothetical species of frog living in, say, equatorial Africa in a single expanse of tropical forest. Suppose that a drought persists for years, drying up the forest such that this single lush range is now broken up into two ranges separated by an expanse of barren terrain (**Figure 17.1**). When the separation occurs, there is still only a single species of frog, but that species is now divided into two *populations* that

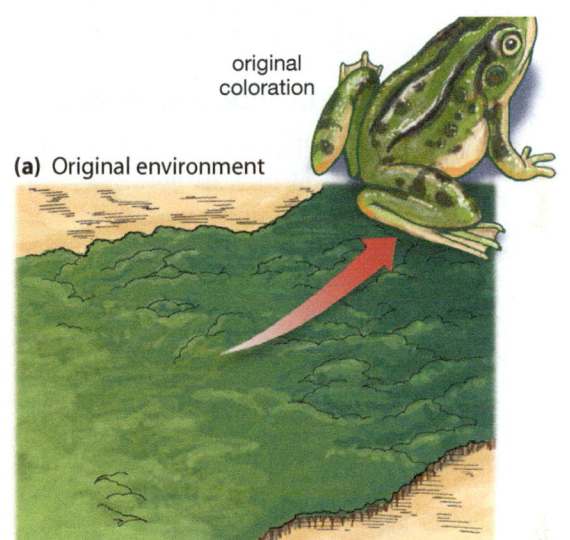

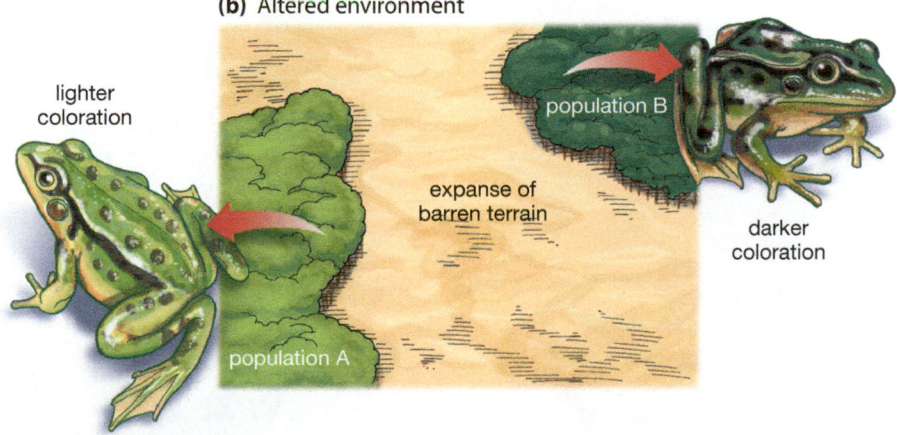

Figure 17.1
Evolution within Populations

(a) A hypothetical species of frog lives and interbreeds in one expanse of tropical forest.

(b) After several years of drought, the forest has been divided by an expanse of barren terrain. The single frog population has thus been divided into two populations, separated by the barren terrain. The two frog populations now have different environmental pressures, such as the kinds of predators each faces, and they no longer interbreed; after many generations their coloration diverges as they adapt to the different pressures.

are geographically isolated and hence no longer inter-breeding. Each population now faces the natural selection pressures of its *own* environment—environments that may differ greatly even though they are nearly adjacent. One area may get more sun, while the other has a larger population of frog predators, for example. Thus, each population stands to be modified individually over time—to evolve. What is it that evolves? The essential unit that does so is a **population,** which can be defined as all the members of a species that live in a defined geographic region at a given time.

The question then becomes: What is the process by which populations evolve? In the population of frogs that has more predators, a darker coloration may evolve over many generations. But how does this darker coloration come about? Through genes. A given frog is unable to change its spots, but over many generations, a *population* of frogs can have its spots changed through the shuffling, addition, and deletion of genetic material.

Gene Pools and Evolution

Recall from Chapter 11 that the genetic makeup of any organism is its **genotype**—and that a genotype provides an underlying basis for an organism's **phenotype,** meaning any observable traits that an organism has, including its physical characteristics and behavior. Many genes are likely to be involved in producing a phenotype such as coloration. In sexually reproducing organisms, *each* of these coloration genes will likely come in variant forms, called **alleles,** with each offspring inheriting one allele for each gene from its

father and one from its mother (**Figure 17.2**). Although both alleles help code for coloration, one may result in slightly lighter or darker coloration than the other.

Genes don't just come in two allelic variants, however; they can come in many. In most species, no one organism can possess more than two alleles of a given gene, but a *population* of organisms can possess many allelic variants of each gene (which is one reason human beings don't just come in one or two colors, but in a continuous range of colors).

When the concept of a population is put together with that of genes and their alleles, the result is the concept of the **gene pool:** all the alleles that exist in a population. This gene pool is very important to the story of evolution, because all evolutionary change ultimately is based on it. If evolution were a card game, the gene pool would be its deck of cards. Individual cards (alleles) are endlessly shuffled and dealt into different "hands" (the genotypes that individuals inherit), with the strength of any given hand dependent on what game is being played. (Survival in a drought? Survival against a new set of predators?)

17.2 Evolution as a Change in the Frequency of Alleles

Thinking of genes in terms of a gene pool gives us a new perspective on what evolution is at root: a change in the frequency of alleles in a population. This may sound a little abstract until you consider the frog coloration example. A frog inherits, from its mother and father, a darker coloration that allows it to evade predators slightly more successfully than other frogs of its generation. It thus lives longer and leaves more offspring than the other frogs. It is successful in this way because of the advantageous set of alleles it inherited. Because this frog is more successful at breeding, its alleles are being passed on to the *next* generation of frogs in greater numbers than the alternative alleles carried by less-successful frogs. As such, this frog's alleles have increased in frequency in the frog population—and this population is now slightly darker in coloration as a result. Put another way, through a change in its allele frequencies, the frog population has evolved in the direction of a slightly darker coloration. In looking at any example of evolution in a population over time, you would find this kind of change in allele frequency as its basis, assuming a stable environment.

With this perspective in mind, you're ready for a definition of **microevolution:** a change of allele frequencies in a population over a relatively short period of time. Why the *micro* in microevolution? Because evolution *within* a population is evolution at its smallest scale. This concept of microevolution allows you to understand the formal definition of evolution

Figure 17.2
Genetic Basis of Evolution

Many genes can produce a trait such as body coloration, and each gene often has many alleles or variants. Each individual in a population, however, can possess only two alleles for each gene, one allele inherited from its mother and one inherited from its father. The two frogs in the figure both have maternal and paternal copies of chromosome 3 that house genes for coloration. The chromosomes of the two frogs may differ, however, in the allelic variants they have of these genes. The frog with dark coloration may possess alleles a_1 and a_2, while the light-colored frog may possess alleles a_2 and a_4 of this same gene.

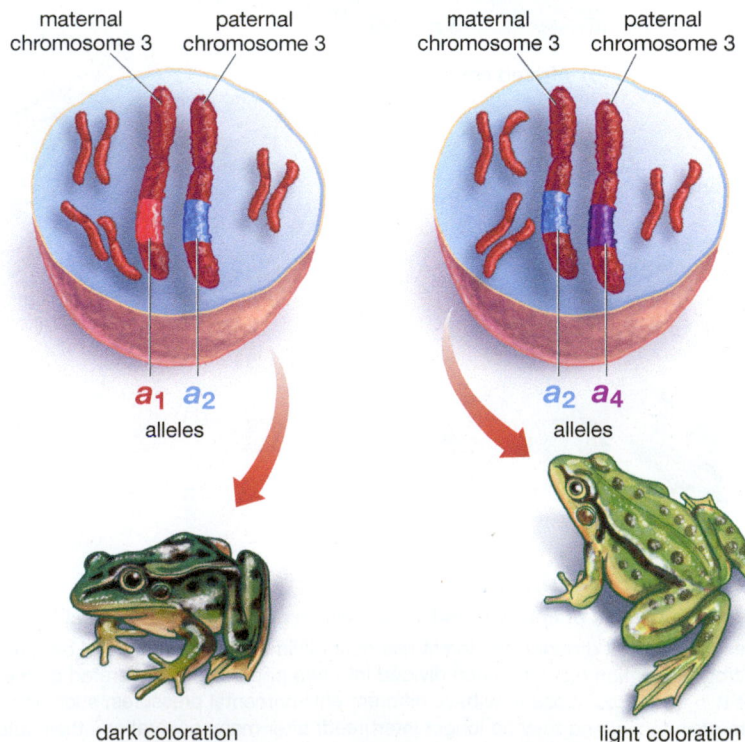

maternal chromosome 3   paternal chromosome 3

a_1 a_2
alleles

dark coloration

maternal chromosome 3   paternal chromosome 3

a_2 a_4
alleles

light coloration

promised to you last chapter. **Evolution** can be defined as any genetically based phenotypic change in a population of organisms over successive generations. Taking a step back, the large-scale patterns produced by microevolution eventually become visible, as with the evolution of, say, mammals from reptiles. This is **macroevolution,** defined as evolution that results in the formation of new species or other large groupings of living things.

SO FAR . . .

1. The essential unit that evolves in nature is the _____, which can be defined as all the members of a _____ living in a specific geographic region at a given time.

2. Genes come in variant forms called _____. A gene pool is defined as all the _____ that exist in a _____.

3. Microevolution is defined as a change in _____ in a population over a relatively short period of time.

17.3 Five Agents of Microevolution

So, what is it that causes microevolution? Put another way, what causes the frequency of alleles to change in a population? In the frog-coloration example, a familiar process was at work: natural selection. The frog's coloration allowed the frog to evade predators, thus helping the frog to live and produce more offspring than did other frogs. This particular coloration was thus *selected* for greater transmission to future generations. From a microevolutionary perspective, this selection brought about a change in allele frequencies in the population: Thanks to the frog's greater reproductive success, the alleles it possessed increased relative to other sets of alleles.

Natural selection is not the only process that can change allele frequencies, however. There are five "agents" of microevolution, meaning forces that can alter allele frequencies in populations. These are mutation, gene flow, genetic drift, sexual selection, and natural selection. You can see these processes summarized in **Table 17.1**. (For a review of how scientists tell whether one of these processes is actually at work changing allele frequencies, see "Detecting Evolution: The Hardy-Weinberg Principle," on page 314.) We'll now look at each of these agents of change in turn.

Table 17.1

Agents of Change: Five Forces That Can Bring about Change in Allele Frequencies in a Population

Agent	Description
Mutation	Alteration in an organism's DNA; generally has no effect or a harmful effect. But beneficial or "adaptive" mutations are indispensable to evolution.
Gene flow	The movement of alleles from one population to another. Occurs when individuals move between populations or when one population of a species joins another, assuming the second population has different allele frequencies than the first.
Genetic drift	Chance alteration of gene frequencies in a population. Most strongly affects small populations. Can occur when populations are reduced to small numbers (the bottleneck effect) or when a few individuals from a population migrate to a new, isolated location and start a new population (the founder effect).
Sexual selection	Occurs when some members of a population mate more often than other members.
Natural selection	Some individuals will be more successful than others in surviving and hence reproducing, owing to traits that give them a better "fit" with their environment. The alleles of those who reproduce more will increase in frequency in a population.

Evolution and Genetics

Mutations: Alterations in the Makeup of DNA

As you saw in the genetics unit, a mutation is any permanent alteration in an organism's DNA. Some of these accidental alterations are *heritable,* meaning they can be passed on to future generations. Mutations can be as small as a change in a single base pair in the DNA chain (a point mutation) or as large as the addition or deletion of a whole chromosome or parts of it. Whatever the case, a mutation is a change in the informational set an organism possesses (**Figure 17.3**). Looked at one way, it is a change in one or more alleles.

The rate of mutation is very low in most organisms; during cell division in humans, it might be just one DNA base pair per billion. And of the mutations that do arise, very few are beneficial or "adaptive." Most do nothing, and many are harmful to organisms. So mutations usually are not working to *further* survival and reproduction, as the frog-coloration alleles did. Given this, any trait produced by a mutation is not likely to become more common over successive generations. The upshot is that mutations are not likely to account for much of the change in allele frequency that is observed in any population.

But a few mutations occur that *are* adaptive. These genetic alterations are something like creative thinkers in a society: They are rare but very important. Such mutations are the only means by which new genetic information comes into being—by which new proteins are produced that can modify the form or capabilities of an organism. The evolution of eyes or wings had to involve mutations. No amount of shuffling of existing genes could get the living world from no eyes to eyes. Of course, no mutation can bring about a feature such as eyes in a single step; such changes are the result of many mutations, followed by rounds of genetic shuffling and natural selection, generally over millions of years.

Gene Flow: When One Population Joins Another

Allele frequencies in a population can also change with the mating that can occur after the arrival of members from a *different* population. This is the second microevolutionary agent, **gene flow,** meaning the movement of genes from one population to another. Such movement takes place through **migration,** which is the movement of individuals from one population into the territory of another. Some populations of a species may truly be isolated, such as those on remote islands, but migration and the gene flow that goes with it are the rule rather than the exception in nature. It may seem at first glance that migration would be limited to animal species, but this isn't so. Mature plants may not move, but plant seeds and pollen do; they are carried to often-distant locations by wind, sea currents, and animals (**Figure 17.4**). Of course, for a migrating population to alter allele frequencies of another population, its gene pool must be different from that of the population it is joining.

Genetic Drift: The Instability of Small Populations

To an extent that may surprise you, evolution turns out to be a matter of chance. You can almost see the dice rolling in the third microevolutionary agent, genetic drift. Imagine a hypothetical population of 10,000 individuals. An allele in this gene pool is carried by 1 out of 10 of them, meaning that 1,000 individuals carry it. Now imagine that some disease sweeps over the population, killing half of it. Say that this allele had nothing to do with the disease, so the illness might be expected to decimate the allele carriers in rough accordance with their proportion in the population. If this were the case, 5,000 individuals in

Figure 17.3

Basis of New Genetic Information

A mutation is any permanent alteration in an organism's DNA. Examples of mutations include:

(a) point mutations, in which the nucleotide sequence is incorrect, and

(b) deletions, in which part of a chromosome is missing.

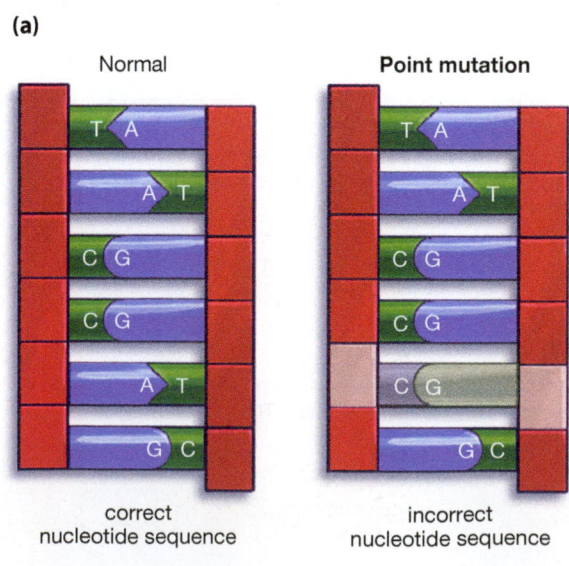

(a)

Normal   **Point mutation**

correct nucleotide sequence   incorrect nucleotide sequence

(b)

Normal   **Deletion**

complete chromosome 5   incomplete chromosome 5

the population would survive, and 1 in 10—or about 500 of them—could be expected to be carriers of this allele. Let us say, however, that just by chance, *550* of the allele carriers were killed, thus leaving the surviving population of 5,000 with only 450 allele carriers. In that scenario, the frequency of the allele in this population would drop from 10 percent to 9 percent (**Figure 17.5a**).

Now for the critical step. Imagine the same allele, with the same 1-in-10 frequency, only now in a population of *10*. There is now but a single carrier of the allele, and that individual may not be one of the 5 members of the population to survive the disease. In this case, the frequency of this allele drops from 10 percent to zero (**Figure 17.5b**). It can be replaced only by a mutation (which is extremely unlikely) or by migration from another population. Assuming that neither happens, no matter how this population grows in the future, in genetic terms it will be a different population than the original in that it lost this allele. The allele might be helpful or harmful, but its adaptive value doesn't matter. It has been eliminated strictly through chance. This is an example of **genetic drift**: the chance alteration of allele frequencies in a population, with such alterations having the greatest impact on small populations. It is true that some genetic drift has taken place in the larger population, but consider how small the effect is. The allele simply went from a 10 percent to a 9 percent frequency, with no loss of allele. Chance events can have much greater effects on small populations than on large ones.

Genetic drift can have large effects on small populations through two common scenarios: Populations can be greatly reduced through disease or natural catastrophe, or a small subset of a population can migrate elsewhere and start a new population. The first of these scenarios is called the *bottleneck effect*;

(a) Hawaiian silversword   **(b)** Tarweeds in California

Figure 17.4
Migration in Plants
Brought about by volcanic eruptions, the Hawaiian Islands have always been surrounded by the Pacific Ocean. Therefore, all the plant species existing on the islands today are descended from species that were introduced to the islands through one means or another—human activity, wind currents, water currents, or animal dispersal of seeds and pollen.
(a) Hawaiian silverswords are derived from
(b) a lineage of California plants commonly known as tarweeds.

(a) Large population = 10,000
(allele carriers in red)

allele frequency = $\frac{1,000}{10,000}$ = 10%

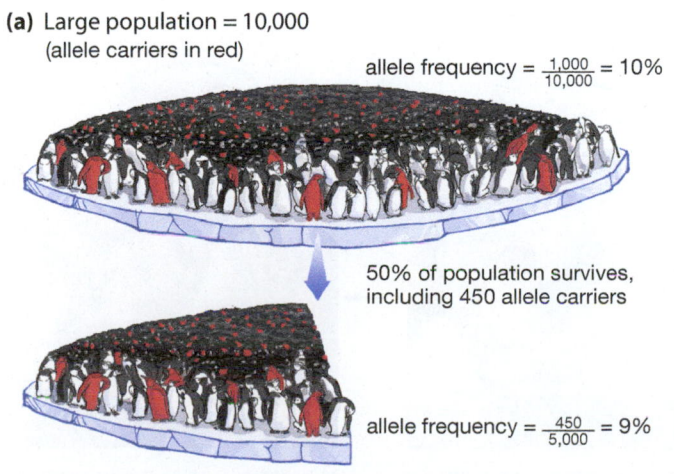

50% of population survives, including 450 allele carriers

allele frequency = $\frac{450}{5,000}$ = 9%

little change in allele frequency (no alleles lost)

(b) Small population = 10
(allele carriers in red)

allele frequency = $\frac{1}{10}$ = 10%

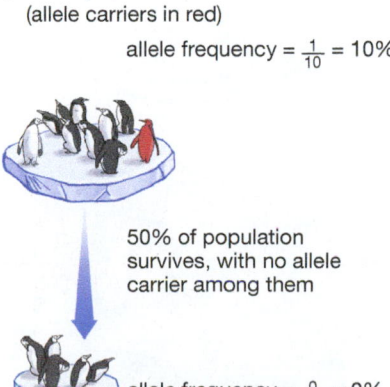

50% of population survives, with no allele carrier among them

allele frequency = $\frac{0}{5}$ = 0%

dramatic change in allele frequency (potential to lose one allele)

Figure 17.5
Genetic Drift

(a) In a hypothetical population of 10,000 individuals, 1 in 10 carries a given allele. The population loses half its members to a disease, including 550 individuals who carried the allele. The frequency of the allele in the population thus drops from 10 percent to 9 percent.

(b) A population of 10 with the same allele frequency likewise loses half its members to a disease. Because the one member of the population who carried the allele is not a survivor, the frequency of the allele in the population drops from 10 percent to zero.

Bioflix: Mechanisms of Evolution

The Bottleneck Effect

Figure 17.6

The Bottleneck Effect

Northern elephant seals were hunted so heavily by humans that by the 1890s, fewer than 50 animals remained. The population has grown from the few survivors (represented here by the three seals in the second frame), but the resulting genetic diversity of this population is very low. (Seal coloration for illustrative purposes only.)

the second is called the *founder effect*. Let's have a look at both of them in turn.

The Bottleneck Effect and Genetic Drift

The **bottleneck effect** can be defined as a change in allele frequencies in a population due to chance following a sharp reduction in the population's size. Real populations, or even species, can go through dramatic reductions in numbers. For example, northern elephant seals, which can be found off the Pacific coast of North America, were prized for the oil their blubber yielded and thus were hunted so heavily that, by the 1890s, fewer than 50 animals remained. Thanks to species protection measures, the seals' numbers have rebounded somewhat in recent decades. But genetically, all the members of this species are very similar today because they all descended from the few seals that made it through the nineteenth-century bottleneck. What occurs in these reductions is a "sampling" of the original population—the "sample" being those who survived the devastation (**Figure 17.6**).

The reason allele frequencies change in such an event has to do with the nature of probability and sample size. Imagine a box filled with M&Ms candies, with equal numbers of red, green, and yellow M&Ms inside. You close your eyes and grab a handful of M&Ms and pull out 12. With such a small sample, you might get, say, 6 reds, 4 greens, and 2 yellows, rather than the 4 of each kind that would be expected from probability. If you pulled out *120* M&Ms, conversely, your reds, greens, and yellows would be much more likely to approach the 40 of each kind that would be expected. In just such a way, a small sample of *alleles* is likely to yield a gene pool that's different from the distribution found in the larger population.

The Founder Effect and Genetic Drift

Genetic drift can also result from the **founder effect**, which is simply a way of stating that when a small subpopulation *migrates* to a new area to start a new population, it is likely to bring with it only a portion of the original population's gene pool. This is another kind of sampling of the gene pool, in other words, but in this case it's caused by the migration of a few individuals rather than the survival of only a few. This sample of the gene pool now becomes the founding gene pool of a new population. As such, it can have a great effect; whichever genes exist in it become the genetic set that is passed on to all future generations as long as this population stays isolated.

The genetic drift that comes with the founder effect can be seen most clearly when a founding population brings with it the alleles for rare genetic diseases. There is, for example, a very rare genetic affliction of the eyes called *cornea plana* that results in a misshapen cornea—the first structure of the eye through which light passes. The result can be impaired close-range vision and a general clouding of eyesight. Cornea plana is known to affect only 113 people worldwide. The strange thing is that 78 of these people are in Finland, most of them in an area in northern Finland. Current research indicates that about 400 years ago, a small population arrived in this isolated area, with at least one member of this population carrying the recessive allele that causes this affliction. Since then, this allele has continued to profoundly affect subsequent generations in the area. This will always be the case if the descendants of a founder population stay relatively isolated over time—that is, if the descendants breed mostly among themselves.

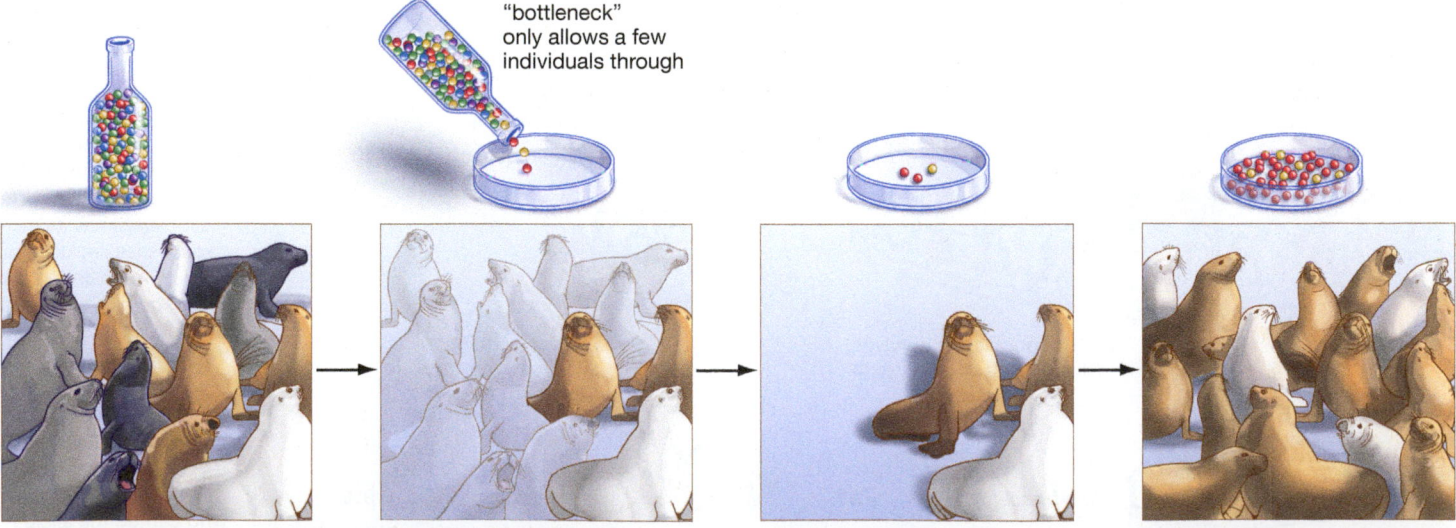

"bottleneck" only allows a few individuals through

Original population, original allele frequency.

Hunting of seals in late 1800s greatly reduced population size.

Surviving population had different allele frequency and little genetic diversity.

This different allele frequency is reflected in today's population.

From this and the earlier examples, you may have perceived by now that there is an inherent value in genetic diversity; you can read about this in "The Price of Inbreeding," on page 308.

Sexual Selection: When Mating Is Uneven across a Population

You've now looked at mutation, gene flow, and genetic drift as agents of change with respect to allele frequencies in a population. Now let's look at a fourth agent, **sexual selection,** defined as a form of natural selection that produces differential reproductive success based on differential success in obtaining mating partners. In practice, this is mating based on *phenotype*, which you may recall is any observable trait in an organism, including differences in appearance and behavior. It is the appearance of particularly nice plumage on a male peacock that makes female peacocks choose it for mating, rather than another male, while it is the behavior of strutting and chest-swelling in a male sage grouse that causes a succession of female grouse to mate with it rather than nearby competitors (**Figure 17.7**). It's easy to see that if one male mates four times as much as the average male of his generation, his alleles stand to increase proportionately in the next generation.

Differential mating success among members of one sex in a species often is based on choices made by members of the opposite sex in that species. It generally is females who are doing the choosing in these situations. Differential mating success can also, however, be based on differences in combative abilities that give individuals of one sex (generally males) greater access to members of the opposite sex. Sexual selection is a form of natural selection in that some alleles—those that help bring about attractiveness, for example—are being preferentially selected for transmission to future generations. But while natural selection has to do with differential survival and reproductive capacity (and hence reproduction), sexual selection has to do with differential *mating* (and hence reproduction).

Natural Selection: Evolution's Adaptive Mechanism

You've already gone over a good deal about natural selection in this review of evolution, but it's time now to look at it as the last agent of change in microevolution. First, what is meant by natural selection? Here is a short definition: **Natural selection** is a process through which traits that confer a reproductive advantage to individual organisms grow more common in populations of organisms over successive generations. In our frog example, the trait of darker coloration conferred a reproductive advantage to an individual frog, in that this coloration helped this frog evade predators—

something that allowed the frog to live longer and leave more offspring than other frogs of its generation. The fact that this frog's more numerous offspring *likewise* bore the darker coloration meant that this trait was becoming more common in this population of frogs. And to the extent that darker coloration continued to carry a reproductive advantage, in one frog generation after another, then this trait would continue to become more common in the frog population over time. Through a process of being selected for transmission, then, a darker coloration was evolving within this population of frogs.

A key concept in natural selection is that of **adaptation,** meaning a modification in the form, physical functioning, or behavior of organisms in a population over generations in response to environmental change. Environmental change may come to a population (through streams drying up or new predators arriving, for example), or a population may come to environmental change (through migration). Either way, natural selection is a process through which populations can adapt to new environmental conditions. Indeed, among the five agents of microevolution, natural selection is the *only* process that consistently works to adapt populations to their environment. Genetic drift is random; it could as easily make an adaptive trait less common in a population. Although mutations can have an adaptive effect, they more often have no effect or even a negative effect. Gene flow doesn't necessarily bring in genes that are better suited to a given environment. And sexual selection has to do with the ways mating partners are selected, not with matching individuals to environment.

Natural selection is, however, constantly working to modify populations in accordance with the

Figure 17.7
Sexual Selection

Individuals in some species choose their mates based on appearance or behavior. Female sage grouse (*Centrocercus urophasianus*) prefer to mate with males who put on superior courtship displays—physical demonstrations that include sounds, a kind of strutting, and a puffing up of their chests. Here a male grouse takes part in such a display by inflating two air sacs on his chest.

ESSAY

The Price of Inbreeding

"Marry afar" is advice given by the pygmy men of Africa, by which they mean, "marry a woman who lives far from your home." The pygmies point to the hunting rights a man acquires in distant territory by marrying a woman from a remote region, but there is another benefit to this practice as well: It helps ensure genetic diversity among the pygmies. When pygmies from remote locations marry, it cuts down on the chances of *inbreeding*, meaning mating in which close relatives produce offspring. Inbreeding can have harmful effects not only in humans but in any species.

Inbreeding brings together the recessive alleles that are required for many genetic diseases.

What are these effects? Well, consider what has happened in the United States with "purebred" dogs (**Figure 1**), which is to say dogs that over many generations have been bred solely with members of their own breed (cocker spaniels mating only with cocker spaniels, and so forth). In the mid-1990s it was estimated that up to one-fourth of all U.S. purebred dogs had some sort of serious genetic defect—ranging from improper joint formation, to heart defects, to deafness. We also have a good deal of experimental evidence from other animals on the effect of inbreeding. In one instance, a group of white-footed mice that had been inbred and then let out into the natural environment survived at only 56 percent the rate of a group of genetically diverse mice. In humans, meanwhile, evidence of the harm that can come when husbands and wives come from the same isolated community can be found in some of the Amish and Mennonite groups of Pennsylvania and Ohio, which for generations have been plagued by numerous cases of genetic conditions that are either exceedingly rare in society in general or that are not found at all in it. There is, for example, a condition called Amish Lethal Microcephaly that results in infants with very small brains and a life expectancy of between five and six months. In the 40 years

Figure 1
Result of Generations of Controlled Breeding

environments around them. As such, it is regarded as the most important agent in having shaped the natural world—in having given zebras their stripes and flowers their fragrance. Because it is so important, we'll now look at it in a little more detail.

SO FAR . . .

1. What the agents of microevolution all have in common is that each of them can alter _____ in a population.

2. Genetic drift is a _____ alteration of _____ in a population that has its greatest effect on _____ populations.

3. In sexual selection, differential reproductive success comes about because of differential success in obtaining _____.

17.4 Natural Selection and Evolutionary Fitness

Even the strongest supporters of natural selection as a shaping force would not maintain that it has the effect of producing perfect organisms. The concept of "fitness" is helpful here, if only to clear up some misconceptions. To a biologist, **fitness** means the success of an organism in passing on its genes to offspring *relative to* other members of its population at a particular time. An organism cannot be deemed "fit," even if it has 1,000 offspring. It can only have *more or less* fitness relative to other members of its population (who might have 900 or 1,100 offspring). This has to do, once again, with allele frequencies in a gene pool. No matter how many offspring an individual has, its allele frequencies will increase in a population only if it has *more* offspring than other members of its generation. Thus, fitness is a measure of impact on allele frequencies in a population.

This concept then gets us to the notion of "survival of the fittest" and the misunderstandings that arise

that ended in 2005, at least 61 infants from 33 Pennsylvania Amish families were born with this condition, whereas outside this community not a single case of the condition has ever been documented.

But why should reproduction within a limited community cause such trouble? As you've seen, every organism has a whole series of genes, most of which come in variant forms, or alleles. In any individual, one allele for a given gene will be inherited from the father and one from the mother. In all organisms, a very small proportion of these alleles is potentially harmful or even lethal. But in most instances, such alleles are recessive. *One* defective allele is not enough to cause trouble because the second differs from it and provides sufficient information for the organism to function properly. Only an individual who has two copies of the defective allele—an individual who is homozygous for it—would suffer from the condition. Meanwhile, an individual who has one "good" and one harmful allele would have no trouble.

So how do these biological facts play out in inbreeding? Imagine two seals, brother and sister, that have mated together because their population has been so reduced.

Imagine that their mother had a recessive allele that brings about a heart defect and that this allele occurs in only 1 of every 100 seals. There is a 1-in-2 chance that the mother will pass this allele on to any one of her offspring, and let's say it happened with the male and female here: Both brother and sister have inherited a single defective allele. When *they* mate, the chance that either of them will pass along the defective allele to their offspring is likewise 1 in 2. More important, however, the chance that *both* of them will pass the allele along—producing an offspring with two defective alleles—is 1 in 4. If, on the other hand, the brother had mated with an unrelated member of the population, the chances of this union passing along two copies of this defective allele are 1 in 400.* Why the difference? Because the unrelated female had only a 1-in-100 chance of having the allele in the first place.

The essence of this lesson, as you may have realized, is that inbreeding has brought together rare, identical alleles—the recessive alleles that are required for many genetic diseases. Random breeding, meanwhile, tends to keep such alleles apart. It tends to preserve genetic diversity, to put it another way.

You might think from this that there are no uses for genetic *similarity*, but consider what human beings have been doing for centuries with dog breeding. Cocker spaniels did not come about by accident. Dogs that had cocker spaniel features were bred together for many generations, ultimately giving us today's dog. It was the *novel features* of the spaniel that people wanted, and novel features are often the result of recessive alleles coming together. While the breeding done to produce American cocker spaniels has resulted in dogs with cute, floppy ears, the unintended effect of this practice is a breed that is also prone to ear infections and a "rage syndrome" that can result in unprovoked aggression, even against familiar humans. In short, inbreeding produces a range of "novel features," some of which are desirable, but others of which are more aptly referred to as genetic defects.

* Remembering the rule of multiplication from Chapter 11, to calculate the odds of passing on two defective alleles in the random mating example, take the son's chances of passing along the allele (1 in 2, or 50 percent, or 0.5) $\times$ 0.01 (the chance of his female partner *having* the allele) $\times$ 0.5 (the chance of her passing the allele along). This works out to 0.0025, which equals 1 in 400. For the brother-and-sister mating, the figures are 0.5 (the chance of the brother passing it along) $\times$ 0.5 (the chance of the sister passing it along), which equals 0.25—or 1 in 4.

from it. The phrase can be taken to imply the existence of superior beings; that is, organisms that are simply "better" than their counterparts, with images of being faster, more muscular, or smarter coming to mind. In fact, however, evolutionary fitness tells us nothing about organisms being *generally* superior, and it certainly tells us nothing about the value of any particular capacity, be it brawn or brain. All it tells us about are organisms who are better than others in their population at passing along their genes in a given environment at a given time. An accurate phrase, as others have pointed out, would not be "survival of the fittest," but rather "survival of those who fit—for now." Let's look at a real-life example of natural selection to see why this is true.

Galapagos Finches: The Studies of Peter and Rosemary Grant

When Charles Darwin stopped at the Galapagos Islands in 1835, some of the animal varieties he collected were various species of finches. Over the years,

biologists kept coming back to "Darwin's finches" because of the very qualities Darwin found in them: They seemed to present a textbook case of evolution, with their 14 species having evolved from a single ancestral species on the South American mainland. Yet for more than 100 years after Darwin, it was a puzzle for scientists to figure out how Darwin's posited mechanism of evolution, natural selection, could have been at work with the finches. This changed beginning in the 1970s, when the husband-and-wife team of Peter and Rosemary Grant began a painstakingly detailed study of the birds.

Natural selection in the finches came into sharp focus in 1977, when a tiny Galapagos island, Daphne Major, suffered a severe drought. Rain that normally begins in January and lasts through July scarcely came at all that year. This was a disaster for the island's two species of finches; in January 1977 there were 1,300 of them, but by December the number had plunged to fewer than 300. Daphne's medium-size ground finch, *Geospiza fortis,* lost 85 percent of its population in this calamity. The staple of this

Figure 17.8
A Product of Evolution on the Galapagos
A male of the finch species *Geospiza fortis,* which is native to the Galapagos Islands.

difference were the survivors able to get into large, tough seeds and make it through the catastrophe, eventually to reproduce (**Figure 17.9**).

This is natural selection in action, but there is more to be learned from the Grant study, which is to say *evolution* made visible. The Grant team knew that beak depth had a high "heritability" in the finches, meaning that beak depth is largely under genetic control. As it turned out, the offspring of the drought survivors had beaks that were 4 to 5 percent deeper than the average of the population before the drought. In other words, natural selection had preferentially preserved those alleles from the starting population that brought about deeper beaks, and the result was a population that evolved in this direction.

But the Grant study yielded one more lesson. In 1984–1985 there was pressure in the opposite direction: Few large seeds and an abundance of small seeds provided an advantage to smaller birds, and it was these birds that survived this event in disproportionate numbers.

bird's diet is plant seeds. When times get tough, as in the drought, the size and shape of *G. fortis* beaks— their beak "morphology"—begins to define what one bird can eat as opposed to another (**Figure 17.8**). In *G. fortis,* larger body size and deeper beaks turned out to make all the difference between life and death in the drought of 1977. Measuring the beaks of *G. fortis* that survived the drought, the Grant team found they were larger than the beaks of the population before the drought by an average of some 6 percent. This was a difference of about half a millimeter, or roughly two-hundredths of an inch; by such a

Lessons from *G. fortis*

So, where is the "fittest" bird in all this? There isn't any. Evolution among the finches was not marching toward some generally superior bird. Different traits were simply favored under different environmental conditions. Second, there is no evolutionary movement toward combativeness or general intelligence here. Survival had to do with size—and not necessarily *larger* size at that. Looking around in nature, it's true that some showcase species, such as lions and mountain gorillas, gain success in reproduction by being aggressive. And it's true that our own

Figure 17.9
Who Survives in a Drought?
Peter and Rosemary Grant observed that individuals of the finch species *Geospiza fortis* who survived an island drought in 1977 had a greater average beak depth than individuals surveyed before the drought, in 1976. The offspring of the survivors likewise had a larger average beak size than did the population before the drought. Thus, evolution through natural selection was observed in just a few years on the island.

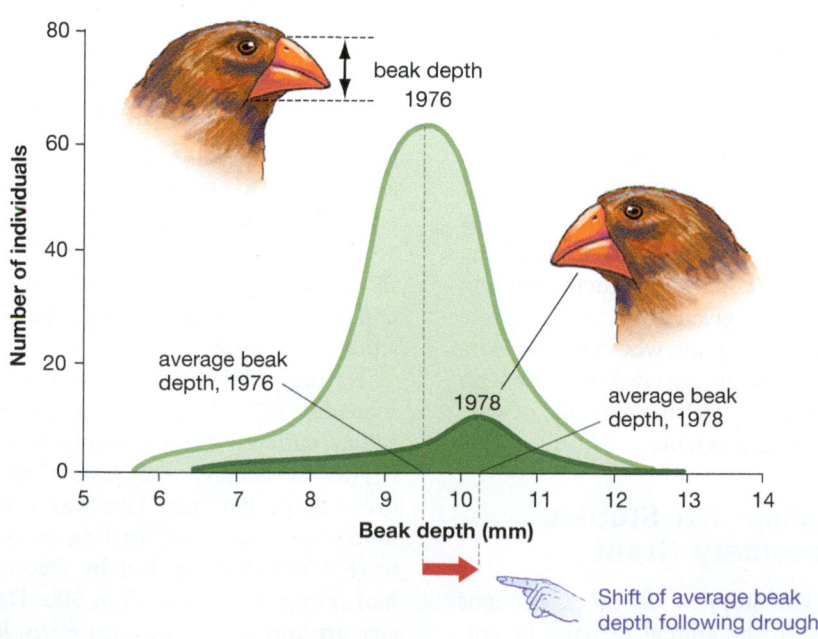

species owes such success as it has had to intelligence. But in most instances, it is not brawn or brain that make the difference; it is something as seemingly benign as beak depth and its fit with the environment.

Finally, consider how imperfect natural selection is at the genetic level. Suppose the smaller *G. fortis* that disproportionately died off in 1977 also disproportionately carried an allele that would have aided just slightly in, say, long-distance flying. In the long run this might have been an adaptive trait, but it wouldn't matter. The flight allele would have been reduced in frequency in the population (as the smaller birds died off) because flight distance didn't matter in 1977; beak depth did. Evolution operates on the phenotypes of whole organisms, not individual genes. As such, it does not work to spread *all* adaptive traits more broadly. Instead, the destiny of each trait is tied to the constellation of traits the organism possesses. Genes are "team players," in other words, that can only do as well as the team they came with.

17.5 Three Modes of Natural Selection

In what directions can natural selection push evolution? As noted in Chapter 11, a character such as human height is under the control of many different genes and is thus **polygenic.** Such polygenic characters tend to be "continuously variable." There are not one or two or three human heights, but an innumerable number of them in a range. (See "Multiple Alleles and Polygenic Inheritance" on page 204.) When natural selection operates on characters that are polygenic and continuously variable, it can proceed in any of three ways. The essential question here is: Does natural selection favor what is average in a given character or what is extreme (**Figure 17.10**)?

Three Modes of Selection

Figure 17.10
Three Modes of Natural Selection

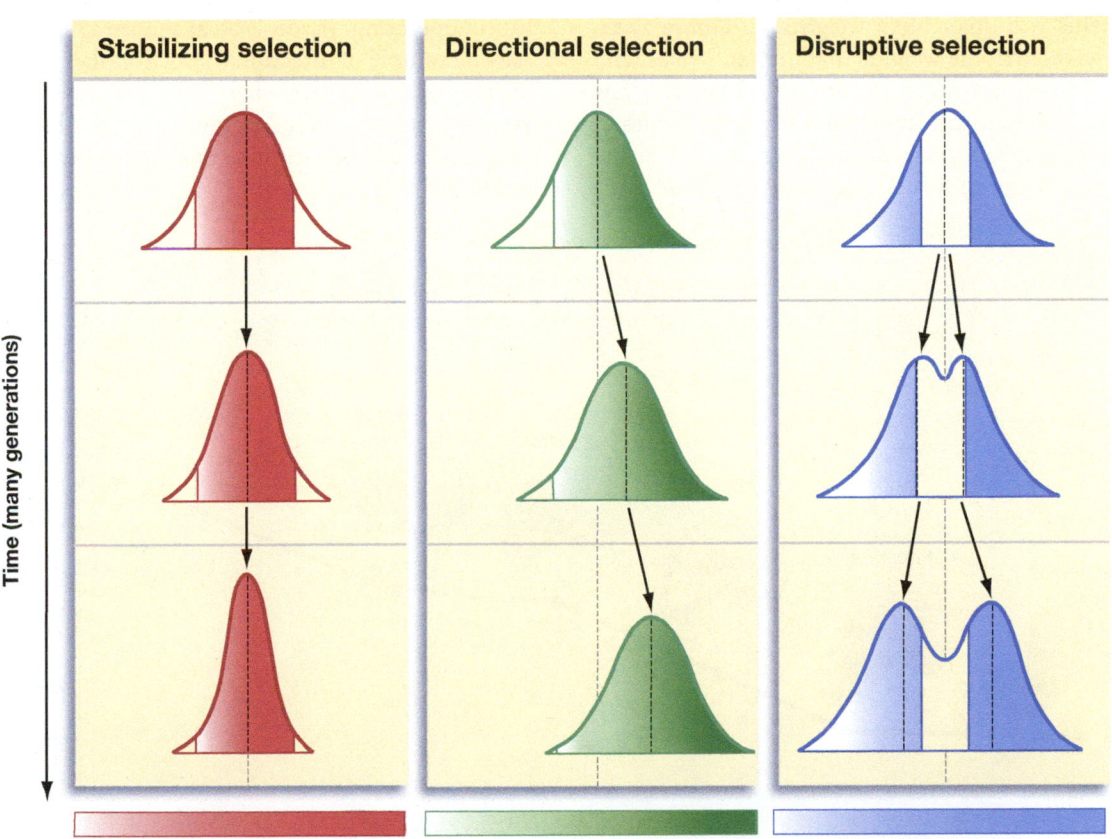

Range of a particular characteristic (in this instance, lightness or darkness of coloration)

In stabilizing selection, individuals that possess extreme values of a characteristic are selected against and die or fail to reproduce. Over succeeding generations, an increasing proportion of the population becomes average in coloration.

In directional selection, one of the extremes of a characteristic is better suited to the environment, meaning that individuals at the other extreme are selected against. Over succeeding generations, the coloration of the population moves in a direcron–in this case toward darker coloration.

In disruptive selection, individuals with average coloration are selected against and die. Over succeeding generations, part of the population becomes lighter, while part becomes darker, meaning the range of color variation in the population has increased.

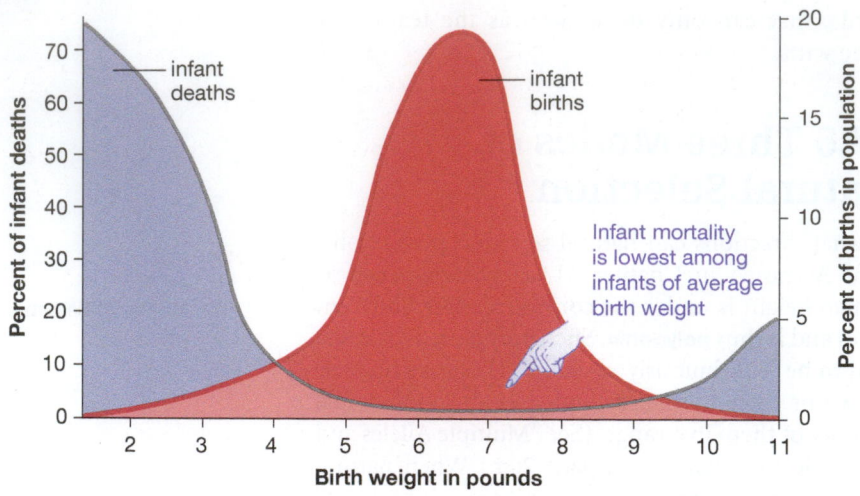

look at the infant mortality curve. Infant deaths are highest at both extremes of birth weight; low-birth-weight babies *and* high-birth-weight babies are more at risk than are average-birth-weight babies (although low birth weight poses the greater risk). Put another way, the children most likely to survive (and reproduce) are those carrying alleles for intermediate birth weights. Thus, natural selection is making intermediate weights even more common. Through it, birth weights are not moving toward the extremes of higher or lower weights; they are being stabilized at intermediate levels. Stabilizing selection is assumed to be the most common type of selection operating in the natural world. This should not be surprising because most organisms are well adapted to their environments.

Figure 17.11
Stabilizing Selection: Human Birth Weights and Infant Mortality
Note that infant deaths are more prevalent at the upper and lower extremes of infant birth weights.

Stabilizing Selection

In **stabilizing selection,** intermediate forms of a given character are favored over extreme forms. An example of this is human birth weights. In **Figure 17.11**, you can see, first, the weights that human babies tend to be. Notice that there are not only relatively few 3-pound babies but relatively few 9-pound babies as well. A great proportion of birth weights fall at the average or mean of a little less than 7 pounds. Now

Directional Selection

When natural selection moves a character toward *one* of its extremes, **directional selection** is in operation—the mode in which we most commonly think of evolution operating. If you look at **Figure 17.12**, you can see an example of directional selection that took place over a very long period of time—about 4 million years—involving evolution toward larger brain size in some close, but now extinct, relatives of ours, a group called the hominins.

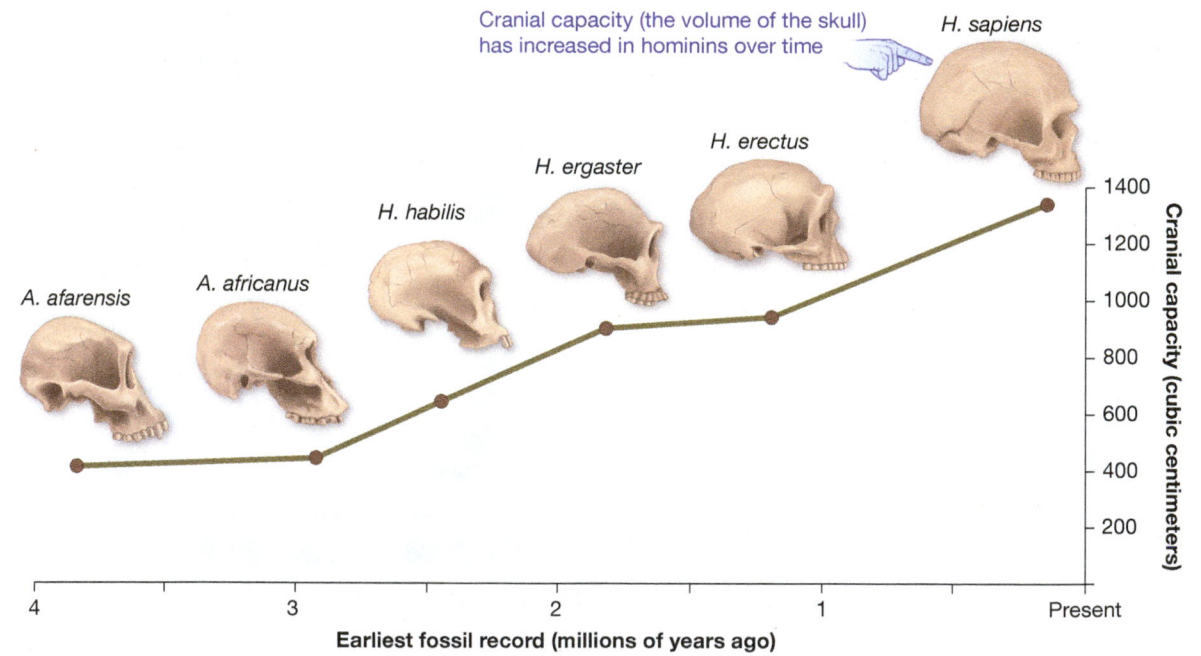

Figure 17.12
Directional Selection: Evolving toward Increased Brain Size
Modern humans are part of an evolutionary group, called the hominins, that includes all the upright-walking or "bipedal" primate species. Cranial capacity is a measure of the volume of brain tissue that can fit inside a skull. Directional selection for cranial capacity can clearly be seen in the ever-increasing capacities of this group of hominins (all of whom are now extinct, except for ourselves). The line in the figure does not trace ancestry, but merely shows the trend toward increased brain size. Going from left to right, the full names of the species shown are *Australopithecus afarensis, Australopithecus africanus, Homo habilis, Homo ergaster, Homo erectus,* and *Homo sapiens* (ourselves).

Disruptive Selection

When natural selection moves a character toward *both* of its extremes, the result is **disruptive selection,** which appears to occur much less frequently in nature than the other two modes of natural selection. This mode of selection is visible in the beaks of yet another kind of finch. A species of these birds found in West Africa (*Pyrenestes ostrinus*) has a beak that comes in only two sizes. Thomas Bates Smith, who has studied the birds, has observed that if human height followed this pattern, there would be some Americans who are 4 to 5 feet tall and some who are 6 to 7 feet tall, but no one who is 5 to 6 feet tall (**Figure 17.13**).

The environmental condition that leads to this mode of selection in the finches is, once again, diet.

Figure 17.13
Disruptive Selection: Evolution That Favors Extremes in Two Directions
Finches of the species *Pyrenestes ostrinus,* found in West Africa. Birds of this species have beaks that come in two distinct sizes—large and small.

When food gets scarce, large-billed birds specialize in cracking a very hard seed, while small-billed birds begin feeding on several soft varieties of seed, with each type of bird being able to out-compete the other for its special variety. Given how these birds have evolved, it's probably safe to assume that a bird with an intermediate-size bill would get less food than one with a bill of either extreme type. Bill size seems to be under the control of a single genetic factor in these birds, so that bills are able to come out either large or small. This genotype was presumably shaped by natural selection over generations, such that any alleles for intermediate-size bills were weeded out.

On to the Origin of Species

Stabilizing selection does what it says: It stabilizes given traits in a population, thereby preserving the similarity this population has with other populations in its species that likewise are undergoing stabilizing selection. However, both disruptive and directional selection can serve as the basis for speciation—for bringing about the transformation of a single species into two or more *different* species. How is it, though, that such speciation works? And how do we classify the huge number of species that are the outcome of evolution's myriad branchings? These are the subjects of the next chapter.

SO FAR. . .

1. True or false: If an insect has 1,000 offspring, we can label it as "fit" in an evolutionary sense.

2. In stabilizing selection, forms that are _____ are favored over forms that are _____.

3. We normally think of evolution working through _____, in which a character moves toward _____.

ESSAY

Detecting Evolution: The Hardy-Weinberg Principle

How can scientists tell when evolution is taking place? As you've seen, evolution always entails a change in allele frequencies in a population. And there are five primary "agents" of allele frequency change—processes such as genetic drift and natural selection. So, how can scientists tell when one of these agents actually is at work? They can employ a procedure—a kind of test—devised early in the twentieth century by a British mathematician, Godfrey Hardy, and a German physician, Wilhelm Weinberg. Working independently, Hardy and Weinberg arrived at an insight, now called the *Hardy-Weinberg principle*, that today is used routinely not just in evolutionary biology, but in fields like public health as well.

The starting point for the Hardy-Weinberg principle is its assertion that allele frequencies will *not* change unless one of the five agents is at work. A population whose alleles are not being altered by one of the agents is said to be at *Hardy-Weinberg equilibrium*—a state in which the population's allele frequencies remain constant from one generation to the next. Let's walk through an example now that shows how a population at Hardy-Weinberg equilibrium stays the same. Along the way, you'll see the kinds of insights that are made possible by the Hardy-Weinberg principle.

Imagine a hypothetical species of plant whose flowers come in two colors, red and white, and say that we are looking at a population of 100 of these individuals. Further, we know from breeding these plants that red results from a dominant allele for a gene, while white results from a recessive allele for that gene. In the tradition of

Mendel, we signify the red allele as *R* and white as *r*. Finally, we have sampled this population's DNA and found that *R* has a frequency of 78 percent, while *r* has a frequency of 22 percent. What does "frequency" mean in this context? It means that if we looked at the gametes of these plants—their eggs and sperm—78 percent of them would contain a dominant *R* allele while 22 percent would contain a recessive *r* allele. Now, keep your eye on this 78:22 allele frequency because that is what we are primarily interested in. The essential question is: When we look at the *next* generation of plants, will the allele frequencies in it remain 78:22?

To see what happens in the next generation, we need to calculate the *genotypes* of our starting generation. What is a genotype in this context? Each egg or sperm may contain only *one* allele for a gene—in this case *R* or *r*—but in diploid organisms such as flowering plants, egg and sperm fuse in fertilization, thus bringing alleles together to create an organism that has *two* alleles for each gene—one from the male, one from the female. In this context, then, genotype means the combination of *R* and *r* alleles that an individual possesses. Now, what possible combinations could a given individual have? An easy way to visualize them is to return once again to the Punnett squares that were used in Chapter 11. Imagine, in the parental generation, a male and female that are both heterozygous for the color gene—

Possible genotypes in the population

each of them has one *R* allele and one *r* allele. In this case, our Punnett square looks like the one you can see above.

So, the possible genotypes in the next generation are *RR, Rr,* or *rr.* (Further, note that because the *r* allele is recessive, only the *rr* genotype yields white flowers, signified by the white flower in the square.) But how do we calculate the genotype frequencies that will come about from this range of possibilities? How do we calculate what percentage of the population will have the *RR* genotype, for example, based on the allele frequencies we started with? The heart of the Hardy-Weinberg principle, the Hardy-Weinberg equation, allows us to do this. To generalize beyond the case of our red- or white-flowered plants, let's use a couple of conventional symbols in the equation: *p* will stand for the frequency of the more common *R* allele, and *q* will stand for the frequency of the less common *r* allele. The Hardy-Weinberg equation then looks like this:

The Hardy-Weinberg equation ...	p^2	+	$2pq$	+	q^2
Applied to this example means ...	R^2	+	$2Rr$	+	r^2
And this yields the ...	frequency of the *RR* genotype	+	frequency of the *Rr* genotype	+	frequency of the *rr* genotype

This formula says: If we take the frequency of the *R* allele and square it (R^2), we will get the frequency of the *RR* genotype in the population. Well, we know that the frequency of the *R* allele in the population is 0.78, so this means 0.78 × 0.78, which equals 0.6084. This is another way of saying that 60.84 percent of the population should have the genotype *RR*. Next, doubling *R* × *r* (2*Rr*) yields the genotype frequency of individuals that are heterozygous for the gene, meaning the plants that are the *Rr* genotype. This works out to 34.32 percent of the population. Finally, taking *r* times itself (r^2) yields the frequency of the *rr* genotype, which works out to be 4.84 percent of the population.

Although this formula has allowed us to work out genotype frequencies for the population, we have, in a sense, merely taken Mr. Hardy and Mr. Weinberg's word that things work like this. A question you may be asking yourself at this point is: *Why* do these particular mathematical steps reveal this information? Why, for example, should squaring the percentage of *R* alleles in the population tell us what the percentage of *RR* genotypes is in this population? To grasp the logic that's at work, it's important to understand that the Hardy-Weinberg equation deals with *probabilities.* It starts with a set of allele frequencies in a population and then says, given these, what are the probabilities of getting various genotypes? The particular steps that the formula uses to answer this question can perhaps best be understood by looking at a more familiar set of probabilities, those that come with flipping a coin.

When we flip a coin, what is the probability that heads or tails will come up? Fifty–fifty, or 0.5, right? Now, what are the chances that when we flip two coins, *both* will come up heads? The answer is contained in something covered in Chapter 11, the rule of multiplication (see page 198). This rule states that the probability of any two events happening is the *product* of their respective probabilities. In the case of our coin toss, this means the respective probabilities of each of our tosses, 0.5 × 0.5, which equals 0.25—so we have a 25 percent chance of getting two heads.

Now, it's important to recognize that the 50/50 odds that exist in coin tossing actually stem from an initial set of *frequencies.* There are heads, which have a frequency of 0.5 in our population of coins, and there are tails, which likewise have a 0.5 frequency. But there are other frequencies besides 0.5—and other probability exercises to run besides flipping coins. In our *flowering plant* population, we have frequencies of *alleles.* These are 0.78 for *R* and 0.22 for *r*. If the probability of getting one *R* allele is 0.78, then the rule of multiplication tells us that the probability of getting two *R* alleles in an organism is 0.78 × 0.78, or to put it another way, 0.78^2. This last number is simply a specific value for p^2 in the Hardy-Weinberg equation of $p^2 + 2pq + q^2$. You can easily see that the *rr* genotype frequency is similarly derived from q^2, which in this case means 0.22^2.

But what about the middle term in the equation, the 2*pq* that gives us the percentage of *Rr* genotypes in the plant population? Look at the Punnett square again; you can see that there are *two* ways a plant could end up with the *Rr* genotype—male *R* and female *r* or female *R* and male *r*. When an outcome can occur in two or more different ways, something called the rule of addition comes into play. It states that the probability of a given outcome is the *sum* of its respective probabilities. In this case, the probability of getting one *R* and one *r* is 0.78 × 0.22. But this same probability can happen in two different ways (*Rr* and *rR*), so the formula for it is (0.78 × 0.22) + (0.78 × 0.22), which can be stated as 2 × (0.78 × 0.22). With this, we can see the basis of the whole equation, which plays out this way in our example: The Hardy-Weinberg equation of $p^2 + 2pq + q^2$ in this case means 0.78 × 0.78 + 2 × (0.78 × 0.22) + 0.22 × 0.22. Doing the calculations on this yields the answers: 0.6084 + 0.3432 + 0.0484, which provides us with our final probabilities: 60.84 percent of the population is likely to be *RR*, 34.32 percent *Rr*, and 4.84 percent *rr*.

Now we have only one more step to go, which is to calculate expected allele frequencies in the *second* generation based on the first-generation genotype frequen-

cies that we got. This is just a matter of unpacking, we might say, the genotype frequencies. To calculate what the *R* allele frequencies will be, we start by dividing the 0.3432 *Rr* genotype in half—yielding 0.1716—because half the alleles in this genotype are, by definition, *R* (and the other half *r*). Then we add to this number all of the 0.6084 alleles in the *RR* genotype because all these alleles are *R*. To get the frequency of *r* alleles, we take the 0.1716 figure and add to it the 0.0484 figure for the *rr* genotype. Thus:

$$R = 0.6084 + 0.1716 = 0.78$$
$$r = 0.0484 + 0.1716 = 0.22$$

Hopefully, these last two figures look familiar, because they are, in fact, the allele frequencies we started with in the first generation. Thus the Hardy-Weinberg equation has demonstrated that allele frequencies do not change from one generation to the next simply as a matter of reproduction.

To grasp the importance of this lack of change, imagine that you are a geneticist who has sampled a human population for the frequency of an allele suspected of contributing to disease in young adults. With your allele frequencies in hand, you run your Hardy-Weinberg equation to get the genotype frequencies for this population. Then you sample the population for its *actual* genotype frequencies and find that they don't match the Hardy-Weinberg prediction. You discover that there are fewer people in the population with the homozygous recessive (*qq*) genotype frequency than predicted by Hardy-Weinberg. This means that at least one of the agents noted earlier has to be at work—one of them must be disturbing Hardy-Weinberg equilibrium. Among the possibilities is that the agent at work is natural selection. The *qq* genotype may in fact contribute to people dying younger and hence having fewer children. This could be why there are fewer people with this genotype in the population than predicted by the formula. The Hardy-Weinberg principle, then, provides scientists with insights not just about evolution, but about public health as well.

Summary

17.1 What Is It That Evolves?

- The smallest unit that evolves is a population, defined as all the members of a single species living in a defined area at one time. (p. 301)

- Genes exist in variant forms called alleles. In most species, no individual will possess more than two alleles for a given gene—one allele coming from the individual's father, the other allele coming from the mother. A population, however, is likely to possess many different alleles for a given gene. The total of all alleles in a population is that population's gene pool. (p. 302)

17.2 Evolution as a Change in the Frequency of Alleles

- If a set of alleles increases in frequency from one generation to the next within a population, the phenotypes, or observable characteristics, produced by those alleles will be exhibited to a greater extent within the population. Through this, a population can be said to have evolved. Evolution at this level is referred to as microevolution: a change of allele frequencies within a population over a relatively short time. Macroevolution is evolution on a larger scale, which results in the formation of new species or other large groupings of living things. (p. 302)

17.3 Five Agents of Microevolution

- Five evolutionary forces can produce changes in allele frequencies within a population: mutation, gene flow, genetic drift, sexual selection, and natural selection. (p. 303)

- A mutation is any permanent alteration in an organism's DNA; some mutations are heritable. Mutations appear infrequently and most either have no effect or

are harmful; rare adaptive mutations are vital to evolution, however, because they are the only means by which entirely new genetic information comes into being. (p. 304)

- Gene flow, the movement of genes from one population to another, takes place through migration. (p. 304)

- Genetic drift is the chance alteration of allele frequencies in a population. It has its greatest effects on small populations via two common scenarios: the bottleneck effect, a change in allele frequencies due to chance during a sharp reduction in a population's size; and the founder effect, a change in frequencies brought about when a small subpopulation migrates to a new area, thus founding a new population that possesses only a portion of the original population's gene pool. (p. 304)

- Sexual selection occurs when differences in reproductive success arise because of differential success in mating. (p. 307)

- Natural selection is a process through which traits that confer a reproductive advantage to individual organisms grow more common in populations over successive generations. A given organism that has a trait that provides it with a reproductive advantage will leave more offspring than other members of its generation. This trait will then be seen with greater frequency in the succeeding generation. (p. 307)

- Natural selection is the only agent of microevolution that consistently works to adapt populations of organisms to their environments; as such, it is regarded as the most powerful force underlying evolution. (p. 307)

17.4 Natural Selection and Evolutionary Fitness

- The phrase "survival of the fittest" is misleading because it implies that evolution works to produce beings who would be

superior in any environment. Evolutionary fitness refers only to the relative reproductive success of individuals in a given environment at a given time. Fitness can change within an individual's lifetime in accordance with changes in the surrounding environment. (p. 308)

17.5 Three Modes of Natural Selection

- Natural selection has three modes: stabilizing selection, directional selection, and disruptive selection. Stabilizing selection moves a given character in a population toward an intermediate form and hence tends to preserve the status quo; directional selection moves a given character toward one of its extreme forms; and disruptive selection moves a given character toward two extreme forms. (p. 311)

Key Terms

Understanding the Basics

Multiple-Choice Questions

(Answers are in the back of the book.)

1. In most populations of living things, each individual has __ copies of each gene, which may be the same or different allelic variants. A population may have _____ allelic variants for a given gene.
   - **a.** 1, 2
   - **b.** 2, 4
   - **c.** 2, many
   - **d.** 1, many
   - **e.** 2, 2

2. Match the terms with their meanings:
   - **a.** gene pool   a variant form of a gene
   - **b.** allele   the set of all alleles in a population
   - **c.** allele frequency   the genetic makeup of an organism
   - **d.** genotype   exchange of genes between populations
   - **e.** gene flow   the relative representation of a given form of a gene in a population

3. Which of the following statements is true of microevolution? (Select all that apply.)
   - **a.** It is a change in allele frequency within populations.
   - **b.** It is caused by inheritance of characters acquired during the life of an organism.
   - **c.** It is only seen when some allelic variants cause individuals to have increased reproductive success.
   - **d.** It always leads to adaptation.
   - **e.** It is caused only by natural selection.

4. Agents of allele frequency change in populations include (select all that apply):
   - **a.** genetic drift.
   - **b.** meiosis.
   - **c.** mutation.
   - **d.** segregation.
   - **e.** natural selection.

5. Agents that consistently produce adaptive evolution (adaptation) include (select all that apply):
   - **a.** genetic drift.
   - **b.** mutation.
   - **c.** meiosis.
   - **d.** natural selection.
   - **e.** gene flow.

6. Mutations are (select all that apply):
   - **a.** generally beneficial.
   - **b.** sometimes heritable
   - **c.** always caused by changes in single base pairs of DNA.
   - **d.** the ultimate source of new genetic information.
   - **e.** prone to occur with great frequency.

7. Match the agent of microevolution to its description:
   - **a.** mutation   chance alterations of allele frequencies, having the greatest effect on small populations
   - **b.** gene flow   a process through which traits that confer a reproductive advantage to individuals are seen with greater frequency in succeeding generations of a population
   - **c.** natural selection   movement of alleles between populations by migration
   - **d.** genetic drift   genetic drift brought about by migration of a subpopulation
   - **e.** founder effect   changes in the genetic material

8. Finches in the Galapagos experienced _____ selection on beak depth following a drought; the African finch *P. ostrinus* experienced _____ selection on beak size because it eats either very small, soft seeds or much larger, hard seeds.
   - **a.** stabilizing, directional
   - **b.** sexual, disruptive
   - **c.** directional, disruptive
   - **d.** disruptive, directional
   - **e.** stabilizing, disruptive

Brief Review

(Answers are in the back of the book.)

1. Explain the statement "Individuals are selected, populations evolve."

2. What is a gene pool, and why do changes in gene pools lie at the root of evolution?

3. What are the five causes of allele frequency changes (microevolution), and how does each work?

4. Why is evolutionary fitness always a measure of relative reproduction in a population?

5. Why is natural selection the only agent of evolution that consistently produces adaptation?

6. Is the evolutionary fitness of an individual expected to be the same or different in different environments? How did the Grants' study of beak depth of finches after the drought and then again in 1984–1985 provide a real example of this?

Applying Your Knowledge

1. What common misconceptions about nature result from the phrase "survival of the fittest," as opposed to the phrase that accurately describes what happens in nature, "survival of those who fit—for now"?

2. Explain why genetic drift may be important when captive populations of animals or plants are started with just a few individuals.

MasteringBiology®

1. MasteringBiology Assignments
Activities: Evolution and Genetics; The Bottleneck Effect; Three Modes of Selection; **Bioflix:** Mechanisms of Evolution; **ABC News Videos:** Herbicide-resistant Pigweed; **Current Events:** New Fossils Indicate Early Branching of Human Family Tree; A Colorful Way to Watch Evolution in Nebraska's Sand Dunes; Mammoth Hemoglobin Offers More

Clues to Its Arctic Evolution; **Building Vocabulary; Questions:** Bioflix Quiz, Reading Questions, Test Bank

2. eText
Read your book online, search, take notes, highlight text, and study more effectively.

3. The Study Area
Self-study tools to test comprehension.

18

The Outcomes of Evolution:
Macroevolution

New species come about when populations of existing species cease to breed with one another. Each new species that arises constitutes another branch on life's family tree. Scientists are trying hard to figure out what this tree looks like.

ESSAYS

Available on MasteringBiology®

Activities: Allopatric Speciation; Speciation and Polyploidy; **Current Events:** Genetic Data and Fossil Evidence Tell Differing Tales of Human Origins; Is Sugar Toxic?; **Building Vocabulary; Questions:** Reading Questions, Test Bank

Male and female little green bee-eaters (*Merops orientalis*) engage in a courtship ritual in which the male feeds the female. Such rituals are one of the mechanisms that help produce new species.

How many types of living things are there on Earth? How many varieties of life-forms are there that we can recognize as being fundamentally different from one another? It may surprise you to learn that we haven't the foggiest idea. The lowest estimate is about 3 million species, but higher-end estimates often come in at 10 to 15 million, and the highest of them all is 100 million. Scientists simply have been unable to catalogue the vast diversity that exists on our planet. In 250 years of watching, digging, netting, and bagging, they have identified about 1.8 million species—a large number, to be sure, but only a fraction of the species that actually exist, even if our lowest estimates are correct.

Since no one knows how many species there are, let's suppose that there are "only" 3 million on Earth. Isn't this number staggering? This seems particularly true given its starting point. A single type of organism arose perhaps 3.8 billion years ago, branched into two types of organisms, and then the process continued—branches forming on branches—until at least 3 million different types of living things came to exist on Earth. Now here's the real surprise: These are just the *survivors*. The fossil record indicates that more than 99 percent of all species that have ever lived on Earth are now extinct. Branching indeed.

The questions to be answered in this chapter are: How does this branching work? How do we go from the microevolutionary mechanisms explored in Chapter 17 to the actual divergence of one species into two? And how do we classify the creatures that result from this branching?

18.1 What Is a Species?

The category of life that's been mentioned so far, the species, has a great importance not only to biology but to society in general. If we allow that some bacteria are harmful to human beings while others are helpful, that some varieties of mosquitoes transmit malaria while others do not, that some species of birds are endangered while others are not, then it follows that there must be some means of distinguishing one of these groupings from another (**Figure 18.1**).

(a) Endangered species

(b) Not endangered

Figure 18.1
Separate Entities

The concept of a species can have very practical effects.

(a) This bird is a northern spotted owl (*Strix occidentalis*), which is an endangered species, protected by federal law.

(b) This bird is a barred owl (*Strix varia*), which is not endangered.

The whole notion of knowing something about the natural world begins to break down if we can't say *which* grouping it is that is endangered or harmful.

Over the centuries, human beings have devised various ways of defining the groups now called species. But in science today, the most commonly accepted definition stems from what is known as the **biological species concept,** which uses breeding among organisms to make classifications. We actually looked at a version of the biological species concept in Chapter 17. Here it is again, as formulated by the evolutionary biologist Ernst Mayr:

> Species are groups of actually or potentially interbreeding natural populations which are reproductively isolated from other such groups.

Note that the breeding Mayr talks about can be real or potential. Two populations of finch may be separated from one another by geography, but if, on being reunited in the wild, they begin breeding again, they are a single species. Note also that Mayr stipulates that species are groups of *natural* populations. This is important because breeding may take place in captivity that would not in nature. No one doubts that lions and tigers are separate species, yet they will mate in zoos, producing little tiglons or ligers, depending on whether the father was a tiger or a lion (**Figure 18.2**). In natural surroundings, however, they apparently have never interbred, even when their ranges overlapped centuries ago.

You might think that, with the biological species concept in hand, scientists would be able to study any organism and, by discerning its breeding activity, pronounce it to be a member of this species or that. Nature is so vast and varied, however, that this doesn't always work. We can't look at the breeding activity of single-celled bacteria or archaea, for example, because they don't *have* any breeding activity; they multiply instead by simple cell division. To complicate matters further, different varieties of these microbes transfer genes among themselves (a phenomenon that can be appreciated if you imagine, say, an adult mouse being able to exchange genes with an adult squirrel). Given this, if one bacterial "species" exchanges a substantial number of genes with another, do we even have two distinct species? Microbiologists define bacterial and archaeal species today by sequencing their DNA or RNA and looking for identifying patterns among these sequences, but most specialists would say that, so far, we do not have a satisfactory means of placing bacteria and archaea into clearly defined species categories.

Given this limitation, the biological species concept applies only to organisms who reproduce *sexually*—that is, to organisms who reproduce by means of generating individual reproductive cells, such as sperm and eggs, that come together to produce a new generation of organism. Almost all the plants and animals with which we're familiar reproduce this way, as do many fungi and even a good number of the generally microscopic organisms called protists. Even among these sexual reproducers, however, there are "separate" species that sometimes interbreed in nature, producing so-called hybrid offspring as a result. Such mixing between species happens more in the plant world than the animal, but it does take place in both. If species are supposedly "reproductively isolated" from one another, then what are we to make of these crossings? (Ernst Mayr pointed out that it is *populations* in his concept that are reproductively isolated, not individuals, and that whole populations do tend to stay within their species confines.)

Despite these difficulties, the biological species concept provides a meaningful way of defining a

Figure 18.2
Hybrid Animal

Pictured is a liger—a big cat whose father was a lion and whose mother was a tiger. These animals are produced only in captivity, never in nature.

fundamental category of Earth's living things. This is so because the concept is rooted in a critical activity of organisms: their mating, which controls the flow of genes among them. As you saw last chapter, it is the flow of genes across generations that lies at the root of microevolution. Let's look now at how the flow of genes between populations controls **speciation,** which we can define as the development of new species through evolution.

18.2 How Do New Species Arise?

When we begin to think about speciation, it's helpful to recall the gene variants or alleles reviewed in Chapter 17. As you saw in that chapter, evolution within a population means a change in that population's allele frequencies. Let's shift now, however, to thinking about *two* populations of a single species—say, a species of bird in which one population has become separated from another by means of having flown to a nearby location. To the extent that these two populations continue to breed with one another, with individuals moving between the locations, each population will *share* in whatever allele frequency changes are going on with the other population. Hence, the two populations will evolve together, remaining a single species. A critical step would come, however, if migration should *stop* between the two populations. Through the various agents of change we looked at last chapter—natural selection, genetic drift, and so forth—each population would continue to undergo allele frequency changes. But now neither population would be sharing these changes with the other. And, over time, such genetic changes may bring about alterations in form and behavior. Coloration or bill lengths may change; feeding habits may be transformed; mating preferences may be modified. After enough time, should the two populations find themselves geographically reunited, they may no longer freely interbreed. At that point, they are separate species. Speciation has occurred.

The critical change here came when the two populations quit interbreeding. For scientists, the key question thus becomes: Why would this happen? What could drastically reduce interbreeding, and hence gene flow, between two populations of the same species?

The Role of Geographic Separation: Allopatric Speciation

In the preceding example, a very clear factor set the stage for reduced gene flow between the bird populations: They were separated geographically. And geographical separation turns out to be the most important starting point in speciation.

Populations can become separated in lots of ways. On a large scale, glaciers can move into new territory, cutting a previously undivided population into two. Rivers can change course, with the result that what was a single population on one side of the river may now be two populations on different sides of it. On a smaller scale, a pond may partially dry up, leaving a strip of exposed land between what are now two ponds with separate populations. Such environmental changes are not the only ways that populations can be separated, however. Part of a population might *migrate* to a remote area, as did the bird population, and in time be cut off from its larger population. For a real-life example of how migration can help bring about speciation, see **Figure 18.3**.

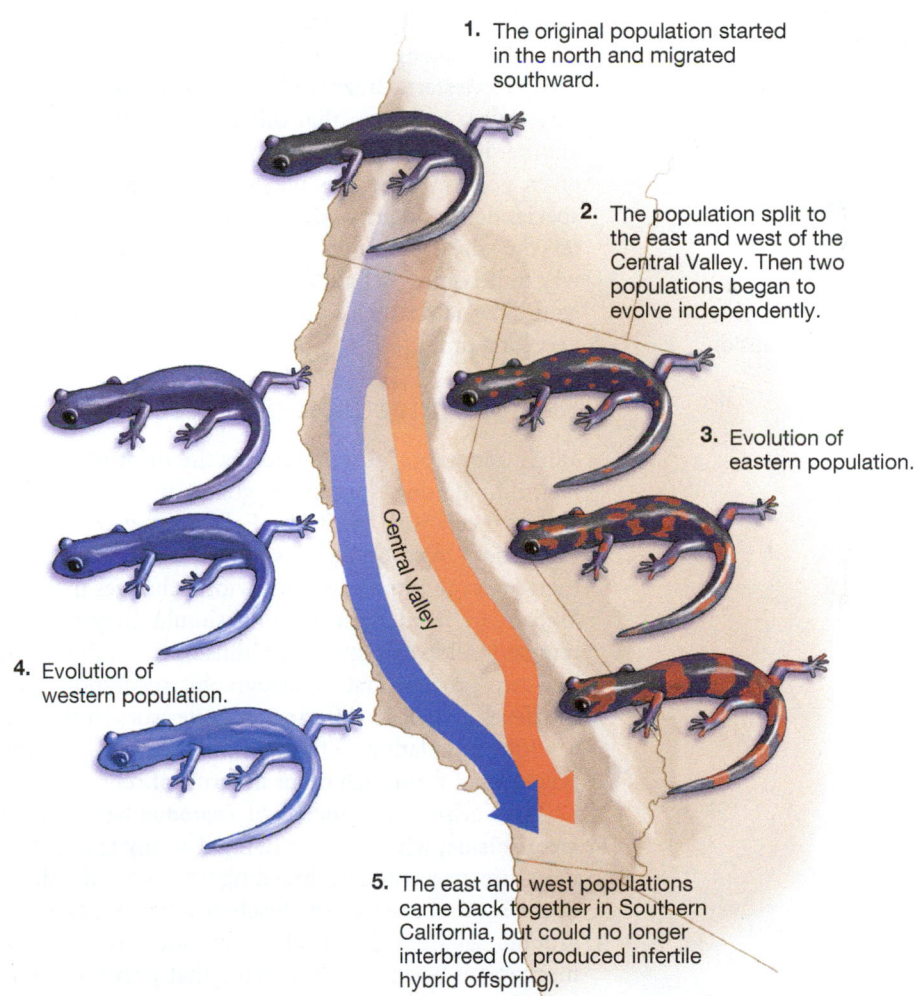

1. The original population started in the north and migrated southward.

2. The population split to the east and west of the Central Valley. Then two populations began to evolve independently.

3. Evolution of eastern population.

4. Evolution of western population.

Central Valley

5. The east and west populations came back together in Southern California, but could no longer interbreed (or produced infertile hybrid offspring).

Figure 18.3
Speciation in Action

Millions of years ago, the salamander *Ensatina eschscholtzii* began migrating southward from the Pacific Northwest. When it reached the Central Valley of California—an uninhabitable territory for it—populations branched west (to the coastal range) and east (to the foothills of the Sierra Nevada). Over time, the populations took on different colorations as they moved southward. By the time the two populations were united in Southern California, they differed enough, genetically and physically, that they either did not interbreed or produced infertile hybrid offspring when they did. (Salamanders are falsely colored in the drawing for illustrative purposes.)

Figure 18.4
Geographical Separation—Leading to Speciation?

These two varieties of squirrel had a common ancestor that at one time lived in a single range of territory. Then the Grand Canyon was carved out of land in northern Arizona, leaving populations of this species separated from one another, about 10,000 years ago. In the area of the Grand Canyon, **(a)** the Abert's squirrel lives only on the canyon's south rim, while **(b)** the Kaibab squirrel lives on the north rim. It's unclear whether these two varieties of squirrel would be reproductively isolated if reunited today—that is, whether they are now separate species—but the geographical separation they have experienced is the first step on the way to allopatric speciation.

Allopatric Speciation

(a) Abert's squirrel, south rim of Grand Canyon

(b) Kaibab squirrel, north rim of Grand Canyon

When geographical barriers divide a population and the resulting populations then go on to become separate species, what has occurred is **allopatric speciation.** *Allopatric* literally means "of other countries" (**Figure 18.4**). Look along the banks on either side of the Rio Juruá in western Brazil and you will find small monkeys, called tamarins, that differ genetically from one another in accordance with how wide the river is at any given point. Where it is widest, the members of this species do not interbreed, while at the narrow headwaters they do. Are the non-breeding populations on their way to becoming separate species? Perhaps, but only if gene flow is drastically reduced between them for a very long time.

Reproductive Isolating Mechanisms

While geographical separation is the most important factor in getting speciation going, it cannot bring about speciation by itself. Following geographical separation, two populations of the same species must then undergo physical or behavioral changes that will *keep* them from interbreeding should they ever be reunited. Allopatric speciation thus operates through a one–two process: first the geographic separation, then the development of differing characteristics in the two resulting populations. These are characteristics that will *isolate* them from each other in terms of reproduction.

Thus arises the concept of **reproductive isolating mechanisms,** which can be defined as any factor that, in nature, prevents interbreeding between individuals of the same species or of closely related species. Geographic separation is itself a reproductive isolating mechanism because it is a factor that prevents interbreeding. But because the mountains or rivers that are the actual barriers to interbreeding are outside of or *extrinsic* to the organisms in question, geographic separation is called an **extrinsic isolating mechanism.** For allopatric speciation to take place, what also must occur is the second in the one–two series of events: the evolution of *internal* characteristics that keep organisms from interbreeding. Such characteristics are referred to as **intrinsic isolating mechanisms:** evolved differences in anatomy, physiology, or behavior that

prevent interbreeding between individuals of the same species or of closely related species.

SO FAR . . .

1. Under the biological species concept, species members are defined by their ability to _____ with each other in nature.

2. The critical factor leading to speciation is reduced _____ between two _____ of the same species.

3. Two factors always play a part in allopatric speciation: first, the development of _____ and second, the development of one or more _____.

Six Intrinsic Reproductive Isolating Mechanisms

So, what are these intrinsic isolating mechanisms? A list of the most important ones is presented in **Table 18.1**. An easy way to remember them is to think about what sequence of events would be required for fertile offspring to be produced in any sexually reproducing organism. First, organisms that live in the same area must encounter one another; if they don't, the "ecological isolation" mechanism is in place. If they do encounter one another, they must then mate in the same time frame; if they don't, then temporal isolation is in place, and so on.

Ecological Isolation

Two closely related species of animals may overlap in their ranges and yet feed, mate, and grow in separate areas, which are called *habitats*. If they use different habitats, this means they may rarely meet up. If so, gene flow will be greatly restricted between them. Lions and tigers *can* interbreed, but they never have in nature, even when their ranges overlapped in the past. One reason for this is their largely separate habitats: Lions prefer the open grasslands, tigers the deep forests.

Temporal Isolation

Even when two populations share the same habitat, if they do not mate within the same time frame, gene flow will be limited between them. Two populations of the same species of flowering plant may begin releasing pollen at slightly different times of the year. Should their reproductive periods cease to overlap altogether, gene flow would be cut off between them.

Behavioral Isolation

Even when populations are in contact and breed at the same time, they must choose to mate with one another for interbreeding to occur. Such choice is often based on specific courtship and mating displays, which can be thought of as passwords between members of the same species. Birds must hear the proper song, spiders must perform the proper dance, and fiddler crabs must wave their claws in the proper way for mating to occur.

Mechanical Isolation

Reproductive organs may come to differ in size, shape, or some other feature, such that organisms of the same or closely related species can no longer mate. Different species of alpine butterfly look very similar, but their genital organs are different enough that one species cannot mate with another.

Gametic Isolation

Even if mating occurs, offspring may not result if there are incompatibilities between sperm and egg or between sperm and the female reproductive tract. In plants, the sperm borne by pollen may be unable to reach the egg lying within the plant's ovary. In animals, sperm may be killed by the chemical nature of a given reproductive tract or may be unable to bind with receptors on the egg.

There is one special form of gametic isolation, called *polyploidy*, that is very important in plants and that may have played a part in giving rise to the vertebrate evolutionary line that we humans are part of. You can read about it in "New Species through Genetic Accidents: Polyploidy" on page 326.

Hybrid Inviability or Infertility

Even if offspring result, they may develop poorly—they may be stunted or malformed in some way—or they may be *infertile*, meaning unable to bear offspring of their own. A well-known example of such infertility is the mule, which is the infertile offspring of a female horse and a male donkey (**Figure 18.5** on the next page).

Sympatric Speciation

Thus far, you have looked at the development of intrinsic reproductive isolating mechanisms strictly as a second step in speciation—one that follows a

Table 18.1		
Reproductive Isolating Mechanisms		
Extrinsic isolating mechanism		**Geographic isolation** Individuals of two populations cannot interbreed if they live in different places (the first step in allopatric speciation).
Intrinsic isolating mechanisms		**Ecological isolation** Even if they live in the same place, they can't mate if they don't come in contact with one another.
		**Temporal isolation** Even if they come in contact, they can't mate if they breed at different times.
		**Behavioral isolation** Even if they breed at the same time, they will not mate if they are not attracted to one another.
		**Mechanical isolation** Even if they attract one another, they cannot mate if they are not physically compatible.
	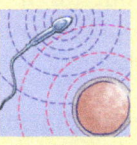	**Gametic isolation** Even if they are physically compatible, an embryo will not form if the egg and sperm do not fuse properly.
		**Hybrid inviability or infertility** Even if fertilization occurs successfully, the offspring may not survive, or if it survives, may not reproduce (e.g., mule).

geographic separation of populations. As it happens, however, intrinsic reproductive isolating mechanisms can develop between two populations in the *absence* of any geographic separation of them. If these isolating mechanisms reduce interbreeding between the populations sufficiently, speciation can take place. What occurs in this situation is not allopatric speciation, however; it is **sympatric speciation,** which can be defined as any speciation that does not involve geographic separation. (*Sympatric* literally means "of the same country.")

Sympatric speciation has been a contentious subject in biology for decades. Scientists have long recognized one of its forms: the polyploidy reviewed in "New Species through Genetic Accidents." But for years, most biologists were skeptical that it existed in

Speciation and Polyploidy

Figure 18.5
Mules Are Infertile Hybrids

Even though mules cannot themselves reproduce, humans frequently cross horses and donkeys to produce them to take advantage of their exceptional strength and endurance. Chromosomal incompatibilities between horses and donkeys leave the mules sterile.

Figure 18.6
A Species Undergoing Sympatric Speciation?

The fruit-fly species *Rhagoletis pomonella*, pictured here on the skin of a green apple, may be undergoing speciation.

any other form. Over the past decade, however, new evidence has convinced most experts that sympatric speciation has operated in a number of instances. No one claims that it has the importance of allopatric speciation, but for most biologists today, the key question about sympatric speciation is not whether it has taken place but how frequently it has occurred.

Sympatric Speciation in a Fruit Fly

To see how sympatric speciation works, consider one of its best-studied examples, a species of fruit fly named *Rhagoletis pomonella*. Prior to the European colonization of North America, *R. pomonella* existed solely on the small, red fruit of hawthorn trees (**Figure 18.6**). The

Europeans brought *apple* trees with them, however, and by 1862 some *R. pomonella* had moved over to them. It seemed to at least one mid-nineteenth-century observer, however, that with the introduction of apples there had arisen separate varieties of the flies—what are sometimes called apple flies and haw flies—with each variety courting, mating, and laying eggs almost exclusively on its own type of tree. A modern researcher named Guy Bush led the way in investigating whether this was the case, and the answer turned out to be yes. The two varieties of flies are separated from each other in all these ways. In one study of them, only 6 percent of the apple and haw flies interbred with one another. The apple and haw *R. pomonella* are not separate species yet, but they certainly give indication of being in transition to that status.

For our purposes, the first thing to note is that this separation has not come about because of *geographic* division. The apple and hawthorn trees that the two varieties of flies live on may scarcely be separated in space at all. Given this, how did this single species move toward becoming two separate species? Bush offers a likely scenario, based on a critical difference between the hosts the two species live on. Apples tend to ripen in August and September, while hawthorn fruit ripens in September and October. In the summer, all fruit flies emerge as adults after wintering underground as larvae, after which they fly to their host tree to mate and lay eggs in the fruit.

Bush believes that about 150 years ago (when hawthorns were hosts to all the flies), some individuals in a population of *R. pomonella* experienced either a mutation or perhaps a chance combination of rare existing alleles. In either event, this change did two things: It caused these flies to emerge slightly *earlier* from their underground state than did most flies, and it drew these flies to the smell of *apples* as well as hawthorns. Because apples mature slightly earlier than the hawthorn fruit, these flies had a suitable host waiting for them. More important, these flies would have bred *among themselves* to a high degree. Recall that the other flies were emerging later, and the adult fly only lives for about a month. Thus, the variant alleles were passed on to a selected population in the next generation. Today, the apple *R. pomonella* flies indisputably do emerge earlier than the hawthorn flies. This is one of the things that ensures reproductive isolation between the two groups; their periods of mating don't fully overlap.

Such a lack of overlapping mating periods may sound familiar because it is one of the intrinsic reproductive isolating mechanisms you looked at earlier. (It is temporal isolation.) You also can see ecological isolation in operation with these flies. Although the two types of flies live in the same area, they meet up with relatively little frequency because they have different habitats (hawthorn vs. apple trees). In short, these populations have developed intrinsic reproductive isolating mechanisms without ever having been separated from one another geographically. They are headed toward speciation, but in this case it is sympatric speciation.

SO FAR . . .

1. Name as many intrinsic isolating mechanisms as you can.

2. The mule provides an example of the intrinsic isolating mechanism of hybrid _____.

3. Sympatric speciation is speciation that does not involve _____.

18.3 Adaptive Radiation and the Pace of Speciation

The Hawaiian Islands are places of such lush natural diversity that it can be a surprise to learn that they harbor no native species of reptiles or amphibians and only one native variety of land mammal: the bat. On the other hand, Hawaii is home to more than 500 native species of the fruit flies known as drosophilids—a number so large it represents a quarter of all the drosophilid species found in the world (**Figure 18.7**). This story becomes more puzzling with knowledge that genetic analysis has revealed that every one of these 500 species evolved from a single drosophilid species that made it to the islands millions of years ago. Hence we have mystery: A small geographic area has been home to a tremendous amount of speciation within one group of living things. What can explain this kind of speciation? A phenomenon known as adaptive radiation, which turns out to have had a far-reaching importance in the diversification of Earth's organisms.

New Environments and Speciation

As most people know, Hawaii's islands were formed by volcanic activity. The oldest of the islands, Kauai, arose through this activity about 5 million years ago, while the youngest island, Hawaii, formed about 375,000 years ago. If you've ever seen a landscape recently coated by lava, you know the meaning of the word *barren*. Each Hawaiian island originally was barren in this way, which meant that, for a brief time, each island was sterile, with no life on it. Very quickly, however, life did come to the islands, with bacteria, fungi, plant seeds, and tiny animals landing on them, all being borne by ocean currents, wind, or migrating animals. In time, numerous large plant species became established.

By the time the drosophilid flies arrived, it's likely they encountered two things that were critical to their speciation: an environment that *had* plenty of food and habitat but that *lacked* any flies of the drosophilid variety. To understand the importance of these factors, imagine that you are a graphic artist, working in a big city with lots of other graphic artists, most of whom specialize in this or that (magazines, Internet websites, etc.). Now you and a few other graphic artists move to a new city in which there are few graphic artists but a good number of *possibilities* for graphic arts work. You would thus be able to specialize fairly easily—filling a *niche* or working role in this new environment—because many of these niches would not yet have been taken. Just so did the drosophilids rush in to fill previously unoccupied niches on the Hawaiian Islands—this mating environment, this food source, this place on the forest floor for laying eggs.

Such a situation is ripe with possibilities for change (meaning speciation) because, while niches are in flux, there is a good deal of shaping of species to environment through natural selection. But more of this occurred on Hawaii because this niche-filling was taking place on separate *islands*. These islands represented a geographic barrier to interbreeding, and you know what follows from this: allopatric speciation.

Taking a step back from these steps, we can see that the drosophilid species of Hawaii proliferated through

(a) *Drosophila plantibia*

(b) *Drosophila cyrtoloma*

Figure 18.7
Two of Many

The Hawaiian islands are home to more than 500 species of the flies known as drosophilids. Pictured are two varieties of these flies, *Drosophila plantibia*, on the left, and *D. cyrtoloma* on the right. Both flies live in rainforest habitats on the island of Maui.

ESSAY

New Species through Genetic Accidents: Polyploidy

In this chapter so far, speciation has been portrayed as something that takes place over many generations. But plants have a means of speciating in a single generation. It is called *polyploidy,* and it is very important in the plant world, as it is thought to have figured in the evolutionary history of more than 70 percent of all flowering plants. Beyond this, polyploidy may have had great importance in the evolution of animals. Evolutionary biologists currently are debating whether a polyploidy "event" helped speed the rise of animals with backbones—the vertebrates, such as ourselves. As you'll see, whenever polyploidy comes about, it helps speed up evolution.

How does polyploidy work? Here is one of the ways. From the genetics unit, you may recall that human beings have 23 pairs of chromosomes, *Drosophila* flies 4 pairs, and Mendel's peas 7. Whatever the *number* of chromosomes, the commonality among all these species is that their chromosomes come in *pairs,* which is the general rule for species that reproduce sexually (**Figure 1**).

As noted at the start of the chapter, plants are more adept than animals at producing "hybrids," with the gametes from two separate species coming together to create an offspring. Often these offspring are sterile, however, because when it comes time for them to produce *their* gametes (eggs and sperm) the sets of chromosomes they inherited from their different parental species may not "pair up" correctly in meiosis owing to differences in chromosome number or structure.

Now comes the accident. Suppose that, back when it was first created as a single-celled zygote, the hybrid offspring carried out the usual practice of doubling its chromosome number in preparing for its initial cell division. Now, however, the cell fails to actually divide; whereas it was supposed to put half its complement of chromosomes into one daughter cell and half

Did polyploidy speed the evolution of our own vertebrate line?

into another, it doesn't do this. It doubles the chromosomes but keeps them all in one cell. Equipped with *twice* the usual number of chromosomes, this single cell then proceeds to undergo regular cell division, meaning that every cell in the plant that follows will have this doubled number of chromosomes. Critically, this plant will have doubled both its *sets* of chromosomes—the set it got from parental species A and the set it got from species B. If you double a set, by definition every member of that set now has a partner to pair with in gamete formation. Thus, the roadblock to a hybrid producing offspring has been removed; the chromosomes can all pair up.

This pairing up yields eggs and sperm that can then come together and fuse thanks to another capability of plants. Recall that many plants can *self-fertilize.* A single plant contains male gametes that can fertilize that same plant's female gametes. Thus, the plant with the doubled set of chromosomes can fertilize itself, theoretically beginning an unending line of fertile offspring. This line of organisms is "reproductively isolated" from either of its parental species because it has a different number of chromosomes than either parental species. With reproductive isolation, we have a *separate* species, and with self-fertilization, we have a species that can perpetuate itself.

A multiplication of the normal two sets of chromosomes to some other set number is known as **polyploidy**; here we have speciation by polyploidy, which is one type of sympatric speciation. The importance of this in the plant world is immense. Many of our most important food crops are polyploid, including oats, wheat, cotton, potatoes, and coffee. Indeed, the type of polyploid speciation we've looked at—which begins with a hybrid offspring—often produces bigger, healthier plants. As a result, breeders have developed ways to artificially induce polyploidy.

But what about animals? The self-fertilization that some plants are capable of does not exist among more complex animals. As a result, polyploidy that can be maintained over generations is extremely rare in the animal kingdom. But in the course of hundreds of millions of

adaptive radiation: the rapid evolution of many species from a single species that has been introduced to a new environment. Though this mode of speciation can be perceived most readily on islands, it has an importance that goes way beyond the emergence of limited groups of living things in small geographical areas. Indeed, a number of experts in speciation believe that much of the diversity we see in the living world has come about through episodes of adaptive radia-tion. The basic factors in play with the Hawaiian drosophilids stand to have been repeated, in less extreme form, with one species after another over the long course of Earth's evolutionary history. All that's necessary for an adaptive radiation to take place anywhere is that a given environment present a species with a set of unfilled niches. Critically, species can come to such environments, through migration, but such environments can also come to species, through

years of evolution, it clearly has been maintained several times. An extremely important polyploidy event may have occurred about 500 million years ago, just as vertebrate animals were evolving from invertebrates. A controversial hypothesis holds that this event doubled the chromosomal set of our invertebrate ancestors, thus greatly accelerating the pace of their evolution.

How can polyploidy speed up evolution? Think of it this way: If you added an extra room to a house, that would give you the flexibility to transform, say, an existing bedroom into a den. Now, if you add genes to a genome, as polyploidy does, that provides the *genome* with a kind of flexibility: the flexibility to have one kind of gene transformed into another. Where once an organism would have but two copies of a given gene—one allele from each parent—with polyploidy it has four. With this change, members of one pair of these alleles can mutate without causing harm to the organism because the *other* pair of alleles can carry out their function all by themselves. And in a few cases, alleles that mutate will produce new proteins that can help transform an organism—that can help it evolve down different lines. Polyploidy, then, provides for a tremendous increase in the genetic capacity to evolve. And it provides this capacity in a single generation.

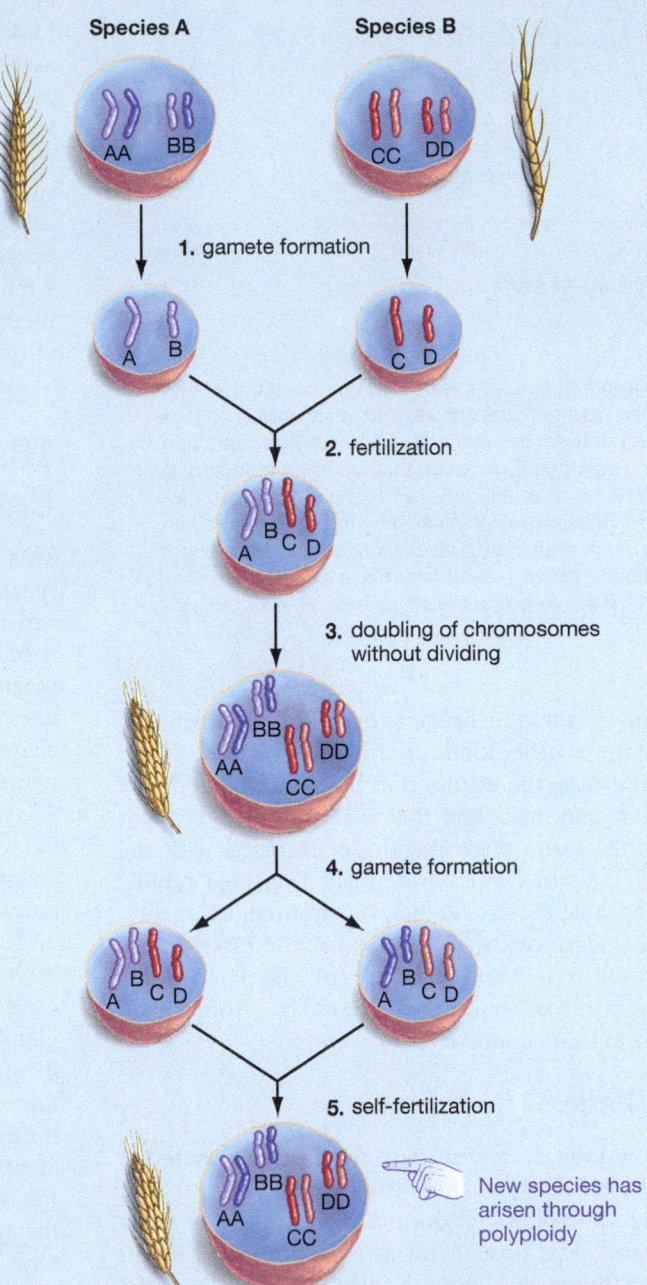

Species A   Species B

1. gamete formation

2. fertilization

3. doubling of chromosomes without dividing

4. gamete formation

5. self-fertilization

New species has arisen through polyploidy

Figure 1
Polyploidy in Wheat

Two different species of wheat exist in nature, with slightly different "genomes" or complements of DNA.

1. Gametes (eggs and sperm) are formed in the different species.

2. These gametes fuse, in fertilization, to form a zygote—a single cell that will develop into a new plant. Such a mixed-species or "hybrid" zygote generally develops into a sterile plant because its chromosomes cannot pair up correctly when the plant produces its own gametes during meiosis.

3. In this case, however, the zygote doubles its chromosomes in preparation for cell division, but then fails to divide. With this doubling, each chromosome now has a compatible homologous chromosome to pair with during meiosis. This is the polyploidy event.

4. Gamete formation then takes place in the plant.

5. These gametes from the same plant then fuse because this is a self-fertilizing plant. With this, a new generation of wheat plant has been produced—one that is a different species from either parent generation because each of the parent species has two pairs of chromosomes, while this new hybrid species has four pairs.

such means as climate change. Either way, the result is a changing relationship between environment and species. When this occurs, speciation becomes much more likely.

To get a sense of how this process can play out on a large scale, look at **Figure 18.8** on the next page, which traces speciation in the evolutionary group we ourselves are part of, mammals. Note that mammalian speciation did not take place in a steady way over time,

but instead came about in two waves, the first and most dramatic of which ran from 100 million to 85 million years ago. What occurred to spark this diversification is unknown. Mammals may have come into new environments, bringing with them traits, such as warm-bloodedness, that were *novel* traits within those environments. Conversely, it's possible that new environments came to mammals, in the form of, for example, the cooling temperatures of the period. A third

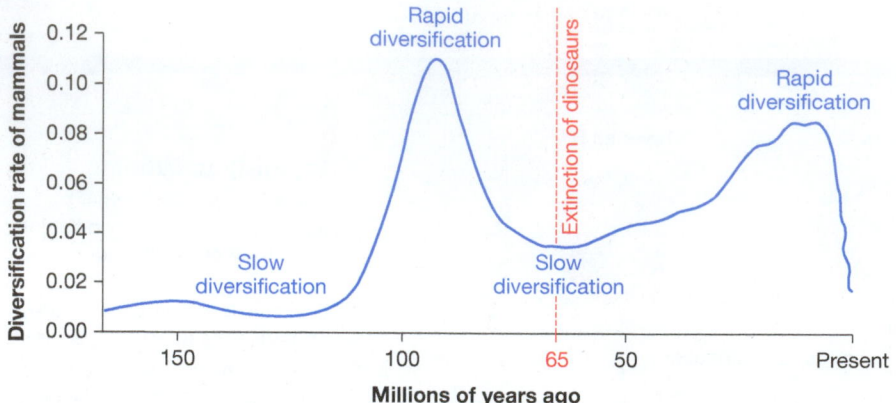

Figure 18.8
The Diversification of Mammals

Mammals evolved from reptiles beginning about 220 million years ago but did not diversify in a steady way from that time forward. Instead, as the figure shows, mammals appear to have undergone two bursts of speciation, the first of them beginning about 100 million years ago, and the second beginning about 50 million years ago. Speciation that follows a trajectory such as this is likely to have come about through a series of adaptive radiations. Contrary to the long-held hypothesis that mammalian diversification was unleashed by the extinction of the dinosaurs 65 million years ago, the first round of mammalian diversification occurred long before the extinction of the dinosaurs, while the second round took place long after.

(Adapted with permission from: Bininda-Emonds, Olaf R.P. et al., "The Delayed Rise of Present-Day Mammals," *Nature*: 29 March 2007, 446:507–512.)

Figure 18.9
Speedy Speciation

Through sympatric speciation, the fish on the right, *Amphilophus zaliosus*, evolved from the fish on the left, *A. citrinellus*, within less than 10,000 years in a relatively small lake in Nicaragua. The two species are kept separate by both ecological and behavioral isolating mechanisms. Ecologically, *A. zaliosus* is a surface-feeder, while *A. citrinellus* is a bottom-feeder. Behaviorally, differences in courtship signals keep the two species from mating, even when males and females from the two species come together in nature or in the laboratory.

possibility is that niches were opened up to mammals by means of other kinds of animals dying out. Whatever the case, the assumption is that an altered "fit" between mammals and their environments brought about the same phenomenon as occurred with the drosophilids in Hawaii: Mammals diversified rapidly as they adapted to fill new environmental niches. When this process was repeated over and over again, the result was the emergence of the fantastically diverse group of mammals we see today—from whales to bats to bears to ourselves.

The Pace of Speciation

So, how long does speciation take? Setting aside the instantaneous speciation that can come about through polyploidy, it can occur in as little as a few thousand years or as many as several million. If you look at **Figure 18.9**, you can see an example from the speedy side of this spectrum. Pictured are two species

of fish from a small lake in Nicaragua, Lake Apoyo. The species on the right (*Amphilophus zaliosus*) evolved from the species on the left (*Amphilophus citrinellus*) in a sympatric speciation that certainly took no more than 10,000 years and may have taken as few as 2,000. (One of the ways we know who evolved from whom is that *A. citrinellus* is widespread in the region, while *A. zaliosus* exists nowhere in the world except Lake Apoyo.) On the other end of the spectrum, species within a group of freshwater fish called the characins—which include piranhas and many aquarium fish—have been shown to take up to 9 million years to speciate. In between these extremes, there are species that appear to have taken hundreds of thousands of years to give rise to sister species.

18.4 The Categorization of Earth's Living Things

This chapter began by noting the importance of the species concept—of being able to say that this organism is fundamentally separate from that. To have a species concept means that there has to be some means of *naming* separate species. You have probably noticed that most of the species considered in this chapter have been referred to by their scientific names; you didn't just look at a fruit fly, but rather at *Rhagoletis pomonella*. To the average person, such names may be regarded as evidence that scientists are awfully exacting—or just plain fussy. Why the two names? Why the Latin? Can't we just say fruit fly?

To take these questions one at a time, scientists can't just say "fruit fly" for the same reason a person can't say "the guy in the white shirt" while trying to identify a player at a tennis tournament. There are lots of guys in white shirts at tennis tournaments, and there are lots of different kinds of flies in the world. It is important to be able to say *which* fly we are talking about. The importance of this can be seen with *Rhagoletis pomonella* itself, which is a major pest in the apple industry. Tell some apple growers that you have spotted flies on their apples, and you may get a shrug of the shoulders; tell them you have spotted *R. pomonella*, and you may get a very different reaction.

The Latin that is used stems from the fact that this naming convention was standardized by the Swedish scientist Carl von Linné in the eighteenth century, at a time when Latin was still used in the Western world in scientific naming. (In fact, Linné is better known by the Latinized form of his name—Carolus Linnaeus.) Linnaeus recognized the confusion that can result from having several common names for the same creature and thus devoted himself to giving specific names to some 4,200 species of animals and 7,700 species of plants—all that were known

(a) *Amphilophus citrinellus*

(b) *Amphilophus zaliosus*

to exist in his time. Many of the names he conferred are still in use today.

The fact that there are two parts to scientific names—a **binomial nomenclature**—points to the central question of groupings of organisms on Earth. Consider the domestic cat, which has the scientific name *Felis domestica*. The first part of this name specifies a genus—*Felis* in this case—which is to say a group of closely related but still separate species. Meanwhile, the second, "specific," part of this name designates a *particular* member of the *Felis* genus. Together, these two names constitute the name of a species: *Felis domestica*. The practical importance of making genus names a part of species names is that, if we know that two organisms are part of the same genus, then to know something about one of the organisms is to know a good deal about the other. There are six species of small cat within the *Felis* genus, for example, and all of these species share a very similar set of physical characteristics.

Taxonomic Classification and the Degree of Relatedness

But what is the basis for placing organisms in a category such as genus? Modern science classifies organisms largely—although not entirely—in accordance with how closely they are *related*. In this context, "related" has the same meaning as it does when the subject is extended human families. If people have the same mother, they are more closely related than if they shared only a common *grandmother*, which in turn makes them more closely related than if they had only a common great-great grandmother, and so on.

In the same way, species can be thought of as being related. Domestic dogs (*Canis familiaris*) are very closely related to gray wolves (*Canis lupus*), with all dogs being descended from these wolves. Indeed, by some reckonings dogs and gray wolves are the same species (*Canis lupus*) because they will interbreed in nature. Such differences as exist between them are the product of a mere 15,000 years or so of the human domestication of dogs. It may be, then, that 15,000 years ago there was but one species (the gray wolf) that gave rise to both today's wolf and today's dog—a close relation indeed.

Domestic dogs are also related, however, to domestic cats; if we look far back enough in time, we can find a single group of animals that gave rise to both the dog and cat lines. This does not mean going back 10,000 years, however; it means going back perhaps 60 million years. So there is a big difference in how closely related dogs and wolves are as opposed to dogs and cats. Establishing such *degrees* of relatedness is the most important task of the scientific classification system. There is a field of biology, called **systematics**, that is concerned with the diversity and relatedness of organisms; part of what systematists do is try to establish the truth about who is more closely related to whom. They study the evolutionary history of groups of organisms.

Setting aside for the moment the difficulty in *determining* what such evolutionary histories are, there obviously is a tremendous cataloguing job here, in light of the number of living and extinct species mentioned earlier. Given this diversity, a method of classifying organisms, called a *taxonomic system,* is employed to classify every species of living thing on Earth. There are eight basic categories in use in the modern taxonomic system: species, genus, family, order, class, phylum, kingdom, and domain. The organisms in any of these categories make up a grouping of living things, a **taxon.**

A Taxonomic Example: The Common House Cat

If you look at **Figure 18.10**, you can see how this taxonomic system works in connection with the domestic cat. As noted, this cat is only one species in a genus

Figure 18.10
Classifying Living Things

The classification of the house cat, *Felis domestica,* based on the Linnaean system.

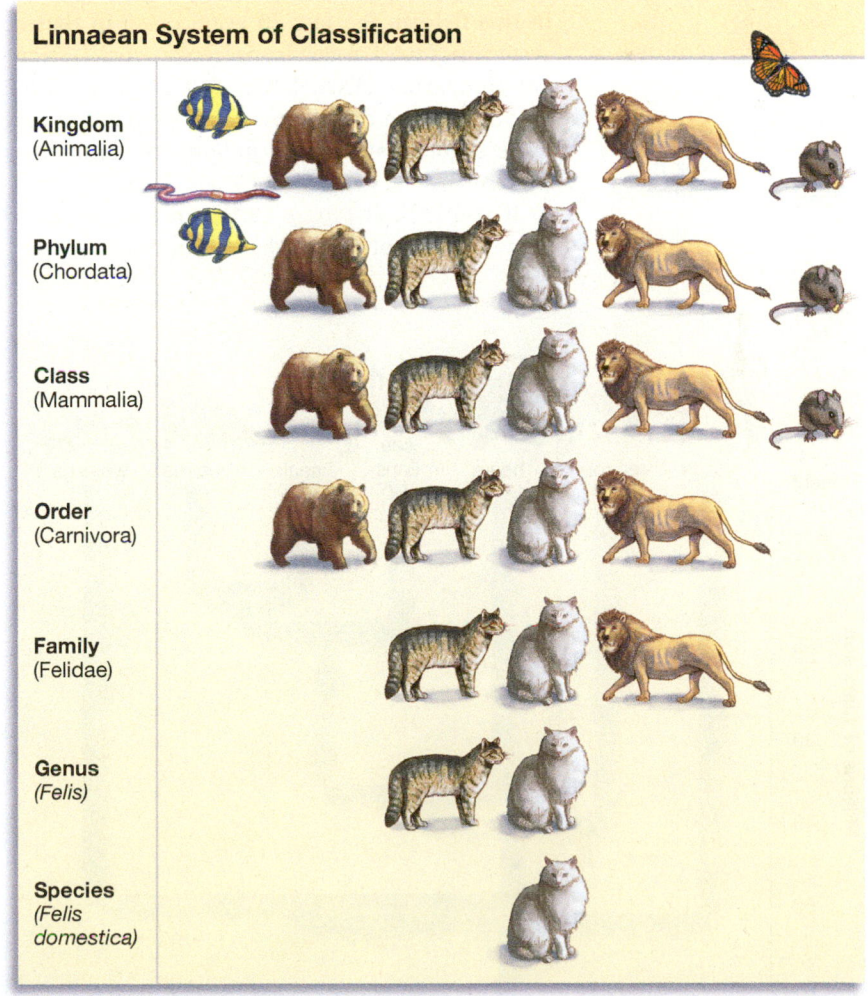

Linnaean System of Classification

Kingdom (Animalia)

Phylum (Chordata)

Class (Mammalia)

Order (Carnivora)

Family (Felidae)

Genus (Felis)

Species (Felis domestica)

(*Felis*) that has six living species in it. The genus then is a small part of a family (Felidae) that has 17 other genera in it (panthers, snow leopards, and others). The family is then part of an order (Carnivora) that includes not only big and small "cats" but other carnivores, such as bears and dogs. On up the taxa we go, with each taxon being more inclusive than the one beneath it until we get to the highest category in this figure, the kingdom Animalia, which includes all animals. (In Chapter 19, you'll look at the "supercategory" above kingdom, called domain, but for simplicity's sake, kingdom is the highest level taxon considered here.)

Constructing Evolutionary Histories

So how do systematists go about putting organisms into these various groups? What evidence do they use in constructing their evolutionary histories? They rely on radiometric dating, the fossil record, DNA sequence comparisons—all the things reviewed in Chapter 16 that are used to chart the history of life on Earth. In practice, there has been a revolution in this field in the last 20 years or so in that most systematics work today is molecular work—scientists are comparing DNA, RNA, and protein sequences among modern organisms and using patterns among these sequences to make judgments about lines of descent.

If you look at **Figure 18.11**, you can see the outcome of some of this work, an evolutionary "tree" for

one group of organisms, in this case one of the major groups of mammalian carnivores. You can see that the tree is "rooted," about 60 million years ago, with an ancestral carnivore whose lineage split two ways: to dogs on the one hand and everything else on the other. One interesting facet of this evolutionary history is how closely related bears are to the aquatic carnivores. All such histories are hypotheses about evolutionary relationships, with each such hypothesis known as a **phylogeny.**

Obscuring the Trail of Evidence: Convergent Evolution

One of the things that makes the interpretation of evolutionary evidence difficult is that similar features may arise *independently* in several evolutionary lines. A bedrock of systematic classification is the existence of **homologies,** which can be defined as common structures in different species that result from a shared ancestry. In Chapter 16, a strong homology was noted in forelimb structure that exists in organisms as different as a gorilla, a bat, and a whale. If you look at **Figure 18.12**, you can see the gorilla and the bat again.

Now, however, consider an extinct group, the litopterns, that once lived in what is now Argentina. Like the modern-day horse, they roamed on open grasslands. Over the course of evolutionary time, they developed legs that were extraordinarily similar to the horse's, right down to having an undivided hoof. In fact, litopterns are only very distantly related to the horse, but the leg similarities were enough to convince one nineteenth-century expert that he had found in litopterns the ancestor to the modern horse.

What is exhibited in the legs of the litoptern and horse is an **analogy**: a feature in different organisms that is the same in function and superficial appearance. When nature has shaped two separate evolutionary lines in analogous ways, what has occurred is **convergent evolution.** (Both the litoptern and horse lines converged in their development of similar legs.) But analogous features have nothing to do with common descent; they merely show that the same kinds of environmental pressures lead to the same kinds of physical features. (There are only so many leg structures that work well for grass-eating mammals that roam over long distances, so it's not surprising that natural selection would have channeled two separate evolutionary lines toward a similar leg shape.) The problem is that analogy can be confused with homology; analogies can make us think that species share common ancestry when in fact they don't. (Convergent evolution can take place *within* a species—indeed, it has in our own species in the relatively recent past, as you can see in "Converging on an Ability to Drink Milk" on page 332.)

Figure 18.11
A Family Tree for a Selected Group of Mammalian Carnivores

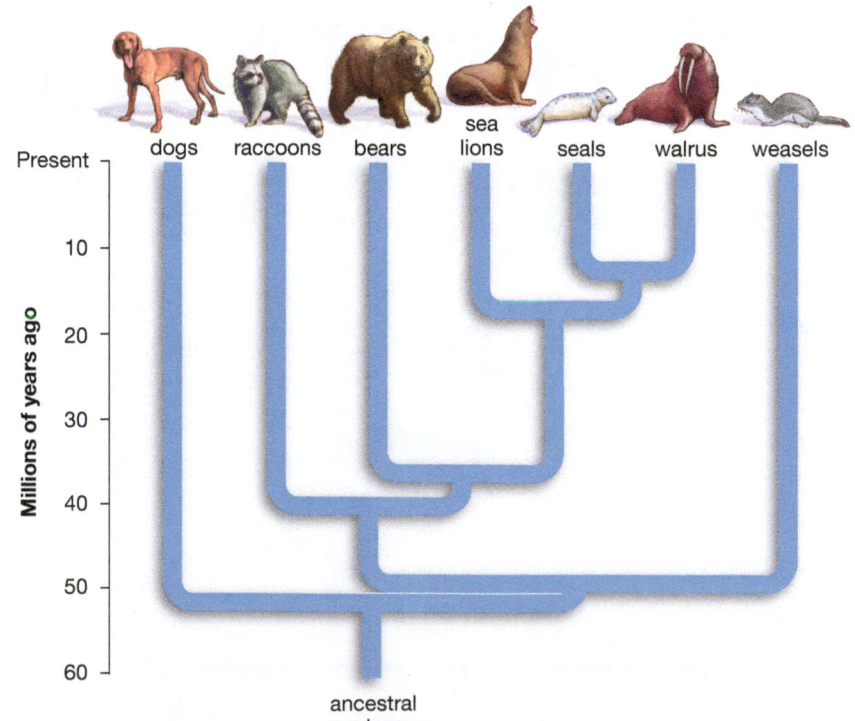

(a) Homology: Common structures in different organisms that result from common ancestry

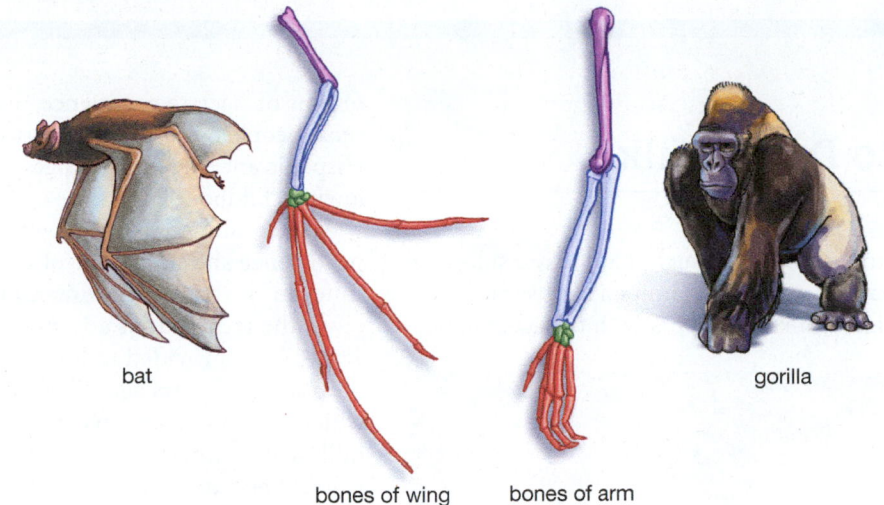

bat

bones of wing   bones of arm

gorilla

(b) Analogy: Characters of similar function and superficial structure that have *not* arisen from common ancestry

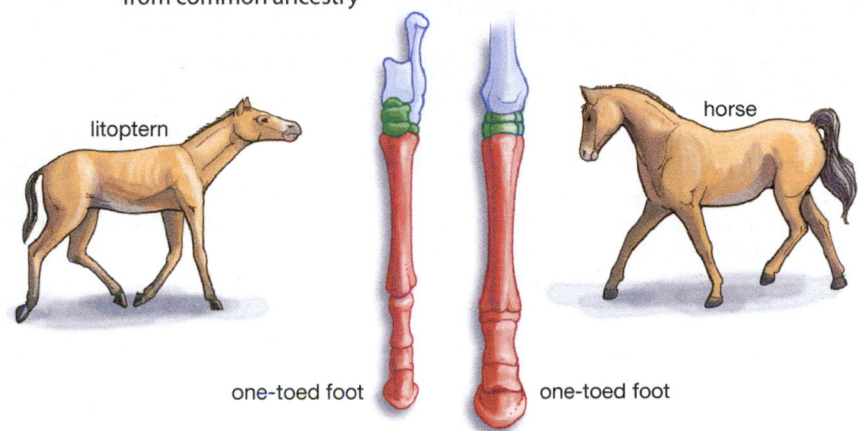

litoptern

one-toed foot   one-toed foot

horse

Figure 18.12

(a) Bats and gorillas have the same bone composition in their forelimbs—one upper bone, joined to two intermediate bones, joined to five digits—because they share a common ancestor. Of course, these forelimbs are used today for very different functions.

(b) Horses and the extinct litopterns have long legs and one-toed feet that serve the same function, but these features were derived independently in the two creatures—in separate lines of descent.

SO FAR . . .

1. Adaptive radiation is the rapid emergence of many species from a single species that has been introduced to _____.

2. Living things are classified largely in accordance with degrees of _____.

3. All species are referred to by a two-part name, the first part of which is a species' _____.

18.5 Classical Taxonomy and Cladistics

Given difficulties such as convergent evolution and a less-than-perfect fossil record, systematists over the years have been driven to devise several method-ologies for producing phylogenies. The essential question they have asked is: What means of *interpreting* the evidence will give us the most accurate picture of how Earth's living and extinct organisms are related?

One of the methodologies for interpreting evidence actually predates the field of systematics and most of the techniques used in it. Handed down from the time of Darwin, this methodology is sometimes called *classical taxonomy* (or descriptive taxonomy). In thinking about animals in Earth's family tree, classical taxonomists would look first and foremost at the physical form or morphology of the animals in question, as preserved in the fossil record, and compare it to the morphology of modern animals. Skull shape, "dentition" (teeth patterns), limb structure, and much more would be considered. They would also look at *where* the ancient forms existed compared to modern forms. More recently, they have been using molecular techniques, such as comparing DNA or protein sequences

ESSAY

Converging on an Ability to Drink Milk

Convergent evolution may sound like an exotic concept with little relevance to everyday life, but we human beings have experienced an important convergent evolution of our own: The human ability to digest milk throughout life evolved at least three separate times over the past 7,000 years—once in East Africa, once in Northern Europe, and once in the Middle East. In each case, this ability originated by means of a mutation that occurred within a population of humans who had begun to domesticate milk-yielding animals such as cows or camels (**Figure 1**). Each of these populations experienced a different mutation, but all of the mutations resulted in alleles that had the same effect: They allowed individuals to keep producing an enzyme called lactase once past infancy, a change that gave these individuals a lifelong ability to digest the lactose in milk. Once these alleles appeared, they spread through each of the three local populations and then spread to ever-larger groups of people, such that today a substantial minority of human adults can digest milk. In keeping with the geographic

Figure 1
An Ability that Evolved Several Times
The ability to digest milk throughout life has evolved at least three times in the human species over the last 7,000 years. In each case, the ability arose in a geographical region in which human beings had begun to domesticate herd animals. One such region was the Middle East, where camels were domesticated. Here a man drinks milk provided by his camels near Suweibah, Kuwait.

origins of "lactase persistence," however, most people of East Asian, West African, or Hispanic ancestry cannot digest milk following childhood.

It's actually not surprising that lactase persistence should have evolved several times in the course of human history, given the tremendous advantage this trait would have provided to humans who, thousands of years ago, lived in groups in which food in general was scarce but milk from animals was available. Anyone who inherited an allele for lactase persistence was effectively granted a tremendous survival and reproduction advantage compared to someone whose ability to digest milk ended in childhood. The result was that lactase-persistent individuals left more offspring, which meant that lactase persistence quickly became prevalent in populations that possessed the allele for it. In short, natural selection acted very strongly to spread lactase persistence, which is why this trait traveled so far in a time span that, in evolutionary terms, was the mere blink of an eye. (For more on human milk digestion, see "An Evolving Ability to Digest Milk" on page 294.)

in different living species to determine the relatedness among them. In essence, classical taxonomists would look at how many similarities one group has with another and, on that basis, try to judge relatedness among them. The word *judge* is used advisedly here because, at the end of the day in the classical system, subjective judgments usually need to be made about who is more closely related to whom. It is a matter of weighing one piece of evidence against another, but who's to say which piece of evidence is more important?

Another System for Interpreting the Evidence: Cladistics

The subjectivity inherent in classical taxonomy helped bring about the formation of another system for establishing relatedness. Developed in the 1950s by the German biologist Willi Henning, it is called **cladistics** (from the Greek *klados*, meaning "branch"). In prac-

tice, cladistics has become the core of most phylogenetic work going on today.

In **Figure 18.13** you can see a **cladogram,** which is an evolutionary tree constructed within the cladistic system. Note that there is a very simple branched line and that no time scale is attached to it, as was the case with the evolutionary tree of carnivores. Cladistics concerns itself first with lines of descent—with the *order* of branching events. Once this order has been established, efforts can be made to fix events in time, but that is a secondary concern. First and foremost, this cladogram is a proposed answer to the question: Among the animals lizard, deer, lion, and seal, which two groups of animals have the most recent common ancestor? Then, which other two have the *next* most recent common ancestor, and so on. By extension, the question is: Who is more closely related to whom? Note that these questions are being asked about only four of the five animals listed. You'll get to the role of the hagfish in a moment.

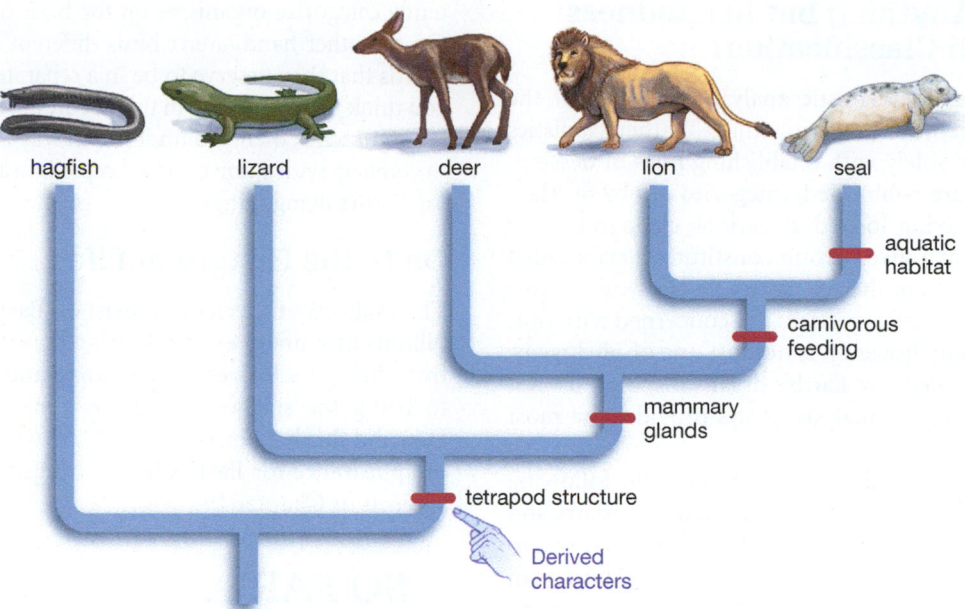

Figure 18.13
A Simple Cladogram
Among the lizard, deer, lion, and seal, which two are most closely related? The tentative answer displayed in this cladogram is the lion and the seal. The basis for this assertion is the number of unique or "shared derived" characters these two creatures have compared to any other two organisms under consideration. (They share three characters: tetrapod structure, mammary glands, and carnivorous feeding.) Such a system pays no attention to superficial similarities among organisms. Like fish, seals swim in the water, but they share far more derived characters with lions than they do with fish—enough to persuade scientists that seals have returned to the water in relatively recent times, following a period in which their ancestors lived on land.

Shared Ancestral and Derived Characters

The starting point for cladistic analysis is the difference between so-called ancestral characters and derived characters. If you look at any group of species (any taxon), its **ancestral characters** are those that existed in an ancestor common to them all. There are about 50,000 species of animals on Earth that possess a dorsal vertebral column (better known as a backbone). The ancestor to all fishes, reptiles, and mammals had such a feature, which makes the vertebral column an ancestral character for all these groups. On the other hand, only *some* vertebrates then went on to become animals that have four limbs (and are thus tetrapods). The tetrapod feature is thus a **derived character** of the vertebrate state; it is a character *unique* to taxa descended from a common ancestor.

In Figure 18.13, you can see a continuation of this process of ever-more selective grouping in that only some mammals are carnivorous feeders, and only some members of this taxon are aquatic mammalian carnivores. The presumption in cladistics is that shared derived characters are evidence of common ancestry and that the *more* derived characters any two organisms share, the more recently they will have shared a common ancestor compared to other organisms under study. At its most fundamental level, cladistics becomes a matter of counting shared derived characters. Seals and lions share three derived characters (tetrapod structure, mammary glands, and carnivorous feeding), while the lizard has only one derived character that it shares with all the other animals in Figure 18.13 (tetrapod structure). Cladistics infers from this that the lion and the seal have a more recent common ancestor than do any other two groups in the cladogram—that their branching from this common line of descent came later than any other branching in the line. And the hagfish? It is a starting-point organism for the mammalian carnivores, known as an *out-group*. If we are to say that the animals under consideration have derived characters, they must be derived *compared to* some other organism—in this case, the hagfish.

The examples here employ visible, physical traits, but in practice modern cladistics relies heavily on comparisons among molecular sequences, such as the number of shared, derived base pairs in lengths of DNA or RNA. Furthermore, since cladistics often compares many species, its phylogenies cannot be produced through simple counting; instead, they are the result of multiple rounds of computer-aided calculations.

Taking a step back, the important thing to note about the cladistic method is that it has done away with one of the problems of classical taxonomy: It has removed the element of subjectivity. There is a firm rule in cladistics for inferring relatedness—the number of shared derived characters—which means that judgment need not play a part in its analyses.

Should Anything but Relatedness Matter in Classification?

Strictly speaking, cladistic analysis is not about the classification of Earth's living things. Rather, cladistics is concerned solely with establishing lines of descent. Once these are established, categories can be overlaid on them. Having looked at various cladograms, we could say that a given group constitutes a class called Reptilia, while another constitutes a class called Amphibia. But cladistics per se is not concerned with this. If the question, however, is not just one of phylogeny, but how to *categorize* Earth's organisms, then it's not clear that cladistic analysis always provides the most sensible categories.

To appreciate this point, consider the following question. Who is more closely related: dinosaurs and lizards or dinosaurs and birds? It may surprise you to learn that it is dinosaurs and birds. Birds split off from a dinosaur line within the last 200 million years or so while the split between the lizard and dinosaur lines came much earlier. Despite this, conventional taxonomy says that dinosaurs and lizards belong in one class (Reptilia) while birds belong in another (Aves; **Figure 18.14**). The assumption here is that, in classifying creatures, something should count *besides* their relatedness—in this case, the special qualities of birds (feathers, flight).

To some scientists, a categorization such as this makes no sense. Why should relatedness dictate taxonomy in some instances but not in others? Shouldn't we consis-

tently categorize organisms on the basis of relatedness? On the other hand, aren't birds different enough from lizards that they deserve to be in a separate grouping? If you think that both sides in this debate are making sensible arguments, then you understand why it is so difficult to come up with a universally accepted means of classifying Earth's living things.

On to the History of Life

The millions of species that exist on Earth today took billions of years to evolve. During that time, life went from being exclusively microscopic and single-celled to being the staggeringly diverse entity it is today. How did this happen over time? And what is the order of appearance for Earth's living things? These are the subjects of Chapter 19.

SO FAR . . .

1. Cladistics determines relatedness among organisms by counting shared, derived characters: characters that are _____ to taxa descended from a _____.

2. One school of thought holds that Earth's organisms ought to be categorized in strict accordance with _____, while another holds that other factors, such as shared physical qualities, ought to be considered as well.

Figure 18.14
Are Birds Reptiles?

(a) Classical taxonomy employs three classes, Aves, Mammalia, and Reptilia, for birds, mammals, and reptiles, respectively. Thus, dinosaurs and birds are in separate classes—a recognition on the part of classical taxonomists of the unique features of birds, such as feathers.

(b) In cladistic analysis, by contrast, birds are grouped very closely with dinosaurs in recognition of the recent shared ancestry of the two taxa. The only consideration in this system of classification is who is more closely related to whom.

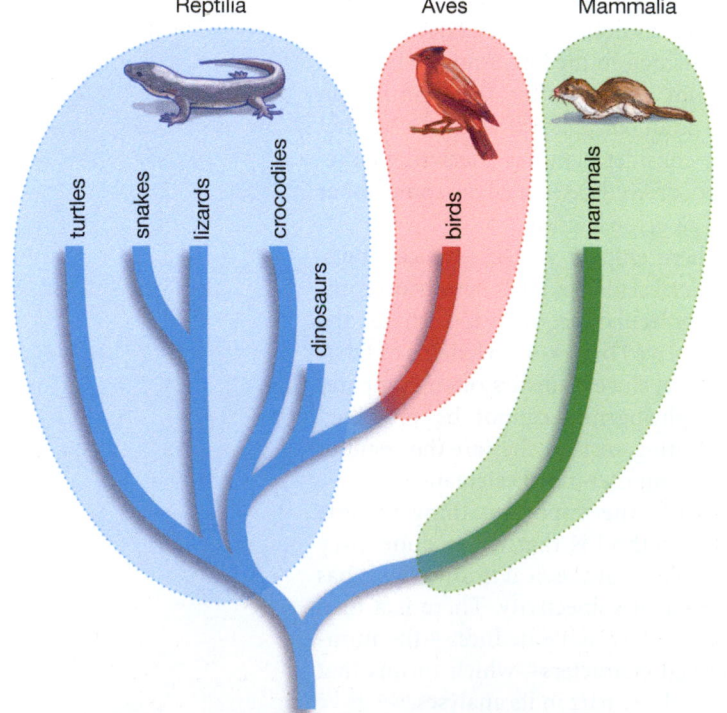

(a) Classical view of relationships among tetrapods

Reptilia — Aves — Mammalia

turtles snakes lizards crocodiles dinosaurs birds mammals

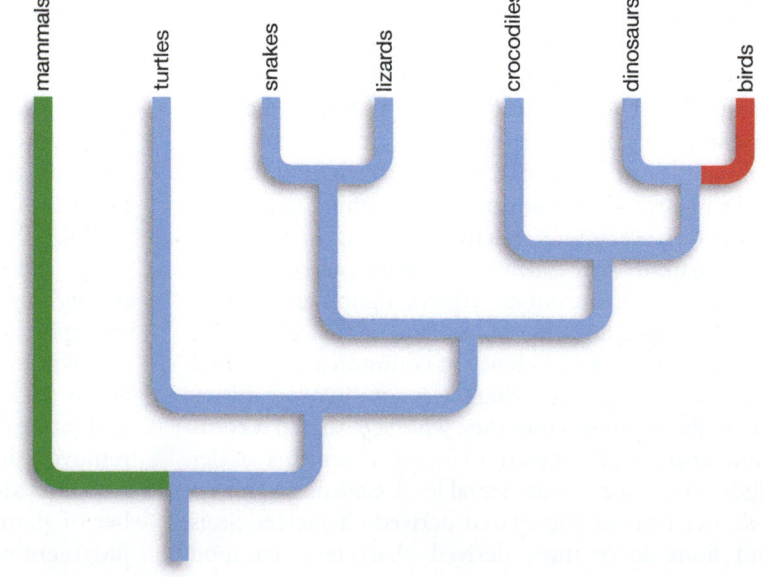

(b) Cladistic view of relationships among tetrapods

mammals turtles snakes lizards crocodiles dinosaurs birds

Summary

18.1 What Is a Species?

- The biological species concept defines species as groups of actually or potentially interbreeding natural populations that are reproductively isolated from other such groups. This definition does not apply to bacteria and archaea (which reproduce through simple cell division), but does apply to all organisms that multiply through sexual reproduction. **(p. 319)**

18.2 How Do New Species Arise?

- Branching evolution, which occurs when a single "parent" species diverges into two species, is based on reductions in gene flow between populations of the same species. When evolution has altered the characteristics of separated populations of the same species enough that they will no longer interbreed, even if reunited, speciation has occurred. **(p. 321)**

- Geographic separation is the most important factor in reducing gene flow between populations of the same species. When geographic separation plays a part in the evolution of a species, allopatric speciation has occurred. **(p. 321)**

- Geographic separation cannot bring about speciation by itself. Following geographic separation, separate populations of the same species must develop internal characteristics, called intrinsic isolating mechanisms, that will keep them from interbreeding should they reunite. The six most important of these mechanisms are ecological, temporal, behavioral, mechanical, and gametic isolation and hybrid inviability or infertility. **(p. 322)**

- Speciation can occur in the absence of geographical separation, but such sympatric speciation is thought to occur much less frequently than allopatric speciation. **(p. 323)**

18.3 Adaptive Radiation and the Pace of Speciation

- Speciation is more likely to occur when a species comes to a new environment or a new environment comes to a species (through climate change, for example). Environmental shifts of this sort make niches available to a species, a condition that fosters the adaptation of the species to its environment through natural selection and that hence promotes the evolution of new species from it. This is known as adaptive radiation: the rapid emergence of many species from a single species that has been introduced to a new environment. Much of the living world's diversity may have been produced through episodes of adaptive radiation. **(p. 325)**

- Polyploidy is sympatric speciation that can give rise to a new species in a single generation. It is most important in plants, but it has occurred in animal species as well. Polyploidy speeds evolution by providing a redundancy in alleles that allows some to mutate without causing harm to the organism. **(p. 326)**

18.4 The Categorization of Earth's Living Things

- A binomial nomenclature is used for each of Earth's species. The first name designates the organism's genus and the second designates its species. Genus and species are categories that fit into a larger framework of taxonomy, meaning the classification of species. Going from least to most inclusive, the eight most important categories in the taxonomy of Earth's organisms are species, genus, family, order, class, phylum, kingdom, and domain. **(p. 328)**

- Species that are closely related are in the same genus while species that are distantly related may only be in the same phylum or kingdom. The scientific discipline called systematics establishes degrees of relatedness among both living and extinct species. **(p. 329)**

- Systematists establish phylogenies, or hypotheses about evolutionary family trees, by reviewing radiometric dating, the fossil record, and DNA sequence comparisons, among other kinds of evidence. Most of the work in systematics today is molecular. **(p. 330)**

- Systematists look for homologous structures (common structures in different species that result from a shared ancestry). One problem with the use of these structures is that they can be confused with analogous structures (similar features that developed independently in separate lines of organisms). **(p. 330)**

18.5 Classical Taxonomy and Cladistics

- Classical taxonomy employs subjective judgments in deciding what weight to give one piece of evidence versus another in proposing phylogenies. An alternate methodology, cladistics, does away with this subjectivity by following a firm rule for inferring relatedness. It counts the number of shared derived characters two organisms have. **(p. 331)**

- Cladistics is chiefly concerned with establishing phylogenies, although the results it produces can be used to put species into taxonomic categories. Classical taxonomy holds that, in putting organisms into various categories, factors other than phylogeny ought to be taken into account. **(p. 334)**

Key Terms

Understanding the Basics

Multiple-Choice Questions

(Answers are in the back of the book.)

1. In the biological species concept, the factor that defines a species is:
   a. geographical separation.
   b. reproductive isolation.
   c. hybridization.
   d. cladogenesis.
   e. its behavior.

2. In allopatric speciation (select all that apply):
   a. allele frequency changes are shared between populations of the same species.
   b. populations are geographically isolated.
   c. gene flow between populations is greatly reduced because of geographic separation.

   d. populations accumulate genetic changes over time that result in physical or behavioral changes that eventually block successful interbreeding.
   e. populations are directly adjacent.

3. Possible intrinsic reproductive isolating mechanisms include (select all that apply):
   a. mating at different times of year or times of day.
   b. geographical isolation.
   c. incompatibility between eggs and sperm that prevent fertilization.
   d. singing a different mating song than another species.
   e. producing fertile hybrids.

4. Match the reproductive isolating mechanism with the means by which it works:
   a. temporal isolation — populations utilize different habitats
   b. ecological isolation — populations employ different courtship or mating displays
   c. behavioral isolation — mating occurs at different times of day or year
   d. gametic isolation — progeny of a cross are unable to reproduce
   e. hybrid sterility — egg and sperm do not fuse

5. In ocean-dwelling creatures such as sea urchins or corals, sperm and eggs are released into the water by males and females and fertilization occurs externally. If several closely related species of coral live in the same location, what reproductive isolating mechanisms could potentially prevent interbreeding? (Select all that apply.)
   a. behavioral isolation
   b. gametic isolation
   c. temporal isolation
   d. ecological isolation
   e. hybrid infertility

6. Order the following categories into the appropriate biological hierarchy, with the least inclusive category first
   a. family
   b. species
   c. phylum
   d. genus
   e. class

7. Factors that can be problems in determining relatedness among organisms may include (select all that apply):
   a. homologous traits.
   b. analogous traits.
   c. convergent evolution.
   d. poor fossil record.
   e. shared derived traits.

8. Adaptive radiations (select all that apply):
   a. can be seen most clearly in mainland environments.
   b. are episodes of rapid speciation that occur when the "fit" between environment and species is in flux.
   c. concern the evolution of heat loss in vertebrates experiencing stressful climates.
   d. are exemplified by the Hawaiian drosophilid flies.
   e. can be seen most clearly in island environments.

Brief Review

(Answers are in the back of the book.)

1. State the biological species concept.

2. Why is gene flow central to both microevolution and the macroevolutionary process of speciation?

3. Why is the evolution of intrinsic reproductive isolation mechanisms required for two groups to be called separate species? Why isn't simple geographical isolation sufficient?

4. Describe one major difference between allopatric and sympatric speciation. What is one thing that they have in common?

5. What is convergent evolution, and why is it a problem for phylogenetic analyses?

Applying Your Knowledge

1. Living things are classified into a set of hierarchical categories. Why is this more useful to biologists than simply giving everything a one-word name (or a number) and eliminating some of the categories?

2. Populations of some kinds of organisms may be isolated by a barrier as small as a roadway, while other organisms require a much larger physical barrier to block gene flow. What features of organisms might determine how readily they become geographically isolated? What impact might these features have on how often or how rapidly speciation is likely to occur in these groups?

3. In an adaptive radiation, one species may colonize a previously empty environment, such as an island, and diversify evolutionarily into a set of closely related species that occupy a very wide range of habitats. For example, with Darwin's finches on the Galapagos Islands, one species fills the role of woodpecker by using a stick to tap on trees and dig out insects. Do you think that such a species would have been likely to evolve in an area that already had woodpeckers? Why or why not?

MasteringBiology®

1. MasteringBiology Assignments

Activities: Allopatric Speciation; Speciation and Polyploidy; **Current Events:** Genetic Data and Fossil Evidence Tell Differing Tales of Human Origins; Is Sugar Toxic?; **Building Vocabulary; Questions:** Reading Questions, Test Bank

2. eText

Read your book online, search, take notes, highlight text, and study more effectively.

3. The Study Area

Self-study tools to test comprehension.

A Slow Unfolding:
The History of Life on Earth

Life started out small, perhaps 3.8 billion years ago, and stayed that way for a long time. Eventually it exploded into a multitude of forms that included human beings.

Available on MasteringBiology®

Activities: RNA-Like Self Replication; **Current Events:** Life in the Sea Found Its Fate in a Paroxysm of Extinction; A Lesson of Genealogy; Looks Can Be Deceiving; Could Conjoined Twins Share a Mind; Signs of Neanderthals Mating with Humans; **Building Vocabulary; Questions:** Reading Questions, Test Bank

The remains of a group of ocean-dwelling animals called crinoids have been preserved in this piece of rock dating from Earth's Carboniferous period, which began 359 million years ago.

Within 5 minutes of its birth, a newborn wildebeest on the plains of central Africa can be off and running with its herd (**Figure 19.1**). By contrast, human infants probably will not be able to take a step by themselves until they are about a year old. Female lions reach sexual maturity at about 3 years, and killer whales at about 7, but in human hunter-gatherer societies, reproduction generally begins at age 18 or 19. Even in comparison to our fellow mammals, we humans mature at a notably slow rate.

At first glance, it might seem puzzling that our evolution took this turn. From what we have learned of natural selection, wouldn't it make sense that survival would *decrease* for offspring that developed more slowly? What is the fitness payoff in having young who are essentially helpless for several years and then immature for many more? Here is one possible answer: As a species, we have survived not so much because of our physical capabilities but because of our wits. Our success is the product of our remarkable capacity to learn. But learning takes time, and it is arguably best undertaken by minds that remain for a long time in what might be called a state of flexible immaturity. (Think of how easily young children acquire a second language in comparison with adults.) Among our human ancestors, it may be that those who survived tended to be those who could learn the most. And who could learn the most? Those who took the longest to mature.

Under this view, we carry our evolutionary heritage with us in the form of delayed maturity. We did not leave this heritage behind in the African savanna, where human beings first evolved; we can see it every time we look at a 2-year-old. In the same way, we can see evolutionary vestiges in us from a much earlier time as well. Note the way we walk, our left arm swinging forward as our right leg does the same. Almost all four-limbed animals walk this way. One view is that we inherited this from the elongated, four-finned fish we evolved from. They came onto the muddy land by slithering in an S-pattern, the left-front fin going back as the left-rear fin went forward, then the reverse (**Figure 19.2**).

Note the sweep of evolution in all this. Maturation is with us from one period, walking style from another. If we look at the genetic code—this DNA sequence specifying that amino acid—we have within us a heritage that stretches back to the beginning of life on Earth because this code is shared by every living thing.

Figure 19.1
No Time to Lose
A newborn wildebeest struggles to its feet immediately after its birth on the plains of Africa. Wildebeest predators, such as hyenas, often concentrate their efforts on the newborn.

Figure 19.2
Our Evolutionary Heritage

The gait exhibited by almost all four-limbed animals, including human beings, may stem from the reversing S-pattern that our evolutionary ancestors used in getting around when they were making the transition from aquatic to terrestrial life. Here a salamander employs this same motion in getting from one place to another.

Under this view, evolution is not some abstract process that lies separate from us; on the contrary, it is with us in every step we take. But how has evolution proceeded in its long sweep down the ages? Starting with the spark of life itself, how did it produce the range of creatures that inhabit the Earth? In this chapter and the one that follows, you'll begin to find out.

19.1 The Geological Timescale

We can start this inquiry by getting a sense of life's timeline. The Earth, which is home to all the life we know of, came into being about 4.6 billion years ago. Given the human life span of 80-plus years, 4.6 billion years is an unimaginably long time. Indeed, it's a long period of time relative to any time-frame we know of. The universe as a whole is thought to be 13.7 billion years old, so the Earth's history stretches back a third of the way to the beginning of time.

In measuring something as long as 4.6 billion years, it obviously won't do to think in terms of individual years, so scientists use something called the geological timescale, which you can see in **Figure 19.3**. It divides earthly history into broad eras and shorter periods. The scale begins with the formation of the Earth (at the bottom of the figure) and runs to "historic time," meaning time in the last 10,000 years.

But what do the demarcations in this timescale indicate? What, for example, is the distinction between the Permian period you can see at the middle-left of Figure 19.3 and the Triassic period that followed it? To understand what separates them, it's important to know *when* the timescale was developed: in the late eighteenth and early nineteenth centuries, at which point there was no radiometric dating of materials—no using the decay of uranium, for example, to provide absolute dates for any fossils that might be found. Hence, no one could say whether a fossil was 10 million or 100 million years old. Geologists of the time did know, however, that sedimentary rocks had been put down in layers (or strata), with the oldest strata on the bottom and the youngest strata on top (**Figure 19.4** on page 342). And as it turned out, each stratum tended to have within it a group (or "assemblage") of fossils that was unique to it. Thus, the strata could be divided up in accordance with the transitions the scientists saw in the life-forms. The Permian period is considered different from the Triassic period because, with movement from one to the other, a different set of life-forms appears, as recorded in the fossil record.

Features of the Timescale

Now a couple of notes on the timescale. First, it may surprise you to learn what the basis is for many of the transitions in it: death on a grand scale, in the form of "major extinction events." Five of these events are noted in Figure 19.3 (over on its right), with all of them defining the end of an era or period. The most famous extinction event, the **Cretaceous Extinction,** occurred at the boundary between the Cretaceous and Tertiary periods. This was the extinction, aided by the impact of a giant asteroid, that brought about the end of the dinosaurs along with many other life-forms. The greatest extinction event of all, however, occurred earlier, in the boundary between the Permian and Triassic periods. This was the **Permian Extinction,** in which as many as 96 percent of all species on Earth were wiped out.

This is not to say, however, that all Earth's eras or periods are marked off by extinctions. The transition between the Precambrian and Paleozoic eras marks not a major extinction, but a seeming surge in the diversity of animal forms. Nevertheless, this Cambrian Explosion, which you'll read about later, simply marks a different kind of transition in living things.

Now, when did some of these transitions come about? You might think that a transition such as a sudden increase in animal forms would have occurred very early in life's history, but the Cambrian Explosion took place 542 million years ago. (Another way of writing this is 542 Mya.) This is a long time ago, to be sure, but life is thought to have begun 3.8 *billion* years ago. There may not have been an animal on the planet for 3.2 billion years after life got going—nor any plants, either. Life was single-celled to begin with, and 2 billion years later, it still consisted of nothing but microbes. As you'll see, these tiny organisms were undergoing important evolutionary changes during Earth's long Precambrian era (a segment of time so extended it technically is referred to as a "supereon"). But to find changes in the *forms* of life's organisms, we have to go very far forward indeed. If the fossil record is correct, all the birds, reptiles, fish, plants, and mammals that exist today only came about in the last 16 percent of evolutionary time—the last 600 million years of life's 3.8-billion-year span. As you'll see, our own species, *Homo sapiens,* is an extreme latecomer in this series of events.

Notable Evolutionary Events and Value Judgments

One final word about the timescale and the path you will follow in this chapter and the one that follows. Picking out the "notable events" in evolution is inevitably an exercise in making value judgments. The notable events covered in this book will lead you along a line that begins with microscopic sea creatures and ends with human beings. The value judgment guiding this path is that students have a great interest in knowing about their own evolution and that the path to humans presents a broad sweep of evolutionary development. Such a course of study may leave the impression, however,

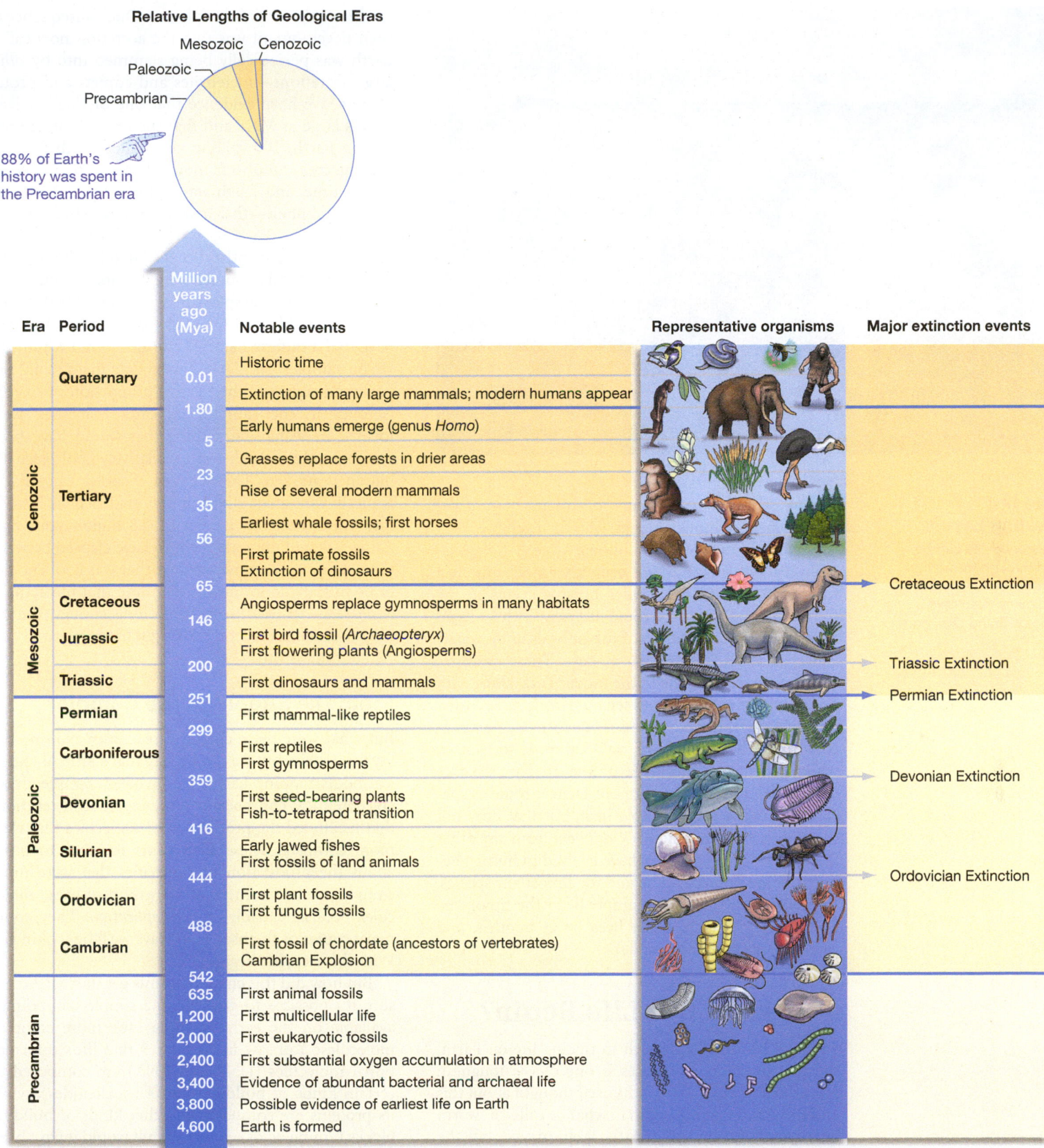

Relative Lengths of Geological Eras

88% of Earth's history was spent in the Precambrian era

Era	Period	Million years ago (Mya)	Notable events	Representative organisms	Major extinction events
Cenozoic	Quaternary	0.01	Historic time		
		1.80	Extinction of many large mammals; modern humans appear		
	Tertiary	5	Early humans emerge (genus *Homo*)		
		23	Grasses replace forests in drier areas		
		35	Rise of several modern mammals		
		56	Earliest whale fossils; first horses		
		65	First primate fossils Extinction of dinosaurs		Cretaceous Extinction
Mesozoic	Cretaceous	146	Angiosperms replace gymnosperms in many habitats		
	Jurassic	200	First bird fossil (*Archaeopteryx*) First flowering plants (Angiosperms)		Triassic Extinction
	Triassic	251	First dinosaurs and mammals		Permian Extinction
Paleozoic	Permian	299	First mammal-like reptiles		
	Carboniferous	359	First reptiles First gymnosperms		Devonian Extinction
	Devonian	416	First seed-bearing plants Fish-to-tetrapod transition		
	Silurian	444	Early jawed fishes First fossils of land animals		Ordovician Extinction
	Ordovician	488	First plant fossils First fungus fossils		
	Cambrian	542	First fossil of chordate (ancestors of vertebrates) Cambrian Explosion		
Precambrian		635	First animal fossils		
		1,200	First multicellular life		
		2,000	First eukaryotic fossils		
		2,400	First substantial oxygen accumulation in atmosphere		
		3,400	Evidence of abundant bacterial and archaeal life		
		3,800	Possible evidence of earliest life on Earth		
		4,600	Earth is formed		

Figure 19.3
Earth's Geologic Timescale

Figure 19.4
Revealing Layers
The history of life can be traced through fossilized life-forms found in layers of sediment, seen here in America's Grand Canyon.

that all of evolution amounts to a march toward the development of human beings, who then get to occupy the highest branch of an evolutionary tree.

In reality, we occupy one ordinary branch among a multitude of branches. Under notable events in the Cenozoic era, Figure 19.3 could just as easily have listed milestones in the evolution of fish or birds or the emergence of social insects such as bees. These creatures have continued evolving in the modern era, along with our own species, but in lines separate from us. Focusing on a line that leads to humans is like focusing on a railroad route that leads from, say, Baltimore to Denver, while ignoring the multitude of lines that are separate from it. We regard the Baltimore-Denver route as special because we are *interested* in it, but that does not mean that it is fundamentally *different* from any other line. Some special qualities have evolved in human beings, but the same could be said for almost any species. In phylogenetic terms, we simply lie at the tip of one evolutionary branch, while bees lie at another, and cactus plants lie at still another.

19.2 How Did Life Begin?

Taking a historical approach to tracing life on Earth, the first question that arises is one of the toughest: How did life begin? Darwin himself thought about this and imagined that life began in what he called a "warm little pond" in the early Earth. Well, maybe; but the *very* early Earth had no warm little ponds. What it had was an environment more akin to hell on Earth.

You have seen that Earth was formed about 4.6 billion years ago. This took place in a process of "accretion," in which ever-larger particles clumped together: cosmic dust to gravel, gravel to larger balls, larger balls

to objects the size of tiny planets. One consequence of such development was that the accretion now called Earth was periodically being slammed into by *other* large accretions—meteorites and comets and protoplanets. We have evidence that one of these objects was as large as Mars and that the result of its impact on the Earth, 100 million years after Earth formed, was the creation of our moon. There were no oceans at this time, and Earth was so hot—up to 1,800 degrees Fahrenheit—that it was covered with a layer of molten rock.

At one time, scientists believed that Earth remained in this state until about 3.8 billion years ago, but in recent years, tiny samples of a mineral called zircon found embedded in the rocks of Western Australia have told a different story. Uranium found within the zircon samples has allowed us to date them to 4.2 billion years ago, and the relative proportion of oxygen isotopes in the samples indicates that they were formed in *water*—something that would have been possible only if Earth were a relatively cool place 4.2 billion years ago. This view of a cooler early Earth is nicely consistent with the first evidence we have of life, which remember dates to 3.8 billion years ago. Some 400 million years seem to have elapsed between the time Earth became a planet that was cool enough to be hospitable to life and the time that life actually appeared—an expanse of time long enough for the kind of chemical reactions that got life going to be carried out many times.

From the Simple to the Complex

But what were these chemical reactions? No one can say for certain, but we can be sure that a kind of chemical bootstrapping had to be involved. Simple, available chemical compounds came together in reactions that facilitated the production of molecules that were more complex. These then came together to bring about the production of molecules that were more complex yet. At some point, a group of these molecules became capable of *self-replication:* They could make copies of themselves in ways we'll be looking at. With this step, life began.

But how did this chain of events get started? Are we sure it's possible to go from the simple compounds available on the early Earth to life's more complex molecules? You saw in Chapter 3 that life's informational molecules (DNA and RNA) are composed of certain kinds of building blocks (nucleotides), while its proteins are composed of other kinds of building blocks (amino acids). How difficult would it have been for these sorts of molecules to have been produced from the starting chemical ingredients available on the early Earth?

In 1953, a young graduate student named Stanley Miller was asking himself this very question. In seeking to answer it, he performed what is undoubtedly the most famous experiment in origin-of-life studies. In essence, he tried to re-create the conditions of the young Earth in the set of glass flasks and tubes you can see in **Figure 19.5**. Working with his mentor, Harold Urey, Miller put together an "atmosphere" inside the glassware that he thought would approximate the atmosphere of the early Earth—water vapor combined with the gases methane, ammonia, and hydrogen. He then ran an electric current through the system, thus simulating the lightning he believed would have been flashing through Earth's ancient skies. After a couple of days, he got a brown goo in the apparatus that, on analysis, turned out to contain . . . amino acids! Some of the prime building blocks of life had been created out of simple compounds in just a few days. This result had an enormous impact on origin-of-life research for it meant that it was *not* difficult to go from some very simple substances to one variety of life's building blocks under the conditions believed to exist on the young Earth. With this experiment, origin-of-life studies were turned into an experimental science.

Figure 19.5
Origins-of-Life Researcher
In 1994, Stanley Miller posed with part of the laboratory apparatus he had used 41 years earlier in attempting to re-create the atmosphere that surrounded the young Earth. Miller ran an electric current through the gases in the apparatus and discovered that, in only a few days time, the system generated amino acids, some of the key building blocks of life.

Since the time of Miller's work, generations of researchers have essentially been asking the same kinds of questions he did: What was Earth's atmosphere like at the time life got going? What was the climate like? What compounds might have been made available through the action of ocean waves or volcanic activity? The idea is to test plausible scenarios for Earth's early conditions against the likelihood of life being produced in those conditions.

Modern Origin-of-Life Theories

In recent years, this research has produced two fundamentally different conceptual frameworks for how life may have gotten started. One of these is referred to as the "replicator-first" model, while the other is called the "metabolism-first" model.

Replicator-first takes as its starting point the fact that reproduction is a central feature of life and that reproduction today is essentially a matter of the "replication" or copying of a single molecule—the DNA that became so familiar to readers of the genetics unit. By means of base pairing, a length of DNA can specify the structure of a virtually identical "daughter" DNA molecule. Replicator-first adherents believe that the first component of life was likely to have been a primitive replicator molecule, similar in function to today's DNA (though with a crucial added ability we'll be looking at).

To this, metabolism-first advocates reply that any imaginable replicator molecule would necessarily have been complex—too complex, in their view, to come into being as a *first* element of life. A more likely scenario, they believe, is that a metabolism, or set of self-sustaining chemical reactions, got going inside a protective compartment through the interactions of a set of small, uncomplicated molecules.

If we ask how the validity of either of these models could be demonstrated, the answer, oddly enough, is through experiments carried out in modern chemistry laboratories. Let's look now at what contemporary research has revealed about the plausibility of replicator-first as opposed to metabolism-first.

Replicator-First: The RNA World

In modern organisms, DNA contains information not only for its own replication but also for the production of the molecular workers known as proteins. In this latter role, we could think of an organism's full complement of DNA as a kind of cookbook containing recipes (individual DNA sequences) that lead to products (proteins) that undertake a wide variety of tasks. Yet, while DNA is the cookbook, it is not the cook. When it comes time for DNA to be copied, for example, the double helix cannot unwind itself, nor pair its bases up, nor do anything else. Instead, all these tasks are carried out by the class of proteins called enzymes. Hence, there is a chicken-and-egg dilemma: Enzymes

RNA-Like Self Replication

ESSAY

Physical Forces and Evolution

Climate Change

The living world is very much affected by the physical forces operating on the Earth. You can see this in connection with the mass extinctions discussed in the main text. What has caused these extinctions? Evidence links two of those listed in Figure 19.3 to episodes of global climate change, specifically global cooling. (The two are the Ordovician and Devonian Extinctions.) All kinds of things happen with global cooling. Glaciers develop, wiping out everything in their path and tying up huge amounts of water in ice. This reduces sea levels, thereby killing many organisms that live on the continental shelves. The oceans themselves get colder, particularly at greater depths, which is its own kind of killing event.

Against this, the biggest extinction event of them all, the Permian Extinction of 251 million years ago (Mya), seems likely to have been caused by global *warming*, though this warming appears to have been triggered by another kind of physical force: a set of massive volcanic eruptions that spewed, into the atmosphere, huge quantities of the gases carbon dioxide and methane (the same gases that are causing global warming on the planet today). This series of events lowered the levels of oxygen in the Earth's seas and, with this, the stage was set for the direct cause of the Permian Extinction: Bacteria that produced the deadly gas hydrogen sulfide as part of their normal metabolism multiplied and rose toward the surface of the ancient oceans in great numbers, producing hydrogen sulfide that now went into Earth's atmosphere as well as its oceans. With this, both atmosphere and oceans essentially were turned into poisonous gas chambers and the result was the elimination of up to 96 percent of Earth's species.

Extraterrestrial Objects and Catastrophism

Volcanic eruptions may also have played a part in the famous Cretaceous Extinction

that brought about the end of the dinosaurs. We know that massive eruptions occurred in what is now India from 67 to 65 Mya, and the effect of these may have been a poisoning of Earth's atmosphere and a blockage of sunlight, which would have decreased production of the food that plants put together in photosynthesis. On the other hand, the Cretaceous Extinction appears to have been very abrupt relative to other extinction events, in that, to judge by the fossil record, organisms both large and small simply disappeared from the face of the Earth near the precise date of 65 Mya. Moreover, we know that a cataclysmic event took place right at this date: An enormous asteroid, 6.5 miles in diameter, struck the Earth, crashing into the coastal waters off Mexico's Yucatan Peninsula. If there is a consensus among contemporary researchers, it's that this event is largely responsible for the Cretaceous Extinction. The asteroid's impact would have produced tsunamis 130 feet tall, pushed massive clouds of sunlight-blocking dust into the atmosphere, and could have unleashed giant firestorms that would have blocked sunlight even further. From an evolutionary point of view, a sequence of events such as this is an example of what scientists call catastrophism: a type of environmental disaster, unrelated to Earth's normal physical processes, that alters the course of evolution.

Continental Drift

As noted in Chapter 18, the primary engine driving speciation is the geographic separation of populations. Geographic separation was described there as taking place in scales that were both small (a stream changing course) and large (a glacier's movement). It turns out that geographic separation can take place on a much larger scale yet, in that whole *continents* have come together and separated again during Earth's history.

If you look at **Figure 1**, you can see the landmasses of the world as they have ex-

isted at four separate times: 250 Mya, 150 Mya, 70 Mya, and the present. In the first of these, nearly all the Earth's continental mass was part of a single supercontinent, Pangaea, which stretched from the South to the North Pole. By the mid-Mesozoic era, Pangaea was separating along the lines you see in the figure. This in turn gave way to our current division, which like its predecessors, is a configuration in flux.

These maps are only four "snapshots" in a process of **continental drift,** or the movement of continents in geological time, that has produced lots of other alignments. How does such movement come about? In essence, the Earth's crust and part of its upper mantle are divided into a series of plates that move over the globe. The movement of some of these plates can be thought of in terms of a conveyor belt. Halfway between North America and Europe there is a long underwater mountain chain, the Mid-Atlantic Ridge. This ridge has in its middle a deep valley that forms the border between the North American and Eurasian Plates. Hot material in the Earth's interior is constantly being moved up into the valley from either plate, after which, as new crust, it moves eastward on one side of the ridge but westward on the other. What we have, then, is a spreading of the ocean floor in opposite directions. *Continents* sit atop this spreading material, and they are being moved in opposite directions— Europe to the east, North America to the west—as if they were objects on conveyor belts (see **Figure 2**).

The Impact of Continental Drift on Evolution

Continental drift has had profound effects on evolution. If we ask, for example, what causes global cooling, one factor is the development of glaciers, which as it turns out can grow only when continents are located at or near the Earth's poles. This *was* the case 300 Mya, then it was *not* the case 200 Mya, and now it is again. By moving landmasses on and off the poles, continental drift has helped change the climate, and climate greatly affects Earth's living things.

Continental drift has had a second profound effect on evolution in that it

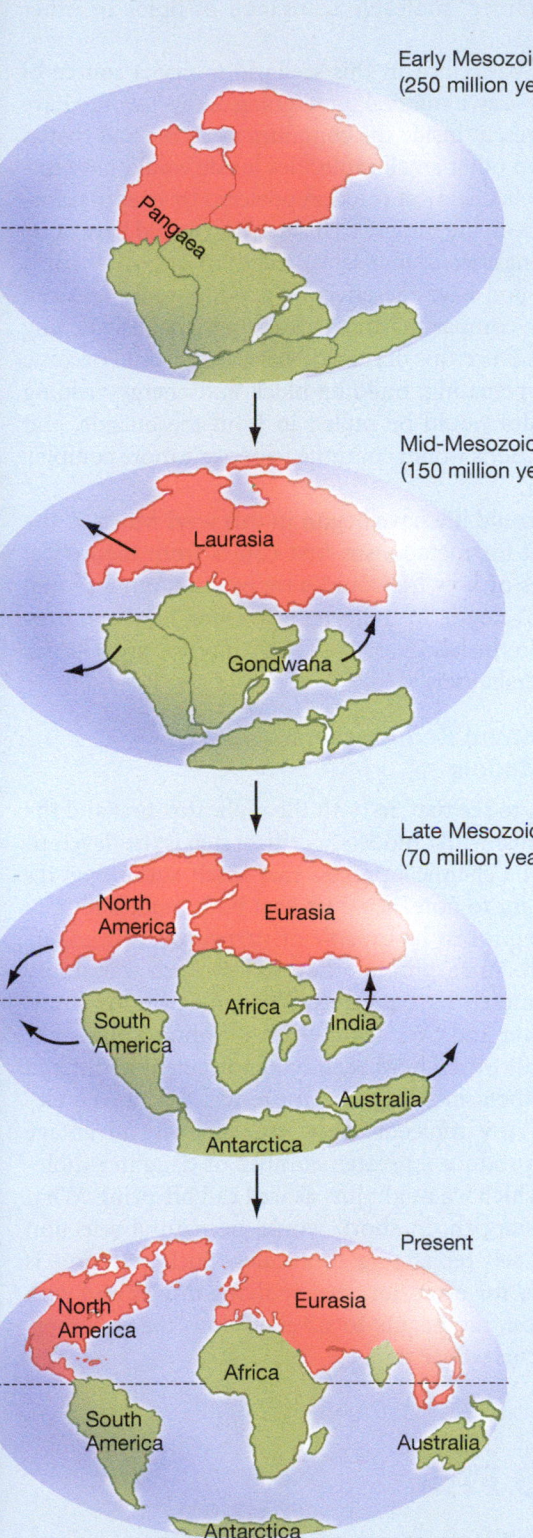

Early Mesozoic era
(250 million years ago)

Mid-Mesozoic era
(150 million years ago)

Late Mesozoic era
(70 million years ago)

Present

Figure 1
Continental Drift
During Earth's history, the tectonic plates that make up the surface layer of Earth have moved with respect to one another, causing the continents to spread apart and move together.

(a) Mesozoic era Pangaea (250 million years ago)

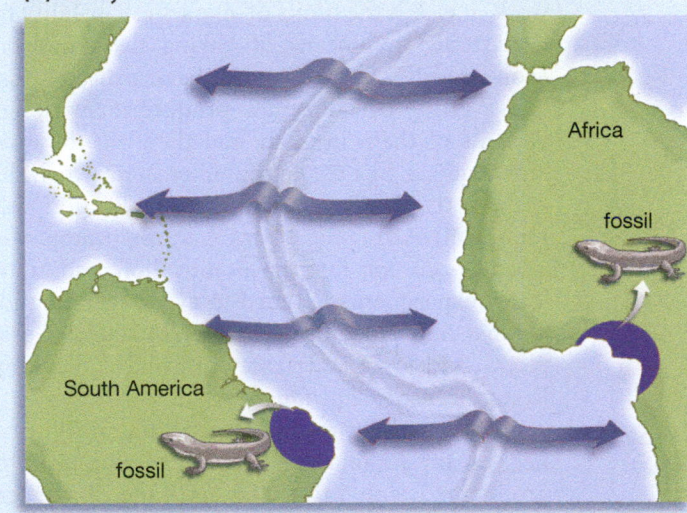

(b) Today

Figure 2
Similar fossils are found on the east coast of South America and the west coast of Africa because:
(a) 250 million years ago, the two continents were contiguous and formed part of the giant continent Pangaea.
(b) Animals from that period are now found as fossils on opposite sides of the ocean because the ocean floor has been spreading westward and eastward.

has brought about the separation and mixing of living *populations*. For example, the breakup of the supercontinent Pangaea eventually meant the creation of island continents such as Australia and Antarctica. Until the coming of humans, Australia had scarcely any placental mammals—mammals with embryos that are nourished by a placenta (see Chapter 23). Since prehistoric times, however, Australia has had an abundance of marsupial (or "pouched") mammals, such as kangaroos and koalas. We can read evolutionary history in this distribution. As traditional theory has it, marsupials developed in Pangaea before its breakup, in what is now North America and western Europe, and spread to Australia across land bridges to what is now Antarctica. (See the first and second drawings, Figure 1.) Meanwhile, placental mammals were spreading, too, but few were able to reach the Australia-Antarctic landmass before it broke away from Pangaea. In effect, placental mammals missed the Antarctic-Australian boat.

can't be produced without the information contained in DNA, but DNA can't copy itself without the activity of enzymes. So, how did life get going if neither thing can function without the other?

In the late 1960s, researchers hit on an idea that has come to have a good deal of support: In early life-forms, a single molecule performed both the DNA and enzyme roles. This molecule was RNA (ribonucleic acid), which was portrayed back in Chapter 14 as a kind of genetic middleman, ferrying DNA's information to the sites where proteins are put together. A critical piece of evidence about RNA's role in early life came in 1983, when researchers discovered the existence of enzymes that are composed of RNA instead of protein: **ribozymes.** These molecules can encode information *and* act as enzymes (by, for example, facilitating bonds that bind RNA units together).

This evidence has led researchers to the presumed existence of an "RNA world"—an early living world that consisted solely of self-replicating RNA molecules. In recent years, the existence of such a world has been bolstered by a great deal of experimental work in which the limits of ribozyme capability have been tested by means of having these molecules evolve inside a test tube. The basic procedure is to take a group of ribozymes, produce many designer variants of them in a lab, and then put these variants together in a single container, after which those variants that best perform a given replication task—joining RNA fragments together, for example—are selected out for further modification and testing. These experiments have led to ever-expanding ideas of what ribozymes are capable of. In 2009, for example, Gerald Joyce and Tracey Lincoln of the Scripps Research Institute produced a pair of ribozymes that could continually replicate each other without the help of any proteins. This is not the same thing as producing life in a test tube, since the ribozymes in question originated as RNA fragments that were designed by scientists. But it does show that, by itself, RNA is capable of carrying out some fundamental replication processes.

Even if we assume, however, that RNA is versatile enough to have been life's main component at some point, replicator-first advocates still have another daunting question to answer: How did RNA come into being in the first place? In the laboratory, researchers have now mapped out an almost complete chemical route by which RNA could have been put together from simpler building blocks, but this kind of evidence leaves metabolism-first advocates unimpressed.

Metabolism-First

Metabolism-first researchers argue that any replicating molecule with an RNA-like structure would have been so large and complex that the chances of it ever *spontaneously* forming are vanishingly small. A more promising route to life, they argue, runs through small molecules that could have come together inside a "container" molecule composed of lipids or other materials.

The basic idea in this scenario is that a source of energy—say, a mineral undergoing decay—would have been present inside the container molecule and that at least two other small molecules inside would have undergone a chemical reaction powered by energy transferred from the mineral. This reaction would have done two things: result in a set of reactions of greater complexity and have the effect of drawing more material into the compartment than was spontaneously exiting from it. Once this occurred, the system would become self-perpetuating; building-block and energy-yielding molecules would be pulled in from the outside, and the reactions would continue to become more complex in form.

So, could life have begun in this way? Much of the support for metabolism-first comes from theoretical analyses of how life could have evolved, but a partial metabolic cycle of the type just noted has been produced in the laboratory of German researcher Günter Wächtershäuser.

A Common Requirement of the Two Models

It's easy to see that, in both the replicator-first and the metabolism-first models, a critical step is the development of a chemical process that is self-sustaining. It's important to note, however, that both models require that the process be capable of *varying* over time. Any replicator-first process that lacked such variability would merely have produced the same primitive molecule over and over again. With variation, conversely, the result would have been one molecule that differed from others in being, say, more resistant to the elements. Any molecule that had such an advantage would produce a greater number of daughter molecules, which we might just as well call offspring. What would happen, in short, would be natural selection among self-replicating molecules. Such selection is the basis for evolution among living things, and with evolution, life begins to diversify into the innumerable forms we see today.

SO FAR . . .

1. Each of the eras or periods within the geological timescale is characterized by a particular grouping of _____ preserved today as _____.

2. True or false: The first animals and plants evolved early in life's history.

3. Life is generally regarded as having begun with the development of _____ molecules or chemical systems.

19.3 The Tree of Life

Once life was established, how did it evolve? If you look at **Figure 19.6**, you can see the biggest picture of all. In the upper part of the drawing, you can see life as it currently exists on Earth, divided up into its major categories. As noted last chapter, the broadest of these categories are the three "domains" of life—**Bacteria, Archaea,** and **Eukarya.** Then there are four "kingdoms" within Eukarya: Protista, Plantae, Animalia, and Fungi. The figure gives you some idea of the kinds of organisms that fall into each of these categories. Domains Bacteria and Archaea are made up strictly of single-celled microbes, while Domain Eukarya has within it all the organisms that are most familiar to us—plants,

animals, and fungi—along with a group that is less familiar, the protists, which are mostly microscopic and water-dwelling.

If you then look at the lower part of the drawing, you can see how these groups evolved. At the bottom you see a "universal ancestor," which is the organism—or perhaps group of organisms—that gave rise to all current life. The evolutionary line that leads from the universal ancestor then branches out to yield, on the one hand, Domain Bacteria and, on the other, a line that leads to both Domain Archaea and Domain Eukarya. The archael microbes were long thought of as simply a different kind of bacteria, but we now know that these organisms are distinct from bacteria in fundamental ways. Nevertheless, there are some important similarities. All bacteria and archaea are essentially single-celled, and no archael or bacterial cell has a nucleus. This latter fact makes the archaea and bacteria **prokaryotes.** Conversely, all the organisms in Domain Eukarya have nucleated cells and are thus **eukaryotes,** which do come in single-celled varieties, but which can be organisms that are made up of trillions of cells.

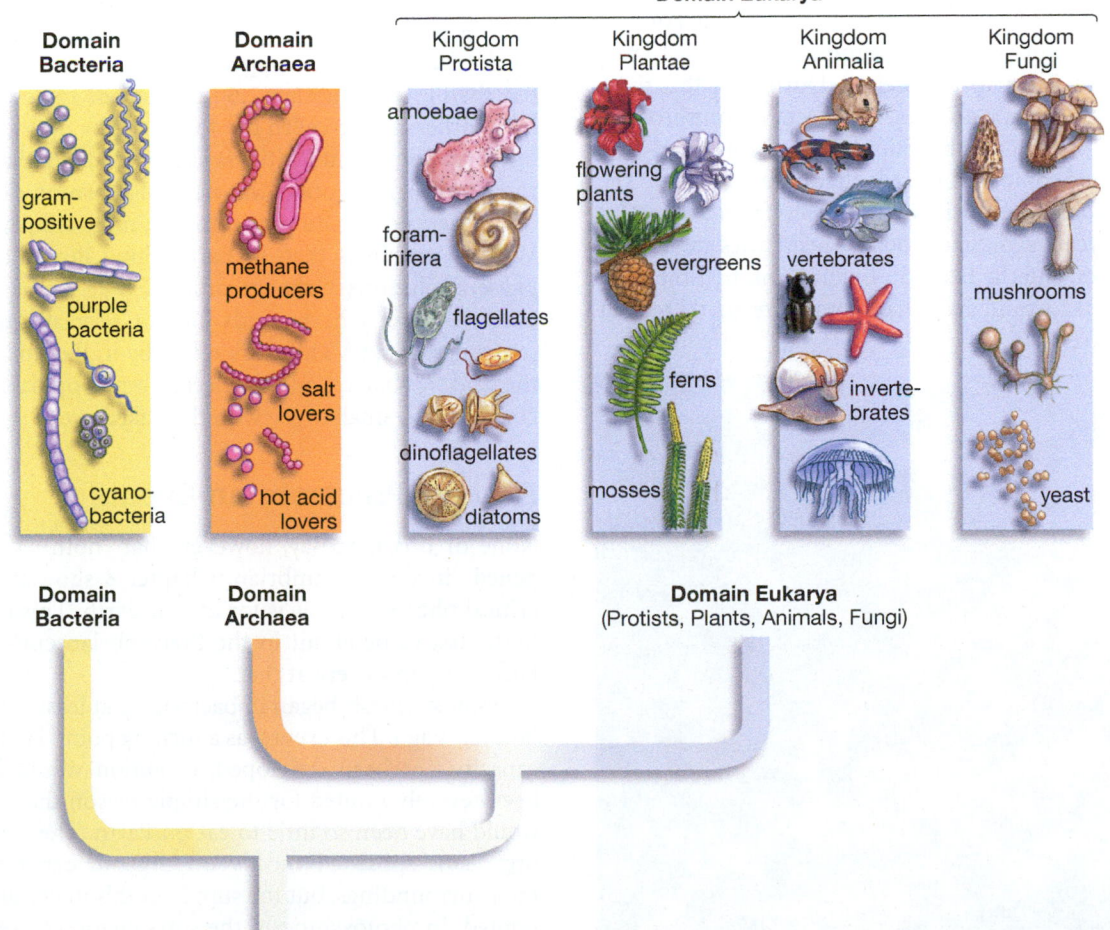

Domain Eukarya

Domain Bacteria — gram-positive, purple bacteria, cyano-bacteria

Domain Archaea — methane producers, salt lovers, hot acid lovers

Kingdom Protista — amoebae, foram-inifera, flagellates, dinoflagellates, diatoms

Kingdom Plantae — flowering plants, evergreens, ferns, mosses

Kingdom Animalia — vertebrates, inverte-brates

Kingdom Fungi — mushrooms, yeast

Domain Bacteria

Domain Archaea

Domain Eukarya
(Protists, Plants, Animals, Fungi)

Universal ancestor

Figure 19.6
Earth's Organisms and How They Evolved

Scientists divide Earth's living organisms into three major categories: the Domains Bacteria, Archaea, and Eukarya. There are then four kingdoms within Domain Eukarya: Protista, Plantae, Animalia, and Fungi, which represent the plants, animals, and fungi of the world, along with a group of mostly microscopic, water-dwelling organisms, the protists. The lower drawing shows the evolutionary relationships among life's three domains.

19.4 A Long First Era: The Precambrian

With this picture in mind, let's start tracing life from its cellular beginnings in the first of Earth's geologic time spans, the Precambrian era. Recall that the Earth was formed about 4.6 billion years ago. The earliest evidence we have of life is not a fossil, but instead a kind of chemical signature of life left in the barren, frozen Isua rocks of Greenland (**Figure 19.7**). Carbon trapped in these 3.8-billion-year-old rocks is of a type produced only by living things. The evidence here is sparse, but most authorities think we can date life from this point. Shift forward another 400 million years, and we don't have to wonder whether life existed; we have abundant evidence of it, much of it found in the rock deposits of Western Australia. These signs of life are not fossils, but rather evidence of living things having altered the rock deposits we see today. If you look at **Figure 19.8**, you can see a 3.4-billion-year-old rock deposit from Australia containing structures, called stromatolites, whose internal patterns are so complex it's likely they could only have been created by living organisms.

What were these organisms? All of them were microbes. To put a finer point on it, all of them were microbes that were either bacteria or archaea. For perhaps 2 billion years, these simple, single-celled creatures had Earth to themselves. During all this time, there were no fish, no trees, no insects, no mushrooms. Indeed, there were no eukaryotes of any kind—no organisms composed of cells that have a nucleus. The oldest eukaryote fossils we have date from a mere 1.8 billion years ago. Since today's eukaryotes *do* include fish, trees, and the like, you might think that the early eukaryotes would have had some size to them, but

Figure 19.8
Traces of Ancient Organisms

A geologic structure called a stromatolite in the Strelley Pool Chert of Western Australia shows evidence of microbes that flourished in a shallow-water ocean environment there some 3.4 billion years ago. Note the inverted V-shaped structures in the formation (one of which has been bracketed). Scientists believe that such features could only have been produced by living organisms.

they were single-celled, too—and they stayed that way for a long time. The important feature of multicellularity didn't evolve in eukaryotes until about 1.2 billion years ago.

Taking a step back from these various dates, note the big picture in early evolution. It took about 800 million years for life to appear after the Earth formed, but then it took another 2.6 billion years to get from life's earliest single-celled life-forms to any multicelled life-forms. Some authorities say this is evidence that the real hurdle in evolution was not the initiation of life, but rather the initiation of complex, multicellular life. What is certain is that, in Earth's long Precambrian phase, evolutionary change proceeded at such a slow pace we scarcely have words for it. ("Glacial" greatly overstates things.) The years rolled by in the millions, the land was barren, and the oceans were populated by creatures too small to see with the naked eye.

Notable Precambrian Events

None of this is to say, however, that "nothing happened" in the Precambrian. Chapter 8 showed how critical photosynthesis is for life on Earth. This capability first came about in the Precambrian era—and fairly early in the era at that.

Photosynthesis began in bacteria by at least 3.4 billion years ago. This event was a turning point. Had this capacity not been developed, evolution would have been severely limited for the simple reason that there would have been so little to eat on Earth. The earliest organisms subsisted mostly on organic material in their surroundings, but the supply of this material was limited. In photosynthesis, the sun's energy is used to *produce* organic material—in our own era, the leaves and grasses and grains on which all animal life depends.

Figure 19.7
Home to the First Signs of Life

The rocks at Isua, Greenland contain possible chemical signatures of life that have been dated to nearly 3.8 billion years ago.

Through photosynthesis, a massive quantity of energy-rich food was made available.

One particular kind of photosynthesis, again beginning in the Precambrian, had a second dramatic impact on evolution. A single type of bacteria, the cyanobacteria, were the first organisms to produce *oxygen* as a by-product of photosynthesis. For us, the word *oxygen* seems nearly synonymous with the word *life,* but until 2.4 billion years ago, very little of it existed in the atmosphere. When it did arrive, through the work of the cyanobacteria, it had the effect of greatly increasing the capacity of Earth to support a diverse group of organisms because it provided a huge infusion of *energy* into the living world. As Chapter 7 made clear, organisms that obtain their energy through oxygen-aided or "aerobic" means can extract a great deal more energy from a given quantity of food than can organisms that lack this aerobic capacity. Just as cars need oxygen to burn gasoline efficiently, so living things need oxygen to burn food efficiently. As such, the appearance of atmospheric oxygen set the stage for the emergence of life-forms larger and more complex than the early microbes. Even so, the early eukaryotes that benefited from this oxygenic change only managed to do so through a means that sounds rather unpromising: They were invaded.

Those who read Chapter 7 will recall that most of the energy that eukaryotes get from food is extracted within special cellular structures called mitochondria. Although mitochondria are tiny organelles within eukaryotic cells, they have characteristics of free-living *cells,* specifically free-living bacterial cells. As it turns out, that is what they once were. Ancient bacteria that could metabolize oxygen took up residence in early eukaryotic cells and eventually struck up a mutually beneficial or "symbiotic" relationship with them. The bacteria benefited from the eukaryotic cellular machinery, and the eukaryotes got to survive in an oxygenated world (**Figure 19.9**). A later bacterial invasion (by cyanobacteria) allowed one group of eukaryotes, the algae, to start carrying out photosynthesis. The tiny oblong organelles called chloroplasts that are the sites of photosynthesis in plant and algal cells today are the descendants of this second set of bacterial invaders.

The Malnourished Earth

Surviving in an oxygenated world is not the same thing as thriving in it and for a long time the early eukaryotes didn't seem to manage much more than survival: For the first billion years after they appeared, the vast majority of them remained single-celled and fairly simple. Eukaryote evolution was not quite dead in the water, but it was uneventful. Given the astounding variety of forms that eukaryotes eventually assumed, we might wonder why they experienced such a long holding pattern early in their evolution. A fascinating idea put forward is that they essentially were starving.

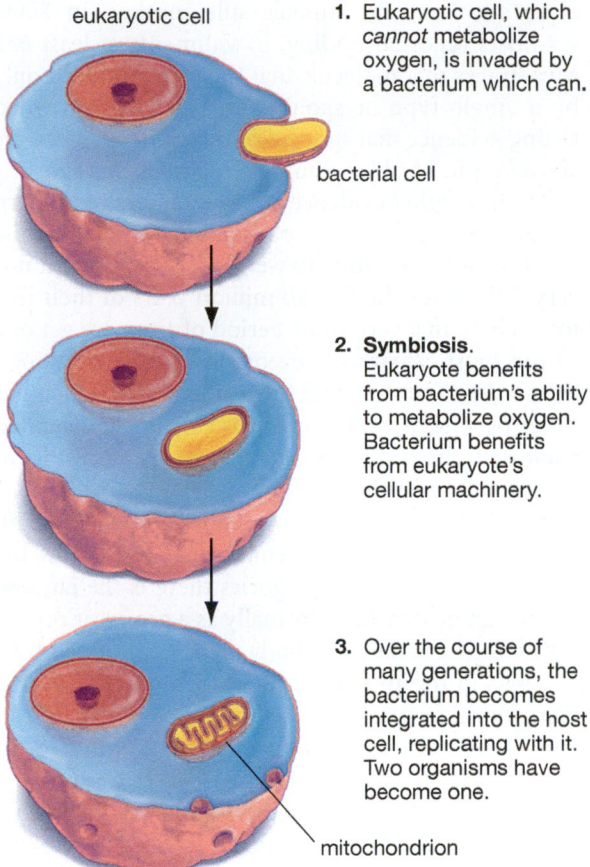

eukaryotic cell

bacterial cell

1. Eukaryotic cell, which *cannot* metabolize oxygen, is invaded by a bacterium which can.

2. **Symbiosis.** Eukaryote benefits from bacterium's ability to metabolize oxygen. Bacterium benefits from eukaryote's cellular machinery.

3. Over the course of many generations, the bacterium becomes integrated into the host cell, replicating with it. Two organisms have become one.

mitochondrion

Figure 19.9
Adapting to Oxygen
The "powerhouses" of our cells—the organelles called mitochondria—originated as free-standing bacteria that invaded ancient eukaryotic cells. Over time, the two organisms developed a symbiotic relationship: The bacterium utilized the eukaryote's cellular machinery while the eukaryote utilized the bacterium's ability to metabolize oxygen. Over the course of many generations, the bacterium made a transition from separate organism to organelle within the larger eukaryotic cell.

According to what's called the Malnourished Earth hypothesis, the oxygen that started building up in Earth's atmosphere 2.4 billion years ago had the effect of leaching sulfur off the land and dumping it into the ancient oceans. This was, in essence, a global toxic spill that effectively removed elements from the oceans that the photosynthesizing eukaryotes—the algae—needed in order to take up nutrients. The end result was a group of organisms so malnourished that they could not diversify very far through the kinds of evolutionary processes we looked at last chapter. The first notable increase in eukaryote diversity did not arrive until about 635 million years ago (635 Mya), and when it did come, it was not algae that were at the forefront.

19.5 The Cambrian Explosion

As the Precambrian was coming to a close, life consisted of bacteria, archaea, and the eukaryote protists, almost all of which lived in the oceans. By 635 Mya, however, we get the first signs of a new form of life in the sea: animals. The evidence in this case is not a set of fossils per se, but instead a set of "biomarker" fossils, meaning a set of fatty molecules that were produced by a living thing. Such molecules bear the chemical imprint of the type of organism that generated them: Bacteria produce one type of molecule, algae another, and animals still another. In 2009, scientists reported finding, in sediments at least 635 Mya, traces of a molecule that today is produced only by a single type of sponge. This imprint provided strong evidence that sponges of this same variety were abundant in Earth's oceans 635 million years ago.

We have other evidence of the ancestors of modern animals dating from this period, but to judge by the fossil record, such animals were neither abundant nor very diverse for the first 90 million years of their history. Then, in a very short period of time, we get one of the most remarkable events in all of evolution: a tidal wave of new animal forms the like of which has never been seen before or since. This is the **Cambrian Explosion**: the sudden appearance of the ancestors of most modern animals beginning 542 Mya.

Recalling the taxonomic system you went over in Chapter 18, you may remember that just below the domain and kingdom categories there is the **phylum,** which can be defined informally as a group of organisms that share the same body plan. There are at least 36 phyla in the animal kingdom today, some of these being Porifera (sponges), Arthropoda (insects, crabs), and Chordata (ourselves, salamanders). With two exceptions, the fossil record indicates, every one of these phyla came into being in the Cambrian Explosion. This seeming riot of sea-floor evolution was actually more extensive than even this implies in that a good number of phyla that *don't* exist today—phyla that have become extinct—also seem to have first appeared in the Cambrian. Some of these creatures are so bizarre it's surprising that Hollywood hasn't mined the Cambrian archives for new ideas on monsters (**Figure 19.10**). As if all this weren't enough, the Cambrian Explosion appears to have taken place in a very short period of time relative to evolution's normal pace—a few tens of millions of years at most.

But did it? The Cambrian Explosion is so extreme that one of the primary challenges to evolutionary biology has been to explain why such a thing came about—or why it came about only once. Note the series of events: only a few modern animal forms before the Cambrian, then the explosion, then almost no fundamentally new animal forms since. One possibility is that, as with sponges, other modern animal phyla actually evolved prior to the Cambrian but that the individuals in these phyla were so small they failed to leave traces of themselves. Consistent with this idea, most molecular clock studies indicate that modern animal phyla emerged much earlier than the supposed explosion—perhaps a billion years ago. Nevertheless, most experts believe that the Cambrian Explosion was a real event; they believe that the diversity and size of animals did increase tremendously in a relatively brief time span during the Cambrian.

We might ask, however: Why did the starter gun for this event go off when it did—at 542 Mya, rather than earlier or later? A factor that we've talked about before may have played a part. Recall that significant quantities of oxygen first appeared in Earth's atmosphere about 2.4 billion years ago, thanks to the photosynthesis carried out by bacteria. It turns out, however, that even with this increase, atmospheric oxygen levels remained relatively low until about 600 Mya when, for reasons unknown to us, they climbed to something approaching current levels. This second oxygen increase occurred near the point at which we get the first animals, which is to say near the point at which we get the first large organisms of any sort. Large organisms need large amounts of energy, and only oxygen-driven or "aerobic" respiration is capable of delivering this quantity of energy. It may be that the oxygen that had done so much to change life early on ended up altering it again as the long Precambrian era came to a close.

Figure 19.10
Life on the Ancient Sea Floor

The Cambrian Explosion resulted in a multitude of new animal forms. In this artist's interpretation of the fossil evidence, a number of now-extinct Cambrian animals are shown in their ancient sea-floor habitat. Two of these are Anomalocaris, raising up with the hooked claws, and Hallucigenia, on the seafloor with the spikes extending from it.

SO FAR . . .

1. Life has three domains, and one of these domains has four kingdoms within it. Name both the domains and the kingdoms.

2. Life is thought to have begun as early as _____ years ago. One important event that took place in the long, early time span called the Precambrian was the appearance of ____ in Earth's atmosphere, all of it produced as a by-product of photosynthesis carried out by a type of _____.

3. The Cambrian Explosion was an explosion in the number of _____ life-forms.

19.6 Plants Move onto Land

The teeming seas of the early Cambrian stand in sharp contrast to what existed on land at the time, which was no life at all except for some hardy bacteria. Earth was simply barren—no greenery, no birds, no insects. When multicelled life did come to the land, the first intrepid travelers were plant-fungi combinations. (Even today, about 80 percent of plants have a symbiotic relationship with fungi, the plants providing fungi with food through photosynthesis, the fungi providing plants with water and mineral absorption through their filament extensions.) Exactly when this transition onto land took place is a matter of debate. The earliest undisputed fossil evidence we have of plants and fungi on land dates from 460 Mya, meaning these organisms had become abundant by that time. But when did they first appear? Our best estimates are sometime before 500 Mya. The general transition here began with algae in the ocean, then continued with freshwater algae, then freshwater algae that came to exist in *shallow* water, living partially above the waterline. It was one variety of this algae, the green algae, that gave rise to all of Earth's plants. When these plants made the transition to full-time living on land, it was to damp environments.

Adaptations of Plants to the Land

Such a change could not have come about without a lot of adaptation. Aquatic algae did not have to deal with water loss or the crushing effects of gravity, but land plants did. Not surprisingly, one of the adaptations that evolved in plants was an outer covering, called a cuticle, that serves to retain moisture. (When you polish an apple by rubbing it, what you're actually polishing is this waxy, protective material.) Meanwhile, the force of gravity meant that the initial land plants were almost all *short* plants, as they lacked any means of transporting water up through themselves, against the force of gravity. Some of the most primitive land plants are mosses that often hug the ground like so much green carpet (**Figure 19.11**).

Then there were the changes that life on land brought to reproduction. When green algae reproduce sexually, their gametes (eggs and sperm) float off from them as individual cells, after which they are brought together by ocean currents or the cells' own movement. Once this happens, the resulting embryos develop completely on their own. An embryo left to mature by itself on *land,* however, would dry out and perish. Plant evolution, however, produced the adaptive feature of an internal housing for embryos—in plants, embryos develop *within* a parent.

Another Plant Innovation: A Vascular System

If you look at **Figure 19.12**, you can see how the major divisions of plants evolved. The most primitive

Figure 19.11
Lying Low
Moss covering rocks in the Columbia River Gorge's Fairy Creek.

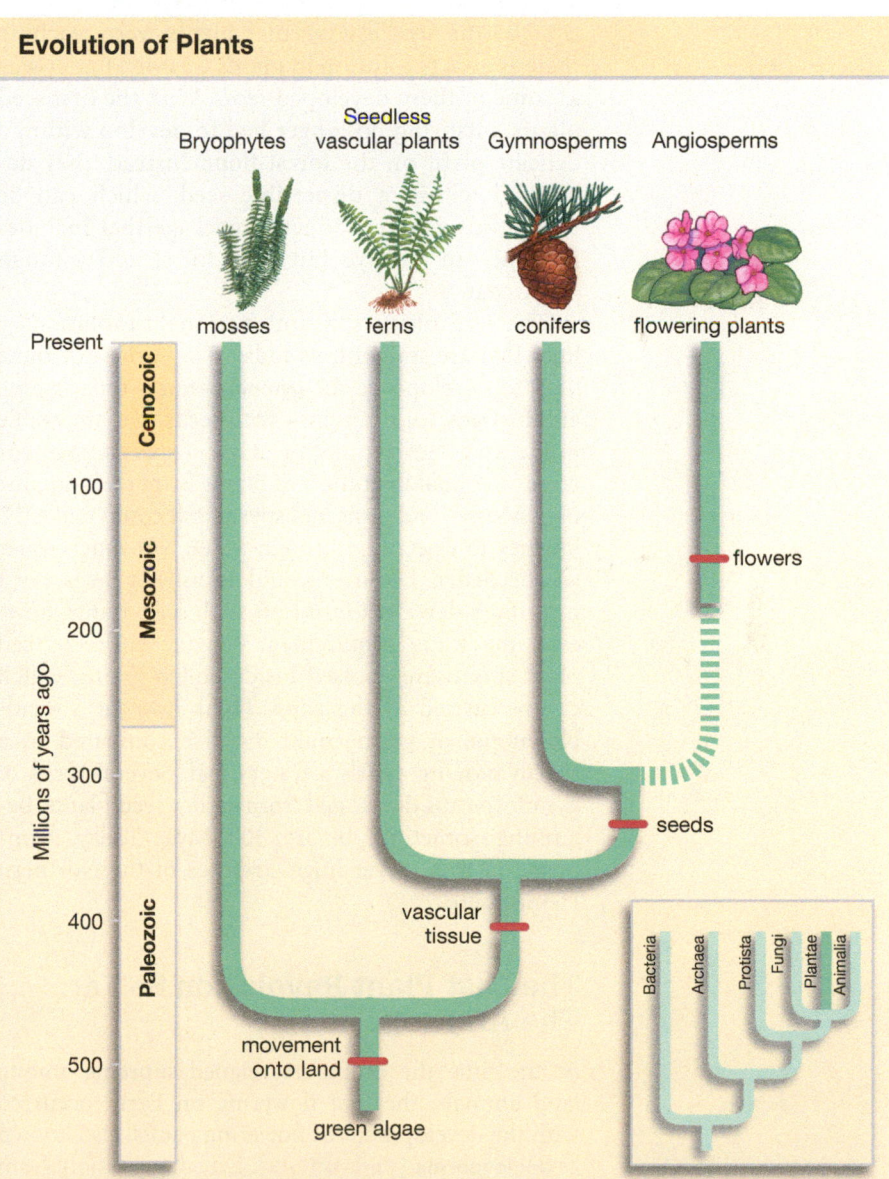

Figure 19.12
How Plants Evolved
The tree of life for bryophytes, seedless vascular plants, gymnosperms, and angiosperms. The dashed line in the angiosperm lineage represents scientific uncertainty about the evolutionary origins of these plants.

land plants that are still with us, the **bryophytes** (represented by today's mosses), had no vascular structure—meaning a system of tubes that transports water and nutrients. Plants without such a system have very limited structural possibilities. They can grow out, but they lack ability to grow *up* very far, against gravity. By contrast, the ancestors of today's ferns developed a vascular system; with this, over the ensuing 100 million years, a variety of **seedless vascular plants** evolved, including huge seedless trees such as club mosses, some of them 130 feet tall.

Plants with Seeds: The Gymnosperms and Angiosperms

Even as the seedless vascular plants were reaching their apex, a revolution in plants was well under way as some of them developed *seeds*. With the first seed plants, offspring no longer had to develop within a delicate plant on the forest floor. Instead, they developed within a dispersible seed, which can be thought of as a reproductive package that includes not only an embryo but food for it and a tough outer coat.

The seed plants split into two main evolutionary lines that are still with us today. The earliest of these lines to develop was the **gymnosperms**—today's pine and fir trees, for example—with seeds that are visible in the well-known pinecone. Gymnosperms also represent the final liberation of plants from their aquatic past. Mosses and ferns had sperm that could make the journey to eggs only through water. Not much water was needed; a thin layer would do, as with the last of a morning's dew on a fern leaf. With seed plants, however, the water requirement is gone entirely. Seed plant sperm are encased inside pollen grains, which can be carried by the *wind*. Think how far a wind-blown pollen grain could disperse compared to a sperm moving across a watery leaf. Several types of gymnosperms developed from earlier seed plants, beginning sometime before 300 Mya. Today, gymnosperm trees cover huge stretches of the Northern Hemisphere.

The Last Plant Revolution So Far: The Angiosperms

At the time the dinosaurs reigned supreme among land animals, the first flowering on Earth occurred with the development of flowering plants, also known as **angiosperms** (**Figure 19.13**). Evolving possibly from gymnosperms between 180 and 140 Mya, the angiosperms eventually succeeded the gymnosperms as the most dominant plants on Earth. Today, there are about 1,000 gymnosperm species, but some 300,000 angiosperm species, with more being identified all the

Figure 19.13
An Early Flowering Plant

A fossilized impression of the 125-million-year-old plant *Archaefructus sinensis* can be seen at left. The plant clearly had angiosperm-like traits—for example, the male reproductive structures known as anthers—yet it had no petals. One hypothesis is that it represents a transitional form in the evolution of angiosperms. An artist's conception of it is at right.

time. Angiosperms are not just more numerous than gymnosperms; they are vastly more diverse as well. They include not only magnolias and roses but cacti and common grasses, wheat and rice, lima beans and sunflowers, almost all the leafy trees—the list is very long.

One of the reasons for the angiosperms' success was their utilization of a more efficient means of pollination. Gymnosperms used the wind, but angiosperms used *animals*—birds and bees and bats—which traveled from one angiosperm to another, picking up nectar and inadvertently delivering pollen, thus providing a pollination that did not depend on the happenstance of wind currents, but that instead took the form of a direct transfer of pollen from one plant to another. A second key to angiosperm success was their more efficient use of resources. Gymnosperms produce food for every seed they generate, but angiosperms produce food for a given seed only if the egg within it has become fertilized by sperm. This difference means that angiosperms do not waste precious resources on seeds that have no chance of developing into offspring. In the living world, success usually comes to those organisms that are more efficient. For proof of this, just look out your window; chances are you'll see a profusion of angiosperms growing all around you.

19.7 Animals Move onto Land

The movement of plants onto the land made it possible for animals to follow, as plants brought not only food but shelter from the sun's rays. Recalling the division of the animal kingdom into various major groups called phyla, it was the phylum of arthropods that first moved to land. About 440 Mya—20 million years after the first plant fossils were laid down—a creature similar to a modern centipede put down the oldest set of terrestrial animal tracks we know of. It makes sense that arthropods were the first animals to come onto land because the hallmark of all of them is a tough external skeleton, or *exoskeleton,* which can prevent water loss and guard against the sun's rays.

Other kinds of arthropods, the insects, came on land some 50 million years after the first arthropod immigrants. The insect pioneers were wingless, perhaps being something like modern silverfish. Insect flight would not develop for tens of millions of years more, but when it did insects had the skies to themselves for 100 million years. The more primitive flying insects are represented today by dragonflies, whose wings cannot be folded back on their body. The compactness that came with foldable wings gave later insects the advantage of being able to get into small spaces. (Think of a common housefly.) In both winged and wingless forms, the insects took to life on land with a vengeance. About 1.8 million species of living things have been identified on Earth to date, of which more than *half* are insects. For millions of years, the insects and their fellow arthropods were the only animals on land—but this would change.

Vertebrates Move onto Land

You'll recall that back at the time of the Cambrian Explosion almost all animals existed on the ancient ocean floor. Within just a few million years, however, animals with backbones, known as vertebrates, were moving throughout the ocean; several of them were primitive, now extinct, orders of fishes. About 450 Mya, we get the rise of the family of fishes, the gnathostomes, that are the ancestors of nearly all the fish species still alive today, from sharks to gars to goldfish. Critically, gnathostomes were the first creatures to have *jaws.* With the development of jaws, these vertebrates and their descendants (one of whom is us) could securely grasp all kinds of things—big chunks of food, to be sure, but also offspring and inanimate objects. The new foods that became available to those with jaws allowed the gnathostomes to outgrow their jawless vertebrate relatives.

Fins to Limbs

A particular, later-arriving type of gnathostome, the lobe-finned fish, had an anatomical feature that turned out to be critical in vertebrate evolution. The lobed *fins* these fish possessed were the precursors to the four *limbs* that are found in all tetrapods, including ourselves. With fins, the fish swam in water, but with limbs the tetrapods moved onto land (**Figure 19.14**).

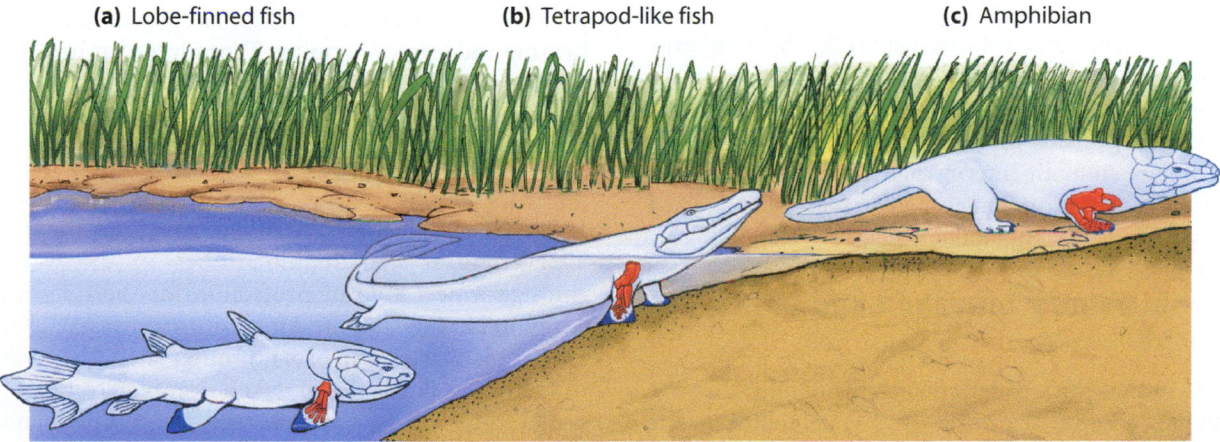

(a) Lobe-finned fish   **(b)** Tetrapod-like fish   **(c)** Amphibian

Figure 19.14
Vertebrates onto Land

(a) The lobe-finned fish had small bones in their fins that enabled them to pull out of the water and support their weight on tidal mudflats and sandbars.

(b) Numerous species of tetrapod-like fish evolved from the lobe-finned fish. These animals displayed a mixture of fish and tetrapod features—some had both the fins of fish and the developed rib cages of tetrapods, which facilitated breathing air when these animals pushed themselves up out of water.

(c) Finally, the bony fins of the tetrapod-like fish evolved into the limbs of the first true tetrapods, the amphibians, which lived partially on land while still spending considerable time in water.

Figure 19.15
Transitional Animal
An artist's conception of the tetrapod *Ichthyostega,* whose fossilized remains, found in present-day Greenland, date from about 370 million years ago. This 4-foot-long animal probably spent a good deal of time in the water, as its ears were of a type that would have functioned best there. Yet its shoulders, forearms, and ribs were of a type consistent with an animal that moved across dry land, perhaps in the manner of modern seals.

This momentous transition is known to have occurred in a period of about 10 million years, from 380 to 370 Mya. How can we be sure of the dates? We now have fossils that span the entire period. By searching from nearly the North Pole to Australia, scientists have retrieved the fossils of 15 species whose forms run the gamut between fully aquatic fish and partially terrestrial tetrapods. **Figure 19.15** is an artist's interpretation of the appearance of a creature, called *Ichthyostega,* that lay closer to the terrestrial end of this spectrum. And in "Going after the Fossils," on page 356, you can get a view of a more recently discovered creature, named Tiktaalik, that lay closer to the fish end. It's worth noting that one of the arguments sometimes advanced against the theory of evolution is that there are no "intermediate forms" in the fossil record—no forms that show us the "in-between" organisms in an evolutionary transition. We now have so many intermediate forms in the fish-to-tetrapod transition that even experts are hard-pressed to say where fish end and tetrapods begin.

Evolutionary Lines of Land Vertebrates

It is several lines of tetrapods—the amphibians, the reptiles, and the mammals—that you'll follow for the rest of your walk through evolution. Here is the order of emergence among these forms. The early tetrapods gave rise to amphibians, and early, salamander-like amphibians gave rise to reptiles. Early reptiles then diverged into several lines, one leading to the dinosaurs and another leading to mammals. Dinosaurs eventually died out, of course, but before they did, a lineage that lived on—the birds—diverged from them (**Figure 19.16**).

Amphibians and Reptiles

Amphibian literally means "double life," which is appropriate because in these creatures we can see the pull of both the watery world from which they came and the land onto which they moved. Modern-day frogs can serve as an example here. Many species of frog spend their adult lives as air-breathing land dwellers but must return to the water to reproduce. Female frogs deposit eggs directly into the water, and the males then deposit sperm on top of them. The young that survive from this clutch go on to live as swimming tadpoles, complete with gills and a tail. After a time, however, both these features disappear, even as lungs start to develop, along with ears for hearing and legs for hopping. The tadpole has become a frog—a double life indeed.

A Critical Reptilian Innovation: The Amniotic Egg

In the transition from amphibian to *reptile,* there is a severing of the amphibian ties to the water through a new kind of protection for the unborn. Amphibian eggs require an environment that is at least moist; lacking any sort of outer shell, they will dry out or "desiccate" if taken completely out of water, thus killing the embryos inside. Reptilian evolution, however, produced a significant innovation, the **amniotic egg:** an egg that has not only a hard outer casing but an inner padding in the form of egg "whites" and a series of supportive membranes around the growing embryo. These membranes help supply nutrients and get rid of waste for the embryonic reptile. With this hardy egg, the tie to the water had been broken; reptiles could move *inland.*

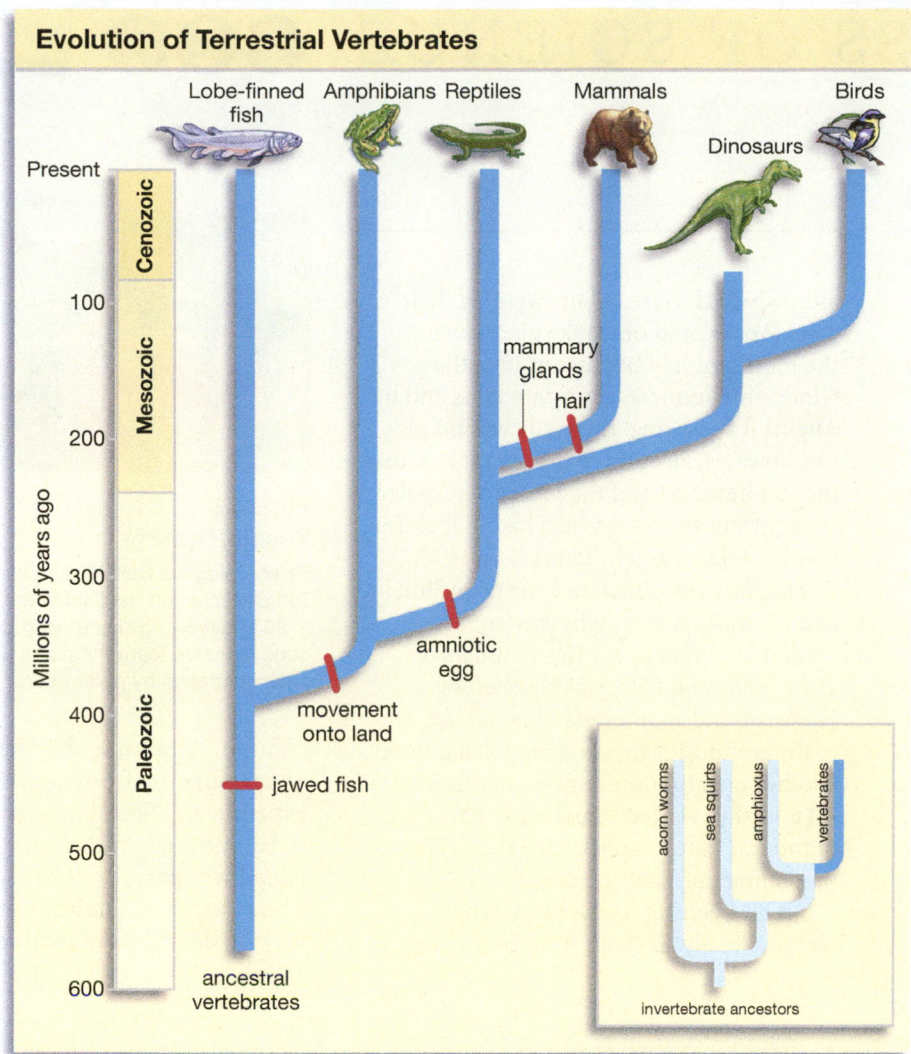

Evolution of Terrestrial Vertebrates

Figure 19.16
How Terrestrial Vertebrates Evolved
A tree of life for lobe-finned fish, amphibians, reptiles, mammals, and birds.

Amphibians were the sole terrestrial vertebrates for about 30 million years, beginning about 350 Mya. We think of these creatures now in terms of small frogs and salamanders, but the amphibians that roamed the Earth in the Carboniferous period were often the size of modern-day pigs or crocodiles. The carnivores among them probably fed on small early reptiles, but reptiles ultimately became the hunters rather than the hunted. Thriving everywhere from water to dry highlands, reptiles grew in number and diversity.

One line of reptiles evolved, about 225 Mya, into the most fearsome creatures ever to walk the Earth—the dinosaurs. For 160 million years, no other land creatures dared challenge them for food or habitat. Not all dinosaurs were huge, and many did not eat meat. Yet even a leaf eater such as *Apatosaurus* must have commanded a good deal of respect because it measured more than 85 feet from head to tail and weighed more than 30 tons. By comparison, a modern African elephant may weigh about 7 tons.

From Reptiles to Mammals

Scurrying about in the underbrush when the dinosaurs reigned supreme were numerous species of insect-eating animals, few of them bigger than a rat. These animals fed their young on milk derived from special female mammary glands, however. In addition, they probably possessed a warm coat of fur over their skins—an important feature for animals restricted to feeding at night, when the dinosaurs were less active. These were the mammals, an evolutionary line of organisms that split off from reptiles 225 Mya.

The conventional wisdom has long been that it took the disappearance of the dinosaurs, in the Cretaceous Extinction, to bring the early mammals into more diverse environments, thus allowing them to evolve into the forms seen today. A major analysis of fossil and molecular evidence completed in 2007 told a different tale, however. According to it, all the basic groups of modern mammals (all of today's mammalian "orders") had appeared by 75 Mya,

THE PROCESS OF SCIENCE

Going after the Fossils

For the University of Chicago's Neil Shubin and his colleagues, it would have been nice if the ancient animal they dubbed *Tiktaalik roseae* had made its home in what is now Tahiti or Hawaii, but that's not the way things happened. Instead, this startling creature made its home on what is now Canada's Ellesmere Island. Back when Tiktaalik was sloshing around in streambeds, Ellesmere actually was a balmy place. But that was 375 million years ago, when Canada's Hudson's Bay sat on the equator. In the eons that followed, North America moved north, and this movement put Ellesmere Island *way* north—you have to travel 600 miles south from it just to reach the Arctic Circle.

But fossil hunters have to go where the fossils are, and the evidence told Shubin and his colleague Ted Daeschler that the Ellesmere region might well contain the kind of fossils they were looking for: those of ancient animals—you could probably call them fish—that lived mostly in water but were starting to make the transition to living on land. So, beginning in the summer of 1999, the two researchers, along with colleague Farish Jenkins, Jr. and a changing group of graduate students, began spending summers walking the barren tundra of northern Canada looking for 375-million-year-old fish fossils encased in rock.

In undertaking this task, the researchers were practicing a time-honored discipline known as expeditionary paleontology. Scientists who study evolution by examining fossils are called paleontologists, and *finding* these fossils often means mounting full-scale expeditions. Certainly this is what Shubin's team had to do in connection with the Ellesmere fossils. For starters, the fossil sites lay 240 miles from the nearest civilization. As such, the researchers and their supplies had to be helicoptered in,

after which the researchers were on their own. Work could only take place during the month of July because in June the winds on Ellesmere can rip up tents and by August it's snowing. Even July was no picnic, however. Sleeping was difficult because the sun never set and the presence of polar bears meant the researchers had to look for fossils while toting shotguns (**Figure 1**).

Despite these difficulties, the prize Shubin and his colleagues sought was important enough that they spent the summers of 1999, 2000, and 2002 trekking across Ellesmere and another island to its west, eyes to the ground, digging in one spot and then another, only to come away empty-handed. All told, they walked across some 400 miles of tundra before finally hitting the jackpot in the summer of 2004 (**Figure 2**).

"It was getting to the point where we were ready to chalk it up in the loss

Figure 1
Fossil Hunters
Paleontologists Neil Shubin, on the left, and Ted Daeschler in the field in the Canadian north in 2006. Daeschler is carrying one of the shotguns research team members used to guard against attacks by polar bears.

column," Shubin said by telephone, days after returning from another Ellesmere expedition. "The 2004 year was going to be our last." And although that year ended wonderfully, it had started badly. The researchers' arrival had been delayed for two days by foul weather. And once

Figure 2
A Long, Difficult Search
Research team members trek across the valley in Ellesmere Island where they made their discovery.

they did get to work, a morning rain washed out their second day of digging on a fossil site.

"So, I said 'Well, let's just walk for the second half of the day,'" Shubin recalls. "And on that walk I was sitting down having a snack. And next to where I was sitting was the side of a Tiktaalik, exposed—sitting right out on the surface. And I was like 'Holy cow! That rings a bell.'" The researchers started digging right away, then returned to the site the next day and found a second exposed side of a Tiktaalik. Then on the third day, they made the biggest discovery of all.

"My colleague Steve Gatesy was at the top of the hill by the site, and he's flipping rocks, and he says 'Hey, what's this?' So we go over there and look. And it was the snout of a Tiktaalik sticking out of the hill."

This was the kind of specimen every paleontologist dreams of finding: an important creature in evolutionary history whose most important body part—in this case, its head—is well preserved. (See **Figure 3** for a picture of it.) In time, biologists around the world would agree it was a terrific find, for it revealed, in the words of paleontologist Jennifer Clack, "What our ancestors looked like when they began to leave the water." Tiktaalik had the scales of the fish it evolved from, but the neck and flexible forelimbs of the land dwellers that came later. It had the gills of a fish, but also had lungs that it probably filled by gulping air the way frogs do. Overall, it sat midway in the evolutionary transition between fish and the four-limbed land-walkers called tetrapods. This was an important transition because tetrapods are the ancestors of all land vertebrates, including ourselves. Tiktaalik is thus a candidate for what *human* ancestors looked like when they started moving onto land.

None of this was lost on the researchers in the summer of 2004 as they looked at the creature—soon to be named Tiktaalik—with its snout sticking out of

Figure 3
Transitional Creature

A cast of the fossilized remains of *Tiktaalik roseae,* on the right, accompanied by a model of *Tiktaalik* as it may have looked in its shallow-water habitat 375 million years ago. Note the combination of fish-like and tetrapod-like features in this single animal: fin-like forelimbs, but a flattened head more similar to that of a crocodile than a fish.

the Ellesmere rock formation. Shubin and his colleagues couldn't wait to find out what the whole fossil would reveal. Their problem was that, having discovered this specimen and the others, they then had to get them out of the rocks and safely home. And they had to do this quickly, before the August snows set in. The first challenge facing them was to leave the right amount of rock around the fossils: enough so that they wouldn't be damaged, but not so much as to make them too heavy to transport. The second trick was to figure out how to get a protective layer of plaster to dry around them, even as rains were washing over Ellesmere every day.

In the end, all the specimens made it safely to the United States. Over the course of the next several months, as museum experts carefully separated rock from fossil, it became apparent that the researchers had discovered a creature rich in evolutionary information. They decided to give it a name based in the Inuktitut language used on Ellesmere. They called it Tiktaalik (tik-TA-lik), which means large, shallow-water fish. In April 2006, the wider world got its first look at Tiktaalik when Shubin and his colleagues published two scientific papers on the creature in the journal *Nature.* Newspaper, radio, and television stories followed. It was a busy time for Shubin because he was simultaneously doing two things: working away in Chicago, fielding press calls all the while, and laying plans to be 3,000 miles away in two months' time—in Ellesmere, looking for fossils.

Figure 19.17
Primate Characteristics
These modern-day lemurs in Madagascar exhibit several characteristics common to most primates: opposable digits that enable grasping, front-facing eyes that allow binocular vision, and a tree-dwelling existence.

made it through this asteroid-aided extinction, while not one group of dinosaurs did. It's not clear, however, which factors account for the difference. Pure size may have had something to do with it; small creatures seem to survive extinction events better than large ones.

The Primate Mammals

To judge by the fossil record, the mammalian order Primates didn't emerge until about 10 million years after the dinosaurs died out—about 55 Mya. It will not surprise you to learn, however, that molecular work places primate origins much further back, at about 85 Mya. If you look at **Figure 19.17**, you can see, in the lemur of Madagascar, a modern-day descendant of the earliest primates. Three things characteristic of most primates are apparent in this picture: large, front-facing eyes that allow for binocular vision (which enhances depth perception); limbs that have an opposable first digit, like our thumb (which makes grasping possible); and a tree-dwelling existence. There are about 230 species of primates living today (**Figure 19.18**), which is a small portion of the 4,554 species of mammals. It's a tiny number of species indeed compared to, say, the 60,000 species of molluscs or the 1 million species of insects. **Figure 19.19** sets forth a primate family tree.

which is some 10 million years *before* the dinosaurs disappeared. A second round of speciation, this one within the mammalian orders, did take place after the demise of the dinosaurs, but it took place so *long* after the dinosaurs died out—from 55 to 10 Mya—that it's clear the extinction of the dinosaurs had nothing to do with the evolution of modern mammals. An obvious question here is: Why did the mammals survive the Cretaceous Extinction while the dinosaurs didn't? Forty-three modern mammalian lineages

On to Human Evolution

If you look to the upper right in the primate tree, you can see a fork that leads to chimpanzees on the one hand and human beings on the other. The simplicity of the fork might give you the impression that a couple of speciation events, occurring about 6 Mya, led

(a) A prosimian

(b) A New World monkey

(c) A great ape

Figure 19.18
Several Types of Primates
(a) A loris from India (*Loris tardigradus*) is representative of an early primate lineage, the prosimians, that includes lemurs.
(b) A woolly spider monkey (*Brachyteles arachnoids*) from the Atlantic rain forest in Brazil. Note the tail, which is capable of grasping objects. Only New World monkeys have such tails.
(c) An eastern lowland gorilla (*Gorilla beringei*) in Central Africa.

Figure 19.19
How Primates Evolved

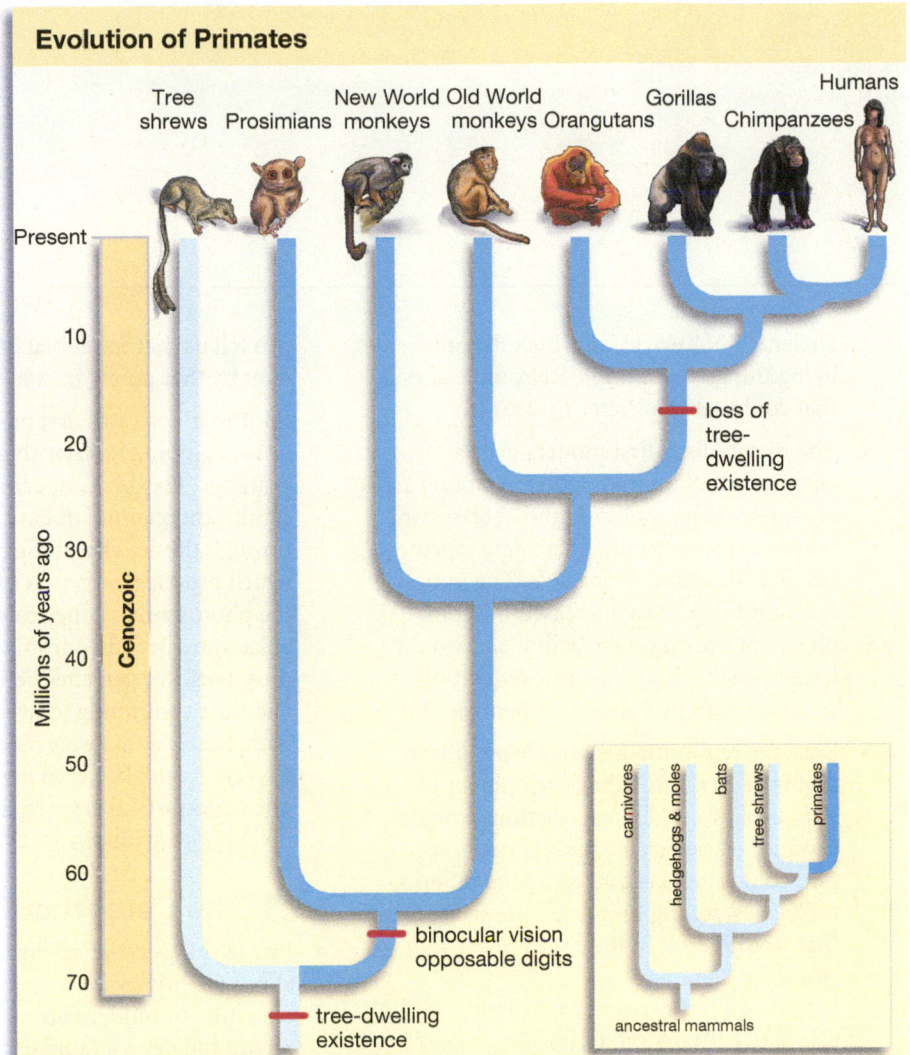

Evolution of Primates

to today's chimps and today's humans. But looking at an evolutionary tree of this sort is something like looking at a river from an airplane: It's valuable in that you get the big picture, but it's misleading in that you miss the details. In reality, the evolution of human beings was a complicated thing—a process that produced many species of primates that were somewhat like us. Where are they now? They're all gone except for us. Among the primates, we're the last two-legged walkers still standing. The story of how we evolved is coming up.

SO FAR . . .

1. All plants are thought to have evolved from water-dwelling _____.

2. All land-dwelling vertebrates are _____, which evolved from _____.

3. The amniotic egg that came about in reptile evolution was important because it could be laid _____.

CHAPTER 19 REVIEW

Summary

19.1 The Geological Timescale

- Earth's 4.6 billion years of history are measured in the geological timescale, which is divided into broad eras and shorter periods. These time-frames are defined in large part by the different life-forms that evolved within them, as reflected in the fossil record. Many of the transitions evident in the fossil record came about because of major extinction events. **(p. 340)**

- Life is thought to have begun on Earth about 3.8 billion years ago; from that time until about 1.2 billion years ago, all life was microscopic and single-celled. To judge by the fossil record, all the animal and plant life-forms that exist today came about in the last 16 percent of life's 3.8-billion-year history. **(p. 340)**

19.2 How Did Life Begin?

- Life arose through a chemical process: Simple elements and compounds available on early Earth came together to produce more complex molecules, which eventually became capable of self-replication. **(p. 342)**

- In a famous 1953 experiment, researcher Stanley Miller attempted to re-create the environmental conditions that he believed existed on early Earth and found that one variety of life's chemical building blocks, amino acids, could be produced in a short period of time. **(p. 342)**

- The replicator-first model of the origin of life holds that the first life-form probably was a single molecule, similar to today's DNA, which was capable of replicating itself. The metabolism-first model holds that life probably originated through the initiation of a self-sustaining chemical reaction that involved a set of small, uncomplicated molecules. **(p. 343)**

- The discovery of RNA molecules called ribozymes provided evidence for an ancient "RNA world" in which the only living things were simple RNA molecules that could self-replicate. **(p. 343)**

- The metabolism-first model posits a chemical reaction that occurred among an energy-yielding molecule and at least two other simple molecules that came together inside a "container" molecule. The initial reaction would have increased the complexity of the original reaction and would have brought about a net movement of new materials into the container. **(p. 346)**

- Both the replicator-first and metabolism-first models assume the development of self-sustaining chemical reactions whose products varied over time. Natural selection could then have acted on the differing molecules, setting in motion the evolution that would result in today's diverse living world. **(p. 346)**

19.3 The Tree of Life

- Every living thing on Earth belongs to one of three domains of life: Bacteria, Archaea, or Eukarya. All members of Domains Bacteria and Archaea are single-celled microbes, and no bacterial or archael cell has a nucleus. All organisms in Domain Eukarya are composed of one or more cells that have a nucleus. Domain Eukarya is composed of four kingdoms: plants, animals, fungi, and protists. **(p. 347)**

- Life can be conceptualized as beginning with a universal ancestor that produced an evolutionary line leading to Domain Bacteria on the one hand and to Domains Archaea and Eukarya on the other. **(p. 347)**

19.4 A Long First Era: The Precambrian

- The earliest evidence of life is a chemical signature of it: carbon utilization by a living thing, as recorded in rocks found in Greenland and dated to about 3.8 billion years ago. Patterns in sedimentary rock formations dating from 3.4 billion years ago tell us that microbial life was abundant by that point. **(p. 348)**

- Photosynthesis was first performed by bacteria beginning no later than 3.4 billion years ago. Oxygen later came to exist in significant quantity in Earth's atmosphere through the activity of the cyanobacteria, which produce oxygen as a by-product of the photosynthesis they carry out. Early eukaryotes benefited from the rise in atmospheric oxygen after being invaded by, and then continuing to live in symbiosis with, bacteria that were able to metabolize oxygen. A later bacterial invasion allowed one variety of eukaryotes, the algae, to perform photosynthesis. **(p. 348)**

19.5 The Cambrian Explosion

- Living things began to diversify in form about 635 Mya with the appearance of the ancestors of modern animals. The fossil record indicates a rapid expansion in the number of animal forms in a "Cambrian Explosion" that began 542 Mya. **(p. 349)**

19.6 Plants Move onto Land

- Plants, which evolved from green algae, made a gradual transition to land in tandem with their symbiotic partners, the fungi. The earliest plant–fungi fossils date from about 460 Mya. During the course of their evolution, plants developed a water-retaining cuticle, embryos that matured inside parent plants, and fluid-transport systems. **(p. 351)**

- Later plants developed seeds. The closest living descendants of the first seed plants are the gymnosperms, represented by today's conifers. Angiosperms, or flowering plants, evolved between 180 and 140 Mya and eventually succeeded the gymnosperms as the most dominant plants on Earth. **(p. 352)**

19.7 Animals Move onto Land

- The first land animals were arthropods, who appear to have moved onto land no later than 440 Mya. Eventually, an abundance of one variety of arthropod, the insects, came onto land. **(p. 353)**

- One group of fish, the lobe-finned fishes, gave rise to the four-limbed vertebrates, called tetrapods, that moved onto land between 380 and 370 Mya. **(p. 353)**

- Early tetrapods gave rise to amphibians, which in turn gave rise to reptiles. All mammals then evolved from reptiles. Birds are an offshoot of the dinosaur branch of the reptile family. **(p. 354)**

- Amphibians inhabit both water and land, in keeping with their close evolutionary relationship with animals that lived strictly in water. Reptilian evolution produced the amniotic egg, which provided reptilian embryos with superior protective housing, thus allowing reptiles to move inland. **(p. 354)**

- Mammals first appeared as a group of small, insect-eating animals about 220 Mya. Contemporary research indicates that all of today's basic forms of animals had evolved by 75 Mya, some 10 million years before the demise of the dinosaurs in the Cretaceous Extinction. **(p. 355)**

- The mammals called primates, whose earliest fossils date from 55 Mya, generally are characterized by large, front-facing eyes, limbs with an opposable first digit, and a tree-dwelling existence. **(p. 358)**

Key Terms

amniotic egg   354

angiosperm   352

Archaea   347

Bacteria   347

bryophytes   352

Cambrian Explosion   350

continental drift   344

Cretaceous Extinction   340

Eukarya   347

eukaryote   347

gymnosperm   352

Permian Extinction   340

prokaryote   347

phylum   350

ribozyme   346

seedless vascular plant   352

Understanding the Basics

Multiple-Choice Questions

(Answers are in the back of the book.)

1. Put the following in the proper time sequence, earliest first:
   a. first dinosaurs   **b.** first bacteria
   c. first amphibians   **d.** first primates
   e. oxygen accumulation in the atmosphere   **f.** first land plants   **g.** first birds

2. The time-frames in the geological timescale generally denote _____.
   a. 10,000-year increments   **b.** 10-million-year increments   **c.** the reappearance of certain life-forms   **d.** transitions in life-forms   **e.** the appearance of larger life-forms

3. The first life on Earth may have been:
   a. eukaryotic cells   **d.** formed on
   b. RNA molecules   land
   c. bacteria   **e.** algae

4. Which of the following characteristics of early plants were adaptations to life on land? (Select all that apply.)
   a. multicellularity   **b.** outer cuticle
   c. photosynthesis   **d.** lack of vertical growth   **e.** internal embryonic development

5. What is the order of evolution among the following organisms, earliest first?
   a. amphibians   **b.** lobe-finned fish
   c. mammals   **d.** reptiles   **e.** primates

6. Photosynthesis was an important innovation in the history of life because (select all that apply):
   a. it allowed RNA to replicate itself.
   b. it put large quantities of carbon dioxide into the atmosphere.   **c.** it allowed living things to capture energy from the sun.   **d.** it brought about the

production of a large amount of food
   e. it helped the Earth to warm.

7. Mitochondria and chloroplasts (select all that apply):
   a. are organelles within cells.   **b.** have a bacterial structure.   **c.** were early protists.
   d. resulted from symbiosis.   **e.** appeared late in evolutionary history.

8. What is the order of the evolution among the following organisms, earliest first:
   a. gymnosperms   **d.** bryophtes
   b. green algae   **e.** angiosperms
   c. seedless vascular plants

Brief Review

(Answers are in the back of the book.)

1. What is the best date we have for the origination of life? What forms of life existed for the next 1.8 billion years thereafter?

2. Which conceptual framework for the origin of life is linked to the "RNA world?" What two kinds of capabilities would a self-replicating RNA molecule need to possess?

3. What group of organisms did all plants evolve from? What group of organisms did all land-dwelling vertebrates evolve from?

4. Why was the amniotic egg a breakthrough for terrestrial animals? In which group of animals did it evolve?

5. Describe two characteristics that evolved in mammals that had not been present in earlier organisms.

Applying Your Knowledge

1. In what way are the bryophyte plants and the amphibian animals alike?

2. In many areas of human endeavor, consistency is an important quality. Why was inconsistency a critical quality in the initiation of life as we know it?

MasteringBiology®

1. MasteringBiology Assignments
Activities: RNA-Like Self Replication;
Current Events: Life in the Sea Found Its Fate in a Paroxysm of Extinction; A Lesson of Genealogy; Looks Can Be Deceiving; Could Conjoined Twins Share a Mind; Signs of Neanderthals Mating with Humans; **Building Vocabulary; Questions:** Reading Questions, Test Bank

2. eText
Read your book online, search, take notes, highlight text, and study more effectively.

3. The Study Area
Self-study tools to test comprehension.

20

Arriving Late, Traveling Far:

The Evolution of Human Beings

Human beings evolved as one species within an evolutionary group called the hominins whose origins stretch back 6 or 7 million years.

Available on MasteringBiology®

ABC News Video: Modern Humans and Neanderthal DNA; **Current Events:** Family Tree of Languages has Roots in Anatolia, Biologists Say; A Masterpiece of Nature? Yuck!; Lucy's Kin Carved Up a Meaty Meal, Scientists Say; **Building Vocabulary; Questions:** Reading Questions, Test Bank

A skull of *Homo habilis* or "handy man," surrounded by additional fragments of its fossilized skeleton. This close human relative lived about 1.8 million years ago in East Africa.

In August of 1856, workers at a limestone quarry in Germany's Neander Valley blasted away the entrance to a cave and discovered within it some bones they recognized as unusual. Thinking that they had perhaps come upon the bones of a cave bear, they set some aside and asked a local schoolteacher and naturalist, Johann Fuhlrott, to come take a look at them.

On seeing them, Fuhlrott knew immediately that these were not the bones of a bear. But what were they? Although they were much like the bones of a human being, they were not *exactly* human in form. There was a partial, human-like skull, but its browridge was strangely thick and prominent. And there were limb bones, but they were too curved and massive to belong to an ordinary human.

Together with Hermann Schaaffhausen, a professor of anatomy at the University of Bonn, Fuhlrott presented both the bones and drawings of them to a natural history society in 1857. News about the presentation followed, and with this, humanity got a wake-up call about itself. Although the notion was resisted at first, over time it became clear that there once had been beings on Earth who were much more human in form than, say, chimpanzees or gorillas, but who nevertheless were not *fully* human. So, how should these beings be classified? Schaaffhausen's view was that the Neander individual was a member of a "savage race" of Europeans who had been supplanted by modern Germans. Although it took a while, scientists eventually rejected this view in favor of another: that there once had been separate *species* of human-like beings on the Earth. The bones of the Neander individual belonged to one of these species, which in time came to be called *Homo neanderthalensis,* or Neanderthal Man.

20.1 The Human Family Tree

Like a child who has matured to the point that he or she can start to recognize family members in photos, humanity had matured enough by the nineteenth century that it could start to recognize its ancient family members in fossils. In 1891, Dutch physician Eugène Dubois found, in present-day Indonesia, a partial skull of a being who had both ape- and human-like qualities. Today, we know this 400,000-year-old individual as a member of the relatively modern species *Homo erectus*. In 1924, Raymond Dart, an Australian physician then working in South Africa, identified the skull of a child who came to be known as Taung baby. Today, we know this child as a member of the relatively primitive species *Australopithecus africanus,* who lived between 2 and 3 million years ago (Mya). By the early 1960s, at least three more human-family species members had been found, but as it turned out, things were just getting started in the 1960s. Discoveries kept coming in the 1970s, and a trove of them were added in the 1990s and early 2000s. Today, there are at least two dozen separate species who are candidates for membership in the taxonomic category known as **Hominini,** which is to say the taxon of human-like primates.

The Neander find that started this chain of discovery came 3 years before Charles Darwin published his path-breaking work, *On the Origin of Species*. This work, and the research on evolution that followed, provided researchers with a framework for making sense of the human-family fossils. Over time, a growing understanding of evolution converged with the fossil evidence to produce a startling realization: Some members of Hominini were not just human relatives, but human ancestors—we humans had evolved from them. All the fossil evidence told us that human beings were extreme *latecomers* among the Hominini. At the time Taung baby lived, there were no modern human beings. And the fossil evidence told us that, with movement forward in time from species like the Taung baby's, hominin species became progressively more human-like in form. The obvious implication was that we humans formed a single, late-budding branch on a Hominini family tree.

Throughout these years of discovery, scientists were asking a question that is still with us today: What does this Hominini tree look like? **Figure 20.1** presents one attempt to answer this question. What's pictured is a proposed phylogeny, or tentative family tree, for hominins—species that are members of Hominini. You'll notice that it contains not only a convoluted line that leads to our own species, **Homo sapiens,** but several lines that turned out to be evolutionary dead ends. Note also that our own genus (*Homo*) is only one of several within Hominini. At the center of the figure, for example, you can see members of the genus *Australopithecus,* whose species are, in general, more ape-like than any of the *Homo* species.

If you look at the figure's lower right, you'll also get an idea of who's *not* included in the Hominini. Note that there is a primate ancestor who gave rise to all the primate lines you can see—monkey, gorilla, chimpanzee—but that the final split at the right comes between the chimpanzee and hominin lines. By working backward through a set of primate molecular clocks, scientists have determined that sometime between 6 and 7 Mya, there was a single primate "common ancestor" that gave rise to the chimpanzee line on the one hand and the hominin line on the other. Figure 20.1's large tree is simply one idea of what the hominin line looks like in detail.

So, who was the common ancestor of the chimps and the hominins? No one knows. We know that it was not the equivalent of a modern-day chimpanzee; today's chimpanzees did not evolve until about the same time modern humans did. All we can say is that it was an ape-like creature whose closest living relatives are chimpanzees and ourselves.

Connecting the Species

Figure 20.1 was drawn up by a respected expert in human evolution, Ian Tattersall, but if we asked six experts of Tattersall's stature to provide hominin family trees, we would almost surely get six different trees. Why the uncertainty? We just don't have enough evidence to *connect* the species with confidence. We don't have enough fossils to be sure about who evolved from whom. This is the case even though our inventory of fossils is now quite large—we have at least one good fossil from every species you see on the tree and scores of fossils from some species. Moreover, we have good dates for most of these species. See the species called *Australopithecus anamensis* toward the center? We're certain that the fossils we have for it date to between 3.9 and 4.2 Mya. Meanwhile, the fossils of the *Ardipithecus ramidus* species you can see just to the lower left of the main tree date from about 4.4 Mya. But did *Ar. ramidus* evolve into *Au. anamensis*? A number of experts think the answer is yes; they think, in other words, that the two *Ardipithecus* species in the tree ought to be moved over to the right, and that a line from them ought to lead straight up to *Au. anamensis*. Tattersall isn't con-

vinced, however, so he's left *Ardipithecus* with no known evolutionary heirs and *Australopithecus* with no known evolutionary predecessors. If we were to review the entire tree, we'd find many such points of disagreement. Even with such uncertainty about linkage, however, we still know a great deal about human evolution, as a little more analysis of the tree will confirm.

20.2 Human Evolution in Overview

One of the notable features of the hominin tree is something we've touched on already. If you look right at the top of it, you'll notice that among all the hominins, only one species made it to the line marked "present." Numerous hominin species arose over the past 6 or 7 million years, but all of them are extinct now except for us. If you could take a time machine back 50,000 years, you would find, in Africa and elsewhere, fully modern human beings, which is to say hominins who physically are indistinguishable from us. But in Europe you would also find stocky Neanderthals, and over in the Far East you would find the dwindling remnants of the *Homo erectus* species. Jump forward another 10,000 years, and all the *H. erectus* are gone; jump forward another 12,000, and all the Neanderthals are gone. Today, only we *Homo sapiens* remain among the hominins.

Another way to think of this tree is in terms of relative time frames. An "ancient" civilization, such as that of Egypt, stretches back perhaps 5,000 years. Yet the Neanderthals disappeared from the face of the Earth 23,000 years before Egyptian civilization began, and they had been in existence for at least 100,000 years by the time they died out. Even so, the Neanderthals were latecomers among the hominins. Recall that the hominin line stretches back as far as 7 million years. Thus, human evolution took a great deal of time, even when dated just from the last primate ancestor. On the other hand, when we consider the entire sweep of evolution, all hominins are latecomers. If we assume that life on Earth began 3.8 billion years ago and think of all evolution as taking place in one 24-hour day, then the most ancient of the hominins evolved in the last minutes of that day. And modern human beings? We arose about 200,000 years ago, which means we have only been around for the last 5 seconds of Earth's evolutionary day.

Yet another way to think of human evolution is in terms of *where* it took place. The message here is straightforward: For perhaps 70 percent of the time we hominins have been around, we lived and died solely in Africa. This is clear because the earliest hominin fossils found *outside* of Africa date to only

A Hominin Family Tree

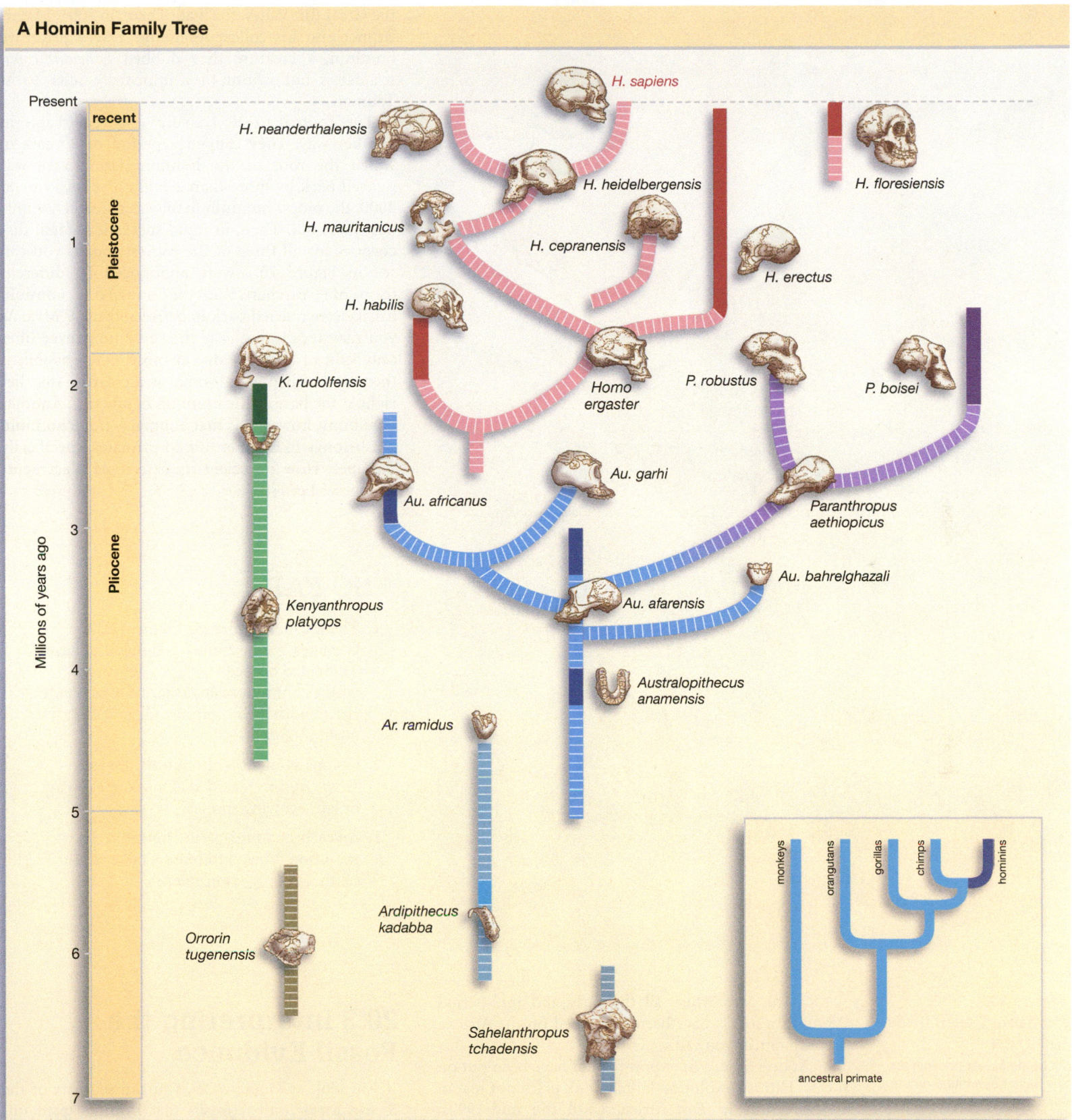

Figure 20.1
A Possible Hominin Family Tree

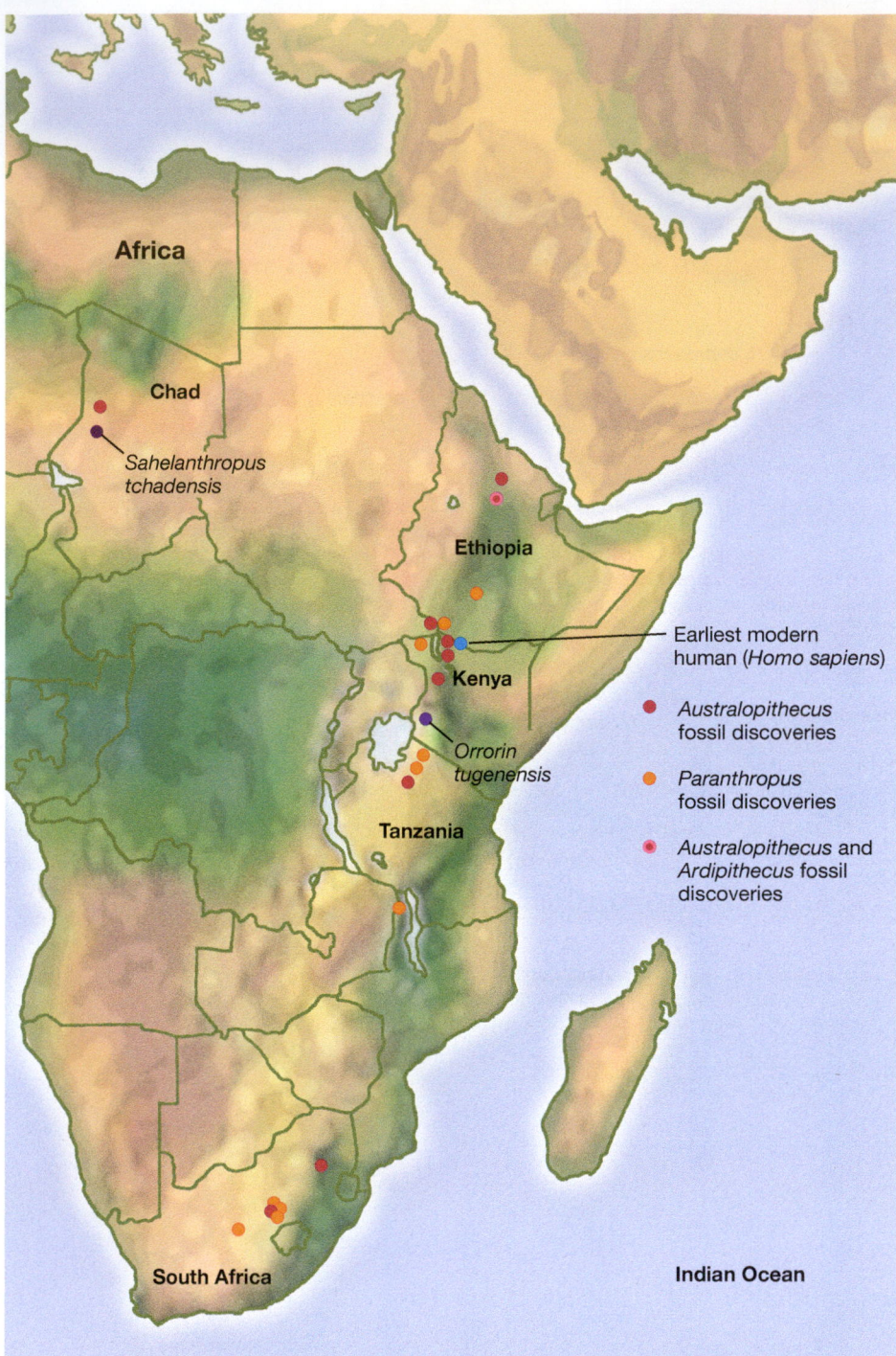

Figure 20.2
Hominin Fossil Sites in Africa

All of the oldest hominin fossils have been found on the African continent. In addition, the oldest known remains of fully modern human beings (*Homo sapiens*) have been found in an area known as the Omo Kibish site in southern Ethiopia, which has been dated to 195,000 years ago.

the Great Rift Valley, to Chad. There, in 2001, Michel Brunet and his colleagues found a very primitive hominin, a creature they dubbed *Sahelanthropus tchadensis* but whom they informally refer to as Toumaï.

The Toumaï fossils actually were remarkable in another way. They capped a period of 2 years in which the root of the hominin family tree was pushed back by more than 2 million years. Prior to 2000, the oldest hominin fossil was dated at 4.4 million years old. Then, in rapid succession, fossil discoveries for all three of the earliest species you can see in Figure 20.1 were announced by different teams of researchers. With the Toumaï find, hominin fossils were pushed back to between 6 and 7 Mya. As you may recall, the hominin and chimpanzee lines only split at about 7 Mya at most. One possibility, then, is that Brunet's Toumaï is a creature that lies right at the base of the hominin family tree. Another possibility, however, is that Toumaï is not a hominin but belongs in another line of primates, specifically the apes. How do scientists determine placements like these? Let's see.

1.8 Mya. Meanwhile, all the early and mid-period hominin fossils have been unearthed on the African continent, as you can see in **Figure 20.2**.

At one time, a finer point could have been put on this: Human evolution largely took place not just in Africa, but in *East* Africa. Until the mid-1990s, nearly all early and mid-period hominin fossils were found in a swath of Africa that runs from the eastern portion of South Africa up through the Great Rift Valley that traverses modern Kenya, Tanzania, and Ethiopia. But the range of likely hominin fossils has now been expanded some 2,500 kilometers west of

SO FAR . . .

1. The taxonomic group of human-like primates, the hominins, is thought to have originated between _____ years ago with a common primate ancestor who gave rise to both the hominin and _____ evolutionary lines.

2. Compared to other hominins, are we *Homo sapiens* regarded as an early evolving or late-evolving species?

3. All early and mid-period hominin evolution is believed to have taken place on the _____ continent.

20.3 Interpreting the Fossil Evidence

If you look at **Figure 20.3a**, you can see part of the *S. tchadensis*/Toumaï fossils that Brunet's team unearthed in the Chadian desert (**Figure 20.4**). This remarkably complete cranium of Toumaï was complemented by additional artifacts the researchers found: two lower jaw segments and some individual teeth. Fragmentary as this evidence may seem, it actually is more extensive than most hominin fossil finds. We tend to think that the term *fossils* means complete skeletons, but this is rarely the case. The claim for the primitive *Orrorin tugenensis* species seen in Figure 20.1 originally was based mostly on three thigh bones and several teeth.

The researchers who look at this evidence have a professional name that is a mouthful. Any scientist who studies evolution by analyzing fossils is a paleontologist. Meanwhile, a scientist who studies human culture or development is an anthropologist. Putting these two terms together, a scientist who studies fossils in the human lineage is a **paleoanthropologist.** These researchers are, to be sure, not the only variety of scientist who investigates human evolution. As noted, the dates we have for the hominin–ape split do not come from fossil evidence, but from *molecular* evidence—from analyses of primate DNA sequences. The importance of molecular evidence has increased greatly in recent years (see "Sequencing the Neanderthal Genome" on page 374). Nevertheless, the central evidence in human-origins research has always been fossils, and this is as true today as it was when the first hominin fossils were being dug up in Germany's Neander Valley in the nineteenth century.

So, what are paleoanthropologists looking for in such fossils? What features help them decide, for example, whether a given species should be placed in the hominin family tree or be put over with the apes? When Brunet first looked at Toumaï, three features stood as evidence that it was a hominin. One was its canine teeth; they are smaller and less sharp than those of the apes. Another was the enamel on its teeth—it is thicker than the enamel apes have. A third was what might be called the slope of the face; the lower face doesn't jut out beyond the eyes in the way ape faces do. Against this, Toumaï clearly had some ape-like features. Note its widely spaced eyes and massive brow, for example. Moreover, Brunet's fellow paleoanthropologist, Brigitte Senut, interpreted Toumaï's canine

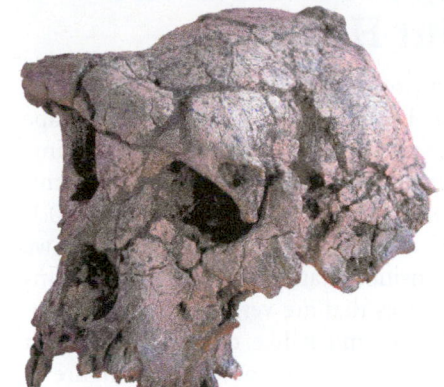

(a) Cranium of Toumaï.

(b) Model of Toumaï.

teeth not as evidence of a male hominin but as evidence of a female ape.

These various arguments were set forth at the time Brunet unveiled Toumaï, in 2002. In 2005, however, two experts reported on a painstaking computer analysis they did of Toumaï's skull that yielded an additional, important piece of information: Toumaï appears to have walked upright, on two legs. This is critical because, along with tooth structure, upright walking, or **bipedalism,** is the single most important defining feature of a hominin. Our closest living relatives, the chimpanzees, are "knuckle walkers" who only occasionally rise up to walk on two legs. But all the hominins you see in Figure 20.1 are either known to have walked on two legs or arguably did so. The evidence that Toumaï did is one of the factors paleoanthropologists would take into account in deciding whether to place it among the hominins or the apes. The computer analysis of Toumaï's skull that yielded the evidence of its upright walking also yielded another benefit, which is an idea of what this creature may have looked like in the flesh (**Figure 20.3b**).

Figure 20.3

An Ancient Human Ancestor?

(a) This cranium of *Sahelanthropus tchadensis,* or Toumaï, was discovered in the deserts of Chad and has been determined to be 6 to 7 million years old. Toumaï's discoverers believe it is a very ancient hominin, but other researchers believe it should be grouped with apes.

(b) A procedure called computed tomography yielded this reconstruction of the appearance of Toumaï.

Figure 20.4

The Hard Work of Finding Fossils

Hominin fossils can be located in some inhospitable places. Here Michel Brunet (in the foreground) looks for hominin fossils with his crew in Chad's Djurab Desert in 2006. It was Brunet's team that discovered the Toumaï fossils in Chad in 2001.

20.4 Snapshots from the Past: Four Hominins

It shouldn't surprise us that a controversy exists over whether Toumaï belongs in the hominin or the ape family tree. After all, Toumaï lived right about the time that the hominin and ape lines were parting company. If we looked at each of the species noted in Figure 20.1, moving from most ancient to most recent, what we would see is a transition that reflects this split—a transition from creatures that are very much like apes to those that look very much like ourselves. We don't need to examine all the species noted in the figure to get a sense of this transition, however. Here are four snapshots from the past that will provide some idea of the nature of our hominin relatives over time. An artist's interpretation of the appearance of three of these relatives can be seen in **Figure 20.5**.

Australopithecus afarensis: **Lucy and Her Kin.** If we jump forward about 3 million years from Toumaï, we find a hominin whose features illustrate a partial transition from ape-like to human-like. The species, *Australopithecus afarensis,* is best known through its most famous individual, dubbed Lucy, whose remains were found in Ethiopia in the 1970s. We have a very good idea of when Lucy lived; her fossils have been dated to 3.18 Mya.

There is no doubt that, like us, Lucy was bipedal; the structure of her pelvis makes this a certainty. Nevertheless, she possessed long arms, short legs, and feet that were built for grasping—features consistent with

a species that no longer lived exclusively in the trees but that may have still spent considerable time in them. Meanwhile, our own species is thought not to have been "arboreal" or tree-dwelling at all. Lucy also differed from us in the size of her brain; she had a cranial volume of about 450 cubic centimeters—about that of a chimpanzee—while ours is about 1,400 cubic centimeters.

When we think about these various features together, we can see that hominin features evolved in what is called a *mosaic pattern*—different features developed at different points in time in different species. Lucy didn't have upright walking, a big brain, and modern limb lengths; she only had upright walking. To get some idea of Lucy's anatomy compared with ours, see **Figure 20.6**.

One of the notable things about Lucy's species is that there is good agreement about its location in the hominin family tree. Most authorities would give *A. afarensis* the central place you can see in Figure 20.1—an important place because, as you can see, *A. afarensis* is thought to have given rise to our own genus (*Homo*). In Lucy, then, we probably are seeing not just a relative but an ancestor.

Homo ergaster and Homo erectus: **More Like Us.** Changes to physical forms that are more like ours come at several steps in the hominin line, but the most dramatic change comes with the rise of *Homo ergaster,* whose best-preserved remains come from a boy who died 1.6 Mya on the shores of Lake Turkana in Kenya. Experts estimate that, had this 8-year-old

(a) *Australopithecus afarensis*

(b) *Homo ergaster*

(c) *Homo neanderthalensis*

Figure 20.5
Three Hominins

What did now-extinct members of the hominin group look like? Working from fossil and other evidence, artist Viktor Deak has produced images that provide some idea of their appearance.

(a) *Australopithecus afarensis,* the "Lucy" species, lived a little more than 3 million years ago. *A. afarensis* possessed so many ape-like features that some researchers have dubbed it a "bipedal chimpanzee."

(b) *Homo ergaster* clearly is a hominin with a much more human-like form. Evolving about 1.7 million years after Lucy and her relatives, *H. ergaster* is generally believed to be a human ancestor.

(c) *Homo neanderthalensis* lived in Europe and elsewhere in a period running from about 130,000 years ago to 28,000 years ago. These were the Neanderthals, a species that lived in proximity to modern humans for thousands of years.

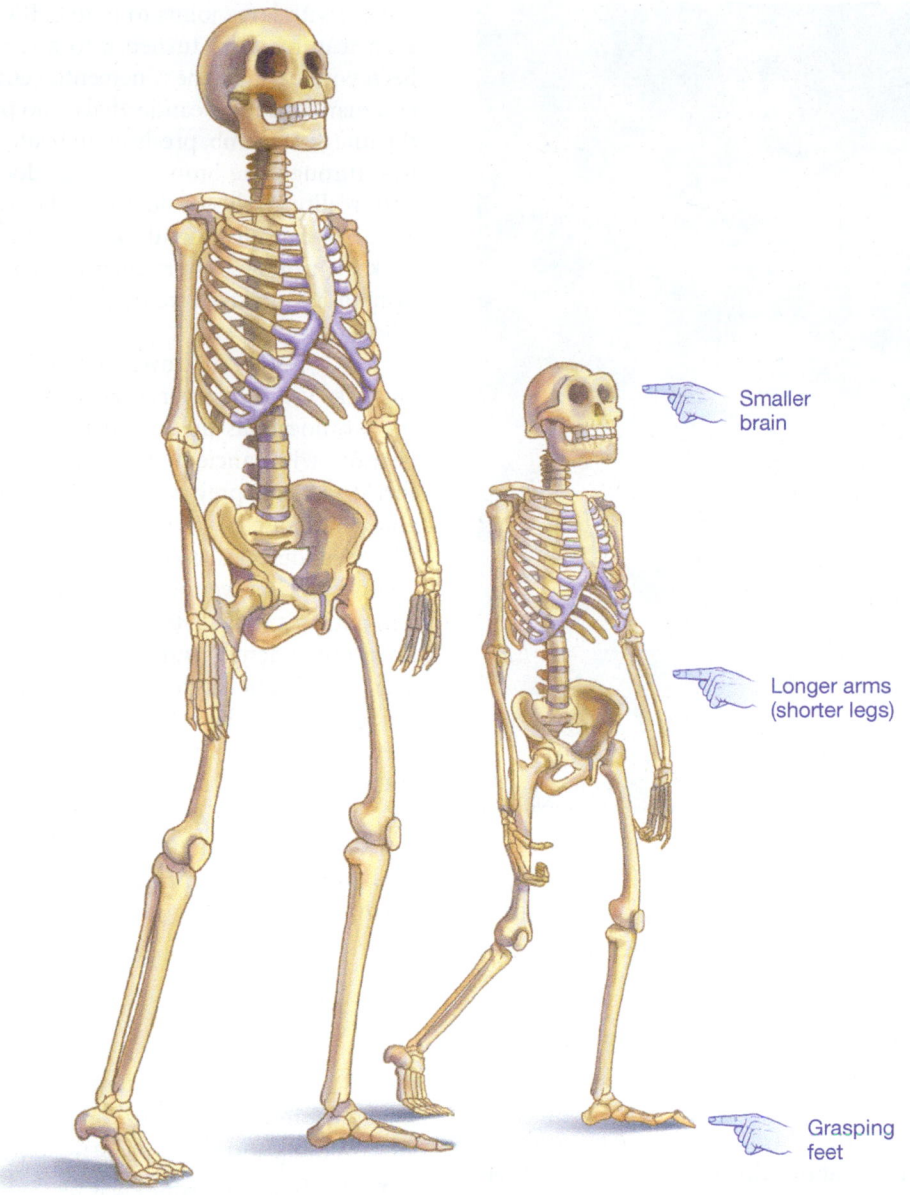

Smaller brain

Longer arms (shorter legs)

Grasping feet

Figure 20.6
"Lucy" and Modern Humans Compared
What was the hominin Lucy like compared to a modern human female? The figure gives an idea. Lucy stood about three-foot-seven and had a much smaller brain than modern humans, even allowing for her smaller stature. Lucy's hip and pelvic bones make clear that she was bipedal, but note the longer arms and grasping feet common to tree-dwelling primates.

(Adapted from Boyd, R., and Silk, J.B., *How Humans Evolved.* New York: W.W. Norton & Company, 2000, p. 329.)

"Turkana Boy" grown to maturity, he would have reached a height of 6 feet. His brain was more than half the size of the average modern *Homo sapiens* brain. He had a much more modern face, long limbs typical of those humans who dwell today in arid climates in Africa, and advanced tool technology. If you look at **Figure 20.7** on page 370, you can see Turkana Boy's skeleton, which was found in amazingly complete form.

Homo ergaster is regarded by many researchers as being ancestral to our own species. (If you look back at Figure 20.1, you can see the place that Ian Tattersall believes *H. ergaster* holds in our evolutionary line.) Moreover, this long-legged hominin is generally regarded to have directly given rise to another species, *Homo erectus,* which has the honor of being the first hominin to leave Africa. By 1.8 Mya, it had reached the former Soviet Republic of Georgia, but at that point its migration was just getting started.

Eventually, *H. erectus* groups made it all the way to far-eastern China and the Indonesian archipelago. Why did *H. erectus* travel so much farther than any of its predecessors? Perhaps because it could. In 2009, scientists announced the discovery of a series of footprints that were laid down 1.5 Mya near Lake Turkana, probably by individuals who were either *H. erectus* or *H. ergaster*. Analysis of the prints showed that these hominins not only had long legs that would have been good for striding but a foot structure that allowed each step to end with a propulsive push off the toes. This may sound familiar, because it's exactly the kind of stride we have; it may be that the way we walk had become established in our ancestors by 1.5 Mya.

Relative to most hominin species, *H. erectus* ended up having a terrifically long existence—nearly 2 million years from the time of its origins in Africa to the time its last individuals died out in

Figure 20.7
Much More Like Us
In the remarkably complete skeleton of Turkana Boy we can see the evolution of hominins to a form more like our own. This member of the species *Homo ergaster* was tall and had a much larger brain capacity than the earlier hominin Lucy.

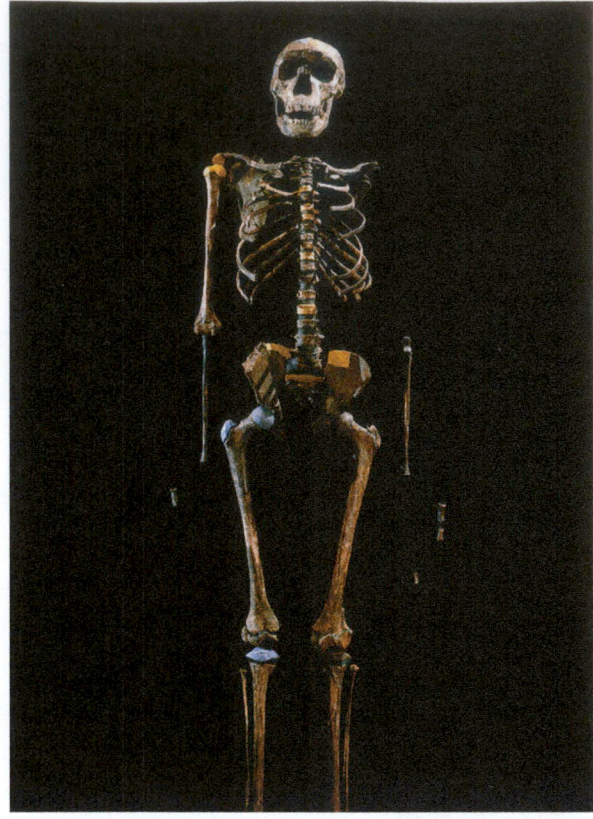

Indonesia, 40,000 years ago. (By way of comparison, our own species has been in existence for only a tenth this amount of time.) Pure intelligence may have played a part in *H. erectus'* long run, but why did increased intelligence arrive with such seeming speed in this species? A Harvard primate expert named Richard Wrangham thinks the reason may have been that *H. erectus* was the first hominin to undertake the critical human activity of cooking. You can read more about this in "Cooking Up Intelligence" on page 371.

Homo neanderthalensis: Modern and Far-flung. In the **Neanderthals**, we see a modern species of hominin—one that nobody ever confused with an ape. Various dates are assigned to their first appearance; we have signs of them as much as 350,000 years ago, but they didn't evolve into their most modern form until about 130,000 years ago. Either way, the Neanderthals appeared recently indeed compared to the 1.6 Mya when *H. ergaster* flourished, to say nothing of the 3.18 Mya when Lucy was alive. The Neanderthals evolved in Europe, but they migrated great distances; fossils of them have now been recovered all the way from England to Siberia.

So, what were these hominins like? Figure 20.5c shows you an artist's interpretation of their physical stature: short, stout bodies—the average male was about five foot six—powerfully built, with a heavy "double-arched" brow and a receding chin; big-

boned, with large joints to match. It's easy to see how such features were turned into a *caricature* that has been with us since the nineteenth century: that of the caveman. It is the Neanderthals who provided us with the image of dumb, pre-human brutes, grunting their way through the Stone Age. But does this image fit with reality? Well, consider that the Neanderthals had a cranial capacity slightly *larger* than ours; that they took the trouble to bury their dead; and that they appear to have taken care of those within their groups who were old or sick.

This is not to say, however, that Neanderthals were "just like us." Whenever direct comparisons are made, the Neanderthals do indeed look primitive in comparison with ancient *Homo sapiens*. Neanderthals thrust spears at prey, but humans developed the valuable technique of *throwing* spears from a distance. The Neanderthals used nothing but stone for their relatively primitive tools, while *H. sapiens* used bone and antler as well as stone in constructing their much finer implements. And the Neanderthals were "foragers," while ancient humans were "collectors." What's the difference? Humans monitored their environments and used "forward planning" by, for example, placing their campsites near animal migration paths. By contrast, Neanderthals do not appear to have timed their migrations in this way. Such differences may have been important to the fate of the Neanderthals, as you'll see.

SO FAR . . .

1. The two most important features in distinguishing hominins from apes are _____ walking and _____ structure.

2. The "Lucy" hominin was like modern human beings in that she was _____, but unlike them with respect to _____ size.

3. Lucy shows us that hominin features evolved in a _____ pattern, which is to say that different features evolved at different points in time and in different species.

20.5 The Appearance of Modern Human Beings

The Neanderthals had Europe to themselves for tens of thousands of years, but beginning about 40,000 years ago, they had company, in the form of ourselves—modern *Homo sapiens*. Where had we come from? The short answer is Africa. Beginning sometime between

ESSAY

Cooking Up Intelligence

When our evolutionary ancestors first learned to cook food, tasty meals may have been the least of the benefits they pulled from the fire. Under a theory championed by Harvard primate expert Richard Wrangham, a big increase in intelligence that our ancestors experienced about 2 Mya was brought about by their discovery of cooking.

What makes us think that hominin intelligence ramped up 2 Mya? The earliest hominins lived 6 to 7 Mya, but 3 million years later the cranial capacity of the hominin "Lucy" was still only 450 cubic centimeters—about that of a chimpanzee. Go forward another 1.3 million years, however, and the cranial capacity of the species *Homo erectus* was about 1,000 cubic centimeters—more than twice that of Lucy (**Figure 1**). So, why the big increase in such a short amount of time? Wrangham thinks that *H. erectus* learned to cook food over a fire and that this made all the difference.

Cooking had this effect, Wrangham believes, because of the sheer amount of energy it made available for brain growth and maintenance. Present-day chimpanzees spend 5 hours a day just chewing food, whereas people in today's hunter-gatherer societies—all of which utilize cooking—spend just an hour a day doing the same thing. And once food is swallowed, cooking yields another energy benefit. In experiments on pythons, Wrangham and physiologist Stephen Secor found that snakes that were fed cooked beef spent 13 percent less energy digesting this food than snakes who were fed raw meat.

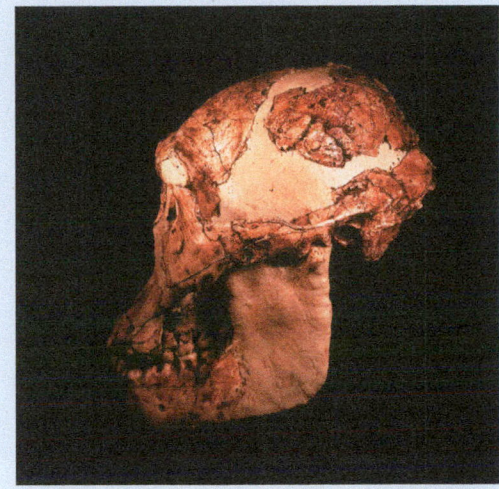

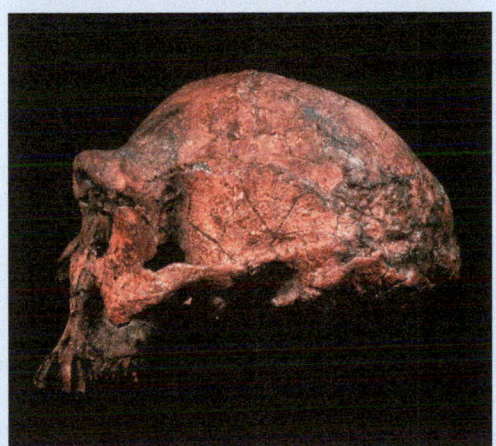

Figure 1
Big Increase in Brain Power
The cranial capacity of the Lucy species *Australopithecus afarensis,* in the top photo, was less than half that of the later-evolving *Homo erectus,* in the bottom photo. Primatologist Richard Wrangham believes that a larger brain evolved in *H. erectus* over a relatively short period of time thanks to its discovery of cooking.

All this matters to brain power because brains are voracious consumers of energy. Up to 25 percent of the calories that a human body expends while at rest go to the maintenance of brain tissue. Wrangham has calculated that, even if *H. erectus* had access to raw meat, a hominin of its brain size and physical stature would have needed to spend at least 5.7 hours a day just chewing the meat to satisfy its energy requirements. We are certain, however, that something happened to reduce *H. erectus'* reliance on raw, tough foods. Compared to earlier hominins, it had smaller teeth and a smaller digestive tract—traits that are consistent with a species that had either learned to cook food, or that at least had gained access to a large quantity of soft, highly nutritious food.

Wrangham has won praise for developing a hypothesis that is internally consistent and that is supported by a large amount of circumstantial evidence. All parties are agreed, however, that something critical is missing from his argument: clear evidence that *H. erectus* actually cooked food. The earliest uncontested evidence we have of hominin cooking—stone hearths that surround charred earth—date from a mere 250,000 years ago. Meanwhile, Wrangham's hypothesis would have *H. erectus* employing cooking some 1.8 Mya. Fires that may have had a human origin have been dated to between 1.4 and 1.6 Mya, but each of these could plausibly have been set by a natural force such as lightning, rather than by early humans preparing a meal. Wrangham is convinced he's right, however, and no one doubts that he'll leave no stone unturned in trying to determine whether human intelligence got a big boost from human cooking.

50,000 and 60,000 years ago, we left our ancestral African home and began a migration that eventually took us all over the globe. Our travels initially took us not west into Europe, however, but east into Asia, Indonesia, and Australia. It was a second round of migration, thought to have originated in the region between modern-day India and the Arabian Peninsula, that took us both westward into Europe and northeastward into Russia.

Critically, the first round of our migration seems to have begun after we humans evolved into our present anatomical form. Put another way, *H. sapiens* fully

Figure 20.8
**Out of Africa, then
Out of Western Asia**

The first wave of modern
human migration out of Africa
began 50,000 to 60,000 years
ago and followed an easterly
route—from Africa through
the Arabian Peninsula, down
through South Asia, and
eventually all the way into
Australia. In a second wave
of migration, begun in West
Asia 35,000 to 45,000 years
ago, modern humans went
northwest into Europe,
northeast into Russia, and
southeast into North Africa.

(Adapted with permission from Ted
Goebel, "The Missing Years for
Modern Humans," *Science,* 12 January
2007, 315: 194–196.)

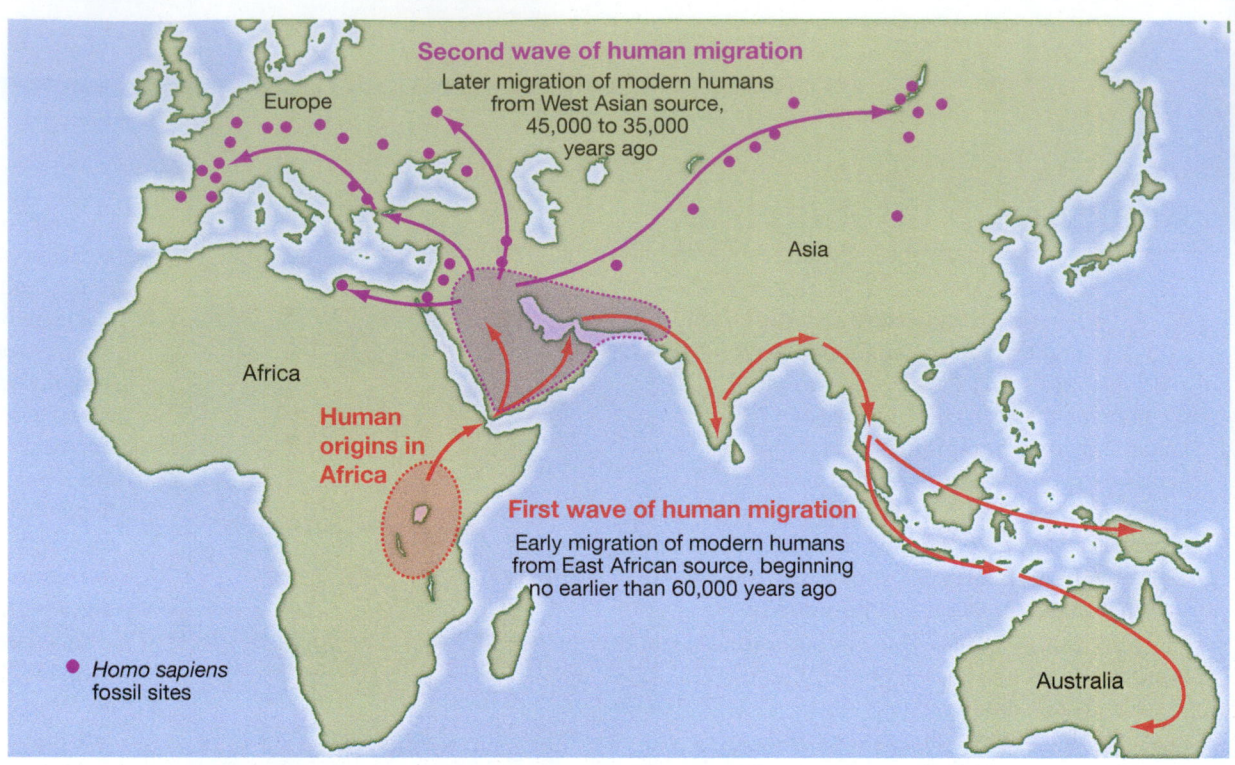

evolved *in Africa* before migrating out to the rest of the world (**Figure 20.8**). But a fascinating twist was added to this story in 2010 with news that some modern humans who migrated out of Africa mated with Neanderthals, with the result that an estimated 1-4 percent of the DNA of today's European and Asian human beings is of Neanderthal origin. Meanwhile, the DNA of today's Africans contains no Neanderthal DNA sequence at all, which is not surprising since Neanderthals are not known ever to have inhabited Africa proper.

The amount of interbreeding between modern humans and Neanderthals clearly was small but the fact that it occurred at all has upset the strongest version of what has been called the "out-of-Africa" hypothesis, which holds that fully modern humans migrated out of Africa and remained a self-contained species—mating with no others—right down through modern times. This hypothesis stands in contrast to another long-standing school of thought, called multiregionalism, which holds that several late-stage hominins left Africa at different times and then interbred in Europe and Asia. Over time, the hypothesis goes, these species evolved into both modern *H. sapiens* and the late-stage Neanderthals. The small amount of mating now known to have taken place between Neanderthals and modern humans does not mean that multiregionalism is correct, but it does mean that the out-of-Africa hypothesis is not true in its strictest sense, since we know that modern human beings did interbreed with at least one other species.

The idea that modern humans interbred with several species, however, is not supported by the evidence we have—specifically by evidence that looks at the genetic patterns found in the DNA of *living* human beings. In recent years, scientists have been collecting DNA samples from sets of human populations that have deep ancestral roots in different geographic regions around the world—Europe, the Middle East, and South Central Asia, for example. A recent analysis of these samples confirmed that human genetic variation is greatest in sub-Saharan Africa and then declines in a steady way with geographic *distance* from Africa. Thus, present-day African populations are the most genetically diverse on Earth, European populations are less genetically diverse, and Native American populations are less diverse yet. This is exactly what we would expect if human beings had lived in Africa for tens of thousands of years—accumulating genetic diversity there—after which they began migrating throughout the world through a series of what are called "founder" events: One human population would break away from another and found a new human settlement at a distant location. As readers of Chapter 17 know, each time this happens, the migrating population takes with it only a portion of the gene pool of the original population. Thus, each migrating population is less diverse genetically than the group it came from. If this occurred within a single species, the result should be a steady drop-off in genetic diversity with distance from the ancestral homeland. If you look at **Figure 20.9**, you can see how remarkably steady this drop-off is in actual fact. In every major group of contemporary human beings studied, the farther a human population is from Africa, the less diverse it is genetically.

Evidence of this sort has convinced most human-origins researchers that modern humans had a purely African origin, with the understanding that non-African human beings then acquired a small amount of DNA from Neanderthals. But when did we first appear in Africa? Molecular evidence indicates that we evolved there no earlier than 200,000 years ago, and we now have fossil evidence that dovetails nicely with this date. The oldest known *H. sapiens* fossils, found in present-day Ethiopia, have been dated to 195,000 years ago. (You can see the location of these remains in Figure 20.2 on page 366.)

Who Lives, Who Doesn't?

When our ancestors left Africa, it's possible that they encountered not only Neanderthals but individuals from the species *Homo erectus* as well. Now let's think about extinction and survival among these three species. Recall that about 40,000 years ago, *H. sapiens* arrived in Europe, where *H. neanderthalensis* was living, but that by 28,000 years ago, all the Neanderthals were gone. Then recall that we humans had left Africa, headed east initially, by no later than 50,000 years ago, but that by 40,000 years ago the last *H. erectus* were gone, having died out in Indonesia.

You can probably see the implication in this chain of events. If there is a consensus account of why we humans are the last surviving hominins, it is that we "replaced" our fellow hominins throughout the world once we encountered them. This is not to say that we simply killed them outright. We have no evidence of violent encounters between ourselves and the Neanderthals or *H. erectus*. But similar species are bound to compete for the same kinds of resources, such as food and shelter. And we have every reason to believe that modern humans could *out*-compete both the Neanderthals and *H. erectus*, thanks to our superior brain power. (Remember the comparisons noted earlier between Neanderthals and modern humans?) So, one possibility is that the hominin family is now a family of one—ourselves—in part because our success brought about the extinction of our near relatives.

20.6 Next-to-Last Standing? The Hobbit People

In 2004, a fascinating twist was added to this story. Researchers working on the Indonesian island of Flores reported finding fossils of a hominin species whose members stood just 3 feet tall and who lived on Flores as recently as 12,000 years ago (**Figure 20.10**). This discovery was stunning for three reasons. First,

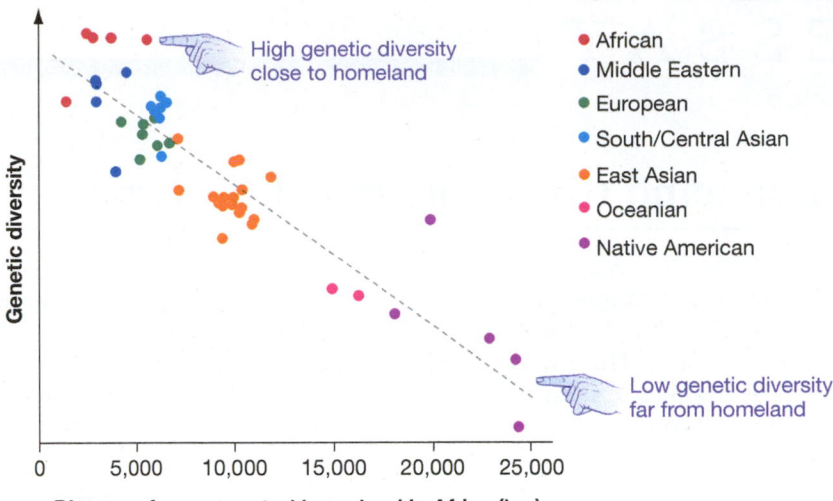

Figure 20.9
Worldwide Human Genetic Diversity
The level of genetic diversity of 51 modern-day human populations, each of which had deep ancestral roots in its current geographic location. The farther away each population was from Addis Ababa, Africa, the less genetically diverse it was.
(Adapted with permission from Jun Z. Li, et al., "Worldwide Human Relationships Inferred from Genome-Wide Patterns of Variation," *Science* 22 February 2008, 319: 1100–1104.)

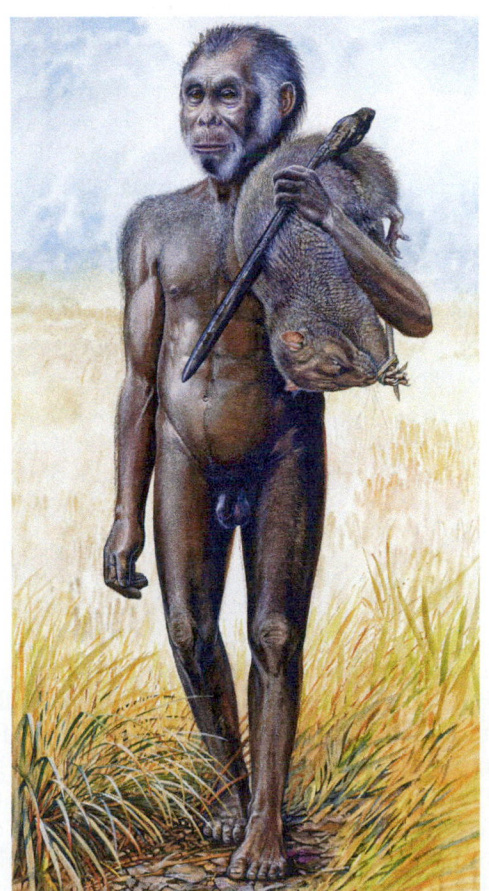

Figure 20.10
Small Hominin, Possibly Big Discovery
An artist's interpretation of the appearance of the 3-foot-tall hominin dubbed *Homo floresiensis*, whose skeletal remains were found on the Indonesian island of Flores.

ESSAY

Sequencing the Neanderthal Genome

Sequence the genome of a living species? It's done all the time. Since 2003, when the Human Genome Project was completed, scientists have managed to get complete DNA sequences for chimpanzees, sheep, fruit flies, a mustard plant, several fungi, and many bacteria—the genomes of hundreds of species are now in hand. But how about sequencing the genome of an *extinct* species? This is much more difficult because of what might be called the DNA access problem. DNA degrades over time, and of course, many extinct species have been gone a *long* time. Moreover, any DNA that remains in ancient bones is likely to be not only fragmented but also contaminated: It will be mixed with the DNA of both bacteria and any scientists who have handled the bones it came from.

Given such obstacles, the scientific world cheered in 2006 when a team of scientists based in Germany announced that they intended to sequence the genome of a long-extinct relative of ours, *Homo neanderthalensis,* or Neanderthal Man. Led by geneticist Svante Pääbo of the Max Planck Institute for Evolutionary Anthropology, the team decided to begin its work by extracting DNA from the bones of two Neanderthal women who lived 38,000 years ago in a cave in what is now Croatia. Pääbo and other scientists actually had sequenced Neanderthal DNA several times before. But in each case, the starting genetic material was so-called mitochondrial DNA, meaning DNA that comes from the tiny organelles within cells (the mitochondria) that carry their own small complement of DNA. In the new project, Pääbo has been sequencing the Neanderthal's "genomic DNA," meaning its primary complement of genetic material, derived from all its chromosomes.

In 2009, the world got a peek at what this sequencing had revealed, but at that time the news coming from Germany simply confirmed what scientists had long believed: That Neanderthals and modern human beings had never interbred. In 2010, however, with a more complete analysis of the genome in hand, Pääbo and scientists working with his data revealed that Neanderthals and modern human beings had in fact interbred. The practice was not widespread, but at least 1 percent of the DNA of today's non-African human beings is of Neanderthal origin.

At least 1 percent of the DNA of today's non-African human beings is of Neanderthal origin.

Since human beings and Neanderthals lived in physical proximity in Europe for at least 12,000 years, beginning about 40,000 years ago, experts assumed that any interbreeding the two species undertook would have taken place in Europe during this time. But since both Europeans and *Asians* show the same amount of Neanderthal sequence in their DNA, it seem likely that such interbreeding as Neanderthals and humans undertook occurred at least 60,000 years ago—before modern humans migrated to Asia—with the location for the interbreeding being the Middle East rather than Europe. We have evidence of both modern humans and Neanderthals occupying the same region in what is now Israel for a period of up to 10,000 years during this period but pinpointing the time and place of Neanderthal-human interbreeding is work for the future.

Apart from telling us what Neanderthals and modern humans share in the way of DNA, the sequencing of the Neanderthal genome has also told us how the two species differ genetically. Since the amount of interbreeding between them was so small, they largely accumulated DNA mutations independently of one another from the time they shared common ancestor in Africa between 270,000 and 440,000 years ago. When the Pääbo team compared the Neanderthal genome to that of five modern human beings of diverse ethnicities, it found that the genetic differences between humans and Neanderthals were few and far between. In the A's, T's, G's, and C's that make up genes, there were only 78 significant differences between the genes of the two species. The challenge for researchers now is to identify the exact functions of these differing genes. We have a general idea about some of these genes already, but this information only deepens the mystery about the differences between Neanderthals and ourselves. Two of the genes that differ most significantly between ourselves and Neanderthals, for example, have to do with skin—with the sweat glands in it, with the deep structure of the hair that extends over it, and with its pigmentation. Fascinatingly, three other genes have to do with mental development. The reason we have a good idea of the general function of these genes is that in a mutated state they can contribute to three well-known modern mental afflictions: schizophrenia, autism, and down syndrome.

Finally, analysis of the Neanderthal genome has revealed that Neanderthals possessed a variant of a pigmentation gene, called *mc1r,* that can result in red hair and pale skin. In one sense, this is not surprising. Since the Neanderthals lived to a large extent in colder climates, they would have benefited from a skin tone that could absorb more sunlight and produce more vitamin D as a result. This kind of physical detail is, however, just one of many that can be expected to be revealed over the next few years as Pääbo and his team continue to report on the genome of our long-gone, but very close, relative.

anthropologists have long believed that the last hominins other than ourselves died out not 12,000 years ago but 28,000 years ago, with the passing of the Neanderthals. Second, if modern humans really were spanning the globe replacing other hominin species 50,000 years ago, how did the Flores people, dubbed *Homo floresiensis,* manage to avoid this fate for some 38,000 years? Third, the Flores people used fire and had sophisticated tools. Yet, they had a cranial capacity of only 400 cubic centimeters—less than that of a chimpanzee! How could they do so much in the way of culture with so little in the way of brain size?

The answers to these questions may be a while in coming. One possibility is that much of the hoopla about *H. floresiensis* has been undeserved. Several teams of researchers who have analyzed the Flores evidence believe *H. floresiensis* represents not a new hominin species but a modern human who suffered from a medical condition. Indeed, no fewer than three medical conditions have now been proposed to account for the means by which a modern human could have ended up with *H. floresiensis'* combination of short stature and small brain size. The scientists who made the initial Flores discovery have rejected all these ideas, however, and their position has been supported by other research teams. The upshot of this back-and-forth is that the *H. floresiensis* find has turned into a long-running, acrimonious debate about what actually was discovered on Flores. The only thing the opposing parties seem to agree on is the desirability of finding an additional *H. floresiensis* skull, along with some of its DNA, so that the issue might be settled.

Assuming that the Flores "Hobbit people" really were a separate species, how might they fit into the greater hominin family? The hypothesis that seems to have the most support is that *H. floresiensis* was a later-evolving offshoot of the first hominin world traveler, *Homo erectus,* which you'll recall survived in the Indonesian archipelago up until about 40,000 years ago. Of course, *H. erectus* was a *tall* hominin,

but there may be an explanation for how one of its descendant species ended up being so short: island dwarfism. Once isolated on islands, mammals can evolve into unusually small or large forms because of the unique environment an island can provide—a complete lack of some predators, for example, and an abundance of others. If the Hobbit people were isolated for long enough on Flores, it may be that the island's environment worked with natural selection to produce a group of extremely small hominins. More research will hopefully sort out the truth about this. For now, we're left with a fascinating question: Did a hominin species other than our own—a tiny one at that—survive until just 12,000 years ago?

On to the Diversity of Life

In reviewing the history of life in the last two chapters, you necessarily looked, in some small way, at almost all of Earth's major life-forms—at primates in this chapter and at other animals, bacteria, plants, and so forth in Chapter 19. In the four chapters coming up, you'll take a more detailed look at each of these life-forms. This is a tour that will make plain the astounding diversity that exists in the living world today.

SO FAR . . .

1. Human beings appear to have evolved into their present anatomical form in _____.

2. Were modern human beings the first hominin species to reach Europe and Asia?

3. The "Hobbit" people, *Homo floresiensis,* are thought to have survived until as recently as _____ years ago on an island that is part of present-day _____.

20 REVIEW

Summary

20.1 The Human Family Tree

- Human evolution is the study of the taxonomic grouping called the Hominini. Every member of this group is referred to as a hominin, including human beings. (p. 363)

- A common primate ancestor is believed to have given rise to both the chimpanzee and the hominin evolutionary lines between 6 and 7 Mya, but the structure of the hominin tree is a matter of considerable debate. (p. 364)

20.2 Human Evolution in Overview

- All hominins are extinct except for *Homo sapiens*, the modern human species. *H. sapiens* was late in arriving among the hominins, but all hominins are late arrivers when the entire sweep of evolution is considered. All early and mid-period hominin evolution took place in Africa. (p. 364)

- In 2001 the range of likely hominin fossils was expanded from east Africa to Chad with the discovery of the 6- to 7-Mya remains of a primitive hominin named *Sahelanthropus tchadensis*. (p. 366)

20.3 Interpreting the Fossil Evidence

- Molecular evidence is increasingly important in human evolution studies, but the discipline's primary evidence remains fossil evidence. (p. 366)

- Paleoanthropologists interpret fossils to make judgments about where a given fossil form lies in the hominin family tree, but interpretations can differ. The two most important defining characteristics of a hominin are tooth structure and bipedal walking. (p. 367)

20.4 Snapshots from the Past: Four Hominins

- The bipedalism seen in our own species existed in the hominin species *Australopithecus afarensis*, whose most famous individual, Lucy, lived 3.18 Mya in what is now Ethiopia. Lucy had a much smaller brain than modern humans do, however, and she probably was partly arboreal. Lucy makes clear that hominin features developed in a mosaic pattern. (p. 368)

- A change to a physical form and mental capacity much closer to ours comes with the evolution of *Homo ergaster*, exemplified by Turkana Boy, who lived 1.6 Mya. A direct descendant of *H. ergaster*, *H. erectus* was the first hominin to leave Africa. Its migratory capability may have resulted in part from its anatomy, which likely allowed it to walk much the way modern humans do. (p. 368)

- In their most modern form, the Neanderthals (*Homo neanderthalensis*) appeared in Europe by about 130,000 years ago. Evidence of their settlements exists as far west as present-day England and as far east as Siberia. The last of them died in Europe 28,000 years ago. (p. 370)

20.5 The Appearance of Modern Human Beings

- Modern humans appear to have evolved into their current physical form prior to the time they began migrating out of Africa, though evidence from the sequencing of the Neanderthal genome has revealed that a small amount of interbreeding did take place between Neanderthals and human beings no later than 60,000 years ago in the Middle East. This interbreeding undercuts the strongest version of the "out-of-Africa" hypothesis, which holds that modern humans remained reproductively isolated from all other hominin species from the time they first evolved in Africa. (p. 370)

- Molecular evidence indicates that modern human beings evolved no earlier than 200,000 years ago. The earliest human fossils we have date from 195,000 years ago and were found in Ethiopia. (p. 373)

- The initial wave of modern human migration out of Africa appears to have begun sometime between 50,000 and 60,000 years ago, and to have taken modern humans through south Asia all the way into Australia. A second wave of human migration, initiated in western Asia, took modern human beings into Europe and Russia. (p. 370)

- The arrival of modern human beings in Europe 40,000 years ago was followed by the extinction of the Neanderthals 12,000 years later, while the arrival of human beings in the Far East was followed by the extinction of the *Homo erectus* species no later than 40,000 years ago. If there is a consensus about why modern human beings are the only living species of hominin, it is that they "replaced" such species as *H. erectus* and *H. neanderthalensis* by out-competing them after having migrated to Asia and Europe. (p. 373)

20.6 Next-to-Last Standing? The Hobbit People

- In 2004, researchers reported finding, on the Indonesian island of Flores, fossils of a previously unknown hominin, *Homo floresiensis*, who stood only 3 feet tall and who may have survived until 12,000 years ago. However, several teams of researchers have concluded that *H. floresiensis* actually was a small, modern human being suffering from one of several possible medical conditions. (p. 373)

Key Terms

bipedalism 367

Hominini 363

Homo sapiens (*H. sapiens*) 364

Neanderthal 370

paleoanthropologist 367

Understanding the Basics

Multiple-Choice Questions

(Answers are in the back of the book.)

1. The first fossils that were recognized as belonging to an extinct hominin species (select all that apply):
   a. were found in the nineteenth century in Germany.
   b. belonged to the species *Australopithecus africanus*.
   c. were found in the twenty-first century in Indonesia.
   d. belonged to the species *Homo neanderthalensis*.
   e. belonged to the species *Homo erectus*.

2. The taxonomic group Hominini is defined as the group of:
   a. modern humans and chimpanzees.
   b. modern humans and *Homo neanderthalensis*.
   c. modern humans, chimpanzees, and gorillas.
   d. human-like primates.
   e. primates lacking a tree-dwelling existence.

3. The Hominini (select all that apply):
   a. originated between 135,000 and 350,000 years ago.
   b. includes one living species.
   c. originated between 6 and 7 million years ago.
   d. includes two living species.
   e. shares a common ancestor with the chimpanzee taxonomic group.

4. To judge by fossil remains and other evidence, the "Lucy" hominin *Australopithecus afarensis* appears to (select all that apply):
   a. have been bipedal.
   b. have been about as intelligent as modern humans.
   c. have existed roughly 3.2 million years ago.
   d. have been replaced by modern humans.
   e. have existed roughly 1.6 million years ago.

5. The first hominin species to leave Africa:
   a. was *Homo sapiens*, 50,000–60,000 years ago.
   b. was *Homo neanderthalensis* 350,000 years ago.
   c. was *Australopithecus afarensis* roughly 3.2 million years ago.
   d. was *Homo erectus* roughly 1.8 million years ago.
   e. was *Homo sapiens* roughly 1.8 million years ago.

6. Neanderthals (select all that apply):
   a. first evolved in Africa.
   b. first evolved in Europe.
   c. died out in Indonesia.
   d. became extinct after a period of living in proximity to *Homo sapiens*.
   e. died out 28,000 years ago.

7. *Homo sapiens* or modern humans (select all that apply):
   a. evolved into their present anatomical form in Africa.
   b. first migrated out of Africa 1.8 million years ago.
   c. never interbred with Neanderthals.
   d. went north and west, into Europe, on their first migration out of Africa.
   e. first appeared in Africa about 200,000 years ago.

8. The "Hobbit" species that has been named *Homo floresiensis* (select all that apply):
   a. lived in present-day Indonesia.
   b. was arguably a Neanderthal.
   c. was arguably a modern human suffering from a physical condition.
   d. is thought to have died out 28,000 years ago.
   e. may represent a previously unknown species of hominin.

Brief Review

(Answers are in the back of the book.)

1. Did human beings evolve from chimpanzees?

2. All phylogenies or family trees are tentative, but why is the hominin family tree particularly tentative?

3. East Africa has sometimes been called the cradle of humanity. Why has it gotten this title?

4. Arrange the following hominin species by order of their chronological appearance: *Homo erectus*, *Sahelanthropus tchadensis* (Toumaï), *Homo neanderthalensis*, *Australopithecus afarensis* (Lucy).

5. How was it possible for modern humans to "replace" Neanderthals in locations such as Europe without killing them outright?

Applying Your Knowledge

1. A common misconception in society is that scientists are hunting for a "missing link" that falls in an evolutionary line between human beings and apes. Why does our current knowledge of human evolution make the concept of a missing link meaningless?

2. With the sequencing of the Neanderthal genome, some scientists have argued that it is now theoretically possible to produce a Neanderthal through a combination of genetic engineering and cloning. If such a thing were done, how do you think society would treat any Neanderthal that resulted? As a species more akin to chimpanzees, which is to say subject to being kept in zoos? Or as regular, if somewhat different, human beings?

3. In his 1903 essay "Was the World Made for Man?" Mark Twain wrote, "If the Eiffel tower were now representing the world's age, the skin of paint on the pinnacle-knob at its summit would represent man's share of that age; and anybody would perceive that that skin was what the tower was built for. I reckon they would. I dunno." Evaluate Twain's words in light of what you have learned about the history of life on Earth.

MasteringBiology®

1. MasteringBiology Assignments
ABC News Video: Modern Humans and Neanderthal DNA; **Current Events:** Family Tree of Languages has Roots in Anatolia, Biologists Say; A Masterpiece of Nature? Yuck!; Lucy's Kin Carved Up a Meaty Meal, Scientists Say; **Building Vocabulary; Questions:** Reading Questions, Test Bank

2. eText
Read your book online, search, take notes, highlight text, and study more effectively.

3. The Study Area
Self-study tools to test comprehension.

Viruses, Bacteria, Archaea, and Protists:

The Diversity of Life 1

Across its categories, life is stunningly diverse. Some of the most interesting living things are microscopic.

Available on MasteringBiology®

Activities: Retrovirus (HIV) Reproductive Cycle; Phage Lysogenic and Lytic Cycles; Phage Lytic Cycle; Replication in Detail; Simplified Viral Reproductive Cycle; Bacterial Conjugation; Structure and Reproduction of Bacteria; **ABC News Video:** Turning Algae into Biofuel; **Current Events:** Battle Brewing Over Labeling of Genetically Modified Food; Scientists Move Closer to a Lasting Flu Vaccine; Can Answers to Evolution Be Found in Slime?; The New Generation of Microbe Hunters; Heart of Iowa as Fault Line of Egg Recall; How Microbes Defend and Define Us; Hunting Fossil Viruses in Human DNA; Virus Ravages Cassava Plants in Africa; **Building Vocabulary; Questions:** Reading Questions, Test Bank

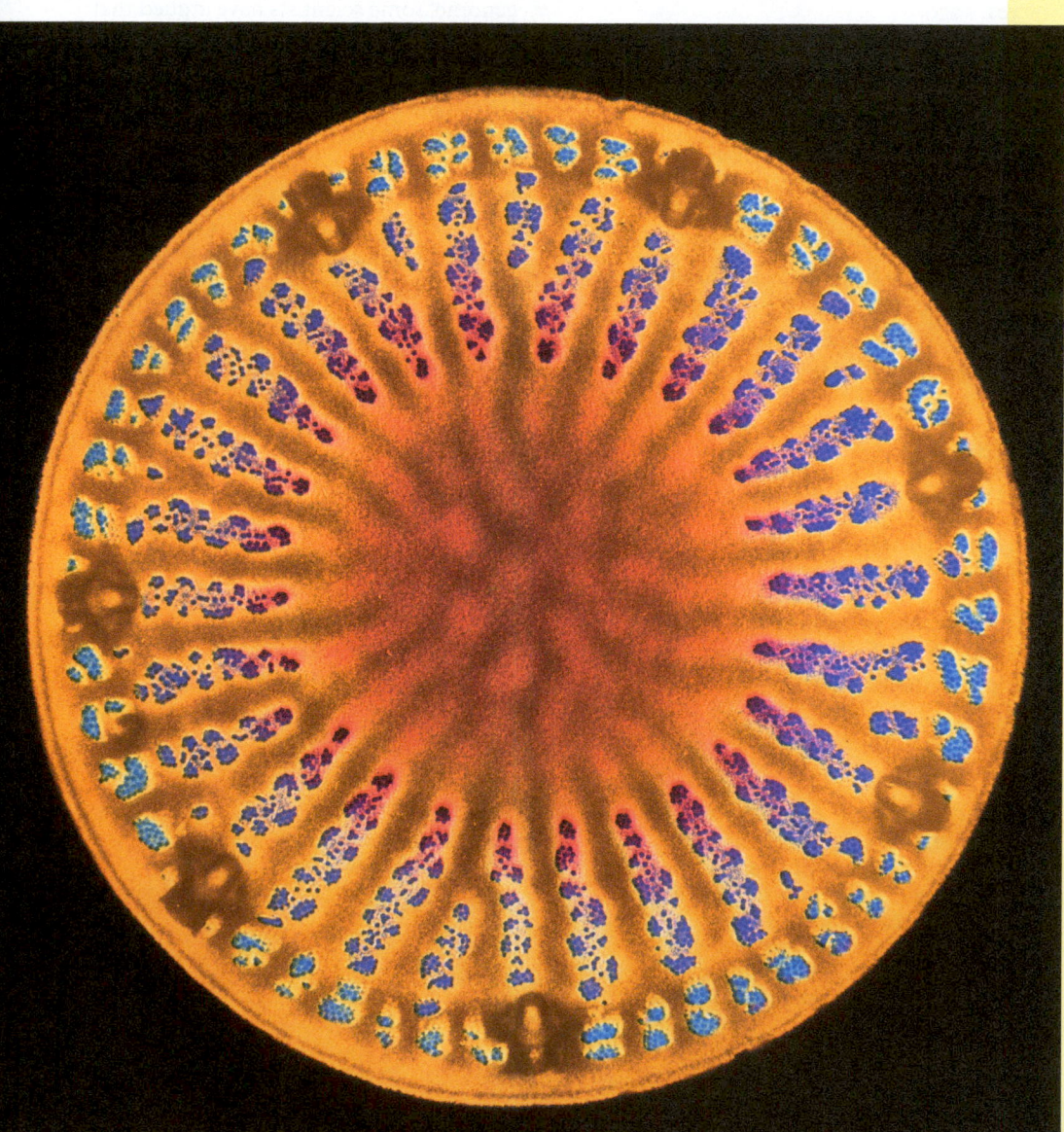

The single-celled microscopic algae known as diatoms possess outer cell walls, made of silica, that come in a fantastic variety of geometric forms. This color-enhanced micrograph is of the diatom *Cyclotella pseudostelligera*, which lives in both fresh and brackish water.

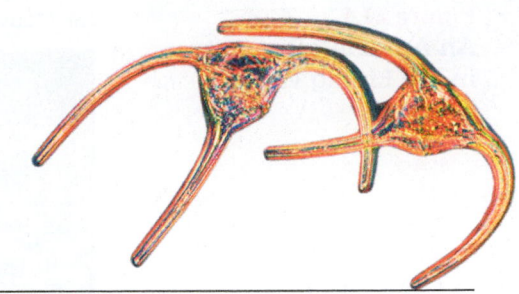

The smallest living thing discovered to date is a type of bacteria that measures about two-tenths of a micrometer in diameter—that is, two-tenths of one-millionth of a meter in diameter. When we turn to the biggest living thing, elephants or whales might come to mind, but it turns out that animals aren't even in the running for this title. At a maximum length of about 89 feet, the blue whale is the largest animal that has ever existed, but it's small compared to California's coastal redwood trees, which might reach a height of 330 feet. This is large indeed, but consider that a giant sea kelp once was found that was about 900 feet long. It may be, though, that in terms of size, both kelp and redwoods have to take a backseat to a life-form normally thought of as quite small. In 1992, researchers reported finding a fungus growing in Washington state that runs underground through an area of 1,500 acres and that arguably is a single organism.

Now, how about the *deepest* living things? The current record holders are bacteria that were found in the early 1990s—in a combination of oil drilling and scientific exploration—living almost 1.9 miles beneath Earth's surface. Food and water are so scarce at such depths that these bacterial cells may live in a kind of suspended animation, dividing perhaps once a year or even once a century. Turning to the highest living things, 12 species of bacteria have been found living in the Himalayas at 27,000 feet above sea level, which is just a couple thousand feet lower than the peak of Mount Everest. A species of chickweed survives in the Himalayas at more than 20,000 feet.

All kinds of life-forms—clams, tube worms, crabs—live kilometers beneath the surface of the oceans, near the "hot-water vents" that spew out water and minerals from the Earth's interior onto the ocean floor. Microbes from a group called archaea, living at these deep-sea vents, set the pace for life in a hot environment. An average, midday temperature in the Sahara Desert would be so *cold* it would end the reproduction of the archaeon *Pyrolobus fumarii*, which lives within hot-water vents at temperatures of up to 113°C, which is hotter than the temperature at which water boils. At the other extreme, the well-named bacterium *Phormidium frigidum* carries out photosynthesis in lakes in Antarctica whose temperatures sit right at the point at which water freezes.

Finally, what about the oldest living things? Humans have been known to live to 120, but this is an eyeblink compared to the lives of some plants. There is a bristlecone pine tree (*Pinus longaeva*) that has been living in the dry slopes of eastern California for 4,900 years, meaning it has been alive since the pharaohs ruled Egypt. For many years, this "Methuselah" tree was believed to be the oldest in the world, but in 2008 scientists performed carbon-14 dating on a set of spruce trees growing in the mountains of western Sweden and found that they are 8,000 years old. All trees, however, are mere youngsters compared to some microbes, which can survive in a form called spores for immensely long periods of time. In the 1990s, bacterial spores were found living within a bee that had been encased in amber for 25 to 40 million years. Even this eye-popping number did not prepare scientists for what came next, however. In 2000, researchers announced that they had found, encased within salt crystals at the bottom of a New Mexico air shaft, archaeal spores that date from 250 million years ago. These spores were revived, and billions of their offspring now exist. This claim has been controversial from the time it was made; one possibility is that the crystal sample the researchers were looking at was contaminated with modern DNA. But if true, it raises the question of whether some microbes are "effectively immortal," as one researcher has put it.

Looking at life like this—at its biggest or deepest or oldest extremes—is one way to get at its incredible diversity. But it does not begin to do justice to life's variety because life is diverse in so many ways. Honeybees do a "dance" to let their hive-mates know the location of a food supply. Corn plants, when attacked by army worms, can call in an air force: The plants release an airborne substance that attracts parasitic wasps, which then prey on the worms. Pacific salmon live for years in the ocean, only to make one arduous, upstream journey to spawn in the waters where they were born, after which they die. Look closely at almost

Figure 21.1
An Amazing Diversity in the Living World

(a) Mouse beneath the foot of an elephant

(b) A lion nudibranch moving across a giant kelp blade

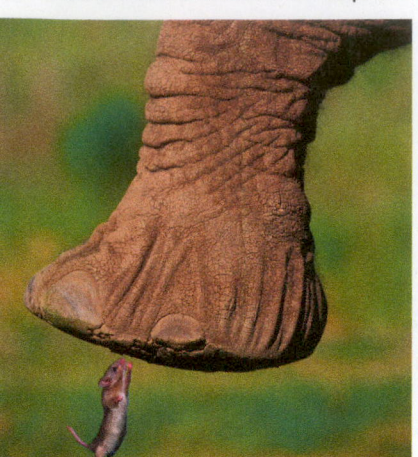

(c) Fungus growing on a fallen log

(d) Water flea swimming near green algae

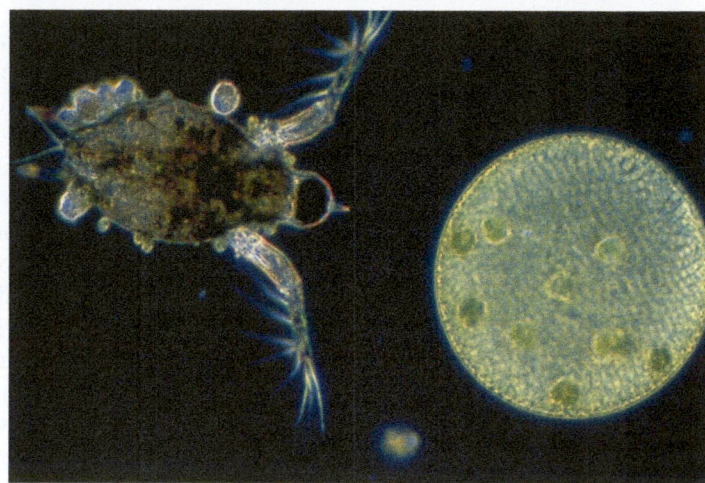

any creature, and you are likely to find something only slightly less dramatic (**Figure 21.1**).

21.1 Life's Categories and the Importance of Microbes

The purpose of this chapter and the three that follow is to introduce you to life as it exists across all its large-scale categories. As you may recall from Chapter 19, all living things can be placed into one of three *domains* of life. Two of these domains are Archaea and Bacteria, whose members are all single-celled and microscopic. The third domain, Eukarya, encompasses such incredible diversity that it is further divided into four separate *kingdoms*. These are the kingdoms of plants, animals, fungi, and a group of mostly microscopic organisms called protists. We will begin our tour of the living world by starting small, looking in this chapter at Domains Bacteria and Archaea and at the protists within Domain Eukarya. Then, we'll look at fungi in Chapter 22, animals in Chapter 23, and plants in Chapter 24.

Figure 21.2 shows the evolutionary relationships among all three domains of life and the kinds of organisms that are part of each domain.

Because this chapter is devoted to mostly unfamiliar *microbes*—meaning living things so small they can't be seen with the naked eye—it might be helpful to say a word at the start about their significance. Microbes are something like the foundation of a house: seldom thought of but critically important. To the extent that they cross our minds at all, we are aware of them mostly for the diseases they cause. This is an important topic—one we'll go into extensively—but thinking of microbes strictly as disease-causing organisms is like thinking of cars strictly as wreck-causing machines. Where does the oxygen we breathe come from? If your answer is "plants," that's partly correct. But the photosynthesis carried out by plants supplies less than half our oxygen; the balance comes from microscopic algae and bacteria—mostly drifters on the ocean's surface. How about the nitrogen that is indispensable to all plants and animals? Among living things, only bacteria and archaea are capable of

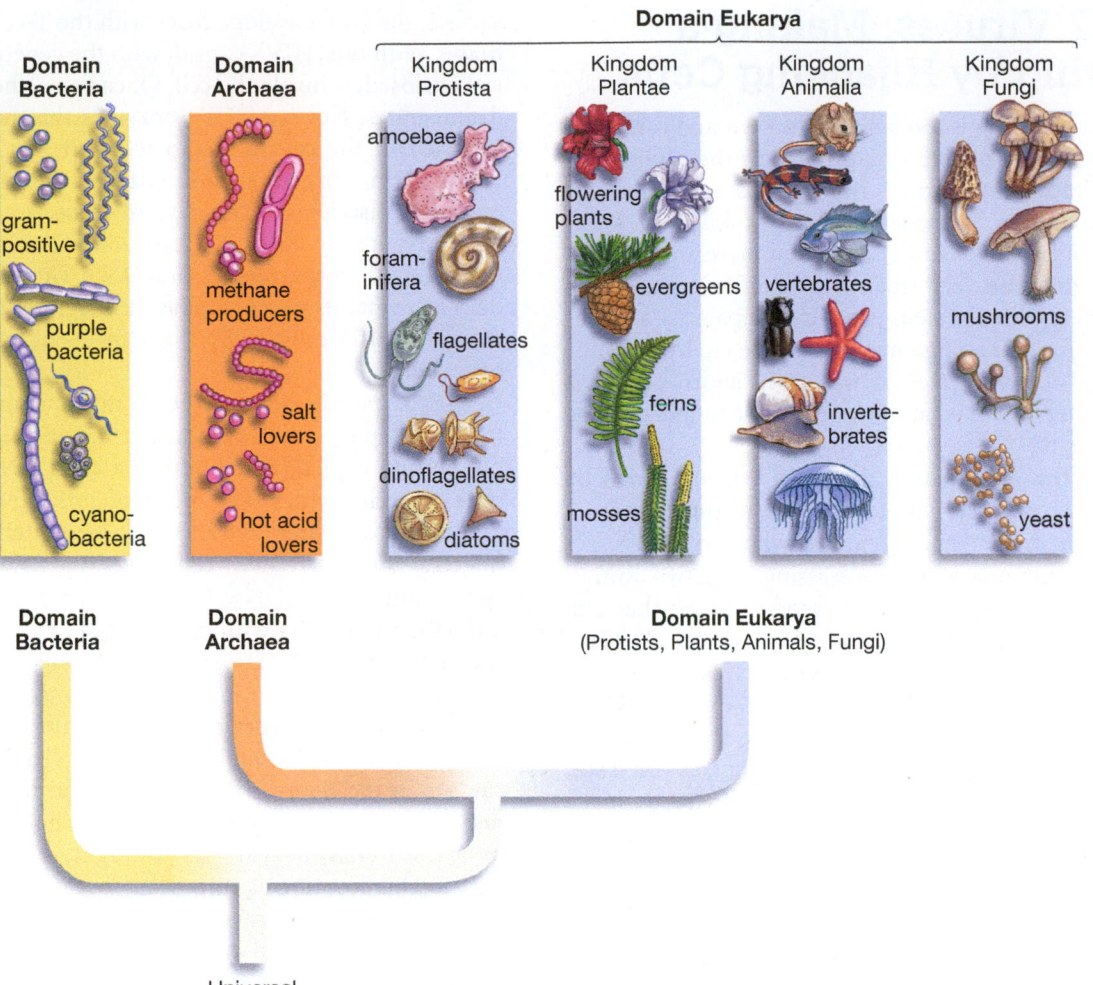

Figure 21.2
The Tree of Life

Life is presumed to have developed from a single common ancestor (represented by the base of the tree), which gave rise to the three domains of the living world: Bacteria, Archaea, and Eukarya.

taking nitrogen that exists in the air and transforming it into a form that plants can use. If there were no bacteria and archaea, there would be no plants—and consequently no land animals. And how is it that dead tree branches, orange peels, or the bones of animals are broken down and recycled into the Earth? Almost entirely through the work of bacteria and fungi. Without these organisms, Earth would long ago have become a garbage heap of dead, organic matter. Animals cannot decompose a dead tree branch, but bacteria and fungi can.

The upshot of all this is that life on Earth would grind to a halt without microbes; they are an essential underpinning to all forms of life. They come to this status in part because of the environments they live in, which is to say all environments on Earth that support life of any kind. It probably didn't escape your attention that the coldest-, hottest-, highest-, and deepest-living things were all bacteria or archaea. Wherever a plant or animal can live, microbes can be found, too.

Just as impressive is the sheer number and tonnage of these tiny organisms. A common denominator for measuring the amount of living material is something called *biomass,* defined as the total dry weight of material produced by a given organism. Each bacterium or microscopic alga has an infinitesimally small weight, of course—but then again, there are so many of them! It has been estimated that there are more bacteria living in your mouth now than the number of people who have ever lived, and there are far more bacterial cells in the human body than there are human cells (a topic we'll return to). A typical gram of fertile soil—about a quarter teaspoon's worth of material—has about 100 million bacteria living in it, and a mere drop of water near the ocean's surface contains thousands of the tiny algae called phytoplankton. All this adds up. When the calculations are done, we find that microbes probably make up more than half of Earth's biomass. Put another way, the microbes of the world outweigh all the plants, animals, and visible fungi of the world combined.

In sum, microbes matter, and not just in ways that are harmful. We'll now start to look at the portion of the living world that is made up of them. Our first stop on this tour, however, will be a category of tiny replicating entities that most biologists would say lie just outside of life, although they have a great effect on living things.

Retrovirus (HIV)
Reproductive Cycle

Phage Lysogenic and Lytic
Cycle

Phage Lytic Cycle

Replication in Detail

21.2 Viruses: Making a Living by Hijacking Cells

If a living thing is too small to be seen and can cause disease, most people would put it in the vague category known as "germs." But there is a fundamental distinction to be made between *two* kinds of infectious organisms. On the one hand there are bacteria, which may be very small but nevertheless are *cells,* complete with a protein-producing apparatus, a mechanism for extracting energy from the environment, a means of getting rid of waste—in short, complete with everything it takes to be a self-contained living thing.

On the other hand, there are *viruses,* which by themselves possess none of these features. Indeed, viruses can be likened to a thief who arrives at a factory he intends to rob possessing only two things: the tools to get inside and some software that will make the factory turn out items he can use. The factory being broken into is a living cell, and the software that the virus brings is its DNA or RNA, which it puts inside this "host" cell. Once this is accomplished, viruses employ different tactics, as you'll see, but in most cases the result is that viruses make more copies of themselves. Viruses are an integral part of the living world, and not all of them cause harm. But unlike the case with bacteria, it's right to think of them first and foremost in terms of the diseases they cause, not only in humans but in every variety of living thing. **Viruses** are noncellular replicating entities that must invade living cells to carry out their replication.

HIV: The AIDS Virus

If you look at **Figure 21.3a**, you can see a simplified rendering of a particular virus, the human immunodeficiency virus (HIV), which is the cause of AIDS. All viruses have two of the three large-scale structures you can see in HIV: the genetic material at the core (in HIV's case, two strands of RNA) and a protein coat, called a **capsid,** that surrounds the viral genetic material. Many viruses then go on to have a third major element—a fatty membrane, called an *envelope,* which you can see surrounding the HIV capsid. Protruding from the HIV envelope are a series of receptors, often referred to as "spikes," which are proteins capped with carbohydrate chains. These serve the critical role of binding with receptors that protrude from the target cell, thus giving the virus a way to get in.

If you look at **Figure 21.3b**, you can see how the life cycle of HIV proceeds from this initial binding. The target cell in this case is the one most commonly invaded by HIV, an immune system cell called a helper T-cell (also known as a CD-4 cell). As you can see, once HIV binds with two receptors on the T-cell's surface, the viral envelope fuses with the T-cell membrane. With this, HIV's capsid, with the genetic material enclosed, is inside the cell. Once there, the capsid disintegrates. Now two HIV enzymes that had been enclosed in the capsid get to work. If you look at Figure 21.3a, you can see these enzymes, integrase and reverse transcriptase. The reverse transcriptase gets busy first, using HIV's RNA strands as a template to produce double-stranded viral DNA. Then integrase does just what its name implies: It integrates the viral DNA into the *cell's* DNA through a cut-and-paste operation.

At this point, the viral DNA might simply stay integrated in the T-cell's DNA, causing no great harm but getting copied, along with the cell's DNA, each time the cell divides. The effect of this, however, is that every "daughter" cell of this infected cell will be infected too; each new cell will have viral DNA spliced into its own DNA. Each of these cells then has a time bomb ticking within it because at some point the HIV DNA will change course. It will no longer simply go on replicating within the cell's DNA. Instead, it will start turning out the materials necessary to make whole new copies of the virus. As you can see in the figure, virus construction takes place just inside the cell's outer membrane (helped along by a third enzyme that HIV brought with it, protease). This location is important because the cell's membrane ends up serving as the final structural part of each new viral copy or "particle." When the capsid is completed, it buds off from the cell, taking with it part of the cell's membrane, which now becomes the capsid's envelope. The completed viral particle that emerges from the cell then goes on to infect other cells.

Helper T-cells that produce new viral particles in this way may be targeted for destruction by the immune system (since they are infected cells). Beyond this, HIV receptors that have become detached from the virus can land on the surface of *un*infected helper T-cells, thus causing them to initiate a kind of cellular suicide known as programmed cell death. Once both these processes have been set in motion, the result is a greatly reduced population of helper T-cells—a potentially calamitous event since these cells are central to the body's immune system response. A person whose helper-T cell-count has dropped below a certain minimum level loses the ability to fight off everyday infections caused by bacteria, fungi, and viruses other than HIV.

Viral Diversity

In the HIV life cycle, you can see four steps that are common to most viruses: Get genetic material inside the host cell, turn out viral component parts, construct new particles from these parts, and move these particles out of the cell. HIV uses RNA as its genetic

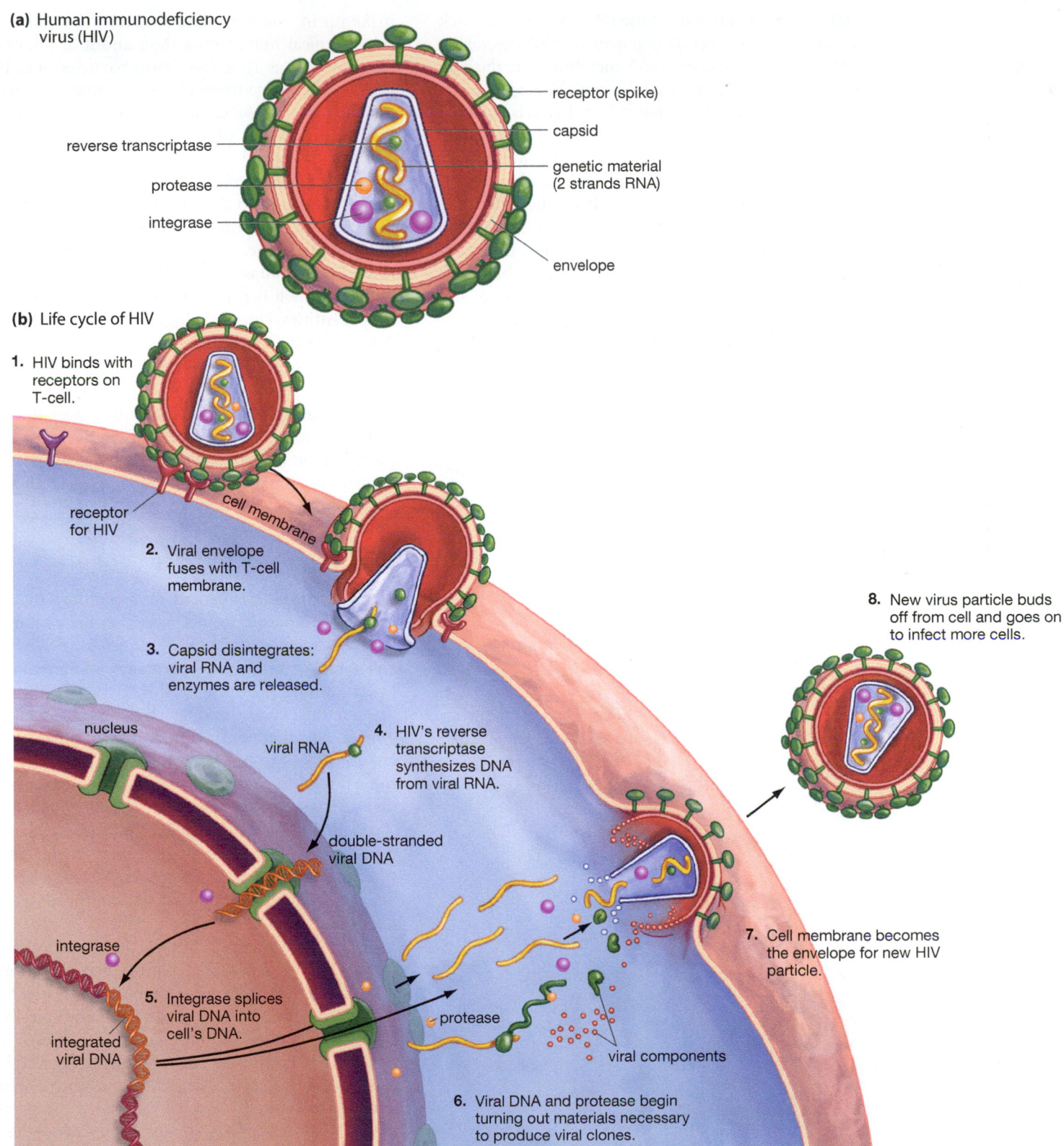

(a) Human immunodeficiency virus (HIV)

receptor (spike)

capsid

reverse transcriptase

genetic material (2 strands RNA)

protease

integrase

envelope

(b) Life cycle of HIV

1. HIV binds with receptors on T-cell.

receptor for HIV

cell membrane

2. Viral envelope fuses with T-cell membrane.

3. Capsid disintegrates: viral RNA and enzymes are released.

nucleus

viral RNA

4. HIV's reverse transcriptase synthesizes DNA from viral RNA.

double-stranded viral DNA

integrase

5. Integrase splices viral DNA into cell's DNA.

integrated viral DNA

protease

6. Viral DNA and protease begin turning out materials necessary to produce viral clones.

viral components

7. Cell membrane becomes the envelope for new HIV particle.

8. New virus particle buds off from cell and goes on to infect more cells.

Figure 21.3
The Virus That Causes AIDS
The anatomy and life cycle of HIV, the human immunodeficiency virus.

Simplified Viral
Reproductive Cycle

material, but many viruses use DNA. Viruses that lack an envelope don't get their genetic material inside the cell by fusing with the cell's membrane in the way HIV does. In **Figure 21.4**, you can see a spacecraft-shaped virus called T4 that infects bacteria through an alternate means. T4 keeps its capsid outside the target cell but gets its DNA inside by injecting it through the bacterium's cell wall and membrane.

As you might be able to tell from the size of T4 compared to its target cell, viruses are very small things. Thousands of them can typically fit within a bacterial cell. Some amount to nothing but their genetic material and the capsid that surrounds it, and there are even virus-like entities, called *viroids,* that lack a capsid; they are simply small strands of infectious RNA. What viruses lack in size, however, they

make up in numbers. One researcher has calculated that a typical milliliter—a thousandth of a liter—of ocean water has 10 million virus particles in it. If so, then ocean-going viruses contain as much as 270 million metric tons of carbon, which is more than 20 times the amount of carbon contained in all the whales in the world.

In every case, viruses are not only small but *simple* things compared to the organisms they infect. First, they lack the usual machinery that cells have—a protein assembly line, a waste disposal system, and so forth. Even when we look at the one thing all of them do have, which is genetic material, they are still very simple entities. Human beings are thought to have between 20,000 and 25,000 genes, and even a primitive organism like baker's yeast has 6,000 genes. But the AIDS virus? It has nine genes. To be sure, some viruses are more complex than HIV, but in no case is a virus complex enough to replicate by itself. Because viruses can carry on scarcely any of life's basic functions by themselves, most scientists don't classify them as living things.

The Influenza Viruses

But what trouble is caused by something that isn't even alive! Apart from AIDS, viruses are responsible for polio, measles, chickenpox, rabies, herpes, rubella, and some forms of cancer, hepatitis, and pneumonia, to say nothing of common colds.

As if this weren't enough, there are the influenza viruses, which come in common forms that cause a fair amount of trouble each year—in the form of seasonal flus—but which also come in uncommon forms that can throw societies into full-blown states of alarm. In 2009, for example, countries around the world took extraordinary measures to prevent the spread of an influenza virus properly known as H1N1, but referred to in the early days of its outbreak as swine flu (**Figure 21.5**).

In the United States, about 11,700 Americans died from H1N1 from its emergence in April 2009 through mid-January 2010—a large number of deaths to be sure, but about 24,000 *fewer* deaths than typically result from seasonal flu in the United States each year. So why did H1N1 cause such widespread anxiety? Because some classes of viruses carry more potential danger than others and H1N1 was a threatening member of a very dangerous viral class—the influenza type A viruses.

The twentieth century saw three *pandemics,* or worldwide outbreaks of fatal disease, that were caused by influenza A viruses. One of these, in 1957, killed between 1 and 4 million people, while another, in 1968, killed about a million people. Both paled in comparison, however, to an outbreak that occurred in

Figure 21.4
**Another Variety
of Virus**

(a) The T4 virus looks like a spacecraft that has landed on the surface of a bacterium.

(b) A viral invader lands. An artificially colored T4 virus injects its DNA into an *Escherichia coli* bacterium (in blue).

(a)

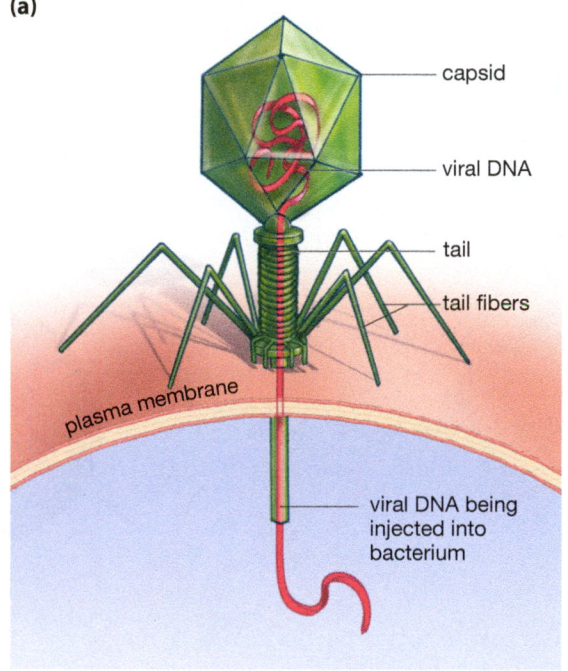

capsid

viral DNA

tail

tail fibers

plasma membrane

viral DNA being injected into bacterium

(b)

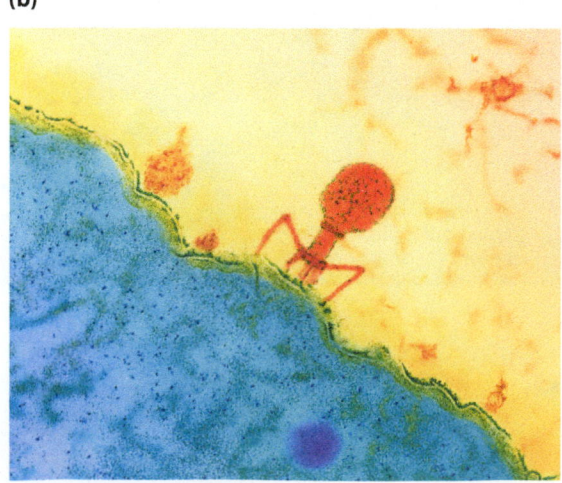

1918–1919. That pandemic is estimated to have infected a third of the people then living in the world and to have killed 50 million of them. (By way of comparison, AIDS took a quarter century to kill 25 million people.) When H1N1 first surfaced in Mexico and the United States, it was immediately recognized as an influenza A virus that shared several characteristics of the twentieth-century's killer influenzas: It contained a mixture of genes so novel that people were unlikely to have any natural immunity to it; it spread easily from one person to another; and its effects could be lethal—of the first 26 people known to have contracted it in Mexico, 7 died from it.

Given these characteristics, the initial reactions of public health officials to the H1N1 outbreak now seem understandable. Even so, we might ask how these experts managed to be taken by surprise by the appearance of a virus as potentially catastrophic as this one. The answer is that many influenza A viruses are likely to come as a surprise to experts because of the way these viruses evolve. Though all of them are assumed to originate in aquatic birds, they can move between several animal species and they can combine with each other in new ways *within* animal species. Two different viruses that commonly circulate among, say, human beings and birds will infect a population of pigs and exchange genetic sequences with one another during this infection. The result? A new virus that is different from either of the parental varieties. This process of *reassortment,* as it's called, was on full display in the evolution of the H1N1 virus. It contained two genes that came originally from an "avian" or bird influenza virus that infected swine in 1979, three genes from a swine virus that was first isolated in 1930, two genes from human and avian viruses that reassorted in swine in the 1990s, and one gene that came from humans after being transmitted to them from birds in 1968.

It would be nice to report that a mind-bogglingly complex pattern of evolution such as this is only seen in viruses like H1N1, but that's not the case. Influenza A viruses in general evolve in this way, which is why, even in their less dangerous forms, these infectious agents join the influenza B viruses in giving societies a predictable problem to deal with each year: The need for annual flu vaccinations. Last year's flu vaccine will not protect anyone against this year's viruses because this year's threat is a set of recently evolved viral "strains" or subtypes. Ultimately, society is helpless to prevent the emergence of harmful influenzas because new strains of them are generated so quickly in so many different ways. The best we can do is to keep close tabs on the animals that commonly serve as influenza hosts and to put in place containment responses that can be ramped up quickly when the time comes.

SO FAR . . .

1. The three domains of life are _____, _____, and Eukarya. The four kingdoms within Eukarya are composed of _____, _____, _____, and _____.
2. Microbes perform several activities that are critical to the maintenance of other life-forms on Earth. Name three of these activities.
3. In order to replicate, viruses must invade _____.

Figure 21.5
Trying to Prevent a Pandemic

An employee of the Richardson, Texas school district sprays an antimicrobial solution on the Canyon Creek Elementary School in April 2009. The school closed for several days after two of its students contracted H1N1 flu.

21.3 Bacteria: Masters of Every Environment

Domain Bacteria is made up entirely of the microbes called bacteria. Keeping in mind that almost all the organisms we're looking at are microbes, what separates bacteria from any of the others? Recall from Chapter 4 that if we set a bacterial cell and, say, a fungal cell side by side, anyone peering through a microscope could see a fundamental difference between the two. The fungal cell would have a large, circular-shaped structure near its center, while the bacterial cell would not. The circular structure is the fungal cell's nucleus, which contains almost all of its DNA. But bacteria are

ESSAY

Good News in the Fight Against Herpes

The first outbreak usually is the worst, but the real problem is that the first outbreak usually isn't the last. Herpes comes back, as herpes sufferers well know. A few weeks or months after the initial outbreak, everything seems fine, but then there may be a tingling—nothing that can be seen yet, but instead just a feeling. Not much later, however, it's back: the same unsightly "cold sore" on the mouth or lesions on the genitals, in the same place that the first outbreak occurred.

This sequence of events is depressingly familiar to millions of Americans. And at present, the best medical treatment available to any of them is a drug called Acyclovir, which can shorten the duration and frequency of herpes outbreaks but which cannot eliminate herpes from the body altogether. In 2008, however, researchers at Duke University made a fundamental breakthrough against herpes: They discovered the means by which the virus that causes it hides from the human immune system. With this knowledge in hand, pharmaceutical researchers began working on a drug that essentially will force the herpes virus to show itself in all its hiding places. Any virus that is flushed out in this way is vulnerable to being eradicated from the body.

The term "herpes virus" is shorthand for a pair of viruses, the *herpes simplex viruses,* which in turn are part of a group of 100 known viruses in the herpes family. (Other viruses in the family cause chicken pox, shingles, and the Epstein-Barr disease.) Herpes simplex viruses come in type 1 and type 2 varieties. The type 1 variety supposedly is "oral," while the type 2 is "genital." But either variety can end up in either location because of oral sex.

The herpes sores that can be so distressing are the result of newly formed virus particles breaking out of skin cells, thereby destroying them. Fever, swollen glands, and other symptoms often accompany these lesions, particularly in an initial infection. Within a couple of weeks or so, however, the immune system (or a drug like Acyclovir) has killed all the virus particles in and around the skin cells, and the symptoms are gone. The problem is that, by this time, the virus has also infected *nerve* cells in the area. Such nerve cells (or neurons) have long extensions of themselves called axons. With the initial infection, herpes virus particles start making a slow migration up these axons to the *body* of the nerve cells, specifically to the nucleus inside the cell body. In the case of genital herpes, this means a journey from the vagina or penis to cell bodies that sit right at the base of the spinal cord. And these cell bodies are where the virus stays—permanently. Once there, it does next to nothing; it's not causing any nerve damage or replicating. And the nerve cells it is in never divide, so it isn't passed on to a larger group of cells.

This lack of activity is, however, the essence of the problem with herpes. Most viruses are *active* inside the cells they infect; they produce viral proteins, fragments of which end up being displayed on the surfaces of infected cells. These fragments serve as signals to the immune system, conveying the message "Infected cell here; begin attack." The herpes virus, however, avoids all this by staying silent—it produces no proteins at all. It is this lack of viral productivity that the researchers from Duke figured out. They discovered that, in an infected cell, the herpes virus actually is active in one major way: It produces a length of RNA that then is processed into four *micro*-RNAs, which is to say RNAs that *prevent* the production of proteins.

Indeed, micro-RNAs often end up cutting up the messenger RNAs that are required for the production of any protein. Thus, the herpes virus silences itself inside cells; it undercuts its own production of viral proteins.

The only change in this series of events comes when the virus senses that the cell it's part of is threatened in some way—say, by the body having acquired a cold. Now the virus' production of proteins overwhelms the micro-RNA control system. The result? Newly made viral particles leave the infected cell—in a manner that one researcher has compared to rats deserting a sinking ship—and migrate back up the cell's axon and end up at the same old place in skin cells on the lips or the genitals. With this, the body has a new herpes outbreak. (Since a cold is capable of bringing this sequence of events on, you can see why herpes lesions are sometimes referred to as cold sores.) A drug like Acyclovir can kill all the virus particles in any of these outbreaks. The problem is that not all of the body's infected cells go through this cycle at the same *time;* the virus remains latent in some cells even while it has become active in others. Thus, there is no point at which the virus is showing itself in all of the body's infected cells. As a result, in any given person, the herpes virus always perseveres to break out another day—at least for now, but perhaps not in the future.

With the discovery made at Duke, pharmaceutical researchers began trying to develop a drug that will stop the production of herpes micro-RNAs. Any drug that could do this should theoretically *initiate* the production of herpes proteins. Once this happens, each infected cell could be identified as infected by the immune system and Acyclovir, and that could spell the end of herpes throughout the body. Only time will tell whether this sequence of events actually will unfold, but at the moment, promising anti-herpes compounds are being tested for safety in animals.

prokaryotes: organisms whose DNA is not contained within a cell nucleus. Only bacteria and archaea lack a nucleus. All other organisms—all plants, animals, fungi, and protists—are composed of cells that have a nucleus, which makes them **eukaryotes.** The lack of a cell nucleus is, however, just one defining feature of bacteria. Here is a list of others.

- **No membrane-bound organelles.** A nucleus is an example of an *organelle:* a highly organized structure within a cell that carries out specific cellular functions. You may remember from Chapter 4 that, in addition to a nucleus, eukaryotic cells have such organelles as mitochondria and lysosomes, which are surrounded by a membrane. Bacteria have only a single kind of organelle, the ribosome, which differs from other organelles in that it is not surrounded by a membrane.

- **Single-celled organisms with small cell size.** Bacteria often come together in groupings that have an organization to them, but each bacterial cell is a self-contained living thing—each one can live and reproduce without the aid of other cells. Most bacterial cells are also very small, relative to eukaryotic cells; thousands of them could typically fit into a single eukaryotic cell. (If you look at **Figure 21.6**, you can see a size comparison of bacteria, a T4 virus, and a protist you'll be looking at called a paramecium.)

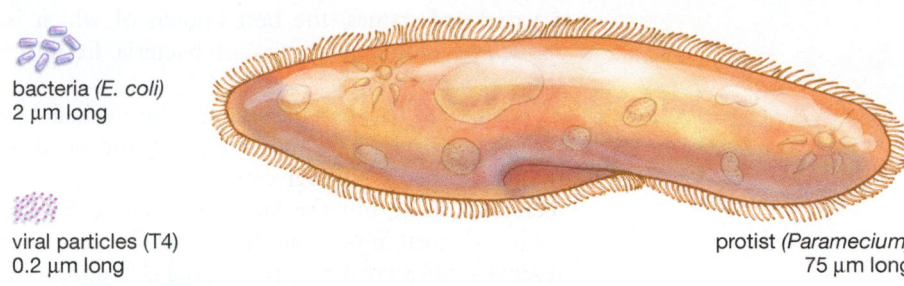

bacteria (*E. coli*)
2 µm long

viral particles (T4)
0.2 µm long

protist (*Paramecium*)
75 µm long

Figure 21.6
Small Is a Relative Thing
The bacteria, protist, and viral particles shown in the figure are all microscopic. Nevertheless, they differ greatly among themselves in terms of size. A typical human cell might be about a third the size of the *Paramecium*.

- **Asexual reproduction.** Bacteria reproduce through a form of simple cell splitting called **binary fission:** one cell splits into two, with both "daughter" cells being genetic replicas of the parental cell. In contrast, most eukaryotes are capable of sexual reproduction, in which reproductive cells from two separate organisms come together to produce offspring (as with egg and sperm coming together in human reproduction).

MB

Bacterial Conjugation

If you look at **Figure 21.7**, you can see micrographs of three forms that bacterial cells commonly take: round, rod shaped, and spiral shaped. A round bacterium is known as a *coccus;* a rod-shaped bacterium is a *bacillus;* and a spiral-shaped bacterium goes under

(a) Round bacteria (cocci)

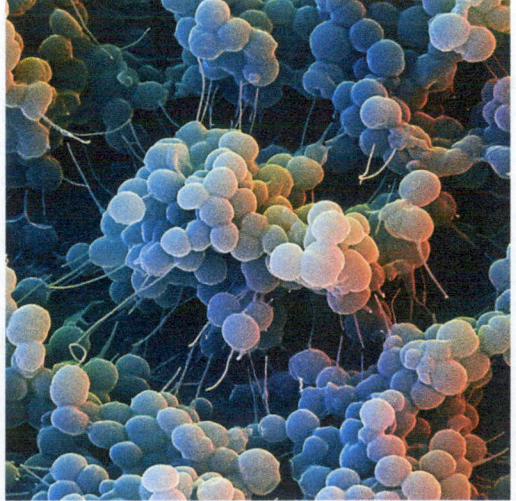

(b) Rod-shaped bacteria (bacilli)

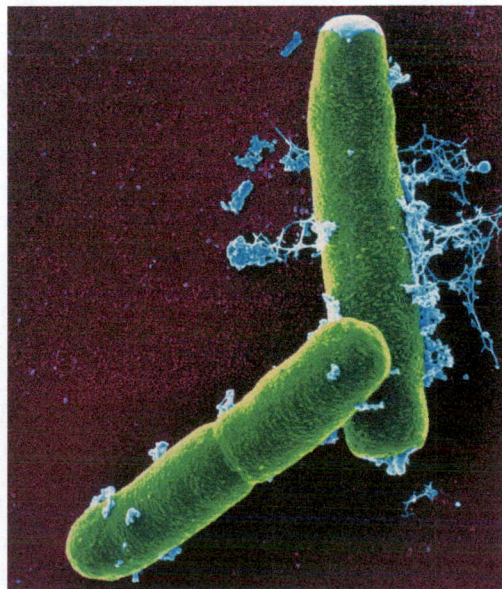

(c) Spiral-shaped bacterium (spirochete)

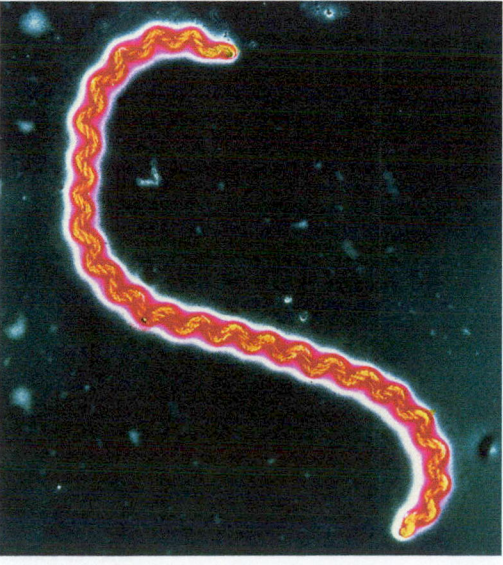

Figure 21.7
Different Shapes of Bacteria

(a) The round bacterium *Staphylococcus epidermis* is a normal part of the bacteria that cover our skin.

(b) The rod-shaped bacterium *Bacillus anthracis* is the cause of the deadly disease anthrax.

(c) The spiral-shaped bacterium *Leptospira interrogans* often infects rodents, sometimes infects dogs, and occasionally infects human beings.

a variety of names, the best known of which is a *spirochete*. Different species of bacteria fit into all three groups.

Almost all bacteria have a structure called a cell wall at their periphery that is largely composed of a material called peptidoglycan. This may seem like a technical detail, but the fact that bacteria have cell walls is of great importance to humans because it represents a *difference* between bacterial cells and human cells. As you'll see, we humans have exploited this difference in our development of the medical drugs known as antibiotics.

Bacterial Numbers and Diversity

In 2006, scientists reported on a bacterial census they undertook by means of looking for bacterial DNA sequences within tiny quantities of soil from Alaska. The result? Each gram of soil contained about 4,000 different species of bacteria. Meanwhile, a comparative analysis indicated that a gram of Minnesota farm soil contained about 10,000 bacterial species. Fascinatingly, only about 20 percent of the species the researchers identified were present in the soils of *both* states. You might think that surveys such as these could give us a handle on how many bacterial species there are in total, but so far that's not the case. We really have no idea how many different kinds of bacteria there are in the world; the most we can say is that the number appears to be very large.

When we begin to examine these various types of bacteria, we find that an incredible diversity exists among them. It may seem strange to think of bacteria as "diverse," given that all of them are microscopic and single-celled. But physical form is only one kind of diversity. Imagine that only some human beings got their nutrition in the conventional way—by ingesting food—while others could make their own food by carrying out photosynthesis. Bacteria do things both ways. Imagine that only some human beings needed oxygen to live, while others could take it or leave it, and still others were poisoned by it. Bacteria fit into all three categories. Animals and plants have diverse forms, but bacteria have diverse metabolisms.

21.4 Intimate Strangers: Humans and Bacteria

In the chapter introduction, you saw how bacteria are indispensable to life in general because they can capture atmospheric nitrogen, decompose dead organic material, and so forth. As it turns out, however, bacteria may be indispensable to *human* life in an even more direct way. From the time we travel down our

Structure and Reproduction of Bacteria

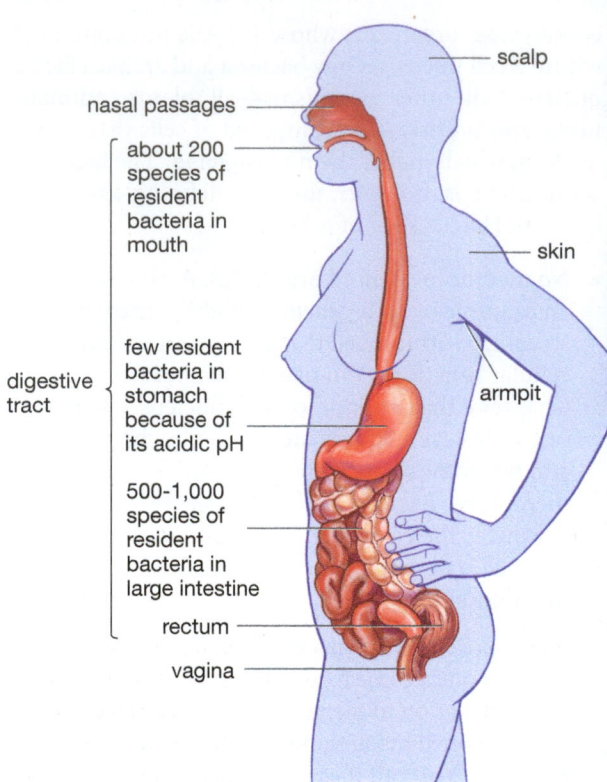

Figure 21.8
At Home in Us
We human beings live with trillions of "resident" bacteria, meaning bacteria that are always present on us or within us. Bacteria on the scalp or skin might temporarily be washed away, but they quickly return.

mother's birth canal, we enter into a lifelong interdependence with these microorganisms. If you look at **Figure 21.8**, you can see the places bacteria are found in quantity outside and inside the human body. Outside, we are covered in bacteria from head to toe, with heavy concentrations in the armpits and scalp. Inside, they exist in our nasal passages, in the vaginas of females, and most especially, in our digestive tracts. All told, perhaps 100 trillion bacterial cells live in or on the human body. Meanwhile, the number of *human* cells in a human body is only a tenth of this number—about 10 trillion cells. (So why aren't we bloated with bacterial cells? They're so small that their collective weight is only about 3 pounds.) This is not to say that bacteria are everywhere within us; most of our tissues are kept bacteria free, except for occasional invaders. But about half the contents of our colon are bacteria, as are perhaps a fourth of our feces by weight. These are not transient varieties of bacteria that come for brief periods and then are gone. They are so-called resident bacteria whose permanent home is the human body.

This account may make you feel like you've been colonized, but the relationship between us and many of the bacteria we harbor actually is one of **mutualism,** which is to say a relationship between two organisms that benefits both of them. If we could shrink ourselves down to the size of a bacterium and enter, say, the human large intestine, we would see a community of competing but *interdependent* organisms, most of which are bacteria, but one of which is us. What bacteria get from participating in this community is food and habitat. What humans get is a fully functional digestive system. How do we know that bacteria are helping maintain this system? Consider that rats whose bacteria have been wiped out through laboratory techniques require 30 percent more calories to maintain their body weight than do normal rats. Without bacteria, the structure of these rats' intestines is altered in a way that's unhealthy: They have low numbers of the cells that move nutrients from the digestive tract to the bloodstream. Researchers have made mice "germ free," then reintroduced bacteria into them and watched as mouse genes were turned on that help metabolize sugars and fats. Bacteria even metabolize sugars that we cannot digest, and they produce some vitamins as well. The essential message here is that, within the digestive tract, mammals and bacteria have had such a long working relationship that each has become dependent on the other.

21.5 Bacteria and Human Disease

Many of our resident bacteria are not beneficial to us but instead are *commensal* with us—they benefit from the relationship, while we are unaffected by it. And, of course, there are bacteria that are **pathogenic,** which is to say disease-causing. These are not bacteria that routinely inhabit any part of our body in large numbers. Instead, they are opportunists that invade specific tissues. Among the diseases they cause are tuberculosis, syphilis, gonorrhea, cholera, tetanus, leprosy, typhoid fever, and diphtheria, not to mention all manner of food poisonings and blood-borne infections. In the fourteenth century, the bacterially caused bubonic plague wiped out a *third* of Europe's population in just 4 years. Closer to home, you may recall a substance referred to as *anthrax* that was maliciously mailed to various offices after the September 11, 2001, attacks. Anthrax actually is a bacterium (*Bacillus anthracis*) that, in nature, occasionally infects farm animals.

How do pathogenic bacteria cause illness? A few invade human cells and reproduce there. More often, however, bacterial damage comes from substances bacteria either secrete or leave behind when they die. These compounds are *toxins,* which is to say substances that harmfully alter living tissue or interfere with biological processes. In the case of respiratory anthrax, *Bacillus anthracis* secretes a trio of toxins that

cause blood vessels in the lungs and brain to "hemorrhage" or leak. The botulism bacterium, *Clostridium botulinum,* secretes a toxin that blocks the signals that go from nerves to muscles. The end result can be paralysis of the muscles that control the diaphragm, in which case victims lose the ability to breathe. Of course, all pathogenic bacteria must have a way to get into the body in the first place. The bacterium that causes bubonic plague, *Yersinia pestis,* enters on the bite of a flea, while the tetanus bacterium, *Clostridium tetani,* may hitch a ride in on a rusty nail. Meanwhile, the tuberculosis bacterium, *Mycobacterium tuberculosis,* can simply be inhaled.

Killing Pathogenic Bacteria: Antibiotics

Prior to the 1930s, doctors were nearly powerless to stop any of these bacterial invaders. Some progress was made in the late 1930s with the development of the synthetic chemicals called sulfa drugs. But modern medicine was born in the 1940s with the development of **antibiotics,** meaning chemical compounds produced by one microorganism that are toxic to another microorganism. The first antibiotic, penicillin, was a substance produced by a fungus. Like all antibiotics, it had two critical qualities: It could kill microorganisms within human beings while leaving the humans themselves unharmed. (For an account of penicillin's development, see "The Discovery of Penicillin" on page 398.)

How can antibiotics be toxic to bacteria but not to people? They exploit the differences between bacterial cells and human cells. Remember earlier when we noted that almost all bacteria have cell walls, while human cells don't? Penicillin and several other antibiotics work by blocking cell-wall construction, which bacteria have to undertake when they divide. Penicillin confiscates a construction tool; it binds with a bacterial enzyme that helps stitch the cell wall together. A bacterium without a cell wall is doomed: It ruptures, floods, and dies. But the *human* cells around it—lacking cell walls—are left unharmed.

It is hard to overstate how much of human health hinges today on antibiotic processes such as these. If you know anyone who was alive in the 1930s, they may well have had friends or relatives who died of tuberculosis or one of the staphylococcal infections better known as blood poisoning. But do you know of anyone who, during your lifetime, has died of such an illness? Probably not. Antibiotics have made all the difference.

The Threat of Antibiotic Resistance

This power of antibiotics ought to spark alarm in us about something that is happening today, which is that antibiotics are steadily *losing* their ability to kill bacteria. How could this happen? An estimated 30 percent of antibiotic prescriptions written today are thought to be needless, in the sense that the person taking the antibiotic doesn't even have a bacterial infection but rather is suffering from a viral or other illness. (So, why do people get such prescriptions? They ask their doctors for them, and the doctors go along.) Even when people do have bacterial infections, they often take their antibiotics only until they start feeling better, rather than following through with the full course of recommended treatment, which can last weeks or even months.

Those of you who have read through the chapters on evolution may be able to see what this leads to. Normal bacteria may be wiped out by a needlessly short course of antibiotics. But a few bacteria will survive these inadequate treatments because they have sets of genes that allow them to do so. In the *absence* of antibiotics, these survivors might not have had anything special going for them. (They aren't *generally* superior bacteria; they just resist antibiotics better.) But with antibiotics, these survivors have been left without bacterial competitors for food and habitat. As a result, they flourish, giving rise to antibiotic-resistant *lines* of bacteria. Moreover, they will do this rapidly since one generation of bacteria can give rise to another in as little as 20 minutes.

Other factors also enter into this picture. It may surprise you to learn that nearly half the antibiotics used in the United States don't go into people at all; they are put into animal feeds because they serve as growth stimulants. In a similar vein, farmers spray huge tracts of fruit trees with antibiotics to ward off bacterial infections. Then among consumer products there are antibacterial hand-cleaners, which, according to a panel convened by the U.S. Food and Drug Administration, are no more effective than plain soap in reducing the spread of infections.

Methicillin-Resistant Staph

The upshot of this overuse of antibiotics is that bacterial infectious diseases are making a comeback. The problem is most pressing in hospitals, where physicians are seeing one formerly useful drug after another fail to cure infections. But antibiotic-resistant infections aren't limited to hospitals anymore. Indeed, they're now turning up in some settings that usually are brimming with health: high school and college locker rooms.

Students who participate in contact sports have always been subject to skin infections caused by the bacterium *Staphylococcus aureus*. In the past, these infections could easily be cured by a family of antibiotics that includes methicillin. What's now appeared, however, is the bacterium known as methicillin-resistant *Staphylococcus aureus*, or MRSA. Once seen almost entirely in the very old, the very young, or those with weakened immune systems, "methicillin-resistant staph" is now showing up in the general populace. Athletes are particularly prone to MRSA infections because one route to acquiring them is to make contact with an infected individual through a break in the skin—something that could easily happen during sporting events. To be sure, MRSA has mostly caused mild, though persistent, infections among the athletes it has affected. Nevertheless, the fact that an otherwise healthy group of young people is now having to deal with a general threat of bacterial infection ought to give society pause about what it is doing with antibiotics. If we can't stay ahead of bacterial evolution by developing new antibiotics—while cutting back on the needless use of existing ones—we may find ourselves returning to the days of our grandparents, when a simple cut could be life-threatening.

21.6 Archaea: From Marginal Player to Center Stage

At one time, almost all biology textbooks referred to the archaea as archae*bacteria,* and most mentioned these microbes as a kind of marginal life-form, existing in only a few extreme habitats, such as hot-water pools. Today, the archaea are recognized as constituting their own domain in life—standing alongside Domains Bacteria and Eukarya—and they're known to exist in large numbers in many environments. So, why were the archaea lumped together with bacteria for so long, and what kind of environments do they actually live in? Let's take these questions in order.

The Separate Status of Domain Archaea

Even under the microscope, archaea really are similar to bacteria in some ways. All of them are single-celled, and all of them are prokaryotes that reproduce through simple cell splitting. Since we saw these same qualities in bacteria, why do the archaea constitute their own domain of life?

To answer this question, let's think about what is *fundamentally* different in organisms as opposed to what is *superficially* different in them. Remember the cell wall we talked about that forms the outer boundary of most bacterial cells? As it turns out, archaea have cell walls too, but their cell walls do not contain the peptidoglycan found in bacterial cells—nor the cellulose found in plant cells, nor the chitin found in fungal cells. The materials that make up archaeal cell walls are unique to the archaea. To understand the importance of this, imagine that you are a construction worker who is inspecting the outer walls of some houses. You do a little scraping and discover that one house has a wall that is made from a material you've never seen before. You'd rightly regard this as a fundamental difference between this house and any others. So it is with the archaea. In essence, these microbes have a deep, chemical structure that is unique to them. If we look at their plasma membrane—the cellular lining that lies just inside the cell wall—we find that it is so chemically unique that it takes a special set of enzymes just to build it.

Structural differences such as these have to be based on genetic differences, and this is indeed the case with the archaea. In 1996, when scientists sequenced the first genome of an archaeon, they were stunned to find that 56 percent of its genes were completely unknown to science—they were unlike anything seen in either bacteria or in eukaryotes. As for the remaining 44 percent of the genome, some genes worked like those of eukaryotes, while others worked like those of bacteria.

Findings such as these produced the picture you can see in **Figure 21.9**; they convinced scientists that life evolved along three principal evolutionary lines, one of which is the archaea.

Archaea and Their Habitats

So, what about the second confusion we talked about—the misunderstanding about where archaea live? A large number of archaea species do in fact live in "extreme" environments, as we'll see. But archaea also live in large numbers in a variety of common habitats. Scientists divide the ocean into a series of vertical zones, one of which—starting at 200 meters down—is informally dubbed the "twilight zone" because it lies just beneath the level at which sunlight penetrates strongly. An eye-opening survey done in the 1990s found that archaea account for 40 percent of the microbial life in this zone over vast stretches of the ocean. This alone makes the archaea "one of the most abundant organisms on the planet," in the words of researchers Stephen Giovannoni and Ulrich Stingl. More recently, archaea have been found to outnumber bacteria in the sediment that *underlies* the ocean in regions as diverse as coastal Japan and Peru. Beyond this, archaea have recently been found to exist in abundance in common soil; there actually are more archaea than bacteria taking part in one phase of the "nitrogen-fixing" process we talked about earlier.

Extremophiles

All this said, many species of archaea do live in environments that are harsh or forbidding. We can thus think of some archaea as being **extremophiles:** organisms that grow optimally in one or more conditions that would kill most other organisms. Archaea are not

the only extremophiles; many bacteria, some fungi, and even some organisms such as lichens thrive under tough conditions as well. Some of these organisms can withstand high pressure, while others can resist radiation; some flourish in cold temperatures, while others can get by with very little water. Here are three important categories of extremophiles, each of which has many species of archaea within it:

- **Thermophiles.** These are microbes that make their homes in extremely hot environments, such as the hot-water vents on the ocean floor or hot-water pools such as exist in Yellowstone National Park.

- **Halophiles.** These are microbes that live in environments so *salty* that few other types of organisms can survive in them. Utah's Great Salt Lake and the Dead Sea that separates Israel and Jordan are examples of *hypersaline* environments, meaning environments with salt concentrations greater than those found in seawater.

- **Anaerobes.** These are microbes that either can live without oxygen or that actually are poisoned by it and thus can only live in oxygen-free environments. You might think it would be impossible to get away from oxygen on Earth, but oxygenless environments are surprisingly widespread. They exist in ocean sediments, swamps, rice fields, and the digestive tracts of many animals, including ourselves. (The microbial menagerie within the human gut is mostly made up of anaerobic bacteria, but it includes archaea as well.) One group of anaerobic microbes—all of them archaea—produce the gas methane in the course of their metabolism and thus are *methanogens*. Their production of this gas can be substantial; the amount diffusing from rice fields alone has been estimated at a minimum of 20 million tons per year. The problem with this? Methane is a potent greenhouse gas—it contributes to global warming.

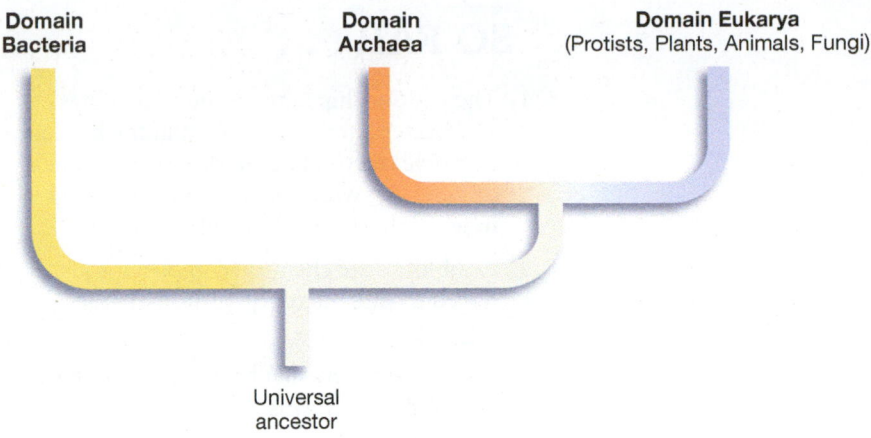

Domain Bacteria **Domain Archaea** **Domain Eukarya** (Protists, Plants, Animals, Fungi)

Universal ancestor

Figure 21.9
Archaea and the Universal Tree of Life
Archaea form one of three main branches on the universal tree of life. Note that they appear to be more closely related to eukaryotes (Domain Eukarya) than to bacteria (Domain Bacteria). The archaea and eukaryotes shared a common ancestor that was not shared by bacteria.

SO FAR . . .

1. The relationship between human beings and bacteria that are resident in the human digestive tract can be characterized as one of _____, which is to say a relationship in which both parties benefit.

2. Antibiotics are chemical compounds produced by one microorganism that are _____ to another.

3. Name three ways that bacteria and archaea are similar life-forms.

21.7 Protists: Pioneers in Diversifying Life

We've so far looked at two of life's domains, Bacteria and Archaea. As noted, the third domain, Eukarya, is so diverse we will look at only one of its four kingdoms in this chapter, the mostly microscopic organisms known as protists.

What's a protist? No one can provide a satisfactory definition. Domain Eukarya has within it three well-defined kingdoms—plants, animals, and fungi—but the protists within Eukarya are defined in terms of what they are *not* relative to these other life-forms: **Protists** are eukaryotic organisms that do not have all the defining characteristics of a plant, an animal, or a

fungus. Why do we have such an unsatisfactory definition for this one group of organisms? The root of the problem is that the term *protist* is a label of convenience rather than a label that reflects evolutionary reality. **Figure 21.10** shows the difference between naming and reality. You can see the three domains of life in the figure, the rightmost of which is Eukarya. Note that, early on in eukaryotic evolution, there was a branch that, toward its top, diverged into red algae on the one hand and plants on the other. Now note that there is a separate branch—once again starting out early in eukaryote evolution—that, toward the top, split into animals on the one hand and a group called the choanoflagellates on the other. Now, nobody calls animals and plants by a single name. But note that the choanoflagellates that are the "sister" group to animals and the red algae that are the sister group to plants *are* called by a single name—they're both "protists." Red algae are as distantly related to choanoflagellates as plants are to animals, yet both the red algae and choanoflagellates go under the catch-all title of protist. It is as if we had multiple family lines—the Garcias, the Lees, the Beesons—and then decided to call all of them the Johnsons.

This naming convention is partly a holdover from the past: Only recently have we understood how early the branching occurred among various evolutionary lines of eukaryotes. But this convention also is with us because it's convenient to have a single name for any living thing that's (a) a eukaryote and (b) not a plant, animal, or fungus. We'll use this convention in this chapter, but keep in mind that the term *protist* actually refers to many different evolutionary lines of organisms.

Figure 21.10
Many Evolutionary Lines, One Name: The Protists

The problem with the term *protist* is that this single name is applied to several different evolutionary lines of organisms. Within Domain Eukarya, on the right, protists make up one kingdom (Kingdom Protista), which has evolutionary lines that exist within the blue cone. Plants, animals, and fungi make up Eukarya's other kingdoms (shown outside the cone). Note that, among the protists, choanoflagellates and red algae evolved along separate lines and are no more closely related to each other than animals are to plants. Yet, plants and animals are regarded as members of separate kingdoms, while red algae and choanoflagellates are both regarded as members of one kingdom, Protista.

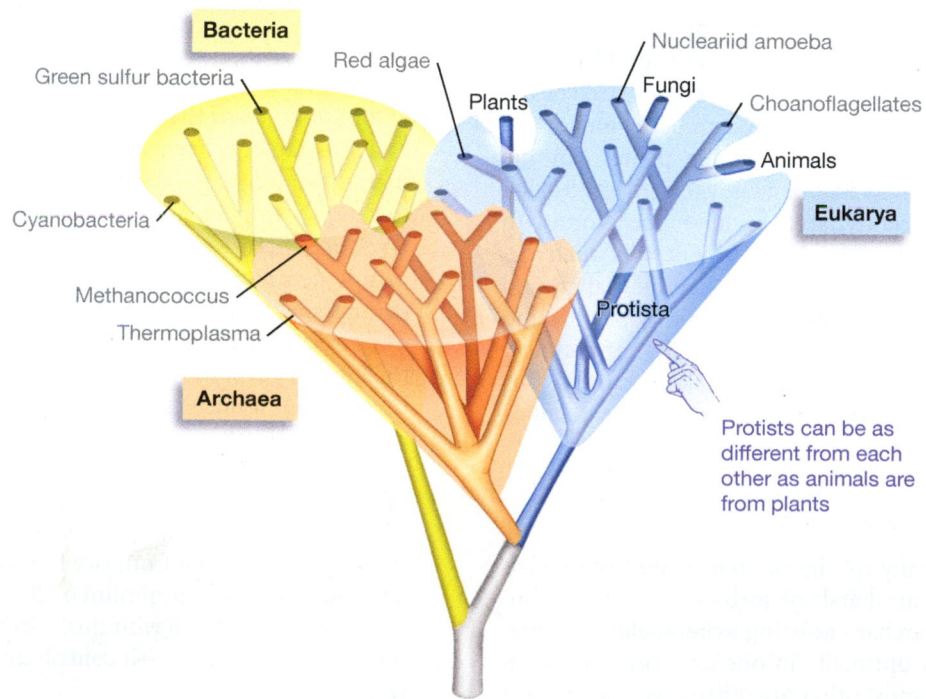

Bacteria
Green sulfur bacteria
Red algae
Nucleariid amoeba
Plants
Fungi
Choanoflagellates
Cyanobacteria
Animals
Eukarya
Methanococcus
Thermoplasma
Protista
Archaea

Protists can be as different from each other as animals are from plants

Given how distantly related many protists are to each other, it's not surprising that, as a group, they don't have much in common. Beyond being eukaryotic, about the only characteristic that all of them share is that they live in environments that are moist, if not fully aquatic—you can find them on damp forest floors as well as in oceans and lakes. Many are single-celled, and most are microscopic, but some are very large, as you'll see. Perhaps the way to think of protists is as organisms that occupy a kind of middle ground of complexity in the living world; they are more complex than bacteria or archaea, but tend to be less complex than plants, animals, or fungi (all of which evolved from protists). About 100,000 species are known to exist.

Although only a small proportion are parasites that affect humans, the rogue's gallery of protists is a formidable one. A microscopic protist, *Plasmodium falciparum,* causes one of the world's most widespread diseases, malaria, by passing between human and mosquito hosts. Campers in the United States are aware of the intestinal parasite *Giardia,* which contaminates water. Other protists cause sleeping sickness and amoebic dysentery. Pathogenic protists also affect humans in less direct ways. The cause of the great Irish Potato Famine of the 1840s was a fungus-like protist, *Phytophthora infestans,* that devastated the Irish potato fields. A close relative of this pest, *Phytophthora ramorum,* is now wiping out stands of California oak trees and bay laurels.

21.8 Protists and Sexual Reproduction

When we look at the living world as a whole, the protists in it are fascinating because they are the organisms that made the break with prokaryotic life—with life as nothing but single-celled bacteria and archaea. After life first appeared on Earth, it consisted solely of bacteria and archaea for almost 2 billion years. The first life-forms to appear other than bacteria and archaea were the protists. And in these organisms, we can see how the living world branched out; we can see how living things made transitions to life as we know it today.

One such transition has to do with reproduction. You may recall that bacteria and archaea reproduce through a simple splitting of one cell into two. And as it turns out, most protists reproduce through this means as well (though they employ a different process than do bacteria or archaea). Many protists, however, can reproduce by mixing genetic material from *two* individuals; they give rise to offspring through *sexual* reproduction, in other words. This change was terrifically important to evolution because it meant that offspring would be different from their parents. Bacterial daughter cells are clones of their parent cell—

they are genetically identical to it. But the offspring of sexually reproducing organisms have a mixture of genes from both parents, meaning these offspring will not be identical to either parent. Once variation started appearing simply as a consequence of reproduction, evolution was off to the races in terms of creating new life-forms because so much variation was being produced.

This does not mean, however, that life went in one step from reproduction by simple cell splitting to reproduction as we now know it today in, say, human beings. Remember, the protists show us transitions. Let's look at what sexual reproduction is like in an ancient, early-evolving protist.

The protist in question is a type of single-celled water-dweller called *Chlamydomonas.* Most of the time, *Chlamydomonas* reproduces through cell splitting. In times of little nutrition, however, *Chlamydomonas* switches over to sexual reproduction. This begins with two normal *Chlamydomonas* cells, which means two cells that are haploid—that contain only a single set of chromosomes. You may recall that bacteria and archaea are always haploid. It makes sense, then, that the usual state for *Chlamydomonas* is haploid since, on life's family tree, it lies so close to the bacteria and archaea. Yet, *Chlamydomonas* has made a switch, as you can tell by looking at **Figure 21.11**. Note that at one point, two haploid *Chlamydomonas* cells fuse to create a zygote—the same thing that takes place with human sperm and egg cells. The result is a cell with *two* sets of chromosomes, one set coming from each parent. This cell then undertakes the meiosis reviewed in Chapter 10. You may recall that meiosis always entails a swapping of pieces between the chromosomes that came from parent A and those that came from parent B. Further, these chromosomes then assort randomly, such that any "daughter" cell will get a different mix of chromosomes from each parent. In *Chlamydomonas,*

Figure 21.11
Sexual Reproduction in a Protist

The simple algae *Chlamydomonas* normally is a haploid organism, meaning one that possesses a single set of chromosomes. In reproduction, however, *Chlamydomonas* cells of opposite mating types (+ and -) can fuse to create a zygote or fertilized egg that is diploid—that has a paired set of chromosomes. Meiosis then ensues in the zygote, and the result is four haploid *Chlamydomonas* cells, each of which differs genetically from the others thanks to the mixing of chromosomes that took place in meiosis.

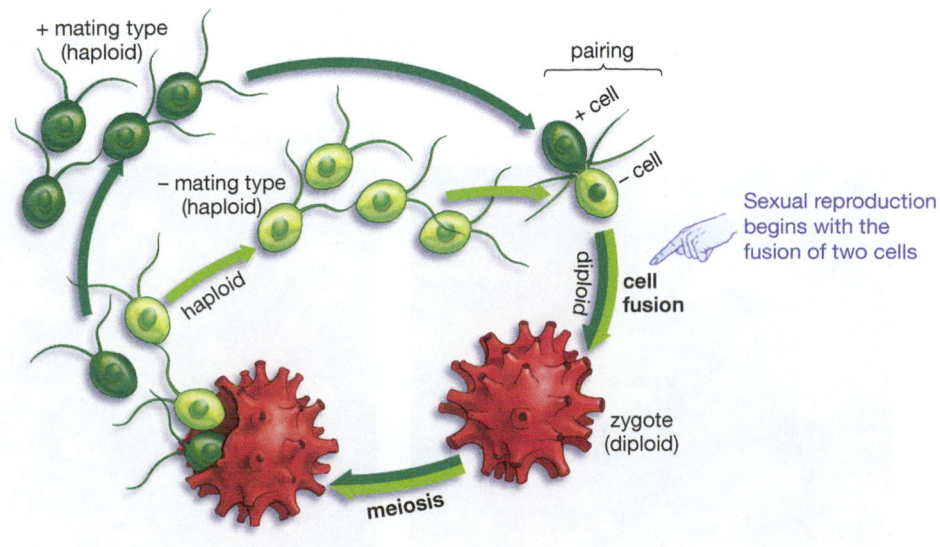

+ mating type (haploid)

− mating type (haploid)

haploid

pairing

+ cell

− cell

diploid

cell fusion

zygote (diploid)

meiosis

Sexual reproduction begins with the fusion of two cells

four new offspring cells are produced from the fusion of the original two cells. The critical thing is that, because of the shuffling done in meiosis, each of these cells has a unique mixture of chromosomes from each parent. The difference between bacteria and *Chlamydomonas* is the difference between clones and diverse offspring.

One more detail about this process is important. You may wonder about the little plus (+) and minus (−) signs next to the *Chlamydomonas* cells in Figure 21.11. This is an early protist version of what it means to be male and female. There are so-called + and − "mating types" of *Chlamydomonas,* and only these opposites can come together in sexual reproduction. (A + cell cannot fuse with another + cell.) The whip-like flagella that extend from + and − mating types differ in their microstructure. As a result, only opposite types of flagella can adhere to one another—something that is necessary to get the fusion process going. What we think of as male–female differences are later-arriving variations on this theme.

With this as background, let's now look at a few representatives of the protists. We'll review them not within their evolutionary groupings, but within categories that have to do with how they obtain their nutrition. We'll look at some protists that perform photosynthesis and some that do not.

21.9 Photosynthesizing Protists

The protists that perform photosynthesis are perhaps the only large group of protists known to the average person. The catchall name for them is **algae.** If you look at **Figure 21.12**, you can see the three varieties we'll be reviewing. Figure 21.12a shows a microscopic golden alga called *Synura scenedesmus* that can be found in freshwater ponds in the United States. In this species, we can see a transition between the single-celled life carried out by bacteria and the multicelled

life carried out by more complex organisms. All varieties of golden algae come in species that exhibit **colonial multicellularity:** a form of life in which individual cells form stable associations with one another but do not take on specialized roles. Each of the three objects you see in the picture of *Synura* is a cluster of *Synura* cells—a colony of them. Each member of the colony keeps its two whip-like flagella pointed to the outside of the group, the flagella beat back and forth, and the colony literally rolls through the water. None of these cells, however, performs a function different from any other cell, which is why this *Synura* is referred to as colonial and not truly multicellular.

If you look at Figure 21.12b, you can see a form of *green* algae, called *Volvox*, that represents an evolutionary step beyond *Synura*. The beautiful, translucent circles that outline *Volvox* are a gelatinous material that initially encloses a single layer of anywhere from 500 to 60,000 *Volvox* cells, all of them spaced around the periphery of the sphere. Almost all of these cells have flagella that they move back and forth in unison, so that *Volvox* moves slowly through the water in a given direction, spinning along an axis as it goes (something like a torpedo). For our purposes, the key thing about *Volvox* is that not all of the cells in this colony will reproduce. Only a select group of cells without flagella, generally at the rear pole of the axis, will divide to produce new colonies. These colonies in turn give rise to new spheres that exist inside the original parent sphere. Eventually, the gelatinous material around the parent sphere ruptures, and the daughter spheres emerge, each of them then continuing the process.

The important point is that *Volvox* has cells that specialize. As such, *Volvox* has arguably achieved **true multicellularity:** a form of life in which individual cells exist in stable groups, with different cells in a group specializing in different functions. In human beings, of course, we can see a great specialization in, for example, nerve and muscle cells. *Volvox* shows us specialization in a rudimentary form.

Figure 21.12
Transitions Visible in Algae

(a) Three colonies of the golden algae *Synura*, each of which is composed of a cluster of individual cells. *Synura* is a colonial protist in that none of the cells in a colony takes on any specialized function.

(b) Colonies of the green algae *Volvox* with daughter colonies growing inside. *Volvox* colonies can be regarded as truly multicellular organisms in that they have specialized reproductive cells. The largest *Volvox* colonies can be seen with the naked eye.

(c) A frond of the brown algae known as sea kelp. These unusually large protists have assemblages of cells that conduct food throughout the organism.

(a) Golden algae

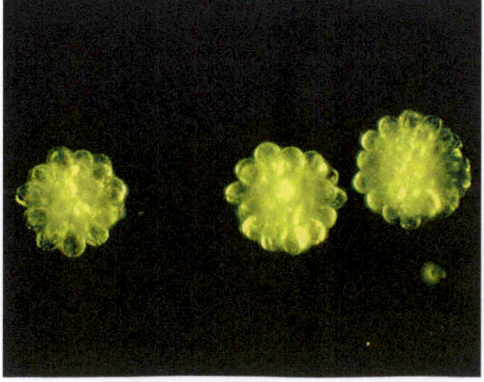

(b) Green algae

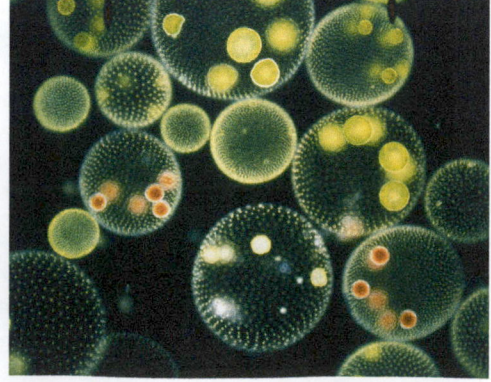

(c) Brown algae

Finally, if you look at Figure 21.12c, you can see some representative brown algae—the giants of the protist world. One variety of brown algae, the sea kelp, has individuals that may be 60 meters long and that often grow in enormous kelp forests. For our purposes, brown algae are notable because their leaf-like "blades" and stem-like "stalks" represent another leap up in complexity: They contain organized groups of cells that conduct food throughout the organism.

One other thing is important about the algae. Note that the golden algae and *Volvox* share three characteristics: They are microscopic, aquatic, and carry out photosynthesis. If we were to go through all the algae, we would find these same qualities in countless species. Creatures in this category, including some bacteria, are known as **phytoplankton:** small photosynthesizing organisms that float near the surface of water. Such organisms are extremely important to life on Earth for two reasons. First, they produce most of the Earth's atmospheric oxygen. Second, they sit right at the base of most aquatic food chains. All large life-forms on Earth ultimately depend on photosynthesizers for their food. On land, lions may eat zebras, but zebras eat grass. In the oceans, whales may eat the shrimp-like creatures called krill, but krill eat phytoplankton.

21.10 Heterotrophic Protists

Lots of species of protists do not get their nutrients from photosynthesis, as algae do, but instead get them from consuming other organisms or bits of organic matter that they find. This makes them heterotrophs (or "other-eaters").

Heterotrophs with Locomotor Extensions

Some of these protists have slender extensions, called cilia and flagella, that allow them to move toward their prey or away from danger. **Figure 21.13** shows two varieties of these highly mobile heterotrophic protists.

Figure 21.13a shows a *Paramecium,* which is known as a ciliate because of all the hair-like cilia that cover its exterior. These cilia beat in unison, allowing paramecia to move forward or back in an exacting fashion—toward or away from objects in its environment. A *Paramecium* may be single-celled, but it is complex. It takes food in through a mouth-like passageway called a gullet, digests it in a special organelle called a food vacuole, and then empties the resulting waste into the outside world through a special pore on its exterior. Some species even have tiny dart-like structures that they can shoot out when threatened. Most animals require different *groups* of cells to do all these things, but a *Paramecium* does them within a single cell. Given this, it may be that paramecia represent the most complex single cells in all of nature.

Figure 21.13b shows a famous human parasite that enters human beings through contaminated water, the "camper's parasite," *Giardia lamblia,* which can live in the intestines of animals. It gets around not by cilia like the *Paramecia,* however, but by pairs of whip-like flagella, as we saw in the algae protists.

Heterotrophs with Limited Mobility

Figure 21.14a, on the next page, shows a so-called amoeboid protist, meaning a protist that gets around by means of a pseudopod or "false foot." The amoeba sends out a slender extension of itself (the foot), and then the rest of its body flows into this extension. It takes in bits of food by encircling them with extensions and then fusing these extensions together. What once had been outside the amoeba is now inside, being digested. Many amoeboid protists are parasites, and one of them, *Entamoeba histolytica,* is a serious

(a) A *Paramecium*

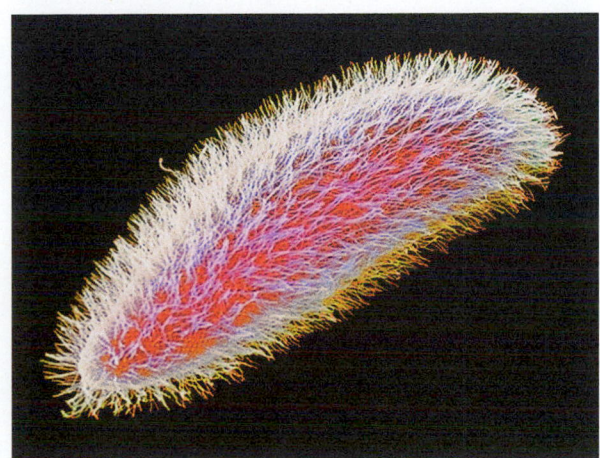

(b) The *Giardia* parasite

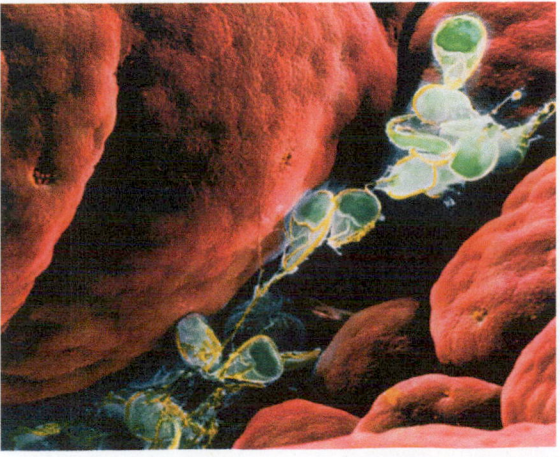

Figure 21.13
Movement through Cellular Extensions

(a) The many hair-like cilia extending from this paramecium allow it both to move through the water and to bring food to its mouth-like gullet.

(b) The parasite *Giardia lamblia* within the human small intestine. Seen here as the pear-shaped green and white objects, *Giardia* use sucking disks to attach themselves to fingerlike extensions of the intestine's inner lining.

Figure 21.14

Life as a Blob

(a) This member of the species *Amoeba proteus* has encircled a much smaller protist by extending two pseudopodia around it. Soon thereafter it ingested its prey, which it then began to digest with enzymes.

(b) This member of the plasmodial slime mold species *Physarum polycephalum* was photographed while moving slowly across a fallen log in southern Canada.

(a) Amoeba

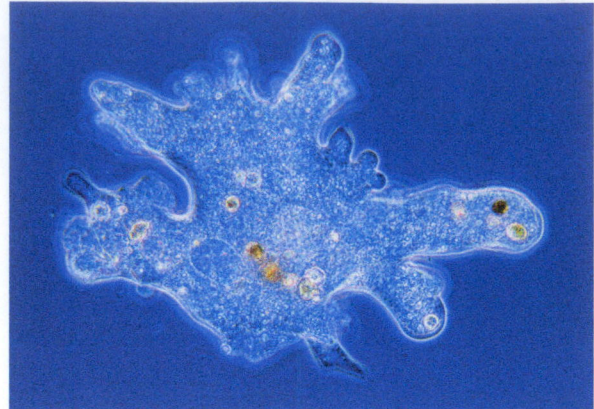

(b) Plasmodial slime mold

human parasite that causes an estimated 100,000 deaths per year, mostly in the tropics, by entering the human digestive system through contaminated water. When a parasite invades the human intestinal wall, the resulting condition is known as *dysentery*. When that parasite is an amoeba, the result is amoebic dysentery.

Figure 21.14b shows a protist known as a plasmodial slime mold. This organism assumes several forms during its life cycle. In the form you see in the picture, its name is appropriate because it is essentially moving slime. If you looked at this organism under a microscope, you would see lots of cell nuclei but none of the *boundaries* (the plasma membranes) that normally separate one cell from another. The result is a free-flowing but slowly moving mass of cytoplasm—the interior of a single cell spread out over a large area, with only one plasma membrane at its periphery. This material moves by the same "cytoplasmic streaming" you saw in the amoeboids. As it rolls over rocks or forest floor, it consumes underlying bacteria, fungi, and bits of organic material. Should an object stand in its path, it simply flows around it, or even through it, if the object is porous enough.

All this makes the plasmodial slime molds sound like very primitive protists, but these single-celled creatures arguably possess an early-evolving form of a trait we normally associate only with animals: intelligence. In 2008, Japanese researchers reported that the slime mold *Physarum polycephalum*, which you can see in Figure 21.14b, has the capacity to remember what it has experienced and to act on the basis of its experience. In nature, *P. polycephalum* moves at differing rates depending on the humidity of the air around it—faster in humid air, slower in drier air. In the laboratory, the Japanese researchers provided three pulses of dry air around *P. polycephalum*, with each pulse spaced an hour apart. They then watched as the mold slowed down at the time when a fourth pulse was due, even though none was provided. Indeed, the mold would sometimes repeat its slow-downs a second and third time, right at the hourly intervals, in the absence of the pulses of dry air.

By now, you may be getting the impression that protists could be star contestants in a show called "Can You Top This for Strangeness?" If you look at **Figure 21.15**, you can see another contestant, a

Figure 21.15

Changing Forms

A *Dictyostelium* in its slug stage is composed of individual cells that have come together to produce an organism capable of moving across the forest floor. Although the slug is composed of tens of thousands of cells, it is no bigger than a grain of sand. Cells in the slug eventually arrange themselves into this tower-like reproductive structure. Reproductive cells called spores will be dispersed from the top of the structure to the forest floor. There they will begin life as individual *Dictyostelium* cells.

much-studied organism, *Dictyostelium discoideum,* that is a so-called social amoeba. This may sound like a contradiction in terms, but consider the varying forms that this protist takes. At one point in its life cycle it exists as a collection of individual amoeba cells that crawl along the forest floor feeding on bacteria. Each "Dicty" is on its own at this stage, but if the bacterial food runs low, these individual cells begin "aggregating," or coming together in an organized group. Eventually, they develop into a barely visible "slug" of up to 100,000 cells that has front and back ends and that migrates to a more fertile area of the forest. Then, at a certain point, these cells change form again. They arrange themselves into a tower-like reproductive structure that has, at its top, spores that will be dispersed and develop into new individual cells that once again crawl along the forest floor. The whole aggregation process will not take place unless starvation is imminent, however; without the danger of starvation as a motivator, the individual cells stay separate.

Now, once you get past the shock of realizing that such a thing could happen—that individual cells could just collect themselves and create a larger, moving organism—you can begin to see what interesting questions it raises. What's the signaling mechanism that brings 100,000 separate cells together? How do they begin to cooperate to form these various structures? (To put this another way, how does multicellularity begin to work?) And, perhaps most intriguingly, why should some cells sacrifice themselves by serving on the *stalk* of the reproductive structure—from which location they will die, without reproducing themselves—while others get to be at the top and disperse to a new life in better hunting grounds? If evolution is all about passing on genes, why should some cells give up the chance to pass on theirs? Scientists study Dicty in hopes of learning more about all these questions.

On to Fungi

The bacteria, archaea, and protists reviewed in this chapter represent only a portion of the living world. Coming up is a life-form that is as close as the toadstool on our front lawn or the athlete's foot between our toes. What we'll be looking at next is fungi.

SO FAR ...

1. In contrast to bacteria and archaea, all protist cells have a _____, which makes protists eukaryotes. While some protists are large, most are _____; all of them live in environments that are at least moist if not fully _____.

2. The organisms known as phytoplankton are important to life in general for two reasons: They produce most of the world's atmospheric _____ and they lie at the base of most aquatic _____.

3. The catchall term for protists that perform photosynthesis is _____.

THE PROCESS OF SCIENCE

The Discovery of Penicillin

In wartime England, late in 1940, as the German bombings raged, a 43-year-old policeman lay dying in the Radcliffe Infirmary in Oxford. Albert Alexander was, however, not a victim of the German air raids. His enemy was smaller and more subtle. A few months earlier, he had scratched his cheek on a rosebush thorn. This minor wound had then become infected, and from that point he steadily went downhill as bacteria worked away inside him. By December his face was a mess of festering wounds, one of which cost him an eye.

Incomprehensible as it may seem, in 1940 a scratch from a rosebush thorn could do this to a person. The notable thing about Albert Alexander was not that he became deathly ill from such a trivial injury, but that he came close to beating his tiny enemies. He happened to live near an Oxford University laboratory where a handful of researchers were hurriedly trying to extract, from a common fungus, a product that the world eventually came to know as penicillin. When Albert Alexander became the first human being to receive a therapeutic dose of penicillin in February 1941, the results were startling. His fever dropped, his infection began to clear; even his appetite returned. But there was to be no happy ending for him, for the researchers had only a 5-day supply of the drug. Trying everything they could think of, they even extracted precious micrograms of penicillin from Alexander's own urine to put back into him. But it was not enough; he died days later.

Nevertheless, Alexander's case had shown that penicillin could work in human beings, and clear-cut successes with the drug were not long in coming. A 15-year-old boy with an infected hip underwent a remarkable turnaround when injected with it; children had their eyesight saved when blindness would have been the expected outcome.

These early trials made clear that, if ever there was a "miracle drug," penicillin was it. Prior to its advent, doctors could only watch and hope for the best as their patients were wracked by infections that ranged from pneumonia to meningitis to diphtheria to blood poisoning. With the advent of penicillin, at last doctors could do something.

How did it happen that a small band of researchers in wartime England ended up realizing the wondrous potential of this substance? The story begins not in the 1940s, but in 1928, when another scientist

The team had to brew up liters of fungal fluid to get mere millionths of a gram of active penicillin.

in England, Alexander Fleming, discovered the existence of penicillin through a combination of perceptiveness and luck. Returning to his hospital laboratory from a summer vacation, he noticed a spot of mold (a type of fungus) in one of the Petri dishes in which he was growing or "culturing" bacteria. Such bacterial cultures can easily become contaminated; this dish probably became tainted by means of fungal spores wafting in from another lab in the building. Something about this particular contamination caught Fleming's eye, however: The bacteria in the vicinity of the mold had clearly burst—they were dead. Realizing that chance had brought to him a substance that had the power to kill bacteria, Fleming grew more of the fungal mold, prepared a broth from it, tested its power against several kinds of deadly bacteria, and even went so far as to inject it into healthy mice to see if it had adverse effects (which it didn't). This last finding was crucial. There are lots of substances that can kill bacteria; the trick is to find one that can kill bacteria while leaving human beings unharmed.

Having advanced his work with penicillin to this point, however, Fleming then abandoned it. He had clearly recognized its therapeutic potential right from the start, but his subsequent research seemed to convince him that penicillin was just one bacteria killer among many that had no future in medicine. He found that it was unstable; that it appeared very difficult to purify from the stew of fungus chemicals it existed in; and that it quickly lost its power to kill bacteria once they had been exposed to it. After 1931, Fleming didn't publish anything on penicillin or encourage his students to work with it. He seemed to regard it as a substance that was useful mostly for killing off unwanted bacteria in Petri dishes. A great deal of credit goes to

him for being perceptive enough to recognize what had been laid in front of him. But had research on penicillin ended with him, there never would have been a drug called penicillin. By 1935, as Trevor Williams has written, "There was not one person in the world who believed in penicillin as a practical aid to medicine."

But this was soon to change, thanks to the work of two immigrants to Britain. One was Howard Florey, an Australian trained as a medical doctor who was then head of a school of pathology at Oxford University. The other was Ernst Chain, a refugee from Hitler's Germany who had been hired by Florey to develop a biochemistry unit in the school. The critical decision the two made in 1938 was to investigate substances in the same class as penicillin—antibiotics, which is to say substances made by one microorganism that are toxic to another. Critically, one of the compounds Florey and Chain decided to investigate was penicillin; they knew of its existence from reading the earlier papers of Fleming and others.

From the beginning, the team's research on penicillin proceeded on two tracks. One was simply seeing if the substance worked. The other was finding out what the substance *was* and trying to get enough of it to work with. Remember that penicillin is the product of a living organism—originally the fungus *Penicillium notatum*—and as such it existed within a mixture of thousands of other compounds the fungus produces. The trick was to isolate and characterize the germ-killing substance within the fungal soup. This biochemical work was the job of Chain. While this was going on, the team had to brew up liters of fungal fluid to get mere millionths of a gram of active penicillin.

Finally armed with enough of the drug to put it to the test, the group carried out the critical experiment. On the morning of May 25, 1940, they injected eight mice with deadly doses of streptococcal bacteria, and an hour later they injected four of the mice with penicillin. Then, a round-the-clock watch began. Normally, the bacterial dose given would have killed all the mice within 24 hours, and indeed by 3:30 a.m. the next day the four unprotected mice were dead. But the four mice given the penicillin were all fine! A long road lay ahead with human subjects such as Albert Alexander, but penicillin had done in mice what had so long been sought: It killed bacterial invaders while leaving their "host" unharmed.

This effort was being carried out against the backdrop of the outbreak of World War II. By the time the mouse experiments were started, much of Europe had fallen to Hitler, and it was feared that England might be next. (The wartime importance of a drug such as penicillin was clear to the Oxford team, which had contingency plans to destroy all its lab equipment and notes in the case of a successful German invasion. Only the precious *P. notatum* spores would be retained, by means of the researchers rubbing them into their clothing.) In this chaotic environment, the scientists were forced to go outside England to get the help they needed to ramp up the production of penicillin to an industrial scale. Howard Florey thus went to America to enlist aid there. Oxford researchers eventually found themselves working with chemical engineers from the U.S. Department of Agriculture in the unlikely location of Peoria, Illinois. Where once the English scientists had fermented penicillin in milk bottles, eventually they and their American colleagues were brewing it up in 25,000-gallon tanks.

Getting to this point meant coming up not only with better fermenting methods but also with better starting ingredients. To this end, military pilots collected soil samples from around the world in an effort to find a *Penicillium* species that would yield more of the precious drug.

As luck would have it, a species that did the trick was found on the project's doorstep. In 1943, lab worker Mary Hunt bought a moldy cantaloupe at a Peoria market. The melon contained a fungus, *Penicillium chrysogenum,* that doubled the amount of penicillin produced; its daughter spores are the ones used to this day to produce penicillin.

Despite some detours, this crash program was a smashing success. The drug that once was restricted to mice, then to a few human subjects, was put into commercial production early in 1944. By the time of the Allied invasion of Europe, in June 1944, enough was being produced that every soldier who needed it could have it. Not much later, all restrictions on it were lifted; penicillin was available by prescription from the corner drugstore.

In 1945, Alexander Fleming, Howard Florey, and Ernst Chain were awarded the Nobel Prize in Physiology or Medicine for their work in discovering this wonder drug. Bestowing the prize just after the war ended, the Nobel committee noted:

In a time when annihilation and destruction through the inventions of man have been greater than ever before in history, the introduction of penicillin is a brilliant demonstration that human genius is just as well able to save life and combat disease.

Penicillin's Discoverers

The three scientists who were awarded the Nobel Prize for their work on the development of penicillin. From left to right, Alexander Fleming, Howard Florey, and Ernst Chain.

21 REVIEW

Summary

21.1 Life's Categories and the Importance of Microbes

- All living things on Earth can be classified as falling into one of three domains: Bacteria, Archaea, or Eukarya. All the members of Domains Bacteria and Archaea are single-celled and microscopic. Domain Eukarya is further divided into four kingdoms: plants, animals, fungi, and protists. (p. 380)

- Microbes are indispensable to life. They produce more than half of Earth's atmospheric oxygen; the bacteria and archaea among them are responsible for putting atmospheric nitrogen into a form plants can use; and bacteria and fungi are the most important decomposers of the natural world. (p. 380)

- Microbes live in all environments in which larger life-forms exist. They are present in numbers so immense that their combined biomass exceeds the biomass of all of Earth's larger life-forms. (p. 381)

21.2 Viruses: Making a Living by Hijacking Cells

- Viruses are noncellular replicating entities that must invade living cells to carry out their replication. Most scientists do not classify them as living things. (p. 382)

- The HIV virus has two structures common to all viruses: genetic material and a protein coat, called a capsid, surrounding this material. HIV also has an fatty membrane, called an envelope, which surrounds the capsid. HIV does its damage by invading immune system cells called helper T-cells, which are then destroyed. (p. 382)

- Most viruses: (1) get their genetic material inside a "host" cell; (2) turn out viral parts; (3) construct new virus particles from these parts; and (4) move the new particles out of the cell, at which point the particles go on to infect more cells. (p. 382)

- Viruses cause a host of human illnesses. Health officials worldwide must constantly be on alert for the emergence of new members of a particularly dangerous class of viruses, the influenza A viruses, one of which was the H1N1 virus that infected people around the world in 2009. (p. 384)

21.3 Bacteria: Masters of Every Environment

- Bacteria are microscopic, single-celled organisms that are prokaryotes. Bacteria have only a single organelle (the ribosome) and reproduce asexually through binary fission. Millions of species of bacteria exist. Some bacteria make their own food through photosynthesis, and others consume organic material in their environment. (p. 385)

21.4 Intimate Strangers: Humans and Bacteria

- Bacteria live on and in human beings in great numbers. In the digestive tract, the relationship between humans and many bacteria is one of mutualism: Bacteria get food and habitat; human beings get an efficiently functioning digestive system. (p. 388)

21.5 Bacteria and Human Disease

- Only a small proportion of bacteria are pathogenic, but these bacteria are responsible for some of humanity's worst diseases. A few pathogenic bacteria cause harm by invading human cells, but bacteria generally do their damage by either releasing toxins or by leaving them behind when they die. (p. 389)

- The primary human medical defense against pathogenic bacteria is antibiotics, which work by exploiting the differences between bacterial and human cells, such that they kill bacteria while leaving human cells unharmed. (p. 389)

- The power of antibiotics is being threatened by the emergence of antibiotic-resistant strains of bacteria. These bacteria are evolving in greater numbers because of an overuse of antibiotics in medicine, agriculture, and consumer products. (p. 389)

21.6 Archaea: From Marginal Player to Center Stage

- Archaea were once thought to be a form of bacteria but are now known to constitute their own domain of life. Archaea are superficially similar to bacteria in that they are single-celled prokaryotes that reproduce through simple cell splitting. However, archaea are unique in the living world at the level of their genes and the chemical structure of their cells. (p. 390)

- Archaea exist in large numbers in some common environments. They make up 40 percent of the microbial life in large portions of the world's oceans, and they are seen in large numbers in soil, where they join bacteria in carrying out one phase of the nitrogen-fixing process. (p. 391)

- Many species of archaea are extremophiles; three large classes of extremophiles are thermophiles, halophiles, and anaerobes. (p. 391)

21.7 Protists: Pioneers in Diversifying Life

- A protist is a eukaryotic organism that does not have all the defining features of a plant, an animal, or a fungus. This unsatisfactory definition stems from the fact that the term *protist* doesn't refer to a single evolutionary grouping but instead is used as a label for several different evolutionary lines of organisms. (p. 392)

- Protists are mostly microscopic, and all of them live in environments that are at least moist, if not aquatic. About 100,000 species are known to exist. (p. 393)

21.8 Protists and Sexual Reproduction

- Protists were the first life-form to evolve other than bacteria or archaea; they were the organisms that made transitions to many of the capabilities and forms seen in larger organisms today. Among these transitions was the change to sexual reproduction, which protists were the first to practice. (p. 393)

21.9 Photosynthesizing Protists

- Protists that get their nutrition by performing photosynthesis are known as algae. Some algal species provide examples of colonial multicellularity. Others provide examples of true multicellularity. (p. 394)

- Microscopic algae are important phytoplankton—small photosynthesizing organisms that float near the surface of water—which are important to life because they produce most of Earth's atmospheric oxygen and because they form the base of so many aquatic food chains. All phytoplankton are either algae or bacteria. (p. 395)

21.10 Heterotrophic Protists

- Heterotrophic protists get their nutrients from consuming either other organisms or bits of organic matter. Some have evolved tiny cilia and flagella, with which they move toward prey or away from danger. (p. 395)

- The protists called amoeba move through use of pseudopodia. Likewise, the protists called plasmodial slime molds and social amoeba move by means of this "cytoplasmic streaming." (p. 395)

- The plasmodial slime mold *Physarum polycephalum* arguably displays a primitive form of intelligence, while the social amoeba called *Dictyostelium discoideum* exists as a collection of individual cells that can come together during times of little food to form a single slug-like organism. (p. 396)

Key Terms

Understanding the Basics

Multiple-Choice Questions

(Answers are in the back of the book.)

1. Microbes:
   a. produce most of the world's atmospheric oxygen. **b.** make nitrogen available to plants. **c.** recycle dead organic matter. **d.** account for most of the world's biomass. **e.** All of the above.

2. Viruses:
   a. are the smallest living things. **b.** have only one organelle. **c.** are beneficial to most organisms. **d.** replicate only by invading living cells. **e.** have circular chromosomes.

3. All viruses possess _____, which they _____.
   a. toxins; kill cells with **b.** cell walls; protect themselves with **c.** DNA or RNA; copy inside themselves **d.** many cells; insert inside a host cell **e.** DNA or RNA; transfer into a host cell

4. Bacteria:
   a. are single-celled eukaryotes. **b.** live outside cells. **c.** reproduce sexually or asexually. **d.** are multicelled prokaryotes. **e.** are single-celled prokaryotes.

5. Within the human body, bacteria exist in greatest numbers in the _____, with most of these bacteria having a _____ relationship with their human host.
   a. lungs; commensal **b.** feet; antagonistic **c.** digestive tract; mutualistic **d.** digestive tract; pathogenic **e.** kidneys; mutualistic

6. Archaea (select all that apply):
   a. are single-celled prokaryotes. **b.** are restricted to marine environments. **c.** are superficially similar to bacteria. **d.** are restricted to extreme environments. **e.** constitute one of the three domains of life.

7. Protists were the first life-forms to (select all that apply):
   a. appear on Earth. **b.** reproduce through binary fission. **c.** reproduce sexually. **d.** evolve following bacteria and archaea. **e.** exhibit true multicellularity.

8. Phytoplankton are important because they (select all that apply):
   a. produce antibiotics. **b.** produce oxygen. **c.** sit at the base of terrestrial food chains. **d.** sit at the base of aquatic food chains. **e.** fight viruses.

Brief Review

(Answers are in the back of the book.)

1. How do plants get the nitrogen they need?

2. Why do most biologists classify viruses as non-living?

3. Antibiotics used in humans must have at least two qualities. What are they?

4. How can we say that archaea are superficially like bacteria and yet fundamentally different from them?

5. Name three features of more complex organisms that first appeared in protists.

6. What is an amoeba? What class of diseases is associated with them?

Applying Your Knowledge

1. What does it mean to be a "successful" organism? Bacteria are certainly the most numerous organisms on Earth, they live in nearly all environments, and they are the least susceptible to being eliminated through environmental catastrophe. Are they Earth's most successful organisms?

MasteringBiology®

1. MasteringBiology Assignments
Activities: Retrovirus (HIV) Reproductive Cycle; Phage Lysogenic and Lytic Cycles; Phage Lytic Cycle; Replication in Detail; Simplified Viral Reproductive Cycle; Bacterial Conjugation; Structure and Reproduction of Bacteria; **ABC News Video:** Turning Algae into Biofuel; **Current Events:** Battle Brewing Over Labeling of Genetically Modified Food; Scientists Move Closer to a Lasting Flu Vaccine; Can Answers to Evolution Be Found in Slime?; The New Generation of Microbe Hunters; Heart of Iowa as Fault Line of Egg Recall; How Microbes Defend and Define Us; Hunting Fossil Viruses in Human DNA; Virus Ravages Cassava Plants in Africa; **Building Vocabulary**

2. eText
Read your book online, search, take notes, highlight text, and study more effectively.

3. The Study Area
Self-study tools to test comprehension.

22

Fungi:
The Diversity of Life 2

Fungi have a powerful ability to break down and consume most of the materials that nature produces—something that has both good and bad consequences for human beings.

Available on MasteringBiology®
Activities: Life Cycle of Fungi; Classification of Fungi; **Current Events:** Kentuckians Take Distilleries to Court Over Black Gunk; Out West, 'Black Fingers of Death' Offer Hope Against an Invader; Unearthing the Sex Secrets of the Perigord Black Truffle; **Building Vocabulary**; **Questions:** Reading Questions, Test Bank

Mushrooms of the fungus *Amanita muscaria* sit side by side with several species of plants in a Russian forest.

Common houseflies can suffer a hostile takeover by a fungus called *Entomophthora muscae*. It's not unusual for one organism to live off another, but in the case of *E. muscae,* the story has a bizarre twist. This fungus gets inside flies either by drilling through their tough outer skeleton or by squeezing between some cracks in it. The strange thing is that,

on entry, the first thing *E. muscae* does is send out extensions of itself to the fly's brain—specifically to parts of the brain that control crawling behavior. Eventually, the fly begins to crawl upward. This usually means upward on a plant stem or leaf, but it may mean upward on the wall of a house, toward the ceiling. Like any fungus, *E. muscae* lives by digesting organic material, and in this case the organic material is the fly's body. By the time the fly begins making its upward trek, it is nearly dead from the internal damage *E. muscae* has done.

If you look at **Figure 22.1**, you can see the end result of all this: a fly that has crawled up on a conifer needle and, in one of its last acts, has put its mouthparts on the needle, to which they are now glued by the fungus. The white growths you see around the fly's abdomen

are fungal spores. Soon they will be showered over everything *beneath* the fly—including, perhaps, some more flies, which then can be parasitized as well. Fungi have various ways of getting their spores up off the ground for dispersal. The structure we call a mushroom is essentially a spore-dispersal tower. But *E. muscae* has taken another tack. It gets its prey to lift it to higher ground, and sticks its prey in place; then it uses the prey as a launching pad for a shower of fungal spores. *E. muscae* shows us once again that, while nature is certainly diverse, it is not always pretty.

22.1 The Fungi: Life as a Web of Slender Threads

Entomophthora muscae is a member of one of the "kingdoms" of the living world, the fungi. There are lots of fungi that make their living as parasites, as *E. muscae* does, but then again there are fungi that live in a beneficial association with plants and fungi that do us the favor of decomposing fallen trees (**Figure 22.2** on the next page.). At first glance, fungi and plants may seem to be alike. After all, they're fixed in one spot and tend to grow in the ground. But as you'll see, the main connection between plants and fungi is that many of them have struck up underground partnerships. When it comes to evolution, the surprising thing is that fungi and *animals* are more closely related than fungi and plants. A single, common protist ancestor gave rise to animals on the one hand and fungi on the other, with plants only distantly related to the members of either kingdom (**Figure 22.3** on the next page.).

But if a fungus is not a plant, what is it? As the fly example shows, fungi do not make their own food as plants do. Instead, fungi are heterotrophs—they consume existing organic material in order to live. They make a living by sending out webs of slender, tube-like

Figure 22.1
Doing a Parasite's Bidding
Pictured is a root maggot fly (genus *Delia*) that has been killed by an *Entomophthora* fungus; the white spores of the fungus can be seen bursting out from the fly's abdomen.

(a)

(b)

(c)

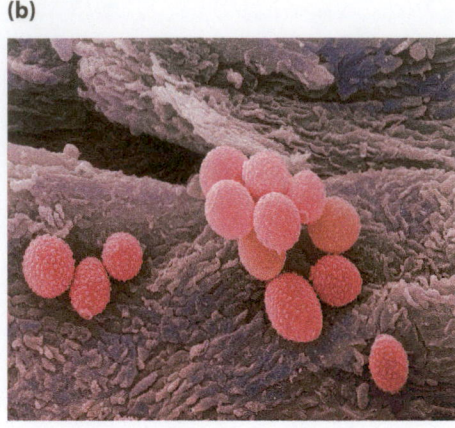

Figure 22.2

Fungal Diversity

(a) What happens to dead branches on a forest floor? They are decomposed, mostly by fungi. The mushrooms sprouting from this branch are outgrowths of a network of fungal fibers that are growing within the branch. These fibers digest and absorb nutrients from the branch, a process that eventually breaks the branch down into its chemical component parts, which are then recycled back into the soil.

(b) Fungi are also human pests. Pictured is a human nail with fungal spores on it that are the cause of athlete's foot.

(c) Fungi distribute their spores in various ways. Pictured are spores being expelled from some puffballs in Costa Rica. What brings about the expulsion? Drops of rain falling on the puffballs.

threads to a food source. This food source could be a fly, as in *E. muscae*'s case, but it could just as easily be bits of organic matter below ground or a fallen log in a forest.

Whatever the case, fungi cannot immediately take in or "ingest" their food in the way that animals do. Fungal cells have cell walls, which means that only relatively small molecules can pass into the fungal

filaments. As such, fungi digest their food *externally*, through digestive enzymes they release. When the food has been broken down sufficiently by these enzymes, the resulting molecules are taken up into the filaments. Almost all fungi obtain nutrition in this manner. Whether we are talking about a microscopic bread mold or extensive webs of fungi that spread underneath the forest floor, this is a defining characteristic of fungi. This characteristic and others can be stated in a more formal way:

- Fungi largely consist of slender, tube-like filaments called **hyphae.**

- Fungi get their nutrition by dissolving their food externally and then absorbing it into their hyphae.

- Collectively, the hyphae make up a branching web, called a **mycelium.** If you look at **Figure 22.4**, you can see the nature of this structure. You may have thought of a mushroom as the main part of a fungus, but in reality it is merely a reproductive structure, called a *fruiting body,* that sprouts up from the larger mycelium below. A fruiting body amounts to a large, folded-up collection of hyphae.

- Most fungi are **sessile,** or fixed in one spot, but their hyphae grow. The fungal mycelium does not move, but its individual hyphae grow toward a food source. This growth comes, however, only at the tips of the hyphae, not throughout their length.

- Fungi are almost always multicellular. The primary exception to this is the yeasts, which are unicellular and thus do not form mycelia.

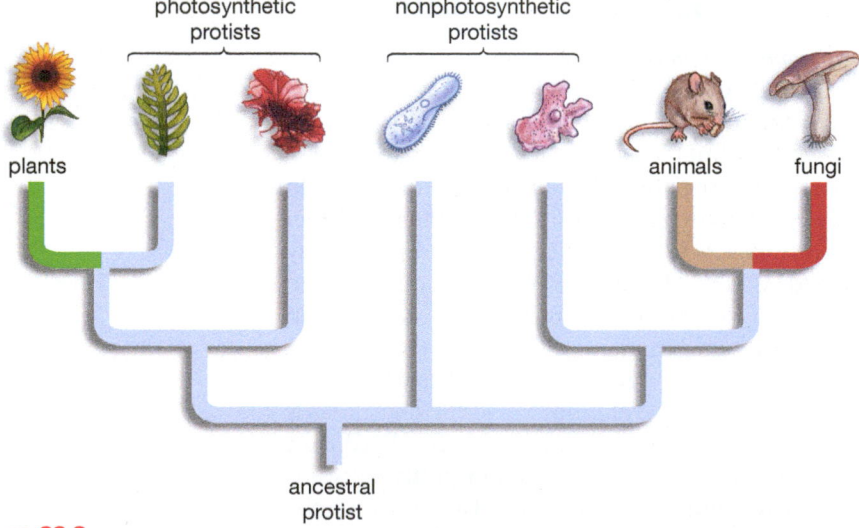

photosynthetic protists

nonphotosynthetic protists

plants

animals

fungi

ancestral protist

Figure 22.3

Fungal Evolution

Fungi may superficially resemble plants, but they actually are more closely related to animals than they are to plants. The evolutionary line that led to plants diverged from the line that led to fungi and animals about a billion years ago, whereas the fungi and animal lines did not diverge until sometime between 35 and 200 million years later. The evolutionary divergence between plants on the one hand and fungi and animals on the other can be thought of as a divergence between organisms that make their own food through photosynthesis (as plants do) and organisms that consume existing organic material (as fungi and animals do).

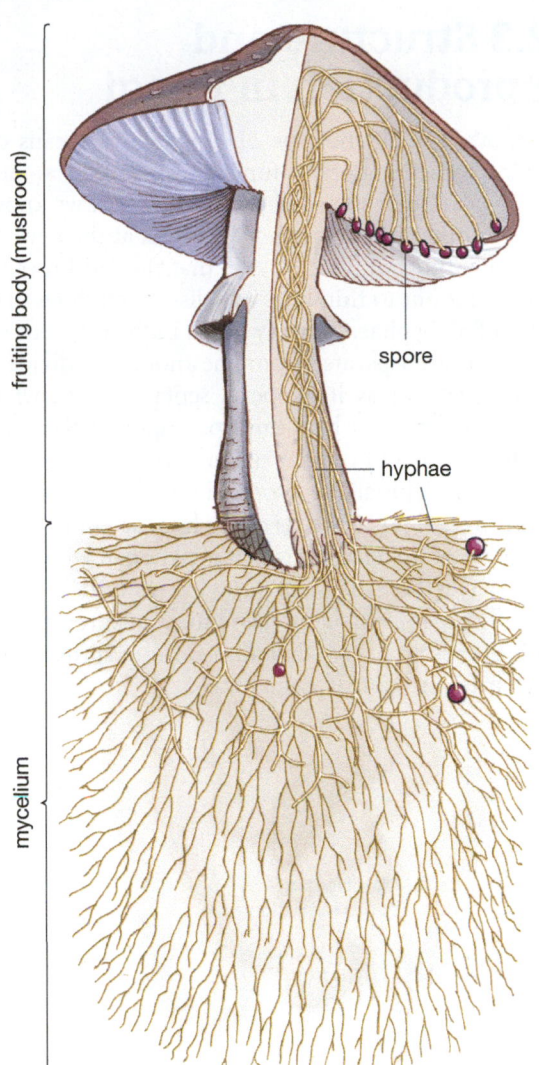

Figure 22.4
Structure of a Fungus
Fungi are composed of tiny slender tubes called hyphae. The hyphae form an elaborate network called a mycelium. The same hyphae also form a reproductive structure called a fruiting body, which in this case is the familiar mushroom.

This last point leads to a more general concept about the living world. You may recall from the last chapter that all bacteria and archaea are single-celled and that protists are usually single-celled. But protists are the limit of single-celled life. With fungi, plants, and animals, life becomes almost completely multicellular. There is no such thing as a single-celled plant or animal, and the one large group of single-celled fungi, the yeasts, seem to have evolved into this state from an earlier multicelled form. This is not to say that all life becomes *big* once we get past the protists. As you'll see, there are plenty of microscopic fungi (and even some microscopic animals). But from here on out, all the life-forms we'll be looking at are made up of more than one cell.

22.2 Roles of Fungi in Society and Nature

Like bacteria, fungi are mostly hidden from us—and mostly known to us by the trouble they cause. This trouble is considerable, as fungi are responsible for mildew, dry rot, vaginal yeast infections, bread molds, general food spoilage, toenail and fingernail infections, and a variety of related skin afflictions, among them athlete's foot, "jock itch," and ringworm. The agricultural blights of corn smut and wheat rust are fungal infections. Indeed, fungi probably are the single worst destroyers of crop plants. One survey in Ohio indicated that only 50 plant diseases there were caused by bacteria and 100 by viruses but 1,000 by fungi. If you look at **Figure 22.5**, you can see an example of what happened to American elm trees in cities across the United States beginning in the 1930s because of a fungus called *Ophiostoma ulmi*—the cause of Dutch elm disease. In recent years, there has been great concern about the fungi commonly called molds and the effects they have on both houses and the people who live in them.

Even given all this fungal pestilence, however, we could say of the fungi what we said of bacteria: While we could do without some of them, we could not live without others. Mold in houses may be a huge problem, but there is another mold that is the source of penicillin. Fungal infections may be a problem, but organ transplants are made possible today by the immune-system-suppressing drug cyclosporin, which is derived from a fungus. If you know someone who has high cholesterol, he or she may control it with the drugs lovastatin or pravastatin, both of which are derived from a fungus. In the realm of food production, brewer's yeast makes bread rise and beer ferment, and blue cheese is blue because of the veins of mold within it. Most soft drinks contain citric acid that is produced from a fungus, and we simply eat fungi outright in the form of mushrooms. Out in nature, fungi join bacteria as the major decomposers of the living world, breaking down organic material such as the wood in fallen logs and turning it into inorganic compounds that are recycled into the soil. In fact, the final breakdown of woody material is almost entirely the work of fungi. And, as noted, fungi are involved in a critical association with plants, about which you'll learn more later.

Figure 22.5
The Effects of Dutch Elm Disease

(a) Elm-lined street, Detroit, 1971

(b) Street after Dutch elm epidemic, 1984

Some 70,000 species of fungi have been identified so far, but this is just a fraction of the number that actually exist. Because fungi tend to grow in inaccessible places, most of them are unknown to us. By one estimate, the total number of fungal species may be about 1.5 million.

SO FAR ...

1. The essential structure of most fungi is a set of slender filaments called _____ that collectively make up a structure called a _____.

2. Which of the following is not a characteristic of fungi: (a) usually fixed in one spot; (b) make their own food; (c) tend to be multicellular?

3. Fungi are indispensable to the natural world because of their role in _____.

Figure 22.6
The Life Cycle of a Fungus

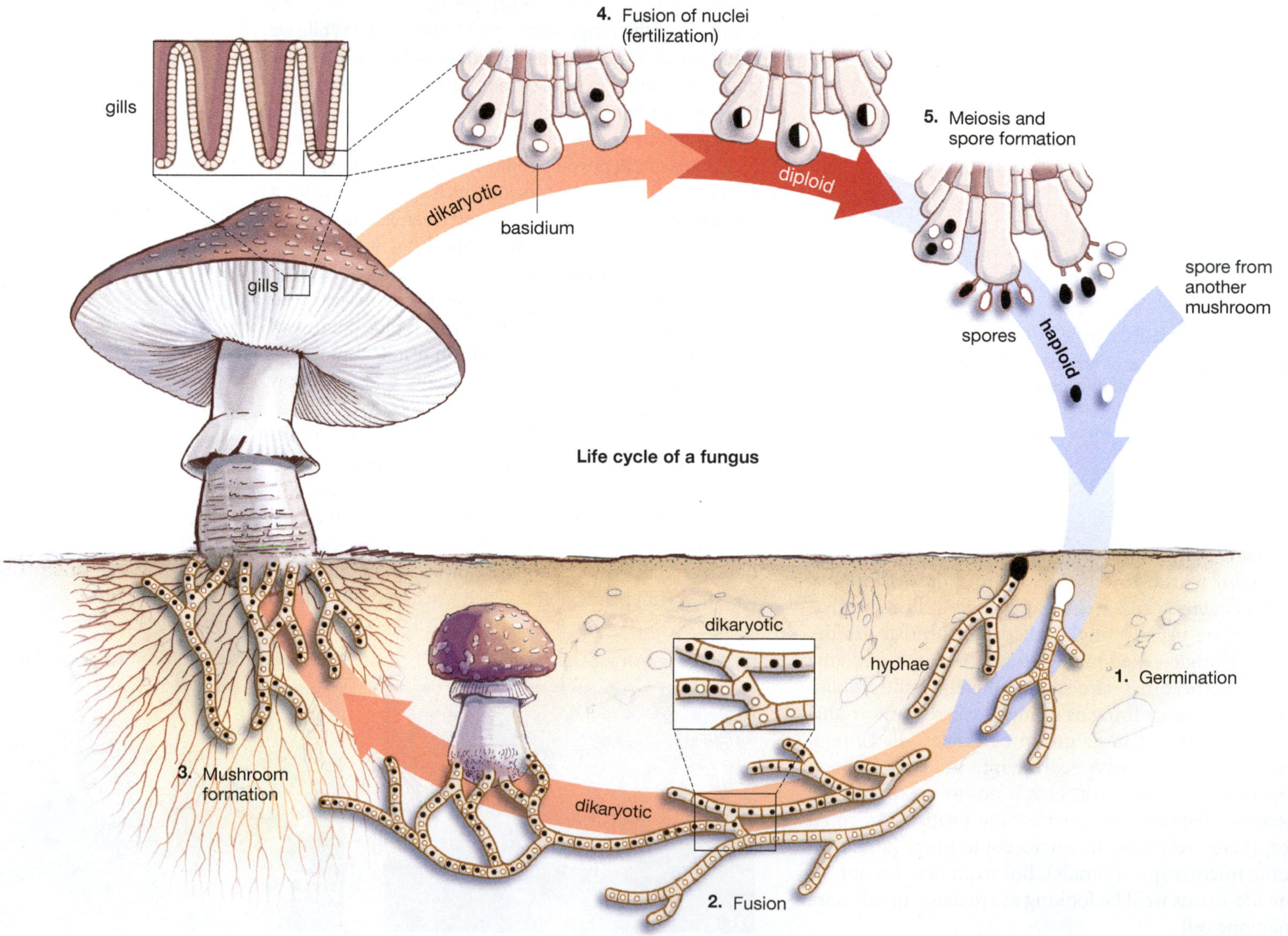

22.3 Structure and Reproduction in Fungi

Let's now look at one type of fungus as a means of learning something about fungi in general. In **Figure 22.6**, you can see a so-called club fungus that we know as the common mushroom. If you look at the figure's lower-right portion, you can see that the mushroom's hyphae amount to thin lines of cells—something that is true of all hyphae. In many cases, individual cells in the hyphae are separated from one another by dividers called septa. But as it happens, septa are somewhat porous dividers; they have tiny openings that allow for a fairly free flow of cellular material between one cell and the next. This allows for rapid movement of cellular resources right to the tip of the hyphae—the site of hyphal growth. The result is that hyphae can grow very quickly, whether they are growing toward a food source or organizing themselves into a mushroom cap. Did it ever seem to you that a mushroom sprouted in your yard overnight? It probably did.

4. Fusion of nuclei (fertilization)

gills

dikaryotic

basidium

diploid

5. Meiosis and spore formation

spore from another mushroom

spores

haploid

Life cycle of a fungus

gills

dikaryotic

hyphae

1. Germination

3. Mushroom formation

dikaryotic

2. Fusion

In the upper left of the figure, you can see how the mushroom cap is structured. The underside of each cap is made up of a profusion of accordion-like folds called gills. Each gill is simply a collection of hyphae. And right at the tip of some of these hyphae is a reproductive structure called a basidium. Reproductive cells called spores are produced within the basidium and then ejected from it. Caught by the wind, these spores are then carried away to new ground on which they can "germinate," or sprout new hyphae. Fungal spores are, however, something like lottery tickets: A multitude are made, but only a few will be winners. Such a tiny fraction of spores end up germinating that huge numbers must be released in order for all nearby environments to receive some. Tens of millions will be ejected each hour from the underside of a fairly small mushroom cap.

The mushroom cap is one example of a large fungal structure, but most fungi don't have this kind of size. Think of a flat circle of blue-gray mold on some spoiled bread. It has its own mycelium, with reproductive structures topped by spores, but magnification is necessary to make out any of this detail, as you can see in **Figure 22.7**. To a great extent, fungi can be regarded as microbes.

The Life Cycle of a Fungus

The mushroom caps and spores we've been looking at are important players in the life cycle of club fungi, which is to say the set of steps by which one generation of these fungi gives rise to the next. If we look at this life cycle from its beginning—in Figure 22.6, over

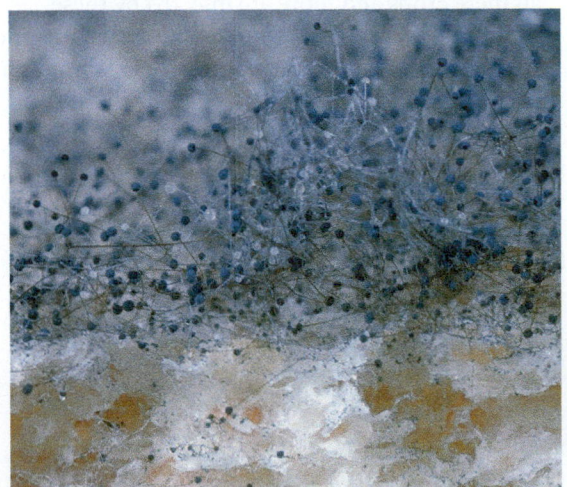

Figure 22.7
Bread Mold up Close
The dark spots on moldy bread are composed of a mycelium that forms stalks tipped by spores, visible in this micrograph as tiny black balls. The spores will be dispersed to produce new mycelia elsewhere—perhaps on more bread. These are spores of the mold *Rhizopus nigricans*, growing on the surface of some bread.

at step 1— you can see that it starts with spores that have been released from fruiting bodies. These spores germinate, generate hyphae, and in step 2, two of these hyphae fuse: Two hyphal cells come together and become one cell, from which a single hypha grows. Note, however, that the cells in this new hypha have *two* nuclei each (symbolized by the little black and white dots within them). This is an unusual condition for a cell; in the normal case, one cell has one nucleus. This condition has come about because, while two cells fused in step 2, their nuclei remained separate. Thus begins a phase of life that is unique to two types of fungi: the **dikaryotic phase,** in which cells in a fungal mycelium have two nuclei.

And what is the nature of the genetic material in these nuclei? Each nucleus exists in what is known as a haploid state—there is but a *single* set of chromosomes in each one. This contrasts with the diploid state in almost all human cells. Our nuclei have *paired* sets of chromosomes within them (one set inherited from our father, the other from our mother). The haploid state of these nuclei may seem like a technical detail, but it's important for what's coming up. Once a mushroom mycelium enters the dikaryotic phase it can stay there for a long time. The phase can go on for years, in fact, and ends only when the most visible manifestation of the mycelium, the mushroom cap, sprouts above ground, as seen in step 3.

Sex in Fungi

The dikaryotic state initially exists in all the cells in the mushroom—even the cells that make up the spore-releasing basidia that line the gills. But in some of these cells, right at the tips of the basidia, this will change. The two nuclei in these cells will fuse into one nucleus, as seen in step 4. This change means these cells have become diploid; they now have a single nucleus that contains *two* sets of chromosomes.

Now, think about where each of these sets of chromosomes came from. One of them came originally from the black spore over at step 1, while the other came from the white spore. They existed separately in a single cell for a while, but now they have fused. With this, we have had a fusion of genetic material from two separate organisms—sex has taken place. It's not the merger of egg and sperm we're used to in human reproduction, but it's sex nevertheless.

Next, as you can see in step 5, each of the cells in which this fusion has taken place undergoes the meiosis that was reviewed in Chapter 10. This means a mixing of chromosomes from the two parents and two rounds of cell division. The result is four haploid spores attached like little bags to the tip of a basidium; these will be among the millions of spores released from the mushroom cap. Each of these spores will be of one "mating type" or another—symbolized by the black and white colors in the figure. Just as only

Life Cycle of Fungi

opposite sexes can come together to create offspring in humans, so only opposite mating types can come together (in hyphae fusion) to create offspring in fungi when they reproduce sexually.

The Nature of Spores

This gets us to the nature of a **spore,** which can be defined as a reproductive cell that can develop into a new organism without fusing with another reproductive cell. Note the difference between the spores of the mushroom and the *gametes* of a human being—eggs or sperm. A human being cannot develop from an egg or a sperm; a fusion of both cells is required. Meanwhile, the spores that our mushroom developed from did not undergo fusion. A fungal spore was produced in meiosis, and once it came into being, it was dispersed and developed into a whole new mycelium by itself.

Reproduction in Other Types of Fungi

The life cycle we've just looked at actually amounts to only one means of fungal reproduction among many that exist. Under some conditions, fungi that are capable of sexual reproduction will instead utilize *asexual* reproduction. The little black balls you can see on top of the bread mold in Figure 22.7 are spores that were produced through mitosis, or simple cell division, as carried out by one organism. There was no fusion of the cells and nuclei from two organisms to produce these spores—there was no sex, in other words. Spores such as the ones on the bread mold develop through cell division, after which they disperse to form new mycelia; then these mycelia develop new spores through cell division and the process continues. Note that the constant within fungal reproduction is spores—they are employed no matter what. But spores can be produced either sexually or asexually. And the sexual processes that fungi employ can vary considerably. In short, fungi exhibit great diversity in the means by which they reproduce themselves.

22.4 Categories of Fungi

We're used to dividing the animal world up into various categories, such as insects, mammals, and reptiles. So, what are the major categories of fungi? It turns out there are four principal groups or "phyla" of fungi, and that three of these phyla are defined by their reproductive structures. As you've seen, toadstool-shaped mushroom fungi have a reproductive structure called a basidium; accordingly, these mushrooms are in the fungal group known as the basidiomycetes. Likewise, there is a group called the ascomycetes and one called the zygomycetes—both named for the reproductive structures they have. Finally, there is a group called the chytrids whose members give us an idea of how close the evolutionary relationship is between fungi and animals. Let's look briefly now at each of these four fungal phyla.

- **Basidiomycetes.** Toadstool-shaped mushrooms are the best-known members of Phylum Basdiomycota, but this group also includes two banes of the agricultural world—the rusts and the smuts—along with the familiar "shelf" fungi that often are seen sprouting from trees (**Figure 22.8**). You may recall that the spore-releasing basidium on a mushroom is shaped something like a club. It is this shape that gives the basidiomycetes the common name of the club fungi. Most of the living world benefits from the work of basidiomycetes as they serve as a major group of decomposers. This is small comfort to farmers worldwide, however, who must live with the incredible agricultural damage caused by the basidiomycete smuts and rusts. Wheat, corn, barley, oats, soybeans, coffee—all these crops and more are infected by these fungi, which can sweep through crops in epidemic waves. In 1993, leaf rust destroyed over 40 million bushels of wheat in Kansas and Nebraska alone.

- **Ascomycetes.** Morel mushrooms are members of Phylum Ascomycota, as are the highly prized

---

Figure 22.8
Basidiomycete Fungi

(a) The basidiomycetes include the well-named shelf fungi, such as this *Ganoderma* fungus, seen growing from the trunk of a beech tree in England.

(b) The agricultural scourges called smuts are also basidiomycetes. Here fungal outgrowths called smut galls can be seen sprouting from an ear of corn that has been infected by basidiomycete fungus *Ustilago zeae.*

(a) Shelf fungus on beech tree trunk

(b) Smut outgrowths on corn ear

(a) Common morel mushroom (*Morchella esculenta*)

(b) Orange peel fungus (*Aleuria aurantia*)

truffles that are used in cooking. On the other hand, the cause of Dutch elm disease is the ascomycete *Ophiostoma ulmi,* and if you see some brown rot on a peach or other stone fruit, it is likely to be the ascomycete *Monilinia fructicola.* All members of this phylum reproduce by using tiny spore-releasing structures whose sac-like shape gives the ascomycetes their common name: the sac fungi. In ascomycete mushrooms, these sacs can be contained in a large fruiting body that is shaped like a cup (**Figure 22.9**). The Ascomycetes are a sister evolutionary group to the Basidiomycetes in that fungi in both of these late-evolving fungal phyla go through the dikaryotic phase in their life cycle, whereas no other fungi do.

- **Zygomycetes.** When you see a patch of blue-gray mold on bread, you're probably looking at a fun-

gus that's part of Phylum Zygomycota. This is so much the case that a common name for the zygomycetes is the bread molds. There are only about 1,050 known species of zygomycetes, which is a small number compared to the more than 22,000 known basidiomycetes or the 32,000 ascomycetes. Nevertheless, the zygomycetes are diverse. Many of them form underground associations with plant roots, but one group of them makes a living by parasitizing insects. (Remember the *Entomophthora* species from the start of the chapter that parasitized the fly? It was a zygomycete.) What all zygomycetes have in common is the use of a spore-releasing structure called a sporangium that is shaped something like a lollypop; you can get a close-up view of this structure in **Figure 22.10**.

Classification of Fungi

(a) Bread mold (*Mucor mucedo*) growing on bread

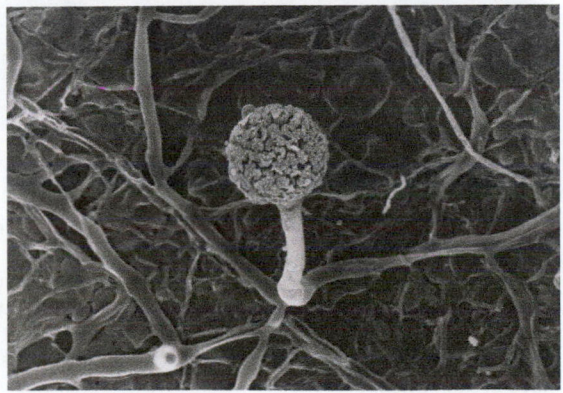

(b) Close-up of spore-laden bread mold sporangium

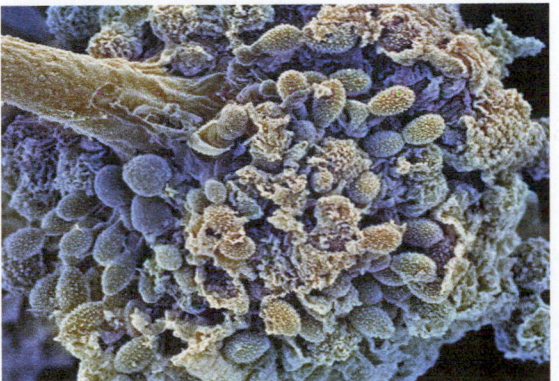

Figure 22.11
The Chytrid Fungi

(a) In this highly magnified micrograph, a Chytrid fungus can be seen releasing a cloud of spores even as it feeds on a pollen grain, visible as the pale brown circle at the photo's center.

(b) Like many of its fellow amphibians, this great barred frog has been infected by the Chytrid fungus *Batrachochytrium dendrobatidis*. Such infections have caused steep, continent-wide declines in some frog species and have brought about the actual extinction of other species. *B. dendrobatidis* invades the skin of frogs, altering it in ways that can be lethal.

(a) Chytrid fungus growing on a pollen grain

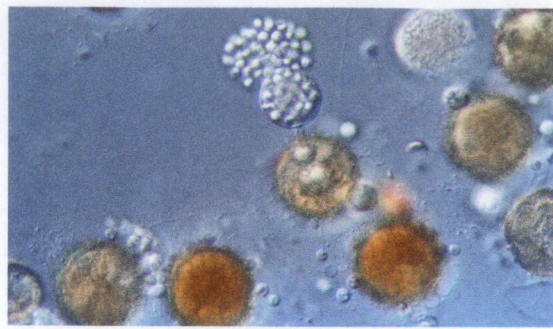

(b) Frog infected with a chytrid fungus

- **Chytrids.** The fungi of Phylum Chytridiomycota are very different from any of the others we've looked at. For one thing, most of them live in water. But beyond this, all of them are *mobile* at one point in their life cycle. Indeed, these fungi move through the water using the kind of tail-like flagellum that we saw last chapter in various species of protists. (To picture a flagellum, just think of the "tail" of a human sperm.) And some chytrids don't develop a mycelium but instead are like the yeasts in that they are single-celled.

Given all this, you might wonder why the chytrids are classified as fungi at all. The answer is that they duplicate their DNA in a manner used by all fungi; their cell walls are partly composed of a material found in most fungal cell walls (chitin); and contemporary sequencing of their DNA shows that their closest relatives are fungi rather than any other life-form. The fascinating thing is that the chytrids probably give us an idea of what fungi were like *originally*—back when fungi first evolved from protists. The sessile fungi we've been looking at seem to have evolved from a group of mobile fungi that resembled today's chytrids. Moreover, the chytrids show us how close the association is between fungi and animals. Think of it: DNA sequencing tells us how close fungi and animals are to each other; the chytrids show us what the early, mobile fungi were like; and a good deal of evidence indicates that early animals had the flagellated form of the chytrids. Given all this, the message is that primitive animals probably were not all that different from primitive fungi. Like most of the fungi we've looked at, the chytrids take a toll on other life-forms. One species of chytrid, *Batrachochytrium dendrobatidis*, has been decimating populations of frogs worldwide in recent years (**Figure 22.11**).

Yeasts: *Saccharomyces cerevisiae*

Several times so far, we've taken note of the single-celled fungi called yeasts. Where do they fit within the four-phylum structure of the fungi? The answer is that, with the exception of the chytrids, each of the fungal phyla has yeast species within it—there are basidiomycete yeast, ascomycete yeast, and zygomycete yeast, in other words. Thus, "yeast" is not a taxonomic category; it merely describes a form that fungi can take. Most yeasts are ascomycetes, but the basidiomycetes are well represented in this category as well. We can define a **yeast** as any single-celled fungus that tends to reproduce by budding. This, of course, raises the question: What's budding? **Figure 22.12** shows it in action. As you can see, budding is not a splitting of one cell down the middle; it is a means of fungal reproduction in which daughter cells are produced as outgrowths of parental cells.

The particular yeast that's shown in the picture deserves special mention since it has been tremendously useful to people for as long as human civilizations

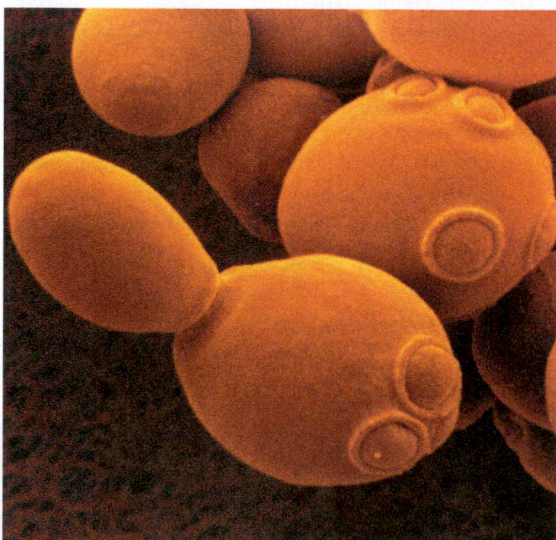

Figure 22.12
Our Partner in Food and Drink

The fungal yeast *Saccharomyces cerevisiae* has been used by human beings for thousands of years in baking and brewing. The *S. cerevisiae* cell at the forefront of the picture is giving rise to a daughter cell through the process of budding. The circular areas on this cell and the one behind it are bud scars—areas from which previous daughter cells have broken away.

ESSAY

A Psychedelic Drug from an Ancient Source

In 1943, chemist Albert Hoffman of Switzerland's Sandoz pharmaceutical company found himself working with a substance that had a long history, most of it tragic. The substance came from a fungus in the genus *Claviceps*, which infects several grain crops, among them rye. The visible manifestation of a *Claviceps* infection is little sickle-shaped growths of black hyphae that pop out among the rye grains (**Figure 1**). Centuries ago, the French perceived these growths to look like an ergot—a little spur that some birds have on their legs—and the name stuck.

In nature, when wild grasses get ergot infections, the effect is not all that notable.

Figure 1
Damaging Intruder
Black ergots of the *Claviceps purpurea* fungus sprout from among the grains of a wheat plant.

The fungus takes some plant nutrition away, the ergot growths fall to the ground in autumn, and spores from it infect the next generation of grass in the spring. The problem for humans comes when the fungus infects *food* crops such as rye and the ergot are ground up along with the rye grains. The result is that ergot becomes a flour ingredient. This is serious because ergot contains a powerful alkaloid; its active ingredient is a member of the family of alkaline substances that also includes cocaine, nicotine, and morphine, along with such poisons as hemlock. In numerous instances, recorded in Europe since the Middle Ages, ergot made its way into the bread that people ate, and the results were disastrous. France in particular had repeated instances of "ergotism," with two of these episodes alone resulting in 50,000 deaths. What did ergot do? Here is a description by Oswald Tippo and William Louis Stern. Ergot's effects included:

> Abortions in pregnant females; burning sensation in extremities; feelings of being consumed by fire; constriction of blood vessels leading to the blockage of circulation and resulting in the appearance of blue discolored areas on the body; onset of gangrene; shriveling and falling off of hands, arms, and legs; experiencing of hallucinations such as wild beasts in pursuit of the victim.

Horrible as these effects were, medical people hearing of them recognized that ergot had the ability to do good as well as harm. Unwanted abortions are a terrible thing, but labor sometimes needs to be *induced* when a fetus is overdue; likewise, the constriction of blood vessels has its uses when properly controlled. Alkaloids often have such double-edged capabilities. Cocaine was first isolated from the coca plant in 1860, but in the 1880s it became a miracle drug for eye surgery when it served as the first local anesthetic. Morphine, which is derived from poppy plants, is used in medicine as a painkiller. Thus, it's not surprising that drugs were also isolated from ergot over time. In the 1930s, for example, a drug called ergonovine—still in use today—was developed that helps stop bleeding after childbirth.

Thus it was that Albert Hoffman found himself working with ergot in his Basel lab in 1943. He was looking for other useful alkaloids that the fungus might yield. Years before, he had isolated from ergot a chemical core—a compound called lysergic acid—and added to it a manufactured compound called diethylamine, but the results seemed uninteresting to him. This time, however, he apparently absorbed a tiny quantity of the substance through his skin. The outcome, as he later wrote, was that:

> I was seized by a peculiar sensation of vertigo and restlessness. Objects appeared to undergo optical changes and I was unable to concentrate on my work. . . . With my eyes closed, fantastic pictures of extraordinary plasticity and intensive color seemed to surge toward me.

Although he had not intended to, Albert Hoffman had discovered one of the world's most powerful psychoactive drugs: lysergic acid diethylamide, which the world would come to know by its initials, LSD.

---

have existed. *Saccharomyces cerevisiae* is the most important yeast by far in making bread rise and in making wine and beer ferment. The alcohol in wine is simply a product of *S. cerevisiae*—it feeds on the sugars in grape juice and produces alcohol in the course of its metabolism. In a similar vein, the rise that we see in yeasted breads usually comes from the carbon dioxide given off by *S. cerevisiae* as it feeds on other bread ingredients. As if all this weren't enough, *S. cerevisiae* is a mainstay of scientific research: An entire field of genetics has been built around the study of it.

Figure 22.13
Life on the Rocks
Lichens growing on sandstone rocks near the Montana–Wyoming border.

22.5 Fungal Associations: Lichens and Mycorrhizae

Fungi obtain their nutrition in three basic ways. Any fungus that gets its nutrition from a living organism is a **parasite**—an organism that feeds off another organism but does not kill it immediately and may not kill it ever. Many fungi are parasites: Think of a basidiomycete rust feeding on a wheat crop, a chytrid fungus living in the skin of a frog, or an ascomycete fungus living between your toes (thus giving you athlete's foot). Against this, if a fungus extracts its nutrition from *dead* organic matter, it is a **saprophyte.** Thus, all the basdiomycetes that make a living by breaking down fallen logs are said to be saprophytic, and this same thing is true of any zygomycete that can be found covering an old piece of bread. Some fungi, however, obtain their nutrition not by being parasitic or saprophytic but by being *cooperative* with other organisms. Indeed, one of the cooperative associations that fungi have entered into is so close that it has resulted in two organisms that function as one.

Lichens

Everyone has seen the thin, sometimes colorful coverings called lichens that seemingly can grow anywhere—on rocks as well as on trees, in Antarctica as well as in a lush forest (**Figure 22.13**). As it turns out, lichens are not a single organism. A **lichen** is a composite organism composed of a fungus and either algae or photosynthesizing bacteria. If you look at **Figure 22.14**, you can see that this association can be structured as a kind of sandwich, with an upper layer of densely packed fungal hyphae on top, then a zone of less-dense hyphae that includes a layer of algal or bacterial cells, then another layer of densely packed hyphae on the bottom that sprout extensions down into the material the lichen is

growing on (such as a rock). The example shown is an association between a fungus and an alga, which is the case in about 90 percent of lichens. In these cases, the fungal hyphae either wrap tightly around the algal cells or actually extend into them.

A lichen is mostly a fungus, then, but it is a fungus that could not grow without its algal partner. Why? Because the alga makes its own food through photosynthesis and then proceeds to supply the fungus with

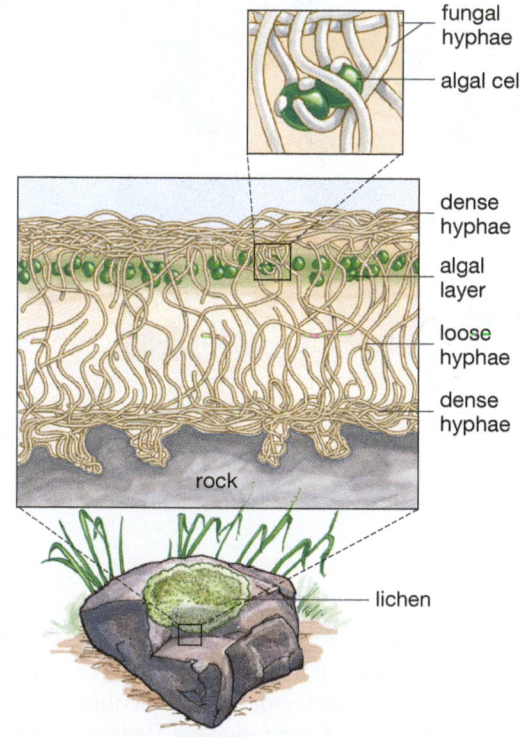

fungal hyphae

algal cell

dense hyphae

algal layer

loose hyphae

dense hyphae

rock

lichen

Figure 22.14
The Structure of a Lichen

most of the nutrients it needs. A debate existed for years about whether the fungus was in turn doing something for the alga, but this has now been settled. The relationship between a fungus and an alga is one of **mutualism**—both parties are benefiting from it. The alga may be supplying food *to* the fungus, but it is getting several things *from* the fungus in return: water, carbon dioxide, minerals, and protection from the elements. This last factor explains why lichens are often so colorful. Their fungi produce brightly colored pigments that can screen out harsh sunlight—something that's important in the exposed environments lichens can be found in.

In nature, the general significance of lichens comes from their role as pioneers in barren habitats where life is just getting established. Because the algae in lichens make their own food, and because the fungi in them can dig into rock itself, lichens can make a living in these tough habitats, whereas most other organisms cannot. Beyond this, lichen fungi produce acids that can help break down rock and build up soil—activities that create habitat for *other* organisms.

Lichens have an ability not only to establish themselves in tough environments but also to persist in them for a long time. If left undisturbed, they can survive for thousands of years in colder climates. The lives of lichens, however, could be characterized not only as long but as slow. At most, they spread out at a rate of about 20 millimeters per year, which is about three-fourths of an inch. Indeed, they have been observed to grow as little as 0.2 millimeters per year, at which pace they would grow by about three-fourths of an inch per century. What accounts for this sluggish rate of expansion? In times of little water, lichen algae stop performing photosynthesis. This puts the entire organism into a state of dormancy—a kind of suspended animation—that persists until water becomes available again.

Mycorrhizae

By some estimates, 90 percent of seed plants live in a cooperative relationship with fungi. The linkage here is one of plant roots and fungal hyphae, with both the plants and the fungi benefiting. What the fungus gets from the association is food, in the form of carbohydrates that come from the photosynthesizing plant; what the plant gets is minerals and water, absorbed by the fungal hyphae. So important is this relationship to plants that some species of trees, such as pine and oak, cannot grow without fungal partners. Other plants will have stunted growth without a fungal partner.

Fungal hyphae generally grow into the plant roots, but in some instances they wrap around the root

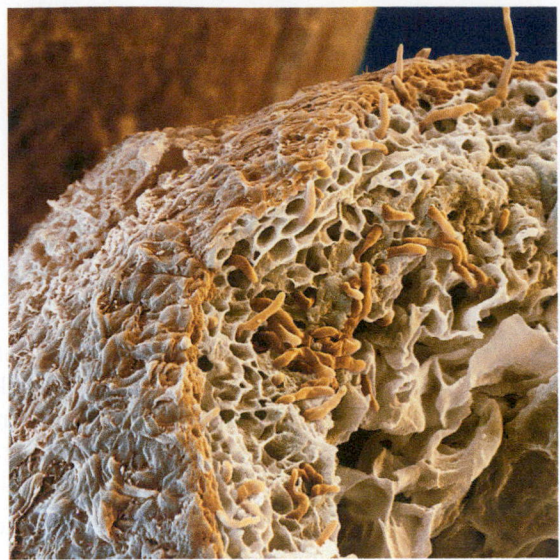

Figure 22.15
Underground Partners
Mycorrhizae are associations between plant roots and fungal hyphae that benefit both the plant and the fungus. Here the association can be seen in a highly magnified image of a root and its surrounding hyphae.

without penetrating it. In either case, the root–hyphae associations of plants and fungi are known as **mycorrhizae** (**Figure 22.15**). The mycorrhizal relationship is one of the oldest and most important in nature. It appears that at least 460 million years ago, when the very first plants made the transition to life on land, they did so with fungal partners.

On to a Look at Animals

In this chapter and the one that preceded it, you've looked at bacteria and archaea and at two of the kingdoms within the eukaryotic domain—protists and fungi. Coming up next is a kingdom that is arguably the most familiar of all because we human beings are part of it. Yet it is a kingdom so diverse that it holds a good many surprises as well. It's the kingdom of animals.

SO FAR . . .

1. A lichen is a composite organism, always composed of a _____ and either _____ or _____.

2. In a lichen, the _____ is providing the _____ or _____ with water, carbon dioxide, minerals, and protection from the elements.

3. The cooperative associations called mycorrhizae provide the _____ with sugars, while providing the _____ with water and minerals.

22 REVIEW

Summary

22.1 The Fungi: Life as a Web of Slender Threads

- Fungi are heterotrophs: They consume existing organic material to live. Fungi consist largely of webs of slender tubes, called hyphae, that grow toward food sources. Fungi cannot immediately ingest the food the hyphae reach since only small molecules can pass through the cell walls of fungal cells. Fungi thus digest their food externally, through release of digestive enzymes, and then bring the resulting small molecules into the hyphae. **(p. 403)**

- Collectively, hyphae make up a structure called a mycelium, which forms the bulk of most fungi. Many fungi have a reproductive structure called a fruiting body that produces and releases reproductive cells called spores. All fruiting bodies are organized collections of hyphae. **(p. 404)**

- Most fungi are sessile, but their hyphae grow toward food sources. Such growth comes only at the tips of the hyphae, not throughout their length. **(p. 404)**

- Most fungi are multicellular. The primary exception to this, the single-celled yeasts, are thought to have evolved into the single-celled state from an earlier multi-celled condition. **(p. 404)**

22.2 Roles of Fungi in Society and Nature

- Fungi can be destructive, but they are also sources of medicines and are used extensively in food processing. In nature, fungi join bacteria as major decomposers of the living world. **(p. 405)**

22.3 Structure and Reproduction in Fungi

- Individual cells that form hyphae often are separated by septa, which are porous enough to allow for a free flow of cellular material between one hyphal cell and the next. Because hyphae can bring resources quickly to the point of growth at the hyphal tips, fungi can grow quickly. **(p. 406)**

- A mushroom cap, one kind of fruiting body, is made up of accordion-like gills that produce and release spores. Only a tiny fraction of fungal spores will successfully germinate, so fruiting bodies release huge quantities to ensure new fungal growth. **(p. 407)**

- Two groups of fungi have a dikaryotic phase of life in which the fusion of cells from different parental hyphae produces a single hypha whose cells each have two haploid nuclei—one nucleus coming from one parental hypha, the other from the second parental hypha. **(p. 407)**

- When the below-ground mycelium in a mushroom sprouts into the above-ground mushroom cap, all the cells in the cap initially are dikaryotic, including those in the reproductive structure called the basidium. The nuclei in certain cells of the basidia then undergo fusion, meaning sex has taken place. The resulting diploid cells then undergo meiosis, thus producing four haploid spores attached to the tip of the basidium. These spores are released and blown by the wind to new locations, where they may germinate. **(p. 407)**

- A spore is a reproductive cell that can develop into a new organism without fusing with another reproductive cell. **(p. 408)**

- Most fungi can reproduce by asexual as well as sexual means. In asexual reproduction, fungal spores are produced through mitosis (or simple cell division), which is carried out by one organism. In sexual reproduction, fungal spores are produced through a fusion of nuclei from two separate fungi. **(p. 408)**

22.4 Categories of Fungi

- There are four phyla of fungi: basidiomycetes, ascomycetes, zygomycetes, and chytrids. The first three are defined by their reproductive structures. **(p. 408)**

- The basidiomycetes, also known as the club fungi, include "toadstool" mushrooms, "shelf" fungi that can be seen sprouting from trees, and the agricultural pathogens called smuts and rusts. The ascomycetes, also known as the sac fungi, include truffles, morel mushrooms, and fungi that attack stone fruits. The zygomycetes, also known as the bread molds, include many fungi that form associations with plant roots but also include one group of fungi that live by parasitizing insects. The chytrids are a primitive group of mostly aquatic fungi, all of which are mobile at one point in their life cycle. The chytrids give us an idea of what the earliest fungi were like, as the sessile fungi that are so common today seem to have evolved from chytrid-like ancestors. Early fungi were probably similar to early animals. **(p. 408)**

- A yeast is any single-celled fungus that tends to reproduce by budding, a process in which fungal daughter cells are produced as outgrowths of parental cells. All fungal phyla except the chytrids have yeast species exist within them. The majority of yeasts are ascomycetes, but many are basidiomycetes. **(p. 410)**

22.5 Fungal Associations: Lichens and Mycorrhizae

- Some fungi obtain their nutrition from living organisms and are thus parasites; some extract nutrition from dead organic matter and are thus saprophytes; and still others obtain nutrition through cooperative associations with other life-forms. One of these associations results in the formation of lichens: composite organisms made up usually of fungi and algae but sometimes of fungi and bacteria. The relationship between fungi and their algal partners in lichens is one of mutualism, in that both parties benefit from the interaction. **(p. 412)**

- Up to 90 percent of seed plants live in a mutualistic association with fungi that links plant roots with fungal hyphae. In this association, called mycorrhizae, plants supply fungi with food produced in photosynthesis, while the fungi supply plants with minerals and water, gathered by the web of fungal hyphae. (p. 413)

Key Terms

dikaryotic phase   407

hyphae   404

lichen   412

mutualism   413

mycelium   404

mycorrhizae   413

parasite   412

saprophyte   412

sessile   404

spore   408

yeast   410

Understanding the Basics

Multiple-Choice Questions

(Answers are in the back of the book.)

1. Fungi (select all that apply):
   a. are more closely related to plants than animals.
   b. make their own food.
   c. use existing organic material as their food.
   d. digest their food externally.
   e. are usually multicellular.

2. Most fungi are composed almost entirely of:
   a. ball-shaped objects called hypha.
   b. leaves and stems.
   c. reproductive cells called spores.
   d. slender filaments called hyphae.
   e. slender filaments called septa.

3. Fungi cause (select all that apply):
   a. tetanus.
   b. athlete's foot.
   c. common colds.
   d. dry rot.
   e. wheat rust.

4. Fungi are important to humans as (select all that apply):
   a. a source of citric acid.
   b. photosynthesizing organisms.
   c. the source of some antibiotics.
   d. decomposers of dead organic material.
   e. producers of oxygen.

5. Fungi reproduce through cells called spores that (select all that apply):
   a. have a single set of chromosomes.
   b. often are released from structures called fruiting bodies.
   c. usually are released in huge numbers.
   d. must fuse with another spore to bring about germination.
   e. germinate without fusing with another cell.

6. Fungi that obtain their nutrition from living organisms are referred to as _____, while fungi that obtain their nutrition from dead organic material are referred to as _____.
   a. carnivores; herbivores
   b. saprophytes; parasites
   c. omnivores; herbivores
   d. parasites; saprophytes
   e. saprophytes; herbivores

7. The association between fungi and algae (or bacteria) in forming lichens is one of _____ in that _____.
   a. parasitism; only one party benefits from the relationship
   b. competition; neither party benefits from the relationship
   c. competition; both parties are vying for the same resources
   d. mutualism; both parties are benefiting from the relationship
   e. saprophytism; one party is feeding on the dead remains of the other

8. The mycorrhizae formed by the association between fungal _____ and plant _____ are found in _____ seed plants.
   a. hyphae; roots; only a few
   b. spores; leaves; only a few
   c. hyphae; roots; the vast majority of
   d. fruiting bodies; roots; the vast majority of
   e. hyphae; stems; only a few

Brief Review

(Answers are in the back of the book.)

1. Explain the phrase "animals ingest then digest, while fungi digest then ingest."

2. Fungi could be said to take a "lottery" approach to reproduction. In what respect?

3. How does a spore differ from a gamete?

4. Match the fungal phylum on the left with the description on the right

   a. Basidiomycota — Bread molds, spore-releasing sporangium
   b. Ascomycota — Mobile, water-dwelling, early-evolving
   c. Zygomycota — Club-fungi, rusts and smuts, toadstools
   d. Chytridiomycota — Sac fungi, morel mushrooms, truffles

5. Name two defining features of yeast fungi.

Applying Your Knowledge

1. Vaginal yeast infections are caused by the fungus *Candida albicans*, which is one of a community of microorganisms that are always present in the vagina in low concentrations. One of the factors that can bring about a yeast infection is the initiation of the use of certain antibiotics. Why would the use of antibiotics foster an unhealthy multiplication of *C. albicans* in the vagina?

MasteringBiology®

1. MasteringBiology Assignments
Activities: Life Cycle of Fungi; Classification of Fungi; **Current Events:** Kentuckians Take Distilleries to Court Over Black Gunk; Out West, 'Black Fingers of Death' Offer Hope Against an Invader; Unearthing the Sex Secrets of the Perigord Black Truffle; **Building Vocabulary; Questions:** Reading Questions, Test Bank

2. eText
Read your book online, search, take notes, highlight text, and study more effectively.

3. The Study Area
Self-study tools to test comprehension.

Animals:

The Diversity of Life 3

The diversity of animals is one of the wonders of the natural world.

Available on MasteringBiology®

Current Events: Frozen Sperm Offer a Lifeline for Coral; One Day, Growing Spare Parts Inside the Body; Warm and Furry, But They Pack a Toxic Punch; It's Complicated: Dragonfly Love Comes Calling; 7,000 Miles Nonstop and No Pretzels; **Building Vocabulary; Questions:** Reading Questions, Test Bank

The blue darner dragonfly (*Aeshna cyanea*). The darner's name seems to have come from a folk belief that they use their slender bodies to sew up, or "darn," the lips of misbehaving children.

We human beings are, in a sense, intimate strangers with the bacteria and fungi that were reviewed in the last two chapters. We are intimate with these organisms in that we carry out our lives right along beside them: We walk through clouds of them, wash off layers of them, and have trillions of bacteria living inside us. That said, except for the

occasional outbreak of athlete's foot or case of upset stomach, these organisms never cross our minds. So small are bacteria and most fungi that, while they may be everywhere, for us they are effectively nowhere. It's a different story with the animals that are the subject of this chapter. Our cats and dogs and neighborhood birds are part of our conscious lives. We pay attention to a new spider web on our porch or to the lizard on our hiking path. We wonder about the circling hawk overhead.

Animals come from the long evolutionary line of *other* feeders: the heterotrophs, who cannot manufacture their own food but must get their nutrition from outside themselves. Many bacteria fit into this category, as do lots of protists and all fungi. But animals, in their quest for food, added something that is unique to them: a nervous system; a system for transmitting complex messages rapidly over long pathways in the body. Once this happened, the living world was off to the races in terms of adding novel abilities. Think of sight, hearing, smell, flight, walking, singing, and reading. Then there are such things as the echolocation of bats and the waggle dance of honey bees. For the average person, to be alive means to sense, to investigate, to respond, to move. In all these areas, animals reign supreme in the living world.

This chapter is about the broad diversity that exists in the animal kingdom (**Figure 23.1** on the next page). Given that animals are part of our daily lives, some of what follows will be familiar, but there are likely to be many surprises as well.

23.1 What Is an Animal?

In coming to an understanding of animals, we first have to understand what separates animals from other varieties of living things. And to do this, the question that needs to be answered is: What characteristics do *all* animals have that other organisms *don't* have? It turns out there is a single, rather technical feature that is sufficient to set animals apart from all other living things.

- Animals pass through something called a blastula stage in their embryonic development. A *blastula* is a hollow, fluid-filled ball of cells that forms soon after an egg is fertilized by sperm. All animals go through a blastula stage, but no other living things do.

Four other characteristics are found in all animals, but they're found in other kinds of organisms as well. All animals:

- Are multicelled; there are no single-celled animals.
- Are heterotrophs; they must get their nutrition from outside themselves.
- Are mobile or "motile" for at least part of their lives. (There are, however, animals that are as "sessile" or immobile as any plant during most of their lives.)
- Are composed of cells that do not have cell walls. (The outer lining of animal cells is the plasma membrane. By contrast, all plants and most other creatures have a relatively thick additional lining—the cell wall—outside their plasma membrane.)

Then there are characteristics of animals that don't fit our preconceived notions. While it's true that animals generally are large relative to bacteria or protists, some animals are microscopic. We often think of animals as creatures with backbones (or "vertebral columns"), but there are far more **invertebrates,** or animals without vertebral columns, than animals with them. Indeed, 99 percent of all animal species are invertebrates. We think of animals primarily as land-living or terrestrial creatures, but there is far more animal diversity in the ocean than on land. This makes sense because animals existed in the sea for at least 100 million years before any of them came onto land.

SO FAR . . .

1. A feature common only to animals is that all animals pass through a _____ stage in their _____ development.

2. Four other features that all animals share are that they are _____ for at least part of their lives, _____, _____, and their cells do not have _____.

3. Vertebrates make up a _____ portion of the animal kingdom.

23.2 Lessons from the Animal Family Tree

One way to get a handle on the range of animal life is to look at the large-scale categories of animals and see how the creatures in these categories differ from one another. The animal kingdom's broadest categories are known as phyla, with each phylum informally defined as a group of organisms that shares a basic body structure. Depending on who is counting, there are between 36 and 41 animal phyla. So, in what ways are the members of these various phyla related to one another? To put this another way, what does the animal family tree look like?

(a) Jellyfish

(b) Feather star

(c) Octopus

(d) Nudibranch

(e) Water flea

(f) Bonobo

Figure 23.1
Animal Diversity

(a) Like all jellyfish, these West Coast sea nettles (*Chrysaora fuscescens*) use stinging extensions to immobilize prey.

(b) This feather star (*Oxycomanthus bennetti*), a relative of sea stars, feeds by catching drifting bits of food.

(c) Octopuses such as this one are marine molluscs, in the same phylum as snails and clams.

(d) Nudibranchs such as this one are also marine molluscs.

(e) Water fleas, from the genus *Daphnia*, are tiny aquatic crustaceans. Note the animal's prominent digestive tract.

(f) Bonobos (*Pan paniscus*) are close relatives of chimpanzees.

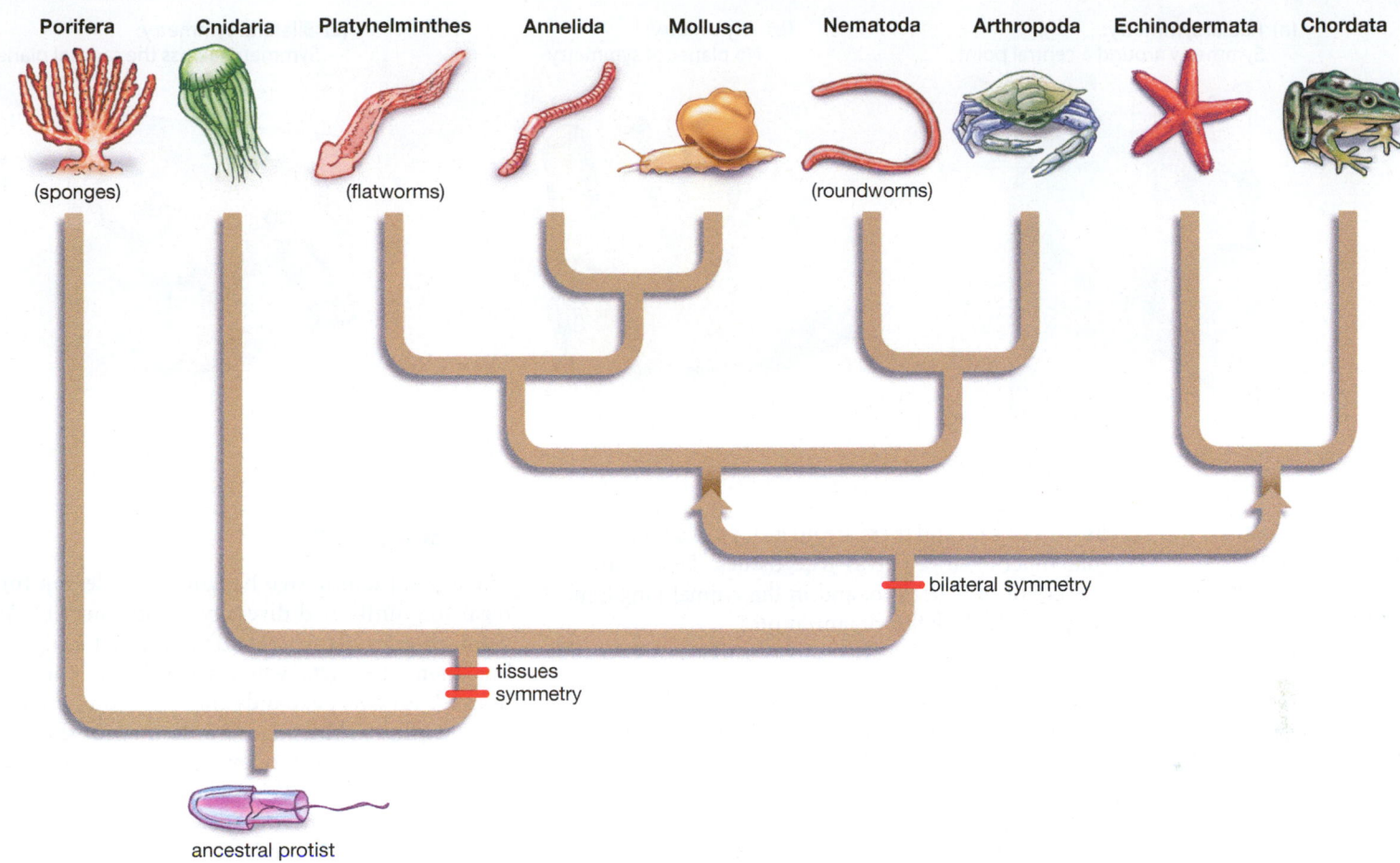

Porifera Cnidaria Platyhelminthes Annelida Mollusca Nematoda Arthropoda Echinodermata Chordata

(sponges) (flatworms) (roundworms)

bilateral symmetry

tissues
symmetry

ancestral protist

Figure 23.2

A Family Tree for Animals

The animal kingdom is divided into groups called phyla. The members of each phylum have in common physical features that are evidence of shared ancestry. Nine of the estimated 36 to 41 animal phyla are shown here. All these phyla are regarded as having evolved from an ancestral species of protist, shown at lower left in the family tree.

If you look at **Figure 23.2,** you can see one version of it. One way to conceptualize this tree is that, over time, animals became more complex through a series of *additions* to the characteristics found in more primitive animals. The twist is that only some varieties of animals evolved to get these additions, while others retained the primitive, ancestral condition. To get a handle on this large-scale evolution, a good place to start is with some little red lines you can see at the bottom of the tree. We're concerned first with one animal trait labeled there as tissues and another labeled as symmetry. Later on we'll look at the significance of the trait of organs.

Tissues

A **tissue** is a group of cells that perform a common function. In your own body, you have cells that perform the common function of contracting. These are muscle cells, and a group of them constitutes a muscle tissue. Likewise, you have nervous tissue and connec-

tive tissue, and other varieties. Looking down at the trunk of the tree in Figure 23.2, you can see that all animals have a common ancestor—probably a single-celled protist that swam through use of a tail-like flagellum. Now notice that the tree splits above this common ancestor and yields, over to the left, the phylum Porifera, which is made up of sponges. Porifera are truly the outliers of the animal world in that, almost alone among animals, they lack true tissues (and symmetry, which we'll get to in a second). If you look again at the split above the common ancestor, this time going to the right, you can see that the animals on this branch did develop tissues. Note also that this right branch leads to all the other animal phyla. Thus, all the animals that stem from this branch will have tissues. If we looked, for example, at the Cnidarian phylum (which includes jellyfish), we'd see that its members have nervous and muscle tissue. But sponges? They are more like a collection of cells that have come together to form an organism. In a given sponge, each cell acquires its own oxygen and eliminates

(a) Radial symmetry:
Symmetry around a central point

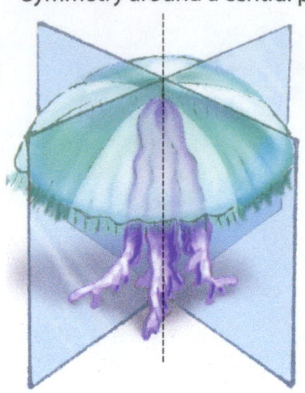

(b) Asymmetry:
No planes of symmetry

(c) Bilateral symmetry:
Symmetry across the sagittal plane

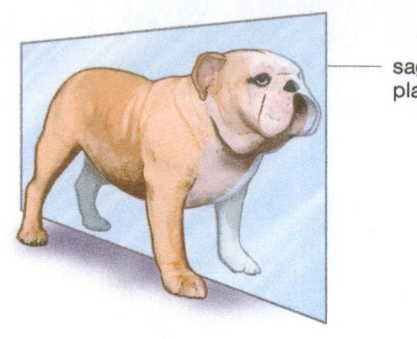

sagittal plane

Figure 23.3
Symmetry Is an Equivalence in Body Sections

If an imaginary plane drawn through an animal can divide that animal into sections that are mirror images of one another, then that animal has symmetry.

a. Radial symmetry. The imaginary planes drawn through the jellyfish show that it has radial symmetry: a symmetry in which body sections are distributed evenly around a central point.

b. Asymmetry. The sponge has no symmetry—no plane drawn through the sponge body would yield sections that are mirror images of one another.

c. Bilateral symmetry. The dog has a kind of symmetry common to most animals: bilateral symmetry, in which the sides of an animal are mirror images of one another.

its own wastes, and there are no groups of sponge cells that function together as true tissues. Tissues are a marker of organization, and in the animal kingdom, only sponges lack this organization.

Symmetry

So what about this second quality noted with a red line, symmetry? If you look at **Figure 23.3a**, you can see that the jellyfish that's pictured can be thought of as being divided by imaginary planes, thus yielding body sections. These sections are mirror images of one another; they have **symmetry,** meaning an equivalence of size, shape, and relative position of parts across a dividing line or around a central point. The jellyfish actually has a particular kind of symmetry. It has **radial symmetry,** meaning a symmetry in which body parts are distributed evenly around a central point. In simpler terms, the symmetry of jellyfish is the symmetry of pie sections.

To appreciate symmetry, consider what a lack of it is like by looking again at our outlier, the sponge, in **Figure 23.3b**. Where is its symmetry? There isn't any, because there is no section of the sponge that is a mirror image of any other.

Symmetry can, however, come in several forms. Look at the dog in **Figure 23.3c**, divided by what is known as a sagittal plane. The dog obviously has symmetry, but it is a different kind of symmetry from that of the jellyfish. As it turns out, the dog has a kind of symmetry that unifies most of the animal kingdom. Animals usually are *different* front-to-back and top-to-bottom, but *symmetrical* side-to-side. Put another way, your head is different than your feet, and your chest is different than your back, but your left *side* is very similar to your right. This is what it means to have **bilateral symmetry:** a bodily symmetry in which opposite sides of a sagittal plane are mirror images of one another. In looking at the animal family tree, you can see that bilateral symmetry was an evolutionary innovation that affected all modern-day animals except for the sponges and the cnidarians.

The Body Cavity

The animal family tree has one other lesson for us regarding unity and diversity among animals. Your stomach expands when you've just had a big meal, but then contracts when you're busy for a few hours. Your heart expands and contracts perhaps 70 times a minute. You bend over to tie your shoes and, unbeknownst to you, many of your internal organs slide out of the way. What allows you to do all this? The answer is an internal space you have: a centrally placed, fluid-filled body cavity called a **coelom.** We humans are not alone in having such a cavity. There are only three animal phyla in Figure 23.2 that *don't* have one: the sponges, the cnidarians, and the members of phylum Platyhelminthes (the flatworms).

What's the value of such a cavity? Well first, it allows for an expandable stomach, which is to say a stomach whose dimensions can vary in accordance with how much food we've ingested, which means it doesn't take up a constant, large amount of space. Along similar lines, it allows reproductive organs to expand to accommodate eggs or offspring. Then there's the fact that if a heart couldn't expand and contract, it couldn't work at all. Finally, a central cavity protects organs from bodily blows and provides a large part of the body with flexibility.

In most instances the coelom surrounds another physical structure, the *digestive tract*—meaning the tube, functioning in digestion, that runs from the mouth to the anus. We can therefore think of the coelom as one tube that encircles another; the coelom is generally tube-shaped, and it surrounds the tube that is the digestive tract. **Figure 23.4** shows you what it means to have no coelom, as in flatworms, and a *true coelom,* as in earthworms. (Not shown is another variation on this theme, a pseudocoel, which roundworms have.) You may wonder why there are no red lines on Figure 23.2 showing where the coelom evolved. So valuable is this internal space that it seems to have evolved independently several times among animals.

Having reviewed a few broad-scale characteristics of the animal kingdom, we'll now look briefly at each of the nine phyla featured in the animal family tree set forth in Figure 23.2. Once we've finished with this review, we'll look at some of the ways animals handle such basic functions as movement, reproduction, and circulation.

SO FAR . . .

1. The sponges in phylum Porifera are outliers in the animal kingdom in that, almost alone among animals, their bodies are not composed of _____ and their body structure lacks the quality of _____.

2. All animal phyla except sponges and cnidarians exhibit _____ symmetry in their body plans, meaning a symmetry in which _____ of their body are mirror images of one another.

3. A coelom is an internal _____ possessed by almost all animals except sponges, jellyfish, and flatworms.

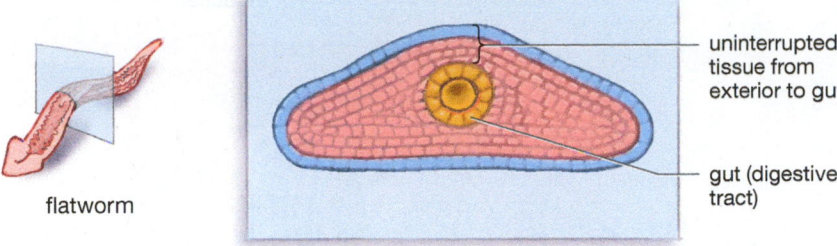

No coelom. Phylum Platyhelminthes, composed of flatworms, is one of the phyla that has no coelom. Note that from its gut to its exterior, the flatworm is composed of uninterrupted tissue.

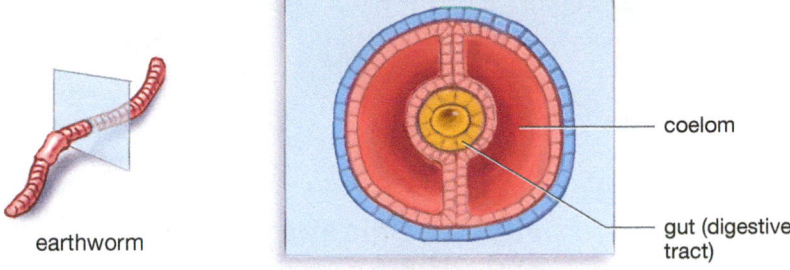

Coelom. The earthworm is one of many animals that has a coelom — a fluid-filled central cavity that usually surrounds the gut.

Figure 23.4
An Important Space
Most animal bodies have an enclosed cavity or coelom—an internal space that surrounds their digestive tract or other internal structures.

23.3 Across the Animal Kingdom: Nine Phyla

Phylum Porifera: The Sponges

The lives of the most primitive members of the animal kingdom hinge entirely on carrying out a single activity: moving water into, through, and then out of themselves. Because adult sponges are fixed in one spot (generally on the seafloor), everything they need in the way of food and oxygen must come to them in water currents while everything they need to get rid of must be washed away from them in these same currents. If you look at **Figure 23.5**, you can see the body plan that allows sponges to carry out this continual transfer of water. The simple variety of sponge pictured there has a layer of outer cells that is pockmarked with thousands of microscopic pores. Millions of so-called collar cells wave tiny, whip-like flagella that keep water moving into these pores. As the water passes through the sponge cells, the food in it is filtered out and the oxygen captured, after which the resulting waste water is expelled through a large opening at the top of the sponge called an *osculum*. Almost all the 5,000 species of sponges identified to date are marine (meaning ocean-dwelling), but a few freshwater varieties exist as well.

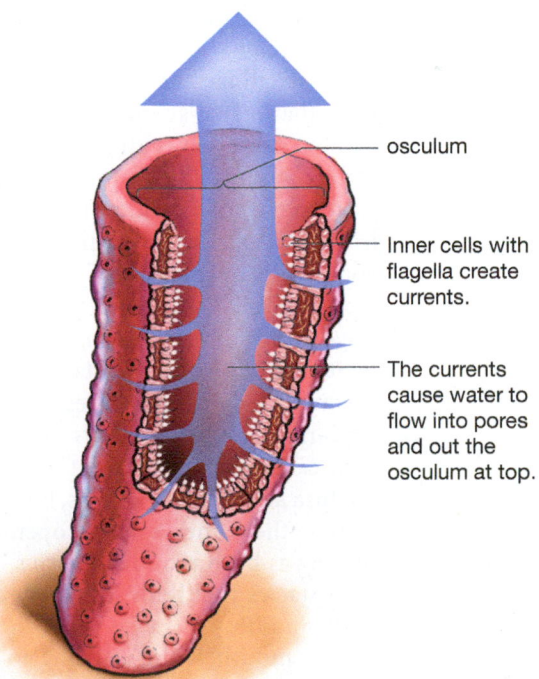

Figure 23.5
A Body Plan for a Simple Lifestyle
Sponges filter water through themselves in order to live. Cells on their interior surface have hair-like extensions, called flagella, whose rapid back-and-forth movement draws water in through pores in the sponge's side walls. After filtering this water for nutrients, the sponge then expels it through a large opening at the top of its body, called an osculum.

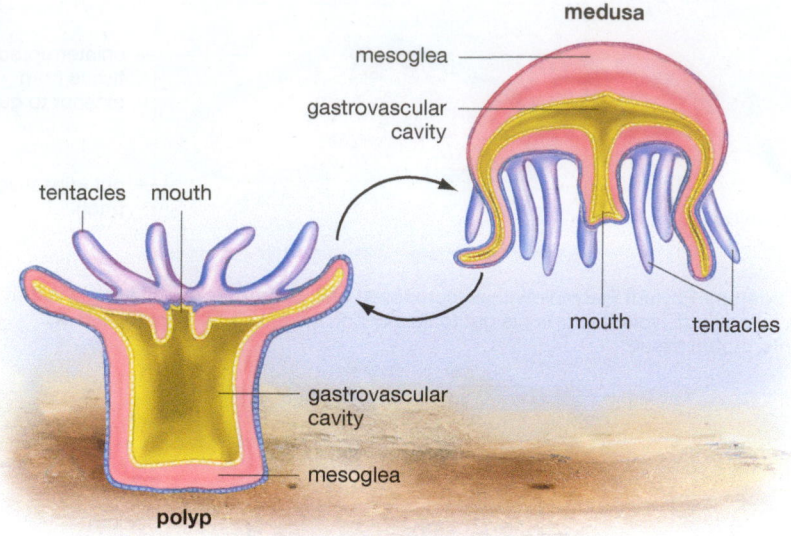

Figure 23.6

Two Stages of Life for Many Cnidarians

Many cnidarians go through both a polyp stage, in which they are attached to a solid surface beneath them, and a medusa stage, in which they swim freely through the water. Note that in the polyp stage, the cnidarian's mouth and tentacles face upward, while in the medusa stage, they face downward. The gastrovascular cavity of the cnidarian functions in both digestion and circulation. The mesoglea noted in the figure is a secreted, gelatinous material that makes up the bulk of the medusa stage in jellyfish cnidarians.

Phylum Cnidaria: Jellyfish and Others

If the signature activity of sponges is filtering water to get food, the signature activity of cnidarians is stinging prey to get it. In the main, cnidarians (knee-DAR-ee-uns) harpoon their prey with extensions that are not only barbed but that may release poisons. Jellyfish are the symbols of this ancient phylum, though sea anemones and the tiny animals that make up coral reefs are also included in it. Almost all the 9,000 identified cnidarian species are marine, though a few live in freshwater.

Most cnidarians have two life stages: a medusa stage that is free-swimming (think of a jellyfish) and a polyp stage that is attached to a surface (think of a sea anemone). The basic body plan in both of these stages is that of a sack, although it is often a sack that is turned upside down. If you look at **Figure 23.6**, you can see how this plays out in connection with an idealized jellyfish. In its familiar adult or medusa stage, its mouth is on the *underside* of its body, while in its immature polyp stage its mouth is on top. Medusa and polyp stages are present in many cnidarians, although corals, sea anemones, and animals called hydra only exist in the polyp stage. In all medusa and in many polyps, however, the basic body plan is the same: a single opening to the outside, serving as both mouth and anus, that is surrounded by tentacles that both sting prey and bring the prey to the mouth.

The mouth is the opening to the jellyfish's gastrovascular cavity, which functions in both digestion and circulation. Surrounding this cavity is the mesoglea, a secreted gelatinous material whose consistency gives jellyfish their name. Cnidarians such as jellyfish can move and exhibit predatory behavior because, unlike sponges, they possess groups of cells that

are organized into the muscle and nervous tissues we talked about earlier.

Phylum Platyhelminthes: Flatworms

The flatworms of phylum Platyhelminthes fit our everyday notions of what animals look like in that they have the bilateral symmetry of most animals, along with a front or "anterior" body portion that is a primitive version of a head (**Figure 23.7**). As we'll see later, flatworms occupy an interesting position in the animal kingdom in that they are the most primitive animals to possess not just the tissues that cnidarians have but organs as well, meaning highly organized structures formed of several kinds of tissues. (Your stomach is an organ that includes nerve and muscle tissues, for example.) Despite this complexity, flatworms lack a good many features that exist in almost all the phyla further up the complexity ladder. They have no coelom or enclosed internal cavity. Thus, except for the tube of their digestive tract, they have very little internal space in them, which limits their flexibility. Further, they have no system of blood circulation. Therefore, every cell in a flatworm must get its oxygen directly from the environment around it, through diffusion from water or air outside the body. This is why flatworms are flat—they have to maximize the number of cells that are near an exterior surface, as opposed to being buried deep inside. (How flat are they? About the width of a sheet of paper.)

Those flatworms that are on land tend to be found under stones or rotting wood; many live at the bottom of the ocean, and a few live in freshwater. Some 25,000 species of flatworm have been identified. Humans are unlikely to notice flatworms because, of the minority

Figure 23.7

Flatworm

Flatworms have the bilateral symmetry seen in most animals, along with a head-like concentration of sensory tissue at their anterior ends. Most flatworms are free-living marine animals like this one, but there are also many parasitic flatworms that live for at least part of their lives inside a host.

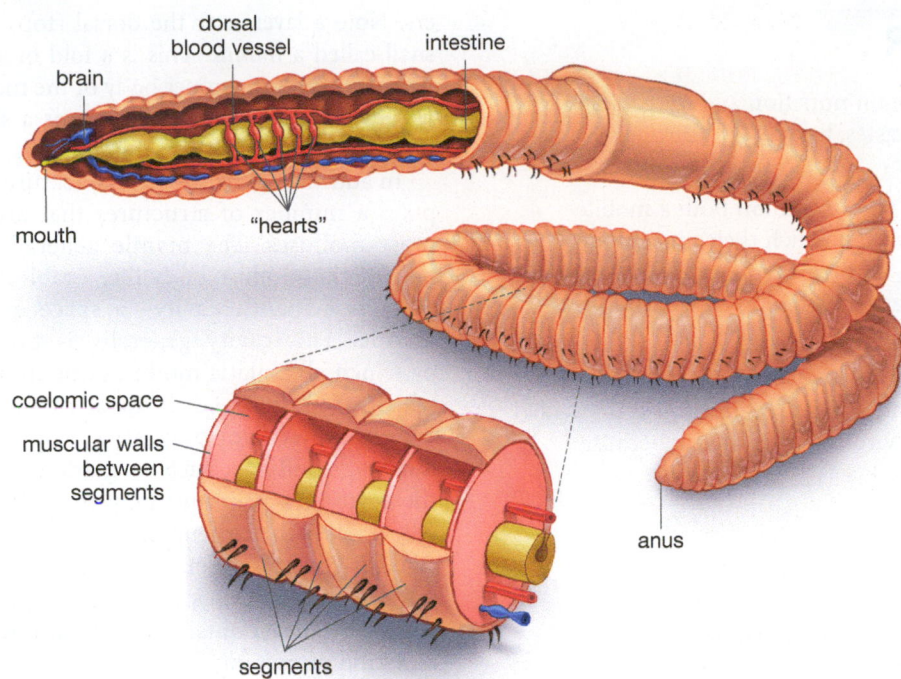

brain
dorsal blood vessel
intestine
mouth
"hearts"
coelomic space
muscular walls between segments
anus
segments

Figure 23.8
The Body Plan of an Earthworm
The segmentation of body parts—a widespread feature in the animal kingdom—is clearly seen in earthworms, whose segments are repeated hundreds of times. The earthworm also expands our notion of what a "heart" amounts to. It has a dorsal blood vessel that contracts, as do human hearts, giving the main impetus to the circulation of red blood that moves through the worm's system. But it also has five pulsating vessels on each side of its intestine that serve as accessory hearts. The coelomic spaces visible in the lower drawing are filled with fluid that serves as a hydrostatic skeleton for the worms.

that live around us, most are very small. When they do come into our consciousness, it is often in the unpleasant role of **parasites,** defined as organisms that feed off prey but do not kill it immediately and may not kill it ever. Most notorious among the flatworm parasites are the animals known as flukes, which are the cause of schistosomiasis, a disease that affects more than 200 million people in tropical areas around the world.

Phylum Annelida: Segmented Worms

The common earthworms that can be seen in such abundance on American lawns after heavy rains are the best-known members of phylum Annelida, the phylum of segmented worms. Annelids provide the clearest example of a feature that exists in numerous animals, including ourselves: **body segmentation,** which can be defined as a repetition of body parts in an animal. In the case of the human body, think of the vertebrae that make up our backbone; 24 of them run in a line from our neck to the base of our back, each of them having a similar structure and a similar function (flexible support and protection). In the case of the earthworm, each of their highly visible "rings" is a marker of a separate segment of their body, as you can see in **Figure 23.8**. Segmentation is valuable to animals in general because it provides a combination of strength and flexibility. (Think of our own backbones, which are highly flexible and yet allow us to push with great force, lift heavy objects, and so forth.)

Earthworms employ this flexible power to burrow in the dirt for hours at a time, ingesting soil for the nutrients it contains and eating leaves or other decaying vegetation.

Earthworms actually represent only the most familiar species in phylum Annelida, which contains about 17,000 named species. Most annelids are ocean dwelling, and some of these colorful creatures, the polychaetes, do not fit in with our idea of drab worms (**Figure 23.9**). The smallest class of annelids is composed entirely of leeches, most of which live up to their reputation of being bloodsucking parasites. Although most leeches live in freshwater, some live in the ocean, and others dwell in damp-land habitats.

Figure 23.9
Annelid Diversity
Not all annelid worms are drab and not all live underground. Polychaetes are ocean-dwelling annelids. Pictured is a Hawaiian Christmas tree worm in the genus *Spirobranchus*. These polychaetes stay in one spot, burrowed into tubes they themselves have constructed, generally within a coral. The bushy "Christmas trees" that extend from the worm's head are structures called radioles that it uses to filter feed from the water.

SO FAR . . .

1. Sponges obtain nutrition and oxygen, and get rid of wastes, by _____ through themselves.

2. Cnidarians tend to exhibit both a mobile _____ stage in which their mouths are on _____ of their bodies, and a sessile _____ stage, in which their mouths are _____.

3. Every cell in a flatworm of phylum Platyhelminthes must get its oxygen directly from the environment around it because flatworms lack a _____ system.

Phylum Mollusca: Snails, Oysters, Squid, and More

Everyone has seen a snail, many of us have eaten clams or mussels, and squid and octopus are probably known to us from television or books. But molluscs are even more varied than these examples would suggest. Some molluscs are slugs and others wormlike; some have highly developed brains and others no brains at all; some are filter feeders and others fierce ocean predators. Depending on the expert consulted, there are between 50,000 and 100,000 described species of molluscs that live in freshwater, saltwater, and on land. Each of these species is a member of one of eight mollusc classes, the most familiar of which are the *gastropods* (snails, slugs), the *bivalves* (oysters, clams, mussels), and the *cephalopods* (octopus, squid, nautilus).

What features unite this huge, diverse group as a single phylum? Outside the realm of technical anatomical details, there really is only one feature common to all molluscs. If you look at **Figure 23.10**, you can see an aquatic snail that displays this feature along with oth-

ers. Note a layer near the dorsal (top) surface of the snail called a *mantle*. This is a fold of skin-like tissue that surrounds the upper body of the mollusc and that usually secretes material that forms a shell (although there are lots of molluscs without shells).

In addition to this universal feature, the snail displays a number of structures that are common to most molluscs. The mantle generally drapes over part of the mollusc body like a tablecloth, meaning there is a mantle cavity—a space underneath the mantle. This cavity generally is the site of gills, by which all aquatic molluscs obtain oxygen and by which the filter-feeding bivalves (clams, mussels, oysters) obtain not only oxygen but food. In land-dwelling molluscs, such as garden snails, much of the cavity is a primitive lung that evolved from the gill. Note, in the blowup of the snail's mouth, there is a tooth-covered membrane called the *radula*. Present in most molluscs except the bivalves, this organ can be extended outside the mouth, where it serves as a kind of rasping file. Molluscs use it for scraping a surface or for cutting small prey, then they retract it back into the mouth, thus moving food into their digestive tracts.

Phylum Nematoda: Roundworms

Nematodes, commonly referred to as roundworms, exist in such enormous numbers that it is impossible to imagine life without them. They fit into numerous ecological roles, such as that of "detritivore," meaning an organism that feeds on dead or cast-off organic material. Nevertheless, many farmers in the United States would *like* to imagine life without them because several species of these animals are a major cause of damage to such crops as soybeans and corn. Farmers and research scientists have been united for decades in an entrenched battle against them. In addition, some are human parasites. The disease trichinosis, usually brought on by eating undercooked pork, is caused by a roundworm; likewise, there is hookworm, which affects hundreds of millions of people in warmer climates throughout the world (**Figure 23.11**).

Figure 23.10
Body Plan of a Mollusc
Aquatic snails such as this one have features that are common to many molluscs. All molluscs possess a skin-like tissue called a mantle, which secretes material that can form a shell, whether it is an external shell (as in the snail) or an internal shell (as in a squid). Snails also possess a tooth-covered rasping structure for feeding, called a radula. The unitary "foot" of the snail which works through wave-like contractions is another common mollusc feature.

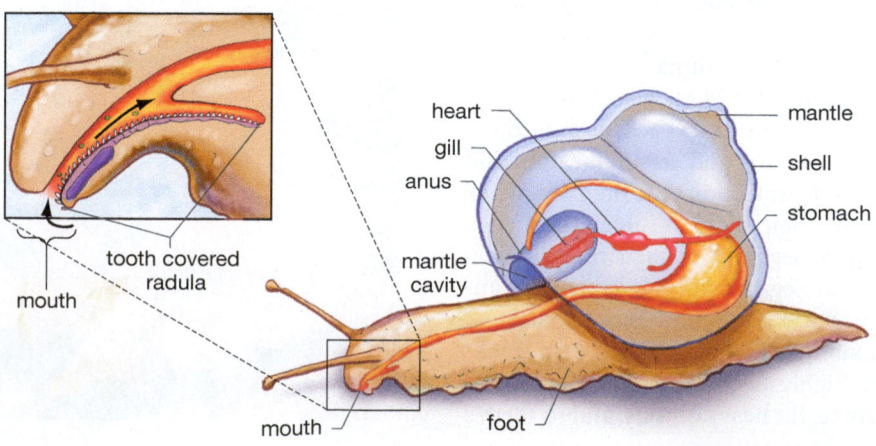

Figure 23.11
Roundworms

The tiny worms called nematodes are a normal part of many ecosystems, including soil ecosystems. However, a small number of roundworm species are harmful to humans, either directly as parasites or indirectly as crop pests. Pictured is a color-enhanced micrograph of the nematode *Toxocara canis*, a parasite that infects dogs. Humans can become infected with *T. canis* eggs if they have contact with an infected dog or contaminated soil.

An acre of prime farmland might contain several hundred billion roundworms, but you don't have to go to farmland to find these animals. They exist in desert sands, polar ice, ocean bottoms, lakes—name a place where life is going on, and nematode worms are likely to be among the living. Indeed, scientists have only the vaguest idea of how many *species* there are; about 15,000 have been catalogued, but 100,000 seems to be the minimum number that actually exist.

Most roundworm species are very small—the crop pests, which feed on roots, generally are microscopic—but some parasitic roundworms are several meters long. The roundworms that have been studied most intensely are those that cause disease and crop damage, and these have a characteristic appearance: transparent, smooth, C-shaped, and cylindrical with tapered ends (hence the term *roundworm*). Across the breadth of roundworm species, however, there is a great diversity in form.

Phylum Arthropoda: Insects, Lobsters, Spiders, and More

Phylum Arthropoda is so large that there are more species in one arthropod group, the insects, than there are in all other animal phyla combined. One way to get a handle on this enormous phylum is to think of it in terms of three subphyla: Uniramia, which includes not only insects but millipedes and centipedes; Crustacea, which includes shrimp, lobsters, crabs, and barnacles; and Chelicerata (chel-is-er-AH-ta), which includes spiders, ticks, mites, and horseshoe crabs.

What characteristics could be shared by a group of animals as varied as these? First, all arthropod animals have a feature we'll be looking at later called an **exoskeleton,** which can be defined as an external material covering the body, providing support and protection. You can see an example of an exoskeleton in the grasshopper pictured in **Figure 23.12**. Second, all arthropods engage in a process of *shedding* their exoskeleton several times during their life—a process known as molting that we'll look at later. Third, all arthropods have what are known as paired, jointed appendages, which are just what they sound like: appendages, such as legs, that come in pairs and that have joints. (The word *arthropod* actually means "jointed leg.") The jointed appendages that arthropods possess come in many forms. Apart from legs, there are claws for predation and defense and wings for flying. These appendages are flexible and often come with sensory attachments, such as sensory "hairs," that keep the arthropod in touch with its environment. The arthropod combination of jointed appendages and an exoskeleton makes for an animal that is often nimble and well defended—one of the reasons arthropods are so numerous.

Phylum Echinodermata: Sea Stars and More

If you look back at the animal family tree in Figure 23.2, you can see that the echinoderms form a "sister" evolutionary lineage to our own phylum, Chordata, which means echinoderms are more closely related to us than they are to, say, arthropods or molluscs. You might expect that, as our close relatives, echinoderms would have the kind of sophisticated features we associate with vertebrate animals—brains, well-developed sense organs, high mobility—but none of

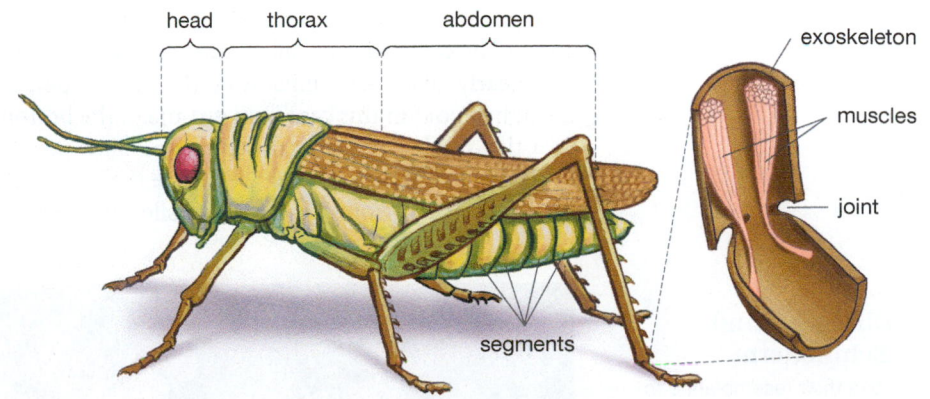

head   thorax   abdomen   exoskeleton

muscles

joint

segments

Figure 23.12
Arthropod Features

Arthropods have paired, jointed appendages, as with the legs of this grasshopper. The insects among the arthropods have bodies made up of three large-scale sections: head, thorax, and abdomen. Like the annelid worms, arthropods are segmented. This segmentation is clearly visible in the abdomen region of the grasshopper but is less pronounced in the head and thorax because there the segments have fused. The entire arthropod body is covered in a rigid exoskeleton that is jointed to allow movement.

Figure 23.13
Sea Star Echinoderms

Sea stars such as this one are fierce predators. Here a sea star is managing to open the shell of a bivalve known as a rock cockle. Note the tubed feet with which the sea star grasps the material beneath it.

this is true of the best-known echinoderm, the sea star (**Figure 23.13**). It has no brain, its only sensory organs are the primitive eyespots found on the tips of its arms, and it's capable only of very slow movement. Even the bilateral symmetry that you've seen in every animal since the jellyfish is absent in the adult sea star: It has radial symmetry, like the jellyfish.

This is not to say, however, that the sea star and its echinoderm relatives are completely primitive. Sea stars have a remarkable system for moving that involves forcing water into a series of suction-tipped tube feet that extend from the underside of their arms. Moreover, sea stars evolved into radial symmetry *from* bilateral symmetry. How do we know? Sea star larvae have bilateral symmetry, a quality that is lost when the larvae develop into adults. Taken together, features such as these make sea stars something of a mystery to scientists, as these animals seem to exhibit characteristics from all over the animal map. Joining them in the echinoderm phylum are sea urchins, with their spine-covered spherical bodies, and the less familiar but well-named sea cucumbers (**Figure 23.14**). There are about 7,000 identified species of echinoderms, all of them ocean dwelling and nearly all of them inhabiting the ocean floor, although "floor" in this case may just mean the bottom of a tide pool. Their shared habitat and slow pace of movement make sense when we realize that the entire phylum evolved from a group of sessile filter feeders.

Figure 23.14
A Different Kind of Echinoderm

Despite their resemblance to a vegetable, sea cucumbers are animals that, like their sea star relatives, possess a series of tubed feet with which they crawl across the ocean floor.

Phylum Chordata: Mostly Animals with Backbones

Say the word *animal* and most people think of a vertebrate animal—be it tiger, elephant, deer, lizard, or some other large and probably toothy creature. Yet vertebrates don't even constitute a phylum; instead, they make up only a subphylum, Vertebrata. The larger phylum into which vertebrates fit, called Chordata, actually has two additional subphyla in it. One of these (Cephalochordata) is made up of only a single kind of animal, a small, eel-like creature called a lancelet, represented by only 25 or so species. The other subphylum (Urochordata) has about 3,000 ocean-going species in it but is represented by only three classes of animals. The largest of these classes is made up of the tunicates, whose practice of expelling water earns them their alternative name of sea squirt.

What distinguishes vertebrates from these other chordates? Only the vertebrates have just what you'd expect: a **vertebral column.** Better known as a backbone, this is a flexible column of bones that extends from the anterior to the posterior ends of an animal's dorsal side. Conversely, what unites all the chordates? Four features, which you can see pictured in a lancelet in **Figure 23.15**. At some point in their lives, all chordates have:

- A *notochord:* a stiff, rod-shaped support structure composed of cells and fluid and surrounded by a lining of fibrous tissue, running from the chordate's head to its tail
- A *dorsal nerve cord:* a rod-shaped structure consisting of nerve cells, running from the chordate's head to its tail
- A series of *pharyngeal slits:* openings to the pharyngeal cavity
- A *post-anal tail:* a tail located posterior to the anus

At this point, as a chordate, you may be wondering where your *own* post-anal tail or pharyngeal slits went, so a little explanation of these features is in order. Although the word *notochord* sounds like "nerve cord," the notochord is a flexible but tough *support* structure, not a nervous system feature. A lancelet retains its notochord throughout life, but most vertebrates lose theirs in embryonic life. (Its outer portion develops into the backbone.) A nerve cord may not seem like a feature unique to chordates, but it is the qualifier *dorsal* that makes the difference. Nerve cords exist in annelids and arthropods but these are ventral cords, meaning they are on the underside of these animals.

The pharyngeal slits you see in the lancelet in Figure 23.15 take on more meaning when you understand that, in both humans and lancelets, the pharynx is a passageway just behind the mouth that can be thought of as the first chamber of the digestive tract. Pharyngeal *slits*, then, are perforations of the pharynx,

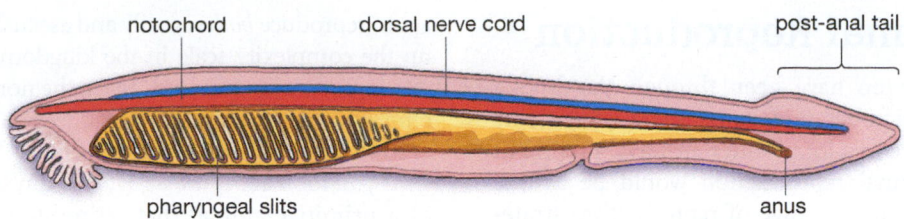

notochord   dorsal nerve cord   post-anal tail

pharyngeal slits   anus

Figure 23.15
Four Universal Chordate Features
All chordates possess four structures at some point in their lives: pharyngeal slits, a support structure called a notochord, a dorsal nerve cord, and a post-anal tail. In the lancelet, all these features are present in the adult. In humans, only the dorsal nerve cord is clearly evident in adults.

and these serve different functions in different chordates. In the filter-feeding lancelets, they serve to trap food. Meanwhile, in embryonic fish, the slits *develop into* gills, which fish use to obtain oxygen. And in people? If you look at a month-old human embryo, you can see a series of pharyngeal clefts that for a time have openings in them. These pharyngeal slits, however, develop not into gills, but into various other structures, including the space inside the middle ear.

Finally, what about that tail? If you look at **Figure 23.16**, you can see an 8-week-old human embryo with a very clearly developed tail. By the time the embryo is 10 weeks old, however, the tail has almost disappeared. All chordates have a tail at some point. In our oceangoing ancestors, it was used to help move them through the water, as is the case with modern fish. In human adults, though, all that remains of our embryonic tails is a small bone, called the coccyx, located at the base of our vertebral column.

About 50,000 vertebrate chordate species have been identified so far—a small number of species compared to the 1 million species of insects, but a fairly large number as animal phyla go. There are more species of vertebrates than there are of sponges, cnidarians, or flatworms, for example, and vertebrate species numbers may match those of molluscs. In other ways, too, the vertebrates have been a big success. All the largest and swiftest animals are vertebrates, whether in water or in air or on land. And the

sheer variety of sensory capabilities that vertebrates have—from echolocation to reading—is unrivaled anywhere in the living world.

Having reviewed some broad-scale features of the various animal phyla, let's now look at some of the ways that varying groups of animals take care of business. How do they reproduce, how do they get oxygen, how do they circulate their bodily fluids, and how do they move?

SO FAR . . .

1. The three most familiar classes of molluscs are the gastropods, one example of which is _____; the bivalves, one example of which is _____; and the cephalopods, one example of which is _____.

2. All arthropods share the features of an _____ or external material that provides support and protection, and _____, _____ appendages.

3. Which of the following traits are not shared by all chordates at some point in their lives: notochord, dorsal nerve cord, exoskeleton, pharyngeal slits, sessile living, post-anal tail?

(a) Human embryo tail at 8 weeks

(b) Human embryo tail at 10 weeks

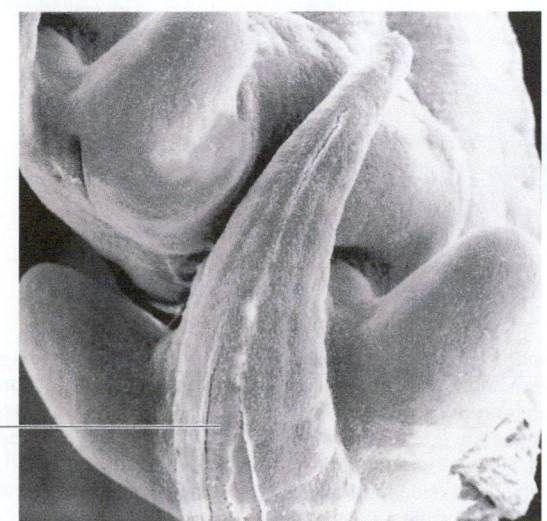

tail

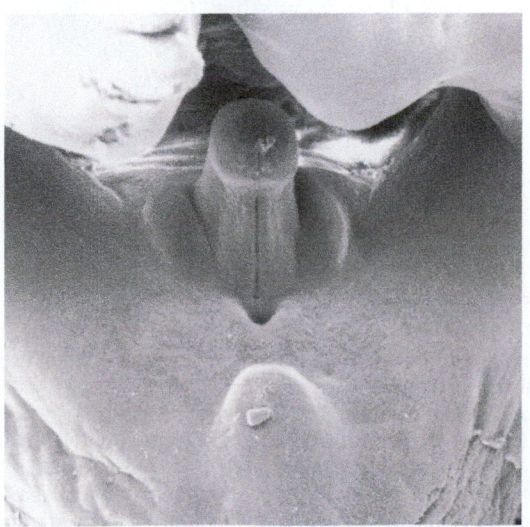

Figure 23.16
The Human Tail
(a) Eight weeks after conception, the human embryo has a very clearly developed tail.
(b) Two weeks later, the tail has almost disappeared. In adult humans, the only vestige of our embryonic tail is a small bone at the base of the vertebral column, the coccyx.

23.4 Animal Reproduction

Those of you who have been through this book's chapters on microbes and fungi know that living things reproduce in a lot of different ways. You might expect that animal reproduction would be a little more familiar, but the range of reproductive "strategies" or practices that animals employ is extremely varied, as you'll see.

Asexual Reproduction

Perhaps the most basic distinction in reproduction has to do with whether it is sexual or asexual—do reproductive cells from two different individuals come together and fuse their genetic material (sexual reproduction), or is there reproduction without this fusion (asexual reproduction)? We're used to thinking of animal reproduction as being sexual because the animals we're most familiar with—the vertebrates—almost always reproduce sexually. But when we look over the animal kingdom as a whole, we find that the more primitive animal phyla are

apt to reproduce *both* sexually and asexually. As we move up the complexity scale in the kingdom, sexual reproduction becomes more and more the norm, but asexual reproduction occasionally exists even in very complex animals, as you'll see.

If you look at **Figure 23.17**, you can see an example of a primitive animal that reproduces both sexually and asexually. The creature is *Obelia*, a tiny jellyfish-like cnidarian, and its sexual reproduction is straightforward: Female medusa-stage *Obelia* produce eggs, the males produce sperm, both kinds of reproductive cells are ejected into the water, sperm fertilize eggs, and a new generation of *Obelia* begins development. But note that the larva that represents this new generation develops into a "polyp" stage *Obelia* that implants on a surface (perhaps on some algae or a rock). Once this happens, the polyp reproduces by **budding**: a form of asexual reproduction in which an offspring is produced as an outgrowth of a parent organism. The polyp that results from this budding could be thought of as a colony that in turn develops polyps of two kinds: those that are specialized for feeding and

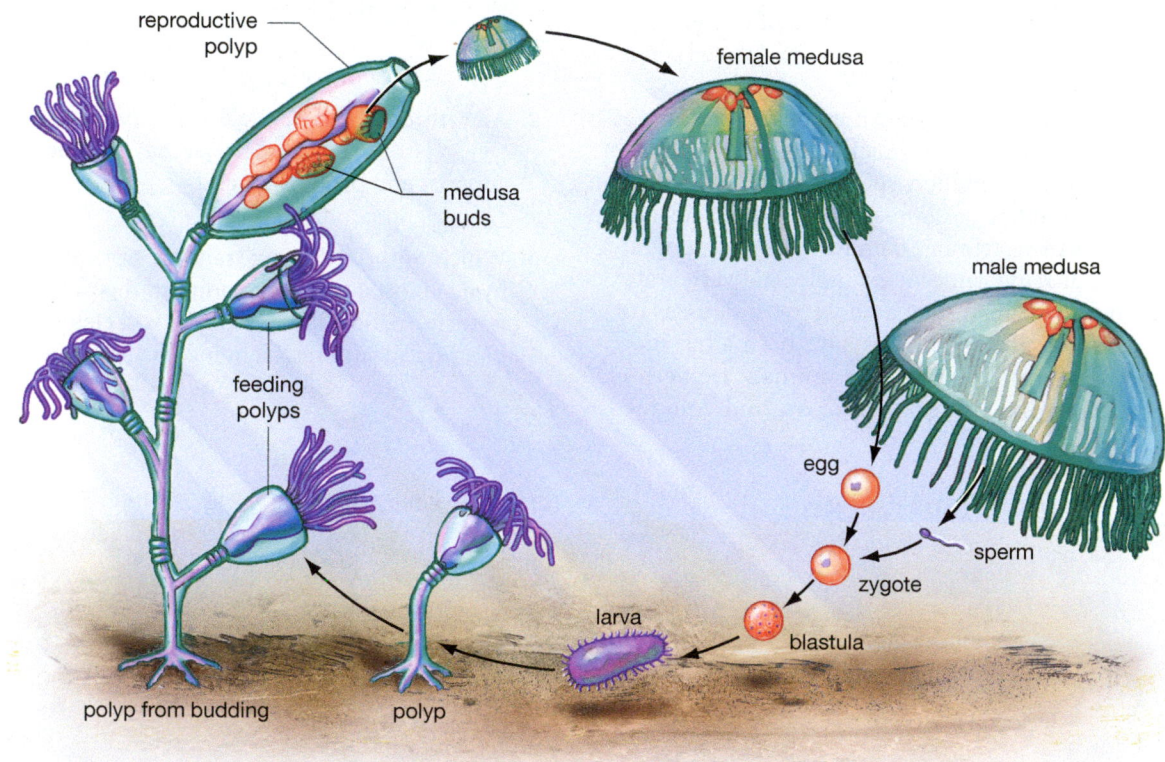

Figure 23.17
Sexual and Asexual Reproducer

This cnidarian, known as *Obelia,* reproduces through both sexual and asexual means. Sexual reproduction can be observed at the right of the figure, where male and female medusae release sperm and eggs, respectively, that fuse to form the new generation of *Obelia.* The polyp that results from this fusion, however, can reproduce by asexual means. Cells bud off from the first polyp (center) and grow into another polyp, shown at left, which may be the first of many that will develop this way. Polyps within these larger colonies specialize. Some are involved in feeding, while others function in reproduction. The reproductive polyps have medusae developing inside them (the medusa buds), which are released to the surrounding water, starting the life cycle over again. The different stages of the life cycle are drawn at different scales.

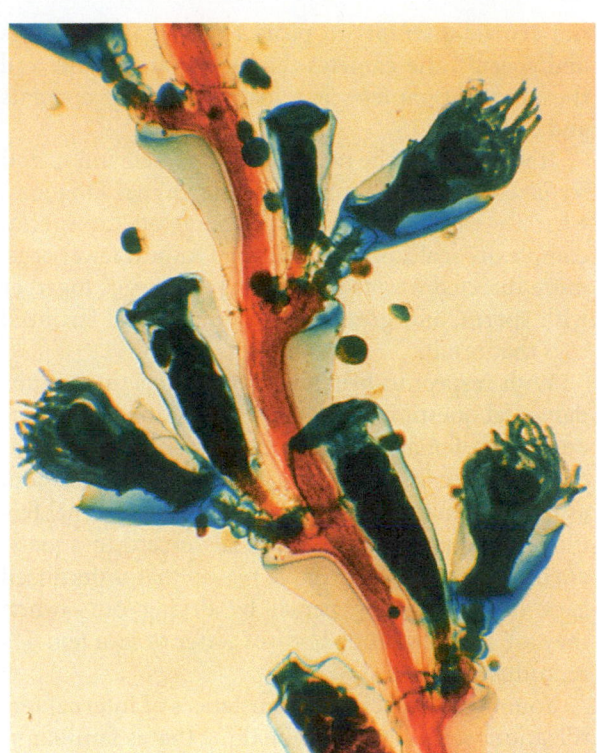

Figure 23.18
Obelia Up Close

A micrograph of the tiny cnidarian *Obelia* in its colonial polyp stage. The red, central stalk attaches to two kinds of polyps: those without tentacles, which are reproductive; and those with tentacles, which function in feeding. The reproductive polyps contain medusa buds that are being released into the water as mature medusae.

those that are specialized for the production of medusa-stage *Obelia*—the very thing our life cycle started with. These medusa *Obelia* then go on to engage in sexual reproduction, and the whole cycle repeats (**Figure 23.18**).

Forms of asexual reproduction other than budding are possible as well. Within the flatworm phylum (Platyhelminthes), certain worms can reproduce by what is known as *fission,* which simply means they break themselves roughly in half. The posterior (tail) portion of the worm grasps the material beneath it, while the anterior (head) portion moves forward. The separated halves then grow whatever tissues they need to make themselves complete worms.

Next, there is a form of asexual reproduction that is found among bees and ants, and that even exists in a small number of reptile species. Among common honeybees (*Apis mellifera*), a queen bee might lay 2,000 eggs per day. Nearly all of these eggs are fertilized by sperm the queen has received during one of her "nuptial" flights, and each one of these eggs will develop into one of the worker bees of the colony—every one of them a female. A queen can, however, choose to let some of her eggs go unfertilized; no

sperm from a male ever fuses with these eggs, yet bees develop within them and hatch from them. Since egg and sperm do not come together in this process, this is not sexual reproduction. Instead, each of these bees has been derived through **parthenogenesis:** a form of asexual reproduction in which an unfertilized egg develops into an adult organism.

Among the honeybees, all the bees derived through parthenogenesis are males—these are the few "drones" of a bee colony. But in other animals, parthenogenesis can produce females. Indeed, if you look at **Figure 23.19** you can see a lizard whose species produces nothing *but* females by employing parthenogenesis. All the members of this species are female, and all reproduction in the species comes through parthenogenesis. Thus, each new lizard develops solely from one of her mother's eggs, meaning that each is a clone of her mother.

Variations in Sexual Reproduction

Having looked at some of the ways animals reproduce asexually, you might think that all sexual reproduction in animals would follow the pattern we saw with *Obelia:* Sperm and eggs are produced by male and female members of a species; these reproductive cells come together and result in offspring. But the animal kingdom is more varied than this. If you look at **Figure 23.20** on the next page, you can see one variation in the form of the flatworm *Dugesia,* which is part of the phylum Platyhelminthes. Note that *Dugesia* has both testes *and* ovaries, along with both a penis and a genital pore (which is analogous to a vagina). *Dugesia* is a **hermaphrodite,** meaning an animal that possesses both male and female sex organs. When two *Dugesia* copulate, each projects its penis and inserts it in the genital pore of the other. (Can a given flatworm fertilize itself? It occasionally happens in the class of parasitic flatworms known as tapeworms.)

Figure 23.19
All-Female Species

Pictured is a whiptail lizard, *Aspidoscelis uniparens,* that reproduces solely through parthenogenesis. Because the eggs these lizards produce are not fertilized by males, they contain only one set of chromosomes, the mother's. As a consequence, each hatchling is genetically identical to her mother—each is a clone of her mother.

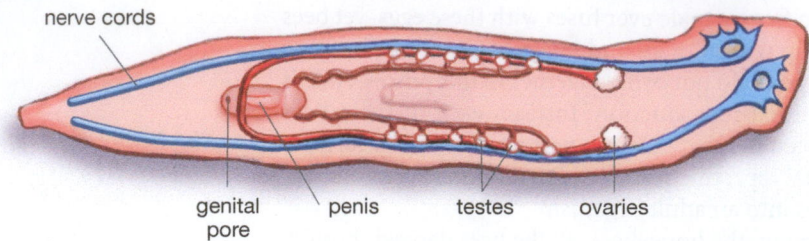

nerve cords

genital pore · penis · testes · ovaries

Figure 23.20
Hermaphrodite
The flatworm *Dugesia* is a hermaphrodite, as it possesses both the sex organs of a male (penis and testes) and those of a female (ovaries and genital pore).

Several variations exist on the theme of hermaphrodism. If we look at the oysters that are part of phylum Mollusca, we find that, while they have separate sexes, a given individual can *change* its sex between mating seasons. A frequent course of events is for an oyster to mature as a male and then change to a female later on. This kind of male-to-female transition is surprisingly common in the animal kingdom; some fish, eels, snails, and shrimp go through it as well. Female-to-male transitions are less common, but they do occur.

SO FAR . . .

1. Sexual reproduction in animals entails a fusion of _____ and _____ from different individuals.

2. Budding, fission, and parthenogenesis are all forms of _____ practiced by animals. In parthenogenesis, an _____ develops into an adult organism.

3. An animal that possesses both male and female sex organs is known as a _____.

23.5 Egg Fertilization and Protection

If we look strictly at sexual reproduction in animals, a central question across the animal kingdom is one we've touched on already: How do eggs get fertilized, and how are they taken care of until such time as offspring result from them? If we think again about the jellyfish-like creature *Obelia* noted earlier, you may recall that it has a very simple strategy on both counts: Eggs are deposited by the female directly into the ocean, after which sperm from the male likewise are deposited; then water currents bring the two kinds of reproductive cells into contact, after which neither parent provides any kind of care to the fertilized eggs. *Obelia* thus provides an example of what is generally regarded as the most primitive and ancestral condition in the animal kingdom. Both males and females simply "broadcast" their eggs and sperm into the environment at the same time. Neither male

nor female attempts to protect these gametes (eggs and sperm), the gametes themselves are fragile, and the union of gametes is a matter of water currents and chance.

If we followed the twin issues of gamete union and fertilized egg protection across the animal kingdom, what we would see, with movement up the species complexity scale, is a general trend toward fewer eggs, more directed means of bringing these eggs together with sperm, and greater protection for the fertilized eggs that result.

With respect to bringing gametes together, a fundamental question is whether fertilization takes place outside the female (external fertilization, as in *Obelia*) or takes place inside the female (internal fertilization, as in the human female). As it turns out, external fertilization is rarely seen in animals except in aquatic environments, for the simple reason that unfertilized eggs desiccate—they perish by drying out—when they are not either in a body of water or in a female's reproductive tract.

You might expect that all instances of internal fertilization would involve something that is familiar to us from the world of mammals: **copulation,** which is to say the release of sperm within a female by a male reproductive organ. But the animal kingdom is more varied than this. The males of many species produce what are known as spermatophores (sperm packets), which in some cases must be introduced into the female reproductive tract by the females themselves. The male of a European amphibian called the great crested newt (*Triturus cristatus*) will do a courtship dance in front of a female, waving his tail, and then deposit a spermatophore on the ground in front of her and block her path in hopes that she will scoop the packet into the opening of her reproductive tract.

Protection for Eggs: The Vertebrates

With respect to the other issue we're concerned with, the protection of fertilized eggs, different strategies are followed across the animal kingdom. But we're going to focus here on just a few reproductive snapshots from one small part of the kingdom, the vertebrates in phylum Chordata.

In **Figure 23.21**, you can see the life cycle of a familiar amphibian vertebrate, the frog, whose reproduction usually involves the kind of contact between the male and female that you can see in the figure: the male clasps the female in an embrace called amplexus, which can go on for hours or even days. This stimulates the female in such a way that she lays her eggs into the water, after which the male deposits his sperm on top of them. These eggs are fragile, however, and predators tend to be plentiful in the environments in which frogs live. A common response of frogs to this situation is to invest time and effort in

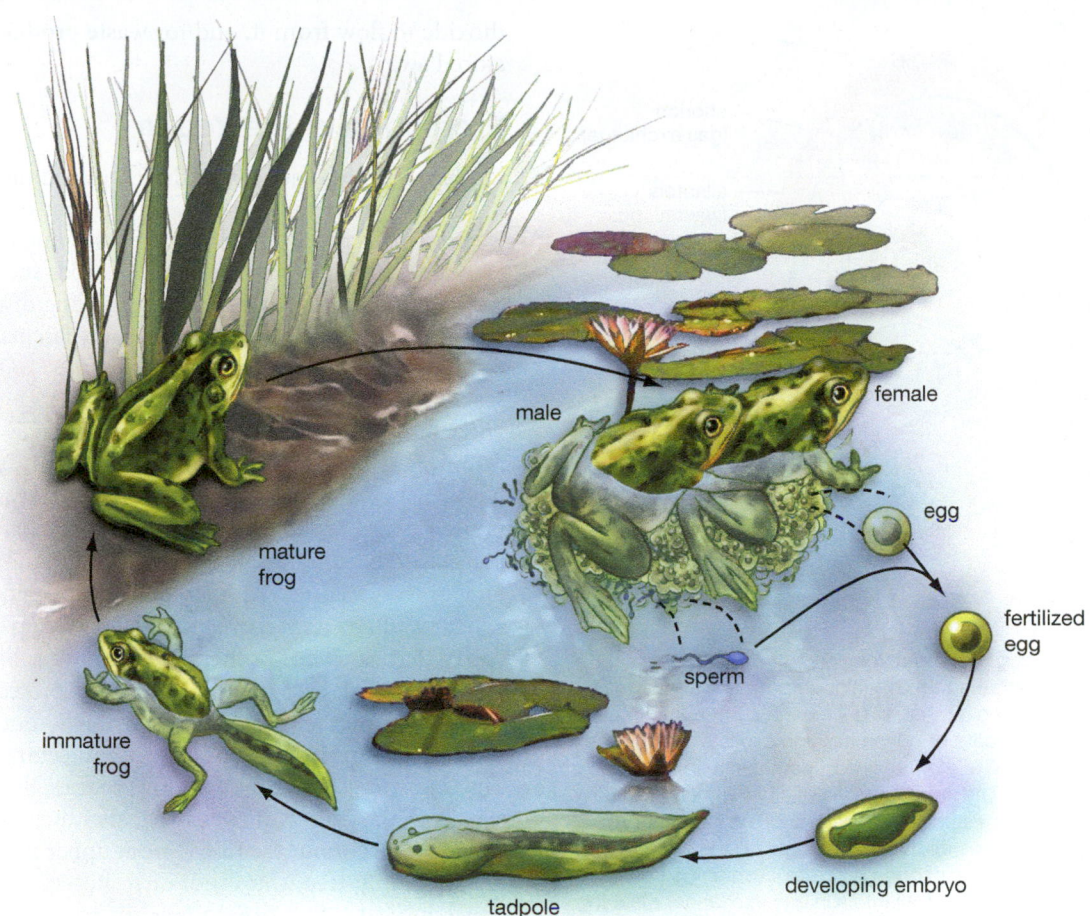

Figure 23.21
Fragile Eggs, Aquatic Environment
Amphibians such as frogs utilize external fertilization, generally releasing both eggs and sperm into water. The eggs must remain in an environment that is at least moist, lest the embryos in them perish from drying out. The tadpoles that hatch out of the eggs in this species are likewise aquatic until such time as they lose their tails, develop lungs, and begin living on land.

keeping the eggs protected. A clear, if unusual, example of this is seen in the Central American frog *Physalaemus pustulosus*, whose females will, while locked in amplexus with the male, secrete a mucus that both frogs then proceed to beat into a froth with their legs. What results is a foam nest, resting on top of the water, that the eggs and sperm are deposited into (**Figure 23.22**). Four days later, little tadpoles drop out of the underside of the nest and start swimming away.

Egg protection efforts of this sort are common across much of the animal kingdom. What interests us is to follow the evolutionary progression in such protection that starts with amphibians. Note that the frogs we looked at laid their eggs in the water. They did so because their eggs were vulnerable not just to predators but to the drying out we talked about earlier. So critical is this issue of desiccation issue to frogs that even those species that are completely terrestrial must lay their eggs in environments that are at least moist if offspring are to result.

Those of you who went through Chapter 19 know, however, that amphibians gave rise to another vertebrate line that was able to completely break the tie between water and reproduction and thus move inland in great numbers. How did these later-evolving creatures preserve their eggs out of water? They

Figure 23.22
Stirring Up Protection
Two tungara frogs (*Physalaemus pustulosus*) stir up foam that will serve as a protective hiding place for eggs that will be deposited by the female and fertilized by the male. These frogs were photographed in the Panamanian rainforest.

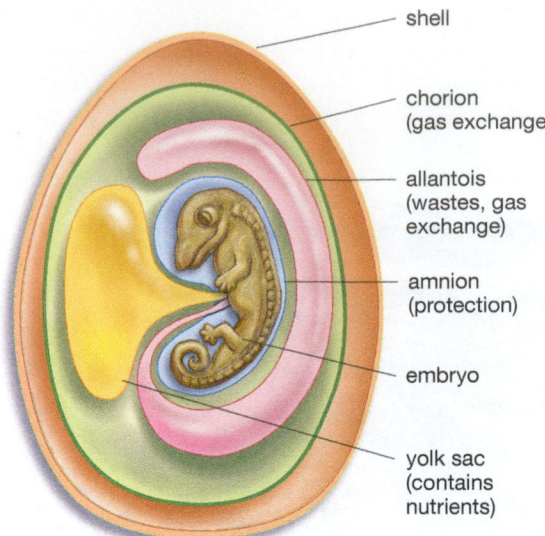

shell

chorion
(gas exchange)

allantois
(wastes, gas
exchange)

amnion
(protection)

embryo

yolk sac
(contains
nutrients)

Figure 23.23
An Egg for Many Environments

The amniotic egg was an evolutionary development that allowed vertebrates to move inland from the water. The egg's features meant that, in dry environments, the embryo would not desiccate or perish through a loss of fluids. The egg's tough shell is the first line of defense; it limits evaporation of fluids. The amnion is one of several membranes that provide some cushioning protection for the embryo. The allantois serves as a repository site for the embryo's waste products. The chorion provides cushioning protection and works with the allantois in gas exchange, meaning the movement of oxygen to the embryo and the movement of carbon dioxide away from it. The yolk sac provides the nutrients the embryo will need as it develops.

employed a different kind of egg. If you look at **Figure 23.23**, you can see the **amniotic egg:** an egg produced primarily by reptiles and birds whose shell and system of membranes provide protection and life support to a developing embryo. Protection? The shell maintains a watery environment *inside* the egg, while several of the egg's membranes provide cushioning. Life support? Other membranes in the egg allow for oxygen to flow to the embryo, for carbon

dioxide to flow from it, and for waste products to be stored away.

Protection of the Young

Modern reptiles generally lay their eggs and then abandon them before the young ever hatch out of them. If we ask why reptiles do this, one answer might be: because they can. The amniotic egg provides such a protective environment that parental guarding of the eggs isn't necessary. There are, however, some fascinating exceptions to this rule. If you look at **Figure 23.24**, you can see a reptile family tree that starts, at the base, with an ancestral reptile. Note that, early on, mammals evolved from this reptile line (over on the left). What really interests us, however, are the three evolutionary lines at upper right: dinosaurs, birds, and crocodilians (meaning both crocodiles and alligators). Now, among contemporary creatures who employ an amniotic egg, which of them also *guard* not only the eggs but the young that hatch out of them? The answer is birds and crocodiles. Birds are well known for their protective behavior toward their eggs and hatchlings, but female crocodiles also turn out to be fierce protectors of their eggs and young. (In some crocodile species, the mother stays with the young for over a year after they hatch.)

Now, what about the third line at upper right in the reptile family tree, the dinosaurs? Recent fossil evidence indicates that dinosaurs may have stayed with their newly hatched young too. The phrase "attentive parents" may not readily come to mind in connection with dinosaurs, but this may have been the case. In 2004, fossils of a plant-eating dinosaur were found in China surrounded by 34 juvenile dinosaurs—a configuration that is hard to explain unless the adult had a close association with the juveniles (**Figure 23.25**). Taking a step back from this, the interesting thing is that protective behavior toward the young may have developed in an evolutionary line that began with the common ancestor of ancient crocodilians and dinosaurs and that then continued in the birds that descended from dinosaurs.

Figure 23.24
The Evolution of Offspring Protection

Most reptiles do not stay with their eggs through the time that hatchlings emerge from them. This is not the case, however, with crocodilians or with birds (who evolved from dinosaurs). They guard both their eggs and, for a considerable time, the young that hatch out of them. Some recent evidence indicates that dinosaurs may have done this as well. Note that mammals evolved from a different lineage of reptiles.

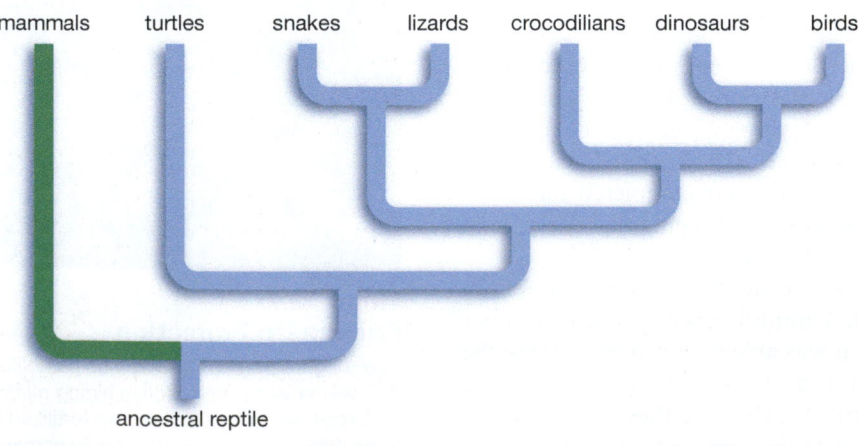

mammals   turtles   snakes   lizards   crocodilians   dinosaurs   birds

ancestral reptile

Figure 23.25
Parent and Offspring?
Researchers in China found the fossilized remains of a parrot-like dinosaur, *Psittacosaurus,* surrounded by 34 juveniles. The 125-million-year-old fossils provide evidence that dinosaurs provided care to their young after they were hatched.

Protection in Mammals

So, thus far we've seen protective behavior on the part of amphibians, a more protective egg that evolved in reptiles, and this egg combined with protective behavior among a select group of reptiles. What else could vertebrate evolution produce in the way of protection for offspring? A dandy of an innovation, as it turned

out. When we look at *mammals,* we find that their embryos are protected and nurtured by the mother's entire body. An egg that is fertilized in a mammalian female's reproductive tract then *stays* in it, implanting in the female's uterus and developing there. Most mammals are thus **viviparous:** Reproduction occurs by means of fertilized eggs developing inside the mother's body. In contrast, most reptiles and all birds are **oviparous:** Reproduction occurs by means of the young developing in fertilized eggs that are laid outside the mother's body.

In most mammals, implantation of the embryo triggers the formation of a network of blood vessels and membranes, called the placenta, through which nutrients and oxygen flow to the embryo and through which wastes and carbon dioxide flow from it. Most mammals are thus said to be placental, but there are some well-known exceptions to this rule. A small group of early evolving mammals lay eggs, just as reptiles do. These are the monotremes, represented by the duck-billed platypus (**Figure 23.26a**). And there are mammals in which the young develop only to a limited extent inside the mother's body. These are the marsupials, which include kangaroos (**Figure 23.26b**). But the vast majority of mammals today are placental (**Figure 23.26c**).

Mammalian care doesn't stop with the birth of the young, of course. Indeed, mammals are the supreme caregivers of the living world. If we look, for example, at elephants, we find that an elephant calf might be nurtured by its mother and a larger kinship group for the first 10 years of its life. To see an even more stellar example of parental care, however, we need look no further than a species that is very familiar to us, the primate known as *Homo sapiens.* We human beings lavish more care on our offspring than any other creature in nature.

Figure 23.26
Three Types of Mammals
Mammals differ in their reproductive strategies.

(a) This duck-billed platypus is a monotreme—a mammal that lays eggs.

(b) Kangaroos are marsupials—they give birth to physically immature young, which then develop further within the mother's pouch. Pictured are a mother and her fairly mature "joey" in eastern Australia.

(c) The young of placental mammals, such as this brown bear (*Ursus arctos*), develop to a relatively advanced state inside the mother, with nutrients supplied not by an egg but by a network of blood vessels and membranes—the placenta.

(b) Marsupial

(a) Monotreme

(c) Placental mammal

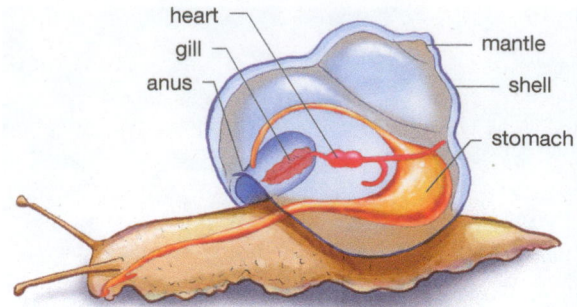

Figure 23.28
Organ Complexity
Gastropods such as this snail have a sophisticated set of organs. For example, the snail has gills to transfer oxygen into its body, a heart to pump its blood, and a stomach to hold and digest its food.

SO FAR . . .

1. With respect to egg fertilization and protection, the primitive, ancestral state in the animal kingdom is believed to be one in which eggs and sperm are _____ into water and brought together by water currents, after which parents provide _____ care to fertilized eggs.

2. The more complex the animal in the animal kingdom, the _____ the number of eggs that are likely to be produced and the _____ amount of care parents are likely to provide to fertilized eggs.

3. The amniotic egg, which evolved in _____, was important because it allowed eggs to be laid _____.

organs *start* in the animal kingdom. Sponges don't even have tissues, and jellyfish have nothing as organized as testes or ovaries. From Platyhelminthes all the way up the ladder of complexity, however, organs are the rule.

You may recall, however, that while a flatworm such as *Dugesia* is organizationally sophisticated compared to some animals, it is organizationally *simple* compared to others. Like all its fellow flatworms, *Dugesia* has no circulatory system, which means that every one of its cells must get its oxygen directly from the environment around it. In addition, it lacks a complete digestive tract, in that it has no anus, which means that waste must be expelled through its mouth.

All this stands in contrast to what you can see in **Figure 23.28**, which is a simplified representation of the aquatic snail we looked at earlier in connection with phylum Mollusca. Notice that the snail has organs that are familiar to us—a heart and stomach—along with gills, which serve to move oxygen into its circulation. Among the items that have helped free the snail from the constraints of shape and size seen in *Dugesia* are its coelom, or internal body cavity, which allows an expandable stomach and a beating heart, and its circulatory system, which allows nutrients to get to tissues that exist far away from its surface.

23.6 Organs and Circulation

Figure 23.27 shows you a creature we talked about earlier in connection with reproductive anatomy, the flatworm *Dugesia*. You can see that, in addition to having the testes and ovaries mentioned earlier, *Dugesia* has two primitive eyes and two collections of nerve cells (called cerebral ganglia), which might be thought of as a primitive brain, along with a highly branched intestine that delivers nutrients to all parts of its body. Meanwhile, the pharynx you can see is a muscular feeding structure that *Dugesia* can use to penetrate the bodies of prey animals.

What all this adds up to is that *Dugesia* has **organs**: highly specialized structures that generally are formed of several kinds of tissues. (You may recall that tissues are formed of collections of similar cells.) We're so used to the idea of animals having organs—such as a heart or a liver—that *Dugesia*'s possession of organs may not seem all that impressive. But as we saw earlier, *Dugesia*'s phylum, Platyhelminthes, actually is where

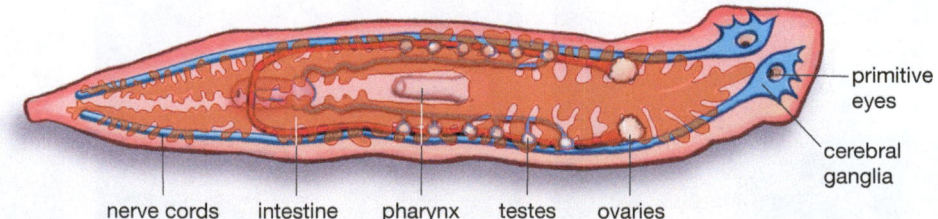

Figure 23.27
Organs in an Early Form
In addition to its sexual organs, the flatworm *Dugesia* has other organs—primitive eyes, for example, branched intestines, a pharynx that functions in feeding, and clusters of nerve cells (the cerebral ganglia) that function like a brain.

Open and Closed Circulation

The snail's circulatory system turns out to offer us another comparative lesson, this time not relative to a flatworm but relative to ourselves. We have what is known as a **closed circulation system**—a system of circulation in which blood *stays within* the blood vessels. There is some leakage of the liquid portion of our blood, but the basic arrangement is that, while oxygen and nutrients diffuse out of our blood vessels, most of the blood itself stays confined. In contrast, the snail and most other molluscs have an **open circulation system**—a circulation system in which blood flows out of blood vessels altogether, into spaces or sinuses where it bathes the surrounding tissues. Veins then collect the blood (or hemolymph) in these sinuses for a return trip to the gills and the heart. The most complex molluscs, the cephalopods (squid, octopus, etc.), have a closed circulation system like ours; but this isn't surprising, since only a closed system has the efficiency to support an animal as large as a squid. The animal kingdom as a whole is mixed with respect to open and closed systems. All vertebrates have closed systems, for example, while all arthropods have open systems.

23.7 Skeletons and Molting

How do we raise our arm? Two things are required: a muscle to exert force and an object to exert this force against—a bone, which is part of a support network called a skeleton. If we look at the human body, we take for granted that our skeleton lies interior to our muscles. But this isn't always the case in the animal kingdom. You may recall that when we looked at arthropods (insects, crustaceans, etc.), we saw that they have an exoskeleton, meaning an external material covering of the body that provides support and protection. Arthropod muscles pull against a skeleton, but this skeleton lies external to the arthropods' muscles.

Having a skeleton on the outside of the body, however, means that the arthropod body can only grow so much before it expands right into the skeleton. When this happens, arthropods are able to continue their growth only after engaging in **molting**: a periodic shedding of an old skeleton followed by the growth of a new one. If you look at **Figure 23.29**, you can see a picture of a crab that has just slipped out of its old skeleton. Molting activity of this sort is not limited to the arthropods. The roundworms of phylum Nematoda also have an exoskeleton, and it turns out that they shed it four times during their lives. Since roundworms and arthropods make up such a large proportion of the animal kingdom, the way to think of molting is as a widespread phenomenon among animals.

Figure 23.29
Old and New

Arthropods have an outer skeleton, or exoskeleton, of a fixed size. They are able to grow by periodically shedding, or molting, this stiff skeleton. The crab in the photograph has just finished molting; its old exoskeleton lies nearby.

The exoskeletons that are part of molting stand in contrast not only to our own *endo*skeleton but to a third variation on animal skeletons, exemplified in the earthworms of phylum Annelida. Now, an earthworm has muscles and everyone knows it can move; yet it has no bones for its muscles to pull against. So *how* does it move? Each of the earthworm's many segments contains a so-called coelomic space (which you can see in Figure 23.8 on page 423). Each space is filled with fluid, and it is this fluid that serves as the earthworm's skeleton. Here's how it works. Imagine squeezing a water-filled balloon. The volume of the water doesn't change, but the *shape* of the balloon does, in response to the pressure of the water. In the same way, the earthworm can squeeze one set of muscles and *lengthen* a set of its segments. Thus, the earthworm employs what is known as a **hydrostatic skeleton,** a skeleton in which the force of muscular contraction is applied through a fluid. Hydrostatic skeletons are common in the animal kingdom. Apart from annelids like the earthworm, they exist in cnidarians (jellyfish, etc.), flatworms, and roundworms.

Table 23.1, on the next page, presents a summary of all the phyla reviewed in this chapter.

On to Plants

The animals we've looked at this chapter set the standard for mobility in the living world, but if you look outside your window, the vast majority of living things you'll see aren't mobile at all, but instead are fixed in one spot from the time they come into being to the time they die. Nevertheless, these organisms are vigorously alive. They are constantly building themselves up with what they take from the Earth and the air; they push their roots into the ground and their stems toward the sky; they produce flowers, fruits, fragrances, wood, and oxygen by the ton. These are the plants, the subject of Chapter 24.

Table 23.1

Animal Phyla

Phylum	Members include	Live in	Characteristics
Porifera	Sponges	Ocean water (marine), with a few living in freshwater	Sessile (fixed in one spot) as adults, no symmetry. System of pores through which water flows serves to capture nutrients. Bodies do not have organs or true tissues. Sexual and asexual reproduction
Cnidaria	Jellyfish, hydrozoans, sea anemones, coral animals	Almost all marine; a few in freshwater	Radial symmetry, medusa (swimming) and polyp (largely sessile) stages of life. Use stinging tentacles to capture prey. Have tissues. Can reproduce sexually or by budding.
Platyhelminthes	Flatworms (including flukes, tapeworms)	Marine, freshwater, or moist terrestrial environments	Bilateral symmetry, hermaphroditic, possess organs, no central cavity. Many are parasites. Can reproduce through fission.
Annelida	Segmented worms	Most are marine, some freshwater and moist terrestrial	Distinct body segmentation in varieties such as earthworms. Leeches are annelids.
Mollusca	Gastropods (snails, slugs), bivalves (oysters, clams, mussels), cephalopods (octopus, squid, nautilus)	Marine, freshwater, moist terrestrial	Mantle that can secrete material that becomes shell; mantle cavity housing gill or lungs.
Nematoda	Roundworms	Almost all environments	Small size in most, but some are several meters long. Many are crop pests, other animal parasites.
Arthropoda	Vast group that includes insects, spiders, mites, crabs, shrimp, and centipedes	Many aquatic and terrestrial environments	External skeleton; paired, jointed appendages; molting; open circulation system.
Echinodermata	Sea stars, sea urchins, sea cucumbers	Always marine, most on ocean floor	Generally have radial symmetry as adults; tube feet in some species, slow ocean floor movement in most.
Chordata	Vertebrates, lancelets, sea squirts	Great variety of aquatic and terrestrial environments	At some point, all possess notochord, dorsal nerve chord, pharyngeal slits, post-anal tail. Vertebrates have vertebral column.

SO FAR . . .

1. In closed circulation systems, blood _____ blood vessels, while in open circulation systems, it _____ to bathe surrounding tissues.

2. Insects and crustaceans have a skeleton different from ours. Theirs is an _____: a skeleton that forms an _____ on an animal body.

3. Animals with an exoskeleton are able to continue to grow by employing the practice of _____, which can be defined as _____.

CHAPTER 23 REVIEW

Summary

23.1 What Is an Animal?

- Only animals pass through a blastula stage in embryonic development. A blastula is a hollow, fluid-filled ball of cells that forms once an egg is fertilized by sperm. **(p. 417)**

- In addition, all animals are multicelled, are motile at some point in life, are heterotrophs, and are composed of cells that do not have cell walls. **(p. 417)**

23.2 Lessons from the Animal Family Tree

- Except for the sponges in Porifera, the members of all the phyla reviewed have bodies composed of tissues. **(p. 419)**

- Except for Porifera and Cnidaria, all phyla reviewed exhibit bilateral symmetry. Cnidaria exhibits radial symmetry while Porifera exhibits no symmetry. **(p. 420)**

- The members of all phyla reviewed other than Porifera, Cnidaria, and Platyhelminthes possess a central body cavity known as a coelom. **(p. 420)**

23.3 Across the Animal Kingdom: Nine Phyla

- Phylum Porifera's sponges live by drawing water into themselves through a series of tiny pores on their exterior and then filtering the food from this water while extracting oxygen from it. **(p. 421)**

- Phylum Cnidaria (jellyfish, sea anemones, hydras, and reef-building coral animals) use stinging tentacles to capture prey. Many cnidarians have both an adult, medusa stage of life, which swims in the water, and an immature, polyp stage of life, which generally remains fixed to a solid surface. **(p. 422)**

- The flatworms of phylum Platyhelminthes are mostly small creatures, dwelling either in aquatic or moist terrestrial environments. Flatworms have bilateral symmetry and organs, but have no coelom or system of blood circulation. **(p. 422)**

- The worms of phylum Annelida provide a clear example of body segmentation. Most annelids are marine, and some dwell in freshwater, though common earthworms are terrestrial. **(p. 423)**

- Three important classes of phylum Mollusca are the gastropods (snails, slugs); the bivalves (oysters, clams, mussels); and the cephalopods (octopus, squid, nautilus). All molluscs possess a fold of skin-like tissue called a mantle that usually secretes material that forms a shell. **(p. 424)**

- The mostly microscopic roundworms of phylum Nematoda exist in enormous numbers in all kinds of habitats on Earth. A number of roundworms are agricultural pests and some are human parasites. **(p. 424)**

- Phylum Arthropoda has three subphyla in it: Uniramia (millipedes, centipedes, insects); Chelicerata (spiders, ticks, mites, horseshoe crabs); and Crustacea (shrimp, lobsters, crabs, barnacles). All arthropods have an exoskeleton and paired, jointed appendages and all molt. **(p. 425)**

- All members of phylum Echinodermata (sea stars, sea urchins, sea cucumbers) are marine and inhabit the ocean floor. **(p. 425)**

- Phylum Chordata has three subphyla in it: Vertebrata, which includes all the vertebrates; Cephalochordata, made up of creatures called lancelets; and Urochordata, made up of three classes of animals, including the tunicates or sea squirts. Only the vertebrates have a vertebral column. All chordates possess, at some point in their lives, a notochord; a nerve cord on their dorsal side; a post-anal tail; and a series of pharyngeal slits. (p. 426)

23.4 Animal Reproduction

- Sexual reproduction in animals always entails the fusion of eggs and sperm from two individuals, while asexual reproduction does not. Members of the less complex animal phyla commonly reproduce both sexually and asexually. Sexual reproduction is the norm with more complex animals, but asexual reproduction is sometimes seen in them. (p. 428)

- Forms of asexual reproduction include budding, fission, and parthenogenesis. (p. 428)

- Sexual reproduction can be practiced by hermaphrodites. Some animals change sex over the course of their lifetimes. (p. 429)

23.5 Egg Fertilization and Protection

- In animals, eggs and sperm are brought together and eggs are protected following fertilization in a number of ways. The most primitive, ancestral method is that of parents broadcasting both eggs and sperm into an aquatic environment and thereafter providing no care for the eggs that become fertilized. With movement up the scale of animal complexity, species tend to produce fewer eggs, to exhibit more directed means of bringing these eggs together with sperm, and to provide greater protection to the fertilized eggs that result. (p. 430)

- Fertilization can be external or internal. Only some species that employ internal fertilization also employ copulation. Other species rely on females to insert sperm packets (spermatophores) into their own reproductive tracts. (p. 430)

- Vertebrate evolution resulted in a range of strategies for protecting fertilized eggs and the offspring that develop from them. Reptile evolution resulted in the amniotic egg, whose shell and system of mem-

branes allows it to be laid in dry environments, whereas mammalian evolution produced the placental system, in which fertilized eggs develop inside a mother's body. (p. 430)

23.6 Organs and Circulation

- Organs are complex bodily structures composed of several types of tissues. Organs are found in all the animal phyla reviewed except Porifera and Cnidaria. (p. 434)

- Vertebrates have closed circulation systems. Arthropods and small molluscs have open circulation systems. (p. 435)

23.7 Skeletons and Molting

- Three kinds of skeletons are found in animals: endoskeletons, exoskeletons, and hydrostatic skeletons. (p. 435)

- To continue their growth, animals with exoskeletons must periodically molt. (p. 435)

Key Terms

amniotic egg   432

bilateral symmetry   420

body segmentation   423

budding   428

closed circulation system   435

coelom   420

copulation   430

exoskeleton   425

hermaphrodite   429

hydrostatic skeleton   435

invertebrate   417

molting   435

organ   434

open circulation system   435

oviparous   433

parasite   423

parthenogenesis   429

radial symmetry   420

symmetry   420

tissue   419

vertebral column   426

viviparous   433

Understanding the Basics

Multiple-Choice Questions

(Answers are in the back of the book.)

1. Animals (select all that apply):
   a. go through a blastula stage in development.
   b. have cells with cell walls.
   c. can be single-celled.
   d. are heterotrophs.
   e. are mostly vertebrates.

2. Bilateral symmetry is a condition in which:
   a. no one animal species can overpower another.
   b. the top and bottom halves of an animal are symmetrical.
   c. two animal species evolve along parallel lines.
   d. one side of an animal is symmetrical to the other.
   e. one side of an animal is as strong as the other.

3. A coelom is:
   a. an enclosed body cavity, not found in sponges, cnidarians, or flatworms.
   b. an enclosed body cavity, found only in vertebrates.
   c. a digestive organ, found in all animals except sponges.
   d. an example of a type of tissue.
   e. an offspring that arises from asexual reproduction.

4. In earthworms, we can see a clear example of the widespread animal feature known as:
   a. filter feeding.
   b. body segmentation.
   c. viviparous reproduction.
   d. a post-anal tail.
   e. a radula.

5. If an animal has an exoskeleton, paired jointed appendages, and goes through the process of molting, it is:
   a. a segmented worm.
   b. a roundworm.
   c. a chordate.
   d. a cnidarian.
   e. an arthropod.

6. Forms of asexual reproduction found in the animal kingdom include (select all that apply):
   a. binary fission.
   b. fission.
   c. budding.
   d. copulation.
   e. parthenogenesis.

7. Larger animals, such as vertebrates and large molluscs (squid, octopus) have _____ circulation systems, while arthropods and smaller molluscs (snails) have _____ circulation systems.
   a. closed; open
   b. closed; no
   c. open; closed
   d. open; no
   e. open; open

8. In a hydrostatic skeleton, the force of muscular contraction is applied through:
   a. a skeleton external to the muscles.
   b. a skeleton that will be shed through molting.
   c. a skeleton internal to the muscles.
   d. a fluid.
   e. a circulation system.

Brief Review

(**Answers are in the back of the book.**)

1. Do all animals have a single ancestor, or do the different animal phyla have different ancestors? Can you think of any evidence to support your answer?

2. Why is it fair to characterize sponges as "primitive" animals?

3. A coelom amounts to internal space in an animal. Why is such a feature valuable?

4. Match the phylum on the left with the description on the right.

5. In animals such as arthropods, what growth issue is solved through the mechanism of molting?

Cnidaria	Sea stars, sea urchins
Nematoda	Fish, amphibians, reptiles
Mollusca	Earthworms, leeches
Annelida	Jellyfish, sea anemones
Arthropoda	Insects, lobsters, spiders
Echinodermata	Roundworms
Chordata	Snails, squid

Applying Your Knowledge

1. One of the distinguishing features of animals is that their cells do not have the thick outer lining known as a cell wall. Plant cells have a cell wall, as do fungi cells. Why would a feature such as this be advantageous to plants but disadvantageous to animals?

2. Plants can have elaborate defense systems, and they can respond in sophisticated ways to their environment—for example, in preserving resources, they can go into a dormant state in winter. Given this, can plants have intelligence? Or is it only animals that have this trait, to one degree or another?

3. The female Atlantic oyster (*Crassostrea virginica*) can produce up to 100 million eggs per year, which she broadcasts into the water and provides no care for thereafter. By contrast, a human female will release 13 eggs per year on average (through ovulation), but in most cases will provide years of care to that small fraction of eggs that become fertilized and develop into babies. Why, with movement up the complexity scale in the animal kingdom, does a trend exist toward fewer eggs but greater parental care for each offspring?

24

Plants:

The Diversity of Life 4

Plants may be silent and fixed in one spot, but they are extremely active and they respond to the world around them in some surprising ways.

Available on MasteringBiology®
Activities: Phototropism; **Building Vocabulary;**
Questions: Reading Questions, Test Bank

Earth's plants come in an amazing variety of forms, including wildflowers and the familiar saguaro cactus (*Carnegiea gigantea*), seen here in Arizona's Sonoran Desert.

T
he optometrist has almost completed the examination but would like a better look at the interior of the patient's eyes. Would she agree to have her pupils dilated with the application of a couple of drops to each eye? The patient tilts her head back, and in her pose, it's possible to imagine her as a lady of the Italian Renaissance, lifting her eyes upward and applying

to them a few drops of a substance she believes will make her a *bella donna*—a beautiful lady. And what does this substance do? It dilates her eyes just slightly, making her pupils large and entrancing.

Spin back further in time to the eleventh century in Scotland where, as legend has it, the Scottish leader Earl Macbeth poisons an invading army of Danish soldiers by lacing their drinks with juice from a plant known as sleepy nightshade. Only later will the English-speaking world call this plant by the name used today: *Atropa belladonna* (**Figure 24.1**).

Skip forward to 1831, when scientists isolate from the belladonna plant one of its active ingredients, a compound called atropine. This is the substance that actually dilated the eyes of the Italian ladies and that, at one time, dilated eyes in optometry clinics. Then, come forward to the summer of 2004, when the federal government announces that it will soon be supplying states with an assortment of antidotes to chemical weapons. As a response to the threat posed by terrorists, the government says, it will distribute sets of "chem-packs" that include the drug atropine.

This makes sense because U.S. armed forces units have been outfitted for years with "auto-injectors" that can pump a dose of atropine into the thigh of any soldier who has been gassed. Now step back from the government officials making their announcement and observe, just a few miles away, some paramedics hard at work. They're making use of atropine, too, although not in connection with a terrorist assault. In ambulances and emergency rooms, atropine is used to revive heart attack victims. An injection of it can bring a pulse back to a person whose heart is failing.

24.1 The Roles and Characteristics of Plants

Like a talented employee, belladonna keeps getting called back to work by human beings. If you were keeping track, we humans have derived, from just this one plant, a poison, an antidote to poison, a cosmetic, an aid to visual diagnosis, and a heart medicine. But that's the way with plants. The number of things they do for human beings is enormous. The best-known thing they do, of course, is provide us with food. But even in this role, their significance is not fully appreciated. In the world today, up to 90 percent of the calories human beings consume come directly from plants. Indeed, as some botanists have noted, there are about a dozen plants that stand between humanity and starvation. (The most important of these plants is rice, with wheat coming in second.) Plants are able to serve this function because they make their own food through photosynthesis. The bounty they produce through this process has made Earth a planet of surplus—a planet of grains and leaves and roots that are available for the taking. Along these same lines, plants also produce much of the oxygen that most living things require. Then there are the products that human beings have

Figure 24.1
A Plant That Has Served Many Purposes
The flowering plant *Atropa belladonna*.

learned to derive from plants, among them lumber, medicines, fabrics, fragrances, and dyes (**Figure 24.2**).

Apart from these uses, plants are a kind of anchoring environmental force. Their roots prevent soil erosion, while their leaves absorb such pollutants as sulfur dioxide and ozone. One of the main greenhouse gases warming Earth is carbon dioxide, and plants absorb carbon dioxide when they perform photosynthesis. Thus, there is not only a *production* side to atmospheric carbon dioxide—in significant part the pollution human beings produce—there is also a *consumption* side to it in the amount plants absorb. One proposed way to fight greenhouse warming, therefore, is to plant trees. Over the course of a summer, a single mature maple tree will absorb about 1,000 pounds of carbon dioxide from the atmosphere.

Plants can take on these varied roles because of a unique set of physical traits they possess. But what are these traits? Put another way, how are plants structured and how do they function? The goal of this chapter is to give you an introduction to these subjects. We'll start by looking at the characteristics that all plants possess and then move on to a consideration of the major types of plants found in the living world. Then we'll finish up by looking at the ways nature's most dominant plants, the angiosperms, interact with the world around them.

The Characteristics of Plants

We all have an intuitive sense of what plants are, although not all plants match our intuitions. A few are parasites (on other plants) and some, like the Venus flytrap, consume animals. Allowing for such exceptions, all plants share the following characteristics:

- Plants are photosynthesizing organisms that are fixed in place, multicelled, and mostly land-dwelling. We know that plants are fixed in one spot (sessile), and that they make their own food through photosynthesis. That they are multicelled follows from the fact that they develop from embryos, which are multicelled by definition. During the course of their evolution from algae, plants made a transition from water to land, and land is where most of them are found today—although some have made the transition back to water, as with water lilies.

- Plant cells have cell walls. Both plant and animal cells are enclosed by a plasma membrane, but plant cells have something just outside the membrane: a **cell wall,** defined as a relatively thick layer of material that forms the periphery of all plant and fungal cells and many types of bacterial, archael, and protist cells (**Figure 24.3**). The plant cell wall is composed in large part of a tough, complex compound called cellulose. In woody plants, cellulose is joined by a strengthening compound called lignin. Why are trees able to grow so tall? Because of the cellulose and lignin found in their cell walls.

- Plant (and algae) cells have organelles called **chloroplasts** that are the sites of photosynthesis. If there is one feature that has fundamentally shaped plant evolution, it is the chloroplast.

Figure 24.2
Uses Galore

Plants are used by human beings in innumerable ways.

(a) Food, here in the form of grapes.

(b) Wood, being harvested from a forest in California.

(c) A foxglove plant, which yields the heart medicine digitalis.

(a)

(b)

(c)

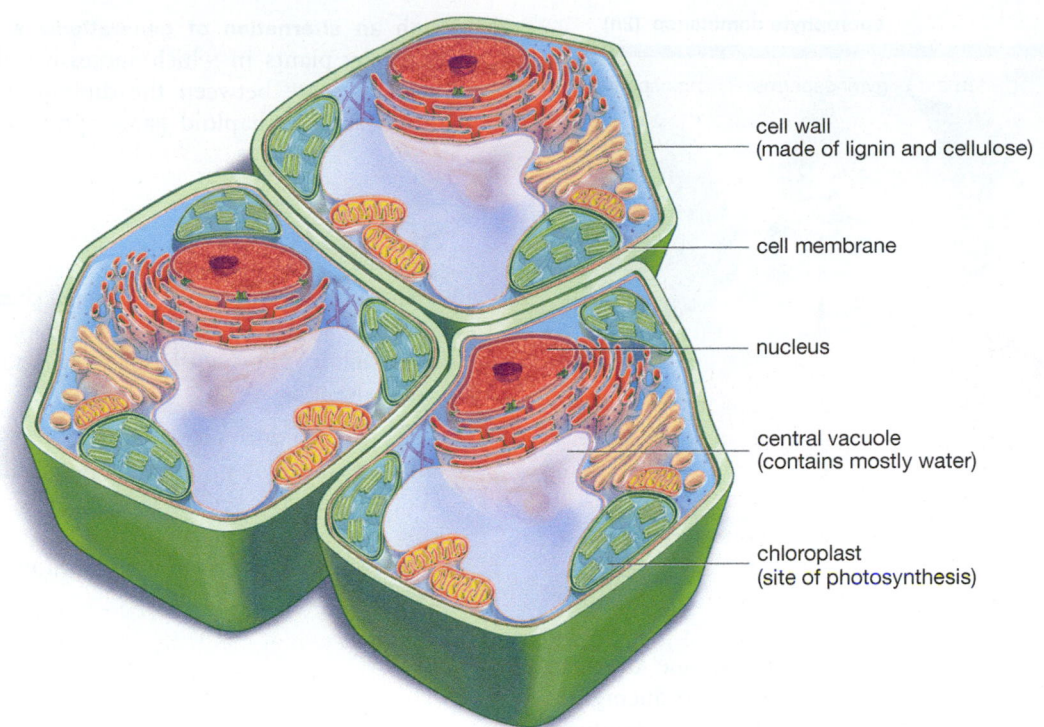

cell wall
(made of lignin and cellulose)

cell membrane

nucleus

central vacuole
(contains mostly water)

chloroplast
(site of photosynthesis)

Figure 24.3
Characteristics of Plants

All cells have an outer membrane, but plant cells have a wall external to this membrane. The compounds cellulose and lignin, which help make up the cell wall, impart strength to it. Plant cells have a higher proportion of water in them than do animal cells, with much of this water located in an organelle called a central vacuole. The sites of photosynthesis in plants are the organelles called chloroplasts.

Why are plants able to be fixed in one spot? Because they make their own food in chloroplasts and thus don't need to move toward food, as animals do.

- Successive generations of plants go through what is known as an alternation of generations.

One way to conceptualize this characteristic is to compare plant reproduction with human reproduction. The eggs and sperm that human beings produce are so-called gametes that are *haploid* cells, meaning they have but a single set of chromosomes in them. When these haploid cells fuse in the moment of conception, what's produced is a *diploid* fertilized egg, meaning an egg with two sets of chromosomes—one set from the father and one from the mother. This egg gives rise to more diploid cells through cell division, and the result is a whole new human being (**Figure 24.4a**).

A typical plant in its diploid phase will, like human beings, produce a specialized set of haploid reproductive cells. Instead of these cells being egg or sperm, however, what's produced is a spore: a reproductive cell that can develop into a new organism without fusing with another reproductive cell. Like the spores we saw in fungi, plant spores have the ability to grow into a new generation of plant all on their own. In some plants, this spore lands on the ground, starts dividing, and after a time a mature plant exists. When *this* plant reproduces, however, it does not do so through more spores. It produces gametes—eggs and sperm. These go on to fuse, and the result is a diploid plant just like the one we started with.

The upshot of all this, as you can see in **Figure 24.4b**, is that plants move back and forth between

(a) Human reproduction

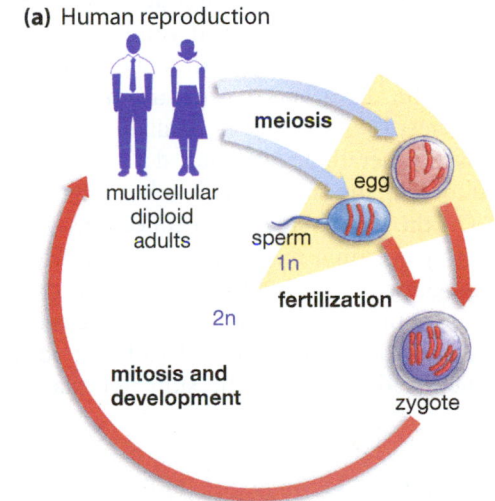

meiosis

multicellular
diploid
adults

egg

sperm
1n

fertilization

2n

mitosis and
development

zygote

(b) Plant alternation of generations

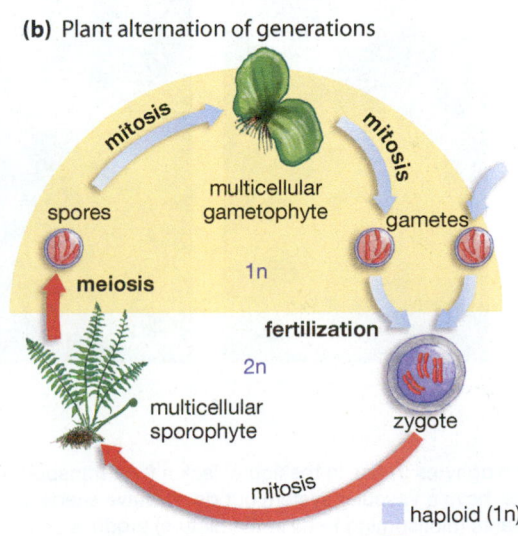

mitosis

mitosis

spores

multicellular
gametophyte

gametes

meiosis

1n

fertilization

2n

multicellular
sporophyte

zygote

mitosis

haploid (1n)

diploid (2n)

Figure 24.4
The Alternation of Generations in Plants, Compared to the Human Life Cycle

(a) Humans. Almost all cells in human beings are diploid, or 2n, meaning they have paired sets of chromosomes in them. The exception to this is human gametes (eggs and sperm), produced through meiosis, which are haploid, or 1n, meaning they have a single set of chromosomes. In the moment of conception, a haploid sperm fuses with a haploid egg to produce a diploid zygote that grows into a complete human being through cell division or mitosis.

(b) Plants. In plants, conversely, diploid (2n) plants—the multicellular sporophyte fern in the figure—go through meiosis and produce individual haploid (1n) spores that, without fusing with any other cells, develop into a separate generation of the plant. This is the multicellular gametophyte shown. This gametophyte-generation plant then produces its own gametes, which are eggs and sperm. Sperm from one plant fertilizes an egg from another, and the result is a diploid zygote that develops into the mature sporophyte generation.

sporophyte dominance (2n)

green algae   mosses   ferns   gymnosperms   angiosperms

gametophyte dominance (n)

through an **alternation of generations:** a life cycle practiced by plants in which successive plant generations alternate between the diploid sporophyte condition and the haploid gametophyte condition.

SO FAR . . .

1. Which of the following traits is *not* characteristic of plants? (a) fixed-in-place, (b) no cell wall, (c) photosynthesizing, (d) mostly land-dwelling.

2. All plants go through an alternation of generations in which one generation produces the reproductive cells called _____, while the alternate generation produces the reproductive cells that fuse in reproduction, _____ and _____.

Figure 24.5
Generational Dominance

In more primitive plants, the gametophyte generation is larger and more dominant. In mosses, for example, the gametophyte generation houses the sporophyte generation. By contrast, in the most recently evolved plants, the angiosperms, the gametophyte generation has been reduced to a microscopic entity that is almost wholly dependent on the sporophyte generation. Although not plants, green algae are included for comparative purposes.

two different kinds of generations, one of them spore-producing, the other gamete-producing. In this cycle, the generation of plant that produces spores is known as the **sporophyte generation,** while the generation of plant that produces the gametes is the **gametophyte generation.** If you look at **Figure 24.5**, you can see how physically different these generations can be. In a massive organism such as a redwood tree, the tree that is familiar to us is the sporophyte generation. But what do its spores develop into? On the male side, barely visible pollen grains. On the female side, small collections of cells, hidden deep within a redwood "pinecone." The fact that these gametophyte individuals are tiny, however, doesn't mean that they are unnecessary. The eggs and sperm they produce are required to bring about a whole new redwood tree. In sum, all plants go

24.2 Types of Plants

The physical characteristics we've been looking at are common to all plants, but within the plant kingdom there are four fundamentally different types of plants, each of which has its own unique set of physical features. These are the bryophytes, which include mosses; the seedless vascular plants, which include ferns; the gymnosperms, which include coniferous ("cone-bearing") trees; and the angiosperms, a vast division of flowering plants—by far the most dominant on Earth today—whose diversity is broad enough to include not only orchids and pear trees but corn and cactus as well (**Figure 24.6**). We'll look at all

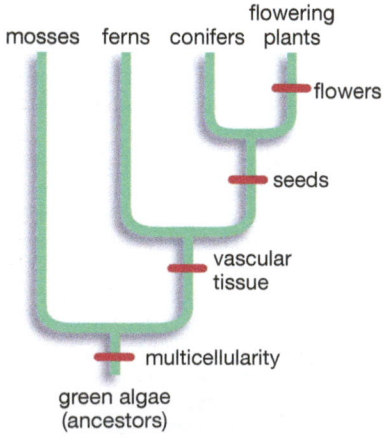

mosses ferns conifers flowering plants

flowers

seeds

vascular tissue

multicellularity

green algae (ancestors)

(a) Moss   (b) Ferns   (c) Conifers   (d) Flowering plants

Figure 24.6
Four Main Varieties of Plants

The most primitive type of plant (a) the bryophytes (moss in the figure) lack a fluid-transporting vascular system. (b) Ferns, representing the seedless vascular plants, have a vascular system but do not have seeds. (c) Gymnosperms (pine trees in the picture) do utilize seeds. (d) Flowering plants (blossoming trees in the picture) produce seeds and are responsible for several other plant innovations, among them fruit.

four types of plants briefly here. Chapter 25 will provide a detailed description of the anatomy and physiology of flowering plants.

Bryophytes: Lying Low in Moist Environments

If you look in low-lying, wet terrain, you are likely to see a carpet-like covering of moss. In countries such as Ireland, whole fields of peat moss grow and are harvested to be used as fuel for heating and cooking. Mosses are the most familiar example of a primitive type of plant called a **bryophyte,** which can be defined as a type of plant lacking a true vascular system (**Figure 24.7**). What's a vascular system? A network of tubes within an organism that serves to transport fluids. We can get a sense of what the bryophytes are like by looking at the mosses.

Like all bryophytes, mosses are close living relatives of the earliest plants that made the transition from living in water to living on land. As such, mosses can be conceptualized as plants that have made only a partial break with the aquatic living of the evolutionary ancestors of all plants, the green algae. In living on land, mosses have to deal with something the algae don't, which is the effect of gravity. Lacking a vascular system, they cannot transport water and other substances very *far* against the force of gravity; instead, they must lie low, hugging the surface to which they are attached while spreading out horizontally to maximize their exposure to sunlight. Nevertheless, mosses are hearty competitors in some tough environments, such as the arctic tundra and the cracks in sidewalks.

In keeping with their aquatic origins, mosses have sperm that usually get to eggs by swimming through water, although not much water is necessary. (A thin film left over from the morning dew will suffice.) For decades, scientists believed that this was the only means by which moss sperm got to eggs, but in 2006 they reported that the bodies of animals also carry out this transport function. Mites and tiny insects called springtails move moss sperm from one kind of moss "tuft" to another. This was a revelation because, up until this point, scientists thought that only one kind of plant used animals as sperm couriers—the flowering plants we'll be looking at shortly.

(a) Mosses

(b) Liverworts

(c) Hornworts

Figure 24.7
Three Kinds of Bryophytes

These plants have no vascular tissue and thus tend to be small. The sperm of bryophytes usually travels through water to get to eggs, although mites and tiny insects can serve this function. Bryophytes are found most often in moist environments.

Given the close ties that mosses have to water, it's not surprising that they can usually be found in moist environments. Like all bryophytes, mosses have no roots at all but instead use single-celled extensions called rhizoids to anchor themselves to their underlying material, which may be soil or rock or wood. You might expect that rhizoids would serve to absorb water, as roots do, but this is true only to a very limited extent. Bryophytes take in water almost entirely through their above-ground exterior surface.

Seedless Vascular Plants: Ferns and Their Relatives

As noted, all plants except the bryophytes have a vascular system, or network of fluid-conducting tubes that transport both food and water. When we begin to look at vascular plants, we find that the most primitive variety of them is a group called the **seedless vascular plants**: plants that have a vascular system but that do not produce seeds as part of reproduction.

Easily the most familiar representatives of the seedless vascular plants are ferns, with their often beautifully shaped leaves, called fronds (**Figure 24.8**). These plants have moved a step further in the direction of separation from an aquatic environment. Their vascular system allows them to grow up as well as out, and it allows for roots that extend into the ground, where they serve their absorptive function. Despite this evolutionary innovation, the sperm of the seedless vascular plants is like that of most of the bryophytes: It needs to move through water to fertilize eggs.

The First Seed Plants: The Gymnosperms

There are two kinds of seed plants, the gymnosperms and the angiosperms, but putting things this way makes it sound as though these plants are on a kind of equal footing with the bryophytes and seedless vascular plants. In reality, the gymnosperms replaced seedless vascular plants as the most dominant plants on Earth, although this event took place a long time ago—about the time the dinosaurs came to dominance. Then the gymnosperms lost this status to the angiosperms, which began flourishing about 80 million years ago. Today, there are only about 1,000 gymnosperm species

Figure 24.8
Three Kinds of Seedless Vascular Plants

Because these plants have vascular tissue, they are able to grow taller than most bryophytes. Like bryophytes, however, they do not produce seeds and are tied to moist environments.

(a) Ferns

(b) Horsetails

(c) Club mosses

compared to 300,000 angiosperm species. Nevertheless, the gymnosperm presence is considerable, especially in the northern latitudes, where they exist as vast bands of coniferous trees, including pine, fir, and spruce. Conifers such as these provide most of the world's lumber (**Figure 24.9**).

(a) Spruce tree

(b) *Ginkgo biloba*

Figure 24.9
Gymnosperms

(a) This spruce tree is a member of the grouping of gymnosperms known as conifers, which account for about three-fourths of all gymnosperm species. The conifers also include redwood, pine, juniper, and cypress trees.

(b) There are other types of gymnosperms as well. Pictured are the leaves and seeds of the maidenhair tree *Ginkgo biloba,* which are used today in herbal preparations.

Reproduction through Pollen and Seeds

Why were the gymnosperms able to out-compete so many of the seedless vascular plants back in the dinosaur days? Part of the answer lies in sperm transport. In gymnosperms, sperm are contained in tiny spheres called pollen grains. Critically, these grains are transported to the female eggs through the *air,* rather than being limited to swimming to them or being carried on the bodies of tiny, crawling animals. Given this innovation, the gymnosperms could propagate over great distances, which gave them a competitive advantage over the seedless plants.

Another advantage the gymnosperms had were their seeds, which can be thought of as tiny packages of food and protection. If you look at **Figure 24.10**, you can see one example of a gymnosperm seed, this one from a pine tree. **Figure 24.11**, on the next page, then shows you how this seed fits into the life cycle of the pine. As you can see, pollen grains are carried by the wind to the female reproductive structure, which resides in the familiar pinecone. Then the sperm within the pollen grain comes together with an egg in the reproductive structure, and the result is a fertilized egg that begins to develop as an embryo. This is no different in principle from human egg and sperm coming together to create a human embryo. In the pine tree, however, the embryo is developing inside a seed, a structure that is unique to plants. Seeds have a tough coat, inside of which is not only the embryo but also a food supply for it—a kind of sack lunch of stored carbohydrates, proteins, and fats. In sum, a **seed** is a plant structure that includes a plant embryo, its food supply, and a tough, protective casing. When seeds reach the ground, they contain all the starting materials necessary to bring about germination of a new generation of plant.

Gymnosperm seeds turn out to differ from the seeds produced by angiosperms. Angiosperm seeds come wrapped in a layer of tissue—called fruit—that gymnosperm seeds do not have. The details of this anatomical feature are presented in Chapter 25. For now, just

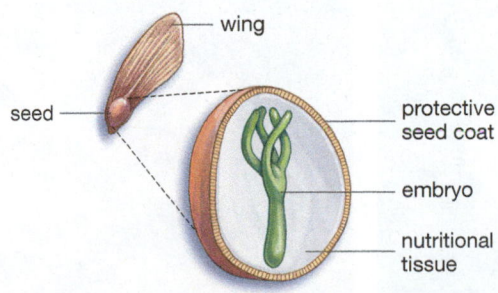

Figure 24.10
A Gymnosperm Seed

Seeds are tiny packages of food and protection. They come in many shapes and sizes, but they all contain an embryo, some food, and a protective seed coat.

Figure 24.11
The Life Cycle of a Gymnosperm

Pine trees have two kinds of cones. Pollen is produced within the smaller male cones, while eggs are produced within the larger female cones. When the wind carries pollen onto the female cone, the sperm within the pollen fertilizes one of the eggs within the cone. An embryo then begins to develop inside a seed, which falls to the ground. Once conditions are suitable, the seed germinates, and a whole new pine tree begins to grow.

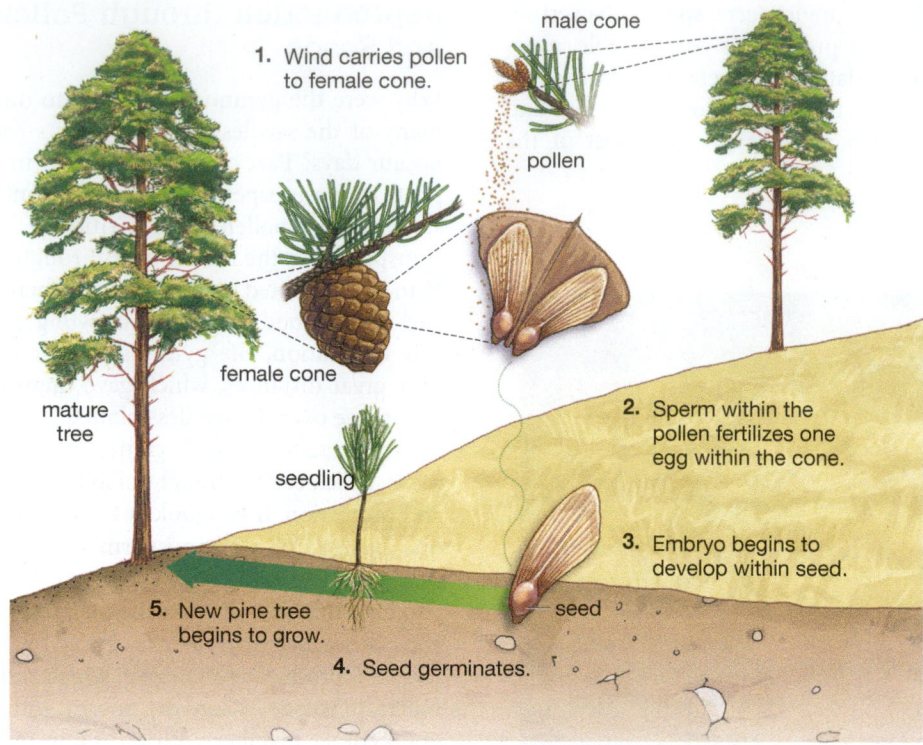

1. Wind carries pollen to female cone.

male cone

pollen

2. Sperm within the pollen fertilizes one egg within the cone.

3. Embryo begins to develop within seed.

mature tree

female cone

seedling

seed

5. New pine tree begins to grow.

4. Seed germinates.

be mindful that a **gymnosperm** can be defined as a seed plant whose seeds are not surrounded by fruit. The very name *gymnosperm* comes from the Greek words *gymnos,* meaning "naked," and *sperma,* meaning "seed."

Angiosperms: Nature's Most Dominant Plants

As noted, there are about 300,000 known species of the flowering plants, or angiosperms. The term *flowering plants* may bring to mind roses or tulips, and indeed, these flowers are angiosperms. But the angiosperm grouping includes all manner of other plants as well—almost all trees except for the conifers, all our important food crops, cactus, shrubs, common grass; the list is very long (**Figure 24.12**). As noted, **angiosperms** are defined by an aspect of their anatomy: Their seeds are surrounded by the tissue called fruit. The details of angiosperm anatomy and physiology are covered in Chapter 25. For the balance of this chapter, we'll focus on the ways these important plants interact with the world around them. First we'll look at their interactions with animals, after which we'll look at the ways they respond to signals they get from their environment. (To learn something about the nutrients that angiosperms need to stay healthy, see "What Is Plant Food?" on page 449.)

Figure 24.12
Angiosperm Variety

The incredible diversity of angiosperms can be seen in the differences among these three types of angiosperm plants.

(a) Calla lily on the California coast

(b) Cholla cactus in California

(c) Corn in a field

What Is Plant Food?

Plants are able to make their own food through photosynthesis if they have access to just four things: water, sunlight, carbon dioxide, and a few nutrients. Water, sunlight, and probably even carbon dioxide are no mystery to you by now, but what exactly are these nutrients?

A nutrient is simply a chemical element that is used by living things to sustain life. Recall from Chapter 2 that elements are the most basic building blocks of the chemical world. Silver is an element, as is uranium, but these are not nutrients because living things do not need them to live. In the broadest sense, carbon, oxygen, and hydrogen are nutrients. And it turns out that almost 96 percent of the weight of the average plant is accounted for by these three elements, which come to plants primarily from water and air.

Plants need at least another 13 elements to live, however, and not all of these are supplied by water and air. When we think of common "plant food," we are generally referring to those nutrients that plants can *use up* in their soil and that thus must be replenished to ensure continued growth and reproduction. Of these, the most important three are nitrogen, phosphorus, and potassium, whose chemical symbols are N, P, and K. When you look at a package of an average plant fertilizer, you will

see three numbers in sequence on it (for example, 10-20-10). What these refer to is the percentage of the fertilizer's weight accounted for by these nutrients (**Figure 1**).

Figure 1
Needed Nutrients
Plants make their own food from sunlight, water, carbon dioxide, and a few nutrients, some of which come from the soil. The most important of these nutrients are nitrogen (N), phosphorus (P), and potassium (K). Because garden plants and houseplants are isolated from natural ecosystems, they must often get these nutrients in the form of fertilizer.

The growth of plants usually is enhanced by a fertilizer that has equal ratios of the three elements, but if increased *blooming* is your goal, an increased phosphorus ratio generally is recommended, as with the 10-20-10 example.

It is not just houseplants or lawns that benefit from fertilizer, but farm crops as well. Indeed, the planting of a crop on a parcel of land generally requires a large investment in fertilizer because a crop such as wheat removes a great deal of N, P, and K from the soil in a single season. Historically, fertilizers used on farms were *organic,* meaning they came from decayed living things, such as the fish that Native Americans taught the Pilgrims to use when planting corn. Nowadays, however, commercial fertilizers generally are *inorganic,* meaning they are mixtures of pure elements within binding materials, produced by chemical processes.

Nature is, of course, perfectly capable of producing a green bounty in such places as forests and marshes without the aid of any human-made fertilizer. Decaying plant and animal matter puts nutrients back into the soil, and bacteria are able to fix nitrogen, meaning to absorb it from the air and transform it into a form that plants can take up. Looked at one way, however, houseplants and farm crops are *isolated* plants whose participation in this web of life is very limited, and thus they require a helping hand from humans in order to remain robust.

SO FAR . . .

1. A seed is a plant structure that includes a plant _____, its _____ supply, and a tough _____.

2. Of the following plants, which produce(s) seeds? (a) moss, (b) fern, (c) spruce tree, (d) orchid.

3. In their role as the male gametophyte generation of seed plants, pollen grains contain plant ____.

24.3 Angiosperm–Animal Interactions

For an angiosperm such as a new honeysuckle plant to come into existence, a two-part process must take place. First, sperm—developing once again inside a pollen grain—must get to an egg and fertilize it, which produces the embryo encased in a seed coat. Next, this seed must land on a patch of soil somewhere and begin to germinate or sprout.

Now, in the angiosperms, how is it that sperm gets to the egg in the first place? Remember that with gymnosperms, pollen grains are carried by the wind. By

Figure 24.13
Animals Help Pollen Get from Here to There
Flowering plants take advantage of the mobility of animals to transport pollen from one plant to another. Relationships between flowers and pollinators can be species specific, which helps to ensure that the pollen will be delivered to the right address.

(a) Pollen-covered honeybee on a dandelion

(b) Ruby-throated hummingbird pollinating a flower

(c) Lesser long-nosed bat pollinating a cactus

contrast, the angiosperms have induced *animals* to carry pollen from one plant to another—to pollinate them. Not all angiosperms are pollinated in this way; some are pollinated by wind, and a few aquatic species rely on water currents. But most are pollinated with the help of animals, be they bees, birds, or even bats (**Figure 24.13**).

Angiosperms use various attractants to encourage animal pollination. The most important of these is food in the form of nectar (which is essentially sugar-water). But angiosperms don't stop there. So critical is pollination to them that evolution produced within them a set of physical characteristics that effectively serve as advertising for pollinators. How about a fragrance that, say, bees are sensitive to? How about colors that they're attracted to? Insect-pollinated flowers generally have both sweet fragrances and specialized coloration patterns—nature's equivalent of homing signals and landing lights. These serve to get an animal to a flower in the first place and then into the right position for feeding and pollination (**Figure 24.14**).

Why do angiosperms spend so much energy attracting animal pollinators while gymnosperms let the wind do the work? Windborne pollination in gymnosperms can be thought of as a kind of scattershot approach: If the wind happens to blow a pollen grain onto the female reproductive structure, then pollination occurs. Think how much more directed things are with animal pollination. It is essentially door-to-door service. Animal pollination is one of the reasons that angiosperms are by far the most dominant plants on Earth.

Seed Endosperm: More Animal Food from Angiosperms

Once pollination has occurred, the resulting seed, with the embryo inside, has to reach the ground and then germinate in it. All seeds contain food reserves

Figure 24.14
Food Lies This Way
Animals may be attracted to a particular flower by its fragrance, color, or pattern of visual elements. Insects are guided into the nutrients in the center of the flower by color patterns called nectar guides, visible to an insect—because it perceives ultraviolet light—but not to us.

(a) This is what we see (normal sunlight).

(b) This is what the bee sees (ultraviolet light).

(c) This is what the bee "thinks."

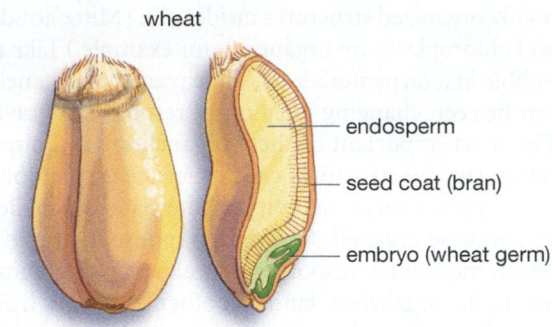

wheat

endosperm

seed coat (bran)

embryo (wheat germ)

Figure 24.15
Food for the Seedling, Food for Humans
Endosperm is a food reserve in angiosperm seeds. In nature, its function is to feed a growing plant embryo in the period before the embryo germinates into a grown plant and starts making its own food through photosynthesis. It is this same endosperm that nourishes humans throughout the world in the form of wheat, corn, rice, and even coconut.

for the growing embryo, but angiosperm seeds develop a special kind of nutritive tissue, called **endosperm,** that often surrounds the embryo (**Figure 24.15**). As it turns out, this endosperm supplies much of the food for *human beings* throughout the world. Rice and wheat grains are seeds consisting in large part of endosperm whose role in nature is to sustain the plant embryo. It's worth noting that whole-grain wheat is wheat that still has its embryo (usually called "wheat germ") and its seed coat (usually called "bran"). By contrast, white bread, regular pasta, and so forth are made from wheat whose wheat germ and bran have been removed through processing. The problem with such processing? Wheat germ is a source of healthy oils and vitamin E, while bran is a terrific source of "fiber" or plant material that doesn't get digested within us and thus provides a host of health benefits. (You can read more about this in Chapter 31.)

Fruit: An Inducement for Seed Dispersal

Some flowering plants also add an inducement for the *dispersal* of their seeds: the tissue called fruit. This is not fruit in the technical sense, but fruit as that term is commonly understood—the flesh of an apricot or cherry, for example. As such, it represents one more piece of bounty that plants provide to animals. To understand how fruit benefits plants as well, imagine a bear that consumes a wild berry, which consists not only of the fruit flesh but of the seeds inside this flesh. The fruit is digested by the bear as food, but the *seeds* are tough; they are passed through its system intact to be deposited, with bear feces as fertilizer, at a location that may be very remote from the place where the bear ate the berry. The result is seed dispersal at what may be a promising new location for a berry plant (**Figure 24.16**).

If we take a step back from processes such as this, it's easy to see that many animals and angiosperms have an interdependent relationship with one another. Angiosperms supply food *to* animals, but they also derive benefits *from* animals in the forms of pollination and seed dispersal. We don't often think of animals and plants as partners in the living world, but that's the case to a high degree with animals and angiosperms.

24.4 Responding to External Signals

Plants may lack a nervous system, but this doesn't mean they can't respond to their environment. They must do so, actually, because their life depends on it. They sense the march of the seasons by measuring available light and the surrounding temperature; they use internal clocks to synchronize their development

(a) Wild berries induce seed dispersal

(b) Burrs help seeds travel

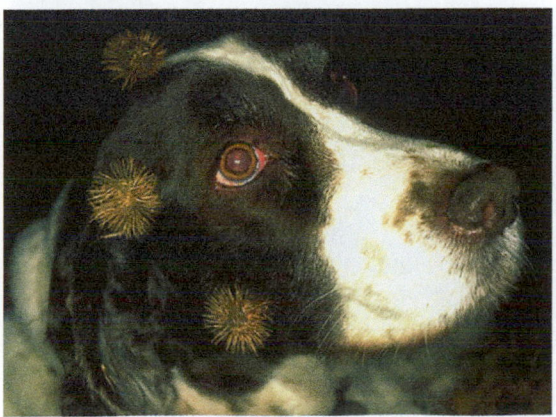

Figure 24.16
Seed Carriers
Angiosperms have taken advantage of the mobility of animals not only to transfer pollen but also to disperse seeds.

(a) Some seeds are wrapped in tasty fruit that is consumed by animals, such as bears.

(b) Other seeds come wrapped in burrs and spines that stick to the fur of animals and are carried away.

and physical functioning; they respond to touch, heat, cold, and pH, among other things. Let's look now at some of the ways the angiosperm plants respond to the world around them. We'll begin with the way they respond to a very familiar force.

Responding to Gravity: Gravitropism

When a seed first starts to sprout, or "germinate," its roots and shoots must go in very specific directions: The root must go down, toward the water and minerals the plant needs, and the shoot must go up, toward the sunlight it requires. Imagine a seed that has simply fallen on the ground; it would not do for either root or shoot to grow *horizontally;* each must grow vertically. If you look at **Figure 24.17**, you can see dramatic evidence of a plant's ability to sense which way is up. The bean plant in Figure 24.17 was placed on its side, and in less than a day its shoot began curving upward, as you can see. What this exemplifies is **gravitropism,** meaning a bending of a plant's shoot or root in response to gravity. But how do shoots or roots "know" which way is up or down, and how do they orchestrate a course correction when they need one? Logically, there must be at least two elements at work here: first, a gravitational *sensing* mechanism and then a means of *responding* to gravitational cues, such that corrective bending points shoot or root the right way.

The consensus among botanists today is that the sensing mechanism in gravitropism is the "sedimentation" that various plant cell organelles perform in response to gravity. Recall that organelles are small but highly organized structures inside cells. (Mitochondria and chloroplasts are organelles, for example.) Like the bubble in a carpenter's level, some plant-cell organelles can be seen changing position in response to gravity. The most important of these organelles is a group of starch-storing structures called *amyloplasts* (although some plants seem to employ alternate organelles). When these organelles move in response to gravity, it sets in motion the responding mechanism in gravitropism. The organelles "land" on other structures inside the cell (such as the cytoskeleton), and the resulting impact triggers a redistribution of a **hormone:** a substance that, when released in one part of an organism, goes on to prompt physiological activity in another part of that organism. In this case, the hormone is known as IAA, and its redistribution to a different portion of the plant has the effect of bringing about *growth* in that portion of the plant—one *side* of the plant stem or root will grow more than the other side. With this growth, a stem that is oriented horizontally will start bending up toward the sky, while a root that is oriented horizontally will start bending down toward the ground.

Responding to Light: Phototropism

Many plants will bend toward a source of light, with the value of this perhaps being obvious: Plants produce their own food, and that production depends on light. Thus, a plant needs to respond when its sunlight becomes blocked by another plant or some other physical object in its surroundings. As it turns out, it is once again the hormone IAA—this time produced in the shoot tips of growing plants—that controls this **phototropism,** defined as a curvature of shoots in response to light. When light strikes one side of the shoot, it causes IAA to migrate to the other side, where the IAA acts to promote the *elongation* of cells on this far side. The effect is to make the shoot curve toward the light (**Figure 24.18**).

Responding to Contact: Thigmotropism

We've all seen plants that manage to climb upward by encircling the stem of another plant. Whole stems can undertake such encircling, but it is often a thin, modified leaflet called a tendril that does so (**Figure 24.19**). But how is a tendril able to wrap around another object? Once again, through differential growth—more rapid growth on one side of a tendril than on the other. Contact with the object is perceived by outer, or epidermal, cells in the tendril. This sets into motion the differential growth, which probably is controlled by the hormones IAA and ethylene. This process is called **thigmotropism,** meaning growth of a plant in

Figure 24.17
Plants Respond to Gravity
A potted bean plant demonstrates the effects of gravitropism.

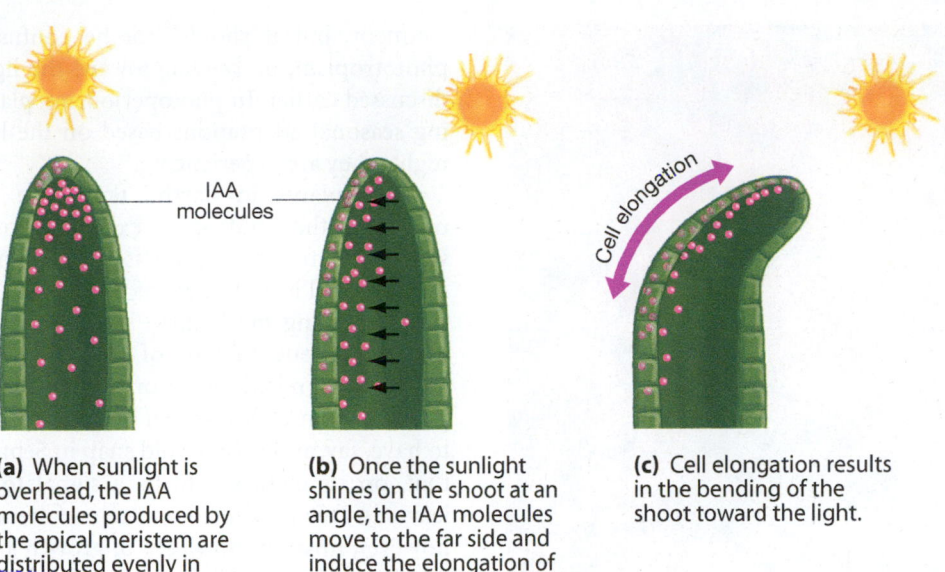

Figure 24.18
Plants Respond to Light

(a) When sunlight is overhead, the IAA molecules produced by the apical meristem are distributed evenly in the shoot.

(b) Once the sunlight shines on the shoot at an angle, the IAA molecules move to the far side and induce the elongation of cells on that side.

(c) Cell elongation results in the bending of the shoot toward the light.

response to touch. Plants with this capability can piggyback on plants or other objects to get more access to sunlight.

Responding to the Passage of the Seasons

Trees that exhibit a coordinated, seasonal loss of leaves are called **deciduous** trees. But why should these broad-leafed trees lose their leaves—their

Figure 24.19
Moving Up by Curling Around
This dodder plant is able to curl around its host plant because of thigmotropism—a plant's ability to grow in response to touch.

primary sites of photosynthesis—while the evergreen pines and firs keep the modified leaves known as needles?

First, there is relatively little water available in winter in cold climates (most of it being frozen in the ground), and trees are just like all vascular plants in that most of the water they absorb through their roots ends up being released into the air through their leaves. Indeed, it's been estimated that 99 percent of the water the average plant takes in through its roots eventually will move out through its leaves (as water vapor) in the process known as transpiration. For our purposes, the important point is that flat-leafed deciduous trees transpire more than evergreens do. In addition, because of the way wind flows over flat leaves, they lose more heat than do the pine needles of evergreens, meaning deciduous leaves could freeze in winter if they were retained—a condition that would damage them and hence result in a tree that was vulnerable to infection by fungi and bacteria.

For deciduous trees, then, any benefits that would come with performing photosynthesis in winter would not be worth the water loss and leaf damage that would result. These trees thus employ a strategy that is a matter of straight economics: Lose the leaves in fall, grow new ones the following spring, and in the meantime perform no photosynthesis at all but exist instead completely on food reserves. This state, in which growth is suspended and metabolic activity is low, is called **dormancy**. Evergreens can exhibit dormancy, too, in low temperatures, but deciduous trees are locked into it until the coming of spring.

Limitation of Loss in the Deciduous Strategy

The dormancy that deciduous trees enter into makes even more sense when we consider one means by which these trees prepare for it: Before losing their

Phototropism

Figure 24.20
Seasonal Marker
Brightly colored maple and birch leaves in Ontario, Canada in fall. Plants have a number of mechanisms that allow them to respond to the passage of the seasons.

leaves, they reclaim the proteins and nutrients that exist in them. Proteins are broken down by enzymes and shipped, along with the nutrients, to storage cells in a tree's trunk and roots. By the time a leaf is ready to detach from the tree, it is little more than an empty shell of cell walls.

Before this happens, however, the well-known change in leaf coloration takes place. It may surprise you to learn there is less adding of new colors in this process than an unmasking of colors that existed in the leaves all along. With the approach of fall, green chlorophyll, which is the main pigment in photosynthesis, begins to break down in the leaf. With it gone, the yellow and orange colors provided by pigments called carotenoids become visible (**Figure 24.20**). New colors are added as well; blue and red come with the synthesis of pigments called anthocyanins. (These pigments are largely responsible for the stunning red color of many maples in the fall.) The combination of carotenoids and anthocyanins gives us the final result, which is multi-hued leaves.

Photoperiodism

But how does a deciduous tree sense when it's time to begin preparing for cold weather? In general, it is an interaction between two factors. First, there is cold itself, which brings about changes in metabolism. In addition, there is a phenomenon known as **photoperiodism,** which is the ability of a plant to respond to changes it is experiencing in the daily duration of darkness, relative to light. This is another example of the plant's ability to respond to its environment, but it should not be confused with the phototropism, or *bending* toward the light, that was discussed earlier. In photoperiodism, plants are making seasonal adaptations based on the length of the nights they are experiencing.

Most plants in Earth's temperate regions and many in the subtropics exhibit photoperiodism, which can affect such processes as the timing of flowering as well as the onset of dormancy. The value of such a sensing mechanism becomes apparent when we think about the kind of trouble plants could get into if they relied solely on temperature signals to provide them with seasonal cues. It's entirely possible to have, say, an unusual cold snap in September—one that, on its own, would be a signal to start losing leaves. A tree that acted on such a false signal would, however, miss out on weeks of growth and food production. In contrast, seasonal dark/light cues are much more reliable—there is no such thing as an unusual *light* condition for a given month. Given this, it makes sense that photoperiodism has an important place in providing plants with a sense of the seasons as they pass.

To get a feel for how photoperiodism works, consider ragweed, a plant that is the bane of many hay-fever sufferers. After sprouting in the spring and reaching a certain level of maturity, a ragweed flowers only when it is in darkness for 10 hours *or more* per day. When would this threshold be crossed? When nights begin to get longer. Ragweed thus joins a group of plants, called long-night plants, whose flowering comes only with an *increased* amount of darkness. As you can imagine, most of these plants are like ragweed in that they flower only in late summer or early fall. Their counterparts are the so-called short-night plants, which flower only when the nights get short enough—that is to say, when nighttime hours are *decreasing* as a proportion of the day. Predictably, these plants tend to flower in spring to midsummer.

The mechanism by which photoperiodism works is now fairly well understood in one short-night plant, a member of the mustard family called *Arabidopsis.* Flowering in *Arabidopsis* is controlled by a protein, known as CO, that can switch on genes by binding to DNA. When enough CO accumulates in *Arabidopsis,* the genes that set flowering in motion are switched on. So, what can bring about CO levels that are high enough to get the flowering started? Shorter nights. Substances called *photoreceptors,* which are sensitive to light, act mostly to allow CO to accumulate during the day, but then predominantly act to destroy it at night. As the nights grow shorter, the balance is shifted toward CO accumulation. When this accumulation reaches a certain threshold, the relevant genes are switched on and flowering begins.

On to a More Detailed Picture of Angiosperms

This chapter has served to introduce you to the basic features of plants—their general traits, their varieties, and the means by which the flowering plants among them interact with animals and respond to environmental signals. We've hardly scratched the surface, however, regarding two important aspects of flowering plant life: the nature of their component parts and the ways these parts interact to produce a functioning organism. Put another way, we've yet to look at the anatomy and physiology of angiosperms. That's just what we'll do in the chapter that's coming up.

SO FAR . . .

1. The pollen of gymnosperms generally is transported by _____, while the pollen of angiosperms generally is transported by _____.

2. Plants respond to gravity, touch, and the direction of sunlight through the _____ growth of roots or stems.

3. Plants in temperate climates make seasonal adjustments based on the relative length of the _____ they are experiencing.

CHAPTER

24 REVIEW

Summary

24.1 The Roles and Characteristics of Plants

- The photosynthesis that plants perform indirectly feeds many life-forms. The oxygen plants produce as a by-product of photosynthesis is vital to many organisms, and many plant products are important to human beings. (p. 441)

- With a few exceptions, all plants are multicelled, fixed in one spot, and carry out photosynthesis. All plant cells have a cell wall and most contain the organelles called chloroplasts, which are the sites of photosynthesis. (p. 442)

- Plants reproduce through an alternation of generations: a life cycle in which one plant generation reproduces through use of the reproductive cells called spores (the sporophyte generation) while the next reproduces through use of the reproductive cells known as gametes (the gametophyte generation). The sporophyte and gametophyte generations

of different plant types can differ greatly in size and structure. (p. 443)

24.2 Types of Plants

- The four principal categories of plants are bryophytes (exemplified by mosses), seedless vascular plants (ferns), gymnosperms (coniferous trees), and angiosperms (a wide array of plants). (p. 444)

- Bryophytes are close relatives of the earliest plants that made the transition from living in water to living on land. They lack a fluid transport, or vascular system and tend to be low-lying. Bryophytes tend to inhabit damp environments and their sperm get to eggs primarily by swimming through water. (p. 445)

- Seedless vascular plants have a vascular system but do not produce seeds in reproduction. Their sperm must move through water to fertilize eggs. (p. 446)

- Gymnosperms are seed-bearing plants whose seeds are not encased in tissue called fruit. There are only about 1,000 gymnosperm species, but their presence is

considerable, particularly in northern latitudes, where gymnosperm trees, such as pine and spruce, often dominate landscapes. In carrying out their reproduction, gymnosperms produce seeds: reproductive structures that include a plant embryo, its food supply, and a tough, protective casing. (p. 446)

- The most dominant group of plants, angiosperms, produce seeds that are encased in fruit. Angiosperm species include plants with flowers and most trees, except for conifers; all important food crops; cactus; shrubs; and common grasses. (p. 448)

24.3 Angiosperm–Animal Interactions

- Angiosperm pollen grains generally are transferred from one plant to another by animals. Flowers produce nectar, colorations, and fragrances that induce animals to carry out pollination. (p. 449)

- Angiosperm seeds contain endosperm, which functions as food for the growing embryo. Endosperm supplies much of the food that human beings eat. (p. 450)

- Angiosperm seeds are unique in the plant world in being wrapped in a layer of tissue called fruit. Edible fruit functions in seed dispersal because animals will eat and digest the fruit but then later excrete the tough seeds inside it, often in a different location. (p. 451)

24.4 Responding to External Signals

- Gravitropism is the ability plants have to sense their orientation and to direct their roots toward the ground and shoots toward the sky. (p. 452)

- Plant shoots can gain additional access to light by bending toward a source of it. This process is called phototropism, meaning a curvature of shoots in response to light. (p. 452)

- Thigmotropism is the growth of a plant shoot in response to touch. Through it, plant shoots can climb upward on other objects by encircling them, thus gaining additional access to light. (p. 452)

- Gravitropism, phototropism, and thigmotropism are made possible by differential growth on one side of a root or stem. (p. 452)

- In temperate climates, deciduous trees exhibit a coordinated, seasonal loss of leaves and enter into a state of dormancy. Dormancy evolved in deciduous trees because any photosynthesis they could perform in winter would not offset the water loss and leaf damage that would result from the retention of leaves. (p. 453)

- Photoperiodism is the process through which plants respond to seasonal changes in the daily duration of darkness relative to light. Some plants that exhibit photoperiodism are long-night plants, meaning those whose flowering comes only with an increased amount of darkness. Others are short-night plants, meaning those whose flowering comes only with a decreased amount of darkness. (p. 454)

Key Terms

alternation of generations   444
angiosperm   448
bryophyte   445
cell wall   442
chloroplast   442
deciduous   453
dormancy   453
endosperm   451
gametophyte generation   444
gymnosperm   448
gravitropism   452
hormone   452
photoperiodism   454
phototropism   452
seed   447
seedless vascular plant   446
sporophyte generation   444
thigmotropism   452

Understanding the Basics

Multiple-Choice Questions

(Answers are in the back of the book.)

1. Plants (select all that apply):
   a. obtain nutrition by consuming existing organic material.
   b. obtain nutrition by performing photosynthesis.
   c. produce much of Earth's oxygen.
   d. absorb atmospheric carbon dioxide.
   e. are a cause of soil erosion.

2. Defining characteristics of plants include (select all that apply):
   a. multicelled
   b. hyphae
   c. cell walls
   d. reproduction through fruiting bodies
   e. land-dwelling

3. In the alternation of generations seen in plants (select all that apply):
   a. one generation produces flowers, the next does not.
   b. one generation produces spores, the next produces eggs and sperm.

   c. succeeding generations tend to be about the same size.
   d. succeeding generations tend to differ in size.
   e. all generations produce spores.

4. Match the plant type on the left with the plant category on the right.
   a. moss   angiosperm
   b. cactus   gymnosperm
   c. pine tree   angiosperm
   d. rose   bryophyte
   e. fern   seedless vascular plant

5. Only angiosperms (select all that apply):
   a. perform photosynthesis.
   b. produce seeds.
   c. encase seeds in fruit.
   d. have seeds containing endosperm.
   e. have eggs encased in pollen grains.

6. Pollen grains contain:
   a. hormones.
   b. food for the new plant generation.
   c. eggs that will be fertilized by sperm.
   d. sperm.
   e. chloroplasts.

7. Phototropism and gravitropism lead to the bending of shoots or roots toward or away from a stimulus (light or gravity, respectively). Such bending is brought about by:
   a. increased growth over the general surface of the root or shoot.
   b. differential growth on opposite sides of the root or shoot.
   c. muscular contractions, leading to a curvature in the root or shoot.
   d. loss of water into the atmosphere.
   e. all of the above.

8. Photoperiodism allows some plants to respond to _____ by timing such things as _____ in accordance with the amount of _____ they are exposed to.
   a. predators; dormancy; moisture
   b. competition; growth; moisture
   c. the seasons; growth; toxins
   d. other plants; flowering; IAA
   e. the seasons; flowering; darkness

Brief Review

(Answers are in the back of the book.)

1. Distinguish between the sporophyte and gametophyte generations of plants.

2. How does a seed differ from a gamete?

3. Describe two means by which angiosperm reproduction is aided by animals.

4. Distinguish among gravitropism, phototropism, and thigmotropism.

5. What factor in a plant's external environment controls photoperiodism?

Applying Your Knowledge

1. As readers of the evolution section of this book know, a lack of genetic diversity can be harmful to a population of living things (see Chapter 17, page 308, for more). Can you think of an aspect of plant functioning that (a) has grown more sophisticated over evolutionary time and that (b) helps ensure genetic diversity among plants?

2. All plant cells have a relatively thick and often strong cell wall, but no animal cells have such a feature. Why does it make sense that plant and animal cells should differ in this way?

3. How are the bryophytes of the plant kingdom like the amphibians of the animal kingdom?

MasteringBiology®

1. MasteringBiology Assignments
Activities: Phototropism; **Building Vocabulary;**
Questions: Reading Questions, Test Bank

2. eText
Read your book online, search, take notes, highlight text, and study more effectively.

3. The Study Area
Self-study tools to test comprehension.

The Angiosperms:
Form and Function in Flowering Plants

Plants are fixed in one spot and have almost no moving parts, yet they manage to reproduce, transport fluids inside themselves, and keep growing for as long as they live.

Available on MasteringBiology®
Activities: Leaves, Reproduction, and Fluid Transport; Plant Reproduction; **Bioflix:** Water Transport in Plants; **Building Vocabulary; Questions:** Bioflix Quiz, Reading Questions, Test Bank

The flowering plant sometimes called the chinche-rinchee (*Ornithogalum thyrsoides*) is native to South Africa. Its flowers grow on a stem that can be up to three feet long.

Insects that eat plants are not much of a surprise, but plants that eat insects? About 500 species of flowering plants are carnivorous, with at least one variety, the tropical pitcher plant *Nepenthes*, able to capture animals up to the size of a small bird. But the most famous plant predator of them all undoubtedly is the Venus flytrap (*Dionaea muscipula*), which is native to the wetlands of North and South Carolina. So, how does this plant, which has no muscles and is fixed in one spot, end up consuming insects?

Each trap is a single leaf folded in two, with the interlocking "fingers" of the two halves able to snap together in less than half a second, imprisoning any small wanderer (**Figure 25.1**). But what keeps the leaf halves from wasting their energy by snapping together in pursuit of wind-borne blades of grass or other objects? The chemical reaction that puts the trap in motion is set off by a series of trigger hairs on the inside of each of the leaf's two sections. The trick is that *two* trigger hairs must be tripped for the halves of the leaf to snap shut. Insects wandering around inside the leaf are likely to trip two hairs, but blades of grass are not. And how does the plant slam its "jaws" shut without muscles? When the trigger hairs are tripped, they set off a chemical reaction that engorges cells at the base of the leaf with water. This rapid movement causes the leaf halves to come together. Enzymes are then released that begin to digest the struggling insect.

It may surprise you to learn that Venus flytraps make their own food through photosynthesis, just like any other plant. So, why do they need these side orders of insects? For nutritional supplements. Like many other carnivorous plants, Venus flytraps grow naturally in mineral-poor soil, and the insects they catch are a good source of nitrogen and phosphate.

The world of plants is filled with species as unique in their own way as the Venus flytrap. Queen Victoria water lilies (*Victoria regia*) have circular leaves that can be nearly 6 feet in diameter and that can serve as a floating platform for a person if pressed into service (**Figure 25.2** on the next page). The seeds of orchids are so tiny that they resemble dust more than seeds, while the seeds of the double coconut come wrapped in a fruit that might be 2 feet across.

Given such diversity, it may surprise you to learn that the orchid and the coconut are united with the water lily and the Venus flytrap in being members of a single group of plants—the flowering plants that are the subject of this chapter. Readers of Chapter 24 may recall that there are four principal types of plants in the living world: the bryophytes, represented by mosses; the seedless vascular plants, represented by ferns; the gymnosperms, represented by coniferous (evergreen) trees; and the flowering plants, which also are known as angiosperms. Chapter 24 introduced all

Figure 25.1
One with Prey, One without
The venus flytrap leaf in the foreground has slammed its two halves together, trapping an insect inside. Meanwhile the leaf in the background will stay open until such time as an insect enters.

Figure 25.2
Floating Platforms

A woman plays a violin while standing atop a Victoria water lily at the Missouri Botanical Garden in St. Louis at the turn of the twentieth century. The lilies are native to South America. Some South American Indians call the lilies *Yrupe*, which can be translated as "big water tray."

One problem in recognizing how dominant these plants are has to do with their name: The term *flowering plant* is likely to bring to mind nothing more than plants that have recognizable flowers on them—say, roses or tulips. These plants are indeed angiosperms, but food crops such as rice and wheat are angiosperms, too, as are all cacti, almost all the leafy trees, innumerable bushes, ice plant, cotton plants—the real challenge is to name familiar plants that *aren't* angiosperms (**Figure 25.3**). In this chapter, you'll get an overview of how angiosperms are structured and how they function. A formal definition of them will be provided once you've learned a little about them.

25.1 The Structure of Angiosperms

Let's begin our tour of the angiosperms by looking at their component parts.

The Basic Division: Roots and Shoots

We'll first look at the larger-scale structures of the angiosperms, starting with a simple, two-part division that rhymes: roots and shoots. Plants live in two worlds, air and soil, with their root system below ground and their shoot system above it (**Figure 25.4**).

The function of the root system is straightforward: Grow toward water and minerals, absorb these substances from the soil, and then begin transporting

Figure 25.3
Angiosperm Variety

(a) A trumpet vine (*Clytostoma callistegiodes*) spreading on a bush.

(b) Saguaro cacti in Arizona's Saguaro National Park.

(c) The cereal grain rye (*Secale*) growing in a field.

four varieties of plants, but in this chapter we'll focus strictly on flowering plants. Why pay so much attention to just one type of plant? Because of the overwhelming dominance of these plants. There are about 16,000 species of bryophytes, 13,000 species of seedless vascular plants, and 1,000 species of gymnosperms in the world today, but there are an estimated 300,000 species of flowering plants.

(a)

(b)

(c)

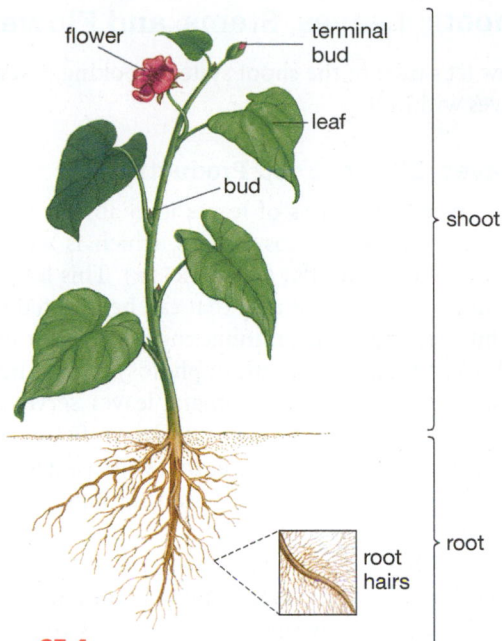

Figure 25.4
A Life-Form in Two Parts
Flowering plants live in two worlds, with their shoots in the air and their roots in the soil. Although flowering plants differ enormously in size and shape, they generally possess all the external features shown.

(a) Taproot system

(b) Fibrous root system

Figure 25.5
Two Root Strategies
(a) Dandelions such as this one employ a taproot system—a large central root and a number of smaller lateral roots.
(b) The chrysanthemum plant employs a fibrous root system—a collection of roots that are all about the same size.

them up through the rest of the plant. Most roots also serve as anchoring devices for plants, and some act as storage sites for food reserves.

The function of the shoot system is more complex. Photosynthesis takes place in this system, primarily in leaves, which means leaves must be positioned to absorb sunlight. Plants are in competition with one another for sunlight, which is a primary reason so many plants are tall. Once food is produced in photosynthesis, it must be distributed throughout the plant, starting with the shoots. But it is not just plant food production that is centered in the shoot system; it is plant reproduction as well. Within flowers and their derivatives, we find all the components of reproduction—seeds, pollen, and so forth. Now let's look in more detail at the components of the root and shoot systems.

Roots: Absorbing the Vital Water

If you look at **Figure 25.5**, you can see pictures of the two basic types of plant root systems, one a **taproot system** consisting of a large central root and a number of smaller lateral roots, and the second a **fibrous root system** consisting of many roots that are all about the same size. **Figure 25.6** then shows you another feature common to root systems: **root hairs,** thread-like extensions of roots that greatly increase their absorptive surface area. Each root hair actually is an elongation of a single outer cell of the root. Between roots and root

hairs, the root structure of a given plant can be extensive. One famous analysis, conducted in the 1930s on a rye plant, concluded that its taproot and lateral roots alone, if laid end to end, would have stretched out for some 386 miles; meanwhile, this same plant had almost 6,600 miles of root hairs! This from a plant whose *shoot* stood 3 inches off the ground.

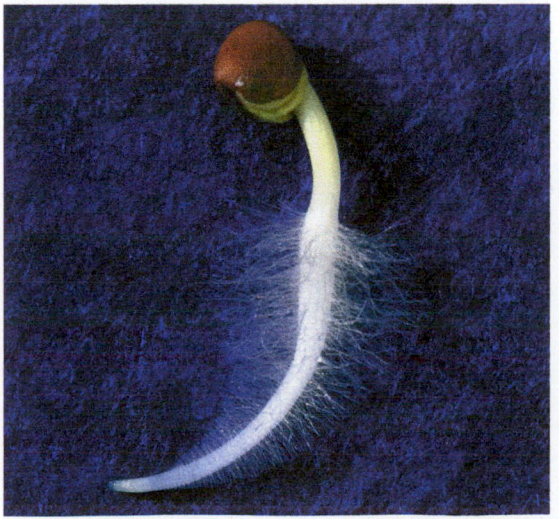

Figure 25.6
Root Hairs
Root hairs enormously increase the surface area of a root. The greater the surface area, the more fluid absorption can occur. Shown here are root hairs on the seedling of a radish.

Leaves, Reproduction, and
Fluid Transport

Why do plants lavish such resources on their roots? Essentially because they have such a great need to take up water, and roots are the structures that allow them to do this. To perform photosynthesis, plants must have a constant supply of carbon dioxide, which enters the plants through microscopic pores, called *stomata*, that exist mostly on the undersides of their leaves. Open stomata, however, don't just let carbon dioxide in; they let the plant's water *out*, as water vapor. To accommodate this loss while still keeping vital tissues moist, plants continually pass water through themselves, from roots, up through the stem, and into the leaves (at which point it exits as water vapor). This process of water movement out of the plant is known as **transpiration;** through it, more than 90 percent of the water that enters a plant through its roots evaporates into the atmosphere. And the scale of this operation can be immense: The roots of a single tall maple tree can absorb nearly 60 gallons of water per *hour* on a hot summer day. Given this, it's easy to see why plants spend so much energy on development of their roots: their basic metabolism makes it a necessity. And the nutrient storage function of roots that was mentioned earlier? We human beings benefit from it along with plants—both sweet potatoes and carrots are roots.

Shoots: Leaves, Stems, and Flowers

Now let's turn to the shoot system, looking first at the leaves within it.

Leaves: Sites of Food Production

The primary business of leaves is to absorb the sunlight that drives photosynthesis, which is why most leaves are thin and flat (**Figure 25.7a**). This leaf shape maximizes the surface area that can be devoted to absorbing sunlight while minimizing the number of cells in leaves that are irrelevant to photosynthesis. Beyond this, through their tiny stomata, leaves serve as the plant's primary entry and exit points for gases. As noted, the most important gas that's entering is carbon dioxide, which is one of the starting ingredients for photosynthesis. What's exiting is the by-product of photosynthesis, oxygen, along with the water vapor.

The broad, flat leaves that are so common in nature have in essence a two-part structure—a **blade** (which we usually think of as the leaf itself) and a **petiole,** more commonly referred to as the leaf stalk. If you look at the idealized cross section of this leaf in **Figure 25.7b**, you can see that the blade can be likened to a kind of cellular sandwich, with layers of cuticle, or waxy outer covering, on the outside, and a layer of epidermal cells just inside them. In the leaf's interior, there are vascular

Figure 25.7
Site of Photosynthesis

(a) Leaves tend to be broad and flat to increase the surface area exposed to sunlight.

(b) Photosynthesis occurs primarily within the mesophyll cells in the interior of the leaf. The vascular bundles carry water to the leaves and carry the product of photosynthesis, sugar, to other parts of the plant. Pores called stomata, mostly on the underside of leaves, allow for the passage of gases in and out of the leaf.

(c) Guard cells of the stomata control the opening and closing of the stomata. When sunlight shines on the leaf during the day, the guard cells engorge with water that makes them bow apart, thus opening the stomata. When sunlight is reduced, water flows out of the guard cells, causing the stomata to close.

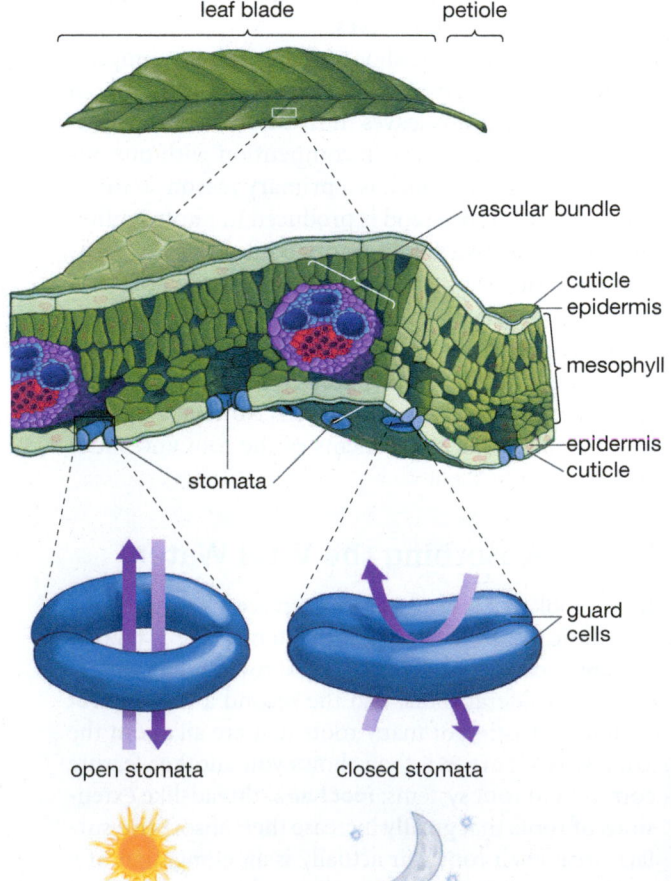

compound
leaves

simple
leaves

leaves modified
as spines

highly reduced leaves
(needles)

leaves modified
as tendrils

Figure 25.8
Leaves Come in Many Sizes, Shapes, and Colors
Simple leaves have just one blade. In compound leaves, the blade is divided into little leaflets. Some leaves are so modified that the average person can hardly recognize them as leaves—for example, the spines of a cactus.

bundles, which bear some relation to animal veins in that they are part of a fluid transport system. Then there are several layers of mesophyll cells. It is these cells that are the sites of most photosynthesis.

Now note, on the underside of the blade, the openings called **stomata.** These are the pores, mentioned earlier, that let water vapor out and carbon dioxide in—but for most plants only during the day. When the sun goes down, photosynthesis can no longer be performed, and it is not cost-effective for a plant to lose water without gaining carbon dioxide. As such, in most plants the stomata close up until photosynthesis begins again the following day. If you look at **Figure 25.7c,** you can see how this opening and closing of stomata is achieved: Two "guard cells," juxtaposed like the sides of a coin purse, are arranged around the stomata. When sunlight strikes the leaf, these cells engorge with water, which makes them bow apart. Then, in the absence of light, water flows out and the door closes again. A given square centimeter of a plant leaf may contain from 1,000 to 100,000 stomata.

Figure 25.8 gives you some idea of the varied forms of leaves. In addition to the so-called simple leaf we've been looking at, there are compound leaves in which individual blades are divided into a series of "leaflets." Pine needles and cactus spines are likewise actually leaves. The cactus, in particular, has reduced the amount of leaf surface it has as a means of conserving water. So, how does it perform enough photosynthesis to get along? It carries out most of its photosynthesis in its stem.

Stems: Structure and Storage

We all generally understand what the stem of a plant is, although it is less generally appreciated that the trunk of a tree is simply one kind of plant stem. The main functions of stems are to give structure to the plant as a whole and to act as storage sites for food reserves. In addition, water, minerals, hormones, and food are constantly shuttling through (and to) the stem, with water on its way up from the roots and food on its way down from the leaves to the rest of the plant.

Figure 25.9
**The Stem and
Its Parts**

Stems provide support to the
rest of the plant, act as nutrient storage sites, and conduct fluids. Vascular plants
have a "plumbing" system
that transports food, water,
minerals, and hormones. The
vascular bundles in the figure
are groups of tubes, running
in parallel, that serve this
transport function.

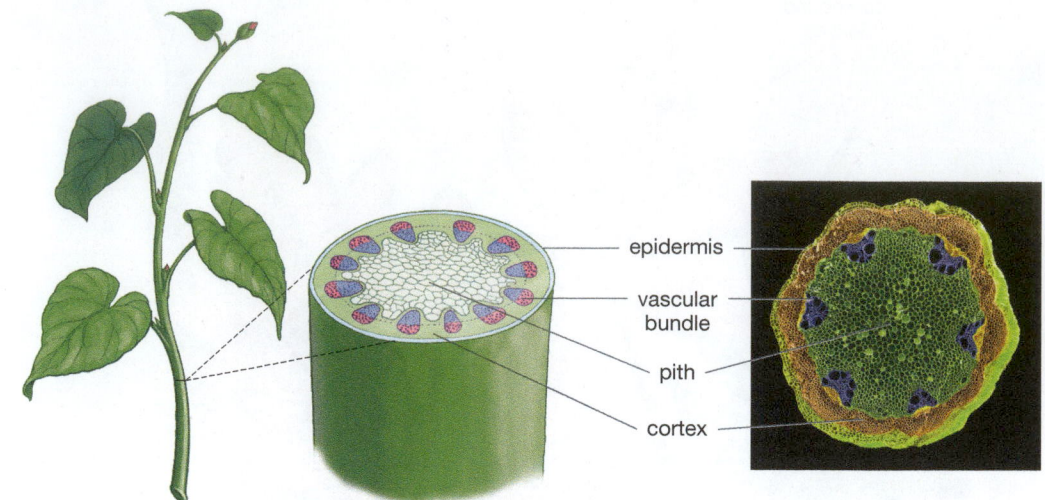

epidermis

vascular
bundle

pith

cortex

If you look at **Figure 25.9**, you can see a cross section of one type of plant stem, showing the vascular
bundles or veins you saw in leaves and the outer, or
epidermal, tissue just as the leaves had. You can also
see so-called ground tissue of two types: an outer cortex and an inner pith, both of which can play a part in
food storage and wound repair and provide structural
strength to the plant. We'll look at all these types of
tissues later.

Flowers: Many Parts in Service of Reproduction

Flowers are the reproductive structures of plants.
A single flower generally has both male and female
reproductive structures on it, which might make
you think that a given plant would fertilize *itself*.
This is indeed the case with some plants, such as
Gregor Mendel's pea plants, reviewed in Chapter 11.
However, because the evolutionary benefit of sexual

reproduction comes with mixing genetic material
from *different* individual organisms, natural selection has worked against self-fertilization by endow-
ing many flowering plants with ways to reduce the
incidence of it. For example, the male pollen of a
given plant might be genetically incompatible with
that plant's female reproductive structures. Some
of the illustrations you'll be seeing show a plant
fertilizing itself, but this is done only for visual
simplicity.

If you look at **Figure 25.10**, you can see the components of a typical flower. Taking things from the bottom, there is a modified stem, called a pedicel, which
widens into a base called a receptacle, from which the
flowers emerge. Flowers themselves can be thought of
as consisting of four parts: sepals, petals, stamens, and
a carpel. The **sepals** are the leaf-like structures that
protect the flower before it opens. (Drying out is a
problem, as are hungry animals.) The function of the

Figure 25.10
Parts of the Flower

Flowers are composed of
four main parts: sepals,
petals, stamens, and a
carpel. The sepals protect
the young bud until it is ready
to bloom. The petals attract
pollinators. The stamens are
the reproductive structures
that produce pollen grains
(which contain sperm cells).
The carpel is the female reproductive structure; it includes an egg-containing
ovary.

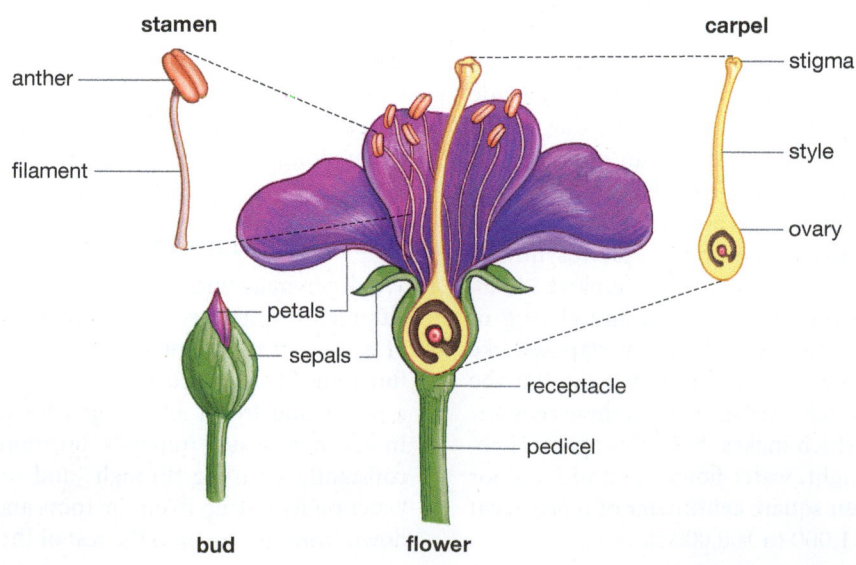

stamen

carpel

anther

stigma

filament

style

ovary

petals

sepals

receptacle

pedicel

bud

flower

colorful **petals** is to announce "food here" to pollinating animals.

The heart of the flower's reproductive structures consists of the stamens and the carpel. If you look at Figure 25.10, you can see that the **stamens** consist of a long, slender **filament** topped by an **anther**. These anthers contain cells that ultimately will yield sperm-bearing pollen grains. Thus, the anthers are the place in the flower where *male* reproductive cells are produced. Pollen grains ultimately will be released from the anther and then carried—perhaps by a pollinating bee or bird—to the carpel of another plant. As Figure 25.10 shows, a **carpel** is a composite structure composed of three main parts: the **stigma**, which is the tip end of the carpel, on which pollen grains are deposited; the **style**, a slender tube that raises the stigma to such a prominent height that it can easily catch the pollen; and the **ovary**, the area in which fertilization of the female egg and then early development of the plant embryo take place. Thus, the ovary is the structure in the flower that houses the *female* reproductive cells, and it is the structure in which the male and female reproductive cells come together.

SO FAR . . .

1. The primary function of roots is to absorb _____ and _____.

2. The primary function of leaves is to serve as sites of _____.

3. Flowers are the _____ structures of plants.

25.2 Monocots and Dicots

The roots, stems, and flowers we've been looking at represent the so-called gross anatomy of angiosperms, which is to say the large-scale physical makeup of these plants. To understand how angiosperms actually function, however, it's necessary to look at their anatomy on a smaller scale. What's the *detailed* makeup of roots, stems, and flowers, and how does this makeup produce a functioning plant? Before we begin looking into this question, it will be helpful to make a distinction between two broad classes of flowering plants. These are the monocotyledons and dicotyledons, which are almost always referred to as monocots and dicots. A **cotyledon** is an embryonic leaf, present in the seed of a plant, as you can see in **Figure 25.11**. **Monocotyledons** are plants that have one embryonic leaf; **dicotyledons** are plants that have two.

	Monocots	Dicots
embryonic leaves	one cotyledon	two cotyledons
mature leaves	narrow leaves / parallel veins	broad leaves / branching veins
roots	fibrous root system	taproot system
vascular bundles	scattered throughout stem	arranged in ring in stem
type of growth	only primary growth	may have secondary woody growth
flower parts	multiples of three	multiples of four or five
examples	orchids, wheat, rice, bananas	oak and maple trees, cacti, sunflowers

This distinction in the embryos is only one of the differences between monocots and dicots. Their roots are different, their leaves are different, and their transport tubes, or vascular bundles, are arranged differently. More than 75 percent of all flowering plants are the broad-leafed dicots. But given the 300,000 known species of flowering plants, this still leaves nearly 75,000 species of narrow-leafed monocots, which

Figure 25.11

Monocots and Dicots Compared

Plants with one embryonic leaf (the monocots) are structured differently from plants with two embryonic leaves (the dicots), as shown in the figure.

(a) Monocot plant

(b) Dicot plant

Figure 25.12
Categorizing Plants by Physical Features

(a) The plant *Dracaena fragrans,* commonly known as a corn plant, is a narrow-leafed monocot.

(b) Geraniums such as this one are broad-leafed dicots.

25.3 Plant Tissue Types

We'll begin our small-scale anatomical review of angiosperms by looking at so-called tissue types in them. Readers of Chapter 23 may recall that a **tissue** is a group of cells that carries out a common function. In our own bodies, we have cells that perform the common function of contracting. These are muscle cells, and a group of them constitutes a muscle tissue. So, what kinds of tissues are there in angiosperms? The short answer is dermal, ground, vascular, and meristematic. Dermal tissue can be thought of as the plant's outer covering, vascular tissue as its transport or "plumbing" tissue, meristematic tissue as its growth tissue, and ground tissue as almost everything else in the plant.

If you look at **Figure 25.13**, you can see the location of all four kinds of primary tissue in a dicot. Looking at the stem in cross section, you can see that ground tissue is well named because visually it forms a kind of background against which you can see the tubes of the vascular tissue. Dermal tissue then is at the periphery of the plant, forming a "skin" layer around the ground tissue. The placement of the meristematic tissue is discussed shortly. Before we get to it, let's look in a little more detail at dermal, ground, and vascular tissue.

Dermal Tissue Is the Plant's Interface with the Outside World

Plants can't move from one location to another, they can't bite, and, as already noted, they must carefully control their water supply. All this makes their dermal tissue, or *epidermis,* very important. In the non-woody plants or "herbaceous" plants, this outer coat is generally only one layer of cells thick, but it serves to

include most of the important food crops (corn, wheat, rice). **Figure 25.12** shows you an example of both a monocot and a dicot.

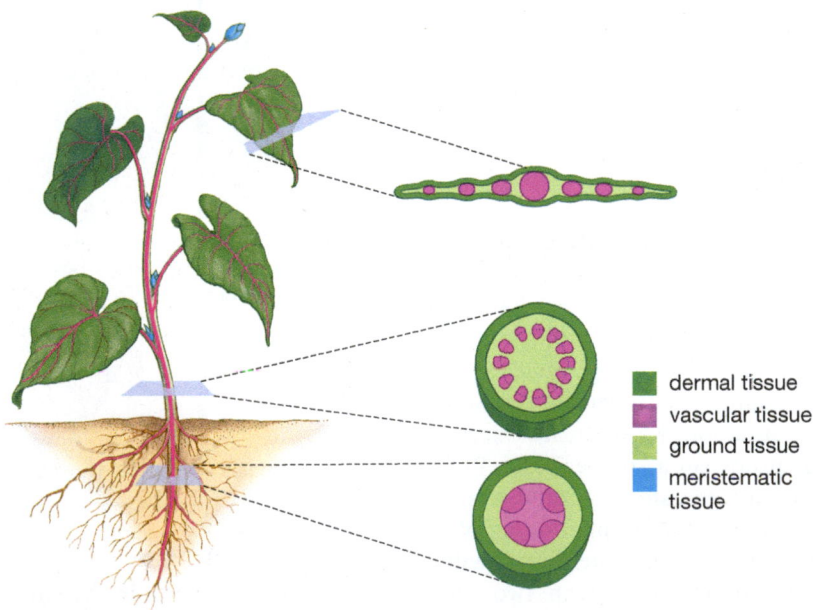

dermal tissue
vascular tissue
ground tissue
meristematic tissue

Figure 25.13
Four Types of Tissue in Primary Plant Growth
Dermal tissue (dark green) is found on the outside of the plant and is the interface between the plant and its environment. Ground tissue (pale green) is found throughout the plant and gives the plant its shape, stores food, and is active in photosynthesis. Vascular tissue (purple) transports water, nutrients, and sugar throughout the plant. Meristematic tissue (blue) occurs at the tips of shoots and roots and at leaf bases and gives rise to the entire plant, through the production of cells that develop into the other three tissue types.

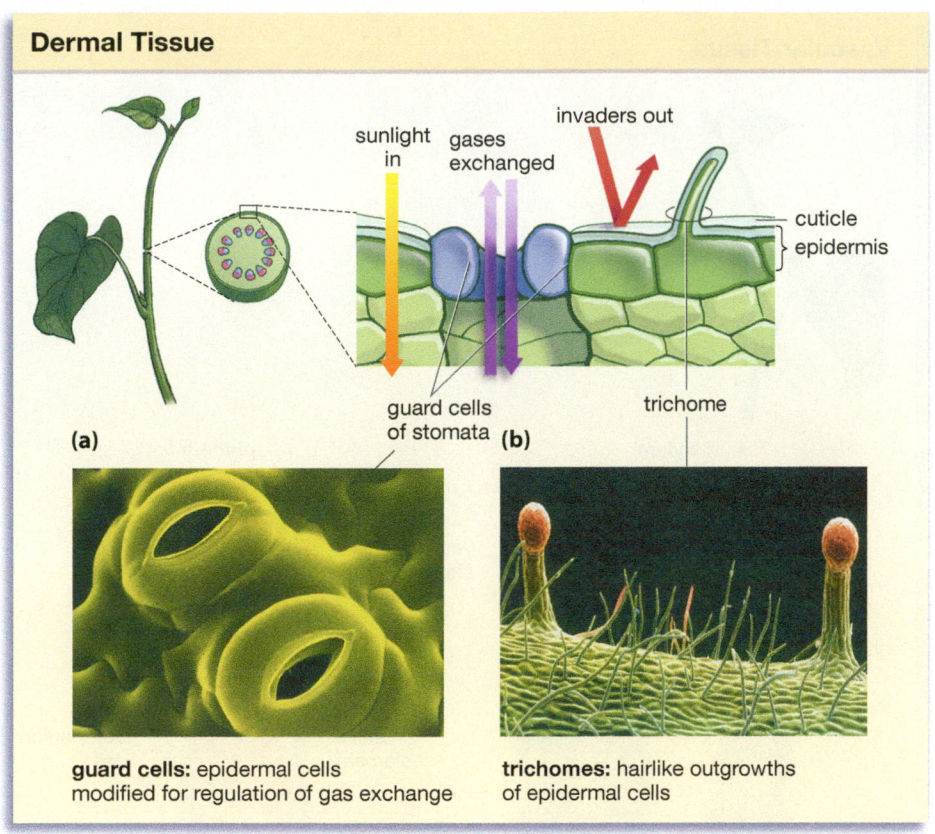

Dermal Tissue

sunlight in

gases exchanged

invaders out

cuticle
epidermis

guard cells of stomata

trichome

(a)

(b)

guard cells: epidermal cells modified for regulation of gas exchange

trichomes: hairlike outgrowths of epidermal cells

Figure 25.14
Where Plant Meets Environment

Dermal tissue serves as the interface between a plant and the environment around it. The waxy cuticle and the single layer of cells in the epidermis work together to protect the plant and to control interactions with the outside world.

(a) Specialized epidermal cells called guard cells regulate the opening of the plant pores called stomata, thus controlling gas exchange between the plant and its environment.

(b) Some epidermal cells have hair-like projections, called trichomes, that serve various functions. Two kinds of trichomes are visible in this rose plant. The larger of the two varieties, with the bulbous tips, help secrete chemicals that guard against plant-eating predators.

protect the plant and control its interaction with the outside world (**Figure 25.14**). A plant's first line of defense against outside predators is a waxy coating called the *cuticle,* which covers the epidermis of the plant's shoot. The cuticle is a kind of waterproofing, serving to keep water in, and an invader-proofing, helping to keep infecting bacteria and fungi out.

Plants need to exchange gases with their environments, however. They take in carbon dioxide and emit oxygen and water vapor through the microscopic stomata noted earlier. Stomata also represent a kind of weak point for plants, however, because they provide a passageway for microbial invaders and a conduit for excessive loss of moisture.

Dermal tissue also forms several kinds of extensions, called trichomes, that protrude from the plant's surface. We actually looked at one variety of these extensions earlier: the root hairs, meaning the thread-like outgrowths of individual epidermal root cells. Other types of trichomes serve not to absorb substances but to secrete them. In some plants, trichomes secrete toxic chemicals that ward off plant-eating animals.

Ground Tissue Forms the Bulk of the Primary Plant

Most of the primary plant is ground tissue, which can play a role in photosynthesis, nutrient storage, and plant structure.

Vascular Tissue Forms the Plant's Transport System

As noted earlier, the plant's vascular system has two main functions. First, it must transport water and minerals; second, it must transport the food made in photosynthesis. In line with this, there are two main components to the plant vascular system. First, there is **xylem,** the tissue through which water and dissolved minerals flow. Second, there is **phloem,** the tissue that conducts the food produced in photosynthesis, along with some hormones and other compounds (**Figure 25.15** on the next page). The movement of water through xylem is directional—from root up through stem and then out through leaves as water vapor. Meanwhile, the food made in photosynthesis must travel through phloem to every part of the plant, although in temperate climates the net flow is downward in summer (from food-producing leaves) but upward in early spring (from root and stem storage sites).

Xylem and phloem are arranged in the *vascular bundles* noted earlier, which is to say collections of xylem and phloem tubes that run together in parallel in the stem—xylem tubes toward the inside of the stem, phloem tubes toward the outside. Such bundles are arranged in different ways within a stem depending on whether the plant is a monocot or a dicot; the circular arrangement seen in Figure 25.15 is the one found in dicots.

Figure 25.15
Materials Movers

Vascular tissue transports fluids throughout the plant. Xylem is composed of cells called tracheids and vessel elements that transport water and dissolved minerals. Phloem is composed of sieve elements (and their companion cells), which transport the food produced during photosynthesis. These cell types are stacked on one another to form long tubes.

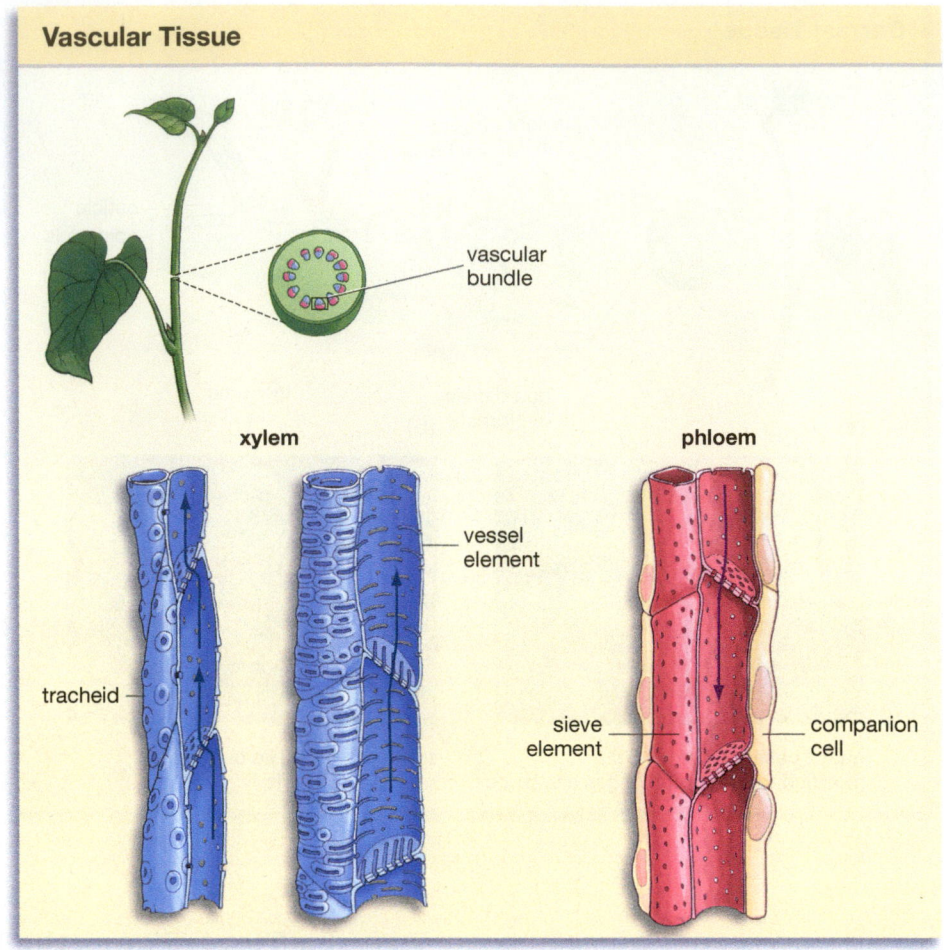

Vascular Tissue

vascular bundle

xylem

phloem

vessel element

tracheid

sieve element

companion cell

SO FAR . . .

1. Most plants fall into the category of the broad-leafed _____, but many of our most important food crops are narrow-leafed _____.

2. Dermal tissue is tissue located at the _____ of a plant, while vascular tissue forms a plant's _____ system.

3. Xylem is the tissue through which _____ and _____ flow, while phloem is the tissue that conducts the _____.

25.4 Primary Growth in Angiosperms

So, you've seen three types of plant tissues—dermal, ground, and vascular. The question now is: How do these get produced? The answer is that the fourth type of tissue mentioned earlier, meristematic tissue, gives rise to all of them. Put another way, meristematic tissue is responsible for all the *growth* that occurs in plants. (We're concerning ourselves here only with a plant's "primary" or vertical growth, but even the so-called secondary or lateral growth that woody plants exhibit comes about through meristematic tissue.)

In coming to understand how this one type of tissue can produce all the others in a plant, it's important to realize that plants do not grow the way people do—throughout their entire length—but instead grow only at the *tips* of their roots and shoots. If you look at **Figure 25.16**, you can see what this means. A boy of, say, eight decides to drive a long nail into the trunk of a secluded tree. Ten years later, he happens to walk by the same tree and, on seeing it, a couple of things occur to him. First, he has grown, but the nail is in the same place; he would have to bend down now to pull it out. Second, to judge by the nail's height, the tree's trunk may not have moved up, but its *top* certainly has; it's much taller than he remembered it. Finally, if the boy continued to visit the tree every 10 years or so, he would get one more lesson about its growth: That there's no point at which it comes to a complete halt. While the boy probably stopped growing at age 18, the tree might keep on growing for decades. Why? Because plant growth is indeterminate—it can go on throughout the life of most plants.

When Billy was eight ...

When Billy was eighteen ...

Figure 25.16

The Basic Characteristics of Plant Growth

If a young boy drives a long nail into a tree and returns 10 years later, two things will be apparent: The nail is at the same height it used to be, but the tree is much taller than it used to be. What does this say about tree growth? Trees do not elongate throughout their whole length, but only at their tips in their primary or vertical growth.

When we put the concept of indeterminate growth together with that of growth at the tips, it follows that there must be cells at the root and shoot tips of plants that remain perpetually young—or more accurately, perpetually embryonic. That is, these cells are able to keep giving rise to both more of themselves and to cells that then differentiate into all the tissue types we just went over. And this is just what takes place; everything in a plant develops ultimately from these cells, which are meristematic cells that form a series of **apical meristems**: groups of plant cells, located at the tips of roots and shoots, that give rise to all the tissues in a plant.

If you look at **Figure 25.17**, you can see the apical meristem locations in a typical plant. Note that in this plant with a taproot, each lateral root tip has its own apical meristematic tissue that is capable of giving rise to yet more roots. In the shoot, plants tend to confine their growth to the shoot apices that lie at the tip of each stem. Why? The better to compete for precious sunlight. It is the shoot apical meristem you can see in the figure that gives rise to all the cells that allow the plant's shoot growth.

Note, however, that there is a second location for meristematic tissue in the shoot—the area nestled between leaf and stem. It's here that we find lateral buds. Any **bud** is an undeveloped shoot, composed mostly of meristematic tissue. A lateral bud is meristematic tissue that may give rise to a branch or a flower (which obviously ends up growing laterally from the stem). Lateral buds are, however, just like any other shoot tissue in that they are the products of the shoot apical meristem. You could think of them as collections of meristematic tissue that are left behind, in a sense, by the apical meristem tissue as it continues its growth. (If you look again at the topmost picture in Figure 25.17, you can see two small collections of darker-colored cells on each side of the apical meristem dome.

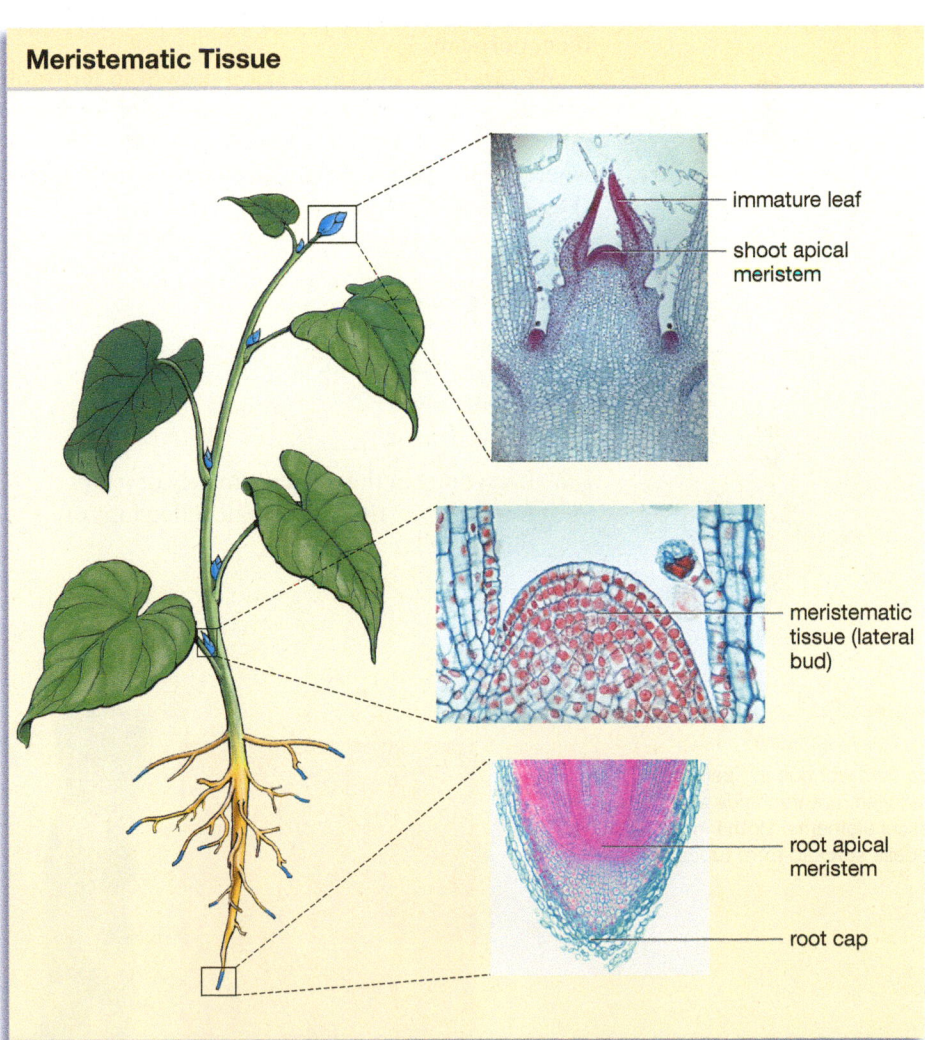

Meristematic Tissue

immature leaf

shoot apical meristem

meristematic tissue (lateral bud)

root apical meristem

root cap

Figure 25.17

Giving Rise to the Whole Plant

Growth in plants originates from meristematic tissue. The shoot and root elongate at the apical meristems, while new branches originate at the lateral buds (which are derived from apical meristem). The root cap noted in the bottom micrograph is tissue that protects the meristematic cells above it while secreting a lubricant that helps the root move into the ground.

week one   week two   week three   week four

leaf
node
internode
} module

Figure 25.18
Vertical Growth Occurs in Modules

Each growth module (indicated here by alternating dark green and light green segments) consists of an internode, a node, and one or more leaves. This plant progresses from one module in week 1 to four modules in week 4.

These are collections of meristem tissue that have been left behind by the shoot apical meristem.) Lateral buds can serve as a kind of insurance policy for the plant in that they can switch roles. Should the plant's apical meristem become damaged, one of the lateral buds steps in to assume the role of the shoot apex, thus allowing the plant to maintain its vertical growth. As long as the original apical meristem is intact, however, it generally will produce hormones that *suppress* the growth of the lateral buds near it, leaving them dormant.

We can think of plant vertical growth in terms of growth modules. Each module consists of an internode (the stem between leaves), a node (the area where a leaf attaches to stem), and one or more leaves (**Figure 25.18**).

SO FAR . . .

1. Unlike the case with animals, plant growth comes only at _____ of roots and shoots and can go on _____.

2. Everything in the plant ultimately develops from _____ tissue, through collections of this tissue known as _____.

25.5 Fluid Movement: The Vascular System

You have had some opportunity already to look at the plant's vascular or "plumbing" system as it relates to plant tissues, but now it's time to take a closer look at how the vascular system actually works, starting with the water-conducting xylem and then continuing to the food-conducting phloem.

How the Xylem Conducts Water

Two types of cells make up the water-conducting portions of plant xylem. These are **tracheids** and **vessel elements.** Both types of cells are peculiar in that they take on their primary function of conducting water only after they die. This happens when they've matured, after which their contents are then cleared out. What's left is a strand of empty, thick-walled cells stacked one on top of another—tubes, in other words (**Figure 25.19**). The walls of these xylem cells remain intact, however, with the upshot that xylem represents not only a transport system but a *load-bearing* system, which is why something as massive as a tree can grow so tall. Indeed, this load-bearing function remains in a given group of xylem cells long after they have ceased to act as a conduit for the water and dissolved nutrients that are sometimes called xylem sap. (To get an idea of one of the uses human beings have for this sap, see "The Syrup for Your Pancakes Comes from Xylem" on page 475.)

Solar Energy and Water Movement

In the tallest trees, water coming from the roots goes through a vertical journey of up to 300 feet before it gets to the topmost leaves of the tree. What kind of force can move water such a long way? As it turns out, plants themselves spend no energy at all on this transport. Instead, it is transpiration—the movement of

Figure 25.19
Water Carriers

The fluid-conducting cells of the xylem are tracheids and vessel elements. Both kinds of cells stack to form tubes.

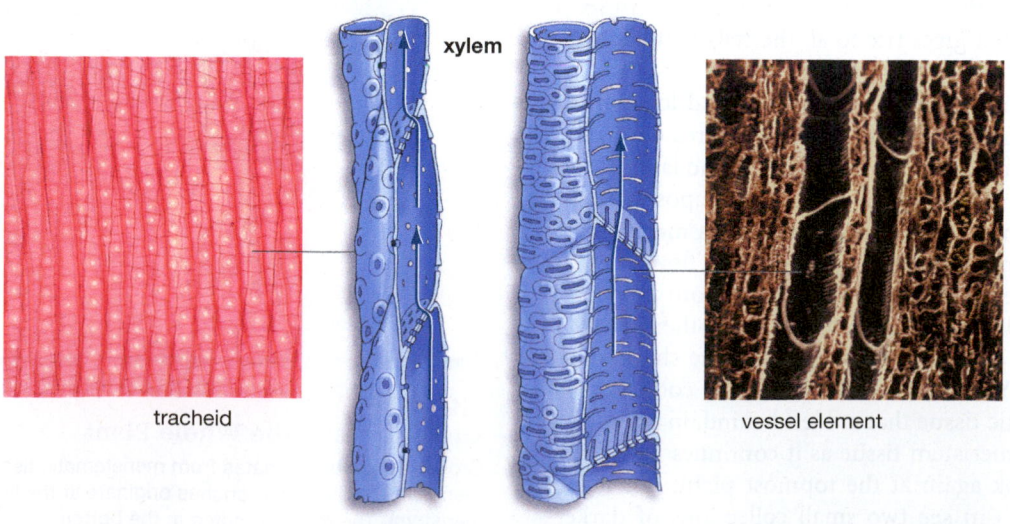

tracheid   xylem   vessel element

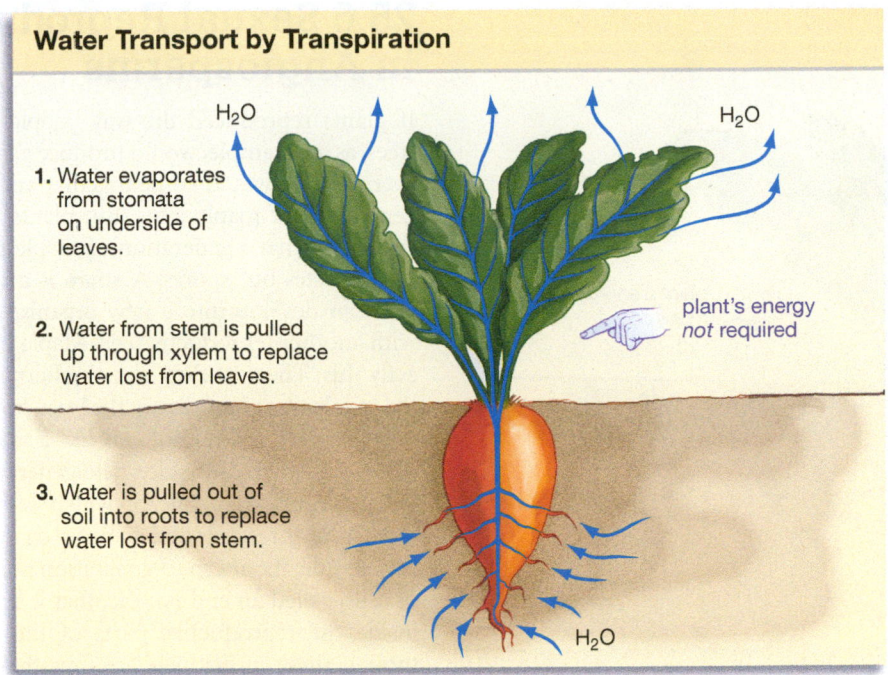

Water Transport by Transpiration

H_2O

1. Water evaporates from stomata on underside of leaves.

2. Water from stem is pulled up through xylem to replace water lost from leaves.

3. Water is pulled out of soil into roots to replace water lost from stem.

H_2O

plant's energy *not* required

H_2O

Figure 25.20
Suction Moves Water through Xylem
When water evaporates from leaves, a low pressure is created that pulls more water up to fill the void. Water is so cohesive that it moves up in a continuous column. The sun's energy, rather than the plant's energy, fuels this process.

(MB)

Bioflix: Water Transport in Plants

water vapor from plant leaves into the atmosphere—that is the primary force that moves water through xylem. As the water vapor evaporates into the air, it creates a region of low pressure, and this pulls a continuous column of water upward (**Figure 25.20**). Ultimately, the energy source for this movement is the sun; its warm rays power the evaporation of water at the leaf surface.

Food Transport through Phloem

Now let's turn from water transport to food transport. What is the nature of the food that flows through phloem? It's mostly sucrose—better known as table sugar—that is dissolved in water. Hormones, amino acids, and other compounds move through phloem as well, but sucrose and water are the main ingredients of this material, which is known as phloem sap.

The phloem cells that this sap moves through are called **sieve elements**. Unlike xylem cells, sieve elements are alive at maturity, though in a curiously altered state. Each one has lost its cell nucleus, which you may remember is the DNA-containing information center of a cell. Sieve elements continue to function as living cells, however, because each one has associated with it one or more *companion cells* that retain their DNA and that seem to take care of all the housekeeping needs of their sister sieve elements (**Figure 25.21**). The association between the two kinds of phloem cells is so intimate that when sieve elements die out, as they often do at the end of a growing season, their companion cells die as well. The reason sieve elements lose their nuclei seems to be to make room for the rapid flow of nutrients through them. The reason they are called *sieve* elements is that the ends of adjoining cells

are studded with small holes, much like a sieve. Stacked on top of each other, these cells form the plant's food-conducting pipes, the *sieve tubes*.

The force that moves food into and through these tubes is different from the one that moves water through xylem. With food, a plant does not get a free ride on transport via transpiration. Instead, it must expend its own energy to take the first step in nutrient transport, which is *loading* sucrose into sieve element cells. Once this happens, a principal player in moving sucrose through the phloem is a force that readers of Chapter 5 are familiar with: osmosis. When sucrose is loaded into the phloem, it sets up an osmotic flow that moves *water* into the phloem—something that is no different in principle than sending water into a

Figure 25.21
Food Carriers
The micrograph at right shows a cross section through a sieve element cell (looking at it from above) and an adjacent companion cell. A plant's phloem sap moves from one sieve element to another through the perforations visible in the cell.

phloem

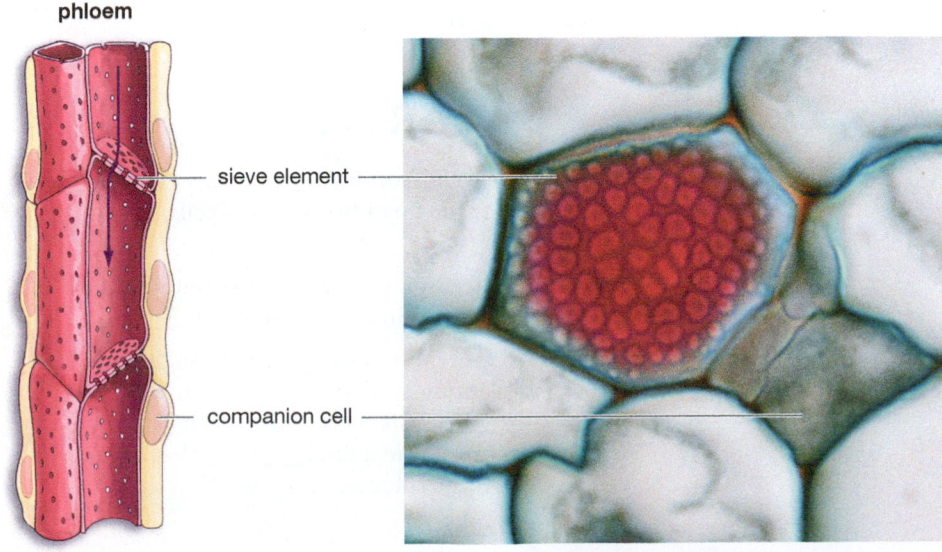

sieve element

companion cell

Sugar Transport by Pressure Flow

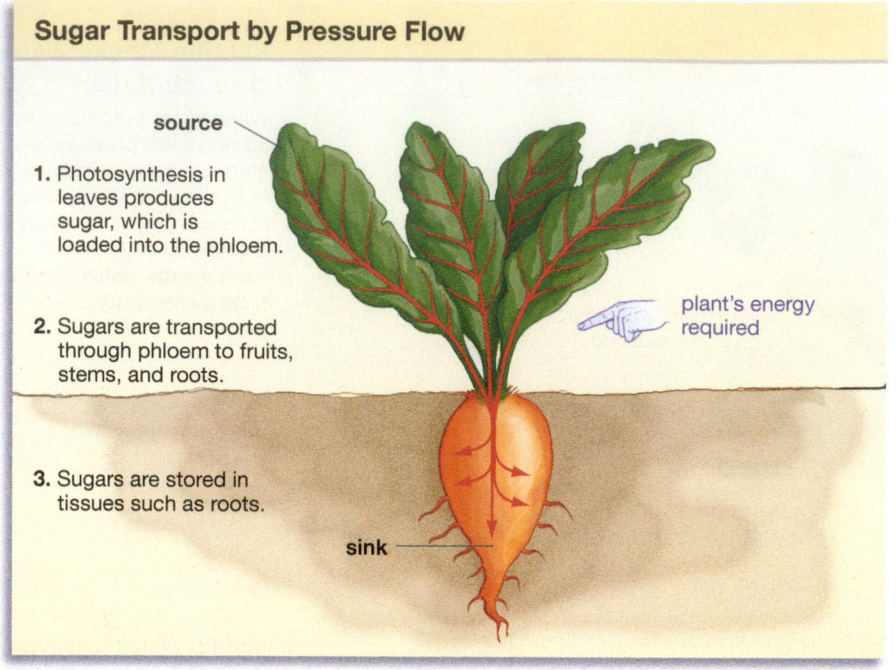

source

1. Photosynthesis in leaves produces sugar, which is loaded into the phloem.

plant's energy required

2. Sugars are transported through phloem to fruits, stems, and roots.

3. Sugars are stored in tissues such as roots.

sink

Figure 25.22

How Food Moves through the Plant

Sucrose is produced in the leaves (source) by photosynthesis, after which the plant expends its own energy to load the sucrose into the phloem. A movement of water into the phloem follows, which produces a fluid pressure that transports the sucrose through the plant, both for use and for storage (sink).

garden hose from an outside source. In both cases, such water is under pressure and it is this pressure that moves the sucrose from one location in the plant to another. The loading site of this system is called a *source* and the unloading site is called a *sink,* but these sites could also be called the "producer" and the "consumer" or perhaps "producer" and "bank" (**Figure 25.22**). Sources tend to be sugar-producing leaves, while sink sites tend to be storage tissue such as roots, along with growing fruits and stems. With this, let's move from a review of transport in angiosperms to a consideration reproduction in them.

SO FAR . . .

1. The water-conducting portions of xylem are made up of two kinds of cells, _____ and _____.

2. The transpiration that moves water through xylem is ultimately powered by _____.

3. To move dissolved sucrose through phloem, a plant must first _____ sucrose into _____ cells. This produces an osmosis that results in a fluid _____ that transports the sucrose.

25.6 Sexual Reproduction in Angiosperms

If plants reproduced the way people do, then maple trees, as an example, would produce gametes—eggs and sperm—that would come together and result in a new generation of maple trees. But plants don't reproduce this way. What a generation of maple trees produces is not gametes but spores. A **spore** is a reproductive cell that can develop into a new organism without fusing with another reproductive cell. Maple tree spores do exactly this: They develop into an alternate generation of the maple tree plant. To say the least, however, this alternate generation does not *look* like a maple tree. On the male side, the mature alternate generation amounts to pollen grains, each one of which is microscopic and consists of three cells and an outer coat, at most. On the female side, the alternate generation is a small collection of cells, called an embryo sac, that is hidden away deep inside the reproductive parts of the maple tree. Although these individuals are tiny, they have a critical role to play: *They* will produce the eggs and sperm that, once fused, will result in more maple trees.

As may be apparent from this account, plants exhibit a back-and-forth between generations. In their reproduction, they go through an *alternation of generations,* which was reviewed in Chapter 24 (see page 443). Because the generation that is familiar to us—the maple tree—produces spores, it is known as the sporophyte generation. Because the generation that is less familiar to us produces gametes, it is called the gametophyte generation. Given how small the gametophyte generation is in angiosperms, most people have trouble seeing it *as* a generation. From a human perspective, pollen grains and embryo sacs look like small component parts of the sporophyte generation. Nevertheless, gametophytes are a separate generation of plant, with their own critical role to play in angiosperm reproduction.

With this in mind, look again, in **Figure 25.23**, at the reproductive parts of an angiosperm—at the flowers that we went over earlier. What you're seeing in the illustration is mostly the *sporophyte* generation of this angiosperm. There is, for example, the stamen mentioned earlier, which consists of a long, slender filament that is topped by an anther. Every anther has chambers within it, however, containing cells that will develop into sperm-containing **pollen grains,** which are the mature *gametophyte* generation of the angiosperm on the male side. Meanwhile, you can also see the carpel, a composite structure composed of three parts: the stigma, style, and ovary. Each of these is a part of the sporophyte generation of plant, but the ovary houses cells that eventually will develop into the egg-containing **embryo sac** you can see in the figure—the mature gametophyte generation of the angiosperm on the *female* side.

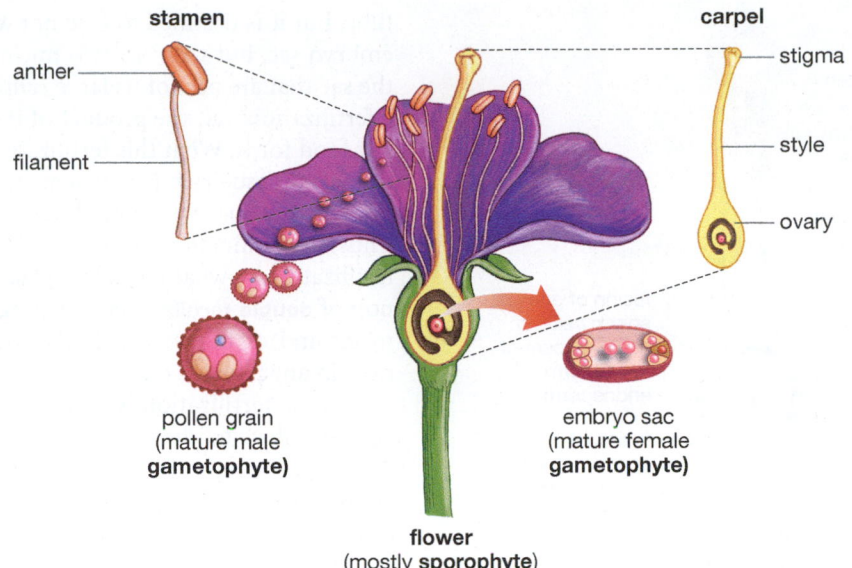

stamen

anther

filament

carpel

stigma

style

ovary

pollen grain
(mature male
gametophyte)

embryo sac
(mature female
gametophyte)

flower
(mostly **sporophyte**)

Figure 25.23
The Sporophyte and Gametophyte Generations
All plants reproduce through an alternation of generations. In angiosperms, this means an alternation between a larger sporophyte generation and a smaller gametophyte generation. The gametophyte generation either develops entirely within the sporophyte plant (the female embryo sac on the right) or largely within it (the male pollen grains on the left).

Now, with these component parts of the two generations in mind, how does reproduction work in angiosperms? If you look at **Figure 25.24**, you can see the process in overview. Note that we're starting with a mature sporophyte plant whose anthers produce the cells, called microspores, that develop into sperm-containing pollen grains. Meanwhile, the ovary of a second sporophyte contains a cell, called a megaspore, that will develop into the egg-containing embryo sac.

Next, a pollen grain from our first plant moves to the stigma of our second, lands on it, thus pollinating it, and a sperm cell from the pollen grain moves down to the egg in the embryo sac and fuses with it, thus fertilizing it. The result is a sporophyte zygote, or early-stage embryo. When this embryo has become surrounded with both nutritive and protective tissue, the result is the reproductive structure known as a seed. When this seed lands on the ground, the embryo

Plant Reproduction

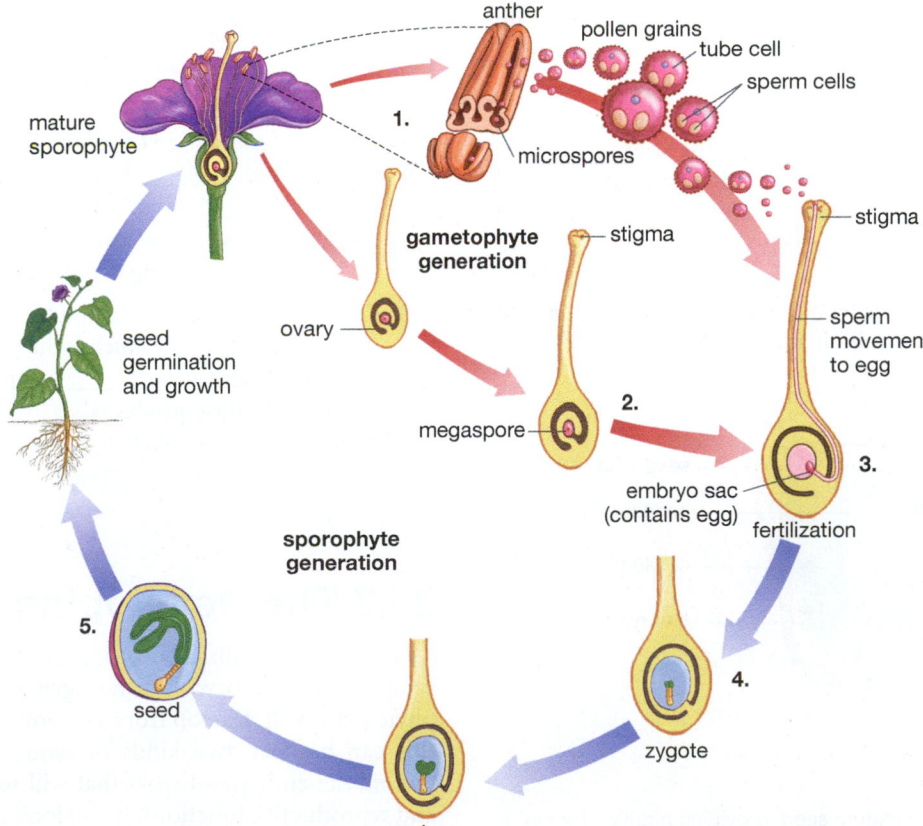

anther

pollen grains

tube cell

sperm cells

microspores

mature
sporophyte

**gametophyte
generation**

stigma

stigma

ovary

sperm
movement
to egg

seed
germination
and growth

megaspore

embryo sac
(contains egg)

fertilization

**sporophyte
generation**

seed

zygote

embryo

1.

2.

3.

4.

5.

Figure 25.24
Overview of Angiosperm Reproduction
1. The mature sporophyte flower produces many male microspores within the anther and a female megaspore within the ovary. The male microspores then develop into pollen grains that contain the male gamete, sperm.
2. Meanwhile, the female megaspore produces the female gamete, the egg. 3. A pollen grain with its sperm inside, moves to the stigma of a plant and a sperm from the grain moves down through the carpel, reaches the egg, and fertilizes it. 4. This results in a zygote that develops into an embryo protected inside a seed. 5. This seed leaves the parent plant and sprouts or "germinates" in the earth, growing eventually into the type of sporophyte plant that began the cycle.

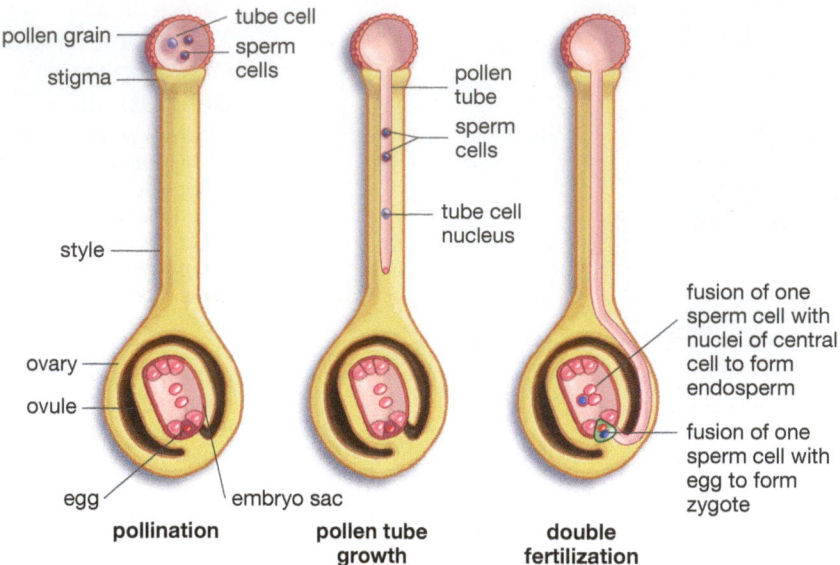

pollen grain
tube cell
sperm cells
stigma
style
ovary
ovule
egg
embryo sac

pollination

pollen tube
sperm cells
tube cell nucleus

pollen tube growth

fusion of one sperm cell with nuclei of central cell to form endosperm

fusion of one sperm cell with egg to form zygote

double fertilization

Figure 25.25

Double Fertilization in Angiosperms

When the male pollen grain (shown greatly enlarged) lands on the stigma, the tube cell within it sprouts a pollen tube, through which the two pollen grain sperm pass in moving to the embryo sac. One sperm cell fertilizes the egg in the sac, forming a zygote—the new sporophyte plant. The other sperm cell fuses with the two nuclei of the central cell in the sac, forming the endosperm that will later nourish the embryonic plant.

it contains will grow into the same kind of sporophyte plant we started with.

A key event in this process obviously comes when the sperm from the first plant fertilizes the egg from the second. But if you look closely at the pollen grains in Figure 25.24, you can see that each one actually contains *two* sperm cells (which are encased inside the larger tube cell that takes up most of the volume of the pollen grain). So, if one egg plus one sperm makes one zygote, why are there two sperm cells inside a pollen grain? And how does a sperm cell get from the stigma at the *top* of the carpel (where the pollen grain lands) to the embryo sac at the *bottom* of the carpel (where the egg lies)?

If you look at **Figure 25.25**, you can see the answer. When the pollen grain lands on the stigma, the tube cell within it then begins to germinate on the stigma—it sprouts a *pollen tube* that grows down through the style. Once this has taken place, one of the sperm in the grain travels through the tube, gets to the female egg, and fertilizes it. And what happens to the other sperm cell? It too travels down the pollen

tube, but it is destined to fuse not with the egg in the embryo sac, but with the two nuclei in the middle of the sac that are part of its large *central cell*. This too is a fertilization, but the product of it is not an embryo, but food for it. What this fertilization produces is endosperm: tissue rich in nutrients that can be used by the embryo that will now develop from the zygote. Thus, the pollination of this plant has resulted in two fertilizations—what has taken place is the phenomenon of **double fertilization,** meaning the fusion of gametes and of nutritive cells that occurs at the same time in angiosperm plants.

Double fertilization is an innovation that belongs almost solely to the angiosperm plants, but it has ended up being a boon to all kinds of *animals,* including people. It is endosperm that supplies much of the food we human beings consume. The rice and wheat grains we're familiar with consist in large part of endosperm meant to sustain the plant embryo (**Figure 25.26**).

SO FAR . . .

1. The plants with which we're familiar, such as maple trees, do not produce gametes (eggs and sperm), but instead produce _____, making them the _____ generation of plants. This generation alternates with the _____ generation, which on the male side consists of _____ and on the female side consists of small collections of cells, each one of which is known as a(n) _____.

2. For a new generation of sporophyte plant to be produced, _____ from one plant must land on the _____ of a plant, a necessary step in allowing the fusion of _____ with _____ that will bring about reproduction.

3. One of the fertilizations in double fertilization produces a _____, which is the original cell in the new generation of _____ plant. The other fertilization produces _____ in the form of the tissue called _____.

corn (monocot)

seed coat

endosperm

cotyledon

embryo

Figure 25.26

Endosperm Is Food for the Embryo

In monocots such as corn, endosperm persists in the mature seed, providing nutrition for the embryo inside. The cotyledon in the figure is the corn plant's embryonic leaf.

25.7 The Developing Plant

As noted, egg fertilization produces a zygote that develops into an embryo. Our new generation of sporophyte plant will develop from this embryo, but before this can happen, two kinds of tissue must develop *around* the embryo—tissues that will serve protective and reproductive functions. If you look at Figure 25.25 again, you can see that the embryo sac is nearly

The Syrup for Your Pancakes Comes from Xylem

In late February or early March, when winter is winding down in New England and eastern Canada, farmers who own sugar maple trees are kept busy in a game of drilling and watching. What they are drilling is a series of 2- to 3-inch holes, about three of them per maple tree. What they are watching is the weather, because they need a series of freezing nights followed by warmer days to get what they want. Such nights prompt the maple trees to convert nutrients they have stored into sugar; the warmer days then bring about a flow of this sugar, upward in sap through the tree's xylem. This material can then be "tapped" through spigots inserted into the drilled holes. The clear white sap that drips out—as much as 40 gallons from a single tree—is the starting material for pure maple syrup. Vermont is the chief maple-syrup-producing state in the United States, with an output of 920,000 gallons in 2009. Canada, however, produces about 80 percent of the world's maple syrup.

The "maple syrup" that most Americans buy in stores actually has very little maple syrup in it (perhaps 2–3 percent, the bulk being made up of corn syrup and syrup from regular sugar). This is understandable because pure maple syrup is expensive; the best varieties cost anywhere between $30 and $45 per gallon. One reason for this expense is that harvesting the syrup is a labor-intensive business; each tap must be put in by hand, just as it was hundreds of years ago (see **Figure 1**). Beyond this, it takes about 40 gallons of sap to make a single gallon of maple syrup. The sap must be boiled down until its sugar concentration reaches 66.5 percent.

Too many taps can kill a tree, but experienced syrup harvesters know how to tap the same tree year after year—potentially for hundreds of years—without harming it. The 40 gallons of sap a tree can yield sounds like a lot, but it may account for only 10 percent of a tree's total supply of sugar.

In the text, you've read much about the plant's food supply flowing through *phloem*, not xylem. But in sugar maples and some other deciduous trees, the food produced in summer's photosynthesis is stored as starch in xylem cells, to be broken down into sugar and released to the growing upper portions of the tree in early spring.

Figure 1
Source of Maple Syrup
Maple trees in Vermont being tapped for their xylem sap, which will become the key ingredient in maple syrup.

surrounded by tissue from the sporophyte plant called an ovule. Once fertilization occurs, tissue from this ovule will develop into a **seed coat** that will protect the embryonic plant until it germinates in the ground. What once was an ovule with a sporophyte embryo inside is now a **seed:** a reproductive structure of a plant that includes an outer seed coat, an embryo, and nutritional tissue for the embryo (the endosperm).

The Development of Fruit

As the seed coat is developing, another layer of tissue is taking shape outside of it. As you can see in Figure 25.25, the ovule we just talked about is surrounded by the *ovary* of the sporophyte plant. The wall of this ovary now develops into something that has a familiar name: fruit. Strictly speaking, **fruit** is the mature ovary of any flowering plant. Given this, *all* flowering plants develop fruit, but it's clear that most of the 300,000 species of flowering plants are not fruits as we commonly understand that term. It follows that the botanical definition of fruit differs from the common definition; under the botanical definition, fruits can take on lots of forms. The pod of a pea plant is a fruit, for example, as is the outer covering of a kernel of corn (although here fruit has fused with the corn seed coat, leaving scarcely anything separate that can be recognized as fruit). On the other hand, fruits can

Figure 25.27
Fruits Come in Many Forms

A fruit, the mature ovary of a flowering plant, surrounds a seed or seeds. Thus, **(a)** the flesh of an apricot fits this definition of fruit, but so does **(b)** the pod of a pea, which has several seeds inside. The structure we commonly think of as **(c)** the strawberry fruit actually is the strawberry plant receptacle, with each receptacle having many fruits on its surface. What are commonly thought of as strawberry "seeds" are tiny strawberry plant carpels, each complete with its own fruit and seed.

(a) Apricot

one carpel, one seed

(b) Pea

one carpel, many seeds

(c) Strawberry

carpels

many carpels, many seeds, one receptacle

receptacle

take exactly the form we're used to, as with the flesh of an apricot or cherry. To add another twist, where is the fruit in a strawberry? The answer, shown in **Figure 25.27**, may surprise you.

Fruit develops in all angiosperms, and it develops *only* in angiosperms. Thus, angiosperms can be defined by this anatomical feature. An **angiosperm** is a plant whose seeds develop inside the tissue called fruit. What is common to many fruits is that they serve a protective function for the underlying seed, and many act as *attractants* for seed dispersal. As noted in Chapter 24 (page 451), the wild berry that a bear eats consists not only of the fruit flesh but of the seeds it surrounds. The fruit is digested by the bear as food, but the seeds pass through the bear's digestive tract. In this way, the seeds get dispersed to what may be a promising new location for a berry plant.

The fruit-encased seed that contains the embryo eventually separates from the sporophyte parent plant and makes its way into a suitable patch of earth, there to sprout both roots and shoots. Once the plant has germinated and matured a little, you could take a cross section of its stem and find concentric circles of the tissue types mentioned near the start of this chapter: dermal, vascular, and ground. Meanwhile, meristematic tissue could be found at the tips of the root and shoot, allowing the plant to grow further. With this, we have come full cycle: Another generation of plant is maturing.

On to the Human Body

In the last two chapters, you've looked at a group of living things that are silent, fixed in one spot, and have almost no moving parts. Now it's time to shift gears; we're going to turn to a group of living things that might be characterized as noisy, highly mobile, and in possession of many moving parts, one of which is so active it contracts and expands ceaselessly within them from the fourth week of their embryonic life to the day they die. These members of the living world are called human beings; the story of how their bodies function is the subject of the next eight chapters.

SO FAR . . .

1. In all angiosperms, a growing embryo is first surrounded by a _____, which will serve to _____ the embryo until it germinates in the ground.

2. Fruit can be defined as the mature _____ of an angiosperm. Fruit in this sense lies outside the _____ that surrounds the angiosperm embryo.

25 REVIEW

Summary

25.1 The Structure of Angiosperms

- Plants consist of a set of above-ground shoots and below-ground roots. Roots absorb water and nutrients, anchor the plant, and often act as nutrient storage sites. Shoots include the plant's leaves, stems, and flowers. **(p. 460)**

- Leaves serve as the primary sites of photosynthesis in most plants. Leaves have a profusion of stomata that open and close in response to light, thus controlling the flow of carbon dioxide into the plant and the flow of oxygen and water vapor out of the plant. **(p. 462)**

- Stems give structure to plants and are storage sites for nutrients. **(p. 463)**

- Flowers are the reproductive structures of plants, with most flowers containing both male and female parts. The male reproductive structure, the stamen, consists of a slender filament topped by an anther, the chambers of which contain the cells that will develop into sperm-containing pollen grains. The female reproductive structure, the carpel, is composed of a stigma, on which pollen grains are deposited; a style, which raises the stigma high enough to catch pollen; and an ovary, in which fertilization of the egg and early development of the embryo take place. **(p. 464)**

25.2 Monocots and Dicots

- Angiosperms are classified according to how many cotyledons, or embryonic leaves, they have: Narrow-leafed monocotyledons have one and the more numerous broad-leafed dicotyledons have two. Monocots and dicots differ in structure in many ways. **(p. 465)**

25.3 Plant Tissue Types

- There are four tissue types in the plant body: Dermal tissue can be thought of as the plant's outer covering, vascular tissue as its "plumbing," meristematic tissue as its growth tissue, and ground tissue as almost everything else. **(p. 466)**

25.4 Primary Growth in Angiosperms

- Plants grow only at the tips of their roots and shoots and plant growth can go on for the life of most plants. **(p. 468)**

- The entire plant develops from meristematic cells located in regions called apical meristems. Meristematic cells remain perpetually embryonic, able to give rise to more of themselves as well as to cells that differentiate into all the plant's tissue types. **(p. 469)**

- In a plant with a taproot, each lateral root tip has its own apical meristematic tissue. In the shoot, meristematic tissue lies at the tip of each stem and forms lateral buds in the area between each leaf and stem. **(p. 469)**

25.5 Fluid Movement: The Vascular System

- The two types of cells that make up the water-conducting portions of xylem tissue are tracheids and vessel elements, both of which are dead in their mature state. **(p. 470)**

- Water movement through xylem is driven by transpiration. As water evaporates into the air, it pulls a continuous column of water upward through the plant. The energy for this process comes from the sun, whose rays power water evaporation at the leaf surface. **(p. 470)**

- The sugar sucrose is the main material that flows through phloem. The fluid-conducting cells of the phloem, sieve elements, lack cell nuclei in maturity. Each sieve element cell has associated with it one or more companion cells, which retain their nuclei and support their related sieve element. **(p. 471)**

- Plants expend their own energy to load the sucrose they produce into the phloem's sieve element cells. This produces an osmotic flow of water into the cells that results in a fluid pressure that serves to transport the sucrose from "sources," such as food-producing leaves, to "sinks," such as nutrient-storing roots. **(p. 471)**

25.6 Sexual Reproduction in Angiosperms

- Plants reproduce through an alternation of generations. A sporophyte generation (a familiar tree or flower) produces spores that develop into their own generation of plant, the gametophyte generation. This gametophyte generation produces the eggs and sperm necessary to give rise to a new sporophyte generation. In angiosperms, the mature male gametophyte is the sperm-containing pollen grain, while the mature female gametophyte is the egg-containing embryo sac, which is housed in the ovary of the sporophyte plant. **(p. 472)**

- In the fertilization of a gametophyte egg by sperm, a sperm-bearing pollen grain from one plant moves to the stigma of a second and lands on it, thus pollinating it, and a sperm cell from the pollen grain then moves down to the egg in the embryo sac and fuses with it, thus fertilizing it. This produces a sporophyte zygote, or early-stage embryo. When this embryo has become surrounded with both nutritive and protective tissue, the result is the reproductive structure known as a seed. When this seed is planted in the ground, the embryo it contains will grow into a new generation of sporophyte plant. (p. 473)

- Each pollen grain contains two sperm cells and a tube cell. When a pollen grain lands on a stigma, the tube cell within it sprouts a pollen tube that grows down to the embryo sac. Both sperm cells from the pollen grain move through the pollen tube to the embryo sac, but only one fuses with the egg there. The second sperm cell fuses with the two nuclei of the central cell in the embryo sac, thus initiating the development of tissue called endosperm, which will provide nutrition to the embryo during its early stages of growth. This second fertilization completes the process of double fertilization. (p. 474)

25.7 The Developing Plant

- With fertilization, ovule tissue that surrounded the embryo sac begins to develop into the protective seed coat that will surround the growing sporophyte embryo. The ovary that surrounded the ovule then starts to develop into a layer of tissue that will surround the seed—fruit, defined as the mature ovary of any flowering plant. An angiosperm can be defined as a plant whose seeds develop inside the tissue called fruit. (p. 474)

Key Terms

angiosperm   476
anther   465
apical meristem   469
blade   462
bud   469
carpel   465
cotyledon   465
dicotyledon (dicot)   465
double fertilization   474
embryo sac   472
fibrous root system   461
filament   465
fruit   475
monocotyledon (monocot)   465
ovary   465
petal   465
petiole   462
phloem   467

pollen grain   472
root hair   461
seed   475
seed coat   475
sepal   464
sieve element   471
spore   472
stamen   465
stigma   465
stomata   463
style   465
taproot system   461
tissue   466
tracheid   470
transpiration   462
vessel element   470
xylem   467

Understanding the Basics

Multiple-Choice Questions

(Answers are in the back of the book.)

1. Plants take in _____ and _____ through their roots, but take in _____ primarily through tiny pores on their leaves.
   a. minerals; carbon dioxide; water
   b. oxygen; carbon dioxide; water
   c. water; carbon dioxide; minerals
   d. oxygen; minerals; carbon dioxide
   e. minerals; water; carbon dioxide

2. Which of the following are not functions of roots? (Select all that apply.)
   a. nutrient storage
   b. loss of water through transpiration
   c. loss of oxygen in photosynthesis
   d. grow to water supplies
   e. absorption of water

3. Dermal tissue _____, while vascular tissue _____.
   a. protects the plant; conducts fluids
   b. provides strength; facilitates growth
   c. facilitates reproduction; protects the plant
   d. facilitates growth; provides strength
   e. protects the plant; facilitates reproduction

4. Which of the following statements about the ascent of water in tall trees is true? (Select all that apply.)
   a. Plants must expend their own metabolic energy to power the ascent of water into treetops.
   b. The evaporation of water from leaf surfaces pulls water from the roots to the canopy of a tall tree.
   c. The ascent of water in the xylem is fastest when the stomata are closed.
   d. The ultimate source of energy for the ascent of water is the sun.
   e. All of the above are true.

5. Pressure develops inside phloem sieve tubes because:
   a. osmotic uptake of water occurs when sugars are loaded into the sieve elements.
   b. sieve elements lack a nucleus.
   c. sieve elements are dead at maturity.
   d. sucrose is a bulky molecule that exerts a pressure against the sieve tube walls.
   e. All of the above are true.

6. Which of the following statements about plant spores is true? (Select all that apply.)
   a. They are another name for gametes.
   b. The multicellular gametophyte generation develops from them.
   c. They produce the eggs and sperm that fuse to produce a new sporophyte generation.
   d. They are part of the sporophyte generation.
   e. They are much smaller than gametes.

7. Which of these is a gametophyte plant? (Select all that apply.)
   a. petal
   b. stigma
   c. embryo sac
   d. sepal
   e. pollen grain

8. Endosperm is formed when:
   a. two sperm from the pollen grain fuse with each other.
   b. two nuclei from the embryo sac fuse with each other.
   c. a sperm nucleus from the pollen fuses with the egg nucleus in the embryo sac.
   d. a pollen grain lands on a stigma.
   e. a sperm nucleus from the pollen grain fuses with two central cell nuclei in the embryo sac.

9. The difference between a seed coat and fruit is that (select all that apply):
   a. fruit is always edible; seed coats are not.
   b. seed coats are always hard; fruits are always soft.
   c. seed coats develop from ovules while fruit tissue develop from ovaries.
   d. dicots produce seed coats while monocots produce fruit.
   e. fruit is tissue that develops outside the seed coat.

Brief Review

(**Answers are in the back of the book.**)

1. The stomata that exist mostly on leaves allow important gas exchanges to take place in plants. What gases are exchanged through the stomata and in which direction? What other substance do plants lose as a result of gas exchange and how do they deal with this loss?

2. What functions do roots perform? What are the main types of roots produced by flowering plants?

3. List the main parts of a flower and describe the functions performed by each part.

4. What is meristematic tissue? Describe its organization.

5. Compare the structure and function of the water-conducting (xylem) and food-conducting (phloem) cells in a plant.

6. Describe the process of double fertilization in plants.

Applying Your Knowledge

1. The "race for sunlight" in plants led to the evolution of the tallest living things in existence, trees. But sunlight is not the only thing that stands to make a tree successful or not in reproducing. Think about how plants function and then answer this question: What are the costs of being taller, as opposed to shorter?

2. It's sometimes said that plants are much more efficient organisms than animals, in that they consume fewer resources and produce fewer waste products. Evaluate this statement in light of what you have learned about the functioning of angiosperms.

3. Gardeners often speak of outdoor plants having become "established," meaning the plants have reached a point at which they need less human care. What threshold does a plant have to cross to become established?

MasteringBiology®

1. MasteringBiology Assignments
Activities: Leaves, Reproduction, and Fluid Transport; Plant Reproduction; **Bioflix:** Water Transport in Plants; **Building Vocabulary; Questions:** Bioflix Quiz, Reading Questions, Test Bank

2. eText
Read your book online, search, take notes, highlight text, and study more effectively.

3. The Study Area
Self-study tools to test comprehension.

26

Body Support and Movement:

The Integumentary, Skeletal, and Muscular Systems

Just as several individuals may make up an office, and several offices a department, so our bodies are organized into a number of working units that go on to make up a functioning human being. Three of the larger units involve skin, bone, and muscle.

Available on MasteringBiology®

Activities: Muscle Structure; **Bioflix:** Muscle Contraction; **Current Events:** A Once-Unthinkable Choice for Amputees; Each Flick of a Digit Is a Job for All Five; Human Muscle, Regrown on Animal Scaffolding; Looking to Save Money, More Places Decide to Stop Fluoridating the Water; A Change in Season and Regimen; Doctors Seek Way to Treat Muscle Loss; **Building Vocabulary; Questions:** Bioflix Quiz, Reading Questions, Test Bank

How is the human body structured and how does it work? The disciplines of human anatomy and physiology are dedicated to finding out.

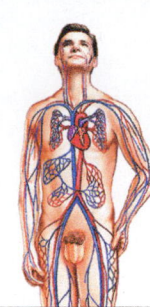

There is an old country saying that goes, "You don't miss your water until your well runs dry." Doesn't this seem applicable to the way we think of our bodies? We go along taking for granted our ability to eat or talk or run—until the day comes when we can't do one of these things. If you've ever, say, broken a leg and ended up with a cast and crutches, you know that just walking around with ease can look like the most wonderful thing in the world. Broken bones heal, of course, and that reassures us—until we run up against a condition that won't heal. While still in his 20s, the actor Michael J. Fox woke up one morning and noticed, as he later remembered, that "the pinky in my left hand was twitching. It was a curiosity as much as anything because I couldn't stop it, no matter how I tried to." About a year later, after many doctor's visits, Fox got definitive word that he had Parkinson's disease, which is incurable. By then, it was not just his little finger that was twitching; it was both hands. Imagine if being able to hold your hands steady seemed as beautiful as the moon and just as far away.

We take our bodies for granted partly because they work so well most of the time, but partly because what goes on inside them is hidden from us and so complicated that our physical functioning just seems mysterious. The next eight chapters of this book are aimed at clearing up some of this mystery. At their conclusion, you will have a basic understanding of how your heart beats, for example, of how your immune system fights off invaders, and of how your eyes and brain provide you with a picture of the world.

If you think about just the first of these things—the workings of the heart—you can get a sense of what's in store. The human heart beats about 100,000 times *each day*, pumping about 8,800 quarts of blood through the body in the process. Moreover, it keeps this rhythmic contraction up without a second's rest from the third week of our embryonic life to the day we die. As if this weren't enough, it is carrying out this feat mostly with its original equipment. Only about 40 percent of the heart's muscle cells will divide over a person's lifetime; as a result, most of the cells that are working for us at age 8 are the cells that are working for us at age 80.

An engineer who could invent such a machine would be considered a genius. But the truly remarkable thing is that there are lots of organs and systems in the body that are just as impressive as the heart. Once you've learned how these things work, it's hard to take your body for granted in the same way because you've come to understand what an incredible machine the human body is.

26.1 How the Body Regulates Itself

When we study the makeup or functioning of any complex organism, we have entered the realms of two overlapping disciplines in biology: anatomy and physiology. All organisms have an anatomy and a physiology. But in these chapters, our primary concern is with the human organism. Within this framework, human anatomy is the study of the body's structure and the relationships between the body's constituent parts. Human physiology is the study of how these parts work. Thus, anatomy tells us where the heart is, for example, how it is structured, and how it is situated in relation to other organs. Physiology describes to us *how* the heart works.

Before we get to specifics regarding either of these subjects, it will be helpful to go over a couple of general concepts about the body's structure and functioning. One of these concepts concerns the question: Who's in charge here? We are in charge of our conscious actions, of course; human beings get to decide what they are going to do from one moment to the next. But what about the huge number of unconscious actions that the body undertakes each second—getting oxygen to cells, getting rid of waste, being prompted to breathe while asleep? You might think the brain would be in charge of all this, and indeed messages from the brain do prompt a lot of physical activity. But there are plenty of physical processes that don't involve brain signals, and the brain *gets* lots of messages that say, in effect, "Get this process going now." Human society is hierarchical to a great degree; we're used to bosses and employees who work in a straight line of authority. But the body doesn't work this way. Instead, it functions through a process of *circular* causation in which A prompts B and B prompts C, but in which C may then prompt A. It is a system of self-regulation in which all parts work together to influence each other.

In understanding self-regulation, it's useful to think about what kind of internal environment the body needs to function properly. The answer is a *stable* environment. The human body must guard against being too hot or too cold; too dehydrated or too hydrated; too stimulated or too relaxed. If it is to exist at all, it must avoid extremes. To put this another way, it must seek **homeostasis,** meaning the maintenance of a relatively stable internal environment. And it turns out that a single process is almost solely responsible for bringing about homeostasis. Let's look at this process by way of an analogy.

Most people are aware of how a home heating system works. Falling temperature causes a thermostat to turn on a furnace. To look at this another way, there is a stimulus (cold air) that brings about a response (furnace operation). The *product* of this response is hot air. When enough hot air circulates to raise the temperature, the thermostat senses this and shuts the furnace down.

Now, let's think of your body, which has, as an example, a certain amount of oxygen circulating through it. All of this oxygen comes in through the lungs and is loaded into red blood cells, which then move through the bloodstream and drop off the oxygen where it's needed. The body requires a great deal

of oxygen in order to function and cells in our kidneys are capable of sensing whether circulating oxygen levels are adequate or not. When oxygen levels fall too low (stimulus), these kidney cells release a hormone, called EPO, which moves through the bloodstream and prompts stem cells in the marrow of our bones to begin producing more red blood cells (response). Thus, the product of *this* stimulus-response chain is more red blood cells and hence a greater amount of oxygen in circulation. When circulating oxygen once again reaches adequate levels, the cells in our kidneys likewise sense this and reduce their production of EPO (**Figure 26.1**).

With both the thermostat and the kidney's sensor cells, we can see that their responses to a stimulus bring about a decrease in their own activity. In other words, their responses feed back on their activity in a negative way—they reduce it. This is **negative feedback,** which can be defined as a process in which the elements that bring about a response have their activity reduced by that response. The EPO example is only one of countless negative-feedback loops that exist in the body. In it, we can see not only how the body maintains homeostasis but how the body is self-regulating. Think of it: There was no central boss giving commands in the regulation of oxygen levels. Instead two elements (oxygen levels and sensor cells in the kidneys) influenced each other to bring about stability. This is the way the body generally works.

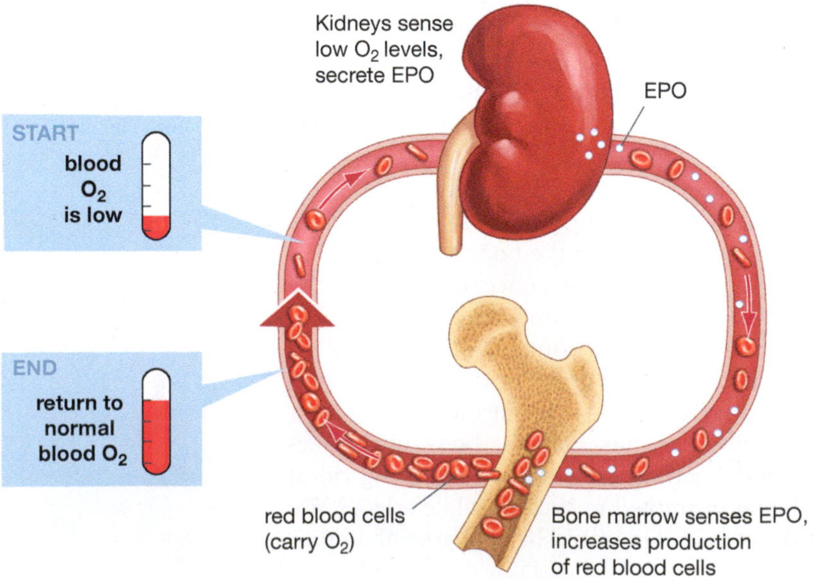

Large-Scale Features of the Body

Many of the large-scale features of a human being—four limbs, one trunk, and so forth—are obvious, but a couple of features are hiding in plain sight, we might say. The first of these is an internal body cavity. All animals more complex than a flatworm have such a cavity. Humans and other mammals actually have two primary body cavities. The first of these is a dorsal (back) cavity that is further divided into cranial and spinal cavities, into which our brain and spinal cord fit (**Figure 26.2**). The second is a ventral (front) cavity that includes a thoracic or "chest" portion containing the lungs and heart as well as a lower abdominopelvic cavity containing such organs as the stomach, liver, and intestines. A muscular sheet called the diaphragm separates the thoracic and abdominopelvic cavities. Cavities provide flexibility and protection. With the space provided by a cavity, an organ like the stomach can expand as we eat and is protected from external blows to some extent.

Humans also have an internal skeleton that supports the body. We take for granted a skeleton that exists inside our skin, but it actually stands in contrast to what most animals have, which is either an *external* skeleton (as with crabs and grasshoppers) or no solid skeleton at all (as with jellyfish).

Figure 26.1

Oxygen Levels: An Example of Negative Feedback

Oxygen levels in the body are regulated in part by cells in the kidneys that are capable of sensing when oxygen levels in the bloodstream have fallen too low. In response, these cells secrete a hormone, called EPO, that causes stem cells in the bone marrow to produce more red blood cells. Production of these oxygen-carrying cells brings about adequate circulating oxygen levels, which the cells in the kidneys can likewise sense. In response, they reduce their production of EPO.

Kidneys sense low O₂ levels, secrete EPO

EPO

START
blood O₂ is low

END
return to normal blood O₂

red blood cells (carry O₂)

Bone marrow senses EPO, increases production of red blood cells

(a) Side view of body cavities

(b) Front view of body cavities

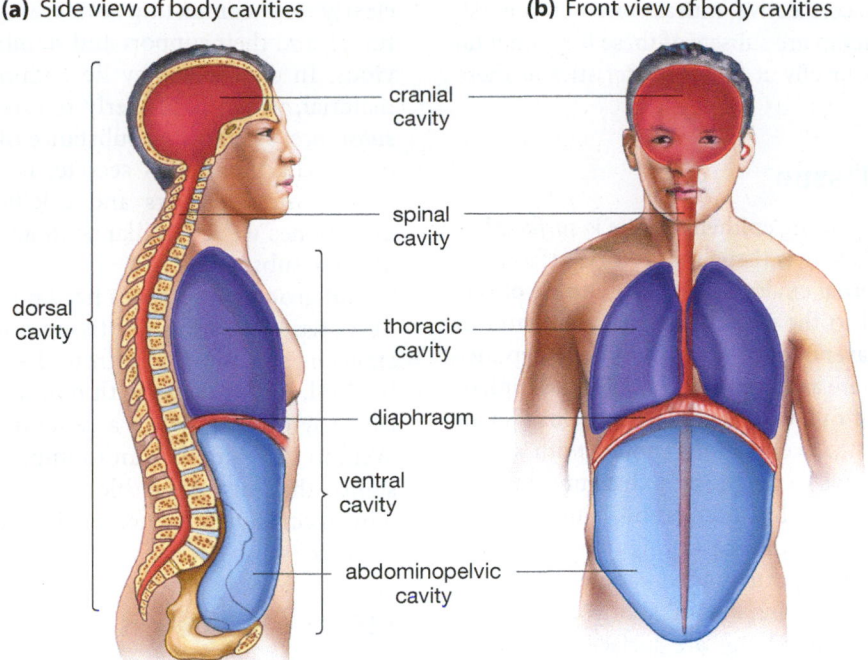

cranial cavity

spinal cavity

dorsal cavity

thoracic cavity

diaphragm

ventral cavity

abdominopelvic cavity

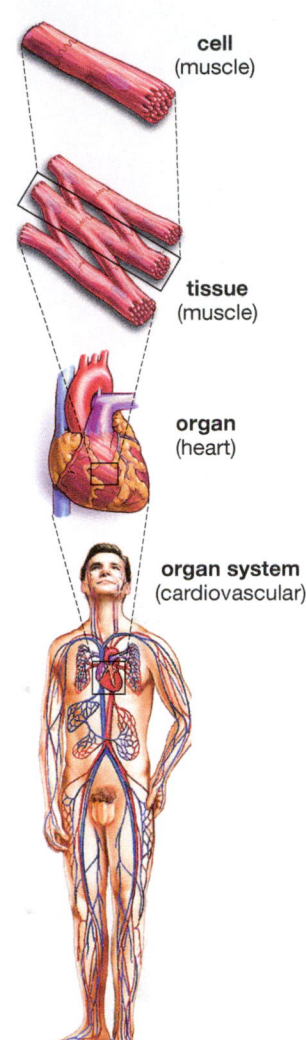

cell (muscle)

tissue (muscle)

organ (heart)

organ system (cardiovascular)

Figure 26.2
Body Cavities
(a) A side or "lateral" view of the two main cavities in the human body—the dorsal (back) and ventral (front) cavities—and the smaller cavities within them.
(b) A front or "ventral" view of the body's cavities.

SO FAR . . .

1. The study of where the component parts of the body are located and how they are structured is called _____, while the study of how the component parts of the body work is called _____.

2. Negative feedback occurs when the elements that bring about a response have their activity _____ by that response.

3. Negative feedback is important because it preserves homeostasis in the body, meaning a _____ internal state.

26.2 Levels of Physical Organization

Features of the human body, such as a skeleton, fit into a larger organizational framework that is common to all living things. In our everyday world, several individuals may make up an office, several offices a department, and several departments a working company. In a similar fashion, each living thing is built up from a series of individual units that go on to make up a working organism. Here's how these units

are organized if we consider just the animals among life's creatures (one of which is us). Recall from Chapter 4 that the basic unit of all life is the cell. Cells and cellular products that work together to perform a common function are known as **tissues.** There is, for example, muscle tissue—a group of cells that perform the common function of contracting. Tissues in turn are arranged into various types of **organs,** meaning complexes of several kinds of tissues that perform a special bodily function. The stomach is an organ that has both muscle and nerve tissue within it. Organs then go on to form **organ systems,** which are groups of interrelated organs and tissues that serve a particular function. Contractions of the heart push blood into a network of blood vessels. The heart, blood, and blood vessels form the *cardiovascular system,* which is one of 11 systems in the human body. **Figure 26.3** shows the various levels of structural organization within the human body, using the cardiovascular system as an example.

26.3 The Four Basic Tissue Types

Looking at the smallest level of organization we're interested in, the tissue, we can then ask: What kinds of tissues are there in humans and other animals? It turns out there are only four fundamental tissue

Figure 26.3
Levels of Organization in the Human Body
A group of cells that performs a common function is a tissue; here, muscle cells have formed heart muscle tissue (whose common function is heart muscle contraction). Two or more tissues combine to form an organ such as the heart. (The heart contains both muscle and connective tissue, for example.) The heart is one component of the cardiovascular system; other parts of the system are blood and blood vessels. All the organ systems combine to create an organism, in this case a human being.

types: epithelial, connective, muscle, and nervous. All other tissue varieties are subsets of these fundamental types. Let's look briefly at the characteristics of these four varieties.

Epithelial Tissue

The key word regarding epithelial tissue is *surfaces* because **epithelial tissue** is tissue that *covers* surfaces exposed to an external environment. The outside of our body is exposed to the external environment, and as such, the outer layer of our skin is composed of epithelial tissue. But there are other surfaces as well. Think of a pipe; it has not only an external surface but also an internal surface that comes into contact with something that is external to it—the fluid that runs through it. Similarly, your body has arteries and veins that have the fluid called blood running through them, and the tissues that line the internal blood vessel surfaces are epithelial tissues. Likewise, the lining of your stomach and the lining of your lungs are surfaces that come into contact with something from outside you; again, this contact is made by epithelial tissues. Given their location, epithelial tissues often serve as a barrier, as in the case of our skin; but many epithelial tissues are "transport" tissues that materials are moved *across*. (The food that enters your bloodstream does so by being transported across the epithelial tissue that lines your small intestines.) Epithelial tissues can be several layers of cells thick or as thin as a single layer, depending on their function (**Figure 26.4**).

Epithelial tissues often produce substances that aid in various bodily processes. The skin has epithelial tissue that produces the water-resistant protein keratin, for example, while the stomach has epithelial tissue that produces digestive juices. This substance-producing function can be carried out by isolated epithelial cells, but it is sometimes undertaken by concentrations of cells. Such concentrations are known as **glands:** organs or groups of cells that are specialized to secrete one or more substances. Later, we'll look at two basic kinds of glands.

Connective Tissue

Connective tissue, defined as tissue that stabilizes and supports other tissue, is very different from epithelial tissue. Whereas epithelial tissue is always in contact with an external environment, connective tissue never is. Whereas epithelial tissue is almost completely composed of cells, connective tissue usually is composed of cells that are separated from each other by an extracellular material—a material that is secreted by the connective tissue cells themselves. Indeed, the prime function of many connective tissues is to produce this kind of material.

We can see these qualities in the connective tissues known as bone and cartilage. These tissues clearly are connecting something (bodily structures), and their support and stabilization role is obvious. In addition, they lie within an extracellular material, which is properly referred to as a *ground substance*. The ground substance of bone, which the bone cells themselves secrete, is a mix of closely packed protein fibers and calcified material. The hard bones we're familiar with are largely made up of these substances.

But ground substances need not be hard. Adipose or fat tissue is connective tissue, and its ground substance is soft. Indeed, a ground substance can even be fluid, as is the case with blood and a bodily fluid called lymph. How can a ground tissue be a fluid? Well, to take blood as an example, the ground substance that surrounds blood cells is mostly water, with electrolytes, proteins, and other materials in the mix.

Muscle Tissue

Muscle tissue is tissue that is specialized in its ability to contract or shorten. There are three kinds of muscle tissue—skeletal, smooth, and cardiac—which differ in their structure and in the way they are prompted to contract. **Skeletal muscle** is the ordinary muscle that is attached to bone and is contained in, for example, our biceps. It is under our conscious control and has a striped or "striated" appearance when looked at under a microscope owing to the parallel orientation of some long, fibrous units that make it up. **Cardiac muscle** exists only in the heart and likewise is striated, but it contracts under the influence of its own pacemaker cells. **Smooth muscle,** which gets its name from its lack of a striated appearance, is muscle responsible for contractions of the uterus, the digestive tract, blood vessels, and the passageways of the lungs. As you might guess from this list of tasks, smooth muscle is not under voluntary control.

Nervous Tissue

Nervous tissue is tissue that is specialized for the rapid conduction of messages, which take the form of electrical impulses. There are two basic types of nervous tissue cells: neurons, which actually carry the nervous system messages, and glial cells, which perform support functions for the neurons.

26.4 Organs Are Made of Several Kinds of Tissues

Having looked at the four types of tissue the body has, let's now see how they combine to form an organ—the human stomach, which is a typical organ in that it is made up of all the kinds of tissue we've been

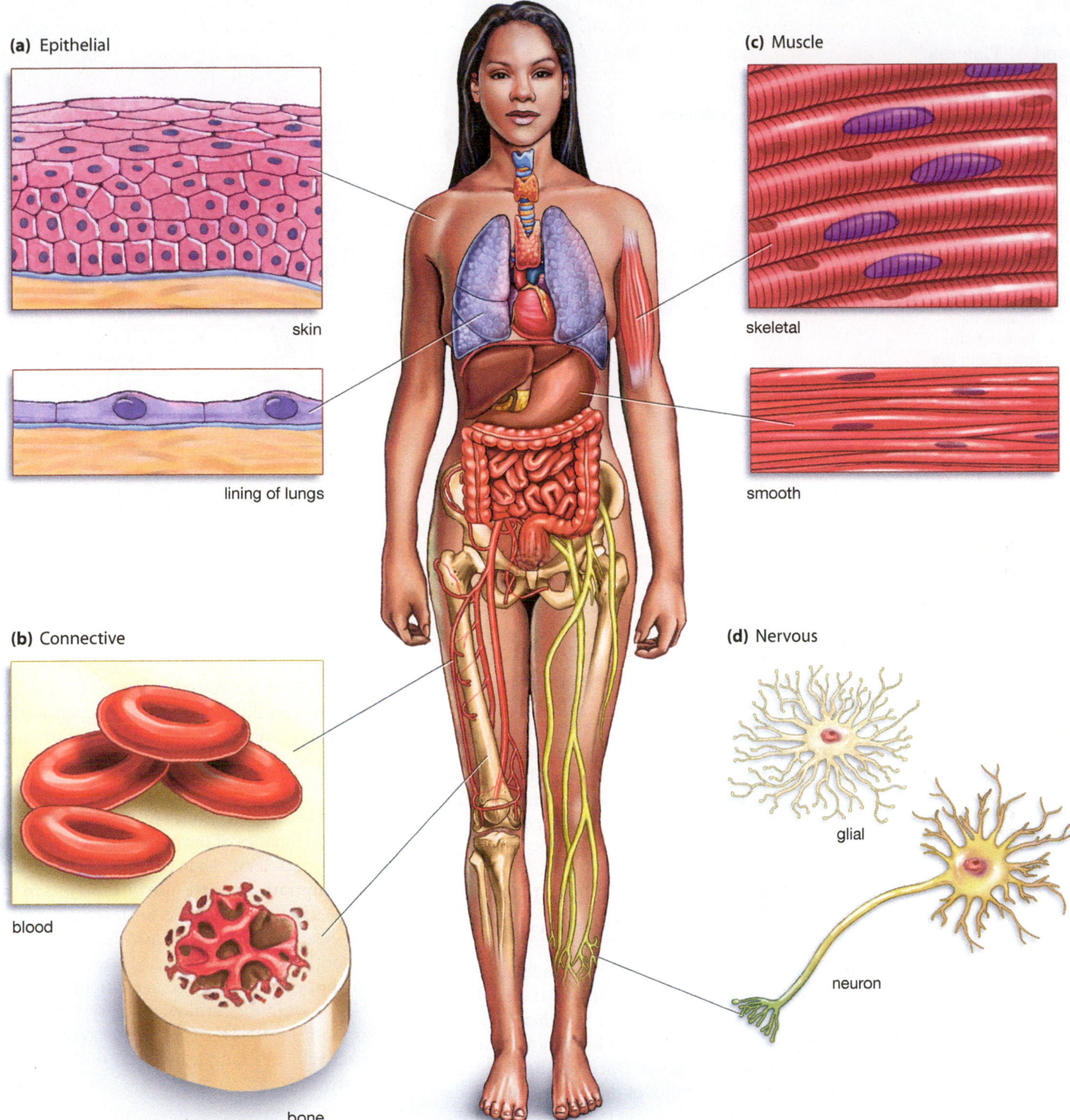

(a) Epithelial

skin

lining of lungs

(b) Connective

blood

bone

(c) Muscle

skeletal

smooth

(d) Nervous

glial

neuron

Figure 26.4
Four Types of Tissue
There are four basic tissue types in the human body: epithelial, muscle, nervous, and connective. Pictured are examples of each of the four types in some representative locations.

(a) Epithelial tissue, which always covers a surface exposed to an external environment, can take the form of many layers of cells (as with the skin) or a single layer of cells (as with the lining of lungs).

(b) Connective tissue, which supports and stabilizes other tissue, can take such diverse forms as bone and blood.

(c) Muscle tissue, which is specialized for contraction, comes in a striped or "striated" form (in both skeletal and cardiac muscle) and in a smooth form that lines the digestive tract, blood vessels, and other structures.

(d) Nervous tissue is specialized for the transmission of electrical signals. The two types of nervous tissue cells are neurons, which conduct nervous system signals, and glial cells, which support the neurons.

Figure 26.5
Organs Are Composed of Tissues
The stomach is a typical organ in that it is composed of various tissue types. Nervous tissue controls the release of digestive juices into the stomach and also controls the contractions of a second type of tissue found in the stomach, its muscle tissue. Connective tissue helps support and stabilize the other tissues around it. Epithelial tissue lines the stomach's surface, which comes into contact with the food we eat. The individual epithelial cells pictured secrete digestive acids and enzymes.

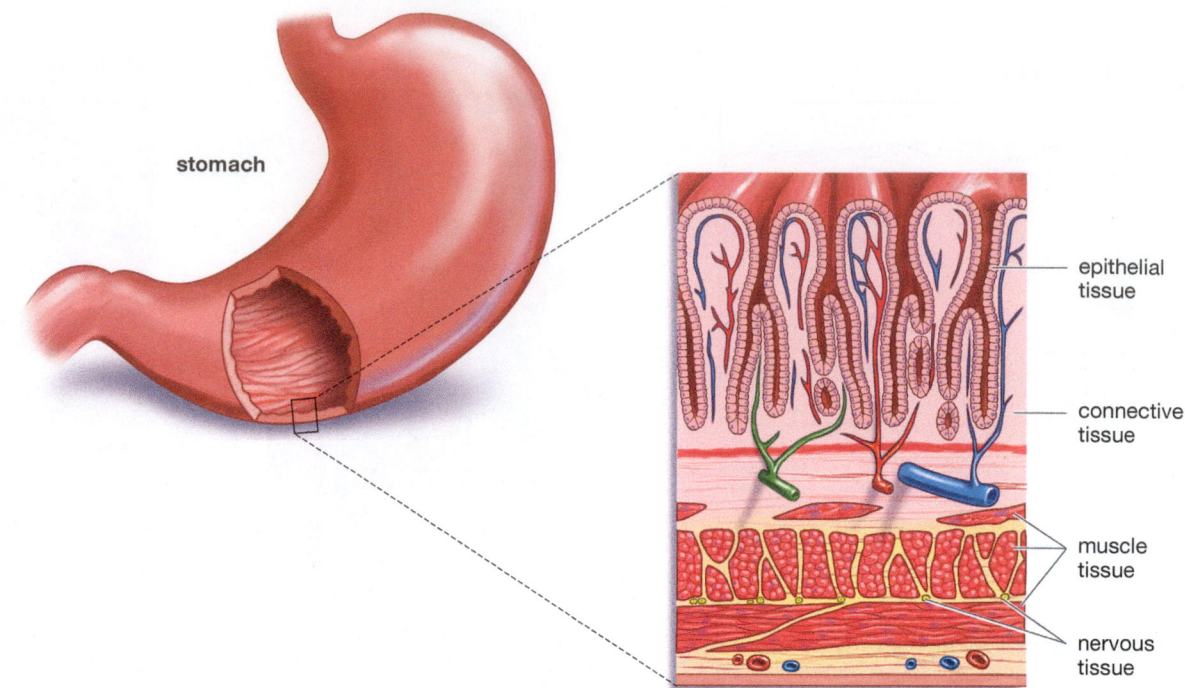

stomach

epithelial tissue

connective tissue

muscle tissue

nervous tissue

discussing. If you look at **Figure 26.5**, you can see that the stomach has several layers of smooth muscle, which bring about contractions that help digest food and move it into the small intestines. Likewise, there is nervous tissue, which controls the release of digestive juices and the contractions of the stomach's muscles. Supportive connective tissue exists as well. And, of course, since the inner lining of the stomach is a surface coming into contact with an external environment, there is epithelial tissue. True to what you saw earlier, some of this epithelial tissue is secretory—that is, it secretes substances that help with digestion.

26.5 Organs and Tissues Make Up Organ Systems

You've now seen how cells make up tissues and how tissues make up organs. Let's now take the final step and see how tissues and organs together make up organ systems. What follows is a brief look at each of the body's 11 organ systems. Once you have finished reviewing them, the balance of this chapter will be devoted to a more detailed look at the first three of these systems. In Chapters 27 through 33, you'll be taking a look at the other eight.

Organ Systems 1: Body Support and Movement—The Integumentary, Skeletal, and Muscular Systems

- The **integumentary system** (**Figure 26.6a**) is sometimes thought of simply as skin, but a number of structures are associated with skin, including hair and nails and the glands we noted

earlier. In addition to serving as an outer, protective wrapping, skin also helps keep body temperature stable.

- The **skeletal system** (**Figure 26.6b**) is composed not only of bones but of cartilages and ligaments. This is a complex framework, as human beings have 206 bones in their body, along with a large number of cartilages. Bones provide support and protection, and they can store fat and minerals.

- The **muscular system** (**Figure 26.6c**) includes all the skeletal muscles of the body—about 600 of them. These muscles provide for movement, posture, and support. They are also vital players in homeostasis in that our bodies are largely warmed by the heat given off from muscle contraction. The smooth and cardiac muscle tissues noted earlier are not part of the muscular system. Cardiac muscle is part of the cardiovascular system, and smooth muscle is part of several systems—in the stomach, for example, it is part of the digestive system.

Organ Systems 2: Coordination, Regulation, and Defense—The Nervous, Endocrine, and Immune Systems

- The nervous tissue reviewed earlier goes on to make up the **nervous system** (**Figure 26.6d**), a rapid communication system in the body that includes the brain, spinal cord, all the nerves outside the brain and spinal cord, and sense organs such as the eye and ear.

- The endocrine system (**Figure 26.6e**) bears comparison with the nervous system in that both systems are in the business of sending signals throughout the body. The **endocrine system,** however, is a communication system that works through chemical messengers called **hormones:** substances secreted by one set of cells that travel through the bloodstream and affect the activities of other cells. Many human hormones are produced in specialized glands. Remember the earlier remark that there are two basic varieties of glands in the body? One of these varieties is named for the endocrine system. **Endocrine glands** are glands that release their materials directly into surrounding extracellular fluid or into the bloodstream. This is what the thyroid gland does, for example. In contrast, **exocrine glands** are glands that secrete their materials through tubes or "ducts." We'll see examples of these glands later.

- The **immune system** is a collection of cells and proteins whose central function is to rid the body of invading microbes. In **Figure 26.6f**, you can see a key component of the immune system, the body's **lymphatic network,** which serves a dual function. First, it acts as a kind of drainage system, collecting fluid that lies outside cells and routing it back to the bloodstream. Second, it acts as an *inspection* system. Lymphatic organs contain heavy concentrations of the body's defenders—its immune system cells—and these cells monitor lymphatic fluid as it passes by for signs of invading microbes. Lymphoid organs include the well-known lymph nodes located under the arms and elsewhere, along with the thymus gland, tonsils, and spleen. Immune system cells and proteins don't reside solely within the lymphatic network, however, but exist throughout the body.

Organ Systems 3: Transport and Exchange with the Environment—The Cardiovascular, Respiratory, Digestive, and Urinary Systems

- The **cardiovascular system** (**Figure 26.6g** on the next page) consists of the heart, blood, and blood vessels and an inner, "marrow" portion of bones where red blood cells are formed. The cardiovascular system is the body's mass transit system. It carries nutrients, dissolved gases, and hormones *to* tissues throughout the body, and it carries waste products *from* tissues to the sites where they are removed: the kidneys and the lungs.

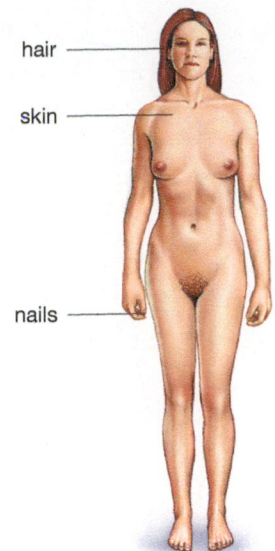

(a) The integumentary system

hair
skin
nails

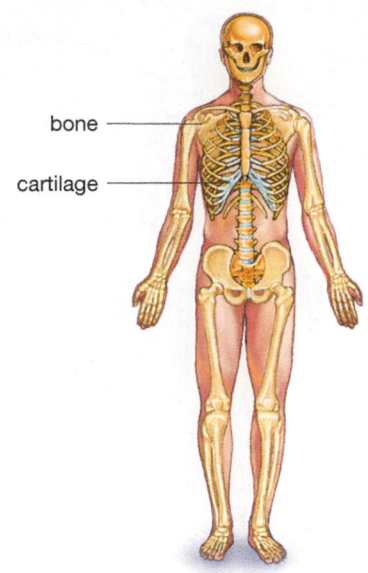

(b) The skeletal system

bone
cartilage

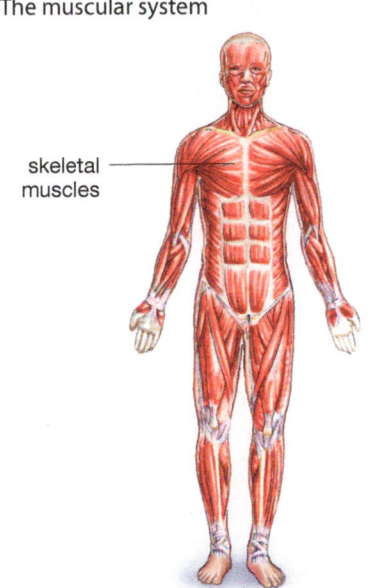

(c) The muscular system

skeletal muscles

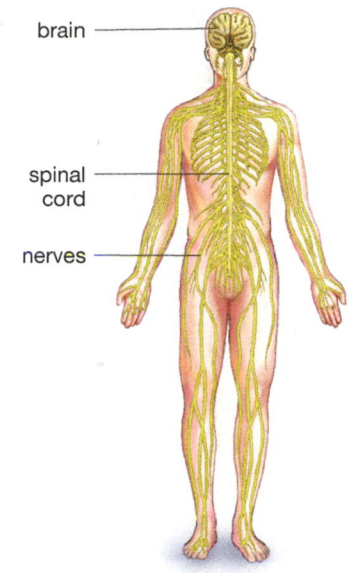

(d) The nervous system

brain
spinal cord
nerves

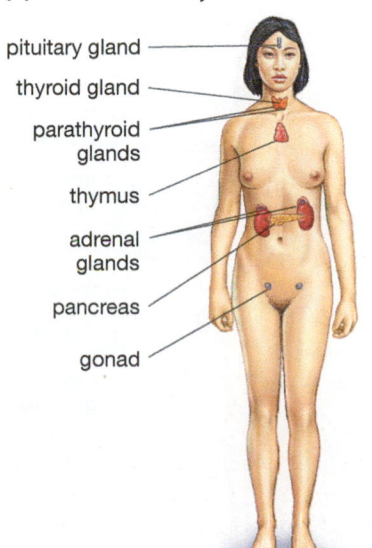

(e) The endocrine system

pituitary gland
thyroid gland
parathyroid glands
thymus
adrenal glands
pancreas
gonad

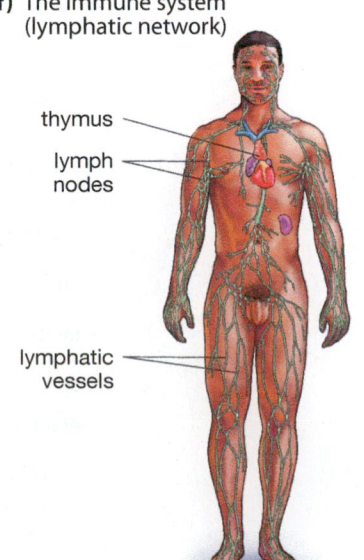

(f) The immune system (lymphatic network)

thymus
lymph nodes
lymphatic vessels

Figure 26.6a–f
The Organ Systems of the Human Body

(g) The cardiovascular system

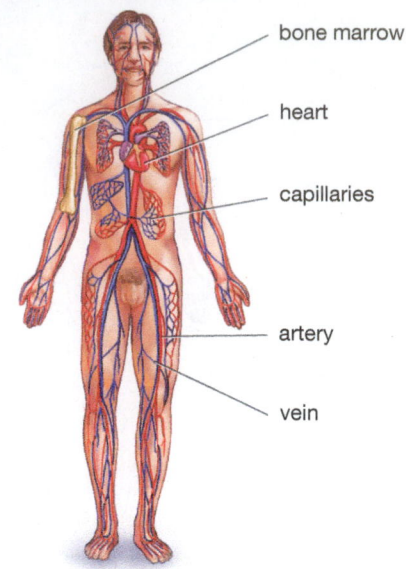

- bone marrow
- heart
- capillaries
- artery
- vein

(h) The respiratory system

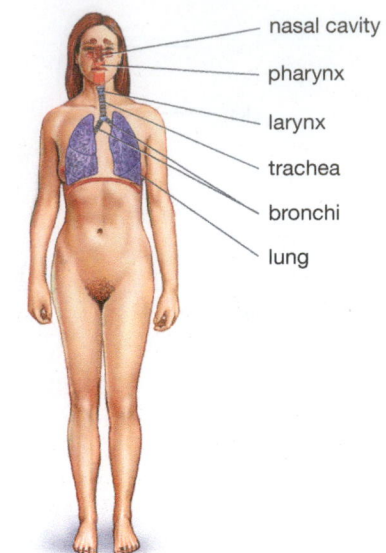

- nasal cavity
- pharynx
- larynx
- trachea
- bronchi
- lung

(i) The digestive system

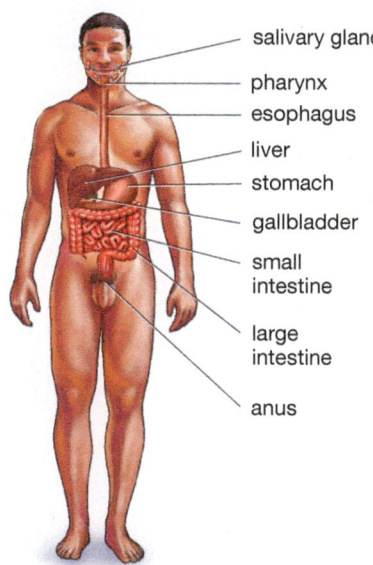

- salivary gland
- pharynx
- esophagus
- liver
- stomach
- gallbladder
- small intestine
- large intestine
- anus

(j) The urinary system

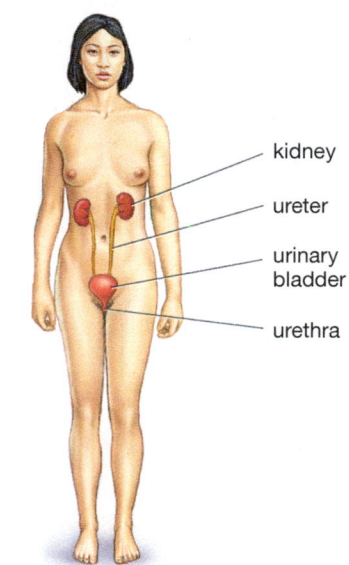

- kidney
- ureter
- urinary bladder
- urethra

(k) The male reproductive system

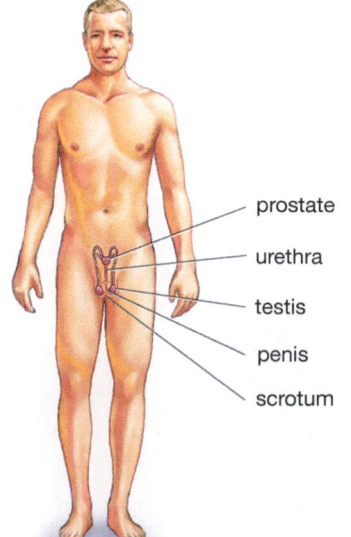

- prostate
- urethra
- testis
- penis
- scrotum

(l) The female reproductive system

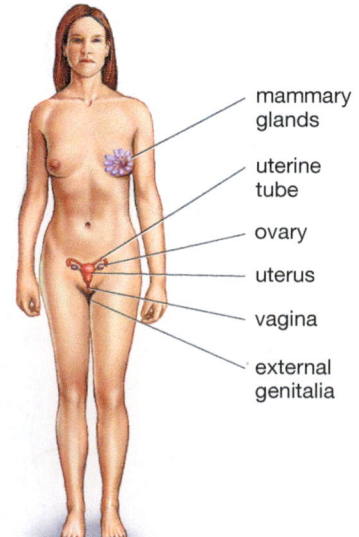

- mammary glands
- uterine tube
- ovary
- uterus
- vagina
- external genitalia

- The **respiratory system** (**Figure 26.6h**) includes the lungs and the passageways that carry air to and from the lungs. Through this system, oxygen comes into the body, and carbon dioxide is expelled from it.

- The central feature of the **digestive system** (**Figure 26.6i**) is the digestive tract, a long tube that begins at the mouth and ends at the anus. Along its course, there are the digestive organs of the system—the stomach, small intestines, and large intestines—and accessory glands, such as the pancreas and liver.

- The **urinary system** (**Figure 26.6j**) has two major functions, one of them well known, the other not so much appreciated. The well-known function is the elimination of waste products from the blood through the production of urine. The underappreciated function is conservation. The system is very good at retaining what is useful to the body—water, proteins, and so forth—even as it eliminates what is useless. The primary organs that carry this selection process out are the kidneys, but the system also includes various other parts, such as the bladder.

- Males and females have different **reproductive systems** (**Figures 26.6k** and **Figures 26.6l**). The two are linked, of course, in that together they produce offspring. Both systems are discussed in detail in Chapter 33.

With this brief review of the organ systems completed, you're ready to look in more detail at the first three organ systems noted earlier—the integumentary, skeletal, and muscular systems. They are batched together in this chapter because each has to do with body support and movement. We'll start with the integumentary system.

SO FAR . . .

1. Going from least inclusive to most inclusive, how would the following levels of organization be ordered in a human being: organ system, cell, tissue, organ?

2. What are the four basic tissue types in a human being?

3. _____ tissue is specialized in its ability to contract, while _____ tissue is specialized in its ability to conduct electrical impulses.

Figure 26.6g–l

26.6 The Integumentary System

The primary component of the integumentary system is the organ known as skin. It may seem strange to think of skin as an organ since it is not a compact, neatly defined structure like the heart or the liver. But remember the definition of an organ: a complex of several kinds of tissues that performs a special bodily function. Clearly, that's what skin is. Joining skin to make up the integumentary system are the related structures of hair, nails, and a variety of exocrine glands.

The primary function of skin is easy to see: It covers the body and seals it off from the outside world. Skin also protects underlying tissues and organs, however, and helps regulate our temperature. In addition, it stores fat and makes vitamin D. Meanwhile, exocrine glands that are accessories to the skin excrete materials such as sweat, water, oils, and milk, and specialized nerve endings in the skin detect such sensations as pressure, pain, and temperature.

The Structure of Skin

Skin is an organ that is organized in two parts: a thin outer covering, the epidermis, and a thicker underlying layer, the dermis, which is composed mostly of connective tissue. Beneath the dermis—and not part of the skin proper—is another layer, called the hypodermis, made up of connective tissue that attaches the skin to deeper structures such as muscles or bones. You can see a cross section of human skin in **Figure 26.7**.

The Outermost Layer of Skin, the Epidermis

The outer layer of our skin, the epithelial tissue called the **epidermis,** seems to face an impossible task. On the one hand, it has to serve as a permanent, protective barrier, keeping out everything from water to invading microorganisms; on the other, it has to continually renew itself, given that it is cut, scraped, and simply worn away each day. How does it manage to do both things? The answer lies in the many *layers* of cells the epidermis is organized into. If we could take a microelevator down to the innermost layer of the epidermis, we would see a group of rounded cells rapidly dividing and, in the process, pushing the cells on top of them *up*, toward the surface of the skin. As we move toward the surface, we notice that the epidermal cells are flattening out. Furthermore, they're getting less active; at a certain point, they cease dividing altogether. What is happening is that they are now so far from the underlying blood vessels (down in the dermis) that they are losing their blood supply. They are still carrying out a task, however, which is the production of **keratin**—a flexible, water-resistant protein, abundant in the outer layers of skin, that also makes up hair and fingernails. By the time epidermal cells reach the surface, they are scarcely cells at all anymore. They are dead and don't even have the remnants of internal organelles within them. Instead, they have become a series of tightly interlocked keratin sacs. In another 2 weeks, they will be scraped or washed away, replaced by the next cells that have pushed up from the inner epidermis.

Figure 26.7
Our Outer Wrap: What Is Skin Made Of?

(a) A micrograph of human skin, showing its epidermal and dermal layers and the subcutaneous hypodermis tissue beneath them. Note the sloughing off of dead cells at the surface of the epidermis. The holes visible in the dermis are spaces through which its blood vessels run.

(b) A cross section of the skin's epidermis and dermis layers and the hypodermis layer beneath them.

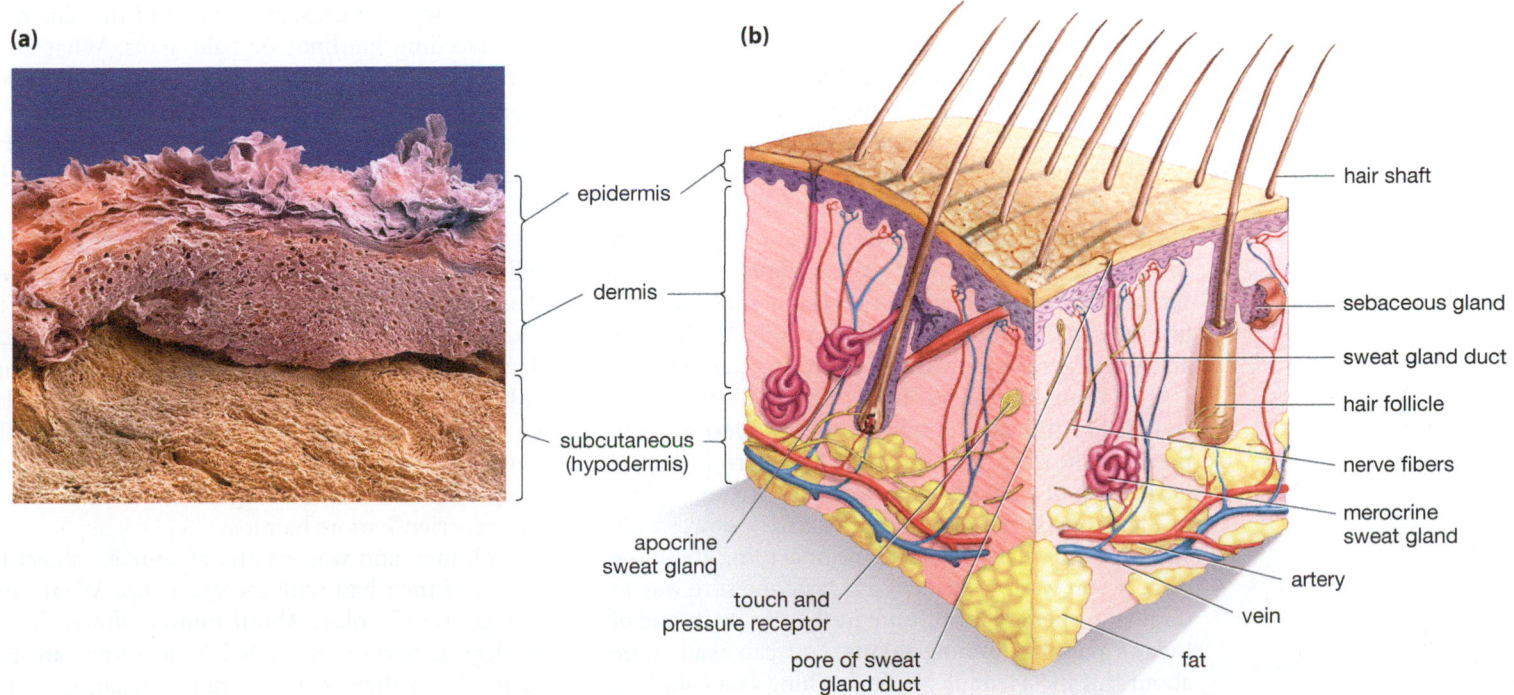

(a)

(b)

- epidermis
- dermis
- subcutaneous (hypodermis)
- apocrine sweat gland
- touch and pressure receptor
- pore of sweat gland duct
- hair shaft
- sebaceous gland
- sweat gland duct
- hair follicle
- nerve fibers
- merocrine sweat gland
- artery
- vein
- fat

Beneath the Epidermis: The Dermis and Hypodermis

Whereas the epidermis is epithelial tissue, the inner layer of skin, the **dermis,** is mostly connective tissue. But if you look again at Figure 26.7, you can see that nerves and muscles exist in the dermis as well, along with sweat glands, hair follicles, and more. This is just another example of how all the various kinds of tissues tend to exist side by side in a typical organ.

Many of the structures in the dermis are serving the epidermis. For example, the nervous system touch and pressure receptors shown in Figure 26.7 allow us to know when something makes contact with our skin, and the blood vessels pictured provide a blood supply to the dividing lower cells of the epidermis. But the dermal blood supply takes on another role as well. It plays a big part in controlling our temperature. If you get too cold, the dermal blood vessels will constrict, thus moving blood away from your surface and retaining the heat that's in the blood. If you get too hot, the dermal blood vessels expand, thereby moving blood toward your surface so that the heat that's in blood can radiate to the outside world (making your skin feel "flushed"). One more thing to note about the dermis is that its cells are relatively fixed, unlike those of the outer epidermis. Ever wonder why a tattoo must be made with a needle sunk into the skin? The tattoo's pigment must be deposited into the fixed cells in the dermis rather than the ever-changing cells of the epidermis.

As noted earlier, the "hypodermis," the layer beneath the dermis, is not actually part of the skin. But it deserves special mention here because of one of the aspects of it you can see in Figure 26.7—its abundant adipose or fat tissue. This tissue makes up the body's *subcutaneous fat,* which can be thought of as a layer of fat located just beneath the skin that varies in thickness throughout the body. The body also has what is known as *visceral fat,* however, which is fat that lies deep within the abdomen (inside the abdominopelvic cavity we talked about). This "belly fat" doesn't exist as a single layer but instead surrounds organs such as the liver. When it exists in excess amounts, it is a more serious health risk than is an excess of subcutaneous fat. You can read more about this in "Why Fat Matters and Why Exercise May Matter More," on page 492.

The term *hypodermis* may ring a bell because it sounds like the more familiar term hypoder*mic,* as in needles. The deeper tissue of the hypodermis doesn't have much of a blood supply and thus is a good place to inject medicines that need a relatively slow, steady entry into circulation. Here's one final thing about the skin. If you want to make it age fast, one sure way to do it is to spend a lot of time in the sun or in one of today's popular tanning salons. You can read more about this in "There Is No Such Thing as a Fabulous Tan" on page 495.

Accessory Structures of the Integumentary System

As noted, skin has a number of accessory structures associated with it, including hair, glands, and nails.

Hair

Five million or so hairs exist on the human body, although only about 100,000 of them are on the human head. Hair clearly serves a major cosmetic role in modern society, but why does it exist on human beings in the first place? You might think that heat retention for our ancestors would have been its main function, but maybe not. Rub your hand with just the lightest stroke over the hair on your head and you can get a sense of how exquisitely sensitive hair is to our sense of *touch.* Like a cat's whiskers, hair may have served primarily to fine-tune our sense of contact with the world around us.

How can hair have this quality since it is not alive but merely a shaft of the protein keratin, which is a product of underlying cells? If you look at Figure 26.7, you can see that each hair grows out of a tube-like structure called a *hair follicle* that begins fairly deep in the dermis. Each follicle is a tissue complex that has not only its own blood vessels and muscles but its own nerve endings as well. Thus, when you touch a hair, you stimulate a hair follicle's nerve endings.

True to what we noted at the start of the chapter, the body's hair is perhaps more missed when it is gone than appreciated when it is here. Many men start losing the hair on their heads in their early 20s, a trend that accelerates as they get older. Women can lose hair, too, although their loss tends to come later in life and usually involves a general thinning of the hair rather than receding hairlines or bald spots. What causes hair loss? Follicles go through growth and resting phases, with a given hair on our heads growing for anywhere from 2 to 6 years until a short resting phase is reached. At that point, the hair drops out, and its follicle's growth phase starts again. Hair loss generally is a condition in which an ever-growing proportion of follicles start to shrink, resulting in a briefer growing phase and hair that comes out either short and fine (like a baby's) or doesn't come out at all. Genes clearly play a part in this process, although contrary to popular belief, it is not just genes from the mother's side of the family that are involved. Hormones called *androgens* are involved as well; men have higher circulating levels of such androgens as testosterone and thus experience more hair loss.

Both men and women are, of course, subject to a *graying* of their hair with increasing age. What causes hair to lose its color? About midway down through the dermal portion of the hair follicle you can see in Figure 26.7b there are cells called melanocyte stem cells. When we are young, these cells produce a steady stream of cells that first migrate to near the bottom of the hair follicle and then differentiate into cells called mature melanocytes. These are the cells that produce the pigments that give hair its color. We thus have a kind of pigment-producing assembly line in our hair follicles, but it's one that aging tends to disrupt in two ways. One is that our melanocyte stem cells can die off altogether; the other is that our mature melanocytes can start maturing too quickly—they differentiate *before* they have migrated to the deepest part of the follicle, which is the only location in which they can function properly.

Glands

You may remember that when we went over the endocrine system, you saw that glands come in two varieties: those that secrete their substances directly into the bloodstream or surrounding tissue (endocrine glands) and those that secrete their substances through the tubes called ducts (exocrine glands). Skin contains two types of exocrine glands. The first of these are known as **sebaceous glands:** glands that produce an oily substance called sebum that is secreted into hair follicles and that then moves through them onto the skin (see Figure 26.7b). Once on the skin, sebum lubricates the hair and inhibits bacterial growth in the surrounding area.

The problem with this process is that, under hormonal influence, sebaceous glands can be *too* active—they can produce too much sebum, particularly in teenagers. When this excess sebum mixes with skin cells that are being shed from the sebaceous follicle, the result can be a blocked follicle that swells with sebum—a whitehead. Once this blockage occurs, bacteria normally found on the skin begin to multiply within the follicle, producing irritating substances that can cause general inflammation and pimples. Taken together, these various skin afflictions are known as *acne*. No matter what you may have heard, acne is not caused by eating fatty foods, stress, or a lack of face scrubbing. As with baldness, the hormones called androgens are the culprits. They prompt the increased sebaceous activity; since males at puberty have more androgens in them than females, it is males who suffer more from acne.

The second type of exocrine gland in the skin is the **sweat gland,** simply defined as a gland that produces the fluid sweat. This can mean a couple of different things, however, because sweat glands come in two varieties. One type is a *merocrine* gland that produces the clear, saltwater-like fluid we normally think of as sweat. We have, at a minimum, 2 million merocrine sweat glands on our bodies. They exist on almost all skin surfaces, but we have an especially dense concentration of them on our feet, forehead, and palms (hence "sweaty palms"). The primary function of the merocrine sweat glands is to bring sweat to the surface of the skin so that it can cool us by evaporating. These glands have their own ducts and follicles, which begin deep in the dermis.

The other sweat glands in the body, called *apocrine* glands, are very different from the merocrine glands. First, they only exist in a few places—notably the groin, the anal area, and the armpits. As such, they have little effect on body cooling. Second, they secrete their products into hair follicles, not up through their own ducts. Most important, apocrine gland secretions contain not only regular sweat but a fatty, milky fluid whose function in humans is unknown. The *effect* of this fluid is very well known, however. Bacteria feed on it, and the resulting bacterial by-products are the cause of body odor. Apocrine glands only begin functioning at puberty, which is why children seldom have body odor.

Nails

Our final accessory structure of the integumentary system is the nails found on our fingers and toes. Like hair, nails are made of a variety of the protein keratin. They are clearly useful for protection and for prying, scratching, and picking up objects. Nails allow us to look, in a sense, at the blood vessels underlying the dermis because it's the reddish color of these vessels that gives nails their pink hue.

SO FAR . . .

1. The primary component of the integumentary system is an organ called _____, which is composed of two primary layers, an outer layer called the _____ and an inner layer called the _____.

2. Fat exists in the human body in two primary locations. First there is subcutaneous fat, which can be thought of as a layer of fat located _____; second there is visceral fat, which is fat that surrounds _____ deep within the _____.

3. Each hair that we have grows out of a tissue complex called a _____.

26.7 The Skeletal System

We now shift gears, going from a consideration of our outer covering to a consideration of the framework that underlies it—our skeletal system. This system has only three structural elements to it: bone, ligaments, and cartilage. Most of us are quite familiar with **bone,** which can be defined as a connective tissue that provides support and storage capacity and that is the site of blood cell production. Ligaments are fairly inaccessible to us;

ESSAY

Why Fat Matters and Why Exercise May Matter More

At one time, scientists thought of the body's billions of fat cells as little more than fuel depots—as storage sites for the calories we ingest but don't burn up (**Figure 1**). This model began to fall apart in 1994, however, with the discovery of a hormone called leptin that plays a role in regulating hunger and that is *produced* by fat cells. Two years later, scientists made an even more puzzling discovery when they found that fat cells also produce a signaling molecule, called tumor necrosis factor (TNF), that helps activate the body's

In the years that followed the TNF discovery, this mystery deepened, when it was discovered that *several* immune signaling molecules are secreted by fat cells and that several specialized varieties of immune system cells take up residence within collections of fat cells. Continuing research revealed, however, that it isn't all fat cells that are involved in these immune system activities, but instead only a select group, located in one part of the body.

As the main text notes, human beings have concentrations of fat cells in two

"belly fat" that draws in immune system cells, that puts significant amounts of TNF into circulation, and so forth. Indeed, the picture that has emerged is that excess visceral fat sets off a low-level immune system response—the inflammation response. When prompted by microbial attack, this response has the virtue of ramping up the body's microbe-fighting capabilities within a limited time frame: Attack-driven

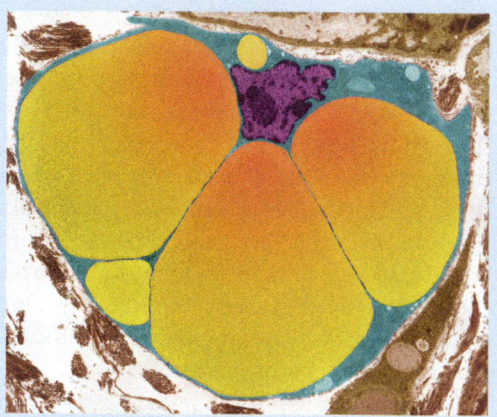

> *It's perfectly possible for a person who is overweight and active to be healthier than a person who is slim but sedentary.*

immune system during periods of viral or bacterial infection. Leptin made sense as a fat-cell hormone, since the message it conveys to the brain—"you're full now"—is one that is consistent with the role fat cells have of storing nutrients. But what did fat cells have to do with the immune system?

separate areas. On the one hand, we have subcutaneous fat, which is the layer of fat that exists just beneath the skin throughout the body. On the other, we have visceral fat, which is fat that surrounds our internal organs deep within the abdomen. Over the years, research has confirmed that it's this

Figure 1
Where Excess Calories Go

Fat cells, such as the one shown here, often contain several lipid droplets whose size dwarfs the cell's other components. In this case, the three large central droplets do just this. The cell's nucleus is the purple object near the top.

Figure 26.8
Elements of the Skeletal System

Bones are the basic element of the skeletal system. Ligaments link one bone to another, and cartilage often serves as padding between bones. Tendons, while not part of the skeletal system, always link bone to muscle.

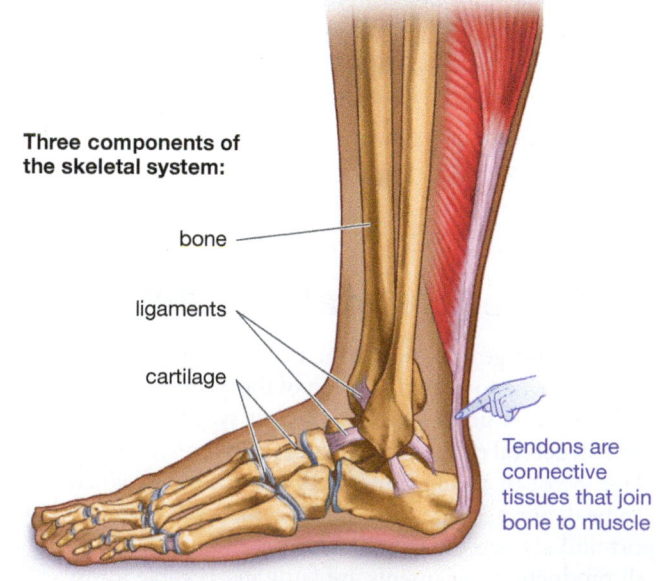

Three components of the skeletal system:

bone

ligaments

cartilage

Tendons are connective tissues that join bone to muscle

but if you look at **Figure 26.8**, you can see some typical ligaments carrying out a typical task—linking bones together. You can also see the third element of the skeletal system, cartilage. Note that cartilage is located *in between* bones in the foot, and this turns out to be one of its common functions. **Cartilage** is a connective tissue that serves as padding in most joints. It is more familiar to us, however, in other parts of the body. It forms our larynx (voice box), the C-shaped "rings" on our trachea (windpipe), and our nose tip and outer ears, and it links our ribs to our breastbone. In all its capacities, cartilage is a very flexible, resilient tissue, which is not surprising given that it is mostly made up of water.

It's a good idea to look at Figure 26.8 to get a sense not only of the difference between bone, ligaments, and cartilage but also of the difference between all of these skeletal elements and another type of tissue that is pictured, tendons. Although not part of the skeletal system, tendons are always associated with

inflammation begins with the microbial invasion and ends when all the microbes have been vanquished. By contrast, an excess of visceral fat engenders *chronic* inflammation, meaning an inflammation that exists day in, day out, for as long as a person carries the fat.

What's the harm in this? Well, consider the signaling protein TNF noted earlier. One of its effects is to make the body's muscle, liver, and fat cells resistant to the effects of another signaling protein, insulin. In a person who is, say, under attack by a virus, this effect may be helpful: The body's immune system cells need extra energy during times of attack and to the extent that insulin resistance is keeping blood sugar from moving into muscle cells, it is making this sugar available to immune system cells. In a state of chronic inflammation, conversely, this effect is anything but helpful. When cells enter into a state of *long-term* resistance to the effects of insulin, the body has started down the path to type 2 diabetes. Blood sugar is persistently kept from entering muscle, liver, and fat cells, with the result that these cells start breaking down their own proteins and lipids just to stay alive.

Since fat-driven inflammation is a systemic, or body-wide, physical condition, it's not surprising that its effects extend beyond those of spurring on diabetes. Consider, for example, a condition called congestive heart failure, in which the heart loses its ability to pump sufficient blood throughout the body. In 2009, researchers released the results of a study of more than 80,000 Swedish men and women whose health status they had tracked for seven years. The scientists found that an excess of visceral fat was a risk factor for heart failure even in persons whose *overall* weight was healthy (as measured by body-mass index, or BMI). Meanwhile, in 2008, researchers in California reported that men and women they studied who were overweight and had large bellies at ages 40–45 were more than three times as likely to be diagnosed with Alzheimer's or other dementias at age 70–75 than were subjects who originally had normal weight and belly size. Critically, subjects who had a large belly size, but who were of *normal* weight in their 40s were still almost twice as likely to have dementia 30 years later than were persons of both normal weight and belly size.

The central message of these research findings is that, while fat matters, the location of that fat seems to matter even more. The good news about this is that something can be done about it. Visceral fat often is the last to be added during weight gain, which means it is likely to be the first to be lost during any program of weight reduction. Beyond this, moderate "endurance" exercise, such as walking or swimming, has been shown to reduce levels of chronic inflammation in the body, irrespective of any weight loss this activity may bring about. Indeed, so strong are the effects of exercise that it's perfectly possible for a person who is overweight and active to be healthier than a person who is slim but sedentary. Evidence of this was provided in 2008 by researchers who looked at a representative sample of 5,400 American adults and ranked their "metabolic fitness" on the basis of such factors as blood pressure and insulin sensitivity. Weight in general, and larger waist size in particular, did tend to make individuals less metabolically fit, but the researchers also found that 51 percent of overweight subjects were metabolically healthy, while 23 percent of normal-weight subjects had two or more indications of poor metabolic health. What factors tended to push slimmer people into the less healthy category? One was the simple fact of getting older, but the other was a lack of regular physical activity.

the bones of the system. Looking at all four elements in the figure, here's one way to think of them. Bones are the basic framework; cartilage often serves as padding between bones; **ligaments** are tissues that join bone to bone; and **tendons** are tissues that join bone to muscle.

Function and Structure of Bones

It's tempting to think of bones as nothing more than support beams, but this conception is wide of the mark in two ways. First, support beams are not alive, but bone is very much a living, dynamic tissue, as you'll see. Second, bones do more than just support us. Once we are past infancy, all of our blood cells (both red and white) are produced in the interior of our bones. Some bones store fat, in the form of yellow marrow, as a kind of energy reserve, and bones serve as the storage sites for important minerals such as calcium and phosphate.

Beyond this, the only reason we can move at all is that our muscles are attached to bones. When we contract our bicep, for example, our forearm is raised by means of the bicep pulling on the bones of the forearm.

The Structure of Bone

Each bone in our body actually is considered a separate organ because each bone is composed of several kinds of tissue. There is the calcified or "osseous" tissue we think of as bone, to be sure, but each bone also has its own blood vessels and nerves. (Cartilage, meanwhile, has neither its own blood vessels nor nerves, which is why torn cartilage takes so long to heal—nutrients must diffuse to it from surrounding fluids, rather than being delivered to it by blood vessels.) Bones are a great example of connective tissue in that the *cells* in bones may account for as little as 2 percent of bone mass. What is the other 98 percent of bone made of? It is overwhelmingly a ground substance, secreted by bone cells, that is composed of hard, calcium-containing crystals and tough, yet flexible collagen fibers. This combination means that bones are hard without being brittle—they can support our weight and yet can bend or twist a little without breaking.

Large-Scale Features of Bone

The typical features of a long bone such as the humerus (in the arm) are shown in **Figure 26.9**. A long bone has a central shaft, or diaphysis, and expanded ends, or epiphyses (singular, epiphysis). There are two types of bone. **Compact bone,** which is relatively solid, forms the outer portion of the bone (all the way around it, as you can see in Figure 26.9b). **Spongy bone** then fills the epiphyses and is well named. It is porous enough that it contains another type of tissue, **red marrow,** which is the tissue in which all the blood cells of the adult body are produced. (These cells include not only the oxygen-carrying red blood cells but the white blood cells that are the central cells of the immune system.) Meanwhile, **yellow marrow,** a tissue largely made up of energy-storing fat cells, is found in the *marrow cavity* of the type of long bones seen in Figure 26.9.

Small-Scale Features of Bone

Dropping down to the microscopic level and looking at bone *cells,* we find that three different types are involved in the growth and maintenance of bone. **Osteoblasts** are immature bone cells that are responsible for the production of new bone. They secrete the material that becomes the bone ground substance. Once osteoblasts are surrounded by this material, they reduce their production of it, and this marks their transition to osteocytes, the second type of bone cell. **Osteocytes** are mature bone cells; they *maintain* the structure and density of normal bone by continually recycling the calcium compounds around themselves. The third type of bone cell, the osteoclasts, could be thought of as the demolition crew of bone tissue. **Osteoclasts** are cells that move along the outside of bones, releasing enzymes that eat away at bone tissue, thus liberating minerals stored in the bone. This may sound harmful, but it actually is essential for bone growth and for the regulation of blood levels of calcium. The upshot of all this is that, at any given moment, osteoblasts are adding to bone, osteocytes are maintaining it, and osteoclasts are removing it.

Within compact bone, the basic functional unit is the osteon seen in Figure 26.9c (which is also known as the Haversian system). The essence of bone growth is that, within each osteon, osteoblasts produce layer after layer of concentric cylinders of bone—much like layers of insulation wrapped around a pipe. These layers surround what could be thought of as the hole in the pipe, which is a central canal (also known as a

Figure 26.9
Large- and Small-Scale Features of a Typical Bone

Note that in **(c)**, the building units of compact bone, the osteons, run parallel to the long axis of the bone. Spongy bone does not contain osteons, but instead is composed of an open network of calcified rods or plates.

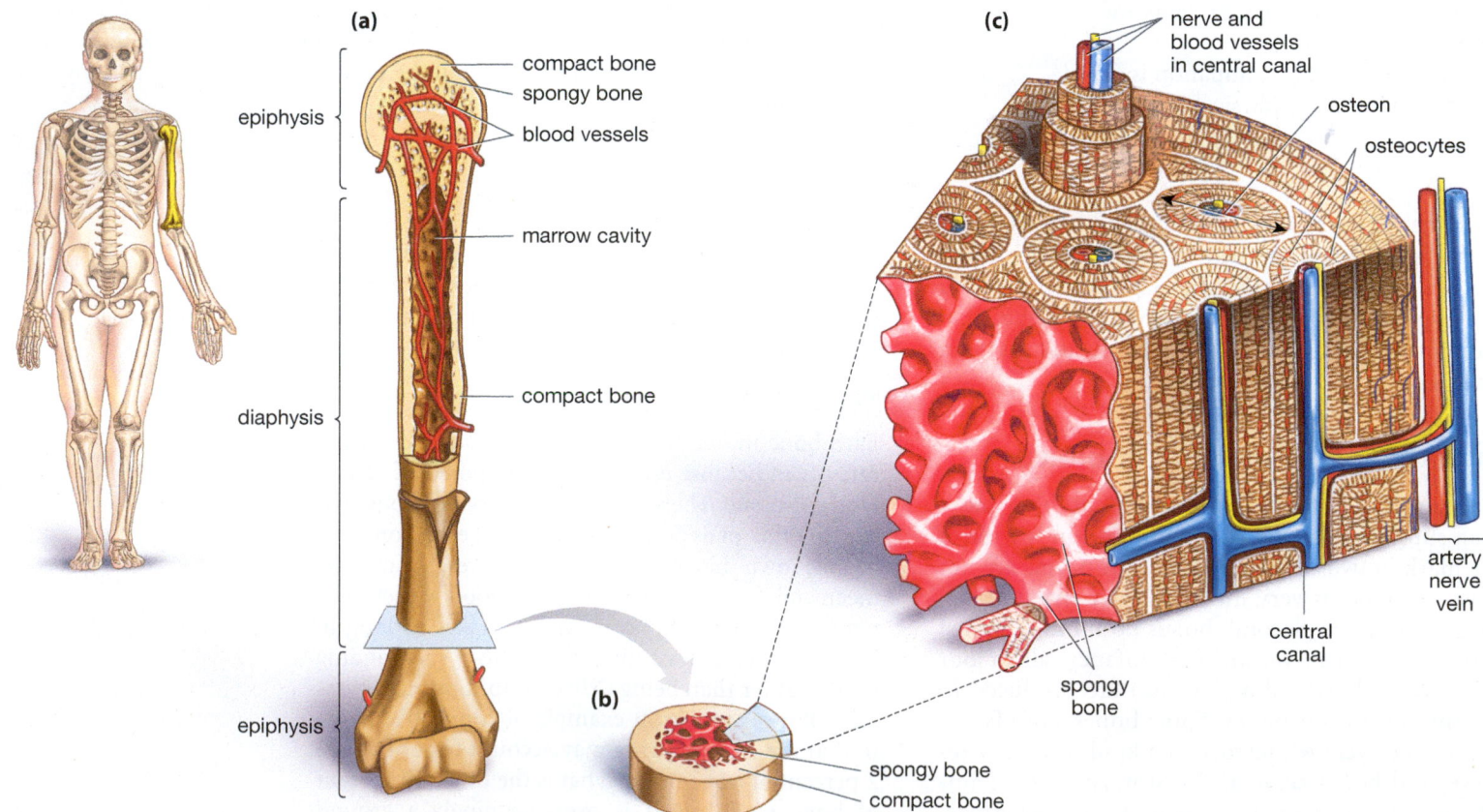

ESSAY

There Is No Such Thing as a Fabulous Tan

The constant message from health professionals these days is to avoid exposure to the sun, but this is a message society has not really taken to heart. Consider that there are an estimated 50,000 businesses in the United States devoted to *giving* people tans through exposure to ultraviolet light—the very thing the sun delivers. On an average day, more than a million Americans will visit tanning salons, with more than 70 percent of these people being women (**Figure 1**). The ongoing attraction that tanning salons hold may in part be attributable to a message the tanning industry tries hard to convey, which is that tanning beds provide a form of "safe tanning." Unfortunately, the scientific evidence says this is not the case.

For starters, the radiation provided by tanning beds is not likely to be less intense, nor of a fundamentally different nature,

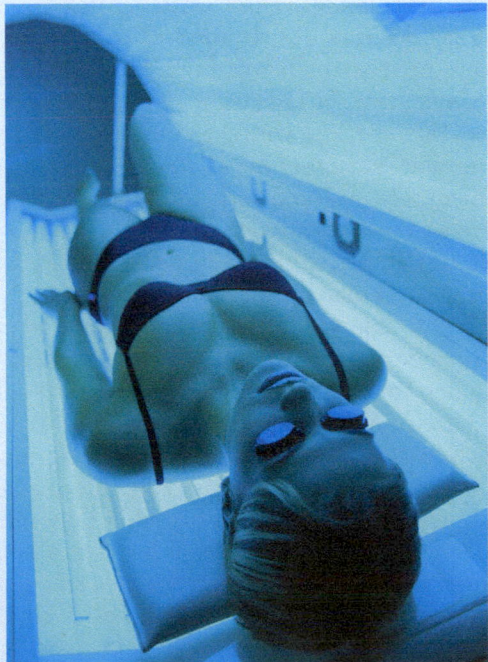

Figure 1
Skin Aging in Progress
A young woman soaks up ultraviolet radiation in a tanning bed.

than the radiation that comes from the sun. The ultraviolet sunlight that reaches Earth during summer months in North America is at least 96 percent UVA radiation, with the balance being the shorter-wave UVB radiation. Meanwhile, the radiation delivered by most tanning beds is

> *The vitamin D produced by UV exposure is chemically identical to the vitamin D that can be obtained by taking a common vitamin pill.*

about 95 percent UVA. It's not clear how much this distinction matters, however, because both UVA and UVB radiation are mutagenic—both cause mutations to the DNA in skin cells—though UVB radiation is more powerful in this respect. Summarizing a great deal of research on this subject, dermatologist Thanh-Nga T. Tran and his colleagues noted in a 2008 review that the chemical pathway that produces a tan in the body is the same pathway that leads to skin cancer. Given this, "safe tanning" through UV exposure "may be a physical impossibility," the researchers wrote. A 2006 analysis of studies that looked at tanning bed exposure and the most serious form of cancer, melanoma, underscored the validity of this message. It found a 15 percent greater risk of contracting melanoma among people who had ever used tanning beds, compared to people who had never used them, and a 75 percent greater risk among people who first used the beds before age 35.

The tanning industry has attempted to counter scientific findings such as these with claims that sun-bed tanning offers health *benefits*—notably a boost in vitamin D levels in the body through UV exposure. Left unsaid is the fact that the vitamin D produced by UV exposure is chemically identical to the vitamin D that can be obtained by taking a common vitamin pill.

The customers who flock to tanning beds may be unaware of the details of this health picture, but they understand in a general way that UV exposure is not healthy. So why do they continue to subject themselves to it? Lots of factors undoubtedly are at play in their decision—they simply think a tan looks good, they minimize the risk they are undertaking, they fall prey to false messages about the benefits of tanning. In the last few years, however, scientists have discovered another factor that might be at work. It turns out that UV-induced DNA damage results not only in the production of the substance melanin (which is what gives you a tan) but in the production of something called beta-endorphin, an opiate compound that brings on sensations of pleasure when released in the brain.

In 2004, Wake Forest University dermatologist Steven Feldman and his colleagues decided to see whether beta-endorphin release made a difference in whether people used tanning beds or not. The researchers had a set of 14 volunteers undertake sessions in one of two "tanning" beds, one of which actually had its UV rays blocked by a filter (a fact known only to the researchers). All the volunteers were regular users of tanning beds, and for six weeks all of them were given access to one research bed on Mondays and the other bed on Wednesdays. They were then given the option of getting additional tanning exposure in *either* bed on Fridays. The result? Volunteers who elected to get the additional Friday exposure selected the UV tanning bed 95 percent of the time. Moreover, volunteers who used the beds on any day reported feeling more relaxed and less tense after having been in the UV bed than in the bed with the UV filter. It may be that people continue to tan not only because of the way it makes them look but because of the way it makes them feel.

Haversian canal). These canals parallel the long axis of the bone and have blood vessels and nerves running through them—a kind of support system for the bone.

Unlike compact bone, spongy bone has no osteons. It consists of an open network of interconnecting calcified rods or plates. The red marrow fills the spaces within this network.

Practical Consequences of Bone Dynamics

Now, let's think about the practical consequences of this structure. Remember the earlier observation that bone is a living tissue? With its three kinds of cells constantly adding, maintaining, and breaking down the tissue, you can see the truth of this. Indeed, the bone "remodeling" that these cells undertake is so extensive that our entire skeleton is replaced through it every 10 years.

But bone is more dynamic than even this would indicate in that bones are responsive to our activities. As physiologist Frederic Martini has noted, when you undertake an exercise like jogging, which stresses your bones, the calcium-containing crystals of the affected bones create small electrical fields that apparently attract osteoblasts. On arrival, these osteoblasts begin producing bone. The result? If you take up jogging, you get denser leg bones. But there's a flip side to this: People who are using crutches to take the weight off a broken leg bone may temporarily lose up to a third of the mass in this bone because it is now scarcely being stressed at all.

Interestingly, although we develop almost all our height by late adolescence, we are still gaining in bone density up until about the age of 30. Then, beginning in middle age, the body may start removing more bone than it adds. This phenomenon particularly afflicts women, in the form of a condition known as *osteoporosis,* meaning a thinning of bone tissue that can result in bone breakage and deformation (**Figure 26.10**). Why should osteoporosis affect more women than men? Hormonal changes that come with

menopause can increase the activity of the bone-depleting osteoclasts relative to the activity of the bone-building osteoblasts. One way to guard against osteoporosis is to reach the age of 30 with as much bone density as possible—the equivalent of putting money in the bank as a hedge against a later withdrawal. And there are a couple of proven ways to get more bone density before 30: Take up programs of "weight-bearing" exercise such as jogging or walking and modestly increase calcium intake.

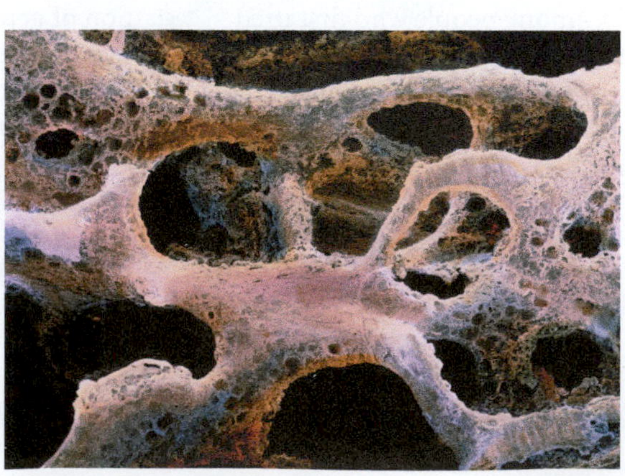

The Human Skeleton

Taking a step back from the microscopic world, you can see in **Figure 26.11** that there are a lot of bones in the skeletal system—206 of them in all. (Only the major bones are visible in the figure.) There are two main divisions to the skeletal system: the **axial skeleton,** whose 80 bones include the skull, the vertebral column, and the rib cage that attaches to it; and the **appendicular skeleton,** whose 126 bones include those of our paired appendages—the arms and the legs, along with the pelvic and pectoral "girdles" to which they are attached. Also note the blue-colored tissue joining ribs to our breastbone and lying in between each of the vertebrae. These are but a few of the cartilages in the body.

Joints

Joints, or articulations, exist wherever two bones meet. What qualities are required of a joint? Well, some only need strength. We think of our skull as being a single bone, but in fact, it's composed of several bones that interlock with each other in joints that are immovable. For other joints, flexibility is the key requirement. The ball-and-socket joint at our shoulder permits a range of motion so extensive that movement of our upper arms is limited more by our shoulder muscles than by our bones.

Highly movable joints are typically found at the ends of long bones, such as those in the legs as well as the shoulder. You can see a view of such a joint—the human knee joint—in **Figure 26.12** on page 498. A joint such as this must serve two competing functions. On the one hand, it must keep the massive femur (thigh bone) and large tibia (shin bone) in a stable arrangement. On the other, it serves as a hinge that allows the tibia to swing as free as a screen door off the femur. How does it do both things? Well, the two bones are partly held in place by four main ligaments, two of which are visible in the figure. Recall that ligaments link bone to bone, but there are also tendons, which attach bone to muscle; the knee joint has these in place as well.

With respect to the motion that such a joint allows, a critical element is that one bone does not actually

Figure 26.10
Bones Weakened by Osteoporosis

The damaging effects of osteoporosis can be seen in this micrograph of a spinal vertebra that has been "thinned" by the condition. Note the many small, circular holes or indentations in the bone; these are areas where bone tissue has been removed at a greater rate than it has been added. The result is a bone that is more susceptible to fractures. The micrograph is of the spongy bone portion of the vertebra; the larger circular holes are thus part of the bone's normal structure.

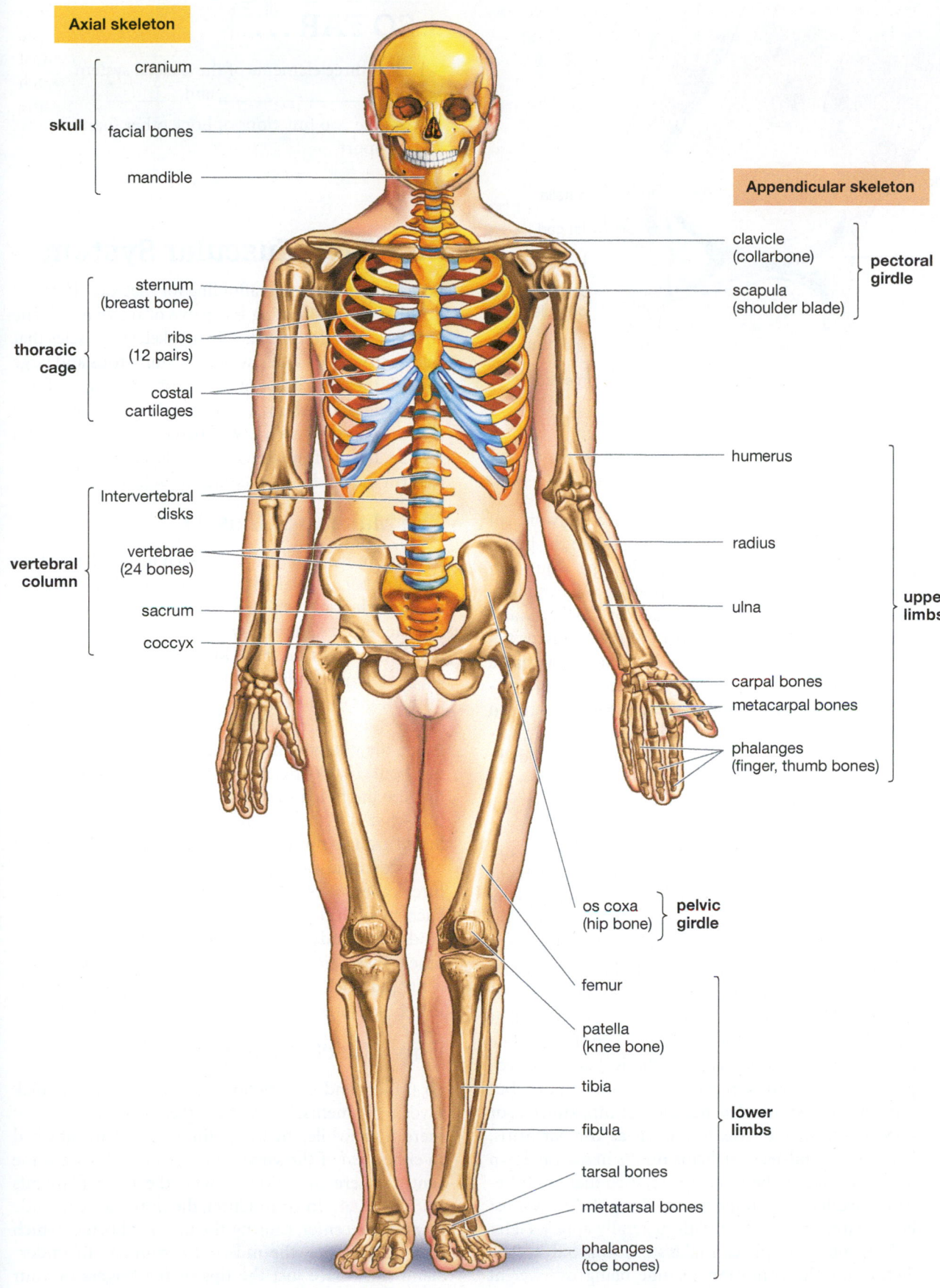

Axial skeleton

skull
- cranium
- facial bones
- mandible

thoracic cage
- sternum (breast bone)
- ribs (12 pairs)
- costal cartilages

vertebral column
- Intervertebral disks
- vertebrae (24 bones)
- sacrum
- coccyx

Appendicular skeleton

clavicle (collarbone)

scapula (shoulder blade)

pectoral girdle

humerus

radius

ulna

upper limbs

carpal bones

metacarpal bones

phalanges (finger, thumb bones)

os coxa (hip bone)

pelvic girdle

femur

patella (knee bone)

tibia

fibula

tarsal bones

metatarsal bones

phalanges (toe bones)

lower limbs

Figure 26.11
Solid Framework
A view of the human skeleton, which has two divisions, axial and appendicular.

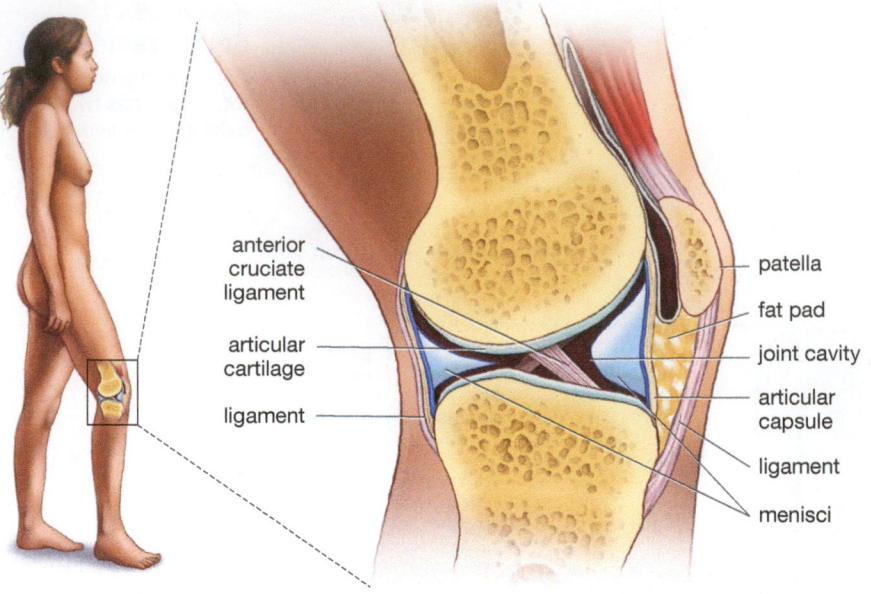

anterior cruciate ligament

articular cartilage

ligament

patella

fat pad

joint cavity

articular capsule

ligament

menisci

Figure 26.12

A Highly Moveable Joint

A simplified view of the human knee joint, which links the femur (thigh bone) to the tibia (shin bone). Note that the two bones never touch each other because they are covered with cartilage. The menisci that you can see are additional cartilage tissue that serves an important shock-absorbing function while also helping to stabilize the joint. The anterior cruciate ligament is the "ACL" we hear about so much in sports injuries. It is one of four main ligaments that keep the femur and tibia in a stable arrangement.

Muscle Structure

touch another. Instead, an extensive network of padding and lubricating tissues in between the bones performs the same function as the "gel" found in running shoes. Note that the top of the tibia and bottom of the femur are surrounded by cartilage, and that more cartilage—in the form of the meniscus tissue—lies on each side of the joint. The yellow tissue shown is fat padding that fills in spaces that are created when the bones move, and the space in between the bones (the joint cavity) is filled with a thick, viscous fluid that not only lubricates joints but aids in their metabolism.

Joints can get injured, of course, and as we age they can wear down altogether. If you follow sports, the name of one of the ligaments shown in Figure 26.12 may ring a bell. The anterior cruciate ligament (ACL), which runs from the back of the femur to the front of the tibia, is a frequent site of injury for athletes who participate in sports that involve rapid "cutting" movements, such as volleyball, soccer, and basketball. An athlete will come to a quick stop, hear a pop, and be down with a torn or ruptured ACL that usually will have to be replaced by grafting tissue onto it from another part of the body. For older people, the usual problem is not a sudden injury, but rather a slow, steady degeneration of hip, knuckle, or knee joints in the disease known as osteoarthritis. Such "wear-and-tear" arthritis results in a breakdown of the cartilage in between bones, such that joints become swollen, with bone sometimes grinding against bone. Although osteoarthritis generally appears later in life, young people can take steps to avoid it. Risk factors for the condition include being overweight and injuring the joints through the traumatic collisions that can take place in sports like football and soccer.

SO FAR . . .

1. The three elements of the skeletal system are _____, _____, and _____.

2. Name two functions of bone other than support.

26.8 The Muscular System

Bones may provide a scaffolding for the body, but how is it that this scaffolding is capable of movement? The answer is muscles, specifically the skeletal muscles that were introduced earlier, which are numerous and large enough that they account for about 40 percent of our body weight. Like bones, skeletal muscles are organs in that they contain a variety of tissues. At each end of a muscle, fibers of the outer muscle layer come together to form the tendons that attach muscle to bone.

The Makeup of Muscle

As you can see in **Figure 26.13**, each muscle has within it a number of oval-shaped bundles called fascicles. Inside each fascicle there is a collection of muscle cells, but the term *cell* here may be misleading because any one of these cells can be as long as the muscle itself—a gigantic length relative to the strictly microscopic dimensions of most cells. Because skeletal muscle cells are so elongated, they are referred to as **muscle fibers.** Along their length, muscle fibers are divided into a set of about 10,000 repeating units, called **sarcomeres,** that are the fundamental units of muscle contraction. But what is it that's contracting in a muscle? A look inside a single muscle fiber reveals that it is composed of more strands—perhaps a thousand long, thin structures called *myofibrils* that run the length of the cell. Each myofibril, in turn, has a large collection of two kinds of strands inside it that alternate with one another; these are thin filaments made of the protein *actin* and thick filaments made of the protein *myosin*. It is these filaments that contract, as you'll now see.

How Muscles Work

To understand contraction, note first that the thick myosin filaments lie in the *center* of a given sarcomere. Meanwhile, the thin actin filaments are attached to either *end* of the sarcomere and extend toward the center, where they overlap with the thick filaments (**Figure 26.14**). In contraction, the thin filaments slide toward the center, causing the unit to shorten, which is to say, causing the muscle to contract. To understand this, place just the tips of the fingers of your right hand in between the tips the fingers of your left hand. Now slide your right hand as far as it will go to the left. Note that neither set of *fingers* shortened in

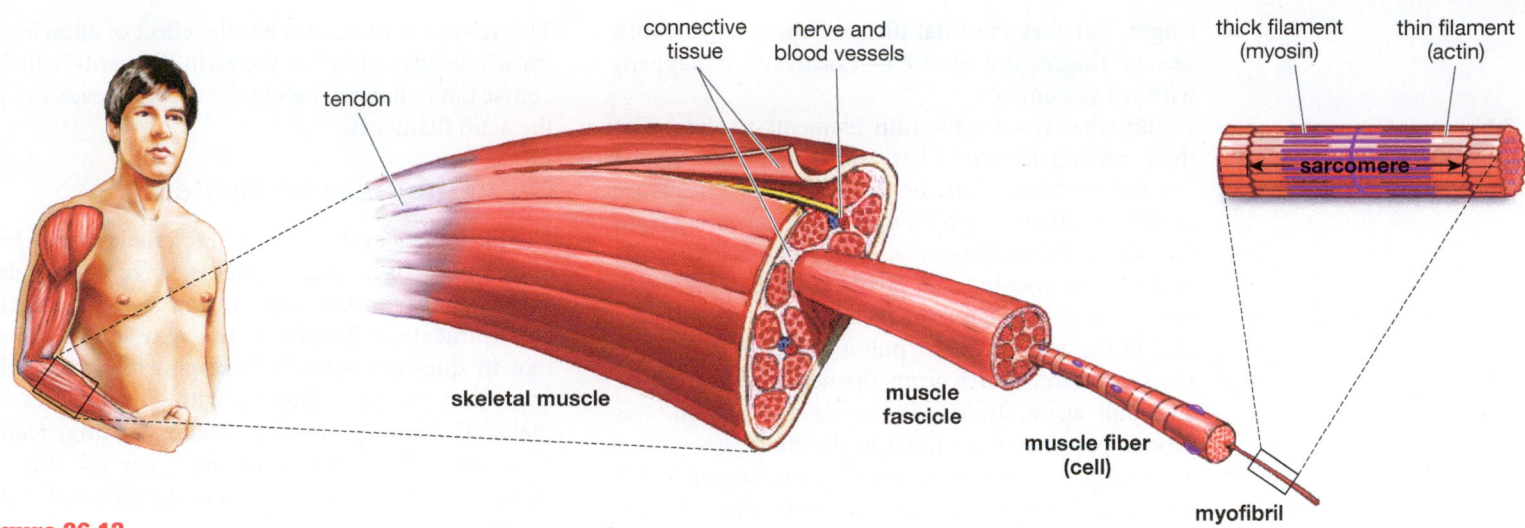

Figure 26.13
Structure of Skeletal Muscle

Skeletal muscles have oval-shaped bundles within them, called fascicles, that are composed of a number of muscle cells, each one of which is so elongated it is referred to as a fiber. Each fiber is composed of many thin myofibrils, and each myofibril has within it a collection of alternating thick and thin protein strands called filaments. The thin filaments are made of the protein actin, the thick filaments of the protein myosin. Each myofibril is divided lengthwise into a series of repeating functional units called sarcomeres.

Figure 26.14
How a Skeletal Muscle Contracts

(a) Muscle at rest

The myofibril contains several sarcomeres, one of which is shown at rest. The thin filaments within the sarcomere, made of actin, overlap with adjacent thick filaments, made of myosin. The thick filaments have club-shaped outgrowths, called heads. The myofibril is also shown in cross section, at a location where thick and thin filaments overlap.

(b) Muscle contraction

Muscle contraction is a matter of the thin actin filaments sliding together within a sarcomere. This comes about when the myosin heads attach to the actin filaments and pull them toward the center of the sarcomere. The myosin heads pivot at their base and are capable of alternately binding with, or detaching from, the thin filaments.

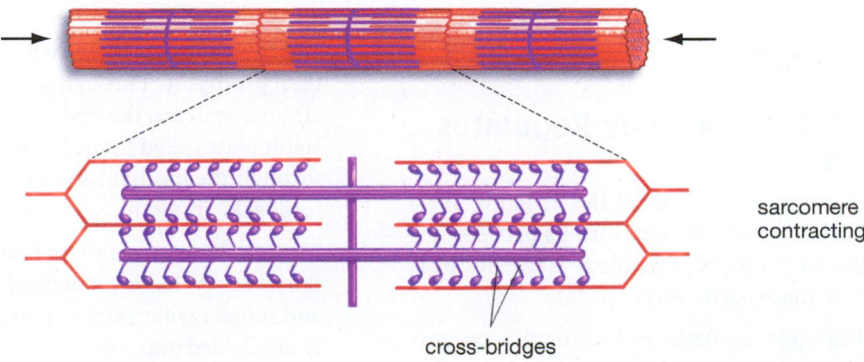

Bioflix: Muscle Contraction

length, but that the total distance made up by both sets of fingers did shorten—exactly what happens within a sarcomere.

But what enables the thin filaments to slide? The thick myosin filaments have numerous club-shaped "heads" extending from them. These myosin heads are capable of alternately binding with or detaching from the adjacent thin filaments, creating so-called cross-bridges. The myosin heads pivot at their base, as if they were on a hinge. Here's the order of their activity: Attach to the actin filament, pull it toward the center of the sarcomere, detach from the actin, reattach to it, and pull again. In actual muscle contraction, this process happens many times in the blink of an eye; at any one point in a contraction, some myosin heads will be pulling an actin filament while others are detaching from it and still others binding to it.

This whole process gets going only when a signal arrives from a nerve telling a muscle to contract—often when a person has decided to move. This is why we speak of skeletal muscles as being under voluntary control, although it's easy to find exceptions to this rule. (The diaphragm that allows you to breathe is a skeletal muscle, but normally there is little that's voluntary in using it.) When a nerve signal arrives, the actual message doesn't go directly from nerve to muscle. Instead, a kind of ferryboat operation takes place in which the nerve end secretes a chemical, called a *neurotransmitter,* that travels across a tiny gap that lies between nerve and muscle. Once the neurotransmitter arrives at the muscle, it binds with it, setting in motion an internal chemical reaction that results in calcium ions (Ca^{2+}) being released inside a sarcomere.

This release of these ions has the effect of allowing the myosin heads to bind to the actin filaments, which of course has to happen before the myosin heads can *pull* the actin filaments.

On to the Nervous System

Having looked at the basic characteristics of the body and at three of its organ systems, we're now ready to move on to another system, one that functions in communication. Routine as this may sound, the system in question actually is responsible, more than any other, for providing us with the qualities that make us human. Hearing, seeing, tasting, feeling, planning, talking, remembering? For all this and more, we're indebted to our nervous system, the subject of Chapter 27.

SO FAR . . .

1. Muscle cells are referred to as muscle _____ because of their unusual _____.

2. Muscle contraction works through the process of a _____ of two kinds of muscle filaments past each other within a muscle unit known as a _____.

3. In this process, relatively thick _____ filaments pull relatively thin _____ filaments toward the center of the _____.

CHAPTER

26 REVIEW

Summary

26.1 How the Body Regulates Itself

- Anatomy is the study of the structure of the body and the relationships between the body's parts. Physiology is the study of how these parts work. **(p. 481)**

- The body regulates itself through a system of circular control in which A prompts B

and B prompts C, but in which C may then prompt A. Through use of such circular systems, the body maintains the stable state called homeostasis, largely through the mechanism of negative feedback. **(p. 481)**

- Humans have two primary body cavities: a dorsal cavity that is subdivided into cranial and spinal cavities; and a ventral cavity that is subdivided into a thoracic cavity, which

contains the lungs and the heart, and an abdominopelvic cavity, which contains such organs as the stomach and liver. **(p. 482)**

26.2 Levels of Physical Organization

- Animal bodies function through a series of working units, whose order, from smaller to larger, runs: cells, tissues, organs, and organ systems. **(p. 483)**

26.3 The Four Basic Tissue Types

- The fundamental tissue types in humans are epithelial, connective, muscle, and nervous. (p. 483)

- Epithelial tissue such as skin covers body surfaces exposed to an external environment and can form glands. (p. 484)

- Connective tissue such as bone supports and protects other tissues. Connective tissue cells are embedded in a "ground" material that they themselves generally secrete. (p. 484)

- Muscle tissue is specialized in its ability to contract. The primary types of muscle tissue are skeletal, cardiac, and smooth. Skeletal muscle is always attached to bone and is under voluntary control. Cardiac muscle exists only in the heart and contracts under the control of its own pacemaker cells. Smooth muscle is not under voluntary control and is responsible for contractions in, for example, blood vessels and the digestive tract. (p. 484)

- Nervous tissue is specialized for the rapid conduction of electrical impulses. It is made up of neurons, which transmit the impulses, and glial cells, which are support cells for neurons. (p. 484)

26.4 Organs Are Made of Several Kinds of Tissues

- The stomach provides a case in point for how various tissue types come together to form a working organ. A typical cross section of the stomach is made up of muscle, nervous, connective, and epithelial tissues. (p. 484)

26.5 Organs and Tissues Make Up Organ Systems

- There are 11 organ systems in the body. The first three—integumentary, skeletal, and muscular—function in body support and movement. The integumentary system is made up of skin and associated structures. The skeletal system is made up of bones, ligaments, and cartilages. The muscular system includes all the skeletal muscles but not smooth or cardiac muscles. (p. 486)

- The nervous, endocrine, and immune systems function in communication, regulation, and defense of the body. The nervous system is a rapid communication system that includes the brain, spinal cord, nerves, and sense organs. The endocrine system is a communication system that works through hormones. The immune system is a network of cells and proteins whose central function is to eliminate microbial invaders. It works in part through the lymphatic network, whose vessels collect extracellular fluid and deliver it to the blood vessels. Immune cells within lymph glands inspect lymph as it passes by to identify invading microbes. (p. 486)

- The cardiovascular, respiratory, digestive, and urinary systems function in transport and in exchange with the environment. The cardiovascular system consists of the heart, blood, blood vessels, and that portion of the bone marrow in which red blood cells are formed. The respiratory system includes the lungs and the passageways that carry air to them. The central feature of the digestive system is the digestive tract, a tube that begins at the mouth and ends at the anus. Along its course, there are digestive organs of the system: the stomach, small intestines, and large intestines and such accessory glands as the liver and pancreas. The urinary system functions to eliminate waste products from the blood, through the formation of urine, and to retain useful substances. It is made up of the kidneys and various other structures, such as the bladder. (p. 487)

26.6 The Integumentary System

- The primary component of the integumentary system is the organ called skin. In addition to covering the body, skin protects underlying tissues and organs, controls the evaporation of body fluids, regulates heat loss, stores fat, and makes vitamin D. Exocrine glands associated with skin excrete such materials as water, oils, and milk. (p. 489)

- Skin has a thin outer epithelial covering, the epidermis, and a thicker underlying dermis. Beneath the dermis—and not part of the skin—is the hypodermis. Epidermal skin cells are constantly worn away and replaced by new epidermal cells being pushed up from the inner epidermis. The dermis is filled with accessory structures and contains blood vessels and nerves that support the surface of the skin. The blood supply of the dermis is important in controlling body temperature. (p. 489)

- The hypodermis is the location for the body's layer of subcutaneous fat. Separate from this tissue is the body's visceral fat, which surrounds organs deep in the abdomen. An excess of visceral fat is more harmful to health than is an excess of subcutaneous fat. (p. 490)

- Hairs originate in tiny organs called hair follicles and may have been important for our evolutionary ancestors in both heat retention and touch sensitivity. (p. 490)

- The skin contains two types of exocrine glands: sebaceous glands, which produce a material called sebum that lubricates hair and limits bacterial growth; and sweat glands, which produce sweat that cools the body by evaporating. Overactive sebaceous glands are the cause of acne. (p. 491)

- Nails are made of the protein keratin. (p. 491)

26.7 The Skeletal System

- The human skeletal system is composed of bone, ligaments, and cartilage. Bone is a connective tissue that provides support and storage capacity and that is the site of blood cell production. Cartilage serves as padding in most joints, forms our larynx and trachea, and links ribs to the breastbone. Ligaments are tissues that link bone to bone. Tendons link bone to muscle but are not part of the skeletal system. (p. 491)

- Each bone of the skeleton is an organ that contains connective tissue, blood vessels, and nervous tissue. (p. 493)

- The typical long bone features a central shaft, called a diaphysis, and expanded ends, called epiphyses. There are two types of bone within this structure: compact bone and spongy bone. In long bones, spongy bone is filled with red marrow, which is the site of the production of all the blood cells in the adult body. Long bones have a central marrow cavity that is filled with yellow marrow, which is composed largely of energy-storing fat cells. (p. 494)

- Three different types of cells are involved in bone growth and maintenance: osteoblasts, which produce new bone; osteocytes, which maintain normal bone; and osteoclasts, which release enzymes that break bone down. (p. 494)

- The basic functional unit of compact bone is the osteon. Osteoblasts produce concentric layers of bone around a central canal of the osteon, which parallels the long axis of the bone and has blood vessels and nerves running through it. (p. 494)

- The human skeleton has 206 bones, in two main divisions: an axial skeleton, whose 80 bones include the skull, the vertebral column, and the rib cage; and an appendicular skeleton, whose 126 bones include those of the arms and legs and the pelvic and pectoral girdles to which they are attached. (p. 496)

- Joints exist wherever two bones meet. Some joints provide a strong linkage between bones, while others provide great flexibility. Normally, the bony surfaces of highly movable joints do not meet directly, but are instead covered with cartilages and pads that absorb shock and provide lubrication. (p. 496)

26.8 The Muscular System

- Because of their tremendous elongation, muscle cells are called muscle fibers. Along their length, they are divided into sarcomeres, which are the basic units of muscle contraction. Each muscle fiber is composed of myofibrils, which are in turn composed of alternating strands of thin filaments made of the protein actin and thicker filaments made of the protein myosin. (p. 498)

- Actin filaments attached to the end of each sarcomere overlap with the myosin filaments that lie in the center of each sarcomere. The myosin filaments bring about contraction by attaching to the actin filaments, pulling them toward the center of the sarcomeres, detaching, and then pulling again. As the actin filaments slide toward the center of the sarcomere, the sarcomere shortens. (p. 498)

- Skeletal muscles contract only when prompted by a signal from a nerve, which initiates a chemical cascade inside muscle tissue that allows the myosin filaments to bind to the actin filaments. (p. 500)

Key Terms

appendicular skeleton 496

axial skeleton 496

bone 491

cardiac muscle 484

cardiovascular system 487

cartilage 492

compact bone 494

connective tissue 484

dermis 490

digestive system 488

endocrine gland 487

endocrine system 487

epidermis 489

epithelial tissue 484

exocrine gland 487

gland 484

homeostasis 482

hormone 487

immune system 487

integumentary system 486

keratin 489

ligaments 493

lymphatic network 487

muscle fiber 498

muscle tissue 484

muscular system 486

negative feedback 482

nervous system 486

nervous tissue 484

organ 483

organ system 483

osteoblast 494

osteoclast 494

osteocyte 494

red marrow 494

reproductive system 488

respiratory system 488

sarcomere 498

sebaceous glands 491

skeletal muscle 484

skeletal system 486

skin 489

smooth muscle 484

spongy bone 494

sweat gland 491

tendons 493

tissue 483

urinary system 488

yellow marrow 494

Understanding the Basics

Multiple-Choice Questions

(Answers are in the back of the book.)

1. To survive, the human body needs to maintain _____ and does so primarily through the means of self-regulation known as _____.
   a. high energy output; positive feedback
   b. adequate fat layers; exocrine secretion
   c. bone density; muscle contraction
   d. homeostasis; negative feedback
   e. nerve signal transmission; hormonal secretion

2. The four basic types of tissue in the human body are:
   a. epithelial, connective, muscle, nervous.
   b. epithelial, blood, muscle, nervous.
   c. skin, bone, muscle, nervous.
   d. skin, connective, heart, nervous.
   e. epithelial, bone, connective, muscle.

3. Muscle fibers that are striated and voluntary are found in:
   a. cardiac muscle.
   b. skeletal muscle.
   c. smooth muscle.
   d. rough muscle.
   e. all four types of muscle.

4. Hormones and the glands or cells that secrete them make up the:
   a. cardiovascular system.
   b. respiratory system.
   c. lymphatic system.
   d. urinary system.
   e. endocrine system.

5. Surface epidermal cells of the skin:
   a. rapidly divide and are filled mostly with keratin.
   b. never divide and remain in place through adult life.
   c. rapidly divide and are in place only a short time.
   d. are dead and filled with keratin.
   e. rapidly divide and are filled with fat.

6. Cells called _____ build bone tissue, _____ maintain it, and _____ break it down.
   a. osteoblasts; osteoclasts; osteocytes
   b. osteoclasts; osteons; osteoblasts
   c. osteoblasts; osteocytes; osteoclasts
   d. osteons; osteocytes; osteoclasts
   e. osteocytes; osteoblasts; osteoclasts

7. The essence of muscle contraction is that thick and thin muscle proteins:
   a. slide past each other.
   b. link irreversibly with each other.
   c. are short enough to respond quickly.
   d. stay completely separate.
   e. pull in opposite directions.

8. No skeletal muscle can contract without first:
   a. receiving a hormone signal.
   b. checking on the status of nearby skeletal muscles.
   c. lining up its actin and myosin filaments.
   d. receiving a nervous system signal.
   e. linking to a bone.

Brief Review

(Answers are in the back of the book.)

1. What is negative about negative feedback?

2. What role does the lymphatic network play in immune function?

3. Why is our skin water-resistant?

4. What are the two principal locations for fat tissue in the body?

5. List three functions of the skeletal system.

6. A sample of bone shows an interconnecting network of calcified rods and plates filled in with red marrow. Is this sample from the shaft (diaphysis) or the end (epiphysis) of a long bone?

Applying Your Knowledge

1. Cells called melanocytes that lie deep in the epidermis produce a pigment, called melanin, that gives skin its color. Exposure to ultraviolet light—from the sun or a tanning bed—causes melanocytes to produce more melanin, which they pass along to the skin cells above them, yielding a suntan. But why would sun exposure prompt this increased melanin production? What is the function of a tan, in other words?

2. The text notes that bone density can increase in response to exercise. Imagine two fairly inactive people who decide to start exercising. One becomes a runner while the other becomes a swimmer. Who experiences the greater increase in bone density: the swimmer (in the arm bones) or the runner (in the leg bones)?

3. Acids secreted by the stomach give the materials within it a pH level that can lie between that of lemon juice and battery acid. Yet our small intestines, which the stomach feeds into, have chemical buffers flowing into them that raise the pH level of the material coming from the stomach. Why should the stomach need such an extreme pH level while the small intestines do not?

MasteringBiology®

1. MasteringBiology Assignments
Activities: Muscle Structure; **Bioflix:** Muscle Contraction; **Current Events:** A Once-Unthinkable Choice for Amputees; Each Flick of a Digit Is a Job for All Five; Human Muscle, Regrown on Animal Scaffolding; Looking to Save Money, More Places Decide to Stop Fluoridating the Water; A Change in Season and Regimen; Doctors Seek Way to Treat Muscle Loss; **Building Vocabulary;** **Questions:** Bioflix Quiz, Reading Questions, Test Bank

2. eText
Read your book online, search, take notes, highlight text, and study more effectively.

3. The Study Area
Self-study tools to test comprehension.

Communication and Control 1:

The Nervous System

To get through a day, our bodies must undertake a huge number of tasks each second, many of them coordinated with each other. This enormous set of activities is controlled in large part by our nervous system.

Available on MasteringBiology®

Activities: The Nervous System; The Human Brain; The Human Ear; The Human Eye; **Bioflix:** How Neurons Work; How Synapses Work; **Current Events:** Whales, Somehow, Are Coping with Humans' Din; Whisper of the Wild; 30 Years In, We Are Still Learning from AIDS; How Far Will Dolphins Go to Relate to Humans?; Stigma Is Toughest Foe in an Epilepsy Fight; What to Do When You Can't Read the Fine Print; Outdoors and Out of Reach, Studying the Brain; Study Says Brain Trauma Can Mimic A.L.S.; **Building Vocabulary; Questions:** Bioflix Quiz, Reading Questions, Test Bank

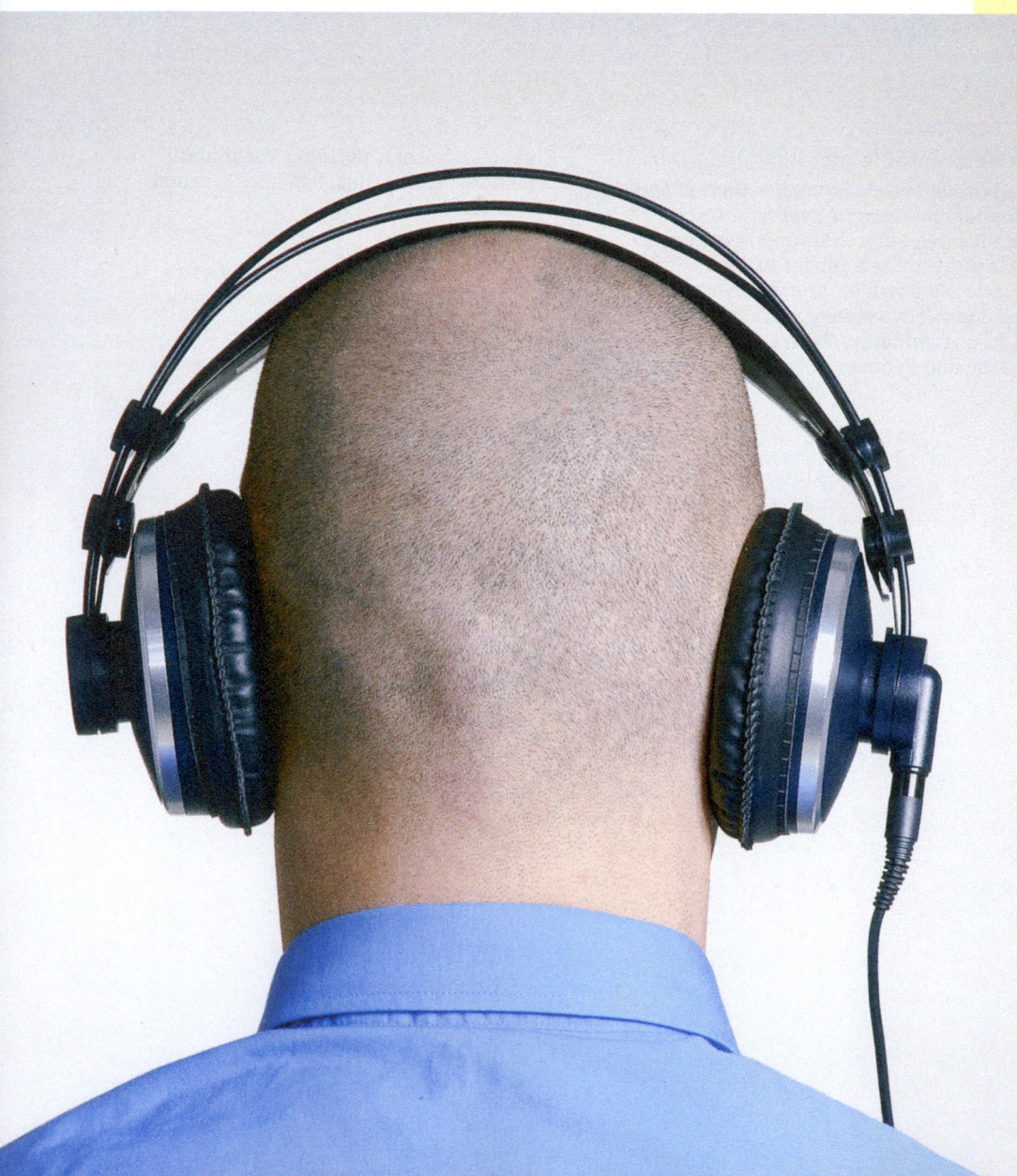

How do we perceive sound or see or move our bodies? Through the communication and control provided by the body's nervous system.

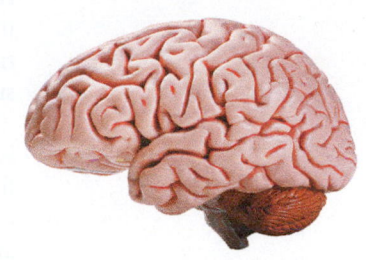

With its depiction of a man who can neither remember his present nor forget his past, the film *Memento* has been a big hit with film buffs over the years. In it, we watch as the fictional Leonard Shelby relentlessly tracks down the murderer of his wife. He does so, however, while taking Polaroid pictures and scribbling notes to himself anywhere he can—some on his own body as tattoos. He does this because he has a big handicap to deal with. Having sustained a head injury in the same break-in in which his wife was killed, he can remember everything in his life up to the moment of his injury but has lost the ability to create new long-term memories. Now, anything that happens more than 15 minutes in the past is lost to him. Hence the constant notes and Polaroids: Where am I staying? Who did I talk to this morning? What did he look like?

Although Leonard Shelby is a movie character, the condition he has actually exists in some people. A version of it was brilliantly described by the physician-writer Oliver Sacks in his book, *The Man Who Mistook His Wife for a Hat.* Sacks introduces us to "Jimmy G.," who was admitted to Sacks' care in the mid-1970s. Although Jimmy G. is a gray-haired, 49-year-old ex-sailor, he perpetually believes himself to be the 19-year-old sailor he was back in 1945. About 1970, he apparently drank so heavily that he damaged his brain, losing not only his memory back to 1945 but his ability to make new memories as well. Hence, in his mind he is always a young man, and it is always the end of World War II. This can lead to brief, horrible moments of disorientation, as Sacks describes:

> "Here," I said, and thrust a mirror toward him. "Look in the mirror and tell me what you see. Is that a 19-year-old looking out from the mirror?" He suddenly turned ashen and gripped the sides of the chair. "Jesus Christ," he whispered. "Christ, what's going on? What's happened to me?"

Sacks would not have done this but for his certainty about what would happen next. Two minutes later, Jimmy G. has no memory of the mirror or of Sacks. When Sacks briefly leaves the room and then reenters it, Jimmy G. greets him as a total stranger.

As Jimmy G. shows, the "self" that we take for granted largely amounts to a set of memories, which exist in the small physical space of the brain. We can thus think of memory as an agent of communication and control—one that allows us to act in the present based on our experience of the past. The human nervous system, which includes the brain, can be thought of as a wider-yet communication-and-control operation—one that constantly monitors our present and issues commands that allow us to function. But how does the nervous system manage this feat? How does it undertake the amazingly detailed task of constantly receiving inputs about the world around us while sending out commands? The goal of this chapter is to provide you with an introduction to this subject. Next chapter, we'll look at another player in communication and control, the human endocrine system.

27.1 Structure of the Nervous System

Taking the broadest view, the nervous system includes all the cells in the body that can be defined as nervous tissue plus the sense organs we have, such as the eye and

In the film *Memento,* the actor Guy Pearce played Leonard Shelby, a man who has lost his ability to create new long-term memories.

ear. Recall from Chapter 26 that two types of cells make up nervous tissue. These are neurons, which actually transmit nervous system messages, and glial cells, which support neurons and modify neuronal communication.

One helpful way to think about the nervous system is to consider three essential tasks it has to perform. It first has to *receive* information, both from outside and inside our bodies. (For example, what taste sensations are coming in?) It next has to *process* this information. (Do I like this?) Then, it has to *send* information out that allows our body to deal with this input. ("Lift your hand for another spoonful.")

The nervous system in which these activities take place can be divided into two fundamental parts. The first of these is the **central nervous system:** that portion of the nervous system consisting of the brain and the spinal cord. The second is the **peripheral nervous system:** that portion of the nervous system outside the brain and spinal cord, plus the sensory organs (**Figure 27.1a**).

The central nervous system's brain and spinal cord are not physically separate entities; the spinal cord simply expands greatly at its top, and we call this flowering the brain. Because the brain and spinal cord are so different in terms of structure and function, however, they are thought of as individual units.

Meanwhile, the peripheral nervous system, or PNS, can be pictured as a group of nerves and related nerve cells that fan out from either the brain or spinal cord. One way to think about these peripheral nerves and related cells is to consider the direction of the messages they carry. Any nerves that help carry messages *to* the brain or spinal cord are said to be part of the **afferent division** of the peripheral nervous system. Any nerves that help carry messages *from* the brain or spinal cord are said to be part of the **efferent division** of the PNS. An easy way to remember the difference between these two terms is to remember that "efferent" sounds like "effect." And the efferent division *effects change* in various organs in the body.

What kinds of change? Many of the activities the nervous system controls are voluntary activities, which often means movement. You decide to lift your finger, you decide to stand up, and so forth. You may remember from Chapter 26 that all movement is handled by skeletal muscles—muscles, like the biceps, that are attached to bones and that are under voluntary control. So, one part of the PNS's efferent division is the **somatic nervous system:** that portion of the peripheral nervous system's efferent division that provides voluntary control over skeletal muscle.

But there are lots of processes in the body that are not under voluntary control. When you walk from bright sunshine into a darkened movie theater, your pupils dilate to let in more light, but you have no control over this dilation. Certain muscles allow our pupils to open up, but these are not skeletal muscles; they are the involuntary "smooth muscles." Likewise, the cardiac muscle of your heart beats in a rhythm that is largely out of your conscious control. And your glands release hormones in a way you have little control over. It is a *second* part of the PNS's efferent division that controls these operations, the **autonomic nervous system:** that part of the peripheral nervous system's efferent division that provides involuntary regulation of smooth muscle, cardiac muscle, and glands (**Figure 27.1b**).

Now let's go down just one more level, this time strictly within the autonomic system. It turns out that the autonomic system is divided into two divisions, the *sympathetic* and the *parasympathetic*. These terms are formally defined later; for now, just be aware of two characteristics they have. First, these divisions differ in that the nerves that make them up stem from different locations in the brain and spinal cord. Second, the sympathetic division generally has stimulatory effects on us—we get adrenaline going through this division, for example—while the parasympathetic division generally facilitates routine maintenance activities, such as digestion.

27.2 Cells of the Nervous System

So, what about the cells that make up these systems? Remember first that the nervous system cells that transmit signals are called neurons. These cells come in three varieties that neatly parallel the idea of a nervous system that receives, processes, and sends information (**Figure 27.2a** on page 508). The first type of neuron is the **sensory neuron,** which does just what its name implies: It senses conditions both inside and outside the body and brings this received information to the central nervous system. (Given the direction of this information, sensory neurons are afferent neurons.) When someone brushes the top of your hand, sensory neurons just beneath the skin sense lots of things about this touch—what direction it came from, what shape the touching object was—and then convey a message containing this information that goes from your hand, into your spinal cord, then up into your brain.

The second type of neuron is the interneuron (or association neuron). Located solely in the central nervous system, **interneurons** interconnect other neurons. These can be very simple connections, but they can be complex as well. The memory we talked about at the start of the chapter amounts to a massive mobilization of interneurons. How do we remember? We process information in complex webs of interneurons.

The third type of neuron is the **motor neuron:** a neuron that sends instructions from the central nervous system (CNS) to such structures as muscles or glands. (Given the direction of this transmission, these are efferent neurons.) The key to understanding motor neurons is to think of them as neurons that transmit messages to organs or tissues *outside* the nervous system, thus prompting some kind of action. If you look at the right side of Figure 27.2a, you can see how this works. A message comes from the cell body of a motor neuron (which lies in the CNS), travels down an extension of it (which lies in the PNS), and then gets transferred to a muscle—causing the muscle to contract. Any organ or group of cells that responds to this kind of nervous system signal is called an *effector*. Thus, a gland prompted to release hormones through a motor neuron signal is also an effector.

(a) The nervous system has two components

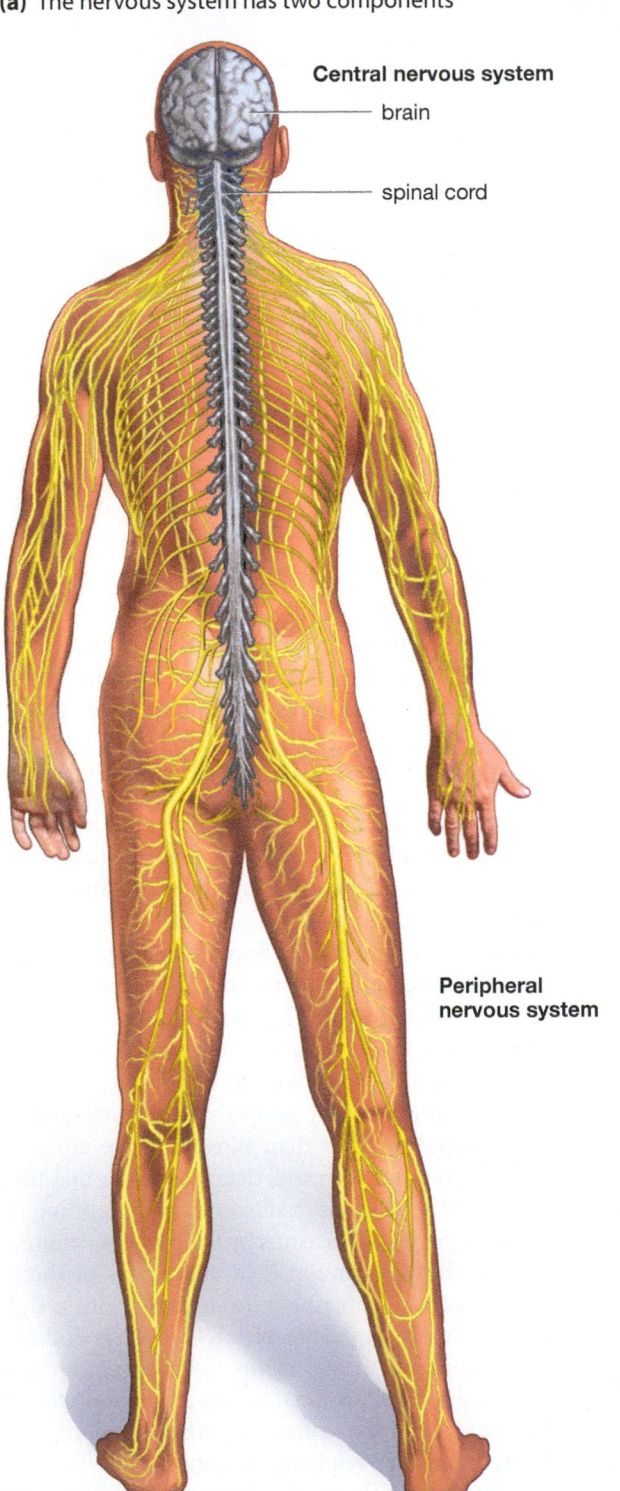

Central nervous system
— brain

— spinal cord

Peripheral
nervous system

(b) How these two components interact

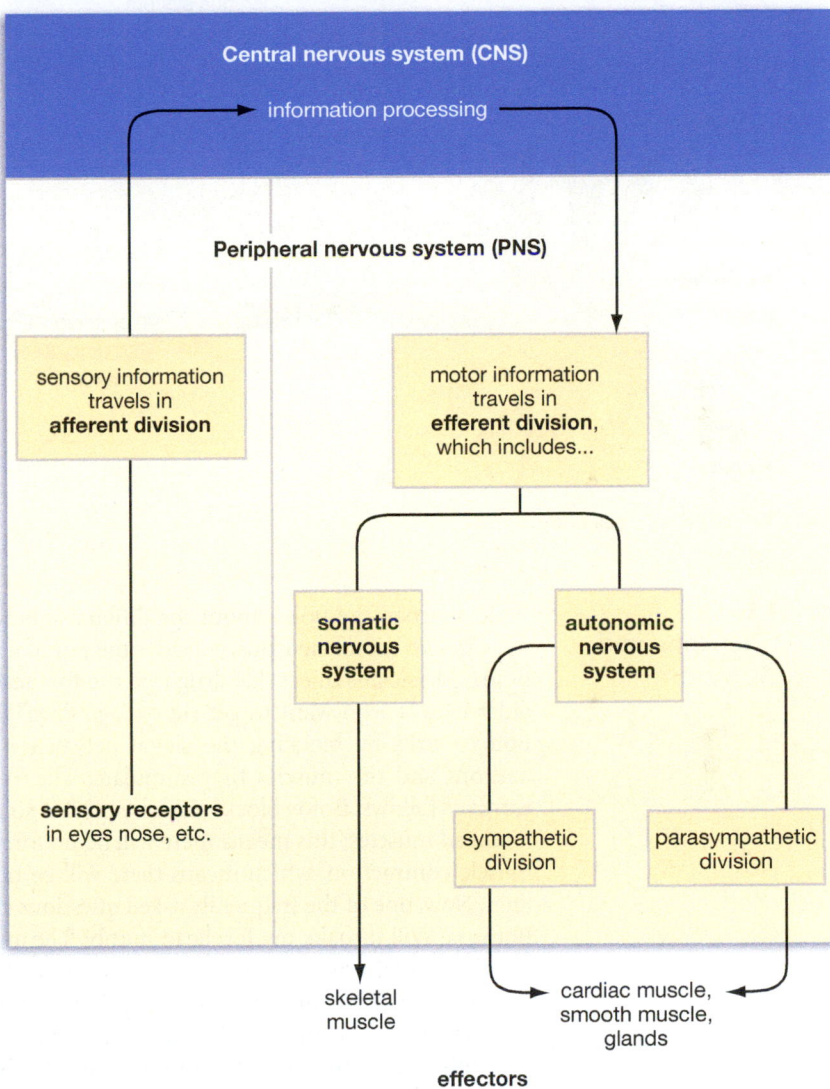

effectors

Figure 27.1
Divisions of the Nervous System

(a) The two main branches of the nervous system are the central nervous system (CNS), which consists of the brain and the spinal cord; and the peripheral nervous system (PNS), which consists of all the nervous tissue outside the brain and spinal cord plus the sensory organs.

(b) Divisions of the nervous system and the paths of signal transmission within them.

(a) Three types of neurons

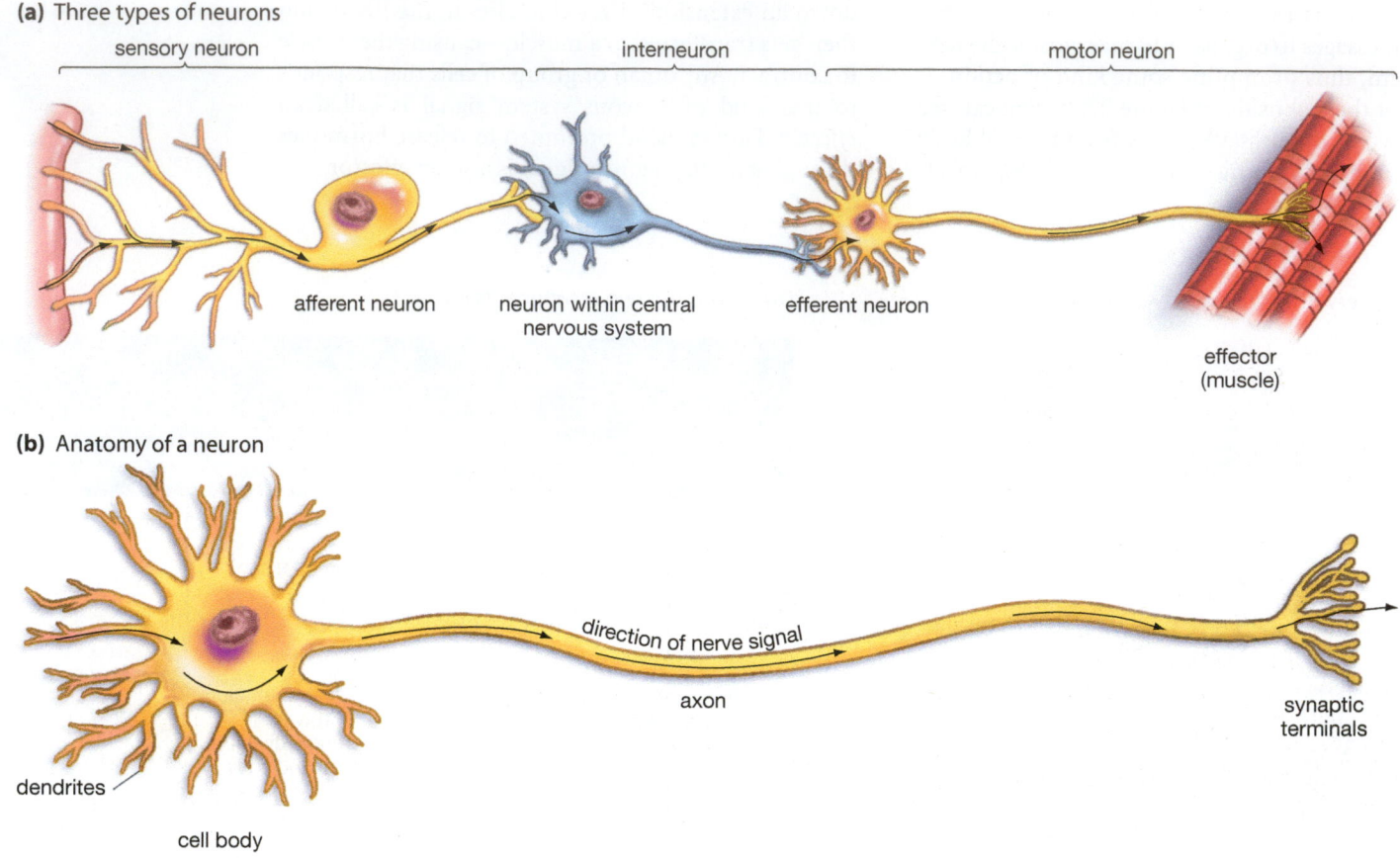

sensory neuron   interneuron   motor neuron

afferent neuron   neuron within central   efferent neuron
                   nervous system

effector
(muscle)

(b) Anatomy of a neuron

direction of nerve signal

axon

synaptic
terminals

dendrites

cell body

Figure 27.2

**Types of Neurons and
Neuron Anatomy**

To sharpen the point about the difference between sensory and motor neurons, consider the popular drug Botox. Physicians inject this drug into the foreheads of older people who want to get rid of "age lines" there. Botox works by blocking the signal between motor neurons and the muscles they stimulate. The linkage works as follows: Botox blocks the nerve signal going to forehead muscles; this means there will be no forehead muscle contraction, which means there will be no age lines. Now, one of the frequently asked questions about Botox is: Will it make my forehead numb? The answer is no because Botox is blocking *motor* neuron transmission, not sensory neuron transmission. The motor signals are going out, the sensory signals are coming in, and only the motor neurons are responsive to Botox.

Anatomy of a Neuron

Now let's look at how a neuron is structured. Like any other cell, a neuron has a nucleus, it makes proteins, it gets rid of waste, and so forth. If you look at **Figure 27.2b**, you can see, however, that the neuron has some extensions sprouting from it that clearly separate it from other cells. Projecting from the cell body are a variable number of **dendrites:** extensions of neurons that carry signals *toward* the neuronal cell body. There is also a single, large **axon:** an extension of the neuron that carries signals *away* from the neuronal cell body. Although you can't see it, the other thing that sepa-

rates the neuron from other cells is its outer or "plasma" membrane, which has an amazing ability to respond to various kinds of stimulation, as you'll see.

The Nature of Glial Cells

Having looked a bit at neurons, let's now consider the second kind of nervous-system cell, the glial cell (also known as glia). Found in both the CNS and PNS, glia usually are described as "support" cells for neurons, and this is true as far as it goes. Some glia, for example, wrap their cell membranes around the axons of neurons in the CNS and PNS. The membranous covering that glia provide to neurons is called **myelin,** and an axon wrapped in this way is said to be myelinated. The importance of this is that axons that are myelinated carry nerve impulses faster than those that are not. (If this sounds like some technical detail, consider that the disease multiple sclerosis results from a dismantling of the myelin covering CNS axons by the body's own immune system.) Because myelin is fat-rich, areas of the brain and spinal cord containing myelinated axons are glossy white. Thus do we get the term *white matter* of the CNS, which contains mostly axons. By contrast, areas containing mostly neuron cell bodies are *gray matter*. If you look at **Figure 27.3a**, you can see what a myelinated axon looks like in the peripheral nervous system.

The support that glial cells provide to neurons goes beyond myelination, however. Some glial cells provide a framework for CNS neurons, acting as a kind of scaffolding for them, and still others serve as a cleanup crew for CNS neurons by taking in cellular debris—from dead or damaged cells—and disposing of it. Recent research has indicated, however, that glial cells are active not just in neuron support but in neuron communication. Glia communicate, by chemical means, both among themselves and with neurons. Indeed, communication between glia and neurons seems to be coordinated in such a systematic way that it appears that glia are modifying neural communication by, for example, increasing the level of signaling that goes on between differing sets of neurons. Another possible role for glia is that they are important in establishing the connections, called synapses, that allow one neuron to communicate with another.

Nerves

All this information about nervous-system cells provides a way to understand a term that's been used extensively thus far but that has not been defined. What is a **nerve**? It is a bundle of axons in the PNS that transmits information to or from the CNS. If you look at **Figure 27.3b**, you can see a representation of a nerve. Note that nerves have support tissue in the form of blood vessels and connective tissue.

SO FAR . . .

1. The brain and spinal cord make up the _____ nervous system, while all the nervous tissue outside the _____ and _____ plus the sensory organs make up the _____ nervous system.

2. The efferent division of the peripheral nervous system is made up of nerves that carry messages _____ the brain or spinal cord, and it is divided up into a portion that provides voluntary control over skeletal muscles—the _____ nervous system—and a portion that provides involuntary regulation over smooth muscle, cardiac muscle, and glands—the _____ nervous system. The afferent division of the peripheral nervous system is made up of nerves that carry messages _____ the brain or spinal cord.

3. The cells of the nervous system that carry nervous system messages are called _____, each of which has a variable number of _____ that carry signals toward the cell body, and a single _____ that carries signals away from the cell body.

(a) A myelinated axon

myelin nodes

glial cells

glial cell nucleus

myelin covering

axon

glial cell cytoplasm

(b) Anatomy of a nerve

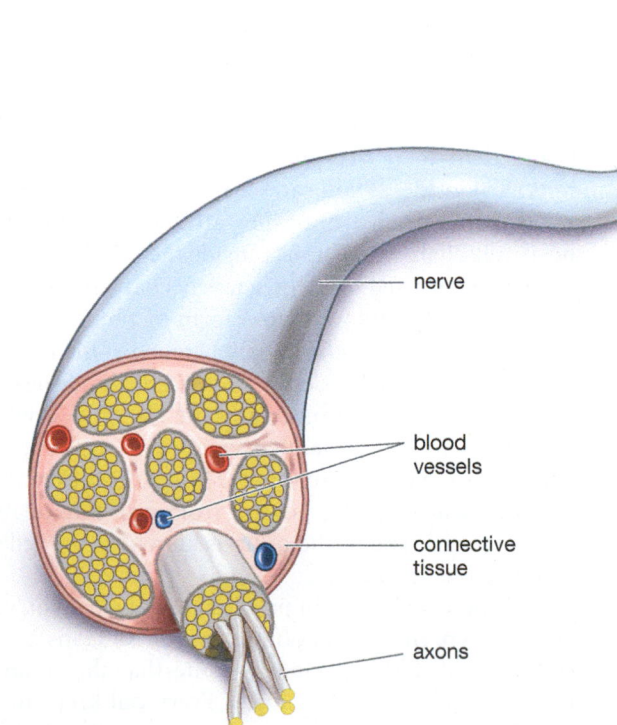

nerve

blood vessels

connective tissue

axons

Figure 27.3
Glial Cells and Nerves

(a) Some glial cells produce a fatty material called myelin that wraps around the axons of neurons, as shown here. Nerve signals travel faster down myelinated axons because the nerve signal skips from one myelin gap or "node" to the next. The micrograph at right shows a cross section of an axon.

(b) A nerve is a bundle of axons in the peripheral nervous system. It includes not only axons themselves but supporting blood vessels and connective tissue.

27.3 Nervous-System Signaling

The Nervous System

Bioflix: How Neurons Work

With this anatomy under your belt, you're ready to see how nervous-system signaling works. We'll look at a two-step process, which we can think of as (1) signal movement down a cell's axon and (2) signal movement from this axon over to a second cell, across a synapse.

Communication within an Axon

All the action in this first step is going to take place on either side of the thin plasma membrane that constitutes the outer border of any animal cell. Inside the plasma membrane is the cell and all its contents; outside is extracellular fluid (**Figure 27.4**). As was noted in Chapter 5, the plasma membrane regulates the passage of substances in and out of the cell. You may remember that some of these substances are the electrically charged particles called ions, meaning atoms that have gained or lost one or more electrons. Because ions are charged, they get the $^+$ or $^-$ signs after their chemical names, as with Na^+ for the positively charged sodium ion. Such ions cannot simply pass through the plasma membrane; rather, they need to move through special passageways called protein channels.

It turns out that a "resting" neural cell—one not transmitting a signal—has a greater positive charge outside itself than inside. This is so mostly because there are more proteins inside the cell than outside, and proteins carry a *negative* charge. The charge difference also exists, however, because there are more of the positively charged Na^+ ions outside the membrane than there are positively charged potassium ions (K^+) inside, as you can see in Figure 27.4. As you know, in electricity, opposites attract, meaning the positively charged Na^+ ions outside the plasma membrane are attracted to the negatively charged interior of the cell. This amounts to a form of potential energy in the same way that a rock perched at the top of a hill does. The rock would roll down the hill (releasing energy) if given a push, and the Na^+ ions would rush into the cell (releasing energy) if the proper channels were opened to them. The charge difference that exists in a resting neuron from one side of its plasma membrane to the other is known as the cell's **membrane potential.**

This potential exists in part because of what you've already seen: the greater abundance of Na^+ ions outside the cell. It's important to note that this abundance doesn't exist by accident. Every cell keeps this imbalance in place by constantly pumping three Na^+ ions out for every two K^+ ions it pumps in. It takes a lot of energy to maintain this *sodium–potassium pump,* but the cell does it for the same reason that you charge

your cell phone battery: to have ready access to a capability. In your case, the capability is talking on the phone; in the cell's case, the capability is that of having Na^+ ions available to rush into it. Indeed, the action of the pump readies a neuron for activity in two ways. In addition to the charge difference it maintains, the pump maintains a *concentration* difference of the ions. The fact that there are more Na^+ ions outside the cell than inside means that the Na^+ ions will have a natural tendency to move down their "concentration gradient"—they will naturally move from where they are more concentrated to where they are less concentrated. Since they are more concentrated outside the cell than inside, there will be a natural net movement of these ions into the cell. (For more on concentration gradients, see Chapter 5.) In sum, the sodium–potassium pump maintains both an electrical charge difference and a concentration difference that can get a nerve signal going, as you'll now see.

When a neuron is stimulated by getting a chemical signal from another neuron, the effect of this stimulation is that, up near where the neuron's axon meets its cell body, Na^+ ion protein channels open up. Now the Na^+ ions can act on their electrical attraction and concentration difference, and they do so, rushing into the cell. After a very brief time, this influx is great enough that the inside of the cell is more positive than the outside. This very state causes the potassium channels to open up more fully, however. Recall that potassium ions exist in more abundance inside the cell than outside it. With the opening of their channels, they act on their electrical and concentration differences, rushing out of the cell, and the positive charge that has briefly existed on the inside of the cell begins to decline. Eventually, enough ions have moved out that the cell returns to its resting state, with a negative charge inside the membrane. This entire process takes place in a few milliseconds, or thousandths of a second.

Movement Down the Axon

All this alters the membrane potential at *one spot* along the neuron's axon. But how does a nerve signal get transmitted down the entire length of the axon? The key to this transmission is that an influx of Na^+ ions at an initial location on the membrane triggers reactions that cause the *neighboring* portion of the membrane to begin an Na^+ influx (see Figure 27.4). What occurs, in other words, is a chain reaction that moves down the entire length of the axon membrane. How does this work? Keep in mind that the initial Na^+ ions that come into a cell are moving into an environment that is negatively charged. On entry, these positively charged Na^+ ions are pulled in both directions along the interior of the mem-

(a) Membrane potential

Electrical energy is stored across the plasma membrane of a resting neuron. There are more negatively charged compounds just inside the membrane than outside of it. As a result, the inside of the cell is negatively charged relative to the outside. The charge difference creates a form of stored energy called a membrane potential. Protein channels (shown in green) that can allow the movement of electrically charged ions across the membrane remain closed in a resting cell, thus maintaining the membrane potential.

(b) Action potential

1. Nerve signal transmission begins when, upon stimulation, some protein channels open up, allowing a movement of positively charged sodium ions (Na⁺) into the cell. For a brief time, the interior of the cell becomes positively charged at this location. On either side of this location, however, the interior of the cell remains negatively charged. Attracted by this negative charge, the Na⁺ ions move laterally in both directions from their point of entry.

2. The Na⁺ gates close and the gates for positively charged potassium ions (K⁺) open up, allowing a movement of K⁺ out of the cell. With this, there is once again a net positive charge outside the membrane. Meanwhile, the arrival of the charged Na⁺ ions "downstream" from their original point of entry triggers an influx of Na⁺ ions at the next Na⁺ channel.

3. Through repetition of this process, the nerve signal is then propagated one way along the axon. Given that Na⁺ ions move laterally in both directions from their original point of entry, why isn't the signal propagated in both directions? Because once an Na⁺ channel has opened, it enters a brief period in which it cannot respond to any additional stimulus. Thus, each "upstream" Na⁺ channel remains briefly closed, while each "downstream" channel is opened in succession.

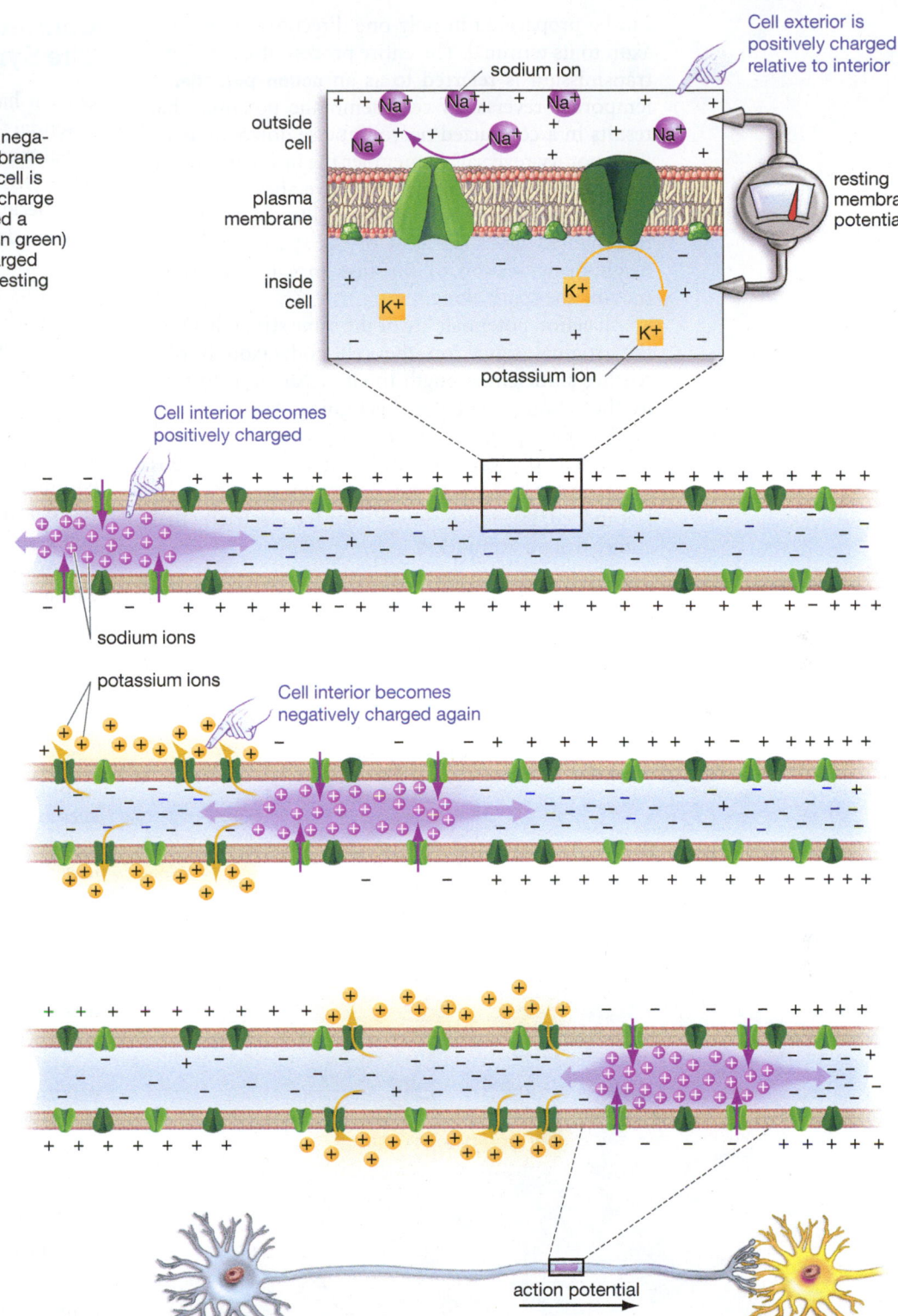

brane by the negative charge they encounter (Figure 27.4). When the Na⁺ ions reach the first "downstream" Na⁺ channel in the axon, their charge causes it to open, letting in Na⁺ ions from outside the cell and getting the whole process going again. Then this same set of steps is repeated all the way down the axon.

You may wonder, however: If the initial Na⁺ ions that enter the cell are pulled in both directions by the negative charge they encounter, then why isn't the *nerve signal* propagated in both directions from this point of entry? The answer is that an Na⁺ gate that has just opened goes through a brief "refractory" period in which it cannot open again. Thus, the nerve signal

Figure 27.4
Nerve Signal Transmission within a Neuron

can be propagated in only one direction—down the axon to its terminal. The entire process of axon signal transmission is referred to as an **action potential:** a temporary reversal of cell membrane potential that results in a conducted nerve impulse down an axon. (Temporary reversal? Remember the inside of the cell goes from negative to briefly positive, then back.) Action potentials have been compared to what happens to a lighted fuse: The heat of the spark causes the neighboring section of the fuse to catch fire, thus moving the spark along.

All action potentials are of the same strength. Once the original signal on the cell body/axon border reaches sufficient strength to allow Na⁺ ions to rush in, the action potential will get going. And thanks to its fuse-like quality, this potential will be the same strength all the way down the axon. How fast can this signal go? At best, about 120 meters per second—very fast indeed, but much slower than the electricity that comes from a wall socket.

Bioflix: How Synapses Work

Communication between Cells: The Synapse

Once it has traveled the length of an axon, an action potential then reaches a tip of the axon. How does the signal then get to the *next* neuron (or muscle, or gland cell)? This question was once very intriguing to biologists because it was clear that, with a few exceptions, an axon does not touch the downstream cell. It comes extremely close, but there is almost always a small intervening space between the two cells. There are thus three entities involved in cell-to-cell transmission: the sending neuron, the receiving cell, and the gap between them. The area where all three come together is called a **synapse** (**Figure 27.5**).

The action potential arrives at a branch of the axon, called a synaptic terminal, which has stored within it small sacs (or vesicles) containing a chemical called a neurotransmitter. The arrival of the action potential causes the synaptic terminal to release neurotransmitter molecules into the gap in the synapse, which is called a **synaptic cleft.** With this, the molecules diffuse over to the receiving cell and bind to receptors in that cell's outer membrane. This binding stimulates the opening of sodium channels there, allowing the now-familiar influx of Na⁺ ions in the receiving neuron, and this keeps the signal transmission going.

But why doesn't the released neurotransmitter just keep on stimulating the receiving cell? Serving as one means of control are the proteins called enzymes. Released into the synaptic cleft, they break down the neurotransmitter, thus inactivating it. The sending cell may also be capable of taking the neurotransmitter back into itself by a process known as "reuptake." In a given nerve impulse, a sending cell typically releases tens of thousands of neurotransmitter molecules, which are stored in hundreds of sacs. With all this in mind, you can understand the definition of a **neurotransmitter:** a chemical, secreted into a synaptic cleft by a neuron, that affects another neuron or an effector by binding with receptors on it.

The Importance of Neurotransmitters

It is difficult to overstate the importance of neurotransmitters. You may recall from Chapter 26 that a neurotransmitter must travel from nerve cell terminals to muscles for any skeletal muscle to work. If this neurotransmitter (called acetylcholine) doesn't make it out of a motor neuron, we can't contract the muscles that allow us to breathe. Likewise, the shaking and stiffness of Parkinson's disease are caused by a lack of the neurotransmitter dopamine, which in turn is brought on by the death of dopamine-producing cells in the brain.

On a different level, remember the reuptake process we talked about, by which a releasing cell will

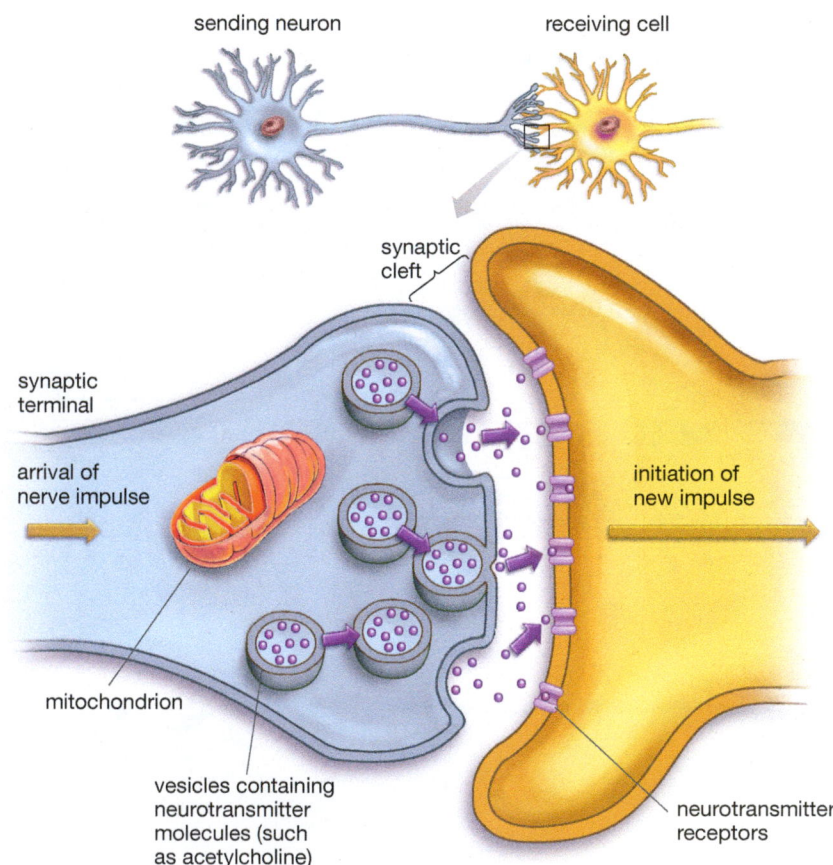

Figure 27.5
Structure and Function of a Synapse

A synapse is made up of a sending neuron, a receiving cell, and the gap between them, called a synaptic cleft. As a nerve impulse reaches the synapse, molecules of a neurotransmitter are released into the synaptic cleft from synaptic terminals of the sending neuron. The neurotransmitter molecules then move across the cleft and stimulate receptors in the receiving cell's membrane, thus initiating a nerve impulse within it.

take back a neurotransmitter that it has secreted into a synaptic cleft? Well, the antidepressant drugs Prozac, Zoloft, and Paxil are all called SSRIs, which stands for selective serotonin reuptake inhibitors. Serotonin is a neurotransmitter found in the brain, and SSRIs are aimed at reducing serotonin reuptake. The result is an increased amount of serotonin in synaptic clefts—and perhaps less depression. By the same token, drugs of abuse such as cocaine and the amphetamine-like ecstasy work by altering neurotransmitter release or reuptake. What makes us feel bad, or OK, or ecstatic? A big part of the answer is: the levels of neurotransmitters in our brains.

SO FAR . . .

1. Transmission of a nerve signal down an axon takes advantage of something called a resting potential, which can be defined as a difference in _____ that exists between the inside and the outside of a neuron in a resting state.

2. The key to signal transmission down the length of an axon is that an influx of Na⁺ ions at an initial location on the axon membrane triggers an influx of _____ on a _____ of the membrane.

3. Communication between neurons takes place by means of substances called _____ moving across a _____ from a sending cell to a receiving cell.

27.4 The Spinal Cord

Now that you've looked at how nerve impulses are transmitted at the cellular level, it's a good time to take a step back and look at the nervous system in its larger dimensions. Let's start by examining the spinal cord, which serves two key functions. First, it can act as a communication center on its own, receiving input from sensory neurons and directing motor neurons in response, with no input from the brain. This is what a reflex amounts to. Second, most sensory impulses that go to the brain don't go *directly* to the brain; they are channeled first through the spinal cord.

Extending from the base of the brain to an area just below the lowest rib, the spinal cord is about 14 millimeters or a half-inch wide and consists of 31 segments, each having both a left and right spinal nerve stemming from it (**Figure 27.6a**). These nerves are grouped into the classes you see named in the figure, which correspond to the region of the body they serve. ("Thoracic" is the chest area; "lumbar" the lower back; and so forth.)

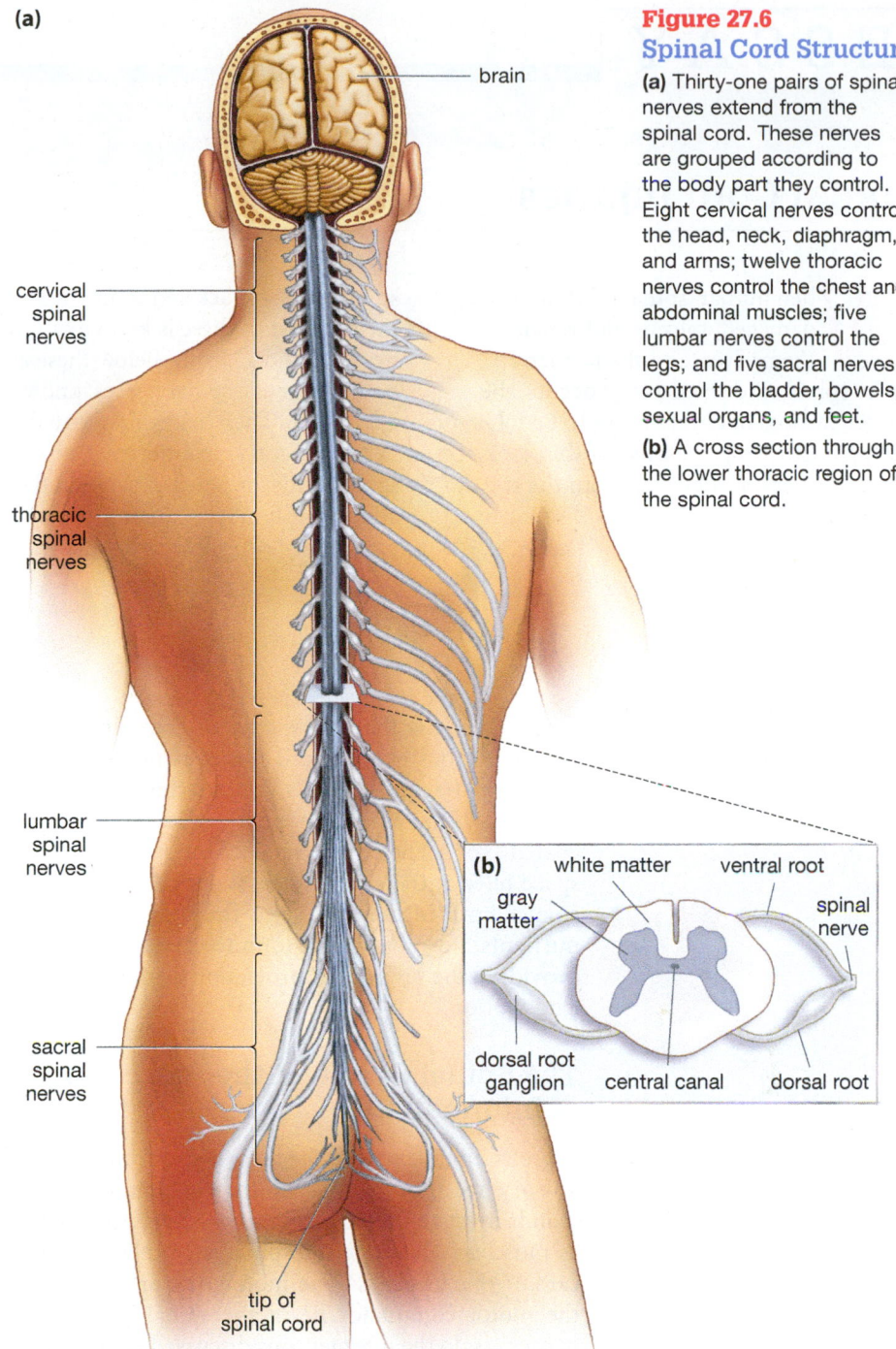

(a) brain

cervical spinal nerves

thoracic spinal nerves

lumbar spinal nerves

sacral spinal nerves

tip of spinal cord

(b) white matter — ventral root
gray matter
spinal nerve
dorsal root ganglion — central canal — dorsal root

Figure 27.6
Spinal Cord Structure
(a) Thirty-one pairs of spinal nerves extend from the spinal cord. These nerves are grouped according to the body part they control. Eight cervical nerves control the head, neck, diaphragm, and arms; twelve thoracic nerves control the chest and abdominal muscles; five lumbar nerves control the legs; and five sacral nerves control the bladder, bowels, sexual organs, and feet.

(b) A cross section through the lower thoracic region of the spinal cord.

Remember how we noted that the peripheral nervous system could be thought of as a group of nerves and related nerve cells that fan out from the brain and spinal cord? In Figure 27.6a, you can see the start of a lot of this fanning, in the form of the spinal nerves. (The other major set of nerves that do this are the so-called cranial nerves, which fan out from the brain.)

The most striking feature of a cross section of the spinal cord, seen in **Figure 27.6b**, is a rough H- or butterfly-shaped area around the narrow central

ESSAY

Spinal Cord Injuries

When human spinal cords are damaged, they do not repair themselves, and the farther up in the spinal cord the damage occurs, the greater the loss of ability tends to be following injury. To get a sense of this, look at the spinal nerves extending from the spinal cord in Figure 27.6 on page 513. You can see that there are eight so-called cervical spinal nerves bracketed in the figure; the topmost of these is known as C1, the next C2, and so forth. Below these are the body's thoracic spinal nerves, known as T1, T2, etc. Then below these are the lumbar and sacral spinal nerves. People whose spinal cords are severed between the T1 and upper lumbar segments become paraplegic—they lose at least partial use of their legs. People whose injury comes between C5 and T1 become quadriplegic—they lose at least partial use of both their legs and arms. But people whose injury comes above C3 lose all this functionality *and* the ability to breathe on their own. Why? The nerve signals that raise our diaphragm—thus allowing us to inhale—travel through the C3–C5 spinal nerves. Given this, a person who sustains a severe enough injury above C3 may be able to breathe only with the assistance of a ventilator.

canal. This H is the gray matter of the spinal cord, composed mostly of the cell bodies of neurons. Surrounding the spinal cord's gray matter is its white matter, which, as you've seen, is largely myelin-coated axons. The central canal of the spinal cord is a space filled with **cerebrospinal fluid,** a fluid that circulates both in the spine and the brain, supplying nutrients, hormones, and immune system cells and providing the brain with protection against any jarring motion.

The Spinal Cord and the Processing of Information

Sensory input comes *to* the spinal cord from various sensory receptors in the body, while motor commands must go *from* the spinal cord to effectors such as muscles. Where are the spinal cord cells that take care of this business? The motor neurons that send the motor commands lie in the spinal cord's gray matter (as do the interneurons discussed earlier). The axons of these motor neurons leave the spinal cord through the ventral roots you can see in Figure 27.6b. Meanwhile, the cell bodies of sensory neurons lie outside the spinal cord in the dorsal root ganglia you can see in Figure 27.6b. (A **ganglion** is any collection of nerve cell bodies in the PNS.) These sensory neurons have dendrites whose endings lie in, say, the hand. When we touch something, these dendrites' terminals sense it and send a message back to the cell body in the dorsal root ganglion. From there, this signal proceeds via the cell's axon into the spinal cord. In sum, sensory information comes to the spinal cord through a dorsal root, while motor commands leave the spinal cord through a ventral root. Note that both dorsal and ventral roots come together, like fibers being joined in a single cable, to form a given spinal nerve.

Quick, Unconscious Action: Reflexes

The sensory/motor division of labor is clearly reflected in simple body **reflexes,** which can be defined as automatic nervous system responses that help us avoid danger or preserve a stable physical state. If you accidentally touch a hot stove, you automatically pull back your hand. No conscious thought of "Oh, I've touched a hot stove; I'd better pull my hand back" is necessary; the reaction is automatic. The neural wiring of a single reflex is called a *reflex arc.* A reflex arc begins at a sensory receptor, runs through the spinal cord, and ends at an effector, such as a muscle or gland.

Figure 27.7 shows one of the best-known examples, the knee jerk, or patellar reflex, in which a properly placed sharp rap on the knee produces a noticeable kick. Note what's at work: A sensory receptor is stimulated, thus prompting a signal to move into the spinal cord. There, the sensory neuron is linked via a synapse to a motor neuron, which issues a command that is carried out by an effector—in this case a set of skeletal muscles. This is an example of the simplest possible reflex arc in that the sensory neuron is linked directly to the motor neuron. Because only one synapse is involved, this kind of simple reflex controls the most rapid motor responses of the nervous system. Many other reflexes have at least one interneuron placed between the sensory receptor and the motor neuron.

27.5 The Autonomic Nervous System

As noted earlier, the part of the peripheral nervous system's efferent or "outgoing" division over which we have no conscious control is the autonomic nervous system. You can grasp its importance by imagining having to consciously control the digestion of the food you eat. Recall that this system controls the involuntary regulation of smooth muscle, cardiac muscle, and glands.

Also recall that, within the autonomic nervous system, there are two "divisions," the sympathetic and parasympathetic. The **sympathetic division,** often called the "fight-or-flight" division of the autonomic nervous system, usually stimulates tissue metabolism, increases alertness, and generally prepares the body to deal with emergencies. What does this mean in practice? Among other things, sympathetic stimulation puts adrenaline into circulation, it diverts blood from our digestive tract to our skeletal muscles, and it opens our lung passages further—a use-

ful set of responses when we're faced with danger. In contrast, the autonomic system's **parasympathetic division** generally conserves energy and promotes what might be called routine maintenance activities. Parasympathetic stimulation slows our heart rate, constricts the pupils in our eyes (thus reducing the amount of light that comes in), and increases muscular activity along our digestive tract. Given effects such as these, the parasympathetic is sometimes called the "rest-and-digest" division of the autonomic nervous system.

In **Figure 27.8** on the next page, you can see which spinal and cranial nerves are part of both the parasympathetic and sympathetic divisions. You can also see a list of the effects each division has on various bodily functions. Although some organs are connected to only one division or the other, most vital organs receive both sympathetic and parasympathetic signals. Where such dual signaling exists, the two divisions often have opposing effects, keeping the body's stability mechanisms working in balance.

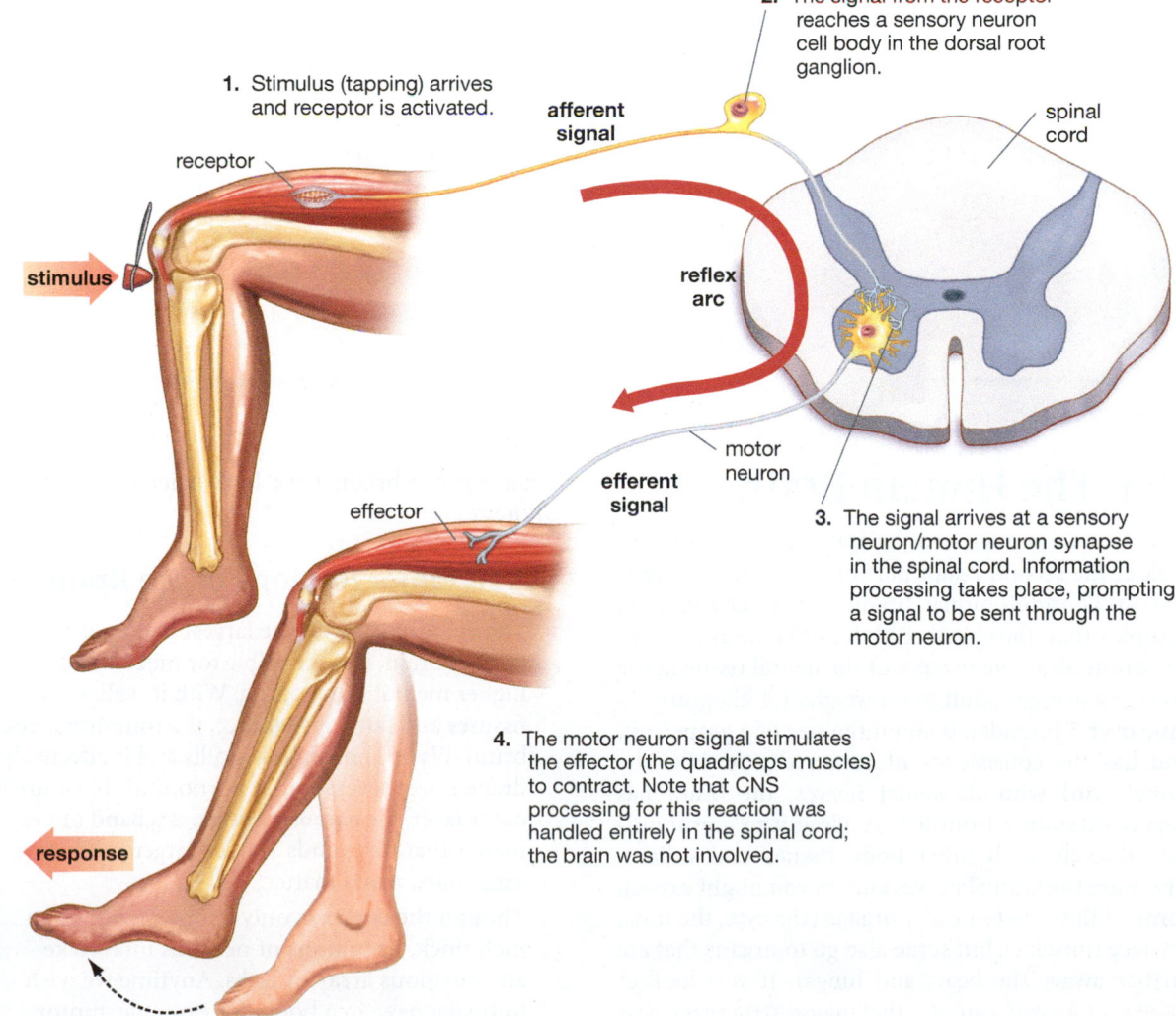

Figure 27.7
Steps in a Reflex Arc

1. Stimulus (tapping) arrives and receptor is activated.

2. The signal from the receptor reaches a sensory neuron cell body in the dorsal root ganglion.

3. The signal arrives at a sensory neuron/motor neuron synapse in the spinal cord. Information processing takes place, prompting a signal to be sent through the motor neuron.

4. The motor neuron signal stimulates the effector (the quadriceps muscles) to contract. Note that CNS processing for this reaction was handled entirely in the spinal cord; the brain was not involved.

receptor

afferent signal

spinal cord

reflex arc

stimulus

effector

efferent signal

motor neuron

response

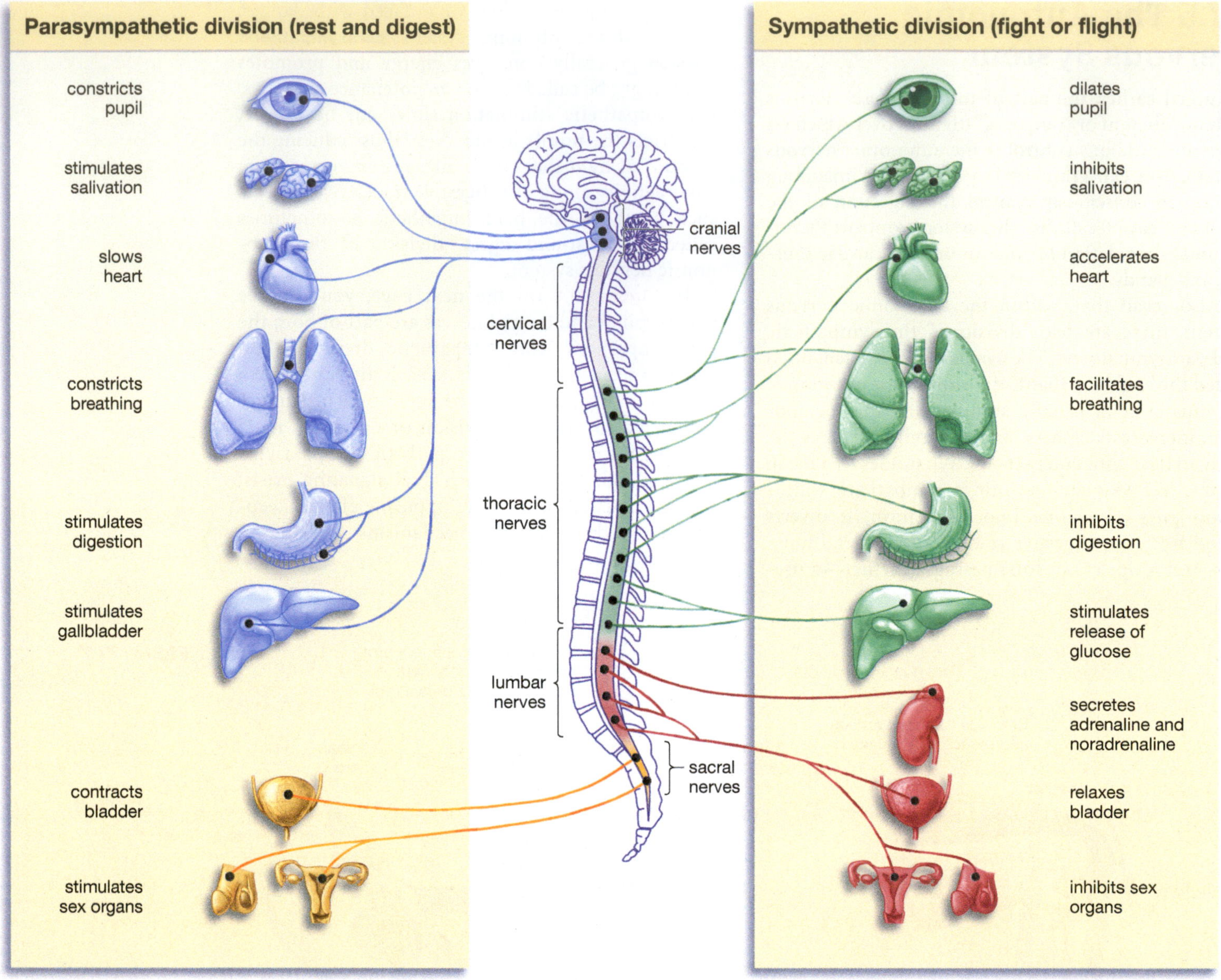

Parasympathetic division (rest and digest)

constricts pupil

stimulates salivation

slows heart

constricts breathing

stimulates digestion

stimulates gallbladder

contracts bladder

stimulates sex organs

cranial nerves

cervical nerves

thoracic nerves

lumbar nerves

sacral nerves

Sympathetic division (fight or flight)

dilates pupil

inhibits salivation

accelerates heart

facilitates breathing

inhibits digestion

stimulates release of glucose

secretes adrenaline and noradrenaline

relaxes bladder

inhibits sex organs

Figure 27.8
Involuntary Control of Bodily Functions

The autonomic nervous system has two divisions, the sympathetic and the parasympathetic, which exercise automatic control over the body's organs, generally in opposing ways. Note how the parasympathetic aids in digestion, while the sympathetic controls "fight-or-flight" responses, such as an increase in heart rate. Axons of the parasympathetic division emerge not only from the spinal cord but from the brain as well.

27.6 The Human Brain

Now that we've reviewed the spinal cord, it's time to look at the second major part of the central nervous system. The adult human brain is far larger and more complex than the spinal cord. Its 100 billion neurons constitute about 98 percent of the neural tissue in the body. An average adult brain weighs 1.4 kilograms (a little over 3 pounds), is about the size of a grapefruit, and has the consistency of cream cheese. Like the spinal cord with its spinal nerves, the brain has nerves extending from it that allow it to communicate directly with other body tissues and organs. These are the cranial nerves and, as you might expect, some of them go to nearby organs (the eyes, the nose, the face muscles), but some also go to organs that are farther away (the heart and lungs). If you look at **Figure 27.9**, you can see the major structures that make up the brain. Let's look briefly now at each of them.

Seven Major Regions of the Brain

- The **cerebrum,** by far the largest region of the human brain, is responsible for much of our higher mental functioning. With its valley-like fissures and fatty appearance, the roundish cerebrum fills up most of our skulls and is effectively draped over several other portions of the brain. Its outer layer, the **cerebral cortex,** is a band of gray matter that surrounds a much larger volume of axon-filled white matter.

Though the cortex is only about a tenth of an inch thick, its billions of neurons undertake an enormous array of tasks. Anytime we wish to turn a page in a book, for example, motor neurons in one area of the cortex must fire to get the muscles of our arms and hands to contract in their proper sequence. To perceive the letters on the book's pages, neurons in a visual association area, located at the back of the cortex, must turn information about units of light entering our eyes into the ordered perception of an array of black marks on a white background. To make sense of this visual information, a language processing area in the cortex will begin to decode the meaning of the words as we scan them. Any reasoning we do about these words will take place in the cortex, and any planning we undertake about what we want to do once we've finished our reading will occur there as well, in a "pre-frontal" area above our temples. In short, the cerebral cortex is central to almost all our voluntary, conscious activities.

To carry out any of these activities, different areas of the cortex must constantly draw on our sense of memory, which is not solely a product of the cortex, but which is almost entirely a product of the cerebrum as a whole. The lower portion of the cerebrum contains a structure called the hippocampus that plays a critical role in the creation and retrieval of long-term memories. Nearby, another cerebral structure, the amygdala, plays a central role in our fear response and our memory of it.

- Though the cerebrum is the seat of most of our voluntary neural activity, if you look again at Figure 27.9a, you can see, nestled beneath the cerebrum, a portion of the brain that's active in a type of involuntary regulatory activity that's critical to our day-to-day functioning. The **cerebellum** is a region of the brain that refines bodily movement and balance based on sensory inputs it receives. A motor command to move our arm may be issued by cells in the cerebral cortex, but for this to be a *smooth* movement, cells in the cerebellum must constantly track the movement while it's in process and issue commands that adjust it. People who have damaged their cerebellum may find that their movements have become jerky, or that they reach too far (or not far enough) for everyday objects, or that they have trouble with balance.

In looking at the brain, if you take away the cerebellum and cerebrum, what remains are (a) the thalamus, (b) the hypothalamus, and (c) the brainstem, which is composed of the midbrain, pons, and medulla oblongata. All of these structures can be seen in Figure 27.9b. Let's take a look at them now.

- When we see, hear, taste, or touch anything, the resulting sensory signals don't go directly to the cerebral cortex; instead, they are routed first to a

region of the brain known as the **thalamus.** If you look at **Figure 27.10** on the next page, you can see how sensory information comes to the thalamus from various sensory organs—the eyes, the ears—and how it is then routed to specific areas of the cortex for processing.

- Lying below the thalamus is the **hypothalamus,** which is small but extremely important in that it is a region of the brain that is critical in regulating drives and in maintaining homeostasis in the body. With respect to drives, the hypothalamus tells us whether we are hungry or thirsty, for example. How does it sense either condition? It monitors the levels of nutrients and water in our bloodstream. Should water levels be too low, it will then respond in two ways: while initiating the thirst drive, it will produce a hormone that has the effect of keeping the body's existing water in circulation (rather than letting it pass out of the body in urine). In numerous ways such as this, the hypothalamus is a key player in preserving homeostasis—the maintenance of a stable internal environment in the body. As you'll see next chapter, hormones are critical to homeostasis and the hypothalamus controls the release of many of the body's hormones.

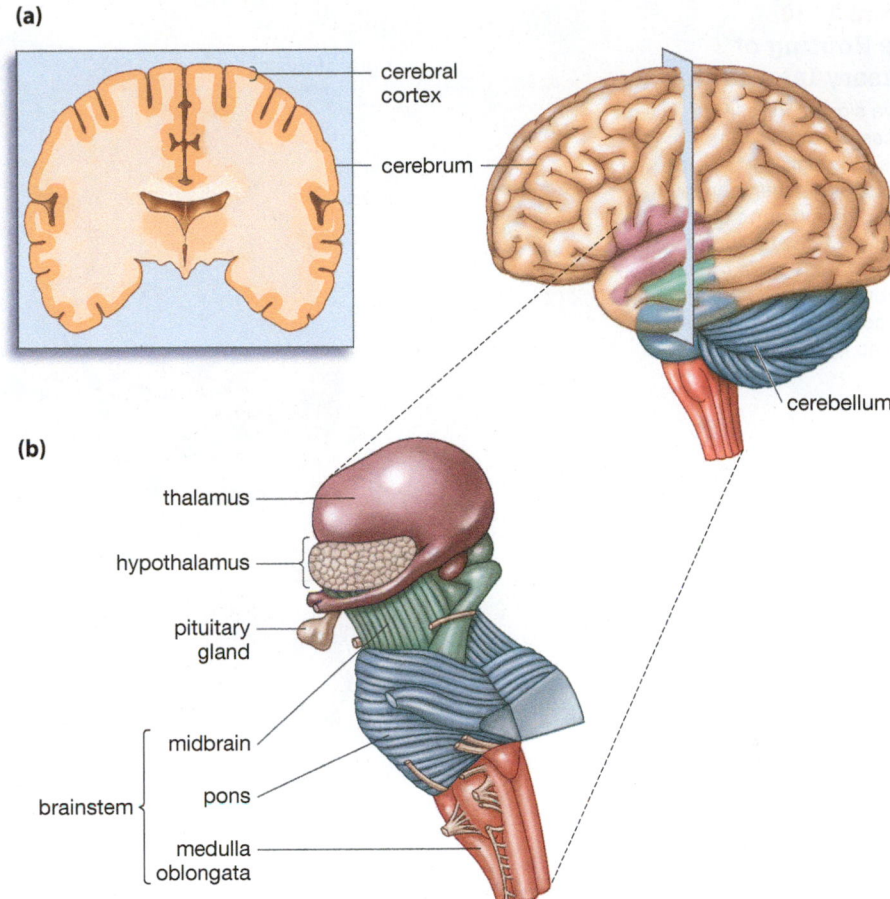

(a)

cerebral cortex

cerebrum

cerebellum

(b)

thalamus

hypothalamus

pituitary gland

midbrain

pons

medulla oblongata

brainstem

Figure 27.9
The Brain

(a) View of a cross section of the brain showing its cerebrum and cerebral cortex and of the left surface of the brain.

(b) Structures of the brain outside the cerebrum and cerebellum. The pituitary gland is connected to the hypothalamus but is a part of the body's hormonal or "endocrine" system, rather than the brain.

The Human Brain

Figure 27.10
The Routing of Sensory Information
Nerve signals for all the senses except smell get routed through the brain's thalamus and then continue on to differing parts of the cerebral cortex for further processing. Signals for smell are routed through the olfactory cortex, the amygdala, and the hypothalamus.

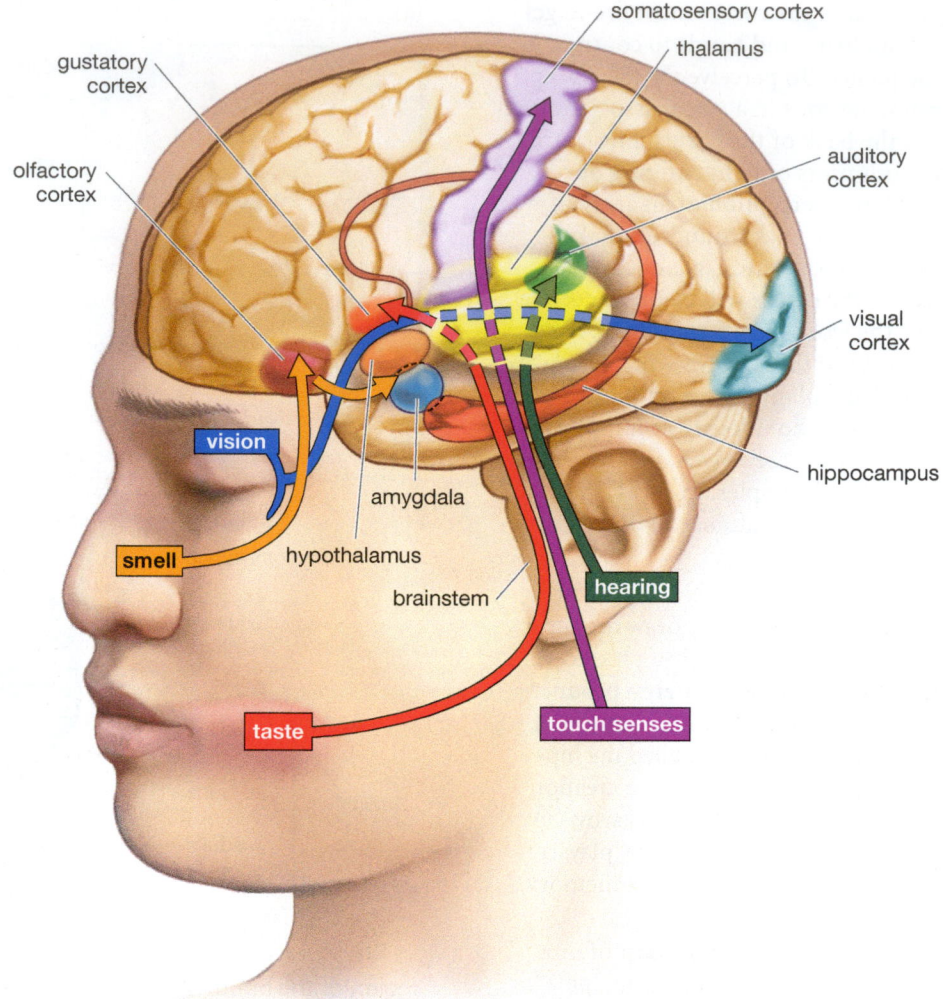

- The **midbrain** is a region of the brain that helps maintain muscle tone and posture through control of involuntary motor responses. Part of this control comes about through the midbrain's production of the neurotransmitter dopamine, noted earlier, which keeps our hands steady and our movements fluid.

- The Latin term *pons* refers to a bridge, and the region of the brain called the **pons** serves as one, in that its primary function is to relay messages between the cerebrum and the cerebellum.

- Located just beneath the pons, and just above the spinal cord, is the **medulla oblongata,** which contains major centers concerned with the regulation of unconscious functions such as breathing, blood pressure, and digestion. No one has to make a conscious decision to breathe, of course, and the medulla oblongata is important in the timing of this involuntary activity, as signals coming from it initiate the contraction of our respiratory muscles. This very function has a connection to a well-known term. What does it mean to be "brain dead"? Under one definition, it means that all the centers of the brain *except* the medulla

oblongata have permanently ceased to function. As long as the medulla is still working, a person continues to breathe despite the loss of all conscious ability.

Table 27.1 summarizes the information just presented on the seven regions of the brain.

SO FAR . . .

1. The spinal cord can receive input from _____ and direct _____ in response, without input from the brain. This entire process is known as a _____.

2. The autonomic nervous system is divided into the sympathetic or "_____" division and the parasympathetic or "_____" division.

3. The portion of the brain that is the site of our highest thinking and processing, the _____, is the outer portion of the brain's _____.

Table 27.1

Major Regions of the Human Brain

Region	Location	Functions In
Cerebrum	Topmost portion of brain; fills most of the skull and is draped over other regions of the brain	Reasoning, planning, memory, processing of sensory information; issuance of voluntary motor commands
Cerebellum	At the base of the skull, between the cerebrum and medulla oblongata	Refinement of motor commands; maintenance of balance
Thalamus	Beneath cerebrum, covered by it	Reception of sensory information and transmission of it to the cerebral cortex for processing
Hypothalamus	Beneath thalamus, above brainstem	Regulation of drives and the maintenance of homeostasis, often by interaction with the hormonal system
Midbrain	Upper brainstem	Maintenance of muscle tone and posture through control of involuntary muscle responses
Pons	Mid-brainstem	Transmission of messages between the cerebrum and cerebellum
Medulla oblongata	Lower brainstem	Regulation of breathing, blood pressure, and digestion

27.7 Our Senses

Having learned some of the basics about the nervous system, we can see how it works in one of its most interesting areas, sensory perception.

It may come as a surprise to learn that we actually have more sensory capabilities than the famous "five senses" of vision, touch, smell, taste, and hearing. We have a sense of balance, for example, but it isn't one of these five senses. We have another little-recognized sensory capability, related to touch, that helps us do such things as dance and simply sit up straight. To demonstrate this capability to yourself, extend one arm out, parallel to the ground, and close your eyes. Now, use the same arm and touch your index finger to the tip of your nose. How were you able to do this? You have sensory neurons that monitor such things as the position of your joints and the tension of the tendons in your body. These neurons provide a sense, called *proprioception,* that in this instance helped guide your finger to your nose. (It sent out signals that said "on the right course" or "correction needed.") It's this sense that the cerebellum relies on to fine-tune the body's movements.

Taking these and other sensory capabilities into account, it becomes apparent that our "five senses" really amount to the senses we are consciously aware of. It is understandable that these senses command our attention, however, because they alone make us not just sensory beings but sensual beings. (If you wanted to evoke the richness of life in as short a space as possible, you could do worse than to write down just five words: seeing, touching, hearing, smelling, and tasting.) As such, we'll briefly review how each of these senses works. But first, it's worth noting what all the senses have in common.

Each of our senses employs the sensory receptor cells noted earlier—cells that can respond to stimulation, such as vibration (for sound) or reflected light (for vision). In addition, all sensory receptor cells *transform* their responses into the language of the nervous system, meaning electrical signals that work through action potentials. Think of this task as similar to the one undertaken by a so-called transducer clipped across the sound hole of an acoustic guitar. It has to turn the vibration of the guitar strings into an electric signal that goes to an amplifier. In a similar way, sensory receptors in your ear must turn the vibrations of your eardrum into an electric signal that goes to your brain. Fittingly enough, sensory receptors are sometimes called *transducers.* All the senses except one also share another commonality. If you look at Figure 27.10, you can see that, except for smell, all sensory perceptions are routed through the brain structure noted earlier called the thalamus, and that all these signals then go on to various parts of the cerebral cortex for processing. Now let's look at the five best-known senses, taking touch first.

27.8 Touch

Humans actually have several senses of touch. One of these is *thermal reception,* which means sensing hot and cold. Another is *tactile reception,* which includes our senses of pressure and the proprioception we

(a) Touch receptors in the skin

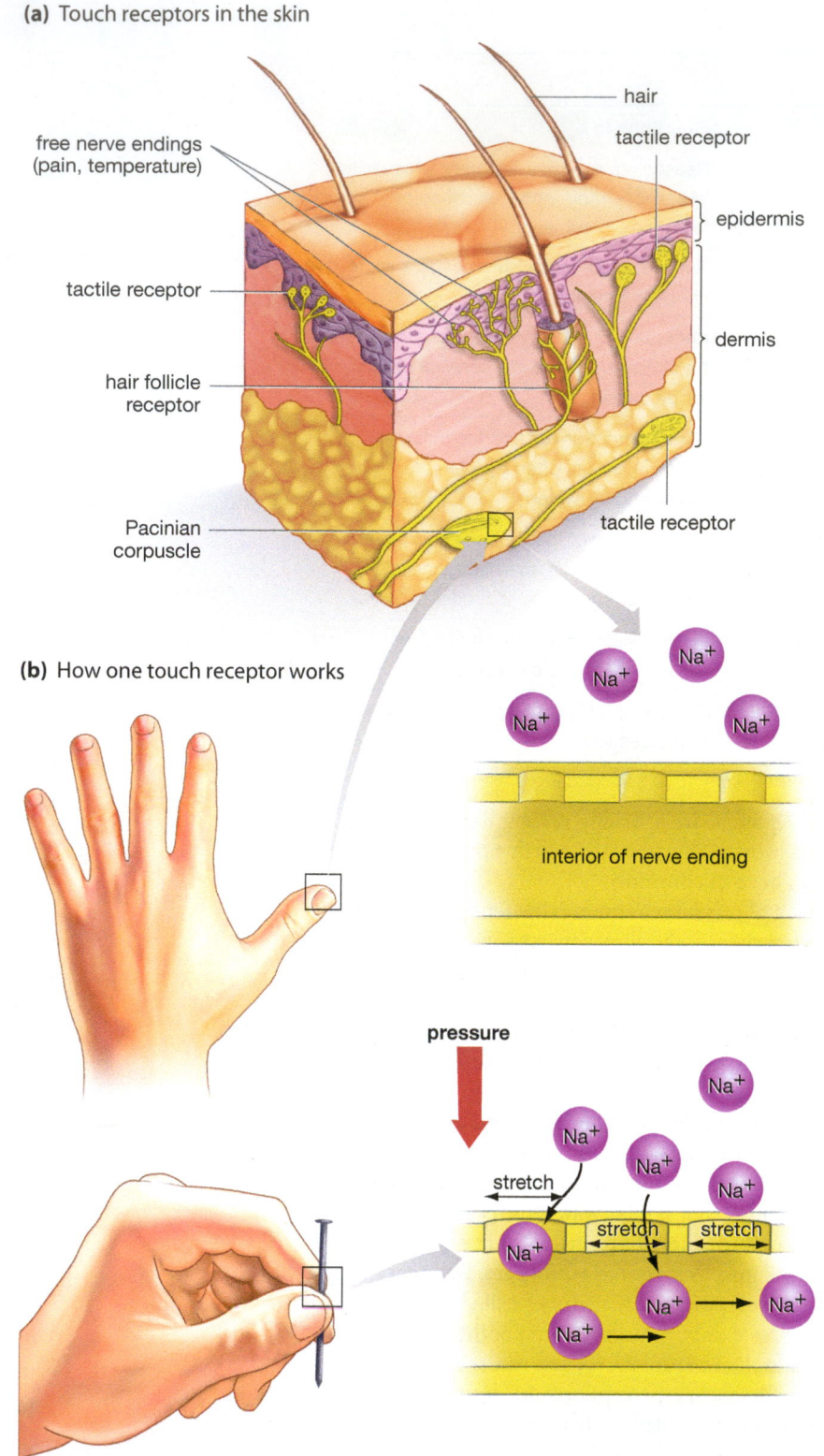

(b) How one touch receptor works

interior of nerve ending

pressure

stretch

stretch   stretch

Figure 27.11
Our Senses of Touch

(a) Free nerve endings near the surface of the skin sense pain and temperature, while five different types of sensory receptors in the skin sense other sorts of tactile sensations.

(b) One of the skin's sensory receptors is the Pacinian corpuscle. When the dendrite within the corpuscle is distorted by touch, its ion channels are stretched in such a way that it allows an influx of sodium ions (Na⁺). This instigates a nerve signal conveying the message "Touch has occurred here."

talked about. A third is *pain reception*, which means just what it sounds like.

Within our skin, we have several kinds of sensory receptors that contribute to these senses. Thermal and pain sensing operate through simple nerve endings near the surface of the skin. But the tactile sense has five special types of receptors in the skin alone. Why so many? Well, pressure can be light or heavy, and we want to know the difference. Likewise, contact can be brief or ongoing, and we need to assess the difference in a sophisticated way. We want the message "your elbows have reached the table" to come to our attention front and center, but we want the message "your elbows are still on the table" to be a *low-level* message, so that we can focus on other things. These various receptors allow for all this and much more.

If you look at **Figure 27.11a**, you can see several varieties of touch receptors, and if you look at **27.11b**, you can see how one particular receptor works. The Pacinian corpuscle receptor consists of a single dendrite that is wrapped within onion-like layers of fluid and supportive tissue (which make up the corpuscle). As a result, this receptor is not likely to respond to gentle pressure. Instead, it responds to somewhat deeper pressure, usually involving vibrations (like pulling your finger across a page). Undisturbed by touch, the ion channels in this receptor's outer membrane are not shaped correctly to let through the sodium (Na⁺) ions that get a nerve signal going. When stretched, however, the ion channels open in such a way that the Na⁺ ions rush in, and a touch message gets going. Note that the Pacinian receptor does the double duty of first responding to a stimulus (pressure) and then transforming this stimulus into the language of the nervous system.

Once the Pacinian's nerve signal is on its way, it comes into the spinal cord through one of the spinal nerves. It then moves up through the spinal cord and brainstem and into the brain's thalamus, from which point it is sent to the brain's somatosensory cortex. It's worth noting that sensory messages such as these always cross over a midpoint in the spinal cord or brainstem, so that anything we feel on the left side of our body is processed on the right side of our brain and vice versa.

27.9 Smell

As noted earlier, our sense of smell stands alone in not routing its messages up through the thalamus. If you look at Figure 27.10, you can see where these messages go: to a part of the cerebral cortex (called the olfactory cortex) and to the amygdala and hypothalamus. These last two structures are active in both emotion and memory. Some scientists believe this linkage is the reason that our sense of smell—or **olfaction,** as it is properly

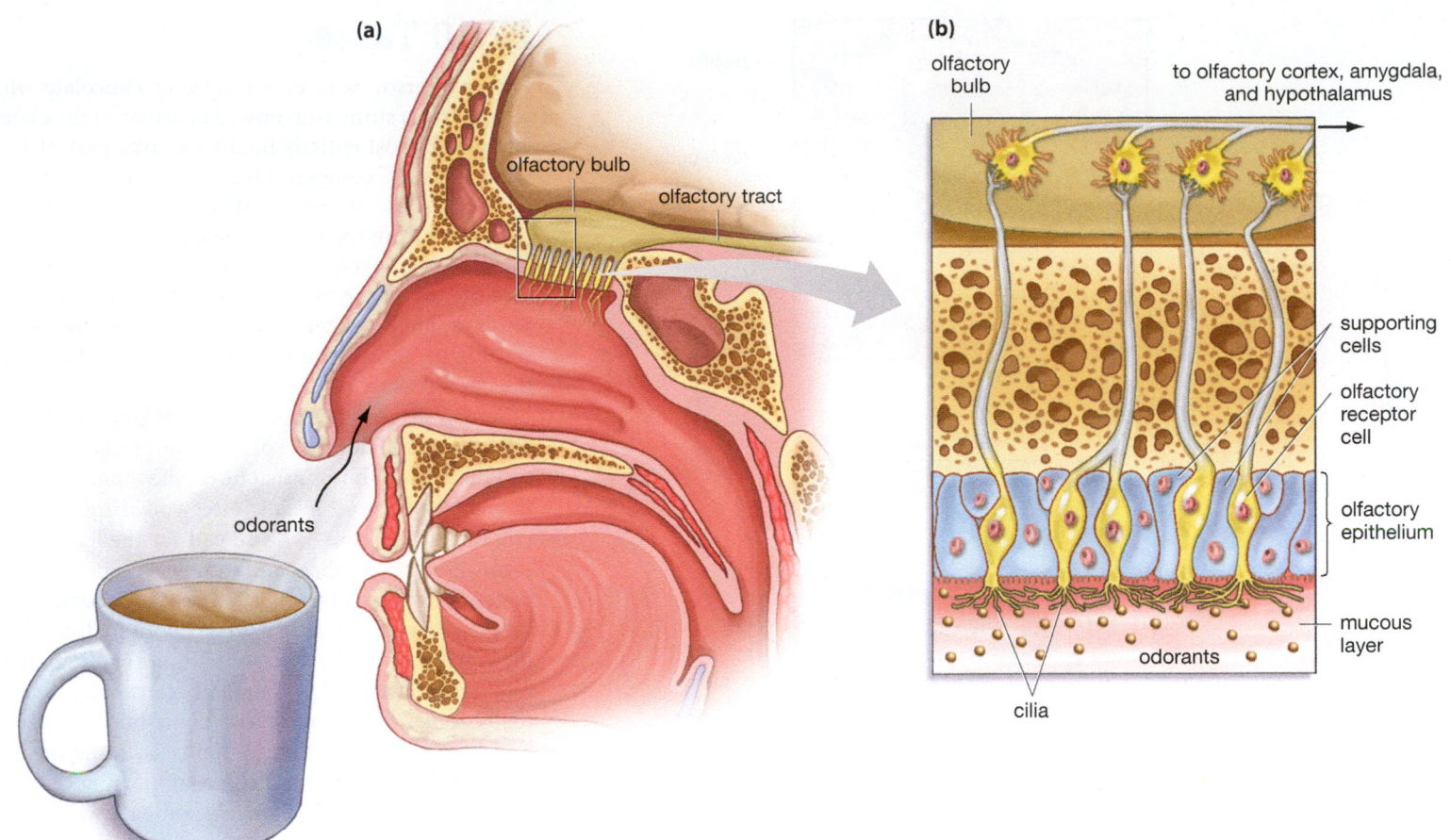

(a)

olfactory bulb

olfactory tract

odorants

(b)

olfactory bulb

to olfactory cortex, amygdala, and hypothalamus

supporting cells

olfactory receptor cell

olfactory epithelium

mucous layer

odorants

cilia

known—is able to evoke such strong memories. People, places, and events can all come back to us in a rush if we happen to encounter the right smell. Olfaction is also thought to be our "oldest" sense in that some of the olfaction mechanisms at work in us are also at work in fruit flies, even though our evolutionary line diverged from theirs about 500 million years ago.

If you look at **Figure 27.12**, you can see how olfaction works. Air that bears fragrant molecules comes into our nasal passages. Right up at the top of these passages, back behind the plane of our forehead and just beneath the level of our eyes, we have a group of neurons, called olfactory receptor cells, whose dendrites extend down into the passages. These dendrites have some hair-like extensions, called cilia, sprouting from their tips. Each of these cilia is covered with receptors capable of binding with the fragrance molecules. These fragrance molecules, which are called "odorants," come in, get trapped in a mucus layer that lines the nasal passages, and soon bind with receptors on the cilia. This binding puts into motion a complicated set of reactions inside the dendrite that results in Na^+ channels being opened up in the dendrite membrane. The result is a neural signal that goes up through the dendrite and then the cell as a whole, which synapses with neurons in

the olfactory bulb—a nerve tract that is part of the brain. From there, the signal gets distributed in the pattern we talked about (to the amygdala, hypothalamus, etc.).

The receptors lying on the surface of the olfactory receptor cilia come in some 340 to 380 different varieties (each of which can bind to many different odorants). This sounds like a lot, but professional perfume creators might be able to discriminate between 2,000 separate scents in the starting ingredients they use, and perfumes are, of course, only one type of scent among many that human beings can recognize. So how can a few hundred different receptor types yield a vastly higher number of scents? Well, as researcher Richard Axel has noted, a mere 26 letters in the English language yield a huge number of words because the letters can be *combined* in different ways. In olfaction, each receptor adds nothing more than a message saying, "receptor here stimulated"—one letter in a word. But the brain is receiving input from all these receptors. When a given combination of receptor signals comes in, the brain senses a "word," meaning the smell of, say, chocolate or lemon. The brain is essentially saying, "When receptors 4, 32, 87, and 220 fire together, that's chocolate." Impressive as human abilities are in this regard, we

Figure 27.12
Our Sense of Smell

Smell or "olfaction" begins when fragrance molecules, called odorants, bind with receptors that cover cilia extending from olfactory receptor cells. This binding causes an influx of Na^+ ions into the cells, resulting in a nerve signal that moves from the cells to the olfactory bulb—a part of the brain.

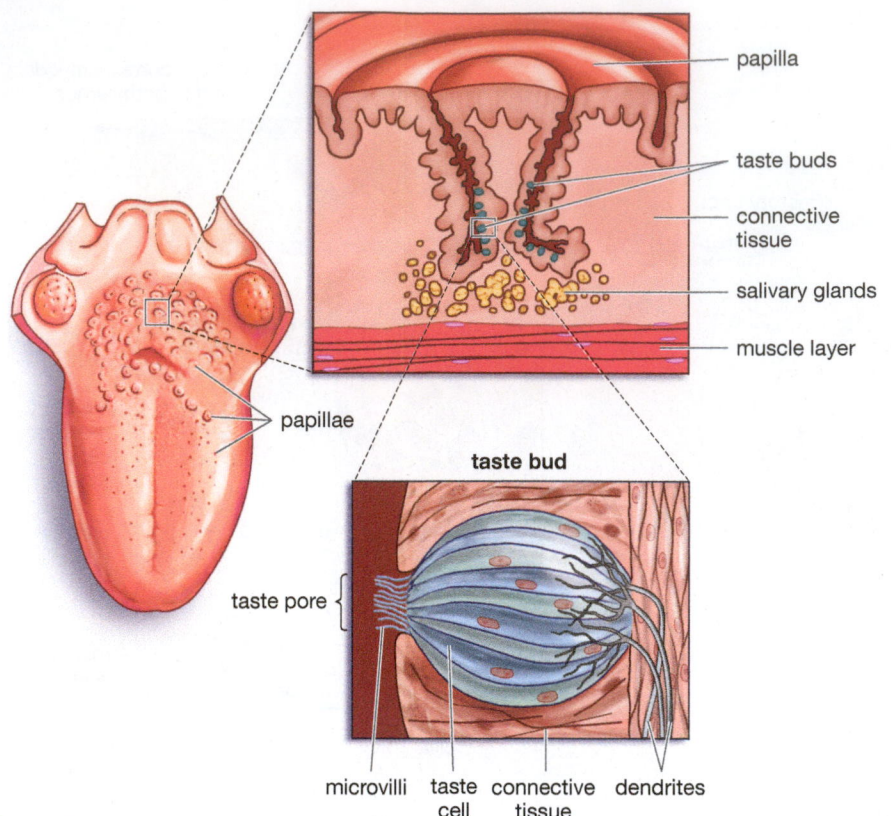

papilla

taste buds

connective tissue

salivary glands

muscle layer

papillae

taste bud

taste pore

microvilli   taste   connective   dendrites
             cell   tissue

Figure 27.13
Our Sense of Taste
Small bumps in the tongue called papillae contain taste buds—collections of tissues and various cells, including taste cells. Microvilli extending from taste cells are capable of binding with food molecules, or "tastants," which they come into contact with through taste pores. This contact causes the taste cells to release a neurotransmitter that stimulates the dendrites of adjacent nerve cells. With this, a nerve signal regarding taste is on its way to the brain.

can only wonder at the finely tuned world of scents inhabited by some of our animal relatives. Dogs are endowed not with a few hundred types of olfactory receptors, but instead with 1,000 of them, and rats have about 1,500.

SO FAR . . .

1. Each of our senses employs the cells known as _____, which can both respond to stimulation and transform their responses into _____.

2. All senses except _____ route their signals through the brain's _____, after which these signals go to differing parts of the _____ for additional processing.

3. The 340 to 380 types of olfactory receptors we have allow us to discriminate among a much greater number of smells because the various receptor types can send signals in different _____.

27.10 Taste

Pity the person who eats a piece of chocolate while also having a stuffed-up nose. The flavor of the chocolate may be lost entirely because a large part of what we call "flavor" comes not from our sense of taste, but from our sense of smell. Earlier, we pictured smells as coming in from the nose to the nasal passages, but remember that these same nasal passages empty out into the top of the throat. Every bite of food you take sends a huge concentration of odorant molecules wafting up the *other* direction—from the throat into the nasal passages.

When we consider pure "taste," as opposed to "flavor," the action does take place on our tongues, within several kinds of bumps that lie on the tongue's surface (some of which are visible after you drink milk). If you look at **Figure 27.13**, you can see the locations of some of these "papillae" and some detail on one of the larger varieties, which exist toward the back of the tongue. As you can see, if you go about halfway down these papillae, you get to something that sounds familiar: the taste bud, so named because each one actually looks something like a plant bud. Each taste bud is a collection of 50 to 100 cells, some of these being touch receptors and some supporting cells. But taste buds also contain *taste cells,* which is what we are interested in. Taste cells are not neurons, but they do have the ability to convey signals. Like the olfactory receptors we looked at, they have tiny hair-like projections extending from them—in this case, projecting out into the cavity that surrounds each papilla. These projections are capable of interacting with "tastants," meaning food molecules that elicit different tastes.

With all this in mind, here's how taste works. You eat some food, and the resulting tastants dissolve in your saliva. Molecules or ions from the tastants then make contact with the projections (called microvilli) sprouting from the taste cells. Through one of several chemical reaction routes, this contact results in a change in the membrane potential of the taste cells, and this change prompts them to release a neurotransmitter. Waiting to receive the neurotransmitter are the dendrites of nearby neurons, and our nerve signal is on its way.

You may have heard that humans have four basic taste sensations: sweet, sour, salty, and bitter. What makes these "basic" is that each one works through a different chemical reaction route in the taste cell. Salty foods spur neurotransmitter release through one set of reactions; sweet foods spur release through another. In recent years, some scientists have become persuaded that there is a fifth basic taste—a savory taste called *umami*—and very recently scientists identified calcium as a basic taste in rodents, which means humans may have this as a basic taste as well. All of the basic tastes discovered so far can be perceived by every taste cell on our tongue, and, contrary to any "tongue maps" you may have seen, no part of the tongue is particularly responsive to any of the basic tastes. Bitter taste is not sensed primarily toward the back of the tongue nor sweet taste toward the front.

At this point, you may be wondering: How can there be just four to six *basic* tastes, but a huge number of tastes we actually experience? The answer is that the neurons that receive the signals from the taste cells vary in their response to specific tastants. If you eat some table sugar, one group of neurons will respond by firing perhaps 10 times over a few seconds, but another group of neurons will fire more than 100 times. Most researchers now agree that it is the *pattern* of firing activity across all taste neurons that the brain uses to tell us whether it is just basil in the soup or a particularly sharp basil. You've seen this pattern processing before, with smell, and it is at work elsewhere. We might say that in pattern processing, the brain is something like a talented news writer: It gets input from lots of individuals and turns these reports into a coherent, highly detailed story—in this instance, our sense of what we are tasting.

27.11 Hearing

Sound is always depicted as coming in "waves," but it does not physically exist as a series of tiny, invisible lines that run in peaks and troughs. Pluck a single string on a harp and then slow the videotape of this action way down. The first thing that happens is that the string moves in the opposite direction from the way it was pulled. This movement *compresses the air molecules* ahead of the string, making this air more dense than that of the ambient air around it. Now the string swings back the other way, and this *lowers* the air pressure ahead of the string—below that of atmospheric air pressure. So what we perceive in hearing are "waves" of air molecules that are, by turns, more and less compressed than those of ambient air. This cycle of high and low compression happens very rapidly. When you hit the A above middle C on a piano, the waves of high and low compression are bumping up against your ear drum 440 times per second—meaning the piano string is going back and forth 440 times per second. It is these cycles that give sound its frequency. The faster the frequency, the higher the sound. The best human ears (usually children's ears) can perceive sounds ranging from about 20 to 20,000 cycles per second.

So, how do we go from a vibrating eardrum to the perception of a sound? If you look at **Figure 27.14a** on the next page, you can see the basic anatomy of the ear, and in **Figure 27.14b**, you can track the steps involved in hearing. (The elongated structure on the right of Figure 27.14b is the cochlea of the ear, uncoiled for illustrative purposes.)

Note that hearing begins when vibrations of the eardrum or *tympanic membrane* in turn vibrate the three smallest bones in the body: the malleus, incus, and stapes (or "hammer, anvil, and stirrup"), which lie in the middle ear. In the same way that a lens serves to take light coming from a large area and focus it on a small one, these three bones take the vibrations coming from the relatively large tympanic membrane and focus them on the smaller oval window. This is important because it *amplifies* the vibration signal for the next step. The oval window's vibrations now shake fluid that lies within the pea-sized structure called the **cochlea:** the coiled, membranous portion of the inner ear in which vibrations are transformed into the nervous-system signals perceived as sound.

In the cross section of the cochlea pictured in **Figure 27.14c**, you can see how vibrations of fluid in the cochlea are transformed into the sensation of sound. The cochlea can be seen to have a couple of ducts (the vestibular and tympanic), which are filled with fluid. In the middle, there is the cochlear duct, likewise filled with fluid, which has at its bottom a "basilar" membrane on which sit the elements that will produce sound. Note that, supported by the basilar membrane, there are a group of so-called hair cells, so named because each of them sprouts a group of hair-like cilia that come close to touching the tectorial membrane that folds over them in the cochlea.

Now, how does all of this yield sound? Think of it this way: If you rested your left hand on a water balloon and then pushed rhythmically on this same balloon with your right hand, your left hand would move up and down slightly. If there were something directly above your left hand, then your hand might make contact with it. In the same way, the vibrations of the fluid in the cochlea move the basilar membrane up and down, and this movement can push the hair cells up against the tectorial membrane that lies above them. If you look at the blowup of Figure 27.14c, you can see what happens next. The cilia on the hair cells have microscopic "trapdoor" channels on them, and when these cilia bend from touching the tectorial membrane, the trap doors open. With this, potassium (K^+) ions flow in, leading to an influx of calcium (Ca^{2+}) ions at the base of the hair cell, and this leads the hair cells to release a neurotransmitter. Waiting to receive the neurotransmitter are adjacent dendrites, and with this, vibration has been turned into a nerve signal.

The details of the hearing process actually are much more complicated than this account would indicate. While we need not go over these details, it's worthwhile to note just one of the ways this system is able to discriminate one sound from another. To locate where a sound is *coming from*, the brain in effect calculates the difference between how intense a sound is in one ear as opposed to the other, and what the time difference is between when sound arrives at one ear and then the other.

Finally, it's easy for hearing to be damaged, and this damage often is to the hair cells we've been looking at, as you can see in "Too Loud" on page 525.

The Human Ear

(a) Anatomy of the ear

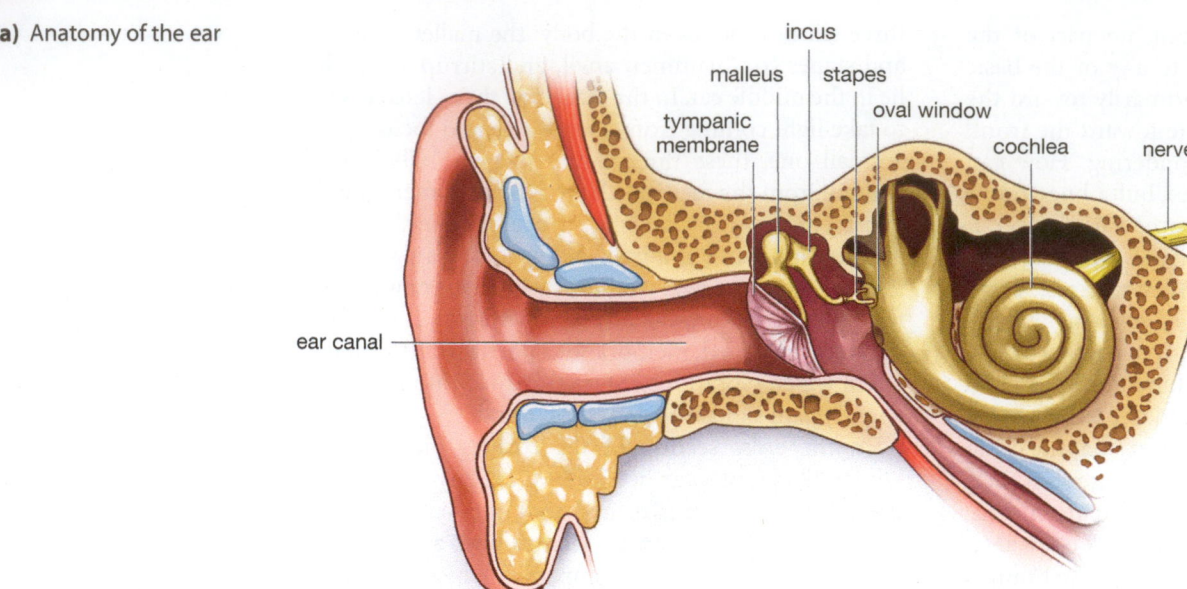

outer ear middle ear inner ear

(b) From air vibration to nerve signal

2. The tympanic membrane vibrates the three bones of the middle ear; the malleus, incus and stapes.

1. Sound waves enter through the ear canal and vibrate the tympanic membrane.

3. The vibration of the stapes focuses the sound-wave vibration on the membrane of the oval window.

4. The oval window's vibrations cause fluid vibrations within the coiled, tubular cochlea (shown elongated here for illustrative purposes).

5. These fluid vibrations cause cells within the cochlea to release a neurotransmitter, which triggers a nerve signal to the brain.

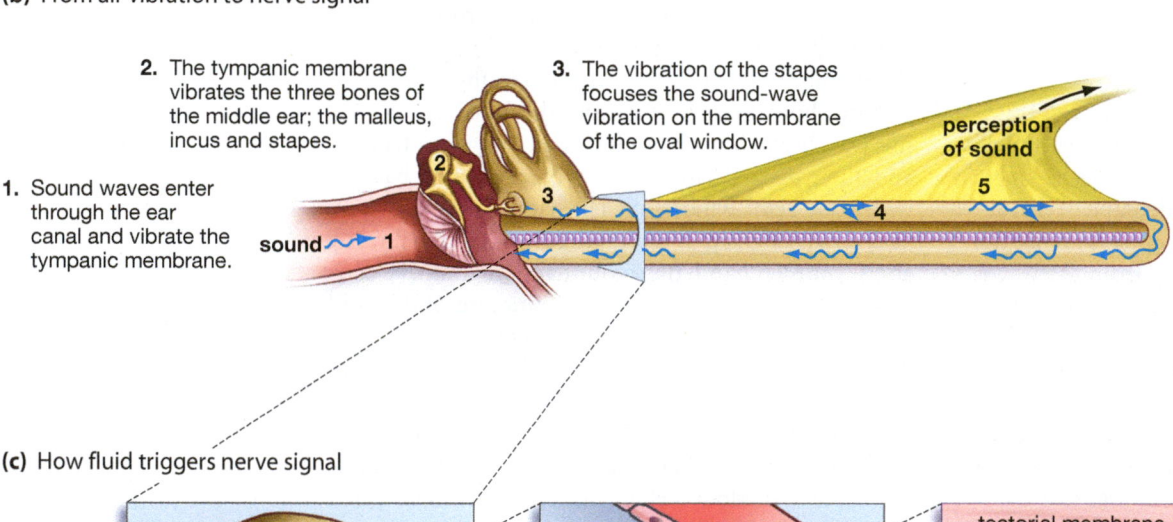

(c) How fluid triggers nerve signal

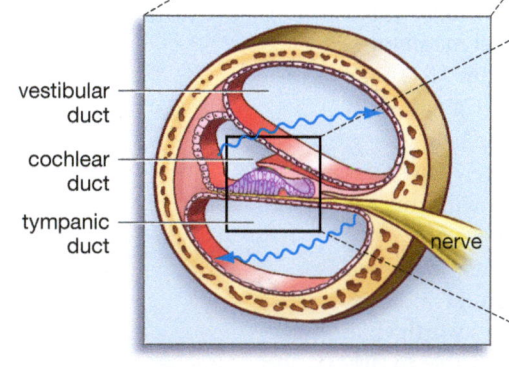

1. Seen in cross section, the cochlea has vestibular and tympanic ducts, in which fluid is vibrating.

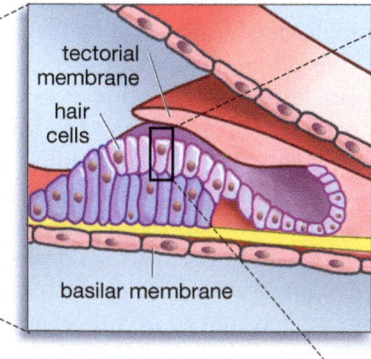

2. This vibration shakes the basilar membrane, pushing hair cells on it up against the overlying tectorial membrane.

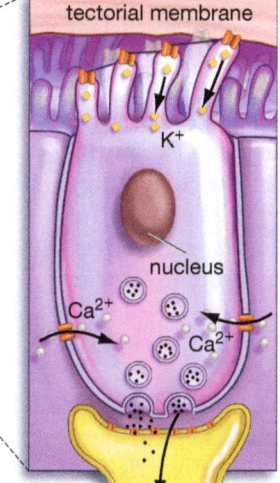

3. As the hair cells contact the tectorial membrane, cilia on them bend. This change in position causes "trap door" channels in the hair cells to open, which allows potassium ions (K^+) to flow into them.

4. This influx triggers an influx of calcium ions (Ca^{2+}) at the base of the hair cells, which in turn causes the cells to release a neurotransmitter.

5. The neurotransmitter is received by adjacent dendrites and a nerve signal is sent to the brain.

Figure 27.14
Our Sense of Hearing

ESSAY

Too Loud: Hair Cell Loss and Hearing

Our sense of hearing can easily be damaged. Eighty percent of the hearing loss in the United States—affecting 28 million people—involves damage to either the auditory hair cells referred to in the main text (**Figure 1**) or to the large nerve into which these cells feed. This type of hearing loss often comes as a consequence of old age, but it can come as a consequence of *young* age, which is to say the tendency of young people to listen to loud music. Damage caused by loud noise may be limited only to the hair cells' cilia—their roots may break off or the links between them may be destroyed—and in most of these instances, these hair cells can recover. When noise is loud or long-lasting enough, however, it can kill hair cells outright. When this happens, the effect may be not only a loss of hearing but a so-called tinnitus, meaning a permanent high-pitched ringing in the ears.

So, how loud is too loud? The rule of thumb is, if you have to shout to be heard, you're in the danger zone. The amount of time a person is exposed to loud noise matters, however. You could be around a typical lawn mower for several hours without harm, but after 15 minutes at a loud rock concert, you are damaging your hair cells. In terms of absolute sound levels, 85 decibels is the limit for indefinite exposure—above that and you could damage hearing if you're exposed for long enough. If you look at **Figure 2**, you can see where various everyday sounds lie on the decibel scale. (Keep in mind that the scale is logarithmic; a 110-decibel sound is 10 times as loud as a 100-decibel sound.) Note that the popular iPod MP3 player can, at maximum volume, deliver up to 120 decibels of sound—an amount equivalent to that produced by a jet airplane when it's taking off.

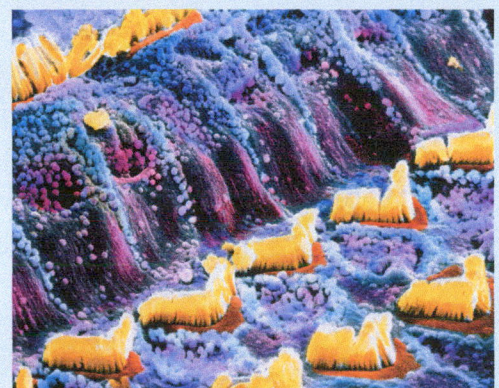

Figure 1
Vulnerable to Damage by Loud Noise
This color-enhanced micrograph shows the ear's hair cells, in yellow. Loud noise can damage the hair-like cilia visible on the cells or can even kill the cells. Two types of hair cells are shown: outer cells, with a V shape at right, and inner cells at left.

Decibel Level	Source
140+	Firearm
140	Jet engine
130	Jackhammer
127	Sporting event
120+	Live music concert
120	Jet plane takeoff
120	Band practice
120	iPods and other MP3 players at maximum volume
120	Health club and aerobics studio
118	Movie theatre
95–120	Motorcycle
100	Chain saw or pneumatic drill
90	Lawnmower
90	Subway
80	Busy street
80	Alarm clock
70	Vacuum cleaner
60	Conversation
50	Moderate rainfall
40	Quiet room
30	Whisper, quiet library

Figure 2
How Loud Is It?
The decibel levels of some everyday sounds. In general, anything above the 85-decibel level can damage hearing if exposure to the sound continues for long enough.
Source: American Speech-Language-Hearing Association.

SO FAR . . .

1. Taste cells are part of organized collections of 50 to 100 cells known as _____. Taste works by means of "tastants" interacting with hair-like projections that extend from the _____.

2. Sound waves are collections of air molecules that are either more or less _____ than molecules of ambient air.

3. The inner ear's cochlea is filled with _____; its vibrations open and shut "trapdoor" channels on _____ cells, thus resulting in the transmission of a nerve signal.

27.12 Vision

The Human Eye

Our visual system has to accomplish three central tasks. First, it has to capture light from the outside world and focus it at a very precise location within our eyes. Second, it has to take this focused light and convert it into a nervous system signal. Third, it has to make sense of the visual information it receives. If you look at **Figure 27.15**, you can see some of the structures of the eye that function in the first two steps. Light comes into our eyes through the transparent cornea and passes through the opening known as the pupil. Surrounding the pupil is the iris, which is composed partly of smooth muscle and thus is capable of contracting or dilating, meaning it can let in less light in a bright environment or more light in a darkened one. The incoming light is bound for the layer of cells at the back of the eye called the **retina:** an inner layer of tissue in the eye containing cells that transform light into nervous system signals. But as the light rays come in, they are bent or *refracted*—first by the cornea and then by the lens that lies behind it—such that they converge on a small area of the retina, as you can see in **Figure 27.16**. This is no different in principle from what happens with a camera. We take a picture of a large object (such as a statue) that ends up as a small image on a frame of film. In a similar way, thanks to refraction, large objects we view with our eyes end up as *very* small images on our retina. How small? At fractions of a millimeter, some of them are microscopic. The eyes do not always focus light rays in such a way that they converge properly on the retina, however. You can see what it means to be nearsighted or farsighted in Figure 27.16.

Now, how do retinal cells convert rays of light into electrical signals? If you look at **Figure 27.17**, you can see that there are three layers of cells in the retina: ganglion cells, connecting cells, and photoreceptors. The **photoreceptors** are the sensory receptors for vision, and they come in two varieties: rods and cones. **Rods** are photoreceptors that function in low-light situations but that provide only black-and-white vision. **Cones** are photoreceptors that respond best to bright light and that provide color vision.

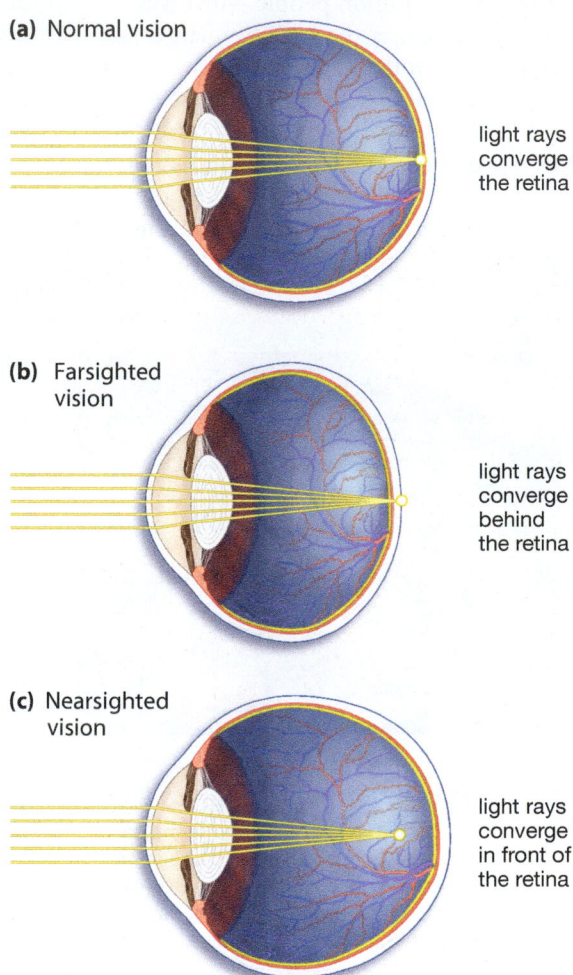

(a) Normal vision

light rays converge on the retina

(b) Farsighted vision

light rays converge behind the retina

(c) Nearsighted vision

light rays converge in front of the retina

Figure 27.16
Focused and Unfocused

Sharp vision requires that incoming light rays converge just as they reach the retina.

(a) In normal vision, the eye's cornea and lens bring about this convergence; they refract the incoming light at an angle that matches the length of the eye.

(b) People who are farsighted have a mismatch between the length of their eye and the amount of refraction provided by their cornea and lens: Their eye is too short for the amount of refraction provided. Thus, light rays converge behind their retina. As a result, these people cannot see nearby objects clearly.

(c) People who are nearsighted have the opposite problem: Their eye is too long for the amount of refraction provided, with the result that light rays converge in front of their retina. These people cannot see distant objects clearly. Both nearsighted and farsighted vision can be corrected.

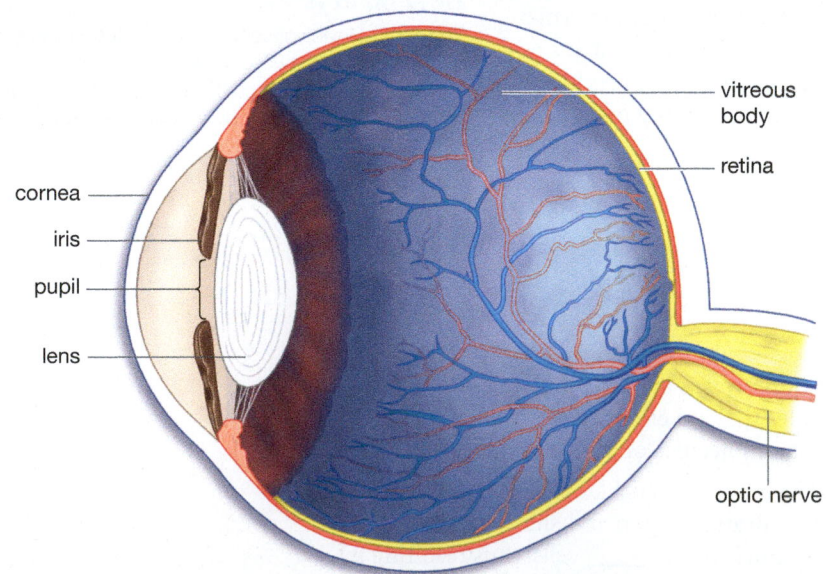

cornea

iris

pupil

lens

vitreous body

retina

optic nerve

Figure 27.15
Anatomy of the Human Eye

The blowup of these cells shows that their posterior ends—the ends toward the back of the head—are filled with pigments, which lie embedded in a series of flattened, membranous disks. A pigment is simply a chemical compound that strongly absorbs light. So, our light comes in and moves to, say, a rod in the retina. There it is absorbed by a pigment, and this absorption causes a normally bent part of the pigment molecule to . . . straighten out. That's it. Our entire sense of vision hinges on this tiny initial effect. Yet, as a shouted "hello" might cause an avalanche, so this small step puts big things in motion. In his book *What Makes You Tick?* vision researcher Thomas Czerner put it this way:

> The infinitesimal energy of a photon [of light] changes the shape of a *single* molecule of pigment, which causes the release of *dozens* of molecules of an enzyme, which cause the rapid breakdown of *hundreds* of molecules of a chemical messenger, which cause the gates of *thousands* of ion channels in the cell membrane to slam shut, which blocks *millions* of sodium ions from entering the cell, which produces the grand finale, a local change in the resting membrane potential.

You might expect that this change in photoreceptor membrane potential would, like the others you've seen, result in the release of a neurotransmitter to an adjacent neuron. The twist in vision, however, is that in their *resting* state rods and cones are releasing neurotransmitter molecules in abundance to their connecting cells. When light strikes the rods and cones, it has the effect of reducing this neurotransmitter release—the blocking of sodium ions that Czerner notes in his account. This lack of neurotransmitter release thus transmits the message "light stimulation here."

The signal that begins with a rod or cone goes first to connecting neurons and then to the ganglion cells you can see in Figure 27.17. These cells have axons that are like tributaries coming together to form a river, only this river is the optic nerve, composed of about a million axons (see Figure 27.15). Because we have two eyes, we thus have a pair of million-fiber optic nerves carrying visual information from the eyes to the brain. The optic nerve takes its visual information to the thalamus, and from there it is transmitted, on other nerve fibers, to an area at the very back surface of the brain, the *primary visual cortex*, where most vision processing is done.

The details of this processing are much too complicated to go into here. But a central lesson of this complexity is that the human visual system does not work like a camera; it does not passively record bits of light and register a collection of these bits as images. Rather,

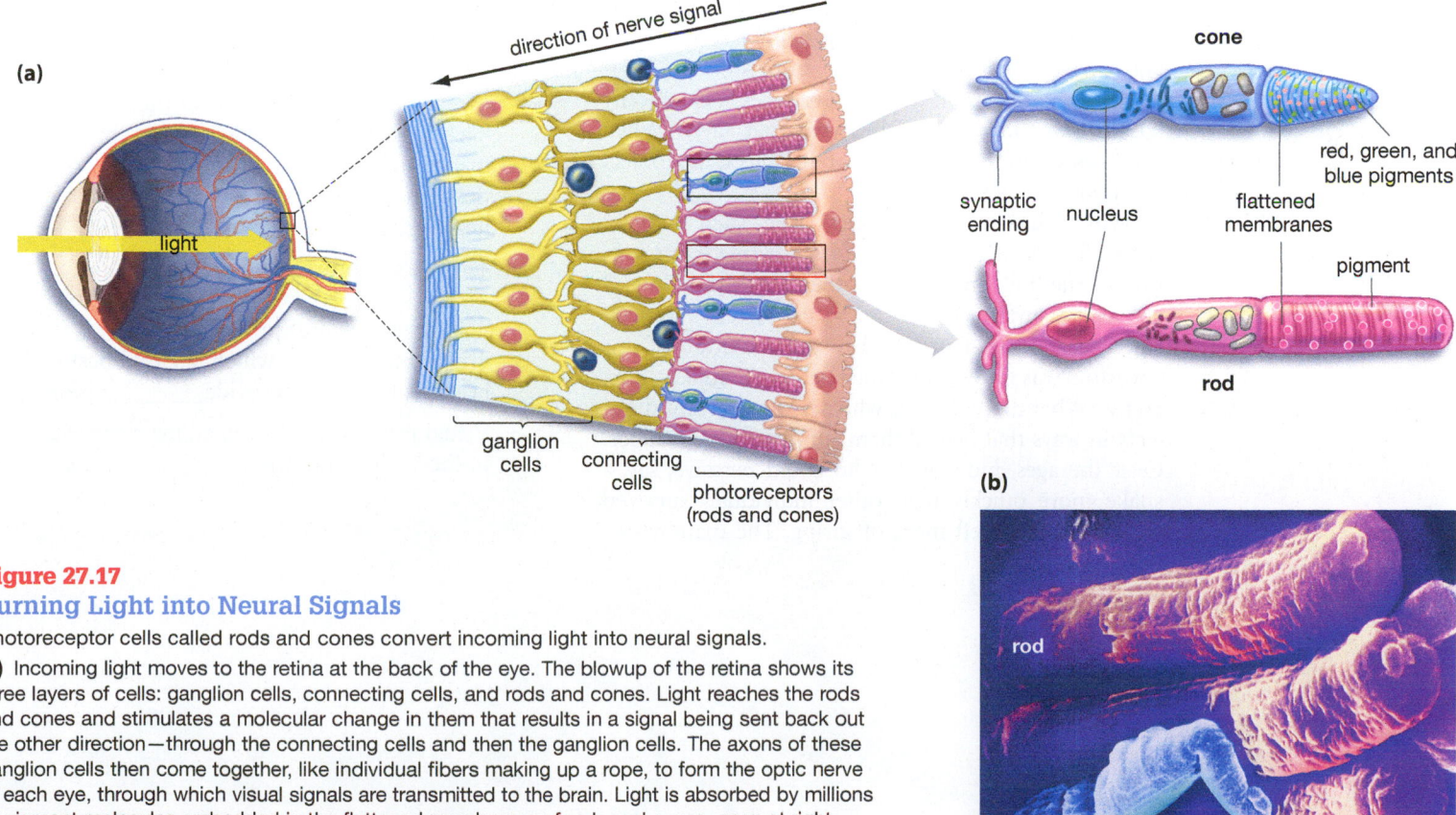

(a) direction of nerve signal

light

ganglion cells — connecting cells — photoreceptors (rods and cones)

cone

red, green, and blue pigments

synaptic ending — nucleus — flattened membranes — pigment

rod

(b)

rod

cone

Figure 27.17
Turning Light into Neural Signals
Photoreceptor cells called rods and cones convert incoming light into neural signals.

(a) Incoming light moves to the retina at the back of the eye. The blowup of the retina shows its three layers of cells: ganglion cells, connecting cells, and rods and cones. Light reaches the rods and cones and stimulates a molecular change in them that results in a signal being sent back out the other direction—through the connecting cells and then the ganglion cells. The axons of these ganglion cells then come together, like individual fibers making up a rope, to form the optic nerve of each eye, through which visual signals are transmitted to the brain. Light is absorbed by millions of pigment molecules embedded in the flattened membranes of rods and cones, seen at right.

(b) A color-enhanced micrograph of rods and cones.

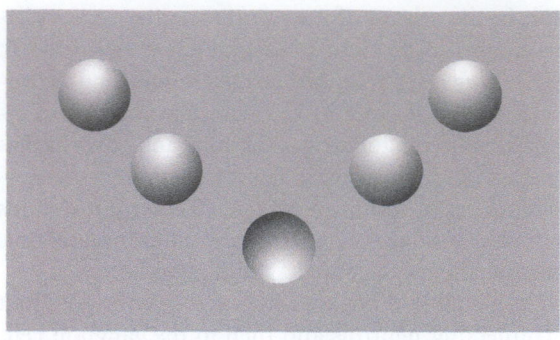

Figure 27.18
Seeing Is Believing
Look at the figure and then turn the book upside down and look at it again. With the shift, dots that once appeared to be raised above the surface of the drawing now appear to be indented, and vice versa. Our visual system has a rule that causes us to perceive objects that are shaded at the bottom to be protruding from a surface, while objects that are shaded at the top appear to be indented. This is but one example of how our visual system creates visual images rather than simply recording them.

the brain *constructs* images as much as it records them. To get an idea of this, look at **Figure 27.18**. In it, you see a V, formed out of bumps and a dot-like indentation. Now turn the book upside down and look again. The bumps have become indentations and vice versa. Why? Note that when the book is right side up, most dots in the V are shaded in such a way that light seems to be coming from above them—there are "shadows" on the lower part of the dots. Light generally comes from above us, whether from a lamp or from the sun. This is so common that evolution has produced a genetically based "rule" in our visual system that says: "When shadows are in the lower part of an object, that object is raised above a surface."

Why would evolution produce rules such as these? After all, this rule is yielding an *inaccurate* interpretation of the dots on the page—they aren't really raised above the surface at all. Our visual system has such rules because what mattered in the evolution of our ancestors was not whether they perceived objects "correctly." What mattered was whether they perceived objects in ways that helped them survive and reproduce. Over the ages, individuals who could perceive, say, a snake more quickly than other individuals survived longer and thus left more offspring. The many visual rules we have amount to rules for survival. They provide a series of "best bets" for quickly telling us what we are looking at. (If an object's lower half is shaded, the best bet is that it's raised above a surface.) The idea that our visual system is often creating perceptions, rather than simply recording them, may seem strange, but that's the reality of visual perception.

On to the Endocrine System

The nervous system you've been reading about handles a particular type of communication within the body—instantaneous communication. Through it, we perceive events as they occur and take action based on information that may be only a fraction of a second old. The human body is so complex, however, that it has communication needs that must be handled through a second, slower means. Is it the time of day when we ought to be sleeping or be alert? Do we have enough water running through our system? Are we full or not? All these questions are dealt with by a system that transmits its signals through a set of chemical messengers. These messengers are known as hormones, and the communication network they're part of is called the endocrine system—the subject of Chapter 28.

SO FAR. . .

1. Light coming in through the eye is focused on a layer of tissue at the back of the eye called the _____ by means of being bent, or refracted, by both the eye's _____ and the _____ that lies behind it.

2. The eye's photoreceptor cells come in two varieties: _____, which function in low-light situations but provide only _____ vision; and _____, which respond best to bright light but that provide _____ vision.

3. Visual information is passed from the eyes to the brain via our two _____.

27 REVIEW

Summary

27.1 Structure of the Nervous System

- The nervous system includes all the nervous tissue in the body plus the body's sensory organs. Nervous tissue is composed of neurons, which transmit nervous system messages, and glial cells, which support neurons and modify their signaling. **(p. 505)**

- The two major divisions of the human nervous system are the central nervous system (CNS), consisting of the brain and spinal cord, and the peripheral nervous system (PNS), which includes all the neural tissue outside the CNS plus the sensory organs. **(p. 506)**

- The PNS has an afferent division, which brings sensory information to the CNS, and an efferent division, which carries action commands to muscles and glands. Within the PNS's efferent division are the somatic nervous system, which provides voluntary control over skeletal muscles, and the autonomic nervous system, which provides involuntary regulation of smooth muscle, cardiac muscle, and glands. The autonomic system is divided into the sympathetic division, which generally has stimulatory effects, and the parasympathetic division, which generally facilitates routine maintenance activities. **(p. 506)**

27.2 Cells of the Nervous System

- There are three types of neurons: sensory neurons, which sense conditions inside and outside the body and convey information about these conditions to neurons inside the CNS; motor neurons, which carry instructions from the CNS to such structures as muscles or glands; and interneurons, which are located entirely within the CNS and which interconnect other neurons. **(p. 506)**

- Each neuron has multiple dendrites through which signals travel to the neuron cell body and a single axon that carries signals away from the cell body. Glial cells produce fat-rich myelin, which can surround neuronal axons and which increases the speed of neural signals. A nerve is a bundle of axons in the PNS that transmits information to or from the CNS. **(p. 508)**

27.3 Nervous-System Signaling

- In the nervous system, a signal moves down a neuron's axon and then moves from this axon to a second cell across a synapse. **(p. 510)**

- An electrical charge difference, called a membrane potential, exists across the plasma membrane of resting neurons because the inside of the neuron is negatively charged relative to the outside. This form of potential energy is used when special protein channels in the neuron's membrane open up on stimulation, thereby allowing ions to flow into the neuron. This influx of ions at an initial point on the axon triggers reactions that cause the adjacent portion of the axonal membrane to initiate the same influx of ions. In this way, a conducted nerve impulse known as an action potential moves down the entire axon. **(p. 510)**

- A nerve signal moves from one neuron to another cell across a synapse, which includes a "sending" neuron, a "receiving" cell, and a tiny gap between the two cells called a synaptic cleft. A chemical called a neurotransmitter diffuses across the synaptic cleft from the sending neuron to the receiving cell. It then binds with receptors on the receiving cell, thus keeping the signal going. **(p. 512)**

27.4 The Spinal Cord

- The spinal cord can receive input from sensory neurons and direct motor neurons in response, with no input from the brain. It also channels sensory impulses to the brain. In cross section, the spinal cord has a darker, H-shaped central area, composed mostly of the cell bodies of neurons (gray matter) and a lighter peripheral area, composed mostly of axons (white matter). Spinal nerves extend from the spinal cord to most areas of the body. **(p. 513)**

- Spinal cord motor neurons have cell bodies that lie within the gray matter of the spinal cord. The axons of these neurons leave the spinal cord through its ventral roots. Sensory neurons, which transmit information to the spinal cord, have their cell bodies outside the spinal cord, in the dorsal root ganglia. Dorsal and ventral roots come together to form a given spinal nerve. **(p. 514)**

- Reflexes are automatic nervous system responses, triggered by specific stimuli, that help us avoid danger or preserve a stable physical state. A reflex arc begins with a sensory receptor, runs through the spinal cord to a motor neuron, and proceeds back out to an effector such as a muscle. **(p. 514)**

27.5 The Autonomic Nervous System

- The sympathetic division of the autonomic nervous system is often called the fight-or-flight system because it generally prepares the body to deal with emergencies. The parasympathetic division is often called the rest-and-digest system because it conserves energy and promotes digestive activities. Most organs receive input from both systems. **(p. 515)**

27.6 The Human Brain

- The major regions in the adult brain are the cerebrum, cerebellum, thalamus, hypothalamus, midbrain, pons, and medulla oblongata. The cerebrum has a thin outer layer of gray matter, the cerebral cortex, that surrounds a much larger volume of cerebral white matter. Differing portions of the cerebral cortex play a central role in processing sensory information and in carrying out nearly all of our conscious mental activities. The cerebellum refines bodily movement and balance based on sensory inputs. The thalamus receives much of the body's sensory information and then transfers it to different regions of the cerebral cortex for processing. The hypothalamus is critical to regulating drives and maintaining homeostasis, in part through its regulation of hormonal release. The brainstem is a collective term for the midbrain, pons, and medulla oblongata. The midbrain helps maintain muscle tone and posture; the pons serves primarily to relay messages between the cerebrum and the cerebellum; and the medulla oblongata helps regulate such involuntary functions as breathing and digestion. **(p. 516)**

27.7 Our Senses

- All human senses operate through cells called sensory receptors that respond to stimuli and then transform these responses into electrical signals that travel through action potentials. Signals from every sense except smell are routed through the brain's thalamus and then to specific areas of the cerebral cortex. **(p. 519)**

27.8 Touch

- Touch works through a variety of sensory receptors that can distinguish among such qualities as light or heavy pressure. In some sensory cells, the stretching of their outer membrane prompts an influx of ions that results in the initiation of a nerve signal. **(p. 519)**

27.9 Smell

- Olfaction works through a set of sensory receptors whose dendrites extend into the nasal passages. "Odorants" bind with hairlike extensions of these dendrites, resulting in a nerve signal to the brain. The higher processing centers of the brain distinguish odorants by sensing unique groups of neurons that fire in connection with given odorants. **(p. 520)**

27.10 Taste

- Taste works through a group of cells, located in taste buds, that have receptors that bind to "tastants." A given taste cell can respond through four to six chemical signaling routes that correspond to the four to six basic tastes of sweet, sour, salty, bitter, and the possible fifth and sixth tastes of umami and calcium. The neurons that receive input from taste cells vary in their response to different tastants. The brain makes sense of the pattern of input it gets from these neurons, yielding the large number of tastes we experience. **(p. 522)**

27.11 Hearing

- Sound vibrations result in "waves" of air molecules that are, by turns, more and less compressed than the ambient air around them. These waves bump against our tympanic membranes, which in turn vibrate, initiating a chain of vibration that ends in the fluid-filled cochlea of the inner ear. "Hair cells" in the cochlea have ion channels that open and close in response to this vibration, resulting in nerve signals to the brain. **(p. 523)**

27.12 Vision

- Light enters the eye through the cornea and then passes through the lens on its way to the retina at the back of the eye. Light is bent or refracted by the cornea and the lens in such a way that it ends up as a tiny, sharply focused image on the retina. **(p. 526)**

- Light signals are converted to nervous system signals by two types of photoreceptor cells in the retina: rods, which function in dim light but provide only black-and-white vision; and cones, which are sensitive to color. Pigments within these photoreceptors are struck by light and change shape, which prompts a cascade of chemical reactions that results in neurotransmitter release being inhibited between a rod or cone and its adjoining connecting cell. This lack of release transmits the signal "photoreceptor stimulated here." Vision signals travel from photoreceptors through two sets of adjoining cells, the latter of which have axons that come together to form the body's optic nerves. **(p. 526)**

- The brain constructs images as much as it records them. Visual perception operates through a series of genetically based "rules" that allow us to quickly make sense of what we perceive. **(p. 527)**

Key Terms

action potential 512

afferent division 506

autonomic nervous system 506

axon 508

central nervous system (CNS) 506

cerebellum 517

cerebral cortex 516

cerebrospinal fluid 514

cerebrum 516

cochlea 523

cones 526

dendrites 508

efferent division 506

ganglion 514

hypothalamus 517

interneuron 506

medulla oblongata 518

membrane potential 510

midbrain 518

motor neuron 506

myelin 508

nerve 509

neurotransmitter 512

olfaction 520

parasympathetic division 515

peripheral nervous system (PNS) 506

photoreceptor 526

pons 518

reflex 514

retina 526

rods 526

sensory neuron 506

somatic nervous system 506

sympathetic division 515

synapse 512

synaptic cleft 512

thalamus 517

Understanding the Basics

Multiple-Choice Questions

(Answers are in the back of the book.)

1. The peripheral nervous system (PNS) is made up of:
   a. the brain and spinal cord.
   b. the brain and cranial nerves.
   c. sensory organs and all of the nervous system outside the brain and spinal cord.
   d. all of the nervous system inside the CNS plus the sensory organs.
   e. all of the above.

2. A nervous system cell capable of responding to light and sending a message about that light to the brain would be both a _____ and _____.
   a. glial cell; an interneuron
   b. glial cell; a motor neuron
   c. sensory neuron; an afferent neuron
   d. motor neuron; an afferent neuron
   e. motor neuron; an efferent neuron

3. An action potential is possible because:
   a. neurons are wrapped in myelin.
   b. the CNS and PNS are separate.
   c. only the spinal cord has to be involved.
   d. there is no charge difference across the outer membranes of resting neurons.
   e. there is a charge difference across the outer membranes of resting neurons.

4. A neurotransmitter:
   a. is stored in sending cell vesicles.
   b. diffuses across the synaptic cleft.
   c. binds with receptors on a receiving cell.
   d. can sometimes be taken back up by a sending cell.
   e. all of the above.

5. The autonomic nervous system (select all that apply):
   a. carries out voluntary decisions.
   b. is part of the central nervous system.
   c. carries out involuntary regulation.
   d. is divided into sympathetic and parasympathetic divisions.
   e. is an efferent system.

6. In the brain, the site of our higher thought processes is the:
   a. cerebellum.
   b. cerebrum.
   c. thalamus.
   d. medulla oblongata.
   e. hypothalamus

7. All sensory receptors ____ and ____.
   a. respond to stimulation; transform that stimulation into a neural signal
   b. respond to all forms of stimulation except smell; send messages through the spinal cord
   c. are located in the spinal cord; send messages through the thalamus
   d. are located in the brain; interconnect with neurons
   e. respond to neurotransmitter stimulation; turn that stimulation into a neural signal

8. Several of the senses work by means of:
   a. the brain distributing portions of its processing to the spinal cord.
   b. the brain analyzing the pattern of input from many sensory receptors.
   c. the brain receiving messages from sensory receptors and then sending messages back to them.
   d. sensory receptors communicating first with each other, then with the brain.
   e. sensory receptors responding to light.

Brief Review

(Answers are in the back of the book.)

1. Three functional types of neurons are found in the nervous system. What are they, and what role does each play?

2. In what way is the neuron a cell that is specialized for rapid communication?

3. Where do neurotransmitters function within the nervous system? Describe two processes that limit the duration of their activity.

4. Describe how a simple reflex arc works.

5. Most of the senses reviewed in the text had these things in common: sensory receptor stimulation, transduction, neurotransmitters, receiving neurons, thalamus, and cerebral cortex processing. Describe what happens in general at each of these steps.

Applying Your Knowledge

1. Given the roles of the various regions of the brain, why is it almost literally true to speak of "higher" brain functions?

2. It is often said that seeing is believing. Is it reasonable to put this much faith in the human visual system?

3. The text notes that humans have a few hundred different types of smell or "olfactory" receptors while dogs have about 1,000. Given this, dogs have a much greater ability to distinguish one scent from another than people do. Meanwhile, humans have sharper daytime vision than dogs and a greater ability to perceive colors in the portion of the visible light spectrum that runs from green through red. With this in mind, in what ways is a daytime walk around the neighborhood likely to be perceptually different for dogs, compared to humans?

CHAPTER

28

Communication and Control 2:
The Endocrine System

The hormones of the endocrine system help keep the human body running smoothly.

Available on MasteringBiology®

Activities: The Endocrine System; Hypothalamic Control of the Pituitary; **Bioflix:** Homeostasis and Regulating Blood Sugar; **Current Events:** The Hormone Surge of Middle Childhood; A Romp Into Theories of the Cradle of Life; **Building Vocabulary; Questions:** Bioflix Quiz, Reading Questions, Test Bank

INSULI

50 Units — 5 C.C.

10 Units per C.C.

CONNAUGHT LABORAT

UNIVERSITY OF TOR

One of the great medical advances of the 20th century was the identification of insulin as the hormone the body uses to move blood-sugar from the bloodstream into the body's cells. Pictured is a bottle that held one of the world's first commercially available doses of insulin. Manufactured by the Connaught Laboratories of the University of Toronto, the bottle dates from 1923.

Oh it certainly is the most interesting thing I've been in the midst of for a long time," said Elizabeth Hughes, age 15, in a letter to her mother in August of 1922. "Why Mumsy when I'm on that diet next year I'm sure I will be able to do a little studying for when I get my strength and weight back I will be eating as much as I would normally anyway. That is a most thrilling

thought to me, and to think too, that I'll be leading a normal, healthy existence is beyond all comprehension."

Indeed, leading a normal, healthy existence had been beyond comprehension for Elizabeth Hughes for a long time. Diagnosed with diabetes at age 11, she then went through an extended period of decline that left her weighing less than 50 pounds and barely able to walk in the weeks before she wrote her letter. That her illness could reduce her to such a state surprised no one, since in 1922 diabetes was nearly always fatal in children. Physicians could extend the lives of young diabetics by putting them on extreme low-calorie diets, but this often meant that they died of starvation rather than of complications brought about by elevated blood-sugar levels.

Over the years, Hughes had been put on diets that at times fell to as little as 300 calories a day, and by age 14, she had become "pitifully depleted " as her mother put it in a letter to Dr. Frederick Banting of the University of Toronto. Mrs. Hughes was inquiring about a then-experimental treatment for diabetes that Dr. Banting and his young assistant, Charles Best, had begun to develop the previous summer. In experiments with dogs, they had managed to isolate an "internal secretion of the pancreas" that had the marvelous capability of lowering blood-sugar levels in diabetic animals. By January 1922, they had injected their first human subject with this substance—which the world would soon know as insulin—and by August of that year Elizabeth Hughes made a journey from Washington, D.C., to Toronto to receive insulin treatments. The results were as spectacular in her case as they had been in others. By October she had gained 15 pounds and by December she was back in Washington with her parents; there she began leading a life that eventually came to include college, marriage, children, and death only at age 73.

In the years since 1922, millions of diabetics have, like Elizabeth Hughes, lived long and productive lives thanks to the discovery of insulin. And during this time, a great deal of scientific work has confirmed that this substance is, as Frederick Banting said, an internal secretion of the pancreas. Small collections of pancreatic cells produce it each day and then release it into the bloodstream, which carries it to muscle, fat, and liver cells throughout the body. Once it arrives at these cells, it latches onto them and sets in motion a chemical process by which they take up the body's excess blood sugar.

All this may make insulin sound like some extraordinary substance, but it turns out that there are dozens of substances that operate in the body much as insulin does and that are just as important to sustaining life as it is. What are these substances? They are the chemical messengers known as hormones. When we hear a noise at night, the hormone adrenaline will course through our body, preparing us to deal with danger. When we're reading before bed, the hormone melatonin will be circulating within us, making us sleepy. And after we eat a big meal, it's the hormone insulin that comes into play, moving into circulation and holding our blood-sugar levels steady.

As may be obvious from these examples, hormones are in the communication and control business: They communicate with many kinds of cells in the body, thereby altering their activity in some way. Note, however, that before hormones can have an effect *on* cells, they must first be released *by* cells, often meaning cells that make up one of the body's endocrine glands, such as the adrenal glands. When we look at these glands, the hormones they release, and the cells these hormones affect, we find that they make up a communication and control *network,* which is known as the endocrine system. The goal of this chapter is to provide you with an introduction to this system, which, as you'll see, works hand-in-hand with the nervous system we looked at last chapter in keeping the body functioning smoothly.

28.1 The Endocrine System in Overview

The **hormones** that the endocrine system uses to regulate bodily processes can be defined as substances secreted by one set of cells that travel through the bloodstream and affect the activities of other cells. Such a definition sets hormones apart from other kinds of signaling molecules in the body. There are, for example, several types of signaling molecules that do not travel through the bloodstream but that instead carry out strictly *local* communication between cells. In contrast, all endocrine hormones have the ability to carry chemical signals throughout the body, given their transportation through the bloodstream.

This means of transportation actually provides a key for understanding how endocrine signaling works. A typical hormone is "broadcast," in a sense, as it moves through the bloodstream. Like a television signal, it can be "picked up" by any cell that has the proper "receiver." What are these receivers? They are receptors that are shaped in such a way that they can latch onto the hormone. With this, we get to the concept of **target cells:** those cell types that can be affected by a given hormone (**Figure 28.1**). The target cells for a hormone called vasopressin are located primarily in the kidneys, while, as we've seen, the target cells for insulin are located throughout the body.

Thus, hormones differ greatly with respect to the number of target cells they affect and the location of these cells.

If hormones are the products of the endocrine system, then the factories that put together or "synthesize" these products are the specialized organs called **glands** noted earlier, which can be defined as any localized group of cells that work together to secrete a substance. **Endocrine glands** are glands that secrete materials they have synthesized directly into the bloodstream or into surrounding tissues, without using ducts. The major endocrine glands are shown in **Figure 28.2**. Although most hormones are secreted by these glands, it's important to note that not all are. The heart, kidneys, stomach, liver, small intestine, placenta, and fatty tissue all secrete hormones, yet they are not glands.

The fact that hormones work through a process of secretion and circulation means that endocrine communication tends to operate more slowly than the nervous system communication we looked at last chapter. The fastest-acting hormones take several seconds to work, while the slowest-acting ones may take several hours. The opposite side of this coin is that the longest-lasting hormones can *keep* exerting their effects for hours after they have been released. Because hormones operate over such extended time-frames, they tend to regulate processes that unfold over minutes, hours, or even years: whether we have enough calcium in our system, whether our blood-sugar levels are right, and whether our growth will continue. If you need to swerve to avoid a collision, look to the nervous system for a response; for the maintenance of

The Endocrine System

Figure 28.1
Hormones and Their Target Cells

For a hormone to affect a target cell, that cell must have receptors that can bind to the hormone. The binding of hormone to receptor then initiates a change in the target cell's activity. The figure shows a peptide hormone that affects skeletal muscle tissue. The hormone does not affect the nerve cell, also shown in the figure, because this cell does not have the appropriate receptors to bind with the hormone.

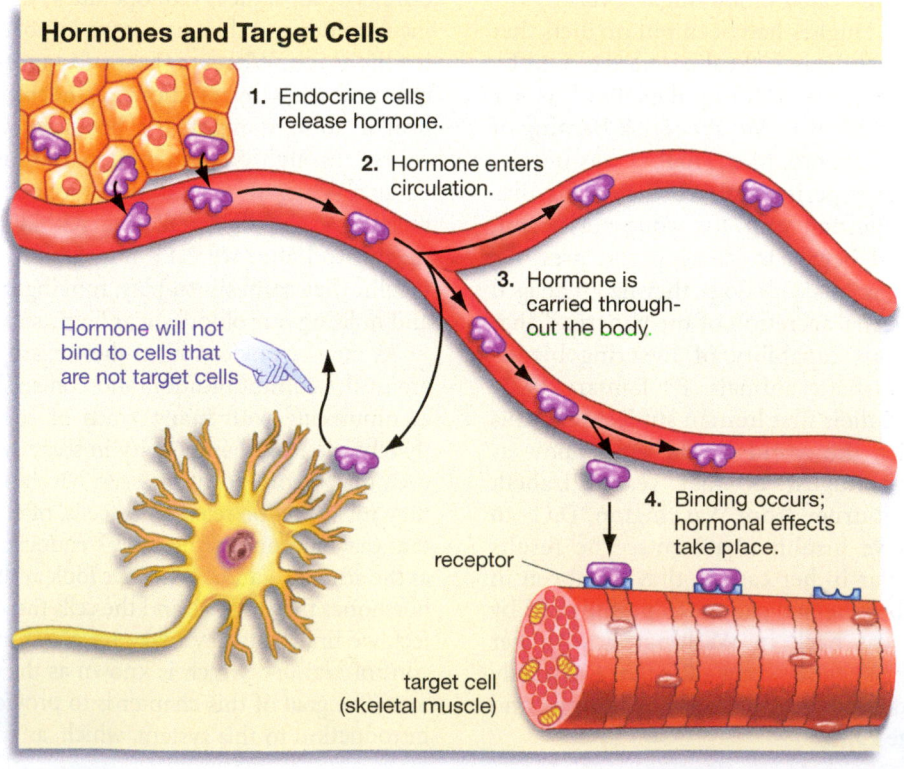

Hormones and Target Cells

1. Endocrine cells release hormone.
2. Hormone enters circulation.
3. Hormone is carried throughout the body.
4. Binding occurs; hormonal effects take place.

Hormone will not bind to cells that are not target cells

receptor

target cell (skeletal muscle)

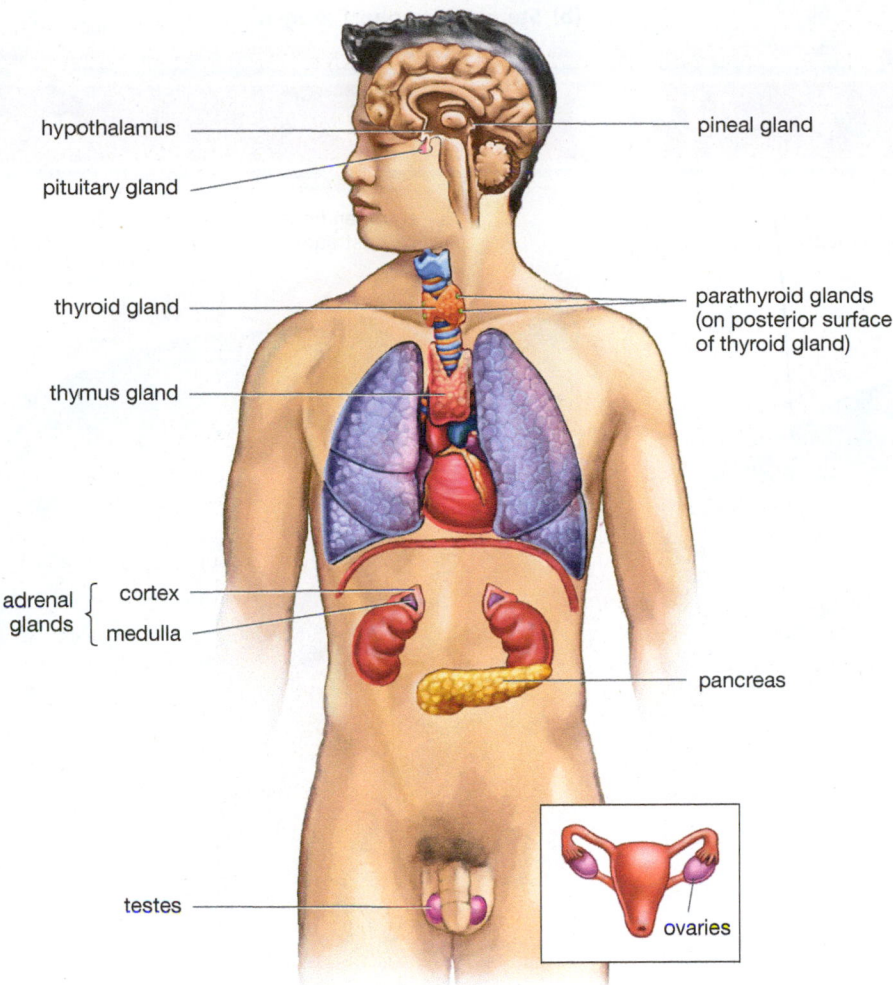

hypothalamus

pituitary gland

pineal gland

thyroid gland

parathyroid glands
(on posterior surface
of thyroid gland)

thymus gland

adrenal
glands {
cortex

medulla

pancreas

testes

ovaries

Figure 28.2
**The Major Hormone-
Secreting Glands of
the Body**
Although it is part of the
brain, the hypothalamus is
shown because it plays a
central role in hormonal regulation and effectively functions as an endocrine gland.

bone density and water balance, look to the endocrine system. In the larger picture, however, *both* the endocrine and nervous systems are communication-and-control operations and it's thus not surprising that the two work closely together, as you'll see.

28.2 Hormone Types and Modes of Action

So, what kinds of hormones are there, and how do they function? Under one classification scheme, there are three primary types of hormones: amino-acid–based, peptide, and steroid. We'll look first at the amino-acid–based and peptide hormones, which can be grouped together because they operate in a similar way.

Amino-acid–based hormones can be defined as hormones that are derived from modification of a single amino acid. The hormone adrenaline is produced through the modification of the amino acid tyrosine, for example, while the hormone melatonin is produced through a modification of the amino acid tryptophan. Meanwhile, **peptide hormones**—by far the largest class of hormones—can be defined as hormones composed of *chains* of amino acids. Such

chains can vary greatly in length. Human growth hormone, for example, is composed of 191 amino acids, while the hormone vasopressin is composed of only 9. Insulin is a peptide hormone, as are most of the hormones secreted by an important endocrine gland we'll be looking at, the pituitary.

Now, how do amino-acid–based and peptide hormones alter the activities of their target cells? As an example, let's consider a peptide hormone called glucagon that, like the insulin we've seen so much of, helps manage the body's blood-sugar or glucose levels. It does so, however, with a twist: After being secreted by cells in the pancreas, glucagon moves to liver cells where it helps transfer glucose *into* circulation by means of breaking up a molecule called glycogen that amounts to a linked chain of glucose units. To accomplish this, glucagon functions as you can see in **Figure 28.3a** on the next page: It binds to a receptor on the outer or "plasma" membrane of a liver cell, never entering the cell itself but by its binding setting off a chemical reaction inside the cell that involves a so-called second messenger. Following another reaction, an enzyme has been activated that begins breaking up glycogen into its glucose units, which then move into circulation and begin supplying cells throughout the body with energy.

(a) Peptide hormone: Glucagon

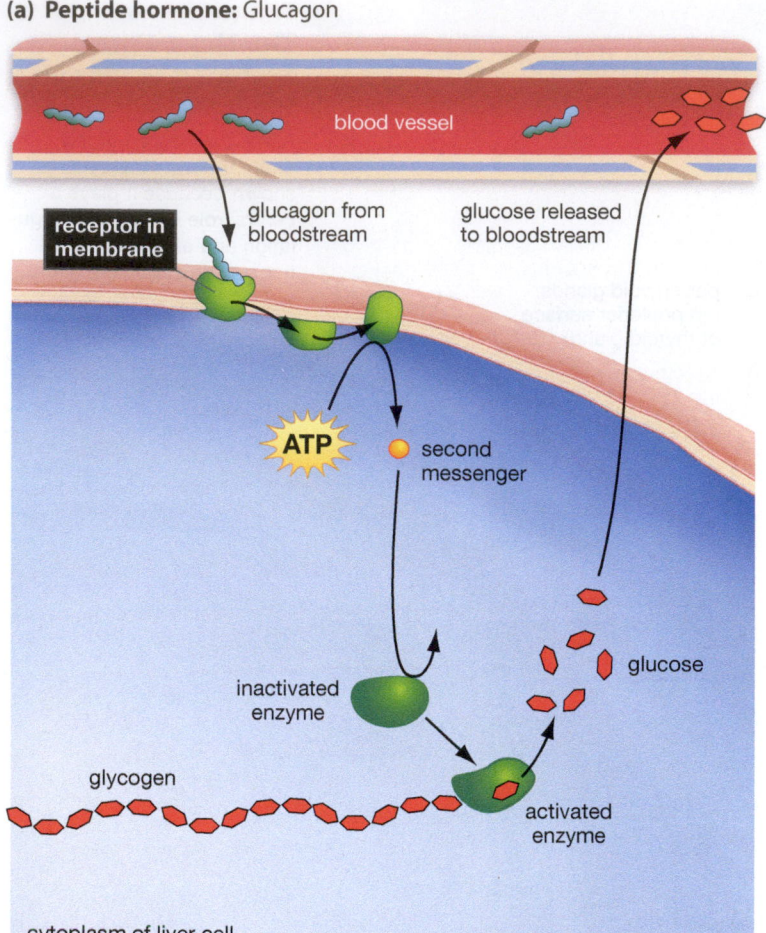

- blood vessel
- **receptor in membrane**
- glucagon from bloodstream
- glucose released to bloodstream
- ATP
- second messenger
- inactivated enzyme
- glucose
- glycogen
- activated enzyme
- cytoplasm of liver cell

(b) Steroid hormone: Estrogen

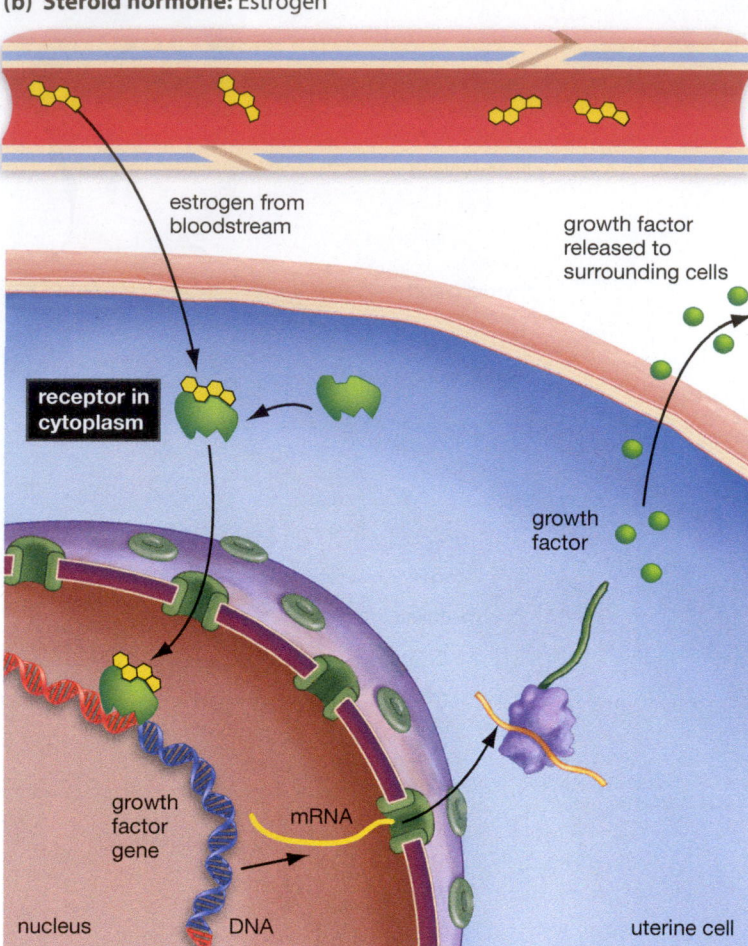

- blood vessel
- estrogen from bloodstream
- growth factor released to surrounding cells
- **receptor in cytoplasm**
- growth factor
- growth factor gene
- mRNA
- nucleus
- DNA
- uterine cell

Figure 28.3

How Peptide and Steroid Hormones Work

(a) Like most peptide hormones, glucagon binds to a receptor located on the outer or plasma membrane of its target cell. This binding sets off a series of chemical reactions inside the cell that ultimately activate an enzyme that clips individual glucose molecules from the energy storage molecule glycogen. These glucose units are then exported from the cell and released into the bloodstream

(b) The steroid hormone estrogen, conversely, works by passing through the plasma membrane of its target cell and binding to a receptor in the cytoplasm of the cell. The combined hormone/receptor molecule then migrates to the nucleus of the cell and binds there to the cell's DNA, thus switching on a gene that codes for a protein known as a growth factor. Once the growth factor proteins are synthesized in the cell's cytoplasm, they are released to surrounding cells.

Most peptide and amino-acid–based hormones operate as glucagon does, in that they begin their work by binding with a cell-surface receptor. The story is different, however, with our third class of hormones, the **steroid hormones,** which can be defined as hormones constructed around the chemical framework of the cholesterol molecule. Most of these hormones pass *through* a cell's plasma membrane and then bind with a receptor protein inside the cell. The resulting combined hormone/receptor molecule then enters the nucleus of the cell and binds with the cell's DNA there—a binding that turns on one or more of the cell's genes, thus bringing about the production of one or more proteins. If you look at **Figure 28.3b,** you can see how this process works in connection with the steroid hormone estrogen, which, in one of its roles, helps bring about the monthly buildup of tissue that lines the female uterus—tissue that can house an embryo, should pregnancy occur. The binding of the combined estrogen/receptor molecule with the DNA of a uterine cell brings about the production of a protein, called a growth factor, that, when exported from the cell, prompts other cells of the uterine lining to begin their monthly buildup.

As may be obvious, steroid hormones operate through a long series of steps when compared to the peptide or amino-acid–based hormones; steroids are bringing about the *production* of proteins through their binding with DNA, while peptide hormones are merely activating *existing* proteins within cells. The upshot of this difference is that steroids are the snails of the hormone world; hours might pass between the time one of them is synthesized and the time it has its effect. Several of the steroids are reproductive hormones that are produced predominately in reproductive organs: Estrogen is produced in large amounts in the ovaries of females (and in smaller amounts in the testes of males); and testosterone is produced in large amounts in the testes of males (and in smaller amounts in the ovaries and adrenal glands of females).

SO FAR . . .

1. Hormones are substances secreted by one set of _____ that move through the _____ and affect the activities of other _____.

2. Hormone production takes place to a significant extent within _____ glands, defined as glands that release their materials directly into the _____ or surrounding tissue, without the use of ducts.

3. Hormones can be thought of as coming in three principal classes: _____ based, _____, and _____ hormones. The largest class of hormones is the _____ hormones, while the slowest acting are the _____ hormones.

28.3 Negative Feedback and Homeostasis

Endocrine glands don't release hormones at a steady rate, day in and day out. Instead, they finely tune their hormonal output in accordance with the body's needs at any given moment. If we ask how their hormonal output is controlled, the answer in almost every case is: through the mechanism of negative feedback. As readers of Chapter 26 know (see page 481), this is the same mechanism that operates in a home heating system. A thermostat senses that a home's temperature is too cold and thus switches on a furnace; the product of this action is hot air, and this very product then feeds back on the thermostat's activity in a negative way—it shuts it down. Likewise, in hormonal secretion, the body has, for example, a set of parathyroid glands that sense when calcium ion levels in the bloodstream have fallen too low. In response, the parathyroids secrete a hormone that prompts the bones in the body to release some of their calcium stores. Higher circulating levels of calcium result, and these levels then feed back on the parathyroid glands, causing them to reduce their hormone secretion.

It's important to note that this process doesn't just *control* circulating calcium levels; it *stabilizes* these levels over the long term—it keeps them within a nar-

row range. With this process in place, calcium levels will neither fall too low nor rise too high. This is conceptually important because the human body requires this kind of stability. To function at all, it cannot be too hot or too cold, too hydrated or too dehydrated, too stimulated or too relaxed. And the endocrine system, with its many negative feedback loops, turns out to be a critical factor in helping the body avoid these kinds of extremes. To state this more formally, the endocrine system and negative feedback are key factors in producing **homeostasis** in the human body, meaning a relatively stable internal environment.

28.4 Hormone Central: The Hypothalamus

The parathyroid glands get input about one condition in the body (its circulating calcium levels) and in response produce one hormone. But human endocrine activity is largely governed by a single structure that gets inputs about *multiple* conditions in the body and in response controls the release of *many* hormones. This Grand Central Station of endocrine activity is, however, not an endocrine gland, but instead is a portion of the *brain* that we looked at last chapter, the hypothalamus. So extensive is the hypothalamus' control over the endocrine system that it is regarded as the most important structure in the body in maintaining homeostasis. If you look at **Figure 28.4a** on the next page, you can see the location of this busy, walnut-sized portion of the brain.

The hypothalamus is well suited to playing its central role in homeostasis because it receives inputs from so many sources and affects the activities of so many cells, either directly or indirectly. On the input side, sensory nerves project to it from, for example, the digestive tract, the brainstem, and the eyes, thus conveying information about how full we are, what our internal temperature is, and whether it is day or night (thus helping set our internal clock). Beyond this, the hypothalamus stands alone among brain structures in not being shielded from the body's general blood circulation by the so-called blood-brain barrier. As such, it can sample the contents of the bloodstream and does so constantly with special receptors it has that are sensitive to such things as the levels of water and nutrients that are circulating within us.

Information from all these sources is then used by the hypothalamus to issue a wide array of endocrine commands through three separate means. One of these is simply the issuance of nerve signals to cells in the inner portion, or medulla, of the adrenal glands (which sit atop the kidneys). When these signals arrive, they cause the adrenal cells to release the hormones adrenaline and noradrenaline.

The hypothalamus exercises most of its endocrine control, however, through the posterior and anterior

Hypothalamic Control of the Pituitary

Figure 28.4
Pivotal Player in Hormone Regulation

(a) The hypothalamus is a region of the brain, but in keeping with its control over much of the endocrine system, it lies adjacent to one endocrine gland, the posterior pituitary, and very near to another, the anterior pituitary.

(b) It controls hormonal release in both of these glands, though in different ways. It exercises control over the posterior pituitary by means of first synthesizing, and then releasing into a posterior capillary bed, two hormones that then move into general circulation. It exercises control over the anterior pituitary by means of synthesizing a number of releasing or inhibiting hormones that diffuse into cells in the anterior pituitary, thus prompting them to release or retain six different hormones.

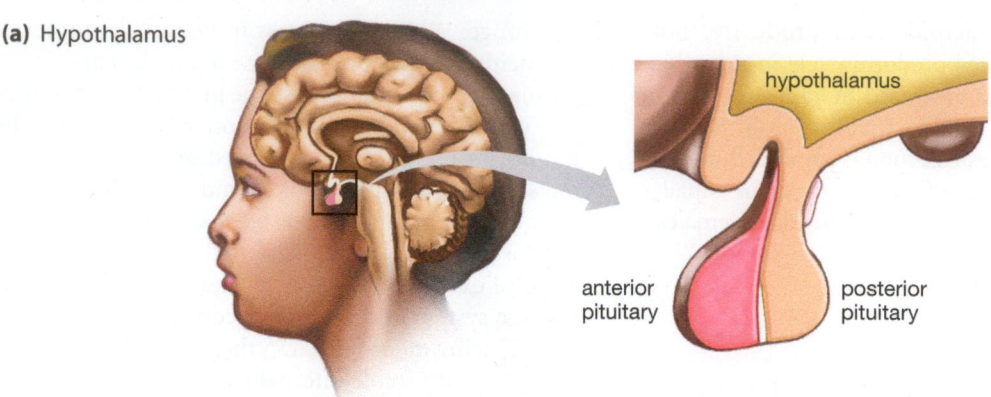

(a) Hypothalamus

hypothalamus

anterior pituitary   posterior pituitary

(b) Two means of endocrine control by the hypothalamus

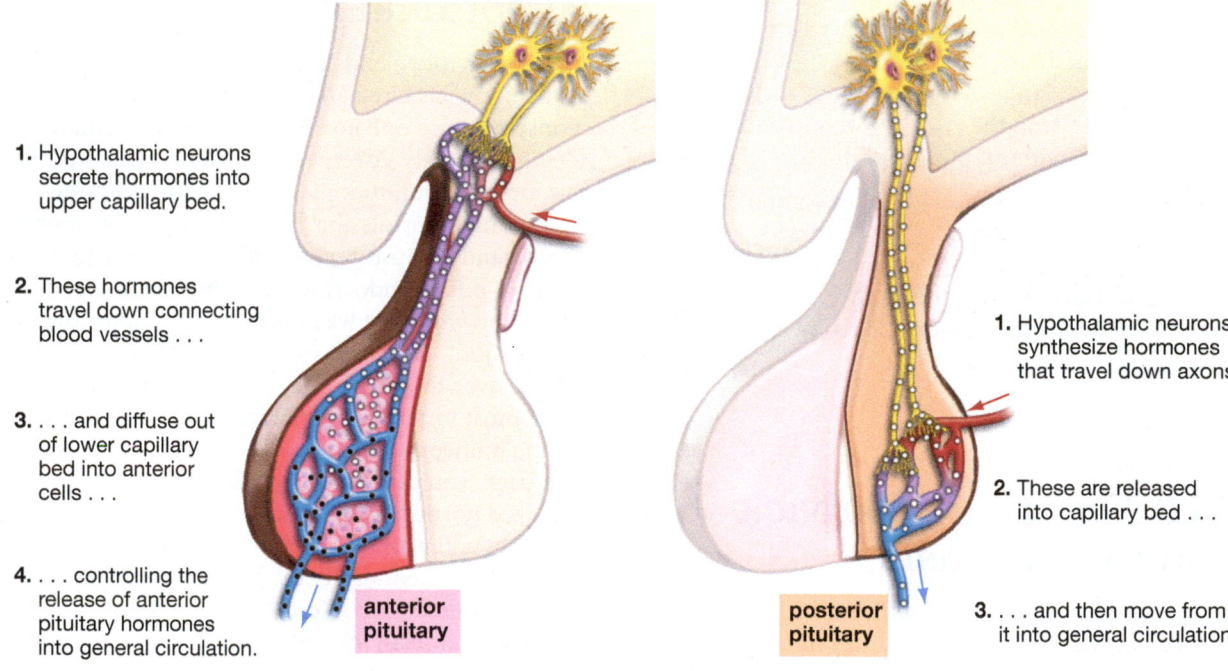

1. Hypothalamic neurons secrete hormones into upper capillary bed.

2. These hormones travel down connecting blood vessels . . .

3. . . . and diffuse out of lower capillary bed into anterior cells . . .

4. . . . controlling the release of anterior pituitary hormones into general circulation.

anterior pituitary

1. Hypothalamic neurons synthesize hormones that travel down axons.

2. These are released into capillary bed . . .

3. . . . and then move from it into general circulation.

posterior pituitary

pituitary glands illustrated in **Figure 28.4b**. On the posterior side, this control is straightforward: Separate groups of neurons in the hypothalamus synthesize two hormones; these hormones move down the neurons' axons, to their very tips, where they will be stored until such time as the neurons issue a release signal, at which point the hormones are secreted into general circulation through the capillary bed you can see in the figure. Thus, the hypothalamus is simply functioning as an endocrine gland in this instance—even though it isn't one—by first synthesizing two hormones and then secreting them within the posterior pituitary.

The two hormones being released in this process could not be more different in their physical effects. One of them, called oxytocin, is best known for hastening childbirth by stimulating uterine contractions, while the other, called vasopressin, plays a critical role in keeping water levels steady in the body. (You can read more about both of these substances, and their newly discovered role in human sociability, in "Social Bonding through Chemistry" on page 542.)

It is when we turn to the hypothalamus and the *anterior* pituitary gland, however, that the full endocrine reach of the hypothalamus becomes apparent. This is so because the anterior pituitary secretes no fewer than six hormones, five of which go on to control hormonal release in *other* endocrine glands. The anterior pituitary secretes its hormones, however, only when prompted into action by signals coming from the hypothalamus. What are these signals? As was the case with the posterior pituitary, they start with neurons in the hypothalamus that synthesize hormones and send them down their axons. These hormones, however, don't then move into general circulation to any appreciable extent; instead they travel down the connecting blood vessels you can see in the figure and diffuse into the anterior pituitary gland itself. There, they will affect several different types of anterior pituitary cells that synthesize the gland's six hormones. In most cases, the hypothalamic hormones will prompt these pituitary cells to *release* their hormones, but in a few cases they will *inhibit* their release of hormones. Either way, the hypothalamus is serving as traffic controller for all the activity of this busy gland. To give you some sense of what this control means in practice, let's follow a path by which a pair of metabolism-regulation hormones are produced and then used.

Metabolic Control through Thyroid Hormones

To function properly, the human body must maintain an optimal metabolic rate, meaning the rate at which it carries out the enormous number of chemical reactions that sustain life. Just as an old-fashioned stove needs to burn wood at the right rate to keep a room at the correct temperature, so the human body needs to burn the calories contained in food at the right rate to keep the *body* at the correct temperature, to supply the body's cells with the energy they need, and so forth. Several hormones are active in metabolic control, but for now we're only interested in two of them—hormones called T_3 and T_4 that are released by the body's thyroid gland, a small, butterfly-shaped organ that wraps around the front of the human windpipe, as you can see in Figure 28.2 on page 535.

So, how do T_3 and T_4 get produced? Their story begins with neurons of the hypothalamus that produce one of the *releasing* hormones we just talked about—a hormone called thyrotropin-releasing hormone, or TRH, which travels down to the anterior pituitary, diffuses into it, and then does just what its name implies: It prompts selected cells in the anterior pituitary to release a hormone *they* have synthesized called thyroid-stimulating hormone, or TSH. This hormone then moves into general circulation, but its target cells exist only in the thyroid gland. It is the binding that takes place between these target cells and TSH that brings about the production of T_3 and T_4. Once these two hormones have been released into general circulation they will end up binding to nearly every cell in the body, and *this* binding will bring about our final hormonal effect, an increase in each cell's metabolic rate through such means as an increased production of the energy transfer molecule ATP. In sum, through its secretion of a tiny amount of releasing hormone, the hypothalamus has ended up altering the activity of nearly all the body's cells.

Even so, the hypothalamus must itself be regulated in this activity. It can't just keep producing its TRH continuously, because this would ultimately produce a sky-high metabolic rate in the body's cells—an extreme condition that these cells could not tolerate. So, how is the activity of the hypothalamus regulated? It will not surprise you to learn that a negative feedback loop is in operation. Both the hypothalamus and the anterior pituitary gland are

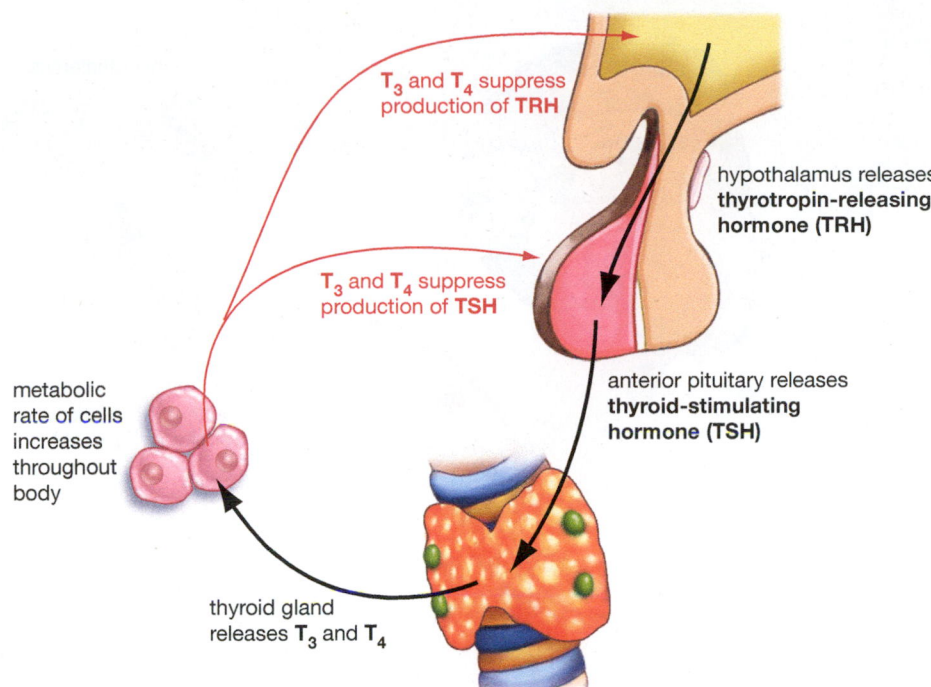

T_3 and T_4 suppress production of TRH

hypothalamus releases thyrotropin-releasing hormone (TRH)

T_3 and T_4 suppress production of TSH

anterior pituitary releases thyroid-stimulating hormone (TSH)

metabolic rate of cells increases throughout body

thyroid gland releases T_3 and T_4

sensitive to circulating levels of T_3 and T_4; as these levels go up, the hypothalamus and pituitary *reduce* their production of TRH and TSH, respectively, which puts a limit on T_3 and T_4 release.

If you look at **Figure 28.5**, you can see a summary of the entire process of T_3–T_4 regulation. Then if you look at **Figure 28.6** on the next page, you can see a summary of everything the hypothalamus does through its three means of endocrine regulation—control over the anterior and posterior pituitary glands and control over the cells of the adrenal medulla. Now we'll continue our review of the endocrine system by looking in a little more depth at three of the hormones it produces.

Figure 28.5
How the Body's Metabolic Rate Is Controlled

The importance of both the hypothalamus and of negative feedback can be seen in the body's regulation of its metabolic rate through use of the hormones T_3 and T_4. Increased secretion of these hormones begins with the hypothalamus' secretion of thyrotropin-releasing hormone, or TRH, but the higher circulating levels of T_3 and T_4 that result from this secretion then feed back on both the hypothalamus and anterior pituitary gland, causing them to reduce their production of thyroid hormones.

SO FAR . . .

1. Almost all hormone secretion is regulated through the mechanism of _____. The endocrine system's use of this mechanism is a key factor in producing homeostasis in the human body, meaning its maintenance of a relatively _____ internal environment.

2. The part of the brain called the _____ controls much of the activity of the endocrine system through its regulation of hormone release in three glands, the _____ and _____ pituitary glands and the medulla of the _____ glands.

3. Five of the six hormones produced by the _____ pituitary gland control the release of hormones produced by _____.

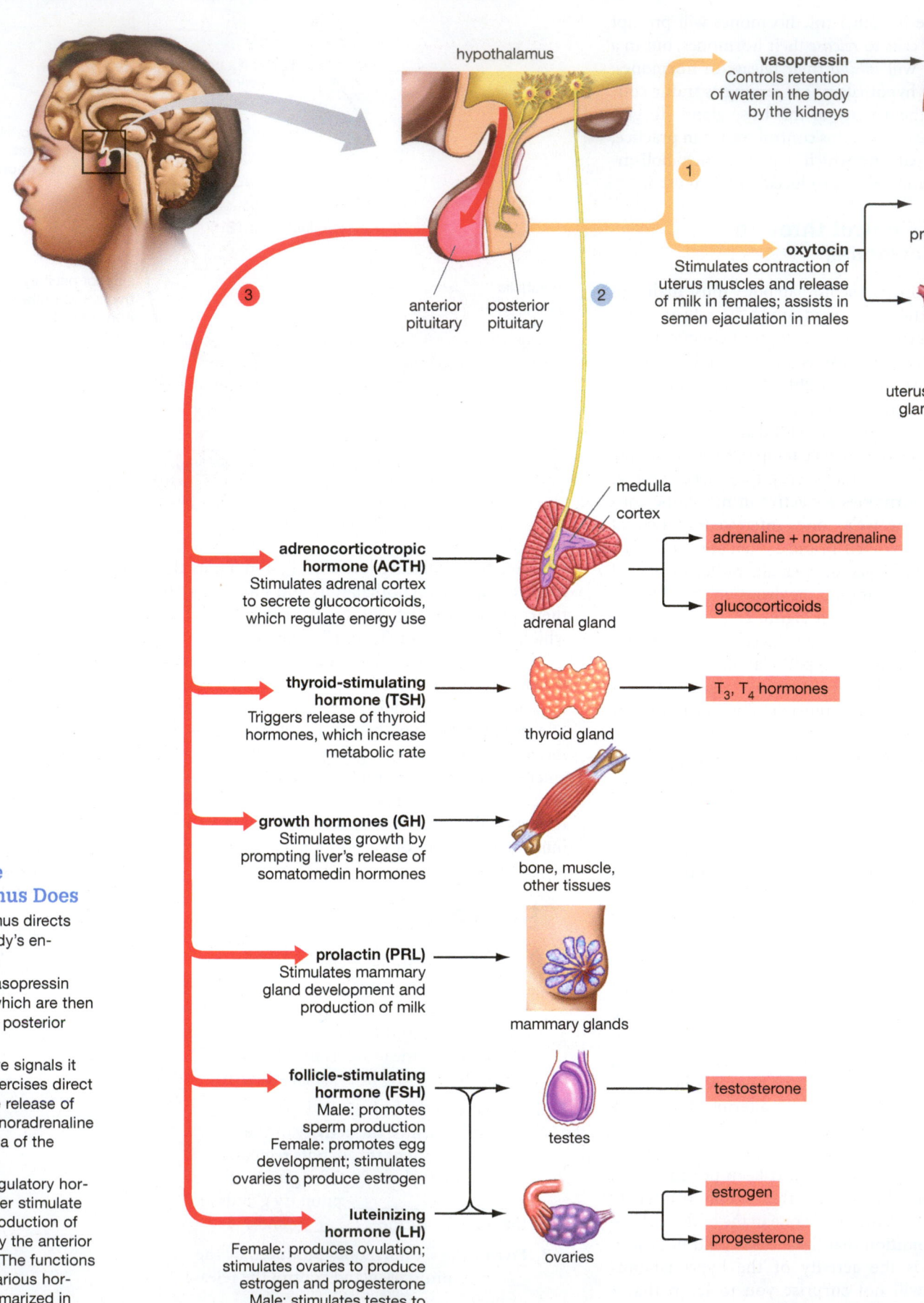

Figure 28.6
All That the Hypothalamus Does

The hypothalamus directs much of the body's endocrine activity.

1. It secretes vasopressin and oxytocin, which are then released by the posterior pituitary gland.

2. Through nerve signals it generates, it exercises direct control over the release of adrenaline and noradrenaline from the medulla of the adrenal glands.

3. It releases regulatory hormones that either stimulate or inhibit the production of six hormones by the anterior pituitary gland. The functions of the body's various hormones are summarized in Table 28.1.

hypothalamus

vasopressin
Controls retention of water in the body by the kidneys

kidneys

oxytocin
Stimulates contraction of uterus muscles and release of milk in females; assists in semen ejaculation in males

prostate gland in males

uterus and mammary glands in females

anterior pituitary   posterior pituitary

medulla
cortex

adrenocorticotropic hormone (ACTH)
Stimulates adrenal cortex to secrete glucocorticoids, which regulate energy use

adrenal gland

adrenaline + noradrenaline

glucocorticoids

thyroid-stimulating hormone (TSH)
Triggers release of thyroid hormones, which increase metabolic rate

thyroid gland

T_3, T_4 hormones

growth hormones (GH)
Stimulates growth by prompting liver's release of somatomedin hormones

bone, muscle, other tissues

prolactin (PRL)
Stimulates mammary gland development and production of milk

mammary glands

follicle-stimulating hormone (FSH)
Male: promotes sperm production
Female: promotes egg development; stimulates ovaries to produce estrogen

testes

testosterone

luteinizing hormone (LH)
Female: produces ovulation; stimulates ovaries to produce estrogen and progesterone
Male: stimulates testes to produce androgens

ovaries

estrogen

progesterone

28.5 Hormones in Action: Three Examples

Insulin and Glucagon: Keeping a Tight Rein on Glucose

We've already had occasion to look at a couple of aspects of the body's hormonal control of its glucose or blood-sugar levels. Now let's see how this control works from start to finish.

Glucose turns out to be the body's single most important source of energy; all human cells use it to power their activities, and it is the sole source of energy for human brain cells. Most of the carbohydrates we eat are broken down (in the digestive tract) into glucose, after which it passes to the liver and then to all the tissues of the body via the bloodstream.

Glucose is, however, a double-edged sword: Our cells require it, but it is harmful when it exists in excess amounts in our bloodstream. The body is thus constantly performing a balancing act; while it must keep some glucose in circulation—to supply cells with energy as needed—it must guard against having too *much* in circulation. So, how does it walk this fine line? It uses two hormones we touched on earlier that have opposite effects. One of these is **insulin:** a hormone that brings about a decrease in blood levels of glucose. The other hormone is **glucagon:** a hormone that brings about an increase in blood levels of glucose.

If you look at **Figure 28.7**, you can see the source of both of these hormones: the organ known as the pancreas, which lies just behind the bottom portion of our stomach. You might think that most of the pancreas would be given over to insulin and glucagon production, since these two hormones are so widely used in the body, but in fact only 1 percent of the volume of the pancreas is devoted to hormone synthesis. (The other 99 percent produces digestive enzymes that empty into the small intestine via the duct you can see in the figure.) Insulin is secreted by "beta" cells in the pancreas, while glucagon is secreted by "alpha" cells. These two kinds of cells sit side by side within small clusters or "islets" of cells in the pancreas called Islets of Langerhans. A profusion of tiny capillaries exists in the islets as well; these blood vessels take up the insulin and glucagon as they are produced, which is the first step in moving them into general circulation.

How do these hormones do their jobs? Here's a simplified account. Insulin binds with receptors on its target cells, and this binding prompts these cells to create channels, called transport proteins, that allow glucose to move from the bloodstream into the interior of the cells. As noted, the primary target cells for insulin are liver, skeletal muscle, and fat cells. (Brain cells, meanwhile, are capable of taking up glucose on their own, without the help of insulin.) Liver and skeletal muscle cells may make quick use of some of the glucose they take in, but

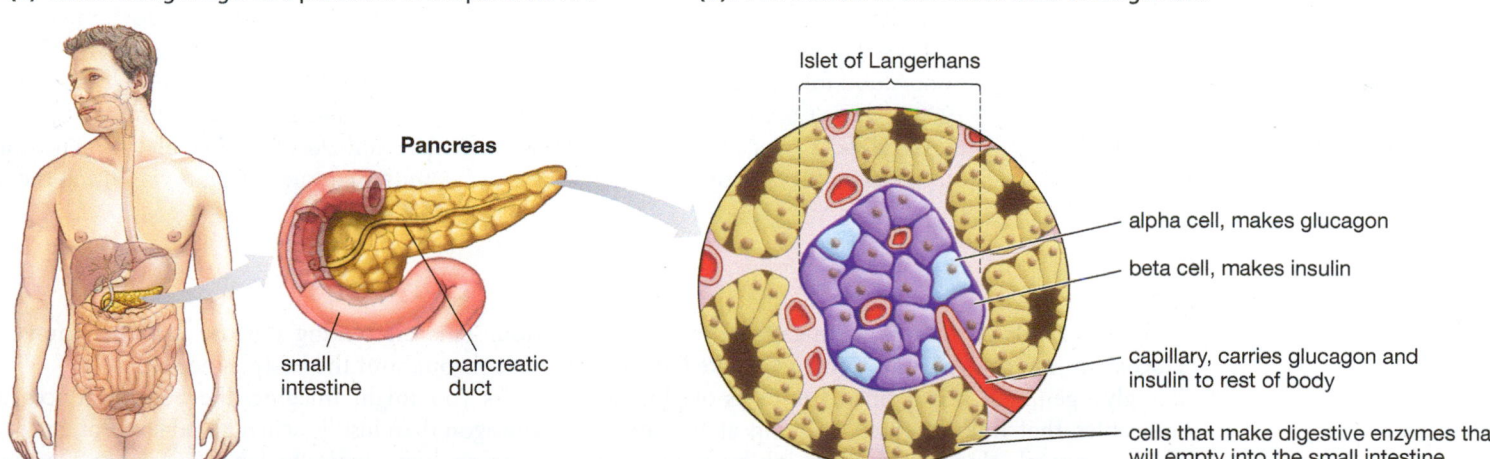

(a) Insulin and glucagon are produced in the pancreas . . .

Pancreas

small intestine

pancreatic duct

(b) . . . in clusters of cells called Islets of Langerhans

Islet of Langerhans

alpha cell, makes glucagon

beta cell, makes insulin

capillary, carries glucagon and insulin to rest of body

cells that make digestive enzymes that will empty into the small intestine

Figure 28.7
Where Insulin and Glucagon Are Produced

(a) The pancreas takes on two very different roles in the body. Most of its cells produce enzymes that help digest the food we eat. (These enzymes move into the small intestine, seen on the left of the figure, through the tube known as the pancreatic duct.) The pancreas is also home, however, to the cells that produce insulin and glucagon—the hormones that control the levels of glucose in the bloodstream. Insulin is produced by beta cells in the pancreas, while glucagon is produced by alpha cells.

(b) Both alpha and beta cells exist in small, roundish clusters of cells, called Islets of Langerhans, that are scattered throughout the pancreas' more abundant digestive cells. Each islet has an abundant set of the blood vessels, called capillaries, that move insulin and glucagon into general circulation.

ESSAY

Social Bonding through Chemistry

The substances vasopressin and oxytocin are both chemical pipsqueaks, and they're both produced in the same part of the brain, but the most intriguing thing about these two compounds is that they both lead a dual existence: They function first as hormones that regulate some basic human physical processes and then as neurotransmitters that influence some complex human *behaviors,* such as the tendency to pair-bond with members of the opposite sex. These behavioral effects, only recently discovered, have raised the possibility that the oxytocin member of this duo might someday be used as a therapeutic agent in combating a debilitating human condition that has a behavioral component to it: autism.

In their hormonal roles, oxytocin and vasopressin take on tasks that, while vitally important, are purely physical. Vasopressin is produced in the brain's hypothalamus and secreted by the posterior pituitary gland during times in which the body has too little water running through it. It then moves through the bloodstream and arrives at the kidneys, where it prompts kidney cells to create water channels that allow water to flow back into blood circulation, rather than being retained in the

fluid that will become urine. The result? More water back into circulation, less water into urination.

Oxytocin is likewise produced in the hypothalamus but is released within the posterior pituitary primarily at two special times: during childbirth and during breastfeeding. When childbirth begins, it stimulates muscle contractions in the mother that push the fetus down further into the uterus, thus stretching it. This stretching then stimulates more oxytocin release, which stimulates more muscle contractions, and the cycle continues. (What's in process here is one of the body's few *positive*-feedback loops. Oxytocin and

Oxytocin and vasopressin help regulate traits as complex as human trust and sociability.

stretching both prompt *more* of each other in a process that ends only when the baby is born.) Oxytocin's role then continues after birth as it triggers the release of milk from the mother's breasts.

These hormonal effects had been understood for decades when, in the mid-1990s, C. Sue Carter, then at the University of Maryland, College Park, began asking questions about how oxytocin functions as

a neurotransmitter in the small rodents known as voles. Why, she wondered, should males and females from one species of vole, the prairie vole (*Microtus ochrogaster*), form long-lasting, monogamous pair bonds when such bonding is practiced by no more than 5 percent of mammal species? As she and other researchers found out, oxytocin and vasopressin make all the difference.

Among vole species, only prairie voles have receptors for these neurotransmitters in midbrain areas that govern sensations of pleasure. Stimulate oxytocin release in a female prairie vole and you also initiate midbrain release of dopamine—the same neurotransmitter that floods the midbrain when drugs of abuse, such as cocaine, are administered. Female prairie voles given doses of oxytocin will start forming a bond with the nearest male around, while a female given a dose of an oxytocin *blocker* will not bond with any male, irrespective of whether she has mated with him. The assumption among neuroscientists is that, for females, oxytocin is providing a linkage between social bonding and sensations of pleasure. Meanwhile, for *male* prairie voles, this linkage seems to be provided by vasopressin. Dose a male with this neurotransmitter and he will start exhibiting the

---

MB

Bioflix: Homeostasis and Regulating Blood Sugar

these two kinds of cells also are capable of *storing* glucose in the form of a molecule we looked at earlier, glycogen, which is formed from sets of glucose molecules that link together. Looking at the other side of control, glucagon does its job by *unpacking* the storage that insulin helps bring about. In muscle cells, glucagon promotes the transformation of glycogen back into individual glucose molecules, and these molecules are then used by the muscle cells themselves. (When we're hard at work exercising, much of the energy we're expending comes from glucose that was originally stored in the muscles as glycogen.) Glucagon also promotes the transformation of glycogen in the *liver,* but in this case, much of the resulting glucose is released into general circula-

tion. This circulating glucose can then move into cells throughout the body as needed.

As you might imagine, the body produces more glucagon than insulin when blood levels of glucose are running low—typically when we haven't eaten for awhile. After we've had a meal, however, blood glucose levels surge, and the body now produces more insulin than glucagon, so that glucose can be moved out of the bloodstream and into cells (**Figure 28.8** on page 544).

All of this sounds like an efficient way for the body to keep a tight rein on its circulating levels of glucose, but as many people know in a personal way, the body's system of glucose control can break down. In a large number of children and adults, the body *fails* to move glucose from the bloodstream into cells, and the

hallmark behaviors of pair-bonding, such as care for offspring and mate-guarding.

Research results such as these sparked a flood of investigations into the effects of vasopressin and oxytocin on human beings, and the result has been one fascinating finding after another. In 2005, for example, a study was published indicating that oxytocin has an ability to affect human trust. Researchers had a group of male university students play a game of financial rewards in which some students were "investors" while others were "trustees" of the money provided by the investors. The investors stood to increase their money by turning it over to trustees (via computer), but there was an element of risk involved: The trustee could violate the investor's trust by simply keeping the money, secure in the knowledge that, under the rules of the game, trustees did not have to interact with investors more than once. Each investor thus had a decision to make before deciding how much money to turn over: How trustworthy was the trustee? The critical variable in the experiment was that, prior to playing the game, some investors were given a single dose of oxytocin, via nasal spray, while others received only a placebo spray. The result? Those who received the oxytocin exhibited much more trust than those who didn't. Those who were dosed with the hormone were, for example, more than twice as likely to invest the maximum

amount with the trustee than those who received the placebo.

Three years after these results were released, vasopressin's effects on human behavior came into sharper focus in a study that looked at genetic variation as it relates to this neurotransmitter. Reviewing DNA and life-history data from several hundred Swedish men and women, researchers asked whether DNA sequence variations found among these people—in a gene that codes for a vasopressin receptor—had any effect on their ability to form lasting pair-bonds. For the women in the study, the answer was no, but for the men it was a different story. Those men who possessed one particular variant of the vasopressin gene were twice as likely as other men to remain unmarried and twice as likely to report a recent crisis in their relationship if they were married. Moreover, the wives of men who possessed this gene variant expressed more dissatisfaction in their relationships than did the wives of men who lacked it.

Research results such as these have sparked an interest among neuroscientists regarding possible therapeutic uses of oxytocin and vasopressin. Might these compounds be used to improve social bonding or "affiliation" among people who have trouble with it? To date, this research has focused most strongly on oxytocin and its possible use in combating the disorder of autism, which generally produces children

and adults who have great difficulty interacting with other people. Autistic individuals have trouble meeting the gaze of other individuals; they lack a sense of empathy and cannot "read" facial expressions, tone of voice, or body language. In recent research, scientists have found that autistic individuals have relatively low levels of oxytocin in the brain and they've learned that oxytocin has the effect of dampening activity in a region of the brain called the amygdala, which is the portion of the brain most active in fear and threat responses. The obvious question is whether people afflicted with autism have, in effect, an overactive amygdala—one that might be calmed to some degree by boosting oxytocin activity in the brain.

A second line of research concerns the ability of autistic individuals to interpret the emotional states of others. Eric Hollander of the Mount Sinai School of Medicine in New York provided autistic adults with a single intravenous dose of oxytocin and found that their ability to interpret some of the emotions expressed through language—happiness, sadness, indifference—was improved over a period that lasted two weeks. Research in this area is still in its infancy, but it is proceeding on many fronts. Only time will tell whether a substance once thought to function solely as a reproductive hormone could eventually be used to bring individuals with autism into a wider social world.

end result of this failure is the disease diabetes, which you can read more about in "When Blood Sugar Stays in the Blood" on page 545.

Cortisol: Energy Management and Stress

One of the hormones that the anterior pituitary gland releases, adrenocorticotropic hormone, or ACTH, prompts cells in the periphery, or cortex, of the adrenal glands to release several so-called glucocorticoid hormones, the most important of which is cortisol. In our everyday lives, cortisol joins the T_3 and T_4 hormones we looked at earlier in being critical to metabolism and hence to energy management. Cortisol

helps bring about the production of our most important energy molecule, glucose, in a couple of ways—it provides the raw materials for glucose by facilitating the breakdown of muscle proteins, and it promotes the synthesis of glucose in the liver. Along similar lines, it promotes the breakdown of fats so that their component-part fatty acids can be used to power our activities.

In general, then, cortisol helps transform energy molecules; it takes them out of storage forms and puts them into forms that can be used. However, cortisol plays more exotic roles as well, as it is a stress hormone. It is part of a group of hormones that are released as part of the body's general "stress response." When you feel the pressure building in the days before a big

(a) After a meal, the role of insulin

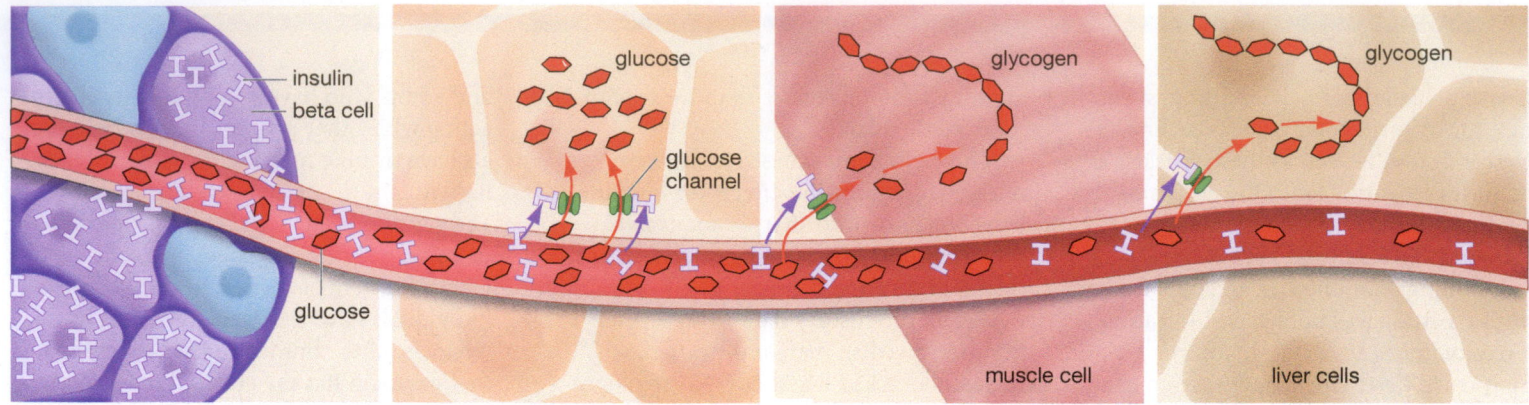

1. **Pancreas:** Stimulated by high levels of glucose in the bloodstream, beta cells in the Islets of Langerhans produce insulin.

2. **Other cells throughout the body:** Insulin enables glucose to move from the bloodstream into cells by triggering the formation of channels in the cell membranes.

3. **Skeletal muscle cells and liver cells:** With insulin's help, glucose can move into these cells and either be used right away or stored in the form of glycogen molecules.

(b) In between meals, the role of glucagon

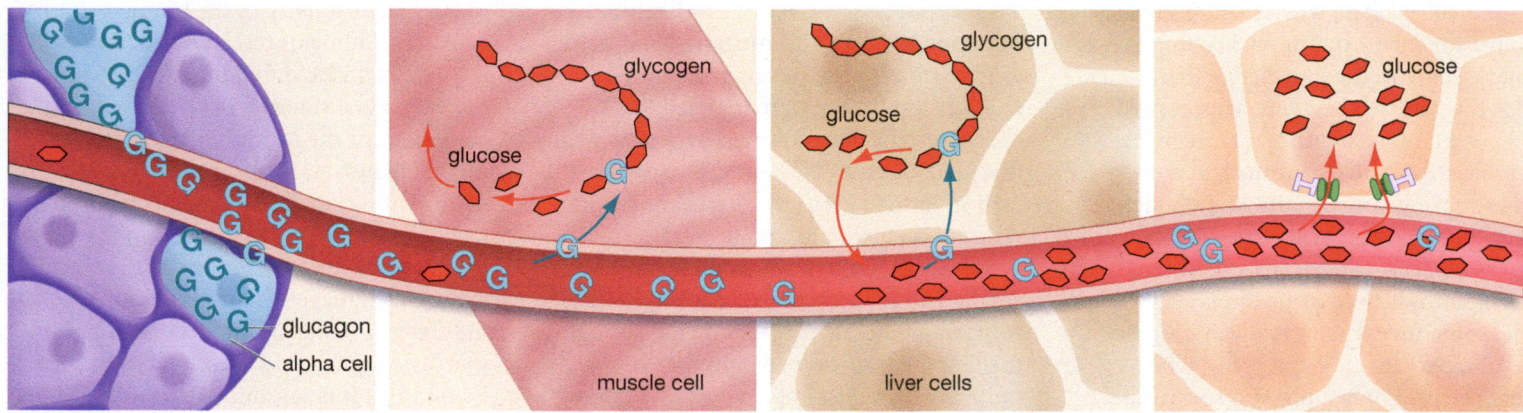

1. **Pancreas:** Stimulated by low levels of glucose in the bloodstream, alpha cells in the Islets of Langerhans produce glucagon.

2. **Skeletal muscle cells and liver cells:** With glucagon's help, glycogen is broken down into glucose. Muscle cells retain all the glucose they derive from this process, using it to power their own activities. Liver cells, meanwhile, move much of the glucose they liberate into general circulation.

3. **Other cells throughout the body:** Glucose released by the liver moves from the bloodstream into cells, supplying them with energy.

Figure 28.8
The Control of Glucose Levels: Insulin and Glucagon

exam and you feel yourself being uncomfortably tense and stimulated, that's cortisol at work. (Its role overlaps that of the more famous stress hormone adrenaline, which can be thought of as functioning more in times of panic rather than ongoing stress.) During prolonged stress, cortisol carries out its normal function, in a sense, in that it's helping to provide energy. But it is also having more sinister effects during these times: It is making you vulnerable to illnesses and temporarily destroying cell connections in the part of the brain called the hippocampus, which is critical in making long-term memories. These effects stem partly from the fact that cortisol weakens the immune system. (Among other things, it actually brings about the death of some immune cells.) But cortisol also plays a part in the narrowing of the arteries that leads

to heart attacks, and it makes cells less sensitive to insulin, leaving people more at risk for developing type 2 diabetes.

All this raises the question: Why should the endocrine system work to make stressed people more vulnerable to illness? In his fine book on the stress response, *Why Zebras Don't Get Ulcers,* researcher Robert Sapolsky has suggested that human beings are stuck with a hormonal stress response that developed, over evolutionary time, in ancestors of ours whose stresses were all *short term.* Such ancestors were not stressed about what would happen next week, but human beings are. Because of our ability to envision the future, our stress responses can go on for weeks, months, or years. And it is long-term stress responses that cause all the problems noted above. In cortisol,

ESSAY

When Blood Sugar Stays in the Blood: Diabetes

The paradox of the disease diabetes is that it amounts to starvation in the midst of plenty. At any given moment in the body of an untreated diabetic, billions of cells are being deprived of the life-sustaining blood sugar known as glucose. Meanwhile, less than a hair's breadth away from each of these cells, ample quantities of glucose are passing by—within the body's blood vessels. In diabetics, then, glucose moves past the body's cells but does not move into them, a state of affairs that deprives cells of the energy they need to function, while leaving harmful levels of glucose circulating in the blood. All kinds of physiological problems can follow, many of them circulatory. In the most serious cases, the result can be blindness, kidney failure, a stroke or heart attack, or a partial loss of limbs. Given such consequences, medical doctors are quick to intervene at the first signs of elevated glucose levels in any patient.

But what brings this elevation about? What causes diabetes, in other words? There actually are two principal varieties of this disease. The variety noted at the start of the chapter, type 1 diabetes, is caused by the death of the body's beta cells—the cells in the pancreas that secrete insulin. Sadly, these cells are victims of the body turning on itself. For reasons we don't understand, the body's immune system can attack beta cells as if they were foreign invaders. With the death of these cells, type 1 diabetics have no natural source of insulin; as a consequence, they must get insulin from outside themselves, in the form of insulin injections (**Figure 1**).

The second variety of diabetes is caused not by a shortage of insulin, but by a resistance to it. In type 2 diabetes, the body's beta cells are still producing insulin, but the body's biggest users of insulin—muscle, liver, and fat cells—become less responsive to it over time. As a result, even with normal levels of insulin in circulation, glucose increasingly stays within the bloodstream. Beta cells

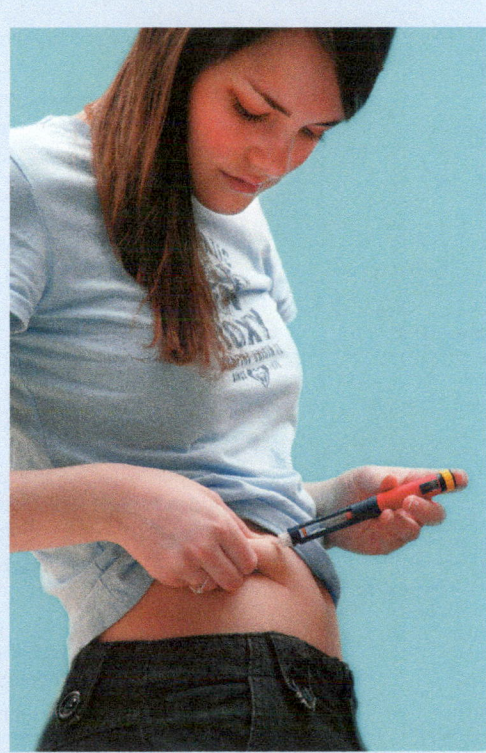

Figure 1
Lifesaving Substance
Type 1 diabetics cannot produce their own insulin and thus must rely on manufactured insulin, which they administer to themselves through means such as the injection seen here.

respond to this change by producing ever-greater amounts of insulin, but beyond a certain point, these cells begin to lose function. When they can no longer produce enough insulin to keep blood levels of glucose within safe limits, the body has moved from insulin resistance to type 2 diabetes.

While type 2 diabetes has always been with us, in recent years the prevalence of this disease has skyrocketed. In 2007, an estimated 7.8 percent of the U.S. population—23.6 million people—had diabetes, with at least 90 percent of these cases being type 2 diabetes. It appears, however, that worse is yet to come. In 2003, researchers from the federal Centers for Disease Control and Prevention estimated that one in three Americans born in the year 2000 will develop diabetes in their lifetime.

What accounts for figures such as these? The single most important risk factor for developing type 2 diabetes is being overweight. Indeed, more than 80 percent of type 2 diabetes patients are overweight. And more Americans are either now overweight or in the still-heavier category of "obese" than at any time in our history. (In 1980, 15 percent of American adults were obese; by 2005–2006, the figure had risen to 34 percent.) Of course, weight can be controlled and therein lies the good news: Most people who are at risk for developing type 2 diabetes can avoid the disease through diet and exercise. Drugs also exist that are aimed at either preventing or controlling type 2 diabetes, but diet and exercise are the first lines of defense against this disease. So, why is weight such a critical factor in its development and why does exercise matter in its prevention? For information on these topics, see "Why Fat Matters and Why Exercise May Matter More" in Chapter 26 on page 492.

Table 28.1

Hormones of the Endocrine System: Their Sources and Effects

Gland/Hormone	Effects
Hypothalamus	
Releasing hormones	Stimulate hormone production in anterior pituitary
Inhibiting hormones	Reduce hormone production in anterior pituitary
Anterior pituitary	
Thyroid-stimulating hormone (TSH)	Triggers release of thyroid hormones
Adrenocorticotropic hormone (ACTH)	Stimulates adrenal cortex cells to secrete glucocorticoids
Follicle-stimulating hormone	Female: promotes egg development: stimulates ovaries to produce estrogen Male: promotes sperm production
Luteinizing hormone (LH)	Female: produces ovulation (egg release): stimulates ovaries to produce estrogen and progesterone Male: stimulates testes to produce androgens (e.g., testosterone)
Prolactin (PRL)	Stimulates mammary gland development and production of milk
Growth hormone (GH)	Stimulates growth by prompting liver's release of somatomedin hormones
Posterior pituitary	
Vasopressin	Also known as antidiuretic hormone; controls retention of water in the body by the kidneys
Oxytocin	Stimulates contraction of the uterus muscles and release of milk in females; assists in semen ejaculation in males
Thyroid	
T_3, T_4	Increase body's metabolic rate
Calcitonin	Reduces calcium ion levels in blood
Parathyroid	
Parathyroid hormone (PTH)	Increases calcium ion levels in blood
Thymus	
Thymosins	Stimulate development of white blood cells (lymphocytes) in early life
Adrenal cortex	
Glucocorticoids	Includes cortisol, which stimulates glucose production and breakdown of fats; a stress-response hormone
Mineralocorticoids	Cause the kidneys to retain sodium ions and water and excrete potassium ions
Adrenal medulla	
Adrenaline	Also known as epinephrine; stimulates release of energy stores; increases heart rate and blood pressure
Noradrenaline	Also known as norepinephrine; effects similar to adrenaline
Pancreas	
Insulin	Decreases glucose level in blood
Glucagon	Increases glucose level in blood
Testes	
Testosterone	Promotes production of sperm and development of male sex characteristics
Ovaries	
Estrogens	Support egg development, growth of uterine lining, and development of female sex characteristics
Progesterones	Prepare uterus for arrival of developing embryo and support of further embryonic development
Pineal gland	
Melatonin	Establishes day/night cycle

then, we can see that the effects of hormones can be both far-reaching and not at all straightforward. **Table 28.1** provides a list of the hormones produced by most of the major endocrine glands in the body along with the effects of these hormones.

On to the Immune System

Having looked at the capabilities of the endocrine system, it's time to move on to a system that's not involved in communication and control, but that instead has the job of defense. Indeed, we might say it is a system that's involved in a never-ending war of defense. On one side in this combat are all kinds of microbes—bacteria, viruses, fungi—that use every opportunity to enter the body and start reproducing in it. On the other side is the body's chief defender, the immune system, whose story is the subject of Chapter 29.

SO FAR . . .

1. _____ is a hormone that promotes the transport of glucose from the bloodstream into cells, while _____ is a hormone that does the reverse. Both hormones are secreted by cells in the _____.

2. In muscle and liver cells, individual glucose molecules can be linked together to form the glucose storage molecule _____, which can later be broken up through the action of the hormone _____.

3. The hormone cortisol is produced in the cortex of the _____ and normally functions to transform energy molecules, taking them out of _____ forms and putting them into forms that can be _____.

CHAPTER 28 REVIEW

Summary

28.1 The Endocrine System in Overview

- The endocrine system regulates bodily processes through hormones: substances secreted by one set of cells that travel through the bloodstream and affect the activities of other cells. Each hormone affects only a select set of cells, known as a hormone's target cells. (p. 534)

- Most hormones are produced and released by endocrine glands, defined as glands that secrete materials they have produced directly into the bloodstream or surrounding tissue. Some hormones are produced, however, not by specialized glands but by organs such as the kidneys or the heart. (p. 534)

- Once secreted, hormones can take from several seconds to several hours to work, but then can continue to exert their effects for extended periods of time. Hormones tend to regulate processes that unfold over minutes, hours, or even years. (p. 534)

28.2 Hormone Types and Modes of Action

- Hormones can be classified as amino-acid–based, peptide, or steroid. All amino-acid–based hormones are derived from a chemical modification of a single amino acid. Peptide hormones—by far the largest class of hormones—are composed of amino-acid chains whose lengths can vary. Amino-acid–based and peptide hormones generally link to their target cells via receptors that protrude from the target cells' outer or plasma membranes. This binding initiates a chain of chemical reactions inside the cell that usually results in the activation of one or more enzymes. (p. 535)

- Steroid hormones generally pass through a cell's plasma membrane and bind with a receptor protein inside the cell. The combined steroid hormone/receptor molecule then binds with the cell's DNA, thus turning on one or more cell genes, which results in the production of one or more proteins. Several steroid hormones, such as testosterone and estrogen, are reproductive hormones. (p. 536)

28.3 Negative Feedback and Homeostasis

- Almost all hormone secretion is controlled by negative feedback. With its many negative-feedback loops, the endocrine system is a key factor in producing homeostasis in the body, meaning a relatively stable environment. (p. 537)

28.4 Hormone Central: The Hypothalamus

- Endocrine activity is largely governed by a portion of the brain, the hypothalamus, which receives multiple sensory inputs and in response controls the release of multiple hormones. (p. 537)

- The hypothalamus exercises endocrine control by regulating hormonal release in three separate organs: the adrenal glands, the posterior pituitary gland, and the anterior pituitary gland. Nerve signals that the hypothalamus issues to cells in the medulla of the adrenal glands bring about the secretion of the hormones adrenaline and noradrenaline. Oxytocin and vasopressin, which the hypothalamus itself synthesizes, are released into circulation within the posterior pituitary gland by axons extending from hypothalamic neurons. Hormones released by the hypothalamus go on to control the release of six hormones produced by the anterior pituitary gland, five of which go on to control the release of hormones by other glands. (p. 537)

28.5 Hormones in Action: Three Examples

- The body controls its circulating levels of glucose through insulin, a hormone that reduces blood levels of glucose, and glucagon, a hormone that increases blood levels of glucose. Insulin is produced by the pancreatic beta cells, and glucagon is produced by pancreatic alpha cells. After moving into circulation, insulin signals cells to create channels through which they take in circulating glucose. In liver and skeletal muscle cells, some of the glucose that is taken in may be stored in the form of the molecule glycogen. Glucagon prompts the transformation of glycogen back into glucose. Diabetes results from a failure of the body to move glucose into cells. (p. 541)

- The hormone cortisol, released by cells in the outer portion, or cortex, of the adrenal

glands, helps bring energy stores into use but also is a "stress" hormone that can have negative effects if stress goes on for extended periods. (p. 543)

Key Terms

amino-acid–based hormones 535

endocrine gland 534

gland 534

glucagon 541

homeostasis 537

hormones 534

insulin 541

peptide hormones 535

steroid hormones 536

target cells 534

Understanding the Basics

Multiple-Choice Questions

(**Answers are in the back of the book.**)

1. As defined in the text, hormones (select all that apply):
   a. tend to be secreted by glands.
   b. affect mostly fat, muscle, and liver cells.
   c. have effects only on target cells.
   d. are distributed by traveling through the bloodstream.
   e. have almost instantaneous effects.

2. Endocrine glands secrete their materials:
   a. into ducts.
   b. into synapses between nerve axons and cell bodies.
   c. directly into cells.
   d. directly into the bloodstream or surrounding tissues.
   e. into the digestive tract.

3. Steroid hormones are all _____ and generally bind to receptors ____.
   a. produced in greater amounts by males; on muscle cells
   b. built around the framework of the cholesterol molecule; on fat cells
   c. composed of chains of amino acids; on a target cell's outer or plasma membrane

   d. built around the framework of the cholesterol molecule; inside target cells
   e. muscle-building hormones; on muscle cells

4. Hormone secretion is generally controlled through the mechanism of _____, which is critical in producing _____, meaning a stable internal state in the body.
   a. hierarchical control; homeostasis
   b. negative feedback; homeostasis
   c. negative feedback; equilibrium
   d. positive feedback; endothermy
   e. hierarchical control; rest

5. The hypothalamus (select all that apply):
   a. is an endocrine gland.
   b. is a part of the brain.
   c. is located in the medulla of the adrenal glands.
   d. produces hormones released in the posterior pituitary.
   e. controls release of adrenaline and noradrenaline.

6. The anterior pituitary gland releases ____ hormones upon receipt of hormonal signals from the ____.
   a. two; posterior pituitary
   b. two; thyroid gland
   c. six; hypothalamus;
   d. six; adrenal glands
   e. two; hypothalamus

7. Insulin ___, while glucagon ____. (Select all that apply.)
   a. is produced in pancreatic alpha cells; is produced in pancreatic beta cells
   b. is produced in pancreatic beta cells; is produced in pancreatic alpha cells
   c. helps move glucose into cells; helps move glucose into circulation
   d. helps move glucose into circulation; helps move glucose into cells
   e. is produced in relatively greater amounts between meals; is produced in relatively greater amounts just after eating

8. Cortisol (select all that apply):
   a. is produced by the anterior pituitary gland.
   b. is produced by t he adrenal cortex.
   c. is active in energy management.
   d. is produced by the adrenal medulla.
   e. is secreted during times of stress.

Brief Review

(Answers are in the back of the book.)

1. Why is the endocrine system better at controlling something like fluid levels in the body than the ability to shoot a basketball?

2. What are the differences between amino-acid–based, peptide, and steroid hormones?

3. Describe the three basic routes through which the hypothalamus is active in hormonal regulation.

4. Describe the process of hormonal control of circulating glucose levels.

5. Why does the hormone cortisol potentially pose a risk to human beings that it does not pose to other animals?

Applying Your Knowledge

1. Pancreatic beta cells can sense high levels of circulating glucose and, in response, secrete insulin, which helps move glucose out of circulation. These cells can likewise sense the resulting low circulating glucose levels, and when they do, they reduce their secretion of insulin. This is negative feedback in action; the product of beta cell activity (insulin) ultimately has the effect of reducing beta cell activity. Why does the body constantly utilize negative feedback to maintain the stable state called homeostasis? Why couldn't it use positive feedback instead?

2. The text makes clear that hormones play critical roles in allowing us to function each day. So why are they commonly thought of as substances that cause us to act in unusual ways?

MasteringBiology®

1. MasteringBiology Assignments
Activities: The Endocrine System; Hypothalamic Control of the Pituitary; **Bioflix:** Homeostasis and Regulating Blood Sugar; **Current Events:** The Hormone Surge of Middle Childhood; A Romp Into Theories of the Cradle of Life; **Building Vocabulary;** **Questions:** Bioflix Quiz, Reading Questions, Test Bank

2. eText
Read your book online, search, take notes, highlight text, and study more effectively.

3. The Study Area
Self-study tools to test comprehension.

29

Defending the Body:

The Immune System

Infectious microbes enter the human body every day. Waiting to meet them is a highly organized network of human cells and proteins. Collectively, these defenders of the body are known as the immune system.

Available on MasteringBiology®
Current Events: Tending the Body's Microbial Garden; An Immune System Trained to Kill Cancer; Do Cellphones Cause Brain Cancer?; Studying Sea Life for a Glue That Mends People; **Building Vocabulary; Questions:** Reading Questions, Test Bank

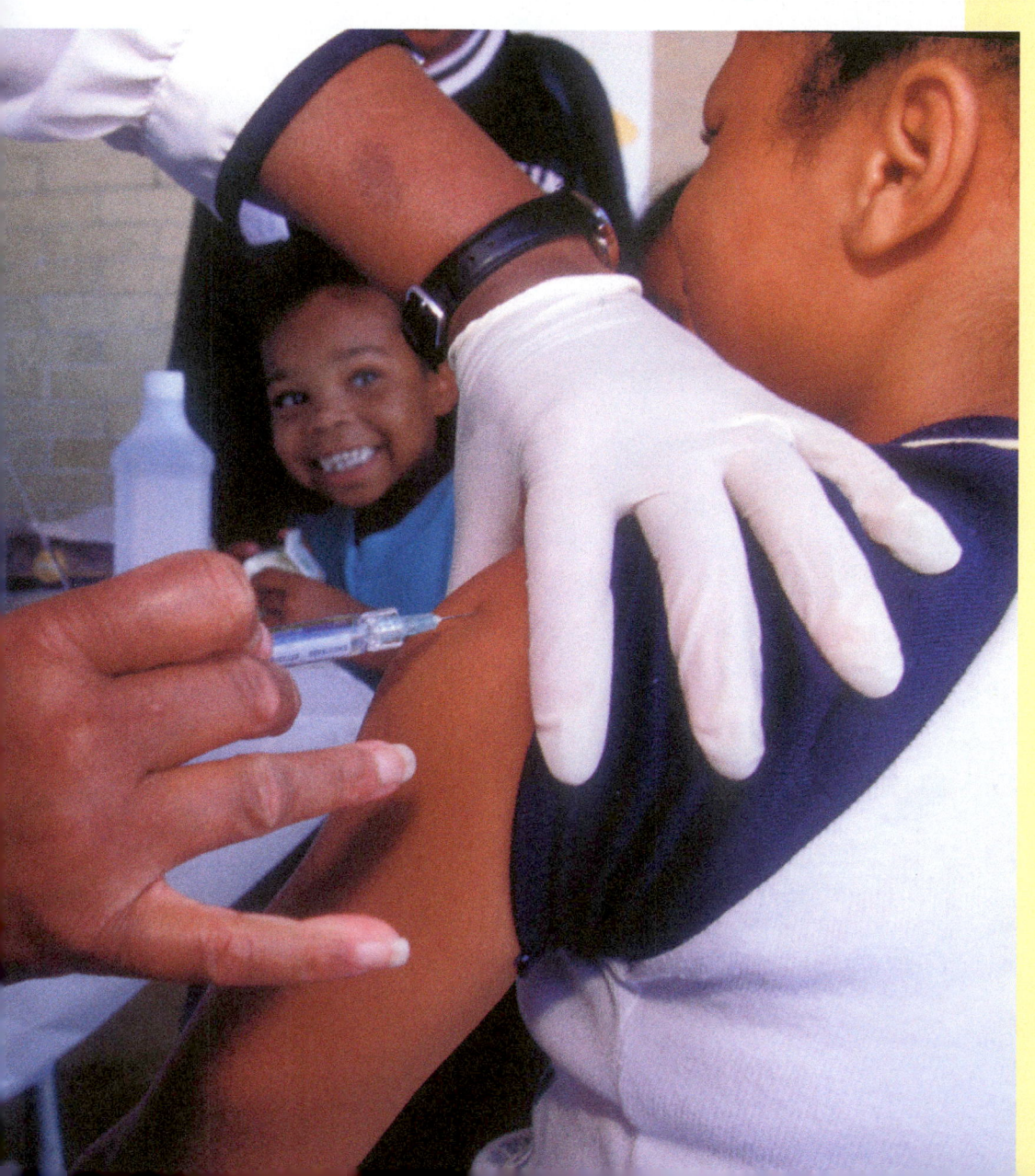

Vaccination puts the body's immune system in a high state of readiness against specific microbial invaders. Here a child in Detroit is getting vaccinated just prior to the first day of school.

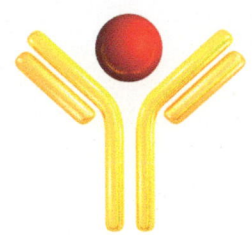

The word "vaccine" comes from the word *vacca,* which is Latin for "cow." What do cows have to do with vaccines? In 1796, an English country doctor, Edward Jenner, decided to test a bit of folk wisdom, which held that farm laborers who caught the mild disease cowpox from cattle could never catch the dreaded disease smallpox from humans. Accordingly,

Jenner took cowpox pus from a sore on the hand of a dairymaid and rubbed it into scratches he made on the arms of an 8-year-old boy, James Phipps. The next step—a chilling one from a modern perspective—was that Jenner applied pus from *smallpox* sores onto Phipps. The boy sailed through this and several subsequent smallpox applications, however, and Jenner's belief was proved correct: A person could be made immune to smallpox by being exposed to cowpox.

With Jenner's demonstration, a revolution was launched—the vaccine revolution. To get an idea of how far-reaching its effects have been, consider this list of killers, cripplers, and worrying childhood diseases: polio, measles, whooping cough, rubella, rotavirus, chicken pox, diphtheria, tetanus, mumps, hepatitis B, invasive pneumococcal disease, and type b meningitis. In developed countries, every one of these infectious illnesses has either been greatly reduced or eliminated altogether thanks to vaccination. And the smallpox that Jenner worked on? It has been eliminated from the face of the Earth as a natural disease. (The last naturally transmitted case of it occurred in 1977.)

But how does vaccination work? To take James Phipps as an example, why should his exposure to cowpox have protected him from smallpox? Two centuries of hard-won scientific advances since Jenner's time have provided the answer. Vaccination activates one part of the *immune system,* a complex network of cells and proteins whose most important function is finding invading organisms and killing them. The immune system is tenacious and has an arsenal of weapons at its disposal. Then again, it needs these very qualities because it faces such a formidable lineup of invaders. In this chapter, you'll learn how the immune system rises to the challenge of protecting the human body.

29.1 The Immune System in Overview

A good place to start our story is by being clear about what the immune system is fighting. Its adversaries are not human-made toxins, such as the poisons that a person might ingest, or carcinogens, such as asbestos. To the extent that the body can eliminate such substances, it does so through the liver and the kidneys. In contrast, the immune system is concerned with *living things.* More specifically, it is concerned with *microbes,* which is to say, living things that are too small to be seen with the naked eye. These tiny organisms are adept at invading the human body, and the damage they can cause by living within us can be severe.

The immune system does sometimes do battle with creatures large enough to be visible—most notably with parasitic worms—but in developed countries, the most

Defenders and Invaders

The large, spherical cells on the left are immune system macrophages that are in the process of attacking invading *E. coli* bacterial cells, at the bottom of the photo. The macrophages will engulf the bacterial cells, ingest them, and dissolve them.

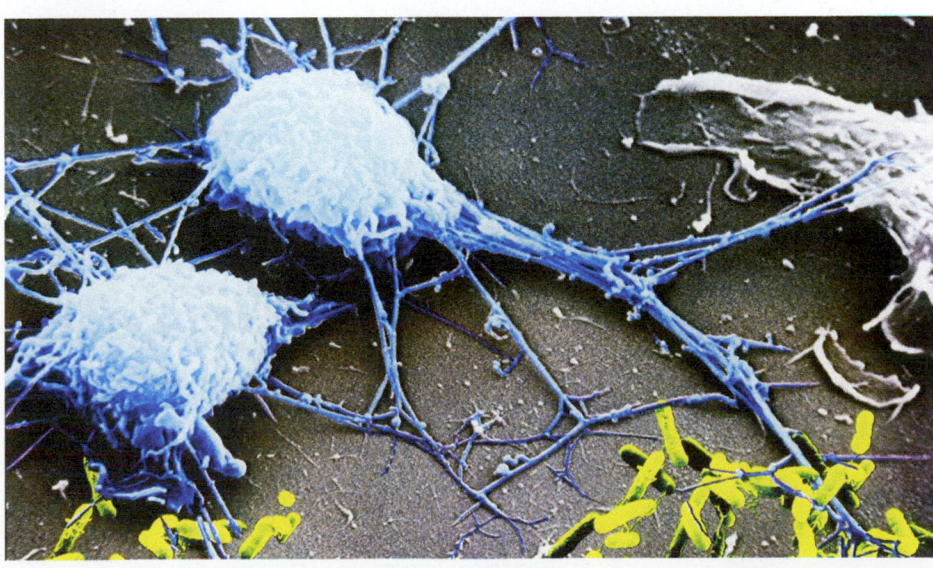

common sources of human infection are microbes, which come in three major invasive classes: bacteria, fungi, and viruses. Bacteria are single-celled microbes that cause illnesses ranging from tuberculosis to today's much-feared "flesh-eating" staph infections. Fungi generally are multicelled microbes that cause afflictions ranging from the relatively harmless athlete's foot to the more serious "valley fever" seen in the southwestern United States or the candida "thrush" infections often seen in AIDS patients. Viruses probably do not qualify as living things, but they have characteristics of living things—including the possession of DNA or RNA—and they are a major class of invader, causing illnesses that range from the common cold to AIDS.

Now, how is it that the immune system recognizes that the body has been attacked by one of these invaders? Think of it this way: If you were asked to close your eyes and run your fingers across, say, a brick wall

and a wooden wall, you would have no trouble telling the difference between the two because the materials brick and wood differ in their basic structure. Just so, the materials that make up human cells on the one hand and bacterial cells on the other differ in *their* structure, and as such, the immune system has no trouble telling one kind of cell from the other. Microbes such as bacteria will display, on their surfaces, proteins and carbohydrates whose micro-structure differs from any protein or carbohydrate found in a human being. Small pieces of these proteins and carbohydrates will be *recognized* as foreign by immune system cells, and when this happens an immune system attack will follow. Any substance that elicits an immune system attack is an **antigen.**

If you look at **Table 29.1**, you can see a partial list of the human cells and proteins that are active in immune system responses. Only a few of the entities you see there are involved in the outright killing of microbes. The immune system functions like an army and as such, it needs not only front-line soldiers but an array of support personnel as well. The mast cells you see in Table 29.1a act, in a sense, as transportation specialists, in that substances released by them enable the rapid movement of immune cells and proteins to the site of an attack. Meanwhile, the cytokine proteins you can see in Table 29.1b are mostly communication specialists—they carry messages from one immune cell to another, though some of them do attack microbial invaders directly. We'll be going over the roles of all these players as the chapter unfolds. It's worth noting that all of the immune cells listed in Table 29.1a are different kinds of white blood cells; any cell that is active in the immune system is likely to be a variety of white blood cell.

First Line of Defense: Barriers to Infection

Immune system attacks will be launched against any disease-causing microbe, or *pathogen,* that has invaded human tissue. But these immune system responses are

---

Table 29.1a

Selected Immune System Cells

Phagocytes (cells that can ingest other cells)

Neutrophils

Eosinophils

Macrophages, which can become antigen-presenting cells (APCs)

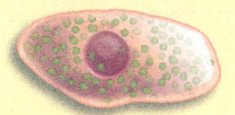

Mast cells (release histamine)

Dendritic cells (antigen presenting cells)

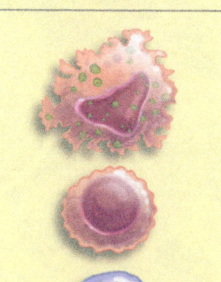

Lymphocytes

Natural killer cells

B cells, which can differentiate into:
Plasma cells
Memory cells

T cells, produced in 3 varieties:
Helper T cells
Cytotoxic (killer) T cells
Regulatory T cells

Table 29.1b

Selected Immune System Proteins

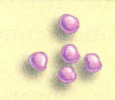

Complement proteins (kill some invaders, bind with others to aid phagocytes)

Antibodies (receptors on surface of B cells; later released by plasma cells derived from B cells)

Cytokines (diverse group of signaling molecules)

costly for the body to mount and, as you'll see, they can sometimes end up damaging human tissue along with invading microbes. Therefore, it makes sense that over evolutionary time, the human body developed a first line of defense that sits *outside* the immune system: We possess a set of barriers that must be breached before an immune system response even gets started. First and foremost among these is human skin, which may seem thin and fragile to us but which actually forms a superb barrier against pathogens. As readers of Chapter 26 know, the top layers of the skin are composed of dead, interlocked cells composed mostly of the tough protein called keratin. Meanwhile, the pores that dot the surface of the skin are regularly flushed with sweat and sebaceous secretions that wash microbes away. Given these qualities, there are only a few microbes that have the ability to make it past an intact layer of skin. Thus, microbes primarily gain access to the body in those parts of it where skin does not exist. What are these parts? The eyes, nose, and mouth and the stomach, lungs, and urinary–genital tract. It is the function of these parts of the body to interact with the outside world, but this very quality allows them to serve as entry points for microbes.

This is not to say, however, that microbes are simply allowed *free* access to the body at these places. Wherever there is a microbial entry point into the body, there are defenses waiting for the microbes. The urinary tract, for example, normally is kept sterile through the flow of urine itself; bacteria may start up the male urethra through the penile opening, but these microbes seldom make it very *far* up the urethra because the flow of urine will wash them back down. Looking at other entry points in the body, the stomach secretes a powerful acid that can destroy most of the pathogens that hitch a ride in on food; the lungs have a set of hair-like cilia that sweep the passageways of the lungs clean like so many tiny brooms; and the female reproductive tract has a microbe-resistant low pH, thanks to lactic acid produced by *helpful* resident bacteria, the lactobacilli.

These defenses are not foolproof, of course, and when they fail, microbes are able to invade human tissues. At this point, the immune system does have to respond and it does so with two primary lines of defense.

The Innate and Adaptive Immune Responses

The immune system's first line of defense is a group of cells and proteins that collectively provide the **innate immune response:** a human immune response that does not target specific microbial invaders. Some immune cells we'll be looking at will target any object in the body perceived to be foreign, and the same is true of a group of immune system proteins we'll be reviewing. These innate immune weapons are in a constant state of readiness to react to any invader, and thus are

the "first responders" to any microbial invasion. They serve, in effect, to counter a microbial invasion until such time as the immune system's slower-acting, but more powerful arm can become fully engaged.

This second arm is the **adaptive immune response:** the human immune response that provides protection against specific microbial invaders. You have immune cells in your body that will latch onto bacterium A but not to bacterium B. Once this happens, a huge number of defensive cells and proteins will be mobilized against this specific invader. In this way, the immune system *adapts* itself to meet the challenges of a particular invader at a particular moment in time. It takes awhile to mobilize weapons that are specialized for one invader, however, and as such, the adaptive immune response is not very effective during the initial stages of an infection. Once it's unleashed, however, it has the ability not only to mount a tenacious attack but to *remember* the adversaries it has faced. This ability allows it to generate a more rapid, and more powerful, response to any subsequent invasion by the same microorganism. The result is **protective immunity:** a state of long-lasting protection that the immune system develops against specific microorganisms.

SO FAR . . .

1. An antigen is any substance that _____.

2. The human body's first line of defense against invading pathogens is not the immune system, but rather a set of physical _____ to infection, the most important of which is _____.

3. The innate immune response will target any object in the body perceived to be _____, while the adaptive immune response targets _____.

29.2 The Innate Immune Response

Let's take a look now at the immune response as it would unfold in time, from the moment the innate immune response is initiated through the time the two kinds of adaptive responses are finished with their work. We'll start with the innate response as it would unfold within a particular context, which is known as **inflammation:** a localized immune system response to tissue injury. Suppose that you were out hiking and that, in getting over a trail fence, an old nail punctured your hand. With this, numerous microbes, including some bacteria, have gotten past the barrier of the skin and are now imbedded in the muscle and connective tissue of your hand. How does the immune system respond?

Figure 29.1
Sensors of Microbial Infection

Human immune system reactions are set off by the binding of an antigen from an invasive microbe with one or more of the sensors called toll-like receptors (TLRs), which in most cases extend from immune system cells such as macrophages. The ultimate result of this binding is the release of several varieties of the proteins called cytokines, which either fight the invader directly or alert other arms of the immune system about the invasion that's now underway.

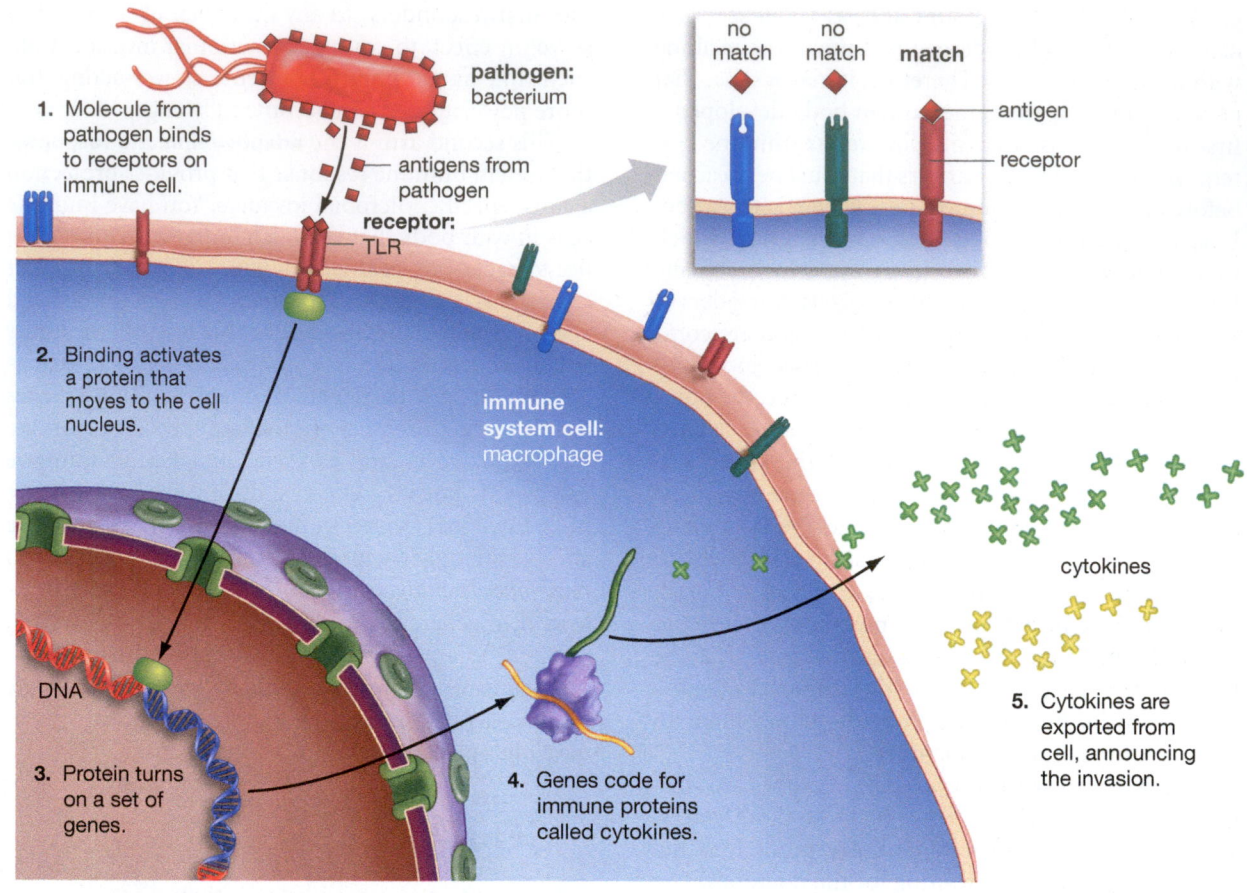

pathogen: bacterium

1. Molecule from pathogen binds to receptors on immune cell.

antigens from pathogen

receptor: TLR

no match   no match   **match**

antigen

receptor

2. Binding activates a protein that moves to the cell nucleus.

immune system cell: macrophage

DNA

3. Protein turns on a set of genes.

4. Genes code for immune proteins called cytokines.

cytokines

5. Cytokines are exported from cell, announcing the invasion.

The very first alarm will go off through the work of the molecular sensors known as cell *receptors*. We'll be seeing numerous examples of receptors, so now might be a good time to get an idea of the nature of some of these cellular extensions by looking at **Figure 29.1**. As you can see there, receptors have the ability to *bind with* material from outside the cell—in the example, with a combined carbohydrate-lipid molecule that is produced by a particular class of bacteria. This binding tends to be fairly specific; the kind of receptor you see in the figure will bind *only* to molecules produced by bacteria in this class while letting all other molecules pass it by. Moreover, note what happens when this binding takes place: A protein is activated within the immune system cell that moves to the nucleus of this cell; its arrival there turns on a set of genes that then produce some of the communication molecules we talked about earlier—the cytokines. These are then exported from the cell in several varieties, some of which may fight the invader directly, but most of which mobilize the larger immune system against the attack that's now underway.

Phagocytes and Toll-Like Receptors

Now, let's associate a few names with these immune system players. The cell in the figure is known as a macrophage—literally a "large eater" owing to its ability to wrap itself around a microbial invader, ingest

it, and then *dig*est it in a series of internal acid baths. You can see the macrophage listed in Table 29.1a as one variety of **phagocyte,** meaning a cell capable of ingesting another cell or cellular materials. Obviously, the macrophage can assume a front-line warrior role in the immune system, but it also can take on a somewhat less-exalted role, which is that of janitor—macrophages engulf and haul away the body's own cells and tissue fragments when they have become worn out or damaged. Like sentries on military bases, some macrophages are stationed permanently in one place. Indeed, they're stationed at exactly those places in the body that are vulnerable to attack: the tiny air sacs at the ends of lung passages, the outside of the digestive tract, and—as with our macrophage in this instance—the tissue just beneath the skin.

Now, what about the most prominent receptors pictured on the macrophage in the figure? These structures are **toll-like receptors (TLRs):** the primary sensors of microbial infection in the human body. These are the most important first-alarm bells of the immune system; the critical sensors that set in motion not only the innate immune response we're looking at but the adaptive immune response as well. Given this status, the amazing thing about these receptors is that their role in the human immune system was completely unknown until 1998. It was only then that scientists realized that human immune system cells had receptors such as these and that the 10 varieties of them discovered so far *channel* the immune response down different pathways, depending on the type of microbe that's invaded the body. The type of TLR pictured in Figure 29.1 binds with a molecule produced by a particular class of bacteria, as we've seen; but a different TLR will bind to a type of *viral* molecule, a third type will bind to a type of protein found in the swimming "tails" of mobile bacteria, and so forth. And the stimulation of, say, a mobile bacteria TLR will result in the release of a suite of cytokines that can optimally mobilize the immune system to fight mobile bacteria.

This clearly is a fine-tuning of the immune response, though it does not have the exquisite specificity that we'll see later in the adaptive immune response. There, the specificity is to unique *species* of invader—to the exact bacterium that causes tuberculosis, for example, or to the single virus that causes polio. With TLRs, conversely, the specificity is to *classes* of invaders (mobile bacteria, etc.). Nevertheless, the ability of TLRs to channel the immune system response down different pathways is something that has scientists very excited about new ways to combat infectious disease, as you'll see later.

Steps in the Innate Response

Having been stimulated by its TLRs, our macrophage has released cytokines that both fight the invader directly and that effectively announce "Infection underway here!" to other players in the innate immune response. One of the first players to get this message is another cell you can see in Table 29.1a, the *mast cell.* It will get involved not by directly killing invading bacteria but by bringing in the resources that will. The immune system's supply lines in this war are blood vessels, and mast cells release several substances that affect them; one of these substances, called histamine, opens blood vessels more widely at the site of infection. This facilitates a greater movement of immune system cells and proteins to the site of attack. In addition, histamine makes blood vessels more permeable—it alters them in such a way that substances can move *out* of them more freely. This is exactly what immune system cells and proteins now do; they move out of the local blood vessels and into the surrounding tissues where the bacteria are now multiplying. In sum, mast cells release **histamine:** a compound that, in the inflammatory response, brings about blood vessel dilation and increased blood vessel permeability.

The first immune system cells to move out of these dilated vessels will probably be one of the smaller phagocytes you can see in Table 29.1a, the neutrophils. Slower in arriving, but longer-lived and more powerful, are so-called free macrophages—wandering macrophages that arrive from distant parts of the body to fight the invader. Both varieties of cells are guided to the site of injury by a form of chemical signaling: Injured cells release compounds that produce a kind of chemical trail that leads to them. Once the neutrophils

and macrophages arrive, they begin ingesting bacterial invaders, but in doing their work, both kinds of cells are prone to leak some of their digestive acids into the surrounding tissue, thus damaging it. The upshot is that sites of infection become war zones filled with dead and dying cells and tissue fragments, along with many still-vigorous phagocytes. All these elements combine to make up the viscous liquid we know as pus.

Also flowing out of the blood vessels at this time are a group of proteins you can see in Table 29.1b, the *complement proteins*—about 30 of them in all. Their name comes from the fact that the actions they undertake are complementary to those taken by other parts of the immune system. In this instance, their role is to cut holes in the cell membranes of the bacteria, thus killing them. (When an invader is a virus, rather than a bacterium, another lethal member of the immune team is the aptly named *natural killer cell,* which will attack any of the body's cells that it perceives as foreign in the sense of having the unusual surface proteins that mark it as having become infected.)

As all this is happening, the site of infection is being limited in size—a critical action, because the greatest threat to the body is a so-called systemic infection, in which the entire body has become infected by invading microbes. To guard against this, the site of attack is sealed off partly through creation of a network of fibers composed of a blood protein that, appropriately enough, is named fibrin (**Figure 29.2**). Finally during an infection, macrophages send out a cytokine that can travel to the brain and reset the body's thermostat, thus bringing about the condition we know as fever (defined as any body temperature over 99°F). Increased body temperature can kill two species of bacteria outright, but scientists believe that fever's main contribution to immunity is to speed up the body's metabolic rate, thus allowing for more cell division, more protein production—in short, allowing for a quicker deployment of all the body's immune forces.

In looking at this account of inflammation, you can see that it brings more blood near a site of injury and more cells and proteins to the site. Blood is warm and red, of course, and this explains why the

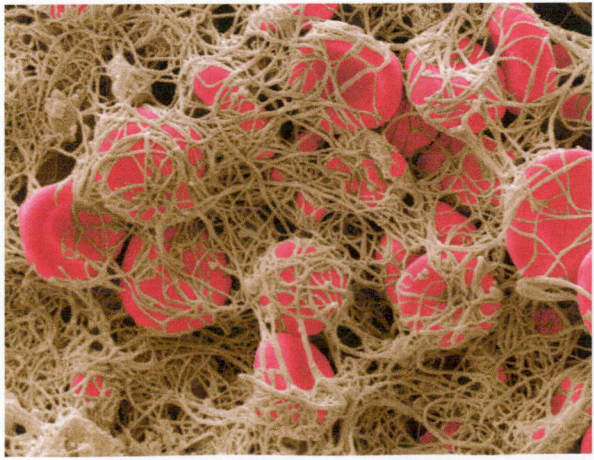

Figure 29.2
Sealing off an Injury

Strands of the protein fibrin begin to surround red blood cells at the site of a wound, thus initiating the formation of a clot that will seal the site off from the rest of the body. Small blood cell fragments called platelets (in green) release factors that help produce the fibrin strands. Meanwhile, a macrophage (in yellow) will help deal with any microbial invaders that may be present.

area around an injury site takes on these same qualities. Meanwhile, the added cells and proteins at the site cause the swelling and pain that we associate with injuries.

The Adaptive Response Begins

We've thus far looked at our bacterial infection as something that would be resisted strictly by the innate immune response. But as noted earlier, the innate response usually is just one step in a two-step process. What's needed to clear any infection that has spread is the slower-acting, but more powerful, adaptive immune response, which engages our invaders in ways we'll now review.

SO FAR . . .

1. Toll-like receptors are the primary _____ of microbial infection in the human body and also serve to _____ both the innate and adaptive immune responses.

2. Invading microbes can be ingested whole by various members of the class of immune system cells known as _____.

Figure 29.3
Development of T Cells and B Cells

Key players in the body's specific defense system are T cells and B cells, both of which start out as a type of white blood cell, called a lymphocyte precursor cell, in the bone marrow. Some of these lymphocytes migrate to the thymus gland, where they differentiate into T cells. Others that remain in the bone marrow develop into B cells.

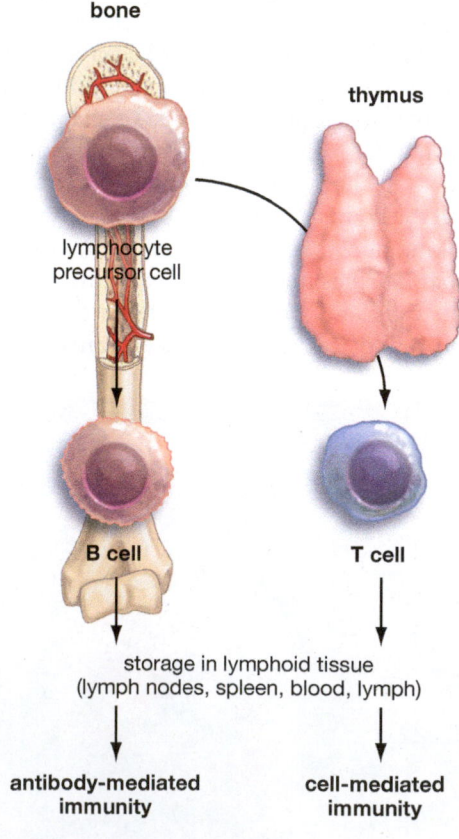

bone

thymus

lymphocyte precursor cell

B cell

T cell

storage in lymphoid tissue
(lymph nodes, spleen, blood, lymph)

antibody-mediated immunity

cell-mediated immunity

29.3 The Adaptive Immune Response

Every living thing on Earth has some kind of defense against pathogenic microbes, but the *vertebrates* on Earth—those animals that have backbones—are alone in having an adaptive immune defense. Vertebrate adaptive immunity operates through two major arms, which makes sense because there are two basic kinds of microbial infections in the body: those in which the invading microbes are free-standing (in the body's tissues), and those in which some of the body's *cells* have become infected and thus must be eliminated. The first arm of adaptive immunity, active primarily against free-standing microbes, is called **antibody-mediated immunity:** an immune system capability that works through the production of proteins called antibodies. The second arm of adaptive immunity, active primarily against infected cells, is called **cell-mediated immunity:** an immune system capability that works through the production of cells that destroy infected cells in the body.

Cells that are central to both antibody-mediated and cell-mediated immunity originate in the same place, which is the marrow of our bones. The cells that are key to adaptive immunity are called *lymphocytes,* which are another group of cells you can see in Table 29.1a. In the bone marrow, some lymphocytes develop into the central cells of antibody-mediated immunity, the **B-lymphocytes** or **B cells** (which is easy to remember if we recall that they are **B**one-marrow-derived cells). Other lymphocytes, however, go on to migrate to the body's thymus (located at the base of the throat), and there they specialize. They develop into the central cells of cell-mediated immunity, the **T-lymphocytes** or just **T cells** (with the T standing for "thymus" cell). Several different types of T cells actually will develop in the thymus; one of them with so-called CD4 receptors on its surface, will become what's called a "helper" T cell; another, with so-called CD8 receptors, will become a "killer" T cell. The development of both T and B cells is tracked for you in **Figure 29.3**.

Eventually, large numbers of both T and B cells will cycle in and out of the organs of an enormous fluid-transport system in the body, the **lymphatic network,** which is laid out for you in **Figure 29.4**. This network actually helps take care of two major tasks in the body, one of which is rather mundane: fluid drainage. Blood pumped by the heart is largely composed of fluid that at one point actually moves *out* of the body's capillary blood vessels—and into surrounding tissues—by virtue of the force with which it has been pumped. Most of this fluid moves back into the capillaries at their far end, but some doesn't get back in; it remains in the surrounding tissues, which gives the body its drainage problem.

(a) Lymphatic network

(b) Lymphatic vessels collect interstitial fluid from tissue.

(c) The fluid, now called lymph, is inspected by immune system cells in lymph nodes.

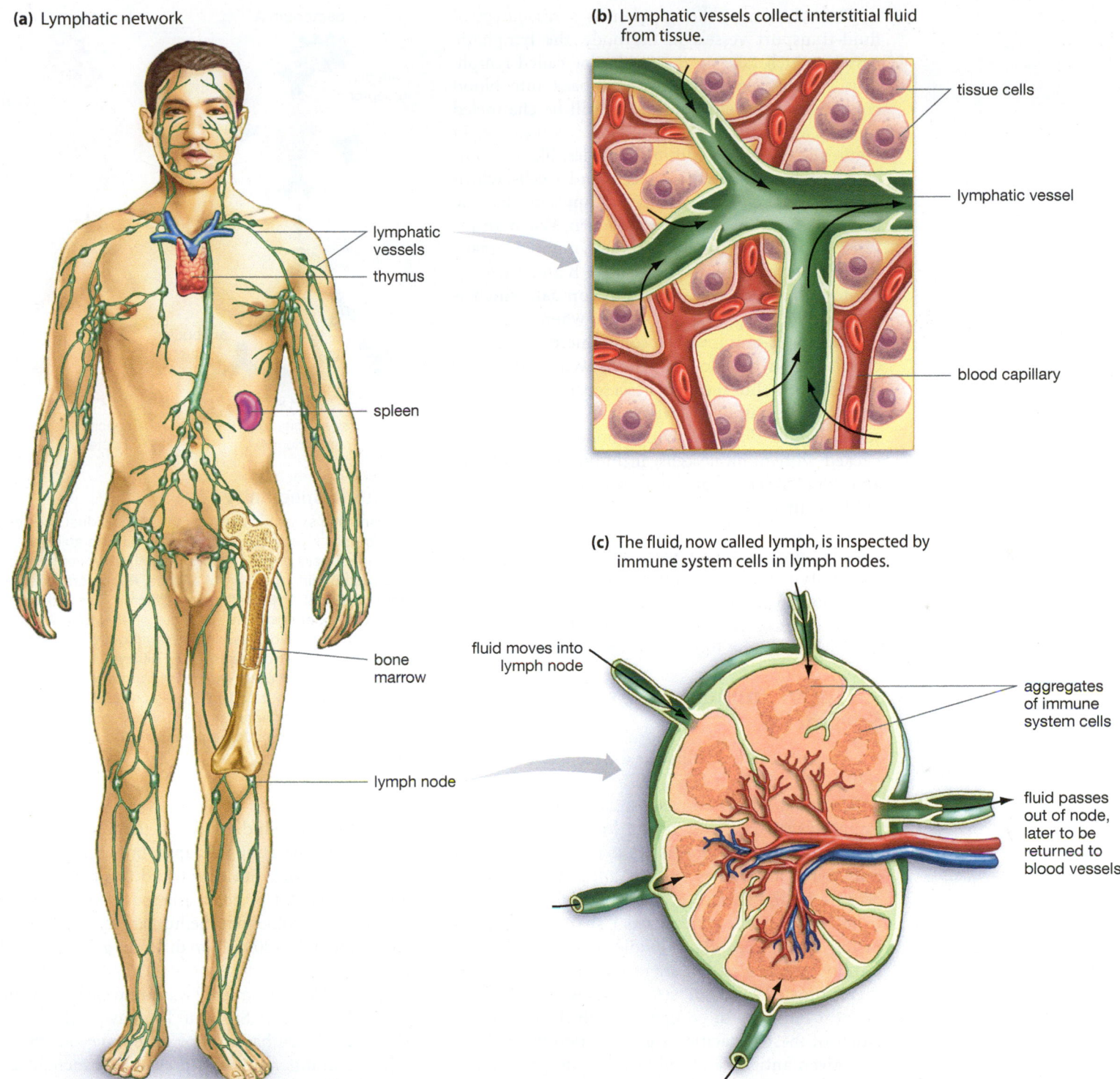

lymphatic vessels

thymus

spleen

bone marrow

lymph node

tissue cells

lymphatic vessel

blood capillary

fluid moves into lymph node

aggregates of immune system cells

fluid passes out of node, later to be returned to blood vessels

Figure 29.4
The Human Lymphatic Network

(a) Fluid moving through lymphatic vessels passes through lymphatic organs, such as lymph nodes, where the fluid is inspected by an array of immune system cells. Note also the location of the body's spleen, where a heavy concentration of immune cells inspects circulating blood for signs of infection.

(b) Detail of interstitial fluid moving into lymphatic vessels.

(c) Detail of fluid moving through a lymph node.

The solution to this problem is a *second* set of fluid-transport vessels in the body, the lymphatic vessels, which take up the fluid, now called lymph. This lymph eventually will move back into blood circulation, but before it does, it will be channeled through a number of the lymph *nodes* you can see in Figure 29.4. Waiting within the nodes, like customs agents at airports, are both B cells and T cells, which will inspect the passing lymph for any sign that the body has been invaded by a pathogen. When an invader *has* entered the body, these lymphoid organs will become even more packed with B and T cells—fighters that now are engaged in combat, which is why our "lymph glands" get swollen when we're sick. Of course, in any invasion some microbes may be circulating in the bloodstream as well as the lymphatic system, but the body has a separate inspection site for these pathogens, which is the *spleen* you can see in Figure 29.4. Like lymph nodes, the spleen is packed with immune cells, including macrophages and B and T cells, that stand ready to recognize and fight any invader.

B cells and T cells are the cells of the immune system that provide the exquisite specificity of adaptive immunity we talked about earlier. Let's take a moment now to see what this specificity means in practice. We'll use B cells in our example, but the lessons from B cells can be applied to T cells as well.

Adaptive Immunity's Fantastic Specificity

An **antibody** can be defined as a circulating immune system protein that binds to a particular antigen, which you'll recall is any substance that sets off an immune system attack. If you look at **Figure 29.5**, you can see why antibodies are so closely linked to B cells. First, antibodies exist as antigen-binding receptors on the surface of B cells, usually extending from B cells in the generalized Y shape you can see in the figure. Second, antibodies also exist as freestanding molecules that are produced by B cells and then exported by them, streaming away from them in great numbers to go fight microbial invaders. In either of these capacities, a given antibody will bind to a given antigen and only to that antigen, be it a few amino acids from the protein coat of a virus or a few carbohydrate molecules from the cell wall of a bacterium.

The *degree* of this specificity is so extreme that, at one time, it presented scientists with quite a mystery. All the antibodies extending from a given B cell were the same, these scientists saw, but every B cell that emerged from the bone marrow was *unique* in terms of its type of antibody. No two emerging B cells displayed exactly the same type of antibody—that is, an antibody that bound to the same antigen. Calcula-

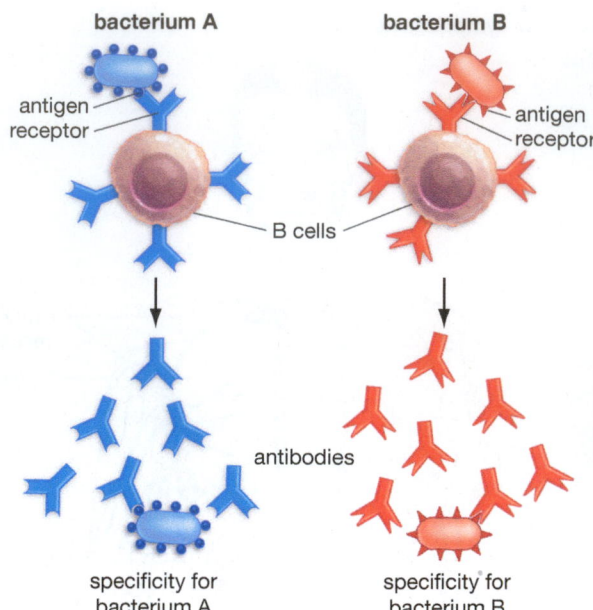

Figure 29.5
Immune Specificity

The immune system's ability to target individual species of microbial invaders hinges on the system's ability to produce B cells and T cells whose antigen receptors come in an astounding diversity of shapes. Receptors extending from the B cell on the left have a different shape than the receptors extending from the B cell on the right, which means that the two cells, and the antibodies produced by them, will bind with antigens from different species of bacteria. Every receptor extending from a given B cell is the same shape, but no two B cells produced in the bone marrow share a receptor shape. This diversity is likewise seen in the receptors that extend from T cells of the immune system.

tions at the time indicated that the human body was capable of generating perhaps a hundred million different antibodies. But how could this be? Antibodies are proteins and proteins are coded for by genes. It seemed impossible that the human genome was large enough to hold 100 million different genes that coded for antibodies.

This mystery eventually was unraveled, in the mid-1970s, by scientist Susumu Tonegawa, who showed that every B cell has a set of gene *fragments* that can *combine* in different ways to create the incredible diversity of human antibodies. We might compare what goes on to one of those machines we sometimes see on television that generate winning lottery numbers. There, a set of numbered balls will be blown around in a glass enclosure, after which perhaps five balls will be pulled out, giving us a unique winning series that goes something like: 13, 24, 39, 51, 52. In the same way, each B cell shuffles its set of gene fragments, ultimately settling on a particular set that will produce a specific antibody. How specific? As immunologist William R. Clark has observed, it's now thought that each of us can produce at least 10 billion different antibodies, but that the true number may be as many as 100 billion. The diversity of T cells, via their antigen receptors, is not this great, but at 100 million receptor types or more, T cells are not too shabby in their diversity either.

Let's now look at how this diversity provides B and T cells with much of the power of the adaptive immune response. We'll start with T cells and cell-mediated immunity and then move on to antibody-mediated immunity.

Cell-mediated Immunity

With a few notable exceptions, pathogenic bacteria are, we might say, exposed to the immune system, in that they live and multiply in the fluids of the body's tissues. Most pathogenic viruses, on the other hand, spend most of their time within the body's *cells*, where they put together new copies of themselves that then go on to infect other cells. For the immune system, then, the challenge posed by viruses comes down to first identifying virally infected cells and then killing them. The key to identification is that nearly every type of cell in the body constantly takes part in a kind of molecular show-and-tell: Each cell puts on display, on its surface, fragments of proteins it has recently produced. Most of the time, these are proteins the cell produces for itself, but a cell that has been infected with a virus will be putting *viral* protein fragments on its surface—proteins the virus has directed the cell to produce so that the virus can make copies of itself. Passing immune system cells take a great interest in these protein fragments, because they serve as the antigens that are the warning signs of infection. You might say that every cell in the body that has been infected has the equivalent of a sign sticking up from its surface that says: "I'm infected; please kill me," and the immune system is able to oblige.

But how does cell-mediated immunity get to the place where it has enough fighters in the field to take care of all the body's infected cells? The road from initial infection to full-blown cell-mediated attack is a drawn-out process of *mobilization*. Once you see what's involved, you'll understand why cell-mediated immunity takes more than a week to get to full throttle following an invasion.

The Cell-Mediated Pathway

The mobilization process typically begins with a type of cell that can be found patrolling parts of the body that are vulnerable to attack. This is the *dendritic cell,* so named because its spiky extensions look like the dendrites on a nerve cell (**Figure 29.6**). Viruses, as we've seen, often reside in human cells, but they can also be found outside cells (when they're moving from

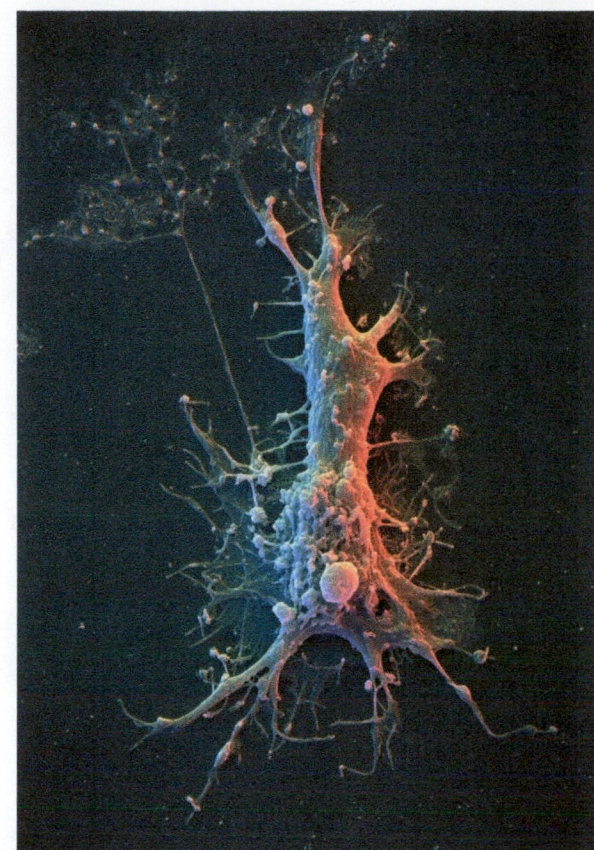

Figure 29.6
A Cell That Activates Other Cells
A human dendritic cell such as this one can ingest microbial pathogens, but its primary function is to activate the T cells of adaptive immunity by means of presenting them with antigens of invaders it has encountered.

one cell to another). When this happens, dendritic cells can *ingest* copies of these viruses (in the way that macrophages can) or they can be *invaded* by these viruses (in the manner that any other cell can). In both cases, the dendritic cell will then display fragments of viral proteins within molecular complexes that extend from its surface. Indeed, the primary function of the dendritic cell is this very display. Dendritic cells are the most important of the immune system's **antigen-presenting cells:** those immune system cells that help initiate the adaptive immune response by displaying microbial antigens on their surface.

Outfitted with its viral antigens on display, our dendritic cell then migrates to the very place you might expect to get the immune system troops mobilized: a lymph node. T cells exist there in abundance, of course, and if you look at **Figure 29.7** on the next page, you can see how scientists believe the next interaction plays out (though there is controversy about the exact sequencing of these events).

You may recall that T cells pour out of the thymus as both CD4 cells, which become "helper" T cells,

Figure 29.7
Bringing T Cells into the Fight

Cell-mediated immunity begins when one of the body's dendritic cells has ingested a virus and been invaded by one of the same type. After moving to a lymph node, the dendritic cell then serves as an antigen-presenting cell: It displays protein fragments from the virus on its surface and docks with both CD4 and CD8 T cells that have receptors specific to the virus. Both kinds of T cells receive information about the invader from this interaction and both then go on to produce invader-fighting clones of themselves—helper T cell clones in the case of the CD4 cell and cytotoxic T cell clones in the case of the CD8 cell.

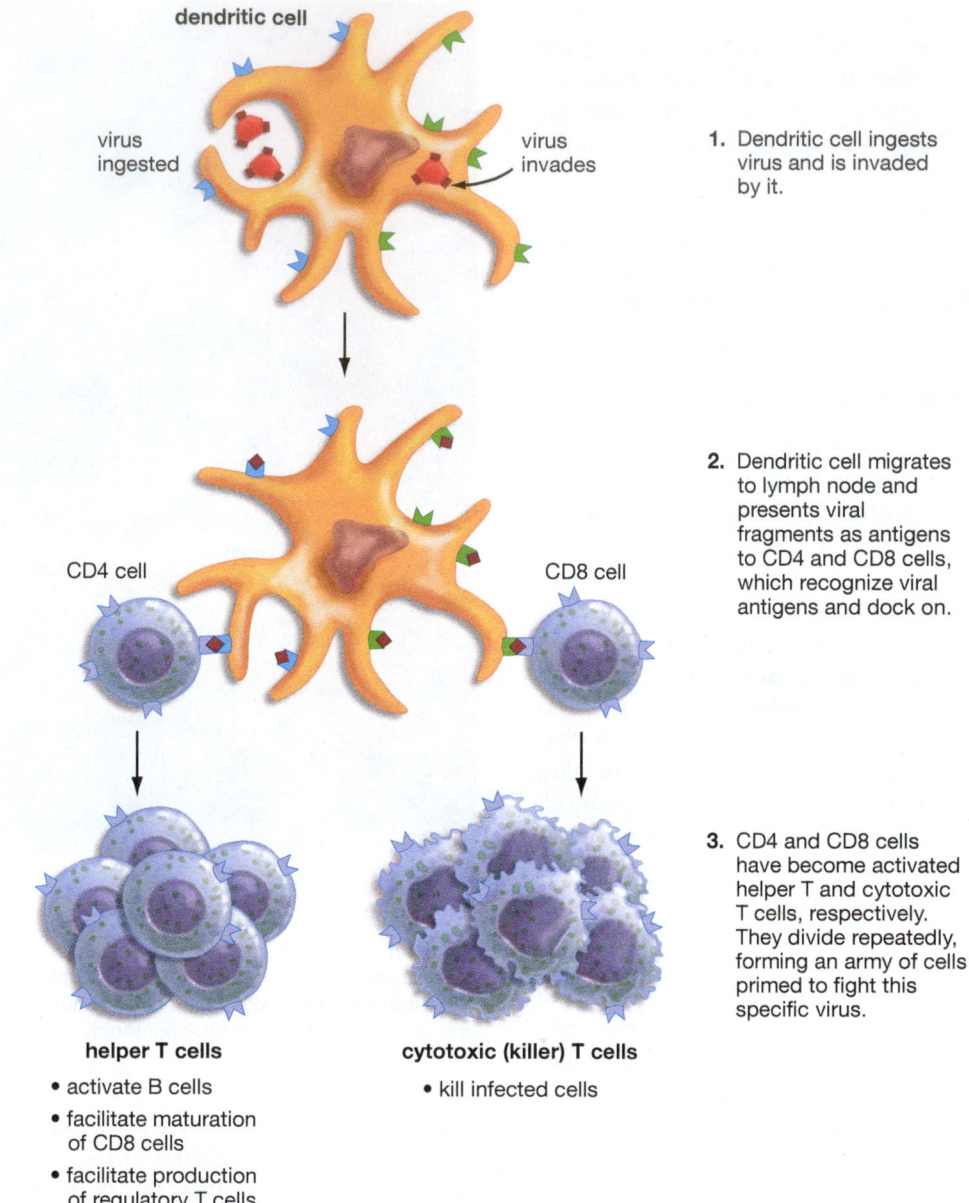

dendritic cell

virus ingested

virus invades

1. Dendritic cell ingests virus and is invaded by it.

CD4 cell

CD8 cell

2. Dendritic cell migrates to lymph node and presents viral fragments as antigens to CD4 and CD8 cells, which recognize viral antigens and dock on.

3. CD4 and CD8 cells have become activated helper T and cytotoxic T cells, respectively. They divide repeatedly, forming an army of cells primed to fight this specific virus.

helper T cells
- activate B cells
- facilitate maturation of CD8 cells
- facilitate production of regulatory T cells

cytotoxic (killer) T cells
- kill infected cells

and CD8 cells, which become "killer" or **cytotoxic T cells:** immune system cells that kill the body's infected cells in the adaptive immune response. Both kinds of cells are referred to as "naïve" prior to their interaction with a dendritic cell, and the term is apt because, upon docking with this dendritic cell, both kinds of T cells will get a schooling; both will learn what kind of infection is underway (bacterial? viral?) and where that infection is in the body. After their interaction with the dendritic cell, the naïve CD4 and CD8 cells have become "activated" helper T and killer T cells, respectively. Their education, however, takes a long time—perhaps 30 hours in all spent interacting with the dendritic cell. Critically, it's not just any CD4 or CD8 cell in a lymph node that can dock with such a cell; it is only those T cells bearing receptors specific to the viral antigen the dendritic cell is displaying.

Once interaction with the dendritic cell is complete, the helper and killer T cells are ready to produce an *army* of soldiers to fight this infection—each produces many clones of itself by dividing repeatedly, one cell becoming two, two becoming four, and so on. The end result is an army of cells primed to fight a specific invader. But the good news doesn't stop there. Each cell division in this process produces not only an "effector" helper or killer T cell—one that will fight the infection that's underway—but what's called a *memory* helper or killer cell, which is to say a cell that will remain in the body long after this infection has passed. Thanks to these cells, should this same virus enter the body again, the immune system will be ready: It will mount an assault on this invader that is much quicker and stronger than this first effort. (If you look at **Figure 29.8**, you can see how much quicker and stronger.) *This* is the basis for the protective

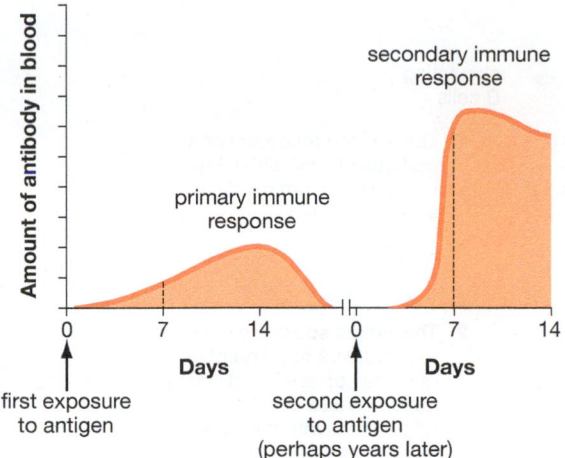

Figure 29.8
Prepared for an Invasion
The memory cells produced by the body during a first attack by an invader allow it to mount a faster, more vigorous defense should the same invader attack a second time.

immunity we talked about at the start of the chapter; the immune system remembers its enemies through this process. The helper T cells also activate a third type of T cell produced back in the thymus, the **regulatory T cell,** whose role is to help limit the immune system response.

Once these steps have taken place, the clones of effector helper and killer T cells take on the enemy directly. Each one of these cells has receptors on it specific to this viral invader, which means that each one is primed to interact with any of the infected cells out in the body that are displaying the "I'm infected; please kill me" antigens. Our helper T cells won't do any of the killing that follows, however. Their role is to release cytokines that more fully prepare killer T cells for battle and to prime the immune reaction in general. Killer T cells will, however, live up to their name. Using their own special receptors, they now bind to these infected cells and kill them, though the way they go about this may be surprising: They get the infected cells to commit suicide. All cells in the body have a built-in suicide or "apoptosis" program, and killer T cells can set this program in motion. Once it's begun, an infected cell will begin to swell up, after which it breaks up, after which the resulting cell fragments are mopped up (by cells such as the macrophages we looked at earlier).

Note that T cells are binding with the body's own infected cells. From this, it may be apparent why they are the main players in *cell-mediated* immunity: In the end, they take part in an orchestrated attack involving several sets of cells. Such an attack is of limited value, however, when the enemy is not a cell-invading virus but instead a bacterium that lives and reproduces outside of cells. For this kind of pathogen, the immune system's main weapons are the proteins called antibodies.

SO FAR . . .

1. The two arms of adaptive immunity are antibody-mediated immunity, which operates through the production of the proteins called _____, and cell-mediated immunity, which works through the production of cells that destroy infected _____ in the body. The central cells of antibody-mediated immunity are known as _____ cells, while the central cells of cell-mediated immunity are known as _____ cells.

2. The cells of both antibody-mediated and cell-mediated immunity exist in great numbers in nodes of the body's fluid-drainage network, the _____.

3. T cells specific to a given invader become activated by binding to a cell whose central function is to initiate adaptive immunity by displaying microbial antigens. These are the _____ cells, the most important of which is the _____ cell.

Antibody-mediated Immunity

As noted earlier, it is B-lymphocytes or B cells that are the primary players in antibody-mediated immunity. Waiting in readiness in the lymph nodes or circulating in the bloodstream, a B cell encounters a bacterium—one that has an antigen specific to this B cell's antigen receptors. To produce numerous clones of itself that will fight this invader, this original B cell must become activated, but it does so only by binding to a player we've been seeing a lot of—a helper T cell (which has become activated against this bacterium by the process we just ran through). This activation then sparks the process you can see in **Figure 29.9** on the next page: The B cell starts dividing rapidly, giving rise to clones of itself. After a time, these B cells then differentiate into two types of cells.

One variety is called a **plasma cell:** an immune system B cell that is specialized to produce antibodies. These antibodies do not function as cell surface receptors, however. They are the free-standing antibodies, noted earlier, that leave the plasma cells by the millions—they are *exported* from the plasma cells—after which they move through the bloodstream and lymphatic system and fight the invader directly.

The other type of B cell produced, the **memory B cell,** functions like the memory T cells we looked at earlier in that it serves as a long-lasting sentry in the body. Should this invader come in a second time, these memory cells will be ready to divide and

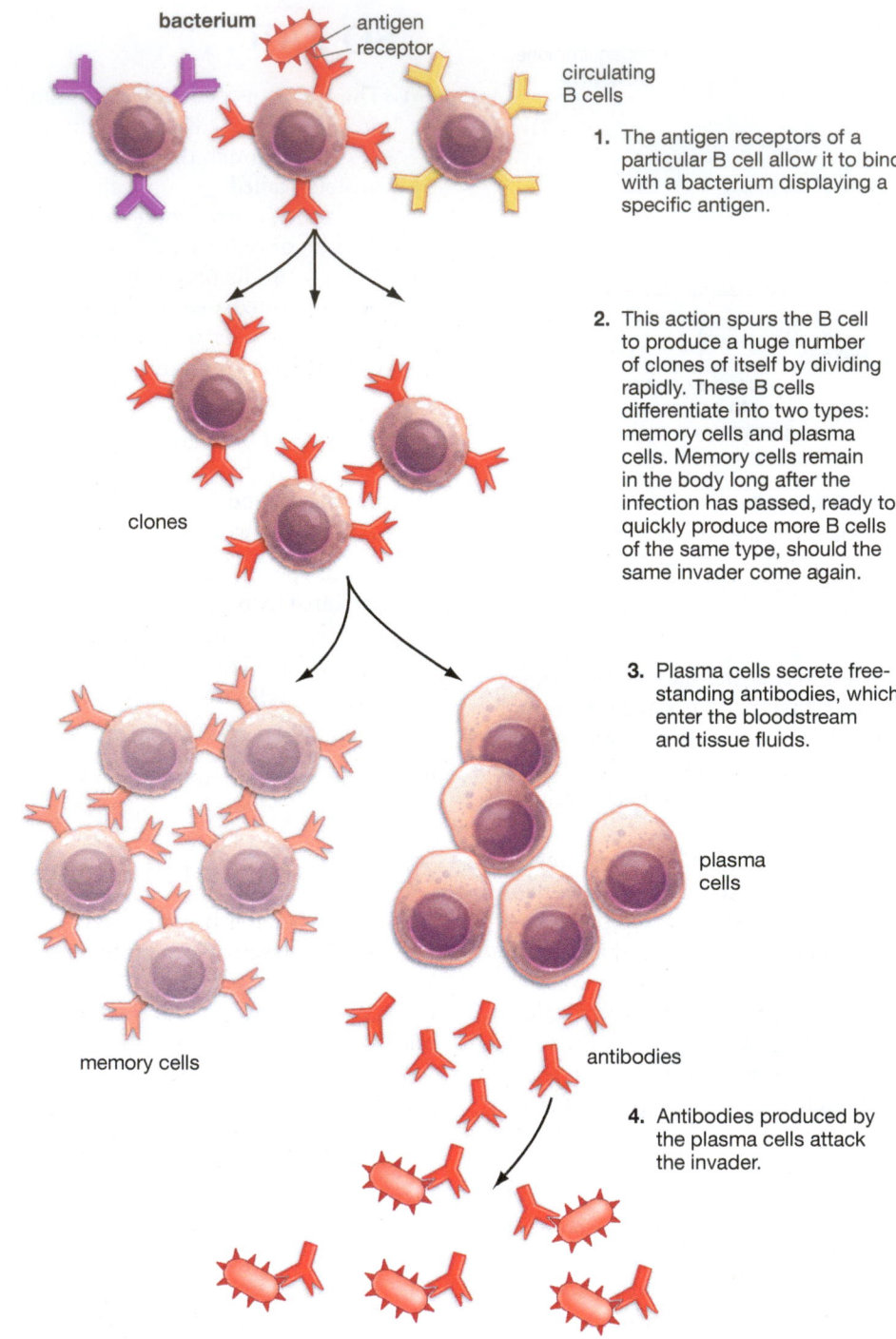

1. The antigen receptors of a particular B cell allow it to bind with a bacterium displaying a specific antigen.

2. This action spurs the B cell to produce a huge number of clones of itself by dividing rapidly. These B cells differentiate into two types: memory cells and plasma cells. Memory cells remain in the body long after the infection has passed, ready to quickly produce more B cells of the same type, should the same invader come again.

3. Plasma cells secrete free-standing antibodies, which enter the bloodstream and tissue fluids.

4. Antibodies produced by the plasma cells attack the invader.

produce more plasma cells to mount a very quick defense

The Action of the Antibodies

But what is it that the exported antibodies do to combat the invader? First, because they are binding to the invader's antigens, the antibodies can often prevent the invader from attaching to anything else. This is referred to as a *neutralization* that stops the invader's spread. Second, antibodies can come together with antigens in a clumping or *agglutination* that renders the invaders inactive. Third, antibodies work in concert with an arm of the immune system you've already reviewed—the innate defenses. Remember the cell-ingesting phagocytes and the complement proteins that you saw in innate defense? Many invading bacterial cells have a slick outer coating that keeps phagocytes from latching onto them. Working together, antibodies and complement proteins often *can* bind to many of these invaders, and this gives the phagocytes something to hold onto so that they can begin ingesting the enemy.

Figure 29.10
An Overview of the
Immune System
Process

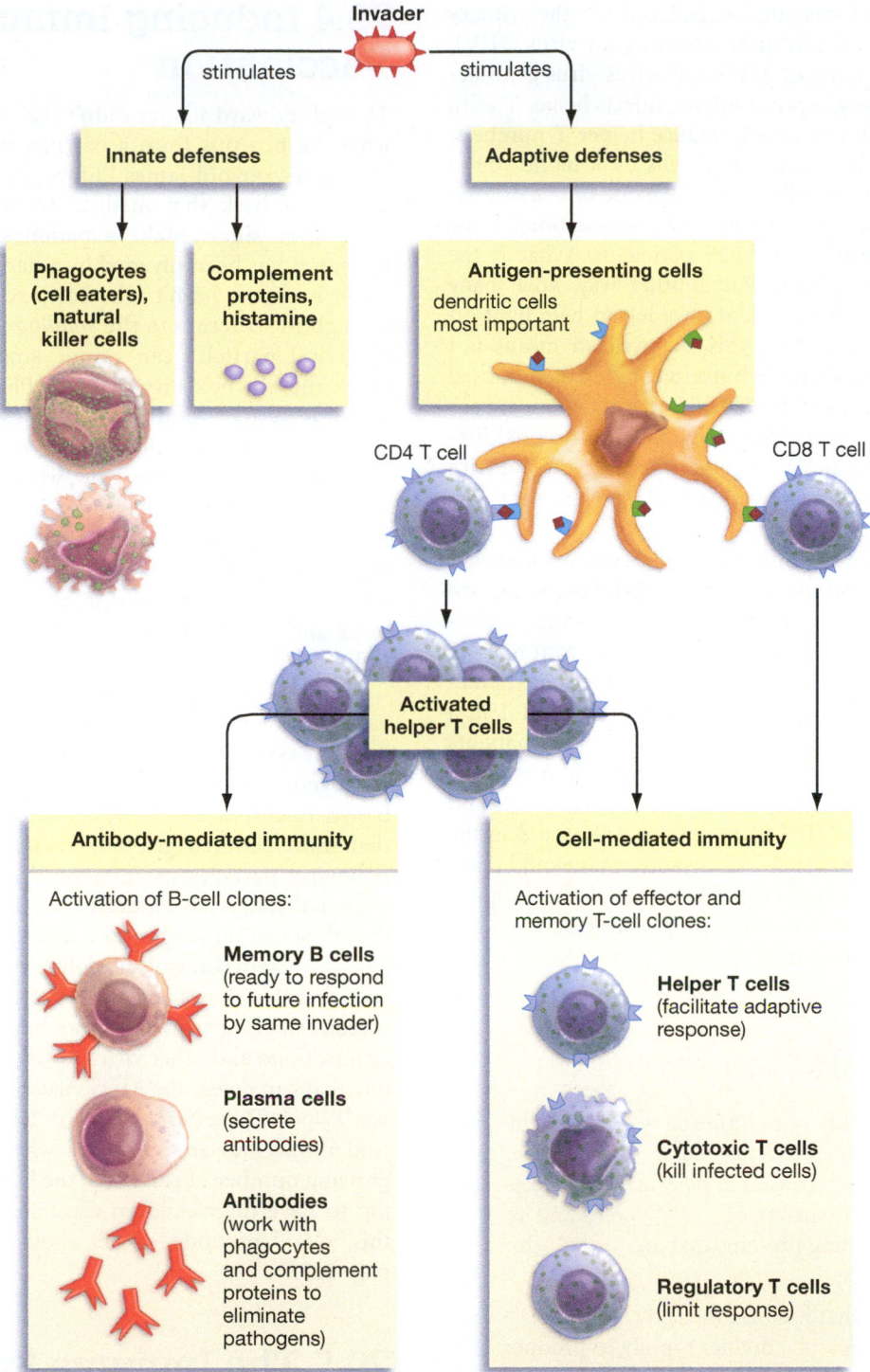

Immunity in Overview and AIDS

Taking a step back from cell- and antibody-mediated immunity, you can see, in **Figure 29.10**, how the two systems work with innate immunity to provide the complete immune response. It's important to note that, in any given microbial invasion, *all* these defenses are likely to be activated. The antibodies produced by B cells take the lead role in fighting off bacterial infections, but antibodies can work quite effectively against viruses when such viruses are not being sheltered within the body's cells. Also note in the figure something we've touched on already: how central **helper T cells** are to the immune system, in that these cells are facilitating both cell- and antibody-mediated immunity.

In recent decades, this pivotal position of helper T cells has been a matter of great practical importance because, as it turns out, helper T cells are the primary target of the human immunodeficiency virus (HIV), which is the cause of AIDS. Once this virus gains access to the body, it preferentially infects helper T cells, a process that can greatly reduce helper T numbers. Physicians refer to helper T cells by a name you've seen before: CD4 cells (from the name of one of their receptors). Not surprisingly, a critical question for any physician treating an AIDS patient is: What is the patient's CD4 count? Put another way, how many CD4 cells does this patient have left to help fight off infectious diseases? By this, a physician means not just one or two diseases, but a long list of them caused by microorganisms. A fungal infection that most people wouldn't even realize they had can become life-threatening to an AIDS patient because the patient's immune system is so weakened, largely because of a lack of CD4 cells.

In developed countries, AIDS became a manageable illness beginning in the mid-1990s, thanks to the discovery of a group of medicines that collectively have a great ability to keep HIV from making copies of itself. The problem is that these medicines cannot *eliminate* HIV from the body. About 1 in a million CD4 cells remain infected even in patients who respond well to AIDS therapy. Given this, researchers have continued to look for a compound that could keep HIV from gaining a foothold in the body in the first place—a substance that would make the body immune to AIDS. But how can a human-made substance provide a long-lasting state of immunity? Let's find out.

SO FAR . . .

1. An antibody is an immune system protein that can bind to a particular _____. Antibodies are seen in two forms: as receptors on the surface of _____ cells and as free-standing proteins that are _____ by these cells.

2. A B cell that has become activated against a given microbe divides rapidly to produce clones of two types of daughter cells: _____ cells, which produce the antibodies that will directly fight the invader, and _____ cells, which will remain in the body long after the infection has passed.

3. AIDS is devastating to the immune system because the virus that causes it has, as its primary target, the _____ cells that are central to both cell- and antibody-mediated immunity.

29.4 Inducing Immunity: Vaccination

Though Edward Jenner didn't realize it at the time, when he put pus from a cowpox patient onto the skin of 8-year-old James Phipps, he was employing a kind of trick that medical scientists have been using ever since: Make a patient's body react as though it has been invaded by a dangerous microbe when in fact it hasn't. The cowpox virus is similar enough in structure to the *smallpox* virus that cowpox viral particles can mimic smallpox antigens. Once the cowpox virus was in Phipps's body, his immune system mounted an attack against it that included the production of the memory B and T cells we looked at earlier. Now, when the *real* killer—the smallpox virus—was introduced into Phipps's body, his immune system was ready: Thanks to the memory B and T cells he had in supply, his body mounted a vigorous, rapid attack against this virus and he sailed through his smallpox exposure unharmed.

In the cowpox example, we can see what all vaccines do: They elicit readiness on the part of the immune system, generally by delivering antigens from pathogenic microbes without delivering the full-blown microbes themselves. The oral polio vaccine that most children get today contains "attenuated" poliovirus particles, meaning particles that have been altered, through lab-induced mutation, in such a way that they can no longer do harm to the body. These particles retain some of the poliovirus *structure,* however, and this allows them to serve as antigens for the actual, crippling virus. (So, if we have potent vaccines against polio and other viruses, why don't we yet have one that can defeat the AIDS virus? For more on this, see "Why Is There No Vaccine for AIDS?" on page 566. And if vaccines are so useful, why are a small but growing number of parents in the United States refusing to have their children vaccinated? For more on this, see "Unfounded Fears about Vaccination" on page 568.)

29.5 The Immune System Can Cause Trouble

In the account you've read so far, the immune system has been portrayed as a kind of multipart weapons system that the body uses to protect itself. The problem with weapons, however, is that they can get pointed in the wrong direction—toward their owners rather than their targets. And this is what happens with the immune system: The price we pay for having it is that it sometimes attacks *us* rather than the organisms that invade us.

Autoimmune Disorders

A critical factor in the immune response is the system's ability to distinguish "self" (the body's own uninfected cells or tissues) from "non-self" (antigens from other organisms). The problem is that the immune system makes mistakes; it sometimes regards self as non-self. The result is an **autoimmune disorder:** an attack by the immune system on the body's own tissues.

T cells learn to distinguish self from non-self as they develop in the thymus—they go through a selection process in which those cells that react against the body's own tissues are destroyed. A similar process of selection takes place with B cells that are developing in the bone marrow. These processes are not foolproof, however; some lines of T cells that pass the selection process in the thymus may nevertheless end up attacking the body's own tissues under special circumstances—when a new virus has entered the body, for example. Similarly, activated B cells can end up making antibodies against the body's own tissues. This is particularly likely when an invading organism presents, for example, an amino acid sequence similar to that found in the body's tissues.

The upshot is that the body can attack itself with devastating consequences. Type 1 diabetes, multiple sclerosis, lupus, and rheumatoid arthritis are just four of more than 80 chronic diseases that have an autoimmune component to them (**Figure 29.11**). In diabetes, macrophages and dendritic cells enter the tiny "islets" in the pancreas where insulin is produced, drawn there, perhaps, by the perception that a pancreatic protein called GAD is a foreign antigen. What follows is the kind of cell-mediated attack we just went over: Helper T and killer T cells migrate to the islets and the killer cells among them start killing the islet's insulin-producing beta cells through apoptosis. The end result is a body whose beta cells have largely been destroyed.

Type 1 diabetes affects equal numbers of men and women, but in general women are far more affected by autoimmune conditions than are men. About 75 percent of all rheumatoid arthritis victims are women, for example, while for lupus the figure is about 90 percent. Why should this difference exist? One hypothesis is that women have more vigilant immune systems than men because the female immune system has to do double-duty: it must protect not only women themselves but the unborn children they carry.

Allergies

The immune system is also responsible for **allergies,** which can be defined as immune system overreactions that result in the release of histamine. The so-called **allergens** that cause these reactions are simply one class of antigens. Allergens need not be living things, but they are almost always at least *derived* from living things. Thus, pollen, foods, fur, insect venom, dust

mite excrement, and penicillin (the product of a fungus), are all potent allergens, in that they all can cause the release of histamine—the compound we looked at earlier that has the effect of allowing fluids to flow more freely out of blood vessels at a site of injury.

Histamine can be released in many areas of the body, however, and when this happens the reactions that ensue can range from merely irritating to potentially life-threatening. Hay fever sufferers might experience nothing more than runny noses or watery eyes, because the pollen they are allergic to brings about a strictly *localized* release of histamine—in their noses and eyes. On the other hand, a person with a severe allergy to, for example, peanuts or insect venom can experience a *body-wide* or "systemic" release of histamine that can produce a body-wide movement of fluid out of blood vessels. When this happens, it brings on the disastrous drop in blood pressure known as shock. As if this weren't enough, systemic histamine release can cause smooth muscle to contract, and it's smooth muscle that surrounds the passages of the lungs. The result can be such an extreme narrowing of lung passages that suffocation can ensue. Taken together, this constellation of reactions is known as anaphylactic shock; thankfully, this condition can usually be brought under control with an injection of adrenaline.

Organ Transplantation

Finally, the immune system is the single most important problem in organ transplantation, for reasons that by now may seem obvious. Imagine a person who receives, say, a donated kidney via transplantation. That person's immune system will regard the new kidney as

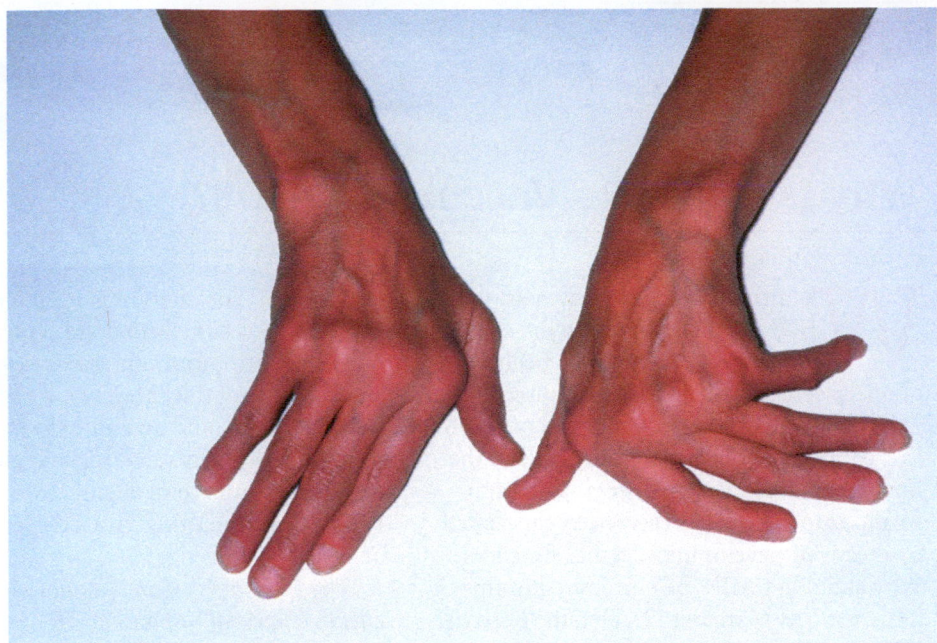

Figure 29.11
When the Body Attacks Itself
The autoimmune condition rheumatoid arthritis has deformed the joints of both hands in this patient.

ESSAY

Why Is There No Vaccine for AIDS?

When the virus now known as HIV was revealed to be the cause of AIDS in 1984, public health officials were optimistic that this discovery would quickly bring about the end of AIDS as a communicable disease. After all, by the mid-1980s, viral illnesses ranging from smallpox to polio had nearly been eliminated by means of developing vaccines for them. Why shouldn't AIDS be eliminated in this same way? As Margaret Heckler, the head of the U.S. Department of Health and Human Services, said at the time: "Yet another terrible disease is about to yield to patience, persistence and outright genius."

More than 25 years later, of course, the world is still waiting for an AIDS vaccine. And it is waiting despite the enormous amount of time and money that societies have poured into AIDS vaccine development from the mid-1980s to the present. In 2008 alone, governments, private funding agencies, and pharmaceutical firms spent an estimated $868 million on AIDS vaccine research. Some of the brightest, most talented research scientists in the world have essentially devoted their professional lives to the search for an AIDS vaccine—all to no avail.

So why have things gone this way? If science can produce vaccines for other viruses, why not one for HIV? The short answer is that HIV is like no other virus we know of. For one thing, it preferentially invades the very cells that are central to the body's adaptive immune response—helper T cells. Beyond this, however, it has a virulence, a stealth, and an ability to morph into different forms that make it an extremely resourceful enemy. How does HIV differ from other viruses? Here's a list of the ways.

The immune system does not mount an effective natural response to HIV. Remember that vaccines themselves don't defeat pathogenic invaders; they merely allow the antibodies and killer T cells of the immune system to take on these invaders more quickly than they would otherwise. In a disease such as polio, the immune system is capable of clearing the poliovirus from all the body's tissues and cells, with or without a vaccine. The only reason a vaccine is needed is that, without it, the virus does damage to the nervous system before the immune attack can be ramped up. When we turn to HIV, however, we find that, while the immune system does mount a defense against it, it never mounts an *effective* defense—one that ends up clearing the virus from the body. We know this because not a single person who has become HIV-positive has ever subsequently become virus-free, even with the aid of the suite of antiviral drugs available today.

This lack of an effective natural response to HIV is a terrific stumbling block to the development of a vaccine for it because scientists don't know which immune responses to try to boost through a vaccine. They cannot say, for example, that a given collection of antibodies is broadly effective in neutralizing HIV particles because they don't know of any group of antibodies that has this effect.

A glimmer of hope appeared late in 2009.

HIV quickly gains a foothold in the body and then can remain hidden from the immune system. In the first month after HIV enters the body, its replication is so intense that each thousandth of a liter of human blood can contain 100 million viral particles. Within weeks after infection, HIV particles have entered a large number of helper T cells and have begun to produce copies of themselves that then bud from the infected T cells in the manner you can see in **Figure 1**. Many cells infected by HIV will not be *perceived* as infected by the immune system, however, because the virus will become dormant within them: It will not carry out replication of itself, but instead will become completely inactive for a period that can last for years. This dormancy results in a host cell that does not put HIV protein fragments on its surface—the fragments that would announce

non-self and immediately begin an attack on it. This is so even though physicians will attempt to get as close a "match" as possible between kidney recipient and kidney donor. (What the physicians are trying to match up is the cell-surface recognition molecules we looked at in cell-mediated immunity.) The challenge for modern medicine has been to walk a fine line: to dampen the recipient's immune response enough so that the donated organ isn't rejected, but not to dampen it so much that the recipient is made vulnerable to serious attacks by microbes.

29.6 New Frontiers in Immune Therapy

Regulatory T Cells

In the world of medicine today, physicians who are trying to damp down overactive immune systems have only a few blunt instruments at their disposal. Drugs such as cyclosporin and prednisone provide a *general* suppression of the immune system that can leave patients vulnerable to microbial infections. Fur-to the immune system that the cell has become infected. This stealth quality of the virus means that large numbers of infected cells will persist even when T cells are going about their work of eliminating such cells. As such, any truly effective HIV vaccine would probably have to produce a collection of antibodies capable of neutralizing HIV particles *before* they infect cells, which is to say: almost immediately after they enter the body, which would be a very tall order for any vaccine.

HIV mutates at a very high rate compared to other viruses. Once it's inside a helper T cell, HIV replicates by turning its viral RNA into DNA through use of an enzyme, called reverse transcriptase, that it brings with it when it infects the cell. (For a view of this process, see Figure 21.3 on page 383.) From the perspective of HIV vaccine development, the problem with reverse transcriptase is that it makes mistakes: The viral DNA it produces in one transcription often differs significantly from the viral DNA it produces in another. This matters because the ultimate product of this viral DNA is viral *proteins*—the very molecules that infected cells put on display on their surface. As far as the immune system goes, then, HIV is constantly changing its identity; this month's virus is not next month's virus. This problem exists to some extent with all viruses. Indeed, this is the reason that different flu vaccines must be administered each year. Viral protein coats change and hence last year's vaccine will not produce enough antibodies to fight this year's virus strains.

HIV, however, presents an extreme version of this morphing problem. As AIDS researcher David I. Watkins has observed, after six years of infection, the diversity of proteins on the surface of HIV particles in a single person is thought to be greater than the diversity of all the human flu virus strains worldwide in a given year. Given this, the only likely route to developing an effective vaccine would be to identify portions of HIV that stay the same from one replication to the next *and* that can be recognized by the immune system—a goal that has so far proven elusive.

So, does all this mean that an AIDS vaccine will never be developed? Actually, a glimmer of hope appeared late in 2009 and since then scientists have been doing all they can to build on it. In a project jointly financed by health agencies in the United States and Thailand, a trial vaccine for AIDS had a positive effect—for the first time ever. About 16,000 Thai residents received either the vaccine or a placebo, after which the HIV status of members of both groups was tracked for six years. Over this time, the vaccinated group had 31 percent fewer infections than the placebo group. Impressive as this may sound, the absolute numbers were very small—51

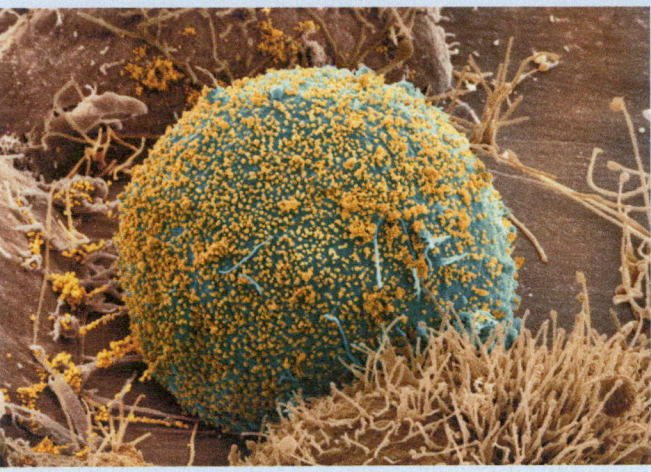

Figure 1
Budding Trouble
HIV particles, in yellow, bud from an infected helper T cell.

vaccine recipients became HIV-positive, compared to 74 placebo recipients—and even the percentage improvement barely reached the level of statistical significance (meaning what could be expected from chance alone). But modest as these results were, they gave scientists something to build on: Through what immune system routes was protection provided? What course did infection take for vaccine recipients who ended up becoming HIV-positive? What behaviors seemed most likely to lead to infection? No one thinks the vaccine used in the Thai trial will ever be employed on a mass scale. But the protection it offered means that the search for an AIDS vaccine is now moving forward with some wind in its sails.

ther, these drugs usually must be taken for long periods of time—often for the lifetime of a patient. In recent years, however, researchers have become excited at the prospect of fighting immune system overreactions through use of one of the immune system's own tools: the regulatory T cells noted earlier that have the natural function of suppressing immune system reactions.

These "T-regs," as they're called, have become one of the hottest research subjects in immunology, and it's not hard to see why. Researchers are hopeful that

T-reg therapy will one day allow physicians to suppress the immune systems of patients in ways that are both selective and long-lasting. Researchers in Italy, for example, are exposing T-regs to specific antigens outside the body and then introducing these cells into bone-marrow transplant patients. If this work goes as hoped, these T-regs will suppress only those immune system cells that are actively involved in rejecting the transplanted tissue. Rheumatoid arthritis sufferers are known to have defective T-reg functioning and there is evidence that this is the case with type 1 diabetes

ESSAY

Unfounded Fears about Vaccination

The average American under the age of 30 today is unlikely ever to have seen a case of the disease measles, but in the not-too-distant past, things were different. In the period 1958–1962, the measles virus infected half a million Americans a year, most of them young children. Cut to the year 2000, however, and only 81 cases were reported in the whole country, according to the U.S. Centers for Disease Control and Prevention (CDC). What brought this dramatic change about? A vaccine for measles; developed in the early 1960s, it was first given to American children in 1963.

Given a success story such as this, it was an unpleasant surprise to the CDC when it looked at measles figures for the first seven months of 2008 and saw that the number of cases had gone up. The numbers were still very small—only 131 cases—but they were nevertheless more than in any full year since 1997. Since a measles vaccine is still routinely given, what accounted for the increase? According to the CDC, the new cases occurred "largely among school-aged children who were eligible for vaccination but whose parents chose not to have them vaccinated."

And this in a nutshell describes a problem that is frustrating to American public health officials and, more important, dangerous to American children: A small proportion of American parents are choosing not to have their children vaccinated against common childhood illnesses. To be sure, the problem is minor in relative terms—according to the CDC vaccination rates for U.S. toddlers were at an all-time high in 2007. The fear, however, is that the measles figures indicate a growing strength in something that has

taken root in America: a movement that sees *vaccination* as a greater threat to children than infectious illness (**Figure 1**). A small proportion of American parents have been persuaded that this is the case, even though all the evidence we have says just the opposite. But evidence is not all-powerful in this issue; what also matters is the ability to understand the difference between evidence and Internet-fed speculation.

The fight over vaccination has been most intense in connection with a single vaccine—the MMR, or combined measles, mumps, and rubella vaccine—and a vaccine preservative, called thimerosal, which until 2002 was added to many childhood vaccines (though

never to MMR). The allegation is that either MMR or thimerosal-containing vaccines, or both, have been a cause of the serious disorder autism, a condition that begins in very young children and that results in lifelong difficulties with language, interactions with other people, and reactions to perceptual sensations.

The opening gun in the supposed link between the MMR vaccine and autism was a 1998 study published by a British medical journal, *The Lancet,* that looked at 12 children, nine of whom had behavioral disorders (including autism) and all of whom had intestinal disorders. The question the paper asked was: Did the MMR vaccine cause a "leaky gut" syndrome that produced circulating toxins that affected the children's brain development? Ultimately, the *Lancet* took the highly unusual step of retracting the paper after the British General Medical Council found that its lead author had

Figure 1
Up in Arms about Vaccines
Protesters at an anti-vaccine rally.

acted "dishonestly and irresponsibly" in carrying out the research that led to it, but in 1998 all that mattered was that scientists had asked a question about a vaccine and autism. As a result, MMR vaccine rates plummeted in Great Britain and even today remain below their 1998 levels. And these lowered rates have persisted even though all the evidence that has accumulated *since* 1998 says that there is no association between MMR and autism. Several studies, for example, have found no connection between autism and intestinal disorders while another found no difference in autism rates between children who had received the MMR vaccine and those who hadn't.

Meanwhile, the alleged link between thimerosal and autism got a boost in 1999 when the American Academy of Pediatrics (AAP) and the U.S. Public Health Service both recommended removing thimerosal from vaccines as a precautionary measure, owing to the chemical similarity between ethyl mercury (a type of mercury found in thimerosal) and the well-known toxin methyl mercury. The AAP later noted that "No scientific data link thimerosal used as a preservative in vaccines with any pediatric neurologic disorder, including autism," but the organization's call for removing thimerosal from vaccines was enough to unnerve parents across the United States.

Faced with this public uncertainty, the prestigious U.S. Institute of Medicine (IOM) convened a panel of experts in 2001 to look at all the evidence regarding vaccine safety. In the years that followed, the panel issued eight reports. By the time the last of these was published, in 2004, the panel had looked at a large body of epidemiological evidence—that is, evidence regarding which children have become autistic and under what circumstances—and on this basis rejected the idea that either the MMR vaccine or thimerosal-containing vaccines

cause autism. In carrying this analysis out, the panel reviewed the findings of the two primary U.S. scientists who claim to have found a link between autism and vaccination, but concluded that the methodology these researchers used is so flawed that their results are "uninterpretable."

None of this has mattered to the anti-vaccine forces, some of whom see, in the IOM and other reports, a government-aided conspiracy aimed in part at protecting the profits of pharmaceutical firms who manufacture vaccines. But in the years since the last IOM report, evidence has continued to pour in, all of it pointing in the same direction. In 2008, for example, researchers from California's Department of Public Health published figures showing that in the 12 years from 1995 to 2007, autism rates increased continuously in children throughout the state, even though thimerosal had been *removed* from all standard childhood vaccines by 2002. If this preservative had been causing autism, the prevalence of this disorder should have declined between 2002 and 2007, but just the opposite was true.

Given all this, why are substantial numbers of parents in the United States and Great Britain still choosing to keep their children away from at least some vaccines? The root of the problem seems to be a misunderstanding about vaccination's risks and rewards.

Many American parents get information about vaccination from exactly the source you'd expect—the Internet—and a study published in 2008 showed that, in more than 90 percent of cases, parents can be expected to encounter messages *against* vaccination among the top 10 hits in any Internet search they conduct on the topic. Faced with the competing claims they see on a computer screen, many parents may find it difficult to separate out scientific evidence from uninformed opin-

ion. As such, some parents may have the idea that, in the vaccination decision, they are being asked to choose between a major risk and a minor reward. The major risk they perceive is that their child stands a substantial chance of becoming autistic through vaccination. The minor reward they perceive is that, by getting vaccinated, their child will avoid some relatively harmless childhood illnesses. If both these things were true, who in their right mind would have a child vaccinated?

In reality, however, *both* perceptions are wrong—not just the risk perception, but the reward perception as well. With respect to risk, there is no substantial risk of vaccination bringing about autism in children; indeed, researchers have been able to identify no risk at all among healthy children. (No one, however, could ever say that the risk is zero; human beings are too biologically diverse for that.) With regard to the rewards of vaccination, the "minor" disease of measles hospitalizes up to 20 percent of its victims today, while the "minor" disease of whooping cough used to strike virtually all children in the United States, and, prior to the advent of a vaccine for it, contributed to as many as 9,000 deaths in a single year, according to the CDC. Add up all the common childhood diseases, and all their effects, and the rewards of getting vaccinated turn out not to be so minor.

Ironically, one of the problems parents have in perceiving these rewards is the very change that vaccines have brought about. Thanks to universal vaccination, a given set of parents in America today may never have seen a single case of measles, with its fever and rashes, or a single case of whooping cough, with its prolonged, wracking coughing spells. You might say that these parents don't know what they're missing. The question public health officials are asking today is whether things can stay this way.

victims as well. In a disease such as diabetes, the hope is that T-reg therapy might be applied early enough in the onset of the condition that a patient's insulin-producing capability could be preserved. Most of the research being done on T-regs today concerns their basic biology—how they develop, how they interact with other immune cells, and so forth. But all of this work has the practical goal of giving physicians a set of tools to use in damping down immune system reactions that are doing harm rather than good.

Toll-like Receptors

A second major avenue of therapeutic immunological research today concerns a set of immune system players we reviewed at the start of the chapter: The toll-like receptors, or TLRs, that serve both to detect microbial invasions and to channel the immune response down different pathways, depending on the class of invader that's been detected. So rich is the potential of TLRs that they have sparked a kind of gold rush among pharmaceutical firms. By 2007—a mere nine years after scientists *discovered* TLRs in humans—some 30 drugs that target TLRs had made it into various stages of testing. And the roster of afflictions these drugs are aimed at fighting is a long one: Asthma, lupus, stroke, hepatitis C, food allergies, and at least seven kinds of cancer are on the list.

TLRs have the ability to affect this wide array of conditions because they can potentially work both sides of the immune system street. Drugs that *stimulate* TLR responses could potentially evoke powerful, directed immune system attacks, while drugs that *reduce* TLR responses could potentially dampen immune system overreactions.

To see how this works, consider three potential uses for drugs that affect just one TLR—TLR-7, which is one of the 10 that human beings have. The normal function of this receptor is to bind to single strands of RNA that help make up a class of viruses that includes HIV and the influenza pathogens. When this happens, signals set in motion by TLR-7 result in the release of a set of cytokines that either kill the cells these viruses are harbored in or mobilize the rest of the immune system to do so. Hence, a drug that evoked a TLR-7 response could have a therapeutic role in fighting viral infections, perhaps by strengthening vaccines against them. Next, the cytokines that are released with TLR-7 binding turn out to be effective in fighting some kinds of cancer; thus, clinical trials are underway aimed at seeing whether TLR-7 activation could be useful in shrinking or eliminating some types of cancerous tumors. Finally, the body's *own* RNA will sometimes gain access to TLR-7 and bind to it, thus setting off a harmful autoimmune reaction. Indeed, we now have reason to believe that lupus sufferers are the victims of this kind of faulty TLR-7 response. Thus, some types of autoimmune conditions might benefit from drugs that can *reduce* TLR-7 binding in affected tissues.

The discovery of TLRs has not yet resulted in a single new drug being put in the hands of physicians, and there is an enormous difference between pharmaceutical potential and a working drug. But the prospects for TLR-based drugs seem bright indeed, given the ability of these receptors to both initiate and channel immune responses. In the next few years, we should know whether it's possible to fight disease by leveraging the power of these first responders of the immune system.

On to Transport and Exchange

Having looked at the complex capabilities of the immune system, it's time to move on to reviews of two other systems that keep the body functioning. Chapter 30, on the circulatory and respiratory systems, is concerned with two subjects that almost seem synonymous with life itself: blood and breathing.

SO FAR . . .

1. Vaccines elicit _____ on the part of the immune system by introducing _____ from a pathogenic microbe without introducing the microbe itself.

2. The immune system can cause trouble in three primary ways: by initiating attacks on the body's own tissues, thus causing the conditions known as _____; by initiating the histamine-releasing overreactions known as _____; and by serving as the primary impediment to _____.

3. Two promising means of using components of the immune system to fight disease have begun to develop in recent years. One of them employs the cells whose natural function is to limit immune system attacks, _____; the other stimulates or suppresses the _____ that are the initial sensors of microbial infection.

Summary

29.1 The Immune System in Overview

- The primary function of the immune system is to find and kill invasive microbes, primarily bacteria, fungi, and viruses (p. 551).

- Any substance that sets off an immune system response is an antigen. Antigens generally are molecules that are part of invading microbes or that are produced by them. (p. 552)

- The body's first line of defense against microbial attack is not the immune system but a set of barriers that keep pathogenic microbes from gaining access to human tissue. Foremost among these is human skin; other barriers include the acidity of the stomach lining, the flow of urine down the urethra, and the low pH of the female reproductive tract. (p. 552)

- The immune system works through two separate but overlapping responses: the innate immune response, which attacks any substance perceived to be foreign, and the adaptive immune response, which produces a set of cells and proteins specific to each invader and which has the ability to remember each pathogen it has faced and hence to mount a rapid response to any subsequent invasion by it. (p. 553)

29.2 The Innate Immune Response

- The innate immune response can operate within the context of inflammation. Each innate response is set off by the binding of a microbial antigen to the receptor of an immune system cell. Such binding often is carried out by a macrophage, which is one member of a class of immune system cells, the phagocytes, that have the ability to ingest invaders. (p. 553)

- The primary sensors of microbial infection are the toll-like receptors (TLRs) that extend from macrophages and other innate immune cells. These receptors channel both the innate and adaptive immune responses down different pathways, depending on the class of microbe that has entered the body. An innate immune cell whose TLRs have bound to an antigen will release proteins called cytokines that either fight the invader directly or that mobilize the immune system against it. (p. 554)

- One of the innate immune cells mobilized by this signaling is the mast cell, which releases the substance histamine, which brings about blood vessel dilation and increased blood vessel permeability. Several varieties of phagocytes will move to the site, along with a group of about 30 complement proteins that can kill invading cells directly. The site of attack will be sealed off during this time, as a means of keeping the infection from spreading. (p. 555)

29.3 The Adaptive Immune Response

- The adaptive immune response operates through two arms. The first is antibody-mediated immunity, which works through proteins, called antibodies, that fight invaders found free-standing in the body's tissues. The second is cell-mediated immunity, which works by producing cells that destroy the body's own cells when they have become infected. The central cells of antibody-mediated immunity are the body's B-lymphocytes or B cells; the central cells of cell-mediated immunity are the body's T-lymphocytes or T cells, which come in several forms, among them CD4 cells, which will become helper T cells, and CD8 cells, which will become killer T cells. (p. 556)

- Both T and B cells can be found in large numbers within the body's lymphatic network, a system of vessels that captures fluids leaked from the bloodstream; that then subjects these fluids to scrutiny by immune system cells located in lymph system nodes; and that then returns the fluids to the circulatory system. (p. 556)

- The fantastic specificity of adaptive immunity rests on the ability of immune system genes to produce B-cell and T-cell receptors specific to nearly every type of pathogenic microbe the body can encounter. (p. 558)

- Cell-mediated immunity begins when a dendritic cell ingests, or is invaded by, a pathogenic virus, after which the dendritic cell will display fragments of the virus on its surface. The dendritic cell then migrates to a lymph node and acts as an antigen-presenting cell: It presents the viral fragments on its surface to CD4 and CD8 T cells that have receptors specific to the invader. The CD4 and CD8 T cells that bind to the dendritic cell then become activated helper T and killer T cells. They divide rapidly, producing clones of themselves that move through the body to fight the invader. These clones include not only effector T cells, which fight the current infection, but memory T cells, which remain in the body long after the infection has passed, thus allowing the body to mount a stronger response to a subsequent attack by the invader. Helper T cells also help activate a third type of T cell, the regulatory T cell, which functions to limit immune responses. (p. 559)

- Antibody-mediated immunity begins when a B cell with receptors specific to a given invader binds to it, after which this B cell will become activated by binding to an activated helper T cell. The B cell will then produce clones of itself of two different types: plasma cells, which are specialized to first produce and then export from themselves free-standing antibodies that will fight the invader; and memory B cells that provide the immune system with a long-lasting ability to mount a rapid defense against this same invader. The antibodies exported by plasma cells can render pathogenic bacteria inactive in several ways. (p. 561)

- The human immunodeficiency virus that is the cause of AIDS is devastating to the immune system because it primarily infects the helper T cells that facilitate both cell-mediated and antibody-mediated immunity. (p. 563)

29.4 Inducing Immunity: Vaccination

- Vaccines work by putting the immune system into a state of readiness. They usually consist of antigens from an infectious microbe without containing the active microbe itself. When injected into the body, these antigens result in the body mounting a full-blown attack on the invader—one that produces memory B and T cells specific to the invader. (p. 564)

29.5 The Immune System Can Cause Trouble

- The immune system can make mistakes in distinguishing "self" from "non-self." Such mistakes can result in autoimmune disorders. More than 80 chronic diseases have an autoimmune component to them. (p. 564)

- An allergy is an immune system over-reaction that results in the release of histamine. Such histamine release can be localized and relatively harmless but can also be systemic, causing anaphylactic shock. (p. 565)

- In organ transplantation, the transplant recipient's immune system may identify a transplanted organ as non-self and thus may attack it as if it were an invading microbe. (p. 565)

29.6 New Frontiers in Immune Therapy

- Scientists are investigating the possibility of suppressing immune system overreactions through the use of the regulatory T cells that function naturally to limit immune system reactions. (p. 566)

- Scientists have begun to develop drugs that work by activating or limiting responses from the toll-like receptors that serve to detect microbial invasions. Drugs that stimulate TLR responses could potentially evoke powerful, directed immune system attacks, and drugs that reduce TLR responses could potentially dampen immune system overreactions. (p. 570)

Key Terms

adaptive immune response   553

allergen   565

allergy   565

antibody   558

antibody-mediated immunity   556

antigen   552

antigen-presenting cell   559

autoimmune disorder   565

B-lymphocyte (B cell)   556

cell-mediated immunity   556

cytotoxic T cell (killer T cell)   560

helper T cell   563

histamine   555

inflammation   553

innate immune response   553

lymphatic network   556

memory B cell   561

phagocyte   554

plasma cell   561

protective immunity   553

regulatory T cell   561

T-lymphocyte (T cell)   556

toll-like receptors (TLRs)   554

Understanding the Basics

Multiple-Choice Questions

(Answers are in the back of the book.)

1. An antigen is:
   a. any immune system protein.
   b. any substance that elicits an immune system response.
   c. any cell that functions in the immune system response.
   d. the ultimate product of a vaccination.
   e. a particular kind of lymphocyte.

2. The pathogens that the immune system fights generally are (select all that apply):
   a. poisons.
   b. living things.
   c. microbes.
   d. produced by the body itself.
   e. fungi, bacteria, or viruses.

3. Associate each immune system component on the left with the description of it on the right.
   a. mast cells — proteins that seal off infection site
   b. cytokines — cells capable of ingesting invading cells
   c. toll-like receptors — cells that release histamine
   d. phagocytes — communication proteins
   e. fibrin proteins — first sensors of infection

4. All T and B cells begin development in the _____, but T cells complete development in the _____.
   a. lymphatic network; bone marrow
   b. lungs; pancreas
   c. pancreas; thymus
   d. bone marrow; thymus
   e. lymphatic network; thymus

5. The body's adaptive immune response differs from its innate immune response in that the adaptive response (select all that apply):
   a. targets specific pathogens.
   b. is faster to ramp up.
   c. has no memory of the pathogens it has faced.
   d. works primarily through T and B cells.
   e. is more powerful than the innate response.

6. Antibodies are seen in two forms in immune system functioning, as _____ and as _____.
   a. a length of DNA in a macrophage; a hormone
   b. an antigen receptor on a T cell; a free-standing protein exported by a T cell
   c. an antigen receptor on a B cell; a free-standing protein exported by a B cell

d. antigen receptors on macrophages; signaling proteins inside macrophages

e. proteins that can kill invaders directly; signaling proteins

7. Cell-mediated immunity works primarily against ___ and primarily through ____.
   a. viruses; mast cells
   b. bacteria; B cells
   c. bacteria; T cells
   d. fungi; B cells
   e. viruses; T cells

8. In autoimmune disorders, the immune system (select all that apply):
   a. attacks some microbes, but not others.
   b. uses only its innate defenses.
   c. fails to distinguish self from non-self.
   d. fails to produce memory cells.
   e. attacks the body's own tissues.

Brief Review

(Answers are in the back of the book.)

1. Describe the main actions that take place as part of inflammation in the innate immune response.

2. In the adaptive immune response, what is the primary function of the dendritic cell?

3. HIV infection can lead to several forms of cancer, including Kaposi's sarcoma and a type of lymphoma; it can lead to a kind of dementia and a physical wasting; it makes its victims prone to several kinds of fungal infections that are rarely seen in the general population. How can one virus have this many effects?

4. Why is the immune system an obstacle to organ transplantation?

5. What element of the adaptive immune response allows vaccines to protect human beings from many serious illnesses?

Applying Your Knowledge

1. When smallpox raged in England in the 18th century, most of its victims were young children. Meanwhile, the people least likely to die from it were senior citizens. Why should this have been so?

2. Public health officials agree that, since the 1960s, asthma has been increasing among children in developed countries such as the United States. Among the tentative explanations for this increase is the "hygiene hypothesis," which holds that the problem is that many children are growing up in environments that are "pristine"—very clean and devoid of household pets. What could the linkage be between an immune system condition such as asthma and growing up in a pristine environment?

3. One of the key treatments for children who lack an immune system is a bone marrow transplant. Why would this procedure help to establish an immune system in a person who does not have one?

Transport and Exchange 1:

Blood and Breath

Our bodies have a respiratory system that acquires oxygen and gets rid of carbon dioxide, a digestive system that breaks down food, and a urinary system that gets rid of some materials while recycling others. Tying all these systems together is another system, the circulatory system.

Available on MasteringBiology®

Activity: The Human Cardiovascular System; Circulatory Systems; The Cardiovascular System; Three-Chambered Hearts; Two-Chambered Hearts; The Respiratory System; **Bioflix:** Gas Exchange; **Current Events:** In Treatment for Leukemia, Glimpses of the Future; **Building Vocabulary; Questions:** Bioflix Quiz, Reading Questions, Test Bank

To carry out any activity, human beings must first acquire oxygen and then move that oxygen to cells throughout the body—tasks that are handled by the human respiratory and cardiovascular systems.

The cells that make up our bodies may be tiny things, but they are still living things, which means they have their own ongoing set of needs. They have to obtain energy and get rid of wastes; they have to obtain oxygen; they have to obtain nutrients such as nitrogen and phosphorus. When life consisted of nothing but single-celled creatures, every cell had access to the world outside itself, which meant that food could enter, waste could be expelled, and gases could be "exchanged" in a fairly simple manner. In a human being, conversely, many of our 10 trillion cells spend their entire existence wrapped deep inside layers of tissue.

During the course of evolution, this access problem, as it might be termed, was solved by the development of physiological *systems*: a special bodily apparatus for extracting nutrients from food, another for delivering oxygen, and another for removing waste. We call the system that extracts nutrients the digestive system, the one that delivers oxygen the respiratory system, and the one that removes waste the urinary system. If you think about it, what all these systems have in common is that they carry out various exchanges of materials between us and our surroundings. They do this, however, only with the help of one additional system, the cardiovascular system, which uses the material called blood to pick up gases, nutrients, and wastes from one location in the body and transport them to another. The cardiovascular system is thus central to our other transport and exchange systems in that it serves as a kind of mass-transit system for them. (How does oxygen get from our lungs to, say, our feet? It travels through the cardiovascular system in blood cells.)

Given the linking role of the cardiovascular system, we'll begin our review of transport and exchange by looking at it, after which we'll review the respiratory system. Then, in Chapter 31, we'll review the digestive and urinary systems. While we're looking at the digestive system, we'll review human nutrition. **Figure 30.1** summarizes the basic functions of the four systems we'll be looking at this chapter and next.

30.1 The Cardiovascular System

The human **cardiovascular system** is a fluid transport system that consists of the heart, all the blood vessels in the body, the blood that flows through these vessels, and the bone marrow tissue in which red blood cells are formed. This system bears a rough comparison with a car's cooling system. A car uses antifreeze and water as its circulating fluids; the cardiovascular system uses

MB

The Human Cardiovascular System

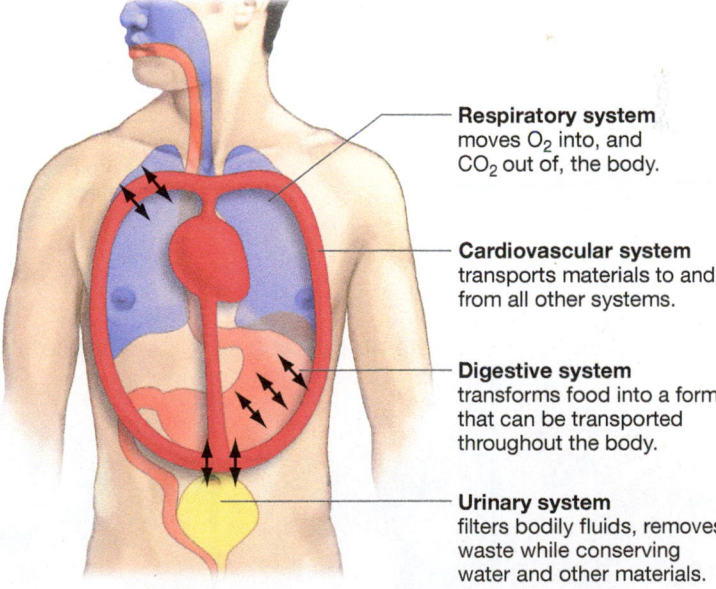

Respiratory system moves O_2 into, and CO_2 out of, the body.

Cardiovascular system transports materials to and from all other systems.

Digestive system transforms food into a form that can be transported throughout the body.

Urinary system filters bodily fluids, removes waste while conserving water and other materials.

Figure 30.1
The Transport and Exchange Systems of the Body
The human body has four systems that are active in transport and exchange: the cardiovascular, digestive, urinary, and respiratory systems. The cardiovascular system is central to all the others in that it transports materials to and from them.

blood. A car may use a water pump to circulate its fluids; the cardiovascular system uses a pump called the heart. A car has an assortment of conducting pipes; the cardiovascular system has blood vessels. The similarities end there, however, because, while a car's cooling system transports only a couple of things, the cardiovascular system transports lots of things. It moves nutrients, vitamins, waste products, hormones, and immune system cells and proteins, not to mention the oxygen we breathe in and the carbon dioxide we breathe out. All these materials are transported in the substance we'll look at first within the cardiovascular system—blood.

30.2 The Composition of Blood

Collect a sample of blood from a human being, add a substance that will prevent it from clotting, and spin it in a centrifuge, and it will separate into two layers.

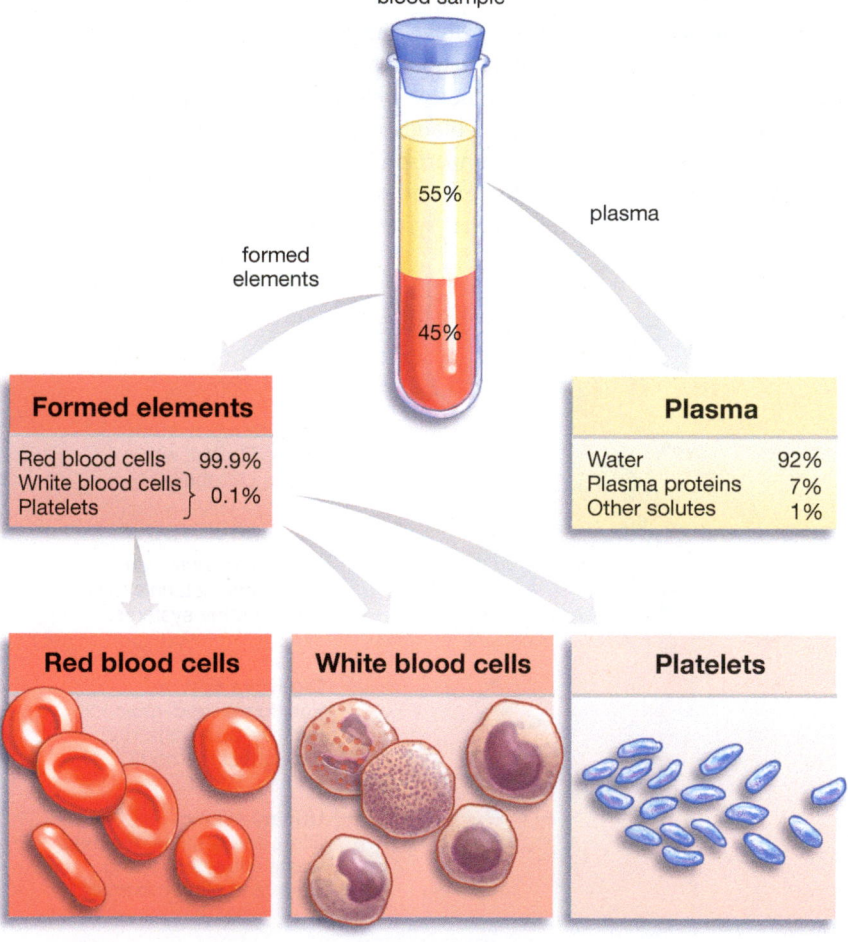

Figure 30.2
The Composition of Blood
Blood that is spun in a centrifuge separates into its two primary constituent parts: formed elements and plasma. Formed elements are cells and cell fragments that help make up the blood—red blood cells, white blood cells, and platelets. Blood plasma, the liquid in which the formed elements are suspended, is composed mostly of water but also contains proteins and other materials.

The heavier, lower layer, accounting for some 45 percent of its volume, consists of blood cells and cell fragments, which collectively are known as the **formed elements** of blood (**Figure 30.2**). The lighter, yellowish, upper layer, accounting for the rest of the volume, is called **plasma,** meaning the fluid portion of blood. Within an actual blood vessel, the formed elements are suspended in the flowing plasma just as tiny bits of sediment might be suspended within a moving stream.

Formed Elements

It turns out there are three kinds of formed elements in the plasma. The most numerous by far are the **red blood cells,** which transport oxygen to, and carbon dioxide from, every part of the body. Then there are **white blood cells,** which are critical players in the immune system reviewed last chapter. Finally, there are **platelets**—small fragments of cells that are important in the blood-clotting process (**Figure 30.3**). Let's now look at each of these elements in more detail.

Red blood cells, or *erythrocytes,* are red because within them is the iron-containing protein **hemoglobin.** Molecules of hemoglobin that lie within a red blood cell (RBC) bind with oxygen as it comes in from the lungs and then hold onto it, releasing it only when the cell reaches a part of the body whose oxygen concentration is relatively low. RBCs then work with blood plasma in picking up the carbon dioxide that cells have produced through their metabolism; they then carry this CO_2 back to the lungs, where it will be exhaled.

Blood is deep red because RBCs constitute 99.9 percent of all the formed elements in it. Indeed, the *number* of RBCs in an average person staggers the imagination. One cubic millimeter of blood—an amount that might scarcely be visible—contains 4.8 to 5.4 million RBCs. So numerous are these cells that they make up about a third of all the cells in the human body.

Apart from their numbers, RBCs are also notable for another quality, which is their odd structure. In their mature state, they essentially consist of a cell membrane that surrounds a mass of hemoglobin. Viewed another way, an RBC has almost none of the standard cellular equipment—a nucleus, other organelles, and so forth. This lack of cellular machinery means that the average RBC life span is only 120 days, and this means that about 180 million new RBCs have to enter the circulation *each minute* for us to have enough of them in our system.

The second formed element in blood, the white blood cells or *leukocytes,* have a completely different function from red blood cells. They have standard cellular equipment, but they are not in the oxygen or

Figure 30.3
Blood's Formed Elements
All three varieties of the formed elements present in blood are visible in this color-enhanced micrograph. The bright-red cells are mature erythrocytes, which carry oxygen to all parts of the body. Three types of the immune system's white blood cells also are visible: a lymphocyte (in blue), a neutrophil (the large yellow cell), and a macrophage (in green). The smaller yellow objects are the cell fragments, called platelets, that help seal off sites of injury by facilitating the formation of blood clots.

CO_2 transport business. Instead, in their many varieties, they are central to the body's immune system operation. A typical cubic millimeter of blood contains 6,000 to 9,000 white blood cells. This is a tiny number relative to the millions of red blood cells in such a volume, but the comparison is somewhat misleading because most white blood cells (WBCs) are found in the body's tissues rather than in the bloodstream. WBCs use the bloodstream primarily to *get to* sites of invasion or injury.

The third formed element, platelets, can be thought of as enzyme-bearing packets. They are not cells but fragments of cells that have broken away from a set of unusually large cells that exist in the bone marrow. The enzymes in them aid in blood clotting at sites of injury. Apart from releasing these enzymes, the platelets clump together at the site of an injury, forming a temporary plug that slows bleeding while the process of clotting continues.

Blood's Other Major Component: Plasma

The formed elements you've just read about are suspended in blood's other major component, plasma, which is 92 percent water. Proteins and a mixture of other materials are dissolved in this fluid as well. Why aren't these proteins and other materials "formed elements"? Well, note that they are not cells, or even cell fragments, but are instead much smaller molecules that go into making up a living thing.

There are three primary kinds of proteins in the plasma. First, there are the albumins, made in the liver, which function in the transport of both hormones and the components of fats known as fatty acids. Second, there is fibrinogen, also made in the liver, which aids in blood clotting. Third, there are globulins, which come in two forms. One is the immunoglobulins or antibodies that you saw so much of in Chapter 29, which attack foreign proteins and disease-causing organisms. The other is the transport proteins that carry small ions, hormones, or other compounds.

Looking strictly at transport proteins, we find that one variety of them has a real importance to our hearts. These *lipoproteins* essentially are capsules of protein that surround globules of fatty material—usually cholesterol. Lipoproteins are best known by the names of their two varieties: **low-density lipoproteins** (or **LDLs**), which carry cholesterol *to* outlying tissues from the liver and small intestine; and **high-density lipoproteins** (or **HDLs**), which carry lipids *from* these tissues to the liver. Keep both LDLs and HDLs in mind because you'll be reading about them later in connection with heart attacks.

Taken together, the plasma proteins plus water make up 99 percent of blood plasma. The other 1 percent is a mixture of hormones, nutrients, wastes, and ions (which are better known as electrolytes). As you can imagine, wastes that are dissolved in the plasma are on their way out of the body altogether. For example, a substance called urea, produced by the liver's breakdown of proteins, is carried within the plasma to the kidneys, is filtered out by them, and is then urinated out of the body.

30.3 Blood Vessels

Having looked at what moves through the cardiovascular system, let's now look at the plumbing that carries these materials—our network of blood vessels.

Blood vessels are tubes, in a sense. To get a real idea of them, however, imagine a regular tube, such as a straw, that could change its internal diameter from second to second, that could take in some materials along its length while keeping others out, and that could repair itself when damaged. Because blood vessels are composed mostly of cells, they are dynamic entities that change their activities in accordance with the body's needs. Blood vessels carrying blood *away* from the heart are **arteries,** while the vessels carrying blood back *to* the heart are **veins.** The farther from the heart these vessels are, the smaller they tend to be. Connecting the arteries to the veins are the smallest blood vessels of all, **capillaries.**

Figure 30.4
The Structure of Blood Vessels
Both arteries and veins have three layers to them: an inner epithelium, a layer of smooth muscle, and a layer of connective tissue. By contracting or relaxing, the smooth muscle can change the diameter of the vessels. Arteries and veins are linked by intervening capillary beds. Each capillary consists of a single layer of cells.

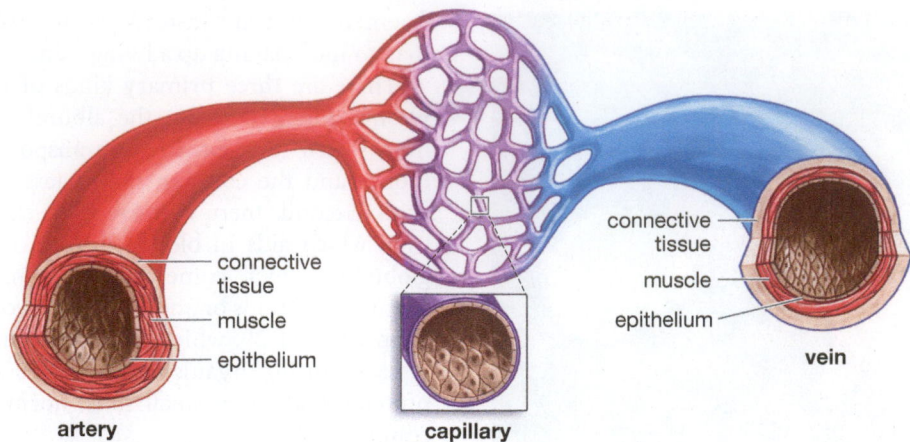

connective tissue
muscle
epithelium
artery

connective tissue
muscle
epithelium
vein

capillary

Circulatory Systems

Arteries and veins are always made up of three distinct layers of tissue (**Figure 30.4**). The innermost layer is composed of a group of flat cells, which you may remember from Chapter 26 are *epithelial cells*—cells that come into contact with materials from outside the body, such as oxygen and nutrients. The middle layer of arteries and veins contains smooth muscle. These are muscles that allow the blood vessel to constrict or expand in diameter. This very ability has health consequences because the dangerous condition called hypertension or high blood pressure is brought about by a *persistent* constriction of the body's smaller arteries. (As the diameter of these arteries decreases, the pressure within them increases.) The outer layer of arteries and veins is a stabilizing sheath of connective tissue. Taken as a whole, these three layers give arteries and veins a great dual capability: strength combined with flexibility. The largest arteries have diameters of up to 2.5 centimeters, or about an inch, but farther out from the heart, the body's smallest arteries, the arterioles, have diameters that average 30 micrometers (μm), which is about a thousandth of an inch.

Tiny as this diameter is, things get a lot smaller yet with the body's other main variety of blood vessel, the capillaries. Their diameters average about three *ten-thousandths* of an inch. Why so minuscule? Their central function is to have substances pass into them and out of them along their length. As such, their walls need to be very thin. Accordingly, each capillary is made of a single layer of cells. So small are capillaries that tiny red blood cells must move through them in single file.

SO FAR . . .

1. Blood is composed of formed elements, which are cells and cell fragments, and plasma, which is the _____ portion of blood.

2. There are three kinds of formed elements: _____, _____, and _____. Plasma, meanwhile, is overwhelmingly composed of _____.

3. Arteries carry blood _____ the heart, while veins carry blood _____ the heart. The tiny blood vessels called _____ connect arteries to veins.

30.4 The Heart and Blood Circulation

What makes blood flow through these vessels? The answer is the muscular pump we call the heart. A small organ, roughly the size of a clenched fist, the heart actually is shaped something like a valentine heart in its major outline. It lies near the back of the chest wall, directly behind the bone known as the sternum.

The heart sends blood out, and then gets it back, through two circulation loops. As noted earlier, one major task of the blood transport system is to get oxygen to all the various bodily tissues. For blood to carry oxygen to tissues, however, it first has to *take up* the oxygen that is brought into the body through the lungs. So imposing is this task of oxygenating blood that one of the heart's circulation loops is devoted solely to getting blood to the lungs and then back to the heart. The oxygenated blood that comes to the heart in this loop is then pumped out to the rest of the body in the heart's second major loop (**Figure 30.5a**). To put these things more formally, there are two general networks of blood vessels in the cardiovascular system. One of these is the **pulmonary circulation,** in which the blood flows between the heart and the lungs. Then there is the **systemic circulation,** in which the blood is moved between the heart and the rest of the body. Let's see how this two-network system works.

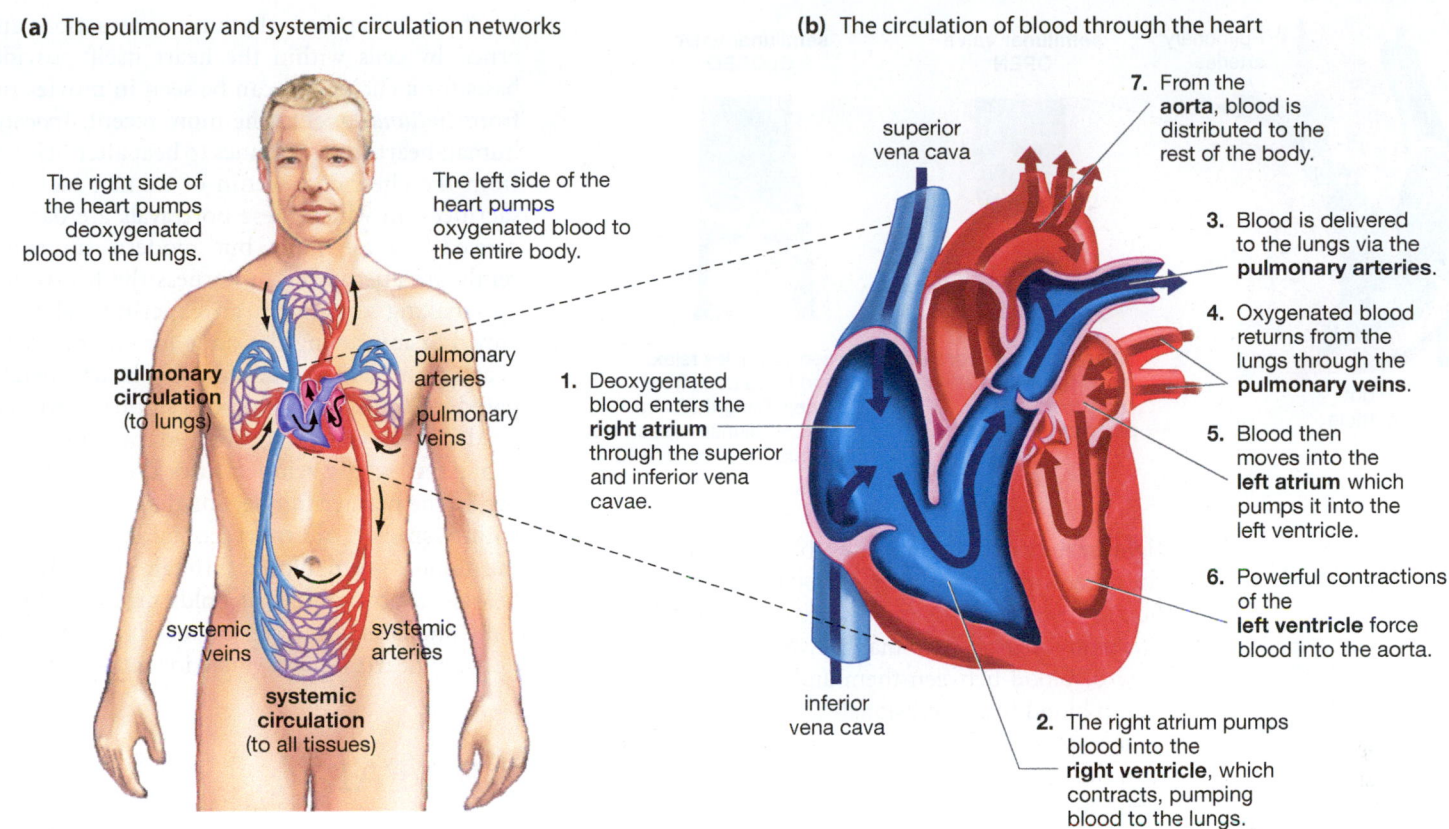

(a) The pulmonary and systemic circulation networks

The right side of the heart pumps deoxygenated blood to the lungs.

The left side of the heart pumps oxygenated blood to the entire body.

pulmonary circulation (to lungs)

pulmonary arteries

pulmonary veins

systemic veins

systemic arteries

systemic circulation (to all tissues)

(b) The circulation of blood through the heart

superior vena cava

inferior vena cava

7. From the **aorta**, blood is distributed to the rest of the body.

3. Blood is delivered to the lungs via the **pulmonary arteries**.

4. Oxygenated blood returns from the lungs through the **pulmonary veins**.

5. Blood then moves into the **left atrium** which pumps it into the left ventricle.

6. Powerful contractions of the **left ventricle** force blood into the aorta.

1. Deoxygenated blood enters the **right atrium** through the superior and inferior vena cavae.

2. The right atrium pumps blood into the **right ventricle**, which contracts, pumping blood to the lungs.

Following the Path of Circulation

The human heart contains four muscular chambers, two associated with the pulmonary circulation and two associated with the systemic circulation. These are the right atrium and right ventricle (pulmonary circulation) and the left atrium and left ventricle (systemic circulation). Let's take as the starting point the blood entering the right atrium from the veins called the superior and inferior venae cavae (on the *left* side of **Figure 30.5b**). These veins are returning blood to the heart from the upper and lower portions of the body, respectively. This is blood that is coming back after distributing oxygen and other blood-borne materials throughout the body—to our legs, our hands, our trunk, and so forth. As such, it is deoxygenated blood, which actually is dark red in color but which is indicated by the blue color of the veins it flows through. After the right atrium receives this blood, it pumps it the short distance to the right ventricle, which contracts, pushing the blood into the pulmonary arteries. (These vessels are arteries instead of veins because blood moves *away* from the heart through them.) This blood is now bound for the lungs, where it will pick up oxygen. Once it does this, it returns to the heart through the two pulmonary veins, which empty into the left atrium. The blood then is pumped down into the left ventricle, which contracts and sends the blood coursing up into the enormous artery called the **aorta**. This vessel has

branches stemming from it that will carry the blood to all the tissues of the body.

Taking a step back from this, you can see that the right side of the heart pumps blood only to the lungs, while the left side of the heart pumps blood to all the body's tissues, as shown in Figure 30.5a. All four chambers of the heart pump blood, but the right and left *atria* pump blood only into another portion of the heart—their adjacent ventricles. The right ventricle is pumping blood to the lungs, so there is some greater contraction power needed there. But the *left* ventricle is pumping to the whole body, and its contractions are by far the most powerful. When we "take our pulse," what we are feeling is the surge of blood produced by the contraction of the left ventricle. (For an account of how our pulse figures in the process of measuring blood pressure, see "Listening in on Blood Pressure" on page 582.)

Valves Control the Flow of Blood

For the heart to work optimally, the flow of blood through it has to be a *one-way* flow—it would not do for blood to back up into, for example, the right ventricle following one of its contractions. But how is this one-way flow insured? The answer is that there are valves between the ventricles and the arteries they pump blood into. These *semilunar valves,* as they're called, open up fully under the force of blood being pumped out of the ventricles, but they make a

Figure 30.5
Two Circulatory Networks and How the Heart Supports Them

(a) Veins in the systemic circulation bring blood into the right side of the heart; it is pumped out via the pulmonary circulation to the lungs, where it is oxygenated, and then returns to the left side of the heart. It is then pumped out from the left side of the heart, through the systemic circulation, to the rest of the body.

(b) The path of the blood's circulation through the four chambers of the heart.

The Cardiovascular System

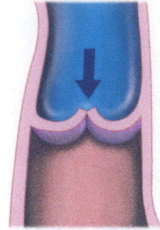

Pulmonary arteries

Right ventricle

semilunar valve **OPEN**

When ventricles contract, blood flows up against semilunar valves, pushing them open.

semilunar valve **CLOSED**

When ventricles relax, blood flows back from arteries, filling the cusps of the semilunar valves and pushing them closed.

Figure 30.6
Controlling Blood Flow

A series of valves keeps blood flowing in one direction through the heart's four chambers. Pictured is the semilunar valve that keeps blood from backing up into the heart's right ventricle from its pulmonary arteries. A second semilunar valve serves this function in the heart's left ventricle, and two atrioventricular valves serve this function in the heart's atrial chambers.

Three-Chambered Hearts

Two-Chambered Hearts

tight seal that will not allow blood to move back the other way (**Figure 30.6**). As you might imagine, the heart's two atrial chambers also have valves—the *atrioventricular valves*—that likewise ensure a one-way flow of blood between them and the ventricles they pump blood into. Occasionally, because of disease or genetic predisposition, people do get some backflow into the atria. The fluid turbulence that results can be heard as a kind of gurgling sound, known as a "heart murmur" (which is not necessarily a sign of a serious heart problem). The heart's two sets of valves are responsible for the familiar "lub-dub" sound of the heartbeat. The "lub" is the sound of the valves closing between the atria and ventricles, while the "dub" is the sound of the valves closing between the ventricles and the arteries.

The Heart's Pacemaker

You might think that the pace of the heart's contractions would be regulated by signals sent from the body's nervous system, but the human heart beats in a rhythm that is fundamentally controlled by a small set of the heart's own cells. The wall of muscular tissue that surrounds the right atrium has a so-called **sinoatrial node** whose cells function as the heart's pacemaker. Though the cells in this node are muscle cells, they function like nerve cells in that they have the ability to generate and transmit electrical signals. The signals they produce are carried throughout the heart along pathways of other muscle cells that likewise are specialized to transmit electrical signals. These conducting cells then transmit their signals to the *contracting* muscle cells of the heart, which do exactly what their name implies in the following order: Cells in the atria contract first, sending blood down into the ventricles that lie beneath them, after which cells in the ventricles carry out a wave of contraction that starts at their base and moves upward, thus sending blood coursing up into the arteries that lie above them.

The fact that these rhythmic contractions are governed by cells within the heart itself provides the basis for a cliché that can be seen in movies ranging from *Indiana Jones* to the more recent *Apocalypto*: a human heart that continues to beat after being pulled from the chest of a victim of human sacrifice. The sacrifices on which these portrayals are based aren't carried out anymore, but modern surgeons currently are attempting to harness the heart's internal pacemaking abilities in an experimental technique aimed at improving the odds for success in human heart transplants. Traditionally, hearts being prepared for transplant have been stopped within brain-dead donors and then removed and transported in containers whose temperatures are kept near freezing. In the new technique, donated hearts are shipped within special transport containers that circulate the donors' own blood through their still-beating hearts—a procedure that holds promise of substantially increasing the period of time during which donated hearts remain viable for transplant.

SO FAR . . .

1. There are two networks of blood vessels in the body. In one of these, the pulmonary circulation, blood flows between the _____. In the other, the systemic circulation, blood flows between the _____.

2. The heart's left atrium and left ventricle function in _____ circulation, while the heart's right atrium and right ventricle function in _____ circulation.

3. Cells in the heart's sinoatrial node function as the heart's pacemaker thanks to their ability to generate _____.

30.5 What Is a Heart Attack?

The heart is a unique organ, but like any other organ, it needs a supply of blood and the oxygen and nutrients that come with it. Indeed, the heart is a voracious user of these materials because it essentially is a set of muscles that never stops working. Blood does not come to the muscles of the heart, however, through any of the chambers you've been reviewing. (Think about the two ventricles: one is sending blood to the lungs, the other to the rest of the body but not directly to the heart itself.) The way blood does come to the heart is through two arteries that branch off from the aorta just after it emerges from the left ventricle.

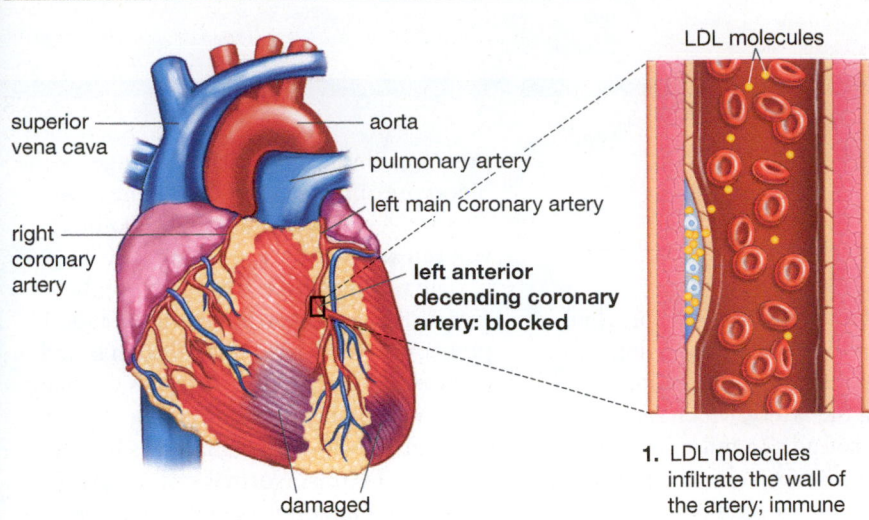

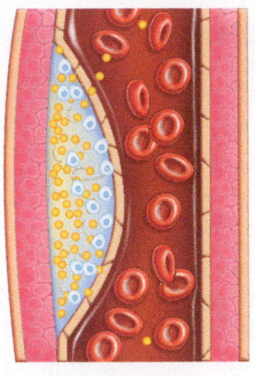

LDL molecules

left main coronary artery

left anterior decending coronary artery: blocked

superior vena cava

aorta

pulmonary artery

right coronary artery

damaged heart muscle

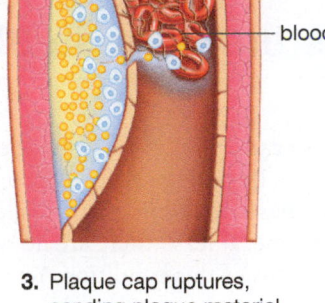

blood clot

1. LDL molecules infiltrate the wall of the artery; immune cells follow.

2. An inflammatory reaction follows; a growing number of cells and cellular debris form a plaque.

3. Plaque cap ruptures, sending plaque material into bloodstream; a blood clot forms that blocks blood flow.

Figure 30.7
Critical Vessels
The coronary arteries begin with a right and left coronary artery that branch off from the aorta. These arteries then undergo branching themselves, thus supplying blood to the whole heart. Blockage of the left anterior descending coronary artery figures in almost half of all heart attacks. The magnification of this artery shows the series of steps that lead to most heart attacks. A buildup of oxidized LDL molecules within the inner artery wall combines with a resulting blood clot to bring about a complete blockage of the artery.

Because these arteries encircle or "crown" the heart before they start branching, they are known as the **coronary arteries** (**Figure 30.7**).

It is difficult to overstate the importance of these arteries because nearly 20 percent of all deaths in the United States today are caused by a blockage of one or more of them. Such blockages generally have, as their starting point, the LDLs or low-density lipoproteins, noted earlier, that carry cholesterol molecules away from the liver and small intestine to varying locations in the body. One place these LDLs go is the coronary arteries. There, they can lodge in a space that develops just outside the layer of epithelial cells that forms the inner lining of all arteries.

Once present in significant numbers, these LDLs can become damaged through oxidation (the same process that causes metal to rust), and this is where the real trouble starts. The immune system now regards these damaged LDLs as *foreign invaders* and thus initiates a complex attack against them. What follows is the inflammatory response that readers of Chapter 29 are familiar with. Inflammation that would take place with, say, a puncture wound causes tissue to become flushed and swollen. The same thing is happening here, only this tissue is swelling into the space through which blood flows in a coronary artery, thus beginning a blockage of it. Some immune cells taking part in this attack become engorged with fatty material from the LDLs, thus transforming themselves into so-called foam cells, and cellular debris from the immune attack collects in this space as

well. The collective term for all these materials is plaque.

Formation of this plaque merely sets the stage, however, for the actual trigger to a heart attack: A "cap" that lies over the plaque ruptures, and with this, plaque material that had been sealed off from the blood flowing through the artery now comes into contact with it. The immune system reacts to this material as it would to a foreign invader and as such takes an action that normally would protect the body but that now does it great harm: It forms a blood clot whose function is to seal the area off. Now there is a blood clot combined with a swollen cap, and the result can be a *complete* blockage of a coronary artery—a **heart attack.** The effect of this blockage is that groups of heart muscle cells no longer have a blood supply, which kills them if it goes on for long enough. This can bring death to the heart attack victim in several ways. The heart muscles may no longer be able to contract with sufficient force to pump blood through the body, or a wild, irregular rhythm, called a ventricular fibrillation, can be set off. Even if the victim survives, the damaged heart muscle tissue will not regenerate; this person will have a weakened heart for life.

The good news about all this is that there are things we can do to lessen the likelihood that this sequence of events will get started. A key concept is to lower the number of LDLs we have circulating within us while raising the number of the HDLs, or high-density lipoproteins, mentioned earlier. The LDLs, as you've

ESSAY

Listening in on Blood Pressure

Because high blood pressure manages to be both harmful and stealthy, it has often been called a "silent killer." For a time, at least, there are no outward symptoms of this condition. Yet as high blood pressure progresses, it is making its victims vulnerable to some very serious illnesses, among them heart attacks and strokes. Is it any wonder, then, that we can expect to get our blood pressure checked almost any time we see a physician?

You may have wondered, however, what physicians are listening for when they take our blood pressure. Here's the story. The tightening of the cuff around our upper arm collapses a major artery—the brachial artery—whose blood pressure is being measured. This ensures that no blood is getting through the artery, which means that there is no sound of a pulse coming through the stethoscope. After the physician starts to release pressure on the cuff and blood starts to flow, the point at which the physician can first hear a pulse marks our systolic blood pressure—the highest pressure we have, brought about by the contraction of our heart's left ventricle. The point at which the sound of the pulse disappears marks our diastolic blood pressure—our lowest blood pressure. Physicians watch the gauge on the blood pressure device and note the blood pressure levels at the two sound points, as you can see in **Figure 1**.

The readings from these two points give us the well-known, two-figure blood pressure levels (as in 120/80). High blood pressure—or hypertension, as it's officially known—can mean that either figure is elevated above a level regarded as healthy. What level is this? Normal systolic pressure for adults is anything below 120, while normal diastolic pressure is anything below 80. People are then regarded as being at risk for high blood pressure if their systolic level stands between 120 and 139, or if their diastolic level stands between 80 and 89. Actual high blood pressure is any systolic reading of 140 or higher or any diastolic reading of 90 or higher.

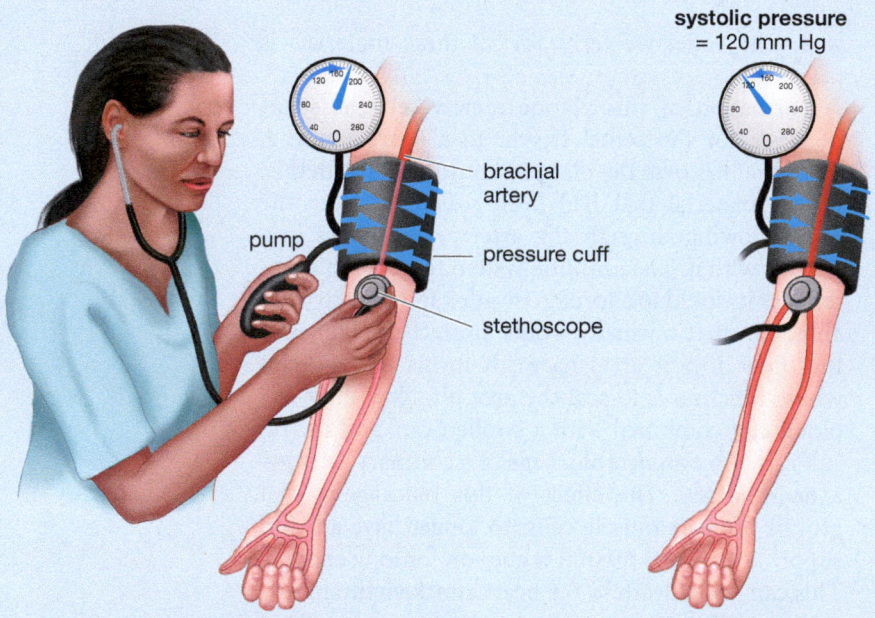

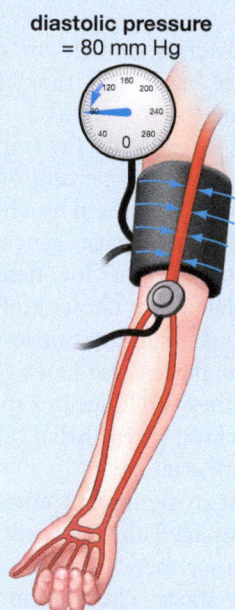

systolic pressure = 120 mm Hg

diastolic pressure = 80 mm Hg

brachial artery

pump

pressure cuff

stethoscope

1. While listening with a stethoscope, the physician inflates the cuff around the patient's arm, tightening it to a level that prevents blood from moving through the patient's brachial artery.

2. While slowly releasing pressure on the cuff, the physician listens for the sound of a pulse. The point at which a pulse can first be heard marks the patient's systolic or highest blood pressure. This is the point at which the patient's blood pressure has overcome the pressure being applied by the cuff.

3. As pressure on the cuff continues to drop, the sound of a pulse begins to fade. The point at which a pulse can no longer be heard marks the patient's diastolic or lowest blood pressure.

Figure 1
How Physicians Measure Blood Pressure

seen, lodge in the coronary arteries, setting off the immune response. The HDLs, conversely, not only remove these LDLs (bringing them back to the liver) but may also carry enzymes that break down the LDLs that have become damaged through oxidation—the LDLs that set off the immune response. So, how do we lower LDLs and raise HDLs? Exercise can do both things, and losing excess weight does both things. Meanwhile, eating saturated (animal) fat raises LDL levels, and smoking may damage the LDLs in ways that set off the immune response. It's worth noting that fatty LDL deposits can begin forming in coronary arteries from childhood forward. Heart attacks may be an affliction of the middle-aged and elderly, but heart disease prevention can be practiced at any age.

30.6 Distributing the Goods: The Capillary Beds

Having looked at the players involved, you're now ready to see how the cardiovascular system actually carries out its central functions of bringing materials to tissues and taking materials away from them. Propelled by the heart through the aorta, blood flows through the tube-like arteries. These large-diameter arteries fork repeatedly, gradually decreasing in size until they become the smaller-diameter arterioles. These then branch into the delivery vehicles of the cardiovascular system, the capillary beds. If you look at **Figure 30.8a**, you can see a representation of one of these beds. Note that at one end, the bed extends from the body's arterial system, but that at the other it feeds into the body's system of veins—the venous system that brings blood *back* to the heart. Looking at the venous system as it begins with the capillaries, you can see that it consists of its smallest vessels, the tiny venules, which then merge into veins. Smaller veins eventually merge into larger ones, which finally come together into the largest veins of all, the two vena cavae that feed directly into the heart.

Thus, blood comes "out" through the arteries and "back" through the veins, but the real action in the circulatory system takes place in the capillary beds that lie between the arteries and the veins. As you've seen, capillaries are very thin. Think now about a line of oxygenated red blood cells, suspended in plasma and moving in single file through a capillary along with hormones, glucose, ions, and other materials needed by the cells *outside* the capillary (**Figure 30.8c**). The oxygen, glucose, and all the rest move out through the capillary wall and into the interstitial fluid—the liquid in which both the capillary and the cells around it are immersed. Once within this fluid, these needed materials will make their way to nearby cells. The distance they must travel to the cells is not far, however. So extensive is the body's capillary network that no cell is farther than two or three cells away from a capillary.

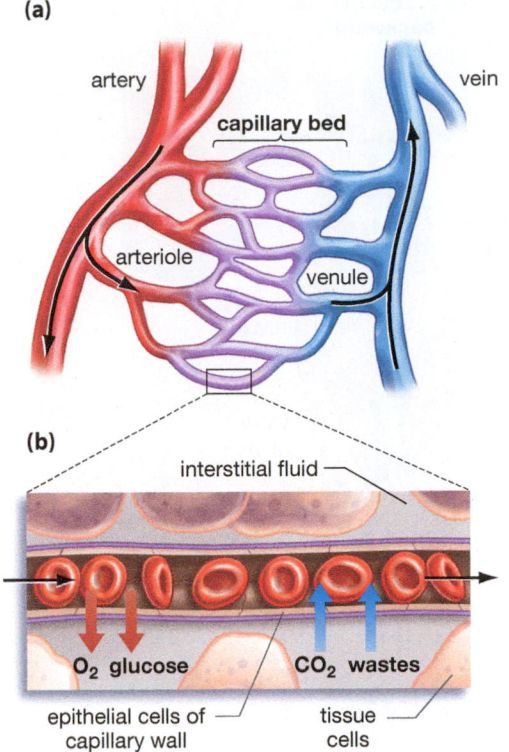

(a)

(b)

(c)

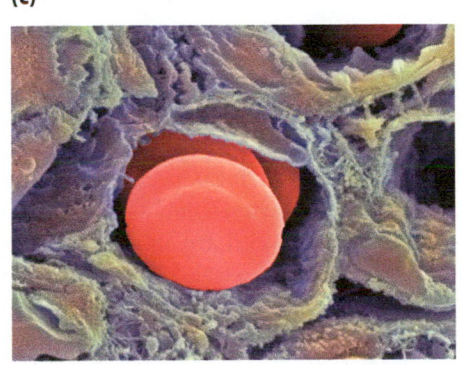

Figure 30.8
Cardiovascular System Delivery Vehicle
Capillary beds are the sites of the exchange of materials between blood vessels and the body's tissues.

(a) An idealized view of how blood flows from arteries to arterioles, through capillary beds, and then into the venules and veins that return it to the heart.

(b) Oxygen and nutrients diffuse out of capillaries and into the surrounding interstitial fluid, on their way to nearby cells, while carbon dioxide and wastes move from these cells, through the fluid, and into the capillaries.

(c) Micrograph of a cross section of a capillary, with red blood cells moving through in single file.

Forces That Work on Exchange Through Capillaries

Red blood cells themselves are too big to pass out of the capillaries, but the oxygen within them diffuses across the capillary lining into the interstitial fluid. Nutrients such as glucose and small ions such as sodium likewise are distributed in this manner. Meanwhile, carbon dioxide and wastes are diffusing *into* the capillaries. The movement of all these substances is aided by concentration gradients, meaning these substances are moving from an area of their higher concentration to an area of their lower concentration. To look at oxygen, for example, there is more of it inside a capillary than outside of it. As a result, it will move "down" its concentration gradient and leave the capillary. (For more on concentration gradients and diffusion, see Chapter 5, page 97.)

Water flows in and out of capillaries as well, propelled by a couple of forces that oppose each other.

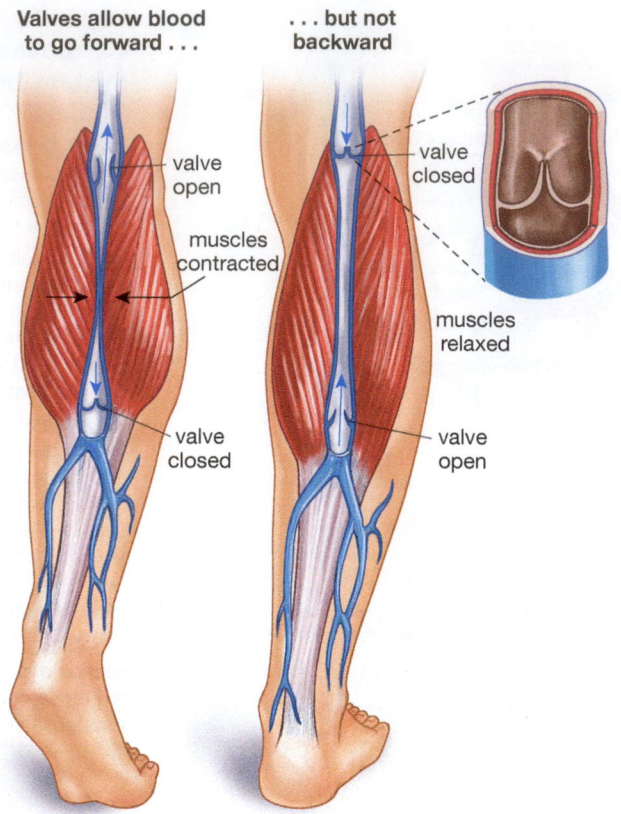

Valves allow blood to go forward . . .

. . . but not backward

valve open

valve closed

muscles contracted

valve closed

muscles relaxed

valve open

Figure 30.9
One-Way Flow to the Heart
Skeletal muscle contraction is the primary force that drives blood back to the heart through the venous circulation. A system of valves guards against the backflow of this blood.

The Respiratory System

It tends to flow out of capillary beds at their arterial end but into the beds at their venous end. The force at work at the arterial end is blood pressure, which results from the heart's contractions. This pressure is relatively strong at the arterial end—strong enough to push water out of the capillary, in a flow that brings with it some suspended materials. But the narrowness of the capillaries means that blood pressure is largely spent by the time blood approaches the venous end of a capillary bed. At that point, another factor, osmotic pressure, tends to move materials back into the capillary. (Proteins too big to leave the capillary result in an osmotic pressure that "pulls" on fluids. You can read more about osmosis in Chapter 5, on page 98.) At the venous end of a capillary, then, osmosis overcomes the force of blood pressure. Most of the water that leaves the capillary beds at the arterial end is returned to them at the venous end. Some remains as interstitial fluid, however, to be picked up by the body's other group of fluid vessels, the lymphatic vessels we looked at last chapter.

Muscles and Valves Work to Return Blood to the Heart

Because blood pressure has dropped to such low levels by the time the blood gets to the venous system, you may wonder how it can get back to the heart. The answer is that our skeletal muscles do double duty. In contracting, they squeeze the veins in a way that moves the venous blood along. Blood can only move

toward the heart in this system because the veins have a series of valves in them that block movement of blood away from the heart (**Figure 30.9**).

30.7 The Respiratory System

As you've seen, one of the things the cardiovascular system transports is oxygen and another is carbon dioxide. Now it's time to look at the bigger picture of how these substances are exchanged between our cells and the air around us.

People can live perhaps a couple of months without food and generally a few days without water. But if they go 5 or 6 minutes without breathing, death is usually the result. This is because oxygen is in the energy transfer business; without oxygen, our cells simply don't have enough energy to function. When oxygen is present, however, cells carry out their work and *generate* carbon dioxide as a result, and this CO_2 has to be disposed of. Thus do we see the two central functions of breathing: The first is capturing and distributing oxygen; the second is disposing of carbon dioxide. Apart from these tasks, breathing also helps balance the pH level in our bloodstream and allows us to talk by providing the air that passes over the vocal cords. All of these tasks are technically part of **respiration,** meaning the exchange of gases between the atmosphere outside the body and the cells within it.

Structure of the Respiratory System

The lungs are but one part of the respiratory system, which you can see in **Figure 30.10a**. This system also includes the nose, nasal cavity, and sinuses, along with the pharynx (upper throat), the larynx (voice box), and the trachea (windpipe). In addition, there are air-conducting passageways: the left bronchus and right bronchus and the many small air passageways they branch into, the **bronchioles,** which deliver air to the lungs.

If we ask what the lungs themselves are, they are made up mostly of tiny, hollow sacs that sprout in grapelike clusters from the end of each bronchiole.

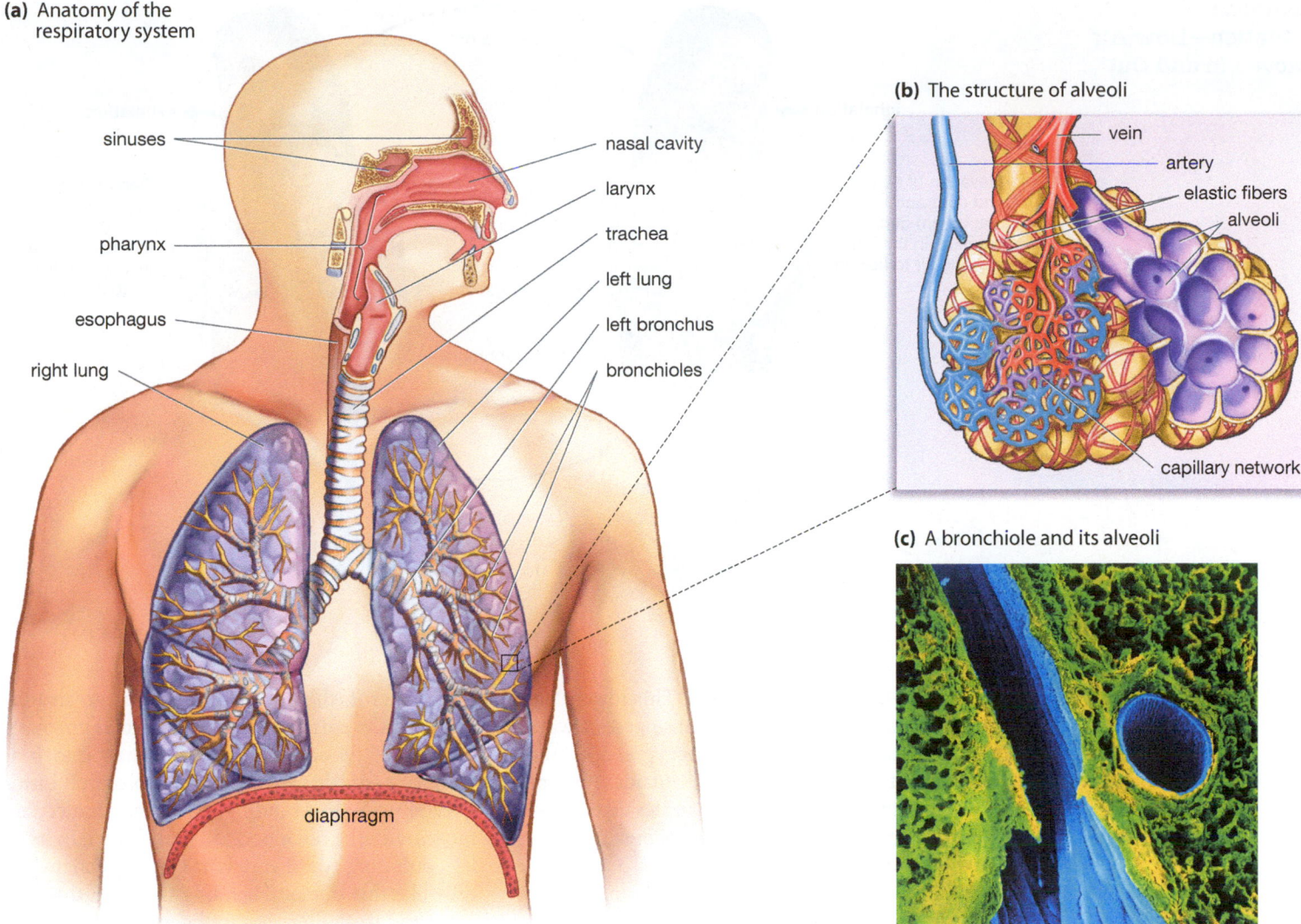

(a) Anatomy of the respiratory system

sinuses
pharynx
esophagus
right lung
diaphragm

nasal cavity
larynx
trachea
left lung
left bronchus
bronchioles

(b) The structure of alveoli

vein
artery
elastic fibers
alveoli
capillary network

(c) A bronchiole and its alveoli

Figure 30.10
Anatomy of the Respiratory System

(a) The large-scale components of the human respiratory system.

(b) In this magnification of the alveoli, the capillary network that all alveoli are surrounded by has been cut away, on the right, to reveal the structure of the alveolar sacs.

(c) A cutaway view of a human lung, artificially colored to highlight its features. The blue tube running from the top is a bronchus, while the blue circle at right is a pulmonary blood vessel. Surrounding both structures is spongy lung tissue (in green) composed of alveolar sacs whose walls are lined with capillaries.

These sacs are the **alveoli,** the air-exchange chambers of the lungs. There are about 300 million of them in a typical set of lungs—a number that gives you an idea of just how tiny these sacs must be. The dual quality of being small and numerous gives alveoli what they need to get their job done, which is *surface area*. What does being small and numerous have to do with surface area? Imagine a cube of Jell-O, 1 inch on each side. Its surface area is 6 square inches. Divide this cube in half, however, and the Jell-O's surface area goes to 8 square inches even though its volume—the amount of space it takes up—stays the same. Repeat this kind of dividing millions of times, as with our alveoli, and the results are startling: The alveoli in a typical set of lungs have a surface area about the size

of a tennis court. If you look at **Figure 30.10b**, you can see how this surface area is used: to give the alveoli and the capillaries that cover them a huge area in which to exchange oxygen and carbon dioxide.

30.8 Steps in Respiration

First Step: Ventilation

The first step in respiration is breathing or **ventilation,** meaning the physical movement of air into and out of the lungs. To understand ventilation, look at the drawing, in **Figure 30.11** on the next page, of the old-fashioned "accordion bellows" that can be used to get a fire going. Pull the bellows handles apart, and the

Figure 30.11
Ventilation—How Air
Is Moved In and Out

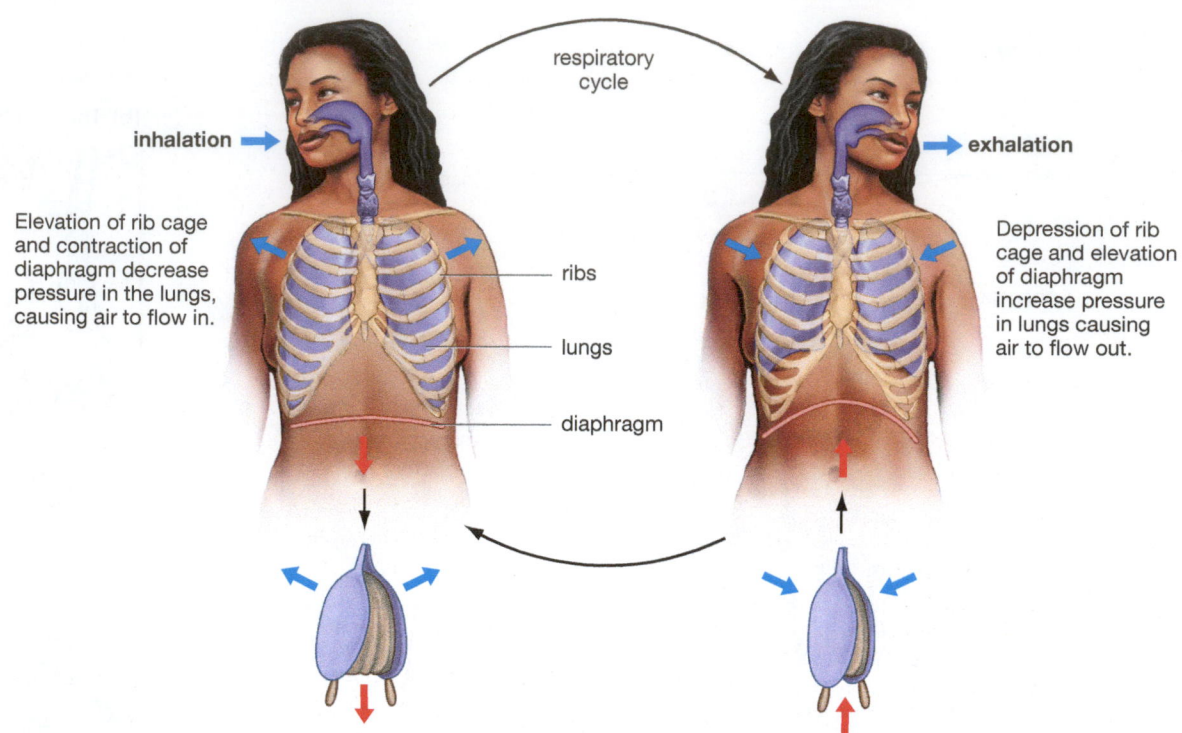

volume inside the bellows increases. This lowers the air pressure inside the bellows. Because gases always flow from areas of higher to lower pressure, the air flows into the bellows from the outside. Reverse this process—squeeze the bellows handles together—and the air pressure inside becomes higher than the pressure outside, meaning the air flows out of the bellows. Similarly, we have a space inside our chests (called the thoracic cavity) whose volume can increase or decrease. The contraction of a muscular sheet called the diaphragm causes the ribs to rise and, with this, the thoracic cavity expands like the inside of the bellows. Air then rushes in. Exhaling in this kind of "quiet" ventilation is a more passive process. When the diaphragm relaxes, the ribs drop due to the combined force of gravity and a natural elasticity the lungs have. As a result, the thoracic cavity shrinks, and air is expelled. During the heavy breathing that takes place during exercise, exhalation is aided by a second set of chest muscles whose contractions help shrink the thoracic cavity. (For information on a disease that decimates the lungs' natural elasticity, see "When Breathing Becomes an Effort: Emphysema" on the next page.

Next Steps: Exchange of Gases

Once oxygen is taken into the lungs, it then moves into the bloodstream. How does it get there? Remember that the sac-like alveoli have capillaries covering them. Think of the interface between a capillary and an alveolus as a border that can be easily crossed. An oxygen molecule is inhaled into an alveolus and then diffuses across the thin alveolar tissue into an adjacent capillary. On entry, this oxygen molecule encounters a passing red blood cell, which it diffuses into, binding with one of the hemoglobin molecules inside it. Locked up by this binding, the oxygen molecule then moves with the red blood cell to the heart. In its ceaseless pumping, the heart is taking oxygenated blood cells like this one and sending them up through the aorta and from there out to the rest of the body. Moving through this route to a tissue somewhere in the body, the red blood cell drops off the oxygen molecule to be used.

This drop-off occurs once again by way of diffusion: The oxygen flows out of a capillary and into the interstitial fluid that surrounds it. Other body cells are immersed nearby in this interstitial fluid, of course, and they can now receive the oxygen. These same cells are, as noted, producing carbon dioxide, which now diffuses *from* them into the interstitial fluid, and then diffuses again into a capillary. This CO_2 is returned to the lungs via the venous system. Then, at the same capillary–alveoli interface noted earlier, the CO_2 moves into the alveoli, from which it is expelled into the environment outside the body (**Figure 30.12**).

As noted, all the oxygen that is loaded into red blood cells binds initially with hemoglobin molecules that exist in these cells. In contrast to this simplicity, the carbon dioxide that is being transferred to the lungs is transported by the blood in three different ways. We need only note here that, in a kind of chemical packing and unpacking, some CO_2 ends up traveling within the hemoglobin in red blood cells, while

MB

Bioflix: Gas Exchange

ESSAY

When Breathing Becomes an Effort: Emphysema

Exhalation may be an easy, passive process for most people, but it's not easy at all for people suffering from the disease emphysema; they must struggle to get air out of their lungs because of what has happened to the elastic fibers, shown in Figure 30.10b, that surround the lungs' alveoli. These fibers normally function like a set of rubber bands: They expand as air is taken into the lungs and then spontaneously contract when the diaphragm relaxes, thus shrinking the alveoli and forcing air out of the lungs. In emphysema victims, however, these elastic fibers have been degraded by an immune system attack upon them. The result is that while the lungs' alveoli can still be filled, they can only be emptied through muscular effort. Overfilled alveoli result despite this effort, however, and in time the walls between adjacent alveoli can be destroyed, resulting in alveoli that merge and become dysfunctional.

But why would the immune system attack the lungs' elastic tissue in the first place? In 80 to 90 percent of emphysema cases, the ultimate answer is: because of cigarette smoking. Irritants from smoke that are deposited in the lungs are attacked by immune system cells, and one result of this attack is the release of an enzyme that degrades the elastic fibers. Unfortunately, this degradation is irreversible; once emphysema has taken hold, there's no cure for it. Because of its strong association with smoking, however, emphysema stands as a crystalline example of a preventable illness: The best way to keep from getting it is to keep away from cigarettes.

---

most of it travels outside of red blood cells, in the blood plasma.

It's also worth noting that the hemoglobin that normally binds with oxygen can also bind with an entirely different gas: carbon monoxide, which is best known as a gas that comes out of the tailpipes of cars. Hemoglobin actually has a greater "affinity" for carbon monoxide than it does for oxygen. As such, most of the hemoglobin molecules in the body can become occupied with carbon monoxide, rather than with oxygen, if exposure to the carbon monoxide is intense enough. When this happens, the body's cells become starved for oxygen, and hence starved for energy. If this goes on for long enough, death can result.

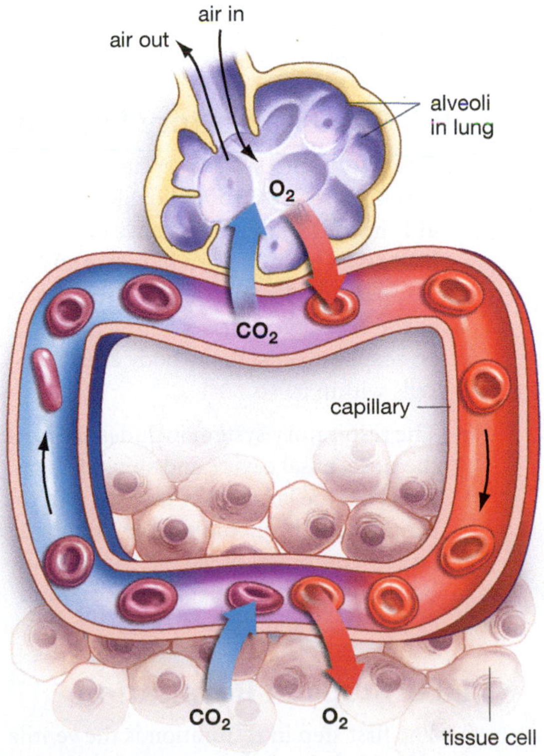

On to the Digestive and Urinary Systems

Oxygen and carbon dioxide are just a couple of the substances that the human body "exchanges" with the outside world. We take in food, of course, and we accumulate—and later rid ourselves of—waste products. Two different transport and exchange systems deal with the digestion of food and the removal of liquid waste. These are the digestive and urinary systems, the subjects of Chapter 31.

Figure 30.12
Gas Exchange in the Body

air in
air out
alveoli in lung
O_2
CO_2
capillary
CO_2 O_2
tissue cell

SO FAR . . .

1. The two central functions of respiration are capturing and distributing _____ and disposing of _____.

2. The air-exchange chambers of the lungs are tiny sacs called _____, which are surrounded by _____.

3. Oxygen binds with the molecule called _____ inside _____.

30 REVIEW

Summary

30.1 The Cardiovascular System

- The human cardiovascular system is a fluid transport system that consists of the heart, blood vessels, blood, and the bone marrow tissue in which red blood cells are formed. This system transports substances both to and from the body's tissues. **(p. 575)**

30.2 The Composition of Blood

- Blood has two primary components: formed elements (blood cells and cell fragments) and blood plasma (the fluid portion of blood in which the formed elements are suspended). The three kinds of formed elements are red blood cells, white blood cells, and platelets. Red blood cells carry oxygen to, and carbon dioxide from, every part of the body; white blood cells are central to the immune system; and platelets are small fragments of cells that are important in blood clotting. **(p. 576)**

- Blood plasma is 92 percent water but also contains other materials, including proteins, nutrients, and hormones. The three primary classes of blood plasma proteins are albumins, which transport hormones and fatty acids; fibrinogen, which aids in blood clotting; and globulins, which aid the immune system and serve as transport proteins. Low-density lipoproteins (LDLs) carry lipids to bodily tissues from the liver and small intestines; high-density lipoproteins (HDLs) carry lipids from these tissues to the liver. **(p. 577)**

30.3 Blood Vessels

- Blood vessels that carry blood away from the heart are arteries; blood vessels that return blood to the heart are veins. Capillaries connect the arteries with the veins. **(p. 577)**

- Arteries and veins include an inner layer of epithelial cells, a middle layer of smooth muscle that can widen or constrict the vessels, and an outer layer of connective tissue. Capillaries are composed of only a single layer of cells; this allows the movement of blood-borne materials into and out of them along their length. **(p. 578)**

30.4 The Heart and Blood Circulation

- Two blood circulation loops exist in the body. The pulmonary circulation loop circulates blood between the heart and the lungs (with the result that blood is oxygenated). The systemic circulation loop circulates blood between the heart and the rest of the body (with the result that needed materials are transported to and from all parts of the body). **(p. 578)**

- The human heart contains four muscular chambers: the right atrium and right ventricle (for pulmonary circulation) and the left atrium and left ventricle (for systemic circulation). Valves ensure that blood flows only one way through the heart. **(p. 579)**

- The pace at which the human heart beats is controlled by a specialized set of muscle cells that are located within the heart itself. These cells, which form the heart's sinoatrial node, generate electrical signals that prompt heart muscles to contract. **(p. 580)**

30.5 What Is a Heart Attack?

- Nearly 20 percent of all deaths in the United States are caused by the blockage of one or more of the coronary arteries that supply heart tissue with blood. Such blockages generally are caused by a buildup of plaque in the wall of a coronary artery, followed by a movement of plaque into the bloodstream, and then formation of a blood clot. A heart attack occurs when a coronary artery is blocked completely, which cuts off the blood supply to cells within the heart, killing them. **(p. 580)**

30.6 Distributing the Goods: The Capillary Beds

- Arteries near the heart branch into smaller arterioles, which feed into capillary beds. The capillary beds then feed into venules, which in turn feed into the system of veins that return blood to the heart. **(p. 583)**

- Materials needed by the body's tissues move out of the capillaries and into the interstitial fluid in which both the capillaries and nearby cells are immersed, after which the materials move into cells. Carbon dioxide and other wastes move from cells into the capillaries. The movement of all these substances is aided by their concentration gradients. **(p. 583)**

- Blood pressure is at low levels by the time blood has moved through the capillaries. Blood returns to the heart through the contraction of skeletal muscles, which squeeze the veins in a way that moves the venous blood toward the heart. **(p. 584)**

30.7 The Respiratory System

- The respiratory system captures oxygen and disposes of carbon dioxide. It also aids in controlling pH balance in the bloodstream and in producing sounds for speaking. Respiration can be defined as the exchange of gases between the atmosphere outside the body and the cells within it. **(p. 584)**

- The respiratory system includes the lungs; the nose, nasal cavity, and sinuses; the pharynx; the larynx; the trachea; and the bronchi and bronchioles. The lungs are largely composed of gas-exchange chambers, called alveoli, that lie at the end of each bronchiole. **(p. 584)**

30.8 Steps in Respiration

- The first step in respiration is the ventilation that brings air into the lungs. Once in the lungs, oxygen diffuses across the thin wall of an alveolus into an adjacent capillary and binds with the hemoglobin protein in red blood cells. Oxygen then moves with the blood cells to the heart, which pumps the blood to body tissues, where the oxygen diffuses into the interstitial fluid and then into nearby cells. The carbon dioxide produced in the body's cells moves into nearby capillaries, to be carried to the lungs. **(p. 585)**

- Oxygen loaded into red blood cells binds initially with the hemoglobin in them. Carbon dioxide is transported both within red blood cells and in blood plasma. Hemoglobin has a great capacity to bind to carbon monoxide as well as to oxygen. **(p. 586)**

Key Terms

alveoli   585
aorta   579
artery   577
bronchiole   584
capillary   577
cardiovascular system   575
coronary artery   581
formed elements   576
heart attack   581
hemoglobin   576
high-density lipoprotein (HDL)   577
low-density lipoprotein (LDL)   577
plasma   576
platelet   576
pulmonary circulation   578
red blood cell (RBC)   576
respiration   584
sinoatrial node   580
systemic circulation   578
vein   577
ventilation   585
white blood cell (WBC)   576

Understanding the Basics

Multiple-Choice Questions

(Answers are in the back of the book.)

1. The formed elements of blood include (select all that apply):
   a. plasma.   b. white blood cells.
   c. fibrinogen.   d. red blood cells.
   e. platelets.

2. What transports oxygen within the blood?
   a. LDLs in plasma   b. albumin in plasma   c. HDLs in plasma
   d. enzymes in platelets   e. hemoglobin in red blood cells

3. The arteries in the body that carry deoxygenated blood are:
   a. arterioles.   b. pulmonary arteries.
   c. coronary arteries.   d. systemic arteries.   e. all small arteries.

4. A heart attack can best be described as:
   a. a sudden, harmful contraction of the heart muscles.   b. an invasion of heart tissue by pathogens such as bacteria.
   c. a death of cells in the heart brought about by blockage of a coronary artery.
   d. a failure of the nervous system to bring about heart contractions.   e. a rapid, harmful expansion of blood volume supplied by the heart.

5. Blood returns to the heart (select all that apply):
   a. through capillaries feeding into venules.   b. through venules feeding into arteries.   c. by force of contractions of the heart's left ventricle.   d. by force of contractions of the heart's right ventricle.   e. by force of skeletal muscle contractions.

6. The two-way exchange of substances between blood cells and body cells occurs only along the length of the blood vessels called:
   a. capillaries.   b. arteries.   c. arterioles.
   d. veins.   e. venules.

7. The small size of alveoli, coupled with their large numbers, provide the lungs with what critical quality for gas exchange?
   a. a resistance to damage caused by pollutants   b. an affinity for oxygen and lack of affinity for carbon dioxide   c. an affinity for carbon dioxide and lack of affinity for oxygen   d. a large surface area   e. a small surface area

8. The movement of oxygen into the interstitial fluid from capillaries, and the movement of carbon dioxide into capillaries from the interstitial fluid is driven by:
   a. ventilation.   b. hemoglobin binding.
   c. the pumping of them.   d. osmotic gradients.   e. concentration gradients.

Brief Review

(Answers are in the back of the book.)

1. What are the primary functions of the three types of formed elements found in blood?

2. Describe the essential difference between the heart's atrial chambers and its ventricular chambers.

3. Why can substances diffuse in and out of capillaries but not arteries or veins?

4. What do eating animal fats, smoking, and exercise have to do with heart attacks?

5. Name as many anatomical structures as you can that are part of the respiratory system.

6. Describe the process of ventilation, in which air is moved into and out of the lungs.

7. Where, specifically, does oxygen enter the bloodstream?

Applying Your Knowledge

1. Heart attacks have sometimes been called a disease of modern civilization because their incidence is much higher in affluent, more-developed societies than in poorer, less-developed ones. Why should this be so?

MasteringBiology®

1. MasteringBiology Assignments
Activity: The Human Cardiovascular System; Circulatory Systems; The Cardiovascular System; Three-Chambered Hearts; Two-Chambered Hearts; The Respiratory System; **Bioflix:** Gas Exchange; **Current Events:** In Treatment for Leukemia, Glimpses of the Future; **Building Vocabulary;**

Questions: Bioflix Quiz, Reading Questions, Test Bank

2. eText
Read your book online, search, take notes, highlight text, and study more effectively.

3. The Study Area
Self-study tools to test comprehension.

Transport and Exchange 2:

Digestion, Nutrition, and Elimination

Our digestive system must break down the foods we eat in order for them to be useful to us. Different varieties of foods must be consumed in order for us to get the nutrients we need. The body conserves what it needs and eliminates what it does not need in large part through the workings of the urinary system.

Available on MasteringBiology®

Activity: The Urinary System; **Current Events:** As Restaurants Cut Salt, Some See Reasons to Pass; Killed by Thousands, Varmint Will Never Quit; Moose Offer Trail of Clues on Arthritis; Told to Eat Its Vegetables, America Orders Fries; **Building Vocabulary; Questions:** Reading Questions, Test Bank

The human digestive system puts the food we eat into a form we can use.

Innovations that work well tend to stay around for a while, but a structure that animals use to digest food has been around for a very long while. The structure is essentially a long tube through which food moves once animals ingest it. This "digestive tract," as it's called, is found almost everywhere in the animal kingdom. Among major animal groups, only the

primitive sponges and jellyfish lack one. Meanwhile, nearly all other animal groups have one. As such, this structure can be viewed as having stood the test of time: Over the course of some 500 million years of evolution, no other structure for digesting food has proved superior to it. Fish may swim and bats may fly, roundworms may burrow and snails may crawl, but in all these creatures, food is taken care of in the same way. To be sure, the digestive tract has had some pieces of auxiliary equipment added to it over time. Flatworms have nothing but a simple tube in a branched form, but more complex animals, such as snails, have the tube plus the digestive organs of a liver and a stomach. Human beings have all this and more—a liver and a stomach plus a gallbladder and a pancreas and no fewer than three sets of salivary glands. Even so, the digestive process that operates in humans is essentially the one that operates in snails: Take food in through a mouth, digest it in the tube, move what's useful out of the tube and into circulation, and move the waste out of the body altogether. In this chapter, we'll follow this whole sequence of events as it occurs in human beings. What happens to the food we eat? In the account that's coming up, you'll find out. Next, we'll look at food from another angle by asking the question: What is it that food does for us, and which foods stand to keep us the healthiest for the longest? The subject in this part of the chapter is nutrition. Finally, we'll look at a filtering operation the body runs that allows it to get rid of blood-borne waste materials while hanging on to important nutrients and fluids. This third subject in the chapter is the urinary system. Let's begin now by following the progress of food as it moves through the human body.

31.1 The Digestive System

The central structure of the human digestive system, the **digestive tract** we've been talking about, can be defined as a muscular passageway for food and food waste that runs through the human body from the mouth to the anus. Along its length, this tube receives input from the various accessory organs we've noted, such as the gallbladder and the pancreas. You saw last chapter that nutrients are delivered to tissues throughout the body by the circulatory system. So the real question for the digestive system is: What does it need to do to get food into a form that can *move into* the circulatory system? For carbohydrates and proteins, the pathway is straightforward. Partway through a portion of the digestive tract called the small intestine, carbohydrates and proteins have been broken down enough that they move out from the inner lining of the intestine into adjacent capillaries. This network of capillaries then feeds into a common vein that goes to the liver. Most nutrients, then, are carried straight from the small intestine to the liver. In a controlled release, the liver then sends these nutrients out (through another vein) to the heart, which pumps them out to all the tissues of the body.

The digestion of fats takes a somewhat more complicated route, as you'll see, but the principle remains the same. Fats must be broken down to such a point that they can leave the digestive tract for the rest of the body. For all foods, "breakdown" means just what it sounds like: a transition from large to small. The transition goes from big, visible bites of food to smaller bits and eventually from big *molecules* of food to the building blocks that make them up. It is only these smaller molecules—simple sugars, amino acids, and fatty acids—that can leave the digestive tract for the rest of the body.

Of course, not everything we eat makes this transition. A good deal of the food we ingest is retained in the digestive tract as a waste product that ultimately is eliminated from the body altogether. It's the digestive system that handles these tasks of waste retention and elimination, even as it's carrying out its other functions of food digestion and nutrient transfer. Now, let's take a look at the components of the digestive system, after which you'll follow the path of digestion.

31.2 Structure of the Digestive System

Figure 31.1 shows the locations and functions of the accessory glands and different subdivisions of the digestive tract. The digestive tract begins with the mouth (or oral cavity) and continues through the pharynx, esophagus, stomach, small intestine, and large intestine before ending at the rectum and anus. The digestive tract is sometimes called the alimentary canal or the gastrointestinal tract.

The Digestive Tract in Cross Section

Thinking again of the digestive tract as a muscular tube, let's look at a cross section of it as it would appear within the small intestine. As you can see in **Figure 31.2,**

this portion of the digestive tract is a four-layered tube. The outermost layer is connective tissue known as the serosa. Inside of it, there is muscle: two sets of smooth muscles, actually, which together are known as the muscularis externa. In a process called **peristalsis,** these muscles take turns contracting, which produces waves of contraction that push materials along the length of the digestive tract. Inside the muscularis externa there is the submucosa, a layer of tissue that contains large blood vessels and a network of nerve tissue that both coordinates the contractions of the smooth muscle layers and regulates the secretions of digestive juices. Finally, there is the epithelial layer that makes up the internal lining of the tract. It is called the mucosa because it is constantly bathed in mucous secretions.

Along much of the length of the digestive tract, the mucosa has a large number of folds that increase

Figure 31.1
Components of the Digestive System

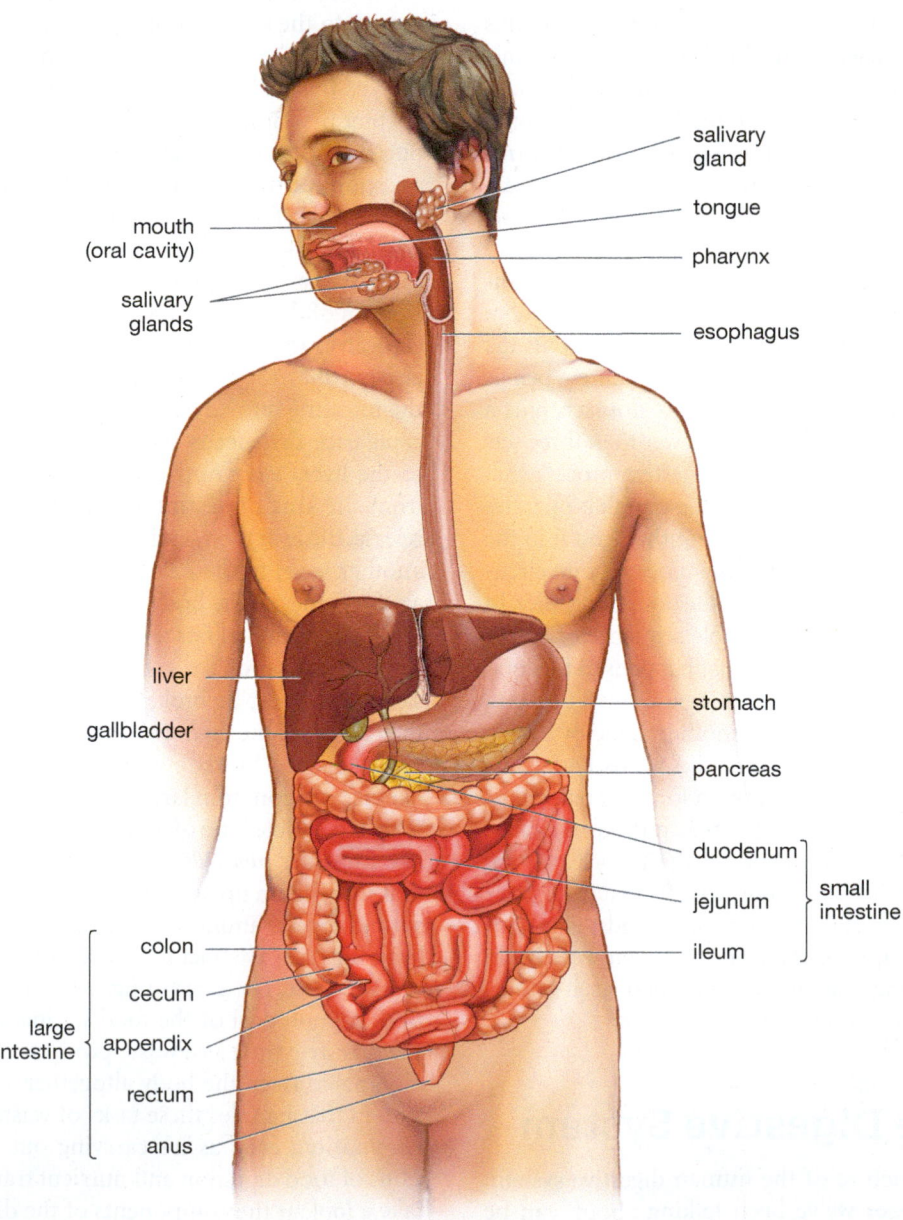

salivary gland

tongue

pharynx

esophagus

mouth (oral cavity)

salivary glands

liver

gallbladder

stomach

pancreas

duodenum ⎤
jejunum ⎬ small intestine
ileum ⎦

colon

cecum

large intestine ⎰ appendix

rectum

anus

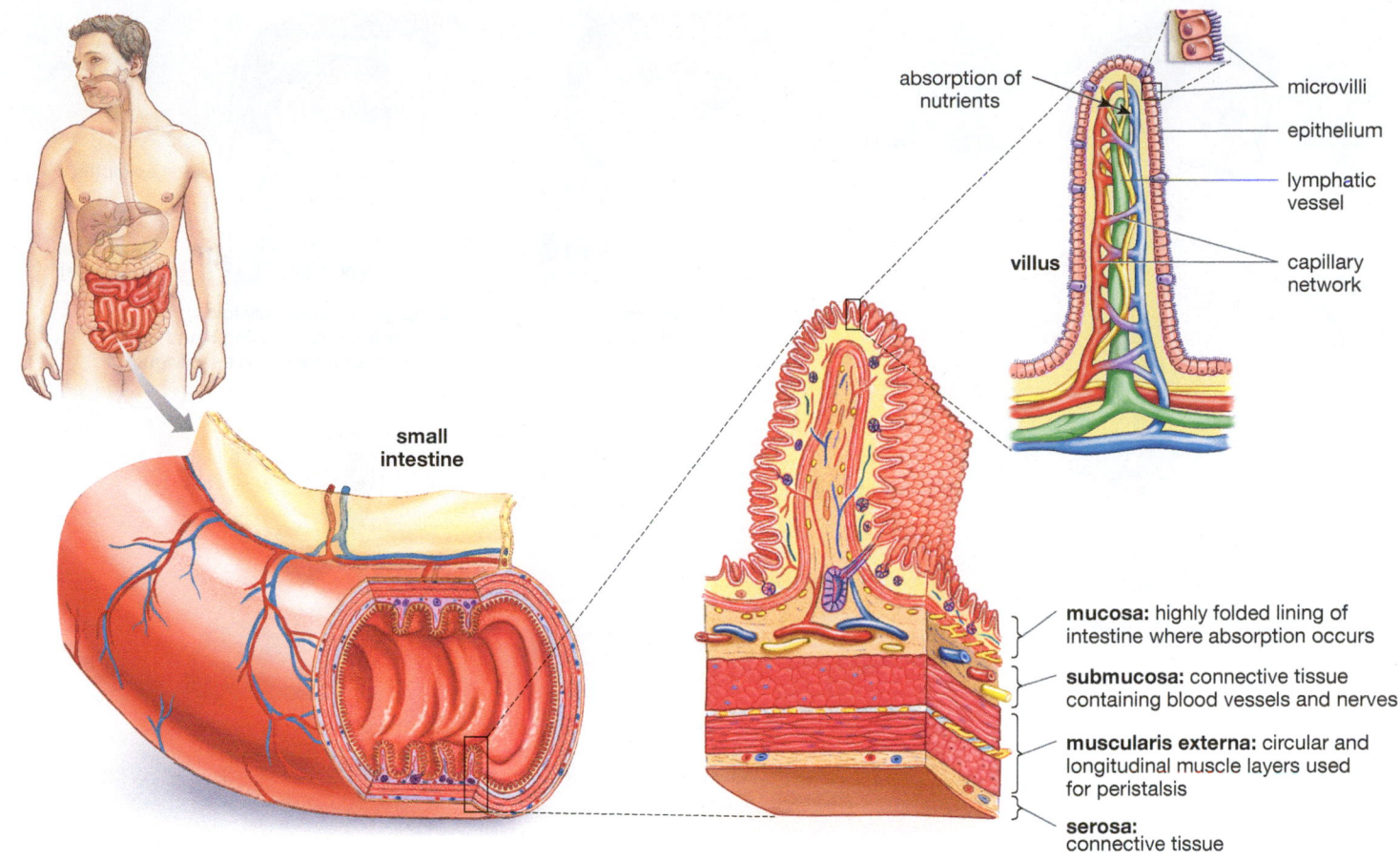

absorption of nutrients

microvilli

epithelium

lymphatic vessel

capillary network

villus

small intestine

mucosa: highly folded lining of intestine where absorption occurs

submucosa: connective tissue containing blood vessels and nerves

muscularis externa: circular and longitudinal muscle layers used for peristalsis

serosa: connective tissue

the surface area available for absorption of digested food; these folds also work like accordion folds in that they permit expansion, which is helpful after a large meal. The mucosa also has on it fingerlike projections called villi that further increase the area for absorption. These villi are the place where most of our digested food actually makes the transition out of the digestive system and into circulation. If you look at the blowup of the villus in Figure 31.2, you can see where nutrients leave the system. Just outside it, they enter either an adjacent capillary or, in the case of fats, a lymphatic vessel whose connecting vessels will move the nutrients out of the lymphatic system and into the bloodstream.

31.3 Steps in Digestion

Let's now look at the process of digestion from start to finish, which means taking it from the top, so to speak. Our mouths are the entry place for food, of course, and we immediately start breaking food down in them by the mechanical means of chewing. Food starts to get digested there by chemical means as well because enzymes in the saliva—secreted by the three pairs of salivary glands noted earlier—have the ability to initiate the breakdown of carbohydrates.

The Pharynx and Esophagus

Food begins to move out of the mouth when the tongue pushes it back toward the pharynx, which we might think of as the upper throat (see Figure 31.1). This action is tricky because the **pharynx** is a passageway that links the mouth with both the food-transporting esophagus and the *air*-transporting trachea. Once the tongue begins pushing food back, tissue called the epiglottis folds down over the larynx to keep the food from entering it, an action that channels the food into the esophagus, as you can see in **Figure 31.3** on the next page. (Sometimes food does go down the "wrong way," of course, at which point we may simply start coughing or, in the worst case, need the Heimlich maneuver to keep us from suffocating.) Next, the esophagus has a set of muscles that constrict in sequence, performing a peristalsis that works with gravity to send food on a 6-second trip to the next stop in the digestive tract.

The Stomach

The stomach digests food partly by the mechanical, muscular action of churning it—in the way that a washing machine churns clothing—thus breaking it into small bits. It also carries out some chemical breakdown of foods, with proteins the primary target of its digestive juices. In addition to performing

Figure 31.2
Structure of the Digestive Tract

Through much of its length, the digestive tract is a tube formed of several layers of tissue, each performing a different function. Note that the absorption of nutrients takes place when food that has been sufficiently digested moves across the epithelium and is taken up by either a capillary or a lymphatic vessel inside a villus.

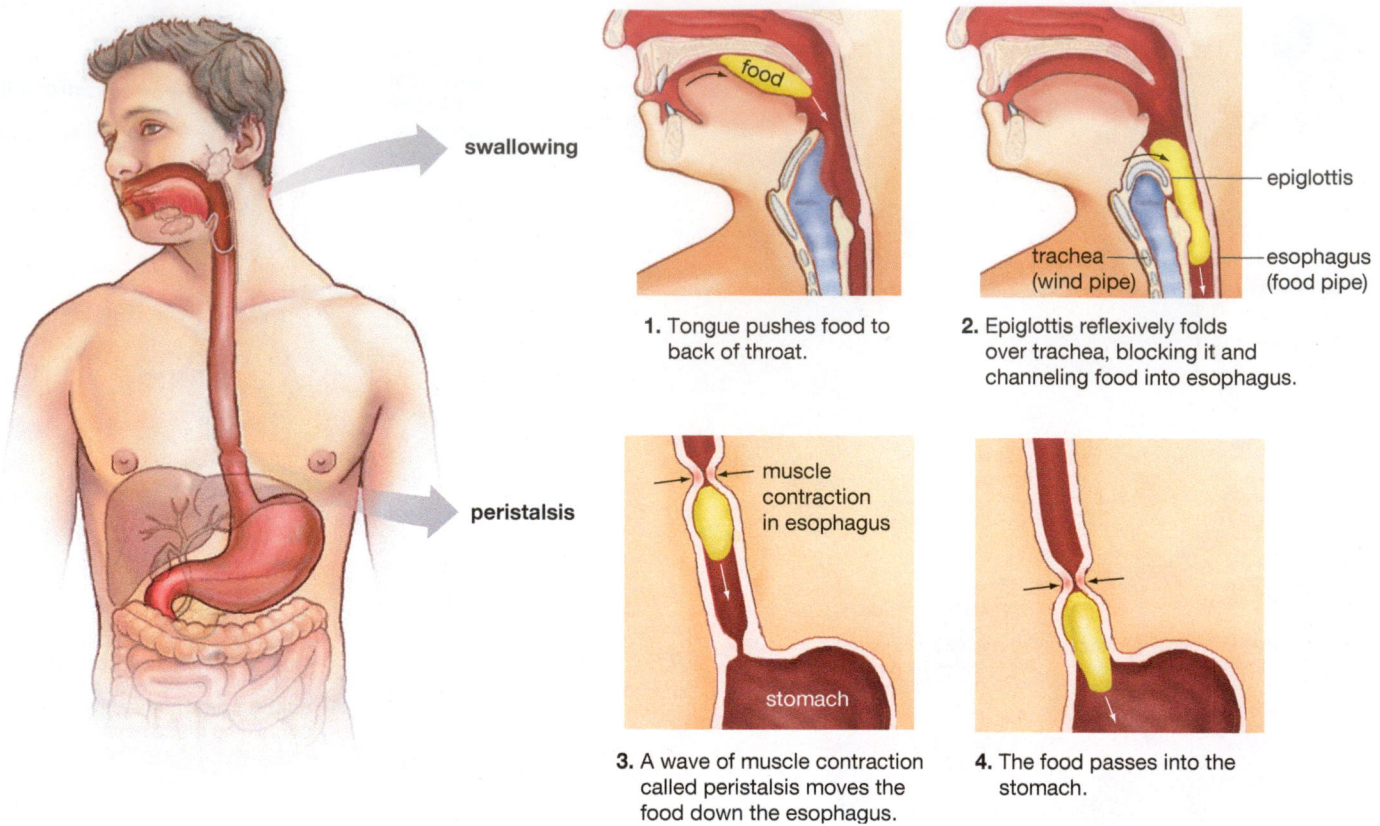

swallowing

peristalsis

epiglottis

trachea (wind pipe)

esophagus (food pipe)

1. Tongue pushes food to back of throat.

2. Epiglottis reflexively folds over trachea, blocking it and channeling food into esophagus.

muscle contraction in esophagus

stomach

3. A wave of muscle contraction called peristalsis moves the food down the esophagus.

4. The food passes into the stomach.

Figure 31.3

Getting Food from Mouth to Stomach

Swallowing moves food from the mouth to the esophagus, after which peristalsis combines with gravity to move food to the stomach.

digestive work, the **stomach** is an organ that has a less-appreciated second role: It serves as a temporary, expandable storage site for food. This capacity is the reason we can go several hours without eating. When empty, the stomach resembles a tube with a narrow cavity, but it can expand to contain up to 1.5 liters, or almost half a gallon, of material. This expansion is possible because the stomach contains numerous folds, called rugae (**Figure 31.4**).

Glands feed into the stomach by way of millions of small depressions, called gastric pits, that open on to its interior. Each day, about 1,500 milliliters, or 45 ounces, of gastric juice pour into the stomach through these pits. This gastric juice is *caustic*—a mixture of hydrochloric acid and the enzyme pepsin. So acidic is the hydrochloric acid that the contents of the stomach can have a pH that lies between that of lemon juice and battery acid. Why is such an extreme environment needed? It serves not only to break down food but also to kill bacteria and other invaders that ride in on the food. You may wonder why the stomach itself wouldn't be eaten away in this kind of environment. The answer is that the cells lining it produce a mucus that protects it.

A few substances, such as alcohol and some drugs, can pass directly from the stomach into circulation, but only a small portion of the nutrients we consume enter circulation in this way. In general, the stomach is only preparing food for a more complete digestion of it that comes later. Food stays in the stomach for 2 to 4 hours,

but on exiting it is no longer just plain food; it is now a soupy mixture of food and gastric juices known as **chyme**. A circular or "sphincter" muscle—the pyloric sphincter—regulates the flow of chyme between the stomach and the next structure in the digestive system.

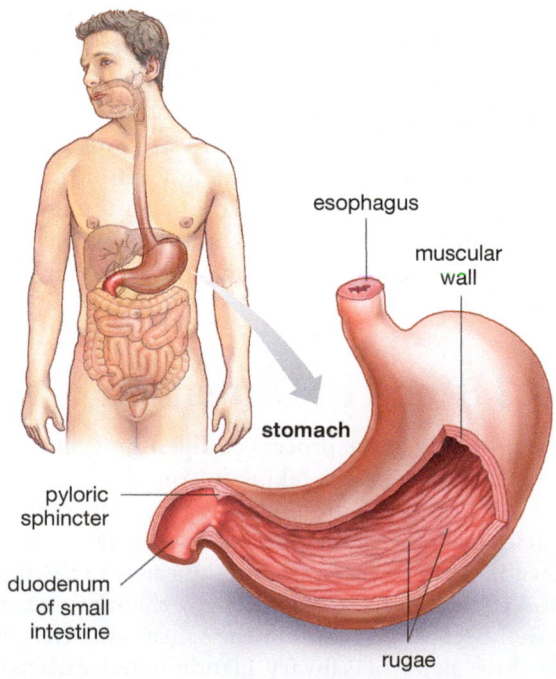

esophagus

muscular wall

stomach

pyloric sphincter

duodenum of small intestine

rugae

Figure 31.4

External and Internal Views of the Stomach

small intestine

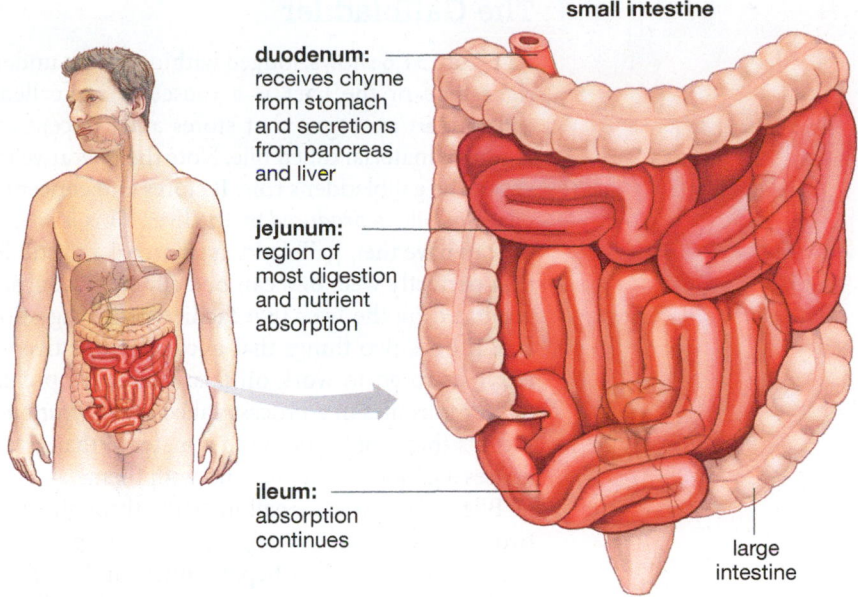

duodenum:
receives chyme
from stomach
and secretions
from pancreas
and liver

jejunum:
region of
most digestion
and nutrient
absorption

ileum:
absorption
continues

large
intestine

SO FAR . . .

1. The central structure of the digestive system is a muscular tube called the _____ that passes through the body from the _____ to the _____.

2. A central function of the digestive system is to get food into a form that can move into the _____.

3. Food is moved along the length of the digestive tract through the action of sets of _____ in a process called _____.

The Small Intestine

The only thing small about the small intestine is its diameter, which is about 4 centimeters, or 1.6 inches, up where it starts (at the stomach) and about 2.5 centimeters, or an inch, where it ends (at the large intestine). In between these two points, the "small" intestine winds on for about 6 meters, or almost 20 feet. Eighty percent of our nutrients make the transition out of the digestive system along the length of the **small intestine,** which can be defined as the portion of the digestive tract that runs between the stomach and the large intestine. (About 10 percent of our nutrients are transferred from the stomach, and the remaining 10 percent are transferred from the large intestine.)

As you can see in **Figure 31.5**, the small intestine is made up of three regions. The first of these, the *duodenum,* is the 25 centimeters, or 10 inches, that are closest to the stomach. Think of the duodenum as that portion of the small intestine that receives not only chyme from the stomach but digestive secretions from three of the specialized digestive organs noted earlier: the pancreas, the liver, and the gallbladder. Next, there is the *jejunum,* which goes on for about 2.5 meters, or 8 feet. This is the site within the small intestine where most chemical digestion and nutrient absorption occur. Absorption continues to some extent in the third segment, the *ileum,* which is the longest, averaging 3.5 meters, or 12 feet, in length. The ileum ends at a sphincter muscle that controls the flow of chyme into the large intestine.

On average, it takes about 5 hours for chyme to pass through the small intestine. Thus, material that began its digestive journey in the mouth takes perhaps 9 hours to reach the end of the small intestine. Now, let's look at what happens at the start of the small intestine, where juices from the digestive glands come into the picture.

The Pancreas

Feeding into the small intestine is the pancreas, which lies behind the stomach and is an elongated, pinkish-gray organ about 15 centimeters, or 6 inches, long. The pancreas is best known as the organ that produces insulin, which it secretes directly into the bloodstream. But our concern is with materials the pancreas secretes through a tube, or duct. The duct in question, as you can see in **Figure 31.6**, is the pancreatic duct, which comes together with the common bile duct (stemming from the liver and gallbladder) to empty into the small intestine's duodenum. In digestion, the **pancreas** can be defined as a gland that secretes, into the small intestine, digestive enzymes and chemical buffers that help raise the pH of the chyme coming from the stomach. (Remember, this chyme is very acidic; when its pH is being "raised," that means it is becoming less acidic.) Pancreatic digestive enzymes are classified according to the kind of food they work on. Lipases break down lipids, proteases break proteins apart, and carbohydrases digest sugars and starches.

Figure 31.5
Structure of the Small Intestine

Figure 31.6
Accessory Glands in Digestion: The Liver, Pancreas, and Gallbladder

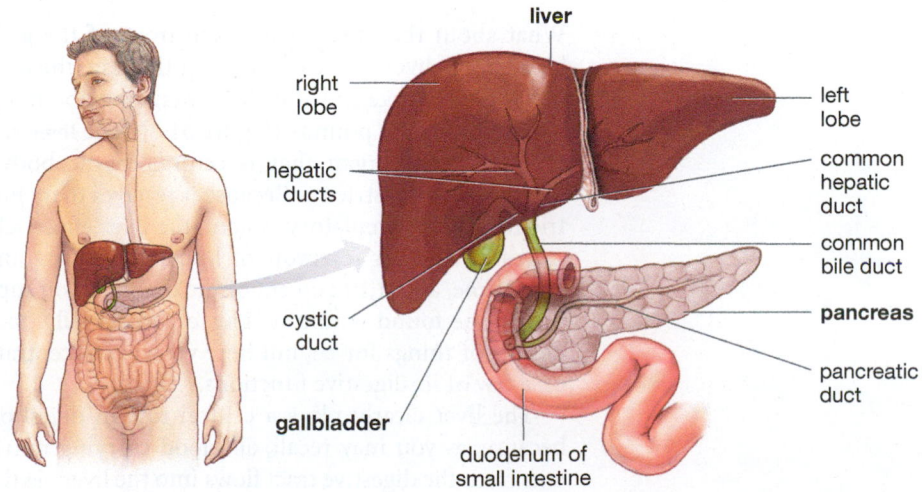

liver

right
lobe

left
lobe

hepatic
ducts

common
hepatic
duct

common
bile duct

pancreas

cystic
duct

pancreatic
duct

gallbladder

duodenum of
small intestine

The Gallbladder

As Figure 31.6 shows, lodged within a recess under the right lobe of the liver is a muscular sac called the **gallbladder:** an organ that stores and concentrates a digestive material called bile. Note the operative terms about the gallbladder's role: It stores and concentrates bile. But bile is *produced* by the liver. This is why people can have their gallbladders removed and still function perfectly well. **Bile** can be defined as a substance produced by the liver that facilitates the digestion of fats. It does two things that allow the digestive lipase enzymes to go to work on fats: It breaks up clusters of fat molecules in a process called *emulsification,* and it coats these molecules with a material that gives the lipases a greater ability to bind with them.

Bile comes to the small intestine through either of two routes. It always travels through passageways within the liver called hepatic ducts and then leaves the liver through the common hepatic duct. This bile may then (1) flow into the small intestine's duodenum through the common bile duct or (2) enter the cystic duct that leads to the gallbladder. If it goes to the gallbladder, only later will it flow into the duodenum through the common bile duct. Why would it take one course or the other? A sphincter at the intestinal end of the common bile duct opens only at mealtimes. At other times, bile backs up through the cystic duct for storage and concentration within the gallbladder before moving to the small intestine.

Bile can become *too* concentrated and harden into the troublesome crystals known as gallstones. Most gallstones are small enough to be flushed away into the duodenum, but they can reach 2 centimeters or more in diameter—about three-quarters of an inch, which is large enough to block either the common bile duct or the cystic duct. The pain that results is so excruciating that gallstones have to be taken care of in one way or another. One modern way of getting rid of them is to pulverize them with high-energy sound waves.

The Liver

What about the organ that lies in front of the gallbladder, the liver? If we don't count the skin, the liver is the largest organ in the body, weighing about 1.5 kilograms, or 3.3 pounds (Figure 31.6). The **liver** is a reddish-brown organ that is central to the body's metabolism of nutrients. Because it receives these nutrients via the circulatory system, the liver effectively serves as a major reservoir of human blood; at any given time, about 10 percent of the body's blood supply can be found within it. This organ actually does dozens of things for us, but here we will concentrate on a few of its digestive functions.

The liver clearly plays a central role in digestion because, as you may recall, all blood carrying nutrients from the digestive tract flows into the liver via the hepatic portal vein. We could thus think of a series of blood vessel "streams"—stemming from the stomach, the small intestine, and so forth—all of them carrying digested nutrients, and all of them flowing like tributaries into the common river of the hepatic portal vein that goes to the liver. This arrangement makes the liver a first stop for most nutrients. From this position, the liver controls which nutrients it will send to the rest of the body and which nutrients it will store. The liver is, however, not only the first stop for nutrients but the first stop for toxins such as alcohol. This is why people who have abused their bodies by drinking too much can end up with damaged livers. In addition to transferring and storing nutrients, the liver produces the bile noted earlier, and it packages waste products for removal by the kidneys.

The Large Intestine

If most nutrients have been transferred out of the digestive system by the time the small intestine ends, what is there left for the large intestine to do? Its main task is to serve as a kind of trash compactor: It holds and compacts material that by now is largely refuse, turning it into the solid waste we know as feces.

If you look at the upper drawing in **Figure 31.7**, you can see that the large intestine neatly frames the small intestine. The small intestine's last section, the ileum, feeds into the large intestine's first section by means of a sphincter (called the ileocecal valve) that you can see at about 7 o'clock in the lower drawing. This sphincter can relax or contract, thus controlling movement of chyme into the large intestine. The large intestine continues for about 1.5 meters, or 5 feet, from this point before ending at the anus. Compared to the 20-foot small intestine, the large intestine is large only in diameter.

The **large intestine** can be defined as that part of the digestive tract that begins at the small intestine and ends at the anus. It is divided into three major regions: the cecum, the colon, and the rectum. Note that the cecum has a pouch called the *appendix* attached to it; the walls of this pouch can become infected, resulting in appendicitis. (When healthy, the appendix is an arm of the immune system, although obviously not an essential one, as it can be removed.) The colon is the large intestine's second section and is by far its longest. The individual pouches that you can see in it allow it to expand and contract, depending on how full it is. The colon then empties into the expandable rectum, which serves as a storage site for feces. The rectal chamber is usually empty except when peristaltic contractions force fecal materials into it from the lower part of the colon. Stretching of the rectal wall then triggers the urge to defecate. Material takes anywhere from 12 to 24 hours to move through the large intestine. Adding this to the other transport times we've noted, you can see that a complete trip through the digestive system takes anywhere from 19 to 33 hours.

Figure 31.7
The Large Intestine

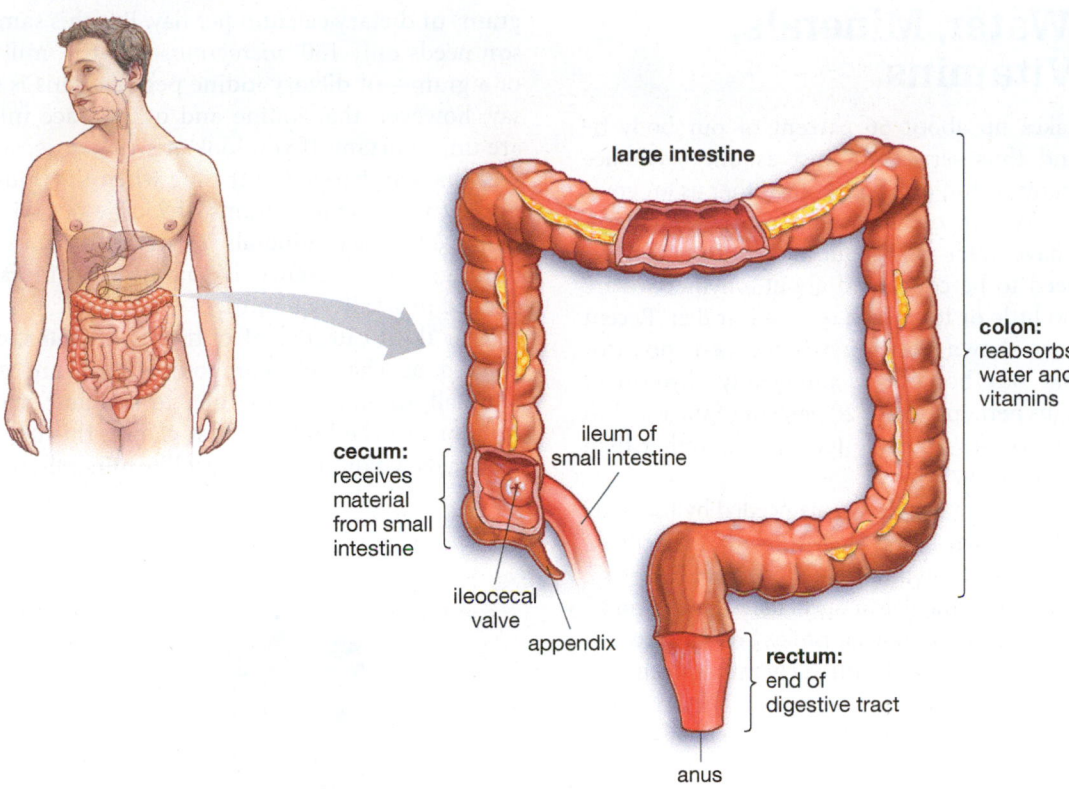

large intestine

colon:
reabsorbs
water and
vitamins

cecum:
receives
material
from small
intestine

ileum of
small intestine

ileocecal
valve

appendix

rectum:
end of
digestive tract

anus

Apart from waste disposal, the large colon also functions in water conservation. Roughly 1,500 milliliters, or 45 ounces, of watery material arrives in the colon each day, but some 40.5 ounces move back out, leaving very little to be ejected with the feces. The large intestine also absorbs vitamin K that is produced by the abundant bacteria that live within the colon. Carbohydrates that cannot be digested arrive in the colon intact and serve as a nutrient source for these bacteria. Their resulting metabolic activities are responsible for intestinal gas. Why do beans often trigger such gas? Because they contain a high concentration of carbohydrates that we cannot digest but that the bacteria can.

SO FAR . . .

1. Eighty percent of our nutrients make the transition out of the digestive system along the length of the _____.

2. Bile is a substance secreted by the _____ and stored and concentrated by the _____. The pancreas secretes _____ and _____ into the small intestine.

3. Arrange the following parts of the digestive system in the order in which food passes through them: stomach, large intestine, esophagus, small intestine.

31.4 Human Nutrition

Having learned something about how the human body processes food, you're now in a better position to understand what kinds of foods the body needs. The subject here is **nutrition,** which can be defined as the study of the relationship between food and health.

What do the foods we eat need to contain in order for us to have a healthy body? Scientists think of foods in terms of substances they contain called *nutrients.* To be a **nutrient,** a substance must do at least one of three things: It must provide us with energy, it must provide us with one of the structural building blocks the body needs, or it must help regulate the body's physical processes. Many of the foods we eat are nutritious in all three senses. While a piece of whole wheat bread is loaded with energy, it also contains some structural building blocks (amino acids) and some substances that help regulate bodily processes (potassium, sodium).

Nutrients can play these diverse roles because nutrients themselves are a diverse group of substances. Every nutrient, however, is a member of a single *class* of nutrients, and it turns out there are only six such classes: water, minerals, and vitamins; and carbohydrates, proteins, and lipids. Given that everything we eat falls into one of these classes, we'll structure our tour of nutrition largely as a walk through these groups. Along the way, you'll see how important it is to choose carefully among foods as a means of getting the nutrients you need.

31.5 Water, Minerals, and Vitamins

Water makes up about 66 percent of our body by weight and thus serves not just as one substance among many within our body, but rather as an environment in which our body tissues are immersed. We must have water to live, but in most instances, we do not need to be concerned about whether we are getting too little or too much of it in our diet. Recent research has shown that thirst is the best indicator of whether our bodies are sufficiently "hydrated." Surprisingly, perhaps, about 20 percent of the water in our systems comes from the food we eat rather than the fluids we drink.

Minerals are chemical elements needed by the body either to help form bodily structures or to facilitate chemical reactions. You may remember from Chapter 2 that a chemical element is a substance that is "pure" in the sense that it cannot be broken down into any set of simpler substances through chemical means. All minerals are elements in this sense. They are not *combinations* of substances; they are elemental substances themselves.

Looking at the functions of minerals in nutrition, we might ask first: What does it mean for a mineral to help form a bodily structure? The mineral calcium provides a good case in point. About 1.5 percent of our body is made up of calcium. Where does this mineral exist? As you can imagine, overwhelmingly in our bones: 99 percent of the calcium in our bodies is found in bone. The 1 percent of calcium *not* found in bone is also critical, but it functions in the second role played by minerals—that of facilitating chemical reactions. Any time we perceive a sound, for example, calcium ions must flow into the cells in our ears that turn sounds into nerve signals. Any time we decide to move a muscle, calcium ions must be released from one variety of muscle fiber in order for another variety of fiber to latch on to it. Calcium thus serves both as a chemical building block and as a partner in chemical reactions. Not all minerals serve both functions—some function only as facilitators of reactions—but as a class of nutrients, minerals are important in both roles.

If you look at **Table 31.1**, you can see a list of 16 of the most important dietary minerals. Note that there are two varieties of them. *Major minerals* are minerals needed in large amounts by the body, while *trace minerals* are minerals needed in small amounts. What's the dividing line? If we need 100 milligrams or more of a mineral per day, it's a major mineral; if we need less than 100 mg per day, it's a trace mineral. What do these differences mean in practice? To take an extreme example, an average adult needs at least 1,000 milligrams (meaning a gram) of dietary calcium per day. But this same person needs only 150 *micrograms*—that is, millionths of a gram—of dietary iodine per day. This is not to say, however, that iodine and other trace minerals are unimportant. If you look at **Figure 31.8**, you can see what can happen over time when the body does not get its 150 micrograms of iodine per day. Both major and trace minerals are critical to our functioning; the difference is that trace minerals are needed in smaller quantities.

Our third category of nutrients, **vitamins**, can be defined as chemical compounds that are needed in small amounts in the diet to facilitate chemical reactions in the body. At first glance, this definition may make vitamins sound just like minerals, but note

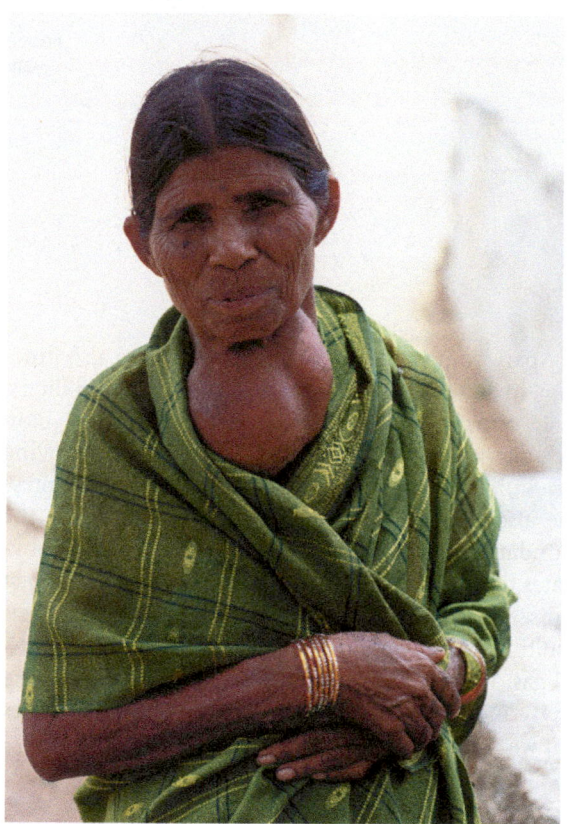

Figure 31.8
Iodine Deficiency

The swelling in this woman's neck is a goiter—an enlargement of the thyroid gland brought about by a lack of iodine in the diet. Goiters of this size are a rare occurrence, but iodine deficiency is a major health problem in the world's less developed countries, where women who are iodine-deficient are at high risk of giving birth to mentally retarded children. In recent years, great progress has been made in combating this problem by means of adding iodine to table salt, a strategy adopted decades ago by the world's more developed countries. The strategy is cost-effective: A mere two ounces of a substance called potassium iodate can iodize a ton of salt at a cost of about $1.15.

Minerals Required by the Human Body

Mineral	Sources	Functions	Symptoms of Deficiency
Major Minerals			
Calcium (Ca)	Dairy products, leafy green vegetables	Maintenance of bones, muscle contraction, nerve signal transmission	Retarded growth in children; osteoporosis in adults
Chloride (Cl)	Table salt	Water balance, digestion	Very rare
Magnesium (Mg)	Broccoli, leafy green vegetables, whole-grain wheat products	Component of bones, teeth; nerve signaling, muscle contraction	Irregular heartbeat, muscle spasms, seizures
Phosphorus (P)	Dairy products, meat, whole grains	Component of bone, nucleic acids, ATP	Bone loss in older women
Potassium (K)	Numerous fruits and vegetables, dairy products	Nerve signal transmission, fluid balance	Muscle cramps, irregular heartbeat, paralysis
Sodium (Na)	Table salt	Nerve signal transmission, fluid balance	In rare instances, muscle cramps, shock
Sulfur (S)	Protein-rich foods (eggs, meat)	Detoxification of drugs; component of some amino acids, vitamins	Unknown when dietary protein is adequate
Trace Minerals			
Chromium (Cr)	Liver, nuts, whole grains	Helps activate insulin	Elevated glucose, insulin levels
Copper (Cu)	Liver, seafood, nuts, legumes	Enzyme activation, hemoglobin synthesis	Ricketts-like skeletal problems, anemia
Fluorine (F)	Fluorinated tap water, toothpaste	Resistance to tooth decay	Increased tooth decay
Iodine (I)	Iodized table salt, seafood, dairy products	Synthesis of thyroid hormones	Goiter, mental retardation in children of iodine-deficient mothers
Iron (Fe)	Whole grains, beef, fish, beans, nuts	Synthesis of hemoglobin	Iron deficiency anemia
Manganese (Mn)	Whole grains, nuts, leafy green vegetables	Enzyme activation	Never observed in humans
Molybdenum (Mo)	Dairy products, grains, legumes	Enzyme activation	Increased heart rate, night blindness, edema
Selenium (Se)	Seafood, liver, eggs	Enzyme activation	Muscle pain, weakness, vascular deterioration
Zinc (Zn)	Meat, poultry, fish, eggs, beans	Enzyme activation in many processes	Stunted growth, diarrhea

the differences between these two kinds of nutrients. First, vitamins are compounds—they are made up of more than one element—whereas each mineral is a *single* element. Thus, a molecule of vitamin B_2 (riboflavin) is made up of the elements carbon, hydrogen, oxygen, and nitrogen. Second, unlike minerals, vitamins serve no structural role in the body; their function is purely to facilitate chemical reactions. As such, vitamins are needed strictly in *small* quantities—sometimes in milligram quantities, but often in microgram quantities. In consequence, the daily diet of a typical American contains only about 300 milligrams of vitamins compared to 20 grams of minerals.

Figure 31.9
One Role for Vitamins
Many of the vitamins we consume serve as coenzymes—substances that bind to enzymes, thus allowing the enzymes to bind to their substrates, meaning the substances they help modify.

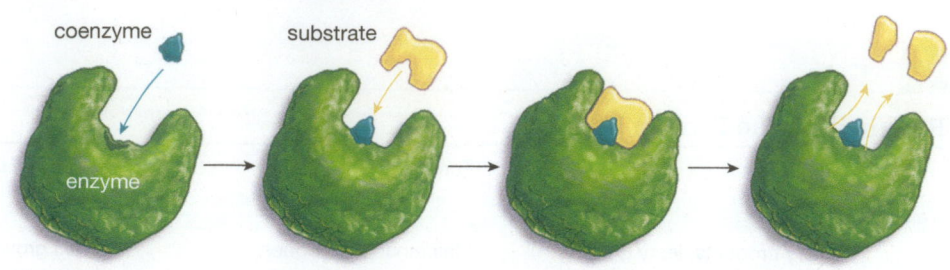

coenzyme   substrate

enzyme

So, in what ways do vitamins facilitate chemical reactions? If you look at **Figure 31.9** you can see an example. From Chapter 6, on energy, you may recall that nearly every chemical reaction that takes place in the body is speeded up—or, in practical terms, *enabled*—by a special kind of protein called an enzyme. Thus, when we drink milk, the lactose (or milk sugar) in it is broken down in a chemical reaction that uses an enzyme called lactase. Like all enzymes, lactase can only do its job by binding with its "substrate," meaning the substance it works on. Now, how do enzymes do this binding? In some cases, they come equipped, we might say, to bind on their own. A few of the amino acids that make up an enzyme will bind to the substrate and help alter it. In other cases, however, enzymes need a chemical partner. In order for them to bind to their substrate, they must first bind with a *coenzyme*. This is the role played by all eight B vitamins: They serve as coenzymes (or parts of coenzymes). To put this another way, the B vitamins are molecules that bind to enzymes, thus allowing them to bind to their substrates.

If we looked at other vitamins—C and A and so forth—we'd find them playing roles other than coenzymes in the facilitation of chemical processes. In each case, however, a vitamin is pivotal in allowing some metabolic process to go forward. Thousands of bodily compounds are important in this same way, but almost all these compounds can be *manufactured* by the body. The reason vitamins are classified as nutrients is because, in most cases, our only source for them is the food we eat or the vitamin supplements we take. (The principal exception to this is vitamin D, which can be manufactured by our skin when it's exposed to sunlight.)

Table 31.2 sets forth a list of all 13 vitamins that the human body must have in order to function properly. Note that vitamins are divided into two classes: water soluble and fat soluble. What's the difference? Water-soluble vitamins dissolve in water, but fat-soluble vitamins dissolve in fat. The practical importance of this distinction is that excess amounts of water-soluble vitamins are simply excreted in urine, with the result that these vitamins do not build up in us. Fat-soluble vitamins, meanwhile, *can* build up in us (particularly in the liver), which means we need to guard against taking in excess quantities of them.

Vitamin and Mineral Sources: Do We Need Supplements?

The foods we eat can be rich sources of both vitamins and minerals. Given this, the question is whether we need to take vitamin and mineral *supplements*—usually pills or capsules—to augment the quantities of vitamins and minerals we get through food. The consensus among nutritionists seems to be that, while a well-balanced diet can supply all the vitamins and minerals the average American needs, a simple one-a-day supplement can't hurt and might be helpful. When it comes, however, to the very large doses of vitamins E or C that have sometimes been recommended in the past, their advice is different. There is no reason to think that "megadoses" of these vitamins will be beneficial to health, and there is even reason to worry that such doses might be harmful. The biggest question mark about vitamin intake today concerns vitamin D, which has traditionally been regarded as important in building strong bones but now is being investigated as a nutrient that may serve to protect human beings from some widespread diseases. You can read more about this in "Vitamin D Moves to Center-Stage" on page 602.

SO FAR . . .

1. To be classified as a nutrient, a substance we ingest must do at least one of three things: It must provide us with _____, it must serve as a _____, or it must help _____ the body's physical processes.

2. The six classes of nutrients are _____, _____, and _____; and _____, _____, and _____.

3. Minerals are chemical _____, while vitamins are chemical _____. Because vitamins do not help form bodily structures, every vitamin is needed in strictly _____ amounts.

Table 31.2

Vitamins in the Human Diet

Vitamins	Sources	Functions	Symptoms of Deficiency
Water-Soluble Vitamins			
Thiamin (B₁)	Whole grains, legumes, nuts	Coenzyme in cellular respiration	Beriberi
Riboflavin (B₂)	Dairy products, leafy green vegetables	Coenzyme in cellular respiration	Skin lesions; blurred vision
Niacin	Peanuts, poultry, beans	Part of cellular respiration enzyme	Pellagra, mental disorders
Vitamin B₆	Whole grains, dairy products, nuts, leafy green vegetables	Part of coenzyme in amino acid synthesis	Anemia, twitching, skin disorders
Pantothenic acid	Nuts, beans, dairy products, eggs	Coenzyme in fat synthesis	Numbness, fatigue
Folic acid	Leafy green vegetables, beans, orange juice	Coenzyme in hemoglobin production, DNA formation	Spina bifida in mothers deficient in folic acid
Vitamin B₁₂	Dairy products, eggs, meat	Coenzyme in nucleic acid formation, red blood cell maturation	Pernicious anemia
Biotin	Eggs, fish, poultry	Coenzyme in synthesis of fat, amino acids	Scaly, inflamed skin, poor growth
Vitamin C	Citrus fruits, green peppers, cauliflower, broccoli	Promotes collagen synthesis in bone, cartilage; antioxidant	Scurvy, poor wound healing
Fat-Soluble Vitamins			
Vitamin A	Liver, fish, eggs, fortified dairy products	Critical to vision	Night blindness, complete blindness in severe cases
Vitamin D	Can be produced by skin exposed to sunlight; fatty fish, fortified dairy products	Regulates bone metabolism; serves as transcription factor	Possible role in many diseases
Vitamin E	Vegetable oils, asparagus, eggs, almonds	Antioxidant; helps protect cell membranes	In infants, hemolytic anemia (rare)
Vitamin K	Liver, leafy green vegetables; some production by bacteria in gut	Production of compounds used in blood clotting	Blood-clotting problems in newborns

31.6 Calories and the Energy-Yielding Nutrients

Thus far, we've looked at three classes of nutrients—water, minerals, and vitamins—that are critical parts of a healthy diet but that have nothing to do with the aspect of food consumption that most people are concerned about: calories. Water contains no calories at all, and minerals and vitamins contain next to none. This is not the case, however, with the carbohy-drates, lipids, and proteins we'll look at now. If you want to know how many calories you're consuming per day, just add up your daily consumption of these three classes of nutrients.

A common assumption, however, is that adding up calories is an exercise in adding up units of *weight,* but this isn't the case. Calories actually are units of *energy.* A single calorie is defined as the amount of energy it takes to raise the temperature of 1 gram of water by 1 degree Celsius. It takes such a small amount of

ESSAY

Vitamin D Moves to Center-Stage

Most Americans understandably regard vitamins as an aspect of nutrition that can cruise along on auto-pilot: The government sees to it that foods are fortified with them; anyone concerned about getting enough of them can take a one-a-day pill; and no one seems to get sick because of a lack of them. Thus it was a notable event in November of 2008 when 17 University of California scientists signed a statement saying that it's likely that Americans *are* getting sick because of an inadequate daily intake of one vitamin—vitamin D. In the scientists' view, a large body of research amassed over the last decade has made clear that vitamin D has the potential to affect vulnerability to a host of diseases, among them breast cancer, diabetes, hypertension, and multiple sclerosis. And in the researchers' view, the level of vitamin D intake necessary to minimize susceptibility to these diseases is far higher than that called for under current government standards. What made the UC scientists believe a lack of vitamin D may be harming the public? Consider some of the evidence relevant to this issue:

- In a study that looked at a representative sample of 13,000 American adults 20 years and older, researchers found that, over a period of six or more years, those subjects whose blood levels of vitamin D were in the bottom 25 percent of the study group had a 26 percent higher risk of death from all causes than those whose levels were in the top 25 percent.

- In a study that looked at military recruits who over time were diagnosed with the disease multiple sclerosis (MS), researchers found that, among non-Hispanic whites, the lower the blood levels of vitamin D, the higher the risk of eventually being diagnosed with MS. The researchers found evidence of what they called a "strong protective effect" of vitamin D among recruits who were younger than 20 at the time their blood levels were sampled.

- In a study of nearly 11,000 people born in Finland in 1966, researchers found that those who received high supplemental doses of vitamin D during their first year of life were 80 percent less likely to develop type 1 diabetes over the course of the next

30 years than subjects not given the supplements.

All of these findings were the result of so-called observational studies, which cannot provide proof of a cause-and-effect relationship. More compelling evidence comes from what are known as clinical trials, in which carefully matched subjects are randomly assigned to either receive an intervention—such as vitamin D supplements—or to receive a placebo instead. Such a trial was conducted beginning in 2000 on a group of nearly 1,200 healthy, older women from rural Nebraska, some of whom were assigned to receive moderate daily supplements of vitamin D and calcium. In a period extending from one year into the study through four years into it, women who received the supplements were 77 percent less likely to be diagnosed with cancer than women who did not receive them.

Findings such as these have challenged the traditional view that vitamin D is important primarily as a nutrient that helps provide the body with strong bones. Only in the last decade have scientists come to recognize the degree to which vitamin D acts, in concert with proteins, as a so-called transcription factor, meaning a substance that has the ability to turn genes on or off. This genetic role of vitamin D appears to be substantial, as it's now believed to affect

energy to do this, however, that the caloric content of food is never measured in individual calories. Instead, each *thousand* calories is referred to as a single calorie (sometimes written as Calorie, with a capital C). This is what is meant by the term "calorie" as we see it in everyday life. Each nutritional **calorie** is the amount of energy it takes to raise the temperature of 1,000 grams of water by 1°C.

Thinking of calories in these terms gives us some insight into the nature of carbohydrates, proteins, and lipids. To say that these nutrients contain calories is the same thing as saying they are capable of yielding energy, which is just what they do. Those of

you who studied Chapter 7, on cellular respiration, are familiar with the means by which these nutrients do this: They yield electrons that are high in energy and that thus can run energetically "downhill," in the same way that the water in a stream can run downhill, yielding energy as it goes. It is the energy given up by carbohydrates, lipids, and proteins that powers every activity we carry out, from moving our muscles to reading words on a page. As such, these three classes of nutrients are sometimes referred to as the *energy-yielding nutrients*. To put a little finer point on this, however, the vast bulk of our everyday energy needs actually are met through carbohydrates

the transcription of least 1,000 human genes. Such broad-based genetic activity provides a basis for understanding how one nutrient could affect so many diseases. The disease multiple sclerosis, for example, stems from an attack by the human immune system on the body's own nerve cells. In one of its roles as a transcription factor, vitamin D has been shown to have the ability to damp down on immune system overreactions by reducing the signaling that goes on between immune system cells.

From the time it was identified in the 1920s, vitamin D was understood to have a strange status among vitamins in that, alone among them, its primary source is not food but sunlight. Even in an era of vitamin-D-fortified milk and cereal, Americans get most of their vitamin D from sun exposure. UVB radiation prompts human skin cells to produce a vitamin D compound that is put into active form through two rounds of chemical modification in the body. And the amounts of vitamin D synthesized in this way can be very large, as measured in the International Units (IUs) that are used to calculate vitamin D intake. An 8-ounce glass of fortified milk provides 100 IUs of vitamin D, but 20 minutes of full-body exposure to bright, midday, summer sun can result in the production of 10,000 IUs. Given this, how can anyone have levels of vitamin D that are too low?

UVB radiation levels plunge during the winter at most U.S. latitudes and people differ greatly in how much sun they expose themselves to year-round and how much radiation they absorb in the process. African Americans, for example, absorb so little UVB radiation, even in summer, that their circulating vitamin D levels are about half those of Caucasian Americans, and sunscreens can stop sun-driven vitamin D production almost entirely in persons of any race. To complicate this issue further, UVB exposure is now a well-documented cause of several kinds of skin cancer and thus is not recommended as a means of boosting vitamin D levels.

The job now before American public health officials is how to deal with both the complexity of vitamin D's multiple sources and the new research on vitamin D's effects in setting standards for how much vitamin D Americans ought to be getting. National dietary standards are set by a federal panel called the Food and Nutrition Board of the Institute of Medicine, which since 1997 has recommended that Americans get 200 IUs per day of vitamin D from childhood through age 50, and then double that between the ages of 51 and 70. For most Americans, the standard that counts is this second one—400 IUs per day—as it is the one used to fix the *percentage* of daily vitamin D that a *given* food or vitamin pill con-

tains. (Look on the side of a milk carton, for example, and you'll see that 8 ounces of fortified milk provides 25 percent of the "daily value" of vitamin D, which works out to 100 of the board's recommended 400 daily IUs.)

In the view of a growing number of researchers, it is this recommended daily intake of 400 IUs that is the critical factor in the vitamin D issue. A coalition of research scientists known as the DAction project has recommended 2,000 IUs per day as the optimal daily intake of vitamin D for adult Americans. An intake of this magnitude is required, the scientists believe, if Americans are not to be left vulnerable to chronic diseases such as diabetes and multiple sclerosis.

Mindful of the views of such scientists, the U.S. Food and Nutrition Board has named a committee to advise it on vitamin D levels for Americans. In carrying out its work, the committee will be weighing an enormous body of evidence (only tiny portions of which are noted above) and will be taking into account the fact that it's possible to ingest too *much* vitamin D, which can lead to the dangerous condition of having excessive blood levels of calcium. Thus, American public health officials now have an issue in front of them whose stakes appear to be high: How much vitamin D should Americans be getting?

and lipids alone; proteins yield only about 5 percent of the energy we use.

Given the importance of energy to our functioning, there's some irony in the fact that the calories used to measure energy generally are thought of as something to be *avoided*. We must have energy to live, and energy is measured in calories. In this sense, calories are a good thing. The problem, of course, is that in modern society it's easy to have too much of a good thing: We can easily consume more calories in, say, a day than we will burn up during that day. When this happens, the grams of food that are not burned up within us are not excreted or dissipated to any significant extent; nearly

all of them are *stored* within us, mostly as fat tissue. This is why we end up thinking of calories as a measure of weight. To see calories as units of energy, however, is the first step in understanding what it means for one food to be more "caloric" than another.

Why is it that lipids (more commonly known as fats) are more caloric than either carbohydrates or proteins? The answer has to do with the *density* of energy in these nutrients—with how much energy is contained in a given portion of them. If we look at carbohydrates and proteins, we find that each gram of them yields 4 calories of energy. Meanwhile, each gram of lipids yields 9 calories of energy. Notice, in

Table 31.3

Daily Calorie Requirements

| Groups, by Age | Calorie Range | |
	Sedentary →	Active
Children		
2–3 years	1,000	1,400
Females		
4–8 years	1,200	1,800
9–13	1,600	2,200
14–18	1,800	2,400
19–30	2,000	2,400
31–50	1,800	2,200
51+	1,600	2,200
Males		
4–8 years	1,400	2,000
9–13	1,800	2,600
14–18	2,200	3,200
19–30	2,400	3,200
31–50	2,200	3,000
51+	2,000	2,800

Sedentary means a lifestyle that includes only the light physical activity associated with typical day-to-day life.

Active means a lifestyle that includes physical activity equivalent to walking more than 3 miles per day at 3–4 miles per hour in addition to the light physical activity associated with typical day-to-day life.

this accounting, that lipids are not *heavier* than carbohydrates or proteins. (We're comparing 1-gram quantities of each kind of nutrient, after all.) Instead, gram per gram, lipids contain more energy than do carbohydrates or proteins. This is why lipids have such power to put pounds on us—they're so packed with energy.

So, how many calories do we need per day to satisfy our energy needs? There actually is no single answer to this question, as calorie requirements vary according to age, gender, and level of physical activity. The U.S. Department of Agriculture has, however, taken these things into account in producing the recommendations you can see in **Table 31.3**. If you then look at **Table 31.4**, you can see the number of calories contained in servings of some common foods.

We've thus far thought of carbohydrates, lipids, and proteins strictly in connection with their role in supplying calories, but we need to think more broadly about these three classes of nutrients. We'll now begin looking at them individually, with an eye toward answering the question: What kinds of proteins, carbohydrates, and lipids should we be consuming to ensure that our diet is as healthy as possible?

Table 31.4

Calories in Portions of Some Selected Foods

Food	Calories	Food	Calories
8 oz. mug unsweetened green tea	2	12 oz. glass of 2% milk	180
1 large slice whole wheat bread	79	1 Krispy Kreme Original Glazed® doughnut	200
1/2 cup cooked white rice	102	1 piece thick-crust, 12″ cheese pizza	256
1/2 cup cooked brown rice	107	1 regular (2.07 oz.) Snickers® bar	280
12 oz. glass nonfat milk	120	1 Starbucks grande Mocha Frappuchino® (with whipped cream)	380
12 oz. cola (Coca-Cola®)	140	1 McDonald's Big Mac®	540
		1 Burger King Whopper®	670

Source: USDA and corporate websites and products.

31.7 Proteins, Carbohydrates, and Lipids

Like all nutrients, proteins are defined by their chemical structure (which you can read about in detail in Chapter 3). The particular structure of proteins is that of a set of linked building blocks called amino acids. In a typical human protein, hundreds of amino acids are strung together, one by one, to create a chain that then folds up into a three-dimensional shape that gives each protein its working ability. For purposes of nutrition, the important thing to note about proteins is that the raw materials for them are amino acids.

When we start to look at the roles that proteins take on in the body, we find that no group of molecules is more diverse in its functions. You saw above that almost every chemical reaction in the body is facilitated by the proteins called enzymes. We might marvel at proteins for this regulatory role alone, but as it turns out, proteins play structural roles as well. They form the solid portion of our muscles, for example, and form most of the connective tissue that holds other tissues together. Beyond this, they are active players in the immune system; they serve as signaling molecules and hormones and transport capsules—the list of their functions is very long.

This versatility notwithstanding, you may remember from the discussion above that proteins are *not* very good at one function of nutrients, which is supplying energy. This is so in part because the body has no reservoir of stored proteins. Lipids can be stored in great quantities within us (think of a big belly), and carbohydrates can be stored to a lesser extent (in muscles and in the liver). But this isn't the case with proteins. All of them we have either are being used already or are being transported toward use somewhere within us. Therefore, for the body to use proteins as an energy source, an *existing* protein has to be broken down somewhere—say, in our muscles. This is an inefficient system for extracting energy, and as such, the body employs it mostly when it has nowhere else to turn—that is, when available supplies of lipids and carbohydrates have started to run low.

The proteins we eat cannot enter the bloodstream "as is" but instead must first be broken down in the digestive tract into their amino acid building blocks—molecules that are small enough to pass out of the digestive tract and into circulation. Thus, protein-laden foods don't supply our bodies with proteins per se but rather with a collection of amino acids that can be turned *back* into proteins through the body's protein-synthesizing machinery.

peanut butter contain almost 10 grams. But not all proteins are equally nutritious; what counts in this regard is the amino acid makeup a protein has. A collection of 20 amino acids is needed to produce every protein in the body. Of these 20, however, 11 can be produced *by* the body and so are called **nonessential amino acids.** Meanwhile, there are 9 amino acids that cannot be produced by the body and that must be obtained from foods; these are the **essential amino acids.** For purposes of nutrition, an important point is that nearly every source of *animal* protein supplies us with a complete, balanced set of essential amino acids. This means that every time we eat fish, poultry, eggs, red meat, or dairy products, we are consuming this set. Meanwhile, among plant proteins, only soy protein contains all nine essential amino acids in their proper proportions. The upshot of this is that meat eaters—or vegetarians who eat dairy products—needn't worry about their sources of protein. Vegetarians who don't eat dairy products, however, may need to pay attention to their diet to make sure they are getting all their essential amino acids.

About 70 percent of the protein in the average American diet comes from animal sources, with beef, poultry, and milk being big suppliers of this class of nutrients. Americans tend to consume much more dietary protein than they need. A 154-pound man needs about 56 grams of protein per day but gets about 100 grams. Meanwhile, a 125-pound woman needs about 46 grams per day but gets about 65 grams.

SO FAR . . .

1. In nutrition, each calorie is defined as the amount of _____ required to raise the temperature of a _____ of water by _____.

2. Each gram of lipids yields _____ calories of _____, while each gram of carbohydrates or proteins yields _____ calories.

3. Proteins move from the digestive system into circulation in the form of their building blocks, _____. Nearly every _____ source of protein supplies us with a complete, balanced set of _____, but this usually is not the case with proteins that come from _____ sources.

Protein Sources and Requirements

Dietary proteins can come to us from both animal and plant sources. Three ounces of salmon contain about 18 grams of protein, but then again 2 tablespoons of

Carbohydrates

If proteins are most important nutritionally in areas other than supplying energy, carbohydrates are the reverse: Their primary importance in the diet is that they furnish the energy that helps power our activities. Like proteins, carbohydrates follow a building-blocks model of chemical organization. Whereas the building blocks for proteins are amino acids, however, the building blocks for carbohydrates are known as monosaccharides. If we take one well-known monosaccharide, glucose, and put it together with a second monosaccharide, fructose, we get a slightly more complex *di*saccharide—sucrose, which is better known as table sugar.

As it turns out, monosaccharides and disaccharides represent one of three principal classes of dietary carbohydrates, the **simple sugars.** When we look at the other two classes of dietary carbohydrates, we find that they are composed not of one or two monosaccharide units but of thousands of linked units of the monosaccharide glucose. These are the so-called complex carbohydrates, which exist in two forms. One of these is the **starches,** which can be defined as complex carbohydrates that are digestible. The other is **fibers,** which can be defined as complex carbohydrates that are indigestible.

To get a sense of what these abstract classes mean in practice, imagine yourself at a picnic with a plate of food in front of you that includes a slice of whole wheat bread on your left and a can of cola over to your right. You eat some of the bread and then wash it down with the cola. What have you consumed? Apart from the water in the cola, what you have consumed is overwhelmingly carbohydrate—in all three of its dietary forms. Two of these forms are found solely in the bread you ate. These are starch and fiber. The starch is a complex carbohydrate in the bread that will be digested; it will be broken down in the digestive system and then put into circulation. The fiber in the bread is different; it is a complex carbohydrate that the body *cannot* break down. As such, it stays within the digestive tract and eventually is excreted. (This actually is a good thing, as you'll see.) Meanwhile, the

third form of carbohydrate you consumed at your meal, the simple sugar, was found primarily in the soft drink. Look on most cans of regular soft drinks today, and you are likely to see "high-fructose corn syrup" listed among the ingredients. Your cola probably contained about 40 grams of this simple sugar, which as the name implies, was derived originally from corn. How did this transformation work? Corn from the fields was processed into cornstarch, which was then broken down to yield a mixture of the two simple sugars fructose and glucose, the final ingredients in high-fructose corn syrup.

Carbohydrate Choices: Think Fresh and Whole Grain

Simple sugars, starches, and fibers all have nutritional uses. But this is not to say that these three forms of carbohydrate are all equally healthy. Nutritionists today uniformly recommend a diet that is not only low in simple sugars that have been added to products—as was the case with our high-fructose corn syrup—but high in particular sorts of *complex* carbohydrates: those that come from fresh fruits, vegetables, and whole-grained cereals. (In a nutritional sense, *cereals* refers to cereal grain crops such as wheat, oats, rice, barley, and corn.) What this advice often means in practice is eating carbohydrates that have been *processed* as little as possible (**Figure 31.10**). To understand this concept, consider the whole wheat bread and the cola that we just looked at in our picnic meal. What made the wheat in the picnic bread "whole wheat" was the fact that it had undergone relatively little processing; each wheat grain used in this bread retained both its bran (outer coat) and its wheat germ (embryonic tissue), whereas the wheat used to make white bread has had both of these components stripped away (**Figure 31.11**). On the other side of the spectrum, if we look at our cola, we can see that a starting ingredient for it was

Figure 31.10
Processed and Unprocessed

Foods that are heavily processed, such as those on the left, tend to be less healthy than those that are fresh and unprocessed, such as those on the right.

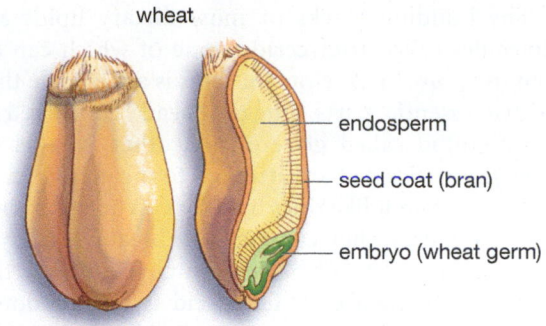

wheat

— endosperm

— seed coat (bran)

— embryo (wheat germ)

Figure 31.11
What "Whole Wheat" Means
Each grain of wheat used in a whole wheat product retains both its seed coat (also known as its bran) and its embryo (also known as its wheat germ). In contrast, grains of wheat used to make white bread or regular pasta have both of these elements removed during processing.

a complex carbohydrate (the starch in whole corn) that, through a good deal of processing, ended up yielding two simple sugars.

The question then becomes: In nutritional terms, how does this cola, with its simple sugars, compare to our slice of whole wheat bread, with its whole grains? When we look at a typical, large slice of whole wheat bread, we find that it contains about 80 calories. Meanwhile, a 12-ounce can of cola might contain 140 calories, all of which come from the corn syrup. Now, *apart* from the calories—which have a value in that they're yielding energy—what do we get from the cola in the way of nutrition? Almost nothing: water and a little sodium, but that's about it. Meanwhile, if we look at the bread, what do we get? Calcium, iron, magnesium, potassium, small quantities of several B vitamins—in short, an array of vitamins and minerals. Comparisons such as these are the reason nutritionists have long referred to soft drinks as foods that provide "empty calories," meaning little in the way of nutrition *except* calories. But the nutritional gap between our simple sugar and our whole-grained bread involves more than just vitamins and minerals. To understand a second factor that's at work, let's think for a moment about how these carbohydrates are processed by the body.

Slower Digestion and Glycemic Load

As you saw earlier in the chapter, only small, simple molecules can pass from the digestive tract into the bloodstream. And as you've seen, the starch that helps make up whole wheat is neither small nor simple. It is a complex carbohydrate whose long chains must be broken down into simple carbohydrates—into the individual glucose units that can move into circulation. It takes time to do this, which means that the glucose that results from the digestion of whole wheat moves *slowly* into circulation. Contrast this pace, however,

with what happens with the carbohydrate from our cola. This carbohydrate is *ingested* as a set of simple sugars—it comes to us in a simple form that requires relatively little further digestion. Consequently, concentrations of simple sugars we ingest can result in a "spike" of glucose in the bloodstream, and this in turn can result in a spike of insulin, released from the pancreas, whose function is to store the glucose away in cells. One problem with these insulin surges is that they seem to bring on hunger sooner than do slow, steady rises in insulin levels. Beyond this issue, however, persistently high blood levels of insulin can raise levels of lipids in circulation while lowering levels of "good" cholesterol (or HDLs). One large, long-standing study of diet and health found that a high intake of calories from refined grains and potatoes is associated with a higher risk of developing both type 2 diabetes and heart disease. Given findings such as these, the take-away message is that the faster a given carbohydrate moves from the digestive tract into circulation, the less healthy it seems to be.

This conclusion is important enough that nutritionists have come up with a couple of ways of measuring the effects of different carbohydrates on blood glucose levels. The best accepted of these is called a food's **glycemic load:** a measure of how blood glucose levels are affected by defined portions of given carbohydrates. Glycemic loads of less than 15 are considered low (which is good), while those of 15 to 20 are considered intermediate, and those of more than 20 are considered high. To give you some perspective on these figures, the single slice of whole wheat bread we've been looking at has a glycemic load of about 9, while a typical 12-ounce cola would have a glycemic load of about 25.

Fiber: The Value of Not Being Digested

Beyond added nutrients and glycemic load, there turns out to be a third reason to prefer unprocessed, fresh, and whole-grained carbohydrates over any other variety. Recall that in addition to the two classes of carbohydrates we've been looking so much at— simple sugars and starches—there is then a third class, which are the various fiber carbohydrates that go undigested when we eat them. You might think that any substance that fails to move into circulation couldn't provide health benefits at all, but this isn't the case with fibers. Among other things, these tough, complex carbohydrates bind in the digestive tract with cholesterol and substances made from it, which has the effect of lowering *blood* levels of cholesterol. In addition, fiber increases the bulk of our stool, in part by absorbing water, and this effect softens our stool, making it easier to pass. (Excessive straining during defecation can produce small pouches, called diverticula, that pop out from the large intestine and that can become sites of inflammation or even infection.)

Moreover, any starch that is bound up with fiber will move more slowly into circulation; thus, fiber helps with the glycemic load problem we just talked about. Beyond these factors, to judge by the long-standing study noted earlier, a high-fiber diet may lower the risk of acquiring type 2 diabetes and heart disease, while as we've seen, a diet high in refined grains and potatoes may raise this risk.

So, where do different foods stand with respect to their fiber content? By now you can probably make some informed guesses. Looking once again at our cola, with its high-fructose corn syrup, we find that it has a fiber content of zero. If we then look not at our slice of whole wheat bread but instead at bread made from more heavily processed *white* flour, we find that a slice of it contains 0.7 grams of fiber. And our whole wheat bread? One slice of it contains 2.2 grams of fiber.

SO FAR . . .

1. There are three principal classes of dietary carbohydrates: _____, _____, and _____.

2. To say that a carbohydrate is a source of "empty calories" is to say that it provides very few _____ or _____.

3. Glycemic load is a measure of how _____ levels are affected by defined portions of given carbohydrates. The _____ the glycemic load, the better, in terms of health.

Lipids

Our final class of nutrients, the lipids, are a terrific source of energy, as we saw earlier, but lipids serve more roles than just providing energy. Those of you who read Chapter 4, on cell structure, may recall that every human cell has an outer, or "plasma," membrane that is largely composed of lipids (the phospholipids and cholesterol) and, beyond this, lipids are used to make hormones such as estrogen and testosterone.

The building blocks of most dietary lipids are molecules called triglycerides, one of which can be seen in **Figure 31.12**. Note that this is a molecule that has two essential parts. One is a "head," composed of a compound called glycerol; the other is a set of three *fatty acids* that are attached to the glycerol and that stem from it like tines on a fork. When our bodies need energy, they can get it by breaking triglycerides apart: The fatty acids are clipped from the glycerol, and then each fatty acid is broken down further. When, however, our bodies have plenty of energy, this process goes the other way: Glycerol and fatty acid units are brought together, and the resulting triglycerides are then stored in cells called adipocytes, which are better known as fat cells. A key point in nutrition, however, is that it doesn't *take* lipids to *make* lipids. Should we have more carbohydrates or proteins in circulation than are necessary to meet our energy needs, these nutrients will go to the liver and be transformed into triglycerides, which will end up being stored in fat cells. In short, we could get fat eating nothing but carbohydrates and proteins.

Oils and Fats

In everyday speech, lipids are invariably referred to as fats, but this common term obscures an important distinction. We actually consume dietary lipids in two different forms. One of these is **oils,** which can be defined as dietary lipids that are liquid at room temperature (olive oil, canola oil). The other is then **fats,** which can be defined as dietary lipids that are solid at room temperature (butter, the fat in a piece of bacon). As we'll see, these two terms will be helpful in making distinctions among different types of lipids with respect to their effects on health. And what is the health concern connected to lipids? As most people know, the central issue is heart disease. Although recent research results have been mixed, a preponderance of evidence indicates that, as our consumption of some lipids goes up, so does our long-term risk for developing heart disease. Meanwhile, other lipids appear to be at least neutral with respect to heart disease, while still others actually seem to protect against it. So, which lipids fall into which category? Let's take a look.

Figure 31.12
Building Block of Dietary Lipids

Most of the lipids we consume are composed of building blocks called triglycerides. Every triglyceride is composed of a single glycerol molecule linked to three fatty acids. This particular triglyceride, called tristearin, has three identical fatty acids linked to its glycerol "head," but in many cases several different fatty acids will be part of a single triglyceride.

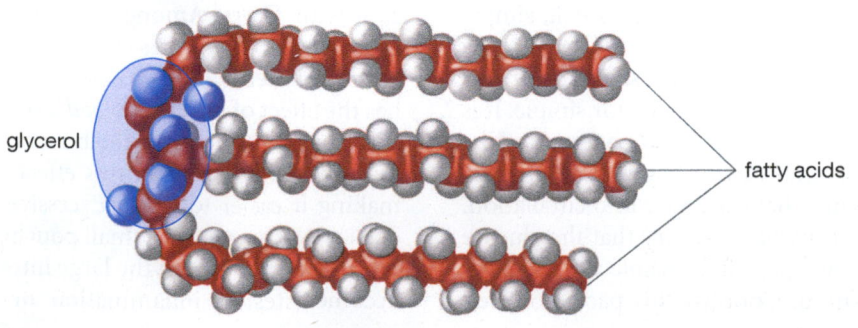

glycerol

fatty acids

Lipid Choices: Think Unsaturated

In discussions of lipids, you've probably heard the terms *saturated* and *unsaturated* fats. What makes a "fat"—meaning a fat or oil—saturated or unsaturated is its fatty acid makeup. About 80 percent of the fatty acids in olive oil, for example, are unsaturated, while only 20 percent of the fatty acids in it are saturated. Hence, olive oil is referred to as an unsaturated fat. Meanwhile, if we look at a stick of butter, we find that 60 percent of the fatty acids in it are saturated, while 40 percent are unsaturated; hence, butter is referred to as a saturated fat. (To learn about the chemical basis for this saturated/unsaturated difference, see Chapter 3, beginning on page 49.)

On one level, the importance of the saturated-to-unsaturated spectrum simply has to do with the *form* of lipids we consume: Lipids that contain a high proportion of unsaturated fatty acids tend to be oils (as in our olive oil example), while those that contain a high proportion of saturated fatty acids tend to be fats (as in our butter example). On another level, however, the transition between unsaturated and saturated lipids is the single most important factor in how *healthy* various lipids are, as defined by their effect on heart disease.

Why should this be so? Saturated fats raise circulating levels of the LDL or "bad cholesterol" molecules we looked at last chapter—the molecules that are instrumental in initiating the most common form of heart disease. Indeed, saturated fats raise LDL levels more than does any other natural class of nutrients, though the cholesterol we consume in eggs and other animal products has this same effect. Meanwhile, two forms of unsaturated fats—the polyunsaturated and monounsaturated fats—are at least neutral with respect to heart disease and may actually help protect against it. Finally, there is a special class of dietary lipids, the trans fats, that are produced through an industrial process and that are even more unhealthy than saturated fats. All this may sound complicated, but it's actually easy to make sense of these various kinds of dietary lipids. Here is a hierarchy of them that runs from least healthy to most healthy:

- Trans fats
- Saturated fats
- Polyunsaturated and monounsaturated fats
- Polyunsaturated and monounsaturated fats rich in omega-3 fatty acids

To put this within the framework we just went over, consumption of trans fats and saturated fats increases our long-term risk of heart disease. Meanwhile, monounsaturated and polyunsaturated fats are at least neutral with respect to heart disease, and the unsaturated fats rich in omega-3 fatty acids actually protect against it.

But how do we know which kinds of fats we're consuming? We can be guided by just a couple of simple rules. First, the monounsaturated and polyunsaturated fats tend to come to us as *oils*: olive, canola, and peanut oil for the monounsaturated fats and safflower and corn oil for the polyunsaturated fats. Thus, a good rule of thumb is that if a lipid is an oil, it is at least neutral with respect to health and may actually be healthy. Which oils fall into the healthy category? Two prime candidates are olive oil and canola oil. Several studies have shown that residents of Mediterranean countries have significantly lower levels of heart disease than people in other parts of the developed world. What these residents have in common, in terms of diet, is consumption of large amounts of olive oil. And canola oil is a good source of omega-3 fatty acids which have been shown to guard against heart disease in several ways: They prevent blood clot formation, they reduce fat levels generally in the bloodstream, and they reduce the growth of the fatty deposits that clog heart arteries. The best source of all for omega-3s isn't oils, however, but instead is certain kinds of fatty fish—notably salmon and albacore tuna—although these lipids can also be found in lesser amounts in plant-based foods such as walnuts and the soybean-based product tofu.

On the unhealthy side of the spectrum, the usual source for *saturated* fats is animal fat, primarily fatty meats and dairy products such as cheese and butter. Finally, the least healthy fats of all, the trans fats, are found in a significant, but decreasing, number of packaged and fast foods: cookies, French fries, cakes, crackers, doughnuts, popcorn, and candy, to name a few foods. Trans fats are created in an industrial process in which naturally occurring oils are turned *into* fats by partially saturating them—by bubbling hydrogen through the oils in a process called hydrogenation. The upside of this is the production of fats that have a longer shelf life, or that have a taste or creamy texture that consumers like. The downside is the production of fats that raise levels of the "bad" LDL cholesterol while lowering levels of "good" HDL cholesterol and boosting circulating fat levels in general. Given these effects, American food producers and fast-food chains have been moving in recent years to remove trans fats from their products. It's actually easy to know whether you're consuming trans fats in packaged foods: Simply look for the "trans fats" line within the Nutrition Facts label. These days, it's likely that the number of grams of trans fats you'll see there will be zero.

31.8 Elements of a Healthy Diet

We've been looking at the six classes of dietary nutrients in isolation from one another, but any real-world diet will, of course, be a mixture of different foods that contain these nutrients. If we think strictly along the lines of nutrition, what would a good combination of

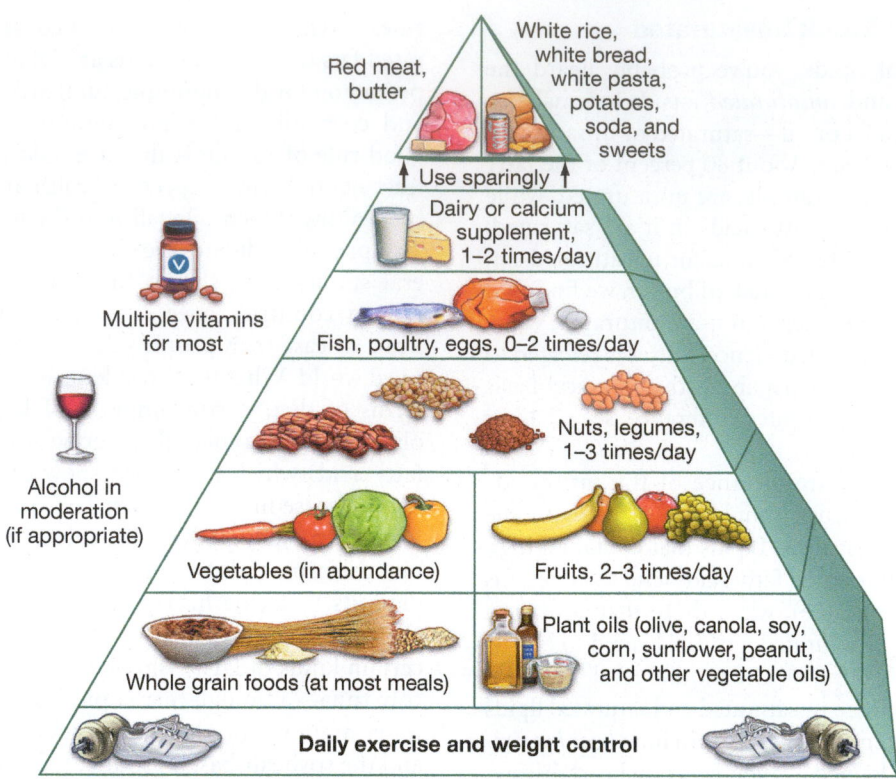

Figure 31.13

Foods in the Proper Proportions

This food pyramid, constructed by researchers at the Harvard School of Public Health, provides one view of the proportions that foods ought to be consumed in to ensure a healthy diet. Note the prominent role recommended not only for whole grains but for plant oils (canola, olive, etc.), which can be thought of as "good" fats. The recommendation for "alcohol in moderation (if appropriate)" stems from evidence indicating that drinking small amounts of alcohol as part of a healthy diet can reduce the risk of heart disease.

foods look like in a diet? More specifically, what *proportions* of different kinds of foods would go into making up a healthy diet?

For years, nutritionists have found the image of a pyramid useful in making recommendations about food proportions. If you look at **Figure 31.13**, you can see one such pyramid—in this case a pyramid constructed by researchers at the Harvard School of Public Health. Many of the recommendations in this figure will by now be familiar. Note that sweets and other refined carbohydrates have been consigned to the narrow "use sparingly" portion of the pyramid, while whole-grained foods are at its base and recommended for "most meals." Meanwhile, fruits and vegetables should make up a substantial portion of a healthy diet, while red meat and butter—both high in saturated fats—are recommended for infrequent consumption.

Although these ideas about food proportions are shared by most nutritionists, they represent only one way of looking at the elements of a healthy diet. In 2005, the U.S. Department of Agriculture abandoned its traditional food pyramid after deciding that no one set of food recommendations could take all Americans into account. The USDA still employs a food pyramid, but only as one element within an in-

teractive website that allows individuals to enter information about themselves as a means of receiving individualized dietary recommendations. You can construct your own set of USDA dietary recommendations by visiting www.MyPyramid.gov.

SO FAR ...

1. Most of the lipids we consume are composed of molecules called triglycerides, which in turn are composed of a molecule called glycerol and three _____.

2. As we move from unsaturated to saturated fatty acids, we move in a general way from _____, which are dietary lipids that are liquid at room temperature, to _____, which are dietary lipids that are solid at room temperature.

3. Most saturated fats have a(n) _____ source, while most monounsaturated and polyunsaturated fats have a(n) _____ source.

31.9 The Urinary System

If nutrition and digestion largely concern what moves into blood circulation, the subject we'll take up now is the reverse: What gets filtered *out* of the blood and then eliminated from the body altogether? Our subject is the urinary system, which filters waste from the blood and then passes this waste on to the bladder for elimination.

Such waste removal obviously is important, but it turns out to be only one of several critical functions the urinary system undertakes. Equally important is the system's regulation of blood volume—how much blood we have within us—a task whose necessity becomes obvious when we think about consuming lots of liquids. Suppose the body simply incorporated all the water in these liquids into circulating blood? We would eventually swell and burst like a water balloon. Instead, the urinary system *regulates* how much water the body retains and how much it excretes, with the kidneys being a critical player in this process. The urinary system also controls ion concentration and maintains pH balance in the body. In short, it is important in preserving the stable state known as homeostasis. We almost never think of the urinary system in this regard, essentially because it works so well. But a person who suffers kidney failure and must rely on a kidney dialysis machine to filter wastes knows the value of the urinary system in a personal way.

In carrying out these activities, the urinary system is very much involved in the *conservation* of bodily resources. Indeed, the kidneys are as much in the conservation business as the waste removal business. They conserve water, amino acids, ions, sugars—all kinds of things that the body needs. Consider that the kidneys process about 180 liters (or almost 48 gallons) of fluid every day. Yet our urination is only about 1.8 liters, or about *half* a gallon, daily. More than 47 gallons recycled back into the body each day! Now let's see how the kidneys and the larger urination system manage this feat.

Structure of the Urinary System

The pivotal structures in the urinary system, shown in **Figure 31.14**, are its two filtering organs, the **kidneys,** which produce urine while conserving useful blood-borne materials. Urine leaving the kidneys travels along two tubes, the left and right **ureters,** to the **urinary bladder,** the hollow, muscular organ that acts as a temporary, expandable storage site for urine. When urination occurs, contraction of the muscular bladder forces the urine through the conducting tube called the **urethra** and out of the body.

The kidneys are located on either side of the vertebral column, at about the level of the eleventh and twelfth ribs. They are shaped like kidney beans, but

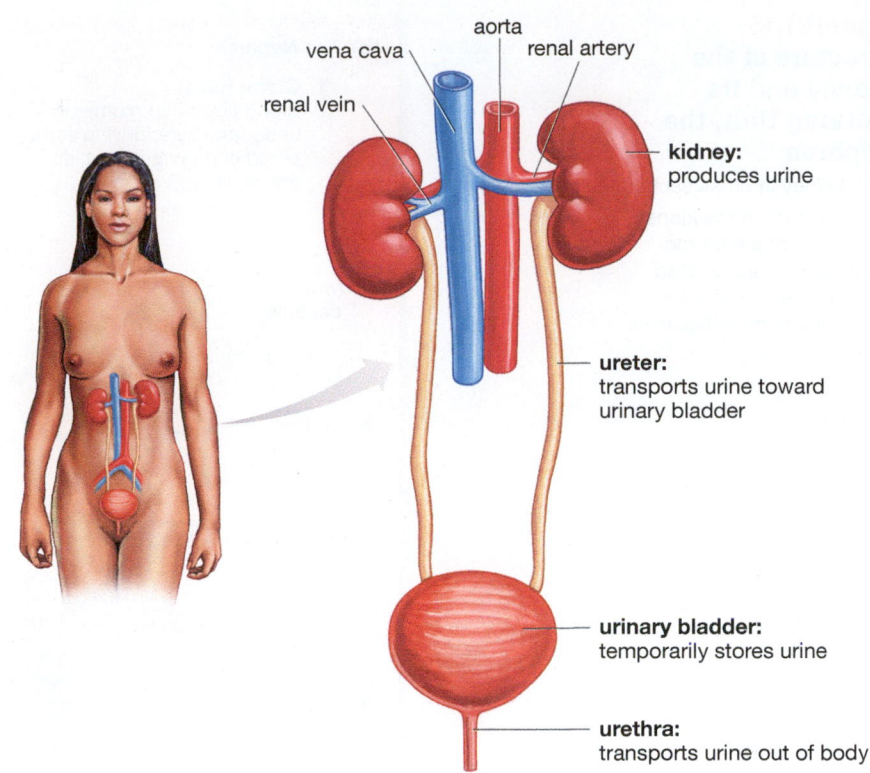

kidney:
produces urine

ureter:
transports urine toward
urinary bladder

urinary bladder:
temporarily stores urine

urethra:
transports urine out of body

aorta
vena cava
renal artery
renal vein

Figure 31.14
The Urinary System
in Overview

since they are each about 10 centimeters, or 4 inches, long, their size is more like that of a small pear.

In considering how the kidneys work, it's helpful to think first in terms of input and output. What comes into the kidney in the way of fluids? Blood from the circulatory system arrives by way of the renal arteries. As you can see in Figure 31.14, these branch off from the aorta, one renal artery going to the left kidney, one going to the right. Now, what comes *out* of the kidney? Two things come out: Blood flows out of the kidneys through the renal *veins* that are pictured, and urine exits each kidney through its ureter. So, each kidney has one input (the renal artery) but *two* outputs. This is a clue to what's happening inside. Some of the material coming into the kidney through the renal arteries is being channeled out of the blood vessels and into a *separate* system of vessels whose output tubes are the ureters. Now let's look more closely at the kidneys.

As seen in the lower left of **Figure 31.15a**, on the next page, each kidney has within it a chamber called the renal pelvis, whose numerous branches collect the urine produced by the kidneys and channel it into the ureter. If you look at **Figure 31.15b**, you can see a blow-up of the basic working unit of the kidney, a structure called a nephron. Each kidney has within it about 1.25 million nephrons, each serviced by a single arteriole that is part of a network of vessels whose blood comes from the entering renal artery.

In the figure, you can see how a nephron begins: An arteriole enters a hollow chamber, called the Bowman's capsule, and then promptly expands into a knotted network of capillaries, called the glomerulus,

Figure 31.15
Structure of the Kidney and Its Working Unit, the Nephron

(a) A kidney in cross section.

(b) A nephron in the kidney, composed of a nephron tubule and its associated blood vessels, which are immersed in interstitial fluid.

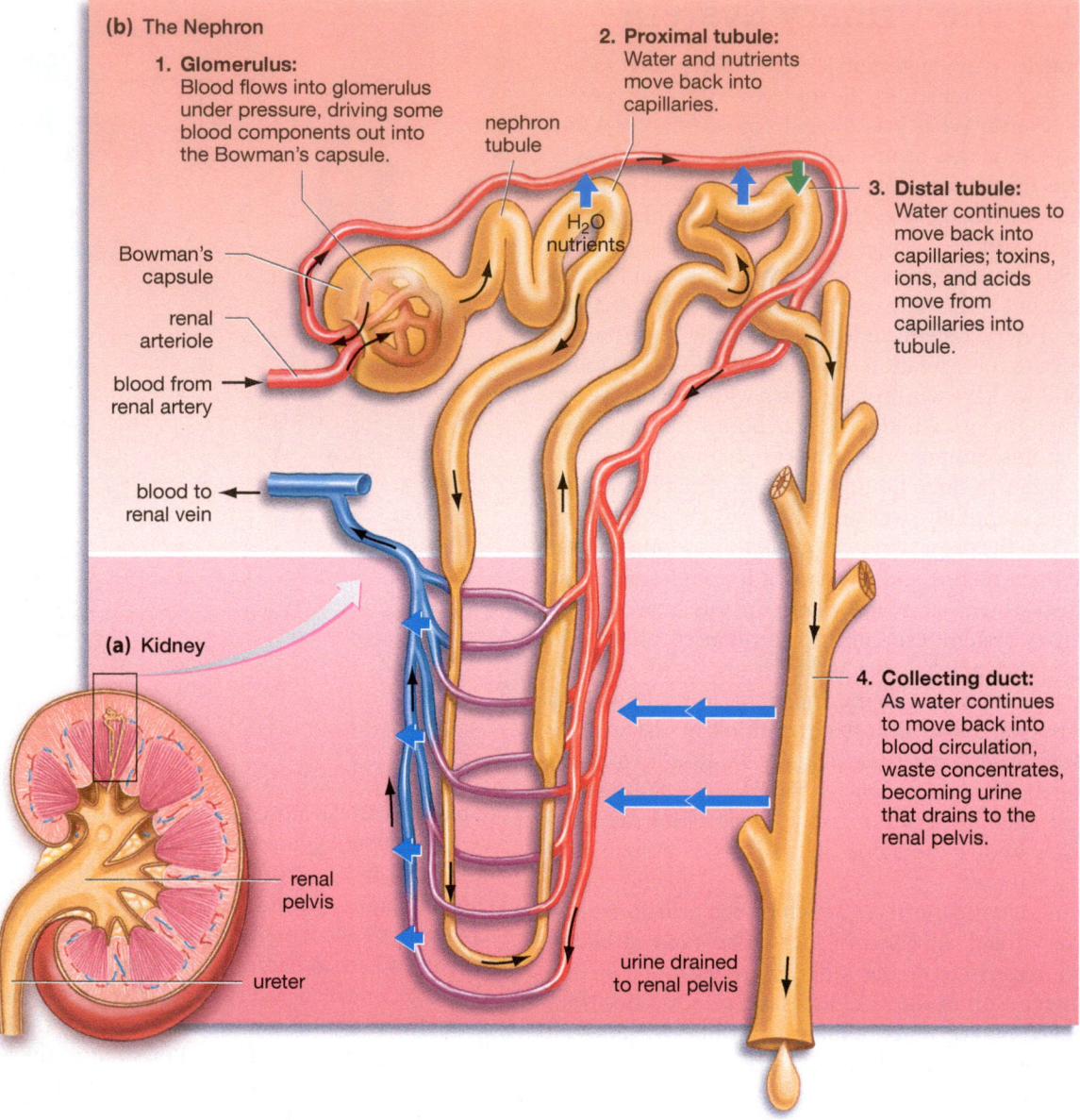

(b) The Nephron

1. Glomerulus: Blood flows into glomerulus under pressure, driving some blood components out into the Bowman's capsule.

2. Proximal tubule: Water and nutrients move back into capillaries.

nephron tubule

3. Distal tubule: Water continues to move back into capillaries; toxins, ions, and acids move from capillaries into tubule.

Bowman's capsule

renal arteriole

blood from renal artery

blood to renal vein

H_2O nutrients

(a) Kidney

4. Collecting duct: As water continues to move back into blood circulation, waste concentrates, becoming urine that drains to the renal pelvis.

renal pelvis

ureter

urine drained to renal pelvis

that is enclosed by the capsule. *Emerging* from this network, there is again an arteriole that goes on to surround something called a nephron tubule. (In reality, this arteriole branches into a network of capillaries that surrounds the tubule, but for simplicity's sake these capillaries are shown here as a single blood vessel.) Thus, we have two sets of tubes—the nephron tubule and the blood vessel—and these will become the two *outputs* just mentioned: The nephron tubule will transmit what the body doesn't need (waste sent to the bladder), while the blood vessel will first lose, and then take back, what the body does need (nutrients, etc.). All this put together adds up to a **nephron:** the functional unit of the kidneys, composed of a nephron tubule, its associated blood vessels, and the interstitial fluid in which both are immersed. At the "output" end of the nephron, you can see that its tubule is only one of several that are emptying into a

common *collecting duct.* Eventually, many of these ducts will themselves merge to feed into the renal pelvis, which feeds into the ureter. Now, let's look at what takes place within a nephron.

31.10 How the Kidneys Function

Imagine holding a washcloth, baglike, underneath a running kitchen faucet. What happens? The water slowly flows out of the washcloth. What if this is a faucet that hasn't been turned on in a while, however, and the water thus contains some sediment? Now the washcloth acts to retain these larger particles while letting smaller substances (the water molecules) flow through. In short, the washcloth acts as a filter. The knotted network of nephron capillaries called the

glomerulus takes on this same filtering role. Receiving blood that is under pressure, it is porous enough to let some smaller blood-borne materials pass out of it—and into the surrounding Bowman's capsule (see Figure 31.15b). What materials pass out? Water, first of all, which makes up such a large portion of the blood, and then many of the materials suspended in the water, such as vitamins, nutrients, and waste materials. All these substances now move from the glomerulus into the Bowman's capsule. What stays in the glomerulus (and hence in the bloodstream) are plasma proteins and blood cells, which are too big to pass out.

We should note at this point the nature of the waste products that are moving out of the glomerulus. Every time you use a muscle, you produce a waste product, called creatinine, that has to be eliminated from the body. Likewise, proteins are broken down mostly in the liver, and a by-product of this breakdown is a substance called urea. Both urea and creatinine are picked up by the bloodstream and move to a glomerulus, where they flow out, along with the water and other substances.

All the materials that have entered the Bowman's capsule constitute a fluid, called filtrate, that flows from the capsule into the nephron's next structure, its proximal tubule. With this tubule, we get to the essence of how the kidneys work as a filtering mechanism—getting rid of waste but retaining what is valuable. The retention process begins immediately after the filtrate moves into the proximal tubule. Some active pumping of materials out of the tubule is coupled to the process of osmosis (reviewed in Chapter 5). The result of both actions is that about two-thirds of the water lost in filtration, as well as almost all the nutrients lost there, move back into blood circulation at the proximal tubule—back into the network of capillaries represented in Figure 31.15b by the single blood vessel. While this "reabsorption" is going on, however, the urea and other waste products have remained in the nephron tubule because, thanks to their chemical makeup, they cannot easily pass out of it.

If you just follow the nephron tubule on around, past the proximal tubule, you can see what comes next: Water continues to flow out of the tubule and back into circulation (blue arrows), while up at the distal tubule, toxins and other waste move the other way: from circulation into the tubule (green arrow). The end result is that, with movement through the tubule, the filtrate becomes an ever-more-concentrated waste product. Water and nutrients are removed from it and sent back to circulation, while urea, creatinine, and other wastes remain in it. By the time the filtrate reaches the collecting duct, it contains such a high concentration of waste that it goes by another name: urine.

Hormonal Control of Water Retention

We noted earlier that the kidneys are important players in the body's control of the volume of fluids we have within us. How does this control work? The key to it is a substance produced by the brain's hypothalamus and released by the adjacent posterior pituitary gland, **antidiuretic hormone** or **ADH,** which is also known as vasopressin. If we get dehydrated, the pituitary releases ADH. When this hormone reaches the kidneys, it has the effect of allowing water to flow more freely out of the distal nephron tubule and the collecting duct. What happens next, of course, is that this water then moves back into blood circulation. Thus, ADH prompts the kidneys to recycle more water back into circulation rather than allowing them to send this water to the bladder as urine. Of course, there are times when we have too much water in the body. In this event, the hypothalamus shuts down its production of ADH, and this prompts the distal tubule and collecting duct to hold on to more water, which means more of it is sent to the bladder. The end result of all this is that the body can tightly regulate the amount of water it retains. Along these lines, if you've ever wondered why 12 ounces of beer make a person have to urinate more than do 12 ounces of lemonade, it's because the alcohol in the beer suppresses the production of ADH. Reduced ADH levels mean that more of the water that enters our systems will be sent to the bladder rather than being moved back into circulation.

31.11 Urine Storage and Excretion

Having seen how the kidneys produce urine, let's now look at the steps by which urine is then excreted from the body. Recall that urine is transported to the urinary bladder by the pair of tubes called the ureters—one stemming from each kidney. The ureters are muscular tubes that extend for a distance of about 30 centimeters, or 12 inches. Starting at the kidney, about every 30 seconds a peristaltic wave of muscle contraction squeezes urine out of the renal pelvis, along the ureter, and into the urinary bladder. Occasionally, solids develop within the collecting tubules, collecting ducts, or ureters. These are kidney stones, and they can be very painful; in addition to obstructing the flow of urine, they may also reduce or eliminate filtration in the affected kidney.

The Urinary System

The Urinary Bladder

The urinary bladder is a hollow muscular organ that stores urine. Its dimensions vary with the volume of stored urine, but a full urinary bladder can contain up to 800 milliliters of urine, which is about 27 ounces. The

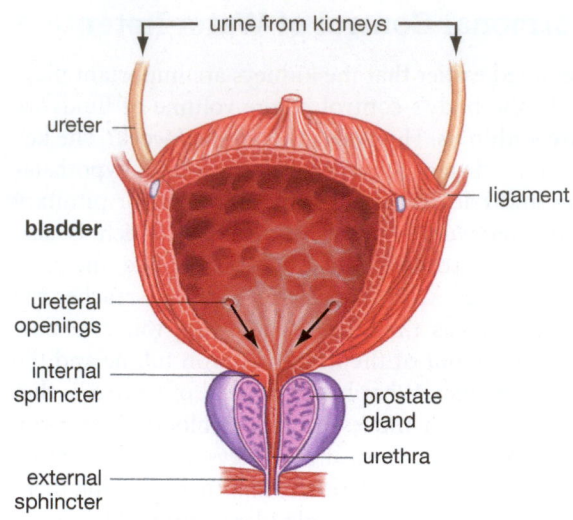

urine from kidneys

ureter

bladder

ureteral openings

internal sphincter

external sphincter

ligament

prostate gland

urethra

Figure 31.16
The Urinary Bladder in a Male

area surrounding the urethral entrance, called the neck of the urinary bladder, contains an internal sphincter muscle that provides involuntary control over the discharge of urine from the bladder (**Figure 31.16**).

The Urethra

The urethra extends from the urinary bladder to the exterior of the body. Its length and functions differ in males and females. In the male (shown in Figure 31.16), the urethra extends to the tip of the penis and is about 18 to 20 centimeters (7 to 8 inches) in length. The initial portion of the male urethra is surrounded by the prostate gland. In contrast, the female urethra is very short, extending 2.5 to 3.0 centimeters (about 1 inch). In both sexes, the urethra contains a circular band of skeletal muscle that forms an external sphincter. Unlike the internal sphincter, the external sphincter is under voluntary control.

Urination

The urge to urinate usually appears when the bladder contains about 200 milliliters, or about 7 ounces, of urine. We become aware of this need through nerve impulses sent to the brain from stretch receptors in the wall of the urinary bladder. The stimulation of these receptors also results in involuntary contractions of the urinary bladder that increase the fluid pressure inside the bladder (which is why we "have to go" to the bathroom). Urine ejection cannot occur,

however, unless both the internal and external sphincters are relaxed. We control the time and place of urination by voluntarily relaxing the external sphincter. When this sphincter relaxes, so does the internal sphincter. If the external sphincter does not relax, the internal sphincter remains closed, and the bladder gradually relaxes. A further increase in bladder volume begins the cycle again, usually within an hour.

On to Development and Reproduction

Over the last six chapters, we've reviewed the ways human beings breathe, think, digest, fight off invaders, remove wastes, and more. While all of these operations are critical to maintaining life, they don't touch on one of the central abilities of living things, which is the ability to pass life on. Each generation of human beings gives rise to another, of course, through the workings of a specialized system, the reproductive system, which comes in two versions: male and female. In essence, these systems must produce reproductive cells—egg and sperm—that come together to produce a single cell that, over the course of 9 months, develops into a 10-trillion-celled baby in full cry. So, how does this process work? How are egg and sperm produced in the first place, how do they come together, and how does the cell that results from their union develop into a new generation of living thing? These are the subjects of the next two chapters.

SO FAR . . .

1. The "input" into each kidney is _____ that flows through one of the body's _____. The "output" from each kidney is _____ that flows into one of the body's _____ and urine that flows into the _____.

2. The working unit of the kidneys is the nephron, composed of a nephron tubule, its associated _____, and the fluid in which both are immersed.

3. Nephrons work by means of retaining _____ within the nephron tubules while returning water and nutrients to the _____.

31 REVIEW

Summary

31.1 The Digestive System

- The digestive system breaks down foods into a form the body can use, transfers nutrients from the foods out of the digestive tract and into blood circulation, retains the waste that remains after nutrients from the food have been transferred, and eliminates this waste from the body. The central structure of the digestive system is the digestive tract, a muscular tube that passes from the mouth to the anus and that receives input along its length from organs such as the gallbladder, liver, and pancreas. (p. 591)

31.2 Structure of the Digestive System

- The digestive tract begins with the mouth (or oral cavity) and continues through the esophagus, stomach, small intestine, and large intestine before ending at the rectum and anus. Through much of its length, it is a tube composed of four layers: the serosa, the muscularis externa, the submucosa, and the mucosa. Nutrients absorbed from the digestive tract are taken up by capillaries (in the case of carbohydrates and proteins) or lymphatic vessels (in the case of fats). (p. 592)

31.3 Steps in Digestion

- Digestion begins in the mouth through both mechanical means and chemical means. Food pushed back by the tongue into the esophagus is transported through it, to the stomach, by peristalsis and gravity. (p. 593)

- The stomach digests food by mechanical means of churning and by chemical means. The mixture of food and digestive juices that then leaves the stomach is chyme. (p. 593)

- Eighty percent of the digestive tract's absorption of nutrients takes place within the small intestine—that portion of the digestive tract that begins at the stomach and ends at the large intestine. (p. 595)

- The pancreas is an organ that secretes, into the small intestine, digestive enzymes that help break down fats, proteins, and carbohydrates. It also secretes "buffers" that raise the pH of the acidic chyme coming from the stomach. (p. 595)

- The gallbladder stores and concentrates bile, a substance produced by the liver that aids in the digestion of fats. (p. 596)

- The liver receives blood-borne nutrients from the digestive tract and then controls which nutrients will be stored and which will be sent into circulation. (p. 596)

- The large intestine holds and compacts material left over from digestion, turning it into feces. It also returns water to general circulation and absorbs vitamins produced by resident bacteria. (p. 596)

31.4 Human Nutrition

- A nutrient is a substance contained in food that provides energy, provides a structural building block, or helps regulate a process. The six classes of nutrients are water, minerals and vitamins; and proteins, carbohydrates, and lipids. (p. 597)

31.5 Water, Minerals, and Vitamins

- People generally do not need to be concerned about getting enough water in their diet; thirst is the best indication of hydration. (p. 598)

- Minerals are chemical elements that are needed either to help form bodily structures or to facilitate chemical reactions. (p. 598)

- Vitamins are chemical compounds that are needed in strictly small quantities to facilitate chemical reactions. (p. 598)

31.6 Calories and the Energy-Yielding Nutrients

- Calories in the human diet are provided by three classes of energy-yielding nutrients: proteins, carbohydrates, and lipids. A nutritional calorie (sometimes written Calorie) is the amount of energy it takes to raise the temperature of 1,000 grams of water by 1°C. (p. 601)

- Quantities of energy-yielding nutrients that we consume but do not burn up end up being stored within us, mostly as fat tissue. Each gram of lipids yields 9 calories of energy while each gram of proteins or carbohydrates yields 4 calories. (p. 603)

31.7 Proteins, Carbohydrates, and Lipids

- Proteins are composed of amino acids. Proteins provide very little energy compared to carbohydrates or lipids but have many structural and regulatory functions in the body. (p. 605)

- Proteins are broken down in the digestive tract into their amino acid building blocks; it is these amino acids that move into circulation. Eleven of the 20 amino acids needed for protein synthesis can be produced by the body and are called nonessential amino acids. Nine amino acids cannot be produced by the body and must be obtained from food; these are called the essential amino acids. (p. 605)

- Monosaccharides and disaccharides form one of the three principal classes of dietary carbohydrates, the simple sugars. The other two classes of dietary carbohydrates, collectively known as complex carbohydrates, are starches, defined as complex carbohydrates that are digestible;

and fibers, defined as complex carbohydrates that are indigestible. **(p. 605)**

- Nutritionists recommend a diet that is low in simple sugars that have been added to processed foods but high in complex carbohydrates. Foods high in added simple sugars often are the source of "empty" calories, meaning calories not accompanied by vitamins or minerals. **(p. 606)**

- Consuming large quantities of simple sugars can result in a surge of glucose into the bloodstream, whereas the consumption of fresh and whole-grained complex carbohydrates results in a healthier slow, steady entry of glucose into the bloodstream. **(p. 607)**

- Fibers lower blood levels of cholesterol and reduce glycemic load. Whole, unrefined grains tend to be high in fiber, while processed, "white" grains have less fiber, and added simple sugars have no fiber. **(p. 607)**

- Lipids are a great source of energy, but they also have significant structural functions in the body. The building blocks of most dietary lipids are molecules called triglycerides that are composed of a "head" of glycerol and three attached fatty acid chains. Triglycerides can be broken down into their components to yield energy or can be stored in fat cells. **(p. 608)**

- Nutritionists distinguish between oils, which are dietary lipids that are liquid at room temperature, and fats, which are dietary lipids that are solid at room temperature. **(p. 608)**

- A fat that is saturated is predominantly composed of saturated fatty acids; a fat that is unsaturated is predominantly composed of unsaturated fatty acids. **(p. 609)**

- Lipids tend to be healthy to the extent that they are unsaturated rather than saturated. Trans fats and saturated fats increase the risk of heart disease; polyunsaturated and monounsaturated fats are at least neutral with respect to it; and unsaturated fats rich in omega-3 fatty acids actually protect against it. Monounsaturated and polyunsaturated fats tend to be found in oils, while omega-3-containing unsaturated fats are found most abundantly in certain kinds of fatty fish. The usual source for saturated fats is animal fat, while trans fats are found in a number of packaged and fast foods. **(p. 609)**

31.8 Elements of a Healthy Diet

- Healthy diets contain a high proportion of healthy foods. Food pyramids are one means of getting a sense of proportionality about the intake of different foods. **(p. 609)**

31.9 The Urinary System

- The urinary system filters waste materials from the blood, regulates blood volume, and conserves useful materials. **(p. 611)**

- The urinary system consists of two kidneys that produce urine; the left and right ureters that the urine travels through on leaving the kidneys; the muscular urinary bladder, which receives the urine from the ureters; and the tube called the urethra through which urine passes from the bladder out of the body. **(p. 611)**

- The working unit of the kidneys is the nephron, composed of a nephron tubule, its associated blood vessels, and the interstitial fluid in which both are immersed. Several nephron tubules empty into a common collecting duct, which will merge with other such ducts to feed into the renal pelvis, which feeds into the ureter. **(p. 611)**

31.10 How the Kidneys Function

- A network of capillaries within a nephron, the glomerulus, receives arterial blood but is porous enough to allow much of the fluid portion of the blood to flow out of it along with smaller molecules such as vitamins, nutrients, and waste products. These materials enter the surrounding Bowman's capsule, thus moving into the nephron tubule as filtrate. At the proximal tubule, much of the original water and almost all the original nutrients are moved back into blood circulation but waste products remain in the nephron tubule. This general process continues over the length of the nephron tubule. By the time the filtrate has reached the collecting duct, it is urine. **(p. 612)**

- Retention of water by the kidneys is regulated by the brain's hypothalamus via its release of antidiuretic hormone, which acts on the kidneys to make them channel more water back into circulation, rather than into urination. When the body is overly hydrated, the hypothalamus will shut down its production of ADH, an action that will channel more water into urination. **(p. 613)**

31.11 Urine Storage and Excretion

- Waves of muscle contraction squeeze urine out of the renal pelvis, thus moving it through the ureters and into temporary storage in the urinary bladder. The urethra then carries the urine from the urinary bladder to the exterior of the body. **(p. 613)**

Key Terms

antidiuretic hormone (ADH)   613

bile   596

calorie (nutritional)   602

chyme   594

digestive tract   591

essential amino acids   605

fat   608

fiber   606

gallbladder   596

glomerulus   613

glycemic load   607

kidneys   611

large intestine   596

liver   596

mineral   598

nephron   612

nonessential amino acids   605

nutrient   597

nutrition   597

oil   608

pancreas   595

peristalsis   592

pharynx   593

simple sugars   606

small intestine   595

starches   606

stomach   594

ureters   611

urethra   611

urinary bladder   611

vitamin   598

Understanding the Basics

Multiple-Choice Questions

(Answers are in the back of the book.)

1. The essential functions of the digestive system are to (select all that apply):
   a. add nutrients to food.
   b. put food into a form that can be transferred into circulation.
   c. filter waste products from the blood.
   d. transfer digested food into circulation.
   e. retain undigested food and eliminate it from the body.

2. Which of the following is not part of the digestive system?
   a. gallbladder
   b. kidneys
   c. liver
   d. small intestine
   e. pancreas

3. Most of the nutrients in the food we eat make the transition out of the digestive system in:
   a. the stomach.
   b. the liver.
   c. the small intestine.
   d. the large intestine.
   e. the pancreas.

4. Which of the following are energy-yielding nutrients?
   a. proteins
   b. minerals
   c. lipids
   d. carbohydrates
   e. vitamins

5. In nutrition, calories are (select all that apply):
   a. units of weight.
   b. a units of energy.
   c. most densely packed in carbohydrates.
   d. most densely packed in lipids.
   e. optional components of human nutrition.

6. Unprocessed carbohydrates generally have the advantage of providing a relatively (select all that apply):
   a. low glycemic load.
   b. high amount of fiber.
   c. high glycemic load.
   d. low amount of fiber.
   e. large array of vitamins and minerals.

7. Saturated fats (select all that apply):
   a. lower circulating LDL levels.
   b. raise circulating LDL levels.
   c. are supplied mostly by animal sources.
   d. are supplied mostly by plant sources.
   e. lower the risk of coronary heart disease.

8. In addition to filtering waste from the body, what are two other critical functions of the kidneys?
   a. storage of blood, distribution of nutrients
   b. production of urea, production of hemoglobin
   c. regulation of blood volume, conservation of useful materials
   d. production of hemoglobin, channeling of lymphatic fluid
   e. storage of nutrients, regulation of digestive system

Brief Review

(Answers are in the back of the book.)

1. Why are most nutrients ready to enter the bloodstream from the small intestine but not ready to enter from the stomach?

2. What are some of the ways that the liver is important in digestion?

3. Apart from waste storage and compaction, what does the large intestine do?

4. What are the differences between vitamins and minerals?

5. Name three nutritional reasons to prefer a food rich in unprocessed whole grains to a food rich in added simple sugars.

6. What kinds of foods are predominantly composed of saturated fatty acids? What kinds of foods are predominantly composed of unsaturated fatty acids?

7. Why don't blood cells and protein molecules get passed into the filtrate that moves through the kidneys' nephrons?

8. What is the small signal of discomfort that tells us when we need to urinate?

Applying Your Knowledge

1. Some weightlifters take protein supplements in the belief that these supplements will provide them with additional energy. Based on what you learned about the dietary role of proteins, are these supplements likely to be doing weightlifters any good?

2. Contrary to popular belief, urine normally has no bacteria in it at all. Indeed, the movement of urine down the urinary tract is one reason the kidneys and the bladder tend to stay free from infection—the flow of urine washes away any encroaching bacteria. Keeping this and the difference between male and female urethras in mind, why is it that women are more prone to bladder infections than men?

3. Suppose that instead of returning nearly all nutrients and 98 percent of fluids back into circulation, the kidneys could return only half these proportions. Could human beings function? How would it change the way they live?

32

An Amazingly Detailed Script:

Animal Development

How can a living thing start out as a single cell and end up as a fully formed animal? It goes through a process called development.

Available on MasteringBiology®

Activity: Embryonic Development; Overview of Animal Development; Role of Bicoid Gene in Drosophila Development; **Current Events:** 4 Decades On, U.S. Starts Cleanup of Agent Orange in Vietnam; **Building Vocabulary; Questions:** Reading Questions, Test Bank

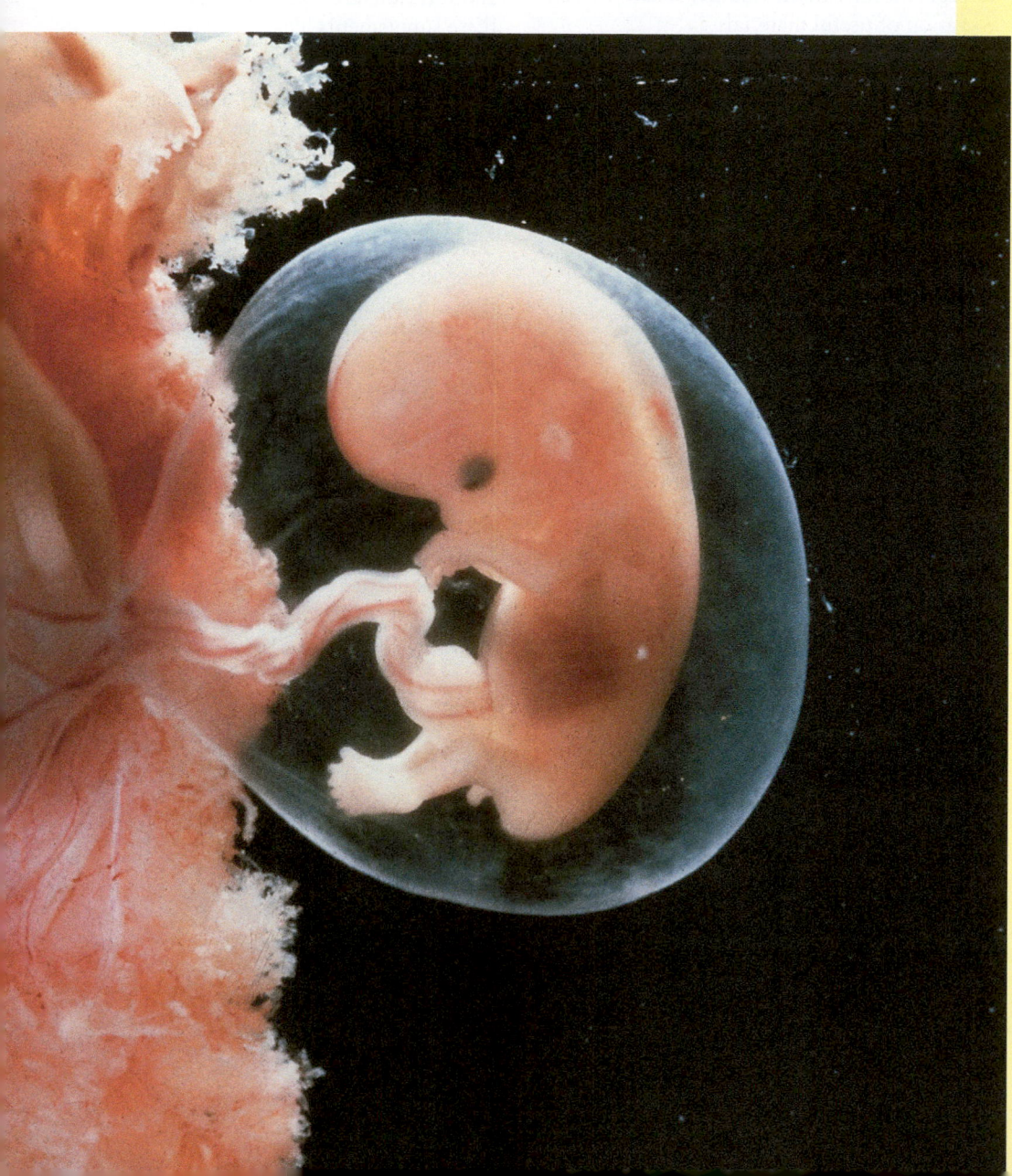

A human embryo at the end of its 8th week of development. It is surrounded by an amniotic sac and attached to a placenta by its umbilical cord.

O f all the marvels in the natural world, the most spectacular may be a complex process that goes under a simple name: development. The set of steps by which, for example, a human embryo goes from being a speck of microscopic cells to a newborn cradled in its mother's arms is so remarkable that even the old adage about "seeing is believing" is turned on its head. People who view pictures of such a work in progress can scarcely believe what they're seeing. It's not uncommon, months after the birth of a child, for parents to find themselves looking at the baby and then at each other in amazement, still wondering how such a thing is possible.

This process is all the more notable because it takes place through self-assembly. From a starting set of instructions supplied by parental DNA, and a starting set of proteins that come in the maternal egg, an embryo fashions itself, aided only by the nurturing environment of the womb.

How can such an event take place? As you'll see, in species as diverse as frogs, insects, and human beings, a common, core set of steps is followed in development. Moreover, nature has provided a number of "tools" that are employed in development, and these likewise tend to be used across different species.

32.1 General Processes in Development

Let's now begin by looking at the developmental progression that is common to much of the animal world, noting first what happens in development and then going over *how* these things happen. We'll be looking at several species, rather than just humans, as this will provide a clearer picture of what takes place.

Two Cells Become One: Fertilization

When a sperm encounters an egg, these two cells fuse in the moment generally referred to as **fertilization,** but more poetically known as conception. The fertilized egg brought about by the fusion of egg and sperm, a **zygote,** is itself a cell, although a very special one in that an entire living thing will develop from it. Development at its simplest level is a process by which this single, original cell divides to become two cells, after which the two become four, the four eight, and so on. Once the zygote starts dividing, it's called an **embryo.**

Three Phases of Early Embryonic Development

Early embryonic development in animals can be divided into three phases:

- *Cleavage.* Cellular division results in a ball of cells with an interior cavity.
- *Gastrulation.* These cells rearrange themselves into three layers that will give rise to specific organs and tissues.
- *Organogenesis.* Organs begin to form.

First Phase: Cleavage

Cleavage is a developmental process in which a zygote is repeatedly divided into smaller individual cells through cell division. The reason the cells get progressively smaller is that none of the usual cell growth occurs between these cell divisions.

The product of this process of cleavage is a **morula** (from the Latin for "mulberry"), a tightly packed ball of early embryonic cells. This configuration soon changes, however, as the cells arrange themselves into a **blastula:** an early embryonic structure composed of one or more layers of cells surrounding a liquid-filled cavity (the blastocoel). The progression from zygote to blastula is illustrated in **Figure 32.1** on the next page. It's worth noting that the human form of a blastula, called a **blastocyst,** is partly composed of cells you may have heard of, *embryonic stem cells*—the cells that hold out promise for treating conditions such as Parkinson's disease and spinal cord injuries. You can read more about these cells in Chapter 15, beginning on page 271.

MB

Embryonic Development

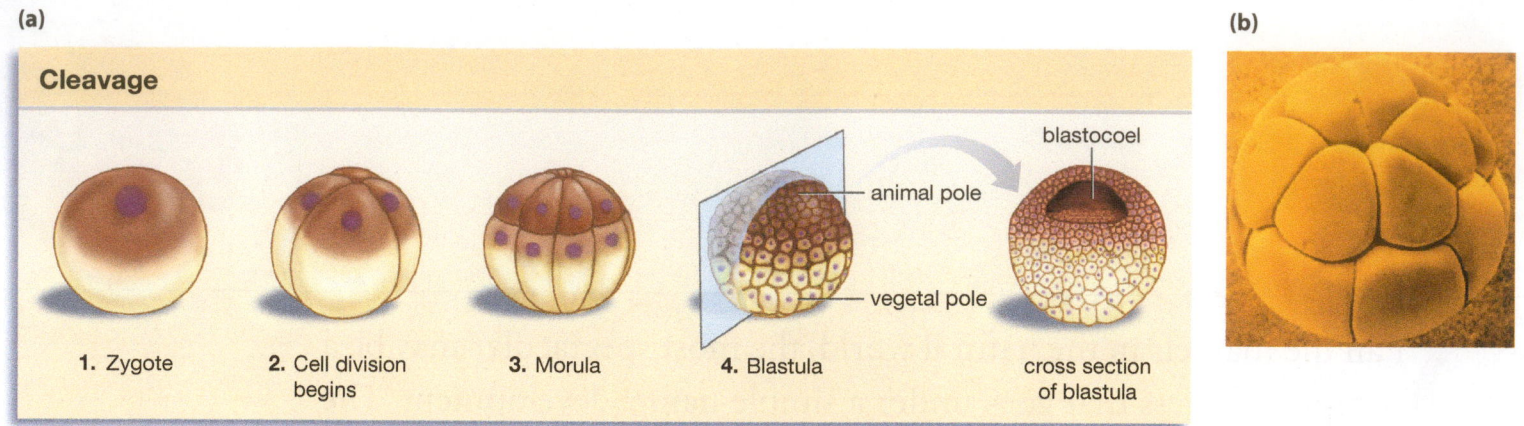

(a)

Cleavage

1. Zygote 2. Cell division begins 3. Morula 4. Blastula cross section of blastula

animal pole, vegetal pole, blastocoel

(b)

Figure 32.1
Cleavage of a Frog Egg

(a) 1. Like all animals, the frog starts out as a single-celled zygote. 2. Cell division begins and the zygote is now called an embryo. 3. Repeated cell division of the embryo results in a tightly packed ball of smaller cells, called a morula. 4. This configuration gives way to a blastula, in which the cells are arranged around a liquid-filled cavity, the blastocoel. The drawing at far right shows a cross section of the blastula, with its blastocoel visible.

(b) An artificially colored micrograph of a frog embryo at the 16-cell stage of development.

An egg that will be fertilized by sperm often contains a quantity of yolk, which helps give the resulting zygotes of many species a *polarity,* meaning a difference between one end and the other. In many zygotes, one end contains a relatively greater proportion of yolk, and this end is called the **vegetal pole.** Meanwhile, the other end has relatively less yolk and lies closer to the cell's nucleus, and this end is called the **animal pole.** As you'll see, this polarity helps define the way the embryo develops.

Second Phase: Gastrulation

Anyone who has ever watched a play in rehearsal probably has witnessed a phenomenon that bears some comparison with the second phase in embryonic development, *gastrulation.* It's the moment in which the director says to the assembled cast, "Take your places." What happens then is a carefully directed movement of actors; some move only a little, whereas others go from one side of the stage to the other. After all this movement is done, the actors—having assumed their places—are ready to take on their assigned roles.

Gastrulation is likewise a process of directed movement, in this case of groups of cells that move to particular places in the developing embryo. Once this movement is completed, the cells will have formed themselves into three different layers that are ready to take on their assigned roles. These layers—the endoderm, mesoderm, and ectoderm—are "germ" layers of cells, meaning layers that *lead to* other kinds of structures (**Table 32.1**). As it turns out, endoderm, mesoderm, and ectoderm have a kind of "inside/middle/outside" quality to them, and this is reflected in the organs and tissues that develop from them. In gastrulation, endoderm takes its place as the innermost layer, mesoderm as the middle layer, and ectoderm as the outer layer. In line with this, endoderm gives rise to interior tissues (the liver and bladder, for example), mesoderm to tissues exterior to these (muscle and skeletal systems), and ectoderm to tissues that are more exterior yet (the skin and nervous system, which first appear in the embryo's outermost layer.) **Gastrulation** can thus be defined as a process in early development in which an embryo's cells migrate to form three layers of tissue: the endoderm, the mesoderm, and the ectoderm.

Let's now look at the process of gastrulation in one organism, the sea urchin, chosen because its gastrulation is so easy to visualize (**Figure 32.2**).

By the time it's ready for gastrulation, the sea urchin blastula amounts to a single layer of about a thousand cells arranged in a hollow ball, the interior of which is the liquid-filled blastocoel. Gastrulation begins at the blastula's vegetal pole. Some cells at the pole, called *mesenchyme cells,* detach from the ring of blastula cells and move into the blastocoel, and with their loss, the vegetal pole begins to pinch inward. This process, called invagination, eventually will carry the vegetal pole as much as halfway toward the blastula's animal pole. A primitive gut, called the archenteron, starts forming around this indentation. Cells growing at the tip of the archenteron begin to send out slender extensions of themselves that adhere to selected cells that lie "above" them on the rim of the blastocoel. These extensions then contract, effectively pulling the archenteron toward the animal pole and eventually moving the archenteron all the way to the pole itself.

Table 32.1

Gastrulation's Three Layers of Cells and Some of the Structures They Give Rise to in Mammals

Cell layer	Organs or types of tissue it gives rise to
Endoderm	Lungs, liver, pancreas, lining of digestive tract, lining of bladder, and several glands
Mesoderm	Much of the body mass of more complex animals, including most of the muscle and skeletal systems; blood, blood vessels, and bone marrow; the uterus and the sperm- and egg-producing gonads; the kidneys; and the cortex of the adrenal gland
Ectoderm	Entire nervous system; lenses of eyes, tooth enamel, and the outer or epidermal layer of skin

Gastrulation

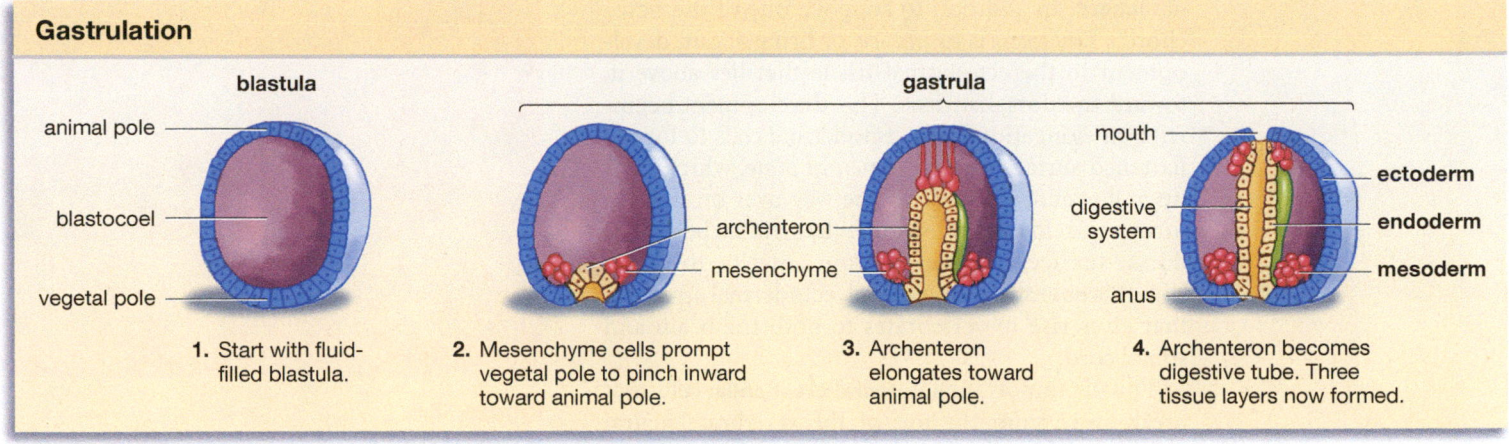

blastula   **gastrula**

animal pole

blastocoel

vegetal pole

archenteron

mesenchyme

mouth

digestive system

anus

ectoderm

endoderm

mesoderm

1. Start with fluid-filled blastula.

2. Mesenchyme cells prompt vegetal pole to pinch inward toward animal pole.

3. Archenteron elongates toward animal pole.

4. Archenteron becomes digestive tube. Three tissue layers now formed.

Now, how does all of this yield the three layers of cells we just talked about? When the archenteron reaches the rim of the blastocoel, a fusion of the two structures takes place that initiates formation of the mouth of the sea urchin. A tube then starts to develop at the mouth that eventually will run the length of the sea urchin, right down to the original vegetal pole. There, the rim of the invagination you saw earlier will develop into an anus. As you might guess, this entire feeding and digestive apparatus is the inner cell layer, or endoderm, of the sea urchin. The cells whose slender extensions reached the blastocoel are also mesenchyme cells; once they get the archenteron to the top of the blastocoel, they break away from the tip of the archenteron and then move back down into the blastocoel cavity. There they will form the mesoderm, or middle layer, of the sea urchin—eventually its skeleton and muscles. Finally, the outer cells of the blastula form the outer cell layer, or ectoderm, of the organism, eventually becoming its skin and related organs.

Third Phase: Organogenesis

In vertebrates, an interplay of two of these three germ layers marks the first steps on the path to the development of organs—in this case, organs of the nervous system.

In talking about development in any organism, an important concept is *body planes*—imagined divisions that pass through the center of an animal, dividing it into two parts of roughly equal size. In **Figure 32.3**, step 1, you can see that a frog embryo has been divided with two such planes. The first yields dorsal and ventral portions of the embryo; the second yields anterior and posterior ends. In the case of a frog, dorsal eventually will mean its back, ventral its belly; anterior will mean its head and posterior its hindquarters.

Looking at steps 2 and 3 in Figure 32.3, you can see what takes place toward the frog embryo's dorsal surface as gastrulation is nearing completion. First, there is the development of a rod-shaped support organ found in all embryonic vertebrates, the

Figure 32.2
Gastrulation in the Sea Urchin

Figure 32.3
Body Planes in a Frog and Formation of the Frog Central Nervous System

Organogenesis

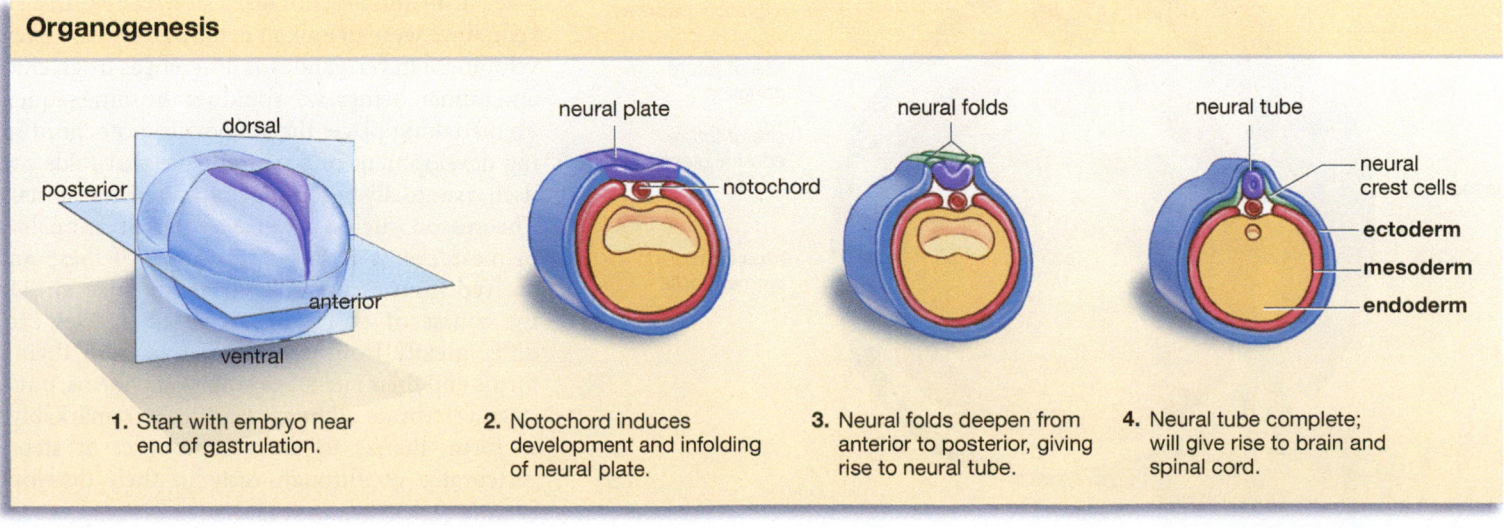

dorsal

posterior

anterior

ventral

neural plate

notochord

neural folds

neural tube

neural crest cells

ectoderm

mesoderm

endoderm

1. Start with embryo near end of gastrulation.

2. Notochord induces development and infolding of neural plate.

3. Neural folds deepen from anterior to posterior, giving rise to neural tube.

4. Neural tube complete; will give rise to brain and spinal cord.

notochord. In addition to support, one of the notochord's functions is to *induce,* or bring about, development in the ectodermal tissue that lies above it, toward the dorsal surface. This development begins with an elongation of the ectodermal cells to form a flattened surface called the *neural plate,* which curls up and eventually folds all the way over on itself to form an enclosed, hollow tube that will lie below the dorsal surface. Developing from anterior to posterior, this **neural tube** is a dorsal, ectodermal structure that gives rise in vertebrates to both the brain and spinal cord.

Equally important are **neural crest cells:** cells that break away from the top of the vertebrate neural tube as it folds together and then migrate to varying parts of the embryo, giving rise to various tissues and organs. For example, the medulla, or inner portion, of the adrenal glands (which in humans sit atop the kidneys) is derived from neural crest cells. You can see the routes of these migrations and some of their outcomes in **Figure 32.4**. One consequence of a failure of neural crest-derived tissue to develop properly is a gap that occasionally occurs between the nose and mouth in humans, which is known as a cleft lip.

As this ectodermal movement is taking place, mesodermal tissue on both sides of the notochord is developing into repeating blocks of tissue, called **somites** (**Figure 32.5**). A good part of the human body is built of repeating units—think of each of the vertebra that make up our spinal column, the muscles that attach to each of these vertebrae, and the ribs that extend from the vertebrae. All these structures develop at least in part from the somites.

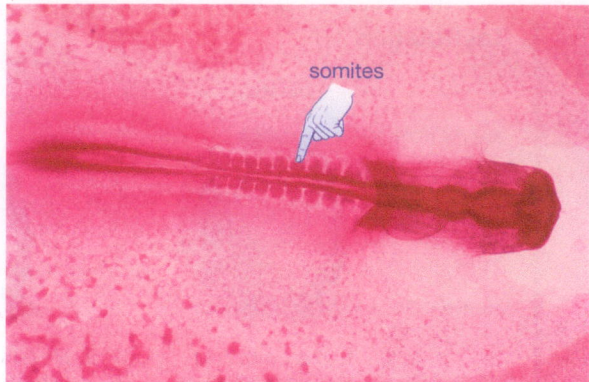

Figure 32.5
Somites
These regular, repeating blocks of mesodermal tissue, visible in this case on either side of the notochord of a chick embryo, will give rise to muscles and the vertebrae that enclose the spinal cord.

Themes in Development

As complicated as all this may seem, it amounts to a brief look at important, but early, periods of development in animals. Looking at the larger picture, notice that in the early stages of development there is a movement from the general to the specific. The animal goes from a small group of superficially similar cells, to three *layers* of cells, to tissues that proceed from these layers (notochord, somites), to actual organs formed from these tissues. To put a little finer point on this, think of the somites. They arise from undifferentiated mesodermal tissue on each side of the notochord, and they take shape as about 40 blocks of mesodermal tissue. These blocks then go on to give rise to several structures that retain the somites' repeating nature but that are much more specialized (vertebrae, ribs).

It's also worth noting that the three most basic developmental processes we looked at—cleavage, gastrulation, and organogenesis—are processes that take place in all animals, not just a selected few. In a similar vein, if we were to look in detail at nervous system development in vertebrates as different as frogs, chickens, and human beings, we would see the same sequence of events taking place: the supportive notochord inducing development of a neural plate that folds over on itself, eventually becoming both brain and spinal cord. The reason such a diverse group of animals share in these events is straightforward: All these animals evolved from a common vertebrate ancestor. During the course of evolution, vertebrate species came to differ greatly from one another in both their adult forms and their late-stage embryonic forms, but *early*-stage vertebrate embryos tend to be remarkably alike in form, thanks to the common set of steps that vertebrates go through early in their development (**Figure 32.6**).

Figure 32.4
Neural Crest Migration in a Frog Embryo
This cross section through a frog embryo shows some of the pathways of neural crest cells as they migrate to various locations, prompting the development of different tissues and organs.

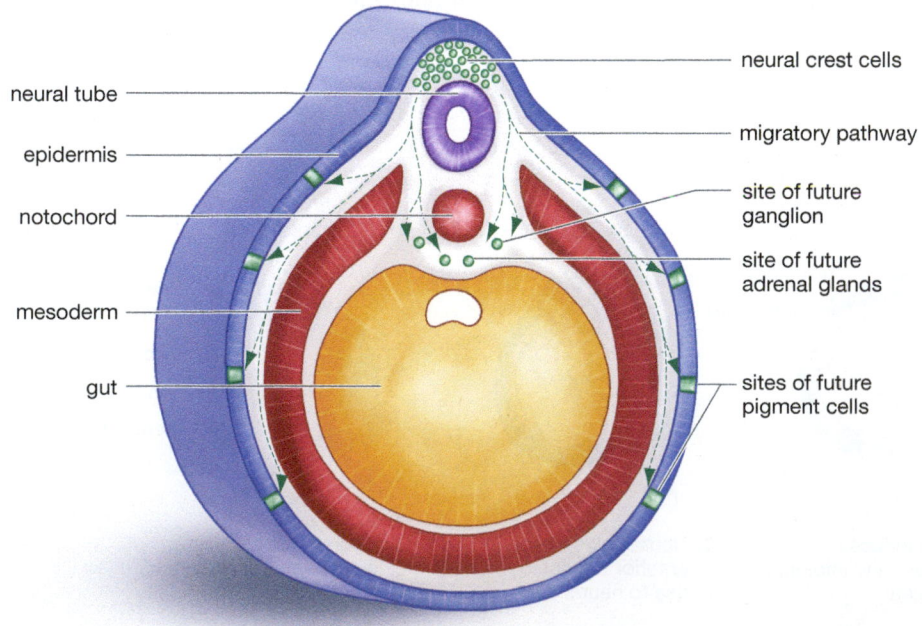

neural tube

epidermis

notochord

mesoderm

gut

neural crest cells

migratory pathway

site of future ganglion

site of future adrenal glands

sites of future pigment cells

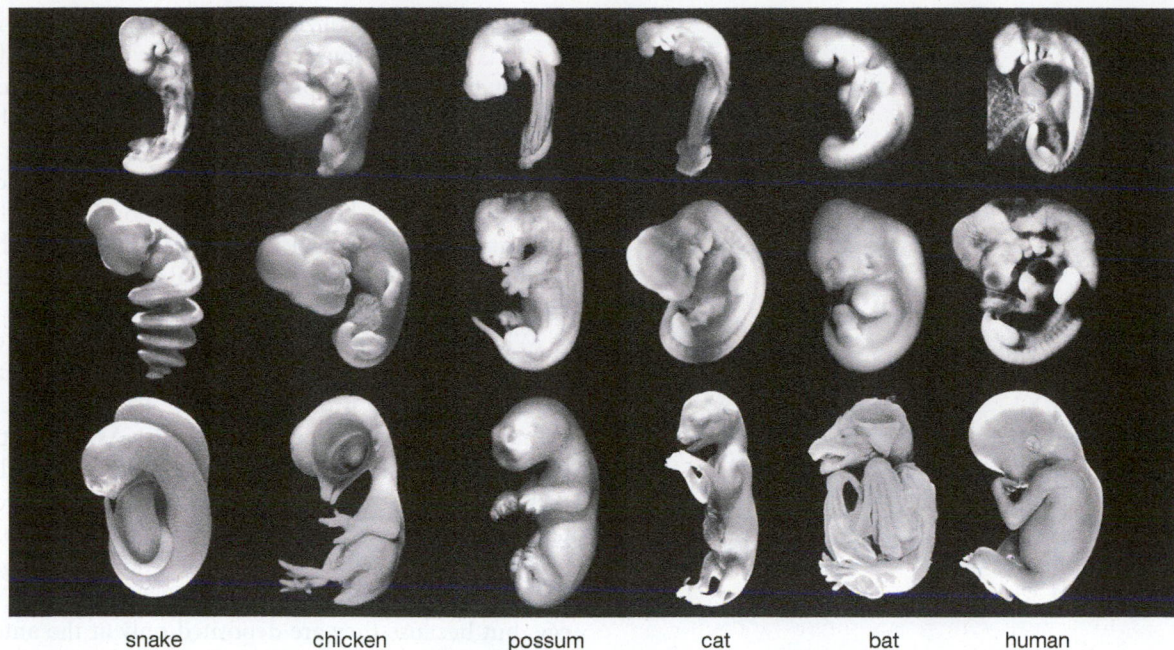

Figure 32.6
Embryonic Development in Six Vertebrates
Pictured are six vertebrates in early, middle, and late stages of their development. Note the general similarity in vertebrate shape in the early stage of development, which roughly corresponds to the period following somite formation. The embryos are not to scale—in reality some are bigger than others.

Photo courtesy of Michael K. Richardson. Human embryo from R. O'Rahilly.

snake   chicken   possum   cat   bat   human

SO FAR . . .

1. Sperm and egg fuse to produce a cell called a _____. Once this cell starts dividing, it is referred to as a(n) _____.

2. In animal development, the process of cleavage produces a ball of cells with an interior cavity. In the succeeding developmental process, called gastrulation, the cells in the embryo migrate to form three layers of cells, the _____, _____, and _____, which in terms of location are the _____, _____, and _____ layers of cells in the organism.

3. In vertebrates, a support structure called a notochord serves to _____, or bring about, the development of a structure called the neural tube, which gives rise to both the _____ and _____.

32.2 Factors that Underlie Development

Now that you've taken a brief look at what happens in early development, the next question is: *How* do these things happen? How do vegetal pole cells "know" they are vegetal pole cells? How do neural crest cells take different routes through an embryo and end up serving as different kinds of cells?

The Process of Induction

As noted, the notochord *induces* or prompts the development of the ectodermal tissue that lies above it. Such **induction** is an important process in development in general. It can be defined as the capacity of some embryonic cells to direct the development of other embryonic cells. If you look at **Figure 32.7** on the next page, you can see how induction works in the well-studied example of chick wings. In essence, there is a relatively small group of cells that gives out information for the development of a portion of the wing. Researchers learned this by working with the "limb bud" that is transformed into the chick wing. They took cells from a posterior "zone of polarizing activity" (ZPA) in one limb bud and transplanted them to the *anterior* portion of a second limb bud. The result was activity in this bud by both the natural and transplanted ZPA; this resulted in two sets of developing digits in wings, spreading out like two sets of human fingers developing from joined hands. In short, the cells in the ZPA induced the development of other cells, sending out a message that said, in effect, "Begin digit development here."

The Interaction of Genes and Proteins

This induction process raises the question of what lies at the root of development. There are lots of cells in an embryo. How can some of them (such as the ZPA cells) control the development of others? It turns out that, at its most basic level, development is controlled by the interaction of genes and proteins—the genetic regulation reviewed in Chapter 14. Biologists'

Overview of Animal Development

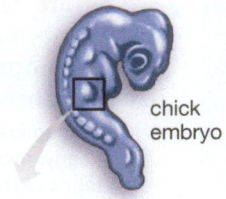

chick embryo

(a) Normal limb bud

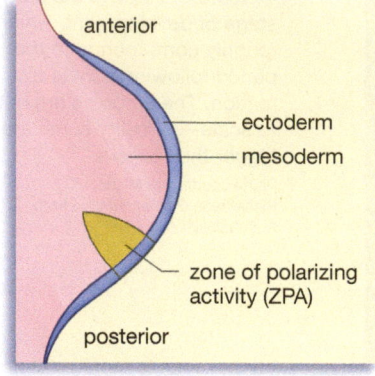

anterior

ectoderm
mesoderm

zone of polarizing activity (ZPA)

posterior

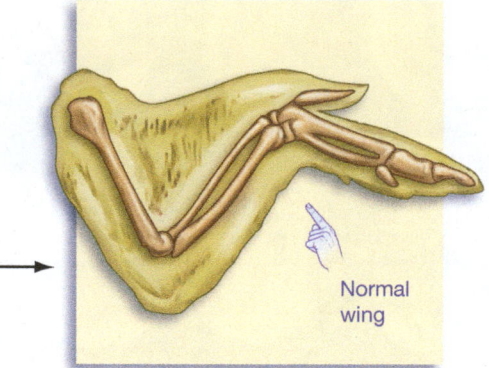

Normal wing

(b) Grafted limb bud

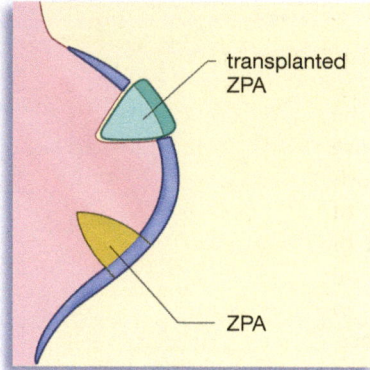

transplanted ZPA

ZPA

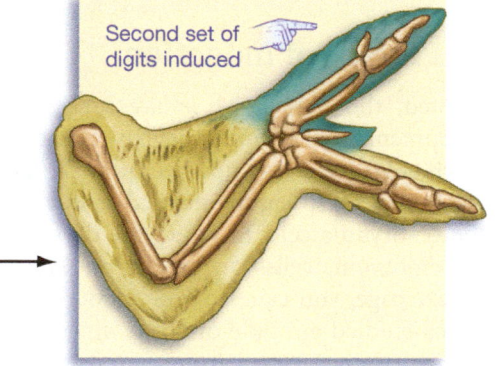

Second set of digits induced

Figure 32.7
Cells Can Induce Development in Other Cells

Transplantation experiments with chick embryos revealed that a group of cells on the posterior portion of the chick's limb bud give out positional information for the development of digits in the chick wing.

(a) When cells from a zone of polarizing activity (ZPA) were left in their normal posterior location, the chick wing developed as usual.

(b) When limb bud ZPA cells were transplanted to the anterior portion of a second limb bud, the result was mirror images of two sets of developing posterior wing digits, much like two sets of human fingers extending from joined hands.

understanding of how this interaction works in development has grown dramatically in the period from the early 1980s to the present. Although there have been several important "model" organisms in this research, the most important of them has been the tiny fruit fly, *Drosophila melanogaster.* Let's look at a little of what the *Drosophila* research has uncovered.

Three Lessons in One Gene

There is a *Drosophila* gene, called *bicoid,* that codes for a protein that is also called bicoid. This single gene offers a clear example of three principles that operate in development:

- A critical early step in development is the establishment of positional information in the embryo.
- Substances called morphogens are important in establishing this positional information.
- Development operates from the general to the specific through a genetic cascade.

Look at the unfertilized *Drosophila* egg pictured in **Figure 32.8** to see how these principles play out with respect to the bicoid protein. Recall first the body planes described earlier. In Figure 32.8a, the top of the egg is its dorsal portion and the bottom its ventral portion, while the left side corresponds to its anterior portion and the right side to its posterior. Eventually, the anterior portion will be the head of the fly, the posterior its hindquarters, the dorsal portion its back, and the ventral portion its belly.

While the egg is still in the mother, a group of "nurse cells" surrounds its anterior portion and pumps molecular products into it. These products include bicoid messenger RNAs (mRNAs)—genetic instructions copied from the *bicoid* gene. Eventually, these mRNA instructions will lead to the bicoid protein. Looking at the figure, you can see that the *bicoid* mRNAs are found at the anterior end of the egg. These mRNAs eventually start to diffuse through the egg, but because they are deposited only at the anterior end, they will exist in a *concentration gradient,* with the greatest concentration being at the anterior end, a lesser concentration toward the center, and none at the posterior end. This leads to the concentration gradient of the bicoid protein you can see in Figure 32.8b.

Positional Information

With this general background, you can now see how *bicoid* exemplifies one of the principles of development noted earlier. What does the bicoid protein do? It imparts positional information to the embryo and is critical in setting in motion the development of structures in the anterior and center portions of the embryo. Recall that the far-anterior portion of the egg eventually becomes the head of the *Drosophila* fly. In experiments, however, scientists have taken bicoid-laden cytoplasm from the anterior portion of one egg, placed it at the *center* of an egg with no bicoid in it, and watched as the head and related structures developed at the center. Meanwhile, an egg with no bicoid protein gives rise to a fly that never develops a head, although its posterior structures start to develop normally.

Morphogens

The bicoid protein is a **morphogen:** a diffusable substance whose local concentration affects the course of local development in an organism. Several other morphogens are operating in *Drosophila* in early development. Indeed, several positional "systems" are at work in the fly. What's important to note is that, with these systems, a three-dimensional grid is set up in which the destiny of various cells is determined by their position within this grid. How? Their exposure to different morphogens varies according to their position.

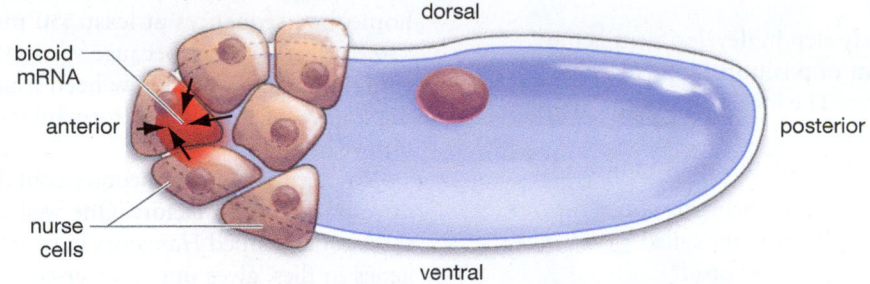

(a) Bicoid mRNAs are deposited in unfertilized egg.

dorsal

bicoid
mRNA

anterior

posterior

nurse
cells

ventral

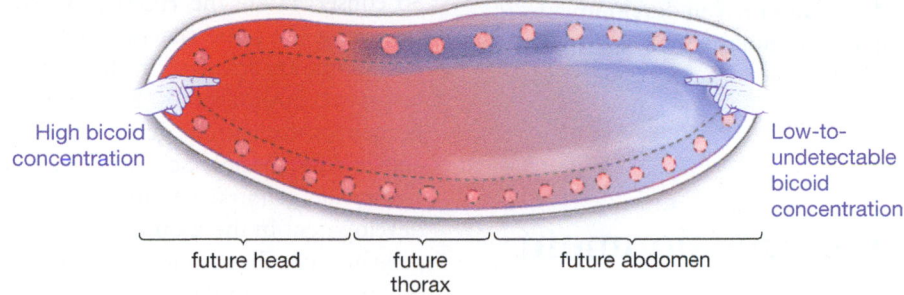

(b) Bicoid proteins diffuse through embryo.

High bicoid
concentration

Low-to-
undetectable
bicoid
concentration

future head   future   future abdomen
                   thorax

Figure 32.8
Drosophila **Egg and**
Bicoid Gradient
A fly begins to develop.
(a) Nurse cells surround
the anterior portion of an
unfertilized *Drosophila* egg.
The cells pump messenger
RNA molecules for the bicoid
protein into the egg.
(b) The messenger RNAs are
translated into bicoid proteins
that diffuse through the now-
fertilized egg. These proteins
exist, however, in a concentration gradient—in high
concentration at the anterior
end of the embryo but in
low-to-undetectable concentration at the posterior end.

Role of Bicoid Gene in
Drosophila Development

But what is it that morphogens are doing to affect cells? You may remember learning in Chapter 14 about proteins that feed back on DNA, binding with it and thus controlling its transcription, which is to say turning its genes on or off. These DNA-regulating proteins are called **transcription factors.** Morphogens are cell-signaling molecules that bring about the production of transcription factors. Diffusing out from one group of cells, morphogens provide a signal to other cells to start producing transcription factors.

We can see an example of this effect in the chick wing development noted earlier. Recall that ZPA cells in the limb buds were able to induce the development of digits in the chick wing. ZPA cells can play their controlling role because they produce a morphogen, called *sonic hedgehog,* that acts on surrounding cells: It diffuses to them and prompts them to begin producing transcription factors. Moreover, like all morphogens, sonic hedgehog is having its effects through a concentration gradient. The order of the digits in the chick wing—the equivalent of the order of our digits from thumb to little finger—is specified by how much sonic hedgehog the receiving cells are exposed to. The chick's posterior digit (the equivalent of our little finger) forms from cells closest to the ZPA—the cells exposed to the greatest amount of sonic hedgehog. Farther out from the ZPA the concentration of sonic hedgehog lessens, and the result is cells that become the chick's *anterior* digit (our thumb). In a similar way, the placement of the head and center

structures of *Drosophila* is determined by the concentration gradient of the bicoid morphogen. Bicoid exists in greatest concentration at the anterior portion of the embryo, and this is where the fly's head develops; it exists in lesser concentration in the middle of the fly, and this is where its center-portion structures develop.

The Genetic Cascade

The developmental effects we've been reviewing work through what is sometimes called a *genetic cascade.* Like water flowing from one level to the next, gene products flow from one developmental level to the next. An initial set of genes is turned on, perhaps by getting a signal from a morphogen, and in time these genes regulate *other* genes, which in turn may regulate still more genes. This explains how development moves from the general to the specific. In *Drosophila,* for example, the first level of genes specifies broad positional patterns across the embryo— anterior and posterior, dorsal and ventral. Then, through the cascade, this positioning gets finer grained; now, other genes are switched on, and development is being specified *within* sections of the embryo. Finally, control is exercised over the development of specific body *structures,* such as wings or antennae. It's worth noting that the genetic cascade turns genes off as well as on. To take one example, once somites have given rise to ribs, vertebrae, and so forth, the genes for building somites will be turned off permanently.

SO FAR . . .

1. A critical early step in development is the establishment of positional information in the embryo. The local concentration of substances called _____ in the embryo helps impart this information.

2. Morphogens work by bringing about the production of the proteins called _____, which turn _____ on or off.

3. Development works from the general to the specific. In *Drosophila,* the morphogen called bicoid operates early in development by specifying the development of _____ and _____ portions of the embryo.

32.3 Unity in Development: Homeobox Sequences

Recall how you saw earlier that species as diverse as fruit flies, frogs, and human beings have similar processes going on in their development. Now that you have some sense of how genes control development, you won't be surprised to learn that different species share in similar developmental processes because they share a similar set of developmental genes. The remarkable thing, however, is *how* similar these genes are across species, how widely they are distributed across the living world, and how long they have existed. As scientists would put it, these genes have been highly conserved—portions of them have been passed on intact, like prized bits of knowledge, down long evolutionary lines.

The best example of this conservation comes with a group of genes whose similarity was discovered in *Drosophila* in the 1980s. Scientists were then tracing development in *Drosophila* by mutating the developmental genes of flies and then watching to see what effect this had on flies when they matured. Some of the changes they got by doing this were dramatic (**Figure 32.9**). What the scientists learned from this exercise was that gene X had developmental effect Y.

In carrying out this work, the scientists were sequencing these genes—they were learning the order of their building-block As, Ts, Cs, and Gs. As they did this, they found that every one of these developmental genes had within it an identical sequence of about 180 letters (or base pairs). The real shock came, however, when this so-called homeobox sequence was found to exist in all animals studied—in frogs, humans, and mice, for example, as well as flies. This was surprising because mice and flies went down separate evolutionary paths about 550 million years ago. The

implication was that a common ancestor of both animals (probably a variety of worm) possessed homeobox sequences at least 550 million years ago. The "at least" comes because versions of homeobox-containing genes have now been found in fungi and in *plants,* which diverged from animals about 1.2 billion years ago.

Not surprisingly, homeobox-containing genes produce transcription factors. One well-studied group of these genes, named *Hox* genes in vertebrates and *HOM* genes in flies, gives out anterior-posterior positioning information—these genes specify where different structures should go along the anterior-posterior axis. So conserved are the *HOM/Hox* genes that they not only have the same general function in humans and in flies but are arranged in the same order along their respective chromosomes. Even more surprising, some homeobox-containing genes are partly interchangeable across species. Scientists have spliced into the *Drosophila* genome a mouse gene that codes for eye development in the mouse. They then watched as this gene brought about normal eye development in the *fly*. Not mouse eyes in the fly, mind you—fly eyes.

What all this means is that there is a general set of genetic instructions for development that are so important they have changed very little over time and across species. Fly eyes and mouse eyes are structurally quite different, but there is a general set of genetic instructions for "eyes" that seems to trigger the development of *specific* kinds of eyes in specific animals.

Figure 32.9
The Power of Developmental Genes
Scientists induced a mutation in one of the developmental genes of this *Drosophila* fruit fly and the result was a fly with normal eyes (the red structures on each side of the head), but with a pair of legs extending from the area where the fly would normally have antennae.

Spurred on by these kinds of findings, scientists today are busy studying how homeobox-containing genes and their proteins differ among the species. This is important work because it stands to tell us a great deal not only about development but about evolution. How does one species evolve from another—how does it branch off? Developmental genes often are involved for the simple reason that one organism will not physically differ much from another without a change in these genes. Insects have six legs, rather than the multitude that many of their fellow arthropods have, because a homeobox-containing gene in one of their ancestors underwent a mutation. In its mutated state, this gene produced a protein that turned *off* another gene—a gene that brings about multiple leg formation. By tracking developmental gene differences in modern animals, scientists can learn how these various animals evolved in the first place.

32.4 Developmental Tools: Sculpting the Body

Genes may lie at the root of development, but the shaping of an organism takes place through genetic instructions that prompt whole cells to take actions. You may wonder how a hand, for example, can be shaped from a seemingly formless mass of cells. The answer lies in capabilities cells have that allow for this sort of development. Three examples of these capabilities are cell movement, cell adhesion, and cell death.

Earlier, you saw examples of cell movement and cell adhesion when reviewing the process of gastrulation in the sea urchin. Recall that cells of the primitive gut were able to move up into the blastocoel, after which cells peripheral to them sent out slender extensions of themselves that adhered to selected cells that lay "above" them in the blastocoel. The first part of this process, cell movement, comes about in various ways. Often, cells use the tent-pole-like structures called microfilaments that grow very rapidly at one end of the cell—in the direction of movement—while decomposing at the other.

Cell adhesion is accomplished through several families of proteins, collectively called *cell adhesion molecules,* that protrude from the surface of cells, allowing them to stick selectively onto other cells. Different groups of cells have different adhesion "affinities" for each other—depending on the proteins they put on their surfaces—and this explains how the *layers* of cells that you've seen come about. Imagine two groups of cells, A and B, that are randomly mixed in with one another. Now imagine that, through genetic control, the A cells start producing a surface protein that makes them strongly adhere to other A cells. By process of

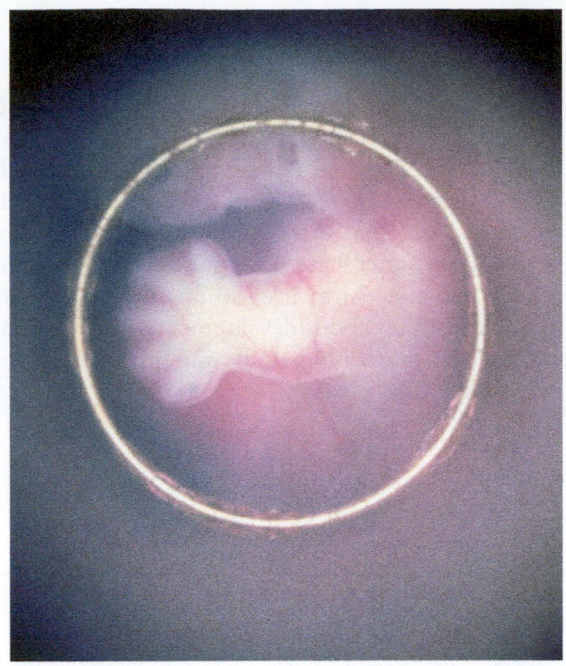

Figure 32.10
Soon to Be Different
The webbed hand of a human embryo 5 weeks after conception. The programmed death of cells in the hand will bring about the emergence of five separate fingers.

random cellular movement, the A and B cells will now start to segregate into two separate groups.

Cell death may not sound like a promising means of developing an organism, but consider that sculptors sometimes work not by adding material but by chipping away at an initial mass of it. Thus it is in development where cell death works to create spaces that define shapes. An example of this is the human hand, which but for the death of large numbers of cells between future fingers would look more like the webbed foot of a duck (**Figure 32.10**). Cells that die in this fashion are not attacked by other cells. Instead, they commit suicide in a process known as *apoptosis* or *programmed cell death.* Apoptosis is a very orderly process, akin to a person dying in a hospital rather than a car crash. The cell undergoing it fades away in a sense; it shuts down its processes, shrinks, and then dies, after which its remains are ingested by neighboring cells.

32.5 Development through Life

It is convenient to think of animal development as something that happens before birth. But if this is the case, what are we to call the loss of "baby" teeth in childhood, the onset of puberty in the teenage years, or the cessation of menstruation (menopause) among middle-aged women? Development never stops throughout the life span. In essence, it is governed by the same factors that control the embryonic development you've looked at so far: the interaction of genes and proteins and the influence of environment on them.

On to Human Reproduction

The steps of development that you have just looked at form a kind of middle phase of the process you'll be looking at next chapter, which is reproduction—specifically human reproduction. Coming before the development you've reviewed are the formation of the egg and sperm that come together to create an embryo. Coming after it are late-stage human embryonic growth and then birth itself. With this chapter, you know something about development; in Chapter 33, you'll see how development fits into the larger scheme of human reproduction.

SO FAR . . .

1. The homeobox-containing genes that are active in development have been highly _____, meaning they have remained relatively unchanged though the course of evolution. These genes contain information for producing _____.

2. Three tools that work in development across animal species are cell _____, cell _____, and programmed cell _____.

3. Human development starts at _____ and continues through _____.

32 REVIEW

Summary

32.1 General Processes in Development

- Development in animals begins with the formation of a zygote, created when sperm fuses with egg. Once a zygote starts to divide, it is called an embryo. (p. 619)

- The three phases of early embryonic development are cleavage, which produces a ball of cells, called a blastula, that has an interior, liquid-filled cavity; gastrulation, in which the blastula's cells rearrange into three layers that give rise to specific organs and tissues; and organogenesis, in which organs begin to form. (p. 619)

- The three layers of cells produced in gastrulation are the endoderm, which gives rise to interior tissues such as the liver and bladder; the mesoderm, which gives rise to tissues exterior to these, such as muscle and skeletal tissue; and the ectoderm, which gives rise to tissues more exterior yet, such as the skin and nervous system. (p. 620)

- In early organogenesis in vertebrates, a notochord develops near the dorsal surface and induces the development of ectodermal tissue that lies above it. A neural plate that develops from this tissue forms a hollow neural tube that gives rise to the brain and the spinal cord. Groups of neural crest cells break away from the top of the neural tube during its development and then migrate to different locations throughout the embryo and eventually develop into different organs and tissues. (p. 621)

- The early stages of development represent a transition from the general to the specific. A small group of superficially similar cells gives rise to three different layers of cells, which give rise to tissues, which in turn give rise to specific organs. (p. 622)

32.2 Factors that Underlie Development

- Induction is a process in which some embryonic cells direct the development of other embryonic cells. (p. 623)

- Development is controlled by the interaction of genes and proteins. A key model organism in scientific understanding of these interactions is the fruit fly *Drosophila melanogaster*. (p. 623)

- A critical early step in development is the establishment of positional information. Positional information in *Drosophila* is provided in part by the protein bicoid, which exists in a concentration gradient in the fly embryo. (p. 624)

- Bicoid is an example of a morphogen: a diffusible substance whose local concentration affects the course of local development. In general, morphogens diffuse from one set of cells to a second set. Binding with these latter cells, morphogens prompt them to begin producing transcription factors. (p. 624)

- Development works through a genetic cascade in which one set of genes regulates a subsequent set, which explains how development proceeds from the general to the specific. A first level of genetic instruction specifies broad positional patterns in an embryo; subsequent levels of instruction specify development within these body sections; and later levels of instruction specify the development of specific structures such as organs. (p. 625)

32.3 Unity in Development: Homeobox Sequences

- Within many of the developmental genes of animals as diverse as flies and humans, there exists an identical sequence of about 180 DNA bases. Called a homeobox, this sequence has remained largely unchanged over hundreds of millions of years of evolution in all animal species studied. As a result of this conservation, similar basic developmental processes are at work in all animals and possibly in fungi and plants. (p. 626)

32.4 Developmental Tools: Sculpting the Body

- Three cell capabilities that help shape the animal body are cell movement; cell adhesion, in which cells produce surface proteins that selectively adhere to other cells; and apoptosis, in which spaces in tissues are created by means of cell death. (p. 627)

32.5 Development through Life

- Development continues through the life span of an organism. (p. 627)

Key Terms

Understanding the Basics

Multiple-Choice Questions

(Answers are in the back of the book.)

1. In what order do the following developmental steps occur?
   a. organogenesis   **b.** fertilization
   c. blastula formation   **d.** morula formation   **e.** gastrulation

2. The process of gastrulation involves:
   a. massive cell death.   **b.** little or no growth of the embryo.   **c.** development of three tissue layers.   **d.** organogenesis.
   e. cell enlargement.

3. Organogenesis is accomplished, in part, by the ability of some cells to bring about developmental change in other cells. This process is referred to as:
   a. reduction.   **b.** transformation.
   c. translation.   **d.** induction.
   e. fusion.

4. The cells that migrate from the developing neural tube and eventually form different organs and tissues throughout the body are:
   a. neural crest cells.   **b.** archenteron cells.   **c.** mesenchyme cells.   **d.** stem cells.   **e.** endoderm cells.

5. Morphogens (select all that apply):
   a. diffuse from one group of cells to another.   **b.** are cell adhesion molecules.   **c.** work through concentration gradients.   **d.** bring about production of transcription factors.   **e.** are active only in nervous system development.

6. The homeobox DNA sequence is said to be ____ because _____.
   a. an induction sequence; it controls general development   **b.** highly unusual; it exists in only a few animals   **c.** a diversifying influence; it brings about differences

in animals   **d.** fast-evolving; it changes every few generations   **e.** highly conserved; it has changed little over evolutionary time

7. The segregation of cells into different layers in development is strongly shaped by _____, while the transformation of the human hand from its webbed to its five-fingered form is strongly shaped by _____.
   a. morphogens; neural crest cells   **b.** cell adhesion; apoptosis   **c.** homeobox sequences; morphogens   **d.** apoptosis; cell adhesion   **e.** induction; neural crest cells

Brief Review

(Answers are in the back of the book.)

1. Briefly describe the order of events that occur after fertilization to produce an embryo with three layers of cells.

2. What is a morphogen concentration gradient, and how does it affect early development of the fruit fly, *Drosophila*?

3. Outline the steps involved in the production of the chick wing.

4. The same homeobox DNA sequence is shared in the developmental genes of mice and fruit flies, even though these two animals diverged from each other evolutionarily about 550 million years ago. What does this high level of conservation tells us about the homeobox sequence?

5. How do cell adhesion molecules influence early events during development?

6. Describe the role cell death plays during the development of an organism.

Applying Your Knowledge

1. In what way is development a good example of a biological process that is self-regulating?

33

How the Baby Came to Be:
Human Reproduction

To produce a child, human males and females must first produce sperm and eggs, which come together in an act of fusion known as conception. The fertilized egg that results will, over the next 38 weeks, develop into a fully formed baby.

Available on MasteringBiology®

Activity: The Female Reproductive System; The Male Reproductive System; **Current Events:** As Algae Bloom Fades, Photosynthesis Hopes Still Shine; **Building Vocabulary; Questions:** Reading Questions, Test Bank

Baby makes three at the end of a long and complicated process.

Memorable events mark every human life, but for parents, one event that is sure to be indelibly etched into memory is the birth of a child. The mother's labor, the final "pushing" that brings the baby forth, the first real look at the infant, and the infant's first look at the world: these are moments that, once experienced, are never forgotten.

The birth of a child is a beginning, of course, and yet it is also a culmination. Human beings must grow to a certain level of biological maturity before being capable of reproduction. After this, finding a partner may be difficult and getting pregnant more difficult yet. Eventually, however, two very special "seeds"—*this* egg and *this* sperm, out of all the millions a man and a woman produce in a lifetime—fuse to create this child.

In this chapter, you'll trace the biology of human reproduction, which might be defined as all the biological steps that must fall into place before a child can be born. The process of development you studied in Chapter 32 represents part of what you'll be looking at, but other elements are involved as well.

33.1 Overview of Human Reproduction and Development

Taking the story of reproduction from the beginning means picking up a narrative that was begun in Chapter 10, when you went over the formation of the reproductive cells called *gametes*—eggs in females, sperm in males. (**Figure 33.1**). In this chapter, the focus widens, taking into account the physical setting in which egg and sperm are formed and the steps that bring these cells together. Finally, you will look at the human development that takes place once egg and sperm have merged. (For information on some of the methods humans use to *prevent* conception, see "Methods of Contraception" on page 642.)

The first part of reproduction, the formation and delivery of egg and sperm, is a straightforward story in outline. The eggs a woman possesses exist in her in a precursor form, called oocytes, that lie within two walnut-shaped structures called ovaries. These exist just right and left of center within the female pelvic area (**Figure 33.2** on the next page). Each oocyte sits in the ovaries within a nurturing complex of cells and fluids called a **follicle.** On average, once every 28 days one follicle-oocyte complex is brought to a state of maturity such that the oocyte and some accessory cells that surround it rupture from their ovary and begin a slow journey to the uterus. This trip takes place inside the structure called the **uterine tube** (or Fallopian tube). The release of the oocyte is called **ovulation.** If, during this trip, the oocyte encounters a male sperm, the two may fuse, in which case conception has occurred.

Sperm in men are produced, in unfinished form, in sets of tubules in the two male testes. They then are transported to a structure called the epididymis (one for each testis) for further development. Then, with

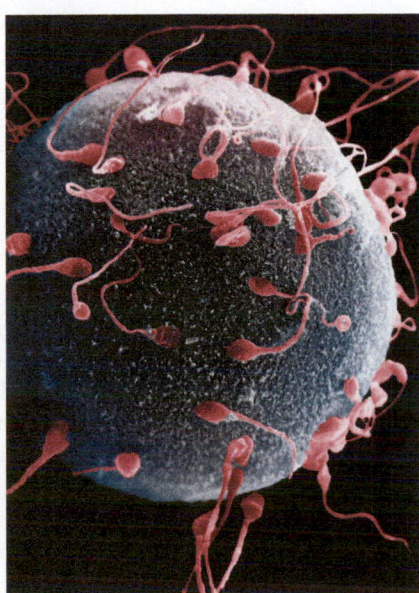

Figure 33.1
Oocyte and Sperm
Numerous sperm surround a single oocyte.

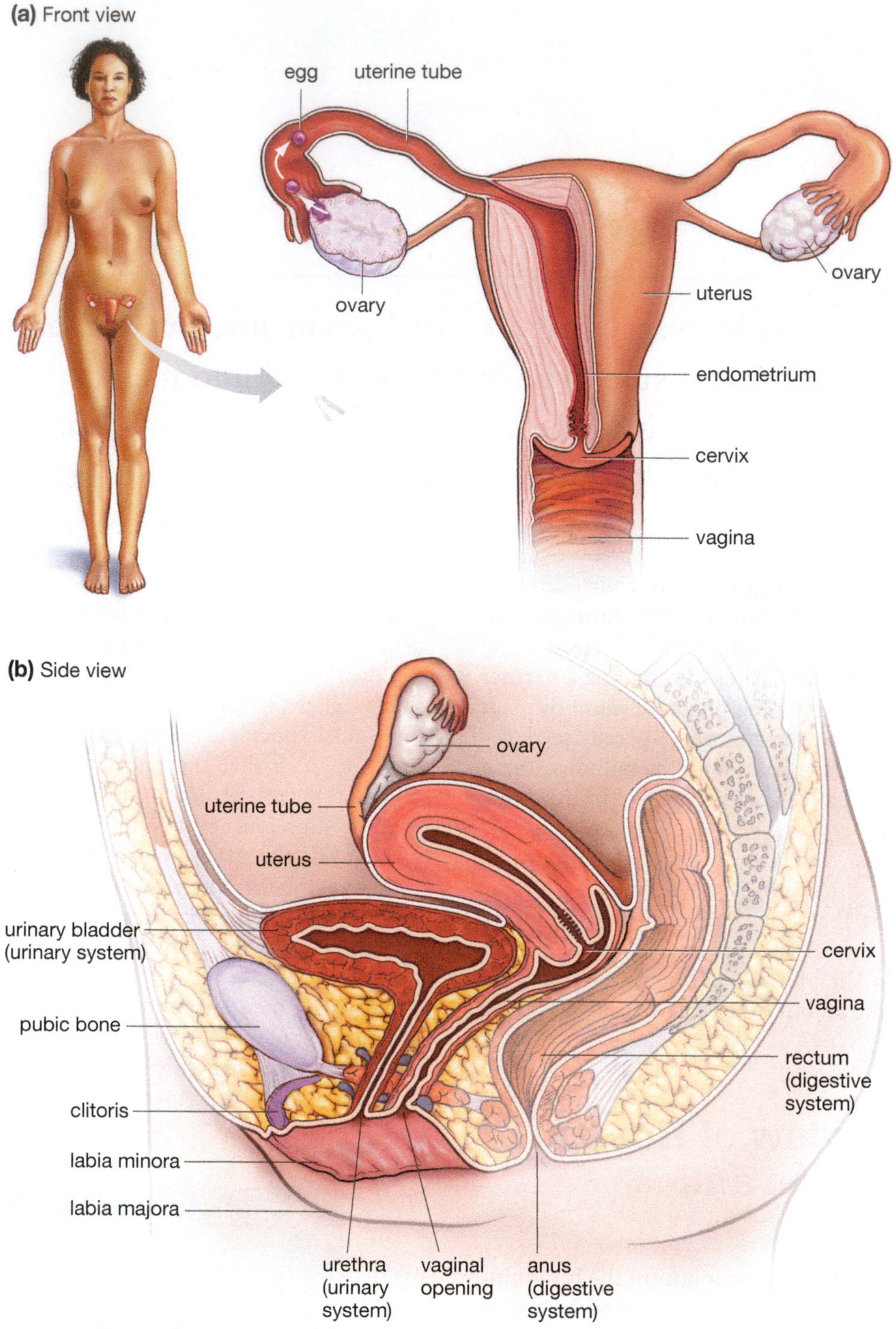

Figure 33.2
Reproductive Anatomy of the Human Female

(a) Eggs develop in the ovaries within nurturing complexes called follicles. Once every 28 days, on average, an egg is expelled from one ovary or the other and journeys through the uterine tube to the uterus. If the egg encounters sperm on the way, pregnancy will result when egg and sperm fuse. The fertilized egg then continues traveling through the uterine tube and implants in the lining of the uterus, the endometrium.

(b) A lateral view of female reproductive anatomy. The labia majora are two folds of skin that enclose two other folds of skin, the labia minora, which in turn enclose a small bud of nerve-dense sensory tissue, the clitoris. The labia minora, vagina, and clitoris all engorge with blood during sex, and stimulation of the clitoris can lead to orgasm.

(a) Front view

egg · uterine tube · ovary · uterus · endometrium · cervix · vagina

(b) Side view

ovary · uterine tube · uterus · urinary bladder (urinary system) · pubic bone · clitoris · labia minora · labia majora · cervix · vagina · rectum (digestive system) · urethra (urinary system) · vaginal opening · anus (digestive system)

The Female Reproductive System

sexual excitation, sperm are transported in a loop: up and over the urinary bladder through two ducts, each called a vas deferens (**Figure 33.3**). The contents of these ducts then empty into a single duct, called the urethra, from which they are ejaculated into the female vagina. Along the way, materials secreted by several glands join the sperm in a process that can be likened to tributaries feeding into a common stream. The resulting mixture of sperm and glandular materials is called **semen.**

Whereas the female releases an average of one oocyte per month, the male releases an average of 200 million sperm per ejaculation. Of the millions of sperm that begin the journey, a few dozen may encounter the female oocyte as it continues its journey from ovary to uterus. The first sperm to breach the coating that surrounds the oocyte fuses with it, while the rest are shut out through a kind of gate-slamming mechanism. When this one sperm has joined with an oocyte, conception has occurred. The now-fertilized egg still floats

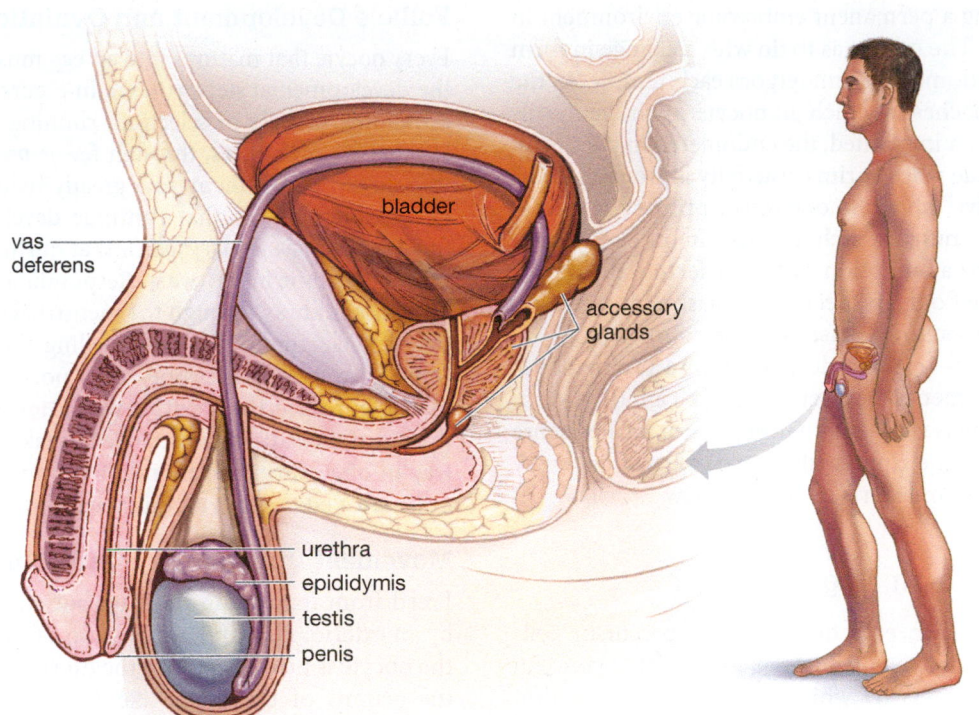

Figure 33.3
Reproductive Anatomy of the Human Male
Sperm are produced in the testes and transported to the epididymis, where they mature and are stored. With orgasm, the sperm are transported through the vas deferens and then the urethra, from which they are ejaculated. Along the way, they are joined by materials secreted by several accessory glands.

freely in the uterine tube, where it begins to develop in the ways described in Chapter 32: One cell becomes two, two become four, and so on. Four days after fertilization, the developing embryo enters the uterus; three days more and it implants itself in the tissue of the uterine wall known as the endometrium and continues to develop.

Both the male and female reproductive systems are greatly influenced by the actions of hormones—chemical signaling molecules that move through the bloodstream. Indeed, the reproductive organs are primary sites of production for some of the most important of these hormones: testosterone in men and estrogen and progesterone in women.

With this outline of human reproduction as background, let's move on now to take a closer look at both the female and male reproductive systems, first as they function separately and then as they function together to produce a baby.

33.2 The Female Reproductive System

The Female Reproductive Cycle

Although hormones operate in both sexes, in females they bring about a reproductive *cycle* that repeats itself about once every 28 days (although it can range from 21 to 42 days). In males, conversely, there is no cycle, but instead a steady production of sperm and a destruction or removal of the sperm that are

not ejaculated. Why a cycle for females and not for males? The answer lies partly in the two roles the female reproductive system must fulfill. It must form an egg *and* create an environment in the uterus in which a fertilized egg can develop. An important part of this uterine environment is essentially built up and then dismantled once every 28 days—except in those rare instances in which an oocyte is fertilized. The end result of the dismantling of the uterine environment is **menstruation:** a cyclical release of blood and the inner portion of a specialized uterine tissue, the **endometrium,** from the female reproductive tract. The endometrium is tissue that lines the interior of the uterus; it is tissue that *would have* housed an embryo had one been implanted there (see Figure 33.1a).

Why Does Menstruation Exist?

From this, it's reasonable to ask: Why do women have this monthly cycle of buildup and destruction rather than just a permanent environment for an embryo? After all, menstruation has some obvious costs attached to it; apart from discomfort, there is a loss of blood and other tissue. (On average, women lose about 40 milliliters—or about 1.4 fluid ounces—of blood in a menstrual cycle and about that much other tissue as well.) So, in the face of these costs, why has menstruation been preserved throughout human evolution? Scientists have no firm answers here, but they do have a couple of favored hypotheses. One is that, costly as menstruation may be, in terms of expended energy it is not *as* costly as the alternative of maintaining a permanent embryonic environment in the uterus. The other has to do with a change in form that the endometrium undergoes each month. As the time approaches in which an oocyte might get fertilized and then implanted, the endometrium goes from being a tissue whose primary activity is growing (in its "proliferative" phase) to one whose primary activity is secreting a nutrient-rich mucus (in its "secretory" phase). The assumption is that a fertilized egg can implant itself only in a secretory-phase endometrium, but that a secretory-phase endometrium cannot perpetuate itself—it must develop from a proliferative-phase endometrium, after which it cannot stay in secretory phase for long. Given this, it makes sense that the endometrium washes a part of itself out every 28 days or so and starts its growth anew.

How Does an Egg Develop?

As noted earlier, eggs develop from precursor cells called **oocytes.** If you look at **Figure 33.4**, you can track the process by which oocytes develop within each ovary. Any oocyte that moves very far through the process of development ends up being surrounded by a sphere of accessory cells, called *follicle cells*, that eventually provide the oocyte with nutrients. The complex of an oocyte and its accessory cells and fluids is called an **ovarian follicle;** it is the basic unit of oocyte development.

Follicle Development and Ovulation

Every oocyte that matures into an egg must go through the developmental steps you see in Figure 33.4. This is a monthly process of selection, running from a large starting set of follicles, through fewer *primary follicles* that develop from them, to a greatly reduced number of *secondary follicles* that continue development, and finally to a single *tertiary follicle* that nurtures an oocyte through its time in the ovary. Responding to hormonal influence midway through the menstrual cycle, the tertiary follicle ruptures, thus expelling the oocyte and some surrounding accessory cells not only from the follicle but from the ovary as well (**Figure 33.5**). With this event, ovulation has occurred. It takes place in this way in one ovary each month, with one ovary or the other selected through an unknown process.

Movement through the Uterine Tube

Freed from its ovarian housing but still surrounded by an exterior coat and a collection of accessory cells, the oocyte is now swept into the uterine tube through the actions of the finger-like *fimbriae* lying at the tube's funnel-shaped end (see **Figure 33.6**). Propelled by liquid currents and the uterine tube's whip-like cilia, the oocyte—technically not yet an egg—makes a journey of about 10 centimeters (or 4 inches) through the uterine tube over the next 4 days. While within the tube, the oocyte may encounter sperm and become fertilized prior to entering the uterus.

Figure 33.4
Steps Leading to an Egg

Eggs develop within ovaries by going through the set of developmental stages seen in the figure. First, the precursor of an egg, a primary oocyte, becomes part of a nurturing complex of cells and fluids called an ovarian follicle. Then, on the periphery of the ovary, one follicle will mature, once every 28 days, by becoming first a primary, then a secondary, and finally a tertiary follicle. The oocyte contained in the tertiary follicle is expelled from the ovary at the end of the process, at which point it begins a journey through the uterine tube. With the oocyte's expulsion, the ruptured tertiary follicle is transformed into the corpus luteum, which for a time produces hormones that facilitate pregnancy. If no pregnancy results from the ovulation, the corpus luteum begins degenerating within about 12 days.

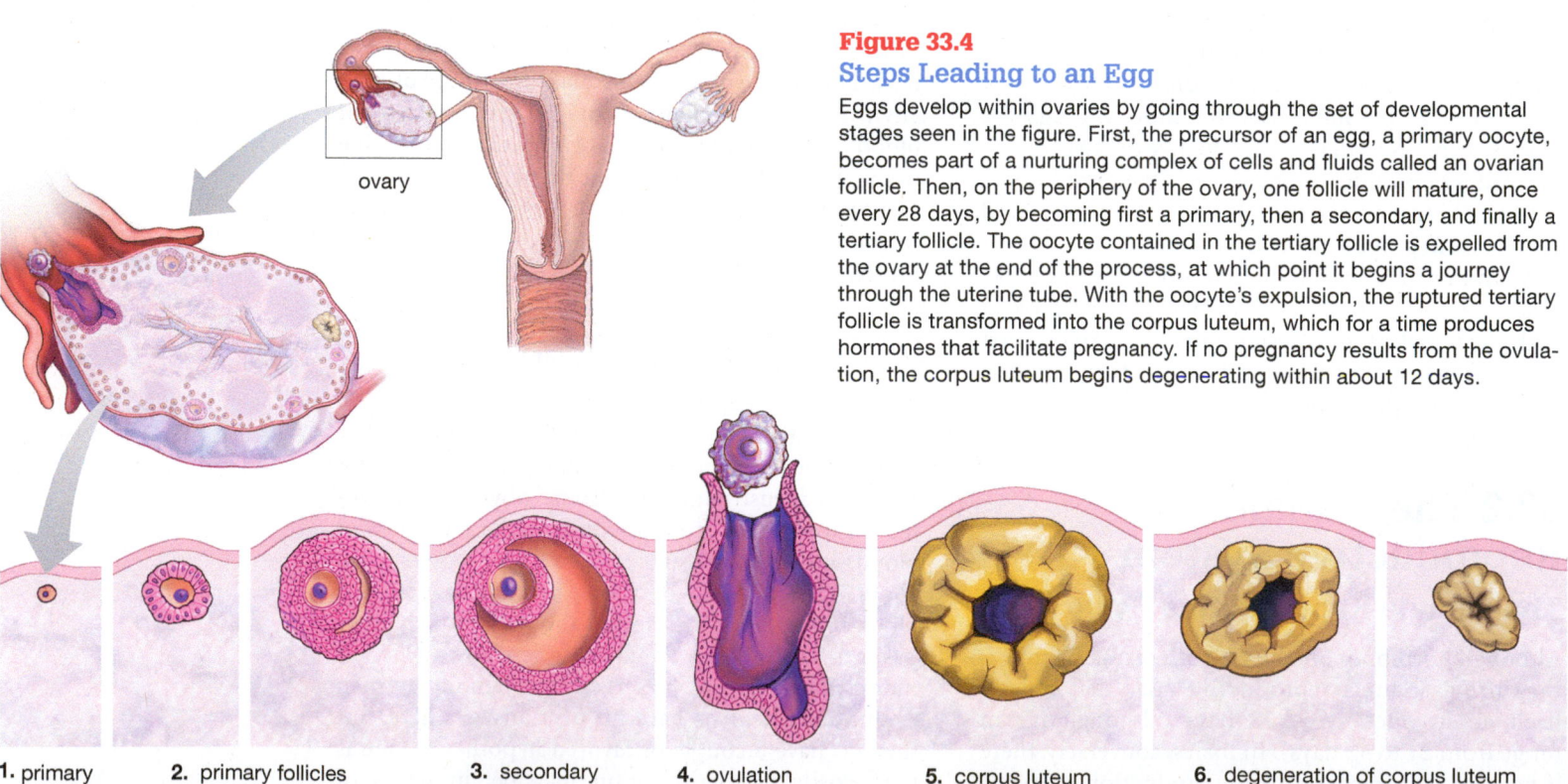

ovary

1. primary oocyte
2. primary follicles
3. secondary follicle
4. ovulation through rupture of tertiary follicle
5. corpus luteum (yellow body)
6. degeneration of corpus luteum

Meanwhile, the ruptured tertiary follicle now takes on a new role: It develops into a body called the **corpus luteum** ("yellow body"), a structure that secretes hormones that first prepare the female reproductive tract for pregnancy and that then maintain the tract during the early phases of pregnancy. If pregnancy does not occur, the corpus luteum begins degenerating after about 12 days. With the corpus luteum, you can see again the part that hormones play in the monthly female reproductive cycle. For details on this hormonal regulation, see "Hormones and the Female Reproductive Cycle," on page 636.

Changes through the Female Life Span

In between the creation of oocytes and the journey of one oocyte down the uterine tube lies a multistep process that could be characterized by the phrase "Many are called, but few are chosen." A 6-month-old female fetus has in its ovaries about 7 million oocytes that have become part of follicle complexes. At birth, however, this number has been reduced to about 800,000, primarily because of the process of *atresia*, or the natural degeneration of follicles. By puberty, the number has declined to about 400,000. A small number of these follicles are "recruited" monthly for further development, but as you've seen, only *one* of these will go on to develop into the oocyte that is expelled from the ovary each month. The only exception to this rule comes in the 1 percent or so of ovarian cycles that result in multiple ovulations, in which case fraternal twins, triplets, and other multiple births are possible. In sum, the oocyte that begins the journey down the uterine tube each month has been selected from more than 7 million initially created and is one of

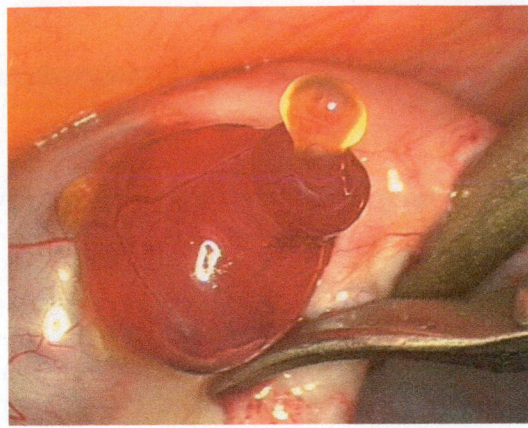

Figure 33.5
Ovulation in Progress
A human oocyte and its surrounding accessory cells emerge from a tertiary follicle. This photo—by far the clearest ever taken of human ovulation—was captured in 2008 through a piece of good luck: A Belgian surgeon was performing a routine hysterectomy on a 45-year-old patient when she began ovulating in mid-procedure. (The gray structure on the right is a surgical instrument.) Images the surgeon captured of the entire process gave reproductive experts reason to doubt a long-held belief about ovulation: that it takes place very rapidly, through a sudden expulsion of an oocyte from its surrounding follicle. This ovulation took a full 15 minutes to complete.

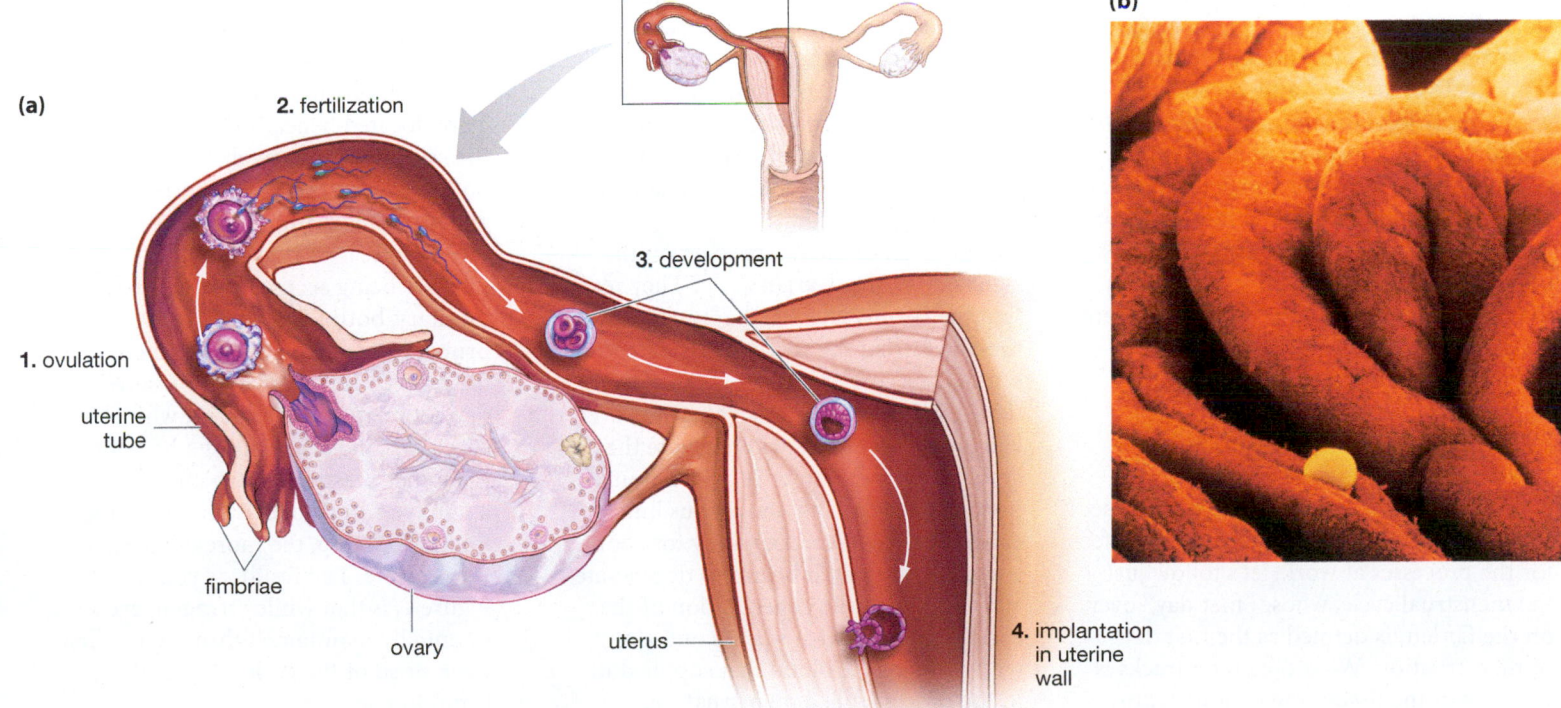

(a)

2. fertilization

3. development

1. ovulation

uterine tube

fimbriae

ovary

uterus

4. implantation in uterine wall

(b)

Figure 33.6
Journey to Pregnancy
(a) Surrounded by accessory cells, an oocyte is expelled from the ovary into the uterine tube, where it encounters sperm and is fertilized by one of them. Now a fertilized egg (or zygote), it begins the process of development and completes its journey with implantation in the endometrium of the uterine wall.
(b) A micrograph of an unfertilized oocyte moving through folds of tissue in a uterine tube.

ESSAY

Hormones and the Female Reproductive Cycle

In females, reproduction is carried out through a *cycle*, which is to say a series of events that repeat themselves regularly—on average, once every 28 days. But how can the female reproductive system "keep time" like this? What is the clock that governs the cycle?

A real clock is actually a good analogy for what happens. If you look at an old-fashioned mechanical watch, what you'll see is a central spring that exerts a force through a series of gears. Given these parts, it takes 12 hours for the physical mechanism of the "small hand" on the watch to make its way through a complete cycle—to end up where it began. In women, the substances called hormones set in motion a group of physical processes that take about 28 days to complete. The last of these processes leaves the body right where it began, at the start of another cycle.

We actually should speak of female reproductive *cycles*, because several are going on in reproduction, all of them interlinked. There is the ovarian cycle, which has to do with the development and release of oocytes. Then there is the menstrual or "uterine" cycle, which concerns the regular buildup and then breakdown of tissue in the inner lining of the uterus. Finally, there are several hormonal cycles, which help govern both menstruation and egg production.

If you look at **Figure 1**, you can see a time line for all these cycles as tracked over a period of 28 days, which you can see marked off at the bottom. To give you a feel for the processes at work, let's follow just the menstrual cycle, whose "first day," over on the far left, is defined as the first day of menstruation. What the graph tracks is the state of the tissue (the endometrium) lining the inside of the uterus. As you can see, beginning on day 1, this tissue goes through a breakdown that lasts about 6 days, after which it begins a steady buildup that continues all the way to about day 21—at

which point it starts to break down again. Now look at what is going on in the ovarian cycle (the second one down from the top). This part of the graph is following the single "successful" oocyte (of many that begin development) through its progression into the oocyte that bursts from the ovary on day 14.

The real lesson, however, is to think about the menstrual and ovarian cycles as they relate to *each other:* The endometrium is built up in the uterus as a means of creating an environment in which an egg could implant itself should one be fertilized. If such an egg begins its journey on day 14 and then takes another 7 days or so to implant itself in the uterus, you can see what stage the *endometrium* will be in at this point: near its maximum thickness (and at a highly productive stage of synthesizing a blood vessel system). Such an environment is just what an implanted embryo would need.

And what of the figures over on the right side of the ovarian cycle? They represent the corpus luteum, the "remains" of the follicle that the successful oocyte broke out of. The corpus luteum is much more than a spent vessel at this point, however. Hormones it secretes will promote the growth of the endometrium. Without a pregnancy occurring, the corpus luteum carries out such secretion only for about a week. After that, it begins to degenerate, and *this* means the degeneration of the endometrium, which washes out, along with some of the blood that supplied it, in the process called menstruation.

Which hormones are secreted by the corpus luteum? Look now at the ovarian hormone cycle, and you'll see a line for "progestins"—a family of hormones whose most important member is progesterone.

As you can see, the level of progestins takes a sharp turn upward just as the corpus luteum is formed—no surprise because it is the corpus luteum that is secreting them.

There is a circularity to all this hormonal control, with the hormones involved influencing each other in different ways as the cycle goes on. Let's arbitrarily take as a starting point, however, first the development of a follicle and then the release of an egg from it.

One important hormone in the cycle is produced in neither the ovaries nor the uterus but in the brain. A brain structure called the hypothalamus releases vanishingly small amounts of a hormone called **gonadotropin-releasing hormone,** whose function is spelled out in its name: It brings about the release of *other* hormones active in reproduction, the gonadotropins.

How can the female reproductive system "keep time"? What is the clock that governs the cycle?

Two gonadotropins are spurred into production in this way, **follicle-stimulating hormone (FSH)** and **luteinizing hormone (LH),** both of which are produced in a gland located beneath the hypothalamus, the anterior pituitary gland. You can see both FSH and LH levels charted in the gonadotropic hormone cycle at the top of Figure 1.

After being secreted by the anterior pituitary, both FSH and LH have their primary early-cycle effects at the site of the ovarian follicles. FSH promotes the development *of* the follicles, while LH stimulates the synthesis of a hormone *by* the follicles, estrogen (which actually is a family of hormones). The more the follicles develop, the more estrogen they secrete. The result, as you can see in Figure 1, is that while estrogen levels are essentially maintained through the first week or so of the cycle, after that they begin to rise.

Estrogen is a hormone that is doing two things in the female reproductive cycle: It is promoting the growth of the endometrium in the uterus, and it is controlling the release of both FSH and

LH. Early on, the relatively low levels of estrogen act to *decrease* FSH—not enough to stop development of the follicles but enough to slightly reduce FSH levels over the first 10 days or so of the cycle, as you can see in Figure 1. In the second week of the cycle, however, estrogen output is greatly increased thanks to the development of the tertiary follicle, whose oocyte will burst forth from the ovary. At this level, estrogen acts to *increase* not only FSH production but LH production as well. The result, as seen in Figure 1, is the "spike" of LH production that begins just before day 14 of the cycle. This hormonal surge is what causes the oocyte to break out from both follicle and ovary.

Now the former tertiary follicle has a new identity, the corpus luteum. In this role, it produces progesterone, which as noted acts on the uterine lining to develop it for embryo implantation. Then, in tandem with estrogen, progesterone has another effect. It acts on the hypothalamus and pituitary to *lower* FSH and LH levels. But it turns out that the corpus luteum needs LH to remain healthy. In the absence of pregnancy, falling LH levels spell the end for the corpus luteum. This in turn causes a decline in progesterone levels, which brings about the degeneration of the endometrium, essentially by depriving it of a blood supply. Now, in the absence of a corpus luteum or developed follicles, the levels of estrogen and progesterone are *low*. Such low levels *stimulate* FSH and LH production in the anterior pituitary, and this is where we started: The cycle begins again as LH and FSH stimulate the growth of another set of follicles.

So, what keeps this cycle from continuing when pregnancy does occur? The implantation of an embryo in the endometrium causes release from the embryo of a hormone (called human chorionic gonadotropin), which acts like LH in that it keeps the corpus luteum robust and generating progesterone and estrogen, which in turn maintains the endometrium.

With this knowledge, you're in a position to understand what birth control pills do. In general, they are compounds that contain both synthetic estrogen and a synthetic progestin. Women usually take them beginning five days after the start of menstruation for three weeks (and during the fourth week take a chemically functionless "spacer" or nothing). Think about what is required at the beginning of a cycle to prompt the development of follicles: *low*

levels of estrogen. (Recall that this low level stimulates the hypothalamus and pituitary to secrete FSH.) The pill provides a *higher* level of estrogen than exists naturally, thus suppressing the release of FSH and the development of follicles. Second, the progestin in the pill suppresses LH release by the pituitary, which prevents the *ovulation* of any follicle that might manage to develop because the LH surge is required for ovulation.

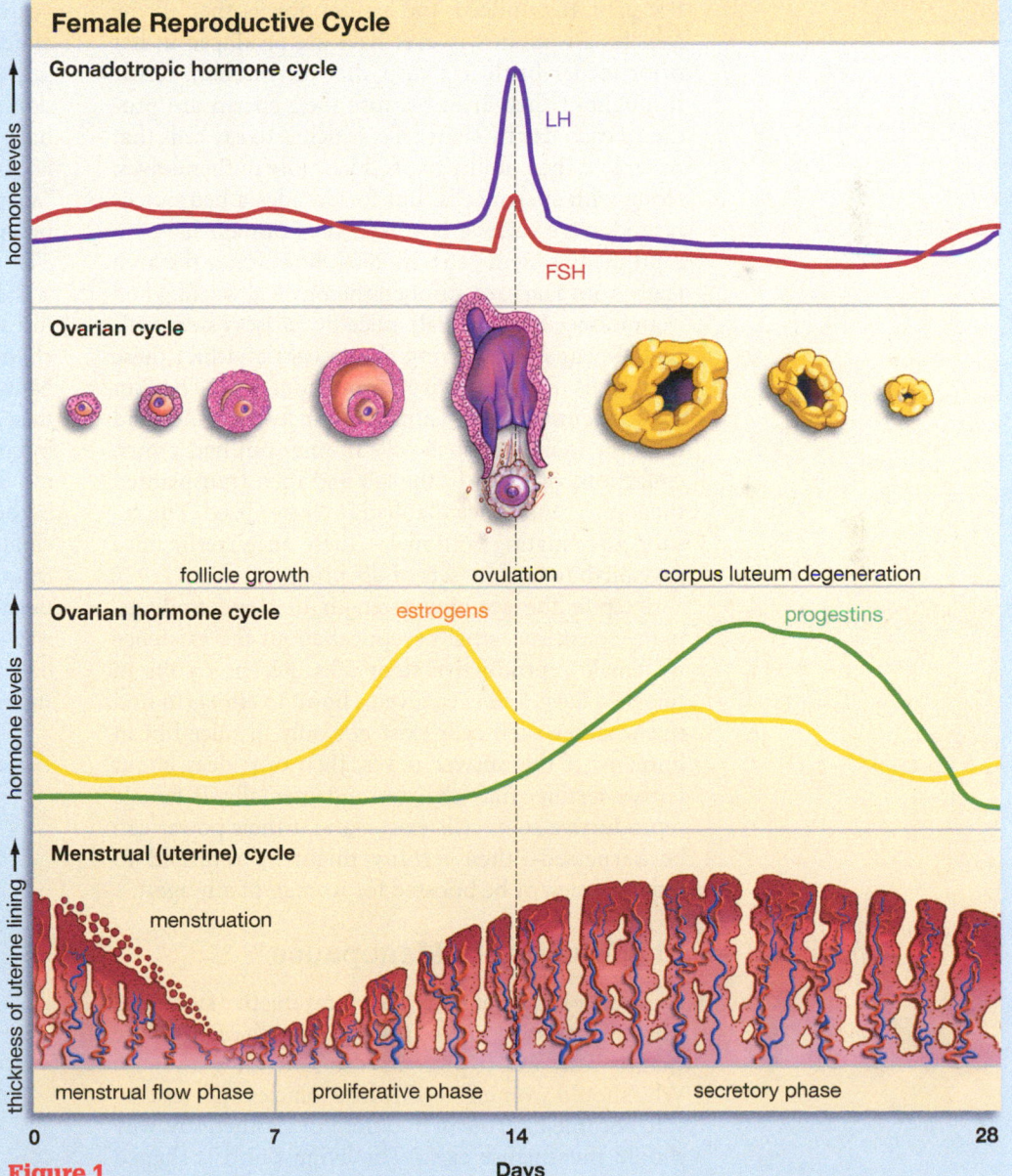

Figure 1
The Female Reproductive Cycle

only 500 or so that will make this trip in a woman's lifetime (about 13 a year for 35 or 40 years).

Follicle Loss and Female Fertility

Given that a woman begins puberty with about 400,000 follicles, you might think that her supply would never run out. This is not the case, however, because the process of atresia continues at a steady pace from puberty through a woman's mid-30s and then accelerates beginning in a woman's late 30s. By the time a woman reaches her early 50s, perhaps 1,100 follicles remain; this scarcity is thought to be the most important factor in bringing about **menopause,** or the cessation of the monthly ovarian cycle.

Taking a step back from this story, you can see that its underlying assumption is that, while a woman will lose follicles over the course of her lifetime, she will not gain any. Indeed, the assumption is that all the follicles a woman will ever have are produced in her prior to her birth. It's clear that men remain fertile throughout their lives because their sperm are produced each day by stem cells, which is to say cells that never lose their ability to produce more of themselves, along with sperm cells. But for decades a bedrock of reproductive science has been that women have no reproductive stem cells. In 2004, however, a research team from Harvard stunned the medical world when it announced that female mice do possess stem cells that are capable of giving rise to eggs and that these stem cells remain active throughout life. Then in 2009, scientists from China announced that they had not only identified such cells in mice but had grown collections of them in the lab and then transplanted them into other mice that had been sterilized. The result? After mating with males, these once-sterile mice gave birth to healthy sets of offspring.

Despite the seeming strength of these findings, many scientists remain skeptical about the existence of female reproductive stem cells. Recent results in this field have, however, given a boost to efforts to find out whether such cells exist not only in mice but in humans. If the answer is yes, then our ideas about female fertility may change fundamentally. If female reproductive stem cells exist—and if their power can be harnessed—then fertility might be returned to older women or be boosted for women of any age.

The Mystery of Menopause

Irrespective of whether female reproductive stem cells exist, the massive loss of follicles over the female life span is real, and it presents science with a mystery: Why should women steadily lose follicles and, as a result, become infertile in midlife? Put another way, why should menopause exist? The living world is shaped by differential success in reproduction. Traits that help a given organism to produce relatively *more* offspring normally are retained over evolutionary time.

Yet the trait of menopause developed in the course of evolution, and it seemingly leads to fewer offspring for women. Through it, contemporary women are taken out of reproduction altogether for the last 20 or 30 years of their lives, during which time they suffer from such menopause-related maladies as osteoporosis. Given these factors, why does menopause exist?

As was the case with menstruation, we have no firm answers, but we do have several candidate hypotheses. One is that menopause is simply a product of women living longer today than their evolutionary ancestors did. Proponents of this view point out that a number of female mammals undergo what amounts to menopause but then live only long enough to raise one additional generation of offspring after this loss of reproductive capacity sets in. Human females may once have followed this pattern and lived to, say, 60 years old—just long enough to give birth to a final child in their late 40s and then raise that child to a self-sustaining age of 10 or so. Today, however, women don't live until 60; they live until 75 or 85 and thus have decades of life following menopause.

A very different hypothesis holds that menopause is "adaptive," as biologists say—it arose through evolution because it actively helps females perpetuate their genes. The starting point for this "grandmother hypothesis" is the observation that what counts in evolution is not just whether you have children; it is whether you have children who themselves go on to have healthy children. Now, how is a given woman most likely to achieve this? Under this hypothesis, she best achieves it not by continuing to give birth throughout life but by leaving direct reproduction behind at a certain point and becoming a grandmother who provides both for her children and her grandchildren. We now have evidence, from eighteenth- and nineteenth-century groups in Finland and Canada, that things actually can work this way—that an increased duration of "postmenopausal" life in women can positively affect the reproductive success of their children.

SO FAR. . .

1. While they are in development, the reproductive cells a woman produces are known as _____ and are produced in the organs called _____.

2. Every 28 days, on average, one of these reproductive cells is released in the process called _____ and then begins a journey down one of the woman's _____.

3. Menstruation, the release of blood and a portion of the uterine tissue called the _____, occurs when there has been no _____ during the reproductive cycle.

33.3 The Male Reproductive System

Switching now to males, if you look at **Figure 33.7**, you can see another view of the male reproductive system. Here you see the large-scale structures of the system. There is the **testis,** in which sperm begin development; the **epididymis,** the tubule in which sperm mature and are stored; and the **vas deferens,** the tube through which sperm move in the process of ejaculation. (All three of these structures exist in pairs, although the second member of each pair is not visible in the figure.) Then there is the single **urethra,** the tube each vas deferens empties into and through which sperm leave the male body. There are also several accessory glands whose contributions are added to sperm as it proceeds: the seminal vesicles, the prostate gland, and the **bulbourethral glands.** (The seminal vesicles and bulbourethral glands also exist in pairs.)

Structure of the Testes

The testes are surrounded by a muscle whose contractions help to regulate the temperature of the sperm inside. Sperm development requires temperatures somewhat cooler than those found in the rest of the body, which is why the testes are *external* to the body's torso. Sperm cannot develop in temperatures that are too cold, however. Thus, when a male steps into, say, a cold swimming pool, the muscle surrounding the testes will contract, drawing the testes up closer to the torso—and warming the sperm.

Figure 33.7a shows that the testes are divided into a small series of lobes, and that within each lobe there are a number of highly convoluted tubes, the **seminiferous tubules:** structures that are the sites of initial sperm development in men. If you now look at a single seminiferous tubule in cross section, in Figure 33.7b, you'll see how this development proceeds: from the outside of each tubule—where sperm precursors are in their *least*-developed state—toward the interior

The Male Reproductive System

(a) Delivery of sperm

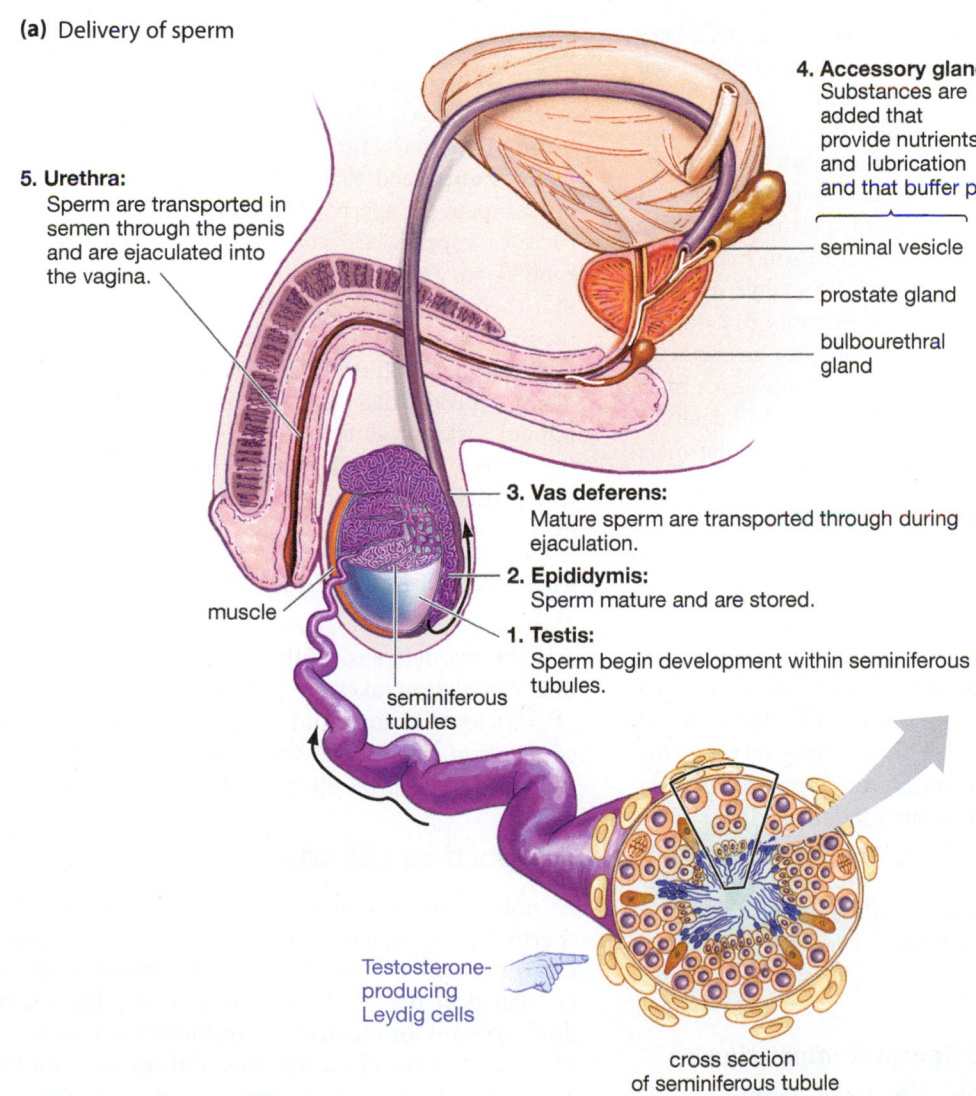

5. Urethra:
Sperm are transported in semen through the penis and are ejaculated into the vagina.

muscle

seminiferous tubules

Testosterone-producing Leydig cells

4. Accessory glands:
Substances are added that provide nutrients and lubrication and that buffer pH.

seminal vesicle

prostate gland

bulbourethral gland

3. Vas deferens:
Mature sperm are transported through during ejaculation.

2. Epididymis:
Sperm mature and are stored.

1. Testis:
Sperm begin development within seminiferous tubules.

cross section of seminiferous tubule

Figure 33.7
Delivery and Development of Sperm
(a) How sperm are delivered.
(b) How sperm mature.

(b) Development of sperm

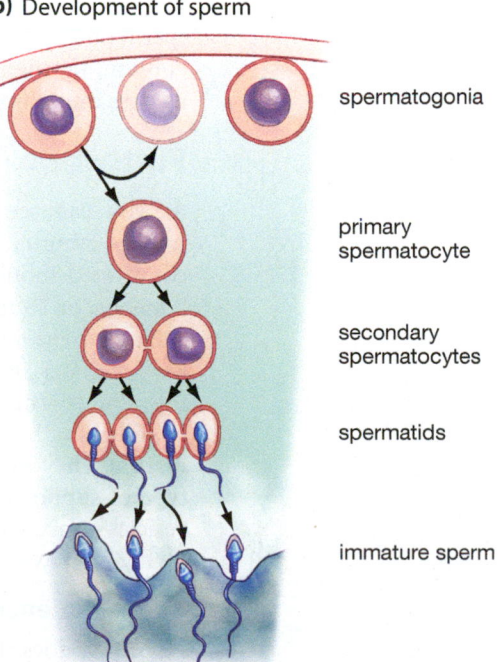

spermatogonia

primary spermatocyte

secondary spermatocytes

spermatids

immature sperm

cavity of each tubule, where sperm are in a more developed, though still immature, state. Eventually, they are released into the interior cavity and are transported up it, moving then to the epididymis for further processing and storage. As you can see, there is an assembly-line quality to sperm production, starting with initial development at the periphery of the tubules, continuing in the interior of the tubules, and ending with storage in the epididymis. Older sperm that are not ejaculated are destroyed (by other cells) or eliminated in the urine. Inside the testes, surrounding the seminiferous tubules, there exist numerous *Leydig cells,* which are the cells that produce testosterone.

Male and Female Gamete Production Compared

The production of sperm bears an interesting comparison with the development of eggs. The cells from which sperm develop are called **spermatocytes.** In going back to the precursors of *these* cells, however, we see a significant difference between males and females. The cells that give rise to spermatocytes are the stem cells called **spermatogonia;** they exist at the periphery of the seminiferous tubules, and through normal cell division, they actually lead to two kinds of cells. One is the primary spermatocyte, which goes on to develop into a sperm; the other, however, is another spermatogonium. Thus, spermatogonia are "sperm factories," if you will, that give rise not only to sperm but to more sperm factories as well. This is the reason males *keep* producing sperm throughout their lives while, as we've noted, females may have a fixed quantity of oocytes. Although male sex hormone production lessens with age and sperm decline in quality, it is possible for male reproductive capability to remain intact until death. If you look at Figure 33.7b, you can see how the male sperm progression works in the testes: from spermatogonium, through spermatocyte to spermatid, and finally to immature sperm.

Further Development of Sperm

As you can see in Figure 33.7a, the seminiferous tubules eventually feed into the epididymis, a single convoluted tubule (one for each testis) that is about 7 meters, or 23 feet, long. This structure serves as the site of storage and final sperm development and as a kind of materials recycling site for damaged sperm. By the time sperm arrive at the bottom portion of the epididymis, they are fairly mature but not yet fully mature. They won't become mobile until fluid from a type of supporting gland noted earlier—the seminal vesicles—mixes with them on ejaculation, thus supplying them with nutrients.

Time Sequence and Sperm Competition

How long does this whole process of sperm development take? About 2.5 months from the start of development to storage in the epididymis. About 200 million sperm are produced in this assembly line each day, which you may recall is about the number that are ejaculated in an average orgasm.

As noted, the watchword for oocyte selection could be "Many are called but few are chosen." How much more extreme is the selection that goes on with sperm! The number expelled with *each ejaculation* vastly exceeds the number of oocytes a woman ever has stored in her ovaries. However, only a single sperm per ejaculation will be able to fertilize an egg—assuming an egg is present in the uterine tube at the right time.

Why is there such an enormous difference between the number of sperm produced and the number that will fertilize eggs? The essential answer here seems pretty clear: Just as males often are in competition with other males to mate with females, so the sperm of males is in competition with the sperm of other males to *fertilize* female eggs. This general phenomenon of *sperm competition* goes on throughout the animal kingdom—in insects, birds, and reptiles as well as in mammals. Its root cause is that there are very few animal species in which females or males are truly monogamous. As a result, mating by itself does not guarantee offspring. If a female who is fertile over a period of several days mates with a number of males during these days, then each male's chance of fatherhood is enhanced to the degree that his sperm can outcompete the sperm of the other males. This is the basis for a sperm "arms race" that has taken place over evolutionary time. Looking at various species, we find that their sperm has gotten longer, or more "motile" (active), or has been produced in greater numbers—qualities that each evolved in various species because of the reproductive edge they provided. This is why human males produce so many sperm. You might say that sperm production in them has followed the lottery principle: The more tickets you have, the more likely you are to win.

From Vas Deferens to Ejaculation

Maturing sperm are stored in both the epididymis and the vas deferens. With sufficient sexual stimulation, ejaculation takes place through the contractions of muscles that surround the penis at its base. This contraction pushes the sperm through the urethra and out the urethral opening at the tip of the penis.

Supporting Glands

As noted, several other kinds of materials join with sperm before ejaculation to form the substance known as semen. Indeed, given the number of sperm ejaculated, it may be surprising to learn that sperm don't account for much of the volume of semen; some 95 percent of the ejaculated material comes from the accessory glands already mentioned: two seminal vesicles, the prostate gland, and two bulbourethral glands. It's helpful to think of sperm as getting "outfitted" for their travels by these glands just as they are exiting from the body. In addition to the nutrients the seminal vesicles supply, accessory glands provide alkaline and lubricating substances (which neutralize the acidic environment of the vagina and facilitate transportation).

One of these supporting glands may sound familiar. The **prostate gland** is a structure that surrounds the male urethra near the bladder and contributes a substantial amount of material to semen. Later in life, a significant percentage of men develop either enlarged prostate glands or, worse, the growth of cancerous cells in the prostate, which is to say, prostate cancer. Given that the prostate encircles the urethra—and that the urethra transmits not only semen but urine as well—you can see why a warning sign of these conditions is reduced urine flow. But with both urine and semen traveling through the urethra, how does the body make sure that one substance doesn't mix with the other? Traffic control is provided by the prostate gland. During ejaculation, a circular muscle in it contracts, thus sealing off the contents of the bladder.

SO FAR . . .

1. Sperm begin development in the _____ and continue development and are stored in the convoluted tubules called the _____.

2. The material that is ejaculated from the penis, _____, is composed of both _____ and fluids secreted by three supporting _____.

3. Males are able to remain fertile throughout their lives because the cells called spermatogonia are reproductive _____ that give rise to more of themselves and to the _____ that develop into sperm.

33.4 The Union of Sperm and Egg

To this point, we've reviewed the process by which male sperm and female egg are first created and then *positioned* so that they can come together. Now let's look at the final steps that lead to the fusion of these two cells.

When the oocyte is expelled from the ovaries, there still is no solitary oocyte in transit, but rather an oocyte surrounded by an extracellular coat that in turn is surrounded by a group of accessory cells (**Figure 33.8b**). When we look at sperm, we find that each one has a "head" that contains not only the nucleus and its vital complement of chromosomes but an outer compartment, called an **acrosome,** which contains enzymes (**Figure 33.8a**). These enzymes, which are released externally, have the ability to dissolve the bonds between the accessory cells that

(a) Structure of a sperm cell

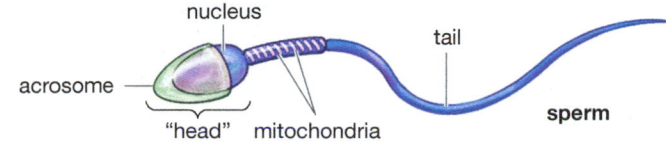

(b) How sperm fertilizes egg

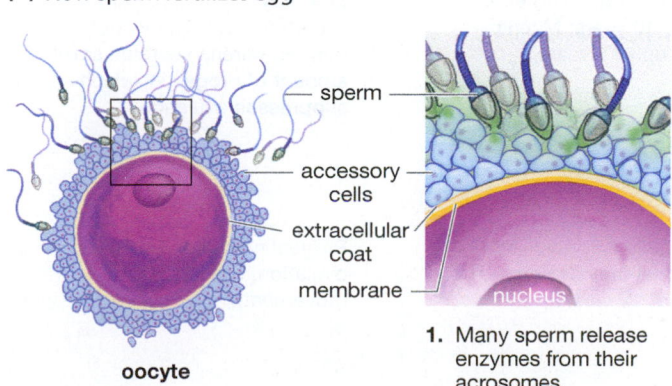

Figure 33.8
Anatomy of a Sperm and Steps in Fertilization

(a) The "head" of each sperm includes not only its complement of chromosomes but also an acrosome, which contains enzymes that, when released externally, break down the bonds between accessory cells that surround the oocyte. Behind the sperm head are the mitochondria that supply the energy for the motion of the tail that propels the sperm.

(b) The process by which a sperm fertilizes an oocyte.

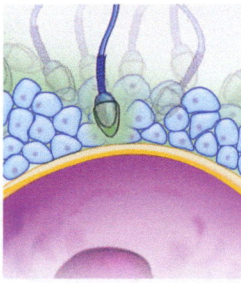
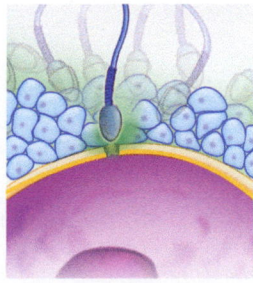
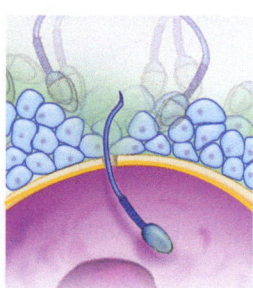

1. Many sperm release enzymes from their acrosomes.

2. The enzymes dissolve bonds between accessory cells, enabling one sperm to pass through.

3. The one sperm uses its own enzymes to clear a path through the extracellular coat.

4. The sperm and oocyte membranes fuse, allowing the sperm to move inside the oocyte.

ESSAY

Methods of Contraception

Selected methods of contraception are presented here, ranked according to how effective each has been shown to be in preventing unwanted pregnancy during an initial year of use. "Typical use" refers to use as practiced by most individuals or couples. "Perfect use" refers to use in which a method is used correctly either during every instance of sexual intercourse (in the case of a method such as condoms) or over the course of consecutive months (in the case of a method such as birth control pills).

Table 33.1

Methods of Contraception

Form	Effectiveness (typical/perfect use)	Description	Works by
Birth control implant (Implanon)	99.9/99.9 percent	Matchstick-sized plastic rod inserted under skin of upper arm, lasts up to 3 years following implantation.	Released hormones prevent ovulation, inhibit sperm movement.
Vasectomy (male sterilization)	99.8/99.9 percent	Surgical procedure for producing permanent infertility, though procedure can sometimes be reversed.	Cutting of vas deferens blocks movement of sperm during ejaculation. Does not affect hormonal production, capacity for erection, or ability to ejaculate.
Tubal ligation (one method of female sterilization)	99.5/99.5 percent (rarely, tubes may reconnect themselves)	Surgical procedure for producing permanent infertility, though procedure can sometimes be reversed.	Cutting and tying of uterine tubes cuts off movement of unfertilized eggs through them. Does not affect feminine characteristics, hormonal production, or sexual performance.
Intrauterine device (IUD), ParaGard (copper-releasing) and Mirena (progestin-releasing)	99.2/99.4 percent (ParaGard) 99.8/99.8 percent (Mirena)	T-shaped plastic device, inserted into uterus by health professional. ParaGard effective up to 10 years; Mirena effective up to 5 years.	Both block sperm movement; may also prevent implantation of embryo in uterus. ParaGard secretes small amounts of copper; Mirena secretes small amount of progestin, which suppresses ovulation.
Birth control shot (Depo-provera)	97/99.7 percent	Shot administered by health professional provides contraception for 12 weeks.	Progestin hormone prevents ovulation, inhibits sperm movement.

Table 33.1

(Continued)

Form	Effectiveness (typical/perfect use)	Description	Works by
Birth control patch (Ortho Evra)	92/99.7 percent	Beige patch placed on skin; new patch once per week, three weeks out of four.	Released hormones prevent ovulation, inhibit sperm movement.
Birth control pill	92/99.7 percent	Estrogen-progestin combination pills (most common) or progestin-only pills.	Released hormones prevent ovulation, inhibit sperm movement.
Birth control vaginal ring (NuvaRing)	92/99.7 percent	Flexible ring inserted into vagina by user; left in place for 3 weeks and then removed for 1 week, after which new ring is inserted.	Released hormones prevent ovulation, inhibit sperm movement.
Emergency contraception (Plan B product)	Reduces chances of pregnancy by up to **89** percent if started within 72 hours of unprotected sex.	Two pills taken after instances of unprotected sex; available over the counter to women 18 and older; rules for availability to younger women in flux.	Progestin hormone prevents ovulation; may inhibit sperm movement and implantation of embryo in uterus.
Condom	85/98 percent	Sheath that covers erect penis; only contraceptive method that offers some STD protection to both partners.	Catches sperm before it can move into vagina and uterine tubes.
Diaphragm	84/94 percent When used with spermicide	Circular rubber barrier inserted by women into vagina before sex. Should be used in tandem with spermicide.	Diaphragm covers cervix, preventing sperm from reaching uterine tubes; spermicide decreases mobility of sperm.
Withdrawal	73/96 percent	Removal of penis from vagina prior to ejaculation.	Prevents ejaculated sperm from reaching uterine tubes; spermicide decreases mobility of sperm.
Spermicide	71/82 percent	Available in foams, creams, gels, vaginal suppositories, and vaginal films that are inserted into the vagina.	Physically block cervix; decrease mobility of sperm.

surround the oocyte. It takes dozens of sperm to release enough enzymes such that *one* sperm can move between the accessory cells and arrive at the external coat that is the last barrier between sperm and oocyte.

When this sperm makes contact with the coat, it binds with receptors on it, an event that triggers the release of this sperm's own complement of digestive enzymes. These enzymes clear a path through the coat that the sperm moves through. With this, the sperm reaches the oocyte, at which point the outer membranes of sperm and oocyte fuse—the two membranes become one, much as two soap bubbles might become one when they come into contact. With this, the contents of the sperm cell move inside the oocyte, which means that the two cells have effectively become one. Once this happens, the nuclei of oocyte and sperm combine, the oocyte at last becomes an egg, and fertilization is complete.

How Latecomers Are Kept Out

But what of the other dozens of sperm surrounding the oocyte? None of them can be allowed entry because an embryo could not survive getting two sets of male chromosomes but only one set of female chromosomes. To guard against this possibility, a kind of gate-slamming process is put into motion. The fusion of the first sperm with the oocyte causes the oocyte to release thousands of tiny granules into the space just outside itself. These granules release substances that harden the membrane immediately outside the oocyte while inactivating the sperm receptors on it. The result is that, with rare exceptions, only one sperm gets in. (The sexual activity that is so important to human reproduction also carries with it the risk of sexually transmitted disease, which you can read about in the essay on page 648.)

33.5 Human Development Prior to Birth

Having seen how sperm and egg form and get together, you have reached a point in human reproduction that should be familiar: *development*, meaning the process by which a fertilized egg is transformed, step-by-step, into a functioning organism. As you'll see, processes you looked at in Chapter 32 in connection with frogs and flies have counterparts in human development. What may be surprising is how *early* in human development many of the critical events of development occur. Here is a list of some of the key developmental processes discussed in Chapter 32:

- Formation of the blastula, or hollow body of cells (called a *blastocyst* in humans and other mammals)

- Gastrulation, in which the three primary layers of cells are formed
- Formation of the neural tube (which gives rise to the brain and nervous system)
- Organogenesis, in which the organs of the body start to take shape

Now, keeping in mind that a human pregnancy lasts an average of 38 weeks, consider the time line for these major developmental events. Blastocyst formation begins 4 or 5 days after conception; gastrulation begins 16 days after conception; neural tube formation begins in the third week after conception; and organogenesis begins in the fourth week. By the end of 12 weeks, immature versions of all the major organ systems have formed. To place some of these events in the context of what the mother experiences, organogenesis is occurring about the time the first menstrual period is missed, and formation of the immature organ systems is being completed at about the same 12-week point at which morning sickness generally ends.

The embryo's development over time can be contrasted with its *growth* over time. By the end of the third week—after blastocyst formation and gastrulation—a human embryo is perhaps 2 millimeters (or less than a tenth of an inch) in diameter. Skip to the twelfth week after conception, when major portions of organogenesis are concluding, and the fetus may still be only about 9 centimeters, or a little over 3.5 inches, in length.

The message here is that development is concentrated early in pregnancy, while the simple growth of what *has* developed comes later. So pronounced are the changes during development that biologists use different terms for the growing organism over time. The original cell—the fertilized egg—is a **zygote.** In humans, about 30 hours after the zygote is formed, it undergoes its first cell division; at this point, the developing organism is referred to as an **embryo,** a name that will be applied to it through the eighth week of development. From the ninth week of development to birth, the organism is referred to as a **fetus.** The typical 9-month pregnancy is divided into periods of roughly 3 months each, called **trimesters.** The first trimester ends at about the time organogenesis does.

Early Development

You saw earlier that the zygote/embryo moves from ovary to uterus during the first 4 days after fertilization. It arrives in the uterus on about day 4 as the *morula,* or tightly packed ball of cells noted in Chapter 32 (see Figure 32.1, page 620). Over the next 2 days, the morula becomes the blastocyst, which implants itself in the endometrial lining. Actually, this "implantation" is more like an invasion. The blastocyst carves out a cavity for itself by releasing enzymes that effectively digest endometrial cells (**Figure 33.9**). Implantation normally occurs in the

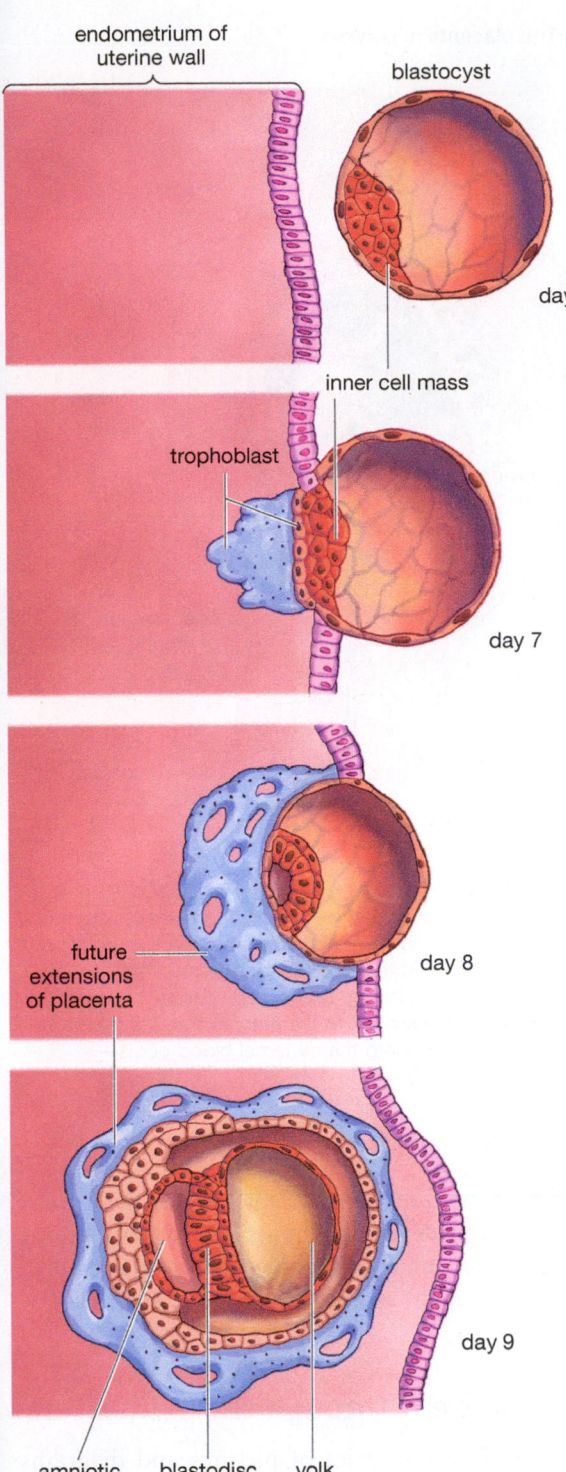

endometrium of
uterine wall

blastocyst

day 6

inner cell mass

trophoblast

day 7

future
extensions
of placenta

day 8

day 9

amniotic   blastodisc   yolk
cavity   (future baby) sac

Figure 33.9
Implantation of the Embryo in the Uterus
Six days after conception, the embryo, now in the
blastocyst stage of development, arrives at the uterine
wall. It is composed of both an inner cell mass, which
develops into the baby, and a group of cells called a
trophoblast, which carves out a cavity for the embryo
in the uterine wall and then begins establishing physical
links with the maternal tissue. The outgrowth of this
linkage is the placenta.

Figure 33.10
Implantation
Accomplished
Twelve days after concep-
tion, the embryo is firmly im-
planted in the uterine wall.

dorsal wall of the uterus (**Figure 33.10**). Occasionally,
however, a blastocyst attaches itself inside the uterine
tube, or sometimes to the **cervix,** which is the lower
part of the uterus that opens onto the vagina. This is
an **ectopic pregnancy:** a pregnancy in which the embryo has implanted in a location other than the uterine wall. An embryo will almost never survive such
an implantation.

Two Parts to the Blastocyst

Figure 33.9 shows you the arrangement of cells in
the blastocyst. On one side, you can see a group of
cells called the **inner cell mass:** the portion of the
blastocyst that develops into the baby. The periphery
of the blastocyst, however, is formed by cells called
the **trophoblast.** These are the cells that etch their way
into the uterine wall; with implantation, they begin
extending farther into the endometrium, establishing indirect links with the maternal blood supply. In
time, these cells grow into the fetal portion of a well-
known structure in pregnancy, the **placenta:** a network of maternal and embryonic blood vessels and
membranes that allows nutrients and oxygen to diffuse *to* the embryo (from the mother), while allowing carbon dioxide and other wastes to diffuse *from*
the embryo (into the maternal bloodstream). See
Figure 33.11 on the next page for a look at how the
placenta functions within a uterine environment as
it exists at about nine weeks.

    The tissue that links the fetus with the placenta is
the **umbilical cord,** which houses fetal blood vessels
that move blood between the fetus and the placenta.
So pressing is the embryo's need for nutrients and
waste removal that its circulatory system, which includes the umbilical cord, becomes functional before
any other organ system.

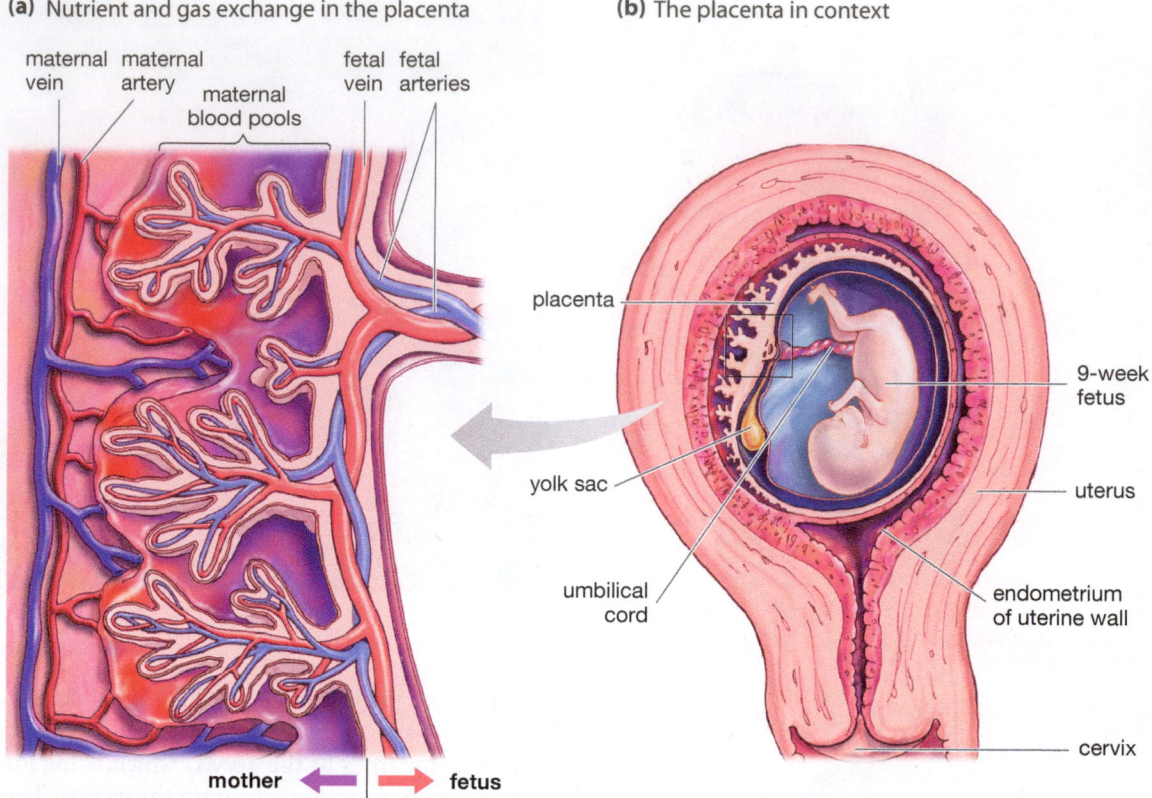

(a) Nutrient and gas exchange in the placenta

(b) The placenta in context

Figure 33.11

Nutrient and Gas Exchange in the Placenta

(a) Oxygen- and nutrient-rich blood comes from the mother through the maternal arteries and moves, through smaller arterial vessels, into the maternal blood pools in the placenta. Oxygen-poor blood from the embryo, transported through the umbilical cord's fetal arteries, does not flow into the maternal blood pools but instead stays within the fetal blood vessels, which project into the maternal blood pools. Oxygen and nutrients in the maternal blood pools move into the fetal blood vessels, however, through the process of diffusion. The resulting oxygen- and nutrient-rich blood is carried back to the embryo through the umbilical cord's fetal vein. Meanwhile, carbon dioxide from the embryo moves the other way—into the maternal blood pools— to be carried away through maternal veins.

(b) Larger structure of the placenta and uterine environment at about 9 weeks.

The Origins of Twins

Having learned a little about fetal development, you're now in a position to understand something about how twins come about. Identical twins can develop at any time in the very early stages in pregnancy—from the time the zygote undergoes its first division through the seventh day after fertilization. In all cases, identical twins result from two *parts* of the embryo developing into separate human beings. This may be a matter of the two cells produced by the zygote's first division each developing as separate embryos, or it may entail the inner cell mass of the blastocyst separating into two masses, with each then giving rise to a different person.

In contrast, **fraternal twins** (or triplets, and so on) are twins who are produced first through multiple ovulations in the mother, followed by multiple fertilizations from separate sperm of the father, and then multiple implantations of the resulting embryos in the uterus. This means that fraternal twins are like any other pair of siblings except that they happen to develop at the *same time*. Meanwhile, **identical twins** are twins who develop from a single zygote. Strictly speaking, they are a single organism at one point in their development and as such have the same genetic makeup.

Development through the Trimesters

If you look at the series of pictures and diagrams that make up **Figure 33.12**, you can see how an embryo and then fetus develop at selected points in the 38 weeks of pregnancy. Here's an overview of what happens over time during the three trimesters of a pregnancy.

First Trimester

As noted, blastocyst formation, gastrulation, and neural tube development all take place within the first month of conception; but much more happens during this period as well. By the twenty-fifth day, a primitive heart has formed in the human embryo and has

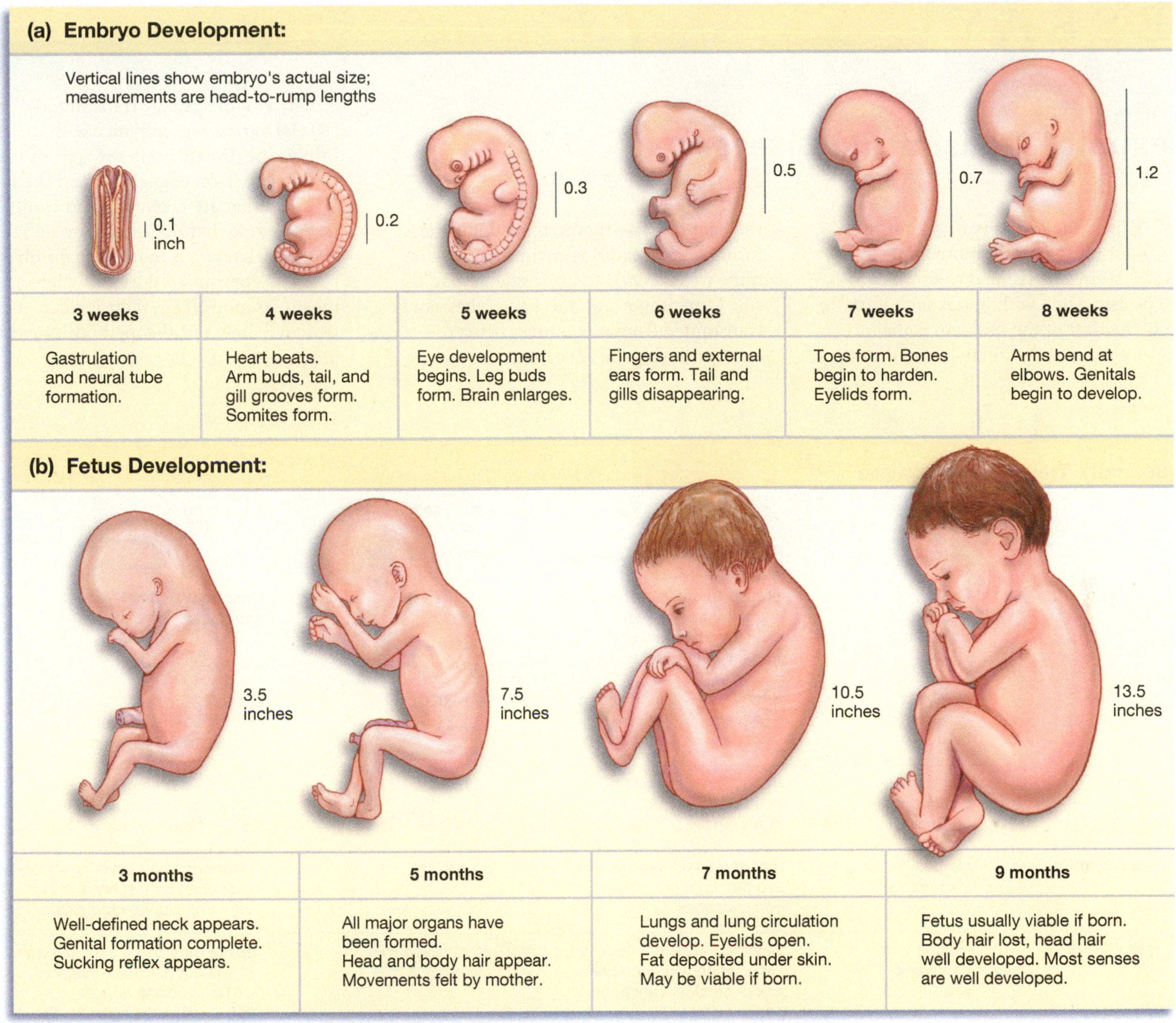

(a) Embryo Development:

Vertical lines show embryo's actual size; measurements are head-to-rump lengths

| | 0.1 inch | | 0.2 | | 0.3 | | 0.5 | | 0.7 | | 1.2 |

3 weeks	4 weeks	5 weeks	6 weeks	7 weeks	8 weeks
Gastrulation and neural tube formation.	Heart beats. Arm buds, tail, and gill grooves form. Somites form.	Eye development begins. Leg buds form. Brain enlarges.	Fingers and external ears form. Tail and gills disappearing.	Toes form. Bones begin to harden. Eyelids form.	Arms bend at elbows. Genitals begin to develop.

(b) Fetus Development:

| | 3.5 inches | | 7.5 inches | | 10.5 inches | | 13.5 inches |

3 months	5 months	7 months	9 months
Well-defined neck appears. Genital formation complete. Sucking reflex appears.	All major organs have been formed. Head and body hair appear. Movements felt by mother.	Lungs and lung circulation develop. Eyelids open. Fat deposited under skin. May be viable if born.	Fetus usually viable if born. Body hair lost, head hair well developed. Most senses are well developed.

(c)

Figure 33.12

Development through the Trimesters

(a) Development of the human embryo (3 weeks to 8 weeks).

(b) Development of the human fetus (3 months to 9 months).

(c) Photos of a human embryo at 6 weeks (left), 11 weeks (middle), and 5 months (right).

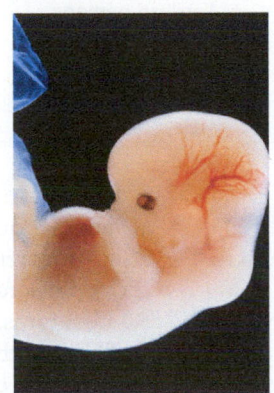

6 weeks

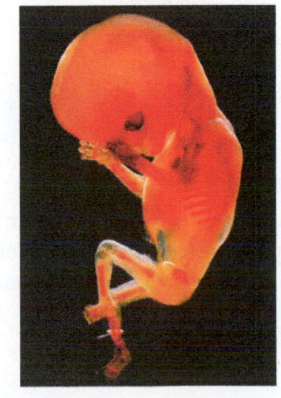

11 weeks

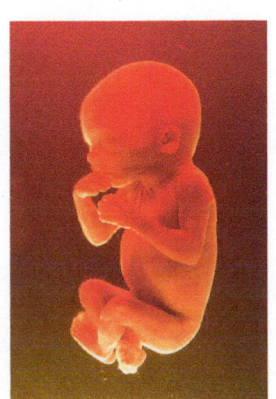

5 months

ESSAY

Sexually Transmitted Disease

Given the necessity of sex for the continuation of human life, it may seem surprising that so much risk can be associated with it, but such is the case. The root of the problem is that microorganisms that literally couldn't stand the light of day—that is, that couldn't exist for long in an exterior environment—are quite well adapted to life inside human tissue. More important, these microbes can be transmitted from one human being to another during the act of sex.

The risk of sexually transmitted diseases (STDs) varies in accordance with lots of things: the disease itself, the type of sexual activity a person engages in, and the precautions that are taken to avoid trouble. But one rule that holds true for all STDs and all behaviors is that as the number of partners goes up, so does risk. Here are the most common STDs, their incidence in the United States, and the methods of treatment for them.

Table 33.2

Sexually Transmitted Disease

Disease	Caused by	Possible consequences	New cases in the United States each year	Treatment
Chlamydia	*Chlamydia trachomatis* bacterium	Spread to uterine tubes in women, causing pelvic inflammatory disease (PID), which can lead to sterility.	2.8 million	Curable with timely use of antibiotics.
Gonorrhea	*Neisseria gonorrhoeae* bacterium	Arthritis, pelvic inflammatory disease.	716,000	Curable with timely use of antibiotics though antibiotic-resistant strains now exist.
Syphilis	*Treponema pallidum* bacterium	Heart, nervous system, bone damage if allowed to progress.	11,500	Curable with timely use of antibiotics though antibiotic-resistant strains now exist.
Genital herpes	Herpes simplex virus	Recurrent skin lesions, eye damage, pregnancy complications.	1 million	Occurrence and severity of outbreaks can be lessened with treatment, but no cure exists.
Genital warts	Human papilloma virus	Cervical infection in women, association with increased risk of cervical cancer and penile cancer.	6.2 million HPV infections	Topical treatments exist for warts, but will not eliminate infection; Gardasil vaccine approved in 2006 for females aged 11–26.
AIDS	Human immuno-deficiency virus (HIV)	Death, dementia, injury from a variety of opportunistic infections.	56,000 new infections; 1 million persons living with HIV	No cure; treatments to reduce symptoms exist.

begun pumping blood; the somites or repeating segments you read about in Chapter 32 have begun to take shape for muscular and skeletal formation.

Second Trimester

The fetus lengthens considerably during this period, but most of its weight gain takes place in the third trimester. Midway through a pregnancy, a fetus has achieved about half of its birth length (which is about 50 centimeters, or 20 inches, on average), but only about 15 percent of its birth weight (which averages 3.3 kilograms, or 7.3 pounds).

The heartbeat of a fetus can be heard through a stethoscope during the second trimester, and the mother can feel the baby begin kicking by the end of the fourth month. Kicking is only one of several

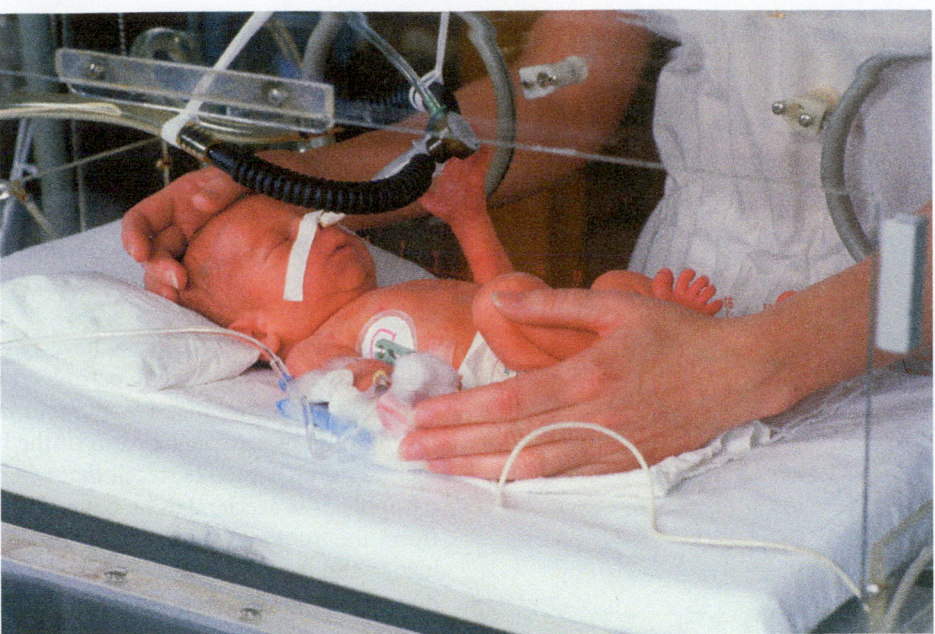

Figure 33.13
Tough Going Early On
A pediatric nurse attends to a premature baby. The incubators such infants are kept in provide them with optimal temperature, humidity, and oxygen levels.

movements the fetus is capable of at this point; others include facial movements and the sucking reflex.

Third Trimester

Along with the terrific growth that takes place during the third trimester, fetuses are running through a series of activities that can be thought of as practice for life outside the uterus (or womb). The fetus is surrounded by a protective and nutritive fluid in the uterus, the **amniotic fluid,** which fills its lungs. It therefore doesn't breathe, but it periodically exercises its breathing muscles by using them to move amniotic fluid in and out of the lungs.

Premature Babies and Lung Function

The end of the second trimester and the start of the third also mark the critical period for a baby being able to survive outside the uterus. Clinically, any baby born before 37 weeks of gestation is defined as premature, but many babies born between 35 and 37 weeks after conception require little special care. The more premature a baby is, however, the greater its risk of having difficulties with physical and cognitive development, at least for the first few years of life. In very premature babies, survival itself is called into question (**Figure 33.13**).

A critical factor in determining whether a premature baby can live outside the uterus is lung function. As it turns out, the respiratory system is the last one to develop in a fetus. Working against the survival of very early "preemies" is the fact that, when they exhale, the alveoli or air sacs in their lungs tend to collapse and then stick together, after which they can be reopened only with great effort. Why doesn't this happen with the lungs of a full-term 38-week fetus? Because, beginning late in pregnancy, fetuses begin synthesizing a lung surfactant—a soap-like fatty substance that keeps lung sacs from collapsing. Babies born before the twenty-sixth week of pregnancy must undergo intensive lung therapy to survive because they have not produced much lung surfactant up to this point. Usually, their therapy includes the administration of a manufactured surfactant.

33.6 The Birth of the Baby

As the fetus grows, the uterus becomes an increasingly cramped place, and fetal movements become more restricted. In late pregnancy, the fetal head normally will lodge near the base of the mother's spine (**Figure 33.14** on the next page). In this "upside-down" position, the fetus is ready to make its entrance into the world. What triggers this journey? The immediate cause is **labor:** the regular contractions of uterine muscles that sweep over the fetus from its legs to its head. The pressure this generates opens or "dilates" the cervix (ultimately to approximately 10 centimeters, or about 4 inches). This creates an opening large enough for the baby to pass through. These things must occur in stages, however. The first phase of the uterine contractions is given over to cervical dilation; then comes the expulsion of the baby, and finally expulsion of the placenta (also called the *afterbirth,* for obvious reasons).

(a) Fetus's position prior to birth

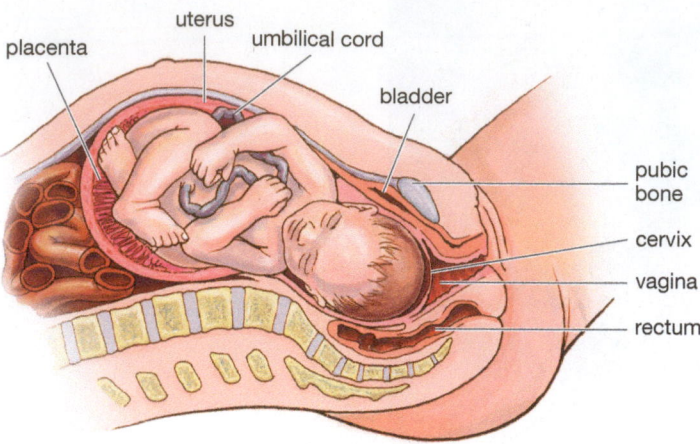

placenta
uterus
umbilical cord
bladder
pubic bone
cervix
vagina
rectum

(b) Movement through birth canal

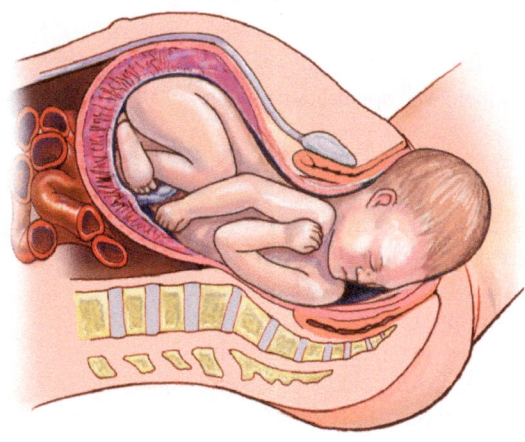

(c) Structure of "afterbirth"

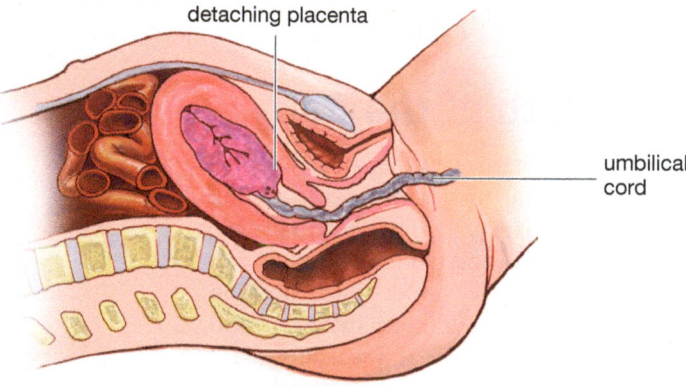

detaching placenta
umbilical cord

Figure 33.14
Birth of the Baby

With uterine contractions, the cervix dilates, allowing room for the baby to pass through the birth canal and to the outside world. The placenta and its associated fluids and tissues are expelled shortly afterward as the "afterbirth."

But what sets this train of events into motion in the first place? Put another way, what is the trigger that initiates labor? Despite a great deal of research on this question, we don't know. One hypothesis is that, as the fetus grows larger, it reaches the point at which even the highly efficient placental system can no longer supply it with enough nutrients. As a result, it secretes hormones that break down nutrients for increased energy, and this secretion triggers a hormonal cascade that brings on labor.

What happens if 38 weeks have passed and labor still doesn't begin? After a period of watching and waiting, a physician may recommend inducing labor through use of a synthetic form of the hormone oxytocin. From Chapter 28, recall that when labor begins naturally, it is accelerated by oxytocin that is released from the mother's pituitary gland. (This oxytocin brings about muscle contractions that push the fetus down further into the uterus, and this pushing brings about the production of more oxytocin.) When a baby is overdue, synthetic oxytocin can help get labor started.

On to Ecology

In this chapter and the seven that preceded it, you followed the workings of the cells and tissues that allow animals to function in their marvelously diverse ways. But animals—and for that matter all the other varieties of living things—are not self-contained units. Rather, they are constantly *interacting* with each other and with the nonliving world around them. In the next three chapters, you'll be looking at life at the level of this integrated web. What you will be studying is ecology.

SO FAR . . .

1. The number of sperm that can fuse with an oocyte is limited to _____. This is vital in order to make sure that a fertilized egg has a balanced number of _____ from the father and the mother.

2. With the implantation of the fertilized egg in the wall of the _____, cells in the egg grow into the fetal portion of a network of blood vessels and membranes that support the fetus. This is the _____.

3. Labor can be defined as regular _____ of uterine _____ that help bring about the birth of the baby.

33

REVIEW

Summary

33.1 Overview of Human Reproduction and Development

- Human male and female reproductive systems produce their respective gametes (sperm and eggs) and then deliver them to a place where they can fuse, in the moment of fertilization. (p. 631)

- The precursors to eggs, called oocytes, exist in the ovaries, which lie just left and right of center in the female pelvic area. Oocytes mature within nurturing complexes of cells and fluids called follicles. On average, once every 28 days, one follicle-oocyte complex is brought to a state of maturity. The oocyte and some of its accessory cells then rupture from the follicle and ovary and begin a journey through the uterine tube to the uterus. Conception can occur if the oocyte encounters a sperm on the way to the uterus. (p. 631)

- Sperm are produced in the male testes and then develop further in a structure called the epididymis. With sexual stimulation, they move through two types of ducts, the vas deferens and the urethra, after which they are ejaculated from the penis into the female vagina. Materials secreted by three separate sets of glands will join with the sperm before ejaculation. The resulting mixture of sperm and glandular materials is called semen. The first of the sperm to reach the oocyte moving through the uterine tube may fertilize it. (p. 631)

33.2 The Female Reproductive System

- A portion of the female reproductive system is built up and then dismantled in a reproductive cycle that takes approxi-

mately 28 days. The result of the dismantling is menstruation, a release of blood and the specialized tissue it infuses, the endometrium, which lines the uterus and which is tissue that an embryo implants itself in if pregnancy occurs. (p. 633)

- The basic unit of oocyte development is the ovarian follicle, a complex that includes the oocyte that develops into an egg and its surrounding accessory cells and fluids. (p. 634)

- On average, a single follicle each month will develop into a tertiary follicle that ruptures, expelling the oocyte and its accessory cells from the ovary. This complex of cells then begins a journey through the uterine tube to the uterus. The ruptured tertiary follicle develops into the corpus luteum, which secretes hormones that prepare the reproductive tract for pregnancy and that maintain it during the early phases of pregnancy. (p. 634)

- The number of follicles a woman possesses decreases throughout her lifetime. The scarcity of follicles a woman has in middle age is the primary factor that brings about menopause. Recent research has called into question the long-standing assumption that women lack reproductive stem cells. (p. 635)

33.3 The Male Reproductive System

- Sperm begin development in the testes and continue development and are stored in the adjacent epididymis. Sperm move from the epididymis through the vas deferens and then the urethra in ejaculation. Accessory glands contribute material to the sperm before ejaculation. (p. 639)

- Within the testes, sperm development takes place in the seminiferous tubules. Developing sperm are transported from the interior of a tubule to the epididymis

for further development and storage. (p. 639)

- Sperm development in the testes begins with spermatogonia, which develop into spermatocytes and then into immature sperm. In cell division, spermatogonia give rise not only to spermatocytes but to more spermatogonia. This self-generation of sperm precursors is the reason men keep producing sperm throughout their lifetimes. (p. 640)

- A human male produces about 200 million sperm each day—about the number of sperm ejaculated with each orgasm. Males produce such a huge number of sperm because of the sperm competition that has taken place over evolutionary time. (p. 640)

- Materials secreted by the seminal vesicles, the prostate gland, and the bulbourethral glands make up 95 percent of semen that is ejaculated; sperm makes up the other 5 percent. (p. 640)

33.4 The Union of Sperm and Egg

- An unfertilized oocyte moving through the uterine tube is surrounded by an external coat that, in turn, is surrounded by a group of accessory cells. Each sperm has an outer compartment, called an acrosome, that can release enzymes capable of dissolving the bonds between the oocyte's accessory cells. Enzymes released by dozens of sperm allow a single sperm to move between the accessory cells and reach the oocyte's external coat. This single sperm binds with receptors on the coat and releases its own complement of enzymes, which clear a path through the coat that the sperm moves through to fuse with the egg. (p. 641)

- Fusion of the sperm and oocyte brings about release of substances from the oocyte that block the entry of any other sperm. (p. 644)

33.5 Human Development Prior to Birth

- Human development proceeds through formation of the blastocyst, gastrulation, formation of the neural tube, and organogenesis. Many of the most important stages in development are completed by the end of 12 weeks of pregnancy. The developing organism is called a zygote at conception, an embryo from the zygote's first division through the eighth week of pregnancy, and a fetus from the ninth week of pregnancy to birth. **(p. 644)**

- The blastocyst implants itself in the endometrial lining of the uterus during the first week of pregnancy. It consists initially of an inner cell mass, which develops into the baby, and a peripheral group of cells called a trophoblast, which grows into the placenta. **(p. 644)**

33.6 The Birth of the Baby

- The immediate cause of birth is labor. The pressure that results from uterine contractions dilates the cervix, creating an opening large enough for the baby to fit through. It is not clear what the triggering mechanism is for the initiation of labor. When labor does not begin naturally following a full-term 38-week pregnancy, it can be chemically induced. **(p. 649)**

Key Terms

Understanding the Basics

Multiple-Choice Questions

(Answers are in the back of the book.)

1. The precursors of the eggs produced by females are (select all that apply):
   a. released from ovaries.
   b. released at the rate of about six per month.
   c. called oocytes.
   d. surrounded by fluids and accessory cells.
   e. put into blood circulation.

2. The structure in which sperm are stored and complete their maturation is the:
   a. epididymis.
   b. oviduct.
   c. penis.
   d. prostate gland.
   e. vas deferens.

3. The tertiary follicle (select all that apply):
   a. contains a maturing spermatocyte.
   b. contains a maturing oocyte.
   c. is transformed into the corpus luteum.
   d. is washed away with the endometrial lining.
   e. expels an oocyte in ovulation.

4. Associate each term on the left with a description on the right.
   a. acrosome — produces fluid found in semen
   b. spermatogonium — enzyme-containing compartment in sperm
   c. spermatocyte — sperm stem cell
   d. prostate gland — duct through which sperm are ejaculated
   e. urethra — maturing sperm cell

5. The natural degeneration of follicles during a woman's lifetime is referred to as:
   a. menopause.
   b. ovulation.
   c. puberty.
   d. atresia.
   e. apoptosis.

6. The portion of the blastocyst called the trophoblast eventually helps form the:
   a. head of the baby.
   b. legs of the baby.
   c. entire baby.
   d. nervous system of the baby.
   e. placenta.

7. Birth of identical twins results from unusual events during:
   a. egg production.
   b. sperm production.
   c. early cell divisions after fertilization.
   d. ovulation.
   e. second trimester.

8. Arrange the following events in their proper sequence:
   a. uterine muscle contractions
   b. birth of baby
   c. assumption of fetal head-down position
   d. expulsion of placenta
   e. dilation of cervix

Brief Review

(Answers are in the back of the book.)

1. Describe the hormonal changes that regulate the female reproductive cycle.

2. How do the testes act to protect developing sperm?

3. Menstruation is costly to females in some ways. Why do scientists believe it occurs?

4. Compare and contrast the tertiary follicle and the corpus luteum.

5. Describe the mechanism that ensures that only one sperm will be able to fuse with a given oocyte.

6. Outline the events in human development that occur up through the time the embryo implants in the uterine wall.

7. Define the term *premature* with respect to human development.

8. How do birth control pills work?

Applying Your Knowledge

1. When a couple is having trouble conceiving a child, the first thing fertility specialists generally will look at is the fertility of the prospective father. What do you think a specialist would be looking for?

2. Most mammals are "placental," which is to say their embryos get nourishment within the mother through the placental system of blood vessels reviewed in the text. By contrast, amphibian and reptilian embryos are nourished by substances contained in eggs the mother lays. Can you think of any advantages mammals derive from the placental means of development? Can you think of any disadvantages?

3. Forty years after birth control pills for women came into widespread use, there is still no birth control pill for men. One reason for this is the basic difference between female production of an oocyte each month and the male production of sperm. Why should it be easier to suppress fertility in women than in men?

MasteringBiology®

1. MasteringBiology Assignments
Activity: The Female Reproductive System; The Male Reproductive System; **Current Events:** As Algae Bloom Fades, Photosynthesis Hopes Still Shine; **Building Vocabulary; Questions:** Reading Questions, Test Bank

2. eText
Read your book online, search, take notes, highlight text, and study more effectively.

3. The Study Area
Self-study tools to test comprehension.

34

An Interactive Living World 1:

Populations in Ecology

Ecology is the study of how living things interact—with each other and with their physical environment. One way of studying these interactions is to focus on populations, meaning all the members of a single species living in one area.

Available on MasteringBiology®

GraphIt!: Age Pyramids and Population Growth; **Bioflix:** Population Ecology; **Current Events:** As Dengue Fever Sweeps India, a Slow Response Stirs Experts' Fears; **Building Vocabulary; Questions:** Bioflix Quiz, Reading Questions, Test Bank

A school of bluecheek butterfly fish (*Chaetodon semilarvatus*) photographed in the Red Sea.

South Mountain Reservation is an oasis of about 2,000 acres of parkland nestled among four suburban New Jersey towns that lie directly west of New York City. Though normally a place for hiking, walking dogs, and enjoying nature, South Mountain briefly became the site of a very different kind of activity in the winters of 2008 and 2009. Volunteer sharpshooters recruited by county government baited the parkland with food, climbed up into park trees, and then waited for white-tailed deer to come collect the food, at which point they shot the deer. In all, almost 300 deer were killed over 18 days of shooting in 2008 and 2009; meat from the animals was distributed to the poor and homeless.

This exercise in animal control was not undertaken lightly or in haste; indeed, a discussion about what to do with South Mountain's deer ran on for years prior to the shootings of 2008. One of the notable aspects of this discussion was that a prominent supporter of killing the deer was the New Jersey Audubon Society, a group normally known for its passionate defense of nature. So why would the Society end up supporting an organized killing of one of nature's creatures? Because this particular creature was threatening to make South Mountain "biologically destitute," in the words of the Society. When deer become numerous enough in a given area, the Society has noted, environmental trouble follows in a fairly predictable way. Some of the first casualties are the flowering plants and herbs that help make up the "understory" of a forest, meaning the plant life that spreads out at low levels beneath large trees. A grown white-tailed deer can consume up to 10 pounds of vegetation per day, and these animals are partial to flowering plants and herbs. Then comes the destruction of the shoots of young shrubs and trees. Large, mature trees aren't being affected, because the deer are powerless to kill them, but a forest that is losing its shoots is losing its future. This decimation of plant life then has an effect on animal life. For some birds, it's difficult to have nests at all in locations that have been heavily browsed by deer, but birds that succeed in hatching out young in these areas may still have their efforts go for nothing, the Society noted in 2005. Early in life, fledglings tend to seek out shrubby forested areas for feeding—areas that only exist far from home when deer are present in high numbers. The problem is that young birds flying long distances are easy marks for predators. As a result, the Society said, "Young birds in a heavily browsed forest are doomed."

Of course, suburban New Jersey is not the only place in the country having a problem with white-tailed deer; just ask any of the estimated 1.5 million Americans involved in car crashes with these animals each year. Why are these accidents happening with such frequency now? The answer is straightforward: The deer population in America has exploded. From a low of about half a million in the early 1900s, deer numbers rose to about 16 million in 1987 and then doubled by 2006, going to an estimated 32 million. How does such an increase come about? Here's the recipe: First, get rid of key deer predators, such as wolves, bears, and bobcats, by

In the Danger Zone

The deer population in the United States has soared in recent decades, creating numerous instances of conflict between deer and human beings.

driving them to extinction in most areas. Then build houses farther and farther out into the sub-urbs, on land adjacent to fields and forests. Now the deer have plenty of land to graze on (suburban lawns) that doesn't even have *human* predators on it. Then couple a growing sensibility about animal rights to a decline in hunting across the country. (Hunters are a potent force in this issue; in Wisconsin alone, they killed 452,000 deer in 2008.) The result? Deer right in your backyard, on your streets, ruining your forest preserve. You might think that the silver lining to this is that the *deer* are at least coming out of it fat and content, but this is not the case at all. An overabundance of deer in a given area produces deer that are emaciated, sick, and beset by parasites; when deer have picked an environment clean, the deer themselves will suffer along with other forms of life.

For our purposes, the story of deer in the United States drives home the message that living things do not live in isolation from one another. What happens with one species affects another. What happens with wolves affects deer, then deer affect plants, and plants affect birds. In short, all these things are part of an *interrelated* web of life. Moreover, this is a web of life that exists in a larger context yet: the *non*living world, which is to say the Earth's climate and atmospheric gases, its sea currents and volcanoes, and the energy it gets from the sun. Life doesn't exist apart from these things. Instead, it is very much conditioned by them—although, as you'll see, some of these things are in turn conditioned by life.

34.1 The Study of Ecology

In your walk through biology so far, you have made it through the molecular world (DNA, for example) and through the world of whole organisms, such as fungi and plants. Now it is time to take the last steps up; in this chapter and the two that follow, you will consider biology's largest realm of all, **ecology,** defined as the study of the interactions that living things have with each other and with their environment.

In common speech, the word *ecology* has become synonymous with the conservation of natural re-sources or with the "environmental movement" in all its political dimensions. Under this view, you might assume that the job of an ecologist is to help preserve the natural world. In fact, however, the job of an ecologist is to *describe* the natural world in its largest scale—to say what is, rather than what should be. Are ecologists environmentalists? Most are, but this shouldn't be surprising. We would no more expect ecologists to be indifferent to the environment than we would expect art historians to be indifferent to paintings. Moreover, there is a branch of biology,

Figure 34.1
An Ecologist on the Job
Ecologists spend a good deal of time going out into nature to study the conditions that exist in various ecosystems—however inaccessible those ecosystems may be. Here, a researcher is ascending into the canopy of a rainforest in Indonesia.

conservation biology, that is directly concerned with preserving natural resources.

But the business of ecology per se is to tell us about the large-scale interactions that go on in the natural world. This work has an important use in that it provides the information base that society can use to make decisions about the environment. Does species diversity make for healthier natural communities? How many gray whales are there in the world? How important are prairie dogs to the prairie? To answer these questions, ecologists end up working in some places you might expect—in tents near the Arctic Circle or at treetop level in tropical rain forests (**Figure 34.1**). They also work in some locations you might not expect, however, such as in offices, hunched over computers, seeing how mathematical models of global warming work out.

Path of Study

In this chapter and the next two, you'll follow a bottom-to-top approach by moving from smaller ecological units to larger ones. What are the scales of life that concern ecology? The smallest is that of the physical functioning, or physiology, of given organisms. At this level, ecologists are examining individual living things, but at all other scales they are looking at groups of organisms or species. For the ecologist, life is conceptualized as being organized into

- Populations
- Communities
- Ecosystems
- The Biosphere

From Individuals to a Population

The smallest level of group organization is a **population,** defined in Chapter 17 as all the members of a single species that live together in a specified geographic region. The North American bullfrog (*Rana catesbeiana*) is a species that can be found from southern Canada to central Florida, but all the *R. catesbeiana* in a given pond constitute a single population of this species. Of course, it's possible to define our population to be all the bullfrogs in *two* ponds, or two states, or in two countries—whatever geographic region is most useful for the question an ecologist is asking.

From Populations to a Community

Going up the scale, if you take the populations of *all* species living in a single region, you have a **community.** Often, the term *community* is used more restrictively to mean populations in a given area that potentially interact with one another.

From a Community to an Ecosystem

If you add to the community all the *non*living elements that interact with it—rainfall, chemical nutrients, soil—you have an ecosystem. More formally, an **ecosystem** is a community of organisms and the physical environment with which they interact. Ecosystems can be of various sizes; you'll be looking at a range that goes from fairly small (a single field, for example) up through the enormous ecosystems called biomes that may take up half a continent.

From Ecosystems to the Biosphere

The largest scale of life is the **biosphere,** which can be thought of as the interactive collection of all the Earth's ecosystems. Given what you've reviewed about ecosystems, this means all life on Earth and all the nonliving elements that interact with life. Sometimes, however, the term *biosphere* is used purely in a territorial sense—to mean that portion of the Earth that supports life. If you look at **Figure 34.2**, you can see a graphic representation of ecology's scales of life. We'll now begin to look at ecology through these levels of biological organization, starting small, with the population.

34.2 Populations: Size and Dynamics

So, what is it ecologists want to know about a population? Well, they need to know how to count it, how and why it is distributed over its geographical area, and how and why its size changes over time—what its *population dynamics* are, to use the term employed by ecologists.

Estimating the Size of a Population

The reason ecologists want to count the members of a given population is straightforward: Without such a count, there is no way to answer a question such as how much territory a group of cheetahs must have to flourish or how fast a population of finches recovers from a drought. (How would you know about the latter unless you could compare the population after the drought with the population before it?)

Figure 34.2
The Scales of Life that Concern Ecology

A population is made up of all the organisms of a single species that live together in a defined region—in the example shown, a population of sea lions living together off the California coast. A community is a more inclusive grouping: all the members of all species that live together in a defined region. In the example, this means not only the sea lions but giant kelp, fish, and a multitude of other species. An ecosystem includes living community members and the nonliving factors that interact with them (such as the tides and wind currents that affect the California coast). The biosphere is then the interactive collection of all the Earth's ecosystems.

organism
(sea lion)

population
(colony)

community
(giant kelp forest)

ecosystem
(Southern California coast)

biosphere
(Earth)

With large, immobile species such as trees, taking a census can sometimes be easy; just mark off an area and count. Things become more difficult when the area under consideration is so large that not all individuals in it can be counted, but rather must be estimated based on a population counted in a smaller, representative area. Estimating becomes more difficult yet when the individuals are numerous and mobile, as with birds. Ecologists employ various means to estimate such populations; they survey bird populations as they migrate, for example.

In some cases, the stakes for coming up with accurate counts of given populations are very high, and in these instances ecologists can spend considerable time simply devising counting techniques that hold promise of yielding meaningful numbers. To take one example, no more than 3,500 tigers are thought to remain on Earth, of which about 2,000 exist on the Indian subcontinent. For years, counts of these elusive animals (*Panthera tigris*) were developed by making plaster casts of their paw or "pug" marks, the idea being that each animal left a unique pug impression that could be identified through the casts. Experts long ago concluded, however, that human beings cannot reliably distinguish one pug mark image from another, which meant that game reserve superintendents in India had only a rough idea of how many tigers they were managing. In the face of this lack of information, ecologists took on the task of devising more accurate counting techniques and managed to come up with three of them. One is computer-driven pug mark interpretation, complete with its own software (Pugmark 1.0); a second is digital camera-trapping identification, in which pairs of remotely placed infrared cameras produce footage of the stripe patterns on tigers (which can distinguish one animal from another); and a third is genetic identification, in which DNA "fingerprints" of tigers are produced through their feces. Over the years, numerous techniques such as these have been devised to get counts of species that are elusive, endangered, or both.

Figure 34.3
Arithmetic and Exponential Growth

When the same number of objects is produced in a given interval of time—in this case, cars from a factory each day—arithmetic growth is at work. By contrast, the populations of most organisms, such as the water flea *Daphnia*, can exhibit exponential growth, at least for a time, meaning the population grows in proportion to its own size.

Growth and Decline of Populations over Time

How is it that populations *change* size? As you begin to think about this question, it's worth going over a more general concept, which is the way a population of anything might increase in number. If you look at, say, the number of cars coming off the end of a production line, there might be 1,000 on Monday, then 1,000 on Tuesday, and so forth. The important thing to note is that the number of cars produced on Tuesday is not related to the number produced on Monday. The increase in the number of cars is thus an **arithmetical increase:** Over each interval of time (a day in this case), an unvarying number of new units (cars) is added to the population.

Now contrast this with what happens to living things. In most cases, each new unit (each living thing) is capable of playing a part in giving rise to more units, which certainly is not the case with cars. Population increases for living things are thus *proportional to* the number of organisms that already exist. Thus, the increase in a population of organisms comes about through a different sort of increase, an **exponential increase,** which occurs when, over an interval of time, the number of new units added to a population is proportional to the number of units that exist. **Figure 34.3** gives an example of the difference between the two kinds of growth over a period of weeks, using cars for arithmetic growth and the tiny water flea *Daphnia* for exponential growth. Let's assume we start out at the end of day 1 with the 1,000 cars produced that day and with 1,000 water fleas existing in an optimal laboratory environment. As you can see, *Daphnia* is a relatively slow starter, but its population quickly overwhelms that of the car population—not surprising, because the *Daphnia* population doubles every 3 days.

Population Growth in the Real World

As noted, the *Daphnia* were growing in an "optimal" laboratory environment—plenty of food, habitat kept clean, no predators. Thanks to human intervention, this was a kind of paradise for water fleas, in other words. But could any population ever grow like this in the real world? In all cases, the answer is "not forever," and in most cases the answer is "not for long," but some population growths can be dramatic while they last.

In 1859, Thomas Austin released 24 wild European rabbits (*Oryctolagus cuniculus*) onto his estate in southern Australia, near Melbourne, to provide more game for sport hunting. His was not the first attempt to introduce rabbits to Australia, but it was by far the most successful—or perhaps the most fateful. Having no natural predators in Australia, this alien species spread north and west like wildfire; within 16 years, it had expanded its range through the entire latitude of

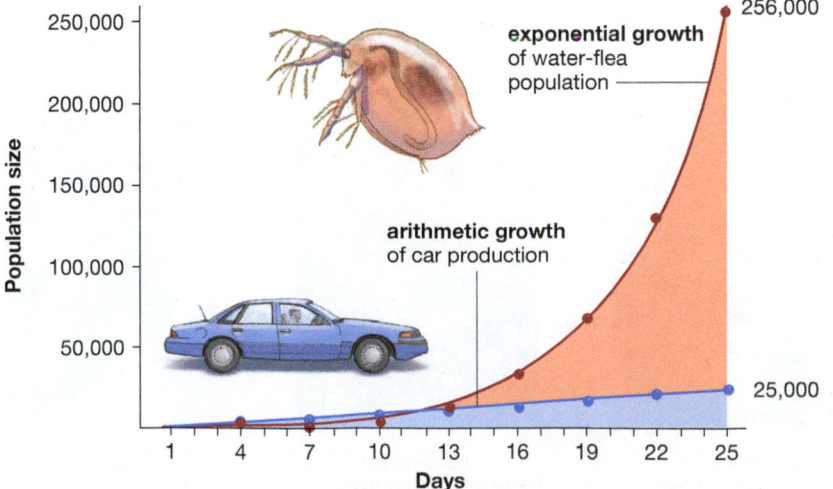

the continent, a distance of almost 1,100 miles. And this invasion was not a matter of a rabbit here and a rabbit there; the European rabbits became a scourge that, even today, has only been brought under partial control through the use of two rabbit-killing viruses. Here is Eric Rolls' description in *They All Ran Wild* of the rabbit problem east of Peterborough in South Australia in 1887:

> For over a hundred miles most trees and plants had been killed. Outalpa head-station was a thousand square miles of rabbits . . . . Rabbits crawled like possums in branches several feet off the ground. They had burrowed under the beautiful granite homestead; they had taken possession of the cellar; they were in the turkeys' coop; they ran in and out of the men's huts apparently without fear; they took no notice of the dogs nor the dogs of them.

There are lots of instances of such seemingly uncontrolled growth. Charles Darwin devoted some thought to the issue as a theoretical matter. He wrote that "There is no exception to the rule that every organic being naturally increases at so high a rate, that, if not destroyed, the Earth would soon be covered by the progeny of a single pair." To put a little finer point on this, a single pair of carrion flies (which are a little larger than a housefly) could in theory give rise to enough flies in a single year to cover the state of New Jersey to a depth of 3 feet.

The Shape of Growth Curves

If you look at **Figure 34.4a**, you can see what growth of this type, **exponential growth,** looks like when plotted on a graph. This is called the *J-shaped growth curve* because it can be so extreme that when plotted it looks like the letter J. It may look familiar because it is the growth form you just saw with *Daphnia*. As with *Daphnia*, there is relatively slow growth at first, then faster growth, then faster yet. This is growth, in other words, in which the *rate* of increase keeps accelerating. In the real world, the curve eventually drops off, often sharply, for reasons we'll go over soon.

Now, look at **Figure 34.4b** to see a second model of growth for natural populations, the *S-shaped growth curve,* which tracks **logistic growth.** This starts like exponential growth, but eventually the rate of growth slows and finally ceases altogether, stabilizing at a certain level, denoted here as *K* (which you'll learn the significance of shortly).

What has intervened to account for the difference between the dizzying increase of the J-shaped curve and the moderate increase of the S-shaped curve? All the forces of the environment that act to limit population growth, which in ecology are known as **environmental resistance.** Organisms will run out of food or have their sunlight blocked; greater numbers of predators will discover the population; the wastes

Figure 34.4
Three Models of Growth for Natural Populations

(a) Exponential growth

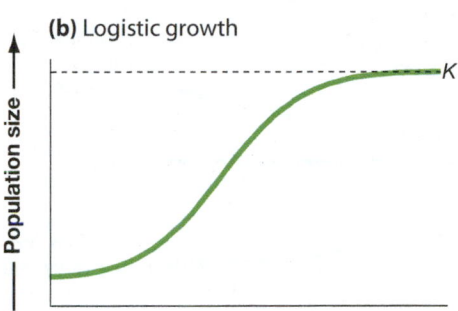

(b) Logistic growth

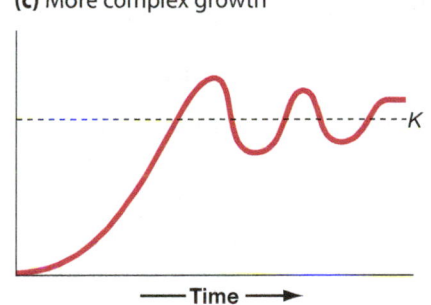

(c) More complex growth

Population size →

← Time →

produced by the organisms will begin to be toxic to the population.

There are populations whose dynamics over time look something like either the J- or S-shaped curve, but in the real world, population size is likely to vary in more complex patterns. In **Figure 34.4c**, you can see one example of such a pattern.

Calculating Exponential Growth in a Population

You've seen that ecologists have ways to count natural populations, but is it possible for them to *predict* the size of populations given certain pieces of starting information? The answer is yes, as you'll now see. Let's look first at a simplified means of calculating exponential growth, meaning the J-shaped curve. For now, assume the population is isolated—that is, that no individuals are moving in or out of it.

What first needs to be done is to pick a time period to evaluate population change. This period may be 40 minutes (for bacterial populations) or 50 years (for elephants), but some unit of time needs to be chosen for the analysis. The next step is to learn the difference between the birth rate and the death rate in the population over the time period you've picked. This, of course, raises the question: What's a birth rate? It is

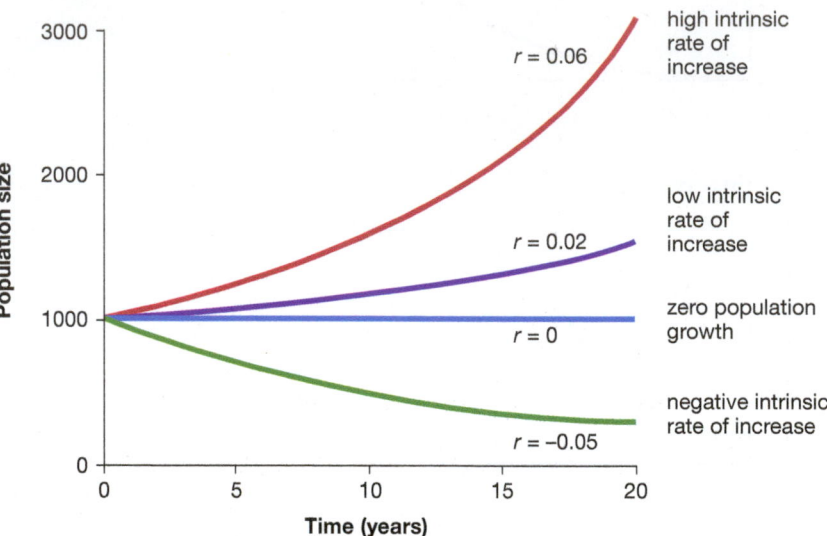

Figure 34.5

A Population's Potential for Growth: The Intrinsic Rate of Increase (r)

If the birth rate exceeds the death rate, then the population size will increase over time (r = 0.02 and 0.06 in the example). If the birth rate equals the death rate, then the population size will stay constant (r = 0). Finally, if the birth rate is lower than the death rate, the population size will decrease over time (r = −0.05). Note what dramatic effects a small change in r can have over time, in this case on a population whose starting size is 1,000 individuals.

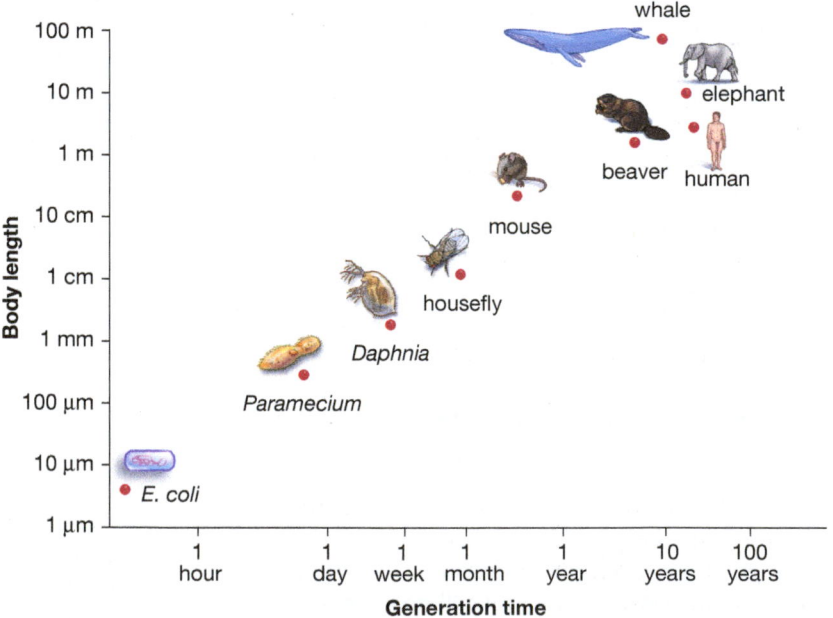

Figure 34.6

How Long between Generations?

An important factor in the population dynamics of any species is its generation time, meaning the time that elapses between the birth of one generation and the birth of the next. Shown on the graph are the minimum generation times for a few selected species, along with a measurement of the length of each organism at the time it reproduces. The larger a species is, the longer its generation time tends to be.

(Redrawn from J. T. Bonner, *Size and Cycle: An Essay on the Structure of Biology*, Princeton University Press, 1965.)

the number of individuals born in a given time period, expressed as a proportion of the whole population. Thus, if a population of 1,000 had 100 births in a year, its birth rate would be 0.10, or 10 percent, per year. Figuring the death rate is similar; if a population of 1,000 had 80 deaths per year, its death rate would be 0.08, or 8 percent, per year.

The *difference* between these birth and death rates, of course, is 2 percent. This number is the population's growth rate in an optimal setting, denoted as *r*—standing for the *intrinsic rate of increase*—and it is very important. With it in hand, you can predict what the population's size will be in the future. Growing at 2 percent per year, the population of 1,000 will increase to 1,020 in the first year and to about 1,480 in 20 years.

Growth rate (*r*) is critical because a small change in it can mean a big change in population over time. Take a population of another species, once again with 1,000 individuals in total, but this time with an *r* of 6 percent rather than 2 percent. In the first year, this population would grow by 60 individuals, giving it 1,060 members as opposed to the first population's 1,020. Go 20 years out, however, and this population has about 3,200 individuals in it—it's now more than twice the size of the first population even though both were the same at the start. Meanwhile, if a population's *r* is less than zero—that is, if the number of deaths exceeds the number of births—the population is shrinking. If *r* is zero on the nose, meaning that births exactly match deaths in a given period, the population is at **zero population growth** (**Figure 34.5**).

As noted, the *r* figure goes by another name, which is a population's **intrinsic rate of increase:** the rate at which a population would grow if there were no external limits on its growth. You could think of it as a population's potential for growth. Importantly, populations of each species have characteristic rates of increase. It won't surprise you to learn that this rate is higher for *Daphnia* populations than it is for whale populations; you can see some examples of an important factor in rate of increase, the time it takes to produce a new generation, in **Figure 34.6**.

Logistical Growth of Populations

You've now observed exponential, or J-shaped, growth, but you've seen that such growth never occurs for long in nature, since environmental resistance will always begin to assert itself. In some instances, there will be a flattening out of the population increase and in time perhaps a complex pattern with the curve moving above and below the line denoted earlier as *K*. But what is *K*? What is this point around which the population hovers? It is the maximum population density of a given species that a defined geographical area can sustain over time. The term for this measure is an area's **carrying capacity (*K*).**

An Example of Carrying Capacity

There is an island in the San Francisco Bay called Angel Island, just a short boat ride from the Golden Gate side of the Bay, that had a small population of deer introduced to it by humans early in the twentieth century. The problem was that the island lacked any natural predators for the deer, with the result that the deer population proliferated way beyond what the island's vegetation could support. Had this taken place in a purely natural environment, the deer population would have shrunk in accordance with the amount of available vegetation. But there were human hikers on Angel Island, and seeing emaciated deer, they did what you might expect: They fed the deer with what they brought along in their backpacks. Surviving through this artificial means, however, *more* deer came to live on the island and stripped it of its vegetation, thus reducing its carrying capacity.

The main lesson here is that there are limits to how many members of a given species an expanse of territory can support. Note also, however, that carrying capacity is not fixed; in the case of Angel Island, it was lowered as the deer, artificially fed by human beings, stripped the island of vegetation. In other instances, carrying capacity might be raised for given periods (perhaps through abundant rainfall). The environments that many species live in tend to be stable over time, however, meaning that their carrying capacities will tend to be stable as well.

The K line can generally be established for a given population by estimating its numbers over an extended period of time. For many species, their population may exceed or "overshoot" K at times, but this excess will be offset by periods during which the population is under K. Once K is known, it is valuable in predicting the change in a population's size because it often serves as a factor that can be used in mathematical equations to calculate the limits to population growth. The higher a population is relative to K, the more severely its growth will be limited.

SO FAR . . .

1. A _____ is all the members of a species that live together in a defined geographical region. A _____ is all the populations of all species living in a single region. An _____ is a community of organisms and the physical environment with which they interact.

2. Environmental resistance can be defined as all the forces of the environment that act to _____.

3. Carrying capacity can be defined as the maximum _____ of a given species that a defined geographical region can _____.

34.3 *r*-Selected and *K*-Selected Species

It's well known that for human beings, about 9 months will elapse between conception and birth, but it's less well known that for an elephant this same process takes almost 2 years. For any given human or elephant female, then, births are relatively few and far between. And once the offspring come, the amount of attention lavished on them is great indeed. It's not unusual for an elephant calf to be nurtured by its mother and a larger kinship group for at least the first 10 years of its life. In the first year of this period, a mother might not ever let her calf wander farther than 20 meters away from her. Meanwhile, humans may lavish 18 years of care and concern on a child before the domestic bonds are finally broken. In contrast, if we look at the common housefly, we find that it can produce a new generation once a month and that the parental generation provides no care whatsoever to the offspring.

Houseflies and elephants lie at opposite ends of a continuum of reproductive strategies, meaning characteristics that have the effect of increasing the number of fertile offspring an organism bears. So, how does the strategy of elephants differ from the strategy of houseflies? This may be apparent from what you've observed already. For elephants, the strategy is to bring forth few offspring but lavish attention on them. For houseflies, it is to bring forth a multitude of offspring but give no attention to any of them. These strategies are in turn related to other characteristics of these species, as you'll now see.

K-Selected, or Equilibrium, Species

Elephants will not seek out a totally new environment; they experience their environment as a relatively stable entity; and they compete among themselves and with other species for resources within it. Given this stability, species like elephants are known as an *equilibrium species*. In line with this, the elephant population stays relatively stable compared to a fly population; its numbers will fluctuate in a relatively narrow range above and below the environment's carrying capacity. This is another way of saying that the population regularly bumps up against carrying capacity (K), which is the density of population that a unit of living space can support. Two things follow from this. First, elephants are said to be a **K-selected species:** one whose population sizes tend to be limited by carrying capacity. The second point, which follows from this definition, is that the pressures on the elephant population are **density dependent:** As the density of the population goes up, the factors that limit the

Figure 34.7
K-Selected and
r-Selected Species
The elephants on the left are a *K*-selected species. Here, a mother stays close to her calf, providing the kind of careful attention typical of *K*-selected species. The pond flies on the right are an *r*-selected species, bearing many young but giving no attention to them after birth.

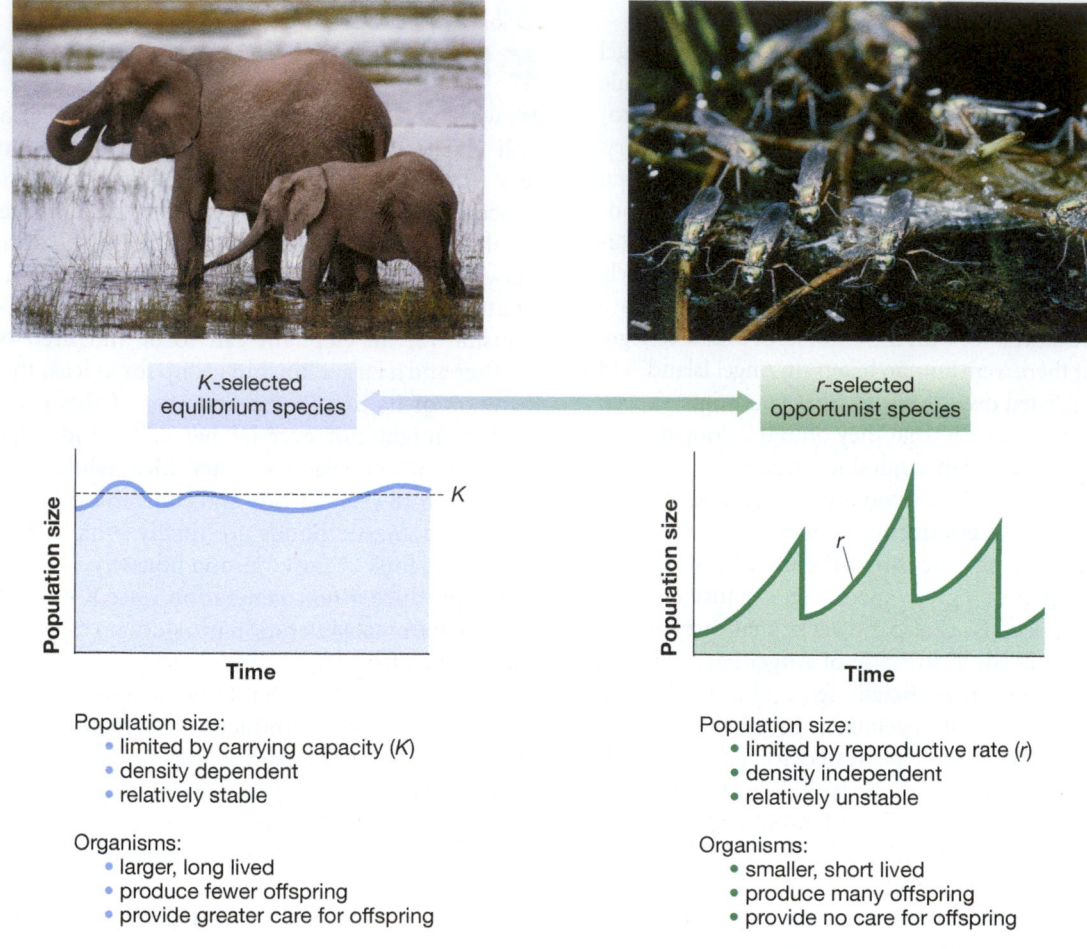

K-selected
equilibrium species ⟷ r-selected
opportunist species

Population size:
• limited by carrying capacity (*K*)
• density dependent
• relatively stable

Organisms:
• larger, long lived
• produce fewer offspring
• provide greater care for offspring

Population size:
• limited by reproductive rate (*r*)
• density independent
• relatively unstable

Organisms:
• smaller, short lived
• produce many offspring
• provide no care for offspring

Bioflix: Population Ecology

population—food supply, living space—assert themselves ever more strongly (**Figure 34.7**).

r-Selected, or Opportunist, Species

In contrast to elephants, houseflies are known as an *opportunist species*—a species whose population size tends to fluctuate greatly in reaction to variations in its environment. Should favorable weather suddenly arrive or a food supply suddenly appear, the fly population in the area will skyrocket. When the food is gone or if the temperature suddenly changes, this same population will plunge in number. In short, environment tends to be highly variable for fly populations, and the flies have a high population growth rate that allows them to take advantage of environmental opportunities. For these animals, the strategy is to produce a multitude of offspring very fast. Flies are therefore said to be an **r-selected species,** meaning one whose population sizes tend to be limited by reproductive rate.

This definition implies that fly population numbers have nothing to do with fly population *density*, at least for a time. The pressures on the population are thus said to be **density independent.** The forces at work in these populations tend to be *physical* forces—

frost, temperature, rain—as opposed to the *biological* forces that operate in density-dependent control (competition for food, disease). Given this, the *r*-selected strategy of many offspring/little attention is understandable. When you exist at the whim of environmental forces, "life is a lottery and it makes sense simply to buy many tickets," as ecologists R. M. May and D. I. Rubinstein have phrased it.

The *r*- and *K*-selected groupings are categories invented by human beings, and as usual, nature refuses to fit neatly into such boxes. It is true that small, short-lived species tend to lie at the *r*-end of the continuum and that large, long-lived species lie at the *K*-end, but sea turtles are long-lived, and they give no care at all to their newborns. Meanwhile, flies can be limited in a density-dependent way for at least part of their population cycle—by predators and by competition among themselves. All sorts of variations exist on this theme. We've looked at animals as examples of *r*- and *K*-selection, but the concept applies to plants as well. A plant such as ragweed is at the *r*-selected extreme in the plant world, while oak trees and cacti are at the *K*-selected extreme. The ragweed grows rapidly and has a short life span, while the oak grows slowly and lives a very long time.

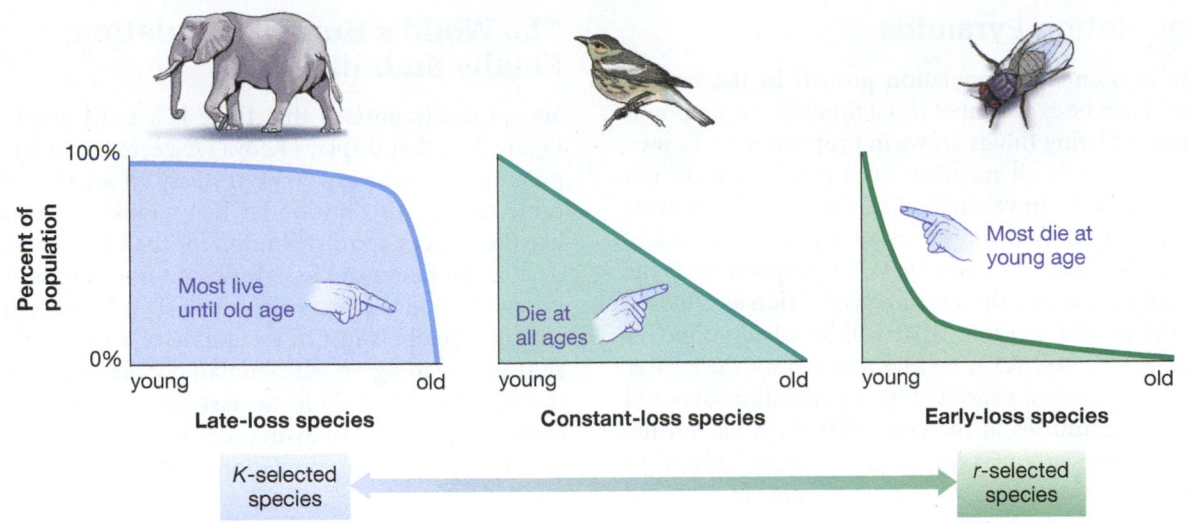

Figure 34.8
When Is Death Likely to Come?

An organism's chance of living a long life is related to the reproductive strategy of its species, as reflected in these survivorship curves. Elephants and humans (late-loss species) produce few young, but most survive until old age. At the other extreme, flies (early-loss species) produce many young, but most die young. In between are constant-loss species; they produce a moderate number of young, which die at a fairly constant rate during a typical life span.

Survivorship Curves

K- and *r*-strategies are related to the concept of "survivorship curves," which as you can see in **Figure 34.8** are thought of in terms of three ideal types: late loss, constant loss, and early loss (sometimes referred to as types I, II, and III). Humans and most of our fellow *K*-selected mammals fall into type I because we tend to survive into old age (our lives are lost late). Insects and many amphibians are type III because their death rates are very high early in life but level out thereafter. Other types of living things, such as birds, fall into type II because they die off at a constant rate through their life span.

34.4 Thinking about Human Populations

We'll look at the final elements in population dynamics in connection with human populations. Some of the concepts involved could be applied to any population of living things. But here, the focus is on human beings, so that you can consider population principles along with the real-world issue of human population growth, which figures so prominently in environmental issues.

Survivorship Curves and Life Tables

The survivorship curves you just looked at are constructed by developing what are known as **life tables** for the species in question. To create these, scientists divide the species' life span into suitable units of time—days for fruit flies, years for human beings—and see how likely it is that an average species member will be alive after a given number of days or years. The technique for doing this was not invented by biologists who wanted to know about the survival of flies, but by a nineteenth-century British actuary—a person trained to calculate probabilities—who knew the information would be useful for life insurance companies. How likely is it that a person will be alive at 10, or 20, or 80 years of age? If you look at **Table 34.1**, you can see the answer for an average group of 100,000 persons born in the United States in the year 2004.

Table 34.1

A Life Table for the United States

At age	Number still living	Average remaining lifetime, in years
10	99,129	68.5
20	98,709	58.8
30	97,776	49.3
40	96,517	39.9
50	93,735	30.9
60	88,038	22.5
70	76,191	15.1
80	53,925	9.1
90	22,219	5.0
100	2,510	2.6

Taking a hypothetical group of 100,000 persons born in the United States in 2004, the table shows the number likely to still be living at the ages indicated and the average remaining lifetime for persons at each age. The numbers are averaged for men and women, a choice that masks significant differences between the sexes in older age cohorts. Of 100,000 women born in 2004, for example, about 61,000 are likely to be alive at age 80, whereas for men the figure is about 46,000.

Source: National Center for Health Statistics

GraphIt!: Age Pyramids and Population Growth

Population Pyramids

You've seen that population growth in the natural world can be exponential (for a time) because a population of living things grows in proportion to its own size. However, all members of a population do not count equally in calculating this growth. If scientists want to peer into the future of a given population, what they want to know is: What proportion of the population is *past* the age of reproduction as opposed to the proportion that is, or will be, of reproductive age? If you look at **Figure 34.9**, you can see the answer to this question, expressed as a population pyramid for two countries in the year 2009. Each bar on the graph represents a 5-year age grouping, or "cohort" of the population (those who are 0–4 years old, 5–9, and so on), with the length of each bar representing the size of the cohort. The reproductive age range for humans generally is considered to be between 15 and 45. If you look at the graph for Kenya, you can see that it has many more individuals of reproductive and pre-reproductive age than of post-reproductive age. What does the future hold for Kenya? In all probability, a large growth in the population. In contrast, the age-structure diagram for the United States shows that a much greater proportion of its population is at or beyond reproductive age.

The World's Human Population: Finally Stabilizing

You probably noticed that there is a third graph in Figure 34.9 that displays Kenya's expected population pyramid for the year 2050. It's easy to see that this projected age distribution for Kenya looks something like the current age distribution for the United States. As a result, Kenya can look forward to slower population growth in the years after 2050. This Kenyan projection actually is just one example of a momentous shift that is going on with human population around the world today. Only in the last few years has it become clear that, following centuries of explosive growth, the human population finally appears to be stabilizing.

To put this change into a larger context, look at **Figure 34.10**, which shows you how human population has grown during "historical time," which is to say the last 12,000 years. For centuries, human numbers grew relatively little but then began an upward climb about 1700. Improved sanitation, better medical care, and increases in the food supply came together to produce the rate of growth you see. The Earth's human population didn't pass the 1-billion mark until 1804; it then took 123 years to double to 2 billion (in 1927), then 48 years to double to 4 billion

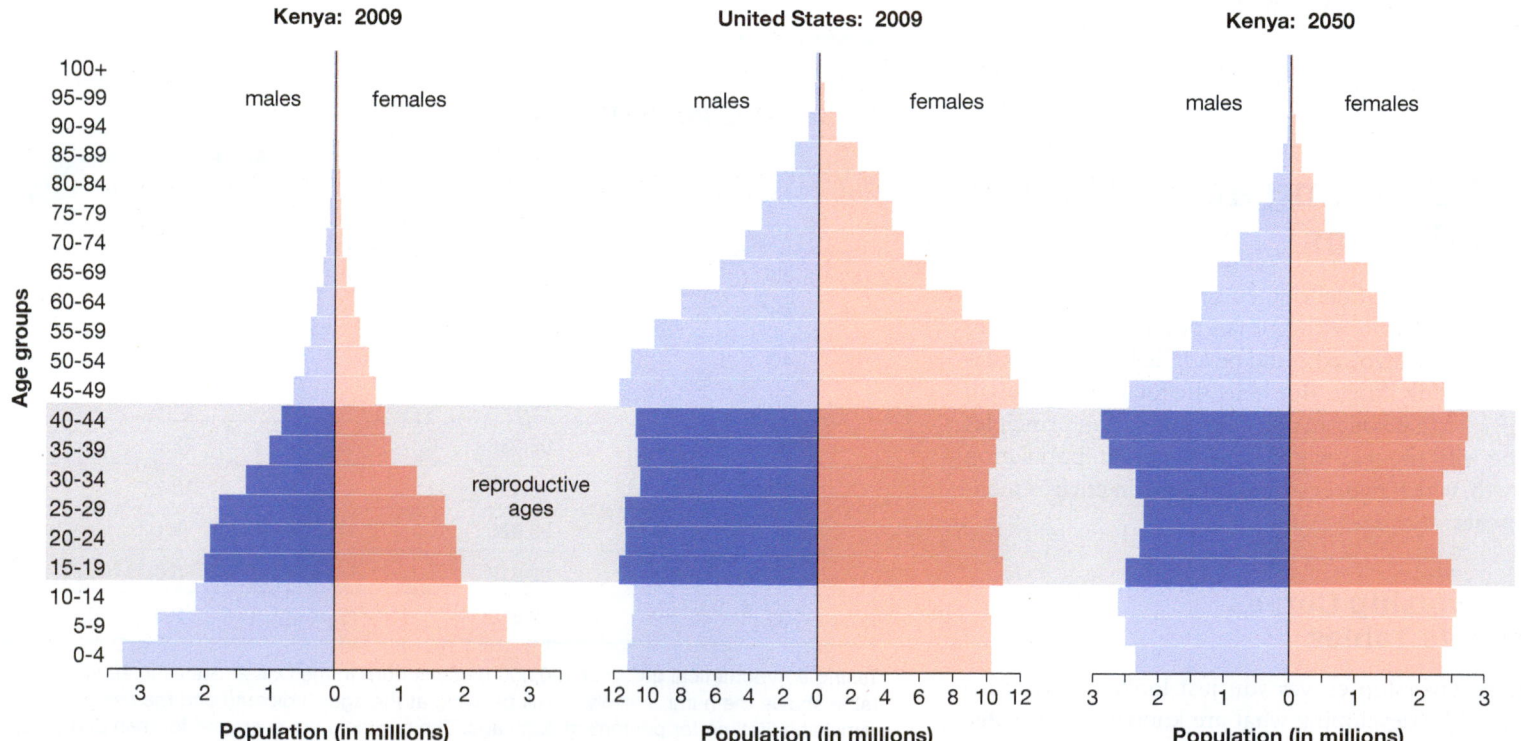

Figure 34.9

Age Structure in Populations: Kenya and the United States

Each bar on the graphs represents a 5-year age grouping of the population. Note the greater proportion of the 2009 Kenya population that is of "pre-reproductive" age—an indication of greater population increases in the future than is the case with the United States. The graph on the right shows the change expected to come about in Kenya's age structure by the middle of the century.

(Data from the U.S. Census Bureau.)

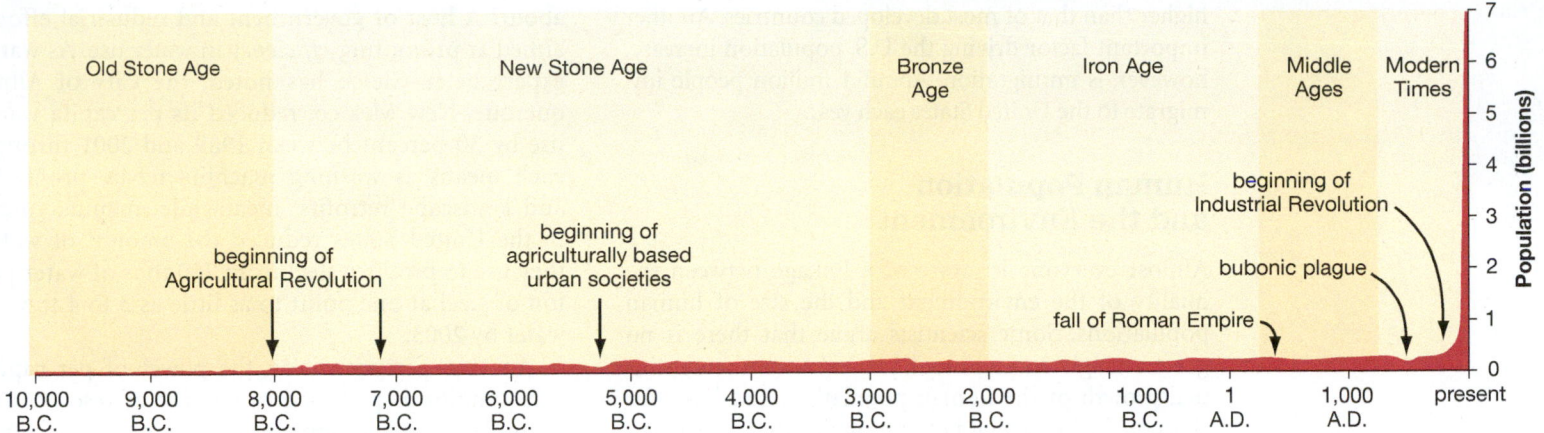

Old Stone Age | New Stone Age | Bronze Age | Iron Age | Middle Ages | Modern Times

beginning of Agricultural Revolution

beginning of agriculturally based urban societies

beginning of Industrial Revolution

bubonic plague

fall of Roman Empire

Population (billions)

10,000 B.C. — 9,000 B.C. — 8,000 B.C. — 7,000 B.C. — 6,000 B.C. — 5,000 B.C. — 4,000 B.C. — 3,000 B.C. — 2,000 B.C. — 1,000 B.C. — 1 A.D. — 1,000 A.D. — present

(in 1974), and is now projected to reach 7 billion in 2012. The curve you see that resulted from this growth may look familiar because it is a more extreme version of the J-shaped curve, although greatly elongated on its left side. Were we to see this kind of growth curve in any species other than our own, we would predict that this kind of increase could not go on for long; that environmental resistance would assert itself, and that the population might even crash in one catastrophic sense or another.

But something has changed. In 1992, demographers with the United Nations were predicting that, by 2050, the Earth would have 12 billion people on it. But in reports issued beginning in 2003, the UN predicted that the human population will be about 9.1 billion by 2050. Further, while the human population will still be growing then, its rate of increase will have dropped considerably. In the period 1990–1995, the world was adding an estimated 82 million people per year; by 2050, that figure should be down to 31 million people per year. In the longer term, it appears the human population may stabilize or even drop to some extent.

What has brought this change about? The key measurement used by demographers to predict human population change is the **total fertility rate,** which can be informally defined as the average number of children born to each woman in a population. In more developed countries, the "replacement" fertility rate is 2.1. If fertility is above 2.1, the population will grow; if it's below 2.1, the population will shrink. (Why 2.1 instead of 2.0? Some children will not live to have children of their own.)

With this in mind, look at **Figure 34.11**. Pictured are fertility levels for a selected group of countries in two periods: 1960–1965 and 2000–2005. The dramatic declines you can see are not being repeated in all countries. Indeed, in some countries fertility is dropping very little. But when all the world's countries are averaged together, we find that global fertility has been cut nearly in half over the past 50 years. Whereas it stood at 5 in the period 1950–1955, it now

stands at 2.56 (and is expected to drop to 2.02 by 2045–2050).

This global average, however, masks some tremendous differences between more-developed and less-developed countries. Think of it this way. In 2005, the world's population was increasing by 154 people per *minute*. And of these 154 additional people, 151 were born in less-developed countries. The South Asian country of Bangladesh, which is about the size of Iowa, is expected to have a population increase of more than 100 million people between 2005 and 2050. Meanwhile, in the countries of Europe, the fertility rate is now so low—at 1.6 on average—that the population is expected to decline by about 36 million people by 2050, even when immigration is factored in.

Where does the United States stand in this spectrum? In the coming decades, it is expected to have more population growth than any other large, developed country. The U.S. population is expected to go from 307 million people today to 439 million people in 2050—a 43 percent increase. What makes the United States different from the other developed countries? Partly the American fertility rate, which at 2.1 is much

Figure 34.10
Human Population Levels through the Centuries

Figure 34.11
Big Changes in Fertility

Pictured are female fertility rates in six countries during two periods: first during the 1960s and then for the period 2000–2005. Changes such as these have brought about a remarkable decline in the worldwide total fertility rate.

(Source: Population Reference Bureau.)

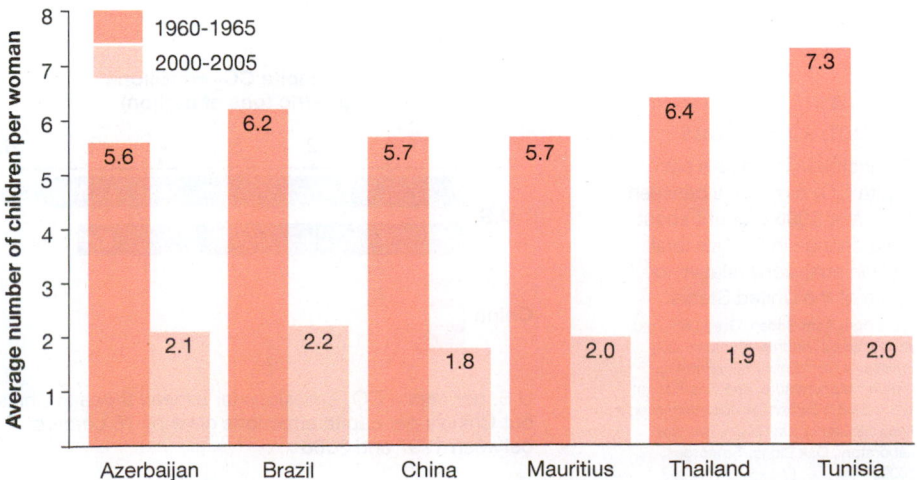

higher than that of most developed countries. Another important factor driving the U.S. population increase, however, is immigration. About 1 million people immigrate to the United States each year.

Human Population and the Environment

Almost everyone is aware of a linkage between the quality of the environment and the size of human populations. Some scientists argue that there is no greater single environmental threat than the continued growth of the human population. The basis for this conclusion is that population affects so many environmental issues: the use of limited natural resources, the amount of waste that is pumped into the environment daily, the reduction of species habitat, the decimation of species through hunting and fishing. Look at almost any environmental problem, and you are likely to find human population growth playing a part in it.

Given that the human population is still growing—but that stabilization of it is on the horizon—it's possible to see the twenty-first century as a crucial moment in time for the environment. The biologist Edward O. Wilson has characterized our century as an environmental *bottleneck*: as a period in which the Earth's ecosystems will be under the greatest assault they are ever likely to face. On the other side of the bottleneck, with human population probably stabilized, the pressures on the environment should ease. The question is: How many species and habitats can make it *through* the bottleneck? How many primate species can be preserved? How many acres of rainforest can be kept intact? How many acres of American wetlands can remain as they are?

Some experts argue that it is not population per se but rather the use of resources *per person* that is of most pressing concern. Consider that the United States used less water in 2000 than it did in 1980 even though the U.S. population grew by 24 percent during the period. What brought this reduction about? A host of government and industrial efforts aimed at promoting *efficiency* in water use. As water expert Peter Gleick has noted, the City of Albuquerque, New Mexico, reduced its per capita water use by 30 percent between 1989 and 2001 through such means as washing machine rebate programs and landscape retrofits; meanwhile, manufacturers in the United States reduced the amount of water they use to produce steel from 200 tons of water per ton of steel at one point to as little as 3 to 4 tons of water by 2003.

Another perspective on the power of per capita consumption is provided by the use of resources in more-developed, as opposed to less-developed, countries. If you look at **Figure 34.12**, you can get a sense of the differences that exist. The carbon tracked in the figure is put into the atmosphere in the form of carbon dioxide (CO_2), a gas that is a by-product of the burning of fossil fuels, such as coal or gas. The CO_2 that is produced by human activities is now regarded as the most significant cause of global warming—the rise in Earth's surface temperature that has occurred over the past century.

The Figure 34.12 graph shows the amount of carbon emitted per capita by human activity in the United States and in China in two different years, 1997 and 2006. A quick glance at the figure shows how little carbon China emitted per person in either year compared to the United States. In 2006, carbon emissions in the United States totaled more than 5 metric tons per person, while in China they came to about 1.27 tons per person. Also note, however, that China's per capita emissions grew by 76 percent between 1997 and 2006, while U.S. emissions declined slightly. To see the effects of this change, look at the graph on the right. As it shows, the carbon emissions for China as a *country* went from 59 percent of those in the United States in 1997 to 106 percent of those in the United States in 2006. What's going on in China? Very little population growth, at least in proportional terms. What is going on is development, which drives the per capita use of resources. China is industrializing at a

Figure 34.12

Per Capita and Total Carbon Emissions

The increase in China's per capita CO_2 emissions between 1997 and 2006 brought about a big change in China's total carbon emissions relative to those of the United States.

(Data from Tom Boden, Gregg Marland, and Robert J. Andres, "National CO_2 Emissions from Fossil-Fuel Burning, Cement Manufacture, and Gas Flaring: 1751–2006," Carbon Dioxide Information Analysis Center, Oak Ridge National Laboratory, Oak Ridge, Tennessee, 2009.)

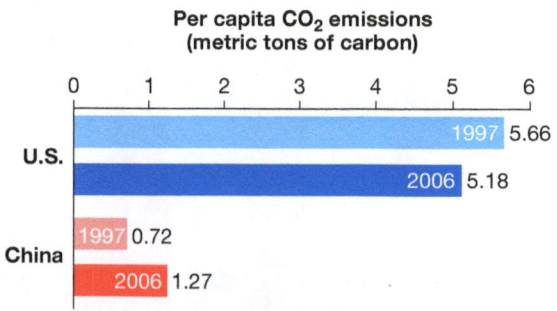

Per capita CO_2 emissions (metric tons of carbon)

U.S. 1997 5.66
U.S. 2006 5.18
China 1997 0.72
China 2006 1.27

U.S. per capita CO_2 emissions far exceed those in China, but China's per capita emissions grew by 76 percent between 1997 and 2006 . . .

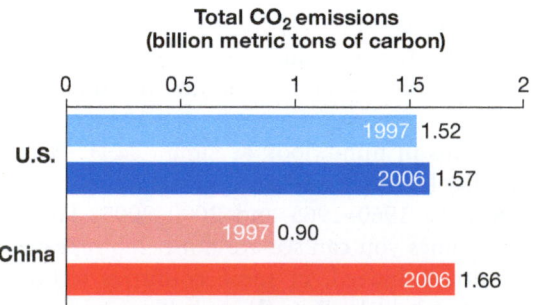

Total CO_2 emissions (billion metric tons of carbon)

U.S. 1997 1.52
U.S. 2006 1.57
China 1997 0.90
China 2006 1.66

. . . When coupled with China's large population, this per capita growth meant that China's total CO_2 emissions went from 59 percent of those in U.S. in 1997 to 106 percent of those in U.S. in 2006.

rapid rate, and it is fueling its industrialization largely with coal. The CO_2 output that is resulting from this industrialization provides a clear, if depressing, lesson: When a large human population increases its per capita use of resources, the result can be big trouble for the environment.

On to Communities

The populations we've looked at this chapter have been abstractions, in a sense, in that we've thought of each of them as existing in a kind of isolation. To be sure, populations do have their own internal dynamics. In an elephant population, for example, only a few young will be born, great care will be lavished on those who are born, and so forth. Yet every elephant population is constantly interacting with populations of other species. Consider that, in the wild, an African elephant can consume up to 200 kilograms, or 440 pounds, of grasses, leaves, and roots each day. Moreover, it's not uncommon for elephants to pull down small trees as a means of getting at some of this material. So African elephant populations can have a significant negative impact on African tree populations. On the other hand, the only way the seeds of some African trees get dispersed is through elephants— *literally* through elephants, in that the seeds pass through the elephant digestive system, after which they're deposited, in elephant dung, at new sites for germination and growth. African elephants and trees thus interact in some complex ways.

When we start thinking about such interactions, we are thinking about life at the level of a community, meaning all the populations of all species that live in a defined geographical area. So, in what ways can community members interact? Beyond this, how do communities get established and then become stabilized? And are some species more important than others in preserving a community's structure? These are some of the questions we'll take up in Chapter 35.

SO FAR ...

1. *K*-selected or equilibrium species tend to have _____ offspring to whom they provide _____ care. Conversely, *r*-selected or opportunist species tend to have _____ offspring to whom they provide _____ care.

2. A country whose population pyramid is widest at the bottom can expect to have _____ population growth in the near future than a country whose population pyramid is of uniform width throughout.

3. A country's human population changes can be predicted by looking at its total fertility rate, informally defined as the _____. The "replacement" level for fertility in more-developed countries is _____.

34 REVIEW

Summary

34.1 The Study of Ecology

- Ecology is the study of the interactions living things have with each other and with their environment. (p. 656)

- The five scales of life that concern ecology are physiology, populations, communities, ecosystems, and the biosphere. (p. 656)

34.2 Populations: Size and Dynamics

- To understand any population, ecologists count the number of individuals in it; various techniques are used to estimate the size of populations whose members can't be counted directly. (p. 657)

- An arithmetical increase occurs when, over a given interval of time, an unvarying number of new units is added to a population. An exponential increase occurs when the number of new units added to a population is proportional to the number of units that exists. (p. 658)

- The rapid growth that sometimes characterizes populations of living things is referred to as exponential growth or the J-shaped growth curve. Populations that initially grow, but whose growth later levels out, have experienced logistic growth, or the S-shaped growth curve. (p. 659)

- The size of populations is limited by environmental resistance, defined as all the forces that limit population growth. (p. 659)

- Exponential growth can be calculated by subtracting a population's death rate from its birth rate, which yields the population's rate of growth. Denoted as *r*, this rate is also known as the population's intrinsic rate of increase. (p. 659)

- Carrying capacity, denoted as *K*, is the maximum population density of a given species that can be sustained within a

defined geographical area over an extended period of time. (p. 660)

34.3 *r*-Selected and *K*-Selected Species

- Different species have different characteristics that affect the number of fertile offspring they bear. *K*-selected species tend to be physically large, experience their environment as relatively stable, and lavish attention on relatively few offspring. The pressures on *K*-selected species tend to be density dependent, meaning that as a population's density goes up, factors that limit its growth assert themselves more strongly. (p. 661)

- *r*-selected species tend to be physically small, experience their environment as relatively unstable, and give little or no attention to the numerous offspring they produce. The pressures on *r*-selected species tend to be density independent. (p. 662)

- Survivorship curves describe how soon species members tend to die within the species' life span. Three idealized types of survivorship curves are late loss, constant loss, and early loss. (p. 663)

34.4 Thinking about Human Populations

- Survivorship curves are created from life tables, which set forth the probabilities of a member of a species being alive after given intervals of time. (p. 663)

- When calculating the growth of human populations, it is important to learn what proportion of the population is at or under reproductive age. (p. 664)

- The world's human population is now growing at a much slower rate than in the past because of a global decrease in the total fertility rate. The global reduction in fertility masks enormous, ongoing differ-

ences between fertility in more-developed and less-developed countries. (p. 664)

- Some scientists believe that there is no greater threat to the environment than the continued growth of the human population. Others argue that a more important concern is the use of natural resources per person. (p. 666)

Key Terms

arithmetical increase   658

biosphere   657

carrying capacity (*K*)   660

community   657

density dependent   661

density independent   662

ecology   656

ecosystem   657

environmental resistance   659

exponential growth   659

exponential increase   658

intrinsic rate of increase (*r*)   660

K-selected species   661

life table   663

logistic growth   659

population   657

r-selected species   662

total fertility rate   665

zero population growth   660

Understanding the Basics

Multiple-Choice Questions

(Answers are in the back of the book.)

1. A(n) _____ consists of all members of a single species that live in a defined geographic region, while a(n) _____ consists of the populations

of all species that live in a defined region.
a. ecosystem; community
b. ecosystem; biosphere
c. community; ecosystem
d. population; community
e. biome; ecosystem

2. A community and all the nonliving elements that interact with it (sunlight, nutrients, soil, etc.) is called
a. a population group.
b. a biosphere.
c. an ecosystem.
d. a biome.
e. a social system.

3. The largest scale of life, which can be thought of as the interaction of all of Earth's ecosystems, is called
a. the biome.
b. the ecosphere.
c. the biosphere.
d. the environment zone.
e. the ionosphere.

4. When population growth increases in proportion to the number of members of a population, _____ growth is said to have occurred.
a. density-dependent
b. density-independent
c. K-selected
d. arithmetical
e. exponential

5. K-selected species tend to (select all that apply):
a. experience their environments as unstable.
b. have stable populations.
c. have their populations limited by carrying capacity.
d. provide little care to their offspring.
e. be most affected by biological forces.

6. r-selected species tend to (select all that apply):
a. have their populations limited by reproductive rate.
b. experience their environments as unstable.
c. provide abundant care to their offspring.
d. have populations that are strongly affected by physical forces.
e. face density-dependent pressures on their populations.

7. The human total fertility rate (select all that apply):
a. is the average number of children born to each family in a population.
b. has been dropping worldwide in recent decades.
c. must be at about 3 for a population to be at replacement level.
d. is generally much lower in less-developed countries.
e. is higher in the United States than in most developed countries.

Brief Review

(Answers are in the back of the book.)

1. Name the four levels of group organization of living things in order of increasing complexity.

2. Compare and contrast r-selected and K-selected species and give an example of each. Are humans r-selected or K-selected?

3. Explain what is meant by environmental resistance and its relationship to population growth.

4. What kinds of forces tend to limit the growth of K-selected species?

5. Which factor has been most important in bringing about the apparent long-term stabilization of Earth's human population? Is this stabilization expected to occur soon?

Applying Your Knowledge

1. Does the concept of carrying capacity apply to human beings? Does the planet as a whole have a human carrying capacity, or do human beings have an infinite capacity to improve life through such means as obtaining greater crop yields and becoming more energy efficient?

2. Is human disease a form of environmental resistance?

3. Some scientists argue that no single factor poses a greater threat to the environment than the continued growth of the human population, while others believe the use of resources per person is a more critical factor. What arguments can you think of for or against either proposition?

MasteringBiology®

1. MasteringBiology Assignments
GraphIt!: Age Pyramids and Population Growth; **Bioflix:** Population Ecology; **Current Events:** As Dengue Fever Sweeps India, a Slow Response Stirs Experts' Fears; **Building Vocabulary; Questions:** Bioflix Quiz, Reading Questions, Test Bank

2. eText
Read your book online, search, take notes, highlight text, and study more effectively.

3. The Study Area
Self-study tools to test comprehension.

35

An Interactive Living World 2:
Communities in Ecology

A community is all the populations of all the species living in one area. Members of different species within a community interact with each other in a variety of ways, some of them competitive, others supportive. Communities establish themselves through a long-term process called succession.

Available on MasteringBiology®

ABC News Video: The Cuttlefish; Eating Behavior in Monkeys; **Video Field Trip:** Invasive Species; Prescribed Fire; **Current Events:** Dead Dolphins and Birds Causing Alarm in Peru; Gulf Coast Towns Brace as Huge Oil Slick Nears Marshes; **Building Vocabulary; Questions:** Reading Questions, Test Bank

A watering hole is a natural congregation spot for members of a community in Botswana's Okavango Delta.

Most people are aware that tropical rainforests are places of great biological diversity, but some years ago Smithsonian biologist Terry Erwin was able to attach a startling number to this diversity. While doing research in Panama, Erwin examined a single tree and found an estimated 1,700 species of beetle living in it. Not 1,700 beetles,

mind you—1,700 *species* of beetle. Of course, these beetles were only one part of a larger group of insects that were living in the tree, and the insects in turn were only one variety of animal living in it. Joining these animals were the fungi that did their best to extend their filaments into the tree's bark each day and the bacteria that lived by the billions in or on all the other organisms that inhabited the tree.

We saw last chapter that a **community** can be defined as all the populations of all the species that inhabit a defined area. This means all the plants, animals, fungi, bacteria, protists—every living thing, in other words. If we think of our tree in Panama as being a defined area, we can get a sense of the mind-boggling diversity that can exist within a single community. And yet, by examining communities from the tropics to the poles, ecologists have found that there is a standard set of questions that can be asked about any community. How diverse is it? Which species in it are dependent on which others? Which species in it are the predators and which are the prey?

If we look for a common thread among these questions, it's not hard to find: The study of communities is first and foremost the study of *relationships* among species. When ecologists ask about predators and prey in a community, it's not the species themselves they're chiefly concerned with; it's the relationships between these varieties of species that interests them. Accordingly, relationships are what we'll primarily be looking at this chapter. How do species relate to each other when living in a common area? Put another way, what makes a community tick? Let's find out.

35.1 Structure in Communities

Large Numbers of a Few Species: Ecological Dominants

If we think about the differences between a tropical rainforest, a prairie, and a desert, it's easy to see that there's a tremendous variability in the mix of species found in different communities. A common feature of many types of communities, however, is that they are dominated by only a few species. Forests tend to be populated by certain kinds of trees, stretches of prairie by certain kinds of grasses, deserts by certain kinds of shrubs or cacti. The small number of species that are abundant in a given community are called **ecological dominants** (**Figure 35.1** on the next page). These generally are plants, but they can be other lifeforms, as is the case with the tiny animals known as corals that are the ecological dominants in coral reefs.

Importance beyond Numbers: Keystone Species

Ecologists have long recognized that there are also species that may not be numerous in a given area but that play a role in a community that cannot be assumed by any other community member. Each of these is a **keystone species**: a species whose absence from a community would bring about significant change in that community. The concept of the keystone species was introduced in the 1960s by marine ecologist Robert Paine, who went with his students to a shallow-water zone of the Pacific Ocean in Washington State and for 6 years regularly removed all sea stars in

Figure 35.1
Ecological Dominants

(a) A Colorado forest floor dominated by aspen trees

(b) A Kansas prairie field dominated by tallgrass

the genus *Pisaster* from a small area. Such a sea star may sound harmless enough, but *Pisaster* was, in fact, the **top predator** in the area: It preyed on other species, but no species preyed on it (**Figure 35.2**). The impact of the *Pisaster* removal was big. Before the change, there had been 15 species of marine animals in the area; after it, there were 8. One species of mussel, freed from its former predator's control, took over much of the attachment space in the area, crowding out other animals, such as barnacles.

Over time, the keystone species concept has undergone some modification. Where once scientists thought of keystones as always sitting at the top of food chains, they now recognize that organisms in other positions can take on a keystone role—beavers building dams, for example, or even lichens that are critical in getting communities going in the desert. Beyond this, it turns out that there are communities without keystones; remove any one species from such a community, and its role will be taken over by another species.

Variety in Communities: What Is Biodiversity?

Apart from ecological dominants and keystones, a third element important to any community is the range of species found within it. This touches on the more general concept of biodiversity, which in the broadest sense simply means variety among living things. In everyday speech, **biodiversity** is thought of as being the same thing as species diversity—to the average person, biodiversity means a diversity *of species* in a given area. This definition of biodiversity is too limited, however. While species diversity is indeed an important type of biodiversity, it is only one of three principal types that exist. To understand

Figure 35.2
Keystone Species
When the predatory sea star *Pisaster ochraceus* was removed from a small area of rocky shore along the Pacific coast of the United States, the species composition of the community changed drastically—more so than if one of the other species had been removed. Pictured are several *Pisaster* sea stars on the Pacific Northwest coast.

the importance of a second type, consider an imaginary experiment that has been put forward by the ecologist Paul Ehrlich.

Suppose that you could get a few members of every species on Earth, but that you restricted each species to a single population housed somewhere (in a single zoo or an aquarium or botanical garden). Species diversity would not drop at all in such a scenario—you'd still have some of each kind of creature, after all—but the Earth quickly would become barren because what's needed is a rich distribution of species in populations across the planet. This geographical distribution of species populations is the second type of biodiversity.

The third type exists *within* populations of different species. It is genetic diversity, meaning a diversity of alleles or variants of genes in a population. Without such diversity, populations are vulnerable to disease; their members may die young or suffer from a variety of inherited mental and physical afflictions (see "The Price of Inbreeding" in Chapter 17, page 308). In summary, there are three principal types of biodiversity among living things: species diversity, geographic diversity, and genetic diversity (Figure 35.3).

The Effects of Species Diversity

In the last 10 years, two intensively researched questions in ecology have concerned species diversity. The first of these questions is: Do diverse communities tend to be more productive communities? In ecology, productivity is a measure of how much energy plants and other photosynthesizers are able to "capture" from the sun's rays and then turn into living material or "biomass." A really productive community is one that is rich with biomass—rich in leaves, seeds, and roots and the animals that feed on them. So, does a diverse community tend to be more productive than one that is not diverse? The answer seems to be yes. The reason for this may be that, in diverse communities, each community member is able to exploit a different portion of the resources available. Some plants will bloom early in a season, while others bloom late; some plants will do well with only a little water, while other plants will flourish with more. In this respect, communities may be something like Jack Sprat and his wife from the old nursery rhyme. Recall that Jack could eat no fat and that his wife could eat no lean, but that between them they licked the platter clean.

A second question about species diversity is whether it leads to greater stability in communities—to a greater ability to recover from a drought, for example, or to a greater ability to resist invasion by new species. Here again, the answer seems to be yes: more diverse communities tend to be more stable communities.

High biodiversity   **Low biodiversity**

(a) Species diversity

many different species   few species

(b) Geographic diversity

broad distribution of species   narrow distribution of species

(c) Genetic diversity

high genetic diversity within population   low genetic diversity within population

Figure 35.3
Three Types of Biodiversity

35.2 Types of Interaction among Community Members

Having considered some structural issues regarding communities, let's now turn to the subject of how members of a community might interact with each other. Here are four primary modes of community interaction:

- Competition
- Predation and parasitism
- Mutualism
- Commensalism

Before continuing with the exploration of these modes, let's go over a couple of concepts that will apply to them all.

Habitat and Niche

A **habitat** is the physical surroundings in which a species can normally be found. Although two populations of a species may be widely separated, they can generally be found dwelling in similar types of natural surroundings. Thus, while the bass-like fish known as groupers can be found in warm-water oceans around the world, they tend to spend their adult lives in a particular type of physical surroundings within the oceans—the coral reef. Habitats such as coral reefs are sometimes described as a species' "address," but a more accurate metaphor might be a species' preferred type of neighborhood.

The word **niche** has been defined in several ways, but it is useful to conceptualize it in terms of a simple metaphor: A niche can be thought of as an organism's occupation. How and where does the organism make a living? What does it do to obtain resources? How does it deal with competition for these resources? The horseshoe crab has found a niche walking on the bottom of shallow coastal waters, feeding on food items that range from algae to small invertebrates. Note that this is about more than what the horseshoe crab eats. It includes specific surroundings (shallow ocean waters), specific behaviors (ocean-floor crawling), and perhaps seasonal or daily feeding times, among other things. If you were to specify all the things that define a horseshoe crab's niche, the odds are that no other organism would exactly fit into it.

With these definitions of habitat and niche in mind, let's look at the ways organisms interact in communities, starting with a familiar type of interaction: competition.

SO FAR ...

1. A keystone species can be defined as any species whose _____ a community would bring about significant change in it.

2. There are three principal types of biodiversity: _____, _____, and _____.

3. A habitat is the _____ in which a species can _____, while a niche can be thought of as an organism's _____.

35.3 Interaction through Competition

Even though niches tend to be specific to given organisms, some species—particularly closely related species—have niches that *overlap* to some degree in a community. A large proportion of both species' diet may be made up of a given organism, or both species may occupy similar kinds of spaces on rocks, branches, or pond surfaces. What arises in such instances is **interspecific competition,** meaning competition between two or more species.

What may come to mind with interspecific competition is a never-ending series of physical battles between two species, but things seldom work like this. For one thing, competition tends to be indirect. It often is a competition for *resources*, in which the winner generally triumphs not by fighting, but by being

more efficient at doing something, such as acquiring food. Lots of animals, including birds and many large mammals, are territorial—meaning they will attempt to keep other creatures out of a territory they have laid claim to—but this usually is a matter of them trying to repel members of their own species.

When the niches of two species greatly overlap, it's unlikely that you'd see long-standing competition among them, for the simple reason that one species or the other is likely to win such a competition in fairly short order. Because the competition is for nothing less than vital resources, the result is that the losing species will be driven to local extinction.

It was a laboratory experiment, performed by the Russian biologist G. F. Gause in the 1930s, that pointed the way on this latter principle. Gause grew two species of *Paramecium* protists, *P. caudatum* and *P. aurelia*, in culture and found that in time, *P. aurelia* was always the sole survivor. True to what was just noted, *P. aurelia* wasn't eating or wounding *P. caudatum*; instead, it grew faster and thus used more of the surrounding resources. Nevertheless, the outcome was always the death of all the *P. caudatum*. This led Gause to formulate what came to be called the **competitive exclusion principle:** When two species compete for the same limited, vital resource, one will always outcompete the other and thus bring about the latter's local extinction (**Figure 35.4**).

Ecologists wouldn't expect to witness the competitive exclusion principle operating much in nature, however. This is because, as noted, the *Paramecium* scenario is likely to be played out quickly. Nevertheless, competitive exclusion has been observed in nature, often when humans have a hand in things. For example, humans introduced a Southeast Asian vine, the kudzu, to the American South on a large scale in the 1930s. Growing at up to 30 centimeters, or 1 foot, per day, kudzu has now taken over millions of acres in the South, locally eliminating many plants in its path in a process of competitive exclusion (**Figure 35.5**).

Resource Partitioning

Competitive exclusion notwithstanding, there are instances in nature in which two related species will use the same kinds of resources from the same habitat over a long period of time. So why isn't one of them eliminated? The answer is contained in another experiment conducted by Gause. He took the successful species from the earlier experiment, *P. aurelia*, and placed it in a test tube with a different paramecium, *P. bursaria*. This time, neither species was eliminated (see Figure 35.4b). Instead, the two species divided up the habitat, *P. aurelia* feeding in the upper part of the test tube, and *P. bursaria* flourishing in the lower part. (*P. bursaria* had an advantage in the lower, oxygen-depleted water because it has symbiotic algae that grow with it, producing oxygen.) In a nutshell, this result describes a situation that often

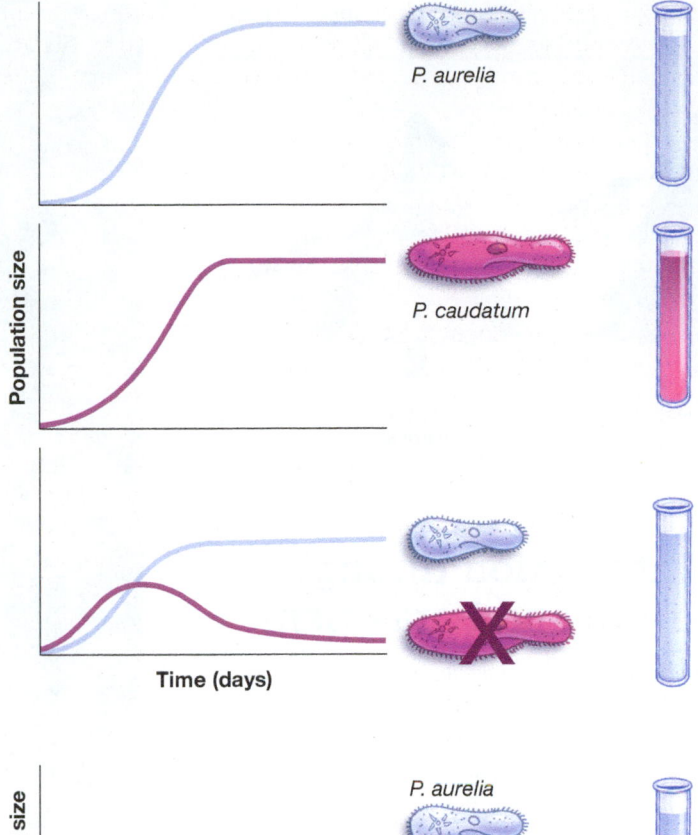

(a) Competitive exclusion

When two species compete for the same limited, vital resource, one will always drive the other to local extinction—as the paramecium *P. aurelia* did to the paramecium *P. caudatum*. This is the competitive exclusion principle at work.

P. aurelia

P. caudatum

Population size

Time (days)

(b) Resource partitioning

Conversely, when Gause put *P. aurelia* together with another paramecium, *P. bursaria*, the two species divided up the habitat, and both survived. This is a demonstration of resource partitioning.

P. aurelia

P. bursaria

Population size

Time (days)

Figure 35.4
Competition for Resources among Species

Laboratory experiments by G. F. Gause showed that competition for resources between two species can have two possible outcomes.

exists in nature: **coexistence** (meaning a sharing of habitat) through a practice called **resource partitioning,** which can be defined as a dividing up of scarce resources among species that have similar requirements. If you look at **Figure 35.6** on the next page, you can see how this works among some species that are a little more familiar—several varieties of warbler. The main message here is that species often flourish by specializing; they derive resources by feeding from different locations (or during different times) within the same habitat.

SO FAR . . .

1. The competitive exclusion principle states that when two populations compete for the same ____, one will always outcompete the other and thus bring about the latter's ____.

2. Habitat often is shared within a community by means of ____, defined as a dividing up of scarce resources among species that have similar requirements.

Figure 35.5
Kudzu Vines Run Wild

The kudzu plant was introduced in the American South in the 1930s and has since spread at a rapid rate, locally eliminating many plants in its path. This is competitive exclusion in action. Here, kudzu has overgrown an abandoned house in Mississippi.

Resource Partitioning

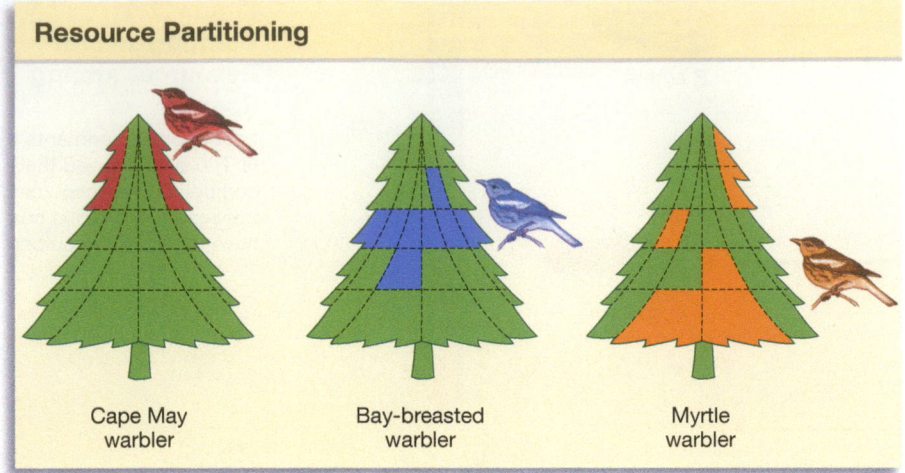

Cape May warbler

Bay-breasted warbler

Myrtle warbler

Figure 35.6
Resource Partitioning
Ecologist Robert MacArthur spent long stretches of time over several years in the 1950s observing the feeding patterns of several species of warblers. All of them ate caterpillars, but from sub-stantially different, though overlapping, parts of the tree.

35.4 Interaction through Predation and Parasitism

It is one thing for two species to compete for re-sources; it is another for one species to *be* a resource for another—to be eaten or used by another species. Thus do ecologists distinguish one mode of interac-tion in communities, competition, from a second mode, predation.

Predation can be defined as one organism feeding on parts or all of a second organism. Note that the prey here can be plants, protists, animals—whatever is preyed on. Predation is generally thought of in terms of animals killing other animals, but note that the definition given includes such things as animals consuming whole plants or their seeds. **Parasitism** is a variety of predation in which the predator feeds on prey but does not kill it immediately—and may not kill it ever. Parasites that come to mind tend to be an-imals, but there are an estimated 3,000 species of par-asitic plants, one of which can be seen in **Figure 35.7**.

Predation and parasitism are the features of nature that many nature lovers can't stand, perhaps because these practices seem cruel or unfair from a human point of view. These modes of interaction have arguably been valuable, however, in that they have stimulated the "arms race" that has resulted in evolutionary adaptations such as vision and flight. (One organism's predatory adaptation spurs the de-velopment of another organism's defensive adapta-tion and vice versa.) Whatever we may think of it, predation is simply a fact of nature, and it is impossi-ble to imagine life without it. But it's also true that to admire the beauty of nature is in significant part to admire the handiwork of predation since it has done so much to shape the living world. You can read about a very familiar predator, the house cat, in "Purring Predators: House Cats and Their Prey" on page 677.

Figure 35.7
A Parasitizing Plant
The strangler fig tree (*Ficus aurea*) is well-named in that it wraps its roots and trunk around a host tree, taking nutri-ents from it, depriving it of sunlight, and eventually killing it. In the United States, strangler figs are native to South Florida. This tree was growing on a host in Queensland, Australia.

Predator–Prey Dynamics

It may be obvious that although predators attack prey, predators are also *dependent* on prey as a food source. An ongoing question in ecological research is how tight this linkage is. To what extent do preda-tor and prey population sizes tend to move up and down together? It's clear that the population dynamics of some predator and prey species are very tightly linked.

The small lemmings (*Dicrostonyx groenlandicus*) of northeast Greenland are fed on by only four preda-tors: the stoat, the arctic fox, the snowy owl, and an-other bird called the long-tailed skua. Of these, the stoat is a "specialist" predator in that it feeds almost entirely on lemmings. If you look at **Figure 35.8**, you can see that both the lemming and stoat populations predictably go through 4-year up-and-down popula-tion cycles. Moreover, note that the movement up or down in stoat population *lags* the change in lemming

ESSAY

Purring Predators: House Cats and Their Prey

House cats are famous for bringing home animals they've killed or captured, of course; intrigued by the carnage brought into their own homes in the 1980s, Peter Churcher and John Lawton decided to make a scientific study of the predatory behavior of the domestic cat. To do so, they enlisted the help of 172 households in the small English village of Bedfordshire, where Churcher lived. The two researchers asked the locals to "bag the remains of any animal the cat caught" and turn the evidence over to them once a week. This process went on for a year with a high degree of cooperation from the cat owners.

Some of the results were not surprising. Young cats hunt more than older cats; small mammals, such as mice, are the favored prey; and cat hunting is not based purely on hunger—all the cats were fed by their owners and yet hunting was widespread.

What was remarkable, however, was the *scale* of the killing. Concentrating on the village's house sparrow population, Churcher and Lawton found that cats were responsible for between one-third and one-half of all sparrow deaths in the village, a figure they believed no other single predator could match. When they looked at all animals killed and projected the village figures onto the whole of Great Britain, the researchers calculated that cats were responsible for about 70 million deaths a year.

Given that birds account for somewhere between 30 and 50 percent of these kills, this means that cats kill at least 20 million birds a year in Britain. The "at least" here may be an important qualifier. An American biologist, the researchers noted, found that cats bring home only about half of the food they catch.

Predator on the Loose
A domestic cat carrying its prey.

population, generally by a few months. What happens is that a higher lemming population means more food for the stoats, which then increase in population. This increase, of course, then means more predation of the lemmings, but the fall you can see in lemming population is not brought about solely by the increase in stoat numbers. When the lemmings' density reaches high levels, the arctic fox, snowy owl, and long-tailed skua begin to prey on them. It is this predation, in combination with the increased stoat predation, that then drives the lemming numbers down. Once the lemmings exist in low densities again, the stoat becomes the only predator, lemming numbers go up, and the cycle repeats.

The Finnish and German researchers who reported these findings determined that declines in lemming numbers could be explained *entirely* by predation; neither the food nor the space available to the lemmings, for example, played any part in reductions of their population. (And, what about the famous mass suicides of the lemmings? Don't they help keep lemming numbers down? The answer is no, because there are no lemming mass suicides; the whole idea is a myth.)

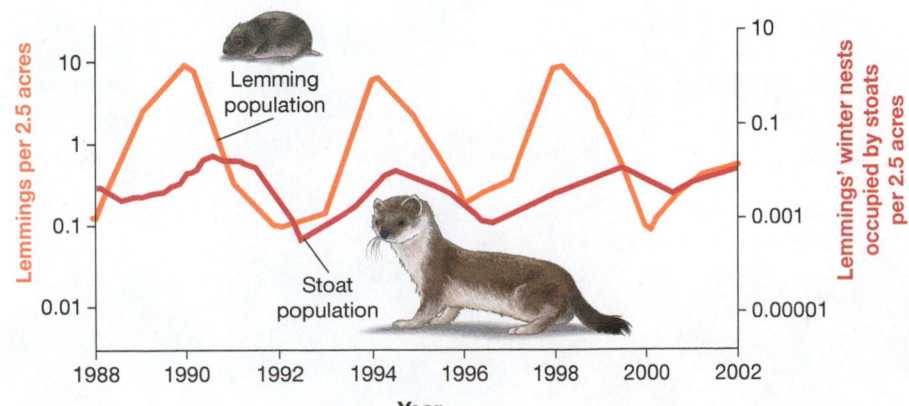

Figure 35.8
Predator and Prey Populations

The population dynamics of a prey species (the lemming) and one of its predators (the stoat) in northeast Greenland over a period of 14 years.

(Reprinted with permission from "Cyclic Dynamics in a Simple Vertebrate Predator-Prey Community," by O. Glig et al., *Science*, 302:867, 2003.)

The crystal-clear connection between predation and the size of the lemming population naturally raises the question of how *common* such a linkage is in nature. What we can say for certain is that, in most species, population dynamics involve more than just predator–prey relationships. Environmental resistance can take many forms, of which increased predation is only one. You might say that the connection between lemmings and their predators serves as an example of how strong predator–prey linkage *can* be, but it is not representative of how strong such linkage is likely to be.

Parasites: Making a Living from the Living

Let's look now at the special variety of predation mentioned earlier, parasitism. Some familiar parasites, such as leeches and ticks, have a straightforward strategy: Get on, hold on, and consume material from the prey, known as the **host.**

But the relationship between parasites and hosts can be much more sophisticated than this. Take, for example, the barely visible roundworm *Strongyloides stercoralis,* which lives throughout the tropics and has, as one of its hosts, human beings in the southeastern United States. The worm comes into the human body from the soil in a thread-like larval form, then makes its way into the veins, which it travels through to get to the lungs. From there, it moves up the trachea and then down into the digestive system, ending up in the small intestine. There it lays eggs that develop into larvae that this time move out, with feces, back into the Earth to start the whole life cycle over again.

Strongyloides is not just using humans as an incubator, however; it can worm its way, you might say, directly from the small intestines into the bloodstream, bringing about a continuous cycle of infection whose symptoms range from severe diarrhea to lung problems. (Lest all of this seem too disturbing, take heart—there is a medicine that can clear this pest from human bodies.) So, note the complexity here: *Strongyloides* not only feeds on a host but uses the host in its reproduction cycle, something that actually is fairly common among parasitic worms and protozoans. Were you to start examining the range of host–parasite dynamics, you would find lots of interactions as elaborate—and chilling—as this. As you can see in "Why Do Rabid Animals Go Crazy?" on 684, parasites sometimes alter the behavior of their hosts.

The Effect of Predator–Prey Interactions on Evolution

Earlier, we noted the "arms race" spurred on by predator–prey interactions: One organism's predatory adaptation spurs the development of another organism's defensive adaptation and vice versa. So, what forms do these evolutionary adaptations take? To see an example on the predator side, look at the warm-water "frogfish" in **Figure 35.9**. These fish have evolved a spine on their dorsal fins that is tipped with a piece of flesh that looks for all the world like a small worm floating free in the ocean. When a would-be predator tries to snatch up this "worm," however, it finds *itself* the prey. On the prey side, look at **Figure 35.10** to see the amazing kinds of camouflage that animals use to keep themselves hidden from the eyes of predators. Another example of prey adaptation may be as close as your own skin. Why are human beings the only primates whose bodies are not covered with a coat of hair? One hypothesis holds that, while hair

Figure 35.9
Fooling Predators about Prey

The predatory frogfish uses a modified spine resembling a worm to lure its prey. Here, a striped frogfish (*Antennarius striatus*) clearly shows its lure while lying in wait in the coastal waters off the Indonesian island of Sulawesi.

(a) Buff-tip moths

(b) Casque-headed chameleon

Figure 35.10
Avoiding Predation through Camouflage

(a) A couple of odd-looking pieces of twig on a branch? No, these are moths—two buff-tip moths (*Phalera bucephala*), native to Europe, whose color and shape make them almost invisible when they are at rest on a forest floor.

(b) The casque-headed chameleon likewise can be very well camouflaged. This one was resting against a tree in Kenya.

warmed our evolutionary ancestors, it also provided a home for many parasites that preyed on them— lice, fleas, and ticks. The adaptation of losing our hair allowed us to lose these parasites to a great extent.

Needless to say, predator–prey evolution goes on within all the kingdoms of life. Plants have developed not only spines, thorns, and other protective structures but also a vast array of chemical compounds—more than 15,000 have been characterized so far—that protect them from predators.

Mimicry Is a Theme in Predator–Prey Evolution

One special form of evolutionary adaptation brought on by predator–prey interaction is **mimicry,** a phenomenon through which one species has evolved to assume the appearance of another. One type of mimicry involves three participants: a model, a mimic, and a dupe. If you look at **Figure 35.11**, you can see a model on the left, the yellow jacket wasp, which obviously can provide a painful lesson to any animal that tries to eat it. On the right you can see the mimic, the clearwing moth, which is harmless but looks a great deal like a wasp. For the dupe, you could select any predator species that sees a clearwing moth and passes it by, believing it to be a dangerous yellow jacket. This is called **Batesian mimicry:** the evolution of one species to resemble a species that has a superior protective capability.

In a second type of mimicry, **Müllerian mimicry,** several species that *have* protection against predators come to resemble each other (**Figure 35.12** on the next page). This creates a visual warning that becomes known to an array of predators. Given this, a would-be predator learns, by interacting with an individual from one species, to keep away from all individuals in any look-alike species. The result is that fewer individuals in these species will be disturbed or killed.

Figure 35.11
Harmless, but Looking Dangerous

In an example of Batesian mimicry, the clearwing moth (on the right) has no sting but has evolved to look like an insect that does, the yellow jacket wasp (on the left). The yellow jacket is the model, and the moth is the mimic.

Figure 35.12
Müllerian Mimicry

The South American butterflies *Heliconius cydno* (top) and *Heliconius sapho* are different in some ways, but they share the characteristic of tasting bad to predators. Over time, they evolved to look like each other, with their appearance serving as a warning to would-be predators. Natural selection favors this Müllerian mimicry because predators can learn about the butterflies' unpalatable taste from both species. The result is that fewer members of either species are likely to be killed or bothered by predators.

35.5 Interaction through Mutualism and Commensalism

The competitive, predatory, and parasitic modes of interaction we've been reviewing have an each-organism-against-all quality to them, but other kinds of relationships also exist within communities. Consider **mutualism**, meaning an interaction between individuals of two species that is beneficial to both individuals. There's a good example of this in Chapter 22 (page 413), with the linkage between plant roots and the slender, below-ground extensions of fungi called hyphae. What the fungi get from growing into the plant roots is food that comes from the photosynthesizing plant; what the plants get is minerals and water, absorbed by the network of hyphae. By some estimates, 90 percent of seed plants have this relationship with fungi. Perhaps the most famous example of mutualism, however, is between the rhinoceros and oxpecker birds. Sitting on the back of the rhino, the oxpecker removes parasites and pests (ticks and flies), thus getting food and a safe place to perch, while the rhino gets relief from its tiny adversaries (**Figure 35.13**).

Our final variety of community interaction is **commensalism,** meaning an interaction between individuals in two species in which one benefits while the other is neither harmed nor helped. Birds can make their nests in trees, and benefit from this, but generally don't affect the trees in any way. Along similar lines, one particular type of bird, the cattle egret (*Bubulcus ibis*), commonly derives a benefit from doing its hunting in fields populated by grazing animals such as cattle and elephants (**Figure 35.14**). The hoof movements of the grazing animals turn over the soil in such fields, thus exposing ground-dwelling insects to the egrets, but the grazing animals are neither harmed nor helped by their interaction with the birds.

Coevolution: Species Driving Each Other's Evolution

Earlier, you read about the effect that predator–prey interactions have had on the evolutionary arms race among species. Yet it's clear that organisms have

Figure 35.13
Mutually Beneficial

(a) Rhinoceros and oxpecker birds

Several oxpecker birds sit atop a black rhinoceros, ridding the rhino of ticks and other pests, while the rhino provides a safe habitat for the birds. This is a demonstration of mutualism—an interaction between two species that is beneficial to both.

(b) Randall's shrimp and flag-tail shrimp-gobi

The randall's shrimp (*Alpheus randalli*), on the left, and the flag-tail shrimp-gobi (*Amblyeleotris yanoi*) also exhibit mutualism. Here the gobi stands guard next to the nearly-blind shrimp, which burrows a hole into the sand into which both animals retreat when the gobi senses danger.

affected each other's evolution through mutualism as well. Indeed, species that are tightly linked in any way over a long period of time are likely to shape one another's evolution. Perceiving this, scientists Paul Ehrlich and Peter Raven developed the concept of **coevolution,** meaning the interdependent evolution of two or more species.

Some of the clearest examples of coevolution involve flowering plants and the animals that pollinate them. Recall from Chapter 22 that many flowers have both fragrances that attract insects and ultraviolet color patterns (invisible to us) that guide the insects to the proper spot for pollination. These features are nature's own homing signals and landing lights, respectively communicating the messages "Food lies this way" and "Land here" (**Figure 35.15**). Now, one of the organisms linked with flowering plants in this way is the honeybee. And it turns out that the color vision of honeybees is most sensitive to the colors that exist in the very flowering plants they pollinate. What seems likely is that honeybee eyesight evolved in response to plant coloration, and that plant coloration evolved to be maximally attractive to the pollinating insects. These groups of organisms coevolved, in other words.

Having considered how community members interact with one another, let's now go on to consider how communities change over time. **Table 35.1** summarizes all the kinds of interactions you've gone over in this chapter.

Figure 35.14
Following in the Footsteps
A group of cattle egrets follow a bull elephant across a field in Botswana, Africa, capturing insects that are exposed in the soil the elephant turns over as it walks.

(a) What we see

(b) What a bee sees

Figure 35.15
Coevolution of Plants and Their Pollinators
This flower has an ultraviolet color pattern that is not normally visible to humans, but that attracts bees. It is likely that these color patterns evolved in the flowers because they aided in attracting the bees. Meanwhile, the bees developed vision that was sensitive to the colors exhibited by these same plants.

SO FAR ...

1. In the interaction known as _____, one species feeds on all or parts of a second organism; in the interaction known as _____, a predator feeds on prey but does not kill it immediately and may not ever kill it.

2. In predator–prey interactions, _____ mimicry has occurred when one species has evolved to resemble another species that has superior protective ability. When several species that have protective ability have evolved to resemble each other _____ mimicry has occurred.

3. Mutualism is an interaction between individuals of two species that is _____. Commensalism is an interaction between individuals of two species in which one is _____ while the other is _____.

Table 35.1		

Community Interactions

In this type of interaction ...	One organism ...	While the other ...
Competition	Is harmed	Is harmed
Predation and parasitism	Gains	Is harmed
Mutualism	Gains	Gains
Commensalism	Gains	Is unaffected

35.6 Succession in Communities

When Washington State's Mount St. Helens volcano erupted in May 1980, it first collapsed inward, thus sending most of the mountain's north face sliding downhill in the largest avalanche in recorded history. Then came the actual eruption, which sent a huge volume of rock and ash hurtling not straight up but out at an angle, toward the north. Some stands of forest in the path of this blast were instantly incinerated down to bare rock; another 35,000 hectares (86,000 acres) of trees were snapped in two like so many twigs. Then came the mudslides caused by the vast expanse of snow and ice melted in the explosion; then came the fall of hundreds of millions of tons of ash, some of it landing as far away as Wyoming. After viewing the devastated area around the blast, President Jimmy Carter said that it "makes the surface of the moon look like a golf course."

But today? No one would claim that the Mount St. Helens area has returned to anything like its former state, but look at the pictures in **Figure 35.16** to get an idea of the transition that has occurred at one location around Mount St. Helens.

The exact form of this rejuvenation may surprise us, but don't we intuitively expect something like this? Just from our everyday experience of watching, say, an abandoned urban lot becoming progressively weed filled, don't we expect this would generally be the case? As it turns out, our intuition is right because almost any parcel of land or water that has been either abandoned by humans or devastated by physical forces will be "reclaimed" by nature, at least to some degree.

The question is, how does this reclamation proceed? Areas that are rebounding don't just get one type of vegetation and then retain it. Instead, the vegetation changes through time—one type of growth *succeeds* another—with a general movement toward more and larger greenery as time goes by. Ecologists call this phenomenon **succession,** meaning a series of replacements of community members at a given location until a relatively stable final state is reached. Within this framework, there are two kinds of succession. The first of these, generally caused by some sort of major ecological disturbance, is **primary succession,** meaning succession in which the starting state is one of little or no life and a soil that lacks nutrients. This is the type of succession that began in parts of the Mount St. Helens environment in 1980. Then there is **secondary succession,** which is succession in which the final state of a habitat has been disturbed by some force, but life remains and the soil has nutrients. The classic example of this is land that has first been cleared for farming but later abandoned by the farmer.

Figure 35.16
Rebound from Disaster
Mount St. Helens, with its crater in the distance, as it appeared in June 1980, one month after the blast, and as it appeared in 1998. This is an example of ecological succession in action.

The relatively stable community that develops at the end of any process of succession is called a **climax community.** There will be some shifts over time in a climax community, just as there are in any other. A prolonged drought will cause some change in the mix of the community's animal and plant life, for example. But, with some exceptions, what does not happen is a shift to a *fundamentally* different mix of life-forms, barring a major change in climate. Grassland stays grassland, and forest stays forest.

Succession can be fairly predictable within small geographical areas. Two abandoned farm plots that lie close together will have very similar kinds of succession—so much so that local farmers or naturalists can sometimes tell to within a few years how long it's been since a field was abandoned. As you may be able to tell, succession generally is thought of in terms of the various *plant* communities that succeed one another, although modern ecology is attempting to bring animals into the picture as well.

An Example of Primary Succession: Alaska's Glacier Bay

One of the best examples of primary succession comes from Alaska's 65-mile-long Glacier Bay, whose lower-lying areas were covered by a glacier until about 240 years ago, when the glacier began retreating. Such withdrawal initially leaves behind a rocky terrain that is completely devoid of organic matter—no fallen leaves or branches, no decomposing animals. Indeed, what's left is not soil but a pulverized rock (called till) that has a high pH and no available nitrogen—not a promising environment for life to establish itself. But establish itself it does. In the case of Glacier Bay, its succession particulars—what species succeeded what—are not important for our purposes. What is important is the general process that occurred. Here's what happened in the eastern arm of Glacier Bay after the glacier retreated (**Figure 35.17**).

In the 20 years immediately following the retreat, the first "pioneer" species to come were lichens and photosynthesizing bacteria, which essentially have the ability to flourish on rock. In addition, there were some primitive plants called horsetails and liverworts.

Over the next 10 years, a gradual transition took place in which an Arctic dwarf shrub, called *Dryas*, took over from the pioneers, growing in a mat that was broken up only by occasional trees. *Dryas* and other early-stage plants have a capability that is critical not only to their own growth in these environments but to the growth of plants that come after them. These plants harbor bacteria in their roots that can *fix nitrogen*—that can take nitrogen from the air and convert it into a form that plants can use. This is how the larger plant community could get started on the till in Glacier Bay, which *lacked* nitrogen.

The *Dryas* community was succeeded by a plant called the alder bush that excluded almost everything else—it was a highly successful dominant, in other words, and its success in the environment meant the end of the *Dryas*. Fifty years after the glacier's retreat, alders were growing in stands that were 10 meters, or about 33 feet, tall. With their growth, the terrain got a great deal of nitrogen and, through acidic alder leaves that fell to the ground, a rapid lowering of pH.

The alder sowed the seeds of its own elimination, however. With the lowered soil pH, an extensive population of Sitka spruce trees could develop; when they did, they "shaded out" the smaller alders by blocking

Figure 35.17
One Example of Primary Succession: Glacier Bay, Alaska

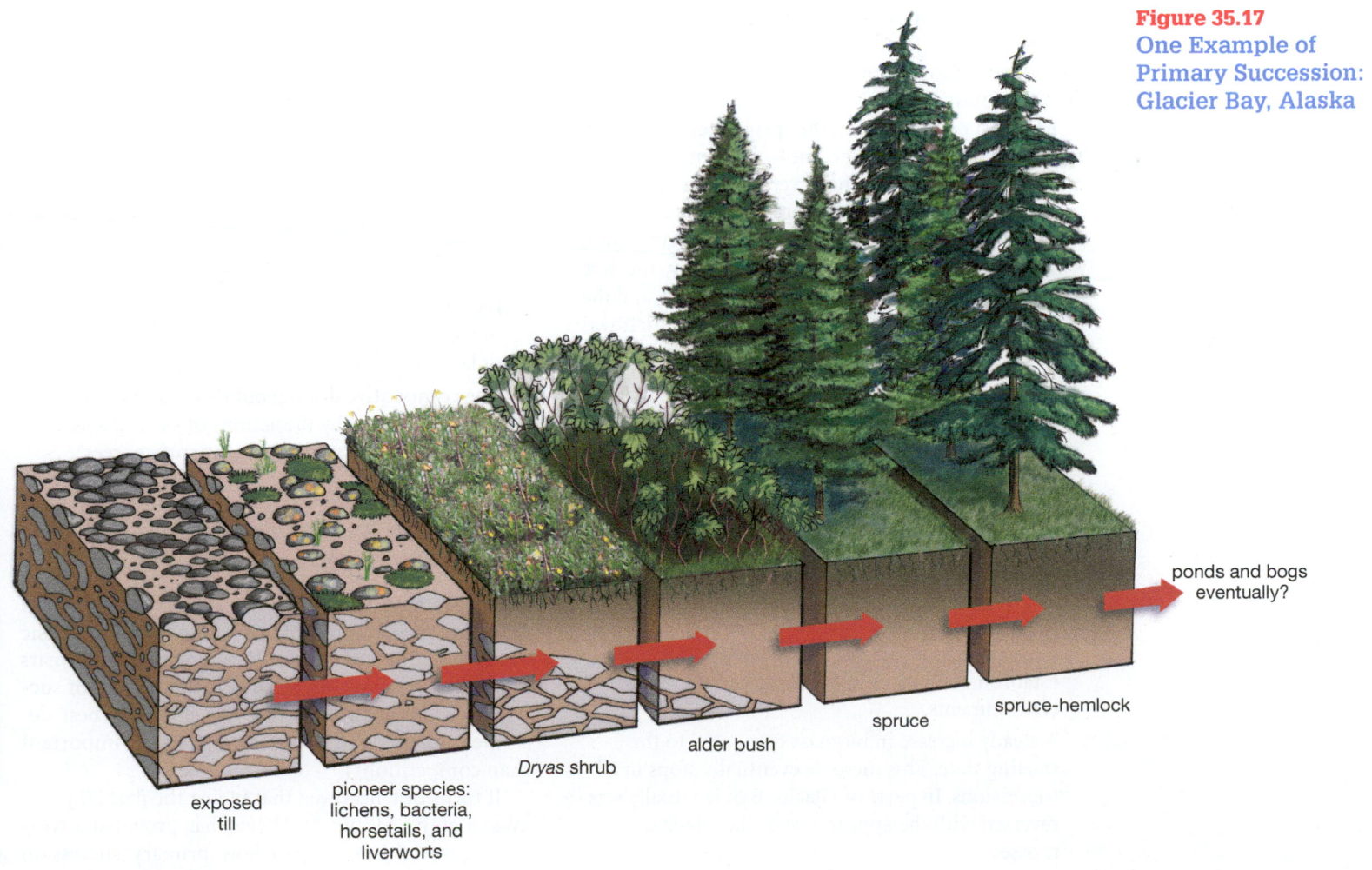

exposed till

pioneer species: lichens, bacteria, horsetails, and liverworts

Dryas shrub

alder bush

spruce

spruce-hemlock

ponds and bogs eventually?

ESSAY

Why Do Rabid Animals Go Crazy?

There is a tiny worm, called *Plagiorhynchus cylindraceus,* that lives its adult life as a parasite in the intestines of starlings and other songbirds. While in a starling, this worm produces eggs that pass out of the bird in its feces. After this, both eggs and feces may be eaten by pillbugs, which are the familiar little arthropods that "roll up into a ball." The worm eggs, however, are not digested by the pillbugs. As a result, a worm *hatches* inside a pillbug, which makes the bug the second "host" for this parasite. And here is where things get interesting, as researcher Janice Moore has shown.

The pillbug is dark, but once infected by a worm, it starts spending more time on light-colored surfaces. What's more, it stops sheltering itself under overhanging objects such as leaves. And the females among the bugs start moving around more and resting less. The effect of all this is that worm-infected pillbugs are more visible to their predators and thus are more likely to be eaten by them. And who are these predators? Some of them are starlings.

Now, when a starling eats a pillbug, the pillbug dies, but the worm inside the bug remains unharmed. Indeed, this is the way the worm gets back inside a starling host

(where it will develop into an adult, lay eggs, and so forth). And that's just the point: It appears that the worm brings about the self-destructive behavior of the pillbug. Through some unknown mechanism, this parasite prompts the pillbug to become an accomplice in its own death. When the pillbug is infected, its behaviors are such that it might as well have a sign on it saying, "Please eat me." The starling is happy to oblige, but the real winner is the *P. cylindraceus* worm, which gets to move from one of its hosts to another.

It would be nice to report that such behavioral takeovers by parasites are rare, but in fact they are common. "Horsehair" worms need to mate in the water, but they spend most of their lives inside arthropods such as ants, whose abdomens they feed on. Just before an ant dies from this,

their sunlight. About 100 years after the glacier's retreat, a spruce forest existed—one that, within 30 more years, started becoming a combined spruce-hemlock forest.

For most of Glacier Bay, the spruce-hemlock forest appears to be the end of the line—the climax community. It's worth noting, however, that on the outer coast of the area there has been one more successional stage. Water-absorbing mosses started to grow and, with them, the forest floor became watery, the tree roots couldn't absorb the oxygen they needed, and the trees died. What eventually was left was land that was filled with ponds and watery bogs. Ecologists have speculated that all of Glacier Bay may be headed in this direction, although this transition could take thousands of years.

Common Elements in Primary Succession

In the Glacier Bay succession, you can see developments that apply to many primary successions, namely:

- An arrival of photosynthesizing pioneer species—in this case, the bacteria and lichens—that can establish themselves in the most barren environments.
- A steady increase in biomass compared to the starting state. This increase eventually stops in all successions. In parts of Glacier Bay, it actually was reversed with the appearance of the late-stage mosses.

- A general movement toward longer-lived species. *Dryas* can live for 50 years, but alders can live for 100 years and spruce trees for 700.
- A general trend toward an increase in species diversity, although not without some reversals (the alder bushes, for example).
- The facilitation of the growth of some later species (the Sitka spruce here) through the actions of earlier species (the shedding of leaves by the alders). More formally, **facilitation** is a process within ecological succession in which the qualities or actions of some early-arriving species enable the success of later-arriving species.
- The competitive driving out of some species (the alder bush) by the actions of later species (the blocking of sunlight by the spruce trees).

Lessons in Succession from Mount St. Helens

Ecologists have been studying succession for about 100 years now, but it remains a field filled with basic questions. Spirited debate has gone on for years about which factors hold true in all instances of succession and which models of development best describe succession. (Is facilitation more important than competition?)

It turns out, however, that in just the past 30 years, Washington's Mount St. Helens has provided a treasure trove of insight into how primary succession

however, it aids the worm in one final way—it makes a journey to water. Indeed, a worm-infected ant seemingly has an uncontrollable urge to get wet as it will return repeatedly to water if blocked. Once it wades into a pool or puddle, however, things proceed quickly. In a matter of minutes, the worm inside it will emerge, ready to mate with another worm.

Now, think about all this in connection with one of the things children are taught to fear from an early age: a rabid animal. Rabies is caused by a virus that spreads through animal saliva. In the "furious" form of the disease, the virus is passed on by means of one animal biting another, thus injecting its saliva into the second animal. Not surprisingly, one of the things the virus does is affect salivary tissue in an infected animal's head and neck.

This ensures that there is plenty of saliva to be transmitted. The virus also inhibits swallowing in an infected animal, which ensures more saliva. Finally, the virus affects the animal's central nervous system in such a way that it brings about the famous *behavioral* change in infected animals—frenzied, unprovoked attacks on other animals. Thus, rabid raccoons or dogs are not simply "going crazy" because of their disease. Their behavior has a function. To be sure, it's a *perverse* function from a human perspective, but then again, aren't the behaviors of infected pillbugs or ants disturbing in a similar sort of way? The general message here is straightforward but hard to take: Living things can be forced to act in ways that benefit the very organisms that have infected them.

Accomplice to Its Own Death
The common pillbug *Armadillidium vulgare*, whose behavior can be manipulated by the parasitical worm *P. cylindraceus*.

works. If there is a silver lining behind the Mount St. Helens explosion, it is that it has given ecologists an unparalleled opportunity to understand succession. This is so because the Mount St. Helens succession began on a crystal-clear starting date (May 18, 1980) and proceeded from such well-defined starting states—obliteration of some habitats and major or minor damage to others.

So, what have we learned from Mount St. Helens? Perhaps the most important lesson is this: What is left from the past is crucial to the future. When trees were knocked over in the blast, their roots were yanked above ground. In this moment of destruction, however, these roots literally carried the seeds of rejuvenation with them. Soil, small plants, and seeds were pulled up with the roots, and they stayed on them—above the deadening layer of ash below. Dead wood, left strewn about after the blast, provided habitat and food for wood-boring insects. These insects in turn attracted birds, which were able to build nests in "snags," meaning trees that were dead but standing. Gophers survived the blast in large numbers because they live below ground, where they feed on roots. They actually flourished in the years following the explosion, and their activity not only turned over the soil—mixing fertile soil below with ash on top—but also provided a kind of subway system for other animals, which were migrating among patches of viable habitat (**Figure 35.18**). So important were the gophers, snags, and surviving seeds to renewal at Mount St. Helens that a collective term for them has now

entered the ecological lexicon. They are examples of **biological legacies:** living things, or products of living things, that survive a major ecological disturbance.

Succession at Mount St. Helens has yielded some other surprises as well. "Classical" primary succession, as outlined in the Glacier Bay example, assumes that life will come to a devastated area from the outside. Seeds and spores will be carried in on the wind, and succession will proceed, in a general way, from the periphery of the barren area to its interior. Although this happened to some degree at Mount St. Helens, in the main recovery was internal: Thousands

Figure 35.18
Important Survivor at Mount St. Helens
The pocket gopher (*Thomomys talpoides*) played a role in the restoration of the Mount St. Helens environment. The gopher's underground habitat allowed it to survive the blast, after which it continued to feed on roots, turning over soil in the process. This gopher is in Wyoming.

of small patches of plant habitat rejuvenated without being seeded from the outside. When organisms did come in from the outside, many were the "pioneer" species predicted by classical succession theory. However, ecologists also found solitary tree seedlings—supposedly latecomers in succession—growing right out of some of the most devastated terrain in the blast area, with no pioneer species setting the stage for them. And some of the external help that did arrive took an unexpected form. In the years immediately following the blast, insects and spiders were blown in on the wind in significant quantity. Once these creatures landed, most of them died, but they themselves went on to serve as food for indigenous ants and other species.

Given all this, another lesson from Mount St. Helens seems to be that succession is more varied and habitat-specific than our theories had predicted. Put another way, we still have a lot to learn about succession.

On to Ecosystems and Biomes

In the past two chapters, you have looked at populations and communities. In populations, you saw the building blocks of communities. In communities, you looked through a lens of the interactions of different species. But it's obvious that community members interact with *nonliving* forces and factors, such as weather and soil composition. When living things and nonliving factors are considered together, the resulting whole is called an *ecosystem*—the subject of the first part of Chapter 36. Very large-scale ecosystems are called *biomes*; a review of them forms the concluding portion of Chapter 36 and will bring to a close this book's coverage of ecology.

SO FAR . . .

1. Succession can be defined as a series of _____ of community members at a given location until a relatively _____ is reached.

2. In primary succession, over time, biomass tends to (pick the correct term from the two) decrease/increase; species tend to be shorter-lived/longer-lived; and species diversity tends to decrease/increase.

3. Succession at Mount St. Helens revealed the importance of biological legacies, meaning living things or products of living things that _____.

CHAPTER 35 REVIEW

Summary

35.1 Structure in Communities

- An ecological community is all the populations of all species that inhabit a given area. Many communities are dominated by only a few species; abundant species in a given area are ecological dominants. A keystone species is one whose absence from a community would bring about significant change in it. (p. 671)

- Biodiversity takes three forms: a diversity of species in a given area, a geographic distribution of species populations, and genetic diversity within species populations. (p. 672)

- Species diversity tends to enhance community productivity and community stability. (p. 673)

35.2 Types of Interaction among Community Members

- There are four primary types of interaction among community members: competition, predation (and a special variety of it, parasitism), mutualism, and commensalism. (p. 673)

- Habitat is the physical surroundings in which a species normally can be found. Niche can be defined as the means by which a species obtains the resources it needs. (p. 673)

35.3 Interaction through Competition

- The competitive exclusion principle states that when two species compete for the same limited, vital resource, one always outcompetes the other and thus brings about the latter's local extinction. (p. 674)

- Coexistence through resource partitioning refers to instances in which two similar species use the same kinds of resources from the same habitat over an extended period of time but divide the resources up such that neither species undergoes local extinction. (p. 674)

35.4 Interaction through Predation and Parasitism

- Predation is defined as one free-standing organism feeding on parts or all of a second organism. Parasitism is a variety of predation in which the predator feeds on prey (known as the host) but does not kill it immediately and may not kill it ever. (p. 676)

- Predator and prey population sizes can move up and down together in a fairly tight linkage, but predator–prey interaction generally is only one of several factors that control the population levels of either kind of community member. (p. 676)

- Predator–prey interactions have spurred the evolution of physical modifications in both predator and prey species. One example of such modification is mimicry, in which one species has evolved to assume the appearance of another. Batesian mimicry occurs when one species evolves to resemble a species that has superior protective capability. Müllerian mimicry occurs when several species that have protection against predators evolve to resemble each other. (p. 679)

35.5 Interaction through Mutualism and Commensalism

- Mutualism is an interaction between individuals of two species that is beneficial to both individuals. Commensalism is an interaction in which an individual from one species benefits while an individual from another species is neither harmed nor helped. (p. 680)

- Coevolution is the interdependent evolution of two or more species. (p. 680)

35.6 Succession in Communities

- Parcels of land that have been abandoned by humans or devastated by physical forces will almost always be reclaimed by nature via succession, a series of replacements of community members at a given location until a relatively stable final state is reached. (p. 682)

- Primary succession proceeds from an original state of little or no life and soil that lacks nutrients. Secondary succession occurs when a final state of habitat is first disturbed by some outside force, but life remains and the soil has nutrients. The final community in any process of succession is known as the climax community. (p. 682)

- Primary succession usually includes the arrival of "pioneer" photosynthesizers, facilitation of the growth of some later species through the actions of earlier species, and the competitive driving out of some earlier species by the actions of later species. As succession proceeds, species diversity tends to increase within communities, and smaller, shorter-lived species tend to be replaced by larger, longer-lived species. (p. 684)

- The rejuvenation of the Mount St. Helens area that has occurred since 1980 has provided ecologists with a wealth of information regarding both primary and secondary succession. (p. 684)

Key Terms

Understanding the Basics

Multiple-Choice Questions

(Answers are in the back of the book.)

1. The Everglades in south Florida contain vast expanses of the sawgrass plant *Cladium jamaicense*. Therefore, in the south Florida Everglades, sawgrass is:
   a. a host.
   b. a model.
   c. a ecological pioneer.
   d. an ecological dominant.
   e. a mimic.

2. If a species preys on others within a community but is not itself preyed upon, and if its absence would bring about significant change within the community, that species is (select all that apply):
   a. an ecological dominant.
   b. a parasite.
   c. a top predator.
   d. a host.
   e. a keystone species.

3. A parasite will often feed on its _____, while using it as part of its _____.
   a. host; reproduction cycle
   b. dupe; camouflage
   c. model; resource partitioning
   d. dupe; reproduction cycle
   e. model; coevolution

4. Plants such as soybeans house bacteria in their roots that receive food from them, while the bacteria provide the soybeans with a usable form of nitrogen. This is an example of:
   a. parasitism.
   b. mutualism.
   c. fertilization.
   d. mimicry.
   e. commensalism.

5. The process by which one species evolves to resemble another species is called:
   a. mimicry.
   b. mutualism.
   c. behaviorism.
   d. parasitism.
   e. natural deception.

6. In communities undergoing primary succession, there are general trends toward (select all that apply):
   a. shorter-lived species.
   b. longer-lived species.
   c. smaller species.
   d. greater species diversity.
   e. less species diversity.

7. In primary succession, pioneer species will predominately be _____ because they ____.
   a. migratory animals; can go elsewhere for their food
   b. cold-weather species; can withstand harsh conditions
   c. insects; can live in so many environments
   d. photosynthesizing organisms; can make their own food
   e. trees; can run roots deep into the soil for water

Brief Review

(Answers are in the back of the book.)

1. In architecture, a keystone is a stone at the top of an arch that holds the entire arch in place. In what way can certain species be thought of as keystones for their communities?

2. Name the four primary modes of interaction in a community.

3. A wasp from the group known as the ichneumons sometimes will paralyze a caterpillar by injecting a toxin into it, after which the wasp deposits its eggs into or onto the caterpillar's body. When the eggs hatch, the resulting wasp larvae live by consuming the caterpillar, which remains alive while paralyzed. What mode of interaction are the wasp and the caterpillar practicing? What is the role of the caterpillar in this interaction?

4. Lichens are composite organisms, usually made up of upper and lower layers of fungi, with a layer of algae in between. Fungi benefit from their association with the photosynthesizing algae in that they receive food from them. At one time, however, scientists questioned whether the algae were in turn benefiting from their relationship with fungi. It now appears, however, that algae do benefit from their association with fungi in that they receive water, carbon dioxide, and protection from the elements from them. For the sake of argument, assume that the fungi are not providing any benefit to the algae; which mode of interaction would the two types of organisms be practicing? In light of the benefits the fungi are providing to the algae, which mode of interaction are they actually practicing?

5. The red-spotted newt *Notophthalmus viridescens* secretes toxins from its skin that make predators avoid it. The red salamander *Pseudotriton ruber* has no such secretions but resembles the red-spotted newt and gains protection from this resemblance. What is this an example of? Which is the model, the mimic, and the dupe?

6. A volcano erupted in Hawaii, and lava covered land that had been farmed for centuries. Eventually, new lichens, and

then plants, grew on the lava. Is this an example of primary or secondary succession?

Applying Your Knowledge

1. In the movie *The Godfather,* Michael Corleone and Enzo the baker stand on the steps outside the hospital where Michael's father is recovering from gunshot wounds and pretend to be holding guns inside their coats as a means of keeping assassins from entering the hospital and killing Michael's father. Allowing for some differences between the natural world and human society, what form of mimicry were Michael Corleone and Enzo the baker practicing?

2. The world's first farmers were not human beings but the leaf-cutter ants of the genus *Atta.* These ants do not eat the leaves they carry off. Instead, they bring them down into their nests, where they process the leaves into a pulp and then spread the pulp onto a fungus that grows only in their nests. The fungus then feeds on the food that has been brought to it and grows, after which the ants harvest the knob-like stalks of the fungus and consume them. What is the mode of interaction between the ants and the fungus in this community?

3. The clearwing moth pictured on page 679 has no protective capability, but has evolved to look like a species—the yellow jacket wasp—that does have such capability. As such, what mode of interaction are the clearwing and the yellow jacket practicing? What mode of interaction is being practiced between two species that have protective capability and that have evolved to look alike?

MasteringBiology®

1. MasteringBiology Assignments
ABC News Video: The Cuttlefish; Eating Behavior in Monkeys; **Video Field Trip:** Invasive Species; Prescribed Fire; **Current Events:** Dead Dolphins and Birds Causing Alarm in Peru; Gulf Coast Towns Brace as Huge Oil Slick Nears Marshes; **Building Vocabulary; Questions:** Reading Questions, Test Bank

2. eText
Read your book online, search, take notes, highlight text, and study more effectively.

3. The Study Area
Self-study tools to test comprehension.

An Interactive Living World 3:

Ecosystems and Biomes

The living world is affected by the larger physical world around it—and vice versa. The interconnected web of living things is being damaged in many ways by human activity.

Available on MasteringBiology®

Activities: The Carbon Cycle; The Nitrogen Cycle; The Hydrologic Cycle; Tropical Atmospheric Circulation and Global Climate; **Bioflix:** The Carbon Cycle; **Current Events:** An Unlikely Way to Save a Species: Serve It for Dinner; As More Eat Meat, a Bid to Cut Emissions; New Jungles Prompt a Debate on Rain Forests; Soil-Free Farming, as Practiced on Board; To Cut Global Warming, Swedes Study Their Plates; **GraphIt!:** Global Fresh Water Resources; Animal Food Production Efficiency and Food Policy; Municipal Solid Waste Trends in the U.S.; Atmospheric CO2 and Temperature Changes; Forestation Change; Prospects for Renewable Energy; Global Soil Degradation; Global Fisheries and Overfishing; **Video Field Trip:** Carbon Neutrality on Campus; Coal-Fired Power Plant; Electronic Waste; Landfill and Recycling Transfer Station; Sustainability on Campus; Sustainable Agriculture; Waste Water Treatment Facility; Water Utility Facility; Wind Power; **ABC News Video:** Ocean Acidification; **Building Vocabulary; Questions:** Bioflix Quizzes, Reading Questions, Test Bank

A toco toucan (*Ramphastos toco*) rests in a tree in the world's largest freshwater wetland, the Panatal of Brazil, Bolivia, and Paraguay.

The word *system* is much in use today. We hear about a given country's political system or its economic system. The term was first used in this context by academics who needed a way to refer to complex entities composed of many interacting parts. This way of looking at things is, however, as applicable to ecology as it is to economics or politics. In Chapter 35,

when you went over ecological communities, you actually were studying something in isolation: the interactions of all the living things in a given area. It goes without saying, however, that these living things also interact with many *non*living entities. To get a complete picture of how the living world functions, then, it's necessary to understand how living things interact not only with each other but with these nonliving elements.

36.1 The Ecosystem

What do nonliving elements add to life? All communities need an original source of energy, generally meaning the sun; they all need a supply of water and a supply of nutrients, such as nitrogen and phosphorus; and all organisms need to exchange gases with their environment (think of your own breathing). The sun's energy is making a one-way trip through the community, ultimately being transformed into heat, but water and nutrients are being *cycled*—from the Earth into organisms and then back again into the Earth. Looking at this, we can begin to perceive the reality of ecological *systems*: of functional units that tightly link both living or **biotic** factors and nonliving or **abiotic** factors to form a whole.

Given this, an important unit in ecology is the **ecosystem,** which can be defined as a community of organisms and the physical environment with which they interact. This chapter begins with a look at small-scale ecosystems and concludes with a look at the very large-scale ecosystems called biomes. In today's world, it's impossible to note the ways in which ecosystems function without also noting the ways in which human activity is damaging them.

36.2 Nutrient and Water Cycling in Ecosystems

In Chapters 34 and 35, you learned a good deal about two of the biotic factors of the ecosystem, namely, individual populations of living things and the communities that are made up of these populations. This chapter focuses on some of the abiotic factors in an ecosystem. In overview, these abiotic factors fall into two categories. First are the *resources* that exist in the ecosystem, such as water and nutrients; second are the *conditions* in which an ecosystem exists, such as average temperature. Resources are considered first in this chapter and conditions later.

The Cycling of Ecosystem Resources

Those of you who went through Chapter 2, on chemistry, know what a chemical **element** is: a substance that is "pure" in that it cannot be broken down into any set of component substances by chemical means. The 92 stable elements include such familiar substances as gold and helium, but only 30 or so of these elements are vital to life and are thus called **nutrients.** Some nutrients, such as iron or iodine, are needed only in small or "trace" quantities by living things, but others are needed in large quantities, among them carbon, oxygen, nitrogen, and phosphorus. Living things also have a great need for water, which is not an element but a molecule composed of two elements, hydrogen and oxygen. Water and nutrients move back and forth between the biotic and abiotic realms, with the term for this movement being a mouthful: **biogeochemical cycling.** Let's look now at the cycling of two elements, carbon and nitrogen, and then at the cycling of water.

Bioflix: The Carbon Cycle

Figure 36.1
The Carbon Cycle

Carbon as One Example of Ecosystem Cycling

A certain amount of cosmic debris makes its way to Earth's surface in the form of meteorites, but this material is pretty sparse. What else comes to Earth from space? The sun's rays, to be sure, as well as light from distant stars, and some other forms of radiation. But when we're talking about chemical elements, the Earth really is a spaceship: It carries a fixed amount of resources with it. Nothing much comes to Earth *from* the outside, and relatively little leaves Earth *for* the outside. For an ecologist, this self-containment simplifies things: What we possess in terms of elements is all we'll ever possess. Thus, the question becomes not so much what we have, but where we have it.

In recent years, there has been considerable concern about the buildup of the heat-trapping gas carbon dioxide (CO_2) in the Earth's atmosphere as this increase is now known to be the primary cause of global warming. Well, if Earth has this fixed quantity of elements—in this case, the carbon and oxygen that make up CO_2—how could there be a buildup of carbon dioxide in the atmosphere? The answer is that carbon has been *transferred* from one place on Earth to another. It has moved from the "fossil fuels" of coal and oil, where it was stored, into the atmosphere.

Storage and transfer: These two concepts are fundamental to biogeochemical cycling. When we look a little deeper into carbon cycling, we find other players involved in carbon's storage and transfer. Plants need CO_2 to perform photosynthesis, and they take in great quantities of it for that purpose, producing their own food as

a result. While the plants are alive, they use some of the carbon they take in to grow—it becomes part of their leaves and stems and roots. And as long as these structures exist, they will store carbon within them, just as coal and oil do. (Thus, one of the proposed solutions to global warming is to grow a huge number of trees, each of which could be thought of as a kind of piggy bank for carbon.) Eventually, the plants die, of course, after which their decomposition by bacteria and fungi releases carbon into the soil and atmosphere, again as CO_2. With this, you can begin to see the whole of the carbon cycle in the natural world (**Figure 36.1**).

Some plants will, of course, be eaten by *animals,* who need carbon-based molecules for tissues and energy, just as plants do. The difference is that animals get these molecules *from plants*—in the form of seeds and leaves and roots—rather than from the air. These animals will likewise lock up carbon for a time, only to return it to the soil with their death and decomposition. Note, however, the critical difference between animals and plants: Carbon comes into the living world only through plants (and their fellow photosynthesizers algae and some bacteria). An animal might take in some carbon dioxide in breathing, but nothing life-sustaining happens as a result. Conversely, a plant that takes in carbon dioxide can *incorporate* it into a life-sustaining molecule, in this case a sugar that is the initial product of photosynthesis. With this sugar as a food source, plants grow, and this bounty ultimately feeds everything else. Look at this simplified cycle:

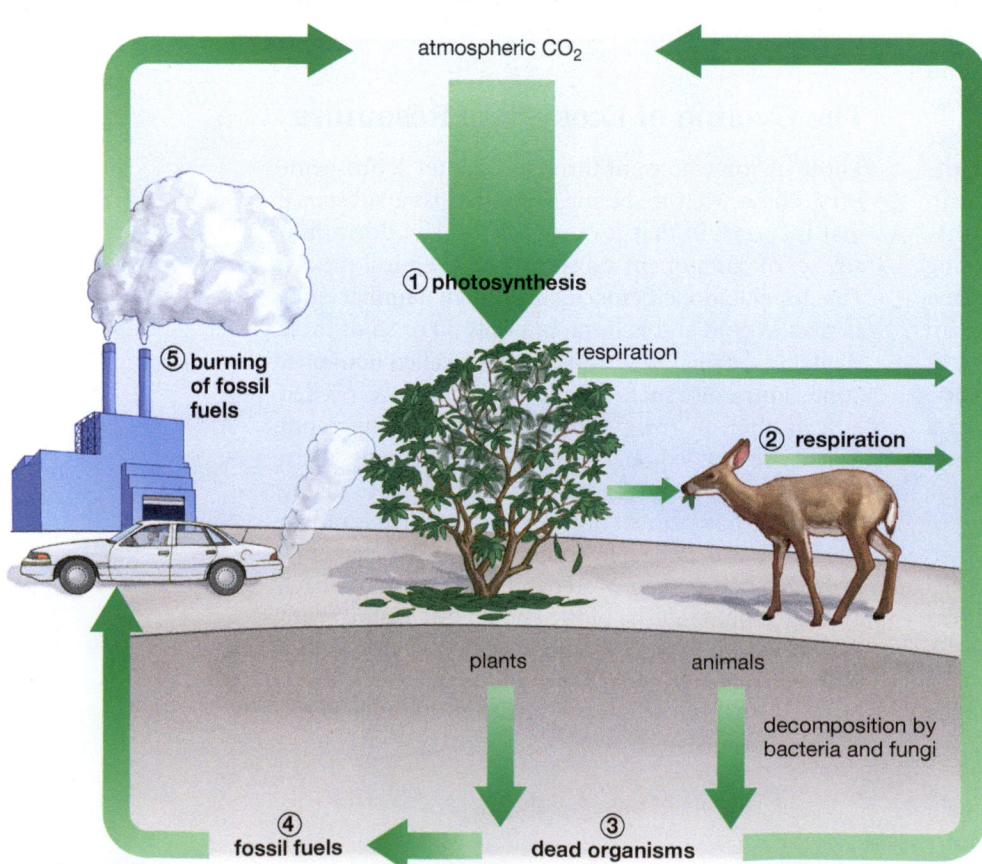

The carbon cycle

1. Plants and other photosynthesizing organisms take in atmospheric carbon dioxide (CO_2) and convert or "fix" it into molecules that become part of the plant.

2. The physical functioning or respiration of organisms converts the carbon in their tissues back into CO_2.

3. Plants and animals die and are decomposed by fungi and bacteria. Some CO_2 results, which moves back into the atmosphere.

4. Some of the carbon in the remains of dead organisms becomes locked up in carbon-based compounds such as coal or oil.

5. The burning of these fossil fuels puts this carbon into the atmosphere in the form of CO_2.

atmospheric CO_2 : plants : animals : decomposers : atmospheric CO_2

What does it track, the path of carbon or the path of food? The answer is both. This is one of the reasons that biologists refer to life on Earth as "carbon based."

How Much Atmospheric CO_2?

In following the carbon trail, it's probably apparent that, just as money is a medium of exchange in our economy, CO_2 is the living world's medium of exchange for carbon; photosynthesizers take *in* carbon as CO_2 and decomposers yield *up* carbon as CO_2. (All organisms also release CO_2 as a by-product of carrying out life's basic processes, as with your own exhalations in breathing.) Given how much **biomass,** or material produced by living things, there is in the world, you might think there would have to be a great deal of CO_2 in the atmosphere. But as a proportion of atmospheric gases, CO_2 is a bit player, making up about 0.035 percent of the atmosphere. But what an important player it is! It's critical in feeding us all, and relatively small changes in its atmospheric concentration are now leading to significant changes in global temperature (something we'll look into in detail later).

The Nitrogen Cycle

Like carbon, nitrogen is an element; and like carbon, nitrogen cycles between the biotic and abiotic domains. Nitrogen makes up only a small proportion of the tissues of living things, but it is a critical proportion because it is required for DNA, RNA, and all proteins. All living things need this element, then. The question is, how do they get it?

The source for all the nitrogen that enters the living world is atmospheric nitrogen, which exists in great abundance. You've seen that carbon dioxide makes up less than 1 percent of the atmosphere, and it turns out that oxygen makes up about 21 percent. But nitrogen makes up 78 percent of the atmosphere. So rich is the atmosphere in this element that a typical garden plot has tons hovering above it. The problem? All this nitrogen is in the form of pairs of nitrogen atoms (N_2) that have a great tendency to stay together, rather than combining with anything else. Thus, in a case of so near and yet so far, plants have no direct access to atmospheric nitrogen. So, how do they obtain it? In the natural world, the answer essentially is, through the actions of bacteria. It is bacteria that are carrying out the process of **nitrogen fixation,** meaning the conversion of atmospheric nitrogen into a form that can be taken up and used by living things.

Bacterial Fixation of Nitrogen

You can see nitrogen fixation and the rest of the nitrogen cycle diagrammed in **Figure 36.2**. The essence of the cycle is that several types of nitrogen-fixing bacteria take in atmospheric nitrogen (N_2) and convert it into

The Nitrogen Cycle

Figure 36.2
The Nitrogen Cycle

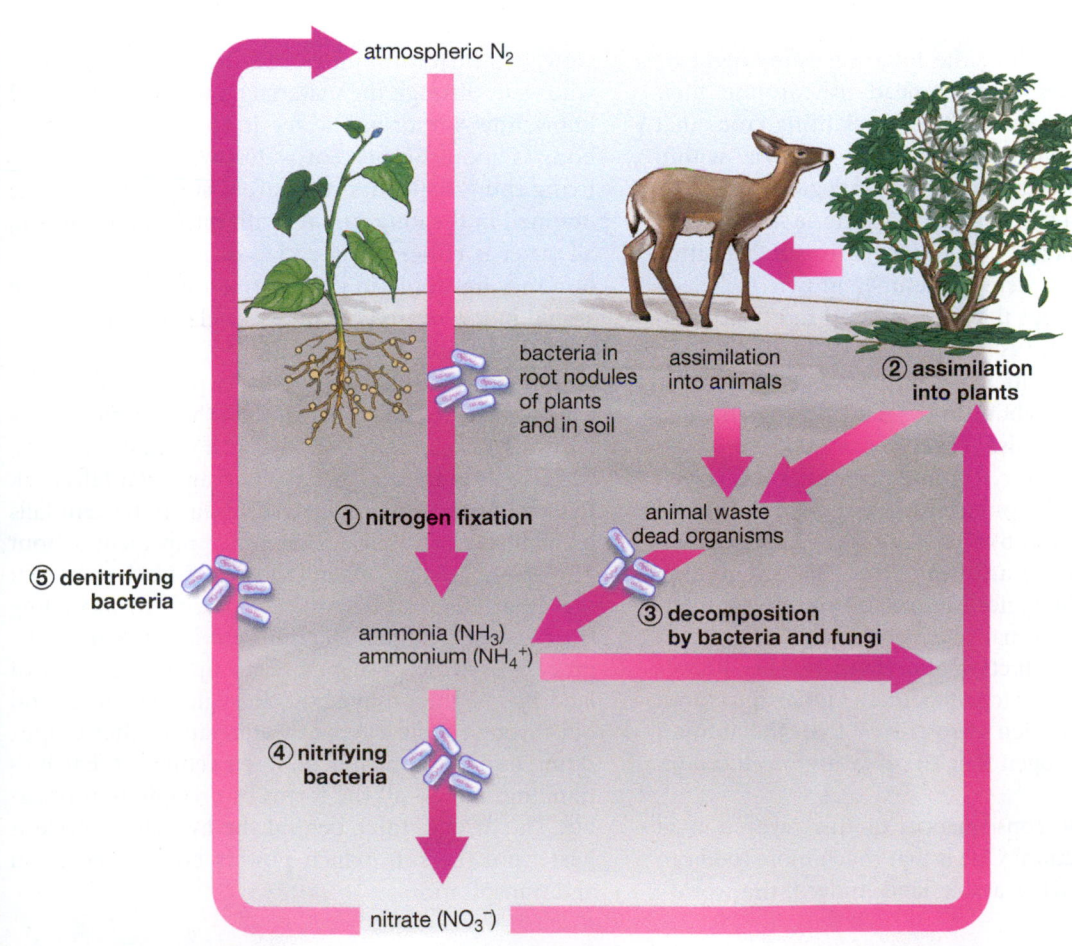

The nitrogen cycle

1. Nitrogen-fixing bacteria convert N_2 into ammonia (NH_3), which converts in water into the ammonium ion (NH_4^+). The latter is a compound that plants can assimilate into tissues. In the diagram, bacteria living symbiotically in plant root nodules have produced NH_4^+, which their plant partners have taken up and used. Meanwhile, free-standing bacteria living in the soil have likewise produced NH_4^+.

2. Other plants take up NH_4^+ that has been produced by soil-dwelling bacteria and assimilate it. Animals eat plants and assimilate the nitrogen from the plants.

3. Animal waste and the tissues of dead animals are decomposed by fungi and by other bacteria, which turn organic nitrogen back into NH_4^+.

4. Other "nitrifying" bacteria convert NH_4^+ into nitrate (NO_3^-), which likewise can be assimilated by plants.

5. Some nitrate, however, is converted by "denitrifying" bacteria back into atmospheric nitrogen, completing the cycle.

ammonia (NH_3). This ammonia then is converted (either in water or by other microbes) into two types of nitrogen-containing compounds that plants can *assimilate*—can take up and use. And, as you've seen, what comes into the plant world will come into the animal world when animals eat plants. The two usable nitrogen compounds produced in the cycle are the ammonium ion (NH_4^+) and nitrate (NO_3^-). Some of the nitrate that results from this process is used by yet another kind of bacteria, denitrifying bacteria, which can convert nitrate back into atmospheric nitrogen, completing the cycle.

Nitrogen as a Limiting Factor in Food Production

If you were to look at the nitrogen story 100 years ago, there would be little more to it than what you've just gone over. In other words, through most of human history, nitrogen was fixed solely by bacteria (with a small assist from lightning, which can produce a usable form of nitrogen that falls to the Earth in rain). Human beings had a problem with this single route to nitrogen, however, because nitrogen is critical for all plant growth, including *agricultural* plant growth. Indeed, for centuries a lack of nitrogen was a primary limiting factor in getting more food from a given amount of land. Long before anyone knew what nitrogen was, farmers were trying to get more of it to crops in two ways. One was by applying organic fertilizer, such as rotting organic material, to their fields. (Remember how the Pilgrims were taught by Native Americans to bury dead fish around their corn plants?) The second was by planting crops that carried their own nitrogen-fixing bacteria within them. This is the case with certain legumes, such as soybeans, which carry symbiotic bacteria in their root nodules. So great was the need for nitrogen that nitrogen-fixing plants were sometimes grown as "green manures"—as crops that were plowed right back into the ground to serve as fertilizers.

Early in the twentieth century, however, a momentous change came about in the use of nitrogen when the German chemist Fritz Haber developed an *industrial* process for turning atmospheric nitrogen into ammonia. In essence, human beings became nitrogen fixers, and by the 1960s, they had taken up this activity on a grand scale. In 2005, 121 million metric tons of biologically active nitrogen were manufactured, and human beings planted crops that resulted in the production of perhaps another 40 million metric tons. Put together, these human-produced quantities of nitrogen were greater than the amount of terrestrial nitrogen that all of nature produced on its own.

What are the consequences of this revolution in nitrogen production? One is that much more food can be grown on Earth's arable land. Indeed, the world's human population could not be fed without the nitrogen fixation that human beings carry out. This massive human intervention in the nitrogen cycle has a downside as well, however. Remember that human-manufactured nitrogen ends up being used in a very *concentrated* way—it is poured onto farmland. A good deal of this nitrogen does not end up in the crops it was intended for, however, but instead departs as runoff. The amount of runoff generated by each farm adds up, and the end result can be large amounts of nitrogen flowing into creeks, ponds, and rivers. On a regional basis, such nitrogen flows can be enormous. The Mississippi River carries an estimated 1.5 million metric tons of nitrogen into the Gulf of Mexico each year.

You might think that since nitrogen is a nutrient, this would be a good thing for the Gulf waters, but the opposite is true. The nitrogen brings on a seasonal "bloom" of algae, the decaying algae feed a huge bacterial population, and the bacteria use up most of the available oxygen at the water's lower depths. The result is a giant "dead zone" that drives away mobile fish and that kills immobile, bottom-dwelling sea creatures (**Figure 36.3**). This problem of *nutrient pollution* is not limited to the Gulf of Mexico, however; it is played out in coastal waters around the world. An analysis delivered in 2008 identified more than 400 coastal regions that are oxygen-depleted or "hypoxic," either seasonally (during summer) or in episodes that may occur once every few years.

The Cycling of Water

Now, let's turn from nutrient to water cycling. Those who went through the material on water in Chapter 2 know how important water is to life. The human body is about 66 percent water by weight, and any living thing that does not have a supply of water is doomed in the long run. As with nitrogen or carbon, all water is either being cycled or is locked up. Carbon may be stored in coal or trees, but water can be stored in ice located on land—in glaciers or in polar ice, for example.

The oceans hold more than 95 percent of Earth's water, but thankfully for us, when the oceans' saltwater evaporates, it falls back to Earth as freshwater. We can also be grateful that not all ocean water falls back to Earth on the oceans; instead, about 10 percent falls on land (**Figure 36.4**). This water represents about 40 percent of the precipitation that land areas get; the other 60 percent comes from a form of cycling reviewed in Chapter 25, *transpiration*, meaning the process by which water is taken up by the roots of land plants, then moved up through their stems and out through their leaves as water vapor. After evaporating into the atmosphere, water returns to Earth as rain, fog, snow—all the forms of precipitation possible. The driving force behind the hydrologic cycle is heat from the sun, which powers both evaporation and transpiration.

(a) Runoff from the Atchafalaya and Mississippi Rivers...

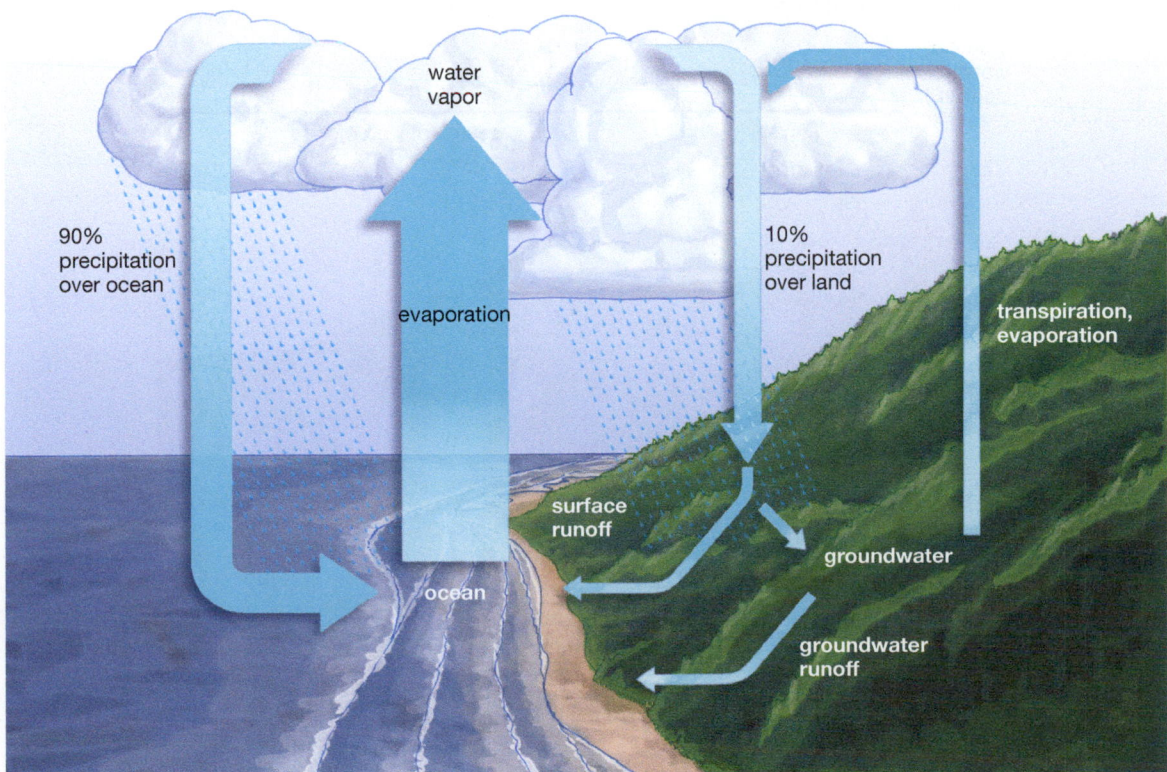

Nitrogen runoff from the enormous Mississippi watershed runs down the river to the Gulf of Mexico.

Figure 36.3
The Gulf of Mexico "Dead Zone"

(a) In 2008, the seasonal "dead zone" in the Gulf of Mexico stretched from the mouth of the Mississippi River to eastern Texas. The waters in the zone are hypoxic, or oxygen depleted, particularly at their lower depths. The primary cause is the tremendous amount of nitrogen carried into the Gulf from both the Mississippi and the Atchafalaya rivers. This nitrogen originates as fertilizer that is applied to farmland all the way up the Mississippi River basin.

(b) Why is this hypoxic zone so extensive? One factor at work is prevailing winds, which carry the discharge of the Mississippi and Atchafalaya Rivers westward, toward Texas. This satellite image of the region was taken in 2007.
(Image courtesy of the Louisiana State University Earth Scan Laboratory.)

(b) ...is given a wide distribution westward in the Gulf of Mexico.

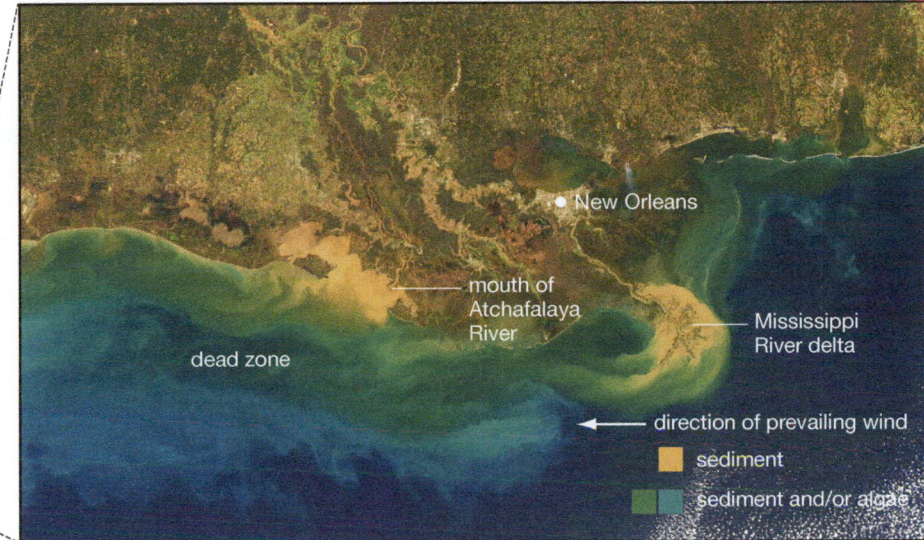

Figure 36.4
The Hydrologic Cycle

The Hydrologic Cycle

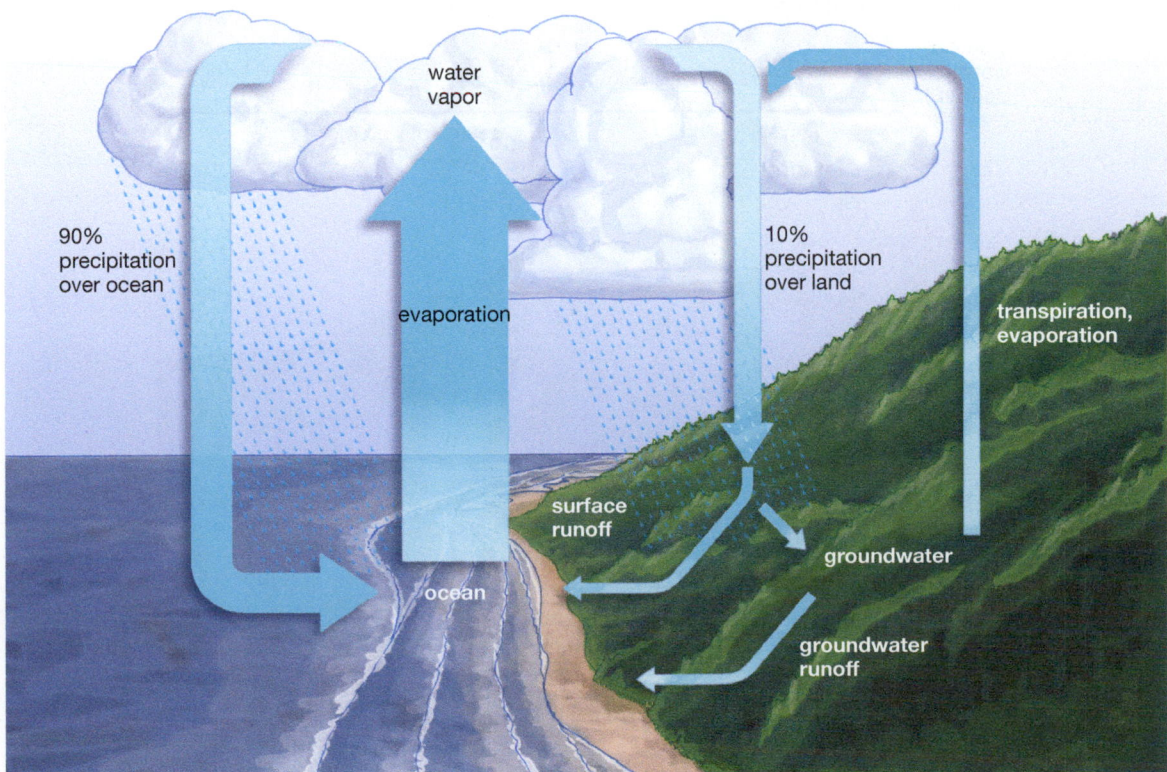

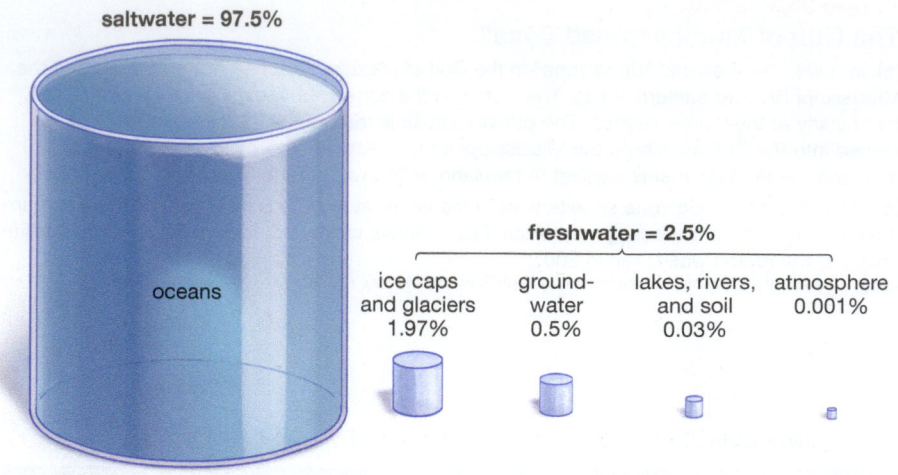

saltwater = 97.5%

oceans

freshwater = 2.5%

| ice caps and glaciers 1.97% | ground-water 0.5% | lakes, rivers, and soil 0.03% | atmosphere 0.001% |

Figure 36.5
Earth's Water

Although there is a tremendous amount of water on our planet, only a small fraction of it is available to us as freshwater.

With more than 95 percent of Earth's water in the oceans, perhaps as little as 2.5 percent of the Earth's water is freshwater at any given time. Of this freshwater, more than 75 percent is locked up in glaciers and other forms of ice (**Figure 36.5**). Thus, any snapshot we would take of Earth would show that as little as 0.5 percent of its water is available as liquid freshwater, but even here we have to qualify things. About 25 percent of this freshwater is **groundwater,** meaning water that moves down through the soil until it reaches porous rock that is saturated with water. In line with this, about

20 percent of the water used in the United States comes from groundwater, and at least a quarter of the world's population depends on groundwater to satisfy basic needs. Groundwater is stored in porous rock called an **aquifer,** an example of which is pictured in **Figure 36.6**. Major aquifers can be enormous, as you can see from the map of the High Plains Aquifer in **Figure 36.7**. Even an aquifer of this size, however, can lose more water through pumping than it gains through replenishment from rain. In 2007, the High Plains Aquifer stored about 2.9 billion acre-feet of water—the equivalent of 2.9 billion acres of water a foot deep. This is a lot of water to be sure, but it is some 270 million acre-feet less than the aquifer contained in 1950.

Global Human Use of Water

Given the sheer number of people on Earth and the importance of water to each of them, it will not surprise you to learn that human beings are laying claim to a great deal of Earth's freshwater; civilization now uses more than half the world's accessible supply. But from what you've seen so far, think of what this means. There is no less *water* than there ever was. Like nitrogen or carbon, all the water we have is either being recycled or is stored. What's changed is that the human population is using an ever-greater proportion of the water that's available. Since 1960, for example, the amount of water countries have impounded behind dams has quadrupled. Even so, the relentless

Figure 36.6
Aquifers Store Groundwater

When water seeps into the ground, it moves freely through layers of sand and porous rock but moves either extremely slowly or not at all through layers of impermeable rock. Water thus becomes trapped in different underground layers, from which humans can withdraw it. Unconfined aquifers receive water directly from a large surface area and therefore tend to be more vulnerable to pollution by chemicals such as fertilizers and pesticides. Confined aquifers are located between layers of impermeable rock. They are replenished more slowly but tend to contain purer water.

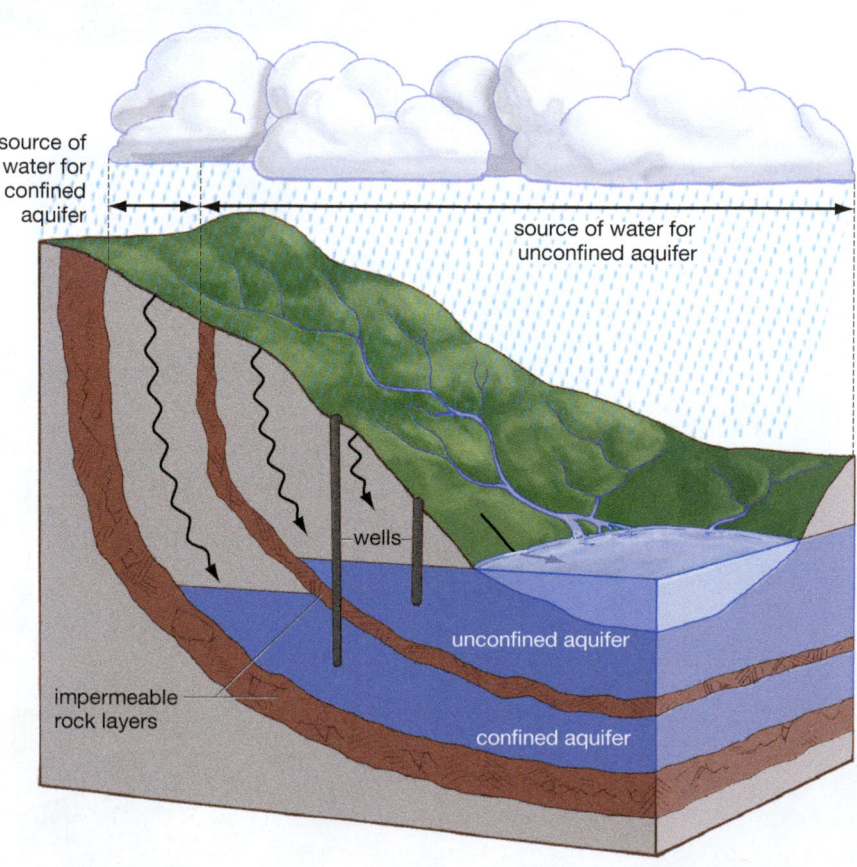

source of water for confined aquifer

source of water for unconfined aquifer

wells

unconfined aquifer

impermeable rock layers

confined aquifer

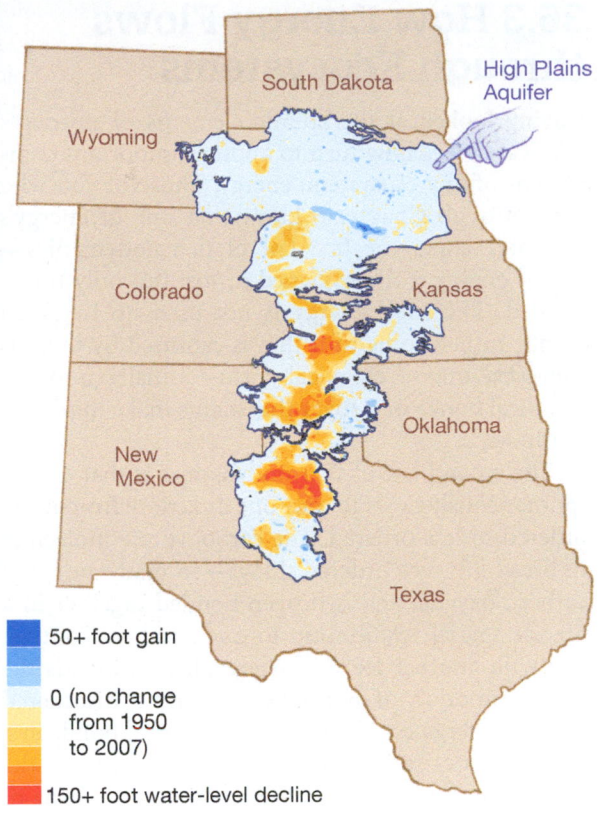

Figure 36.7
Enormous Store of Underground Water

The High Plains Aquifer, perhaps the largest aquifer in the world, underlies 174,000 square miles of land stretching across eight states. Billions of gallons of water are withdrawn from it each day, mostly for agriculture. As the figure shows, large portions of the aquifer have experienced little or no change in depth from 1950 to 2007. In some areas, however, water depth has declined greatly; in one part of Texas, it has been reduced by 234 feet.

(Modified from U.S. Geological Survey Figure 1, "Changes in Water Levels and Storage in the High Plains Aquifer, Predevelopment to 2007," Fact Sheet 2009–3005, February 2009.)

Figure 36.8
Freshwater Is a Limited Resource

Women carry drinking water through a polluted slum in Port-au-Prince, Haiti.

growth of the human population has meant that water remains a scarce commodity for human beings in locations throughout the world. Indeed, sanitary water is so scarce that an estimated 1 billion people worldwide lack access to it (**Figure 36.8**). A large part of the problem is that there is a mismatch between the location of freshwater and the location of human populations. Sixty percent of the world's population lives in Asia, while only 36 percent of Earth's precipitation falls there. Meanwhile, 6 percent of the world's people live in South America, but 26 percent of the world's precipitation falls there. (The Amazon River alone discharges 200,000 cubic meters of freshwater per *second* into the Atlantic Ocean, an amount so large that it constitutes 20 percent of the entire planet's freshwater runoff.)

Another part of the human water problem, water expert Peter Glieck has noted, is that humans make

such poor use of the water they capture. Mexico City's leaky water system *loses* enough water to meet the needs of a city the size of Rome. If the toilet in your house is more than 16 years old, chances are it uses 3.5 gallons of water with each flush, and may use as much as 5.5 gallons. Meanwhile, if it was manufactured since 1995, following imposition of new federal standards for low-flow toilets, it probably uses a mere 1.6 gallons.

When we think of societal use of water, however, household use actually is relatively small. About 70 percent of the water we humans withdraw from rivers, lakes, and groundwater goes to agricultural irrigation. People need food, of course, but the *way* irrigation is carried out in much of the world is extremely wasteful: Water moves from, say, a river to crops by means of an open channel. En route, much of this water either evaporates or is taken up by land that isn't being irrigated. An efficient alternative is "micro-irrigation," in which water gets piped to crops.

Human beings are, of course, capable of diverting water from species that have no say in this division of resources. More than 20 percent of all freshwater fish species are estimated to be threatened or endangered precisely because their water has been dammed or diverted for human use. California's natural populations of coho salmon have been reduced by some 94 percent since the 1940s, in significant part through diversions of the freshwater streams that the salmon swim up in order to spawn. If you look at **Figure 36.9** on the next page, you can get a quick idea of humanity's impact on water and on the two elements we just reviewed, carbon and nitrogen.

GraphIt!: Global Fresh Water Resources

(a) Atmospheric CO₂ concentration

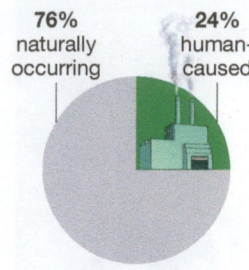

76% naturally occurring   **24%** human-caused

(b) Terrestrial nitrogen fixation

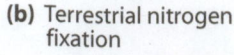
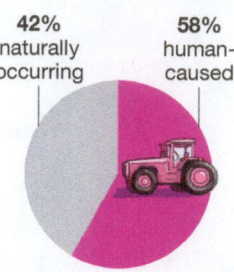

42% naturally occurring   **58%** human-caused

(c) Accessible surface water

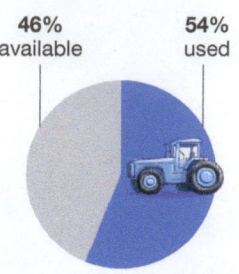

46% available   **54%** used

Figure 36.9
Human Impact on Resources

(a) Nearly a quarter of the current atmospheric carbon dioxide concentration is produced by human activity, primarily through the burning of fossil fuels.

(b) Over half of terrestrial nitrogen fixation comes about because of human activity, including the manufacture of fertilizer for agriculture.

(c) Over half of the Earth's accessible surface water is now diverted for use by humans, mostly for agriculture. Percentages are approximate.

Human Beings Are Not Separate from the Earth on Which They Live

One final thing about nutrient and water cycling: The average person, who has no knowledge of these cycling processes, may regard human beings as a kind of independent line that stretches back in time. Under this view, water and food may pass through us, but they are as separate from us as gasoline is from a car engine.

From your reading of this section, however, it may now be apparent to you that this is not true. Some of the carbon in the bread you ate this morning could have been contained in a tree that decomposed in China 100 years ago. After moving into the atmosphere as CO₂ and staying there for decades, this carbon was taken up by a wheat plant in America whose seeds were ground up, yielding flour that ended in the bread on your shelf. This carbon may be burned by your body to supply the energy it needs, it's true, but it may also become part of your bones or your nerves. Then, when you die, the material you are made of will go *back* to the Earth to be recycled in the ways you've been reading about, perhaps to be part of a tree or a person in a future generation. We are intertwined with the Earth in some basic, physical ways.

SO FAR . . .

1. An ecosystem includes a _____ of organisms and the _____ with which they interact.

2. The carbon that helps make up the living world moves into it by means of organisms that perform _____. Carbon then returns to the soil and atmosphere through means of the _____ of organic material by the organisms called _____ and _____.

3. Almost all the nitrogen that enters the living world by natural means does so through the actions of _____. Nitrogen is then returned to the atmosphere by the actions of _____.

36.3 How Energy Flows through Ecosystems

Having looked at the abiotic elements of nutrients and water, let's now turn to another important component of any ecosystem, energy. Those of you who read through Chapter 6 know that one of energy's inflexible laws—the first law of thermodynamics—is that energy is never gained or lost but only transformed. The sun's energy is not used up by green plants; rather, some of it that is captured by them is *converted* into a chemical form—initially into the chemical bonds of a sugar that plants make in photosynthesis.

The second law of thermodynamics is that energy spontaneously flows in only one direction: from more ordered to less ordered. The carbohydrate molecules in bread are very ordered things—so many atoms of carbon, oxygen, and hydrogen bonded together in a precise spatial relationship to one another. Contrast this with another form of energy, heat, which is the *random* motion of molecules—clearly a disordered form of energy compared to the chemical bonds in a carbohydrate. Indeed, heat is the last stop on the energy line because it is the least-ordered form of energy. And because all energy spontaneously moves toward less order, heat is the ultimate *fate* of all energy. Energy relentlessly ratchets down toward the form of heat because, like a casino that must get a "cut" of each bet laid down, heat gets a cut of each energy transaction. The chemical bonds in gasoline can be broken through combustion and thus be used to power a car engine, but not all the energy released from the combustion drives the engine. Some dissipates as heat, and this happens *every* time energy is used, whether we're talking about a piston firing or a cell dividing.

All of this provides a framework to conceptualize an important part of ecosystems, which is how energy flows through them. Think of the sun, with the energetic rays that leave it as a starting point; now, think of the ultimate destiny of all this energy, which is heat, randomly dispersed in the universe. Looked at one way, life on Earth *intervenes* in this flow. Life is an enormous energy collection and storage enterprise, gathering some of the sun's energy and locking it up for a time in the form of chemical bonds. These bonds can then be broken—think of digesting a muffin—which *releases* this stored energy so that an organism can grow and reproduce. Life does not stop the march of the sun's energy into heat, but it does intervene in it by transforming it into chemical bonds that have order and stability.

Once we see life in these terms, we can begin to think of the *flow* of energy through it. That's what you'll be looking at in the sections that follow.

Producers, Consumers, and Trophic Levels

One way to look at food and energy is in terms of production and consumption. Plants and other photosynthesizers (algae and some bacteria) are an ecosystem's **producers**, while the organisms that eat plants are one kind of **consumer**: an organism that eats other organisms, rather than producing its own food. But as you know, there are consumers *of* consumers—grass is eaten by zebras, but zebras are eaten by lions. Thus do we get to the concept of feeding levels or, as ecologists put it, trophic levels. More formally, a **trophic level** is a position in an ecosystem's food chain or web, with each level defined by a transfer of energy between one kind of organism and another. Producers are one level, and then there are several levels of consumers. Here's a list of trophic levels through four stages:

First trophic level:	**Producers**— (photosynthesizers)
Second trophic level:	**Primary consumers**— plant predators (herbivores)
Third trophic level:	**Secondary consumers**— herbivore predators (carnivores)
Fourth trophic level:	**Tertiary consumers**— organisms that feed on secondary consumers (carnivores)

It is the sun's energy that is locked up at the first trophic level, in leaves and stems, and it is this energy that is being passed along through all the other trophic levels. Note that the levels are not just categories that could be put in any order. We must have primary consumers to have secondary consumers; sharks cannot exist on algae (**Figure 36.10**). You may remember that an animal that eats only plants or algae is a **herbivore;** then there are **carnivores,** which eat only meat, and **omnivores,** which eat both plants and meat. Of course, many organisms cannot be assigned to just a single trophic level. Most human beings, for example, are primary, secondary, and tertiary consumers.

You have thus far thought in terms of food chains, which is to say trophic *lines* in which a single organism follows another. In reality, of course, nature is much more complex than this; what really exists are feeding patterns called *food webs,* one example of which you can see in **Figure 36.11** on the next page.

A Special Class of Consumers: Detritivores

To this scheme of trophic levels, ecologists add a special class of consumers, the **detritivores.** These are con-

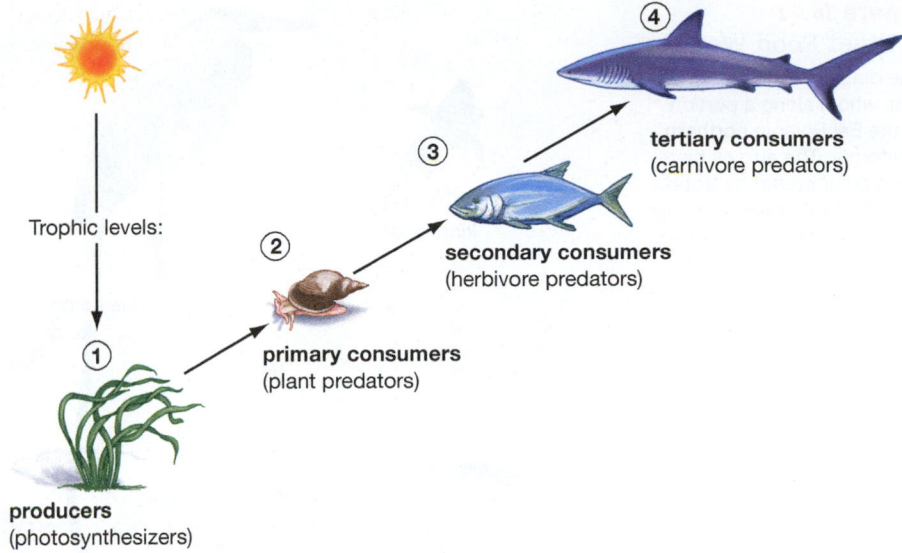

Trophic levels:

① producers (photosynthesizers)

② primary consumers (plant predators)

③ secondary consumers (herbivore predators)

④ tertiary consumers (carnivore predators)

Figure 36.10
Trophic Levels
Plants, algae, and some bacteria are called producers because they use the sun's energy to produce their own food through photosynthesis. Organisms at all other trophic levels ultimately derive their energy from these photosynthesizers.

sumers that feed on detritus, which in normal usage simply means a collection of debris. In ecology, however, detritus is the remains of dead organisms or cast-off material from living organisms (a fallen branch from a living tree, for example). A worm or a dung beetle feeding on such material would fit this description of a detritivore.

Of particular interest, however, is a special kind of detritivore: a **decomposer,** which is an organism that, in feeding on dead or cast-off organic material, breaks it down into its inorganic components, which then can be recycled through the ecosystem. *Inorganic* here can be thought of as building blocks. (More technically, it means a compound that does not contain carbon bound covalently to hydrogen.) The most important decomposers are fungi and bacteria. A number of detritivores usually are responsible for breaking up organic material, first into pieces and then into the inorganic components that are recycled into the Earth and the atmosphere (**Figure 36.12** on the next page).

Stepping back from this, it's worth noting what complete use the living world makes of its own materials. Think of it this way: What does nature produce that it does not eventually decompose and recycle? The answer is, nothing at all. Every twig and beetle and bone that comes into existence eventually will be broken down into its component materials, which then will be used again. This is all to the good, actually, because if there were any natural garbage, spaceship Earth would sooner or later be filled up with this useless material. This very thought ought to give us some pause about the compounds human beings currently are putting into the environment

Figure 36.11
A River Food Web
The diagram indicates who eats whom along a portion of the Eel River in northern California. The arrows have been color coded by trophic level. Note that there are up to five trophic levels in the web.

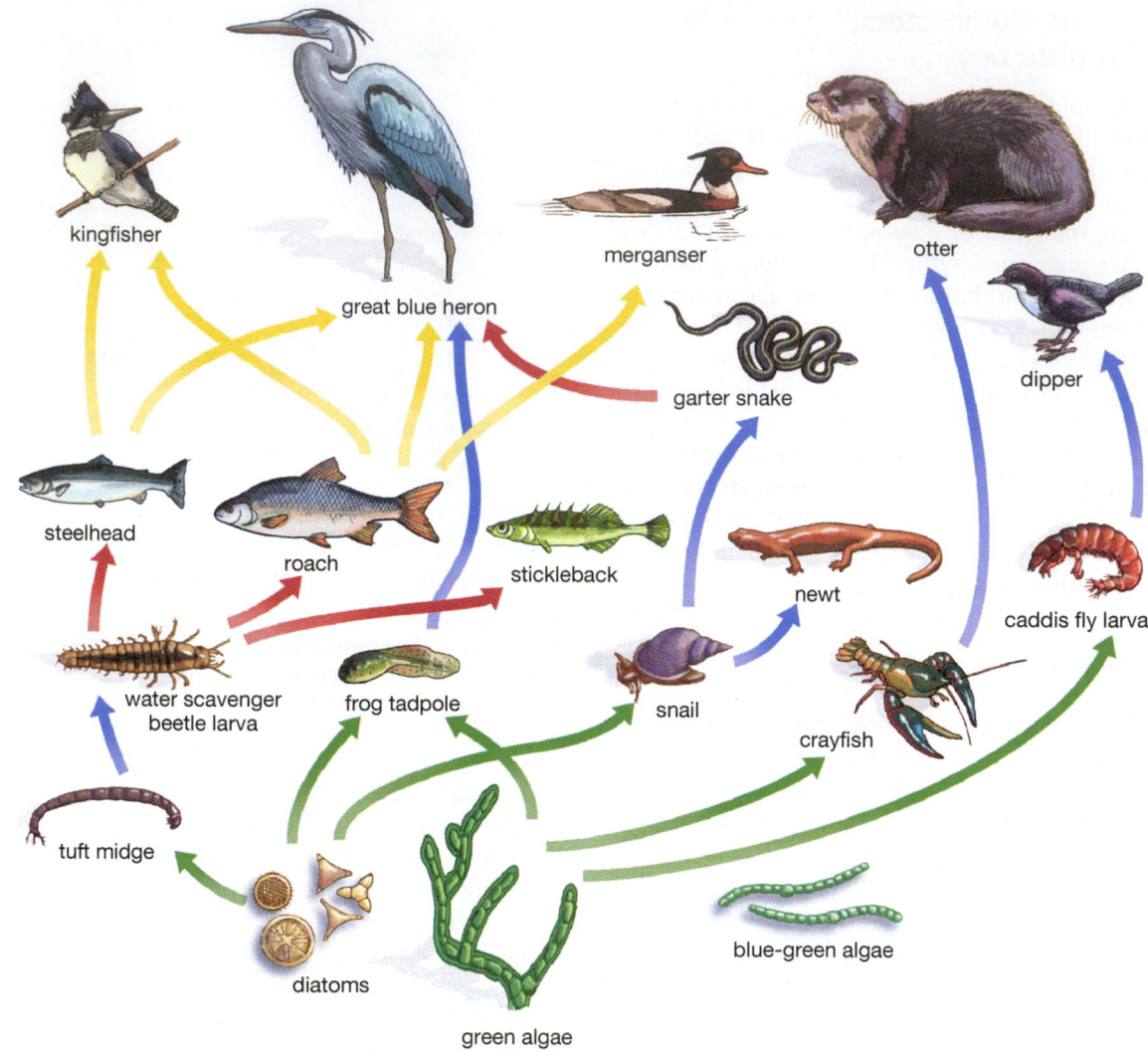

kingfisher

merganser

otter

great blue heron

garter snake

dipper

steelhead

roach

stickleback

newt

caddis fly larva

water scavenger beetle larva

frog tadpole

snail

crayfish

tuft midge

diatoms

blue-green algae

green algae

Figure 36.12
How Organic Material Is Broken Down
Different kinds of detritivores consume the organic material in a branch that has fallen from a tree. Most of these animals are consuming pieces of the log, but the detritivore called a decomposer (a fungus in the figure) is breaking organic materials into their inorganic components, thus returning nutrients to the soil.

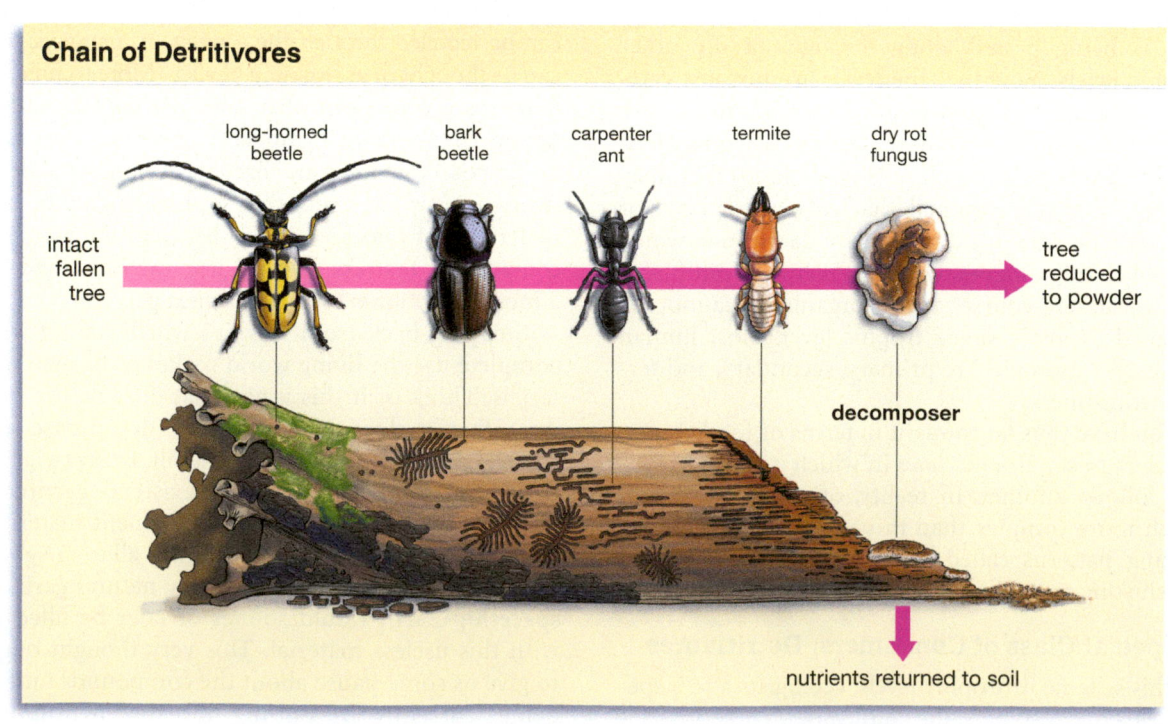

Chain of Detritivores

long-horned beetle

bark beetle

carpenter ant

termite

dry rot fungus

intact fallen tree

tree reduced to powder

decomposer

nutrients returned to soil

that are not **biodegradable,** meaning capable of being decomposed by living organisms. Of particular concern here are plastics, many of which will remain intact for hundreds of years following their production.

Accounting for Energy Flow through the Trophic Levels

All natural substances may be recycled in ecosystems, but this is not true of energy, which is just passing through, on its way to becoming heat. Two interesting questions arise from this: How *much* of the sun's energy does the living world collect in the first place, and then how much of the collected energy makes it through to each successive trophic level?

This may seem like a routine set of questions, but the first ecologists who thought in these terms brought about a revolution in ecology. Working in the 1940s at Minnesota's Cedar Bog Lake, ecologist Raymond Lindeman was the first to conceptualize ecosystems as units in which energy is first captured by given organisms and then transferred to others. This conceptualization is today known as the **energy-flow model** of ecosystems. Part of the impetus for Lindeman's research was to answer a very down-to-earth question, succinctly expressed by ecologist Paul Colinvaux: Why are big, fierce animals rare? (If you doubt that they are, think of this. It's utterly routine to see plants, insects, or small birds, but it's often the experience of a lifetime to see a bear or a shark.) The energy-flow model explained the small numbers of big, fierce animals, as you'll see, but it then went on to shed light on all kinds of ecological relationships. Why? Because energy is a kind of currency in the natural world; all organisms use it, and it is transferred among them. Moreover, it can be measured, usually in units called **kilocalories** (the amount of heat it takes to raise 1 kilogram of water 1 degree Celsius). If we know that a field mouse must use 68 percent of the kilocalories it consumes just to stay alive—to move, to keep a constant body temperature—but that a weasel must use 93 percent of its kilocalories for this purpose, that gives us a meaningful basis for comparison between these two animals. Beyond this, weasels *eat* field mice, and with the energy-flow model we have some basis for tracking this important commodity as it moves through an ecosystem.

The Capture of Energy by Plants

So, what has energy-flow research told us? To take things from the beginning, how much of the sun's energy do plants capture? As you can see in **Figure 36.13,** very little; only about 2 percent of the sun's energy that falls around plants is taken in for photosynthesis. This raises an important point, which is noting the

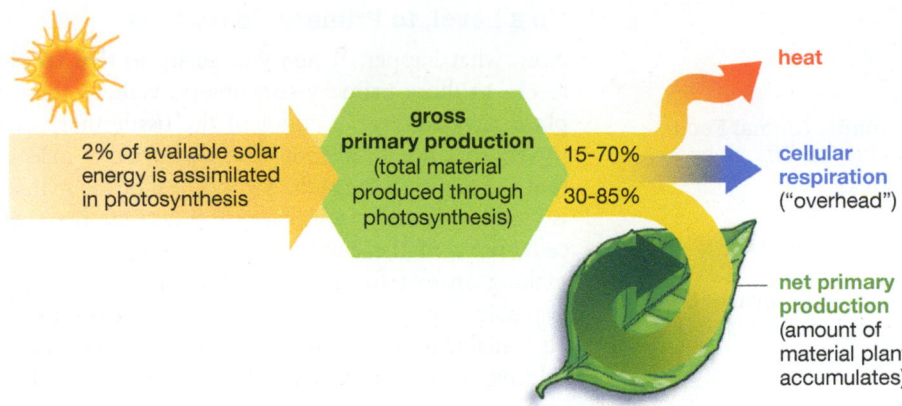

Figure 36.13
Absorption and Use of Solar Energy by Plants
Only 2 percent of the solar energy reaching the Earth's surface is taken in by plants for photosynthesis. Of this energy, only a fraction ends up being contained in plant material, such as leaves. It is this material (the plant's net primary production) that is available to the next trophic level in a food chain. The remaining energy is either dissipated as heat or used by photosynthesizers to power their own physical processes.

difference between the energy that *comes to* an organism and what the organism *does with* that energy.

The amount of material that a plant produces as a result of photosynthesis is known as its **gross primary production.** Those of you who have been through Chapter 8 know that the initial product of photosynthesis is a sugar that serves as a starting "building block" for all the material in the plant. Every time a molecule of this sugar is produced in photosynthesis, the plant has added to its gross primary production. But plants don't get to retain all the material they produce in this way. Why? Because plants incur energy *costs* in just staying alive. It takes energy to perform photosynthesis, to move sap from one end of the plant to the other, and so forth. Such costs are subsumed under the heading of "cellular respiration," but you could also think of them as the plant's overhead; they are the price of doing business. Apart from this, we know that, in every energy transaction, some of the energy is lost as heat. After subtracting these various costs, the plant is left with its **net primary production:** the amount of material a plant accumulates as a result of photosynthesis. This is the material that serves to build the plant up—its stems, leaves, and roots.

One question that arises from this framework is: What fraction of the energy a plant receives from the sun ends up as net primary production? How efficient are plants, in other words, at turning solar energy into stems, leaves, and roots? Of the solar energy that plants receive, somewhere between 30 and 85 percent is transferred into net primary production. In other words, at the low end of this range, only 30 percent of the solar energy a plant assimilates is transferred into material that is even potentially available to the next trophic level up.

GraphIt!: Animal Food
Production Efficiency and
Food Policy

GraphIt!: Municipal Solid
Waste Trends in the U.S.

Up a Level, to Primary Consumers

Now, what happens when you go up to that trophic level—to those primary consumers? Well, in the first place, they can't *get to* a lot of the tissue that's been produced, since it's in roots that are buried or in leaves that are too high to reach. A good deal of the material is inedible in any event, except by detritivores. Critically, most of the energy these scavengers take in is making an exit from the trophic chain; it won't be available to primary consumers, which means it won't be available to secondary or tertiary consumers either. Adding all this up, you can begin to see that, of the energy locked up by plants, very little is assimilated by organisms at the next step up. And this pattern of energy reduction is then repeated in the successive trophic levels. Indeed, the reductions tend to get more severe because animals have higher energy costs than plants have. (The warm-blooded animals we're most familiar with spend a great deal of energy keeping a constant body temperature—producing heat, which obviously is energy lost to the food chain.)

All of this leads to what you see in **Figure 36.14**: a reduction in available energy at each trophic level. The amount that is passed on between levels varies by community, but there is a rule of thumb in ecology that for each jump up in trophic level, the amount of available energy drops by 90 percent. Take all the energy converted into tissue by, say, second-level herbivores (rabbits, mice), and of this, only 10 percent ends up being converted into tissue by third-level carnivores (foxes, weasels). There is a great variability in conversion rates, however, and 10 percent probably is more like a maximum passed on.

The Effects of Energy Loss through Trophic Levels

So, what are the consequences of these reductions in energy at the various trophic levels? To a large extent, they are responsible for the makeup of the living world. Walk through a forest and what do you see? Innumerable small plants and trees, a fair number of small animals—birds, insects, squirrels—and, on very rare occasions, a predator such as a bobcat or a fox. The energy reduction that comes with each tropic level largely dictates these proportions. If you had a field that could feed 100 field mice, how many weasels could the field support if those weasels were existing solely on the mice? Remembering that, at most, only 10 percent of a trophic level's energy is likely to be passed along to the next level, you might guess that 10 weasels could make a living in the field, but even this number would be too high. For 10 predators to make a living from 100 mice, these predators would have to be no *bigger* than the mice, which is certainly not true of weasels. Extending this further, not even a single hawk could survive on the available weasels in the field. Why are big, fierce animals rare? Because of the loss of available energy at each step in the trophic chain. Because weasels are not about to start eating the same things as mice, weasels are locked into a trophic level with a more limited energy supply. When there's only so much energy, there can be only so many weasels.

Even so, do trophic energy reductions *entirely* explain the proportions of different life-forms in a given area? Probably not. More than 50 years ago, a trio of ecologists led by Nelson Hairston proposed that the overwhelming domination of plants within ecosystems comes about not solely because of trophic energy reductions but also because of a form of top-down predator control: the degree to which second-level herbivores are consumed by third-level carnivores. In 2006, Duke University's John Terborgh and his colleagues provided strong evidence that this kind of control does in fact help determine the makeup of ecosystems. The researchers looked at 4,300 square kilometers in Venezuela that had been flooded by government action in 1986, thus creating not only a human-made lake but a series of islands within it, all of which were large enough to support a collection of herbivores (leaf-cutter ants, rodents) but only some of which were large enough to support both herbivores *and* carnivores that preyed upon them (armadillos, large snakes). The result? Islands that lacked carnivores were stripped of vegetation by their now superabundant herbivores, while islands that retained carnivores were able to retain their pre-flood greenery. The take-home message from this is that, while the makeup of ecosystems is powerfully conditioned by the bottom-to-top reductions of available energy, control is also exerted in the other direction—from predators down to producers. (The issue of energy loss by trophic level has relevance to a human issue, which is eating meat that comes from grain-fed cattle, as you can see in "A Cut for the Middleman: Livestock and Food," on page 706.)

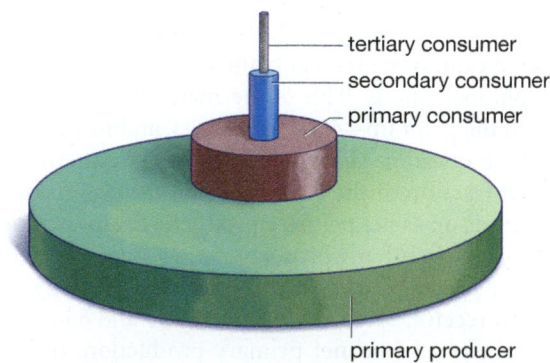

tertiary consumer
secondary consumer
primary consumer

primary producer

Figure 36.14
Energy Pyramid
Only a small fraction (10 percent or less) of energy at each trophic level is available to the next higher trophic level.

Primary Productivity Varies across the Earth by Region

Primary productivity concerns *biomass,* or material produced by first-level photosynthesizers, as measured by weight. It will come as no surprise that some areas of the Earth are more productive than others in this sense. If you look at **Figure 36.15**, you can get an idea of the differences around the globe. The reasons for these differences are explored in this chapter's section on the large ecosystems called biomes. For now, suffice it to say that much of what you see in Figure 36.15 is intuitive. Productivity is very low in such desert areas as North Africa and very high in such tropical rainforest areas as equatorial Africa. Yet why should one part of the Earth be a rainforest and another part a desert? The answer has to do with Earth's large-scale physical environment.

SO FAR . . .

1. All the organisms at the first trophic level perform _____ and thus are referred to as _____. Any organism that feeds solely on first-trophic-level organisms would have to be a (choose one) carnivore/herbivore/omnivore.

2. All material produced by living things is capable of being broken down into its inorganic components by the class of living things called _____. This material is thus said to be _____.

3. There are fewer predator than prey animals due to drastic reductions of _____ at each _____.

36.4 Earth's Physical Environment

Earth itself exists within an environment, called the **atmosphere,** which is the layer of gases surrounding the planet. One of the main things to realize about Earth's atmosphere is how it differs from outer space. There's something *to* Earth's atmosphere (a mix of gases), while outer space really is *space*; there's almost nothing in it.

The atmosphere is divided into several layers, but here you need be concerned with only two of them, which you can see illustrated in **Figure 36.16** on the next page. The lowest layer of the atmosphere, called the **troposphere,** starts at sea level and extends upward about 12 kilometers (or 7.4 miles), and it contains the bulk of the gases in the atmosphere. Nitrogen and oxygen make up 99 percent of the troposphere, but as you've seen, carbon dioxide exists there in small amounts, as do some other gases, such as argon and methane. After a transitional zone, the next layer above the troposphere is the **stratosphere.** Of greatest concern in it is a gas that reaches its greatest density at about 20 to 35 kilometers (13 to 21 miles) above sea level.

The oxygen we breathe comes in the form of two oxygen molecules bonded together (O_2). When ultraviolet sunlight strikes O_2 in the stratosphere, however, it can convert it into a form of three oxygen molecules bonded together (O_3). This is the gas known as **ozone.** Created initially as a *product* of life—since all atmospheric oxygen came from photosynthesis—this ozone layer ultimately came to *protect* life by blocking some 99 percent of the ultraviolet (UV) radiation the sun showers on the Earth. It was this blockage that allowed large life-forms to come on to land some 460 million years ago, and it is this blockage that protects life now. The UV radiation that pours from the sun can wreak havoc on living tissue, bringing about cancer and immune-system problems in people, and damaging vegetation as well.

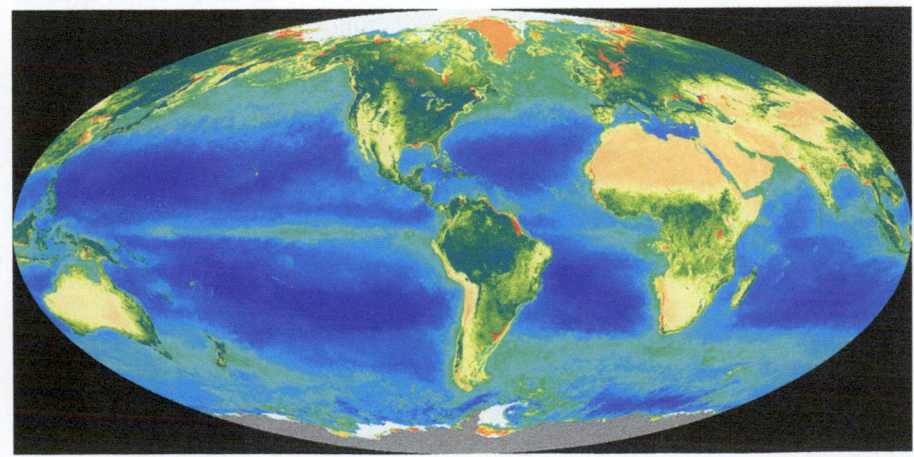

Figure 36.15
Primary Productivity by Geographic Region

Primary productivity around the globe. This composite satellite image shows how the concentration of plant life varies on both land and sea. On land, the areas of lowest productivity are tan colored, as with the enormous swath of land that begins with the eastern Sahara Desert in North Africa. More productive land areas are yellow, more productive yet are light green, and the most productive of all are dark green. On the oceans, the gradient runs from dark blue (the least productive), through lighter blue, green, yellow, and orange-red (the most productive). The ocean measurements are of concentrations of phytoplankton—the tiny, photosynthesizing organisms that drift on the water and form the base of the marine food web. Note the high productivity of the oceans in both the far northern and southern latitudes. Antarctica itself is gray because the imaging technique did not work well for it.

Figure 36.16
Earth's Atmosphere
The two lowest layers of the atmosphere are shown, with Mount Everest (the tallest mountain) included for vertical scale. The troposphere contains most of the atmosphere's gases—mostly nitrogen and oxygen, with some carbon dioxide, methane, and other gases. The stratosphere contains the ozone layer, which blocks 99 percent of the sun's harmful ultraviolet radiation.

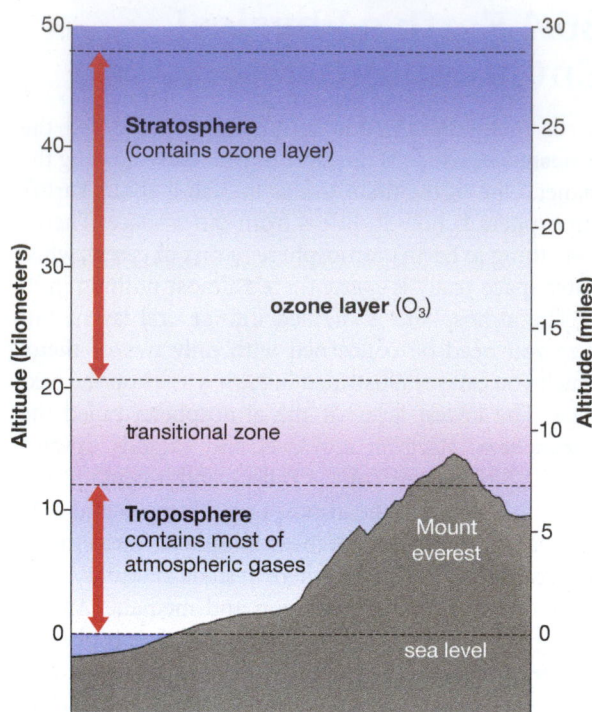

The Worrisome Issue of Ozone Depletion

Given the importance of the ozone layer, it is sobering to contemplate how fragile it is. Some human-made chemical compounds have the effect of destroying stratospheric ozone, and such destruction went on unchecked for years until atmospheric chemists Sherwood Rowland and Mario Molina revealed, in 1974, that compounds called chlorofluorocarbons posed a direct threat to the ozone layer. Chlorofluorocarbons, or **CFCs**—found at one time in spray cans, refrigerators, and plastic foams—are undoubtedly the most famous of the ozone-depleting compounds, but they are by no means the only chemicals that have this effect. Other harmful compounds include methyl bromide, which is used as a pesticide and herbicide, notably on tomato and strawberry crops; and bromine, which is found in a group of fire-extinguishing chemicals. The damage that these compounds have done to stratospheric ozone has brought about the spectacle of annual reports on how big the "ozone hole" is over the South Pole, along with a concern about a general thinning of the ozone layer.

Despite the ominous nature of this news, the long-term outlook on ozone depletion actually is good. In an agreement signed in Montreal in 1987, many of the world's nations pledged to phase out production of ozone-depleting chemicals, and this agreement is working. Production of CFCs was banned in the United States even before the Montreal agreement was signed and has dropped dramatically throughout the world in recent years. This reduction has brought about a reduction in the levels of ozone-depleting compounds

in Earth's atmosphere. As a consequence, the thickness of Earth's ozone layer stabilized beginning in the mid-1990s, according to a United Nations assessment published in 2006. If current trends continue, the Earth's ozone layer outside the polar regions should return to its pre-1980 levels about the middle of this century. The annual Antarctic ozone hole will be a feature of life for decades, but even the Antarctic ozone layer should return to its pre-1980 thickness by 2060 to 2075, according to the U.N.

36.5 Global Warming

Scientists are not nearly so hopeful about a second environmental issue that concerns Earth's atmosphere—the issue of global warming. Identified as a concern as far back as the 1950s, global warming has emerged as perhaps the most pressing environmental problem of our time, a status that's not surprising given that its reach really is global and that its effects stand to be profound: Residents of low-lying areas from Florida to Bangladesh must worry about its ability to raise sea levels; residents of coastal areas from Texas to the Philippines must worry about its ability to increase the intensity of tropical storms; and farmers from Nebraska to Argentina must worry about its ability to change rainfall patterns.

The seriousness of these issues has prompted an intense worldwide research effort that has managed to answer some questions about global warming while leaving other questions open. Uncertainty is a given in almost any analysis of global warming, however, because the most crucial questions about it concern not the effects it's having now, but rather the effects it stands to have in the future.

Warming and Its Causes

Our most authoritative judgments on global warming come from a United Nations group called the Intergovernmental Panel on Climate Change, or IPCC, which has issued four reports since 1990, each of them the work of hundreds of scientists who have reviewed the findings of thousands of their fellow researchers. In its latest report, released in 2007, the panel was able to provide some detail on one of the most basic questions about global warming: How much of it is taking place? In the 100 years that elapsed between 1906 and 2005, the panel said, the Earth's surface temperature increased by about 0.74 degrees Celsius, or about 1.3 degrees Fahrenheit. Moreover, it noted, the warming trend over the 50 years before 2005 was nearly twice that for the previous 100 years, and 11 of the 12 years between 1995 and 2006 were the warmest that have been recorded since such measurements started being taken in the middle of the nineteenth century. A more recent

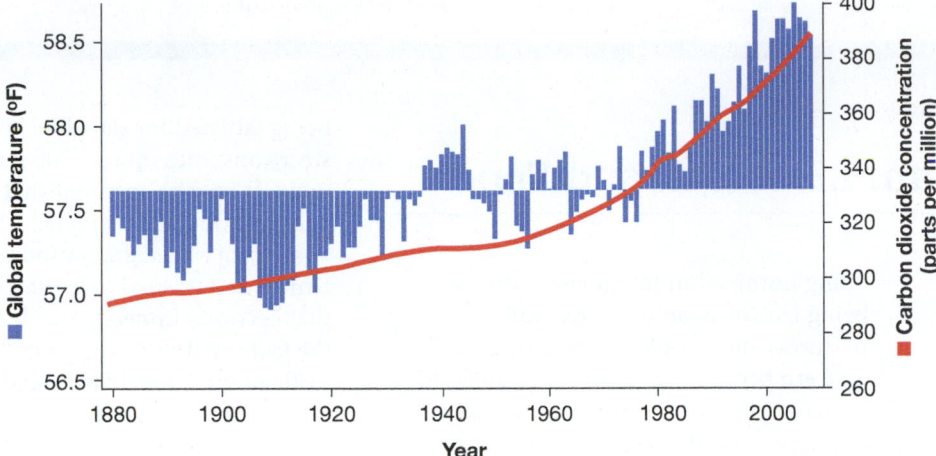

Figure 36.17

With a Rise in CO₂ Levels, a Rise in Temperature

How did global temperatures for each of the years between 1880 and 2000 vary against the average temperature for the entire twentieth century? As the figure shows, yearly temperatures before the late 1930s were uniformly below the century's average, while yearly temperatures after 1980 were uniformly above it. The red line tracks the period's rise in atmospheric CO_2 concentrations, which correlates strongly with the global temperature increases.

(Adapted from "Global Temperature and Carbon Dioxide," *Global Climate Change Impacts in the United States,* Thomas R. Karl, Jerry M. Melillo, and Thomas C. Peterson (eds.), Cambridge University Press 2009, p. 17.)

GraphIt!: Atmospheric CO_2 and Temperature Changes

GraphIt!: Forestation Change

analysis performed by NASA found that the 10 years between 2000 and 2009 were the warmest in the historical record.

So, what's causing this increase? We humans are transferring large amounts of heat-trapping gases into Earth's atmosphere through several means. First and foremost, we are burning fossil fuels, such as oil and coal, which results in a massive transfer of carbon dioxide (CO_2) into the atmosphere. Next, the cattle we raise and the rice we grow put large amounts of the heat-trapping gas methane into the atmosphere. Third, the deforestation we are carrying out—particularly in the tropics—either puts CO_2 directly into the atmosphere, when trees are burned, or results in less CO_2 being taken *out* of the atmosphere, since trees that are killed no longer take up CO_2 to perform photosynthesis.

So, how significantly have these activities altered Earth's atmosphere? To look just at CO_2, it turns out that a greater concentration of it exists in the atmosphere now than at any time in the last 650,000 years. In 2008, Earth's atmospheric CO_2 concentration was 385 *parts per million*—for every million molecules of dry air, there were 385 molecules of CO_2. In contrast, the highest CO_2 concentration ever recorded over the previous 650,000 years was 300 parts per million. (Scientists can look this far back in time by analyzing the CO_2 in ancient ice cores.) If you look at **Figure 36.17**, you can see how CO_2 concentrations have changed from 1880 forward, and how this change correlates with the rise in global temperatures. **Figure 36.18** then shows you how unlikely it is that anything other than human activity could be responsible for the rise in Earth's temperature.

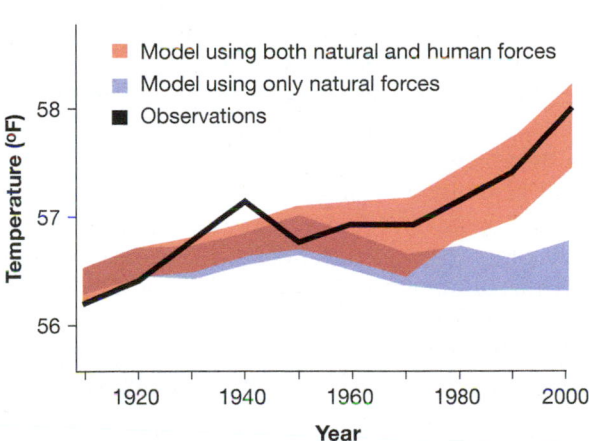

Figure 36.18

A Warming Caused by Human Activity

Scientists who worked on the 2007 IPCC report produced climate models—the blue and red lines in the figure—showing how Earth's temperature would have been expected to change over the course of the twentieth century under two different scenarios: with human inputs (such as CO_2 emissions) and without them. They then overlaid Earth's actual temperature change on these models and found there is a very good fit between what has actually occurred on Earth and what was predicted for it when the human inputs were factored in. Findings such as these have convinced scientists that human activity is responsible for most of the warming Earth has experienced over the last half-century. Were it not for human activity, in fact, Earth probably would have experienced a slight cooling over this period, as the blue line in the figure shows.

(Adapted from "Global Temperature and Carbon Dioxide," *Global Climate Change Impacts in the United States,* Thomas R. Karl, Jerry M. Melillo, and Thomas C. Peterson (eds.), Cambridge University Press 2009, p. 20.)

ESSAY

A Cut for the Middleman: Livestock and Food

It may be difficult to think of cows in a field as occupants of one of the trophic levels you've been looking at, but that's just what they are, with some provocative consequences for human beings and food. People can't survive on grass the way cows and sheep can, and it follows that these

It takes 9 pounds of feed to produce 1 pound of beef ready for human consumption.

animals are turning food that is useless to human beings into products that are use*ful* to human beings—meat, wool, and milk. In this sense, then, these animals are a boon to the efficiency of worldwide agriculture in that they allow us to get more human food and other products from the same amount of land.

There is another side to this story, however, having to do with the *loss* of food that occurs as we move up through the trophic levels. Recall that, as a rough rule of thumb, 90 percent of the energy that is locked up in one trophic level is lost to the organisms in the next trophic level. If cattle and other domesticated animals were

doing nothing but foraging—if they were being fed solely on the grass available in pastures—this would not be a matter of concern because the grass is not useful to humans as a food source anyway. But in developed countries such as the United States, much of the diet of domesticated

animals comes from *grain* that is grown on farms. A large portion of this animal feed could go directly to human beings, but it instead passes through the trophic level of animals, with a predictable result: It takes 9 pounds of feed to produce 1 pound of beef ready for human consumption. What does this mean in practice? A harvest of 2,000 pounds of grain will feed five people for a year. Take roughly this same harvest, feed it to livestock, and then feed the livestock products to people, and you will feed only *one* person for a year.

This calculation takes on added importance when considering the scale of the grain-fed operation. In the feedlot, while

being fattened for slaughter, an average steer consumes up to 2,700 pounds of grain. It may be apparent that such consumption adds up quickly; some 70 percent of the grain produced in the United States is fed to livestock, a figure that becomes understandable in light of the fact that there are about 70 million cattle in the United States and Canada alone.

Fattening at the Feedlot
About 40,000 head of cattle are fed for meat production at this commercial feedlot in Imperial Valley, California.

Figure 36.19
The Greenhouse Effect

The high-energy rays of the sun can easily penetrate the layer of gases in the troposphere. However, the lower-energy radiation (heat) that reflects from Earth's surface cannot penetrate the layer of gases as easily. Carbon dioxide and methane thus take on the role of glass panes in a greenhouse: They let solar energy in, but they retain a good deal of heat.

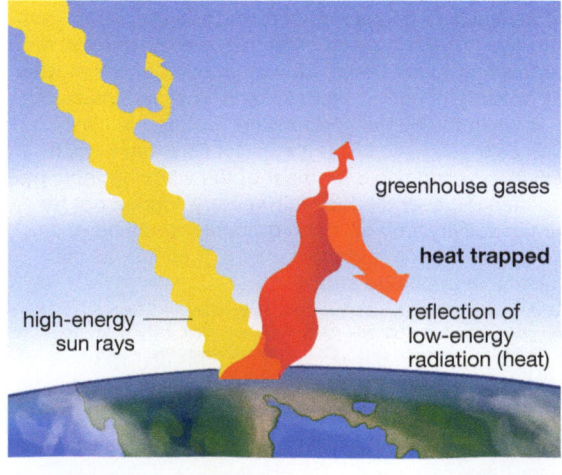

greenhouse gases

heat trapped

high-energy sun rays

reflection of low-energy radiation (heat)

What Is the Greenhouse Effect?

But why should the atmospheric concentration of gases such as CO_2 have anything to do with a warmer planet? Sunlight that is not filtered out by ozone comes to Earth in the form of very energetic, short waves that can easily pass through the atmosphere (**Figure 36.19**). Once this energy reaches the land and ocean, most of it is quickly transformed into heat that does all the things you've just read about: warms the planet, drives the water cycle, and so forth. This heat ultimately is radiated back toward space. But heat is not short-wave radiation; it is *long*-wave radiation, and it can be trapped by certain compounds, among them carbon dioxide and methane.

Global Warming's Consequences

All of this leads to the most critical question about global warming: What will its consequences be? The answer to this question hinges in large part on what the world's industrialized nations do in coming years with respect to their emissions of greenhouse gases. If they put higher levels of these gases into the atmosphere, then the world can expect higher global temperatures, higher sea levels, and so forth. To allow policy makers and the public to understand the linkage between given levels of emissions and the effects they would have, the 2007 IPCC report provided a set of "scenarios" that envisioned humanity either reducing its future emissions by a substantial amount (lower emissions scenarios) or reducing them very little (higher emissions scenarios). In the United States, a federal panel called the U.S. Global Change Research Program then used these alternate scenarios to try to predict what global warming stands to mean in the United States over the course of this century. If you look at **Figure 36.20**, you can see three of the panel's predictions, culled from a list of scores it produced. Looking at the bigger picture,

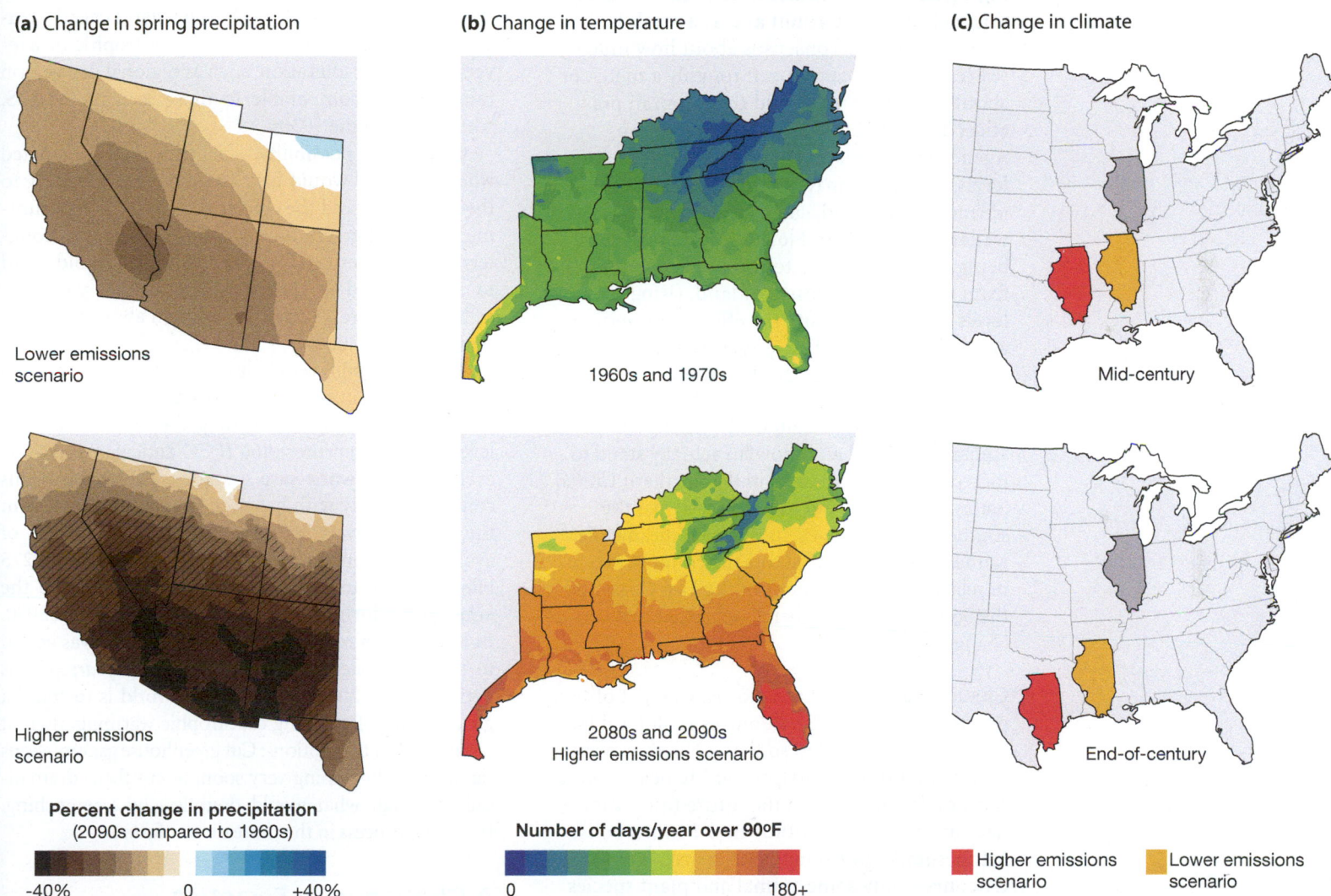

(a) Change in spring precipitation

Lower emissions scenario

Higher emissions scenario

Percent change in precipitation
(2090s compared to 1960s)

-40%   0   +40%

(b) Change in temperature

1960s and 1970s

2080s and 2090s
Higher emissions scenario

Number of days/year over 90°F

0   180+

(c) Change in climate

Mid-century

End-of-century

Higher emissions scenario   Lower emissions scenario

Figure 36.20
Drier West, Warming East, Warming Midwest
Three likely effects of global warming in the United States during the twenty-first century, as predicted by the U.S. Global Change Research Program.

(a) Under either lower or higher emissions scenarios, the Southwest of the late twenty-first century will be a drier place in spring—much drier under the higher emissions scenario.

(b) Under a higher emissions scenario, the United States from the Atlantic seashore to East Texas is likely to experience a great increase in the number of days each year in which the temperature climbs above 90. By the end of the century, North Florida will have more than 165 such days annually, whereas in the 1960s and 1970s, it had about 60.

(c) Under both lower emissions scenarios (orange icons) and higher ones (red icons), the Illinois climate of the mid- and late twenty-first century will resemble the Alabama and Texas climates of today.
(Adapted from "Global Temperature and Carbon Dioxide," *Global Climate Change Impacts in the United States*, Thomas R. Karl, Jerry M. Melillo, and Thomas C. Peterson (eds.), Cambridge University Press 2009, pp. 112, 117, 130.)

GraphIt!: Prospects for Renewable Energy

global warming is certain to produce some large-scale alterations in Earth's physical environment. Among these are:

- **Changes in sea levels.** In 2007, the IPCC predicted that, by century's end, global sea levels could be expected to rise by somewhere between 7 inches and 1.9 feet, depending on levels of greenhouse gas emissions, but it noted that it could not factor one important variable into its calculations: how much the massive ice sheets of Greenland and Antarctica stand to melt over the course of this century because of global warming. Since 2007, scientists have been able to take this melting into account and as a result have come to a revised consensus about how much sea-level rise we can expect: roughly a meter, or about 3.2 feet. What would this mean in practice? The U.S. Environmental Protection Agency has estimated that a 2-foot rise in sea levels would eliminate some 10,000 square miles of land in the United States, including undeveloped wetlands, with south Florida and Louisiana being most vulnerable to these effects. Apart from such outright losses of land, rising sea levels will also intensify the effects of flooding and will increase the salinity of freshwater sources, such as groundwater tables.

- **Changes in precipitation.** Under higher emissions scenarios, rainfall and snowfall actually stand to increase in winter months in the northern United States and Canada but to decrease across the southern United States. In summer, meanwhile, rainfall will decline all across the country, except in relatively small portions of the East Coast and the Southwest, where there is expected to be no change.

- **Changes in habitat.** Each 1.8° Fahrenheit of temperature change on Earth moves ecological habitat by about 100 miles, so that a North American species that today is adapted to life near, say, Des Moines, Iowa, would in the future find its ideal habitat shifted to a latitude near Minneapolis, Minnesota, if global temperatures rise by 4.0°F. Of course, only some animal and plant species will be capable of making such "poleward" migration, or of changing their behavior in accordance with warming—by, for example, flowering earlier in the year—so that the mix of species in different habitats stands to be altered in coming decades. Temperatures are not increasing at uniform rates throughout the world, however; Earth's arctic regions are warming at two to three times the global rate, and many arctic species have nowhere colder to go than their current habitat.

A Narrow Window to Take Action

Predictions such as these often are looked at by the general public as outcomes that are merely hypothetical or that can be avoided by future action. But scientific evidence is increasingly indicating that society has a very narrow time frame in which to take significant action on global warming. To get a sense of what's in play, consider that, in the years since the last IPCC report was issued, scientists have been focused intently on ways to hold warming to no more than 3.6°F above the temperature levels that existed in the pre-industrial era. This is so because there is growing scientific consensus that any increases beyond 3.6°F stand a good chance of producing what is called "dangerous climate change," meaning catastrophic or irreversible climate alteration, such as regional dry-season rainfall levels comparable to those seen in the U.S. "dust bowl" in the 1930s.

Given this possibility, scientists have calculated what the world would have to do to hold warming to the 3.6°F level, and the challenge appears to be daunting. When all greenhouse gases are considered, long-term atmospheric CO_2 concentrations would need to be stabilized at the equivalent of 400 parts per million—a mere 15 parts per million above the levels that existed in 2008.

This would still seem to leave societies some room to maneuver, but in recent years worldwide carbon emissions have not merely been rising; they've been rising at levels higher than those the IPCC anticipated in 2007 even under its worst-case scenarios. The problem this confronts society with has been framed by scientists in the following way: To stand a 75 percent chance of avoiding warming in excess of 3.6°F, no more than 275 billion metric tons of carbon could be added to the atmosphere during the 50 years between 2000 and 2050. But nearly a third of this amount of carbon was added to the atmosphere during the first *nine years* of this period—from 2000 to 2009. If the world is to stand a good chance of avoiding catastrophic warming, then, it essentially has two options: Cut greenhouse gas emissions significantly beginning very soon, or cut them dramatically, through what would doubtless be a wrenching, traumatic process in the more distant future.

A Challenge to Societies

But can emissions be reduced significantly in the near future? Certainly there is no shortage of ideas about how to accomplish this; economists, scientists, and politicians have put forward a long list of proposals. And there is broad agreement about the kinds of steps societies must take to achieve this reduction: They have to use less energy, reduce the degree to which they use fossil fuels as energy sources, find a way to safely store or "sequester" some of the carbon dioxide that continues to be produced, and slow the destruction of

Earth's tropical rainforests. All these ideas, however, involve economic trade-offs. Since the mid-nineteenth century, human societies have been powered mostly by fossil fuels because these have been the cheapest fuels available. To change now means turning to alternatives that are more costly in an economic sense but less costly in an environmental sense. In coming years, we'll see if human societies are up to the task of paying an economic price for getting a healthier planet.

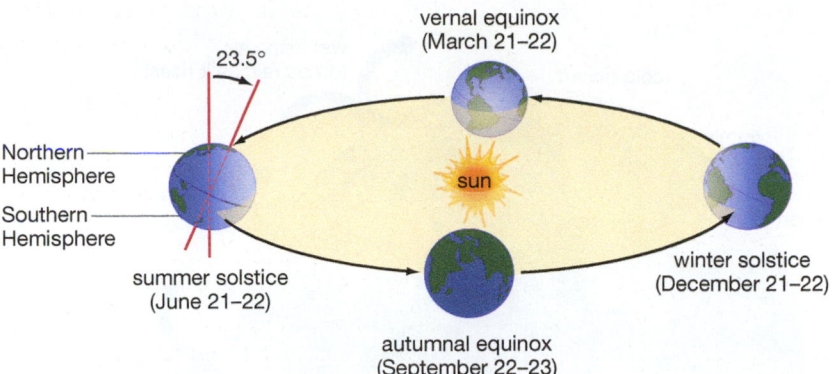

SO FAR . . .

1. Ozone is a gas composed entirely of the element _____ that reaches its greatest density in the layer of the atmosphere known as the _____ and that has the effect of keeping _____ from reaching Earth's surface.

2. Most of the energetic short-wave radiation from the sun is transformed into _____, which can be trapped by greenhouse gases such as _____ and _____.

3. Global warming is certain to produce a _____ in global sea levels and a(n) _____ in precipitation in the United States.

36.6 Earth's Climate

Why Are Some Areas Wet and Some Dry, Some Hot and Some Cold?

Did you ever look at a globe representing the Earth and wonder why it is tilted? Earth exists in space, after all, so what could it be tilted against? When we say that a rod stuck in the ground is tilted, we mean that it is not perpendicular to the ground. Here, of course, we're thinking of the ground as a flat surface—a plane. Earth, too, can be viewed as existing on a plane, in this case the plane of its orbit around the sun. (We generally think of this orbit as looking like a large hula hoop with the sun as a yellow ball in the middle.) Now think of the spherical Earth on this plane as having a rod sticking through it in the form of its north–south axis—the imaginary line that runs from the North Pole straight through to the South Pole. The critical thing is that this rod does not stick straight up and down with respect to the plane of the Earth's orbit; it is tilted, at an angle of 23.5 degrees. This tilt dictates a good deal about Earth's climate, and Earth's climate dictates a great deal about life on Earth. Once again, in other words, the sun determines the basic conditions for life.

You can see the effects of Earth's tilt in **Figure 36.21**: In June, our Northern Hemisphere tilts toward the sun, while in January it tilts away from it. Thus, the sun's rays strike us more directly in June, and the days are warmer (and longer) than days in January. The angle at which the sun's rays strike a given portion of the Earth is very important. Relative to, say, Brazil, sunlight strikes the Arctic at an indirect or "oblique" angle. Thus, the far north gets less of the solar energy that powers photosynthesis.

The Circulation of the Atmosphere and Its Relation to Rain

The variation in temperature caused by these sunlight differences is the most important factor in the circulation of Earth's atmosphere. Near the equator, Earth simply gets more warmth, and, critically, warm air rises. Warm air also can retain more *moisture* than cold air. You've seen an example of this whenever you've looked at the beads of "sweat" that form on the outside of a cold glass in summer. Moisture-laden warm air comes into contact with the cooler air immediately around the glass; thus cooled, the air cannot hold as much moisture and releases it onto the glass.

This same thing happens with the warm air that rises on both sides of the equator. Because of the heat of the tropics, air is rising in quantity from the tropical oceans; because this air is warm, it is carrying with it a great deal of moisture. The air cools as it rises, however, and then drops much of its moisture on the tropics, which is why they're so wet.

Following this a step further, this volume of air is now cooler, drier, and moving toward the poles in both directions from the equator (**Figure 36.22** on the next page). At about 30° N and 30° S of the equator, it descends, warming as it drops and actually absorbing moisture *from* the land. The land will be dry at these latitudes because this is where the dry, hot air descends.

The air that has descended now flows in two directions from the 30° point. In the Northern Hemisphere, this means north toward the North Pole

Figure 36.21
Earth's Tilt and the Seasons

The Northern Hemisphere gets more sunlight in June and less in January, while the reverse is true for the Southern Hemisphere. This is the reason for seasonal climate variations over large portions of the globe.

Tropical Atmospheric Circulation and Global Climate

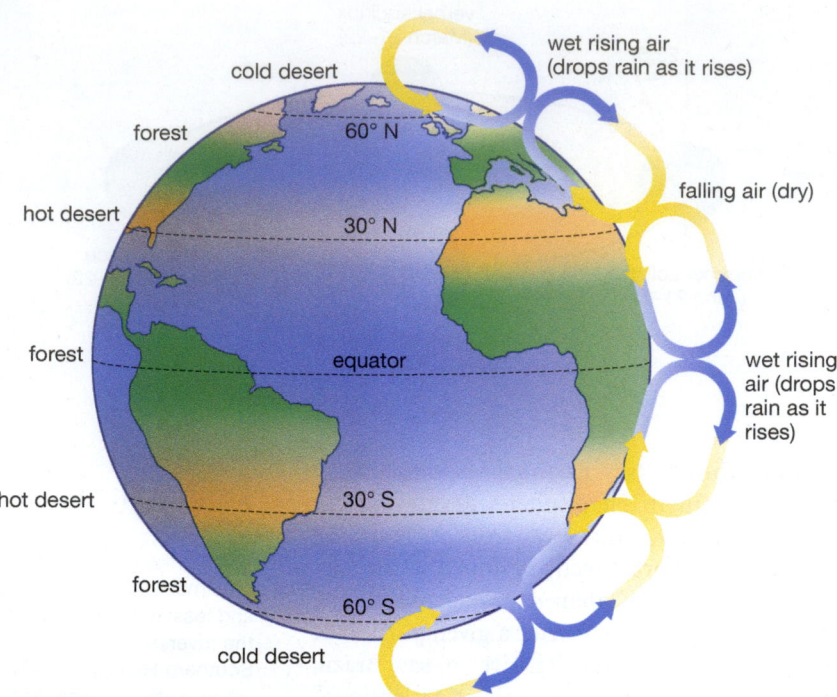

Figure 36.22

Earth's Atmospheric Circulation Cells

Because the Earth receives its most direct sunlight at the equator, this region is warm—and warm air holds moisture and rises. This air moves north and south from the equator, cooling as it rises, and then loses its moisture as rain, with the result that the land and ocean immediately north and south of the equator get lots of rain. As the now-dry air moves farther north and south, it cools and sinks again at about 30° in both hemispheres. Thus, the land at these latitudes tends to be dry. Two other bands of circulation cells exist at higher latitudes, operating under the same principles.

and south toward the equator. Traveling at a fairly low altitude, this air eventually picks up moisture, rises, and deposits its moisture, this time at about 60° N and near the equator, with the same events happening at the same locations in the Southern Hemisphere.

What you get from this rising and falling is what you see in Figure 36.22: a set of interrelated "circu-

lation cells" of moving air, each existing all the way around the planet at its latitude, and each acting like a conveyor belt that is dropping rain on the Earth where it rises but drying the Earth where it descends.

The Impact of Earth's Circulation Cells

The full meaning of this becomes obvious only when you look at a map such as the one in **Figure 36.23** (or better yet, a globe). Draw your finger across the equatorial latitude on the map and look at how much green there is—Southeast Asia, equatorial Africa, the Amazon. This is no surprise because we expect the equator to be wet and green. Now, however, do the same thing at the dry 30° N latitude. Look at how much desert there is at this latitude all around the globe—it's where we find the Sahara, the Arabian, and the Sonoran deserts, among others. And there's more desert territory at 30° S, with the Australian desert and Africa's Namib. From this, you can see that the climatic fate of Earth's various regions is largely written in our globe's large-scale wind patterns.

Mountain Chains Affect Precipitation Patterns

Latitude doesn't tell the whole story about precipitation on Earth, however. Mountain ranges force air upward, and this air also cools as it rises, dropping its moisture on the windward side of the range. Then the air descends on the opposite (leeward) side of the range, only this time it is dry air and is thus picking up moisture from the ground rather than depositing it (**Figure 36.24**). A dramatic example of this "rainshadow" effect can be seen on the Pacific Coast of the United States in southern Oregon, where the western side of the Cascade Mountains is lush with greenery all year round, while the basin below the eastern slopes is a desert.

Figure 36.23

Earth's Atmospheric Circulation Cells and Precipitation

The wettest regions of the Earth lie along the equator, while the driest regions lie along the latitudes 30° N and 30° S. Compare these regions to the circulation cells shown in Figure 36.22. The variations of rainfall patterns within cells are explained by factors such as the presence of mountain ranges.

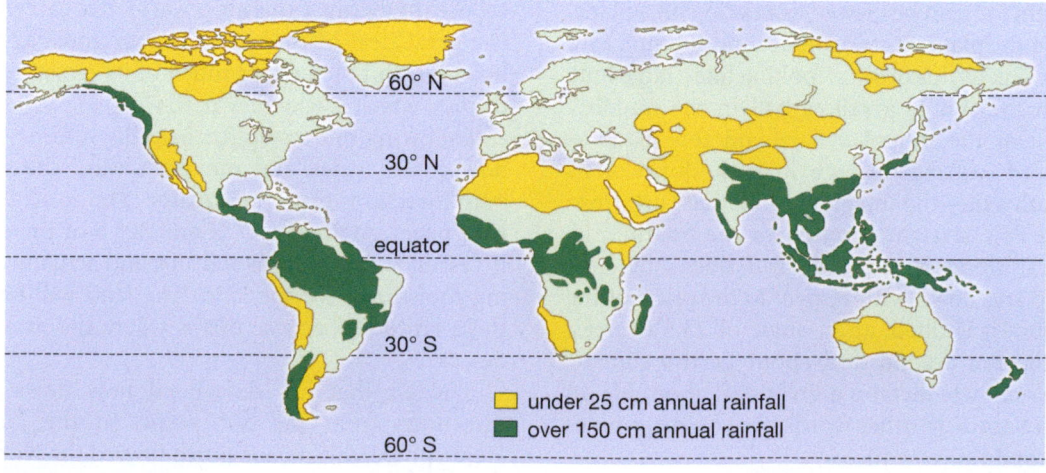

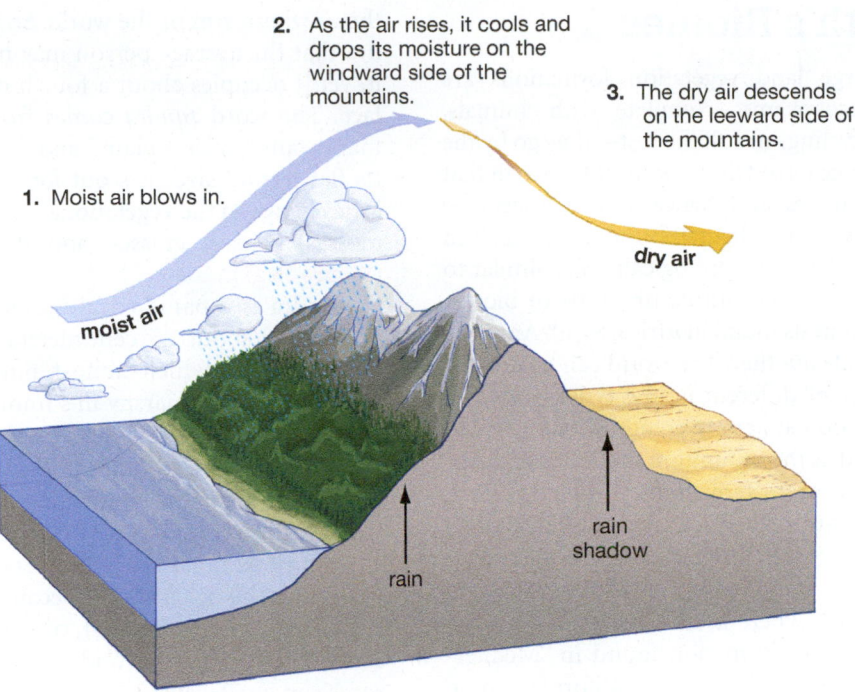

1. Moist air blows in.

2. As the air rises, it cools and drops its moisture on the windward side of the mountains.

3. The dry air descends on the leeward side of the mountains.

dry air

moist air

rain

rain shadow

Figure 36.24
Rain Shadows
Mountain ranges force moving air to rise. This cools the air, causing it to lose its moisture on the windward side of the range. This creates a rain shadow, meaning a lack of precipitation, on the leeward side of the range.

Beyond this, we know that it's possible to be in a warm latitude, such as that of Ecuador in South America, and still be very cold—if you go to the top of the Ecuadorian Andes. As altitude increases, temperature drops, in other words. The changes in climate that occur as you move from flatland to mountain peak are similar to the changes that occur as you move from the equator to one of the poles.

The Importance of Climate to Life

The factors you have just reviewed are among the most important in shaping the large-scale **climates,** or average weather conditions, we find on Earth. It's obvious that climate has a great deal of influence on life since we never see polar bears in Panama or monkeys in Montana, but this influence is even greater than you might think. Looking over the globe, you see different forms of dominant vegetation in different areas—grassland in America's Great Plains, rainforest in northern Brazil. So strongly does climate affect this pattern that Earth's vegetation regions essentially are defined by Earth's climate regions. Far in Canada's north, there is a fairly distinct boundary at which Canada's vast coniferous forest gives way to the mat-like vegetation called tundra. North of this line there is tundra; south of it are the forests. What is this line? In summer, it is the far southern reach of a mass of cold air called the arctic frontal zone (**Figure 36.25**).

Vegetative formations generally overlap one another to a greater extent than is seen with Canada's tundra and coniferous forest, so that the closer we get to a vegetation boundary, the less distinct it looks. Nevertheless, there are discrete regions of vegetation on Earth whose boundaries are largely determined by climate—particularly by temperature and rainfall. Vegetative exceptions within these regions tend to be mountains, whose climates can vary greatly from base to peak.

Figure 36.25
Line of Transition
In Alaska's Denali National Park, a region of forested taiga vegetation gives way to a region of grassy tundra vegetation in a fairly abrupt way.

GraphIt!: Global Soil
Degradation

36.7 Earth's Biomes

When these large land vegetation formations are looked at as ecosystems—complete with animals, fungi, nutrient cycling, and all the rest—they go by the name **biomes:** large terrestrial regions of the Earth that have similar climates and hence similar vegetative formations. Grasslands, whether found in Russia or in the American Midwest, are ecologically very similar to one another and thus constitute one type of biome. The tropical rainforests found in Africa, South America, and Asia constitute another. The world can be divided into any number of different biome types—this is a human classification system overlaid on nature—but six are recognized as the minimum: tundra, taiga, temperate deciduous forest, temperate grassland, desert, and tropical rainforest. Polar ice and mountains often are recognized as separate biomes as well, as is another grassland variation, the savanna, found most famously in equatorial Africa. There is also chaparral, a shrub-dominated vegetation formation found in "Mediterranean" climates such as those on California's coast. See **Figure 36.26** for an overview of locations of these biome types throughout the globe.

The fact that two biomes are of the same type does not mean that they are identical. Two areas that are broadly the same in climate can end up with some significant differences in life-forms. Let's look briefly now at each of Earth's major biome types and then at its aquatic ecosystems.

Cold and Lying Low: Tundra

Along with polar ice, **tundra** is the biome of the far north, stretching in a vast, mostly frozen, ring around the northern rim of the world. So inaccessible is tundra that the average person may never have heard of it, yet it occupies about a fourth of Earth's land surface. The word *tundra* comes from a Finnish word that means "treeless plain," and the description is apt. Its flat terrain stretches out for mile after mile with little change in the vegetational pattern of low shrubs, mosses, lichens, grasses, and the grass-like sedges (**Figure 36.27**).

Tundra is a paradoxical biome in that it gets an average of about 25 centimeters (or 10 inches) of rain per year, which almost puts it in the desert category. Yet it is marshy in summer, with numerous shallow lakes dotting its surface. How can both things be true? Once the summer rain falls, it essentially has no place to go. No more than a meter below the surface, the ground is permanently frozen down to a depth of as much as 500 meters, or a quarter mile (although this is changing because of global warming). This is the **permafrost,** or permanent ice, which acts as a boundary beyond which neither water nor roots can go. Because water cannot drain far into the soil, it sits on top of the tundra in summer, which makes for the bogs and small lakes. The low-lying tundra vegetation makes the most of a short growing season—lasting as little as 50 days—by producing a year's worth of food through photosynthesis in only a few weeks.

The tundra has a good deal of animal life, particularly in the summer when migrating animals move north to feed on the vegetation. Large mammals such as caribou and musk oxen live there. Arctic hares and foxes are year-round residents, as are arctic lemmings—herbivores that are key to much of the tundra's animal food web.

Figure 36.26
Distribution of Biomes across the Earth

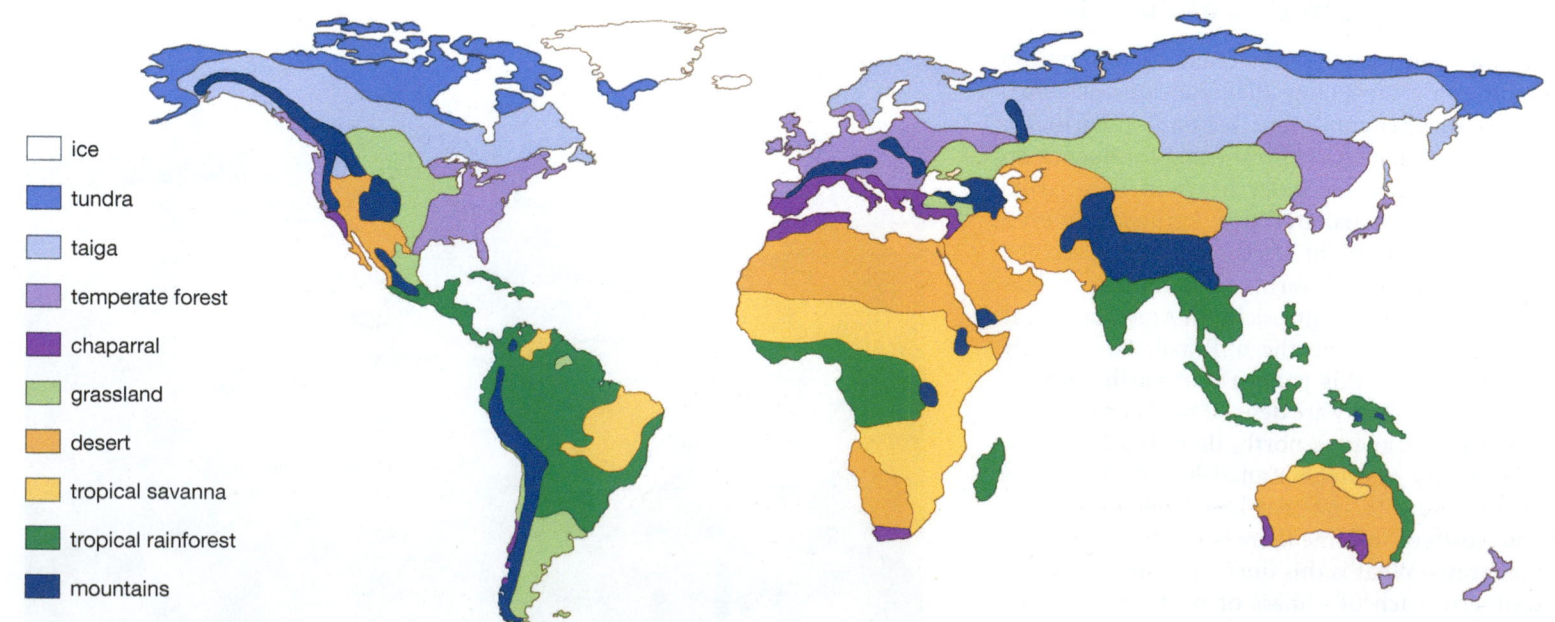

- ice
- tundra
- taiga
- temperate forest
- chaparral
- grassland
- desert
- tropical savanna
- tropical rainforest
- mountains

Figure 36.27
Tundra
An Alaska-Yukon moose (*Alces alces*) in the Alaskan tundra in fall.

Northern Forests: Taiga

The **taiga,** or boreal (northern) forest, includes the enormous expanse of coniferous trees mentioned earlier that juts up against the tundra. Conditions are still fairly dry and cold, and the growing season is short in this northern biome, but in all three respects the taiga is less severe than the tundra. Note in Figure 36.26 that, as with tundra, taiga is found almost solely in the Northern Hemisphere. This is so because there is little land in the Southern Hemisphere at the extreme latitudes of these biomes.

The taiga is a model of species uniformity, as only a few types of trees—spruce, fir, and pine—are the ecological dominants, but they are present in great numbers. The taiga's vegetation supports lots of animals, including large mammals such as moose, bear, and caribou (**Figure 36.28**). When thinking of the fur trade, you might have in mind some freezing, coniferous territory in the far north. The taiga turns out to be it. Sables, minks, beavers, arctic hares, wolverines—this harsh biome has an abundance of these furred creatures.

Figure 36.28
Taiga
Alaskan caribou in winter in the northern-forest biome known as taiga.

Hot in Summer, Cold in Winter: Temperate Deciduous Forest

Temperate deciduous forest is a biome familiar throughout much of the United States, as it exists roughly from the Great Lakes nearly to the Gulf of Mexico and from the western Great Lakes to the Atlantic Ocean. The pine forests that cover large parts of the southeastern United States generally are considered to be part of the temperate deciduous biome because the pines are a successional stage *leading to* deciduous forest. Meanwhile, the warm Gulf Coast is not temperate deciduous forest per se but a variant on

it called temperate evergreen forest. Much of Europe and eastern China fit within the temperate deciduous classification, although large parts of Europe have few forests left, and the same is true of China.

Temperate deciduous forests are the forests of maple, beech, oak, and hickory (**Figure 36.29** on the next page). The *deciduous* in the name refers to trees that exhibit a pattern of loss of leaves in the fall and regrowth of them in the spring. Trees do not dominate these forests to the extent they do the taiga, however. Making use of the good soil, a robust "understory" of woody and herbaceous plants will spring up on the floor of the deciduous forest. Some of this activity occurs early in the growing season,

Figure 36.29

Temperate Deciduous Forest

Plenty of water and seasonally warm temperatures lead to abundant plant and animal life in the biome known as temperate deciduous forest. This landscape is part of Pennsylvania's Allegheny National Forest.

before the trees have "leafed out" and blocked the sunlight. Temperate forests grow where there is plenty of water—between 75 and 200 centimeters (or 30 to 78 inches) per year, with the lower amount being three times what the tundra gets. It's worth noting that while the tundra has no reptiles or amphibians and the taiga has only a few, the temperate deciduous forest has an environment that is wet and warm enough to support a variety of both animal forms.

When Europeans first came to North America in significant numbers, in the seventeenth century, there probably were more than a billion acres of forest in what would become the United States. By 1907, about 250 million acres of this forest had been cleared, primarily for agricultural use. The temperate deciduous forest cover in the eastern United States reached its lowest level by about 1860, after which it began to rebound because of the westward movement of farming into the Midwest. Between 1997 and 2002, forested land in the United States as a whole actually increased by about 10 million acres.

Dry but Very Fertile: Temperate Grassland

There is an irregular line in western Indiana that forms the boundary between deciduous forest and prairie. What is "prairie" in the United States is known as "steppes" in Russia, "pampas" in Argentina, and "veldt" in South Africa. These are all names for the biome that ecologists call *temperate grassland* (**Figure 36.30**).

Looked at one way, grasslands are what often literally lie between forests and deserts. Whereas the precipitation range for deciduous forests is from 75 to 200 centimeters per year (30 to 78 inches), for grassland the range is from 25 to 100 centimeters per year (10 to 39 inches). The North American grassland has its own internal division based on the amount of rainfall. It ranges from tallgrass prairie in Illinois to a drier mixed-grass prairie that begins at about the middle of Kansas and Nebraska—this is where the Great Plains begin—to a short-grass prairie that lies in the rain shadow of the Rocky Mountains. (This same tallgrass/short-grass gradient exists in the steppes of Asia, only there the gradient runs from north to south.) Other American grassland areas can be found in California's Central Valley and eastern Washington and Oregon.

A type of tropical grassland, the **tropical savanna,** exists in Australia, Africa, and South America. Tropical savanna is characterized by seasonal drought, small changes in the generally warm temperatures throughout the year, and stands of naturally occurring trees that punctuate the grassland. This is the homeland of many of the great mammals we associate with Africa, such as lions and zebras.

When Europeans first beheld the American grasslands, they could scarcely believe what they saw: a territory that had essentially no trees but instead a sea of grasses that could grow up to 10 feet tall in the wetter tallgrass prairie. Prairie soil turned out to be some of the most fertile on Earth, however, and that fact eventually spelled the end of natural prairie in the United States and most of the rest of the world. Almost all the tallgrass and mixed-grass prairie in the United States eventually was plowed under for cultivation, while the short-grass prairie is now used for crops and cattle grazing. Prior to the coming of Europeans, 22 million acres of Illinois were prairie; today, that figure stands at slightly more than 2,000 acres. From

Figure 36.30

Temperate Grassland

The Samuel H. Ordway Prairie in South Dakota provides some idea of what America's prairies looked like prior to the arrival of Europeans, as most of the land in it was never plowed under for crops.

the former prairie, however, the United States feeds itself and some of the world outside its borders. Illinois, Kansas, and Nebraska are still grasslands, but the grasses that grow there now are not the native big bluestem or needlegrass; they are corn and wheat.

Chaparral: Rainy Winters, Dry Summers

Chaparral is not grassland but instead is a biome dominated by evergreen shrub vegetation that is dotted with pine and scrub oak trees. Chaparral is found in areas that have a so-called Mediterranean climate. There are five of these areas in the world, all very similar in appearance. One of them is the Mediterranean basin itself; then there is southwestern Australia, central Chile, a small part of western South Africa, and parts of coastal California. What these areas have in common is that they are on the west coast of a landmass. Thus, they lie in the path of ocean winds that bring wet, mild winters and very dry summers.

The Challenge of Water: Deserts

Deserts can be defined as biomes in which rainfall is less than 25 centimeters, or 10 inches, per year, and water evaporation rates are high relative to rainfall (**Figure 36.31**). Note that deserts are not defined by temperature. There can be temperate deserts (the Mojave in Southern California), cold deserts (the Gobi in China), and hot-weather deserts (the Sahara in Africa). Around the globe, some 20 major deserts collectively cover about 30 percent of Earth's land surface. What we think of as desert can transition imperceptibly into dry grassland.

Precipitation is not only low in the desert, it is sporadic; that is, most of the moisture a desert gets will come in just a few days during the year. Given this, desert life has relentlessly been shaped by one overriding requirement: Collect water and then conserve it. The flowering plant grouping called bromeliads have a species, *Tillandsia straminea*, that is blown around on the deserts of coastal Peru much like a dead tumbleweed, only this is a living plant, nearly rootless, that gets all its moisture from *fog* that it is able to absorb.

Desert animals are no less remarkable in this respect. The kangaroo rat (genus *Dipodomys*) can be found in many parts of the American Southwest. It comes out of its underground burrow only at night when temperatures fall, it has no sweat glands, and it generally does not drink standing water (there being almost none available). Instead, the kangaroo rat gets by on the water produced by metabolizing the seeds it eats.

Lush Life, Now Threatened: Tropical Rainforests

Many people have an idea that the Earth's **tropical rainforests** are important, but they're not quite sure why. There are two main reasons. First, along with coral reefs and algal beds, they are the most productive biome type in the world, meaning they produce more biomass per square meter of territory than any other. Second, no other ecosystem can touch them in terms of species diversity. As noted earlier, for mile after mile in the taiga, trees might be limited to one or two species. In the tropical forests of the Amazon, meanwhile, more than 300 species of trees have been identified in a single hectare of land, which is about 2.5 acres. In Chapter 35, you read about a researcher who found an estimated 1,700 species of beetle in one tree in the Panamanian rainforest. In a sense, this isn't surprising because half of all the identified plant and animal species on Earth reside in tropical rainforests. This despite the fact that, compared to the other biomes you've looked at, they don't occupy a lot of territory: about 7 percent of the Earth's surface.

The problem, as most people know, is that the rainforests are disappearing—significant portions of them are being either burned down or cut down. The period of great temperate forest loss came before the twentieth century, but we are living in the period of great tropical forest loss. How significant is the reduction? The United Nations estimates that, in Brazil alone, rainforest losses from 2000 to 2006 totaled about 50,000 square miles—an area about the size of Alabama. The Brazilian forest makes up the lion's share of what is by far the world's largest tropical rainforest, the nearly two-million-square-mile Amazon Basin Forest, which runs from Venezuela south to Bolivia and at one point nearly from the Atlantic to the Pacific. Significant rainforests also can be found in Central America, equatorial Africa, and regions of Southeast Asia near the equator.

Figure 36.31
Desert
When water gets sparse, life does, too. Shown are the Mesquite Flat Sand Dunes in California's Death Valley National Park.

Figure 36.32
Tropical Rainforest
Abundant rain and warm weather mean abundant growth and a great diversity in life-forms in the biome known as the tropical rainforest. Shown is a lowland rainforest on the Segama River in Borneo.

GraphIt!: Global Fisheries and Overfishing

Figure 36.33
Ocean Zones

Rainforests are lands of rain and stable, warm temperatures (**Figure 36.32**). The norm for rainfall is between 200 and 450 centimeters, or about 78 to 175 inches, a year, but some areas get up to 1,000 centimeters annually, which is 390 inches. You've seen that much of the richness of the temperate grasslands can be found in their soil. In the tropical rainforest, however, almost all the richness is on display above ground (much of it 20 to 40 meters above ground in the extensive forest "canopy"). The rainforest has such a large and efficient group of detritivores that organic material dropping to the forest floor is decomposed by them and absorbed almost immediately by plant roots rather than by the surrounding soil. This means the soil never gets built up; it is nutrient poor and often acidic. How can this be squared with this biome's fantastic productivity? The nutrients are simply bypassing the soil, going from plant to detritivore and back to plant.

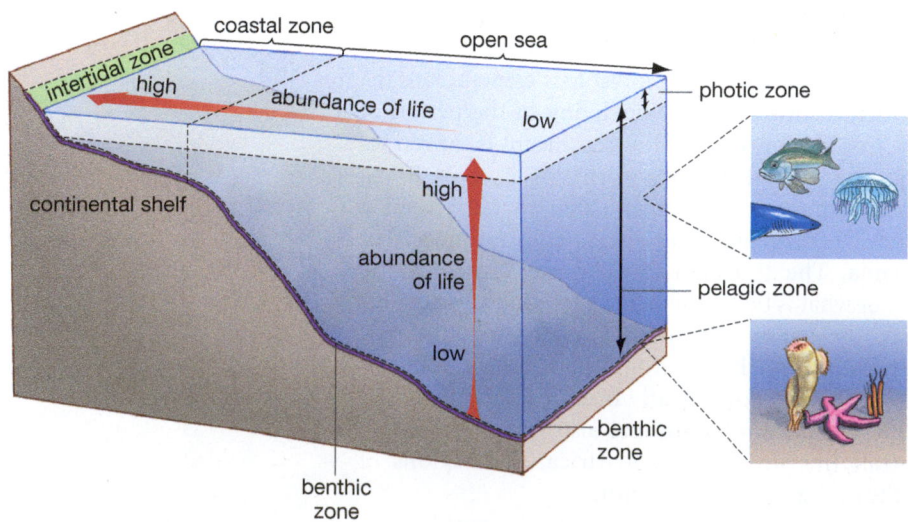

SO FAR...

1. The warm surface air of the tropics _____ as it rises, causing it to _____ moisture. At about 30° N and 30° S of the tropics, however, air descends and _____ moisture from the land.

2. Earth's large vegetation regions are essentially defined by _____.

3. Starting at the North Pole and going south, list the order in which you would be likely to find the following biomes: temperate deciduous forest, tundra, tropical rainforest, taiga.

36.8 Life in the Water: Aquatic Ecosystems

We turn now from life on land to life in the water. Because we humans are land creatures, it might be good to get our bearings by thinking about how one kind of aquatic environment, the ocean, can be divided up in terms of the life in it.

Marine Ecosystems

The ocean's **coastal zone** extends from the point on the shore where the ocean's waves reach at high tide to a point out at sea where an ocean-floor formation called the continental shelf drops off (**Figure 36.33**). Beyond this point, there is the *open sea*. Within the coastal zone, there is an area bordered on one side by the ocean's low-tide mark and on the other by its high-tide mark. This is the **intertidal zone.**

If you look next at the ocean strictly in its vertical dimension, all of the water from the ocean's surface to its floor is called the *pelagic zone*, while the ocean floor itself is called the *benthic zone*. Within the pelagic zone, from sea level down to a depth of at least 100 meters (about 330 feet), the sun's rays can penetrate strongly enough to drive photosynthesis. This zone of photosynthesis is known as the *photic zone*.

More Productive Near the Coasts, Less Productive on the Open Sea

Each of these zones marks off an area that is meaningful in relation to life. Start first at the intertidal zone. This is a world *in between,* we might say, because the creatures in it live part of the time submerged in the ocean and part of the time exposed to the air. It turns out that the intertidal zone is extremely productive for a reason that you might think would make it unproductive: The ocean waves, which bring in nutrients,

ESSAY

Good News about the Environment

A sense of gloom about the environment may arise from between the lines of these pages as the list of current or potential environmental problems piles up: global warming, nitrogen buildup, overuse of water resources, tropical rainforest destruction. These are real problems, but their existence should not blind us to the fact that there has been a great deal of good news about the environment in the past 40 years, much of it showing how effective government can be when prodded into action by its citizens.

If there was a time for unrestrained gloom about the environment in the United States, it probably would have been in the late 1960s, when the country had plenty of endangered species but no Endangered Species Act; when levels of smog in Los Angeles were almost twice what they are now; when there was plenty of household trash to be recycled, but almost no recycling going on; when only a third of the bodies of water in the country were safe for fishing and swimming, whereas today two-thirds are.

As a symbol of change, consider Cleveland, Ohio, whose Cuyahoga River was so polluted in 1969 that it *caught fire,* while southwestern Lake Erie, into which the Cuyahoga flows, was essentially a dead zone for fish during the late summer. The lake came to this low state in part because of the nearly 80 metric tons of phosphorus—then regularly added to washing detergents— that were dumped into it each day. That figure has now been reduced by perhaps 85 percent, and Lake Erie is biologically very active and a major freshwater fishery. Meanwhile, the Cuyahoga River, once virtually dead to all forms of life, now has 20 species of fish swimming in it near Cleveland and eagles and blue herons now nest along its banks farther upstream. How did this change come about? After the 1969 fire, the region worked tirelessly to clean up its waterway, spending $3.5 billion on the effort.

Similar stories could be told about other regions across the United States. Simply put, there has been a profound change with respect to the environment over the past 40 years. From pollution controls to species protection to personal recycling, environmentalism has entered not only our law books but, more impor-

Environmentalism has entered not only our law books but, more important, our consciousness.

tant, our consciousness. When populations of Atlantic swordfish were observed to be declining in the late 1990s, a campaign titled "Give Swordfish a Break" was organized that quickly enlisted 27 top New York chefs. What could chefs do about

environment—in this case, the sensitivity of a group of American chefs. And sensitivity to the environment has grown not just in America but in most developed countries. The cleanup of Lake Erie required the cooperation of the United States and Canada; the heartening trend in preservation of the Earth's ozone layer is the product of an international agreement.

None of this is to say that mechanisms are in place such that we need not worry about the environment. Far from it: Largely because of the growth in human population and the greater per capita use of natural resources, the environment is threatened on thousands of fronts and will remain so as far out as we can see into the future. But the record of the past 40 years is not just one of environmental loss; it is

Back to Healthy Numbers
Populations of Atlantic swordfish (*Xiaphas gladius*) were severely depleted by the late 1990s. A campaign spearheaded by environmentalists spurred governmental action that brought the swordfish numbers back to healthy levels by 2002—just three years after an international conservation program was put in place.

swordfish populations? They pledged to take swordfish off their restaurants' menus. It was government action that brought the Atlantic swordfish population back to healthy levels, but initial public awareness of the swordfish campaign came about because of *personal* sensitivity to the

one of great environmental accomplishment as well. Perhaps the period's most important lesson is that there is nothing inevitable about environmental degradation. Concerned, hardworking people have made a great deal of difference in this issue.

(a) A man and two children investigate tide pools

(b) Penguins swimming underwater

(c) Fish swimming near coral reef

carry away wastes and expose more of the surface area of photosynthesizers to sunlight.

Ocean life continues to be abundant out past the intertidal zone and in general remains so all the way to the end of the coastal zone. In short, ocean life is most productive in its shallower depths (**Figure 36.34**). Once past the continental shelf and into the open ocean, productivity can drop off to levels that are less than those of a desert. Why should this be so? Remember that sunlight strong enough to power photosynthesis penetrates ocean waters only to a depth of 100 meters or so. Meanwhile, the average depth of the ocean is about 3,000 meters, or 1.8 miles. Thus there is a vast depth of open ocean that simply isn't very productive. Indeed, life at greater ocean depths essentially depends on organic particles, referred to as "marine snow," that rain down from above. One measure of the coastal zone's importance is that it is the location of all the world's great fisheries, some of which are now beginning to show signs of returning to health after years of being degraded—something you can read more about in "Hopeful Signs for the World's Fish" on page 720.

More Productive Near the Poles, Less Productive Away from Them

Apart from the shallow-to-deep transition, the ocean also has one other gradient with respect to the amount of life it harbors. In the pelagic zone, there is relatively more ocean life toward the poles, with this concentration decreasing with movement toward the equator—the exact opposite of what we find with land life. The oceans that surround Antarctica are filled with life. **Phytoplankton,** meaning floating, microscopic photosynthesizers such as algae and cyanobacteria, are the base producers for it all. These are consumed by the tiny floating animals known as **zooplankton.** Shrimp-like crustaceans called krill feed mostly on the phytoplankton and serve as the principal source of food for the whales and seals we associate with this area.

Cities of Productivity: Coral Reefs

Despite the richness of the colder oceans, the tropical oceans do boast marine communities that are as productive as any in the ocean—indeed, as any on

Figure 36.34
Diversity in Ocean Life

(a) A man and two children investigate tide pools, a part of the ocean's intertidal zone. The deeper water beyond them can be seen both above and below the surface.

(b) The seas off Antarctica are teeming with aquatic life. Here, emperor penguins (*Aptenodytes forsteri*) dive in Antarctic waters.

(c) Coral reefs are sites of high productivity and diversity. Here, a group of raccoon butterfly fish (*Chaetodon fasciatus*) stand out against the background of a coral reef in the Red Sea.

Earth. These are the coral reefs, which lie in shallow waters at the edge of continents or islands at tropical latitudes. **Coral reefs** are warm-water ocean structures composed of the remains of coral animals. Each animal, known as a polyp, is a tiny relative of the more familiar sea anemones. Coral polyps secrete calcium carbonate, better known as limestone, a practice they share with a type of symbiotic red algae that live inside them. The limestone forms an external skeleton for the corals, and when each coral dies, its limestone skeleton remains. Thus, coral reefs are composed largely of the stacked-up remains of countless generations of coral polyps along with the remains of their symbiotic algae. But the thin outer veneer of each reef is composed of *living* algae- and the latest generation of pinhead-sized corals.

You can get an idea of how many generations of coral are piled up by considering that most reefs are between 5,000 and 8,000 years old—and some are millions of years old. What the reef as a whole creates is *habitat*: nooks and crannies and tunnels and surfaces that are home to an amazing array of living things. So rich is this habitat that it covers only 2 percent of the ocean's floor but is home to perhaps 25 percent of the ocean's species. One coral reef, Australia's Great Barrier Reef, also is one of the largest structures created by any living thing, including humans. It measures some 2,000 kilometers, or 1,200 miles, in length.

Sadly, coral reefs are now imperiled cities of biological productivity. By one estimate, 27 percent of the world's coral reefs have now been lost as functioning ecosystems, although with coral reefs, loss of activity is not always permanent. The greatest single source of this destruction was the El Niño-caused ocean warming of 1997–1998, which brought about a phenomenon known as *coral bleaching*, meaning a discoloration of the reefs due to the death of symbiotic algae that live within the coral animals. Temporary weather events aside, the longer-term threat to coral reefs is human activity. Oil is dumped into the ocean, sewage and runoff from cities harms the reefs, divers may trample them.

The biggest threat that coral reefs face, however, may be another product of the worldwide change we looked at earlier, global warming. Some of the excess atmospheric carbon that drives global warming has been absorbed by the oceans in recent decades, and this carbon is making the oceans more acidic. The problem is that ocean acidification may interfere with the ability of corals and their algal partners to pull dissolved calcium out of ocean water—something these organisms must do to produce their skeletons. We know that coral calcium absorption declined in Australia's Great Barrier Reef by about 14 percent from 1990 to 2005 and that a decline greater than this occurred among some corals in Bermuda over the last few decades. The question now before scientists is

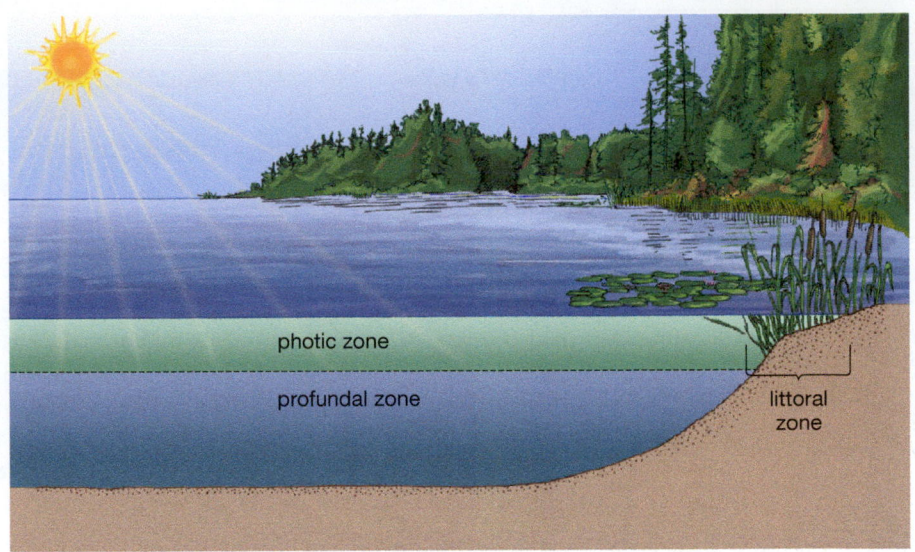

whether these declines have come about because of ocean acidification.

Freshwater Systems

Whereas the oceans cover almost three-quarters of Earth's surface, freshwater ecosystems—inland lakes, rivers, and other running water—together cover only about 2.1 percent of Earth's surface. Like the ocean, lakes can be divided into biological zones. The most productive zone in a lake is the shallow-water area along its edge, the *littoral zone*, whose outer boundary is defined as the point at which the water is so deep that rooted plants can no longer grow. Then there is the *photic zone*, which starts at the surface of the lake and extends down to the point at which sunlight no longer penetrates strongly enough to drive photosynthesis. Finally, there is the area beneath the photic zone, the *profundal zone*, in which photosynthesis can't be performed (**Figure 36.35**).

Nutrients in Lakes

Lakes can be naturally *eutrophic*, meaning nutrient rich, or *oligotrophic*, meaning nutrient poor. A lake that has few nutrients will have relatively few photosynthesizers, and this in turn means few animals. Oligotrophic lakes generally are clear (and deep), while eutrophic lakes often have an abundant algal cover.

Nutrient enrichment can happen naturally over time, or it can take place because of human intervention. The latter variety is known as *artificial eutrophication*, a process that sometimes is undertaken intentionally just to increase the yield of fish from a lake or pond. There is such a thing as having too *many* human-added nutrients in a body of water, however. We touched on this earlier in connection with the Gulf of Mexico's "dead zone," which results from the tons of nitrogen that flow into it. But

Figure 36.35
Zones in a Lake

The littoral zone starts at water's edge and extends out to the point at which rooted plants can no longer grow. The photic zone starts at the lake's surface and extends down to the point at which sunlight no longer drives photosynthesis. The profundal zone is the area beneath the photic zone in which photosynthesis cannot be performed.

ESSAY

Hopeful Signs for the World's Fish

At one time, marine wildlife experts thought that the bounty of fish in the world's oceans was so immense it could never be exhausted. But that idea began to be challenged in the late 1980s, and by the mid-1990s it was effectively dead as it became clear that the numbers of certain kinds of fish had not just dwindled but had dropped to alarmingly low levels. The enormous populations of Atlantic cod once found off the Canadian coast were gone by the early 1990s, for example, and in 2003 researchers reported that the global ocean has lost 90 percent of the fish in its big predatory species—bluefin tuna, blue marlin, Antarctic cod, and sharks among them.

In 2009, however, one of the most comprehensive assessments ever conducted of the world's fisheries yielded a more hopeful message. While 63 percent of the world's ocean fish stocks have been depleted to levels that require them to be rebuilt, fish populations in well-managed fishing areas are recovering and they are doing so through methods that can be applied to fisheries in general: Reducing allowable catches, putting restrictions on the kinds of gear that can be used in fishing, closing off certain areas to fishing altogether, and giving fishermen shares of ownership in regional fish harvests. Such management techniques have stabilized fish populations in five of the world's Large Marine Ecosystems, or LMEs—coastal fishing areas so enormous that one of them includes the entire northeast continental shelf of the United States. Three other LMEs are still being over-fished, however, while two others have never been depleted. Historically, the problem in most fisheries is that rebuilding efforts in them "only begin after there is drastic and undeniable evidence of overexploitation" in the words of the assessment, which was authored by a team of scientists led by ecologist Boris Worm of Dalhousie University in Canada and fisheries scientist Ray Hilborn of the University of Washington, Seattle.

The overexploitation examined in the assessment essentially has come about because, in recent decades, the world's fishing fleets have gotten more efficient at scouring the Earth for large populations of fish. By using ever-more-sophisticated technology, the fleets have managed to find fish no matter what kind of terrain they are located in. Ultimately, however, such fishing practices cannot be sustained because, beyond a certain point of exploitation, fish populations cannot renew themselves. The good news is that proper management techniques can seemingly allow fish populations to remain robust indefinitely.

Figure 36.36
Too Many Nutrients

A pond with an overabundance of nutrients has experienced an overgrowth of algae—an algal bloom—that may be detrimental to other life-forms in the pond.

smaller bodies of water can undergo eutrophication as well. An overabundance of such nutrients as phosphorus or nitrogen brings on an overabundance of algae—a so-called **algal bloom**—and this eventually means an overabundance of dead algae falling to the bottom of the lake (**Figure 36.36**). When this happens, decomposing bacteria flourish, and they are using oxygen—so much of it that the fish in a lake can suffocate. How do these harmful levels of phosphorus or nitrogen get into freshwater aquatic ecosystems? Fertilizers from lawns or agriculture and construction that disturbs bedrock, sewage, and detergents are some of the sources (although in many states phosphate-containing detergents have been banned for decades). This is one form of human environmental impact that can be reversible in a straightforward way: Reduce the nutrients that are flowing into the ecosystem.

Estuaries and Wetlands: Two Very Productive Bodies of Water

There is one important type of aquatic water system that always straddles the line between a freshwater

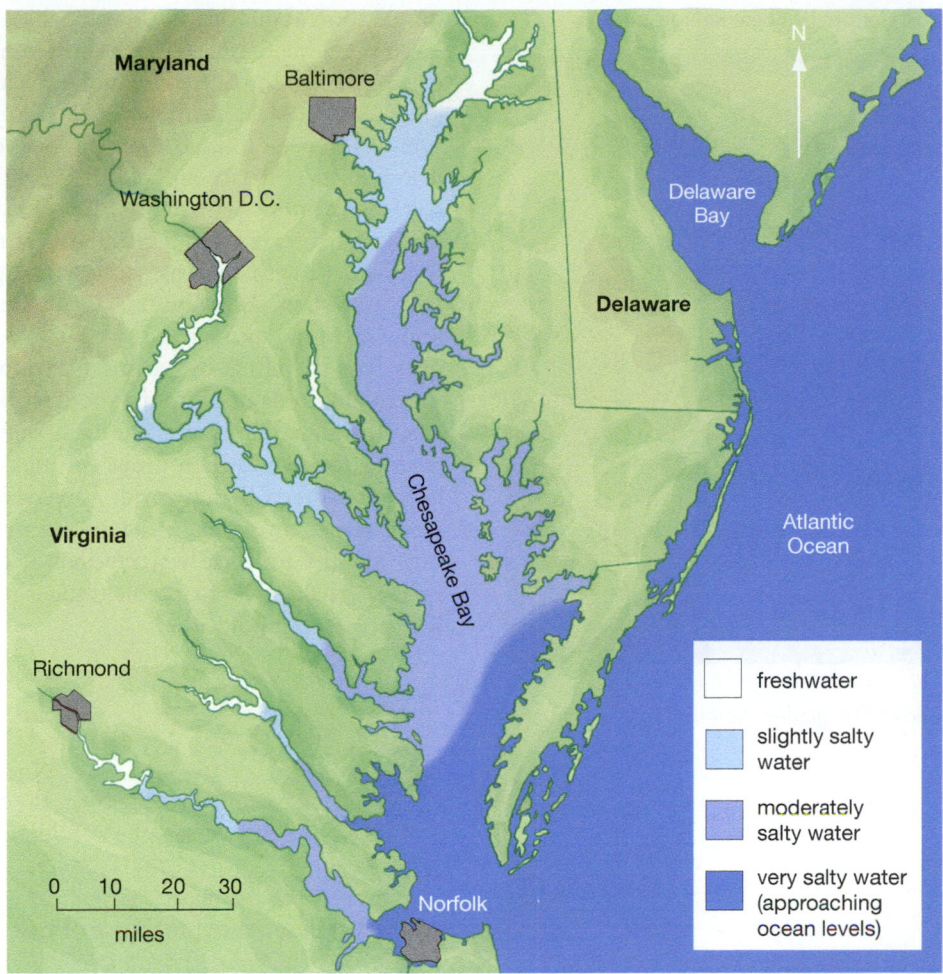

Figure 36.37
Where Saltwater and Freshwater Meet
The Chesapeake Bay is one of the largest and most productive estuaries in the world. In it, the freshwater of 19 principal rivers meets the saltwater of the ocean, forming zones of different degrees of saltiness. Each zone supports specific communities of organisms.

(Redrawn from A. J. Lippson and R. L. Lippson, *Life in the Chesapeake Bay,* Johns Hopkins University Press, Baltimore, 1984.)

Map legend:
- freshwater
- slightly salty water
- moderately salty water
- very salty water (approaching ocean levels)

and saltwater habitat. This is the **estuary,** an area where a stream or river flows into the ocean. Estuaries rank with tropical forests and coral reefs in being the most productive ecosystems on Earth. The cause of this productivity is the constant movement of water—the same force at work in the ocean's intertidal zone. Ocean tides and river flow are constantly stirring up estuary silt that is rich in nutrients. Plants and algae thus get an abundance of these substances, grow in abundance, and pass their bounty up the food chain. In **Figure 36.37**, you can see an image of an important estuary in the United States, the Chesapeake Bay.

Another important type of aquatic ecosystem is **wetlands,** which are just what they sound like—lands that are wet for at least part of the year. "Wet" here covers a variety of conditions, from soil that is merely temporarily waterlogged—a prairie "pothole" in Minnesota, for example—to a cover of permanent water that may be several feet in depth (the bayous of Louisiana). Wetlands go by such specific names as swamps, bogs, marshes, and tidal marshes; the overwhelming majority of them are freshwater inland wetlands rather than the ocean-abutting coastal wetlands, which can be freshwater or saltwater (**Figure 36.38** on the next page). Wetlands are sites of great biological productivity, and they serve as vital habitat for migratory birds. It's been estimated that the amount of wetlands in the United States has been reduced by 55 percent since the arrival of Europeans—120 million of the 215 million acres that were originally wetlands have been drained or paved over. Currently, the state of Florida and the federal government are working on one of the most far-reaching and expensive habitat restoration efforts in history as they try to bring the largest expanse of wetlands in the nation, the Florida Everglades, to a healthier condition over the next 20 years.

Life's Largest Scale: The Biosphere

In this chapter, you've been reviewing life on a very large scale, which is to say the aquatic ecosystems and biomes of the Earth. But there is one larger scale of life yet, which is the collection of all the world's ecosystems, the **biosphere,** which is also thought of as that portion of the Earth that supports life. Much could be said about the biosphere, but here's just one

(a) Prairie pothole ponds in North Dakota's Lostwood National Wildlife Refuge

(b) An overhead view of the sawgrass prairie in Florida's Everglades National Park

Figure 36.38
Inland Wetland and Coastal Wetland

observation about life on this global scale. The Earth is about 24,000 miles in circumference, with life existing along any line you could draw all around the globe. If we look at life in its *vertical* dimension, however, it has been found to exist only about 1.8 miles below Earth's surface and about 5.2 miles above sea level (in the upper Himalayas). This means that all of life on Earth exists in a band that is at most about 7 miles thick, with the vast bulk existing in an even narrower span. Actually, this is all the life that we know of in the *universe,* so that if we could imagine ourselves traveling by cosmic elevator, we could potentially come up from the molten center of Earth, pass through the bacteria lodged between underground rocks, and then move through a tissue-thin layer of a green living world, after which we might never see anything like it again, no matter how far out into the universe we traveled. Taking this perspective, the really remarkable thing is not how much life there is, but how little.

SO FAR . . .

1. With increasing distance from the shoreline toward open sea, the abundance of marine life _____. With increasing distance from the ocean's surface toward the ocean floor, the abundance of marine life _____.

2. The tiny organisms collectively called _____ lie at the base of most marine food webs, thanks to their ability to _____.

3. An estuary is an area where a stream or river _____.

CHAPTER

36 REVIEW

Summary

36.1 The Ecosystem

- An ecosystem is a community of organisms and the physical environment with which they interact. (p. 691)

36.2 Nutrient and Water Cycling in Ecosystems

- Chemical elements that are vital to life are known as nutrients. Along with water, nutrients move back and forth between abiotic (nonliving) and biotic (living) domains on Earth in a process called biogeochemical cycling. Nutrients can be stored in living things, transferred between them, or transferred between them and the abiotic domain. (p. 691)

- Plants, algae, and some bacteria take in atmospheric carbon dioxide to perform photosynthesis. Animals obtain their carbon by consuming these photosynthesizing organisms. Carbon then moves back into the atmosphere in the form of carbon dioxide, which is produced naturally through the respiration of living things and through their decomposition following death. Carbon dioxide, which is vital to life and greatly affects global temperature, makes up a small but critical proportion of the Earth's atmosphere. (p. 692)

- Prior to the twentieth century, nitrogen entered the biotic domain mostly through the action of certain bacteria that have the ability to convert atmospheric nitrogen into forms that can be taken up and used by plants. Other bacteria can convert this organic nitrogen back into atmospheric nitrogen, thus completing the nitrogen cycle. Early in the twentieth century, an industrial process was invented for producing a biologically useful form of nitrogen that can be applied to crops. Nitrogen runoff from agriculture is a form of nutrient pollution that can harm aquatic ecosystems. (p. 693)

- All of Earth's water either is being cycled or is being stored—in such forms as glaciers or polar ice. As little as 0.5 percent of Earth's water is available as fresh, liquid water, with about 25 percent of this being groundwater, which generally is stored in aquifers. (p. 694)

- Fresh, sanitary water is a scarce commodity even for human beings, despite the fact that civilization now uses more than half the world's accessible water. This scarcity of water can be traced in significant part to the inefficient ways in which humans use it. Human diversion of water from natural environments is having harmful effects on other species. (p. 696)

36.3 How Energy Flows through Ecosystems

- Plants and other photosynthesizers are an ecosystem's producers, while the organisms that do not perform photosynthesis are its consumers. Every ecosystem has a number of feeding or trophic levels, with producers forming the first trophic level and consumers forming several additional levels. (p. 699)

- A detritivore is a class of consumer that feeds on the remains of dead organisms or cast-off material from living organisms. A decomposer is a special kind of detritivore that breaks down dead or cast-off organic material into its inorganic components, which can then be recycled. (p. 699)

- The energy-flow model of ecosystems measures energy as it is used by and transferred among members of an ecosystem. (p. 701)

- In general, for each jump up in trophic level, the amount of available energy drops by 90 percent. This reduction in energy by trophic level largely explains the superabundance of plants, the smaller abundance of herbivores, and the smaller-

yet abundance of carnivores. The makeup of given ecosystems can also be affected by the consumption of second-level herbivores by third-level carnivores. (p. 702)

36.4 Earth's Physical Environment

- The atmosphere is the layer of gases surrounding the Earth. Nitrogen and oxygen make up 99 percent of the lowest layer of the atmosphere, the troposphere, which also contains small amounts of carbon dioxide and gases such as argon and methane. (p. 703)

- Ozone, which exists primarily in the stratosphere, screens out 99 percent of the sun's potentially harmful ultraviolet radiation. Human-made compounds such as chlorofluorocarbons (CFCs) can destroy ozone. International action taken on this issue in the 1980s is bringing the ozone layer back to health. (p. 703)

36.5 Global Warming

- Global warming refers to the increasing temperature of the Earth's atmosphere. Between 1906 and 2005, Earth's surface temperature increased by about 1.3°F, and the planet may warm by more than this amount in coming decades, as a result of human activities such as deforestation and the emission of carbon dioxide (CO_2) and methane. (p. 704)

- All of the long-term consequences of global warming cannot be predicted because such consequences will vary with the amount of greenhouse gases emitted in coming years. Among the certain consequences are a rise in sea levels, a change in rainfall patterns, and an alteration in the mix of species in different geographical regions. (p. 707)

- Recent scientific calculations indicate that societies may have a short time frame in which to take action if the phenomenon of dangerous climate change is to be avoided. (p. 708)

36.6 Earth's Climate

- Earth is tilted at an angle of 23.5° relative to the plane of its orbit around the sun, which dictates much about climate on Earth, which in turn dictates much about life. (p. 709)

- Sunlight strikes the equatorial region more directly than the polar regions. The differential warming that results produces a set of enormous interrelated circulation cells of moving air, each of which drops rain on the Earth where the moving air rises but dries Earth where it descends. (p. 709)

- A climate is an average weather condition in a given area. Large vegetative formations essentially are defined by climate regions. (p. 711)

36.7 Earth's Biomes

- Biomes are large terrestrial regions that have similar climates and hence similar vegetative formations. Six types of biomes are recognized at a minimum: tundra, taiga, temperate deciduous forest, temperate grassland, desert, and tropical rainforest. Polar ice, mountains, savannas, and chaparral often are recognized as separate biomes. (p. 712)

- Tundra, the biome of far northern latitudes, is frozen much of the year but has a seasonal vegetation formation of low shrubs, mosses, lichens, grasses, and the grass-like sedges. (p. 712)

- Taiga is found at northern latitudes and exhibits a great deal of species uniformity, with only a few types of trees serving as ecological dominants. (p. 713)

- Temperate deciduous forests, which grow in regions of greater warmth and rainfall than is the case with tundra or taiga, are composed of an abundance of trees, complemented by a robust understory of woody and herbaceous plants. (p. 713)

- Temperate grassland (also known as prairie and steppes) is characterized by less rainfall than that of temperate forest and by grasses as the dominant vegetation. Such regions can be fertile agricultural land. (p. 714)

- Chaparral is a biome dominated by evergreen shrub vegetation. Such regions have mild, rainy winters and very dry summers. (p. 715)

- Deserts are characterized by both low rainfall and water evaporation rates that are high relative to rainfall. Deserts may be hot, cold, or temperate, but all desert life is shaped by the need to collect and conserve water. (p. 715)

- Tropical rainforest is characterized by warm, stable temperatures, abundant moisture, great biological productivity, and great species diversity. Rainforest productivity is concentrated above the forest floor, often in the canopy high above ground. Tropical rainforests are being significantly reduced in size today through cutting and burning. (p. 715)

36.8 Life in the Water: Aquatic Ecosystems

- Marine ecosystems are most biologically productive near the coasts. Coastal areas benefit from wave actions that bring in nutrients, carry away wastes, and expose more of the surface area of photosynthesizers to sunlight. (p. 716)

- There is relatively more ocean life toward the poles, with this concentration decreasing with movement toward the equator. The food webs in the oceans surrounding Antarctica are based on photosynthesizing phytoplankton. (p. 718)

- Coral reefs provide a habitat that results in a rich species diversity. The health of many of the world's coral reefs is being imperiled by human activity. (p. 718)

- Freshwater ecosystems cover only about 2.1 percent of Earth's surface. Freshwater lakes are most productive near their shores and near their surface. Lakes can be naturally eutrophic, meaning nutrient rich, or oligotrophic, meaning nutrient poor. (p. 719)

- Estuaries are areas where streams or rivers flow into the ocean. They are characterized by high biological productivity because of the constant movement of water, which stirs up nutrients. (p. 720)

- Wetlands are lands that are wet for at least part of the year. Wetland soil may merely be waterlogged for part of the year or under a permanent, relatively deep cover of water. Wetlands are very productive and are important habitats for migratory birds. (p. 721)

Key Terms

Understanding the Basics

Multiple-Choice Questions

(Answers are in the back of the book.)

1. Carbon (select all that apply):
   a. is an element
   b. is a nutrient.

c. moves into the living world through animals.

d. moves into the atmosphere through the work of decomposers.

e. makes up about 21 percent of Earth's atmosphere.

2. Nitrogen (select all that apply):
a. is not a nutrient.
b. is transferred into the living world in nature through the actions of plants.
c. is transferred into the living world in nature through the actions of bacteria.
d. has been manufactured by humans since the early twentieth century.
e. exists in such low concentrations in some coastal waters that "dead zones" are the result.

3. Associate each term on the left with one on the right.

a. herbivore   photosynthesizing organism
b. net primary production   material produced through photosynthesis
c. gross primary production   position in an ecosystem's food chain
d. trophic level   organism that eats plants
e. producer   material accumulated through photosynthesis

4. There is a simple food chain in an eastern Oregon desert grassland: Mice are consumed by several species of snakes, which are consumed by hawks. The hawk is a _____ and occupies the _____ trophic level in this ecosystem.
a. tertiary consumer; fourth
b. secondary consumer; fourth
c. tertiary consumer; third
d. secondary consumer; third
e. primary consumer; third

5. A dung beetle (can you guess what it eats?) is an example of a:
a. detritivore.
b. omnivore.
c. carnivore.
d. herbivore.
e. producer.

6. Concentrations of heat-trapping "greenhouse" gases such as carbon dioxide are measured in:
a. kilograms.
b. tons.
c. parts per billion.
d. parts per million.
e. millions of molecules.

7. The greenhouse effect is caused by:
a. incoming long-wave radiation that is trapped as short-wave radiation by the atmosphere.
b. short-wavelength light that is re-radiated as long-wave radiation and is trapped by certain gases.
c. heat that is stored as short-wave radiation by rough surfaces of the Earth and then re-radiated as long-wavelength light.
d. cold air masses that mix with the warm gases and produce warm atmospheric gases.
e. reactions in the stratosphere between these greenhouses gases and ozone, producing long-wave heat.

8. You walk in on the middle of a nature program on television and see aerial shots of huge expanses of one or two types of coniferous trees. The type of biome you are looking at probably is:
a. chaparral.
b. taiga.
c. tundra.
d. savanna.
e. temperate forest.

9. Donna is measuring rainfall over a large elevation gradient in the Cascade Mountains of Oregon. As she walks up the west side, she notices that rainfall increases and that when she reaches the top and walks down the east side, it is much drier. This is because:
a. dry air from the equator flows east at this latitude, dropping moisture on the west side.
b. wet air from the tropical equator flows east at this latitude, dropping

moisture on the west, and then becomes dry by the time it reaches the eastern side.
c. air is flowing from the west and drops moisture as it rises and cools, then absorbs moisture from the land as it descends and warms.
d. warm air from 30° latitudes picks up moisture as it travels to 60° north, then drops this moisture as it cools.
e. the rain-shadow effect means that cool, moist air drops rainfall as it warms with a rise in elevation.

10. A lake affected by high levels of artificial eutrophication will have:
a. high nutrient levels, large phytoplankton populations, and low oxygen levels at depth.
b. high levels of nutrients, low phytoplankton levels, and high oxygen levels in surface waters.
c. low nutrient levels, large phytoplankton populations, and low oxygen levels at depth.
d. low nutrient levels, low phytoplankton populations, and high oxygen levels at depth.
e. low nutrient levels, large zooplankton populations, and low phytoplankton levels.

Brief Review

(Answers are in the back of the book.)

1. In a large reserve in Idaho, Forest Service wildlife scientists measured wolf populations as well as elk populations and found that the wolf population was much lower than that of the elk

population. If wolves feed on elk, why aren't the population sizes similar?

2. Explain why most deserts across the globe occur at about 30° N and 30° S of the equator.

3. What relationship does climate have to the large-scale vegetation patterns that are seen on Earth?

4. A great many dead fish are found floating on the surface of an algae-covered pond. What might be the cause of such a fish kill?

5. Name several ways in which the tundra and taiga biomes differ.

Applying Your Knowledge

1. The filmmakers in the movie *King Kong* land on the imaginary Skull Island, which is the home not only of the giant ape Kong but of a host of meat-eating dinosaurs as well. Setting aside the impossibility of an ape as big as Kong, from an ecological perspective, why could a real Skull Island never have existed, even when dinosaurs roamed the Earth?

2. Bacteria and fungi are vitally important to the living world. If all bacteria and fungi were somehow instantly eliminated, almost no life-forms would survive for long. This is so because of the role bacteria and fungi play in ecological processes. Can you name two of these processes?

3. Tundra and desert are in some ways more fragile biomes than are grasslands or forests. Why should this be so?

Appendix 1

Metric–English System Conversions

Length

1 inch (in.) = 2.54 centimeters (cm)

1 centimeter (cm) = 0.3937 inch (in.)

1 foot (ft) = 0.3048 meter (m)

1 meter (m) = 3.2808 feet (ft) = 1.0936 (yd)

1 mile (mi) = 1.6904 kilometer

1 kilometer (km) = 0.6214 mile (mi)

Volume

1 cubic inch (in.^{3}) = 16.39 cubic centimeters (cm^3 or cc)

1 cubic centimeter (cm^3 or cc) = 0.06 cubic inch (in.^{3})

1 cubic foot (ft^3) = 0.028 cubic meter (m^3)

1 cubic meter (m^3) = 35.30 cubic feet (ft^3) = 1.3079 cubic yards (yd^3)

1 fluid ounce (oz) = 29.6 milliliters (mL) = 0.03 liter (L)

1 milliliter (mL) = 0.03 fluid ounce (oz) = 1/4 teaspoon (approximate)

1 pint (pt) = 473 milliliters (mL) = 0.47 liter (L)

1 quart (qt) = 946 milliliters (mL) = 0.9463 liter (L)

1 gallon (gal) = 3.79 liters (L)

1 liter (L) = 1.0567 quarts (qt) = 0.26 gallon (gal)

Area

1 square inch (in.^{2}) = 6.45 square centimeters (cm^2)

1 square centimeter (cm^2) = 0.155 square inch (in.^{2})

1 square foot (ft^2) = 0.0929 square meter (m^2)

1 square meter (m^2) = 10.7639 square feet (ft^2) = 1.1960 yards (yd^2)

1 square mile (mi^2) = 2.5900 square kilometers (km^2)

1 acre (a) = 0.4047 hectare (ha)

1 hectare (ha) = 2.4710 acres (a) = 10,000 square meters (m^2)

Mass

1 ounce (oz) = 28.3496 grams (g)

1 gram (g) = 0.03527 ounce (oz)

1 pound (lb) = 0.4536 kilogram (kg)

1 kilogram = 2.2046 pounds (lb)

1 ton (tn), U.S. = 0.91 metric ton (t or tonne)

1 metric ton = 1.10 tons (tn), U.S.

Metric Prefixes

Prefix	Abbreviation	Meaning
giga-	G	$10^9 = 1,000,000,000$
mega-	M	$10^6 = 1,000,000$
kilo-	k	$10^3 = 1,000$
hecto-	h	$10^2 = 100$
deka-	da	$10^1 = 10$
		$10^0 = 1$
deci-	d	$10^{-1} = 0.1$
centi-	c	$10^{-2} = 0.01$
milli-	m	$10^{-3} = 0.001$
micro-	μ	$10^{-6} = 0.000001$

Appendix 2

PERIODIC TABLE OF THE ELEMENTS

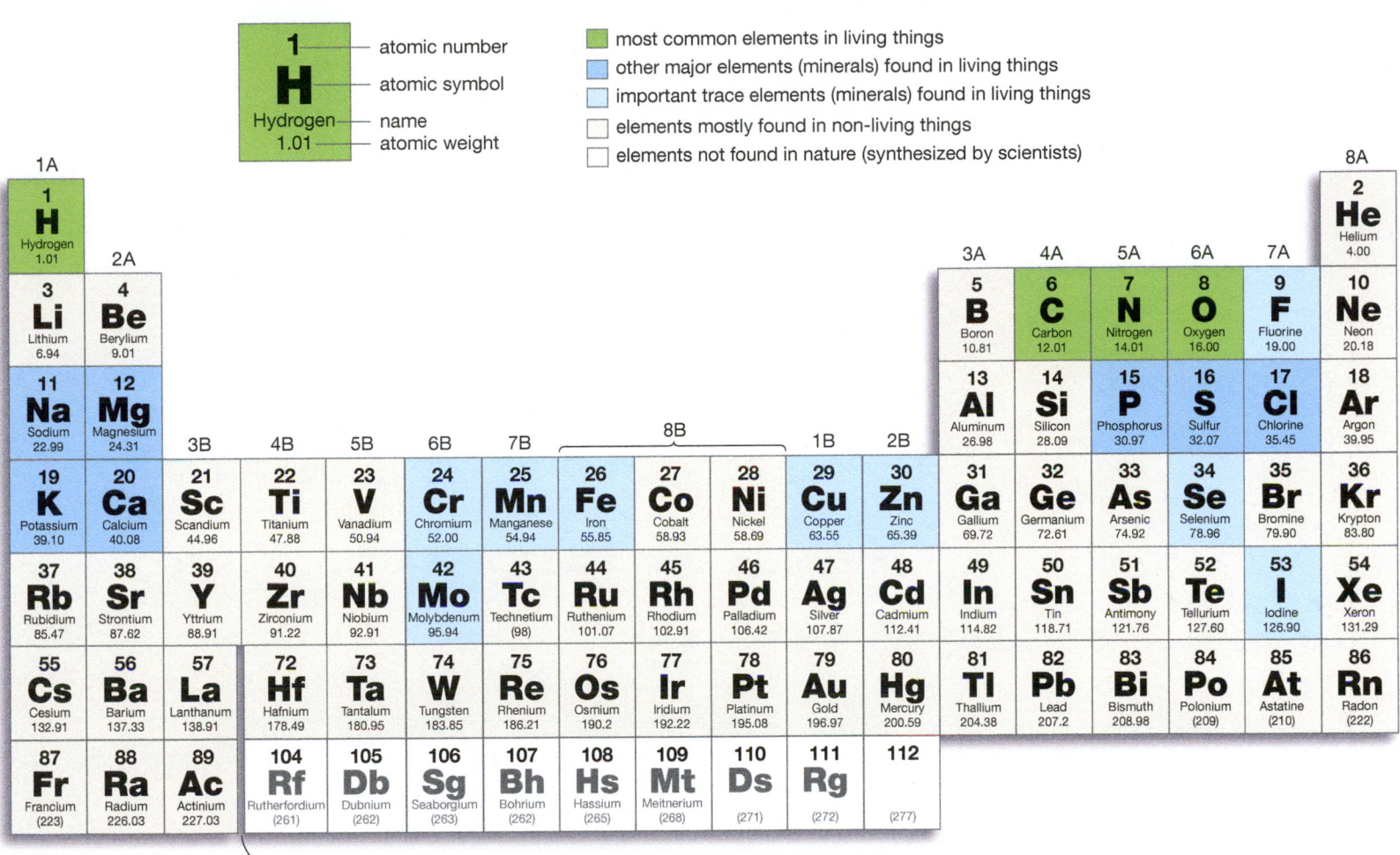

Answers

CHAPTER 1

So Far . . .

page 8

1. guess; principles; nature
2. hypothesis
3. constant; variable

page 9

1. revision; evidence
2. negation
3. natural; supernatural

page 15

1. molecule, cell, organ, organism, community, ecosystem
2. modification; species
3. life

Multiple-Choice

1. c 2. d 3. d 4. a, d, e 5. c 6. e

Brief Review

1. Science is a body of knowledge regarding nature and a formal process for acquiring that knowledge. It is thus a collection of unified insights and a way of learning about the physical world. Unlike religion, science does not acknowledge any immutable truths or unquestioned authority figures. Instead, all scientific principles are open to modification based on the discovery of new evidence.

2. An experiment is said to be controlled when the experimenter controls both the factors in it that will change and those that will be held constant. Suppose a scientist wanted to test the effect of the insecticide DDT on the reproductive development of lab rats. In this case, the experiment might consist of feeding DDT-laced food to one group of rats, while another similar group—called the control group—would receive the same food without DDT. It is important that the two groups of rats be as closely matched as possible in all other factors. For instance, both groups must consist of the same strain of lab mice; both must be the same gender or mix of genders; they must be given the same amount of food and water and housed in identical environments. Suppose differences were observed in, for example, the size of the offspring born to one group of rats as opposed to the other. If the starting conditions were not controlled in the experiment, then there would be no way to tell whether the DDT or some other factor (such as different housing) was responsible for these differences.

3. Pasteur investigated the notion, widely held during his time, that living things can arise spontaneously from basic chemicals. He hypothesized that instances in which life appeared to arise in a seemingly lifeless medium could be explained as the result of contamination of the medium by unseen airborne microscopic organisms. He sterilized meat broth in a flask with a long, S-shaped neck by heating both the broth and the flask. Although the mouth of the long tube remained open, any incoming dust particles laden with microscopic organisms would be trapped in the bent neck and could not gain access to the meat broth. Under these conditions, the broth remained free of any visible signs of life. In other tests, the S-shaped neck was broken off or the flask tilted so that broth made contact with the neck of the flask; in these instances, the broth came "alive" with teeming bacteria and fungi. Pasteur inferred from these experiments that the appearance of life-forms depended on whether the broth was contaminated by airborne microscopic organisms. When the airborne organisms had no access to the broth, however, it remained sterile. There was no evidence, in other words, of life coming about through spontaneous generation.

4. All living things, as we know them on this planet, acquire, store, and use energy; they respond to stimuli in their environment; they maintain a relatively constant internal environment (homeostasis); they reproduce, using DNA to transmit hereditary information; they exhibit a high level of organization; and they show evidence of having evolved from other organisms.

5. Atoms, molecules, organelles, cells, tissues, organs, organ systems, organisms, populations, communities, ecosystems, the biosphere.

CHAPTER 2

So Far . . .

page 22

1. proton; neutron; electron
2. element; protons
3. neutrons

page 25

1. filled; eight; two
2. covalent
3. polar covalent; electrical charge

page 31

1. loses; electrons; charges
2. electronegative; oxygen; nitrogen
3. bind

page 35

1. solute; solvent; solution
2. hydrogen bonds; cohesion; specific heat
3. hydrophobic; hydrophilic

page 39

1. yields; ions; accepts
2. 7; 0; 14
3. neutral (7)

Multiple-Choice

1. b, d 2. b, c 3. a 4. d 5. b 6. e 7. e 8. c

Brief Review

1. The forms are called isotopes. Isotopes of an element differ from one another in accordance with the number of neutrons they have. A regular carbon atom has six neutrons in its nucleus, while a carbon-14 atom has eight neutrons.

2. An atom's electrons move through volumes of space outside the atom's nucleus. An entire atom is about 100,000 times larger than its nucleus. Thus, most of an atom is the space through which electrons move.

3.

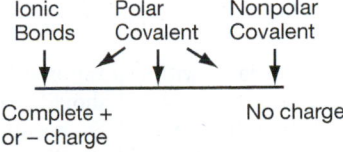

4. When its valence (outer) shell is filled, an atom is at a lower, more stable energy state.

5. Water absorbs tremendous amounts of heat from the sun and releases this heat slowly with the onset of night's colder temperature. The perspiration human beings throw off carries with it a great deal of heat, which has been absorbed by the water.

6. Living things must control their internal pH, keeping it in most instances near the neutral point of 7, because an extreme pH can begin to interfere with such critical

structures as membranes and such critical processes as the work of enzymes. Put another way, the normal structures and operations of organisms can begin to break down if pH is allowed to fluctuate very far from near-neutral levels.

CHAPTER 3

So Far . . .

page 45

1. carbon; bonds with other elements
2. atoms; carbon-based

page 48

1. monomers; polymers
2. carbon; oxygen; hydrogen; monosaccharides; polysaccharides (or complex carbohydrates)
3. starch; glycogen; cellulose; chitin

page 55

1. readily dissolve; water
2. fatty acids; glycerol; carbon atoms; hydrocarbon
3. carbon rings; estrogen; testosterone; phospholipids; phosphate

page 61

1. amino acids; amino acids; folded up
2. lipoproteins; lipids; proteins
3. nucleotides; sugar; phosphate group; nitrogen-containing bases

Multiple-Choice

1. c 2. e 3. a 4. e 5. b 6. b 7. d 8. a

Brief Review

1. Carbon is a mere trace element in Earth's atmosphere and crust and yet makes up nearly 50 percent of the dry weight of living things. It has this status because of its chemical bonding abilities: It has four outer or "valence" electrons, meaning it achieves maximum chemical stability by linking with four more electrons. This bonding capacity means it is able to form strong bonds with many other elements, which is important in creating the kind of complex molecules that are conducive to life.
2. LDL is associated with deposition of cholesterol in the arteries, so you should be concerned about high levels. HDL is associated with carrying cholesterol to the liver, away from the coronary arteries where it does damage, so if your HDL count is high, that's positive.
3. Steroids are a class of lipid molecules that have four carbon rings as a central element in their structure. Under this definition, both cholesterol and estrogen are steroids. In everyday speech, "steroids" refers to a subset of the steroid molecules—those that serve to enhance the ability of muscle to grow stronger. Testosterone is a steroid in this sense, while cholesterol and estrogen are not.

4. Proteins can take on the roles of enzymes, hormones, transport proteins, structural proteins, contractile proteins, storage proteins, and communication proteins. Proteins can do many different kinds of jobs because they take on so many different shapes. Each shape allows the protein to do a different job.
5. DNA contains or "encodes" the information for producing proteins. For a protein to be produced, the instructions for the order of its amino acids must be obtained from a length of DNA.

CHAPTER 4

So Far . . .

page 70

1. cell; cells; cell
2. eukaryotic; nucleus; prokaryotic
3. plasma membrane; nucleus; cytosol; organelles

page 76

1. DNA; proteins
2. ribosomes; rough endoplasmic reticulum (RER); Golgi complex
3. transport vesicles; endomembrane

page 83

1. smooth endoplasmic reticulum (smooth ER); lysosomes
2. energy; ATP
3. cytoskeleton

page 87

1. chloroplasts; central vacuole; cell wall
2. lysosome
3. plasmodesmata; gap junctions

Multiple-Choice

1. e 2. d 3. a 4. b 5. e 6. d 7. b 8. e

Brief Review

1. The cell would have difficulty obtaining enough materials from its external environment because its surface-to-volume ratio would be too low.
2. Eukaryotic cells have a nucleus, bound within a double membrane, that contains almost all their DNA. The DNA in prokaryotic cells is localized, but it is not bound within a membrane. Eukaryotes are usually much larger than prokaryotes. Eukaryotes are also often multicelled organisms, while prokaryotes are always single-celled. Eukaryotic cells have many types of organelles, whereas prokaryotic cells have only one type of organelle, the ribosome.
3. The information for the makeup of insulin is encoded in DNA that is copied onto an mRNA sequence, which leaves the nucleus through the pores in the nuclear envelope. The length of mRNA then arrives at a ribosome, which reads part of it before migrating to, and embedding in, the rough endoplasmic reticulum (rough ER or RER). The ribosome then continues to read the

RNA chain, and the resulting polypeptide chain drops into the cisternal space of the rough ER. There, the polypeptide chain folds up into the insulin protein, which then undergoes editing. Once this editing has been completed in the rough ER, the insulin is encased in a transport vesicle, which buds off from the rough ER and moves to the Golgi complex for further editing and for routing to its proper location. The finished insulin protein then moves through the cytoplasm in a transport vesicle that fuses with the cell's outer or "plasma" membrane, thereby exporting it from the cell.

4. Cilia and flagella are cell extensions formed by microtubules that are either a means of propulsion or, in the case of some cilia, a means of moving material around a cell. Cilia exist as a profusion of hair-like growths extending from cells; they move back and forth very rapidly, while only a few flagella sprout from a given cell—indeed, there is often just a single, tail-like flagellum. In the human body, cilia are found in the nasal passages and in the lungs. The only animal cell that has a flagellum is a sperm cell.
5. Plant cells would be isolated from each other by their thick cell walls. Because communication among cells is critical for the functioning of a plant, the result would be that the plant could not live.
6. A gap junction consists of clusters of protein assemblages that shoot through the plasma membrane of an animal cell from one side to the other, forming a tube. When these protein assemblages line up in two adjacent cells, the result is a communication channel between them. Gap junctions allow for the passage of small molecules and electrical signals between animal cells.

CHAPTER 5

So Far . . .

page 97

1. phospholipid bilayer; cholesterol; proteins; glycocalyx
2. hydrophilic; water; hydrophobic; water
3. receptors; transport proteins

page 100

1. higher concentration; lower concentration; energy
2. semipermeable; solute concentration; solute concentration
3. out of; into

page 104

1. active transport; passive transport; protein channel; protein channel
2. concentration gradient
3. transport vesicle; plasma membrane; phagocytosis; pinocytosis

Multiple-Choice

1. c 2. d 3. a 4. c 5. e 6. b 7. b 8. a

Brief Review

1. Receptor proteins usually are specific to a single communication molecule. This is so because each receptor has a shape that severely restricts the type of communication molecule it can bind with. The receptor in question has a shape that allows it to bind with glucose, but not to insulin; thus, the insulin molecules it comes into contact with simply pass it by.

2. Because steroids are lipids, they pass easily across the phospholipid bilayer by dissolving in it. Thus, they can affect most cells in the body.

3. Because the phospholipid's phosphate "heads" are charged, hydrophilic groups, they are attracted to the polar water molecules that lie on either side of the plasma membrane, and they orient themselves accordingly. Meanwhile, the phospholipid's "tails" are hydrophobic fatty acid chains that do not bond with water and thus end up oriented away from it, toward each other in the interior of the plasma membrane.

4. In clathrin-mediated endocytosis, a group of receptors extending from the surface of the plasma membrane bind with a given target substance (for example, cholesterol) and then migrate laterally to a depression on the membrane surface—a pit whose underside is coated with the protein clathrin. This pit deepens, or invaginates, eventually becoming a transport vesicle that buds off from the plasma membrane, carrying the target substance into the cell.

5. You would not expect that recognition and transport would be as important inside the cell as outside. Thus, the glycocalyx seen on the plasma membrane is unnecessary inside the cell, and proteins on the surface of interior membranes serve different functions than those on the surface of the plasma membrane.

CHAPTER 6

So Far . . .

page 112
1. potential; kinetic
2. created; destroyed; transformed; disorder (or entropy)
3. heat

page 114
1. products; reactants; reactants; products
2. ATP; phosphate group
3. ADP; phosphate group

page 119
1. protein; accelerates
2. substrate; product; substrate
3. initiate; lowering

Multiple-Choice
1. a 2. c 3. d 4. e 5. e 6. c 7. d 8. e

Brief Review

1. Plants abide by the second law of thermodynamics in that, in building themselves up, they are contributing to the total amount of entropy in the universe: They release heat as a by-product of the process by which they build leaves from carbon dioxide, water, and minerals. A plant leaf represents only a local increase in order.

2. The cocking of the extensions was an uphill or endergonic reaction because the extensions were in a higher energetic state when the reaction was completed than when it started. Any such endergonic reaction must be powered by an energy-releasing exergonic reaction. Thus, the two kinds of reactions are linked together—they are coupled. In this case, the binding of ATP to the extensions, and the resulting release of the ATP molecule's third phosphate group, was the exergonic reaction that powered the change in the extension's position.

3. In animals, the energy stored and released by ATP comes from food. ATP is an energy transfer molecule in that it receives energy from food and then passes that energy on when it powers chemical reactions in the body.

4. Enzymes are involved with every aspect of cellular life. They are necessary for breaking things down but they also are used to rearrange molecules or to make larger molecules out of smaller ones.

5. Allosteric regulation of an enzyme occurs when a molecule binds with an enzyme at a site other than its active site, thereby either decreasing or increasing the rate at which the enzyme will bind with its substrate. Such binding has the effect of either decreasing or increasing the amount of product turned out by the enzyme.

CHAPTER 7

So Far . . .

page 126
1. food; phosphate group; ATP
2. oxidation; reduction; redox
3. electron carriers; NAD^+; NADH

page 128
1. glycolysis; Krebs; electron transport chain (ETC); cytosol; mitochondria
2. energetic electrons; the ETC
3. ATP; NADH; pyruvic acid

page 132
1. acetyl CoA (acetyl coenzyme A)
2. energetic electrons; electron transport chain (ETC); carbon dioxide (CO_2)
3. NAD^+/NADH; FAD/$FADH_2$

page 137
1. lose (donate); membrane; hydrogen ions (H^+ ions)
2. H^+ ions; ATP synthase; phosphate groups; ATP

3. oxygen

Multiple-Choice
1. b 2. a 3. a 4. c 5. d 6. b 7. d 8. d

Brief Review

1. The food that we eat contains energetic molecules, such as glucose. In cellular respiration, foods are oxidized—they have electrons removed from them—and the energy released by the "fall" of these electrons from a higher to a lower energetic state powers phosphate groups onto ADP molecules, making them energetic ATP molecules.

2. We would expect people to have more mitochondria in their muscle cells. This is so because muscle cells use a great deal more energy than skin cells, and the organelles that supply that energy are mitochondria.

3. Hydrogen ions (H^+ ions or protons) are pumped across the mitochondrial inner membrane—against their concentration and charge gradients—and their energetic fall back through the ATP synthase in the membrane releases enough energy to move phosphate groups onto ADP.

4. The electron carriers NADH and $FADH_2$ are oxidized at the ETC; the energetic fall of the electrons that are removed from them powers the movement of the H^+ ions across the membrane.

5. No. In short bursts of activity, the energy we are using comes primarily from stored ATP and from glycolysis.

CHAPTER 8

So Far . . .

page 146
1. electrons; carbon dioxide; carbohydrate (a sugar)
2. chloroplasts; leaves; chlorophyll a
3. light reactions; primary electron acceptor; energy state

page 149
1. electrons; NADPH; ATP; oxygen
2. Calvin or C_3
3. sugar; atmosphere; electrons: ATP

page 153
1. low-energy sugar; carbon; oxygen; photorespiration
2. warm-weather; C_4
3. CAM; at night; during the day

Multiple-Choice
1. c 2. e 3. a 4. d 5. e 6. c 7. b 8. c

Brief Review

1. Yes, plants are as dependent on cellular respiration as we are. Plants make food—carbohydrates—in photosynthesis. But just as we have to break down the food we eat to get the energy transferred to ATP, so plants have to break down the food they make to move the energy into ATP.

2. The sunlight that makes it to the Earth's surface is composed of a spectrum of energetic rays (measured by their "wavelengths") that range from very short ultraviolet rays, through visible light rays, to the longer and less-energetic infrared rays. Photosynthesis is driven by part of the visible light spectrum—mainly by blue and red light of certain wavelengths.

3. A photosystem includes, first, several hundred antennae pigments consisting of both chlorophyll *a* and accessory pigments. These molecules collect solar energy and pass it along to another part of the photosystem, the reaction center. The reaction center consists of a pair of special chlorophyll *a* molecules and an initial electron acceptor molecule. These reaction center compounds receive both the solar energy from the antennae pigments and electrons derived from the splitting of water.

4. The C_4 plant would be likely to grow more rapidly in a warm-weather climate. Warm weather causes all plants to close their stomata for longer periods of time as a means of preserving water. When their stomata are closed, however, plants lose access to atmospheric carbon dioxide. In C_3 plants, this series of events leads to an increase in photorespiration, since C_3 plants utilize an enzyme, rubisco, that can bind to internally produced oxygen as well as to atmospheric CO_2. Hence, photorespiration—which undercuts plant growth—is increased in C_3 plants in warm weather. Meanwhile, C_4 never experience photorespiration since they utilize an initial CO_2-fixing enzyme that will not bind to oxygen.

5. The ATP is used to power the steps of the Calvin cycle. ATP is used in two separate steps during the cycle: once when 3-PGA (3-phosphoglyceric acid) is being energized by electrons produced by the light reactions and again when G3P is being transformed into the RuBP that begins the Calvin cycle.

CHAPTER 9

So Far . . .

page 160

1. proteins; bases; A, T, G, C
2. amino acids; bases; gene
3. messenger RNA (mRNA); ribosome; amino acids; mRNA bases; protein

page 164

1. replicate; duplicate
2. proteins; sister chromatids
3. homologous; X; X; Y

page 169

1. chromosomes; daughter cells
2. mitotic: interphase
3. sister chromatids; chromosome

page 171

1. cytokinesis; cell wall; plasma membrane
2. nucleus; plasma membrane; binary fission

Multiple-Choice

1. e 2. d 3. d 4. e 5. a 6. c 7. b 8. c

Brief Review

1. DNA contains information for the production of proteins. These proteins are "workers" and building blocks that serve a vast array of functions in organisms such as ourselves. Through the work of proteins, our bodies develop and carry out everyday functions.

2. The cell is in metaphase. There are eight chromatids present in this stage, and each daughter cell will have four chromosomes.

3. Mitosis serves to equally divide a cell's genetic material into two portions of a parent cell that will become daughter cells. Cytokinesis is the physical separation of a parent cell into two daughter cells.

4. Prophase—Duplicated chromosomes condense and become visible; the centrosome duplicates and centrosomes migrate to the poles; the nuclear membrane breaks down; and the mitotic spindle is formed. Metaphase—Chromosome pairs align along the cell midline or metaphase plate and attach to the microtubules extending from the cellular poles. Anaphase—The sister chromatids paired along the metaphase plate separate and then migrate to the two poles of the cell. Telophase—The condensed chromosomes unwind at their respective poles, and new nuclear membranes are formed around the chromosomes. At this point, division of genetic material is complete. The cell is ready for cytokinesis.

5. Plant cells have cell walls, while animal cells do not. In plant cells, membrane-lined vesicles, filled with a complex sugar, migrate to an area near the metaphase plate and begin fusing together, eventually forming a cell plate that runs from one side of the parent cell to the other. The membrane portion of the plate fuses with the original parent-cell plasma membrane, thus dividing the parent cell in half via formation of two new plasma membranes. The material inside the cell-plate membrane then forms the foundation of the new cell walls of the daughter cells.

6. The 23 pairs of human chromosomes are matched (or homologous) in the sense that the two members of each pair contain information about similar functions, such as hair color, metabolic processes, and so forth. The one exception to the matched-pair rule comes in human males, who have 22 pairs of matched chromosomes, but then one X chromosome and one Y chromosome, which are not matched. Human females have 23 matched pairs, including two X chromosomes.

7. (a) Four different bases—adenine, thymine, guanine, and cytosine, or A, T, G, and C, respectively; (b) in the nucleus; (c) messenger; (d) mRNA is used as a cellular messenger; DNA's information is copied onto it in the nucleus, after which it carries this information to the cytoplasm, where the information is used within ribosomes to put together sequences of amino acids that form proteins; (e) amino acids; (f) in ribosomes residing in the cytoplasm.

CHAPTER 10

So Far . . .

page 178

1. eggs; sperm; diploid; haploid
2. homologous chromosomes; sister chromatids
3. 46; 23

page 182

1. homologous chromosomes; homologous chromosome pairs
2. differ
3. X; X; Y

page 187

1. stem; prior to birth
2. polar bodies
3. identical

Multiple-Choice

1. b 2. c 3. a 4. b 5. b 6. d 7. d 8. a 9. d

Brief Review

1. In mitosis, a single round of chromosome duplication is followed by a single round of cell division. In this cell division, it is sister chromatids (of the duplicated chromosomes) that line up on opposite sides of the metaphase plate and then separate from each other. In meiosis, in contrast, a single round of chromosome duplication is followed by two rounds of cell division. In the first of these rounds, in meiosis I, it is duplicated homologous chromosomes, rather than sister chromatids, that line up on opposite sides of the metaphase plate and then separate from one another. In the second of these rounds, in meiosis II, sister chromatids separate, going to separate cells.

2. Chromosome reduction takes place during meiosis I.

3. The cells in males that go through meiosis, the primary spermatocytes, are themselves the product of cells called spermatogonia. These spermatogonia are stem cells in that, with each cellular division, they give rise not only to primary spermatocytes but to more spermatogonia as well. It is this dual capability in spermatogonia that allows males to keep producing sperm throughout their lifetimes.

4. Female meiotic cytokinesis differs from mitotic cytokinesis in the selective shunting of cytoplasm. In mitosis, cytokinesis divides the cytoplasmic material of a parent cell equally among daughter cells. In female meiosis, cytokinesis preferentially shunts cytoplasmic material to one daughter cell—the cell destined to become the ovum—and

away from the other cells, which become polar bodies.

5. Somatic cells are diploid cells that undergo mitosis. They include every cell in the body except the body's reproductive cells, its gametes, which are sperm in males and eggs in females. Gametes are haploid cells that are derived from diploid cells through the process of meiosis.

6. (**a**) metaphase I, meiosis I; (**b**) anaphase II, meiosis II; (**c**) anaphase, mitosis; (**d**) metaphase II, meiosis II; (**e**) metaphase, mitosis.

7. Meiosis ensures genetic diversity in two ways. In the process called *crossing over* or *recombination*, homologous chromosomes swap reciprocal portions of themselves with each other. Then, in the process called *independent assortment*, homologous chromosome pairs line up in a random way, relative to one another, along the metaphase plate.

8. A sperm has 23 chromosomes and an egg has 23 chromosomes; when sperm fertilizes egg, the result is a zygote with 46 chromosomes—22 homologous pairs of chromosomes and either two X chromosomes (if the zygote is female) or an X and a Y chromosome (if the zygote is male).

CHAPTER 11

So Far . . .

page 196

1. phenotype; genotype
2. dominant; dominant; recessive
3. alleles; homologous chromosomes

page 201

1. alleles; separate
2. identical; differing
3. dominant; recessive

page 204

1. heterozygous; homozygous
2. independent; heterozygous
3. expressed; a second recessive allele

page 208

1. many; polygenic; false; polygenic inheritance tends to produce continuous variation among organisms.
2. bell curve; average
3. environmental

Multiple-Choice

1. a 2. a 3. c 4. b 5. b 6. e 7. a,d 8. e

Brief Review

1. (**a**) Phenotype is a physiological feature, behavior, or physical feature of an organism, while genotype is the genetic composition responsible, at least in part, for the phenotype. In an example from the text, the genotypes *YY* or *Yy* resulted in the phenotype of yellow peas, while the genotype *yy* resulted in the phenotype of green peas. (**b**) The term *dominant* is used to refer to an allele that is expressed in the heterozygous genotype. In the heterozygous genotype *Yy*,

for example, *Y* is expressed (as yellow color) and thus is said to be dominant over *y*. The term *recessive* is used to describe an allele that is not expressed in the heterozygous genotype. Looking again at the heterozygous genotype *Yy*, the green *y* allele is not expressed, since any *Yy* seed will be yellow. The *y* allele is thus is said to be recessive to *Y*. (**c**) Codominance refers to the condition in which a heterozygote simultaneously exhibits the phenotypes associated with each of the alleles in its genotype. This is seen in blood types, where an AB individual produces both A and B cell surface molecules. Incomplete dominance is displayed when the heterozygote displays a phenotype that is intermediate between the dominant and recessive phenotypes. An example is the generation of pink flowers resulting from a parental cross of one red-flowering and one white-flowering snapdragon.

2. True.
3. Each person is likely to possess two alleles for any given gene; one allele is inherited from the individual's father, the other from the person's mother.
4. In Mendelian genetics, two different genotypes may bring about a dominant phenotype; for example, *YY* and *Yy* bring about yellow seed color.
5. About 50 percent of the progeny will be yellow peas.
6. Mendel's "element of inheritance" is now referred to as the gene.
7. Mendel's first law (law of segregation): Members of a gene pair will segregate from each other during gamete formation such that each gamete receives only one of the original two genes from the gene pair. Mendel's second law (law of independent assortment): During gamete formation, gene pairs assort independently of one another.
8. (**a**) Aa × aa; (**b**) Aa × Aa; (**c**) Aabb × aaBb; (**d**) AaBb × AaBb.

Genetics Problems

1. a 2. b 3. b 4. b 5. c 6. e 7. c 8. c
9. Since the purple allele is dominant, flowers will be purple whether the allele is present in two identical copies (homozygous state) or in one copy (heterozygous state). In contrast, expression of the recessive white allele is masked if a purple allele is present in the genotype. Therefore, it must be in homozygous form for the development of white flowers.
10. b
11. (**a**) 3. (**b**) The rule of multiplication. (**c**) 18.75 percent of the offspring should be purple flowered with green seeds. This answer is the product of the joint occurrence of two independent events: the probability of purple flower color (0.75) and the probability of green seed color (0.25).

12. It is possible to unambiguously determine genotype for snapdragons because incomplete dominance governs their color; every genotype manifests as a different phenotype. In the case of flower color in peas, phenotype cannot always unambiguously tell you about genotype. For white flowers, genotype can be assigned from phenotype, but for purple flowers, there are two possible genotypes, *PP* and *Pp*.

13. d
14. c

CHAPTER 12

So Far . . .

page 217

1. functional allele; faulty allele; functional allele
2. True; her second X chromosome contains functional alleles, which are enough to keep her from being color-blind.
3. heterozygous; one faulty allele

page 220

1. genetic; families
2. chromosomes; full set
3. separate; one chromosome too many; one too few

page 228

1. a, b, and c
2. mitosis; meiosis; line
3. segment; chromosome

Multiple-Choice

1. c 2. a 3. d 4. e 5. d 6. e 7. c 8. a

Brief Review

1. Male. Sex-linked diseases usually are associated with the X chromosome, which females have two copies of. A female can carry one faulty version (or allele) of an X chromosome gene and still be protected by a functional allele on her second X chromosome. Males, however, have only a single copy of the X chromosome. Should an allele on it be faulty, they have no second, protective copy of it.

2. Aneuploidy goes widely unrecognized because, in most instances, it is directly affecting embryos rather than children or adults. The most common result of aneuploidy is a miscarriage of the embryo. In many instances, prospective parents will not realize that a miscarriage has taken place, and even when a miscarriage is recognized as such, prospective parents would rarely be able to identify aneuploidy as its cause.

3. An aneuploidy that results in cancer would take place during mitosis, or the division of a regular, nonsex cell in the body. In consequence, not every cell in the body would be affected by this aneuploidy; only the line of cells stemming from the original aneuploid cells would be affected. Conversely, an ane-

uploidy that gives rise to a Down syndrome child takes place during meiosis—the formation of eggs or sperm—and hence affects every cell in the body.

4. Deletions—A portion of the chromosome is completely lost.

Inversions—A portion of the chromosome is broken off and rejoins the original chromosome, but in a reversed order.

Translocations—Occur when two chromosomes that are not homologous exchange pieces, leaving both with improper gene sequences.

Duplications—A portion of a chromosome is duplicated, in some instances by gaining a chromosome piece during abnormal crossing over.

5. (**a**) nondisjunction; (**b**) polyploidy.

6. Because people who have Klinefelter syndrome have a Y chromosome, which confers the male sex, but they also have two copies of the X chromosome, rather than the one copy that males normally possess. Females normally have two copies of the X chromosome. Thus, it makes sense that those with Klinefelter syndrome would exhibit some female characteristics.

7. A heterozygous genotype for hemoglobin would be advantageous in areas of mosquito infestation, while neither homozygous genotype would be advantageous in such areas. Those who are homozygous for hemoglobin A would not have the increased resistance to malaria; those who are homozygous for hemoglobin S would inherit sickle-cell anemia. Those who are heterozygous would have malaria resistance and yet would not inherit sickle-cell anemia.

Genetics Problems

1. None will be affected (although half, on average, will be carriers).

2. One-half of her sons can be expected to be affected. The chance that her first child is affected is 1/4 since there is a 1/2 chance the first child is a boy and a 1/2 chance that any boy is affected (another application of the rule of multiplication).

3. Recessive. Neither of the parents shows the autosomal trait, yet one of their offspring does.

4. His genotype is *aa*.

5. Her genotype must be *Aa*. (Reasoning: if she were *AA*, none of her children would be affected, whereas one is; if she were *aa*, she would manifest the trait herself, which she does not.)

6. One-half can be expected to be affected.

7. d

8. Her genotype is X^+X^p. (Reasoning: She is not affected, her mate is not affected, yet they have affected sons. This is possible only if she is a carrier.)

9. One-quarter of children and one half of their sons will show this trait (no daughters will have the trait).

10.
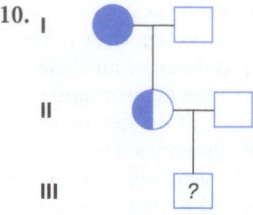

The chance that the son is color blind is 1/2. This is because the woman in question must be a carrier.

11. d

12. One-half of sons and daughters will have hyperphosphatemia. This is in sharp contrast to what would be observed if the trait were X-linked recessive. In this case, half the sons and none of the daughters would be affected.

CHAPTER 13

So Far . . .

page 238

1. structure of DNA
2. A; C
3. bases; protein

page 241

1. new, complementary strands
2. DNA polymerases
3. mutation

Multiple-Choice

1. e 2. e 3. d 4. e 5. c 6. e 7. b 8. b

Brief Review

1. "Something old, something new" is a short-hand means of conveying that, in DNA replication, each newly formed DNA double helix is a mixture of the old and new. Each one is made of an existing "template" strand of DNA and a new strand that is complementary to it.

2. First, the structure of DNA suggested an answer to the question of how genetic information can be passed on from one cell to the next, or from one generation to the next. The answer was: through base-pairing between DNA bases on an existing strand of DNA and bases that are complementary to them. Second, the varied way that bases can be arranged along the "rails" of the DNA helix suggested the means by which DNA could be a versatile enough molecule to contain information for the production of the huge number of proteins found in cells.

3. True—DNA polymerases can remove a mismatched nucleotide and replace it with a correct nucleotide.

4. A mutation is a permanent alteration in a DNA base sequence. A change in a DNA base sequence only becomes permanent when it goes uncorrected by a cell's DNA error-correcting mechanisms, meaning the change will be copied over and over during DNA replication. Not all mutations are harmful. Most have no effect at all, and a

very small proportion actually are useful to organisms as they result in the production of new, functional proteins.

5. Huntington disease results from an abnormal number of repeats of the bases CAG in the gene that codes for the huntingtin protein. The faulty protein clogs up nerve cells, eventually killing them.

CHAPTER 14

So Far . . .

page 246

1. amino acids
2. messenger RNA (mRNA); ribosome
3. ribosome; transfer RNA (tRNA)

page 252

1. complementary base pairing
2. DNA; RNA; amino acid
3. amino acid; messenger RNA codon (triplet)

page 259

1. DNA; transcription
2. micro-RNAs; messenger RNAs
3. editing; messenger RNAs; introns; exons

Multiple-Choice

1. e 2. a 3. c 4. d 5. e 6. d 7. c 8. a

Brief Review

1. The replication of DNA and the production of proteins from DNA.

2. (**a**) mRNA (messenger RNA), which takes DNA information to ribosomes; (**b**) rRNA (ribosomal RNA), which helps make up the ribosomes that act as the sites of protein synthesis; and (**c**) tRNA (transfer RNA), which brings amino acids to the ribosomes and binds with mRNA codons there.

3. Transcription is the process by which the information encoded in DNA sequences is put into the form of mobile mRNA sequences. Translation is the process by which the information encoded in mRNA sequences is put into the form of amino acid sequences; this results in the production of a protein.

4. The enzyme complex RNA polymerase both unwinds DNA that is to be transcribed and pairs DNA bases with their complementary RNA bases.

5. Three; one.

6. The genetic operation is self-regulating in that some of the proteins produced through it go on to regulate the production of other proteins. This is accomplished by means of the regulating proteins (called *transcription factors*) binding to non-coding sequences of DNA, thus facilitating or impeding the work of RNA polymerase, which transcribes genes.

7. Some of the evidence providing reason to doubt the importance of genes, relative to the importance of the regulation of genes is: (**1**) There appears to be no correlation between the number of genes an organism has

and the complexity of that organism; (2) There is a clear correlation between the proportion of non-coding DNA in an organism's genome and the complexity of that organism; and (3) Scientists who study evolution by looking at the development of animals have uncovered evidence indicating that mutations to non-coding DNA sequences have been more important than mutations to coding sequences in bringing about the evolution of differing features in animals.

CHAPTER 15

So Far . . .

page 268
1. genes; another species
2. restriction enzymes
3. human; protein

page 270
1. genetic copy
2. adult mammals
3. provided the donor cell containing DNA

page 275
1. (a)
2. all; ordinary
3. own cells

page 276
1. DNA
2. small
3. short tandem repeats (STRs)

Multiple-Choice
1. c 2. d 3. e 4. e 5. a 6. c 7. c 8. a

Brief Review
1. It is useful because some human beings cannot turn out sufficient quantities of their own supply of needed human proteins; this is so with some diabetics and the protein insulin, for example.
2. The production of sticky ends is useful because this type of end allows different restriction fragments to be joined together easily to create recombinant DNA.

   Fragment sequences:

   ATCG   GATCCTCCG
   TAGCCTAG   GAGGC

3. Dolly looked like the sheep that donated the udder cell, rather than the sheep that donated the egg, because the udder cell was the source of Dolly's DNA, while the DNA in the egg was removed. It was therefore the DNA in the donor cell that provided the instructions for Dolly's development.
4. PCR is valuable because it allows scientists to quickly make a large quantity of a given segment of DNA from a very small starting sample.
5. Because iPS cells are derived from ordinary adult cells, there is no moral controversy surrounding them, as there is with embryo-derived ESCs. In addition, through iPS cells it may be possible for patients to be sources of their own therapeutic cells. Cells from a patient could be reprogrammed into iPS cells, which could then yield quantities of cells of the desired type. These cells would then be implanted in the patient. Should this prove unfeasible because of cost, it may be possible to develop banks of cells, derived from iPS cells, that are specialized for both cell type and immune-system type. The result would be readily available collections of cells that would not set off immune system reactions when transplanted into patients.

CHAPTER 16

So Far . . .

page 284
1. species; common ancestor
2. reproductive advantage; common
3. descended from; separate

page 291
1. False; common descent with modification was an idea that predated Darwin. Darwin was, however, the first person to perceive natural selection as an evolutionary mechanism.
2. Alfred Russel Wallace
3. Common descent with modification; natural selection

page 297
1. radiometric dating; proportions
2. Homologous; common ancestor
3. vestigial

Multiple-Choice
1. b 2. a, b, d 3. a, c, d, e 4. a. Lyell—geological forces observable today caused changes in the Earth; b. Wallace—natural selection as a mechanism underlying evolution; c. Lamarck—one form of organism can evolve into another; d. Cuvier—evidence of extinction of species 5. a, c, and d are homologous; b and e do not belong 6. c 7. a, b, d, e 8. a, c, e

Brief Review
1. The term homology describes structures in different species that are similar due to common ancestry. Homology provides evidence for evolution because it is consistent with the idea of common descent with modification.
2. Figure 16.13 shows that, in a gene shared by pigs, moths, and yeast, there are fewer DNA base differences between the forms of the gene found in humans and pigs than between the forms found in humans and moths. This is what you would expect if humans and pigs shared a common ancestor more recently than did humans and moths.
3. Congruent lines of evidence that are consistent with the theory of evolution: (a) Fossils of creatures believed to exist only in different evolutionary periods are not found together in the same geologic samples, and the ages of these samples, as determined by radiometric dating, are consistent with the fossils found in them. (b) Evidence from molecular clocks is consistent with fossil and radiometric dating evidence regarding the degree of relatedness among species.
4. Darwin brought back many birds that he thought were the blackbirds, wrens, and warblers he was familiar with in Britain, but he found they were in fact all species of finches. Darwin perceived this as an example of descent with modification. Without taking the Beagle voyage, Darwin would not have been exposed to such extravagant examples of branching evolution, or have made various other biological observations that spurred his thinking about evolution by natural selection.
5. Fifteen years or so following the publication of On the Origin of Species, almost all scientists accepted the theory of evolution's principle of common descent with modification—the idea that all living things on Earth are descended from other living things, and that all living things ultimately share a single common ancestor. Not so readily accepted was the theory of evolution's principle of natural selection, which holds that evolutionary modification is driven primarily by differences in reproductive success among individuals.

CHAPTER 17

So Far . . .

page 303
1. population; single species
2. alleles; alleles; population
3. allele frequencies

page 308
1. allele frequencies
2. chance; allele frequencies; small
3. mating partners

page 313
1. False: Evolutionary fitness exists only in a relative sense; in a population, one organism is more or less fit than another depending on whether it has more or fewer offspring than another.
2. intermediate; extreme
3. directional selection; one extreme

Multiple-Choice
1. c 2. a. gene pool—the set of all alleles in a population; b. allele—a variant form of a gene; c. allele frequency—the relative representation of a given form of a gene in a population; d. genotype—the genetic makeup of an organism; e. gene flow—exchange of genes between populations 3. a 4. a, c, e 5. d 6. b, d 7. a. mutation—changes in the genetic material; b. gene flow—movement of alleles between populations by migration; c. natural selection—a process through which traits that confer a reproductive advantage to individuals are seen with greater frequency in succeeding generations of a population; d. genetic

drift—chance alterations of allele frequencies, having the greatest effect on small populations; **e.** founder effect—genetic drift brought about by migration of a subpopulation **8. c**

Brief Review

1. An individual may be selected for survival—and hence greater reproduction—because of the qualities that individual has, such as coloration or height. But this individual's characteristics do not change because it possesses these qualities. An individual does not evolve, in other words. An individual can pass on more of its alleles relative to other individuals in a population, however, and this causes the population's allele frequencies to change, meaning the population has evolved.

2. A gene pool is the sum total of alleles that exist in a population. The inheritable characteristics of living things are based on the alleles they possess. Thus, for a population to change in a way that is passed on from one generation to the next—for the population to evolve—the mixture of alleles in that population must change. Thus the evolution of a population is based at root on changes in that population's gene pool.

3. The five causes of microevolution are (**1**) Mutation—Alterations in the DNA that result in new alleles. (**2**) Gene flow—Brings alleles from other populations into the target population. (**3**) Genetic drift—The chance alteration of allele frequencies in a population. Random events can differentially increase or decrease the frequency of alleles in a population. Such events tend to have much greater effects on small populations than on large ones. Genetic drift can have large effects on small populations through two common scenarios: if a new population is founded by only a few individuals (the founder effect) or if the size of a population is decreased suddenly (the bottleneck effect). (**4**) Sexual selection—Allele frequency change based on the differential success that some members of a population have in securing mating partners. (**5**) Natural selection—Traits that confer a reproductive advantage to individuals in a population will become more common in that population over successive generations.

4. An individual in a population has to have greater reproductive success relative to other members of the population in order to contribute more to the gene pool of the succeeding generation.

5. Mutation and genetic drift alter allele frequencies in random directions and hence have nothing necessarily to do with adaptation. Gene flow doesn't necessarily bring in alleles that will be suited to the new environment, and sexual selection concerns mate choice rather than an organism's fit

with its environment. Through natural selection, however, individuals who are best suited to their environments pass on more alleles to the succeeding generation; in this way, natural selection works to adapt populations to their environment.

6. Evolutionary fitness—the relative success in contributing alleles to the next generation—depends crucially on the environment. It is the fit between environment and organism that determines which organisms will have the greatest success in survival and reproduction. Thus, the evolutionary fitness of any given individual would be expected to vary greatly from one environment to another. In the Grants' study of the Galapagos finches, a deeper bill was favored after the drought, but this situation later reversed.

CHAPTER 18

So Far . . .

page 322

1. breed
2. gene flow; populations
3. geographic separation; intrinsic isolating mechanisms

page 325

1. ecological, temporal, behavioral, mechanical, and gametic isolation and hybrid inviability or infertility
2. infertility
3. geographic separation

page 332

1. a new environment
2. relatedness
3. genus

page 335

1. unique; common ancestor
2. relatedness

Multiple-Choice

1. b 2. b, c, d 3. a, c, d
4. **a.** temporal isolation—mating occurs at different times of day or year; **b.** ecological isolation—populations utilize different habitats; **c.** behavioral isolation—populations employ different courtship or mating displays; **d.** gametic isolation—egg and sperm do not fuse; **e.** hybrid sterility—progeny of a cross are unable to reproduce
5. b, c, e 6. b–d–a–e–c 7. b, c, d 8. b, d, e

Brief Review

1. In the biological species concept, two organisms are in the same species if they actually or potentially interbreed successfully in nature.

2. In microevolution, changes in the frequency of the gene-variants called alleles determine the evolution of individual populations. To the extent that a given set of alleles increases in frequency from one generation to the next within a population, the characteristics produced by those alleles will be exhibited to a greater extent

within the population. In macroevolution, it is a drastic reduction of gene flow between populations of the same species that is the first step in bringing about speciation.

3. Without the evolution of intrinsic isolating mechanisms, two populations that have become geographically separated would simply resume interbreeding, remaining one species, if any physical barriers to interbreeding were removed. Thus, the evolution of intrinsic isolating mechanisms must take place if geographically separated populations are to become two separate species.

4. A first requirement of allopatric speciation is that two populations of the same species must become separated geographically, such that gene flow between them is restricted. This allows the forces operating in microevolution—natural selection, genetic drift, mutation, and so forth—to work separately on the two populations. Eventually, the populations may develop intrinsic isolating mechanisms such that, should they be reunited, they would no longer interbreed. In sympatric speciation, intrinsic isolating mechanisms evolve in the absence of geographical separation. The quality that both sympatric and allopatric speciation have in common is the development of intrinsic reproductive isolating mechanisms.

5. Convergent evolution is the evolution of a similar character in two unrelated organisms, caused when the organisms experience similar types of natural selection pressures in their environments. For example, many unrelated species of desert plants have succulent leaves that store water. But these similar characters are not homologous (they are analogous), so they cannot be used to determine relationships in a phylogenetic study.

CHAPTER 19

So Far . . .

page 346

1. life-forms; fossils
2. False. To judge by the fossil record, animals and plants don't appear until the last 16 percent of life's history.
3. self-replicating

page 350

1. Domains: Bacteria, Archaea, and Eukarya. Within Eukarya, the kingdoms are Protista, Plantae, Animalia, and Fungi.
2. 3.8 billion; oxygen; bacteria.
3. animal

page 359

1. green algae
2. tetrapods; fish
3. out of water

Multiple-Choice

1. b, e, f, c, a, g, d 2. d 3. b. 4. b, d, e
5. b, a, d, c, e 6. c, d 7. a, b, d 8. b, d, c, a, e

Brief Review

1. The oldest evidence we have of life dates from 3.8 billion years ago. The only forms of life that existed for 1.8 billion years following that were ocean-going microbes from Domains Bacteria and Archaea.
2. The "RNA world" forms part of the conceptual framework for the origin of life known as replicator-first. Any self-replicating RNA molecule would have had to possess the information-bearing qualities of today's DNA and the enzymatic capabilities of today's proteins.
3. All plants evolved from green algae. All land-dwelling vertebrates evolved from fish, specifically from lobed-finned fish.
4. The amniotic egg allowed terrestrial animals to lay eggs outside an aquatic environment, thus allowing such animals to live away from water. The amniotic egg evolved in reptiles.
5. Mammals have hair, which helps in heat retention, and the females among them have mammary glands that allow them to feed their offspring with milk.

CHAPTER 20

So Far . . .

page 366

1. six and seven million; chimpanzee
2. Late-evolving. The hominin line is dated to as much as 7 Mya, but modern human beings appeared only about 200,000 years ago.
3. African

page 370

1. bipedal; tooth
2. bipedal; brain
3. mosaic

page 375

1. Africa
2. No; the Neanderthals lived in Europe long before human beings migrated there, and the species *Homo erectus* reached Asia by 1.8 Mya.
3. 12,000; Indonesia

Multiple-Choice

1. a, d 2. d 3. b, c, e 4. a, c 5. d 6. b, d, e
7. a, e 8. a, c, e

Brief Review

1. No. A single common primate ancestor gave rise to both the chimpanzee evolutionary line and the hominin evolutionary line (which includes humans), but present-day chimpanzees evolved relatively recently—about the same time as modern human beings did.
2. The fossil evidence we have is not sufficient to allow us to be confident about who de-

scended from whom among the hominins. Placements on the tree vary in accordance with interpretations of the evidence we have.
3. Hominin evolution is thought to have begun 6 or 7 Mya, but the oldest hominin fossils found outside Africa date from only 1.8 Mya. Nearly all the early- and mid-period hominin fossils have been found in a swath of East Africa that runs from South Africa up through Tanzania, Kenya, and Ethiopia, although some early-period fossils may now have been identified much farther to the west.
4. *Sahelanthropus tchadensis* (Toumaï), *Australopithecus afarensis* (Lucy), *Homo erectus*, *Homo neanderthalensis*.
5. Neanderthals and modern human beings were very similar species and as such were highly likely to compete for the same kinds of resources, such as prey animals and shelter. If humans frequently outcompeted Neanderthals for such resources, over the course of many thousands of years, the Neanderthals may simply have lacked the resources to survive.

CHAPTER 21

So Far . . .

page 385

1. Bacteria, Archaea; protists, plants, animals, fungi
2. production of oxygen; fixation of nitrogen; breakdown of organic matter
3. living cells

page 392

1. mutualism
2. toxic
3. Both are strictly microscopic, single-celled, and prokaryotic (their cells do not have nuclei)

page 397

1. nucleus; microscopic; aquatic
2. oxygen; food chains
3. algae

Multiple-Choice

1. e 2. d 3. e 4. e 5. c 6. a, c, e 7. c, d, e
8. b, d

Brief Review

1. In nature, plants get their nitrogen from certain bacteria and archaea that are able to take in atmospheric nitrogen and turn it into a form plants can use.
2. Viruses cannot replicate, or carry out many of life's other basic functions, without using the machinery of the cells they invade. Most biologists would say that an entity that lacks such abilities should not be regarded as a living thing.
3. They must kill microorganisms while leaving human cells unharmed.
4. Like bacteria, archaea are single-celled and prokaryotic—their cells lack a nucleus.

Archaeal cells are, however, fundamentally different from bacterial cells at the level of deep chemical structure. The molecules that make up archaeal cells are chemically quite different from the molecules that make up bacterial cells (or any other kind of cell). The structural differences seen in archaeal cells are based on genetic differences between them and other life-forms.
5. Cells with nuclei, sexual reproduction, and multicellularity.
6. An amoeba is a protist that moves through use of pseudopodia or "false feet," meaning an extension of one part of the cell and a flowing of the rest of the cell into that extension. Some amoebae are parasites, and some of these are human parasites that cause the class of diseases known as amoebic dysentery.

CHAPTER 22

So Far . . .

page 406

1. hyphae; mycelium
2. (b) make their own food
3. breaking down and recycling organic matter

page 412

1. spore
2. asexual
3. yeasts

page 413

1. fungus; algae; bacteria
2. fungus; algae or bacteria
3. fungus; plant

Multiple-Choice

1. c, d, e 2. d 3. b, d, e 4. a, c, d 5. a, b, c, e
6. d 7. d 8. c

Brief Review

1. Animals take in their food prior to digesting it. Fungi cannot do this because food is made of relatively large molecules and such molecules cannot pass through fungal cell walls. Thus, fungi digest their food externally—using digestive enzymes they secrete—after which they ingest the smaller molecules that this food has been broken into.
2. Only a tiny fraction of the spores that fungi release will be able to germinate, or grow into a new mycelium. The fungi's adaptation to this fact has been to "buy many tickets"—to create a huge number of spores, which scatter to diverse environments strictly because of their numbers.
3. A spore is a reproductive cell that can give rise to a complete organism without fusing with another reproductive cell. A gamete (an egg or sperm) is a reproductive cell that must fuse with another reproductive cell for a complete organism to be produced—egg must fuse with sperm.
4. Basidiomycota—club fungi, rusts and smuts, toadstools; Ascomycota—sac fungi,

morel mushrooms, truffles; Zygomycota—bread molds, spore-releasing sporangium; Chytridiomycota—mobile, water-dwelling, early-evolving.

5. Yeast are single-celled fungi that tend to reproduce through the process of budding.

CHAPTER 23

So Far . . .

page 418

1. blastula; embryonic
2. mobile or motile; multicelled; heterotrophic; cell walls
3. small

page 421

1. tissues; symmetry
2. bilateral; opposite sides
3. body cavity

page 424

1. moving water
2. medusa; the undersides; polyp; on top
3. circulatory

page 427

1. snails or slugs; oyster, clams, or mussels; octopus, squid, or nautilus
2. exoskeleton; paired, jointed
3. exoskeleton, sessile living

page 430

1. eggs; sperm
2. asexual reproduction; unfertilized egg
3. hermaphrodite

page 434

1. broadcast; no
2. fewer; greater
3. reptiles; out of water

page 437

1. stays within; leaves the vessels
2. exoskeleton; external covering
3. molting; a periodic shedding of an old skeleton followed by the growth of a new one

Multiple-Choice

1. a, d 2. d 3. a 4. b 5. e 6. b, c, e 7. a 8. d

Brief Review

1. All animals are believed to have a single ancestor, probably an ocean-dwelling single-celled protist. All animals go through a blastula stage of development, whereas no other life-forms do. It's difficult to imagine a feature as specific as a blastula stage of development arising more than once in evolution.

2. Sponges have no organs, or even true tissues, and each of their cells functions with a great deal of independence. Thus, the cells in a sponge have a lower level of specialization and integration with one another than is the case in more complex animals.

3. A coelom provides an animal with flexibility, allows for the expansion and contraction of organs such as the heart and stomach,

and helps protect an animal's organs from being damaged by external blows.

4. Cnidaria—jellyfish, sea anemones; nematoda—roundworms; mollusca—snails, squid; annelida—earthworms, leeches; arthropoda—insects, lobsters, spiders; echinodermata—sea stars, sea urchins; chordata—fish, amphibians, reptiles.

5. Arthropods have an exoskeleton; were it not for shedding an existing exoskeleton through the process of molting, an arthropod's growth would be limited by the size of its original exoskeleton.

CHAPTER 24

So Far . . .

page 444

1. (b) no cell wall
2. spores; eggs; sperm

page 449

1. embryo; food; protective casing
2. (c) spruce tree and (d) orchid
3. sperm

page 455

1. wind; animals
2. differential
3. nights

Multiple-Choice

1. b, c, d 2. a, c, e 3. b, d 4. a. moss—bryophyte; b. cactus—angiosperm; c. pine tree—gymnosperm; d. rose—angiosperm; e. fern—seedless vascular plant 5. c, d 6. d 7. b 8. e

Brief Review

1. The gametophyte generation is the multicellular, gamete-producing generation of plant, meaning the generation that produces the reproductive cells called eggs and sperm. The sporophyte generation is the multicellular, spore-producing generation of plant, meaning the generation that produces the reproductive cells called spores. The cells of the sporophyte generation are diploid, and thus contain two sets of chromosomes each, while the cells of the gametophyte generation are haploid, and thus contain a single set of chromosomes each.

2. Seeds, which are produced only by plants, result from the fusion of two gametes—an egg and a sperm. Seeds amount to the embryo produced by sperm and egg fusion, food for this embryo, and a tough outer coating that serves to protect the embryo.

3. The pollen of most angiosperms is carried from one plant to another by animals. Such pollination allows the sperm from one plant—contained inside a pollen grain—to fertilize the egg in a second plant. The seeds of some angiosperms are dispersed by animals that ingest plant fruits.

4. Gravitropism is the bending of a plant root or shoot in response to gravity; it allows plants to control the direction of the growth of both their roots and shoots—

roots into the ground and shoots into the air. Phototropism is the curvature of shoots in response to light; it allows plants to maximize their capture of sunlight, which drives photosynthesis. Thigmotropism is the growth of plant shoots in response to touch; it allows plant shoots to grow upwards on other objects, thus maximizing access to sunlight.

5. Photoperiodism is controlled by the daily duration of darkness a plant is experiencing, relative to the daily duration of light.

CHAPTER 25

So Far . . .

page 465

1. water; minerals
2. photosynthesis
3. reproductive

page 468

1. dicots; monocots
2. periphery; fluid transport
3. water; minerals; food produced in photosynthesis

page 470

1. the tips; indefinitely (throughout the life of the plant)
2. meristematic; apical meristems

page 472

1. tracheids; vessel elements
2. solar energy
3. load; sieve element; pressure

page 474

1. spores; sporophyte; gametophyte; pollen grains; embryo sac
2. pollen; stigma; sperm; egg
3. zygote; sporophyte; food; endosperm

page 476

1. seed coat; protect
2. ovary; seed coat

Multiple-Choice

1. e 2. b, c 3. a 4. b, d 5. a 6. b, c 7. c, e 8. e 9. c, e

Brief Review

1. The stomata allow the movement of carbon dioxide into the plant and oxygen out of it. These same openings also allow water to escape from the plant as water vapor. The plant deals with the resulting problem of water loss by moving large quantities of water up through the roots and stem and out through the stomata.

2. Roots serve to anchor the plant and they are the main absorptive structures for the plant. They absorb water and minerals from their surroundings and make them available to the above-ground parts of the plant. In some species, roots also store substantial reserves of food. Taproots and fibrous roots are the two main types of roots produced by flowering plants.

3.

Flower Part	Function
pedicel	Flower stalk; lifts flower up to make it noticeable to pollinators.
sepals	Leaf-like structures that protect flower buds before they open.
petals	Generally most conspicuous part of the flower; often brightly colored; sometimes fragrant; attract pollinators.
stamens	Male reproductive structures; consist of a stalk called a filament, which holds aloft the pollen-producing sacs called anthers.
pollen	The male gametophyte; the vehicle that conducts male sperm to the female reproductive structures; composed at maturity of an outer coat, one tube cell, and two sperm cells.
carpel	Female reproductive structure; consists of the ovary, stigma, and style—the ovary produces and protects the egg-containing female gametophyte, and after fertilization turns into the fruit; the stigma is the receptive surface that captures pollen and promotes germination of pollen; and the slender style raises the stigma to make it accessible to pollinators.

4. Meristematic tissue gives rise to all other tissue types in the plant. Meristematic cells remain perpetually embryonic: They are able to keep giving rise to more of themselves as well as to cells that differentiate into all the plant's various tissue types. Meristematic tissue is concentrated in apical meristems—tissue located at the tips of roots and shoots that gives angiosperms their primary or vertical growth. There is a second location for meristematic tissue in the shoot—the area nestled between leaf and stem known as a lateral bud.

5. The water-conducting xylem cells come in two types: tracheids and vessel elements. Both are thick walled, and both die at maturity, after which their contents are cleared out. Because they are stacked on top of one another, lines of these cells become water-conducting tubes. The food-conducting phloem cells, called sieve elements, are alive at maturity but lack nuclei. Their housekeeping needs are taken care of by one or more adjacent companion cells. A collection of sieve element cells makes up a sieve tube, which transports sucrose, along with hormones and other compounds.

6. When the two sperm cells from a pollen grain enter the embryo sac of a plant, two fertilization events take place. In the first of these, a sperm cell fuses with the embryo sac's egg cell, thus setting in motion development of a new plant embryo. In the second fertilization event, the second sperm cell fuses with the two nuclei of the embryo sac's central cell. This event sets in motion the development of endosperm—food for the growing embryo.

CHAPTER 26

So Far . . .

page 483

1. anatomy; physiology
2. reduced
3. stable

page 488

1. cell, tissue, organ, organ system
2. epithelial, connective, muscle, nervous
3. muscle; nervous

page 491

1. skin; epidermis; dermis
2. just beneath the skin; organs; abdomen
3. follicle

page 498

1. bones; cartilage; ligaments
2. production of blood cells; storage of energy reserves as fat

page 500

1. fibers; length
2. sliding; sarcomere
3. myosin; actin; sarcomere

Multiple-Choice

1. d 2. a 3. b 4. e 5. d 6. c 7. a 8. d

Brief Review

1. In negative feedback, an element that brings about a response is affected negatively by that response. Put another way, the activity of the element is reduced by the response. In the example in the text, the kidneys released a hormone (EPO) that produced red blood cells whose oxygen content fed back on the kidneys, causing them to reduce their output of EPO.

2. Lymphatic organs such as lymph nodes have high concentrations of immune cells within them that monitor passing lymphatic fluid for signs of invading microorganisms.

3. Our skin is water resistant because the surface epidermal layer of the skin is formed of tightly interlocking cells made almost entirely of the protein keratin, which is water resistant.

4. Subcutaneous fat is found in a layer, varying in thickness throughout the body, that exists just beneath the skin. Visceral fat is found surrounding organs deep within the abdomen.

5. Three functions of the skeletal system are support, storage of minerals and fats, and blood cell production.

6. The sample is from the end (epiphysis) of a long bone.

CHAPTER 27

So Far . . .

page 509

1. central; brain; spinal cord; peripheral
2. from; somatic; autonomic; to
3. neurons; dendrites; axon

page 513

1. electrical charge
2. Na$^+$ ions; neighboring portion
3. neurotransmitters; synaptic cleft

page 518

1. sensory neurons; motor neurons; reflex
2. fight-or-flight; rest-and-digest
3. cerebral cortex; cerebrum

page 522

1. sensory receptors; electrical signals
2. smell (olfaction); thalamus; cerebral cortex
3. combinations

page 525

1. taste buds; taste cells
2. compressed
3. fluid; hair

page 528

1. retina; cornea; lens
2. rods; black-and-white; cones; color
3. optic nerves

Multiple-Choice

1. c 2. c 3. e 4. e 5. c, d, e 6. b 7. a 8. b

Brief Review

1. The sensory neuron receives input about conditions inside and outside the body and conveys this received information to the central nervous system (CNS). The motor neuron cell body exists in the CNS, but its axon extends into the peripheral nervous system (PNS); it thus can send instructions from the CNS to such structures as muscles or glands. The interneuron exists solely in the CNS and serves to interconnect neurons and integrate signals from them.

2. Neurons have special extensions of themselves, called axons and dendrites, that respectively carry nervous system signals away from and toward the neuron cell body. In addition, the outer or plasma membrane of each neuron has a charge difference, from one side to the other, that can change rapidly with the proper stimulation, going from negative on the membrane's interior to briefly positive, then back to negative. This change in membrane potential can be propagated down the length of the neuron's axon, after which stimulation of an adjacent neuron can take place through release of neurotransmitters that diffuse from the sending neuron to the receiving neuron across a synaptic cleft.

3. Neurotransmitters function in the synapses of the nervous system—the physical areas that include the tip of one neuronal axon and the cell body of a second neuron (or effector cell). The activity of neurotransmitters can be limited in time by the process of reuptake, in which a sending neuron takes a secreted neurotransmitter back into itself. Conversely, neurotransmitters can be degraded by enzymes that are secreted into the synaptic cleft by the sending cell.

4. A simple reflex arc works by means of a sensory neuron sending an afferent message about some stimulation to the spinal cord. There, the sensory neuron will synapse with a motor neuron (or perhaps an interneuron), after which the motor neuron sends a signal that brings about some action at an effector such as a muscle. The brain is not involved in such an arc.

5. In the senses of hearing, touch, taste, and vision, various types of sensory receptors respond to their respective forms of stimulation and then act as transducers, transforming the stimulation they receive into the electrical impulses of the nervous system. They did the latter by releasing neurotransmitters to receiving neurons (or by not releasing neurotransmitters, in the case of vision). In each case, the resulting nerve signal went first to the brain's thalamus and then to different areas of its cerebral cortex for processing.

CHAPTER 28

So Far . . .

page 537

1. cells; bloodstream; cells
2. endocrine; bloodstream
3. amino acid; peptide; steroid; peptide; steroid

page 539

1. negative feedback; stable
2. hypothalamus; anterior; posterior; adrenal
3. anterior; other glands

page 547

1. insulin; glucagon; pancreas
2. glycogen; glucagon
3. adrenal glands; storage; used

Multiple-Choice

1. a, c, d 2. d 3. d 4. b 5. b, d, e 6. c 7. b, c 8. b, c, e

Brief Review

1. The hormones of the endocrine system take at least several seconds to have any effect. This time-frame is fine for maintaining fluid levels, but of no use in carrying out the rapid, complex movements required to shoot a basketball—the nervous system handles those.

2. Amino-acid-based hormones are derived from a chemical modification of a single amino acid. Peptide hormones are built from chains of amino acids. The building block of all steroid hormones is the cholesterol molecule. Amino-acid-based and peptide hormones generally bind with receptors that are on the surface of target cells. Steroid hormones generally pass through the outer membrane of a target cell and bind with a receptor inside the cell.

3. The hypothalamus produces hormones oxytocin and vasopressin, which it releases from the posterior pituitary gland; through nervous system signals, it directly controls the release of the hormones adrenaline and noradrenaline from the adrenal medulla; and it controls release of six hormones from the anterior pituitary through its secretion of releasing or inhibiting hormones.

4. Two hormones control circulating glucose levels in the body. Insulin, released from beta cells in the body's pancreas, has the effect of moving glucose into cells, where it may be stored or used. Glucagon, released from alpha cells in the pancreas, has the effect of moving glucose out of storage in both muscle and liver cells. Muscle cells retain all the glucose that has been liberated in them—using it to power their own activities—while liver cells release into the bloodstream much of the glucose liberated within them.

5. Cortisol helps liberate stores of energy in the body, particularly during times of stress. When it stays in circulation over extended periods of time, it can suppress immune system activity, reduce the insulin sensitivity of cells, and produce a narrowing of arteries. Human beings are thought to be uniquely vulnerable to these effects because human beings alone have the capacity to experience stress continuously over long periods because of their ability to envision the future.

CHAPTER 29

So Far . . .

page 553

1. elicits an immune system response
2. barriers; skin
3. foreign; specific microbial invaders

page 556

1. sensors; channel
2. phagocytes

page 561

1. antibodies; cells; B; T
2. lymphatic network
3. antigen-presenting; dendritic

page 564

1. antigen; B; exported
2. plasma; memory B
3. helper T

page 570

1. readiness; antigens
2. autoimmune disorders; allergies; organ transplantation
3. regulatory T cells (T-regs); toll-like receptors (TLRs)

Multiple-Choice

1. b 2. b, c, e 3. a. mast cells—cells that release histamine; b. cytokines—communication proteins; c. toll-like receptors—first sensors of infection; d. phagocytes—cells capable of ingesting invading cells; e. fibrin proteins—proteins that seal off infection site 4. d 5. a, d, e 6. c 7. e 8. c, e

Brief Review

1. The immune system increases blood flow near the site of injury and increases the permeability of capillaries near the site (so that more substances can be transported out); it delivers cells and proteins to the site that kill the invaders; and it releases substances that help seal off the site of injury, thus limiting the area of the infection.

2. A dendritic cell is the most important of the immune system's antigen-presenting cells (APCs). Dendritic cells ingest and are invaded by pathogens and then display antigens from the invaders on their surface. They then migrate to a lymph node and present the invader's antigens to CD4 and CD8 cells that have receptors specific to the invader. In the interaction that follows with the dendritic cell, the CD4 and CD8 cells are transformed into activated helper T and killer T cells, respectively.

3. HIV has so many effects because its ability to eliminate helper T cells allows it to weaken the immune system, which normally fights off so many kinds of invaders.

4. The immune system regards any transplanted organ as nonself and thus will initiate an attack against it.

5. Vaccines are able to provide immunity to infectious illness because of the immune system's memory B and T cells. These cells remain in the body long after an infection has passed and allow the immune system to mount a more rapid and effective attack against any second attack by a given invader. Memory B and T cells can be produced by the microbial antigens contained in vaccines; thus vaccines can elicit a state of immune-system readiness against a given invader even in the absence of an actual attack by that invader.

CHAPTER 30

So Far . . .

page 578

1. fluid
2. red blood cells; white blood cells; platelets; water
3. away from; toward; capillaries

page 580

1. heart and lungs; heart and the rest of the body
2. systemic; pulmonary
3. electrical signals

page 584

1. blockage; coronary artery
2. away from the heart
3. Capillaries

page 587

1. oxygen; carbon dioxide
2. alveoli; capillaries
3. hemoglobin; red blood cells

Multiple-Choice

1. b, d, e 2. e 3. b 4. c 5. a, e 6. a 7. d 8. e

Brief Review

1. Red blood cells transport oxygen and carbon dioxide; white blood cells are central to the body's immune system function; and platelets are active in the process of blood clotting.

2. The left and right atrial chambers of the heart pump blood only into their respective ventricles, while the ventricles pump blood to locations outside the heart—the right ventricle to the lungs, the left ventricle to all the tissues of the body.

3. Capillaries are the only blood vessels whose walls are thin enough to permit diffusion of substances in and out along their length; they are composed of but a single layer of cells, whereas arteries and veins have a three-layered structure.

4. Eating animal fats raises the levels of the low-density lipoproteins (LDLs) that lodge within the coronary artery tissue; smoking may damage LDLs in a way that triggers the immune response against them; exercise lowers LDL levels while raising the levels of high-density lipoproteins or HDLs, which help clear or neutralize LDLs.

5. Anatomical structures of the respiratory system include the lungs, the nose, the nasal cavity, the sinuses, the pharynx (upper throat), the larynx (voice box), the trachea (windpipe), the left and right bronchus (the body's two large air-conducting passageways), and the bronchioles (the smaller air passageways, branching off the left and right bronchus, that deliver air to the lungs).

6. A muscular sheet in our chest, the diaphragm, contracts, causing the ribs to rise, which expands the body's thoracic cavity, thus decreasing air pressure within the lungs. Since gases always move from areas of higher pressure to areas of lower pressure, air moves into the lungs. In exhalation, the diaphragm relaxes, and the thoracic cavity shrinks in response. This increases the air pressure inside the lungs, and air is expelled as a result.

7. Oxygen enters the bloodstream at a capillary that covers an alveolus in the lungs. Oxygen is inhaled into the sac-like alveolus and then diffuses across the alveolar tissue into an adjacent capillary.

CHAPTER 31

So Far . . .

page 595

1. digestive tract; mouth; anus
2. circulatory system
3. muscles; peristalsis

page 597

1. small intestine
2. liver; gallbladder; digestive enzymes; chemical buffers

3. esophagus, stomach, small intestine, large intestine

page 600

1. energy; structural building block; regulate
2. water, minerals, vitamins; proteins, carbohydrates, lipids
3. elements; compounds; small

page 605

1. energy; thousand grams; 1 degree Celsius
2. 9; energy; 4
3. amino acids; animal; essential amino acids; plant

page 608

1. simple sugars; starches; fibers
2. vitamins; minerals
3. blood glucose; lower

page 610

1. fatty acids
2. oils; fats
3. animal; plant

page 614

1. blood; renal arteries; blood; renal veins; bladder
2. blood vessels
3. waste; water; circulatory system

Multiple-Choice

1. b, d, e 2. b 3. c 4. a, c, d 5. b, d 6. a, b, e 7. b, c 8. c

Brief Review

1. Food has not been broken down in the stomach to an extent that it can pass out of the digestive system into adjacent capillaries (in the case of carbohydrates and proteins) or into lymphatic vessels (in the case of lipids). Partway through the small intestine, foods have been broken down to this extent, thanks in large part to the digestive enzymes and bile that move into the small intestine from the pancreas, liver, and gallbladder.

2. All carbohydrates and proteins that move out of the digestive tract are first transported to the liver through the hepatic portal vein. The liver controls which of these nutrients it will store and which it will release into general circulation. The liver also is the sole producer of bile, which facilitates the digestion of fats, and it packages waste products for removal by the kidneys.

3. Apart from waste storage and compaction, the large intestine conserves water by moving it back into circulation, and it absorbs vitamins produced by resident bacteria.

4. Vitamins are compounds—they are composed of several different elements—whereas minerals are individual elements. Vitamins function only to facilitate chemical reactions in the body, whereas minerals facilitate chemical reactions and provide structural building blocks for the body. A consequence of this difference is that our diets must contain greater quantities of minerals than vitamins.

5. The food rich in unprocessed whole grains would be preferred nutritionally because it is likely to (**a**) provide vitamins and minerals in addition to calories, (**b**) have a lower glycemic load, and (**c**) have a higher fiber content.

6. Animal fats, such as those found in bacon or beef, are predominately composed of saturated fatty acids. Vegetable oils, such as olive and canola oils, are predominately composed of unsaturated fatty acids.

7. Blood cells and most proteins are too large to pass out of the glomerulus capillaries and into the filtrate; they remain in the circulatory system.

8. The urination signal we are aware of is an involuntary contraction of the urinary bladder. This increases the fluid pressure inside the bladder, which can be uncomfortable.

CHAPTER 32

So Far . . .

page 623

1. zygote; embryo
2. endoderm; mesoderm; ectoderm; inner; middle; outer
3. induce; brain; spinal cord

page 626

1. morphogens
2. transcription factors; genes
3. anterior; center

page 628

1. conserved; transcription factors
2. movement; adhesion; death
3. fertilization (conception); the human life

Multiple-Choice

1. b, d, c, e, a 2. c 3. d 4. a 5. a, c, d 6. e 7. b

Brief Review

1. After fertilization, the cleavage stage of development begins. This first results in the formation of a tightly packed ball of cells, the morula, which is then transformed into a hollow, fluid-filled ball of cells with a central cavity, the blastula. Next, cells in the blastula undergo rearrangement and movement in the process called gastrulation, which results in an embryo composed of three layers of cells—the endoderm, mesoderm, and ectoderm.

2. The concentration of a given morphogen will decrease as distance increases from the cells that secrete the morphogen. Such a concentration gradient exists for all morphogens. In the fruit fly *Drosophila*, a concentration gradient of the morphogen bicoid affects the development of anterior and center-section structures of the fly. Bicoid concentrations are highest at the anterior end of the *Drosophila* embryo, and this high concentration has the effect of bringing about anterior structures of the fly, such as its head. Lower bicoid

concentrations at the center of the embryo help bring about center-section structures.

3. After gastrulation in the chick embryo, a small group of cells is induced to form the wing of the chick. This group of cells forms a structure called the wing bud, which eventually develops into the wing. A group of the cells in the wing bud are called the zone of polarizing activity (ZPA). The ZPA sends out a morphogen that directs the proper order of digit production in the chick's wing.

4. The conservation of the homeobox sequence tells us that this sequence is extremely important. If it were less important, the mutations that naturally occur to DNA over evolutionary time would have left contemporary mice and fruit flies with different DNA sequences in the homeobox portions of their developmental genes.

5. Cell adhesion molecules allow different varieties of cells to adhere selectively to one another. This allows for all kinds of cell organization, including the organization of the embryo into different layers of cells.

6. Cell death works to create spaces that define shapes in a developing organism. Without this death, the human hand would not have separate fingers but would have a webbed structure similar to that of a duck.

CHAPTER 33

So Far . . .

page 638

1. oocytes; ovaries
2. ovulation; uterine (or Fallopian) tubes
3. endometrium; fertilization of an egg

page 641

1. testes; epididymis
2. semen; sperm; glands
3. stem cells; spermatocytes

page 650

1. one; chromosomes
2. uterus; placenta
3. contractions; muscles

Multiple-Choice

1. a, c, d 2. a 3. b, c, e 4. acrosome— enzyme-containing compartment in sperm; spermatogonium—sperm stem cell; spermatocyte—maturing sperm cell; prostate gland—produces fluid found in semen; urethra—duct through which sperm are ejaculated 5. d 6. e 7. c 8. c, a, e, b, d

Brief Review

1. Gonadotropin-releasing hormone, which is released by the hypothalamus, stimulates the release of follicle-stimulating hormone (FSH) and luteinizing hormone (LH). Both of these hormones are released by the anterior pituitary gland. FSH promotes the development of the follicles, and LH stimulates the production of estrogen. As estrogen levels increase, FSH levels at first begin to decrease, but as estrogen levels begin to peak due to the development of the tertiary follicle, both FSH and LH levels spike. This results in ovulation. The corpus luteum then begins to synthesize progesterone, which acts on the hypothalamus to reduce FSH and LH to original levels, resulting in degeneration of the endometrium and menstruation.

2. Sperm development is impaired by temperatures that are either too hot or too cold. The anatomy and physiology of the testes act to maintain a correct temperature for sperm development. Sperm are protected from the high temperatures that exist elsewhere in the body by virtue of the location of the testes outside the body's torso. In addition, a muscle that surrounds the testes contracts when temperatures drop, so that the testes are pulled in closer to the warm body.

3. Two hypotheses currently are favored to explain the existence of menstruation. One is that the menstrual cycle of tissue buildup and breakdown is less costly, in terms of energy expenditure, than the alternative of keeping the endometrium in a permanent state of readiness for embryo implantation. The second hypothesis is that, when it is ready for embryo implantation, the endometrium is in a form that cannot be maintained permanently and that must arise from an earlier form. Thus, it makes sense that the inner endometrium would wash away every 28 days and then enter a new cycle of growth and development.

4. The tertiary follicle first houses a developing oocyte and then releases that oocyte from the ovary. Following this action, the tertiary follicle becomes the corpus luteum, a structure that releases hormones that first prepare the female reproductive tract for pregnancy and that then maintain the tract during the early phases of pregnancy.

5. Fusion of a sperm with an oocyte prompts the oocyte to release granules whose secretions harden a membrane surrounding the oocyte while inactivating sperm receptors on the surface of the membrane.

6. Prior to implantation, the fertilized egg divides to form a morula. This tightly packed ball of cells continues to divide, forming a hollow ball of cells, called the blastocyst, which is composed of two primary structures. One of these is a group of cells called the inner cell mass—the portion of the blastocyst that develops into the baby. The other is a group of cells called the trophoblast—cells in the blastocyst that first attach to, then extend into the maternal endometrium. In time, these cells grow into the fetal portion of the placenta.

7. Any baby born before 37 weeks of gestation is regarded as premature.

8. Birth control pills work using a combination of synthetic estrogen and progestin. Normally, low estrogen levels trigger the beginning of follicle development. The pill creates high estrogen levels that suppress the release of FSH from the hypothalamus. The progestin in the pill suppresses the release of LH to prevent ovulation of any follicle that might develop in the presence of high estrogen levels.

CHAPTER 34

So Far . . .

page 661

1. population; community; ecosystem
2. limit the growth of a population
3. density; support

page 667

1. few; abundant; many; little or no
2. greater
3. average number of children born to each woman; 2.1 births per woman

Multiple-Choice

1. d 2. c 3. c 4. e 5. b, c, e 6. a, b, d 7. b, e

Brief Review

1. population, community, ecosystem, biosphere

2. K-selected species experience their environment as relatively stable, spend most of their lives in populations that are close to carrying capacity (K), and provide a relatively large amount of care to a relatively small number of offspring. In contrast, r-selected species experience their environment as relatively unstable, have populations whose numbers are limited largely by reproductive rate, and provide little or no parental care to a relatively large number of offspring. Humans and oak trees are K-selected species; flies and ragweed are r-selected species.

3. Environmental resistance is the term used to describe all the environmental forces that act to limit population growth. Organisms will run out of food or have their sunlight blocked; temperatures will plunge, or rain will cease; greater numbers of predators will discover the population; the wastes produced by the organisms will begin to be toxic to the population.

4. Populations of K-selected species tend to have their growth limited by the carrying capacity (K) of the environments in which they live. A given environment can provide only so much food and habitat; as the density of a K-selected population increases within an environment, a lack of such resources will tend to limit the population's growth. The limitations on a K-selected population are thus said to be density dependent.

5. The Earth's human population is experiencing a drop in its total fertility rate—the average number of children born to each woman. Whereas this rate stood at 5 during the 1950s, it now stands at about 2.56. For

now, the Earth's human population is still growing; in fact, it is projected to grow from about 7 billion people in 2012 to about 9.1 billion in 2050. However, it is growing at a slower rate than it did through much of the twentieth century. By 2050, it will be growing by an estimated 31 million people per year, and may stabilize or even drop to some extent thereafter.

CHAPTER 35

So Far . . .

page 674
1. absence from
2. species diversity; geographic diversity of species populations; genetic diversity within species' populations
3. physical surroundings; normally be found; occupation

page 675
1. limited, vital resource; local extinction
2. resource partitioning

page 681
1. predation; parasitism
2. Batesian; Müllerian
3. beneficial to both; helped; unaffected

page 686
1. replacements; stable state
2. increase; longer-lived; increase
3. survive a major ecological disturbance

Multiple-Choice

1. d 2. c, e 3. a 4. b 5. a 6. b, d 7. d

Brief Review

1. In ecology, a keystone species is one whose absence from a community would bring about significant change in that community. Just as no other stone within an arch can assume the role of a keystone, so no other species within a community can assume the role of a keystone species. When the sea star *Pisaster* was removed from its community in coastal Washington, the community underwent significant change, in terms of its species makeup, because no other predator in the community could assume *Pisaster's* role in maintaining the populations of other species.
2. Competition; predation (and a special variety of it, parasitism); mutualism; and commensalism.

3. This interaction is one of parasitism of the caterpillar by the ichneumon wasp; in this interaction, the caterpillar is the host.
4. If the fungi were not providing any benefits to the algae, the mode of interaction between them would be parasitism—the fungi would be parasites on the algae. Since both the algae and the fungi are benefiting from their relationship, the mode of interaction between them is mutualism.
5. This is an example of Batesian mimicry, in which one species (the red salamander) has evolved to acquire the appearance of a species with superior protective capabilities (the red-spotted newt). The red-spotted newt is the model, the red salamander is the mimic, and the dupe is any predator species that would avoid the red salamander, believing it to be a red newt.
6. Because this succession begins from a state of little or no life and no soil, it is an example of primary succession.

CHAPTER 36

So Far . . .

page 698
1. community; physical environment
2. photosynthesis; decomposition; bacteria; fungi
3. bacteria; bacteria

page 703
1. photosynthesis; producers; herbivore
2. decomposers; biodegradable
3. energy; trophic level

page 709
1. oxygen; stratosphere; ultraviolet radiation
2. heat; carbon dioxide; methane
3. rise; decrease

page 716
1. cools; lose; absorbs
2. climate
3. tundra, taiga, temperate deciduous forest, tropical rainforest

page 722
1. decreases; decreases
2. phytoplankton; perform photosynthesis
3. flows into an ocean

Multiple-Choice

1. a, b, d 2. c, d 3. a. herbivore—organism that eats plants; b. net primary production—

material accumulated through photosynthesis; c. gross primary production—material produced through photosynthesis; d. trophic level—position in an ecosystem's food chain; e. producer—photosynthesizing organism
4. a 5. a 6. d 7. b 8. b 9. c 10. a

Brief Review

1. Within any ecosystem, the amount of available energy drops significantly at each trophic level. This is so because much of the energy locked up in organisms at each trophic level is used for the organisms' metabolic processes (building up tissue, staying warm), while additional energy is lost as heat. The drop in available energy at each trophic level means there can be fewer predator than prey animals.
2. Warm air (**a**) rises and (**b**) carries more moisture than cold air. Near the equator, Earth gets warm due to the direct rays coming from the sun. As this warm air rises, it cools and drops moisture. This cooler air moves toward the poles in both directions and descends at about 30°N and 30°S, warming as it goes, thus absorbing moisture from the land, which is dry as a result.
3. Climate essentially defines the large-scale vegetation patterns that are seen on Earth. Variations within large-scale climate regions tend to be mountains, whose climates can vary greatly from base to peak.
4. It may be that excessive eutrophication has occurred in the pond. In such occurrences, large algal blooms result from excess nutrients that come into the body of water. As the algae die and their dead organic matter falls to the lower depths, decomposing bacteria consume the organic matter but consume so much oxygen that they deplete the water of oxygen, thus killing fish.
5. Permafrost makes the tundra wet during summer months, even with low rainfall, and it is too cold for trees to grow. Taiga has a less-severe climate, a longer growing season and no permafrost; coniferous trees grow there and support a larger animal community.

Glossary

2n In a living cell, the condition of being diploid: of having two sets of chromosomes. (Contrast with **n:** In a living cell, the condition of being haploid; of having a single set of chromosomes.)

A

abiotic Pertaining to nonliving things.

acid Any substance that yields hydrogen ions in solution. An acid has a number lower than 7 on the pH scale.

acid rain Rain whose pH level has been skewed toward the acidic side of the pH scale, with air pollution the cause of this shift.

acrosome A structure located on the front end of a vertebrate sperm cell; contains enzymes that help the sperm penetrate the accessory cells surrounding the oocyte.

action potential A temporary reversal of neuronal cell membrane potential at one location that results in a conducted nerve impulse down an axon.

activation energy The energy required to initiate a chemical reaction. Enzymes lower the activation energy of a reaction, thereby greatly speeding up the rate of the reaction.

active site The portion of an enzyme that binds with a substrate, thus helping transform it.

active transport Transport of materials across the plasma membrane in which energy is expended. Through active transport, solutes can be moved against their concentration and electrical gradients. The sodium-potassium pump is an example of active transport.

adaptation A modification in the form, physical functioning, or behavior of organisms in a population over generations in response to environmental change.

adaptive immune response The human immune response that provides protection against specific microbial invaders; compare to innate immune response.

adaptive radiation The rapid evolution of many species from a single species that has been introduced to a new environment.

adenosine triphosphate (ATP) A nucleotide that serves as the most important energy-transfer molecule in living things. ATP powers a broad range of chemical reactions by donating one of its three phosphate groups to these reactions. In the process, it becomes adenosine diphosphate (ADP), which reverts to being ATP when a third phosphate is added to it.

afferent division The division of the peripheral nervous system that carries sensory information toward the central nervous system, having gathered information about the body or environment.

algae Protists that perform photosynthesis.

algal bloom An overabundance of algae in a body of water, resulting from an excess of nutrients. The many dead algae that fall to the bottom allow decomposing bacteria to flourish, using up so much oxygen that fish can suffocate.

alkaline Basic, as in solutions. Alkaline (basic) solutions have numbers above 7 on the pH scale.

allele One of the alternative forms of a single gene. In pea plants, a single gene codes for seed color, and it comes in two alleles—one codes for yellow seeds, the other for green seeds.

allergen A foreign substance that triggers an allergic reaction. These substances are usually derived from living things, including pollen, dust mites, foods, and fur.

allergy An immune system overreaction to an antigen that results in the release of histamine.

allopatric speciation Speciation that involves the geographic separation of populations. Most speciation involves geographic separation, followed by the development of intrinsic isolating mechanisms in the separated populations.

allosteric regulation The regulation of an enzyme's activity by means of a molecule binding to a site on the enzyme other than its active site.

alternation of generations A life cycle practiced by plants in which successive plant generations alternate between the diploid sporophyte condition and the haploid gametophyte condition.

alternative splicing A process in genetics in which a single primary transcript can be edited in different ways to yield multiple messenger RNAs, which in turn yield multiple proteins.

alveoli Tiny, hollow, air-exchange sacs that exist in clusters at the end of each of the air-conducting passageways in the lungs, the bronchioles.

amino-acid–based hormones Hormones that are derived from a single amino acid. One of three principal classes of hormones, the other two being peptide and steroid hormones.

amniotic egg An egg with a hard outer casing and an inner series of membranes and fluids that provide protection, nutrients, and waste disposal for a growing embryo. The evolution of the amniotic egg, in reptiles, freed them from the constraint of having to reproduce near water.

amniotic fluid A protective and nutritive fluid that surrounds the fetus of mammals, including filling the lungs.

analogy A structure found in different organisms that is similar in function and appearance but is not the result of shared ancestry. Analogies must be distinguished from homologies to get a true picture of evolutionary relationships.

ancestral character A character that existed in the common ancestor of a group of organisms. Cladistics distinguishes ancestral from derived characters and uses these characters to determine evolutionary relationships.

aneuploidy A condition in which an individual organism has either more or fewer chromosomes than is normally found in its species' full set. Down syndrome is the result of aneuploidy—generally three copies of chromosome 21, rather than the standard two.

angiosperm A flowering seed plant whose seeds are enclosed within the tissue called fruit. Angiosperms are the most dominant and diverse of the four principal types of plants. Examples include roses, cacti, corn, and deciduous trees.

animal pole The end of a zygote with relatively less yolk and lying closer to the cell's nucleus. The location of the egg's poles defines the orientation in which the embryo develops.

anther The part of a flower within which pollen grains are produced. The anther is on top of a filament, and together they make up the flower's stamen.

antibiotics Chemical compounds produced by one microorganism that are toxic to another microorganism.

antibody A protein of the immune system found on the surface of B cells, or exported by them, that is able to bind to a specific antigen, thus playing a role in eliminating it from the body.

antibody-mediated immunity An immune system capability that works through the production of proteins called antibodies.

anticodon The end of the transfer RNA molecule that can bind with a particular codon on the mRNA transcript.

antidiuretic hormone (ADH) Also known as vasopressin, a substance produced by the hypothalamus and released by the posterior pituitary gland that functions as a hormone within the kidneys and as a neurotransmitter within the brain. In the kidneys, ADH acts to conserve water by increasing the amount of it sent back into blood circulation. In the brain, vasopressin acts to facilitate social affiliation in both humans and at least one other species of mammal.

antigen Any foreign substance that elicits a response by the immune system. Certain proteins or carbohydrates on the surface of an invading bacterial cell, for instance, act as antigens that trigger an immune response.

antigen-presenting cell (APC) Any immune system cell that presents, on its surface, fragments of an antigen that it has ingested. Dendritic cells are the immune system's most important type of antigen-presenting cell.

aorta The enormous artery extending from the heart that receives all the blood pumped by the heart's left ventricle. Branches stemming from the aorta supply oxygenated blood to all the tissues in the body.

apical meristem Groups of plant cells, located at the tips of the roots and shoots, that give rise to all tissues in the plant.

appendicular skeleton The division of the skeletal system consisting of the bones of the paired appendages, including the pelvic and pectoral girdles to which they are attached.

aquifer Porous underground rock in which groundwater is stored.

Archaea With Bacteria and Eukarya, one of three domains of the living world, composed solely of microscopic, single-celled organisms superficially similar to bacteria but genetically quite different. Many of the Archaea live in extreme environments, such as boiling-hot vents on the ocean floor.

arithmetical increase An increase in numbers by an addition of a fixed number in each time period.

artery A blood vessel that carries blood away from the heart.

asexual reproduction Reproduction that occurs without the union of two reproductive cells (sexual reproduction). Offspring produced through asexual reproduction are genetically identical to their parent organism.

atmosphere The layer of gases that surrounds the Earth.

atomic number The number of protons in the nucleus of an atom. Gold has 79 protons in its nucleus and thus has the atomic number 79. All elements are ordered on the periodic table according to atomic number.

ATP synthase An enzyme functioning in cellular respiration that brings together ADP and inorganic phosphate molecules to produce ATP.

autoimmune disorder An attack by the immune system on the body's own tissues.

autonomic nervous system That portion of the peripheral nervous system's efferent division that provides involuntary regulation of smooth muscle, cardiac muscle, and glands.

autosomal dominant disorder A genetic disorder caused by a single faulty allele located on an autosomal (non-sex chromosome). Huntington disease is one example.

autosomal recessive disorder A recessive dysfunction caused by a faulty allele on an autosome (non-sex chromosome). Sickle-cell anemia is one example.

axial skeleton The division of the skeletal system that forms the central column, including the skull, vertebral column, and rib cage.

axon A single, large extension of the cell body of a neuron that carries signals away from the cell body toward other cells.

B

Bacteria With Archaea and Eukarya, one of three domains of the living world, composed solely of single-celled, microscopic organisms that superficially resemble Archaea but that are genetically quite different.

ball-and-stick model A diagram showing the three-dimensional structure of a molecule, with the atoms drawn as balls and the bonds between them drawn as sticks. This type of representation clearly shows the spatial relationship of the atoms to each other.

base Any substance that accepts hydrogen ions in solution. A base has a number higher than 7 on the pH scale.

Batesian mimicry A type of mimicry in which one species evolves to resemble a species that has superior protection against predators.

bell curve A distribution of values, found commonly in nature, that is symmetrically largest around the average.

bilateral symmetry A bodily symmetry in which opposite sides of a sagittal plane are mirror images of one another. Animals generally are bilaterally symmetrical.

bile Substance produced by the liver that facilitates the digestion of fats. Bile can be released either directly by the liver or by the gallbladder, which stores and concentrates this substance.

binary fission The form of reproduction carried out by prokaryotic cells in which the chromosome replicates and the cell pinches between the attachment points of the two resulting chromosomes to form two new cells.

binomial nomenclature The system of naming species that uses two names (genus and species) for each species. This system helps identify groupings among living things.

biodegradable Capable of being broken down by living organisms.

biodiversity Variety among living things. There are three principal types of biodiversity: species diversity, geographic distribution of species populations, and genetic diversity within species populations.

biogeochemical cycling The movement of water and nutrients back and forth between biotic (living) and abiotic (nonliving) realms.

biological legacies A living thing, or product of a living thing, that survives a major ecological disturbance.

biological species concept A definition of species that relies on the breeding behavior of populations in nature. It defines species as groups of actually or potentially interbreeding natural populations that are reproductively isolated from other such groups.

biology The study of life.

biomass Material produced by living things, generally measured by dry weight.

biome Large, terrestrial regions of the Earth that have similar climates and hence similar vegetative formations. Six biome types are recognized at a minimum: tundra, taiga, temperate deciduous forest, temperate grassland, desert, and tropical rainforest.

biosphere The interactive collection of all the world's ecosystems. Also thought of as that portion of the Earth that supports life.

biotechnology The use of technology to control biological processes as a means of meeting societal needs.

biotic Pertaining to living things.

bipedalism Among primates, the trait of walking upright, on two legs. Along with tooth structure, the most important trait in defining species that are classified as hominins, meaning members of the taxon of human-like primates.

bivalves A class of molluscs that includes mussels, clams, and oysters, among other organisms.

blade In plants, the major, broad part of a leaf.

blastocyst Hollow, fluid-filled ball of cells that is formed in the early stages of the embryonic development of humans and other mammals. In non-mammalian animals, the blastocyst is known as the blastula.

blastula An early-stage animal embryo composed of one or more layers of cells surrounding a liquid-filled cavity known as a blastocoel. A key feature in defining animals as all animals go through a blastula stage in development, but no other organisms do.

B-lymphocyte cell (B cell) The central cells of antibody-mediated immune system function. B cells produce antibodies, which can function as antigen receptors on the surface of B cells. Conversely, antibodies may be exported from B cells as freestanding entities to fight antigens.

body segmentation A repetition of body parts in an animal. An example can be found in the vertebrae that make up the human vertebral column or backbone.

bone A connective tissue that provides support and structure to the body and that often is the site of fat storage and blood cell production.

bottleneck effect A change in allele frequencies in a population due to chance following a sharp reduction in the population's size. One of the factors that potentiates genetic drift.

bronchiole Tiny air-conducting passageway in the lungs that has at its end several alveoli, the hollow air-exchange sacs of the lungs.

bryophytes A plant that lacks a vascular (fluid transport) structure. Bryophytes are the most primitive of the four principal varieties of living plants. Mosses are the most familiar example.

bud An undeveloped plant shoot, composed mostly of meristematic tissue.

buffering system A physiological system that functions to keep pH within normal limits in an organism. Buffering systems generally utilize weak acids or bases to neutralize any sudden infusion of acid or base.

C

C₄ photosynthesis A form of photosynthesis in which carbon dioxide is first fixed into a four-carbon molecule and then transferred to special bundle sheath cells in which the Calvin cycle is undertaken. C$_4$ photosynthesis is used primarily by plants in warm-weather environments as a means of reducing the photorespiration that undercuts carbohydrate production in C$_3$ plants.

calorie (in nutrition) The amount of energy necessary to raise the temperature of 1,000 grams of water by 1°C. Sometimes written as Calorie to distinguish it from a standard calorie, which is defined as the amount of energy needed to raise the temperature of 1 gram of water by 1°C.

Calvin cycle The set of steps in photosynthesis in which an energy-rich sugar is produced by means of two essential processes—the fixing of atmospheric carbon dioxide into a sugar and the energizing of this sugar with the addition of electrons supplied by the light-dependent reactions of photosynthesis. Also known as the Calvin-Benson and C$_3$ cycle.

CAM photosynthesis (crassulacean acid metabolism photosynthesis) A form of photosynthesis, undertaken by some plants in hot, dry climates, in which carbon fixation takes place at night and the Calvin cycle during the day. Practiced by succulent plants in particular, CAM metabolism allows plants to preserve water by opening their stomata only at night.

Cambrian Explosion The sudden appearance of the ancestors of most modern animals beginning in the Cambrian Period, 542 million years ago.

capillary The smallest type of blood vessel, connecting the arteries and veins in the body's tissues. Gases, nutrients, and wastes are exchanged between the blood and the body's tissues through the thin walls of capillaries.

capsid A protein coat that surrounds the genetic material in almost all viruses.

carbohydrate An organic molecule that always contains carbon, oxygen, and hydrogen and that, in many instances, contains nothing but carbon, oxygen, and hydrogen. Carbohydrates usually contain exactly twice as many hydrogen atoms as oxygen atoms. The building blocks or monomers of carbohydrates are monosaccharides, which combine to create the polymers of carbohydrates, the polysaccharides, such as starch and cellulose.

cardiac muscle The type of striated muscle tissue that forms the muscles of the heart.

cardiovascular system A fluid transport system of the body, consisting of the heart, all the blood vessels in the body, the blood that flows through these vessels, and the bone marrow tissue in which red blood cells are formed.

carnivore An animal that eats meat.

carpel The female reproductive structure of a flower, consisting of an ovary, a style, and a stigma.

carrier A person who does not suffer from a recessive genetic debilitation, but who carries an allele for the condition that can be passed along to offspring.

carrying capacity (K) The maximum population density of a species that can be sustained in a given geographical area over time. In ecology, this is often denoted as K.

cartilage A connective tissue that serves as padding in most joints, forms the human larynx (voicebox) and trachea (windpipe), and links ribs to the breastbone.

catalyst A substance that retains its original chemical composition while bringing about a change in a substrate. Enzymes are catalysts in chemical reactions; one enzyme can carry out hundreds or thousands of chemical transformations without itself being transformed.

cell cycle The repeating pattern of growth, genetic duplication, and division seen in most cells.

cell wall A relatively thick layer of material that forms the periphery of all plant and fungal cells and many types of bacterial, archaeal, and protist cells.

cell-mediated immunity An immune system capability that works through the production of cells that destroy infected cells in the body.

cellular respiration The three-stage, oxygen-dependent harvesting of energy that goes on in most cells. The three stages are glycolysis, the Krebs cycle, and the electron transport chain.

cellulose A structural, complex carbohydrate produced by plants and other organisms.

central nervous system (CNS) The portion of the nervous system consisting of the brain and spinal cord.

central vacuole A large, watery plant organelle whose functions include the maintenance of cell pressure, the storage of nutrients, and the retention and degradation of waste products.

centrosome A cellular structure that acts as an organizing center for the assembly of microtubules. A cell's centrosome duplicates prior to mitosis and plays an important part in the development of the cell's mitotic spindle.

cephalopods A class of molluscs that includes squid, octopus, nautilus, and cuttlefish.

cerebellum A region of the brain that refines movements and the sense of balance based on sensory inputs it receives from various parts of the body.

cerebral cortex Thin outer covering of the region of the brain known as the cerebrum. Responsible for the highest human thinking and processing.

cerebrospinal fluid A fluid that circulates in both the brain and spinal cord, supplying nutrients, hormones, and immune system cells and providing the nervous tissue with protection against jarring injury.

cerebrum The largest region of the human brain, responsible for much of the human capacity for higher mental functioning.

cervix The lower part of the uterus, a narrow neck that opens into the vagina.

CFCs A class of human-made chlorine compounds that destroys the atmospheric ozone that protects life on land from damaging ultraviolet radiation.

chemical bonding General term for a bond created when electrons of two atoms interact and rearrange into a new form that allows the atoms to become attached to each other. Ionic, covalent, and hydrogen bonds are all chemical bonds.

chitin A complex carbohydrate that gives shape and strength to the external skeleton of arthropods, including insects, spiders, and crustaceans.

chlorophyll a The primary pigment of chloroplasts, found embedded in its membranes. Together with the accessory pigments, chlorophyll a absorbs some wavelengths of sunlight in the first step of photosynthesis.

chloroplast The organelle within plant and algae cells that is the site of photosynthesis.

cholesterol A steroid molecule that forms part of the outer membrane of all animal cells and that acts as a precursor for many other steroids, among them the hormones testosterone and estrogen.

chromatid One of the two identical strands of chromatin (DNA plus associated proteins) that make up a chromosome in its duplicated state.

chromatin A molecular complex of DNA and its associated proteins that makes up the chromosomes of eukaryotic organisms.

chromosome Structural unit containing part or all of an organism's genome, consisting of DNA and its associated proteins (chromatin). The human genome is made up of 23 pairs of chromosomes, or 46 chromosomes in all.

chyme The soupy mixture of food and gastric juices that passes from the stomach to the small intestine.

cilia Hair-like extensions of a cell, composed of microtubules. Many cilia protrude from the surface of some cells, moving rapidly back and forth, serving to propel the cells or to move material around it.

citric acid cycle Another name for the Krebs cycle, one of the three main sets of steps in cellular respiration; named for the first product of the cycle, citric acid.

cladistics The branch of systematics that uses shared derived characters to determine the order of branching events in speciation and therefore which species are most closely related. Cladistics is concerned only with evolutionary relationships, not classification.

cladogram An evolutionary tree constructed using the cladistic system.

cleavage The developmental process in which a zygote is repeatedly divided into smaller individual cells through cell division.

climate The average weather conditions, including temperature, precipitation, and wind, in a particular region.

climax community The relatively stable community that develops at the end of any process of ecological succession.

clone An exact genetic copy. Also, used as a verb—to make one of these copies. A single gene or a whole, complex organism can be cloned.

cloning vector A self-replicating agent that, in the cloning process, serves to transfer genetic material. Examples include bacterial plasmids and the viruses known as bacteriophages.

closed circulation system A type of circulatory system, found in all vertebrates and in some invertebrates, in which blood stays within vessels.

coastal zone The region lying between the high-tide point on shore and the point offshore where the continental shelf drops off.

cochlea The coiled, fluid-filled, membranous portion of the inner ear in which fluid vibrations are transformed into the nervous system signals perceived as sound.

codominance A condition in which two alleles of a given gene have different phenotypic effects, with both effects manifesting in organisms that are heterozygous for the gene.

codon An mRNA triplet that codes for a single amino acid or a start or a stop command in the translation stage of protein synthesis.

coelom A central body cavity, found in most animal phyla.

coenzyme A type of accessory molecule that binds to the active site of an enzyme, thus allowing the enzyme to bind to its substrate. Many vitamins are coenzymes.

coevolution The interdependent evolution of two or more species. Coevolution can benefit both species, as in flowering plants and their animal pollinators, or it can be an arms race between species, as in a plant and its predators.

coexistence The condition in which two species live in the same habitat, dividing up resources in a way that allows both to survive.

colonial multicellularity A form of life in which individual cells form stable associations with one another but do not take on specialized roles.

commensalism An interaction between two species in which one benefits while the other is neither harmed nor helped.

common descent with modification The process by which species of living things undergo modification in successive generations, with such modification sometimes resulting in the formation of new, separate species.

community All the populations of all species of living things that inhabit a given area. The term also is used to mean a collection of populations in a given area that potentially interact with each other.

compact bone Dense bone that forms the outer portion of bones, structured as a set of parallel osteons and their associated nerves and blood vessels.

competitive exclusion principle When two species compete for the same limited, vital resource, one will always outcompete the other and thus bring about the latter's local extinction.

competitive inhibition A reduction in the activity of an enzyme by means of a compound other than the enzyme's usual substrate binding with it in its active site.

concentration gradient A gradient within a given medium defined by the difference between the highest and lowest concentrations of a solute. The solute will have a natural tendency to move from the areas of higher concentration to lower, thus diffusing.

cones In human vision, photoreceptors that respond best to bright light and that provide color vision.

connective tissue A tissue, active in the support and protection of other tissues, whose cells are surrounded by a material that they have secreted. In humans, one of the four principal types of tissue.

consumer Any organism that eats other organisms rather than producing its own food.

continental drift The movement of continental plates over the globe, allowing continents to divide and rejoin in different patterns.

convergent evolution Evolution that occurs when similar environmental influences shape two separate evolutionary lines in similar ways.

coral reef Ocean structure, found in shallow, warm waters, that consists primarily of the piled-up remains of many generations of the animals called coral polyps. Such reefs provide habitat for a rich diversity of marine organisms.

coronary artery An artery that delivers oxygenated blood to the muscles of the heart. A heart attack is a complete blockage of a coronary artery.

corpus luteum The structure that develops in the mammalian ovary from the ruptured tertiary follicle following ovulation. The corpus luteum secretes hormones that help prepare the reproductive tract for pregnancy.

cotyledon An embryonic leaf. A major division in plants is between the monocots, which have one embryonic leaf, and the dicots, which have two embryonic leaves.

coupled reaction A chemical reaction in which an endergonic reaction is powered by an exergonic reaction. All "uphill," or endergonic, reactions require an input of energy; this energy is supplied by "downhill," or exergonic, reactions.

covalent bond A type of chemical bond in which two atoms are linked through a sharing of electrons.

Cretaceous Extinction A mass extinction event that occurred at the boundary between the Cretaceous and Tertiary periods. This event, which included an asteroid impact, resulted in the extinction of the dinosaurs along with many other organisms.

crossing over (genetic recombination) A process, occurring during meiosis, in which homologous chromosomes exchange reciprocal portions of themselves.

cross-pollinate To pollinate one plant with pollen of another plant. Mendel used this technique in conducting his experiments to uncover rules of heredity.

cytokinesis The physical separation of one cell into two daughter cells.

cytoplasm The region of a cell inside the plasma membrane and outside the nucleus. Usually, this region is filled with the jelly-like cytosol containing the cell's extranuclear organelles.

cytoskeleton A network of protein filaments that functions in cell structure, cell movement, and the transport of materials within the cell. Microfilaments, intermediate filaments, and microtubules are all parts of the cytoskeleton.

cytosol The protein-rich, jelly-like fluid in which a cell's organelles that are outside the nucleus are immersed.

cytotoxic T cell (killer T cell) Type of T-lymphocyte cell that binds to and kills the body's own cells when they have become infected.

D

deciduous Refers to plants that show a coordinated, seasonal loss of leaves. This strategy allows plants to conserve water during a time they could perform little photosynthesis anyway.

decomposer A type of detritivore that, in feeding on dead or cast-off organic material, breaks it down into its inorganic components. Most decomposers are fungi or bacteria.

deletion A chromosomal condition in which a piece of a chromosome has been lost. Occurs when a chromosomal fragment that breaks off does not rejoin any chromosome.

dendrites Extensions of a neuron that carry signals toward the neuronal cell body.

density dependent In ecology, effects on a population that increase or decrease in accordance with the size of that population. Density-dependent effects tend to involve biological factors.

density independent In ecology, effects on a population that are not related to the size of that population. Density-independent effects tend to involve physical forces, such as temperature and rain.

deoxyribonucleic acid (DNA) The primary information-bearing molecule of life, composed of two chains of nucleotides linked together in the form of a double helix. Proteins are put together in accordance with the information encoded in DNA.

derived character A character unique to groupings of organisms (taxa) descended from a common ancestor.

dermis In certain animals, the thick layer of the skin—composed mostly of connective tissue—that underlies, nourishes, and supports the epidermis.

desert A biome in which rainfall is less than 25 centimeters, or 10 inches, per year and water evaporation rates are high relative to rainfall.

detritivore An organism that feeds on the remains of dead organisms or the cast-off material from living organisms.

dicotyledon (dicot) A type of plant that has two embryonic leaves within its seed. More than three-quarters of all flowering plants are broad-leafed dicotyledons.

diffusion The movement of molecules or ions from areas of their higher concentration to areas of their lower concentration. Over time, the random movement of molecules will result in the even distribution of the material.

digestive system The organ system that transports food into the body, secretes digestive enzymes that help break down food to allow it to be absorbed by the body, and excretes waste products. This system consists of the esophagus, stomach, and small and large intestines plus accessory glands such as the pancreas and gallbladder.

digestive tract A muscular passageway for food and food waste that runs through the human body from the mouth to the anus.

dihybrid cross An experimental cross in which the plants used differ in two of their characters.

dikaryotic Having two haploid nuclei in one cell. A phase of the life cycle of many fungi.

diploid Possessing two sets of chromosomes. All human cells are diploid with the exception of human gametes (eggs and sperm), which are haploid. Such haploid cells possess only a single set of chromosomes.

directional selection In evolution, the type of natural selection that moves a character toward one of its extremes. Compare to stabilizing and disruptive selection.

disruptive selection In evolution, the type of natural selection that moves a character toward both of its extremes, operating against individuals that are average for that character. This type of selection seems to be less common in nature than either stabilizing or directional selection.

DNA polymerase An enzyme that is active in DNA replication, separating strands of DNA, bringing bases to the parental strands, and correcting errors by removing and replacing incorrect base pairs.

dominant Term used to designate an allele that is expressed in the heterozygous condition.

dominant disorder Genetic conditions in which a single faulty allele can cause damage, even when a second, functional allele exists.

dormancy A state in which growth is suspended and there is a prolonged low level of metabolic activity. Dormancy allows organisms to conserve energy during times of unfavorable environmental conditions.

dorsal nerve cord A rod-shaped dorsal structure consisting of nerve cells, running from a chordate animal's head to its tail.

double fertilization A fusion of gametes and of nutritive cells that occurs at the same time in angiosperm plants.

Down syndrome A disorder in humans in which affected individuals usually have three copies of chromosome 21 rather than the standard two. Individuals with this syndrome have short stature, shortened life span, and low IQ.

E

ecological dominants A species that is abundant and obvious in a given community. In any community, a few species, usually plants, will dominate in numbers.

ecology The study of the interactions that living things have with each other and with their environment.

ecosystem A community of living things and the physical environment with which they interact.

ectopic pregnancy An abnormal pregnancy in which the blastocyst has attached to a location other than the uterine wall.

efferent division The division of the peripheral nervous system that carries motor commands from the central nervous system (CNS) toward effectors (muscles and glands).

electron With protons and neutrons, one of three basic constituents of an atom. Electrons carry a negative electrical charge and are distributed in an atom at a distance from the nucleus.

electron carrier A molecule that serves to transfer electrons from one molecule to another in ATP

formation. The most important electron carrier in ATP formation is NAD+ (NADH in its reduced form). The other major carrier is FAD ($FADH_2$).

electron transport chain (ETC) The third stage of cellular respiration, occurring within the inner membrane of the mitochondria, in which most of the ATP is formed.

electronegativity The measure of the strength of attraction an atom has for electrons that are being shared in a covalent bond. An atom with higher electronegativity will tend to pull electrons toward itself, away from atoms with lower electronegativity.

element A substance that cannot be reduced to any simpler set of components through chemical processes. An element is defined by the number of protons in its nucleus.

embryo A developing organism. In humans, the developing organism from the time a zygote undergoes its first division through the end of the eighth week of development.

embryo sac The mature female gametophyte plant. Cells of the embryo sac include the egg that will be fertilized by one sperm cell from the pollen grain, and the central cell, whose two nuclei will fuse with a second sperm cell from the pollen grain, thus producing food for the embryo.

embryonic stem cell A cell from the blastocyst stage of a human embryo that is capable of giving rise to all the types of cells in the adult body.

endergonic reaction A chemical reaction in which the ending set of molecules (the products) contains more energy than the starting set of molecules (the reactants).

endocrine gland A gland that releases its materials directly into surrounding tissues or into the bloodstream, without using ducts.

endocrine system The organ system that sends signals throughout the body through use of the chemical messengers called hormones.

endocytosis The process by which cells bring relatively large materials into themselves through use of transport vesicles.

endomembrane system An interactive group of membrane-lined organelles and transport vesicles that functions within eukaryotic cells.

endometrium The tissue lining the interior of the uterus in mammals, which thickens in response to hormonal secretion during ovulation and is shed during menstruation. If pregnancy occurs, this tissue initially houses the embryo.

endosperm The nutrient tissue that surrounds an angiosperm embryo in the seed. The rice and wheat grains that we eat consist mostly of endosperm.

energy The capacity to bring about movement against an opposing force.

energy-flow model A conceptualization of ecosystems as units in which energy is first captured by given organisms and then transferred to other organisms.

environmental resistance All the forces in the environment that act to limit the size of a population.

enzyme A chemically active protein that speeds up, or in practical terms enables, chemical reactions in living things.

epidermis The outermost layer of skin in animals or the outermost cell layer in plants.

epididymis A collection of tubules near the testis in which sperm complete their development and are stored.

epithelial tissue A tissue that covers surfaces exposed to an external environment. In humans, skin is an epithelial tissue, as is the lining of the digestive tract.

essential amino acid In nutrition, one of nine amino acids that the body cannot make and that hence must be supplied by food.

estrogen A class of hormones, produced primarily by cells of the ovary, that supports egg development, growth of uterine lining, and development of female sex characteristics.

estuary An area where a river or stream flows into the ocean, bringing freshwater and saltwater habitat together. Estuaries are among the most productive ecosystems on Earth.

Eukarya With Bacteria and Archaea, one of three domains of the living world, composed of four subordinate kingdoms: Plantae, Animalia, Fungi, and Protista. All members of domain Eukarya are composed of single cells or groups of cells that have a nucleus.

eukaryote An organism that is a member of domain Eukarya.

eukaryotic cell A cell whose primary complement of DNA is contained within a membrane-lined nucleus. Eukaryotic cells have several other organelles in addition to the nucleus. All organisms except bacteria and archaea either are single eukaryotic cells or are composed of eukaryotic cells.

evolution Any genetically based phenotypic change in a population of organisms over successive generations. Evolution can also be thought of as the process by which species of living things can undergo modification over successive generations, with such modification sometimes resulting in the formation of new species.

exergonic reaction A chemical reaction in which the starting set of molecules (the reactants) contain more energy than the final set of molecules (the products).

exocrine gland A gland that secretes its materials through ducts (tubes). For example, sweat glands conduct perspiration through ducts to the skin.

exocytosis The means by which relatively large volumes of material are moved from the inside of a cell to the outside. In exocytosis, a transport vesicle fuses with a cell's plasma membrane, after which the contents of the vesicle are ejected outside the cell.

exoskeleton An external material covering the animal body, providing support and protection.

exponential growth A form of population growth in which the rate of growth is proportional to the starting size of the population. Exponential growth results in a J-shaped growth curve because, when plotted on a graph, the population's increase resembles the letter J.

exponential increase An increase in numbers that is proportional to the number already in existence.

This type of increase occurs in populations of living things, and it carries the potential for enormous growth of populations.

extremophile An organism that grows optimally in one or more conditions that would kill most other organisms.

extrinsic isolating mechanism A barrier to interbreeding of populations that is not an inherent characteristic of the organisms in the populations. Geographic barriers such as rivers are extrinsic isolating mechanisms.

F

facilitated diffusion A passage of materials through the cell's plasma membrane that is aided by a transport protein.

facilitation In succession, the actions or qualities of earlier-arriving species that in some way assist the establishment of later-arriving species within a community.

fat A dietary lipid that is solid at room temperature (e.g., butter, the fat in a piece of bacon).

fatty acid A molecule, found in many lipids, composed of a hydrocarbon chain bonded to a carboxyl group.

fertilization The fusion of two gametes to form a zygote. In humans (and many other organisms), the gametes known as sperm and egg fuse, resulting in a zygote.

fetus In humans, the developing organism from the start of the ninth week of development to the moment of birth.

fibers In skeletal muscle, a single elongated muscle cell, containing hundreds of long, thin myofibrils that run the length of the cell. In nutrition, one of the three principal classes of dietary carbohydrate, defined as a complex carbohydrate that is indigestible. The other classes of dietary carbohydrate are simple sugars and starches.

fibrous root system A plant root system that consists of many roots, all about the same size.

filament The part of a stamen (male reproductive part of a flower) that is shaped like and functions as a stalk and has an anther at the top.

first filial generation (F_1) The offspring of the parental generation in an experimental genetic cross.

first law of thermodynamics Energy cannot be created or destroyed, but can only be transformed from one form to another.

fitness In evolution, the success of an organism, relative to other members of its population, in passing on its genes to the next generation of offspring.

fixation The process of the incorporation of a gas into an organic molecule. In photosynthesis, atmospheric carbon dioxide (CO_2) is fixed into the sugars of photosynthesizing organisms, such as plants. In ecology, atmospheric nitrogen (N_2) is fixed into ammonia (NH_3) by certain bacteria.

flagella The relatively long, tail-like extensions of some cells, composed of microtubules, that function in cell movement. Often, cells will have but a single flagellum, as with mammalian sperm cells.

fluid-mosaic model A conceptualization of the cell's plasma membrane as a fluid, phospholipid bilayer that has within it a mosaic of both stationary and mobile proteins.

follicle In the vertebrate ovary, the complex of the oocyte (developing egg) and the cells and fluids that surround and nourish it.

follicle-stimulating hormone (FSH) A hormone, secreted by the anterior pituitary, that promotes egg development and stimulates secretion of estrogens in women and supports sperm production and testosterone secretion in men.

formed elements Cells and cell fragments that form the nonfluid portion of blood. Formed elements include red blood cells, white blood cells, and platelets. Contrast with plasma, the fluid portion of blood, consisting mostly of water.

founder effect The phenomenon by which an initial gene pool for a population is established by means of that population migrating to a new area. One of the conditions that potentiates genetic drift.

fraternal twins Twins who are produced first through multiple ovulations in the mother, followed by multiple fertilizations from separate sperm of the father, and then multiple implantations of the resulting embryos in the uterus.

free radical A molecule with an unpaired electron, usually existing for only a very brief time. Although free radicals are natural products of biological processes, these molecules can be destructive in living tissues.

fruit The mature ovary of any flowering plant. Many fruits protect the underlying seeds, and many attract animals that will eat the fruit and disperse the seeds.

functional group A group of atoms that confers a special property on a carbon-based molecule. Functional groups usually are transferred as a unit among carbon-based molecules and often confer an electrical charge or polarity on the molecules they are part of.

G

gallbladder Organ of the body that stores and concentrates the digestive material bile, which is produced by the liver. Bile facilitates the breakdown of fats by digestive enzymes.

gamete A haploid reproductive cell, either egg or sperm.

gametophyte generation The haploid generation in plants that produces gametes (eggs and sperm). In the flowering plants (angiosperms), this generation is microscopic but gives rise to the visible sporophyte generation.

ganglion Any collection of nerve-cell bodies in the peripheral nervous system.

gap junction A protein assemblage that forms a communication channel between adjacent animal cells. Gap junctions open only as necessary, thus allowing small molecules and electrical signals to move between cells.

gastropods A class of molluscs that includes snails and slugs, among other organisms.

gastrulation The process in early animal development in which an embryo's cells migrate to form three layers of tissue: the endoderm, the mesoderm, and the ectoderm. Each of these layers then goes on to give rise to differing tissues and organs in the organism.

gene flow The movement of genes from one population to another.

gene pool The entire collection of alleles, or variant forms of genes, in a population.

genetic code The inventory of linkages between nucleotide triplets and the amino acids they code for. With few exceptions, the genetic code is universal in living things.

genetic drift The chance alteration of allele frequencies in a population, with such alterations having greatest impact on small populations.

genetics The study of physical inheritance among living things.

genome The complete collection of an organism's genetic information. More narrowly, the complete haploid set of an organism's chromosomes.

genotype The genetic makeup of an organism, including all the genes that lie along its chromosomes.

germ-line cell The succession of parent and daughter cells that ultimately produce either eggs or sperm.

gland An organ or group of cells that secretes one or more substances.

glomerulus Knotted network of capillaries in each of the kidneys' nephrons that receives blood and lets some smaller blood-borne materials pass out of it (and into the surrounding Bowman's capsule) while retaining larger materials.

glucagon A hormone, secreted by cells in the pancreas, that brings about an increase in blood levels of glucose.

glycemic load A measure of how blood glucose levels are affected by defined portions of given carbohydrates. Lower glycemic loads are associated with improved human health.

glycocalyx An outer layer of the plasma membrane, composed of short carbohydrate chains that attach to membrane proteins and phospholipid molecules. Such chains serve as the actual binding sites on many membrane proteins, act to lubricate the cell, and can form an adhesion layer that allows one cell to stick to another.

glycogen A complex carbohydrate molecule that serves as the primary form in which carbohydrates are stored in animals.

glycolysis The first stage of cellular respiration, occurring in the cytosol. For some organisms, glycolysis is the sole means of extracting energy from food. In most organisms, it is a means of extracting some energy and a necessary precursor to the other two stages of cellular respiration, the Krebs cycle and the electron transport chain.

glycoprotein A molecule that combines protein and carbohydrate. Glycoproteins play important roles as cell receptors and some types of hormones, among other functions.

Golgi complex A network of membranes, found in the cytoplasm of eukaryotic cells, that processes and distributes proteins that come to it from the rough endoplasmic reticulum.

gonadotropin-releasing hormone A hormone released in tiny amounts from the brain's hypothalamus that stimulates the anterior pituitary to release two other hormones (follicle-stimulating hormone and luteinizing hormone) important in reproduction.

gravitropism The bending of a plant's roots or shoots in response to gravity. This capability helps a plant orient its roots and shoots properly—roots toward the center of the Earth, shoots away from it.

gross primary production The amount of material that a photosynthesizing organism produces through photosynthesis.

groundwater Water contained within underground rock formations.

gymnosperm A seed plant whose seeds do not develop within fruit tissue. One of the four principal varieties of plants, gymnosperms reproduce through wind-aided pollination. Coniferous trees, such as pine and fir, are the most familiar examples

H

habitat The type of surroundings in which individuals of a species are normally found.

haploid Possessing a single set of chromosomes. Human gametes (eggs and sperm) are haploid cells because they have only a single set of chromosomes. All other cells in the human body are diploid, meaning they possess two sets of chromosomes. Some organisms are strictly haploid. Bacteria, for example, have only a single chromosome, making them haploid.

heart attack A complete blockage of one of the heart's coronary arteries, resulting in the death of groups of heart cells from lack of a blood supply.

helper T cell Type of T-lymphocyte cell that stimulates both T-cell and B-cell immunity. Referred to in AIDS therapy as a CD-4 cell.

hemoglobin The iron-containing protein in red blood cells that binds to both oxygen and carbon dioxide, thus assisting in their transportation.

herbivore An animal that eats only plants.

hermaphroditic A state in which one animal possesses both male and female sex organs.

heterozygous Possessing two different alleles of a gene for a given character.

high-density lipoprotein (HDL) A type of protein that transports fat or lipid molecules (usually cholesterol) from various tissues in the body to the liver. Sometimes known as the "good cholesterol," HDLs help remove, and possibly neutralize, the LDLs, or low-density lipoproteins, that can harmfully reside in coronary arteries.

histamine A compound that, in immune system activity, brings about blood vessel dilation and increased blood vessel permeability.

HIV Human immunodeficiency virus, the cause of the disease AIDS.

homeostasis The maintenance of a relatively stable internal environment in living things.

Hominini The taxonomic grouping of human-like primates. Modern human beings, or Homo sapiens, are one of the species within Hominini. All other hominin species are now extinct.

***Homo sapiens* (*H. sapiens*)** The species of modern human beings. The last surviving species within the taxonomic category of Hominini or human-like primates.

homologous In anatomy, having the same structure owing to inheritance from a common ancestor. Forelimb structures in whales, bats, cats, and gorillas are homologous.

homologous chromosomes Chromosomes that are the same in function and hence size. Species that are diploid (have two sets of chromosomes) have matching pairs of homologous chromosomes; one member of each homologous pair is inherited from the male, the second member of each homologous pair is inherited from the female.

homology A structure that is shared in different organisms owing to inheritance from a common ancestor. Homologies are used to help decipher evolutionary relationships.

homozygous Having two identical alleles of a gene for a given character.

hormone A substance that, when released in one part of an organism, goes on to prompt physiological activity in another part of the organism. Both plants and animals have hormones.

host The prey in a parasitic relationship.

human genome The 3.2 billion base pairs of DNA that make up the full complement of DNA found in the nucleus of each human cell.

hydrocarbon A compound made of hydrogen and carbon. Hydrocarbons are nonpolar covalent molecules and therefore are not easily dissolved in water.

hydrogen bond A chemical bond that links an already covalently bonded hydrogen atom with a second, relatively electronegative atom.

hydrophilic The property, possessed by some compounds, of being able to form chemical bonds with water molecules. Table salt ($NaCl$) is hydrophilic and thus will readily dissolve in water.

hydrophobic The property, possessed by some compounds, of being unable to form chemical bonds with water molecules. Oil is hydrophobic and thus will not readily dissolve in water.

hydroxide ion Strongly basic OH^- ion, often used to shift solutions toward neutral or basic on the pH scale.

hypertonic solution A solution that has a high concentration of solutes relative to an adjacent solution.

hyphae (singular, hypha) The slender filaments that make up the bulk of most fungi.

hypothalamus Portion of the brain important in drives and in the maintenance of homeostasis, the latter capacity coming about because of the hypothalamus' control over a good deal of the body's hormonal release.

hypothesis A tentative, testable explanation of an observed phenomenon.

hypotonic solution A solution that has a low concentration of solutes relative to an adjacent solution.

I

identical twins Twins who develop from a single zygote; strictly speaking, identical twins are a single organism at one point in their development and as such have exactly the same genetic makeup.

immune system The collection of cells and proteins that, in vertebrates, function together to kill or neutralize invading microorganisms.

incompletely dominant A genetic condition in which the heterozygote phenotype is intermediate between either of the homozygous phenotypes.

independent assortment The random distribution of homologous chromosome pairs on differing sides of the metaphase plate during meiosis.

induced pluripotent stem cell (iPS cell) A cell that has been induced to a state of pluripotency through the introduction of genes from outside its genome. In humans, a pluripotent cell is a cell that can give rise to all the types of cells found in the body.

induction In animals, the capacity of some embryonic cells to direct the development of other embryonic cells. For example, cells of the notochord induce the tissue above them to form the neural tube.

inflammation A localized immune system response to tissue injury. Inflammation can also be systemic, or body-wide, and chronic, or ongoing. Chronic, systemic inflammation has been linked to illnesses such as type 2 diabetes.

innate immune response The human immune response that does not target specific microbial invaders; compare to adaptive immune response.

inner cell mass In mammalian development, the group of cells in the embryo that will develop into the baby rather than into the placenta.

insulin A hormone, secreted by cells in the pancreas, that brings about a decrease in blood levels of glucose.

integral protein A protein of the plasma membrane that is attached to the membrane's hydrophobic interior.

integumentary system The organ system that protects the body from the external environment and assists in regulation of body temperature. This system consists of the skin and associated structures, such as glands, hair, and nails.

intermediate filament Filaments of the cytoskeleton, intermediate in diameter between microfilaments and microtubules, that help stabilize the positions of the nucleus and other organelles within the cell.

interneuron A type of neuron, located only within the brain or spinal cord, that connects other neurons. These neurons are responsible for the analysis of sensory inputs and the coordination of motor commands.

interphase That portion of the cell cycle in which the cell simultaneously carries out its work and—in preparation for division—duplicates its chromosomes. The other primary phase of the cell cycle is mitotic phase (or M phase), which includes both mitosis (in somatic cells) and cytokinesis.

interspecific competition Competitive interaction between individuals of two different species.

intertidal zone The region within the coastal zone of the ocean that extends from the ocean's low-tide mark to its high-tide mark.

intrinsic isolating mechanism A difference in anatomy, physiology, or behavior that prevents interbreeding between individuals of the same species or of closely related species. One or more intrinsic isolating mechanisms must develop for two populations of the same species to evolve into separate species.

intrinsic rate of increase (*r*) The rate at which a population would grow if there were no external limits on its growth. In ecology, often denoted as *r*.

inversion A chromosomal abnormality that comes about when a chromosomal fragment that rejoins a chromosome does so with an inverted orientation.

invertebrate An animal without a vertebral column.

ion An atom whose number of electrons differs from its number of protons, thus giving it an electrical charge.

ionic bonding A linkage in which two or more ions are bonded to each other by virtue of their opposite charge.

ionic compound A collection of the atoms of two or more elements that have become linked through ionic bonding.

isotonic solution A solution that has the same concentration of solutes as an adjacent solution.

isotope A form of an element as defined by the number of neutrons contained in its nucleus. Different isotopes of an element have the same number of protons but differing numbers of neutrons.

K

karyotype A pictorial arrangement of a full set of an organism's chromosomes.

keratin A flexible, water-resistant protein, abundant in the outer layers of skin, that also makes up hair and fingernails.

keystone species A species whose absence from a community would bring about significant change in that community.

kidneys The filtering organs of the urinary system that produce urine while conserving useful blood-borne materials.

kilocalorie The amount of energy it takes to raise 1,000 grams (1 kilogram) of water 1°C. Food consumption is measured in kilocalories, often written as Calories.

kinetic energy Energy in motion. A rolling rock has kinetic energy.

Krebs cycle The second stage of cellular respiration, occurring in the inner compartment of mitochondria. The Krebs cycle is the major source of electrons that power the third stage of respiration, the electron transport chain. Also known as the citric acid cycle.

***K*-selected species** A species that tends to be relatively long-lived, that tends to have relatively few offspring for whom it provides a good deal of care, and whose population size tends to be relatively stable, remaining at or near its environment's carrying capacity (K). Also known as an equilibrium species.

L

labor The regular contractions of the uterine muscles that sweep over the fetus, creating pressure that opens the cervix and expels the baby and the placenta.

large intestine That portion of the digestive tract that begins at the small intestine and ends at the anus. The large intestine serves largely to compact and store material left over from the digestion of food, turning this material into the solid waste known as feces.

law of conservation of mass In a chemical reaction, matter can neither be created nor destroyed.

law of independent assortment During gamete formation, gene pairs assort independently of one another. Also known as Mendel's Second Law, this is one of the principles of inheritance formulated by Gregor Mendel.

law of segregation Differing characters in organisms result from two genetic elements (alleles) that separate in gamete formation, such that each gamete gets only one of the two alleles. Also known as Mendel's First Law, this is one of the principles of inheritance formulated by Gregor Mendel.

lichen A composite organism composed of a fungus and either algae or photosynthesizing bacteria.

life cycle The repeating series of steps that occur in the reproduction of an organism.

life sciences A set of disciplines that focus on various aspects of the living world. The life sciences include biology and related disciplines such as medicine and forestry.

life table A table showing how likely it is for an average species member to survive a given unit of time.

ligaments In anatomy, a connective tissue that links one bone to another.

lipid A member of a class of biological molecules whose defining characteristic is their relative insolubility in water. Examples include triglycerides, cholesterol, steroids, and phospholipids.

lipoprotein A molecule composed of both lipid and protein. Lipoproteins transport fat molecules through the bloodstream to all parts of the body.

liver An organ that is central to the body's metabolism of nutrients.

logistic growth A form of population growth in which exponential growth slows and then stops in response to environmental resistance. Also known as S-shaped growth because, when plotted on a graph, changes in the population's size resemble the letter S.

low-density lipoprotein (LDL) A type of protein that transports fat or lipid molecules (usually cholesterol) from the liver and small intestine to various tissues throughout the body. Sometimes known as the "bad cholesterol," LDLs can initiate heart disease by coming to reside within coronary arteries.

luteinizing hormone (LH) A hormone secreted by the anterior pituitary; in women, LH induces ovulation and stimulates the ovary to secrete estrogens and progestins to prepare the body for possible pregnancy. In men, LH stimulates the testes to produce androgens such as testosterone.

lymphatic network In humans, the transport network that collects interstitial fluid, transports it as lymph through lymphatic vessels, checks the fluid for infection, and delivers the fluid to blood vessels.

lysosome An organelle found in animal cells that digests worn-out cellular materials and foreign materials that enter the cell.

M

macroevolution Evolution that results in the formation of new species or other large groupings of living things.

mammary glands A set of glands that, in female mammals, provide milk for the young.

marsupial A type of mammal in which the young develop within the mother to a limited extent, inside an egg having a membranous shell. Early in development, the egg's membrane disappears, after which the mother delivers a developmentally immature but active marsupial. Kangaroos are one example of a marsupial.

mass A measure of the quantity of matter in an object. On Earth, equivalent to the weight of an object.

medulla oblongata A region of the brain concerned with the regulation of unconscious functions such as breathing, blood pressure, and digestion.

meiosis A process in which a single diploid cell divides to produce four haploid reproductive cells.

membrane potential The electrical charge difference that exists from one side of a neuron's plasma membrane to the other.

memory B cell One of two types of cells that the B lymphocyte or B cell develops into (the other type being a plasma cell). Memory B cells remain in the system long after a first infection by a microorganism has ended and serve to produce more plasma B cells quickly should the microorganism ever invade again.

menopause The cessation of the monthly ovarian cycle that occurs when women reach about 50 years of age.

menstruation The cyclical release of blood and the specialized uterine lining, the endometrium; occurs about once every 28 days in human females, except when the oocyte is fertilized.

messenger RNA (mRNA) A type of RNA that encodes, and carries to ribosomes, information for the synthesis of proteins.

metabolic pathway A sequential set of enzymatically controlled chemical reactions in which the product of one reaction serves as the substrate for the next.

metabolism The sum of all chemical reactions carried out by a cell or larger organism.

metaphase plate A plane located midway between the poles of a dividing cell.

microevolution A change of allele frequencies in a population over a short period of time. The basis for all large-scale evolution (macroevolution).

microfilaments The most slender of the eukaryotic cell's cytoskeletal fibers, made of the protein actin and functioning in both cell movement and support.

micrograph A picture taken with the aid of a microscope.

micrometer A millionth of a meter, abbreviated as μm.

microtubule The largest of the cytoskeletal filaments, microtubules take the form of hollow tubes composed of the protein tubulin. They help give structure to the cell, serve as the "rails" on which transport vesicles move, and form the cellular extensions known as cilia and flagella.

midbrain A region of the brain that helps maintain muscle tone and posture through control of involuntary motor responses.

mimicry A phenomenon in which one species evolves to resemble another species.

mineral An element essential to the functioning of a living organism.

mitochondria Organelles that are the primary sites of energy conversion within eukaryotic cells.

mitosis The separation of a somatic cell's duplicated chromosomes prior to cytokinesis.

mitotic phase (M phase) That portion of the cell cycle that includes both mitosis and cytokinesis. Mitosis is the separation of a somatic cell's duplicated chromosomes; cytokinesis is the physical separation of one cell into two daughter cells.

mitotic spindle The microtubules active in cell division, including those that align and move the chromosomes.

modern synthesis The unified evolutionary theory that resulted from the convergence of several lines of biological research between 1937 and 1950.

molecular biology The investigation of life at the level of its individual molecules.

molecular formula A notation specifying the elements in a molecule, with the number of atoms of each element in the molecule shown as a subscript. This notation reveals the constituent elements of a molecule, but not the spatial arrangement of those elements.

molecule An entity consisting of a defined number of atoms covalently bonded together in a defined spatial relationship.

molting A periodic shedding of an old exoskeleton followed by the growth of a new one; practiced by animals in the phyla Nematoda and Arthropoda.

monocotyledon (monocot) A type of plant that has one embryonic leaf within the seed. Although comprising only one-quarter of all flowering plants, most important food plants are the narrow-leafed monocotyledons.

monohybrid cross An experimental cross in which organisms are tested for differences in one character.

monomer A small molecule that can be combined with other similar or identical molecules to make a larger polymer.

monosaccharide A building block, or monomer, of carbohydrates. Monosaccharides combine to form complex carbohydrates, or polysaccharides. Glucose is an example of a monosaccharide.

monotremes Egg-laying mammals, represented by the duck-billed platypus and spiny anteaters found in Australia.

monounsaturated fatty acid A fatty acid with one double bond between the carbon atoms of its hydrocarbon chain.

morphogen A diffusible substance whose concentration in a region of an animal embryo affects development in that region.

morphology The physical form of an organism.

morula A tightly packed ball of early embryonic cells.

motor neuron A neuron that sends signals from the central nervous system (CNS) to such structures as muscles or glands. Given the direction of this transmission, motor neurons are efferent neurons.

Müllerian mimicry A type of mimicry in which several species that have protection against predators evolve to look alike.

multiple alleles Three or more alleles—alternative forms of a gene—occurring in a population.

muscle fiber A single skeletal muscle cell; called a fiber because of its extreme length relative to most cells.

muscle tissue Tissue that has the ability to contract. In humans, one of the four principal types of tissue.

muscular system The organ system composed of all the skeletal muscles of the body, which is to say all muscles that are under voluntary control.

mutation A permanent alteration of a DNA base sequence.

mutualism A form of relationship between two organisms in which both organisms benefit.

mycelium A web of fungal hyphae that makes up the major part of a fungus.

mycorrhizae Associations of plant roots and fungal hyphae. The fungal hyphae absorb minerals, growth hormones, and water that are then available to the plant, and the fungus gets carbohydrates from the photosynthesizing plant.

myelin Membranous covering of some neuronal axons, provided by glial cells, that allows faster nerve signal transmission through these axons.

N

n In a living cell, the condition of being haploid: of having a single set of chromosomes.

nanometer A billionth of a meter; abbreviated nm.

natural selection The process through which traits that confer a reproductive advantage to individual organisms grow more common in populations of organisms over successive generations.

Neanderthal A now-extinct species of hominin that lived in Europe and Asia from approximately 350,000 years ago to 28,000 years ago.

negative feedback A system of control in which the product of a process reduces the activity that led to the product.

nephron Functional unit of the kidneys, composed of a nephron tubule, its associated blood vessels, and the interstitial fluid in which both are immersed.

nerve A bundle of axons in the peripheral nervous system that transmits information to or from the central nervous system.

nervous system The organ system that monitors an animal's internal and external environment, integrates the sensory information received, and coordinates the animal's responses. This system consists of all the body's neurons, plus the supporting neuroglia cells, plus the sensory organs.

nervous tissue A tissue specialized for the rapid conduction of electrical impulses. In humans, one of the four principal tissue types.

net primary production The amount of material a plant (or other photosynthesizing organism) accumulates through photosynthesis. Net primary production is gross primary production minus energy lost to heat and the plant's expenditures of energy on its own maintenance.

neural crest cells Cells that break away from the top of the vertebrate neural tube as it folds together and then migrate to varying parts of the embryo, giving rise to various tissues and organs.

neural tube A dorsal, ectodermic structure that gives rise in vertebrates to both the brain and spinal cord.

neurotransmitter A chemical, secreted into a synaptic cleft by a neuron, that affects nervous system signaling by binding with receptors on an adjacent neuron or effector cell.

neutron One of three primary constituents of an atom, possessing no electrical charge and found in the atom's nucleus. Isotopes are defined by the number of neutrons in an atom.

niche A characterization of an organism's way of making a living that includes its habitat, food, and behavior.

nicotinamide adenine dinucleotide (NAD) The most important intermediate electron carrier in cellular respiration.

nitrogen fixation The conversion of atmospheric nitrogen into a form that can be taken up by living things. Bacteria fix nitrogen, which is essential to life.

nondisjunction A failure of homologous chromosomes or sister chromatids to separate during cell division.

nonessential amino acid In nutrition, one of the 11 amino acids that can be produced by the body and that hence does not need to supplied by food.

nonpolar covalent bond A type of covalent bond in which electrons are shared equally between atoms.

notochord A dorsal, rod-shaped support organ that exists in embryonic development in all vertebrates and in the adults of some vertebrates.

nuclear envelope The double membrane that lines the nucleus in eukaryotic cells.

nucleolus The area within the nucleus of a cell devoted to the production of ribosomal RNA.

nucleotide The building block of nucleic acids, including DNA and RNA, consisting of a phosphate group, a sugar, and a nitrogen-containing base.

nucleus A membrane-lined compartment that encloses the primary complement of DNA in eukaryotic cells.

nutrient (in human nutrition) A substance found in food that does at least one of three things: provides energy, provides a structural building block, or regulates a physical process. There are six classes of nutrients: water, minerals and vitamins; and carbohydrates, lipids, and proteins.

nutrient A chemical element that is used by living things to sustain life.

nutrition The study of the relationship between food and health.

O

oil A dietary lipid that is liquid at room temperature (e.g., olive oil, canola oil).

olfaction The sense of smell.

omnivore An animal that eats both plants and animals.

oocyte The precursor cell of eggs in the vertebrate ovary. Diploid primary oocytes develop from oogonia and in turn give rise to haploid secondary oocytes in the process of meiosis.

oogonia (singular, oogonium) The diploid cells that are the starting female cells in gamete (egg) production. Diploid oogonia develop into diploid primary oocytes, which give rise to haploid secondary oocytes.

open circulation A type of circulation, found in many invertebrates, in which arteries carry blood into open spaces called sinuses. The blood then bathes surrounding tissues and is channeled into veins, through which it flows back to the heart.

organ A highly organized unit within an organism, performing one or more functions, that is formed of several kinds of tissue. Kidneys, heart, lungs, and liver are all familiar examples of organs in humans.

organ system A group of interrelated organs and tissues that serve a particular set of functions in the body. For example, the digestive system includes the mouth, stomach, pancreas, and intestines and functions in digestion of food and the elimination of waste.

organelle A highly organized structure within a cell that carries out specific cellular functions. Almost all organelles (meaning "tiny organs") are bound by membranes. The lone exception is the ribosome, which has no membrane and is the single organelle possessed by prokaryote cells (bacteria and archaea). Organelles in eukaryotic cells include the cell nucleus, mitochondria, lysosomes, chloroplasts, and ribosomes.

organic chemistry A branch of chemistry devoted to the study of molecules that have carbon as their central element.

osmosis The net movement of water across a semipermeable membrane from an area of lower solute concentration to an area of higher solute concentration.

osteoblast An immature bone cell that secretes organic material that becomes bone matrix, thus producing new bone.

osteoclast A type of bone cell that dissolves bone matrix, thus liberating the minerals stored in it.

osteocyte Mature bone cell that maintains the structure and density of bone by continually recycling calcium compounds around itself.

ovarian follicle In human reproduction in females, the complex of an oocyte (developing, unfertilized egg) and its accessory cells and fluids.

ovary In flowering plants, the area, located at the base of the carpel, where fertilization of the egg and early development of the embryo occur. In animals, the female reproductive organ in which eggs develop.

oviparous A condition, seen in all birds and many reptiles, in which fertilized eggs are laid outside the mother's body and then develop there.

ovulation The release of an oocyte from the ovary in animals.

oxidation Loss of one or more electrons by an atom or a molecule. Important in energy transfer in living things as part of redox reactions, in which one substance undergoes reduction (a gain in electrons) by oxidizing another.

ozone A gas in the Earth's atmosphere consisting of three oxygen atoms bonded together that serves to protect living things from the sun's ultraviolet radiation.

P

paired, jointed appendages Appendages, such as legs, that come in pairs and have joints. Such appendages are characteristic of all arthropods.

paleoanthropologist A scientist who studies human evolution by means of analyzing fossils.

pancreas In digestion, a gland that secretes, into the small intestine through ducts, digestive enzymes along with buffers that raise the pH of chyme. In nutrient metabolism, a gland that secretes, directly into the bloodstream, the hormones insulin and glucagon, which regulate blood levels of glucose.

parasites Organisms that feed off their prey but do not kill them, at least not immediately and perhaps not ever.

parasitism A type of predation in which the predator gets nutrients from the prey but does not kill the prey immediately and may never kill it.

parasympathetic division The division of the autonomic nervous system that generally has relaxing effects on the body.

parental generation (P) The generation that begins an experimental cross between organisms. Such a cross is used to study genetics and heredity of traits.

passive transport Transport of materials across the cell's plasma membrane that involves no expenditure of energy. Simple and facilitated diffusion are examples of passive transport.

pathogenic Disease-causing. Viruses, bacteria, protists, and fungi are spoken of as being pathogenic or nonpathogenic.

pedigree A familial history of genetically transmissible conditions; generally takes the form of a diagram.

peptide hormones Hormones composed of chains of amino acids. By far the largest class of the three principal classes of hormones, the other two being amino-acid–based and steroid hormones.

peripheral nervous system (PNS) The part of the nervous system that includes all of the neural tissue outside the central nervous system (brain and spinal cord). The PNS brings information to and from the central nervous system; it also provides voluntary control of the skeletal muscles and involuntary control of the smooth muscles, cardiac muscles, and glands.

peripheral protein A protein of the plasma membrane that lies on the inside or outside of the membrane but is not attached to the membrane's hydrophobic interior.

peristalsis Waves of contraction carried out by two sets of muscles in the digestive tract that help digest material and push it through the tract.

permafrost The permanently frozen ground that begins about a meter below the surface in the tundra biome of the far north. Neither roots nor water can penetrate this layer.

Permian Extinction The greatest mass extinction event in Earth's history. This event, in which up to 96 percent of all species on Earth were wiped out, occurred about 251 million years ago.

petal The colorful, leaf-like structure of a flower; petals attract pollinators.

petiole The stalk of a leaf, attaching it to a branch or trunk.

pH scale A scale utilized in measuring the relative acidity or alkalinity of a solution. The scale, ranging from 0 to 14, quantifies the concentration of hydrogen ions in a solution. The lower the pH number, the more acidic the solution; the higher the number, the more basic the solution.

phagocyte An immune system cell capable of ingesting another cell, parts of cells, or other materials. Phagocytes ingest both invading microorganisms and tissue fragments or the body's own cells when they have become damaged.

phagocytosis The movement of large materials into a cell by means of wrapping extensions of the plasma membrane around the materials and fusing the extensions together. One of two primary forms of endocytosis, the other being pinocytosis.

pharyngeal slits In animals, openings to the pharyngeal cavity. All chordates possess pharyngeal slits at some point in their development.

pharynx In humans, the passageway at the back of the mouth that links the mouth with both the food-transporting esophagus and the air-transporting trachea. In some animals, the pharynx can be everted, or turned inside-out, and used to obtain nutrients.

phenotype A physiological feature, bodily characteristic, or behavior of an organism.

phloem In vascular plants, the fluid-transporting tissue that conducts the food produced in photosynthesis along with some hormones and other compounds.

phosphate group A phosphorus atom surrounded by four oxygen atoms.

phospholipid A charged lipid molecule composed of two fatty acids, glycerol, and a phosphate group. The phospholipid's phosphate group is hydrophilic, while its fatty acid chains are hydrophobic. Phospholipids are a major constituent of cell membranes.

phospholipid bilayer One of the chief components of the plasma membrane, composed of two layers of phospholipids, arranged with their fatty acid chains pointing toward each other.

photoperiodism The ability of a plant to respond to changes it experiences in the daily duration of darkness relative to light.

photoreceptor Sensory receptor cell for vision, located in the eye's retina, that transforms light into nervous system signals. Photoreceptors come in two varieties: rods, which function in low-light situations but provide only black-and-white vision; and cones, which respond best to bright light and provide color vision.

photorespiration A process in which the enzyme rubisco undercuts carbon fixation in photosynthesis by binding with oxygen instead of with carbon dioxide. This wasteful reaction takes place most frequently in C_3 plants. The warm-weather C_4 plants have evolved a means of reducing photorespiration.

photosynthesis The process by which certain groups of organisms capture energy from sunlight and convert this solar energy into chemical energy that is initially stored in a carbohydrate.

photosystem An organized complex of molecules within a thylakoid membrane that, in photosynthesis, collects solar energy and transforms it into chemical energy.

phototropism The bending of a plant's shoots in response to light. Generally, this capability helps a plant grow toward the sun to get the most available sunlight.

phylogeny A hypothesis about the evolutionary relationships of a group of organisms.

phylum A category of living things, directly subordinate to the category of kingdom, whose members share traits as a result of shared ancestry. In animals, a group of animals that have a similar structural makeup.

physiology The study of the physical functioning of animals and plants.

phytoplankton Small photosynthesizing organisms that drift in the upper layers of oceans or bodies of freshwater, often forming the base of aquatic food webs.

pinocytosis The movement of relatively large materials into a cell by means of the creation of transport vesicles that are produced through an invagination of the plasma membrane. One of two primary forms of endocytosis, the other being phagocytosis.

placenta A complex network of maternal and embryonic blood vessels and membranes that develops in mammals in pregnancy. The placenta allows nutrients and oxygen to flow to the embryo from the mother, while allowing carbon dioxide and waste to flow from the embryo to the mother.

placental mammal A type of mammal that is nurtured before birth by the placenta, a network of maternal and embryonic blood vessels and membranes. Embryonic placental mammals derive their nutrition not from food stored in an egg, but directly from the mother's circulation.

plasma The fluid portion of blood, consisting mostly of water but also containing proteins and other molecules. Contrast with formed elements—the cells and cell fragments that form the nonfluid portion of blood.

plasma cell One of two types of cells the B lymphocyte develops into (the other type being a memory B cell). Plasma cells are specialized to produce the free-standing antibodies that fight invaders.

plasma membrane A membrane forming the outer boundary of many cells, composed of a phospholipid bilayer interspersed with proteins and cholesterol molecules and coated, on its exterior face, with short carbohydrate chains associated with proteins and lipids.

plasmid A ring of DNA that lies outside the chromosome in bacteria. Plasmids can move into bacterial cells in the process called transformation, thus making them a valuable tool in biotechnology.

plasmodesmata Channels in the plant cell wall that make possible communication between plant cells. The structure of these channels is such that the cytoplasm of one plant cell is continuous with that of another.

platelet One of the three varieties of formed elements within blood, platelets are small fragments of cells that facilitate blood clotting by releasing clotting enzymes and clumping together at the site of an injury.

point mutation A mutation of a single base pair in a genome.

polar body Nonfunctional cell produced during meiosis in females.

polar covalent bond A type of covalent bond in which electrons are shared unequally between atoms, so that one end of the molecule has a slight negative charge and the other end a slight positive charge.

polarity A difference in electrical charge at one end of a molecule compared to the other.

pollen grain A sperm-bearing gametophyte plant, consisting at maturity of three cells and an outer coat. One sperm cell within the pollen grain will fertilize the egg cell in the female gametophyte, thus producing a new generation of sporophyte plant.

pollination The transfer of pollen—by wind, animal, or other means—to a plant's female reproductive structure.

polygenic Having multiple genes affecting a given character, such as height in humans.

polygenic inheritance Inheritance of a genetic character that is determined by the interaction of multiple genes, with each gene having a small additive effect on the character.

polymer A large molecule made up of many similar or identical subunits, called monomers.

polymerase chain reaction (PCR) A technique for generating many copies of a DNA sequence from a small starting sample.

polypeptide A series of amino acids linked in linear fashion. Polypeptide chains fold up to become proteins.

polyploidy A form of sympatric speciation in which one or more sets of chromosomes are added to the genome of an organism. Human beings cannot survive in a polyploid state, but many plants flourish in it. Polyploidy is a means by which speciation can occur (most often in plants) in a single generation.

polysaccharide A polymer of carbohydrates, composed of many monosaccharides. Examples include starch, glycogen, cellulose, and chitin.

polyunsaturated fatty acid A fatty acid with two or more double bonds between the carbon atoms of its hydrocarbon chain.

pons A region of the brain whose primary function is to relay messages between the cerebrum and the cerebellum.

population All the members of a species that live in a defined geographic region at a given time.

postanal tail A tail, existing at some point in the development of all chordates, that is located posterior to the anus.

potential energy Stored energy. A rock perched at the top of a hill has potential energy.

predation One organism feeding on parts or all of a second organism.

primary consumer Any organism that eats producers (organisms that make their own food).

primary oocyte A diploid cell produced in females that may mature into an egg, initially by giving rise to haploid secondary oocytes. After the female reaches puberty, an average of one oocyte per month is selected to continue the process of maturation in the ovary.

primary spermatocyte A diploid cell in a male that will undergo meiosis to produce haploid secondary spermatocytes, which ultimately give rise to mature sperm cells.

primary structure The sequence of amino acids in a protein. This sequence dictates the final shape of the protein because electrochemical bonding and repulsion forces act on the structure to create the folded-up protein.

primary succession In ecology, succession in which the starting state is one of little or no life and a soil that lacks nutrients.

producer Any organism that manufactures its own food. Plants, algae, and certain bacteria are producers. Producers capture solar energy and convert it into biomass that is then passed along in food webs.

product A substance formed in a chemical reaction. The products are written on the right side of a chemical equation.

prokaryote A single-celled organism whose complement of DNA is not contained within a nucleus.

prokaryotic cell A cell whose DNA is not located in the membrane-bound organelle known as a nucleus. Prokaryotes are microscopic forms of life, and all are either bacteria or archaea.

prostate gland In human males, a gland surrounding the urethra near the urinary bladder that contributes fluids to the semen.

protective immunity A state of long-lasting protection that the immune system develops against specific microorganisms.

protein A large polymer of amino acids, composed of one or more polypeptide chains. Proteins come in many forms, including enzymes, structural proteins, and hormones.

protist A eukaryotic organism that does not have all the defining characteristics of a plant, animal, or fungus. The catchall term protist is used to refer to several distinct evolutionary lines of eukaryotic organisms, most of them single-celled and aquatic.

proton A basic constituent of an atom, found in the nucleus of the atom and having positive electrical charge. Elements are defined by the number of protons in their nucleus.

pulmonary circulation The system that circulates blood between the heart and the lungs, with the result that blood is oxygenated.

Q

quaternary structure The way in which two or more polypeptide chains come together to form a protein.

R

radial symmetry A type of animal symmetry in which body parts are distributed evenly about a central axis. Jellyfish are radially symmetrical.

radiometric dating A technique for determining the age of objects by measuring the decay of the radioactive elements within them. The age of fossils can be determined with this technique.

reactant A substance (atoms, molecules, or ions) that goes into a chemical reaction. Reactants interact to form the product(s) of a chemical reaction.

reaction center A molecular complex in a chloroplast that, in photosynthesis, transforms solar energy into chemical energy.

receptor protein A protein that generally protrudes from the plasma membrane of a cell that is active in the transmission of signals between cells. Receptor proteins, often called receptors, function by binding with signaling molecules such as hormones.

recessive Term used to designate an allele that is not expressed in the heterozygous condition.

recessive disorder A medical condition that will not occur when an organism possesses a single functional allele for a given trait. Red-green color blindness is an example of a recessive disorder in that only one set of functional alleles need be present for normal color vision.

recombinant DNA Two or more segments of DNA that have been combined by humans into a sequence that does not exist in nature.

red blood cell (RBC) The blood cells, also known as erythrocytes, that transport oxygen to and carry carbon dioxide from every part of the body.

red marrow A tissue, found in cavities of bones in the human body, within which all of the adult body's blood cells are produced.

redox reaction A combination of a reduction and an oxidation reaction in which the electrons lost from one substance in oxidation are gained by another in reduction.

reduction The gain of one or more electrons by an atom or a molecule. Important in energy transfer in living things as part of redox reactions, in which one substance is reduced by oxidizing (or removing electrons from) another substance.

reflex Automatic nervous system response that helps an organism avoid danger or preserve a stable physical state. The knee-jerk response is a well-known reflex.

regulatory T cell A type of immune system cell that acts to limit the body's immune system response, thus protecting the body's own tissues from attack.

reproductive cloning Cloning intended to produce adult mammals of a defined genotype.

reproductive isolating mechanisms Any factor that, in nature, prevents interbreeding between individuals of the same or closely related species.

reproductive system In humans, the organ system that develops gametes and delivers them to a location where they can fuse with other gametes to produce a new individual.

resource partitioning The dividing of scarce resources among species that have similar requirements. Such partitioning allows species to coexist in the same habitat.

respiratory system The organ system that brings oxygen into the body and expels carbon dioxide from the body. In humans, this system includes the lungs and passageways that carry air to the lungs.

restriction enzyme A type of enzyme, occurring naturally in bacteria, that recognizes a specific sequence of DNA bases and cuts DNA strands at a specific location within the sequence. Restriction enzymes are used in biotechnology to cut DNA in specific places.

retina A layer of tissue at the back of the human eye whose cells convert light signals into neural signals.

ribonucleic acid (RNA) A nucleic acid that is active in the synthesis of proteins and that forms part of the structure of ribosomes. Varieties include messenger RNA (mRNA), transfer RNA (tRNA), ribosomal RNA (rRNA), and micro-RNA.

ribosomal RNA (rRNA) A type of RNA that, along with proteins, forms ribosomes.

ribosome An organelle, located in the cell's cytoplasm, that is the site of protein synthesis. The translation phase of protein synthesis takes place within ribosomes.

ribozyme A molecule composed of RNA that can both encode genetic information and act as an enzyme. Molecules similar to these may have been necessary early in the origin of life.

RNA polymerase In the transcription phase of protein synthesis, the enzyme that unwinds the DNA double helix and puts together a chain of RNA nucleotides complementary to the exposed DNA nucleotides.

rods Photoreceptors that function in low-light situations but that do not provide color vision.

root hair In plants, a threadlike extension of a root cell. Root hairs greatly increase the surface area of roots, thus allowing greater absorption of water and nutrients.

rough endoplasmic reticulum A network of membranes, found in the cytoplasm of eukaryotic cells, that aids in the processing of proteins.

r-selected species A species that tends to be relatively short-lived, that tends to produce relatively many offspring for which it provides little or no care, and whose population size tends to fluctuate widely in reaction to an environment that it experiences as highly variable. Also known as an opportunist species.

rubisco An enzyme that allows plants to incorporate atmospheric carbon dioxide into their own sugars during the process of photosynthesis.

rule of addition In probability theory, the principle that, when an outcome can occur in two or more different ways, the probability of that outcome is the sum of the respective probabilities.

rule of multiplication In probability theory, the principle that the probability of any two events happening is the product of their respective probabilities.

S

sarcomere The functional unit of a striated muscle that contracts when thin filaments slide past thick filaments. The sarcomeres shorten, thus contracting the whole muscle.

saprophyte An organism that obtains its nutrition from dead organic material.

saturated fatty acid A fatty acid with no double bonds between the carbon atoms of its hydrocarbon chain.

science A means of coming to understand the natural world through observation and the testing of hypotheses. Also, a collection of insights about nature, the evidence for which is an array of facts.

scientific method The process by which scientists investigate the natural world. The scientific method involves the testing of hypotheses through observation and experiment.

sebaceous glands A type of gland in the skin that produces a waxy, oily secretion (sebum) that lubricates the hair shaft and inhibits bacterial growth in the surrounding area.

second law of thermodynamics Energy transfer always results in a greater amount of disorder (entropy) in the universe.

secondary consumer Any organism that eats a primary consumer.

secondary structure The structure proteins assume after having folded up.

secondary succession In ecology, succession in which the final state of a habitat has been disturbed by some force, but life remains and the soil has nutrients. A farmer's field that has been abandoned is a site of secondary succession.

seed A reproductive structure in plants that includes a plant embryo, its food supply, and a tough protective casing.

seed coat The tough outer layer of a seed, derived from the integuments of the ovule, which protect the embryo from mechanical damage and water loss.

seedless vascular plant A plant that has a vascular (fluid transport) structure but that does not reproduce through use of seeds. One of the four principal varieties of plants. Ferns are the most familiar example.

semen The mixture of sperm and glandular secretions that is ejaculated from the human male through the urethra.

seminiferous tubule A convoluted tubule inside the testes where sperm development begins. Immature sperm are eventually released into the interior cavity of a tubule and travel to the epididymis, where sperm maturation is completed.

sensory neuron A neuron that senses conditions inside or outside the body and that converts this sensory information into electrical signals that travel through the nervous system.

sepal The leaf-like structure that, with other sepals, protects the flower before it opens.

sessile Fixed in location; organisms such as mushrooms and mature sponges are sessile.

sex chromosome The chromosomes that determine the sex of an organism. The X or Y chromosomes in humans.

sexual reproduction A means of reproduction in which the nuclei of the reproductive cells from two separate organisms fuse to produce offspring.

sexual selection A form of natural selection that produces differential reproductive success based on differential success in obtaining mating partners.

sieve element In plants, a type of cell that conducts phloem sap.

simple diffusion Diffusion through the plasma membrane that requires only concentration gradients, as opposed to concentration gradients and special protein channels. Water, oxygen, carbon dioxide, and steroid hormones can all cross the plasma membrane through simple diffusion.

simple sugars The smallest and simplest form of carbohydrates, which serve as energy-yielding molecules and as the building blocks, or monomers, of complex carbohydrates. In nutrition, one of the three principal classes of carbohydrates. Simple sugars can either be monosaccharides (e.g., glucose or fructose) or disaccharides (e.g., sucrose). The other classes of dietary carbohydrate are starches and fibers.

sinoatrial node A collection of specialized heart muscle cells that function as the heart's pacemaker.

skeletal muscle In humans, muscle that is attached to bone, that is under conscious control, and that microscopically has a striped or "striated" appearance owing to the parallel orientation of the long, fibrous units that make it up.

skeletal system The human organ system that forms an internal supporting framework for the body and protects delicate tissues and organs. This system consists of all the bones and cartilages in the body and the connective tissues and ligaments that connect the bones at the joints.

skin In humans, an organ consisting of two tissue layers, an outer epidermis and inner dermis, and covering the outside of the body. The skin protects the body and receives signals from the environment.

small intestine That portion of the digestive tract that runs between the stomach and large intestine.

smooth endoplasmic reticulum A network of membranes within the eukaryotic cell that is the site of the synthesis of various lipids and site at which potentially harmful substances are detoxified.

smooth muscle In humans, muscle that is not under voluntary control and that lacks a striated appearance. Smooth muscle is responsible for contractions of the uterus, digestive tract, blood vessels, and passageways of the lungs.

solute The substance being dissolved by a solvent to form a solution.

solution A homogeneous mixture of two or more substances. In biology, solutions often consist of a solute dissolved in water, which produces an aqueous solution.

solvent The substance in which a solute is dissolved to form a solution. In an aqueous solution, the solvent is water.

somatic cell Any cell that is not and will not become an egg or sperm cell.

somatic cell nuclear transfer (SCNT) A means of cloning animals through fusion of one somatic (non-sex) cell with an egg cell whose nucleus has been removed (an "enucleated" cell).

somatic nervous system That portion of the peripheral nervous system's efferent division that provides voluntary control over skeletal muscle.

somite One of the blocks of mesodermal tissue in the vertebrate embryo that lies on both sides of the notochord and gives rise to muscles and the vertebrae that enclose the spinal cord.

space-filling model A representation of a molecule showing its three-dimensional structure, particularly showing how big its atoms are in relation to each other and the overall shape of the molecule.

speciation The development of new species through evolution.

species A group of actually or potentially interbreeding natural populations that are reproductively isolated from other such populations.

specific heat The amount of energy needed to raise the temperature of 1 gram of a substance by 1°C.

sperm cell In flowering plants, either of two cells in a pollen grain, one of which fertilizes an egg, the other of which fertilizes the central cell in an embryo sac. In animals, the male gamete, which fertilizes the female gamete (the egg).

spermatocyte An immature sperm cell. Spermatocytes develop from spermatogonia and develop into spermatids, which eventually develop into mature sperm.

spermatogonia Diploid cells that are the starting cells in sperm production in males. Spermatogonia are reproductive stem cells in that, when dividing, each of them produces one primary spermatocyte (which will develop into four mature sperm cells) and one spermatogonium.

spongy bone Type of bone that is porous and less dense than compact bone. Spongy bone fills the expanded ends of long bones.

spore A reproductive cell that can develop into a new organism without fusing with another reproductive cell. (The term spore also refers to a dormant, stress-resistant form of a bacterial or fungal cell.)

sporophyte generation The diploid, spore-producing plant generation. This generation is the dominant, visible generation in flowering plants. Contrast with gametophyte generation.

stabilizing selection In evolution, the type of natural selection in which intermediate forms of a given character are favored over either extreme. This process tends to maintain average traits for a character. Compare to directional and disruptive selection.

stamen The male reproductive structure in a flower, consisting of a stalk-like filament topped by a pollen-producing anther.

starch A complex carbohydrate that serves as the major form of carbohydrate storage in plants.

Starches—found in such forms as potatoes, rice, carrots, and corn—are important sources of food for animals. In human nutrition, one of the three principal classes of dietary carbohydrates, defined as a complex carbohydrate that is digestible. The other classes of dietary carbohydrates are simple sugars and fibers.

steroid A member of the class of lipid molecules that have four carbon rings as a central element in their structure. One steroid differs from another in accordance with the varying side chains that can be attached to these rings. Cholesterol, testosterone, and estrogen are steroid molecules.

steroid hormones Hormones such as testosterone and estrogen that are constructed around the chemical framework of the cholesterol molecule. One of three principal classes of hormones, the other two being peptide and amino-acid–based hormones.

stigma In flowering plants, the tip end of a flower's carpel, where pollen grains are deposited prior to fertilization.

stomach An organ that performs digestion and that serves as a temporary, expandable storage site for food.

stomata Microscopic pores, found in greatest abundance on the undersides of leaves, that allow plants to exchange gases with the atmosphere. Carbon dioxide moves into plants through the stomata, while oxygen and water vapor move out.

stratosphere The layer of the Earth's atmosphere situated above the troposphere, at about 20 to 35 kilometers (13 to 21 miles) above sea level. The ozone layer lies in this level.

stroma In plants and algae, the liquid material of chloroplasts that is the site of the Calvin cycle.

structural formula A two-dimensional representation of a molecule that shows how its atoms are arranged and the types of bonds between them.

style In flowering plants, the stalk-like extension of a carpel that connects its stigma and ovary.

substrate A substance whose chemical alteration is facilitated by an enzyme.

succession In ecology, a series of replacements of community members at a given location until a stable final state is reached.

sweat gland In humans, a type of duct-containing (exocrine) gland that produces perspiration.

symmetry An equivalence of size, shape, and relative position of parts across a dividing line or around a central point.

sympathetic division The division of the autonomic nervous system that generally has stimulatory effects on the body.

sympatric speciation A type of speciation that occurs in the absence of the geographic separation of populations. Polyploidy is a special form of sympatric speciation.

synapse An area in the nervous system consisting of a sending neuron, a receiving neuron or effector cell (i.e., a muscle cell), and the gap between them, called the synaptic cleft.

synaptic cleft The tiny gap that exists between a neuron sending a nervous system signal and a neuron (or effector cell) that is receiving this signal.

systematics The field of biology dealing with the diversity and relatedness of organisms. Systematists study the evolutionary history of groups of organisms.

systemic circulation One of the body's two general networks of blood vessels that serves to transport blood between the heart and the rest of the body following oxygenation of this blood in the body's pulmonary circulation.

T

taiga The biome consisting of boreal (northern) forest, characterized by cold, dry conditions, a relatively short growing season, and large expanses of coniferous trees.

taproot system The type of plant root system that consists of a large central root and many smaller lateral roots.

target cells The set of cells in the body that will be affected by a given hormone. The target cells for each hormone bear receptors capable of binding with that hormone.

taxon Any of the taxonomic categories used in the classification of Earth's organisms, both living and extinct. In order of increasing inclusiveness, these categories are species, genus, family, order, class, phylum, kingdom, and domain.

tendons The connective tissue that attaches a skeletal muscle to a bone.

tertiary consumer Any organism that eats secondary consumers.

tertiary structure The large-scale twists and turns in a protein conformation.

testis The organ in the male reproductive system in which sperm begin development and testosterone is produced.

tetrad The grouping formed by the linkage of two homologous chromosomes in prophase I of meiosis. The four sister chromatids involved in this linkage give it the name tetrad.

tetrapods Any four-limbed vertebrate. All amphibians, reptiles, and mammals are tetrapods.

thalamus The portion of the brain through which most sensory perceptions are channeled before being relayed to the cerebral cortex.

theory A general set of principles, supported by evidence, that explains some aspect of nature.

thigmotropism The growth of a plant in response to touch. This capability allows tendrils to wrap around other objects, thus helping a plant climb upward toward light.

thylakoid A flattened, membrane-bound sac in the interior of a chloroplast that serves as the site for the light reactions in photosynthesis.

tissue An organized assemblage of similar cells that serves a common function. Nervous, epithelial, and muscle tissue are some familiar examples.

T-lymphocyte cell (T cell) A class of lymphocytes that plays a central role in cell-mediated immunity. T cells come in several varieties that play specific roles in recognizing and killing infected cells in the body.

toll-like receptors Receptors, generally existing on the surface of immune system cells, that are the primary sensors of microbial infection in the human body.

top predator A species in a community that preys on other species but is not itself preyed on.

total fertility rate The average number of children born to each woman in a human population. The most important statistic used in predicting human population changes.

tracheid In plants, a slender, tapered cell of the xylem that has perforations (bordered pits) that match up with adjacent cells to allow water to move between them.

transcription In protein synthesis, the process in which DNA's information is copied onto messenger RNA (mRNA).

transcription factor A protein that binds to a regulatory DNA sequence, thus influencing the production of one or more other proteins.

transfer RNA (tRNA) In protein synthesis, a form of RNA that binds with amino acids, transfers them to ribosomes, and then binds with a messenger RNA sequence.

transformation A cell's incorporation of genetic material from outside its boundary. Some bacteria readily undergo this process, and others can be induced to for uses in biotechnology.

transgenic organism An organism whose genome has stably incorporated one or more genes from another species.

translation The process in which a polypeptide chain is produced within a ribosome based on the information encoded in messenger RNA. This process, the second major stage in protein synthesis (after transcription), occurs in the cell's cytoplasm.

translocation The swapping of fragments by nonhomologous chromosomes, resulting in gene sequences that are out of order on both chromosomes.

transpiration The process by which plants lose water when water vapor leaves the plant through open stomata. More than 90 percent of the water that enters a plant evaporates into the atmosphere via transpiration.

transport protein A protein that forms a hydrophilic channel through the hydrophobic interior of the cell's plasma membrane, allowing hydrophilic materials to pass through the membrane.

transport vesicle A membrane-lined sphere that moves within the cell's endomembrane system, carrying within it proteins or other molecules.

triglyceride A lipid molecule formed from three fatty acids bonded to glycerol.

trimester A period lasting about 3 months during a human pregnancy. There are three trimesters during the 9-month pregnancy.

trophic level A position in an ecosystem's food chain or web, with each level defined by a transfer of energy from one kind of organism to another. Plants and other photosynthesizers are producers of food and thus occupy the first trophic level. Organisms that consume producers are primary consumers and occupy the second trophic level, and so on.

trophoblast The cells at the periphery of the developing mammalian embryo that establish physical links with the mother's uterine wall and eventually develop into the fetal portion of the placenta.

tropical rainforest A biome found in Earth's equatorial regions characterized by warm year-round temperatures, abundant rainfall (averaging 200–450 cm per year), and great species diversity.

tropical savanna A grassland biome characterized by seasonal drought, small seasonal changes in the generally warm temperatures, and stands of trees that punctuate the grassland.

troposphere The lowest layer of the Earth's atmosphere, extending from sea level to about 12 kilometers (7.4 miles) above sea level. This layer contains most of the gases in the atmosphere.

true multicellularity A form of life in which individual cells exist in stable groups, with different cells in a group specializing in different functions.

tundra A biome of Earth's far-northern latitudes characterized by very cold temperatures, very little rainfall (averaging 25 cm per year), and a short growing season.

U

umbilical cord In human pregnancy, the tissue linking the fetus with the placenta.

ureters Two tubes in which urine is transported from the kidneys to the urinary bladder.

urethra Tube in which urine flows from the bladder to the outside of the body in both males and females. In males, the urethra also transmits semen.

urinary bladder A hollow, muscular organ that serves as a temporary, expandable storage site for the waste product urine.

urinary system The organ system that eliminates waste products from the blood through formation of urine. In humans, this system consists of the kidneys, where the urine is formed; and the ureters, urinary bladder, and urethra, which transport the urine from the kidneys to the outside of the body.

uterine tube In human females, the tube that transports an ovulated developing egg (an oocyte) from the ovary to the uterus, during which time fertilization may occur. Also called the fallopian tube.

V

variable An element of an experiment that is changed compared to an initial condition.

vas deferens In human males, the tube that carries the sperm from the epididymis to the urethra for ejaculation.

vasopressin See antidiuretic hormone (ADH).

vegetal pole The end of the zygote with relatively more yolk and lying farther from the cell's nucleus. The location of the egg's poles defines the orientation in which the embryo develops.

vegetative reproduction A form of asexual reproduction practiced by plants in which a portion of an initial plant can grow into a second plant. Severed "runners" of aspen trees can grow into separate trees through vegetative reproduction.

vein A blood vessel that carries blood toward the heart.

ventilation The physical movement of air into and out of the lungs, brought about by contractions and relaxations of muscles in the chest.

vertebral column A flexible column of bones extending from the anterior to posterior end of an animal. Also known as a backbone, the vertebral column distinguishes vertebrates from other chordates.

vessel element One of two types of cells that make up xylem tissue, which transports water and dissolved minerals through plants. At maturity, vessel elements are dead and empty of cellular contents, thus facilitating the rapid conduction of water.

vestigial character A structure in an organism whose original function has been lost during the course of evolution.

virus A noncellular replicating entity that must invade a living cell to replicate itself.

vitamin A chemical compound found in foods that is needed in small amounts to facilitate a chemical reaction in the human body.

viviparous A condition in animals in which fertilized eggs develop inside a mother's body.

W

wax A lipid composed of a single fatty acid linked to a long-chain alcohol.

wetlands Lands that are wet for at least part of the year. Wetlands are sites of great biological productivity, and they provide vital habitat for migrating birds.

white blood cell (WBC) The central cells of the immune system. Types of white blood cells include T cells, B cells, neutrophils, eosinophils, and macrophages, each playing a specific role in immune responses.

X

xylem The tissue through which water and dissolved minerals flow in vascular plants.

xylem sap The water and dissolved minerals that are conducted through xylem in plants.

Y

yellow marrow In human beings, a tissue largely made up of energy-storing fat cells found in the marrow cavity of long bones.

Z

zero population growth State of a population in which births exactly equal deaths in a given period.

zooplankton Microscopic aquatic animals that occupy a trophic level above that of phytoplankton, or photosynthesizing aquatic microorganisms.

zygote A fertilized egg. In humans, the developing organism from the time of fertilization through the time of the first cell division (about 30 hours after fertilization).

Credits

CHAPTER 1 **CO-01 verso** Frans Lanting/CORBIS **CO-01 recto** R. Michael Stuckey/Jupiter Images **1-1a** James King-Holmes/SPL/Photo Researchers, Inc. **1-1b** Ethan Miller/Getty Images **1-2** Pearson Science/Kristin Piljay **1-6a** apiguide/Shutterstock **1-ESS-01** John Valls **1-6b** Mark Kostich/iStockphoto **1-6c** George Holton/Photo Researchers, Inc. **1-7a,b** Jose B. Ruiz/naturepl.com

CHAPTER 2 **CO-02 verso** Michael Durham/National Geographic Stock **CO-02 recto** Shutterstock **2-3** HooRoo Graphics/iStockphoto **2-4** Mike Nelson/CORBIS **2-ESS-01** Vienna Report Agency/Sygma/CORBIS **2-13** Stephen C. Harrison; From A.K. Aggarwal, D.W. Rogers, M. Drottar, M. Ptashne, S.C. Harrison, 1988, "Science," 242:899; Courtesy, S.C. Harrison. **2-15** Doug Allen/Photolibrary **2-16a** Herman Eisenbeiss/Photo Researchers, Inc. **2-16b** Pasieka/SPL/Photo Researchers, Inc. **2-17** David Woodfall/Getty Images

CHAPTER 3 **CO-03 verso** David Noton/naturepl.com **CO-03 recto** Arkady/Shutterstock **3-1** Getty Images **3-3** Dorling Kindersley Media Library **3-5** Richard Megna/Fundamental Photographs **3-Table03a** Dr. Jeremy Burgess/SPL/Photo Researchers, Inc. **3-Table03b** edImage/Photo Researchers, Inc. **3-Table03c** Biophoto Associates/Photo Researchers, Inc. **3-Table03d** Eye of Science/Photo Researchers, Inc. **3-13** Eye of Science/Photo Researchers, Inc. **3-14** Quest/SPL/Photo Researchers, Inc. **3-17a,b** Peter M. Colman

CHAPTER 4 **CO-04 verso** David Scharf/Photolibrary **CO-04 recto** LSHTM/Photo Researchers, Inc. **4-01UNa** Prof. P. Motta/SPL/Photo Researchers, Inc. **4-01UNb** Photo Researchers, Inc. **4-01UNc** Andrew Syred/Photo Researchers, Inc. **4-1** NCI/Photo Researchers, Inc. **4-ESS1-01a-c** Tony Brain/David Parker/SPL/Photo Researchers, Inc. **4-6** Biophoto Associates/Photo Researchers, Inc. **4-7** Don W. Fawcett/Photo Researchers, Inc. **4-9** P. Motta & T. Naguro/SPL/Photo Researchers, Inc. **4-11** P. Motta & T. Naguro/SPL/Photo Researchers, Inc. **4-12a** Tomasz Szul/Visuals Unlimited/CORBIS **4-12b** Nancy Kedersha/SPL/Photo Researchers, Inc. **4-12c** Dr. Peter Dawson/SPL/Photo Researchers, Inc. **4-13** Photo Lennart Nilsson/Albert Bonniers Forlag **4-ESS2-01** Scott Camazine/Photo Researchers, Inc. **4-14b** Steve Gschmeissner/SPL/Photo Researchers, Inc. **4-14c** Leroy Francis/SPL/Photo Researchers, Inc. **4-18a** Ed Kanze/Dembinsky Associates **4-18b** Lee W. Wilcox, Univ. of Wisconsin at Madison **4-19** Eldon H. Newcomb & William P. Wergin **4-ESS3-01a** Brian J. Ford

4-ESS3-01b National Library of Medicine

CHAPTER 5 **CO-05 verso** Sinclair Stammers/Photo Researchers, Inc. **CO-05 recto** John Burbidge/Photo Researchers, Inc. **5-4a-c** Pearson Science/Eric Shrader **5-10** Jurgen Bohmer and Linda Hufnagel **5-11a** M.M. Perry **5-11b** Biology Media/Science Source/Photo Researchers, Inc.

CHAPTER 6 **CO-06 verso** Michael Durham/National Geographic Stock **CO-06 recto** Markus Gann/Shutterstock **6-1** Alamy imagebroker/Alamy **6-10** Thomas A. Steitz

CHAPTER 7 **CO-07 verso** Michael & Patricia Fogden/National Geographic Stock **CO-07 recto** Mike Kemp/Photolibrary **7-01UN** Marc Pagani Photography/Shutterstock **7-2** Victor De Schwanberg/SPL/Photo Researchers, Inc. **7-ESS-01** Olaf Doering/Alamy

CHAPTER 8 **CO-08 verso** Roderick Chen/Jupiter Images **CO-08 recto** Kati Molin/Shutterstock **8-1a** Andriy Solovyov/Shutterstock **8-1b** Georgette Douwma/naturepl.com **8-1c** M. I. Walker/Photo Researchers, Inc. **8-2** Catalin Petolea/Alamy **8-7** Eldon H. Newcomb & William P. Wergin **8-11** Jay Directo/AFP/Getty Images

CHAPTER 9 **CO-09 verso** SPL/Photo Researchers, Inc. **CO-09 recto** Dr. Kari Lounatmaa/Photo Researchers, Inc. **9-1** Ariel Skelley/Getty Images **9-7** Biophoto Associates/Photo Researchers, Inc. **9-8a,b** Biophoto Associates/Photo Researchers, Inc. **9-10** Ed Reschke/photolibrary.com **9-11** Don W. Fawcett/Photo Researchers, Inc. **9-14** CNRI/Photo Researchers, Inc.

CHAPTER 10 **CO-10 verso** Hans Dieter Brandl/Foto Natura/National Geographic Stock **CO-10 recto** Xtremer/iStockphoto **10-5** Ron Hainer/Shutterstock **10-6** Andrew Syred/ Photo Researchers, Inc. **10-9** David Phillips/The Population Council/Photo Researchers, Inc. **10-10** Suzanne L. Collins/Photo Researchers, Inc. **10-11** Fred McConnaughey/Photo Researchers, Inc.

CHAPTER 11 **CO-11 verso** Christi Carter/photolibrary.com **CO-11 recto** Miroslava Vasileva Arnaudova/Shutterstock **11-1** Getty Images Inc./Hulton Archive Photos **11-4** Dr. Maban K. Bhattarrya **11-11** Gary Roberts **11-12a** Peter Morenus **11-13** Steffen Hauser/botanikfoto/Alamy

CHAPTER 12 **CO-12 verso** Andrew Syred/Photo Researchers, Inc. **CO-12 recto** Sebastian Kaulitzki/Shutterstock **12-3** Oliver Meckes & Nicole Ottawa/Photo Researchers, Inc. **12-8** PHANIE/Photo

Researchers, Inc. **12-ESS1-01a** Pascal Goetgheluck/Photo Researchers, Inc. **12-ESS1-01b** Dept. of Clinical Cytogenetics, Addenbrookes Hospital/SPL/ Photo Researchers, Inc. **12-9** Sandra Hanks and Nazneen Rahman, Institute of Cancer Research, Sutton, Surrey, UK **12-10a,b** Irene Uchida **12-ESS2-01a** Photo Researchers, Inc.

CHAPTER 13 **CO-13 verso** Zygote Media/3Dscience.com **CO-13 recto** John Harwood/Getty Images **13-1** Omikron/Photo Researchers, Inc. **13-2** Science Source/Photo Researchers, Inc. **13-3** Science Source/Photo Researchers, Inc. **13-4** George Silk/Time Life Pictures/Getty Images

CHAPTER 14 **CO-14 verso** Dr. Torsten Wittman/Photo Researchers, Inc. **CO-14 recto** George Underwood/Getty Images **14-11b** Dr. Elena Kiseleva/Photo Researchers, Inc. **14-14a** Photo Researchers, Inc. **14-14b** Oliver Meckes/Photo Researchers, Inc. **14-14c** James King-Holmes/SPL/ Photo Researchers, Inc. **14-14d** Niall Benvie/naturepl.com **14-17a** Tony Brain/David Parker/ SPL/Photo Researchers, Inc. **14-17b** Photo Researchers, Inc. **14-17c** Niall Benvie/naturepl.com **14-17d** James King-Holmes/SPL/Photo Researchers, Inc. **14-17e** Max Westby **14-17f** Redmond Durrell/ Alamy **14-17g** Shishov Mikhail/Shutterstock

CHAPTER 15 **CO-15 verso** Matthias Schrader/dpa/Landov Media **CO-15 recto** Zirafek/ iStockphoto **15-1** RIKEN Center for Developmental Biology **15-2** GTC Biotherapeutics **15-4** Dr. Gopal Murti/Science Photo Library/Photo Researchers, Inc. **15-6** Jon Christensen/Pearson Education/PH College **15-7** Photograph courtesy of The Roslin Institute **15-12** Pearson Education/PH College

CHAPTER 16 **CO-16 verso** Tony Campbell/Shutterstock **CO-16 recto** Lifeonwhite.com/Getty Images **16-1** Paul Souders/CORBIS **16-2** Painting by John Collier, 1883. London, National Portrait Gallery. ©Archiv/Photo Researchers **16-3** Christopher Ralling **16-5** WaterFrame/Alamy **16-6** Juniors Blildarchiv/age footstock **16-7** Prochasson Frederic/Shutterstock **16-8a** Bruce Coleman/Photoshot Holdings Ltd. **16-8b** Frans Lanting/DanitaDelimont.com **16-8c** © FLPA/Alamy **16-9** Mark Kauffman/Time Life Syndication **16-11** Sinclair Stammers/Science Photo Library/Photo Researchers, Inc. **16-ESS-01** Darrell Gulin/Getty Images

CHAPTER 17 **CO-17 verso** Winfried Wisniewski/Foto Natura/National Geographic Stock **CO-17 recto** Dan Wilton/iStockphoto **17-4a** Greg Vaughn/Alamy **17-4b** Bruce G. Baldwin **17-7** Ray Scott

Index